Lecture Notes in Artificial Intelligence 10535

Subseries of Lecture Notes in Computer Science

More information about this series at http://www.springer.com/series/1244

Michelangelo Ceci · Jaakko Hollmén
Ljupčo Todorovski · Celine Vens
Sašo Džeroski (Eds.)

Machine Learning and Knowledge Discovery in Databases

European Conference, ECML PKDD 2017
Skopje, Macedonia, September 18–22, 2017
Proceedings, Part II

 Springer

Editors
Michelangelo Ceci (iD)
Università degli Studi di Bari Aldo Moro
Bari
Italy

Jaakko Hollmén
Aalto University School of Science
Espoo
Finland

Ljupčo Todorovski
University of Ljubljana
Ljubljana
Slovenia

Celine Vens
KU Leuven Kulak
Kortrijk
Belgium

Sašo Džeroski
Jožef Stefan Institute
Ljubljana
Slovenia

ISSN 0302-9743   ISSN 1611-3349 (electronic)
Lecture Notes in Artificial Intelligence
ISBN 978-3-319-71245-1   ISBN 978-3-319-71246-8 (eBook)
https://doi.org/10.1007/978-3-319-71246-8

Library of Congress Control Number: 2017961799

LNCS Sublibrary: SL7 – Artificial Intelligence

This Springer imprint is published by Springer Nature
The registered company is Springer International Publishing AG
The registered company address is: Gewerbestrasse 11, 6330 Cham, Switzerland

Preface

We are delighted to introduce the proceedings of the 2017 edition of the European Conference on Machine Learning and Principles and Practice of Knowledge Discovery in Databases (ECML PKDD 2017). This year the conference was held in Skopje, Macedonia, during September 18–22, 2017. ECML PKDD is an annual conference that provides an international forum for the discussion of the latest high-quality research results in all areas related to machine learning and knowledge discovery in databases as well as innovative application domains. This event is the premier European machine learning and data mining conference and builds upon a very successful series of 18 ECML, ten PKDD (until 2007, when they merged), and nine ECML PKDD (from 2008). Therefore, this was the tenth edition of ECML PKDD as a single conference.

The scientific program was very rich and consisted of technical presentations of accepted papers, plenary talks by distinguished keynote speakers, workshops, and tutorials. Accepted papers were organized in five different tracks:

- The conference track, featuring regular conference papers, published in the proceedings
- The journal track, featuring papers that satisfy the quality criteria of journal papers and at the same time lend themselves to conference talks (these papers are published separately in the journals *Machine Learning* and *Data Mining and Knowledge Discovery*)
- The applied data science track (formerly industrial track), aiming to bring together participants from academia, industry, government, and non-governmental organizations in a venue that highlights practical and real-world studies of machine learning, knowledge discovery, and data mining
- The demo track, presenting innovative prototype implementations or mature systems that use machine learning techniques and knowledge discovery processes in a real setting
- The nectar track, offering conference attendees a compact overview of recent scientific advances at the frontier of machine learning and data mining with other disciplines, as published in related conferences and journals.

In addition, the PhD Forum provided a friendly environment for junior PhD students to exchange ideas and experiences with peers in an interactive atmosphere and to get constructive feedback from senior researchers. This year we also introduced the EU Projects Forum with the purpose of disseminating EU projects and their results to the targeted scientific audience of the conference participants. Moreover, two discovery challenges, six tutorials, and 17 co-located workshops were organized on related research topics.

Following the successful experience of last year, we stimulated the practices of reproducible research (RR). Authors were encouraged to adhere to such practices by making available data and software tools for reproducing the results reported in their papers. In total, 57% of the accepted papers in the conference have accompanying software and/or data and are flagged as RR papers on the conference website. In the proceedings, each RR paper provides links to such additional material. This year, we took a further step toward data publishing and facilitated the authors in making data and software available on a public repository in a dedicated area branded with the conference name (e.g., figshare, where there is an ECML PKDD area).

The response to our call for papers was very good. We received 364 papers for the main conference track, of which 104 were accepted, yielding an acceptance rate of about 28%. This allowed us to define a very rich program with 101 presentations in the main conference track: The remaining three accepted papers were not presented at the conference and are not included in the proceedings. The program also included six plenary keynotes by invited speakers: Inderjit Dhillon (University of Texas at Austin and Amazon), Alex Graves (Google DeepMind), Frank Hutter (University of Freiburg), Pierre-Philippe Mathieu (ESA/ESRIN), John Quackenbush (Dana-Farber Cancer Institute and Harvard TH Chan School of Public Health), and Cordelia Schmid (Inria).

This year, ECML PKDD attracted over 600 participants from 49 countries. It also attracted substantial attention from industry and end users, both through sponsorship and submission/participation at the applied data science track. This is confirmation that the ECML PKDD community is healthy, strong, and continuously growing.

Many people worked hard together as a superb dedicated team to ensure the successful organization of this conference: the general chairs (Michelangelo Ceci and Sašo Džeroski), the PC chairs (Michelangelo Ceci, Jaakko Hollmén, Ljupčo Todorovski, Celine Vens), the journal track chairs (Kurt Driessens, Dragi Kocev, Marko Robnik-Šikonja, Myra Spiliopoulou), the applied data science chairs (Yasemin Altun, Kamalika Das, Taneli Mielikäinen), the nectar track chairs (Donato Malerba, Jerzy Stefanowski), the demonstration track chairs (Jesse Read, Marinka Žitnik), the workshops and tutorials chairs (Nathalie Japkowicz, Panče Panov), the EU projects forum chairs (Petra Kralj Novak, Nada Lavrač), the PhD forum chairs (Tomislav Šmuc, Bernard Ženko), the Awards Committee (Peter Flach, Rosa Meo, Indrė Žliobaitė), the discovery challenge chair (Dino Ienco), the production and public relations chairs (Dragi Kocev, Nikola Simidjievski), the local organizers (Ivica Dimitrovski, Tina Anžič, Mili Bauer, Gjorgji Madjarov), the sponsorship chairs (Albert Bifet, Panče Panov), the proceedings chairs (Jurica Levatić, Gianvito Pio), the area chairs (listed in this book), the Program Committee members (listed in this book), and members of the MLJ and DMKD Guest Editorial Boards. They made tremendous effort in guaranteeing the quality of the reviewing process and, therefore, the scientific quality of the conference, which is certainly beneficial for the whole community. Our sincere thanks go to all of them.

We would also like to thank the Cankarjev Dom congress agency and the student volunteers. Thanks to Springer for their continuous support and Microsoft for allowing us to use their CMT software for conference management. Special thanks to the sponsors (Deutsche Post DHL Group, Google, AGT, ASML, Deloitte, NEC, Siemens, Cambridge University Press, *IEEE/CAA Journal of Automatica Sinica*, Springer, IBM Research, *Data Mining and Knowledge Discovery*, *Machine Learning*, EurAi, and

GrabIT) and the European project MAESTRA (ICT-2013-612944), as well as the ECML PKDD Steering Committee (for their suggestions and advice). Finally, we would like to thank the organizing institutions: the Jožef Stefan Institute (Slovenia), the Ss. Cyril and Methodius University in Skopje (Macedonia), and the University of Bari Aldo Moro (Italy).

September 2017

Michelangelo Ceci
Jaakko Hollmén
Ljupčo Todorovski
Celine Vens
Sašo Džeroski

Organization

ECML PKDD 2017 Organization

Conference Chairs

Michelangelo Ceci   University of Bari Aldo Moro, Italy
Sašo Džeroski   Jožef Stefan Institute, Slovenia

Program Chairs

Michelangelo Ceci   University of Bari Aldo Moro, Italy
Jaakko Hollmén   Aalto University, Finland
Ljupčo Todorovski   University of Ljubljana, Slovenia
Celine Vens   KU Leuven Kulak, Belgium

Journal Track Chairs

Kurt Driessens   Maastricht University, The Netherlands
Dragi Kocev   Jožef Stefan Institute, Slovenia
Marko Robnik-Šikonja   University of Ljubljana, Slovenia
Myra Spiliopoulu   Magdeburg University, Germany

Applied Data Science Track Chairs

Yasemin Altun   Google Research, Switzerland
Kamalika Das   NASA Ames Research Center, USA
Taneli Mielikäinen   Yahoo! USA

Local Organization Chairs

Ivica Dimitrovski   Ss. Cyril and Methodius University, Macedonia
Tina Anžič   Jožef Stefan Institute, Slovenia
Mili Bauer   Jožef Stefan Institute, Slovenia
Gjorgji Madjarov   Ss. Cyril and Methodius University, Macedonia

Workshops and Tutorials Chairs

Nathalie Japkowicz   American University, USA
Pance Panov   Jožef Stefan Institute, Slovenia

Awards Committee

Peter Flach   University of Bristol, UK
Rosa Meo   University of Turin, Italy
Indrė Žliobaitė   University of Helsinki, Finland

Nectar Track Chairs

Donato Malerba   University of Bari Aldo Moro, Italy
Jerzy Stefanowski   Poznan University of Technology, Poland

Demo Track Chairs

Jesse Read   École Polytechnique, France
Marinka Žitnik   Stanford University, USA

PhD Forum Chairs

Tomislav Šmuc   Rudjer Bošković Institute, Croatia
Bernard Ženko   Jožef Stefan Institute, Slovenia

EU Projects Forum Chairs

Petra Kralj Novak   Jožef Stefan Institute, Slovenia
Nada Lavrač   Jožef Stefan Institute, Slovenia

Proceedings Chairs

Jurica Levatić   Jožef Stefan Institute, Slovenia
Gianvito Pio   University of Bari Aldo Moro, Italy

Discovery Challenge Chair

Dino Ienco   IRSTEA - UMR TETIS, France

Sponsorship Chairs

Albert Bifet   Télécom ParisTech, France
Pance Panov   Jožef Stefan Institute, Slovenia

Production and Public Relations Chairs

Dragi Kocev   Jožef Stefan Institute, Slovenia
Nikola Simidjievski   Jožef Stefan Institute, Slovenia

ECML PKDD Steering Committee

Michele Sebag	Université Paris Sud, France
Francesco Bonchi	ISI Foundation, Italy
Albert Bifet	Télécom ParisTech, France
Hendrik Blockeel	KU Leuven, Belgium and Leiden University, The Netherlands
Katharina Morik	University of Dortmund, Germany
Arno Siebes	Utrecht University, The Netherlands
Siegfried Nijssen	LIACS, Leiden University, The Netherlands
Chedy Raïssi	Inria Nancy Grand-Est, France
Rosa Meo	Università di Torino, Italy
Toon Calders	Eindhoven University of Technology, The Netherlands
João Gama	FCUP, University of Porto/LIAAD, INESC Porto L.A., Portugal
Annalisa Appice	University of Bari Aldo Moro, Italy
Indré Žliobaité	University of Helsinki, Finland
Andrea Passerini	University of Trento, Italy
Paolo Frasconi	University of Florence, Italy
Céline Robardet	National Institute of Applied Science in Lyon, France
Jilles Vreeken	Saarland University, Max Planck Institute for Informatics, Germany

Area Chairs

Michael Berthold	Universität Konstanz, Germany
Hendrik Blockeel	KU Leuven, Belgium
Ulf Brefeld	TU Darmstadt, Germany
Toon Calders	Universiteit Antwerpen, Belgium
Bruno Cremilleux	Universite de Caen Normandie, France
Tapio Elomaa	Tampere University of Technology, Finland
Johannes Fuernkranz	TU Darmstadt, Germany
João Gama	FCUP, University of Porto/LIAAD, INESC Porto L.A., Portugal
Alipio Mario Jorge	FCUP, University of Porto/LIAAD, INESC Porto L.A., Portugal
Arno Knobbe	Leiden University, The Netherlands
Stefan Kramer	Johannes Gutenberg University of Mainz, Germany
Ernestina Menasalvas	Universidad Politecnica de Madrid, Spain
Siegfried Nijssen	Leiden University, The Netherlands
Bernhard Pfahringer	University of Waikato, New Zealand
Jesse Read	Ecole Polytechnique, France
Arno Siebes	Universiteit Utrecht, The Netherlands
Carlos Soares	University of Porto, Portugal
Einoshin Suzuki	Kyushu University, Japan
Luis Torgo	University of Porto, Portugal

Matthijs Van Leeuwen   Leiden Institute of Advanced Computer Science,
                                        The Netherlands
Indré Žliobaité   University of Helsinki, Finland

Program Committee

Niall Adams	Thierry Charnois	Carlos Ferreira
Mohammad Al Hasan	Duen Horng Chau	Cesar Ferri
Carlos Alzate	Keke Chen Wright	Maurizio Filippone
Aijun An	Silvia Chiusano	Asja Fischer
Aris Anagnostopoulos	Arthur Choi	Peter Flach
Fabrizio Angiulli	Frans Coenen	Eibe Frank
Annalisa Appice	Roberto Corizzo	Elisa Fromont
Ira Assent	Vitor Santos Costa	Fabio Fumarola
Martin Atzmueller	Boris Cule	Esther Galbrun
Antonio Bahamonde	Tomaz Curk	Patrick Gallinari
Jose Balcazar	James Cussens	Dragan Gamberger
Nicola Barberi	Alfredo Cuzzocrea	Byron Gao
Christian Bauckhage	Claudia d'Amato	Paolo Garza
Roberto Bayardo	Maria Damiani	Eric Gaussier
Martin Becker	Stepanova Daria	Rainer Gemulla
Srikanta Bedathur	Jesse Davis	Konstantinos Georgatzis
Jessa Bekker	Tijl De Bie	Pierre Geurts
Vaishak Belle	Martine De Cock '	Dorota Glowacka
Andras Benczur	Juan del Coz	Michael Granitzer
Petr Berka	Anne Denton	Caglar Gulcehre
Michele Berlingerio	Christian Desrosiers	Francesco Gullo
Cuissart Bertrand	Nicola Di Mauro	Stephan Gunnemann
Marenglen Biba	Claudia Diamantini	Tias Guns
Silvio Bicciato	Tom Diethe	Sara Hajian
Albert Bifet	Ivica Dimitrovski	Maria Halkidi
Paul Blomstedt	Ying Ding	Jiawei Han
Mario Boley	Stephan Doerfel	Xiao He
Gianluca Bontempi	Carlotta Domeniconi	Denis Helic
Henrik Boström	Frank Dondelinger	Daniel Hernandez
Marc Boulle	Madalina Drugan	Jose Hernandez-Orallo
Pavel Brazdil	Wouter Duivesteijn	Thanh Lam Hoang
Dariusz Brzezinski	Robert Durrant	Arjen Hommersom
Rui Camacho	Ines Dutra	Frank Höppner
Longbing Cao	Dora Erdos	Tamas Horvath
Francisco Casacuberta	Floriana Esposito	Andreas Hotho
Peggy Cellier	Nicola Fanizzi	Yuanhua Huang
Loic Cerf	Fabio Fassetti	Eyke Huellermeier
Tania Cerquitelli	Ad Feelders	Dino Ienco
Edward Chang	Stefano Ferilli	Georgiana Ifrim

Bhattacharya Indrajit
Szymon Jaroszewicz
Klema Jiri
Giuseppe Jurman
Toshihiro Kamishima
Bo Kang
U Kang
Andreas Karwath
George Karypis
Samuel Kaski
Mehdi Kaytoue
Latifur Khan
Frank Klawonn
Dragi Kocev
Levente Kocsis
Yun Sing Koh
Alek Kolcz
Irena Koprinska
Frederic Koriche
Walter Kosters
Lars Kotthoff
Danai Koutra
Georg Krempl
Tomas Krilavicius
Matjaz Kukar
Meelis Kull
Jorma Laaksonen
Nicolas Lachiche
Leo Lahti
Helge Langseth
Thomas Lansdall-Welfare
Christine Largeron
Pedro Larranaga
Silvio Lattanzi
Niklas Lavesson
Nada Lavrac
Florian Lemmerich
Jiuyong Li
Limin Li
Jefrey Lijffijt
Tony Lindgren
Corrado Loglisci
Peter Lucas
Gjorgji Madjarov
Donato Malerba

Giuseppe Manco
Elio Masciari
Andres Masegosa
Wannes Meert
Rosa Meo
Pauli Miettinen
Mehdi Mirza
Dunja Mladenic
Karthika Mohan
Anna Monreale
Joao Moreira
Mohamed Nadif
Ndapa Nakasholc
Mirco Nanni
Amedeo Napoli
Sriraam Natarajan
Benjamin Nguyen
Thomas Nielsen
Xia Ning
Kjetil Norvag
Eirini Ntoutsi
Andreas Nuernberger
Francesco Orsini
Nikunj Oza
Pance Panov
Apostolos Papadopoulos
Evangelos Papalexakis
Panagiotis Papapetrou
Ioannis Partalas
Gabriella Pasi
Andrea Passerini
Dino Pedreschi
Jaakko Peltonen
Jing Peng
Ruggero Pensa
Nico Piatkowksi
Andrea Pietracaprina
Gianvito Pio
Susanna Pirttikangas
Marc Plantevit
Pascal Poncelet
Miguel Prada
L. A. Prashanth
Philippe Preux
Kai Puolamaki

Buyue Qian
Chedy Raissi
Jan Ramon
Huzefa Rangwala
Zbigniew Ras
Chotirat Ratanamahatana
Jan Rauch
Chiara Renso
Achim Rettinger
Fabrizio Riguzzi
Matteo Riondato
Celine Robardet
Pedro Rodrigues
Juan Rodriguez
Fabrice Rossi
Celine Rouveirol
Stefan Rueping
Salvatore Ruggieri
Yvan Saeys
Alan Said
Lorenza Saitta
Ansaf Salleb-Aouissi
Claudio Sartori
Christoph Schommer
Matthias Schubert
Konstantinos Sechidis
Giovanni Semeraro
Sohan Seth
Vinay Setty
Junming Shao
Nikola Simidjievski
Sameer Singh
Andrzej Skowron
Dominik Slezak
Kevin Small
Tomislav Smuc
Yangqiu Song
Mauro Sozio
Papadimitriou Spiros
Jerzy Stefanowski
Gerd Stumme
Mahito Sugiyama
Mika Sulkava
Stephen Swift
Sandor Szedmak

Andrea Tagarelli	Antti Ukkonen	Joerg Wicker
Domenico Talia	Jan Van Haaren	Marco Wiering
Letizia Tanca	Martijn Van Otterlo	Pengtao Xie
Jovan Tanevski	Iraklis Varlamis	Makoto Yamada
Nikolaj Tatti	Julien Velcin	Philip Yu
Maguelonne Teisseire	Shankar Vembu	Bianca Zadrozny
Aika Terada	Deepak Venugopal	Gerson Zaverucha
Georgios Theocharous	Vassilios Verykios	Filip Zelezny
Hannu Toivonen	Ricardo Vigário	Bernard Zenko
Roberto Trasarti	Herna Viktor	Junping Zhang
Volker Tresp	Christel Vrain	Min-Ling Zhang
Isaac Triguero	Jilles Vreeken	Shichao Zhang
Panagiotis Tsaparas	Willem Waegeman	Ying Zhao
Karl Tuyls	Jianyong Wang	Mingjun Zhong
Niall Twomey	Ding Wei	Arthur Zimek
Nikolaos Tziortziotis	Cheng Weiwei	Albrecht Zimmermann
Theodoros Tzouramanis	Zheng Wen	Marinka Zitnik

Additional Reviewers

Anes Bendimerad	Kata Gabor	Marcos de Paula
Asim Karim	Golnoosh Farnadi	Martin Ringsquandl
Antonis Matakos	Tom Hanika	Letizia Milli
Anderson Nascimento	Heri Ramampiaro	Christian Poelitz
Andrea Pagliarani	Vasileios Iosifidis	Riccardo Guidotti
Ricardo Ñanculef	Jeremiah Deng	Andreas Schmidt
Caio Corro	Johannes Jurgovsky	Shi Zhi
Emanuele Frandi	Kemilly Dearo	Yujia Shen
Emanuele Rabosio	Mark Kibanov	Thomas Niebler
Fábio Pinto	Konstantin Ziegler	Albin Zehe
Fernando Martinez-Pluned	Ling Luo	
Francesca Pratesi	Liyuan Liu	
Fangbo Tao	Lorenzo Gabrielli	

Sponsors

Gold Sponsors

Deutsche Post DHL Group   http://www.dpdhl.com/
Google   https://research.google.com/

Silver Sponsors

AGT   http://www.agtinternational.com/
ASML   https://www.workingatasml.com/
Deloitte   https://www2.deloitte.com/global/en.html
NEC Europe Ltd.   http://www.neclab.eu/
Siemens   https://www.siemens.com/

Bronze Sponsors

Cambridge University Press   http://www.cambridge.org/wm-ecommerce-web/
     academic/landingPage/KDD17

IEEE/CAA Journal   http://www.ieee-jas.org/
   of Automatica Sinica

Awards Sponsors

Machine Learning   http://link.springer.com/journal/10994
Data Mining and   http://link.springer.com/journal/10618
   Knowledge Discovery
Deloitte   http://www2.deloitte.com/

Lanyards Sponsor

KNIME   http://www.knime.org/

Publishing Partner and Sponsor

Springer   http://www.springer.com/gp/

PhD Forum Sponsor

IBM Research   http://researchweb.watson.ibm.com/

Invited Talk Sponsors

EurAi   https://www.eurai.org/
GrabIT   https://www.grabit.mk/

Contents – Part II

Transfer and Multi-task Learning

Unsupervised and Semisupervised Learning

Contents – Part III

Nectar Track

Demo Track

Contents – Part I

Ensembles and Meta Learning

Feature Selection and Extraction

Kernel Methods

Networks and Graphs

Neural Networks and Deep Learning

Pattern and Sequence Mining

BeatLex: Summarizing and Forecasting Time Series with Patterns

Bryan Hooi[1,2(✉)], Shenghua Liu[3], Asim Smailagic[1], and Christos Faloutsos[1]

[1] School of Computer Science, Carnegie Mellon University, Pittsburgh, USA
{asim,christos}@cs.cmu.edu
[2] Department of Statistics, Carnegie Mellon University, Pittsburgh, USA
bhooi@andrew.cmu.edu
[3] CAS Key Laboratory of Network Data Science and Technology,
Institute of Computing Technology, Chinese Academy of Sciences, Beijing, China
liushenghua@ict.ac.cn

Abstract. Given time-series data such as electrocardiogram (ECG) readings, or motion capture data, how can we succintly summarize the data in a way that robustly identifies patterns that appear repeatedly? How can we then use such a summary to identify anomalies such as abnormal heartbeats, and also forecast future values of the time series? Our main idea is a vocabulary-based approach, which automatically learns a set of common patterns, or 'beat patterns,' which are used as building blocks to describe the time series in an intuitive and interpretable way. Our summarization algorithm, BeatLex (Beat LEXicons for Summarization) is: (1) fast and online, requiring *linear* time in the data size and *bounded* memory; (2) effective, outperforming competing algorithms in labelling accuracy by *5.3* times, and forecasting accuracy by *1.8* times; (3) principled and parameter-free, as it is based on the Minimum Description Length principle of summarizing the data by compressing it using as few bits as possible, and automatically tunes all its parameters; (4) general: it applies to any domain of time series data, and can make use of multidimensional (i.e. coevolving) time series.

1 Introduction

Consider a medical team who wishes to monitor patients in a large hospital. How can we design an algorithm that monitors multiple time series (blood pressure, ECG, etc.) for a patient, automatically learns and summarizes common patterns, and alerts a doctor when the patient's overall state deviates from the norm?

Time series data has attracted huge interest in countless domains, including medicine [7], social media [12], and sensor data [1]. But as the scale and complexity of time series data has exploded, the human capacity to process data has not changed. This has led to a growing need for scalable algorithms which automatically summarize high-level patterns in data, or alert a user's attention toward anomalies.

M. Ceci et al. (Eds.): ECML PKDD 2017, Part II, LNAI 10535, pp. 3–19, 2017.
https://doi.org/10.1007/978-3-319-71246-8_1

Figure 1 (top) shows an example of an ECG sequence. This ECG sequence contains two distinct types of patterns: indeed, it was manually labelled by cardiologists as shown in Fig. 1 (bottom), who marked a *segmentation* at the start of each heartbeat (shown by the grey vertical lines), and labeled the heartbeats as 'normal beats' and 'premature ventricular contractions,' a type of abnormal heartbeat. A natural way to summarize this sequence, then, would be to exploit patterns that occur multiple times: namely, the two types of heartbeat patterns.

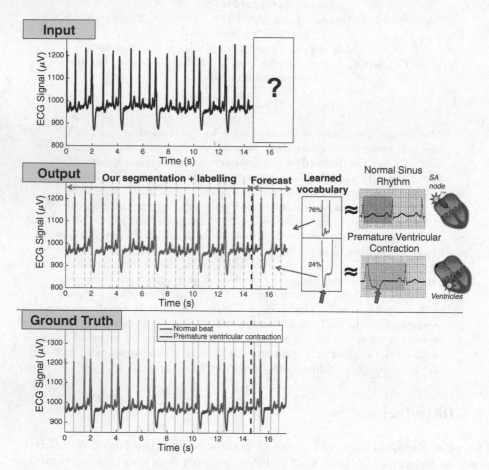

Fig. 1. Accurate segmentation, labelling, and forecasting: BEATLEX learns a vocabulary, segments the data (grey dotted vertical lines), labels heartbeat types based on the closest vocabulary term, and forecasts future data to the right of the black dotted line. Its output matches the ground truth almost exactly, and the vocabulary terms correspond closely to medically relevant patterns: 'normal sinus rhythm' and 'premature ventricular contraction.' (Color figure online)

Thus, our goal is to summarize a time series using **patterns that occur multiple times**. Here patterns refers to any subsequences which are broadly similar to one another. This includes periodic time series, but applies much more generally to allow patterns whose shape or length gets distorted, or changes over time, or even multiple patterns interspersed with one another as in Fig. 1.

Hence, the problem that we focus on is:

Informal Problem 1. *Given a time series X with patterns, find:*

- *Summarization: a model that succintly represents the common patterns in X.*
- *Anomalies: time periods in X during which anomalous events occurred (e.g. abnormal heartbeats).*
- *Forecast: forecast future time ticks of X.*

To robustly handle real-world data such as in Fig. 1, there are several key challenges: (1) patterns can be highly complex and nonlinear, so simple parametric models do not work. (2) Patterns are often distorted in length and shape, so methods that assume that a pattern straightforwardly repeats itself do not work. (3) The correct segmentation of the data is unknown: domain-specific segmentation tools for ECG sequences are not enough as we want be able to handle any type of time series (e.g. motion-capture data in Fig. 2).

To solve this problem, our algorithm adopts a **vocabulary**-based approach, as illustrated in Fig. 1 (middle). It automatically and robustly learns a vocabulary containing the common patterns in the data. At the same time, it finds cut points to break up the data into segments, and describes each segment based on its closest vocabulary term, labelling each segment accordingly. Note in Fig. 1 that both its segmentation and labelling are essentially identical to the ground truth annotation by cardiologists.

This vocabulary-based approach is intuitive and interpretable, since it describes the data exactly as a human would, in terms of its patterns and where they are located. In the ECG case, the learned vocabulary terms are also medically relevant, as shown in Fig. 1 (right): the blue and red beats correspond to known heartbeat patterns.

Figure 1 also shows that BEATLEX also allows accurate forecasting: the part of the ECG sequence to the right of the black dotted line was forecasted by the algorithm, which also matches the ground truth. Note that the algorithm learns from past data that 3 normal (blue) beats tend to be followed by an abnormal (red) beat, then cycling back to normal beats, a known condition called 'quadrigeminy.'

Our method is:

- **Fast and online:** BEATLEX takes *linear* time in the length of the time series. Moreover, it requires *bounded* memory and constant update time per new data point, making it usable even with limited processing power, such as wearable computers or distributed sensor networks.
- **Effective:** BEATLEX outperforms existing algorithms in labelling accuracy by *5.3* times, and forecasting accuracy by *1.8* times.

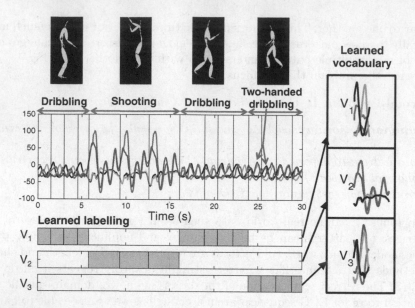

Fig. 2. Generality: BEATLEX accurately segments and labels action types in motion capture data of a basketball player. The 4 coloured lines correspond to the subject's left and right arms and legs. (Color figure online)

- **Principled and parameter-free:** BEATLEX is fit using the Minimum Description Length principle of compressing the data into as few bits as possible, and automatically tunes all its parameters.
- **General:** BEATLEX applies to any type of time series data, and can make use of multidimensional time series, e.g. motion capture data in Fig. 2.

2   Background and Related Work

Time Series Summarization. Methods for summarizing a time series can be divided into model-based and non-model-based methods. Model-based methods estimate a statistical model, and include classic methods such as autoregression (AR), ARIMA [3], and Hidden Markov Models (HMMs) [2]. More recent variants include DynaMMo [11], AutoPlait [14], and RegimeCast [13]. Non model-based methods summarize the data using approximations or feature representations, including SAX [21]. TBATS [5] is a forecasting approach allowing for complex seasonality. SAX [21] discretizes a time series into symbols, and has been used as a preprocessing step before time series clustering [9,17] and anomaly detection [8].

Distance-Based Methods. These methods extract subsequences using sliding windows, and measure distances between the subsequences, using either Euclidean distance or Dynamic Time Warping [6] (DTW). DTW is a distance-like measure that allows elastic shifting of the time axis, which has shown good empirical performance for time series classification [26]. Discord Detection methods [8,27] apply Euclidean distances or DTW between subsequences for anomaly detection, while subsequence clustering methods [9,17] find clusters of similar subsequences.

Table 1 summarizes existing work related to our problem. BEATLEX differs from existing methods as follows: (1) BEATLEX allows for patterns that change over time; (2) BEATLEX is an online algorithm; (3) BEATLEX uses a novel vocabulary-based approach; importantly, this difference allows it to robustly capture arbitrarily complex patterns to summarize the data.

Table 1. Comparison of related approaches. 'AR++' refers to AR and its extensions (ARIMA, etc.). 'Changing Patterns' refer to modelling sequences with patterns that change over time. 'Non-linear' refers to sequences with non-linear dynamics.

	AR++ [3]	HMM++ [2,11,25]	Discord [8,27]	Clustering [9,17]	AutoPlait [14]	RegimeCast [13]	BEATLEX
Segmentation		✓			✓		✔
Anomaly detection			✓	✓			✔
Forecasting	✓	✓				✓	✔
Pattern discovery		✓		✓	✓		✔
Non-linear			✓	✓		✓	✔
Changing patterns							✔
Online							✔

3   Problem Definition

Preliminaries. Table 2 includes the main definitions and symbols used in this paper.

We first introduce the Minimum Description Length (MDL) principle, which will allow us to define what a good summarization is. The MDL principle states that the best representation for some data is the one that leads to the best compression of the data. Formally, given data X, the MDL principle states that we should find the model M to minimize the **description length** of X, which is defined as $\text{Cost}(M) + \text{Cost}(X|M)$, where $\text{Cost}(M)$ is the number of bits needed to encode the model M, and $\text{Cost}(X|M)$ is the number of bits needed to encode the data given model M. The full expression for these costs depends on the type of model used, and will be given later, in Eq. (1).

Based on this cost function, we can formally define our summarization problem:

Problem 1 (Summarization). Given $(X_i)_{i=1}^m$, a real-valued time series of length m, find a model M to minimize the description length, $\text{Cost}(M) + \text{Cost}(X|M)$.

We next define our model M, and explain how it is used to compress the data.

Table 2. Symbols and definitions

Symbol	Interpretation
X	Real-valued input time series
$X_{a:b}$	Subsequence of X from index a to b (inclusive)
m	Length of time series X
V_i	ith vocabulary term
n_i	Length of V_i
k	Number of vocabulary terms
n	Number of segments
a_i	Start of ith segment
b_i	End of ith segment
X^i	ith segment, i.e. $X_{a_i:b_i}$
$z(i)$	Assignment variable for ith segment
N_i	Number of segments assigned to vocabulary term i
$C(\cdot)$	Description cost, i.e. no. of bits needed to describe a parameter
C_F	Number of bits for encoding a floating point number
MDTW	Modified DTW distance (see Definition 1)
s_{min}, s_{max}	Minimum and maximum width of a segment
k_{max}	Maximum vocabulary size
w	Width of Sakoe-Chiba band for DTW [20]

4   Model

Figure 3 illustrates our vocabulary-based model for a time series X. It consists of:

- **Vocabulary:** the 'vocabulary terms' $V_1, \ldots, V_k$ are short time series patterns which will be used to explain segments of the actual data. For example, one pattern may represent normal heartbeats, while another may represent abnormal heartbeats.
- **Segmentation:** this describes how X is split into continuous segments of time. We represent the segments using intervals $[a_1, b_1], \ldots, [a_n, b_n]$, where $a_1 = 1, b_n = m$, and $b_i + 1 = a_{i+1}$ for $i = 1, \ldots, n-1$.
- **Assignment variables:** these describe which vocabulary term is used to encode each segment. For each i, the assignment variable $z(i)$ means that the $z(i)$th vocabulary term (i.e. $V_{z(i)}$) is used to encode segment i.

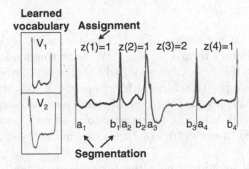

Fig. 3. Illustration of our summarization model. The data is broken into segments, and the assignment describes how each segment is described using the vocabulary.

5 Optimization Objective

In this section we explain our optimization objective, which is based on minimizing the description length of the model, and the data given the model.

5.1 Model Cost

From our model definition in Sect. 4, the parameters in the model are the vocabulary size k, the vocabulary sequences $V_1, \ldots, V_k$, the segmentation intervals $[a_1, b_1], \ldots, [a_n, b_n]$, and the assignment variables $z(1), \ldots, z(n)$. Let $C(\cdot)$ denote the number of bits required to store a parameter. As preliminaries: first, encoding an arbitrary integer requires $\log^* k$ bits.[1] Second, encoding a discrete variable taking N possible values requires $\log_2 N$ bits. In the rest of this paper, all logarithms will be base-2.

The model cost consists of:

- **Vocabulary size:** storing the positive integer k requires $\log^* k$ bits.
- **Vocabulary:** storing V_i requires $n_i \times C_F$ bits, where C_F is the number of bits needed to encode a floating point number.[2]
- **Segmentation:** for the segmentation $[a_1, b_1], \ldots, [a_n, b_n]$, it is sufficient to store $b_1, \ldots, b_{n-1}$, since these completely determine the segmentation. Each b_i takes m possible values. Hence, the total number of bits required is $(n - 1)\log(m)$.
- **Assignment variables:** there are n such variables, each taking k possible values. Hence, the number of bits required is $n \log(k)$.

In total, the number of bits needed to store the model is:

$$\text{Cost}(M) = \underbrace{\log^*(k)}_{\text{cost of } k} + \underbrace{C_F \sum_{i=1}^{k} n_i}_{\text{vocab. cost}} + \underbrace{(n-1)\log(m)}_{\text{segmentation cost}} + \underbrace{n\log(k)}_{\text{assign. cost}}$$

---

[1] Here, $\log^*$ is the universal code length for positive integers [19].
[2] We use $C_F = 8$, following [15].

5.2   Data Cost

How do we encode X given this model? Consider each segment X^i, and its corresponding vocabulary term $V = V_{z(i)}$. We want to encode X^i such that its encoding cost is low if it is similar to V, where this similarity should allow for slight distortions in shape. To achieve this, we encode X^i based on a modification of Dynamic Time Warping (DTW). Recall that given two sequences A and B, DTW aligns them while allowing for **expansions** on either sequence (e.g. if an entry of A is matched to two entries of B, we say that entry of A was 'expanded' once).

We modify DTW by adding penalties for each expansion. Similar penalized variants of DTW exist [23]; however, since we use an MDL framework, our choice of penalties has a natural interpretation as the number of bits needed to describe X^i in terms of V.

Definition 1 (Modified DTW).   *Given two sequences A and B, $MDTW(A, B)$ modifies regular DTW by adding a penalty of $\log n_A$ for each expansion to A and $\log n_B$ for each expansion to B, where n_A and n_B are the lengths of A and B.*

The number of bits needed to describe X^i in terms of V is given by the MDTW cost: i.e. $C(X^i|V) = \text{MDTW}(X^i, V)$. To see this, note that the penalty, $\log n_A$ for expansions to A, exactly describes the number of bits needed to encode an expansion to A, since each expansion is of one of n_A possible entries. Thus the penalties capture the number of bits needed to encode expansions. In addition, we need to encode the remaining errors after warping; assuming these errors are independently Gaussian distributed, based on Huffman coding, their encoding cost in bits is their negative log likelihood under the Gaussian model [4], which is the standard squared-error DTW cost.

5.3   Final Cost Function

Combining our discussion so far, the description length cost under model M is:

$$f(M) = \text{Cost}(M) + \text{Cost}(X|M)$$

$$= \underbrace{\log^*(k)}_{\text{cost of } k} + \underbrace{C_F \sum_{i=1}^{k} n_i}_{\text{vocab. cost}} + \underbrace{(n-1)\log(m)}_{\text{segmentation cost}} + \underbrace{n \log(k)}_{\text{assign. cost}} + \underbrace{\sum_{i=1}^{n} \text{MDTW}(X^i, V_{z(i)})}_{\text{data cost}}$$

$$(1)$$

6   BEATLEX Summarization Algorithm

6.1   Overview

In this section, we describe our algorithm for learning vocabulary terms, a segmentation, and an assignment, to minimize description length. From a high level, we first generate the initial vocabulary term (*New Vocabulary Term Generation*). Then, starting at the beginning of the time series, our algorithm repeatedly finds the closest match between any existing vocabulary term and a prefix of the time series (*Best Vocabulary-Prefix Match*). If a sufficiently good match is found, it assigns this segment to this vocabulary term (*Vocabulary Merge*). Otherwise, it creates a new vocabulary term (*New Vocabulary Term Generation*).

Algorithm 1. BEATLEX *summarization algorithm*

    Input : Time series X
    Output: Vocabulary $V_1, \ldots, V_k$, assignments z, segmentation S.
1 $k = 0$;
2 $i = 1$; // current position
3 **while** $i < m$ **do**
4   ▷Find best vocabulary-prefix match:
5      $j^*, s^* = \underset{j,s}{\arg\min} \; \frac{C(X_{i+1:i+s}|V_j)}{s}$;   ▷see Section 6.2
6   ▷If using existing vocab. has lower cost than creating new vocab. term:
7   **if** $C(X_{i+1:i+s^*}|V_{j^*}) < C_F \cdot s^*$ *or* $k = k_{max}$ **then**
8   ▷Use existing vocabulary term:
9          $V_{j^*} = \text{VOCABMERGE}(X_{i+1:i+s^*}, V_{j^*}))$;   ▷see Section 6.3
10   Append j^* to z;
11   **else**
12   ▷Create new vocabulary term:
13          $s^* = \text{NEWVOCABLENGTH}(X, i, (V_1, \ldots, V_k))$;   ▷see Section 6.4
14          $V_{k+1} = X_{i+1:i+s^*}$;
15   Append $k + 1$ to z;
16          $k = k + 1$;
17   Append $[i + 1 : i + s^*]$ to S;
18      $i = i + s^*$;

6.2 Best Vocabulary-Prefix Match

Assume that the algorithm has processed the time series up to time i so far, and the current vocabulary it has learned is $V_1, \ldots, V_k$. The next key subroutine the algorithm uses is to find the best match between any vocabulary term and any **prefix** of the current time series, i.e. $X_{i+1:i+s}$, for some s with $s_{min} \leq s \leq s_{max}$, where s_{min} and s_{max} are lower and upper bounds of the allowed segment length. We choose the best vocabulary term index j^* and the best prefix length s^* by minimizing **average encoding cost**:

$$j^*, s^* = \underset{j,s}{\arg\min} \; \frac{C(X_{i+1:i+s}|V_j)}{s} = \underset{j,s}{\arg\min} \; \frac{MDTW(X_{i+1:i+s}, V_j)}{s} \qquad (2)$$

Dividing by s allows us to compare fairly between different prefix lengths, by finding the prefix that can be encoded most efficiently per unit length.

How do we efficiently perform the minimization in (2)? Consider a single vocabulary term; we have to compare it against $s_{max} - s_{min} + 1$ choices of prefixes, and running a separate MDTW computation for each would be very slow. It turns out that we can minimize across all these subsequences optimally in a single MDTW computation. Across k vocabulary terms, this gives a total of k MDTW computations:

Theorem 1. *The best j^*, s^* minimizing (2) can be found in k MDTW computations, and requires $O(k \cdot w \cdot s_{max})$ time.*

Proof. For each j, consider computing the MDTW on V_j and $X_{i+1:i+s_{max}}$. Each entry of the DTW matrix computed by the dynamic programming DTW algorithm encodes the minimum DTW cost between each prefix of V_j and each prefix of $X_{i+1:i+s_{max}}$. Hence, the last $s_{max} - s_{min} + 1$ entries of the last row of the DTW matrix contains the minimum MDTW distance between V_j and each of the prefixes $X_{i+1:i+s}$ for $s_{min} \leq s \leq s_{max}$. Our algorithm can extract these values, divide them by their prefix lengths s, and choose the one minimizing average encoding cost. This requires k MDTW computations, one for each vocabulary term; each MDTW computation requires $O(w \cdot s_{max})$ time, using a Sakoe-Chiba band of width w [20]. ∎

6.3 Vocabulary Merge (VOCABMERGE)

Now assume we have computed the best vocabulary-prefix match. If the resulting average encoding cost is less than C_F, then since generating a new vocabulary term would require C_F bits per unit length, the most efficient choice is to encode $X_{i+1:i+s^*}$ using the existing vocabulary V_{j^*}, so our algorithm makes this choice (Line 7). Otherwise, it creates a new vocabulary term (see Sect. 6.4).

After encoding $X_{i+1:i+s^*}$ using V_{j^*}, we would like to update V_{j^*} to make it an 'average' of the subsequences it encodes, to allow our algorithm to keep track of patterns that change over time. We use a *running average* approach, in which we keep track of how many subsequences have been assigned to V_{j^*} so far (call this number N_{j^*}). Intuitively, we should replace V_{j^*} by a weighted average of $X_{i+1:i+s^*}$ and itself, with weight of $\frac{1}{1+N_{j^*}}$ given to $X_{i+1:i+s^*}$, and $\frac{N_{j^*}}{1+N_{j^*}}$ given to the current value of V_{j^*}; this makes sense since the current value of V_{j^*} is an average of the N_{j^*} subsequences currently assigned to it. However, rather than using a straightforward average, we first run the MDTW algorithm in order to align $X_{i+1:i+s^*}$ to V_{j^*} before averaging it with V_{j^*}. Straightforward averaging would adversely affect the sequence shape, e.g. if both sequences have sharp peaks in slightly different locations, the average would have two peaks. Aligning the sequences first avoids this problem.

6.4 New Vocabulary Term Generation (NEWVOCABLENGTH)

In this case our algorithm chooses to generate a new vocabulary term V_{k+1}. To select the ideal length n_{k+1}, we iterate over all possible n_{k+1}. For each, we then use *Best Vocabulary-Prefix Match* to compute the average encoding cost of the *next* prefix after $X_{i+1:i+n_{k+1}}$, matched to any of the vocabulary terms (including V_{k+1}). The final length n_{k+1} chosen is the one that results in the lowest such average encoding cost.

6.5 Computation Time and Memory

BeatLex is linear in the length m of the time series:

Lemma 1. BeatLex *runs in* $O(m \cdot k \cdot w \cdot s_{max}/s_{min} + k^2 \cdot w \cdot s_{max}^2)$ *time.*

Proof. As shown in Theorem 1, each vocabulary-prefix match step takes $O(k \cdot w \cdot s_{max})$ time. We perform this step up to once per segment (not including vocabulary term generation), so at most once every s_{min} steps, taking $O(m \cdot k \cdot w \cdot s_{max}/s_{min})$ in total. The vocabulary term generation step runs k times, each trying at most s_{max} candidates taking $O(k \cdot w \cdot s_{max})$ each time, for a total of $O(k^2 \cdot w \cdot s_{max}^2)$. ∎

Note that typically m greatly exceeds the other variables, the $O(k^2 \cdot w \cdot s_{max}^2)$ term is typically negligible. Figure 5c verifies the linear scalability of our algorithm in practice.

Lemma 2. BeatLex *does not require the past history of the time series, and hence requires only bounded memory.*

Proof. We verify from Algorithm 1 that BeatLex never needs to access the past history of the time series, and operates on incoming time series values while only keeping track of its vocabulary, storing at most its k_{max} vocabulary terms at any time, which requires bounded memory. ∎

6.6 Extensions

Multidimensional Time Series: Figure 6 shows our algorithm applied on a 2-lead (i.e. 2-dimensional) ECG. To handle multidimensional time series, only a small change is needed: now, our vocabulary terms are likewise multidimensional. When encoding a subsequence using a vocabulary term, we use the dth dimension of the vocabulary term to encode the dth dimension of the subsequence. The description length of encoding a subsequence is then the sum of description lengths of encoding each individual dimension separately. The rest of our BeatLex algorithm applies without change.

Anomaly Detection: Figure 1 reports the vocabulary terms with their frequencies in the data; PVCs are the rarer pattern. The most straightforward kind of anomaly are patterns that occur exactly once; we can easily detect these by selecting vocabulary terms that are only used to encode a single segment. More generally, we may also wish to detect *rare* patterns such as the PVCs in Fig. 1. We can easily detect this kind of anomaly (as well as the former type) by ranking the vocabulary terms by how many segments they encode, and returning those encode the fewest.

Forecasting: How do we forecast future values of the time series, as done in Fig. 1? Based on our learned segment labels, we learn Markov Models of orders $0, 1, \ldots, r_{max}$ using the usual maximum likelihood approach [2]. Here the 0th order model simply ignores the past and forecasts the most frequent label. To forecast the next segment label, we first try to use the highest (r_{max}) order model, but repeatedly drop to the next lower order if the sequence of the last r labels (where r is the current order) has not been seen in the past. This process continues to forecast as many segment labels as we need. Our forecasts for the actual data are the associated vocabulary terms.

Automatic Parameter Setting: Our MDL objective (5) provides an easy way to choose parameters automatically, via grid search minimizing the description length cost. To keep the algorithm fast, we only do this once per type of data (e.g. on a single ECG sequence), then fix those values.

7   Experiments

We design experiments to answer the following questions:

- **Q1. Labelling Accuracy:** how accurate are the labels returned by BEAT-LEX?
- **Q2. Forecasting Accuracy:** how accurately does it forecast future data?
- **Q3. Scalability:** how does the algorithm scale with the data size?
- **Q4. Interpretability:** are the results of the algorithm interpretable by the user?

Our code, links to datasets, and experiments are publicly available at www.andrew.cmu.edu/user/bhooi/beatlex. Experiments were done on a 2.4 GHz Intel Core i5 Macbook Pro, 16 GB RAM running OS X 10.11.2.

7.1   Data

We evaluate our algorithm on ECG sequences from the MIT-DB Arrythmia Database[3] (MITDB) [16], as well as motion capture data from the CMU motion capture database[4].

MITDB Dataset. The MITDB dataset contains 48 half-hour ECG recordings from test subjects from Beth Israel Hospital. Each recording consists of two ECG sequences, or 'leads,' at 360 time ticks per second, for a total of 650,000 ticks. Ground-truth annotations by two or more independent cardiologists indicate the position and type of each heartbeat; disagreements were resolved to obtain the annotations.

---

[3] https://physionet.org/cgi-bin/atm/ATM?database=mitdb.
[4] http://mocap.cs.cmu.edu/.

CMU Motion Capture Dataset. The dataset consists of motion-captured subjects performing various actions. Each recording is a 64 dimensional vector (representing the subject's body parts), of which we chose 4 dimensions (left-right arms and legs). The recording lasts 50 s with 120 ticks per second, for a total of 6000 time ticks.

7.2 Q1: Labelling Accuracy

Recall that BEATLEX labels each time tick based on which vocabulary term it was assigned to. We evaluate this by comparing it to the ground truth labelling using standard clustering metrics: Adjusted Rand Index [18] and Normalized Mutual Information [22]. Due to the fairly large number of ECG sequences and trials, we subset time series to 5000 time ticks, and do the same for our forecasting tests in the next subsection.

Baselines. The baselines are Hidden Markov Models (HMMs) [10] and Autoplait [14], which identifies 'regimes' in time series data using a hierarchical HMM-based model.

Figure 4 shows the labelling accuracy of BEATLEX, HMMs and Autoplait, averaged across all 48 ECGs. Under both metrics, BEATLEX clearly outperforms both baselines. The key difference is that HMM-based methods (including Autoplait) have difficulty accurately representing the complex and nonlinear ECG patterns based on a model only using state transitions. In contrast, BEATLEX treats entire vocabulary terms as 'building blocks,' and hence has no particular difficulty handling complex patterns.

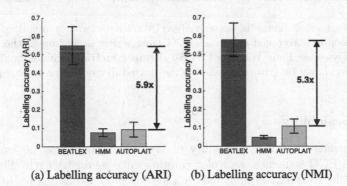

(a) Labelling accuracy (ARI) (b) Labelling accuracy (NMI)

Fig. 4. Labelling accuracy (higher is better): BEATLEX outperforms baselines, according to the Adjusted Rand Index (ARI) and Normalized Mutual Information (NMI) metrics, averaged across ECG sequences. Error bars indicate one standard deviation.

7.3 Q2: Forecasting Accuracy

We now evaluate BEATLEX in terms of forecasting future data. For each ECG sequence, the last 1000 time ticks (approximately 3 s) are hidden from the algorithm, and the algorithm forecasts this data. We compare the forecasts to the true values using Root Mean Squared Error (RMSE), as well as Dynamic Time Warping (DTW) distance.

Baselines. As baselines, we use the classical ARIMA [3] algorithm, as well as the more recent TBATS [5] forecasting algorithm. We select the ARIMA order using AIC.

Figure 5 shows the forecasting error of BEATLEX compared to the baselines (lower is better). BEATLEX outperforms the baselines according to both metrics, but particularly under DTW distance. This difference between metrics is not surprising: RMSE is extremely sensitive to small temporal variations, e.g. of the spikes present in ECG data.

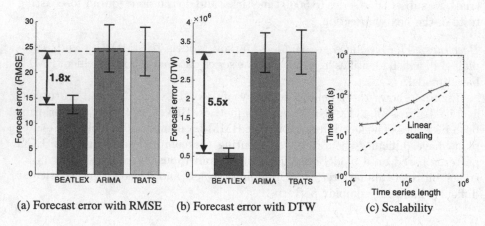

(a) Forecast error with RMSE   (b) Forecast error with DTW   (c) Scalability

Fig. 5. Forecast error (lower is better): BEATLEX outperforms competing baselines in forecast accuracy averaged across all ECG time series, according to the standard RMSE and Dynamic Time Warping (DTW) distance metrics. Scalability: BEATLEX scales linearly, shown by growth parallel to the dotted diagonal on a log-log plot.

7.4   Q3: Scalability

Figure 5c verifies the linear scalability of BEATLEX. We vary the length of a subset of an ECG sequence, and plot running time against length. The plot is around parallel to the main diagonal on a log-log plot, indicating linear growth. BEATLEX scales to large datasets, running on data of length 512, 000 in less than 200 s.

7.5   Q4: Discoveries and Interpratability

In this section, we show that BEATLEX automatically discovers interpretable, medically relevant patterns. In Fig. 1, the learned vocabulary terms can be easily and accurately matched with medically recognized patterns, 'normal sinus rhythm' and 'premature ventricular contraction' (PVC) respectively. Going beyond individual heartbeats, our algorithm also successfully learns a pattern present in Fig. 1 known as 'quadrigeminy' in which PVC beats show up approximately every 4 beats. Medically, these patterns occur because the heart repolarizes after a PVC, during which normal beats occur [24].

Multidimensionality: Section 6.6 explains that our algorithm handles multidimensional time series by learning multidimensional vocabulary terms. Figure 6 illustrates this. The two vocabulary terms still accurately correspond to normal sinus rhythm and PVCs respectively, except that each now has a multidimensional vocabulary term.

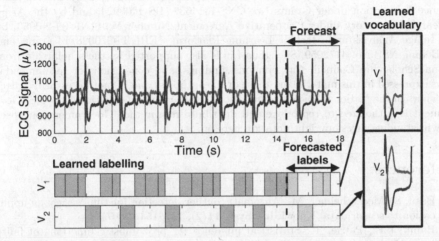

Fig. 6. Accurate multidimensional summarization and forecasting: an ECG sequence, with its two dimensions in red and blue. BEATLEX accurately segments the data, learns the two types of heartbeats, and forecasts future data. (Color figure online)

Motion Capture Dataset: Our algorithm is generalizable to other time series datasets. Figure 2 shows its results on motion capture data of a basketball player. Despite significant variation in the width and shape of a pattern, our algorithm still successfully learns, segments and labels vocabulary terms corresponding to dribbling, shooting and two-handed dribbling.

8   Conclusion

We presented the BEATLEX algorithm, a novel vocabulary-based approach designed to handle time series with patterns. It robustly learns interpretable vocabulary terms, in the face of possibly nonlinear patterns, distortions, and an unknown segmentation. BEATLEX is:

– **Fast and online:** BEATLEX takes *linear* time, *bounded* memory and constant update time per data point.

- **Effective:** BEATLEX outperforms existing algorithms in labelling accuracy by *5.3* times, and forecasting accuracy by *1.8* times.
- **Principled and parameter-free:** BEATLEX is fit using the Minimum Description Length (MDL) principle, and automatically tunes its parameters.
- **General:** BEATLEX applies to any domain of time series data, and multidimensional time series.   .

Reproducibility: Our code, links to datasets, and experiments are publicly available at www.andrew.cmu.edu/user/bhooi/beatlex.

Acknowledgments. This material is based upon work supported by the National Science Foundation under Grants No. CNS-1314632, IIS-1408924, and by the Army Research Laboratory under Cooperative Agreement Number W911NF-09-2-0053, by the Image Analysis and Machine Learning Platform - ERI/TIC/0028/14 grant, and by Beijing NSF No. 4172059. Shenghua Liu is also supported by the scholarship from China Scholarship Council. Any opinions, findings, and conclusions or recommendations expressed in this material are those of the author(s) and do not necessarily reflect the views of the National Science Foundation, or other funding parties. The U.S. Government is authorized to reproduce and distribute reprints for Government purposes notwithstanding any copyright notation here on.

References

1. Basu, S., Meckesheimer, M.: Automatic outlier detection for time series: an application to sensor data. Knowl. Inf. Syst. **11**(2), 137–154 (2007)
2. Baum, L.E., Petrie, T.: Statistical inference for probabilistic functions of finite state Markov chains. Ann. Math. Stat. **37**(6), 1554–1563 (1966)
3. Box, G.E., Jenkins, G.M.: Time Series Analysis: Forecasting and Control, revised edn. Holden-Day Inc., San Francisco (1976)
4. Cover, T.M., Thomas, J.A.: Elements of Information Theory. Wiley, Hoboken (2012)
5. De Livera, A.M., Hyndman, R.J., Snyder, R.D.: Forecasting time series with complex seasonal patterns using exponential smoothing. J. Am. Stat. Assoc. **106**(496), 1513–1527 (2011)
6. Keogh, E.: Exact indexing of dynamic time warping. In: Proceedings of the 28th International Conference on Very Large Data Bases, pp. 406–417. VLDB Endowment (2002)
7. Keogh, E., Chu, S., Hart, D., Pazzani, M.: An online algorithm for segmenting time series. In: Proceedings IEEE International Conference on Data Mining, ICDM 2001, pp. 289–296. IEEE (2001)
8. Keogh, E., Lin, J., Fu, A.: Hot sax: efficiently finding the most unusual time series subsequence. In: Fifth IEEE International Conference on Data Mining, pp. 8–pp. IEEE (2005)
9. Keogh, E., Lin, J., Truppel, W.: Clustering of time series subsequences is meaningless: implications for previous and future research. In: Third IEEE International Conference on Data Mining, ICDM 2003, pp. 115–122. IEEE (2003)

10. Letchner, J., Re, C., Balazinska, M., Philipose, M.: Access methods for markovian streams. In: IEEE 25th International Conference on Data Engineering, ICDE 2009, pp. 246–257. IEEE (2009)
11. Li, L., McCann, J., Pollard, N.S., Faloutsos, C.: Dynammo: mining and summarization of coevolving sequences with missing values. In: Proceedings of the 15th ACM SIGKDD International Conference on Knowledge Discovery and Data Mining, pp. 507–516. ACM (2009)
12. Mathioudakis, M., Koudas, N., Marbach, P.: Early online identification of attention gathering items in social media. In: Proceedings of the Third ACM International Conference on Web Search and Data Mining, pp. 301–310. ACM (2010)
13. Matsubara, Y., Sakurai, Y.: Regime shifts in streams: real-time forecasting of co-evolving time sequences. In: Proceedings of the 22nd ACM SIGKDD International Conference on Knowledge Discovery and Data Mining, pp. 1045–1054. ACM (2016)
14. Matsubara, Y., Sakurai, Y., Faloutsos, C.: Autoplait: automatic mining of co-evolving time sequences. In: Proceedings of the 2014 ACM SIGMOD International Conference on Management of Data, pp. 193–204. ACM (2014)
15. Matsubara, Y., Sakurai, Y., Faloutsos, C.: The web as a jungle: non-linear dynamical systems for co-evolving online activities. In: Proceedings of the 24th International Conference on World Wide Web, pp. 721–731. ACM (2015)
16. Moody, G.B., Mark, R.G.: The impact of the MIT-BIH arrhythmia database. IEEE Eng. Med. Biol. Mag. **20**(3), 45–50 (2001)
17. Rakthanmanon, T., Keogh, E.J., Lonardi, S., Evans, S.: MDL-based time series clustering. Knowl. Inf. Syst. **33**(2), 371–399 (2012)
18. Rand, W.M.: Objective criteria for the evaluation of clustering methods. J. Am. Stat. Assoc. **66**(336), 846–850 (1971)
19. Rissanen, J.: A universal prior for integers and estimation by minimum description length. Ann. Stat. **11**(2), 416–431 (1983)
20. Sakoe, H., Chiba, S.: Dynamic programming algorithm optimization for spoken word recognition. IEEE Trans. Acoust. Speech Sig. Process. **26**(1), 43–49 (1978)
21. Shieh, J., Keogh, E.: iSAX: indexing and mining terabyte sized time series. In: Proceedings of the 14th ACM SIGKDD International Conference on Knowledge Discovery and Data Mining, pp. 623–631. ACM (2008)
22. Strehl, A., Ghosh, J.: Cluster ensembles—a knowledge reuse framework for combining multiple partitions. J. Mach. Learn. Res. **3**(Dec), 583–617 (2002)
23. Sun, H., Lui, J.C., Yau, D.K.: Distributed mechanism in detecting and defending against the low-rate TCP attack. Comput. Netw. **50**(13), 2312–2330 (2006)
24. Wagner, G.S.: Marriott's Practical Electrocardiography. Lippincott Williams & Wilkins, Philadelphia (2001)
25. Wang, P., Wang, H., Wang, W.: Finding semantics in time series. In: Proceedings of the 2011 ACM SIGMOD International Conference on Management of Data, pp. 385–396. ACM (2011)
26. Wang, X., Mueen, A., Ding, H., Trajcevski, G., Scheuermann, P., Keogh, E.: Experimental comparison of representation methods and distance measures for time series data. Data Min. Knowl. Discov. **26**(2), 1–35 (2013)
27. Yankov, D., Keogh, E., Rebbapragada, U.: Disk aware discord discovery: finding unusual time series in terabyte sized datasets. In: Seventh IEEE International Conference on Data Mining, ICDM 2007, pp. 381–390. IEEE (2007)

Behavioral Constraint Template-Based Sequence Classification

Johannes De Smedt[1]([⊠]), Galina Deeva[2], and Jochen De Weerdt[2]

[1] Management Science and Business Economics Group, Business School, University of Edinburgh, Edinburgh, UK
`johannes.desmedt@ed.ac.uk`
[2] Department of Decision Sciences and Information Management, Faculty of Economics and Business, KU Leuven, Leuven, Belgium

Abstract. In this paper we present the interesting Behavioral Constraint Miner (iBCM), a new approach towards classifying sequences. The prevalence of sequential data, i.e., a collection of ordered items such as text, website navigation patterns, traffic management, and so on, has incited a surge in research interest towards sequence classification. Existing approaches mainly focus on retrieving sequences of itemsets and checking their presence in labeled data streams to obtain a classifier. The proposed iBCM approach, rather than focusing on plain sequences, is template-based and draws its inspiration from behavioral patterns used for software verification. These patterns have a broad range of characteristics and go beyond the typical sequence mining representation, allowing for a more precise and concise way of capturing sequential information in a database. Furthermore, it is possible to also mine for negative information, i.e., sequences that do not occur. The technique is benchmarked against other state-of-the-art approaches and exhibits a strong potential towards sequence classification. Code related to this chapter is available at: http://feb.kuleuven.be/public/u0092789/.

Keywords: Sequence mining · Sequence classification Constraint-based mining

1 Introduction

Analyzing sequential data [1] has seen a vast surge in interest during recent years, driven by the growth of typical sources such as DNA databases, text repositories, road analysis [2] and user behavior analysis [3]. Many techniques exist to derive ordered items from temporal databases, focusing on either different techniques for discovery, e.g., using prefix-oriented and constraint-based approaches, or towards different outcomes, e.g., regular expressions or closed sequences. These sequential features can be used for classifying new database entries, a discipline that does not only focus on constructing the most complete set of features, but rather the most discriminating.

In this paper, we propose a new sequence classification technique, called iBCM (interesting Behavioural Constraint Miner), which featurizes sequences

© Springer International Publishing AG 2017
M. Ceci et al. (Eds.): ECML PKDD 2017, Part II, LNAI 10535, pp. 20–36, 2017.
https://doi.org/10.1007/978-3-319-71246-8_2

according to a predefined set of behavioral constraint templates. As such, a fine-granular view of the temporal relations between items can be achieved and applied towards classification. Furthermore, iBCM allows for easy identification of the differences between classes, and gives insight into what types of relations are typically relevant for classification. In the experimental evaluation, it is shown that iBCM is capable of obtaining high discriminative power while minimizing the number of features needed. In addition, only deriving a certain type of constraint templates can already capture the most discriminating features.

This paper is structured as follows. In Sect. 2, an overview of the state-of-the-art of both sequence mining and classification is discussed. In Sect. 3, the backdrop for mining behavioral sequence patterns is introduced, which leads into the discussion of the inference part of iBCM in Sect. 4. Next, Sect. 5 reports on a benchmark with other state-of-the-art techniques. Finally, Sect. 6 summarizes the contributions and provides suggestions for future work.

2   State-of-the-Art

In this section, an overview of existing sequence mining and classification techniques is discussed.

2.1   Sequence Mining

Sequence mining, also referred to as frequent ordered itemset mining or temporal data mining, has been tackled in numerous ways. The original approach was rooted in frequent itemset discovery [4] and based on apriori-concepts. Extensions to this original approach have been proposed to obtain closed sequences [5] and to achieve performance benefits through prefix representation of the dataset [6]. A constraint-based approach was proposed in [7] in the form of cSPADE, and has recently seen a strong interest towards extending it along the declarative constraint programming paradigm. More specifically, several studies investigate how to generically build a knowledge base of constraints covering the sequences in a temporal dataset. For example, in [8], a satisfiability-based technique is devised for enumerating all frequent sequences using cardinalities for the constraints retrieved. In [9] a better prefix representation for sequences mining constraints was introduced, which was later extended for GAP constraints [10]. A similar approach was devised in [11], in which the authors propose an approach that speeds up the retrieval of constraints by precomputing the relations between items in a dataset to avoid reiterating over the sequences. These approaches can also be used to quickly retrieve regular expressions. In [12], a general constraint programming approach that steers away from explicit wildcards is introduced. Finally, a similar vein of research was pursued with Warmer [13], an inductive logic programming pattern discovery algorithm that relies on the Datalog formalization for expressing multi-dimensional patterns. It was elaborated further for sequences in [14]. The proposed work is a special purpose algorithm that mines for a subset of Datalog patterns.

2.2   Sequence Classification

While many insights from sequence mining carry over into sequence classification, the nature of the objective is different. Rather than eliciting the full set of sequences or constraints supported, it is paramount that the feature set exhibits the following characteristics.

- **Compact:** in order to build classifiers in reasonable time, the set of features should be reduced to a minimum,
- **Interesting:** features of sequential patterns should be supported in a database, but their usefulness towards classification, i.e., their discriminative power, also depends on other factors such as confidence and interestingness [15]. In general, there is a need for a balance in the feature set that strikes support values in between extremely high and low values [16],
- **Concise:** the feature set is small though comprehensive, and explains the sequential behavior in an understandable way.

Many sequence classification techniques have been proposed [17–19], each focusing on a different approach ranging from extensions to sequence pattern mining algorithms, to statistical approaches that infer the explanatory power of subsequences. They can be classified as either being direct, i.e., the features are extracted according to their strength towards the classifier, or indirect, i.e., all features are generated and later selected by a classifier. [17] extends the cSPADE algorithm with an interestingness measure that is based on both the support and the window (cohesion) in which the items of the constraint occur. In [18] BIDE-D(C) is introduced which rather incorporates information gain into BIDE to provide a direct sequence classification approach. In [19], the sequence database is split up in smaller parts to be recreated by a sparse knowledge base that punishes for infrequent behavior by constructing a Bayesian network of posteriors that are able to reconstruct the sequence database. A similar approach is used in [20], where a strong emphasis is used towards finding interesting sequences.

In contrast to the previously mentioned techniques, iBCM draws from insights in constraint programming, but rather than constructing a complete constraint base that is able to elicit the sequence database as a whole, highly diverse and informative behavioral patterns are used that incorporate cardinality, alteration, gaps, as well as negative information. By fixing the pattern base, it becomes easy to write a specific and fast algorithm for retrieving them from large databases. The technique employs only binary constraints, however, other studies such as [17] have already revealed that for sequence classification, the length of the patterns does not have to exceed 3, or even 2.

3   The Framework of Behavioral Templates

In this section, the preliminaries are established and an overview of the behavioral constraint templates/patterns and their characteristics is given.

3.1   Sequences and Sequence Databases

The task of sequence classification relies on the principles of both a sequence and a sequence database, as well as the classes or labels needed to discern their behavior.

Definition 1. *A sequence $\sigma = \langle \sigma_1, \sigma_2, ..., \sigma_n \rangle$ is a list of items with length $|\sigma| = n$ out of the alphabet Σ_σ. We denote:*

- *$occ(a, \sigma) = \{i \mid \sigma_i = a, i \in \mathbb{N}\}$ the ordered set of positions of $a \in \Sigma_\sigma$ in σ,*
- *$min(occ(a, \sigma))$ the first occurrence,*
- *$max(occ(a, \sigma))$ the last occurrence, and*
- *$|occ(a, \sigma)|$ the number of occurrences.*

Sequences, or ordered sets of items, are typically bundled in sequence databases, which can be defined as follows.

Definition 2. *A sequence database $\mathcal{SB}$ is a set of sequences with $L : \mathcal{SB} \to \mathbb{N}$ a labeling function assigning a class label to a sequence consisting of the items in $\Sigma_{\mathcal{SB}}$. The number of sequences in the database is $|\mathcal{SB}|$.*

Consider the example sequence database in Table 1, with $\Sigma_{\mathcal{SB}} = \{a, b, c\}$, $|\mathcal{SB}| = 6$, and $|img(L)| = 2$.

Table 1. Example database.

ID	Sequence	Label	ID	Sequence	Label
1	abbcaa	1	4	acbbcaacc	2
2	abbccaa	1	5	acbbcaa	2
3	abbaac	1	6	acbbcaa	2

3.2   Declare Pattern Base

The iBCM approach relies on a set of behavioral constraint templates based on the Declare language [21], which itself is inspired by the formal verification patterns of Dwyer [22]. These are widely used for identifying not only sequential, but overall behavioral characteristics of programs and processes. The Declare template base consists of a number of patterns for modeling flexible business processes, which are typically expressed in linear temporal logic (LTL), or regular expressions and finite state machines (FSMs). The template base is extensible, but the most widely-used entries are listed in Table 2. The patterns contain both unary and binary constraints. The unary constraints focus either on the position (first/last), or the cardinality. The choice constraint can be considered an *existence* constraint over multiple items. The binary constraints exhibit a hierarchy [25]. There are unordered constraints (*responded/co-existence*), simple ordered (*precedence, response, succession*), alternating ordered, and chain

Table 2. An overview of Declare constraint templates with their corresponding LTL formula and regular expression.

Template	LTL formula [23]	Regular expression [24]
Existence(A,n)	$\Diamond(A \wedge \bigcirc(existence(n-1, A)))$	.*(A.*){n}
Absence(A,n)	$\neg existence(n, A)$	[^A]*(A?[^A]*){n-1}
Exactly(A,n)	$existence(n, A) \wedge$ $absence(n+1, A)$	[^A]*(A[^A]*){n}
Init(A)	A	(A.*)?
Last(A)	$\Box(A \implies \neg X \neg A)$	.*A
Responded existence(A,B)	$\Diamond A \implies \Diamond B$	[^A]*((A.*B.*) |(B.*A.*))?
Co-existence(A,B)	$\Diamond A \impliedby \Diamond B$	[^AB]*((A.*B.*) |(B.*A.*))?
Response(A,B)	$\Box(A \implies \Diamond B)$	[^A]*(A.*B)*[^A]*
Precedence(A,B)	$(\neg B\, U\, A) \vee \Box(\neg B)$	[^B]*(A.*B)*[^B]*
Succession(A,B)	$response(A, B) \wedge$ $precedence(A, B)$	[^AB]*(A.*B)*[^AB]*
Alternate response(A,B)	$\Box(A \implies \bigcirc(\neg A\, U\, B))$	[^A]*(A[^A]*B[^A]*)*
Alternate precedence(A,B)	$precedence(A, B) \wedge$ $\Box(B \implies$ $\bigcirc(precedence(A, B))$	[^B]*(A[^B]*B[^B]*)*
Alternate succession(A,B)	$altresponse(A, B) \wedge$ $precedence(A, B)$	[^AB]*(A[^AB]*B[^AB]*)*
Chain response(A,B)	$\Box(A \implies \bigcirc B)$	[^A]*(AB[^A]*)*
Chain precedence(A,B)	$\Box(\bigcirc B \implies A)$	[^B]*(AB[^B]*)*
Chain succession(A,B)	$\Box(A \iff \bigcirc B)$	[^AB]*(AB[^AB]*)*
Not co-existence(A,B)	$\neg(\Diamond A \wedge \Diamond B)$	[^AB]*((A[^B]*) |(B[^A]*))?
Not succession(A,B)	$\Box(A \implies \neg(\Diamond B))$	[^A]*(A[^B]*)*
Not chain succession(A,B)	$\Box(A \implies \neg(\bigcirc B))$	[^A]*(A+[^AB][^A]*)*A*
Choice(A,B)	$\Diamond A \vee \Diamond B$	.*[AB].*
Exclusive choice(A,B)	$(\Diamond A \vee \Diamond B) \wedge \neg(\Diamond A \wedge \Diamond B)$	([^B]*A[^B]*) |.*[AB].*([^A]*B[^A]*)

ordered constraints. Hence, the opportunity exists to express not only the ordering, but also the repeating (alternation) and local (chain) behavior of two items. Furthermore, there are negative constraints, expressing behavior that does not occur. These can prove especially useful in the context of classification, and are typically not generated by sequence classification techniques that only mine for positive patterns.

Definition 3. *A sequence constraint* $\pi = (A, t)$ *is a tuple with* A *a set of items and* t *the type of constraint.*

A binary constraint has an antecedent, implying the constraint, and a consequent. Both can exist out of a set of items, however, in the rest of the paper we will assume both to be singletons. The type of the constraints correspond with the templates that are defined in Table 2. For convenience, the constraints are written in an abbreviated fashion, e.g., $altPrec(a, b)$. They all correspond with a certain regular expression which can be converted into an FSM. We denote the corresponding regular expression as $\S(t)$. We write the FSM $\mathcal{A}$ corresponding with the regular expression as $\mathcal{A} = \S(A, t)$ or $\mathcal{A} = \S(altPrec(a, b))$. An example of $altPrec(a, b)$ is depicted in Fig. 1.

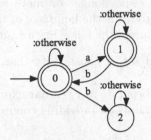

Fig. 1. Automaton of *alternate precedence(a,b)*.

Definition 4. *A sequence σ supports a constraint π iff $\sigma \in \mathcal{L}(\mathcal{A}(\pi))$ where $\mathcal{L}$ denotes the language of the corresponding FSM. The support of the constraint in the database is $sup(\pi)_{SB} = |\{\sigma | \sigma \in \mathcal{L}(\mathcal{A}(\pi)), \forall \sigma \in SB\}|$.*

E.g., in $SB = \{aab, abb\}$, $\sigma_1 \in \mathcal{L}(\S(altPrec(a, b)))$, $\sigma_2 \notin \mathcal{L}(\S(altPrec(a, b)))$, and $sup(\pi)_{SB} = 1$.

3.3   Comparison with Other Sequence Constraint Representation

The iBCM approach does not intend to be able to reproduce the database. Rather it is able to capture the most discerning sequence-based features. Consider for example the database in Table 1. Table 3 lists the constraints that are present for both labels. Notice that for label 1, a does not always precede b. Also, for label 2 c occurs before b. This can be discerned by only 3 constraints which are marked in bold. Hence, with only 3 features, it is possible to classify

Table 3. The behavioral constraints present in the sequence database of Table 1. The constraints that are supported at 100% are left out for 50%.

Support	Label	Supported constraint templates
100%	1	init(a), existence(a,3), exactly(b,2), response(b,a), precedence(a,c), **succession(b,c)**, **not succession(c,b)**, precedence(a,b)
	2	init(a), existence(a,3), exactly(b,2), response(b,a), precedence(a,c), precedence(c,b), response(b,c), **precedence(a,b)**
50%	1	exactly(c,1), last(a), response(c,a), **alternate precedence(a,c)**, **alternate precedence(b,c)**,
	2	last(a), **exactly(c,2)**, response(c,a)

the traces correctly. Lowering the support threshold results in more constraints being different, although the number of constraints does not have to drastically increase, as for example *response(a,b)* will eventually be replaced by *alternate response(a,b)* because of the hierarchy between the constraints. To achieve the same results with typical sequence-based constraints as used in, e.g., SPADE, it is harder to make such concise distinctions, as non-local information present in, e.g., *succession* requires either longer or more sequences to approach the behavior that will converge towards the language of the regular expression.

To summarize, iBCM exhibits the following advantages:

- It employs a rich, varied set of constraints that can be derived in a fast manner,
- It can be extended to incorporate any regular expression,
- It includes negative constraints for providing counter evidence, useful towards classification,
- It includes both unary cardinalities, as well as relational constraints,
- It enables easy comparison of constraint sets,
- It enables understanding what type of behavioral relations are present,
- It can be converted into a global automaton for representing behavior graphically.

4   iBCM: Algorithm Design and Implementation

This section outlines the algorithm for constructing the set of features based on the constraint templates discussed in Sect. 3. iBCM is an indirect sequence classification approach, i.e., the featurization and classification part are separate. Section 5 outlines the performance of the constraints generated by the approach as binary features (present/not present).

4.1   Inferring Constraints

The featurization approach is employed as a 3-step approach and outlined in Algorithm 1.

Step 1: Retain frequent items. First, items that exceed the support threshold are withheld in set A (line 2). Only these items will be used for checking unary constraints, and will be used in pair for checking binary constraints.

Step 2: Generate constraints. Next, every sequence in the database is checked in the following manner (line 4, and Algorithm 2). The sequence is traversed completely, and for every item in the frequent itemset the positions are stored. This allows for easy verification of the binary constraints. For every item $a \in A$, $|occ(a, \sigma)|$ is used for determining the cardinality constraints, i.e., absence/exactly/ existence. It is also checked whether it occurred as the first or last item in the sequence. Next, a is paired with every other $b \in A \backslash a$ to determine the type of behavioral constraint pattern. If a happens before b, the precedence lineage is reviewed. For every next occurrence of b, it is checked whether there was another a preceding it for *alternate precedence*. In the meantime for every occurrence, the exact position is checked for *chain precedence*. Both checks stop when there is no further evidence. If all occurrences of b fit, the constraints are added to the constraint set. If b happens after a, the response hierarchy is scrutinized. Similar to *alternate precedence*, every occurrence of a is checked for a subsequent b before the next occurrence. If every next occurrence of a is b, *chain response* is stored. After every pairwise check, the respective *succession* constraints are added if both *(alternate/chain) response* and *precedence* are present in the sequence. When b is not present in the sequence, there is evidence for *exclusive choice*.

Step 3: Retain frequent constraints. Finally, for every constraint it is checked whether it satisfies the minimum support level for the different labels in the sequence database in line 6 of Algorithm 1. This allows for precise measuring of sequential behavior, as some sequences might support both *response* and *precedence*, and others do not. Still, they can be merged (i.e. the simultaneous presence of *response* and *precedence* forms *succession*) to reduce the size of the number of features.

As can be seen from Algorithm 2, the binary constraints can be derived very efficiently by boolean and string operations. The approach is inspired by both [25, 26]. However, for classification purposes the sequences need to be labeled right away. The former uses DFAs to check constraints for each frequent pair. Doing this on a sequence level is computationally expensive, as it would require running each string many times. The latter builds a knowledge base of occurrence and precedence relations and calculates the support for constraints. This, however, is done on a log level, rather than at entry/sequence level, which requires extra featurization steps afterwards.

Algorithm 1. Mining constraint features per class i

1: **procedure** RETRIEVE_CONSTRAINTS($\mathcal{SB}, minsup$)   ▷ Input: Data and parameters
2:     $A \leftarrow frequentItems(\Sigma_{\mathcal{SB}}, minsup)$ ▷ Retain only frequent items
3:     **for** $l \in img(L)$ **do** $C_l \leftarrow \emptyset$ ▷ C_l a list with the constraints supporting label l
4:         **for** $\sigma \in \mathcal{SB} \wedge L(\sigma) = l$ **do** $C_l \leftarrow mineConstraints(\sigma, A)$
5:         **for** $c \in C_l$ **do**
6:             **if** $|\{c | c \in C_l\}| \geq |C_l| \times minsup$ **then** $C_{\mathcal{SB}} \leftarrow c$
7:     applyHierarchyReduction
8:     **return** $C_{\mathcal{SB}}$

4.2   Considerations on Constraint Template Base

Not all Declare constraint templates are fit to be considered for obtaining features from single sequences. First of all, constraints might suffer from being vacuously satisfied, i.e., they are satisfied because no counterevidence is provided.

Algorithm 2. Mining behavioral constraint templates

```
1:  procedure MINECONSTRAINTS(σ, A)
2:      C ← ∅
3:      for σᵢ ∈ σ do occ(σᵢ, σ) ← i
4:      for a ∈ A ∩ Σσ do
5:          if |occ(a, σ)| = 0 then C ← absence(a, 1)          ▷ Unary constraints
6:          else if |occ(a, σ)| > 2 then C ← existence(a, 3)
7:          else C ← exactly(a, |occ(a, σ)|)
8:          if 1 ∈ occ(a, σ) then C ← init(a)
9:          if |σ| ∈ occ(a, σ) then C ← last(a)
10:         for b ∈ A ∩ Σσ do                                  ▷ Binary constraints
11:             if min(occ(a, σ)) < min(occ(b, σ)) then
12:                 C ← prec(a, b)
13:                 i ← min(occ(b, σ)), chain ← (i − 1) ∈ occ(a, σ), continue ← ⊤
14:                 while ∃n ∈ occ(b, σ), n > i ∧ continue do
15:                     if ∃p ∈ occ(a, σ), i < p < n then i ← n
16:                         if ¬chain ∨ (n − 1) ∉ occ(a, σ) then chain ← ¬
17:                     else continue ← ¬
18:                 if continue then C ← altPrec(a, b)
19:                 if chain then C ← chainPrec(a, b)
20:             if max(occ(a, σ)) < max(occ(b, σ)) then
21:                 C ← resp(a, b)
22:             if max(occ(a, σ)) < min(occ(b, σ)) then C ← notSuc(a, b)
23:                 i ← min(occ(a, σ)), chain ← (i + 1) ∈ occ(b, σ), continue ← ⊤
24:                 while ∃n ∈ occ(a, σ), n > i ∧ continue do
25:                     if ∃p ∈ occ(b, σ), i < p < n then C ← altResp(a, b), i ← n
26:                         if ¬chain ∨ (n + 1) ∉ occ(b, σ) then chain ← ¬
27:                     else continue ← ¬
28:                 if continue then C ← altResp(a, b)
29:                 if chain then C ← chainResp(a, b)
30:             add succession if (alternate/chain) response and precedence
31:             if b ∉ Σσ ∧ b ∈ A then C ← exclChoi(a, b)
32:     return C
```

Hence, only binary pairs that are both present in a sequence are considered. This automatically satisfies the *choice* constraint, as well as *responded* and *co-existence*. Secondly, in a single sequence, *absence, exactly,* and *existence* are not distinguishable. It is opted not to generate all of them, but rather stick with a layered approach of *absence* for no occurrences, *exactly* for 1 to 2 occurrences, and *existence* for more than 3 occurrences. It would be possible to check them separately, and merge them afterwards, however, experiments showed that this does not have an impact on the results. Finally, *exclusive choice* and *not chain succession* both mine for negative behavior that reflects everything that is not present in the sequences. While absence does the same, the magnitude of the number of not existing sequence pairs is vastly larger. Although mining for negative information is one distinctive feature of the proposed approach, the gain in accuracy performance does not outweigh the burden in terms of the number of extra constraints generated. Hence they are not included in the final constraint set. *Not succession* is the only negative constraint used. Note that all constraints are mined with a confidence of 100%.

4.3 Scalability

The computational tractability of the technique relies heavily on two components. First of all, the length of the sequence is an important factor as they are traversed completely. Hence, the performance is bound in the extreme by the length of the longest sequence. Secondly, the minimum support determines the number of activities, hence the number of pairs and constraint templates that need to be checked. In the worst case, all pairs have to be checked for all binary templates. Most constraints can be checked by simple lookups, but in case the templates in the upper part of the hierarchy are checked, the complexity in the worst case is the length of the string for checking alternating and chain behavior. This results in $O(|A|^2 \times |\sigma|)$. As will become clear from experimental evaluation, however, iBCM can achieve good results at high minimum support levels, reducing $|A|$ drastically.

5   Experimental Evaluation

In this section, the technique will be evaluated on widely-used, realistic datasets and compared with 4 other approaches.

5.1 Setup

Below, an overview of the data, implementation, and other approaches is given.

Data and Classification. The datasets that were used are summarized in Table 4 and are a mix with both a large set of distinct items, as well as a large number of data entries. They are discussed in more detail in [17,19]. All techniques were first employed to generate interesting sequences, and next to build a predictive model by using the presence of the sequences as a binary feature. Three classifiers were considered, i.e., naive Bayes (NB), decision trees (DT), and random forests (RF), for which the Weka[1] Java implementation was used. All runs were executed using a Java 8 Virtual Machine on an Intel i7-6700HQ CPU with 16GB DDR4 memory. A 10-fold cross-validation was applied for all the experiments.

Approaches. iBCM is benchmarked against 4 other state-of-the-art techniques, being cSPADE [7], Interesting Sequence Miner (ISM) [19], Sequence Classification based on Interesting Sequences (SCIP) [17], and Mining Sequential Classification Rules (MiSeRe) [20], which all have the clear goal of obtaining discriminative, informative sequences for classification and are compared in Table 5. A comparison with other techniques can be found in the respective works as well. For cSPADE, iBCM, and SCIP, the support levels were set at 0.1–1.0 by 0.1

---

[1] http://www.cs.waikato.ac.nz/ml/weka/.

Table 4. Characteristics of the datalogs used for evaluation.

	$|\mathcal{SB}|$	$|\Sigma_{\mathcal{SB}}|$	$|\mathbf{img(L)}|$	$\overline{|\sigma|}$	$\max(|\sigma|)$
context	240	94	5	88.39	246
Unix	5,472	1,697	4	32.34	1,400
auslan2	200	16	10	5.53	18
aslbu	424	250	7	13.05	54
pioneer	160	178	3	40.14	100
news	4,976	27,884	5	139.96	6,779

intervals. SCIP was used for a minimum interestingness level of 0.05 and a maximum sequence length of 2 (this length was devised by the authors in [17] and a longer length increased computation time and did not return better results). For MiSeRe, 1, 2, 5, and 10 second run times were considered. Finally ISM was used with a maximum number of iterations of 200, and a maximum number of optimization steps of 10,000. No notable differences were reported when using different settings. The implementation of the benchmark can be found online at https://feb.kuleuven.be/public/u0092789.

Table 5. An overview of the techniques used for benchmarking.

Technique	Description	Parameters
cSPADE [7]	Sequence mining approach based on window, gap, length, width, and other constraints	Support
SCIP [17]	Extension to SPADE based on an interesting measure that next to the support of a sequence also consists of the proximity of its items	Support, interestingness, maximum length of sequences
ISM [19]	Technique that interleaves subsequences and infers the most compact set of sequences that can regenerate the database	max # optimization steps, max # iterations
MiSeRe [20]	Randomly generates diverse sequences and applies a Bayesian approach to retain interesting sequences	Maximum runtime (in seconds)
iBCM	The devised approach, based on mining a set of behavioral constraints	Support

5.2   Results

The results in terms of accuracy and the number of generated constraints along the support spectrum are displayed in Figs. 2 and 3. The results for ISM and MiSeRe are reported separately in Table 6. An overview of the share of each constraint template family in the results of iBCM is given in Fig. 4.

Overall, iBCM is capable of achieving a high accuracy, without inducing a big amount of constraints ($|C|$). Especially for the *aslbu* and *auslan* datasets,

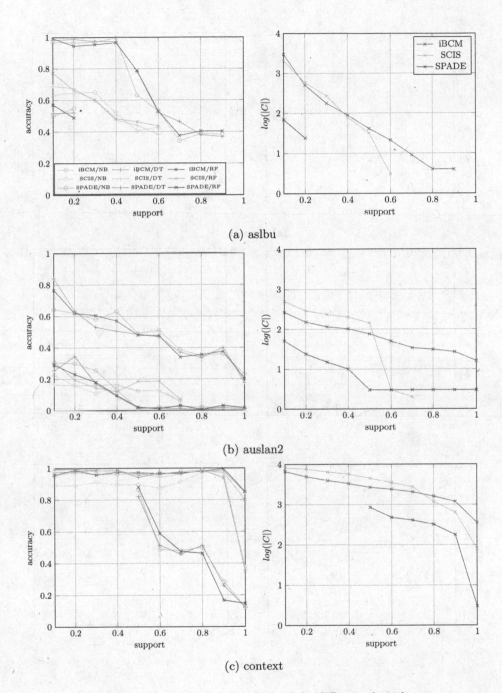

(a) aslbu

(b) auslan2

(c) context

Fig. 2. Overview of the performance of the different algorithms.

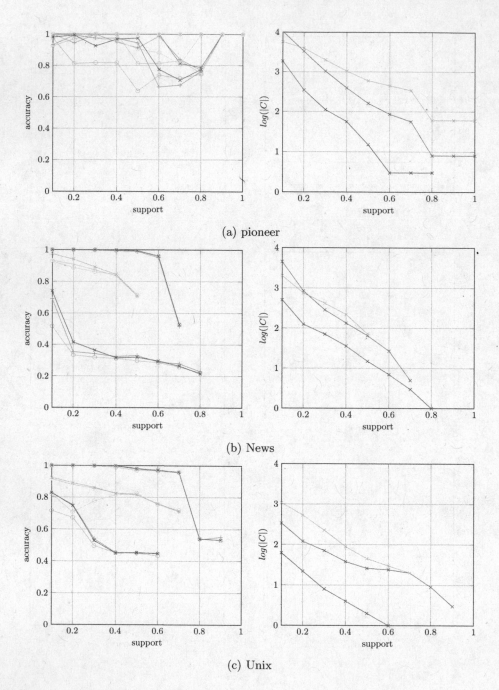

(a) pioneer

(b) News

(c) Unix

Fig. 3. Overview of the performance of the different algorithms.

a higher accuracy is obtained than using the state-of-the-art techniques. Also, iBCM achieves a higher accuracy more rapidly when going down the support spectrum, achieving high accuracy already for 50% to 60% with a small amount of constraints (<100). The differences in terms of the type of sequential behavior present becomes apparent. In the text-based datasets, such as *news*, the *absence* constraint clearly provides a prominent source of information, since rather the presence of items, not relations, are needed for classification. This lies in line with the findings in [19]. In the other datasets, the whole set of constraint patterns is used, except for the very specific *chain* constraints. The inclusion of negative constraints might explain the higher accuracy for *aslbu* and *auslan2*. The more comprehensive *alternating* constraints are indeed often present (note that the hierarchy reduction cuts away all simple/alternating ordered constraints when alternating/chain constraints are found).

cSPADE was not able to finish executing the *context* dataset within 60 min for support values lower than 50%. Similarly, ISM was not able to do generate interesting sequences for the *News* dataset. Besides, the algorithms did not always generate constraints for certain higher support values. In terms of performance, in Fig. 5 the time needed to generate the constraints and label the sequences is plotted. All constraints could be derived in less than 1 second, except for the News dataset due to the bigger size of $|A|$. In this case, the technique clearly scales exponentially with the size of $|A|$. This is probably due to the nature of the data, being plain text. The higher number of items, of which there are no particularly frequent after a certain threshold, increases the runtime. In the other datasets, infrequent

Table 6. Accuracny and $\log(|C|)$ (between brackets) results for MiSeRe and ISM.

Dataset	Classifier	misere (1s)	misere (2s)	misere (5s)	misere (10s)	ISM
aslbu	NB	0.56 (2.086)	0.556 (2.111)	0.542 (2.111)	**0.595 (2.107)**	0.602 (2.274)
	DT	0.542 (2.083)	0.547 (2.1)	0.556 (2.111)	0.544 (2.111)	0.626 (2.274)
	RF	0.53 (2.097)	0.593 (2.111)	0.586 (2.111)	0.595 (2.111)	**0.623 (2.274)**
auslan2	NB	**0.37 (2.4)**	0.31 (2.401)	0.37 (2.401)	0.26 (2.401)	0.225 (1.279)
	DT	0.27 (2.401)	0.305 (2.401)	0.27 (2.401)	0.25 (2.401)	**0.24 (1.279)**
	RF	0.25 (2.401)	0.325 (2.401)	0.305 (2.401)	0.295 (2.401)	**0.24 (1.279)**
context	NB	**0.938 (2.587)**	0.938 (2.892)	0.929 (3.259)	0.888 (3.568)	0.821 (2.025)
	DT	0.888 (2.592)	0.871 (2.854)	0.908 (3.282)	0.867 (3.547)	**0.821 (2.025)**
	RF	0.908 (2.645)	0.888 (2.814)	0.933 (3.235)	0.913 (3.554)	0.725 (2.025)
pioneer	NB	0.963 (1.903)	0.969 (1.982)	0.813 (2.201)	0.863 (2.369)	0.981 (2.093)
	DT	0.994 (1.845)	0.988 (1.959)	0.988 (2.152)	0.988 (2.336)	0.963 (2.093)
	RF	0.994 (1.863)	0.994 (2.021)	**1 (2.188)**	1 (2.342)	**1 (2.093)**
Unix	NB	0.863 (2.423)	0.86 (2.447)	0.815 (2.687)	0.744 (2.707)	0.897 (3.232)
	DT	0.901 (2.42)	0.902 (2.616)	0.898 (2.549)	0.903 (2.702)	**0.927 (3.232)**
	RF	**0.923 (2.511)**	0.91 (2.555)	0.912 (2.716)	0.91 (2.842)	0.908 (3.232)
News	NB	0.929 (3.445)	0.929 (3.445)	0.928 (3.446)	0.919 (3.448)	NA
	DT	0.899 (3.446)	0.901 (3.446)	0.902 (3.446)	0.905 (3.447)	NA
	RF	**0.973 (3.445)**	0.969 (3.445)	0.971 (3.446)	0.971 (3.445)	NA

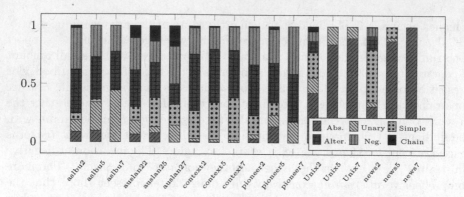

Fig. 4. An overview of the share (in %) of the type of constraint templates mined from the databases. The numbers stand for the minimum support, e.g., 2 stands for 20%.

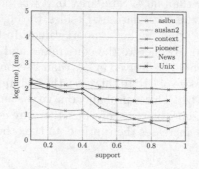

Fig. 5. Time needed to mine constraints.

items are truly infrequent and $|A|$ does not necessarily grow. Nevertheless, support settings as of 0.6 already guarantee a decent level of accuracy.

There is no notable difference in accuracy when using different classifiers, except for the *aslbu* and *auslan2* datasets. Especially the constraints generated by cSPADE seem to have a different impact on the classifiers. In general, the classifiers perform more stably on the datasets with either less labels or with more sequences to learn from. Random forests seem to perform the best overall.

6   Conclusion and Future Work

This work proposed iBCM, a new technique with the ability to discover features for sequence classification. Based on behavioral constraint templates, iBCM is able to concisely distinguish different sequential behaviors in databases. It is capable of achieving results with high accuracy, while minimizing the number of features needed compared with other approaches. Furthermore, the inference technique devised can also be applied towards descriptively interpreting the nature of the patterns present in a sequence database, offering insights into what types of interplay are present between the items in the data.

In future work, a more in-depth comparison of which types of constraints contribute the most to the classifiers will be made. This establishes the base for building direct sequence classification techniques as well. Finally, the data-aware versions of the patterns can be introduced as well. Most patterns are also described in first-order LTL and can be extended to include non-sequential information [27] to bridge the gap with `Datalog` [13]. Also, the target-branched version of Declare [28], i.e., constraints with a consequent being a set rather than a singleton will be investigated.

References

1. Agrawal, R., Srikant, R.: Mining sequential patterns. In: ICDE, pp. 3–14. IEEE Computer Society (1995)
2. Lee, J., Han, J., Li, X., Cheng, H.: Mining discriminative patterns for classifying trajectories on road networks. IEEE Trans. Knowl. Data Eng. **23**(5), 713–726 (2011)
3. Eichinger, F., Nauck, D.D., Klawonn, F.: Sequence mining for customer behaviour predictions in telecommunications. In: Proceedings of the Workshop on Practical Data Mining at ECML/PKDD, pp. 3–10 (2006)
4. Agrawal, R., Imielinski, T., Swami, A.N.: Mining association rules between sets of items in large databases. In: SIGMOD Conference, pp. 207–216. ACM Press (1993)
5. Wang, J., Han, J.: BIDE: efficient mining of frequent closed sequences. In: ICDE, pp. 79–90. IEEE Computer Society (2004)
6. Pei, J., Han, J., Mortazavi-Asl, B., Pinto, H., Chen, Q., Dayal, U., Hsu, M.: Prefixspan: mining sequential patterns by prefix-projected growth. In: ICDE, pp. 215–224. IEEE Computer Society (2001)
7. Zaki, M.J.: Sequence mining in categorical domains: Incorporating constraints. In: CIKM, pp. 422–429. ACM (2000)
8. Coquery, E., Jabbour, S., Saïs, L., Salhi, Y.: A sat-based approach for discovering frequent, closed and maximal patterns in a sequence. In: ECAI. Frontiers in Artificial Intelligence and Applications, vol. 242, pp. 258–263. IOS Press (2012)
9. Kemmar, A., Loudni, S., Lebbah, Y., Boizumault, P., Charnois, T.: PREFIX-PROJECTION global constraint for sequential pattern mining. In: Pesant, G. (ed.) CP 2015. LNCS, vol. 9255, pp. 226–243. Springer, Cham (2015). https://doi.org/10.1007/978-3-319-23219-5_17
10. Kemmar, A., Loudni, S., Lebbah, Y., Boizumault, P., Charnois, T.: A global constraint for mining sequential patterns with GAP constraint. In: Quimper, C.-G. (ed.) CPAIOR 2016. LNCS, vol. 9676, pp. 198–215. Springer, Cham (2016). https://doi.org/10.1007/978-3-319-33954-2_15
11. Aoga, J.O.R., Guns, T., Schaus, P.: An efficient algorithm for mining frequent sequence with constraint programming. In: Frasconi, P., Landwehr, N., Manco, G., Vreeken, J. (eds.) ECML PKDD 2016. LNCS (LNAI), vol. 9852, pp. 315–330. Springer, Cham (2016). https://doi.org/10.1007/978-3-319-46227-1_20
12. Negrevergne, B., Guns, T.: Constraint-based sequence mining using constraint programming. In: Michel, L. (ed.) CPAIOR 2015. LNCS, vol. 9075, pp. 288–305. Springer, Cham (2015). https://doi.org/10.1007/978-3-319-18008-3_20

13. Dehaspe, L., Toivonen, H.: Discovery of frequent DATALOG patterns. Data Min. Knowl. Discov. **3**(1), 7–36 (1999)
14. Esposito, F., Mauro, N.D., Basile, T.M.A., Ferilli, S.: Multi-dimensional relational sequence mining. Fundam. Inform. **89**(1), 23–43 (2008)
15. Cule, B., Goethals, B.: Mining association rules in long sequences. In: Zaki, M.J., Yu, J.X., Ravindran, B., Pudi, V. (eds.) PAKDD 2010. LNCS (LNAI), vol. 6118, pp. 300–309. Springer, Heidelberg (2010). https://doi.org/10.1007/978-3-642-13657-3_33
16. Cheng, H., Yan, X., Han, J., Hsu, C.: Discriminative frequent pattern analysis for effective classification. In: ICDE, pp. 716–725. IEEE Computer Society (2007)
17. Zhou, C., Cule, B., Goethals, B.: Pattern based sequence classification. IEEE Trans. Knowl. Data Eng. **28**(5), 1285–1298 (2016)
18. Fradkin, D., Mörchen, F.: Mining sequential patterns for classification. Knowl. Inf. Syst. **45**(3), 731–749 (2015)
19. Fowkes, J.M., Sutton, C.A.: A subsequence interleaving model for sequential pattern mining. In: KDD, pp. 835–844. ACM (2016)
20. Egho, E., Gay, D., Boullé, M., Voisine, N., Clérot, F.: A parameter-free approach for mining robust sequential classification rules. In: ICDM, pp. 745–750. IEEE Computer Society (2015)
21. Pesic, M., Schonenberg, H., van der Aalst, W.M.P.: DECLARE: full support for loosely-structured processes. In: EDOC, pp. 287–300. IEEE Computer Society (2007)
22. Dwyer, M.B., Avrunin, G.S., Corbett, J.C.: Patterns in property specifications for finite-state verification. In: ICSE, pp. 411–420. ACM (1999)
23. Pesić, M.: Constraint-based work on management systems: shifting control to users. Ph.D. thesis, Eindhoven University of Technology, p. 26 (2008)
24. Westergaard, M., Stahl, C., Reijers, H.A.: Unconstrainedminer: efficient discovery of generalized declarative process models. BPM Center Report BPM-13-28, p. 28. BPMcenter.org (2013)
25. Di Ciccio, C., Mecella, M.: A two-step fast algorithm for the automated discovery of declarative workflows. In: CIDM, pp. 135–142. IEEE (2013)
26. Maggi, F.M., Bose, R.P.J.C., van der Aalst, W.M.P.: Efficient discovery of understandable declarative process models from event logs. In: Ralyté, J., Franch, X., Brinkkemper, S., Wrycza, S. (eds.) CAiSE 2012. LNCS, vol. 7328, pp. 270–285. Springer, Heidelberg (2012). https://doi.org/10.1007/978-3-642-31095-9_18
27. Maggi, F.M., Dumas, M., García-Bañuelos, L., Montali, M.: Discovering data-aware declarative process models from event logs. In: Daniel, F., Wang, J., Weber, B. (eds.) BPM 2013. LNCS, vol. 8094, pp. 81–96. Springer, Heidelberg (2013). https://doi.org/10.1007/978-3-642-40176-3_8
28. Di Ciccio, C., Maggi, F.M., Mendling, J.: Efficient discovery of target-branched declare constraints. Inf. Syst. **56**, 258–283 (2016)

Efficient Sequence Regression by Learning Linear Models in All-Subsequence Space

Severin Gsponer[✉], Barry Smyth, and Georgiana Ifrim

Insight Centre for Data Analytics, University College Dublin, Dublin, Ireland
{severin.gsponer,barry.smyth,georgiana.ifrim}@insight-centre.org

Abstract. We present a new approach for learning a sequence regression function, i.e., a mapping from sequential observations to a numeric score. Our learning algorithm employs coordinate gradient descent with Gauss-Southwell optimization in the feature space of all subsequences. We give a tight upper bound for the coordinate wise gradients of squared error loss which enables efficient Gauss-Southwell selection. The proposed bound is built by separating the positive and the negative gradients of the loss function and exploits the structure of the feature space. Extensive experiments on simulated as well as real-world sequence regression benchmarks show that the bound is effective and our proposed learning algorithm is efficient and accurate. The resulting linear regression model provides the user with a list of the most predictive features selected during the learning stage, adding to the interpretability of the method. Code and data related to this chapter are available at: https://github.com/svgsponer/SqLoss.

1 Introduction

A wide range of applications benefit from methods that can learn a mapping from sequential observations to categorical or numeric scores. For example, a mapping could be learned from a set of labeled DNA sequences, to classify each sequence into subfamilies [10], or assign to it a numeric score, such as a protein binding affinity [18]. Methods aimed at solving such problems typically employ Hidden Markov Models (HMM) [14,19], kernel Support Vector Machines (SVM) [11] or more recently, Convolutional Neural Networks (CNN) [2]. While the accuracy of such techniques is promising, their efficiency and interpretability are still critical challenges. An alternative to the above approaches is to explicitly generate all k-mers (i.e., subsequences of length k) up to a specified k, followed by learning a classification or regression model using all the generated k-mers as features. Such methods are much simpler and achieve an accuracy comparable to the more sophisticated methods above [3,18]. Nevertheless, they are limited by the huge computational burden of explicitly generating all k-mers, and therefore tend to be applied to small datasets with k fixed manually, e.g., up to 6 or 8, and need to use additional filtering to further reduce the large feature space.

In this work, we propose a regression approach that can use the entire space of k-mers, of unlimited length, by learning a linear model using an iterative

© Springer International Publishing AG 2017
M. Ceci et al. (Eds.): ECML PKDD 2017, Part II, LNAI 10535, pp. 37–52, 2017.
https://doi.org/10.1007/978-3-319-71246-8_3

branch-and-bound strategy. The main idea behind this technique is to exploit the nested structure of the feature space via greedy search, thus avoiding the need for explicitly generating all subsequences, and instead focusing on the most discriminative ones during learning. The resulting approach combines feature selection and learning into a simple algorithm and, as shown in our experiments, delivers accuracy similar to the state-of-the-art, with no pre-processing or domain knowledge required. Since during learning we only need to explore a small subset of the feature space, we can employ richer features such as gapped k-mers to allow inexact feature matches. This enables our linear models to achieve high accuracy. Our optimization algorithm relies on greedy coordinate-descent with Gauss-Southwell selection. To enable efficient coordinate selection, we give a tight upper bound for the coordinate-wise gradients of squared error loss.

We test the proposed algorithm on a simulated benchmark and on two real-world applications. First, we compare our algorithm to other regression methods on a synthetic sequence regression dataset, where we vary parameters such as the true motif length and the alphabet size, to study potential gains from using rich features such as all k-mers. Next, we apply our model to a sequence regression problem where the goal is to score the protein binding affinity of DNA sequences. We work with a publicly available dataset of 40,000 DNA sequences prepared by [18] for a popular data challenge[1]. Finally, we study our algorithm on a large sequence classification dataset to compare the effectiveness of our approach to existing methods designed for linear sequence classification [7,8]. In this application the aim is to score software files represented as hexadecimal sequences, in order to categorize malicious software into known families, also known as malware classification [1]. This dataset was released by a recent data challenge[2] organized by Microsoft.

Contribution. We propose a new method for efficient sequence regression by learning linear models with rich subsequence features, e.g., unrestricted-length, contiguous and gapped k-mers. Our algorithm uses an optimization strategy based on coordinate-descent coupled with an upper bound on the coordinate-wise gradients of squared error loss, to enable efficient Gauss-Southwell selection. We evaluate our learning algorithm on simulated data and on two real-world applications and show that our simple linear models are as accurate as more complex state-of-the-art sequence regression methods, while requiring no feature engineering or heavy parameter tuning. We release all our code for producing synthetic sequence regression data, as well as for our learning algorithm[3].

2   Related Work

We discuss a range of approaches for sequence regression and classification, with a focus on the two application domains studied in this paper.

---

[1] https://www.synapse.org/#!Synapse:syn2887863/wiki/72186.
[2] https://www.kaggle.com/c/malware-classification.
[3] Code of our algorithms: https://github.com/svgsponer/SqLoss.

Sequence Regression for DNA. The sequence regression benchmark provided by the DREAM5 challenge aims to advance the state-of-the-art in recognizing DNA-binding proteins. It consists of 40,000 DNA training sequences, each with a numeric score describing the binding specificity of a particular protein, from a class of proteins named Transcription Factors (TF). The task is to learn a regression function for a given TF, then predict the TF binding affinity for unseen sequences. There are 66 TF defining 66 different sequence regression tasks. The interpretability of the learned model is also important as new knowledge can be extracted from knowing the individual k-mer binding specificity. The work by [18] presents a systematic comparison of 26 methods evaluated on this benchmark. The winning method was a linear regression model with squared error loss and optimization by gradient descent on sequences represented in the feature space of all 4 to 8-mers (Team_D) [3]. Among the top-5 methods were a Markov model (Team_F) [9], an HMM trained by Expectation-Maximization combined with a linear model (Team_E) [16], and a linear regression model using contiguous and gapped 6-mers (Team_G)[4].

A second round of the challenge has added new competing techniques, the most notable of which are a new HMM model for regression (RegHMM) [19] and a deep learning method based on CNN (named DeepBind) [2]. The linear regression method of Team_D came second. RegHMM which restricts the k-mers to $k = 6$, obtained results similar to Team_D. DeepBind slightly outperformed Team_D regarding prediction accuracy, but used a CNN architecture designed for this specific task and required a custom implementation for GPUs with extensive parameter calibration over 30 sets of parameters. We test our approach on the same TF-DNA benchmark and show that we can achieve similar accuracy to prior techniques, with a much simpler and more efficient approach.

Sequence Classification for Malware. Microsoft released a malware classification benchmark on Kaggle in 2015, containing about 20,000 files amounting to 500GB of data, to train and evaluate classifiers of malware files. Besides the large number of samples, individual files are also quite large (up to 50 Mb per file). Each file can be interpreted as a discrete sequence of bytes. Several competing methods have exploited the sequence structure of the files. The winning method [17] has explicitly generated all k-mers with $k \in \{2, 3, 4\}$ using a machine with 104 Gb memory, 16 CPUs, and extra 200 Gb disk space for the generated data. They have used these features with a boosting method implemented in the Xgboost library. The k-mer features have proven to be very useful for achieving high classification accuracy, but generating them explicitly requires extensive storage and computational resources. Recent work [1] has studied new feature engineering approaches on the same benchmark. They have decided not to use k-mer features due to the excessive computational requirements, but this has lead to lower accuracy than the challenge winner. In [5] the authors use explicitly

---

[4] http://www.nature.com/nbt/journal/v31/n2/extref/nbt.2486-S1.pdf.

generated 3-mers as features, but use mutual information and random projections to select a feature subset that is manageable for learning a neural network. Malware coders routinely use masking and other code obfuscation techniques. To detect such manipulations, gapped k-mers that allow flexible, rather than exact matching, could improve the classifier. To compare our regression algorithm to prior work, we treat sequence classification as a regression with binary scores. We learn and evaluate our linear regression function using *unrestricted-length k-mer features*.

Linear Sequence Classification. Some of our key intuitions come from the work of [7,8] which proposed the SEQL framework implementing greedy approaches for linear sequence classification. In [8] it was shown that a branch-and-bound approach can be used for a variety of classification loss functions. We build on previous research [8] and propose an algorithm for efficient sequence regression, by exploiting the structure of the feature space and separately bounding the positive and negative coordinate-wise gradients for squared error loss. This bound allows us to guarantee that we iteratively find the best (with respect to a given loss function) k-mer feature from a very large feature space. We compare our methods to the linear sequence classifiers from [7,8] on the malware classification benchmark.

3   Method Proposed

3.1   Basic Notation

Let $D = \{(s_1, y_1), (s_2, y_2), \ldots, (s_N, y_N)\}$ be a training set of instance-label pairs, where $s_i = c_1 c_2 \ldots c_{m_i}$ is a sequence of variable length m_i, with each $c_i \in \Sigma$ a symbol from the alphabet of possible symbols denoted by Σ. For example, in the case of DNA sequences $\Sigma = \{A, C, G, T\}$. Each sequence s_i has an associated score $y_i \in \mathbb{R}$. We define a subsequence as a contiguous part of a sequence, e.g., $s_j = c_2 c_3 c_4$ and write $s_j \subseteq s_i$ if s_j is a subsequence of s_i. Given this definition we can represent a sequence s_i as a binary vector in the space of all subsequences in the training data: $x_i = (x_{i1}, \ldots, x_{ij}, \ldots, x_{in})^T, x_{ij} \in \{0, 1\}, i = 1, \ldots, N$, where $x_{ij} = 1$ means that subsequence s_j occurs in sequence s_i. We denote by n the number of distinct subsequences in the feature space, i.e., the coordinates of the vectors space in which we learn. Although this space is huge and in practice infeasible to generate explicitly, we show how to work with this representation by exploiting its nested structure to develop a lazy search procedure.

The goal is to learn a mapping from sequences to scores, $f : S \to \mathbb{R}$, from the given training set D, so that we can predict a score $y \in \mathbb{R}$ for a new sample $s \in S$. In our framework we want to learn a linear model, i.e., a parameter vector β that allows us to estimate the real score y by setting $\hat{y} = \beta^T x_i$. Although linear models are not powerful enough to capture non-linear relationships, by working in a very complex feature space (e.g., all k-mers) we can learn a powerful model, similar to the kernel trick applied by kernel Support Vector Machines. We compute $\beta = (\beta_1, \ldots, \beta_j, \ldots, \beta_n)$ by minimizing a loss function over the training set:

$$\beta^* = \underset{\beta \in \mathbb{R}^n}{\arg\min}\, L(\beta). \tag{1}$$

In our work $L(\beta)$ is the regularized squared loss:

$$L(\beta) = \sum_{i=1}^{N}(y_i - \beta^T \cdot x_i)^2 + C R_\alpha(\beta). \tag{2}$$

$C \in \mathbb{R}_0^+$ is the weight for the regularizer $R_\alpha(\beta)$. We use the elastic-net regularizer $R_\alpha(\beta) = \alpha \sum_{j=1}^{n} |\beta_j| + (1 - \alpha)\frac{1}{2}\sum_{j=1}^{n} \beta_j^2$ defined in [6] which allows trading-off $l1$ and $l2$ penalties.

3.2   Learning via Coordinate-Descent with Gauss-Southwell Selection

Recent work [12] has shown that for a class of loss functions, which includes the squared loss, learning via coordinate descent is faster than random coordinate descent optimization. Furthermore, for squared loss in particular, coordinate descent via the Gauss-Southwell rule was proven to converge much faster than other coordinate descent methods. In our setting, the feature space of all subsequences is potentially exponential, thus it is not even possible to explicitly compute the full gradient. We first give the generic learning algorithm and then provide an upper bound that makes Gauss-Southwell selection feasible for this complex feature space.

We are interested in solving the convex optimization problem in (1). The coordinate descent method is based on the iteration step:

$$\beta^{(t)} = \beta^{(t-1)} - \eta_{j_t} \frac{\partial L}{\partial \beta_{j_t}}(\beta^{(t-1)})e_{j_t} \tag{3}$$

To determine the descent direction we use the Gauss-Southwell rule [12]:

$$j_t = \underset{j}{\arg\max}\left|\frac{\partial L}{\partial \beta_j}(\beta^{(t-1)})\right| \tag{4}$$

This formulation transforms the learning problem into a search problem since in each iteration we have to find the best coordinate j_t, i.e., the subsequence with the largest absolute gradient value. Algorithm 1 shows the basic mechanics of our method. The crucial part of this algorithm is the search for the best coordinate (line 5), for which we present an efficient algorithm in the next section.

Algorithm 1. Greedy Coordinate Descent with Gauss Southwell Selection

---
1: Set $\beta^{(0)} = 0$
2: **while** not termination condition **do**
3:      Adjust intercept
4:      Calculate objective function $L(\beta^{(t)})$
5:      Find coordinate j_t with maximum gradient value
6:      Find optimal step size η_{j_t} by line search or exact optimization
7:      Update $\beta^{(t)} = \beta^{(t-1)} - \eta_{j_t} \frac{\partial L}{\partial \beta_{j_t}}(\beta^{(t-1)}) e_{j_t}$
8:      Add corresponding feature to feature set
9: **end while**

Step Size. The parameter η_{j_t} is called step size and acts as a scaling factor on the gradient value, to enforce convergence. The work in [12] analyzed a variety of options to set this parameter, from constant step size to exact optimization. As exact optimization was shown to produce much faster convergence, and is feasible to compute for squared loss, we also optimize η_{j_t} exactly.

3.3   Upper Bound for Fast Gauss-Southwell Selection

Formulating a learning algorithm via coordinate descent with Gauss-Southwell rule does not provide a solution for the problem of finding the best subsequence in a huge feature space. Here we give an upper bound on the coordinate-wise gradient value for the squared loss function, that enables us to efficiently search for the best coordinate in each iteration. The theory relies on the following intuitions. First, the subsequence space has a structure that we can exploit to focus the search on parts of the feature space. Namely, we can bound the frequency of a sequence, based on the frequency of any of its subsequences, using an argument similar to that of the Apriori market basket analysis algorithm. Second, we can separate the positive and negative terms of the gradient, to obtain an upper bound on the gradient for squared loss. This allows us to incrementally generate feature candidates[5] and to quickly rule out parts of the feature space, while guaranteeing to find a coordinate with maximum gradient magnitude, as required by the Gauss-Southwell optimization strategy. In the following, $s_j \in x_i$ means that the corresponding vector entry $x_{ij} = 1$. Furthermore, we denote $s_p \subseteq s_j$ if s_p is a subsequence of s_j. Theorem 1 gives an upper bound on the gradient value of any subsequence s_j, using only information about its prefix s_p.

Theorem 1 (Bounding the search for the best coordinate). *Let $L(\beta)$ be the squared loss function and $y_i \in \mathbb{R}$. For any subsequence $s_p \subseteq s_j$, it holds that*

$$\left| \frac{\partial L}{\partial \beta_j}(\beta) \right| \leq \mu(s_p), \text{ where}$$

---

[5] To generate candidate features we start from 1-mers and use breadth-first-expansion to generate k-mers with $k \geq 1$.

$$\mu(s_p) = \max \left\{ \left| \sum_{\{i|x_{ip}=1, y_i - \beta^T x_i \geq 0\}} -2x_{ip}(y_i - \beta^T x_i) + C(\alpha\,\text{sign}(\beta_j) + (1-\alpha)\beta_j) \right|, \right.$$

$$\left. \left| \sum_{\{i|x_{ip}=1, y_i - \beta^T x_i \leq 0\}} -2x_{ip}(y_i - \beta^T x_i) + C(\alpha\,\text{sign}(\beta_j) + (1-\alpha)\beta_j) \right| \right\}$$

Proof. We first focus on bounding the positive terms of the coordinate-wise gradients:

$$\frac{\partial L}{\partial \beta_j}(\beta) = \sum_{i=1}^{N} -2x_{ij}(y_i - \beta^T x_i) + CR'_\alpha(\beta_j) \tag{5}$$

$$= \sum_{\{i|x_{ij}=1\}} -2x_{ij}(y_i - \beta^T x_i) + CR'_\alpha(\beta_j) \tag{6}$$

$$\leq \sum_{\substack{\{i|x_{ij}=1, \\ y_i - \beta^T x_i \leq 0\}}} -2x_{ij}(y_i - \beta^T x_i) + CR'_\alpha(\beta_j) \tag{7}$$

$$\leq \sum_{\substack{\{i|x_{ip}=1, \\ y_i - \beta^T x_i \leq 0\}}} -2x_{ip}(y_i - \beta^T x_i) + CR'_\alpha(\beta_j) \tag{8}$$

$$\leq \sum_{\substack{\{i|x_{ip}=1, \\ y_i - \beta^T x_i \leq 0\}}} -2x_{ip}(y_i - \beta^T x_i) + C(\alpha\,\text{sign}(\beta_j) + (1-\alpha)\beta_j) \tag{9}$$

The step from (7) to (8) moves from coordinate j to coordinate p. The inequality holds since $\{i|x_{ij}=1, y_i - \beta^T x_i \leq 0\} \subseteq \{i|x_{ip}=1, y_i - \beta^T x_i \leq 0\}$ as every sequence which contains s_j also contains its subsequence s_p. Similarly, by separating the negative terms, we get a second bound:

$$\frac{\partial L}{\partial \beta_j}(\beta) = \sum_{i=1}^{N} -2x_{ij}(y_i - \beta^T x_i) + CR'_\alpha(\beta_j) \tag{10}$$

$$\geq \sum_{\substack{\{i|x_{ip}=1, \\ y_i - \beta^T x_i \geq 0\}}} -2x_{ip}(y_i - \beta^T x_i) + C(\alpha\,\text{sign}(\beta_j) + (1-\alpha)\beta_j) \tag{11}$$

The two bounds provide an upper bound on the absolute value of the gradient at coordinate j, using only information about coordinate p:

$$\sum_{\{i|s_p \in x_i, y_i - \beta^T x_i \geq 0\}} -2x_{ip}(y_i - \beta^T x_i) + CR'_\alpha(\beta_j) \leq \frac{\partial L}{\partial \beta_j}(\beta) \tag{12}$$

$$\leq \sum_{\{i|s_p \in x_i, y_i - \beta^T x_i \leq 0\}} -2x_{ip}(y_i - \beta^T x_i) + CR'_\alpha(\beta_j)$$

With the regularization term $CR'_\alpha(\beta_j)$ included, the bound depends on the prefix s_p as well as on the weight β_j of the subsequence s_j. Since in the beginning all β_j are set to zero this does not represent a problem, as the regularizer is zero in this case. The bounds of features that were already selected in a previous iteration are the only ones that have to be adjusted by adding this term.

The bound allows us to efficiently search for the coordinate with the largest gradient. Algorithm 2 shows the search procedure. We start the search by expanding from 1-mers. Throughout the search we keep track of the current best feature (*best_feature*) and in τ we save its absolute gradient value. Before the expansion of any subsequence, we check if its upper bound μ is smaller than τ. If this is the case, we can prune the subtree starting at this node, as no further expansion can improve the current gradient.

Algorithm 2. Fast Gauss-Southwell Coordinate Selection

1: $\tau \leftarrow 0$
2: *best_feature* $\leftarrow NIL$
3: **for all** $s' \in \bigcup_{i=1}^{N} \{s | s \in s_i, |s| = 1\}$ **do**   ▷ For each 1-mer
4:     GROW_SEQUENCE(s')
5: **end for**
6: **return** *best_feature*
7:

1: **function** GROW_SEQUENCE(s)
2:     **if** $\mu(s) \leq \tau$ **then return**   ▷ $\mu(s)$ like in Theorem 1
3:     **else if** $abs(gradient(s)) > \tau$ **then**
4:         *best_feature* $= s$ ▷ Suboptimal solution
5:         $\tau = abs(gradient(s))$
6:     **end if**
7:     **for all** $s'' \in \{s' | s' \supseteq s, s' \in \bigcup_{i=1}^{N} s_i, |s'| = |s| + 1\}$ **do**
8:         GROW_SEQUENCE(s'')
9:     **end for**
10: **end function**

Proposition 1 (Tightness of upper bound). *The upper bound given in Theorem 1 is tight.*

Proof. It suffices to show one example in which the upper bound (12) is reached. The inequality becomes an equality when, e.g., $y_i - \beta^T x_i = 0, \forall i = 1, \ldots, N$ or whenever all $y_i - \beta^T x_i \leq 0, \forall i = 1, \ldots, N$ and the set of occurrences of a subsequence s_j is the same as that of its subsequences $s_p \subseteq s_j$, i.e., $\{i | x_{ij} = 1\} = \{i | x_{ip} = 1\}$.

Proposition 2 (Convergence rate). *The proposed learning algorithm for sequence regression by optimizing the squared loss, converges to the global optimum of the objective function with a convergence rate of*

$$L(\beta^{(t)}) - L(\beta^*) \leq \left[\prod_{r=1}^{t} \left(1 - \frac{\mu}{l_{j_r}} \right) \right] [L(\beta^{(0)}) - L(\beta^*)].$$

Proof. We use recent convergence results for coordinate-descent optimization of functions that are μ-strongly convex, with coordinate-wise l_{j_r}-Lipschitz continuous gradient (e.g., squared loss). In particular, we use coordinate descent with Gauss-Southwell selection for squared loss and exact step size optimization. For a detailed proof see [12].

Algorithm Complexity. The time complexity of the proposed algorithm is $O(fN)$ per iteration, where f is the number of features that need to be investigated for Gauss-Southwell selection and N is the number of training examples. *Implementation.* In practice we use data structures such as inverted indexes and tries to fully take advantage of the sparsity and the nested structure of the feature space. We also investigate empirically the quality of the upper bound by computing the number of distinct features investigated per iteration, and measuring the running time per iteration, and for fully learning a model.

4 Experiments

We evaluate our learning algorithm on synthetic data and two benchmarks from recent data challenges. First, we analyze and compare our method to other linear regression methods on simulated sequence regression data where we vary the data generation parameters. Next, we study a sequence regression problem, to compare our learning algorithm to state-of-the-art sequence regression approaches on real data. Finally, we study a sequence classification problem, in order to compare the squared loss bound effectiveness to related methods developed for sequence classification. We run all experiments on a PC with 132 GB RAM, single Intel Xeon 2.4 GHz CPU and 5.4 TB HDD. All our code and data is available online[6].

4.1 Synthetic Data

In this section we analyze our method (named **SqLoss**) on synthetic data. The controled generation of data allows us to compare SqLoss to the state-of-the-art (SotA) methods in a systematic way. We generate sequence regression datasets according to Algorithm 3. Before the actual sequence generation starts, n motifs have to be generated by drawing m symbols from a given alphabet Σ. For each of these motifs, the influence on the response variable (i.e., the motif weight) is set randomly. The first step of the sequence generation is to define which motifs each sequence contains. The binary indicator variables I_{ij} encode this as in (13). Each of these indicator variables is set to 0 or 1 according to a user set probability. Depending on the value of the indicator variable, an insertion position for the motif in question is determined. Next, the actual generation of the string starts. For each position in the sequence, the algorithm checks if a motif has to be inserted. If not, a random symbol from the alphabet is inserted.

---

[6] Code of our algorithms: https://github.com/svgsponer/SqLoss.

Otherwise the corresponding motif is placed at this position. As soon as the end of the sequence is reached, a score is assigned to the sequence, according to (13), where ϵ is Gaussian noise. This generation process can lead to the case that a motif is present in a sequence by chance. Our implementation checks all generated sequences for unintentionally inserted motifs and replaces them with a random subsequence of the same length as the motif.

$$y_i = \sum_{j=1}^{n} w_j I_{ij} + \epsilon, \text{ where } I_{ij} = \begin{cases} 1, & \text{if sequence } i \text{ contains motif } j \\ 0, & \text{otherwise} \end{cases} \quad (13)$$

Algorithm 3. Generation of Sequence Regression Dataset with $n = 2$ Motifs

---

Input: Number of sequences N, length of sequence L, number of motifs n, motif length m, alphabet Σ.
Output: Dataset with N (sequence, score) pairs.

Generate n motifs by drawing m symbols $\sim U(\Sigma)$
Set weights for each motif
for $i < N$ **do**
    pos1 $\sim U(L)$ if I_{i1}
    pos2 $\sim U(L)$ if I_{i2}
    for $l < L$ **do**
        if $l = pos1$ or $l = pos2$ **then**
            add motif to sequence
        else
            add symbol $c \sim U(\Sigma)$ to sequence
        end if
    end for
    add sequence s_i to data set with $y_i = w_1 I_{i1} - w_2 I_{i2} + \epsilon$
end for

In the following experiment we generate 10,000 sequences of length 5,000 and insert 2 motifs. We compare SqLoss to three regression methods: ordinary least squares (ols), ridge regression (ridge) and linear support vector regression (linsvr). For all these methods we use the implementations in scikit-learn (version 0.17) [13] with default parameters and explicitly generate all k-mers up to $k = 5$. For SqLoss we do not restrict k and use default parameters (see code online). Figures 1 and 2 show the average mean squared error (MSE) and training time over 5 runs, for various alphabet sizes and 4 different motif lengths (3, 5, 7, 10). We note that the SotA methods' performance suffers if the motifs are longer than the maximal extracted k-mer. With increasing alphabet size, this effect vanishes as the density of k-mers feature space decreases. If the k-mer density is low enough, subsequences (e.g., 5-mers) of motifs can already indicate the presence of the whole motif and so be used to learn an appropriate weight. When the motif is shorter than the extracted k-mers, all methods perform similar, even though with increasing alphabet size the SotA methods achieve slightly worse results. We suspect this is caused by overfitting since the feature space becomes huge. Even though this is a simplified regression problem, it is promising to see that SqLoss achieves comparable or better results in different data generation settings, without the need to set k explicitly. This means that our method requires much less feature engineering for achieving good prediction quality.

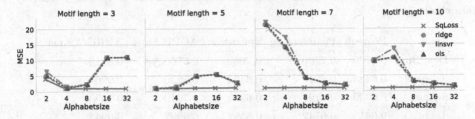

Fig. 1. Average MSE across 5 runs, comparing the impact of varying data generation parameters (alphabet size, motif length) on 4 regression methods. SqLoss has the lowest MSE across all data generation scenarios.

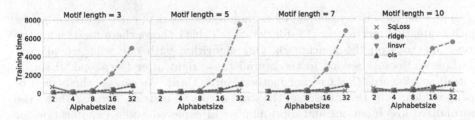

Fig. 2. Average training time (sec) across 5 runs, comparing the impact of varying data generation parameters (alphabet size, motif length) on 4 regression methods.

4.2   TF-DNA Binding Prediction Challenge

The DREAM5 challenge[7] [18] provides 40,000 DNA sequences, each with a numeric score describing the binding affinity of a particular protein called Transcription Factor (TF). The task is to learn a regression function for a given TF. There are 66 TF each defining a different sequence regression task. We use the DREAMTools [4] that allow us to compare our results to the results of the 26 challenge participants [18]. We also show results for DeepBind [2], a recent method that achieved higher scores than previous participants. Of special interest is the comparison to challenge winner (Team_D) [3], a linear regression method using k-mers as features. The authors of [3] pre-process the training data as follows: (1) log2-transform the target scores followed by subtracting the mean; (2) remove all sequences that were flagged as bad by the challenge organizers; (3) remove noisy training sequences; (4) filter low intensity probes and (5) restrict

---

the k-mer length between 4 and 8. We only apply the first two transforms but none of the latter filters, as we want to study our algorithm's effectiveness when using minimal domain knowledge. We also study the influence of using wildcards for allowing more flexible gapped k-mer features (e.g., A*B where * is a extra symbol in Σ which stands for any symbol of the alphabet).

Table 1. TF-DNA: results of top-5 sequence regression methods for the DREAM5 data challenge.

Team	Pearson	Pearson_Log	Spearman	AUPR_8mer	AUROC_8mer
DeepBind	0.6780	0.7260	0.7060	0.6760	0.9910
SqLossWc	0.6483	0.6846	0.6423	0.7236	0.9967
SqLoss	0.6399	0.6791	0.6390	0.7049	0.9953
Team_D	0.6413	0.6742	0.6394	0.6997	0.9942
Team_E	0.6375	0.6936	0.6735	0.5223	0.9524
Team_F	0.6103	0.6732	0.6555	0.5456	0.9766

The final rank is determined by averaging the ranks of each algorithm under each evaluation metric (see [18] for details). Table 1 shows these metrics for the top-5 methods on the benchmark. Our algorithm with one wildcard allowed (SqLossWc) comes second in the overall rank, right after DeepBind. If we do not allow wildcards (SqLoss) our method comes third. We can see the benefit of using wildcards by the increase of the score across all metrics. Figure 3 shows the normalized loss functions and total number of explored nodes per iteration for both models for TF47. Additionally, it shows the Pearson correlation achieved by SqLossWc at each iteration. The flexibility of wildcards clearly increases the number of nodes in the search tree which influences the runtime. SqLoss needs on average 513 s total training time per TF, and SqLossWc takes 2,834 s total training time per TF. Our algorithm reaches similar metrics to Team_D and DeepBind, but the variation without wildcards uses a fraction of the training time of DeepBind and no specific pre-processing as Team_D. The average length of the learned motifs for SqLoss is 5.05 (std across different TFs: 0.16) with a maximum length of 10 bases. For SqLossWc the average motif length is 5.4 (std: 0.09) and the longest learned motif is 11 bases long.

An advantage of our linear model is that it can easily be interpreted, unlike complex non-linear methods such as DeepBind. Table 2 shows some high rank features for TF13 learned with k-mer features with one wildcard allowed. The weights directly reflect the importance of the learned features and can provide important knowledge to domain experts. The learned features are stored in a trie to allow efficient prediction by linearly scanning test sequences. Prediction for 40,000 sequences takes 2 s for SqLoss and 4 s for SqLossWc.

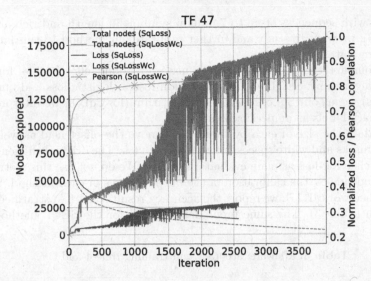

Fig. 3. TF-DNA: loss and explored nodes per iteration, w/o wildcard for protein TF47. In blue SqLoss, in red SqLossWc. Pearson correlation (green) for SqLossWc. (Color figure online)

Table 2. TF-DNA: example positive/negative features from SqLossWc model learned for TF13, where * indicates a wildcard.

Motif	Weight
TAAT*A	0.733985
TAATG*G	0.706344
ATG*AAA	0.674507
⋮	⋮
GGATA	−0.188202
TCAAT	−0.214858
G*ATAG	−0.218132

4.3 Microsoft Malware Classification Challenge

The goal of the Microsoft challenge[8] is to classify files into one of 9 malware families. The training set has 10,868 labeled samples, each with a binary file with hexadecimal representation and a file with the disassembled code. We want to find out if we can build effective classifiers using only the binary (bytes) representation, since disassembling the code requires expensive computation. Even though this dataset poses a classification, rather than a regression task, we use it to compare our method to the challenge winners and to similar approaches developed for linear classification, as implemented in the SEQL framework [8]. In

---

[8] https://www.kaggle.com/c/malware-classification.

addition, with sequences of up to 7 million symbols in length and rich alphabet ($|\Sigma| = 16$), we want to study and further calibrate our method on this large and challenging dataset.

We compare the accuracy, convergence and bound effectiveness for three learning algorithms: SEQL with classification losses (logistic loss and quadratic hinge loss) versus our SqLoss regression algorithm (for SqLoss we interpret one-vs-all binary labels as numeric scores).

To reduce the size of each bytes file we remove the offset field as well as all question marks and white spaces between hexa bytes. The original challenge uses the multiclass log-loss as main evaluation metric. We do not use this metric as it heavily depends on the calibration of the output scores of each method. Similar to published work [1,17], we report the accuracy results on 4-fold stratified cross-validation (Table 3). The same input data is used for the SEQL methods and SqLoss.

Table 3. Malware: accuracy and training time 4-fold CV.

Method	Accuracy	Training time (mins)	
Wang et al. (challenge winner)	0.9983	(multicore, preprocessing only)	2,880
Ahmedi et al.	0.9976	(multicore, preprocessing only)	2,780
SEQL logistic regression loss	0.9958	(singlecore, full training)	603
SEQL quadratic hinge loss	0.9949	(singlecore, full training)	410
SqLoss (our method)	0.9916	(singlecore, full training)	3,183

We note that SqLoss has lower accuracy than the challenge winner [17] and the recent solution of [1]. Nevertheless, both those methods use a variety of hand picked features extracted from both the binary and the disassembled files. Disassembled code is expensive to extract and is inexact, i.e., it is possible for a single program to have two or more disassemblies. To extract features [1,17] have to do costly data preprocessing and heavy feature engineering. In particular [17] have to limit k-mers to a maximum $k = 4$ and need explicit generation of k-mers (for both training and test data) which requires 100 GB memory, 16 CPUs, and 48 h extraction time on the training data alone.

To better analyze our method we compare it to SEQL, a linear classification method that uses branch-and-bound for selecting subsequence features. As we can see in Table 3, the accuracy of the SEQL losses is comparable to that of SqLoss, while training time is better for the classification losses. Figure 4 (left column) shows the value of the three loss functions (normalized by the start loss) per iteration, for classes 1, 2 and 7. As in [15], we also find that the squared error loss decreases slower than the classification loss functions. This is expected given that SqLoss is a regression algorithm used for a classification task. We currently implement the same stopping criterion for all three losses and we believe this may be ill suited for SqLoss, as in practice we could stop the iterations earlier without compromising accuracy. Figure 4 (right column) shows the total number of nodes explored per iteration during the search for the best feature, for each

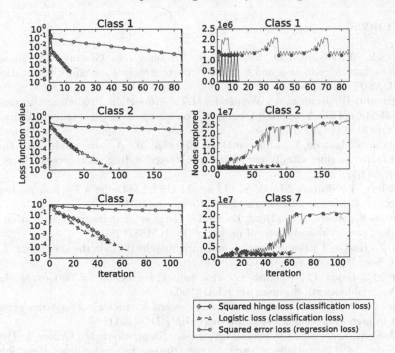

Fig. 4. Malware: comparison of normalized loss function value (left, log scale) and number of nodes explored (right) per iteration for classes 1, 2 and 7. Blue for squared hinge loss, green for logistic regression and red for SqLoss. (Color figure online)

of the three methods. The pruning for SqLoss seems to be less efficient than the one for logistic regression or the hinge loss, on this classification task.

5 Conclusion and Future Work

We present a new method for efficient linear sequence regression in the feature space of all subsequences. The proposed method uses coordinate gradient descent with Gauss-Southwell rule to optimize squared error loss. We propose a branch-and-bound algorithm for efficient Gauss-Southwell selection. Our empirical study shows that we can achieve results comparable to the state-of-the-art, with a simple linear regression model, while employing little to no domain knowledge or pre-processing. In particular, our models can use unrestricted-length, flexible k-mer features (with wildcards), without compromising training and testing efficiency. In the future we want to explore other flexible operators on feature representations of sequences. Further, we want to improve the scalability of our method as well as extend it to make use of multiple cores.

Acknowledgment. This work was funded by Science Foundation Ireland (SFI) under grant number 12/RC/2289.

References

1. Ahmadi, M., Ulyanov, D., Semenov, S., Trofimov, M., Giacinto, G.: Novel feature extraction, selection and fusion for effective malware family classification. In: CODASPY (2016)
2. Alipanahi, B., Delong, A., Weirauch, M.T., Frey, B.J.: Predicting the sequence specificities of DNA- and RNA-binding proteins by deep learning. Nat. Biotechnol. **33**(8), 831–838 (2015)
3. Annala, M., Laurila, K., Lähdesmäki, H., Nykter, M.: A linear model for transcription factor binding affinity prediction in protein binding microarrays. PLoS ONE **6**(5), e20059 (2011)
4. Cokelaer, T., Bansal, M., Bare, C., et al.: DREAMTools: a Python package for scoring collaborative challenges. F1000Research (2016)
5. Dahl, G.E., Stokes, J.W., Deng, L., Yu, D.: Large-scale malware classification using random projections and neural networks. In: ICASSP (2013)
6. Hui, Z., Hastie, T.: Regularization and variable selection via the elastic net. J. Roy. Stat. Soc. B **67**(2), 301–320 (2005)
7. Ifrim, G., Bakir, G., Weikum, G.: Fast logistic regression for text categorization with variable-length n-grams. In: KDD (2008)
8. Ifrim, G., Wiuf, C.: Bounded coordinate-descent for biological sequence classification in high dimensional predictor space. In: KDD (2011)
9. Keilwagen, J., Grau, J., Paponov, I.A., Posch, S., Strickert, M., Grosse, I.: De-novo discovery of differentially abundant transcription factor binding sites including their positional preference. PLoS Comput. Biol. **7**(2), e1001070 (2011)
10. Leslie, C., Eskin, E., Noble, W.S.: The spectrum kernel: a string kernel for SVM protein classification. In: PSB (2002)
11. Leslie, C., Kuang, R.: Fast string kernels using inexact matching for protein sequences. JMLR **5**(Nov), 1435–1455 (2004)
12. Nutini, J., Schmidt, M., Laradji, I.H., Friedlander, M., Koepke, H.: Coordinate descent converges faster with the gauss-southwell rule than random selection. In: ICML (2015)
13. Pedregosa, F., Varoquaux, G., Gramfort, A., Michel, V., Thirion, B., Grisel, O., Blondel, M., Prettenhofer, P., Weiss, R., Dubourg, V., Vanderplas, J., Passos, A., Cournapeau, D., Brucher, M., Perrot, M., Duchesnay, E.: Scikit-learn: machine learning in Python. JMLR **12**(Oct), 2825–2830 (2011)
14. Punta, M., Coggill, P.C., Eberhardt, R.Y., Mistry, J., Tate, J., Boursnell, C., Pang, N., Forslund, K., Ceric, G., Clements, J., Heger, A., Holm, L., Sonnhammer, E.L.L., Eddy, S.R., Bateman, A., Finn, R.D.: The Pfam protein families database. Nucleic Acids Res. **40**(Database issue), D290–D301 (2012)
15. Rosasco, L., De Vito, E., Caponnetto, A., Piana, M., Verri, A.: Are loss functions all the same? Neural Comput. **16**(5), 1063–1076 (2004)
16. Schütz, F., Delorenzi, M.: MAMOT: hidden Markov modeling tool. Bioinformatics **24**(11), 1399–1400 (2008)
17. Wang, X., Liu, J., Chen, X.: Microsoft malware classification challenge (BIG 2015) first place team: say no to overfitting. In: BIG (2015)
18. Weirauch, M.T., Cote, A., Norel, R., Annala, M.: Evaluation of methods for modeling transcription factor sequence specificity. Nat. Biotech. **31**(2), 126–134 (2013)
19. Zhang, Y., Henao, R., Carin, L., Zhong, J., Hartemink, A.: Learning a hybrid architecture for sequence regression and annotation. In: AAAI (2016)

Subjectively Interesting Connecting Trees

Florian Adriaens[(✉)], Jefrey Lijffijt, and Tijl De Bie

IDLab, Department of Electronics and Information Systems,
Ghent University – Imec, Ghent, Belgium
{florian.adriaens,jefrey.lijffijt,tijl.debie}@ugent.be

Abstract. Consider a large network, and a user-provided set of *query* nodes between which the user wishes to explore relations. For example, a researcher may want to connect research papers in a citation network, an analyst may wish to connect organized crime suspects in a communication network, or an internet user may want to organize their bookmarks given their location in the world wide web. A natural way to show how query nodes are related is in the form of a tree in the network that connects them. However, in sufficiently dense networks, most such trees will be large or somehow trivial (e.g. involving high degree nodes) and thus not insightful. In this paper, we define and investigate the new problem of mining *subjectively interesting trees* connecting a set of query nodes in a network, i.e., trees that are highly surprising to the specific user at hand. Using information theoretic principles, we formalize the notion of interestingness of such trees mathematically, taking in account any prior beliefs the user has specified about the network. We then propose heuristic algorithms to find the best trees efficiently, given a specified prior belief model. Modeling the user's prior belief state is however not necessarily computationally tractable. Yet, we show how a highly generic class of prior beliefs, namely about individual node degrees in combination with the density of particular sub-networks, can be dealt with in a tractable manner. Such types of beliefs can be used to model knowledge of a partial or total order of the network nodes, e.g. where the nodes represent events in time (such as papers in a citation network). An empirical validation of our methods on a large real network evaluates the different heuristics and validates the interestingness of the given trees.

Keywords: Exploratory Data Mining · Subjective interestingness Information theory · Graphs · Graph pattern mining

1 Introduction

Given a graph and a set of query nodes, we are interested in connecting these query nodes in a minimal but highly informative manner. *Minimal* in the sense that we are looking for a preferably small subgraph to which the query nodes belong to. *Informative* meaning that our aim is to show a user a subgraph that is highly insightful to them, i.e., the subgraph contains relationships between nodes that are unexpected and surprising to the user. In this paper we consider the case of connecting the query nodes through a subgraph that has a tree structure.

© Springer International Publishing AG 2017
M. Ceci et al. (Eds.): ECML PKDD 2017, Part II, LNAI 10535, pp. 53–69, 2017.
https://doi.org/10.1007/978-3-319-71246-8_4

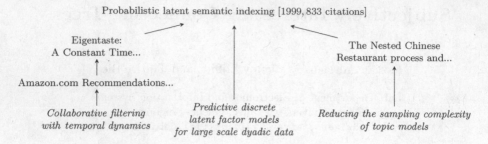

Fig. 1. Tree connecting the three most recent KDD best paper award winners listed at the official ACM SIGKDD webpage (http://www.kdd.org/awards/sigkdd-best-research-paper-awards) that are also present in the Aminer ACM-Citation-network v8 (https://aminer.org/citation, [10]). The result of our algorithm with heuristic s-IR given no background knowledge about the graph. See Sect. 5 for more details.

An example: suppose we have a scientific paper citation network, where edges denote that one paper references another. Given a set of query papers, a directed tree containing these query papers is one possible way to represent interesting citation relationships between these papers. The root of the tree could represent a paper that was (perhaps indirectly) highly influential to all the papers in the query set. Connections between nodes are subjectively interesting if they are surprising. E.g., if a user knows certain papers are widely cited (have high degree), those papers would be less interesting to find in the connecting tree: the user already expects this connection to exist and hence does not learn much.

An example of an informative tree connecting three recent KDD best paper award winners where no prior knowledge about degrees was assumed is given in Fig. 1. Another example application would be to organize your bookmarks by constructing a tree where the bookmarks are the query nodes and the network is the WWW. With a prior containing the degrees of the nodes in the network, an informative tree would partition the bookmarks according to links that are surprising and hence specific to a sub-network (they have low degree). Our method find such trees without doing community detection as an intermediate step.

The main question here is: what makes a certain tree interesting to a given user? We believe that the goal of *Exploratory Data Mining* (EDM) is to increase a user's understanding of his or her data in an efficient way. However, we have to consider that every user is different. It is in this regard that the notion of *subjective interestingness* was formalised [9] and more particularly the creation of the data mining framework FORSIED that we build upon [4,5].

The FORSIED framework specifies in general terms how to model prior beliefs the user has about the data. Given a *background model* representing these prior beliefs, we may find patterns that are highly surprising to the particular user. Hence in our setting, a tree will generally be more interesting if it contains, according to the user's beliefs, more unexpected relationships between the nodes.

This paper contributes the following:

- We define the new problem of finding subjectively interesting trees connecting a set of query nodes in a network. (Sect. 2)
- We show how to formalize a user's knowledge that the graph has a 'DAG'-like structure, for example because the nodes represent events in time. (Sect. 3)
- We propose heuristics for mining the most interesting trees efficiently in the case of directed graphs. (Sect. 4)
- We evaluate and compare the effectiveness of these heuristics on real data and study the utility of the resulting trees, showing that the results are truly and usefully dependent on the assumed prior beliefs of the user. (Sect. 5)

2 Subjectively Interesting Trees in Graphs

We denote a network (aka graph) G as $G = (V, E)$, where V is the set of nodes (aka vertices) and $E \subseteq V \times V$ is the edge set. We denote the adjacency matrix of a graph as $\mathbf{A}$, where $\mathbf{A}_{ij} = 1$ iff there is an edge connecting node i to j, i.e., iff $(i, j) \in E$. The main focus of this paper will be on directed networks. However, our methods directly apply to undirected networks, when considered as a special case where $\mathbf{A}$ is symmetric. We assume that the set of nodes V is fixed and known, and the user is interested the network's connectivity, i.e., the edge set E, especially in relation to a set of so-called *query nodes* $Q \subseteq V$.

2.1 Trees Connecting Query Nodes as Data Mining Patterns

The data mining process we consider is query-driven: the user provides a set of query nodes $Q \subseteq V$ between which they suspect connections exist in the graph that might be of interest to them. In response to this query, the methods proposed in this paper will thus provide the user with a tree-structured sub-network connecting the query nodes. We consider trees because they are easy to interpret. We refer to the presence of a tree as a *pattern* found in the network.

Formally, a tree $T = (V_T, E_T)$ is a network over the nodes $V_T \subseteq V$ with edges $E_T = \{e_1, \ldots, e_{|V_T|-1}\} \subseteq V_T \times V_T$, where $e_i(2) \neq e_j(2)$ for $i \neq j$ (i.e., each node has only one parent). The tree $T = (V_T, E_T)$ is said to be present in the network $G = (V, E)$ iff $V_T \subseteq V$ and $E_T \subseteq E$. The methods proposed below search for interesting trees $T = (V_T, E_T)$ present in the network $G = (V, E)$ with $Q \subseteq V_T$.

Remark 1. *The above description is a special type of tree: a rooted arborescence. This is a tree-structured directed sub-network with a unique directed path between the root and each of the leaves. The edges all point away from the root (out-arborescence), but by reversing all edge directions also in-arborescences can be considered. We will simply refer to the considered patterns as trees.*

2.2 Subjective Interestingness

The FORSIED framework aims to quantify interestingness of a pattern in a subjective manner, dependent on prior beliefs the user holds about the data. To model the user's belief state about the data, the framework proposes to use a so-called *background distribution*, which is a probability distribution P over the data space (in our setting, the set of all possible edge sets E). It was argued that a good choice for the background distribution is the maximum entropy distribution subject to the prior beliefs as constraints [4,5].

The FORSIED framework then prefers patterns that achieve a trade-off between how much information the pattern conveys to the user (considering their belief state), versus the effort required of the user to assimilate the pattern. Specifically, De Bie [4] argued that the *Subjective Interestingness (SI)* of a pattern can be quantified as *the ratio of the Information Content (IC) and the Description Length (DL)* of a pattern. The IC is defined as the negative log probability of the pattern w.r.t. the background distribution P. The DL is quantified as the length of the code needed to communicate the pattern to the user.

The IC of a tree. The background distributions P for all prior belief types discussed in this paper have the property that P factorizes as a product of independent Bernoulli distributions[1], one for each possible edge $e \in V \times V$. Hence the IC of a tree T with edges E_T decomposes as

$$\text{IC}(T) = -\log\left(\prod_{e \in E_T} \Pr(e) \right) = \sum_{e \in E_T} \text{IC}(e), \tag{1}$$

where we defined the IC of an edge e to be $\text{IC}(e) = -\log(\Pr(e))$.

The DL of a tree. A tree can be described by first describing the set of nodes V_T and then the set of edges E_T over this set of nodes. To describe the set $V_T \subseteq V$ efficiently, note that $Q \subseteq V_T$ such that only $V_T \backslash Q$ needs to be described. This can be done using a sequence of $|V_T| - |Q| + 1$ symbols from $V \backslash Q \cup \{\text{'stop'}\}$, where the last one is a stop symbol. This results in a description length of $(|V_T| - |Q| + 1)\log(|V| - |Q| + 1)$ bits for V_T. Given V_T, E_T can be described by listing the parents of all nodes from within $V_T \cup \{\text{'none'}\}$, where the 'none' symbol is used for the root. This requires $|V_T|\log(|V_T|+1)$ bits. Thus:

$$\text{DL}(T) = (|V_T| - |Q| + 1)\log(|V| - |Q| + 1) + |V_T|\log(|V_T|+1). \tag{2}$$

2.3   Finding Subjectively Interesting Trees

The methods presented in this paper aim to solve the following problem:

Problem 1. *Given a graph $G = (V, E)$ and set of query nodes $Q \subseteq V$, we want to find a root $r \in V$ and an out-arborescence rooted at r, such that the arborescence is maximally subjectively interesting. We additionally require that all leaf nodes are query nodes, and we constrain the height of the tree not to be larger than a user-defined parameter k.*

---

[1] This just happens to be true for the studied prior beliefs. This may indeed reduce computational complexity and it surely reduces the complexity of exposition.

Since the SI depends on the background distribution and thus on the user's prior beliefs, the optimal solution to Problem 1 does as well. As stated in Remark 1, by transposing the adjacency matrix $\mathbf{A}$, we can equivalently consider in-arborescences in exactly the same manner.

3 The Background Distribution to Model the User Beliefs

As mentioned, the background distribution is computed as the maximum entropy distribution subject to the prior beliefs as constraints. Here we discuss how this is done in detail for three types of prior beliefs: (1) on the overall edge density; (2) on the individual node degrees; and () for networks with nodes that correspond to timed events, on the tendency of nodes to be connected to nodes *at a specified time difference* (as well as generalization thereof). These prior beliefs can be combined as well. Note that (1) and (2) were introduced before in [6].

3.1 Prior Beliefs on Overall Density, and on Individual Node Degrees

As shown in [6], given prior beliefs on the degrees of the nodes, the maximum entropy distribution factorizes as:

$$P(\mathbf{A}) = \prod_{i,j} \frac{\exp((\lambda_i^r + \lambda_j^c)\mathbf{A}_{ij})}{1 + \exp(\lambda_i^r + \lambda_j^c)},$$

where λ_i^r and λ_j^c are parameters from the resulting optimization problem. [3] showed how these can be computed efficiently. For a prior belief on the overall graph density, every edge probability in the model equals the assumed density.

3.2 Prior Beliefs When Nodes Represent Timed Events

If the nodes in G correspond to events in time, we can partition the nodes into bins according to a time-based criterion. For example, if the nodes are scientific papers in a citation network, we can partition them by publication year. Given these bins, it is possible to express prior beliefs on the number of edges between two bins. This would allow one to express e.g. beliefs on how often papers from year x cite papers from year y. This is useful e.g. if one believes that papers cite recent papers more often than older ones.

We consider the case when our beliefs are in line with a *stationarity* property, i.e. when the beliefs regarding two bins are independent of the absolute value of the time-based criterion of these two bins, but rather only depend on the time difference. Given an adjacency matrix $\mathbf{A}$, this amounts to expressing prior beliefs on the total number of ones in each of the block-diagonals of the resulting block matrix (formed by partitioning the elements into bins), see Fig. 2 for clarification.

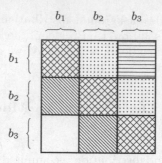

Fig. 2. A resulting block matrix with 3 bins b_1, b_2 and b_3. There are 5 block-diagonals D_k (indicated by the same fill). For each D_k, we express prior beliefs on the sum of all elements in D_k.

We consider the problem of finding the maximum entropy distribution over the set of rectangular binary matrices $\mathcal{A} = \{0,1\}^{n \times n}$, while constraining the expectation of the sum of the elements in each of the block diagonals, as well each of the row and column sums. It is found by solving:

$$\underset{P(\mathbf{A})}{\arg\max} - \sum_{\mathbf{A} \in \mathcal{A}} P(\mathbf{A}) \log P(\mathbf{A}),$$

$$\text{s.t.} \sum_{\mathbf{A} \in \mathcal{A}} P(\mathbf{A}) \sum_{j=1}^{n} \mathbf{A}_{ij} = d_i^r, \quad \sum_{\mathbf{A} \in \mathcal{A}} P(\mathbf{A}) \sum_{i=1}^{n} \mathbf{A}_{ij} = d_j^c,$$

$$\sum_{\mathbf{A} \in \mathcal{A}} P(\mathbf{A}) \sum_{(i,j) \in D_k} \mathbf{A}_{ij} = B_k,$$

$$\sum_{\mathbf{A} \in \mathcal{A}} P(\mathbf{A}) = 1,$$

with $i, j \in \{1, \dots, n\}$ and $k \in \{1, \dots, 2\#\text{bins} - 1\}$, and with d_i^r the expected sum of the i'th row, d_j^c the expected sum of the j'th column, and B_k the expected sum of the k'th block diagonal D_k. The resulting maximum entropy distribution factorizes as a product of independent Bernoulli distributions, one for each random variable $\mathbf{A}_{ij} \in \{0,1\}$:

$$P(\mathbf{A}) = \prod_{i,j} \frac{\exp((\lambda_i^r + \lambda_j^c + \alpha_k)\mathbf{A}_{ij})}{1 + \exp(\lambda_i^r + \lambda_j^c + \alpha_k)}, \tag{3}$$

where λ_i^r, λ_j^c and α_k are the Lagrange multipliers for the corresponding row, column and block-diagonal constraints. These Lagrange multipliers are found by minimizing the Lagrange dual function, as given by:

$$L(\lambda^r, \lambda^c, \alpha) = \sum_{i,j} \log(1 + \exp(\lambda_i^r + \lambda_j^c + \alpha_k)) - \sum_i \lambda_i^r d_i^r - \sum_j \lambda_j^c d_j^c - \sum_k \alpha_k B_k.$$

Standard methods for unconstrained convex optimization such as Newton's method can be used to infer the optimal values. The number of variables to be optimized over is equal to $2(n + \#bins) - 1$, where $1 \leq \#bins \leq n$. Using Newton's method then requires solving a linear system of $O(n)$ equations, with computational complexity $O(n^3)$. For practical problems involving large networks, this quickly becomes infeasable. However, with a similar argument as in [3], we can dramatically reduce the number of variables. Observe that if $d_k^r = d_l^r$ and k and l belong to the same bin, then we have $L(\ldots, \lambda_k^r, \lambda_l^r, \ldots) = L(\ldots, \lambda_l^r, \lambda_k^r, \ldots)$. The convexity of L implies $\lambda_k^r = \lambda_l^r$ at the optimum. A similar argument holds for the λ^c parameters. Thus the number of free variables per bin to be optimized over, is bounded by the number of distinct row and column sums *per bin*.

Let $\widetilde{m}$ be the total number of free row variables, and $\widetilde{n}$ be the total number of free column variables. The following Lemma provides an upper bound on $\widetilde{m} + \widetilde{n}$ in terms of the number of non-zero elements of $\mathbf{A}$ and the number of bins k:

Lemma 1. *Let $\mathbf{A}$ be a binary rectangular matrix and denote $s = \sum_{i,j} \mathbf{A}$. Then it holds that $\widetilde{m} + \widetilde{n} \leq 2\sqrt{2ks}$.*

Proof. Let $\widetilde{m}_i$ be the number of distinct row variables in the i-th bin and similarly for $\widetilde{n}_i$ with $i \in \{1, \ldots, k\}$. Let s_i (s_i') be the total number of ones in all the rows (columns) of the elements in bin i. Then the following inequalities hold [3]:

$$\widetilde{m}_i \leq \sqrt{2s_i}, \quad \text{and} \quad \widetilde{n}_i \leq \sqrt{2s_i'}.$$

Hence $\widetilde{m} + \widetilde{n} \leq \sqrt{2}(\sqrt{s_1} + \ldots + \sqrt{s_k'})$. Clearly also $\sum_i s_i + s_i' = 2s$ and thus by Jensen's inequality $\sqrt{s_1} + \ldots + \sqrt{s_k'} \leq 2\sqrt{ks}$, which proves the lemma. □

Denote $\widetilde{\lambda}_{k,l}^r$ as the l-th unique row parameter in the k-th bin. Denote the corresponding row sum constraint as $\widetilde{d}_{k,l}^r$ having $\widetilde{m}_l^k$ occurences in that bin. Similarly for $\widetilde{\lambda}_{k,l}^c$, $\widetilde{d}_{k,l}^c$ and $\widetilde{n}_l^k$. Denote $\alpha_{kk'}$ as the α parameter of the $\mathbf{A}_{ij}$ elements with $i \in$ bin k and $j \in$ bin k'. The reduced Lagrange dual function then becomes

$$L(\widetilde{\lambda}^r, \widetilde{\lambda}^c, \alpha) = \sum_k \sum_{k'} \sum_l \sum_{l'} \widetilde{m}_l^k \widetilde{n}_{l'}^{k'} \log(1 + \exp(\widetilde{\lambda}_{k,l}^r + \widetilde{\lambda}_{k',l'}^c + \alpha_{kk'}))$$

$$- \sum_k \sum_l \widetilde{m}_l^k \widetilde{d}_{k,l}^r \widetilde{\lambda}_{k,l}^r - \sum_{k'} \sum_{l'} \widetilde{n}_{l'}^{k'} \widetilde{d}_{k',l'}^c \widetilde{\lambda}_{k',l'}^c - \sum_m \alpha_m B_m.$$

The gradient is computed as

$$\frac{\partial L}{\partial \widetilde{\lambda}_{k,l}^r} = \sum_{k'} \sum_{l'} \widetilde{m}_l^k \widetilde{n}_{l'}^{k'} \frac{\exp(\widetilde{\lambda}_{k,l}^r + \widetilde{\lambda}_{k',l'}^c + \alpha_{kk'})}{1 + \exp(\widetilde{\lambda}_{k,l}^r + \widetilde{\lambda}_{k',l'}^c + \alpha_{kk'})} - \widetilde{m}_l^k \widetilde{d}_{k,l}^r, \quad (4)$$

$$\frac{\partial L}{\partial \widetilde{\lambda}_{k',l'}^c} = \sum_k \sum_l \widetilde{m}_l^k \widetilde{n}_{l'}^{k'} \frac{\exp(\widetilde{\lambda}_{k,l}^r + \widetilde{\lambda}_{k',l'}^c + \alpha_{kk'})}{1 + \exp(\widetilde{\lambda}_{k,l}^r + \widetilde{\lambda}_{k',l'}^c + \alpha_{kk'})} - \widetilde{n}_{l'}^{k'} \widetilde{d}_{k',l'}^c, \quad (5)$$

$$\frac{\partial L}{\partial \alpha_k} = \sum_{D_k} \frac{\exp(\widetilde{\lambda}_{k,l}^r + \widetilde{\lambda}_{k',l'}^c + \alpha_k)}{1 + \exp(\widetilde{\lambda}_{k,l}^r + \widetilde{\lambda}_{k',l'}^c + \alpha_k)} - B_k. \quad (6)$$

and a similar expression for the Hessian. In all cases (rows, columns and block diagonal) the corresponding gradient is simply the difference between the expected number of ones and the corresponding parameter as given by the constraints. When applying Newton's method to the reduced model, we need $O(\widetilde{m}\widetilde{n})$ calculations to compute both the gradient and Hessian. After that we need to solve a linear system with $\widetilde{m} + \widetilde{n} + 2k - 1$ equations, with cubic complexity. By Lemma 1, this is $O(\sqrt{k^3 s^3} + k^3)$, making it very efficient in many real life applications (sparse networks and a small number of bins).

Remark 2. *Note that we are not limited to the case of stationarity, nor is it necessary that nodes correspond to timed events. Expressing prior beliefs on the density of* any *particular subset of edges* is possible in a similar manner. We tackled this specific case because it directly applies to the data used in this paper.

4   Algorithms for Finding the Most Interesting Tree

The problem of finding a directed Steiner arborescence (spanning all the query nodes) with maximum SI is NP-hard in general, as can be seen from the case of constant edge weights (e.g., if the prior belief is the overall graph density). In this case the SI of a tree will be a decreasing function of the number of nodes in the tree. Hence the problem is equivalent to the minimum Steiner arborescence problem, with constant edge weights, which is NP-hard. For nonconstant background models it will be a trade-off between the IC and the DL of a tree. In most cases, we are looking for small trees with highly informative edges.

There are a number of algorithms that provide good approximation bounds for the directed Steiner problem [2,7,11], and this problem has also been studied recently in the data mining community, e.g., [1,8]. However, Problem 1 is equivalent to the Steiner problem in the case of a uniform background distribution, i.e., when the IC of the edges is constant and hence irrelevant. In general, we aim to solve a *maximization* problem, while Steiner tree problems aim to *minimize* the cost of the tree. For this reason we propose fast heuristics for large graphs, that perform well on different kinds of background distributions.

A Python implementation of the algorithms and the experiments is available at http://www.interesting-patterns.net/forsied/sict/.

4.1   Proposed Heuristics

Our proposed methods all work in a similar way. We apply a preprocessing step, resulting in a set of candidate roots. Given a candidate root r, we build the tree by iteratively adding edges (parents) to the *frontier*—initialized as $Q\backslash\{r\}$—, until *frontier* is empty. We exhaustively search over all candidate roots and select the best resulting tree. The heuristics differ in the way they select allowable edges. The outline of SteinerBestEdge is given in Algorithm 1.

Preprocessing. All of the proposed heuristics have two common preprocessing steps. First we find the common roots of the nodes in Q up to a certain level k, meaning we look for nodes r, s.t. $\forall q \in Q : \text{SPL}(q, r) \leq k$, with $\text{SPL}(\cdot)$ denoting the shortest path length. This can be done using a BFS expansion on the nodes in Q until the threshold level k is reached. Note that query nodes are also potential candidates for being the root, if they satisfy the above requirement.

Secondly, for each r we create a subgraph $H \subset G$, consisting of all simple paths $q \rightsquigarrow r$ with $\text{SPL}(q, r) \leq k$, for all $q \in Q$. This can be done using a modified DFS-search. The number of simple paths can be large. However, we can prune the search space by only visiting nodes that we encountered in the BFS expansion, making the construction of H quite efficient for small k.

SteinerBestEdge. Given the subgraph H, we construct the arborescence working from the query nodes up to the root. We initialize the frontier as $Q \setminus \{r\}$, and iteratively add the best feasible edge to a partial solution, denoted as *Steiner*, according to a greedy criterion. The greedy criterion is based on the ratio of the IC of that edge to the DL[2] of the partial *Steiner* that would result from adding that edge. This heuristic prefers to pick edges from a parent node that is already in *Steiner*, yielding a more compressed tree and thus a smaller DL.

Algorithm 2 checks if an edge is *feasible* by propagating its potential influence to all the other nodes in H. The check can fail in two ways. First, the addition of an edge could yield a *Steiner* tree with height $> k$, see Fig. 3 for an example. Secondly, the addition of an edge may lead to cycles in *Steiner*. Cycles are avoided by only considering edges (s, t) that do not potentially change $\text{SPL}(t, r)$. If $\text{SPL}(t, r)$ would change, the shortest path –given the current *Steiner*– from s to r is not along the edge (s, t) and hence for all $f \in frontier$ we always have 1 feasible edge to pick (i.e. an edge that is part of a shortest path $f \rightsquigarrow r$). One way to select the best feasible edge is to first sort the edges according to the greedy criterion. Then try the check from Algorithm 2 on this sorted list (starting with the best edge(s)), until the first success, and add the resulting edge to *Steiner*. Algorithm 2 will also return an updated shortest path function *NewSP*, containing all the changes in $\text{SPL}(n, r)$ for $n \in H$ due to the addition of that edge to *Steiner*. After performing the necessary updates on the *SP* function, and the *frontier*, *parents* and *level* sets, we continue to iterate until *frontier* is empty.

SteinerBestIC. Instead of adding 1 edge at a time, this heuristic adds multiple edges at once. We look for the parent node that (potentially) adds the most total information content of allowable edges to the current *Steiner*. However, given such a parent node, it not always possible to add multiple edges, see Fig. 4. Instead we sort the edges coming from such a parent node according to their IC, and iteratively try to add the next best edge to *Steiner*.

SteinerBestIR. A natural extension of SteinerBestIC is to actually take in account the DL of the partial *Steiner* solution, as we did in SteinerBestEdge.

---

[2] Note that during construction, the partial solution *Steiner* is often a forest. However, we compute the DL as if it was an equally sized tree. This makes sense because the end result will in fact be a tree, and we are optimizing towards the IR of that tree.

Algorithm 1. SteinerBestEdge(subgraph H, root r, queryset Q, maxlevel k)

1: $Steiner \leftarrow Q \cup \{r\}$
2: $\forall x \in H : SP(x) \leftarrow$ ShortestPathLength(r, x)
3: $frontier \leftarrow Q$.remove(r)
4: $level(frontier) \leftarrow 0$
5: $parents \leftarrow$ parents$(frontier)$
6: **while** frontier **do**
7:      $n \leftarrow$ length$(Steiner), ranking \leftarrow \emptyset$
8:      **for all** $edges$ from $parents$ to $frontier$ **do**
9:          **if** $parent \in Steiner$ **then**
10:              $ranking$.add$(\{edge,$ IC$(edge)/$DLTree$(n)\})$
11:          **else**
12:              $ranking$.add$(\{edge,$ IC$(edge)/$DLTree$(n + 1)\})$
13:      $ranking \leftarrow$ sort$(ranking)$
14:      **for** $edge \in ranking$ **do**
15:          **if** CheckChildren$(H, edge, Steiner, k, SP)$ is True **then**
16:              $update(Steiner, frontier, parents, level, SP)$
17:              Quit loop
18: **return** $Steiner$

SteinerBestIR favors parent nodes that are already in $Steiner$, steering towards an even more compressed tree.

SteinerBestEdgeBestIR. Our last method simply picks the single best edge coming from the best parent, where the best parent is determined by the same criteria as in SteinerBestIR. In general this will pick a locally less optimal edge than SteinerBestEdge, but it will pick edges from a parent node that has lots of potential to the current $Steiner$ solution.

Correctness of the solutions. The following theorem states that all the heuristics indeed result in a tree with maximal height $\leq k$.

Theorem 1. *Given a non-empty query set Q, a candidate root r and a height $k \geq 1$. In all cases all four heuristics will return a tree with height $\leq k$.*

Proof. In all cases the proposed heuristics return a tree rooted at r with height $\leq k$. We call a partial forest solution $Steiner$ valid, if for all leaf nodes $l \in Steiner$: SPL$(l, r|Steiner) \leq k$, where $SPL(\cdot|Steiner)$ denotes a shortest path length given the partial $Steiner$ solution. Note that the initial $Steiner$ is valid, due to the way the subgraph H was constructed. It is always possible to go from one valid $Steiner$ solution to another valid one, by selecting an edge (incident to a frontier node) along a shortest path—given we have $Steiner$—from r that frontier node. This will result in an unchanged SPL for all other nodes (in particular the leaf nodes), and hence remains a valid $Steiner$. If we have n frontier nodes, we have at least n such valid edges to pick from. Hence, all of the heuristics have at least $n \geq 1$ valid edges to pick from. The process of adding edges is finite, and will eventually result in an arborescence rooted at r with height $\leq k$.

Algorithm 2. CheckChildren($H, edge, Steiner, k, SP$)

1: $frontier \leftarrow$ frontier($Steiner$), $level \leftarrow$ level($Steiner$), ($source, target$) $\leftarrow edge$
2: $NewSP \leftarrow \{source : SP(target) + 1\}$
3: **if** $NewSP(source) = SP(target)$ **then return** True
4: **else if** $NewSP(source) + level(source) > k$ **then return** False
5: **else**
6:     $children \leftarrow$ children($source$)

7: **while** $children$ **do**
8:     $nextChildren \leftarrow \emptyset$
9:     **for** $c \in children$ **do**
10:         **if** $c \notin Steiner \setminus frontier$ **then**
11:             $updatedP =$ parents(c) $\cap NewSP$
12:             $otherP =$ parents(c) $\setminus updatedP$
13:             $cand \leftarrow \min(NewSP(p) : p \in updatcdP) + \min(SP(p) : p \in otherP) + 1$
14:             **if** $cand > SP(c)$ **then**
15:                 $NewSP(c) = cand$
16:                 **if** $c \in query$ and $NewSP(c) > k$ **then return** False
17:                 **if** c is $target$ **then return** False   ▷ Possible cycle avoided
18:                 $nextChildren$.add(children(c))
19:             **else**
20:                 $NewSP(c) = NewSP(SteinerParent(c)) + 1$
21:                 **if** c is $target$ **then return** False   ▷ Possible cycle avoided
22:                 $nextChildren$.add(children(c))
23:     $children \leftarrow nextChildren$
24: **return** True, $NewSP$

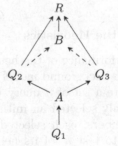

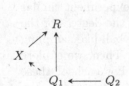

Fig. 3. Example of why look-ahead is needed to ensure the returned tree has depth as most k. If $k = 2$, the only valid tree is $(Q_1, R)(Q_2, Q_1)$. Initially, the *frontier* is $\{Q_1, Q_2\}$ and X is a candidate parent for Q_1 because there is a path $Q \rightsquigarrow R$ of at most length 2. Yet, adding the dashed edge violates the shortest path constraint for Q_2.

Fig. 4. Example of why look-ahead is needed for *sets of edges*. For $k = 3$, neither of the two dashed edges violate the depth constraint—they are part of a valid tree—, but together they indirectly violate the shortest path constraint for Q_1. Regardless of which parent is chosen for A, the path from Q_1 to R has length 4.

5   Experiments

In this section we empirically evaluate our proposed methods on real data. All experiments are based on the ACM-Citation-network v8[3], a scientific paper citation network. This (directed) network contains 2,381,688 papers and 10,476,564 citations. The oldest paper is a seminal paper of C.E. Shannon from 1938. The most recent papers are from 2016. We will use the acronyms s-E, s-IC, s-IR and s-EIR for resp. SteinerBestEdge, SteinerBestIC, SteinerBestIR and SteinerBest-EdgeBestIR. First, we evaluate and compare the performance of the heuristics.

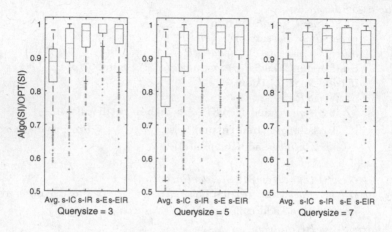

Fig. 5. The interestingness of the heuristics (relative to the optimal interestingness) versus querysize. We also compare with the average interestingness over all trees. Note the decrease in performance of s-E for larger querysizes.

5.1   Comparing the Heuristics

To compare the performance of the heuristics we set up an experiment similar to [1]. We fitted the background model with prior beliefs on the degrees of the network. To generate a set of n query nodes we used a snowball-like sampling scheme. We randomly selected an initial node in the graph. Then, we explore $n' < n$ of its neighbors, each selected with probability s. For each of these nodes we continue to test n' of its neighbors until we have n selected nodes. From this query set we randomly select a valid common root within a maximum distance k. To have a baseline, we find the arborescence with maximal SI using exhaustive search. To keep this comparison feasible, we only consider cases where the number of trees is <200,000. For querysizes = {3, 5} we generated 1000 query sets, for querysize = 7 we have done 250 sets. In all cases k was limited to 3, the beamwidth n' was chosen to be 2 and sampling rates $s \in \{0.1, \ldots, 0.9\}$.

Figure 5 shows a boxplot of the interestingness scores of the tree-building heuristics (relative scores to the optimal arborescence interestingness) versus

---

3 https://aminer.org/citation, [10].

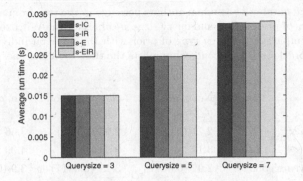

Fig. 6. Average run time of the heuristics. The main bottleneck is finding all the simple paths, which is included here. Hence, the run time differences are small.

query size. All four heuristics clearly are better strategies than randomly selecting an arborescence (the Avg. case). s-IR outperforms s-IC in all cases, which makes sense because s-IC has no regard for the DL of the tree. s-E performs comparatively worse for larger query sizes, and s-IR seems to be the best option for larger querysizes. This result is not definite, it could be due to the fact that we fixed the height at $k = 3$. However, while not reported here, we observed that s-IR is also the best option for larger query sizes and larger k. We also tested the effect of the sampling rate (not shown), and this appeared to affect neither the SI of the resulting trees, nor the ranking of the algorithms.

Figure 6 shows the average run time of our methods. The run time of the heuristics are all negligable compared to the time needed to find all simple paths from the root to the query nodes (see Sect. 4), which in all cases takes up more than 90% of the total running time. We conclude that with prior beliefs on the individual node degrees, s-IR seems to be best option for larger queries, while for smaller queries s-E seems to give a more interesting tree.

5.2 The Effect of Different Prior Beliefs, and a Subjective Evaluation

Here we evaluate the outcome of our heuristics w.r.t. three different kinds of prior beliefs on the ACM citation network. The first prior belief is on the overall graph density. In this case, every edge has the same probability in the background model and hence same information content. The optimal arborescence then is the smallest Steiner arborescence. The second set of prior beliefs is on the individual degree of each node. As for a citation network the number of citations a paper has is easier to estimate than the number of references of that paper (without reading it), we only constrained the expected in-degree of each node. As a result, edges to highly cited nodes are more probable, and a tree will be more interesting if it is not only small, but has a preference for less frequently cited papers. The final type of prior belief is on both the individual in-degree of each node, as well as the dependency of citation probabilities on the difference in publication date.

Table 1. Average number of common authors per edge in the tree from algorithm s-IR for different types of prior beliefs and query sizes. p-values for the Wilcoxon signed-rank test (pairwise comparison) of each type of prior with the prior on individual degrees, shown between brackets. The second column lists the time to fit the background model on the full data.

Prior beliefs	Time (s)	No. of authors, $|Q| = 5$	No. of authors, $|Q| = 7$	No. of authors, $|Q| = 9$
Overall density	-	2.71 (0.0035)	2.94 (0.005)	3.66 (0.0044)
Indiv. degrees	22	3.17	3.34	4.09
Bin every 3 years	380	2.92 (0.007)	3.12 (0.0417)	3.9 (0.0996)
Bin every 5 years	254	2.80 (5.63e−04)	2.84 (9.05e−06)	3.71 (0.0049)

In Sect. 3.2 we showed how to formalize prior beliefs on diagonal block sums. Here it is natural to group papers together according to their publication year (or per 2 years, 5 years, ...). In this way, it is possible to incorporate prior beliefs such as: "The number of papers from year X citing a paper from year $X - 3$ is high". In general, an edge will have a high probability if the corresponding expected block diagonal sum is high, see Eqs. (3) and (6). Note that the citation network should (in theory) be a directed acyclic graph, since no paper can cite a paper with a higher publication year. Yet the data contains 66,772 (<0.01%) violating edges, which our method handles gracefully.

Common authors as external validation. In many scientific fields, self-citations are common practice. We expect the trees to reflect this to differing degrees, depending on the prior beliefs taken into account. To test this, we set up an experiment similar as in Sect. 5.1. The queries are generated in the same way, but with a preference for queries that have some authors in common. If a paper has an author in common with the current query set, it is automatically chosen instead of being sampled with probability s. We generated 200 random queries for each querysize $\{5, 7, 9\}$, with max. height $k = 4$. For each query, we look at the tree generated by s-IR, computed for 4 different types of prior beliefs. Our measure is the total number of common authors per edge in the tree.

Table 1 shows the results for 4 types of prior beliefs. There is a substantial difference between the first and second prior. This makes sense because with a constant background model, s-IR is indifferent to the number of citations of papers. With the second prior, s-IR prefers nodes with fewer citations, penalizing highly cited papers. This means we are also favoring self-citations a bit more, since chances are high that nodes encountered in our experiment do not have authors in common for references to seminal papers. Secondly, there are differences in the self-citation rate between a prior on the time relations (priors 3 and 4) versus prior 2. Most people stop publishing after their PhD, but during that time they will have some references to their own papers. Hence with a background model of type 3 and 4, s-IR will prefer citations between papers with a high difference in publication year, making self-citations less common.

Subjective evaluation. We queried three recent KDD best paper award winners that were present in the network, see Figs. 1, 7 and 8 for results for different prior beliefs. We used $k = 3$, resulting in 33 candidate roots. Notice the number of citations and the publication year of the root in each of the resulting trees, confirming our expectations of the influence of the prior beliefs on the SI of trees.

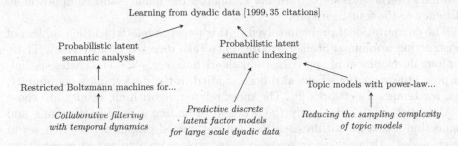

Fig. 7. Like Fig. 1, with as prior knowledge the degree of each node.

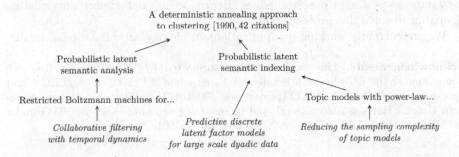

Fig. 8. Like Fig. 1, with as prior knowledge degrees and time constraints.

6   Discussion and Related Work

We studied the problem of finding interesting trees that connect a user-provided set of query nodes in a large network. This is useful for example to, based on citation data, find papers that (indirectly) influenced a set of query papers, perhaps to understand the structure of an organization from communication records, and in many other settings. We defined the problem of finding such trees as an optimization problem to find an optimal balance between the informativeness (the Information Content) and conciseness (the Description Length) of a tree. Additionally, by encoding the prior beliefs of a user, we propose how to find results that are surprising and interesting to a specific user.

We have introduced a general algorithmic strategy to construct such trees along with four heuristics of varying complexity. We have introduced a tractable model to include prior knowledge about the density of sub-networks and more specifically for the case where the nodes appear in time blocks and the probability of edges is expected to be a function of time. Finally, we evaluated the interestingness of the results in several experiments, both subjectively and using external criteria, plus we empirically compared the quality and computational efficiency of the four heuristics.

The computational problem solved in this paper is related to the problem of constructing a minimal Steiner arborescence (aka directed Steiner tree). There is a long development of approximation algorithms, e.g., [2,7,11]. Faster special-purpose approximations have also been studied in the data mining community, e.g., for temporal networks [8]. The most related algorithmic results are those of Akoglu et al. [1], who study the problem of finding a good partitioning and connection structure within each part on undirected graphs for a given set of query nodes. Although their purpose is to explore an undirected graph, they map the problem to graph partitioning plus finding Steiner arborescences.

It should be noted that Problem 1 is *not* equivalent to the Steiner arborescence problem, because in general the subjective interestingness of a tree does not factorize as a sum over the edges. Hence, we do not expect any existing algorithm to solve this problem well.

We are currently working on applications in biology as well as social media.

Acknowledgements. This work has been supported by the European Research Council under the EU's Seventh Framework Programme (FP/2007-2013)/ERC Grant Agreement no. 615517, the FWO (project nos. G091017N, G0F9816N), and the European Union's Horizon 2020 research and innovation programme and the FWO under the Marie Sklodowska-Curie Grant Agreement no. 665501.

References

1. Akoglu, L., Chau, D.H., Faloutsos, C., Tatti, N., Tong, H., Vreeken, J.: Mining connection pathways for marked nodes in large graphs. In: Proceedings of SDM, pp. 37–45 (2013)
2. Charikar, M., Chekuri, C., Cheung, T.Y., Dai, Z., Goel, A., Guha, S., Li, M.: Approximation algorithms for directed Steiner problems. In: Proceedings of SODA, pp. 192–200 (1998)
3. De Bie, T.: Maximum entropy models and subjective interestingness: an application to tiles in binary databases. Data Min. Knowl. Disc. **23**(3), 407–446 (2011)
4. De Bie, T.: An information theoretic framework for data mining. In: Proceedings of KDD, pp. 564–572 (2011)
5. De Bie, T.: Subjective interestingness in exploratory data mining. In: Proceedings of IDA, pp. 19–31 (2013)
6. van Leeuwen, M., De Bie, T., Spyropoulou, E., Mesnage, C.: Subjective interestingness of subgraph patterns. Mach. Learn. **105**(1), 41–75 (2016)
7. Melkonian, V.: New primal-dual algorithms for Steiner tree problems. Comput. Oper. Res. **34**(7), 2147–2167 (2007)

8. Rozenshtein, P., Gionis, A., Prakash, B.A., Vreeken, J.: Reconstructing an epidemic over time. In: Proceedings of KDD, pp. 1835–1844 (2016)
9. Silberschatz, A., Tuzhilin, A.: On subjective measures of interestingness in knowledge discovery. In: Proceedings of KDD, pp. 275–281 (1996)
10. Tang, J., Zhang, J., Yao, L., Li, J., Zhang, L., Su, Z.: ArnetMiner: extraction and mining of academic social networks. In: Proceedings of KDD, pp. 990–998 (2008)
11. Watel, D., Weisser, M.A.: A practical greedy approximation for the directed Steiner tree problem. In: Proceedings of COCOA, pp. 200–215 (2014)

Privacy and Security

Malware Detection by Analysing Encrypted Network Traffic with Neural Networks

Paul Prasse[1], Lukáš Machlica[2], Tomáš Pevný[2], Jiří Havelka[2], and Tobias Scheffer[1(✉)]

[1] Department of Computer Science, University of Potsdam, Potsdam, Germany
{prasse,scheffer}@cs.uni-potsdam.de
[2] Cisco R&D, Prague, Czech Republic
{lumachli,tpevny,jhavelka}@cisco.com

Abstract. We study the problem of detecting malware on client computers based on the analysis of HTTPS traffic. Here, malware has to be detected based on the host address, timestamps, and data volume information of the computer's network traffic. We develop a scalable protocol that allows us to collect network flows of known malicious and benign applications as training data and derive a malware-detection method based on a neural embedding of domain names and a long short-term memory network that processes network flows. We study the method's ability to detect new malware in a large-scale empirical study.

1 Introduction

Malware violates users' privacy, harvests passwords, can encrypt users' files for ransom, is used to commit click-fraud, and to promote political agendas by popularizing specific content in social media [1]. Several different types of analysis are being used to detect malware.

The analysis of an organization's network traffic complements decentralized antivirus software that runs on client computers. It allows organizations to enforce a security policy consistently throughout an entire network and to minimize the management overhead. This approach makes it possible to encapsulate malware detection into network devices or cloud services. Network-traffic analysis can help to detect polymorphic malware [2] as well as new and as-yet unknown malware based on network-traffic patterns [3,4].

When the URL string of HTTP requests is not encrypted, one can extract a wide range of features from it on which the detection of malicious traffic can be based [5]. However, the analysis of the HTTP payload can easily be prevented by using the encrypted *HTTPS* protocol. Google, Facebook, LinkedIn, and many other popular sites encrypt their network traffic by default. In June 2016, an estimated 45% (and growing) fraction of all browser page loads use HTTPS [6]. In order to continue to have an impact, traffic analysis has to work with HTTPS traffic.

On the application layer, HTTPS uses the HTTP protocol, but all messages are encrypted via the Transport Layer Security (TLS) protocol or its predecessor,

© Springer International Publishing AG 2017
M. Ceci et al. (Eds.): ECML PKDD 2017, Part II, LNAI 10535, pp. 73–88, 2017.
https://doi.org/10.1007/978-3-319-71246-8_5

the Secure Sockets Layer (SSL) protocol. An observer can see the client and host IP addresses and ports, and the timestamps and data volume of packets. Network devices aggregate TCP/IP packets exchanged between a pair of IP addresses and ports into a *network flow* for which address, timing, and data-volume information are saved to a log file. Most of the time, an observer can also see the unencrypted host domain name; we will discuss the underlying mechanisms in Sect. 3. The HTTP payload, including the HTTP header fields and the URL, are encrypted. Therefore, malware detection has to be based on properties of the host-domain names that a client contacts, and on statistical patterns in the timing and data-volumes of the sequence of network flows from and to that client. We will pursue this goal in this paper.

In order to extract features from the domain name, we explore a low-dimensional neural embedding [7] of the domain-name string. As a baseline, we will study manually-engineered domain features [5]. In order to model statistical patterns in the sequence of network flows, we will employ *long short-term memory (LSTM)* networks [8]. LSTMs are a neural model for sequential data that account for long-term dependencies in information sequences. As reference method, we will explore random forests.

The effectiveness of any machine-learning approach crucially depends on the availability of large amounts of labeled training data. However, obtaining ground-truth class labels for HTTPS traffic is a difficult problem—when the HTTP payload is encrypted, one generally cannot determine whether it originates from malware by analyzing the network traffic in isolation. We will make use of a VPN client that is able to observe the executable files that use the network interface to send TCP/IP packets. We deploy this VPN client to a large number of client computers. This allows us to observe the associations between executable files and network flows on a large number of client computers. We use antivirus tools to determine, in retrospect, which network flows in our training and evaluation data originate from malware.

This paper is organized as follows. Section 2 discusses related work. Section 3 describes our application environment and our data collection. Section 4 describes the problem setting and Sect. 5 describes our method. Section 6 presents experiments; Sect. 7 concludes.

2   Related Work

Prior work on the analysis of *HTTP logs* [9] has addressed the problems of identifying command-and-control servers [10], unsupervised detection of malware [11], and supervised detection of malware using domain blacklists as labels [5,12]. HTTP log files contain the full URL string, from which a wide array of informative features can be extracted [5]. In addition, each HTTP log file entry corresponds to a single HTTP request which also makes the timing and data volume information more informative than in the case of HTTPS, where the networking equipment cannot identify the boundaries between requests that pass through the same TLS/SSL tunnel.

Prior work on *HTTPS logs* has aimed at identifying the application layer protocol [13–15], identifying applications that are hosted by web servers [16], and identifying servers that are contacted by malware [17]. Some methods process the complete sequence of TCP packets which is not usually recorded by available network devices. Lokoč et al. [17] use similar features to the ones that we use—that can easily be recorded for HTTPS traffic—and a similar method for generating labeled data based on a multitude of antivirus tools. However, they focus on the problem of identifying servers that are contacted by malware.

Prior work on neural networks for network-flow analysis [18] has worked with labels for client computers (infected and not infected)—which leads to a multi-instance learning problem. By contrast, our operating environment allows us to observe the association between flows and executable files. Prasse et al. [19] have explored HTTPS analysis using random forests in a preliminary report. We have presented some of the results of this paper to a computer-security audience in a prior workshop paper [20].

LSTM networks [8] are widely used for translation, speech recognition and other natural-language processing tasks. Their ability to process sequential input and to account for long-term dependencies makes them appealing for the analysis of network flows. In computer security, their use has previously been explored for intrusion detection [21].

3 Operating Environment

This section describes our application and data-collection environment. In order to protect all computers of an organization, a Cloud Web Security (CWS) service provides an interface between the organization's private network and the internet. Client computers establish a VPN connection to the CWS service. The service enforces the organization's security policy; it can block HTTP and HTTPS requests based on the host domain and on the organization's acceptable-use policy. The CWS service can issue warnings when it has detected malware on a client. Since administrators have to process the malware warnings, the proportion of false alarms among all issued warnings has to be small.

The CWS service aggregates all TCP/IP packets between a single client computer, client port, host IP address, and host port that result from a single HTTP request or from the TLS/SSL tunnel of an HTTPS request into a network flow. This information is available for network devices that support the IPFIX [22] and NetFlow [23] formats. For each network flow, a line is written into the log file that includes data volume, timestamp, domain name, and duration information. For unencrypted HTTP traffic, this line also contains the full URL string. For HTTPS traffic, it includes the domain name—if that name can be observed via one of the following mechanisms.

Clients that use the Server Name Indication protocol extension (SNI) publish the unencrypted host-domain name when they establish the connection. SNI is widely used because it is necessary to verify certificates of servers that host multiple domains, as most web servers do. When the network uses a transparent

DNS proxy [24], this server caches DNS request-response pairs and can map IP addresses to previously resolved domain names. To further improve the availability of host-domain names the CWS could—but does not currently—employ passive DNS replication [25] and build a look-up table of observed DNS request-response pairs. The resulting sequence of log-file lines serves as input to the malware-detection model.

3.1   Data Collection

Since the proportion of HTTP versus HTTPS traffic declines continuously, we want to study malware detection in encrypted traffic. In our data collection, we therefore discard the URLs of HTTP requests from the log files, which leaves us only with information that would still be visible if all traffic were encrypted.

In order to obtain training and evaluation data, we have to label network traffic by whether it originated from malicious or benign applications. The CWS service can be configured to inspect HTTPS connections, provided that the service's root certificate has been installed as a trusted certificate on all client computers. This allows the CWS service to act as a man-in-the-middle between client and host, where it can decrypt, inspect, and re-encrypt HTTPS requests. This inspection can only be carried out for a small proportion of clients; we rely on it to collect training data. However, the deployed malware-detection model only uses observable features of HTTPS traffic and therefore does not require this interception. The VPN client that runs on the client computer has access to the process table and the network interface. The VPN client identifies applications by means of a SHA hash code of their executable file, and can be configured to communicate the association between HTTP/HTTPS traffic and applications to the CWS server. This allows us to augment all network flows of clients that operate with this configuration with a hash key of the application that has sent and received the network traffic. It also allows us to observe host-domain names even when SNI and transparent DNS proxy servers are not used. We configure all VPN clients of a number of participating organizations accordingly.

We label the traffic in retrospect, after the malware status of the files has been established by signature-based antivirus tools. Virustotal.com is a web service that allows users to upload executable files or hash keys of executable files. The files are run through 60 antivirus solutions, and the results of this analysis are returned. We upload the hash keys of all executable files that have generated HTTP/HTTPS traffic to Virustotal; we label files as benign when the hash is known—that is, when the file has been run through the scanning tools—and none of the 60 scanning tools recognize the file as malicious. When three or more tools recognize the file as malicious, we label it as malicious. When only one or two virus scanners recognize the file as a virus, we consider the malware status of the file *uncertain*; we do not use uncertain files for training and skip them during evaluation. In order to limit its memory use, Virustotal removes hashes of benign files from its database after some time. Therefore, we label all files whose hashes are unknown to Virustotal as benign. We then label all traffic that has been generated by malicious executables as malicious, and all traffic of benign files as benign.

We collect three different data sets; we will refer to them as *current data*, *future data*, and an independent set of *training data for domain-name features*, based on the roles which these data sets will play in our experiments. The *current data* contains the complete HTTP and HTTPS traffic of 171 small to large computer networks that use the Cisco AnyConnect Secure Mobility Solution for a period of 5 days in July 2016. This data set contains 44,348,879 flows of 133,437 distinct clients. The *future data* contains the complete HTTP and HTTPS traffic of 169 small to large different computer networks that use the Cisco AnyConnect Secure Mobility Solution for a period of 8 days in September 2016. The data set contains 149,005,149 flows of 177,738 distinct clients. The *training data for domain-name features* contains the HTTPS traffic of 21 small to large computer networks, collected over 14 days between February and April 2016. All data sets have been anonymized—information about organizations and user names have been replaced by random keys—and full URLs of HTTP requests have been removed.

3.2 Quantitative Analysis of the Data

We query the malware status of all observed hash keys in the current and future data sets from Virustotal after the collection of the respective data set has been completed, and reiterate the queries to Virustotal in February 2017. We study the stability of this labeling procedure. Table 1 shows that over this period of five months for the *future data* and seven months for the *current data*, 10% of all previously uncertain files (classified as malware by one or two antivirus tools) change their status to benign, while 3.5% change their status to malicious. From all initially unknown files, 2.3% become known as benign while 0.2% become known as malicious. From the files initially labeled as benign, 1.5% become uncertain and only 0.07% become known as malicious. Likewise, only 0.16% of all initially malicious files change their label to benign. We use the labels obtained in February 2017 in the following. We can conclude that once a file has been labeled as benign or malicious by at least 3 antivirus tools, it is very unlikely that this classification will later be reversed.

The current data contains 20,130 unique hashes, 350,220 malicious and 43,150,605 benign flows; the future data 27,263 unique hashes, 955,037 malicious and 142,592,850 benign flows. Table 2 enumerates the types and frequency of different types of malware families, according to the public classification of antivirus vendors. A large proportion of malware files is classified as *potentially unwanted applications (PUAs)*. PUAs are usually bundled to freeware or shareware applications; depending on how a distributor chooses to employ these programs, they can change the browser settings and starting page, install tools bars and browser extensions, display advertisements, and can also reveal passwords to the attacker. They typically cannot be uninstalled without a virus removal tool or detailed technical background knowledge.

Table 1. Virustotal malware status confusion matrix

uncertain	6,507	8	746	260
unknown	61	17,371	411	38
benign	949	38	58,899	40
malicious	150	0	8	4,781
	uncertain	unknown	benign	malicious

Status July/September 2016

Status February 2017

Table 2. Malware families

Malware family	Type	Variations
dealply	adware	506
softcnapp	adware, PUA	119
crossrider	adware, PUA	98
elex	adware, PUA	86
opencandy	adware, PUA	57
conduit	adware, spyware, PUA	56
browsefox	PUA	52
speedingupmypc	PUA	29
kraddare	adware, PUA	28
installcore	adware, PUA	27
mobogenie	PUA	26
pullupdate	PUA	25
iobit downloader	adware, PUA	24
asparnet	trojan, adware	24

4   Client Malware Detection Problem

We will now establish the malware-detection problem setting. Our goal is to flag client computers that are hosting malware. Client computers are identified by a (local) IP address and a VPN user name.

For each interval of 24 h, we count every client computer that establishes at least one network connection as a separate *classification instance*; a client that is active on multiple days constitutes multiple classification instances. Each classification instance has the shape of a *sequence* $x_1, \ldots, x_T$ of *network flows* from and to a particular client on a particular day. This sequence generally blends the network flows of multiple applications that run on the client computer. When at least one malicious application generates any network traffic throughout a day, we label that instance as positive; when at least one benign application but no malicious application generates traffic, the instance is negative.

For each classification instance, malware detection model f has to decide whether to raise an alarm. The model processes a *sequence* $x_1, \ldots, x_T$ of *network flows* from and to a particular client in an online fashion and raises an alarm if the malware-detection score $f(x_1, \ldots, x_t)$ exceeds some threshold τ at any time t during the day; in this case, the instance counts as a (true or false) positive. In each 24-h interval, model f with threshold τ will detect some proportion of malware-infected clients, and may raise some false alarms. The trade-off between the number of detections and false alarms can be controlled by adjusting threshold τ. Increasing the threshold decreases the number of detections as well as the number of false alarms.

We will measure precision-recall curves because they are most directly linked to the merit of a malware detection method from an application point of view. However, since precision recall curves are not invariant in the class ratio, we will additionally use ROC curves to compare the performance of classifiers on data sets with varying class ratios.

We will measure the following performance indicators.

1. *Recall $R = \frac{n_{TP}}{n_{TP}+n_{FN}}$* is the proportion of malicious instances that have been detected, relative to all (detected and undetected) malicious instances.
2. *Precision $P = \frac{n_{TP}}{n_{TP}+n_{FP}}$* is the proportion of malicious instances that have been detected, relative to all (malicious and benign) instances that have been flagged. Note that the absolute number of false alarms equals $(1 - P)$ times the absolute number of alarms; a high precision implies that the number of false alarms is small and the detection method is practical.
3. The recall at a specific precision, $R@x\%P$, quantifies the proportion of malicious instances that are detected when the threshold is adjusted such that at most $1 - x\%$ of all alarms are false alarms.
4. The precision-recall curve of decision function f shows the possible trade-offs that can be achieved by varying decision threshold τ. The precision recall curve of a fixed decision function will deteriorate if the class ratio shifts further toward the negative class.
5. The false-positive rate $R_{FP} = \frac{n_{FP}}{n_{FP}+n_{TN}}$ is the proportion of benign software that is classified as malicious. Note that the absolute number of alarms equals R_{FP} times the number of benign instances. Since the number of benign instances is typically huge, a seemingly small false-positive rate does not necessarily imply that a detection method is practical.
6. The ROC curve displays the possible tradeoffs between false-positive rate and recall that can be achieved by varying decision threshold τ. It is invariant in the class ratio but it conveys no information about the proportion of alarms that are in fact false alarms. To visualize the decision function behavior for small false-positive rates, we use a logarithmic x-axis.
7. The *time to detection* measures the interval from the first network flow sent by a malicious application on a given day to its detection.

Note the difference between *precision* and the *false-positive rate*—the risk that a benign file is mistakenly classified as malware. For instance, at a false-positive rate of 10%, the expected number of false alarms equals 10% of the number of benign instances; hence, false alarms would by far outnumber actual malware detections. By contrast, at a precision of 90%, only 10% of all alarms would be false alarms.

Training data for model f consists of labeled sequences $S = \bigcup_{i=1}^{n} \{((x_{i1}, y_{i1}), \ldots, (x_{iT_i}, y_{iT_i})\}$ in which each flow x_{it} is labeled by whether it has been sent or received by a malicious ($y_{it} = +1$) or benign ($y_{it} = -1$) application.

5   Network-Flow Analysis

This section presents our method that flags malware-infected client computers based on their network traffic.

5.1   Flow Features

The detection model processes *sequences of flows* $x_1, \ldots, x_T$ sent to or received by one particular client computer. This sequence is a blend of the network traffic of multiple applications. When a computer hosts malware, benign applications will generally interfere with any patterns in the malware's communication. For each flow, a client identifier (IP address and VPN user name), host address, ports, a timestamp, inbound and outbound data volume, and a duration are observable. From each flow x_t, we extract a vector $\phi(x_t)$ that includes the log-transformed duration, log-transformed numbers of sent and received bytes, duration, and the time gap from the preceding flows.

5.2   Domain-Name Features

Each flow contains an unencrypted host IP address. For most HTTPS connection requests, the host-domain name is visible. Otherwise, the host domain name string consists of a numeric IP address. We explore several types of features that can be extracted from the host-domain name.

Engineered Features. Franc et al. [12] develop a comprehensive set of 60 features of URL strings for malware detection in unencrypted HTTP traffic—here, the entire URL is visible to third parties. Their features include the ratio of vowel changes, the maximum occurrence ratio of individual characters for the domain and subdomain, the maximum length of substrings without vowels, the presence of non-base-64 characters, the ratio of non-letter characters, and many other characteristics. We extract the vector of these *engineered* domain-name features for all domains.

Character n-gram Features. Character n-gram features decompose the domain string into sets of overlapping substrings; for instance, "example.com" is composed of the 3-g "exa", "xam", "amp", ..., ".co", and "com". The number of n-grams that occur in URLs grows almost exponentially in n; in our data, there are 1,583 character 2-g, 54,436 character 3-g, and 1,243,285 character 4-g. If we added all character 3-g features to the feature representation $\phi(x_t)$ of a flow, then the total number of features of an entire sequence of T flows would impose a heavy computational burden, and cause overfitting of the malware-detection model. In our experiments, we therefore explore character 2-g features.

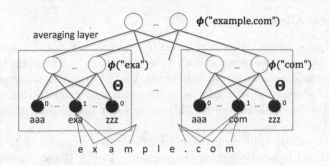

Fig. 1. Continuous bag-of-words architecture

Neural Domain-Name Features. We condense the character n-grams by means of a *neural embedding*. Neural language models [7] derive low-dimensional, continuous-state representations of words which have the property that words which tend to appear in similar contexts have similar representations. Here, "words" are the overlapping character n-grams that constitute a domain name. We apply neural embeddings with the goal of finding a representation such that substrings that tend to co-occur in URL strings have a similar representation.

We use the continuous bag-of-words architecture illustrated in Fig. 1. The input to the network consists of character n-grams that are one-hot coded as a binary vector in which each dimension represents an n-gram; Fig. 1 illustrates the case of $n = 3$. The input layer is fully connected to a hidden layer; we find that 100 hidden units give the best experimental results. The same weight matrix is applied to all input character n-grams. The activation of the hidden units is the vector-space representation of the input n-gram of characters. In order to infer the vector-space representation of an entire domain-name, an "averaging layer" averages the hidden-unit activations of all its character n-grams.

Neural language models are trained to complete an initial part of a sentence, which forces the network to find a vector-space representation that allows to "guess" a word's context from its vector-space representation. The "natural" reading order of a domain string is from the right to the left, because the domain ends with the most general part and starts with the most specific subdomain. Therefore, we use the preceding character n-gram, one-hot coded, as training target.

We train the neural language model using all domain names in our *training data for domain-name features*, and the 500,000 domains that have the highest web traffic according to alexa.com. For each URL and each position of the sliding window, we perform a back-propagation update step using the character n-gram that immediately precedes the input window as prediction target. We use the word2vec software package [26].

5.3   Client Classifier

We develop a client classifier based on long-short term memory networks (LSTMs) and, as reference method, a classifier based on random forests. LSTMs process input sequences iteratively. Each LSTM unit has a memory cell which allows it to store results of inferences; the unit may refer to this memory cell in later time steps. This allows LSTMs to account for long-term dependencies in sequential input.

Both classifiers split each client's traffic into sequences of 10 flows—for the LSTM, a fixed upper-bound on the sequence length is not necessary in theory, but it allows for more efficient training. In preliminary experiments, we have observed the performance of both LSTMs and random forests to deteriorate with longer sequences. After 10 flows $x_1, \ldots, x_{10}$ have been observed for a client, the sequence is processed by the neural network. The feature representation of each flow (duration, bytes sent and received, and the neural host domain-name representation) is processed sequentially by a layer of 32 LSTM units. After processing the input sequence, the LSTM layer passes its output to a layer of 128 cells with ReLU activation function, which is connected to two softmax cells whose activation represents the decision-function scores of the classes malicious and benign. The softmax layer is trained with 50% dropout regularization.

During training, the target label for a sequence of 10 flows is positive if at least one of the 10 flows originates from a malicious application. At application time, the overall decision-function value $f(x_1, \ldots, x_T)$ for the client's full traffic over a 24-h interval that is compared against threshold τ is the maximum activation of the positive output cell over all adjacent sequences of 10 flows.

For the random-forest classifier, the feature representation of 10 flows is stacked into a feature vector $\Phi(x_1, \ldots, x_{10}) = [\phi(x_1) \ldots \phi(x_{10})]^\top$ which serves as input to the random-forest classifier. At training time, the target label of a sequence is positive if at least one flow originates from a malicious application, and at application time the decision-function value of the random forest is maximized over all 10-flow sequences.

6   Experiments

We will first study the capability of the neural domain-name features to discriminate between benign and malicious domains. We will then explore the contribution of different types of features to malware detection, and the relative performance of LSTMs versus random forests. We will study the classifiers' ability to detect malware in current and future data, and will investigate how this performance varies across known and unknown malware families.

6.1   Classification of Host Domains

In our first experiment, we investigate the types of domain-name features with respect to their ability to distinguish between domains that are contacted by

Table 3. Domain classification and feature types

Feature type	R@70%P	R@80%P	R@90%P
Neural	**0.84**	**0.79**	**0.73**
Char 2-grams	0.83	0.76	0.62
Engineered	0.68	0.36	0.0
Neural+engineered	0.75	0.64	0.24

benign and domains that are contacted by malicious applications. In this experiment, domains serve as classification instances; all domains that are contacted more often by malicious than by benign software are positive instances, and all other domains are negative instances. In our training data for domain features, there are 860,268 negative (benign) and 1,927 positive (malicious) domains. A total of 3,490 domains are contacted by both malware and benign applications; many of them are likely used for malicious purposes (*e.g.*, "https://r8---sn-4g57km7e.googlevideo.com/", "https://s.xp1.ru4.com"), while others are standard services ("maps.google.com", "public-api.wordpress.com"). Malware frequently sends requests to legitimate services and uses URL forwarding techniques to reach the actual recipient domain. For 90,445 of the domains, only the IP address string is available.

We infer engineered domain-name features, character 2-g, and the vector-space representation of each domain string using the neural language model, as described in Sect. 5.2. We use 75% of the domains for training and 25% of the domains for testing; no domain is in both the training and the test data. We train a random forest classifier to discriminate positive from negative instances. Table 3 shows precision-recall trade-offs for the different set of feature types. We find that a parameter combination of $n = 6$ (input character 6-g), $m = 4$ (during training, the vector-space representation of 4 adjacent character 6-g is averaged) and $k = 100$ (the vector-space representation of a domain name has 100 dimensions) works best. Comparing the neural domain features to the raw character 2-g and the 60 engineered features in Table 3, we find that the neural features outperform both baselines. A combination of neural and engineered features performs worse than the neural features alone, which indicates that the engineered features inflate the feature space while not adding a substantial amount of additional information.

In order to analyze the domain-name classifier in depth, we look at domain-names that achieve the highest and lowest score from the random-forest classifier that uses the neural domain features. We find that a wide range of domains receive a decision-function value of 0 (confidently benign). These lowest-scoring domains include small and mid-size enterprises, blogs on various topics, university department homepages, games websites, a number of Google subdomains, governmental agencies; sites that can perhaps best be described as "random web sites". Table 4 shows the highest-scoring domains. They include numeric IP addresses, cloud services, subdomains of the YouTube and Facebook content

Table 4. Domain-names most confidently classified as malicious

https://52.84.0.111/
https://139.150.3.78/
https://uswj208.appspot.com/
https://ci-e2f452ea1b-50fe9b43.http.atlas.cdn.yimg.com/
https://pub47.bravenet.com
https://service6.vb-xl.net/
https://sp-autoupdate.conduit-services.com
https://external-yyz1-1.xx.fbcdn.net/
https://doc-14-28-apps-viewer.googleusercontent.com/
https://239-trouter-weu-c.drip.trouter.io

delivery networks, and domains that do not host visible content and have most likely been registered for malicious purposes. Based on these findings, we continue to study the neural domain-name features and exclude the engineered and character 2-g features from the following experiments.

6.2   Client Malware Detection

Learning Methods and Feature Types. We will first report on experiments in which we conduct 10-fold cross validation on the *current data*; we split the data into partitions with disjoint sets of users. We tune the parameters of the random forest using a grid search in an inner loop of two-fold cross validation on the training part of the data. The number of LSTM units is optimized in the first fold of the 10-fold cross validation on the training part of the data and fixed for the remaining folds.

Figure 2 compares the precision-recall curves of LSTMs and random forests using neural domain-name, flow, and combined features. We conclude that LSTMs outperform random forests for this problem, and that the combination of neural domain-name features and numeric flow features outperforms either set of features. We therefore exclude random forests from the remaining experiments.

Evolution of Malware. We will now explore whether a model that has been trained on the *current data* is able to detect malware in the *future data*. We train the LSTM on the entire *current data* and evaluate it on the entire *future data*. The number of LSTM units is left fixed. We compare the resulting model to the LSTM trained by 10-fold cross validation on the *current data*. Figure 3 compares precision-recall and ROC curves. Since the *future data* contains a smaller ratio of malicious instances—the prevalence of malware changes over time and over companies—the difference between the precision-recall curves is not necessarily due to a deterioration of the decision function. But a comparison of the ROC

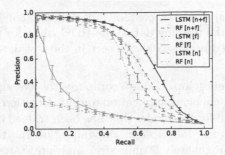

Fig. 2. Precision-recall curve for malware detection on current data. Error bars indicate standard errors.

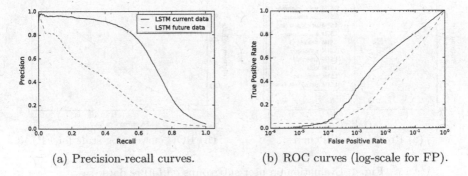

(a) Precision-recall curves.   (b) ROC curves (log-scale for FP).

Fig. 3. Comparison between performance on current and on future data.

curves shows that the decision function does deteriorate to a small extent in the two months that separate the training and test data.

Malware Families. We will now study the detection performance on specific malware families, on previously unseen malware, and on malware that does not contact any previously known domain. We use an LSTM that has been trained on all *current data*. We evaluate its performance on specific subsets of malware in the *future data*. In each experiment, each user who is hosting a malicious application from a selected malware family that has sent at least one network flow within a 24-h interval counts as a positive test instance, and each user who has run at least one benign application with at least one network flow but no malicious application within a 24-h interval counts as negative classification instance. Test users who are hosting malware of different families are skipped.

Figure 4 compares precision-recall and ROC curves. Since each specific subgroup entails only few users, the class ratios are highly skewed towards the negative class and the precision-recall curves cannot be directly compared. Comparing the ROC curves, we observe that the decision function performs similarly well across the entire range of malware families, including malware that does not belong to any known family, malware that does not occur in the training data, and malware that does not contact any domain that has been contacted by any application in the training data. The ROC curve for the malware family asparnet is fairly ragged because only 7 users in the test data are infected with.

Average Time to Detection. Table 5 compares the average intervals between the first flow sent by a malicious application and its detection for cross-validation on the *current data* and for the model that has been trained on all of the *current data* and is applied to the *future data*. Depending on the threshold, detection occurs on average approximately 90 min after malware starts communicating.

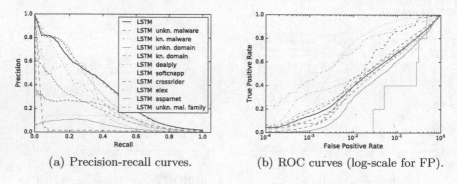

(a) Precision-recall curves.   (b) ROC curves (log-scale for FP).

Fig. 4. Evaluation on user subgroups on future data.

Table 5. Average time to detection

Data set	TTC@80%P	TTC@70%P	TTC@60%P
Current	1.66 h	1.67 h	1.65 h
Future	1.40 h	1.36 h	1.38 h

7   Conclusion

We can draw a number of conclusions from our study. A neural language model can transform a domain name into a low-dimensional feature representation that provides more information about whether the site is malicious than a set of carefully engineered features of domain-name characteristics. Our experimental setting allows us to collect large volumes of malicious and benign network traffic for model training and evaluation. The VPN client records the hash key of the executable file that has generated each flow, and by using VirusTotal we are able to determine, in retrospect, which flows originate from malware.

We have developed and studied a malware-detection model based on LSTMs that uses only observable aspects of HTTPS traffic. We conclude that the LSTM-based model outperforms random forests, and that the combination of neural domain-name features and numeric flow features outperforms either of these feature sets. This mechanism is able to identify a substantial proportion of malware in traffic that was recorded two months after the training data were recorded, including previously unseen malware. Its average time to detection is approximately 90 min after the first HTTPS request; its performance is uniform over many different malware families, including previously unknown malware. It complements signature-based and other behavior-based malware detection.

Acknowledgment. We would like to thank Virustotal.com for their kind support.

References

1. Kogan, R.: Bedep trojan malware spread by the angler exploit kit gets political. SpiderLabs Blog (2015). https://www.trustwave.com/Resources/SpiderLabs-Blog/Bedep-trojan-malware-spread-by-the-Angler-exploit-kit-gets-political/
2. Karim, M.E., Walenstein, A., Lakhotia, A., Parida, L.: Malware phylogeny generation using permutations of code. J. Comput. Virol. **1**, 13–23 (2005)
3. Gu, G., Zhang, J., Lee, W.: BotSniffer: detecting botnet command and control channels in network traffic. In: Proceedings of the Annual Network and Distributed System Security Symposium (2008)
4. Perdisci, R., Lee, W., Feamster, N.: Behavioral clustering of HTTP-based malware and signature generation using malicious network traces. In: Proceedings of the USENIX Conference on Networked Systems Design and Implementation (2010)
5. Bartos, K., Sofka, M.: Robust representation for domain adaptation in network security. In: Bifet, A., May, M., Zadrozny, B., Gavalda, R., Pedreschi, D., Bonchi, F., Cardoso, J., Spiliopoulou, M. (eds.) ECML PKDD 2015. LNCS (LNAI), vol. 9286, pp. 116–132. Springer, Cham (2015). https://doi.org/10.1007/978-3-319-23461-8_8
6. Aas, J.: Progress towards 100% HTTPS. Let's Encrypt (2016)
7. Bengio, Y., Ducharme, R., Vincent, P., Jauvin, C.: A neural probabilistic language model. J. Mach. Learn. Res. **3**, 1137–1155 (2003)
8. Hochreiter, S., Schmidhuber, J.: Long short-term memory. Neural Comput. **9**, 1735–1780 (1997)
9. Nguyen, T., Armitage, G.: A survey of techniques for internet traffic classification using machine learning. IEEE Commun. Surv. Tutor. **10**, 56–76 (2008)
10. Nelms, T., Perdisci, R., Ahamad, M.: ExecScent: mining for new C&C domains in live networks with adaptive control protocol templates. In: Proceedings of the USENIX Security Symposium (2013)
11. Kohout, J., Pevny, T.: Unsupervised detection of malware in persistent web traffic. In: Proceedings of the IEEE International Conference on Acoustics, Speech and Signal Processing (2015)
12. Franc, V., Sofka, M., Bartos, K.: Learning detector of malicious network traffic from weak labels. In: Bifet, A., May, M., Zadrozny, B., Gavalda, R., Pedreschi, D., Bonchi, F., Cardoso, J., Spiliopoulou, M. (eds.) ECML PKDD 2015. LNCS (LNAI), vol. 9286, pp. 85–99. Springer, Cham (2015). https://doi.org/10.1007/978-3-319-23461-8_6
13. Wright, C.V., Monrose, F., Masson, G.M.: On inferring application protocol behaviors in encrypted network traffic. J. Mach. Learn. Res. **7**, 2745–2769 (2006)

14. Crotti, M., Dusi, M., Gringoli, F., Salgarelli, L.: Traffic classification through simple statistical fingerprinting. ACM SIGCOMM Comput. Commun. Rev. **37**, 5–16 (2007)

15. Dusi, M., Crotti, M., Gringoli, F., Salgarelli, L.: Tunnel hunter: detecting application-layer tunnels with statistical fingerprinting. Comput. Netw. **53**, 81–97 (2009)

16. Kohout, J., Pevny, T.: Automatic discovery of web servers hosting similar applications. In: Proceedings of the IFIP/IEEE International Symposium on Integrated Network Management (2015)

17. Lokoč, J., Kohout, J., Čech, P., Skopal, T., Pevný, T.: k-NN classification of malware in HTTPS traffic using the metric space approach. In: Chau, M., Wang, G.A., Chen, H. (eds.) PAISI 2016. LNCS, vol. 9650, pp. 131–145. Springer, Cham (2016). https://doi.org/10.1007/978-3-319-31863-9_10

18. Pevny, T., Somol, P.: Discriminative models for multi-instance problems with tree structure. In: Proceedings of the International Workshop on Artificial Intelligence for Computer Security (2016)

19. Prasse, P., Gruben, G., Machlika, L., Pevny, T., Sofka, M., Scheffer, T.: Malware detection by HTTPS traffic analysis. Technical report. urn: urn:nbn:de:kobv:517-opus4-100942 (2017)

20. Prasse, P., Machlica, L., Pevný, T., Havelka, J., Scheffer, T.: Malware detection by analysing network traffic with neural networks. In: Proceedings of the Workshop on Traffic Measurements for Cybersecurity at the IEEE Symposium on Security and Privacy (2017)

21. Staudemeyer, R., Omlin, C.: Evaluating performance of long short-term memory recurrent neural networks on intrusion detection data. In: Proceedings of the South African Institute for Computer Scientists and Information Technologists Conference (2013)

22. Claise, B., Trammell, B., Aitken, P.: Specification of the IP flow information export (IPFIX) protocol for the exchange of flow information (2013). https://tools.ietf.org/html/rfc7011

23. Cisco Systems: Cisco IOS NetFlow (2016). http://www.cisco.com/c/en/us/products/ios-nx-os-software/ios-netflow/index.html

24. Blum, S.B., Lueker, J.: Transparent proxy server. US Patent 6,182,141 (2001)

25. Weimer, F.: Passive DNS replication. In: Proceedings of the FIRST Conference on Computer Security Incidents, p. 98 (2005)

26. Mikolov, T., Sutskever, I., Chen, K., Corrado, G., Dean, J.: Distributed representations of words and phrases and their compositionality. In: Advances in Neural Information Processing Systems (2013)

PEM: A Practical Differentially Private System for Large-Scale Cross-Institutional Data Mining

Yi Li[1(⊠)], Yitao Duan[2], and Wei Xu[1]

[1] Tsinghua University, Beijing 100084, China
xiaolixiaoyi@gmail.com
[2] Netease Youdao, Beijing 100193, China

Abstract. Privacy has become a serious concern in data mining. Achieving adequate privacy is especially challenging when the scale of the problem is large. Fundamentally, designing a practical privacy-preserving data mining system involves tradeoffs among several factors such as the privacy guarantee, the accuracy or utility of the mining result, the computation efficiency and the generality of the approach. In this paper, we present PEM, a practical system that tries to strike the right balance among these factors. We use a combination of noise-based and noise-free techniques to achieve provable differential privacy at a low computational overhead while obtaining more accurate result than previous approaches. PEM provides an efficient private gradient descent that can be the basis for many practical data mining and machine learning algorithms, like logistic regression, k-means, and Apriori. We evaluate these algorithms on three real-world open datasets in a cloud computing environment. The results show that PEM achieves good accuracy, high scalability, low computation cost while maintaining differential privacy.

1 Introduction

With the increasing digitization of society, more and more data are being collected and analyzed in many industries , including e-commerce, finance, health care and education. At the same time, however, privacy concerns are becoming more and more acute as our ever-increasing ability to extract valuable information from the data may also work against people's fundamental rights to privacy. How to make good use of the data while providing an adequate level of privacy is an urgent problem facing both data mining and security communities.

Fundamentally, the problem boils down to striking a balance among three factors that are often at odds with one another: privacy, utility, and efficiency. Previous works largely focus on one or two of these aspects and, as a result, fail to provide practical solutions that can be used in real-world systems (to be elaborated later). Also, due to the diversity of applications regarding goals, algorithms, and data partition situations, a general solution that supports all the situations appears to be extremely hard, if not entirely impossible.

In this paper we target the cross-institutional mining problem where each institution (referred to as a *client* here) collects data independently and together

© Springer International Publishing AG 2017
M. Ceci et al. (Eds.): ECML PKDD 2017, Part II, LNAI 10535, pp. 89–105, 2017.
https://doi.org/10.1007/978-3-319-71246-8_6

they would like to mine collective data to extract insights. This setup can be found in many real-world situations and typically it is more beneficial to include more data in the analysis. For example, to study the side effects of a drug, it is more accurate and timely if one could pool all the data from multiple relevant institutions (medical research institutes, hospitals, pharmaceutical companies and so on). The same applies to quick public health threats detection, educational data mining etc. Due to privacy concerns, however, institutions are prohibited by law from sharing their data.

This problem is called the horizontally partitioned database model in the privacy-preserving data mining community and has been studied extensively. Many solutions are algorithm-specific, such as clustering [25] and recommendation [3], and they are not scalable to handle large-scale problems. Some solutions are data-specific, e.g. [21]. Other solutions either make impratical assumptions (e.g., [27] assumes all the participants are semi-honest and the exact results do not reveal privacy), or involve heavy computation and communication cost that makes it unacceptable (e.g., [2] is based on secure multi-party computation and is very slow even for vector addition). In addition, early works tend to have a less rigorous notion of privacy.

The privacy we provide is differential privacy [12], a rigorous and strong notion that limits the probability that an adversary can distinguish between two datasets that differ only by one record. A holistic privacy solution needs to protect any information that is released from its owner. In a distributed setting where data is partitioned among multiple institutions, general differential privacy doctrine would add noise in every of these occasions [3, 21, 22, 24]. This approach does poorly for iterative data mining algorithms, as these algorithms would require noise for all intermediate results at each iteration, resulting in a final result too noisy to be useful.

The key idea of our approach is that we can use efficient and noise-free cryptographic primitives to reduce or even eliminate the noise for the intermediate results, preserving final accuracy while achieve the same level of differential privacy. We adopt a secret sharing over small field approach, similar to that of P4P [9] and Sharemind [5]. These frameworks are noise-free and efficient, but lacking explicit mechanisms to protect the intermediate results, and supports only limited operations (e.g. addition). Our noise adding mechanisms compensate the limitation.

Concretely, we make the following contributions:

(1) We combine a natural noise addition mechanism with an efficient noise-free cryptographic primitive, making it differentially private. We show that utilizing our framework significantly reduces the amount of noise necessary for maintaining differential privacy. We can use *orders of magnitude* lower noise and thus improve resulting model accuracy dramatically.
(2) We provide a complete solution, together with easy to use programming APIs and flexible options that enable differential privacy and performance optimizations. PEM automatically determines the noise level. This enables users to easily implement tradeoff between privacy, scalability and accuracy.

(3) We design mechanisms to deal with common distributed system issues such as fault tolerance and load balancing.
(4) The system provides a differentially private gradient method that can be the basis of many machine learning algorithms. We also implement a number of commonly used machine learning algorithms such as k-means, Apriori etc.

Our goal is to provide a practical solution with provable privacy for many real-world analysis. We evaluate the tradeoff between accuracy and privacy using three real-word datasets. Our results show that PEM not only guarantees differential privacy, achieving similar results as the original non-private algorithms, but also introduce very low computation overhead. We are adding more machine learning algorithms into PEM, and we will release PEM open source.

The paper is organized as follows. Section 2 reviews recent research on privacy-preserving computation. Section 3 formulates the problem and introduces import definitions. Section 4 introduces the goals of our system and describes some details about the system design. Section 5 gives some proofs why our system works. Section 6 illustrates three kinds of algorithms implemented in our system. Section 7 shows the experiment results. We conclude in Sect. 8.

2 Related Work

People have proposed many definitions of privacy over the years. Earlier versions, such as k-anonymity and l-diversity [16], are vulnerable to simple attacks [4,12]. Differential privacy is a popular definition that has strong privacy guarantee while still achievable in practice [12]. There are extended versions of differential privacy definitions such as pan privacy [11] and noiseless privacy [4]. These extensions often come with restriction to the datasets or use cases. Thus we adopt the general differential privacy definition.

The general approach to achieving differential privacy is to add noise to the released results [10,22]. Adding noise inevitably affects the accuracy and people have explored several noise-free methods [4,8]. These works make use of the adversary's uncertainty about the data thus eliminating the need for external noise. However, both [8] and [4] make strong assumptions about the data distribution to maintain differential privacy.

Many systems use cryptography to achieve differential or other types of privacy. They are based on either homomorphic encryption (HE) or secure multi-party computation (MPC) [28]. The problem with HE or MPC is that both make expensive use of large integer operations and are orders of magnitude slower than the non-private version, making them impractical. For example, Rastogi et al. proposes an algorithm PASTE, which uses HE for noise generation to mine distributed time-series data without any trusted server [21]. SEIPA performs privacy-preserving aggregation of network statistics using Shamir's secret sharing scheme [23], which involves heavy polynomial calculation, making it slow in large datasets [6]. DstrDP [26] and SDQ [29] use HE and secure equality check to garble the noise from participants. Other approaches, including [2,5,14,19], also use expensive cryptograph techniques.

The other trend is to take advantage of the property of application algorithm or statistics of the dataset to avoid the expensive cryptographic operations. Chaudhuri *et al.* shows that sampling is helpful for increasing the accuracy of principal component analysis while preserving differential privacy [7]. Shokri *et al.* implements a system for privacy-preserving deep learning [24], using distributed selective SGD algorithm. However, the convergence of the algorithm strongly depends on the dataset. PINQ assumes a trusted third party to implement privacy-preserving queries [18] by adding Laplace noise. However, a single trusted third party is not only a strong assumption but also a performance and security bottleneck. P4P relaxes the trust assumption by allowing non-colluding semi-honest servers [9]. However, the noise-free approach in P4P still assumes too much about the dataset.

We combine these methods into a coherent system: we take the relaxed trust model, adding (reduced amount of) Laplace noise to achieve provable differential privacy, and leverage the properties of algorithms (such as sampling) to further reduce the amount of noise.

3 Preliminaries

3.1 Problem Formulation

In PEM, there are n clients. Each client $C_i (i = 1, 2, \ldots, n)$ has a subset of records D_i to contribute to the computation. The goal of our computation is to use the union of all D_i's, which we denote as D, to compute some statistics, $f(D)$. During the computation, each client wants to ensure that no information about the individual records in the dataset is leaked to other clients or any third pary. Specifically, we want to support any iterative computation where each iteration can be expressed in the form of

$$f(D) = g(\sum_{d \in D} h(d)). \tag{1}$$

where both h and g can be nonlinear. This simple form can support a surprisingly large number of widely-used data mining algorithms, including k-means, expectation maximization (EM), singular value decomposition (SVD), etc.

Assume we have m independent *servers*, $S_1, \ldots, S_m$, and an *aggregator* server. We make the following key assumptions about the servers, which usually hold for real-world applications:

1. All servers are *semi-honest*, which means each server follows the protocol but may attempt to learn about all private data when it has a chance.
2. There are at least two servers that do not conspire with other servers to break the protocol. We can achieve this goal by using servers from different administrate domains.
3. All communications are secure so that adversaries cannot see/change any useful information in the communication channels. We can ensure this assumption using encryption and message authentication methods.

In a realistic scenario, the clients can be collaborating institutions (e.g., hospitals, schools) that do not wish their data exposed to any party. The servers and aggregator can be cloud computing facilities that are owned by different service providers. The non-colluding assumption upholds as the service providers are prohibited by law to leak their customer data. The number of servers, m, does not have to be very large. Using more servers increases security but also the cost. $m = 2$ or 3 is enough for many situations.

3.2 Definitions

We summarize important definitions for readers not familiar with the field.

Differential Privacy [12]. An algorithm K gives ϵ-differential privacy if for all *adjacent datasets* D, D' and all $S \subseteq Range(K)$,

$$Pr[K(D) \in S] \leq e^\epsilon \cdot Pr[K(D') \in S] \tag{2}$$

where adjacent datasets are two data sets that differ in at most a single record. Intuitively, with differential privacy, we make the distributions of $K(D)$ and $K(D')$ nearly undistinguishable by perturbing the outputs.

Laplace Distribution [12]. A random variable X follows Laplace distribution $Lap(\lambda)$ if the probability density function (PDF) is

$$Pr\{X = x\} = \frac{1}{2\lambda}e^{-|x|/\lambda} \tag{3}$$

where λ is a parameter that determines the variance. The mean and variance of $Lap(\lambda)$ are 0 and $2\lambda^2$, respectively. We denote $\mathbf{Lap}^d(\lambda)$ as a d-dimensional vector consisting of d independent $Lap(\lambda)$ random variables.

Sensitivity. For a function $f : \mathcal{D} \to \mathbb{R}^n$, the L_1-sensitivity $S(f)$ is defined as:

$$S(f) = \max_{D_1, D_2} \|f(D_1) - f(D_2)\|_1 \tag{4}$$

where D_1 and D_2 are two neighboring datasets differing in at most one row. It can be shown that $f(D) + r$ is ϵ-differentially private if $r \sim \mathbf{Lap}^d(S(f)/\epsilon)$.

4 System Design

Differential privacy provides a rigorous definition and tuneable parameters for us to specify the level of protection that we desire. However, prior works such as [3,13,24] indicate that direct application of the noisy response approach may not allow us to find a spot where acceptable level of privacy and utility coexist. In all the cases a large ϵ must be used, meaning that there is essentially little privacy, if we want the results to have any sensible usage.

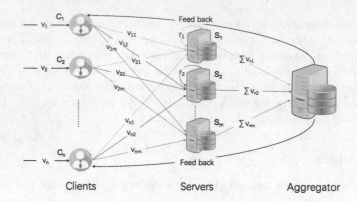

Fig. 1. The overview of PEM architecture.

To address this problem, it is clear that we must reduce noise as much as possible. Considering the distributed setting that we are dealing with, we are motivated to adopt the following design principle: use efficient cryptographic tools whenever possible to eliminate the need for noise. Instead of adding noise at every client, we adopt secret sharing over small field to perform the aggregation. This paradigm allows us to only perturb the final aggregates, as secret sharing itself protects each client's input. Since the number of servers is far less than the number of clients, the final noise is much smaller.

4.1   System Architecture

There are three roles in PEM, and Fig. 1 shows an overview of the system.

Clients. The clients are the owners of the data sources. The PEM client module uses secret sharing protocol to privatize the data while providing useful tools for privacy tracing and model updates.

Servers. The servers compute partial sums and add noise to the sums. Each server is logically a single component, while it can be implemented as a distributed system for scalability and fault tolerance.

Aggregator. The aggregator performs all data aggregation and application-specific computation. It completes an iteration of model update based on the noisy aggregates and sends the model back to the clients.

4.2   Data Processing Protocol

On the start of the job, PEM distributes important parameters and the application executables to all the nodes, such as the privacy parameter ϵ and the list of participants. Then the data processing follows four major steps in PEM:

(1) (Clients-Servers) Using secret sharing to aggregate client data. Each client secret-shares each data vector into m random vectors, one for each server. Specifically, for a vector $\mathbf{v}_i$ held by client C_i (here we assume $\mathbf{v}_i$ is an integer vector, while real-value vectors can be converted by discretization), the client can first generate $m - 1$ random vectors $\mathbf{v}_{ij} \in \mathbb{Z}_\phi^d (j = 1, 2, \ldots, m - 1)$, where ϕ is a 32- or 64-bit prime and $\mathbb{Z}_\phi$ is the additive group of integers module ϕ. Then it computes the m-th vector as $\mathbf{v}_{im} = \mathbf{v}_i - \sum_{j=1}^{m-1} \mathbf{v}_{ij} \mod \phi$. Obviously, the sum of the m vectors equals to $\mathbf{v}_i \mod \phi$. The client C_i sends a random vector $\mathbf{v}_{ij}$ to the server $S_j (j = 1, 2, \ldots, m)$, and S_j receives a random vector from each client. Each server computes a sum vector using vectors from all clients.

(2) (Servers) Generating Laplace noise to achieve privacy. To preserve differential privacy, each server automatically determines the level of noise, generates Laplace noise following the approaches in [10] and adds it to its partial sum. Each server $S_j (j = 1, 2, \ldots, m)$ generates a d-dimensional random vector $\mathbf{r}_j$ following $\mathbf{Lap}^d(\lambda)$, and sends $\sum_{i=1}^{n} \mathbf{v}_{ij} + \mathbf{r}_j$ to the aggregator.

(3) (Servers - Aggregator) Computing the sum of all vectors. At this step, each server sends its perturbed partial sum to the aggregator for final aggregation. On receiving the sums from servers, the aggregator sums up the vectors and obtains the final sum, $\mathbf{v} = \sum_{i=1}^{n} \mathbf{v}_i + \sum_{j=1}^{m} \mathbf{r}_j$. The noise prevents the aggregator from guessing the original sum.

(4) (Aggregator - Clients) Update model and complete an iteration. The aggregator runs the application code to generate or update the model. As many learning algorithms can be written in the form in Eq. 1, the sum obtained from the previous step is sufficient for a single iteration in the model update. Then the aggregator sends back necessary information, e.g., the latest models, back to each client for the next iteration.

4.3 ϵ-Splitting and Dynamic L Setting

Privacy parameter (ϵ) splitting. As mentioned above, the servers add noise $r \sim \mathbf{Lap}^d(\lambda)$ to a d-dimensional sum. This works in many cases. However, sometimes we want to apply different noise levels to each dimension. For example, when we use k-means for clustering, we should calculate the sum of and the count of each class to get the mean of these records. As the sensitivity of the records and the sensitivity of the count of each class are different, it is reasonable to add noise of different levels to them. PEM allows users to specify different privacy parameters to different dimensions. Specifically, users can assign privacy parameter as $[d_1 : \epsilon_1, d_2 : \epsilon_2, \ldots, d_k : \epsilon_k]$ where d_i is the subset of dimensions, as long as $\forall_{i,j} d_i \cap d_j = \emptyset$, $\cup_i d_i = \{1, 2, \ldots, d\}$ and $\sum_{i=1}^{k} \epsilon_i = \epsilon$. Servers automatically compute the amount of noise according to the manner of ϵ splitting.

Dynamic sensitivity (L) setting. PEM splits the ϵ(s) equally to each iteration in advance, and thus the noise levels of different iterations can be calculated independently according to the sensitivity L and other parameters. However, sometimes the vectors to be added up from the clients may change over iterations, leading to a change of the sensitivity. In this case, L is set dynamically over each iteration, making the servers generate noise of different levels. PEM allows users to specify L for each iteration. Then the λ will change accordingly and the servers will update the level of noise automatically.

4.4 Automatically Computing the Noise Level (λ)

A key feature of PEM is the automatical computation of λ based on user settings, such as ϵ and L. As mentioned, each server generates noise following Laplace noise $\mathbf{Lap}^d(\lambda)$, where λ is the noise level. In PEM, the servers automatically compute the noise level according to iteration count T, ϵ and L. As different parameter setting means different ways to generate noise, we consider the following cases:

- In the simplest case, where there is no ϵ-*splitting* or *dynamic L setting*, each server S_j computes $\lambda = T \cdot L/\epsilon$ and generates noise $\mathbf{r}_j = \mathbf{Lap}^d(\lambda)$.
- If ϵ-*splitting* is specified, i.e., the privacy parameter is assigned as $[d_1 : \epsilon_1, d_2 : \epsilon_2, \ldots, d_k : \epsilon_k]$, each server S_j computes $\lambda_k = T \cdot L_k/\epsilon_k$ where L_k is the sensitivity of elements in d_k and generates noise $\mathbf{r}_j = [\mathbf{r}_{j,1}, \mathbf{r}_{j,2}, \ldots, \mathbf{r}_{j,k}] = [\mathbf{Lap}^{|d_1|}(\lambda_1), \mathbf{Lap}^{|d_2|}(\lambda_2), \ldots, \mathbf{Lap}^{|d_k|}(\lambda_k)]$.
- If *dynamic L setting* is specified, i.e., the sensitivity of iteration t is $L(t)$, each server S_j computes $\lambda(t) = T \cdot L(t)/\epsilon$ and generates noise $\mathbf{r}_j = \mathbf{Lap}^d(\lambda(t))$.

4.5 Optimizations

PEM is a federated system across multiple administrative domains. To handle the privacy-related requirements, we make the following extensions.

Extending servers to clusters. Although these operations are simple, we can still accelerate them to make PEM more efficient. We can extend each server to a cluster (e.g., a Spark cluster) to parallize the communications and computations such that the servers are unlikely to be bottlenecks.

Handling client failures. If a server does not hear from a client for an extended period, it informs the aggregator about the client's failure, and the aggregator notifies all servers to remove the client.

Intuitive configurations and programing interfaces. The application developers only need to provide a few intuitive parameters, such as the sensitivity L, privacy strength ϵ, and the switch of sparse vector technique. PEM automatically determines other settings. The programmers only need to implement three functions: *init*, *mapInClient* and *reduceInAggregator* (see Sect. 6 for details).

Online processing. PEM also supports online processing. Clients send the data streams to servers for online processing. We model a data stream as a sequence of *mini-batches*, to reduce the communication overhead. In a mini-batch iteration, each client buffers the data records locally and sends them out periodically.

Algorithm specific optimizations. There are often algorithm specific optimizations to further reduce the required level of noise, such as *sampling* and *sparse vector technique* [12]. Sampling is commonly used to deal with imbalanced data, and can introduce more uncertainty about the raw data, reducing the level of noise required for differential privacy [15]. Sparse vector technique selects the queries above a pre-defined threshold τ and set the unselected elements to 0. Thus we can compress the proposed sparse vector and save communication cost with a small accuracy loss [12]. PEM allows users to enable these optimizations, and PEM automatically performs them and adjusts the noise level accordingly.

5 Analysis

Provable differential privacy. We first prove that we can make the algorithms differentially private to the aggregator and clients, and then prove that the algorithms are also differentially private to the servers. Formally, we determine the parameter λ using the sensitivity method [10]. Assuming the sensitivity is no more than L, i.e. $|\mathbf{v}_i - \mathbf{v}_j| \leq L$, we have the following theorem.

Theorem 1. *For any $v \in \mathbb{R}^d$, $v + \sum_{j=1}^{m} \mathbf{r}_j$ is ϵ-differentially private if, for $j = 1, 2, \ldots, m$, $\mathbf{r}_j$ is drawn independently from $\mathbf{Lap}^d(L/\epsilon)$.*

Proof. According to [10], if we add only one of $\mathbf{r}_j (j = 1, 2, \ldots, m)$, the algorithm is already ϵ-differentially private. As $\mathbf{r}_j (j = 1, 2, \ldots, m)$ are independent, adding another random noise to the result can still achieve ϵ-differential privacy.

We now show that the mechanism in PEM for calculating the sum is differentially private. Specifically, in PEM, as long as there are at least two servers that do not collude with others, the noisy sum calculated by the aggregator is ϵ-differentially private to all roles. The reason is straightforward: both the aggregator and the clients see the true vector sum plus at least two pieces of noise generated independently according to $\mathbf{Lap}^d(L/\epsilon)$. By Theorem 1, the aggregation is ϵ-differentially private to the aggregator and clients. Meanwhile, according to [9], PEM leaks no information beyond the intermediate sums to the servers as the servers' view of the intermediate sums is uniformly random and contains no information about the raw data. On the other hand, as there are at least two semi-honest servers generating noise faithfully, each server sees the result perturbed by at least one piece of noise following $\mathbf{Lap}^d(L/\epsilon)$ even if it excludes its own noise from the sum, which is enough for preserving ϵ-differential privacy.

Cost and complexity. In PEM, there are m servers and n clients. We have $m \ll n$ in general. To preserve the privacy of data, m servers generate independent Laplace noise. As the variance of Laplace distribution $Lap(\lambda)$ is $2\lambda^2$, the variance of the sum of m independent noises is $2m\lambda^2$. Comparing to the methods where all clients add noise (e.g. [3,24]), where the overall variance of the sum is $2n\lambda^2$, our method reduces much noise as $m \ll n$.

As most of operations in PEM are vector addition and noise generation, the computational overhead grows nearly linearly with the number of clients and the data dimensionality. Thus, PEM is suitable for processing high-dimensional vectors. In PEM, noise generation and addition is orders of magnitudes faster, comparing to many homomorphic-encryption-based approaches, especially those related to high-dimensional vectors.

6   Sample Algorithms in PEM

To implement privacy-preserving data mining algorithms in PEM, users only need to implement the following necessary functions:

- *init*: initialize the parameters, including privacy parameter ϵ and the sensitivity L. Note that ϵ can be an array and L may change over the iterations.
- *mapInClient*: map the records of the local database to the vectors.
- *reduceInAggregator*: analyze the sum of the vectors from the clients and update the parameters.

We present three algorithms we have implemented on PEM: logistic regression, k-means and Apriori as examples.

6.1   Gradient Descent

Gradient descent (GD) is a commonly-used algorithm for finding a local minimum of a function. Given an optimization problem, we want to minimize the loss function $\phi(\mathbf{x}, y, \mathbf{w})$, where $\mathbf{x}$ is the input, y is the expected output and $\mathbf{w}$ is the parameter. In GD, to estimate the optimal $\mathbf{w}$, we first calculate the gradient $\nabla\phi(\mathbf{w})$ on the dataset, then update $\mathbf{w}$ as $\mathbf{w}_{t+1} = \mathbf{w}_t - \eta_t \cdot \nabla\phi(\mathbf{w}_t)$ where η_t is the learning rate in iteration t. We set η_t to $O(\frac{1}{\sqrt{t}})$, like [30] does.

In PEM, servers add Laplace noise to the sum of the gradients from the clients. The formula for update becomes $\mathbf{w}_{t+1} = \mathbf{w}_t - \eta_t \frac{1}{\sum_{i=1}^{n} b_i}(\sum_{i=1}^{n} \nabla\phi_i(\mathbf{w}_t) + \sum_{j=1}^{m} \mathbf{r}_j)$ where b_i is the size of sub-dataset D_i held by client C_i, $\nabla\phi(\mathbf{w})$ is the sum of gradients on D_i and $\mathbf{r}_j$ is the Laplace noise. Usually, GD consists of multiple iterations. Assuming the number of iterations is T, $\mathbf{r}_j$ follows $\mathbf{Lap}^d(TL/\epsilon)$. Algorithm 1 shows the user functions for gradient descent.

6.2 k-Means

k-means is a simple yet popular algorithm for clustering. In standard k-means, if we want to partition the d-dimensional records to l clusters, we first initialize l centroids $\mathbf{c}_1, \mathbf{c}_2, \ldots, \mathbf{c}_l$ of different clusters. In each iteration, we assign each record $\mathbf{x}$ to the cluster whose centroid $\mathbf{c}$ is the closest to it. Then we update each centroid to the mean of the records in its cluster: $\mathbf{c}_i = \frac{1}{|S_i|} \sum_{\mathbf{x} \in S_i} \mathbf{x}$ where

---

Function *Init()*:
    Initialize ϵ and L as constants. Initialize the iteration count T, the weight $\mathbf{w}$, loss function ϕ and the learning rate η.

Function *mapInClient*$(\mathbf{x}, y, \mathbf{w}, \phi)$:
    return $\partial\phi(\mathbf{x}, y, \mathbf{w})/\partial\mathbf{w}$.

Function *reduceInAggregator*$(\mathbf{v}, \mathbf{x}, \eta, t)$:
    update $\mathbf{w}$ as $\mathbf{w} = \mathbf{w} - \eta\frac{1}{\sqrt{t}} \cdot \frac{1}{\sum_{i=1}^{n} b_i}\mathbf{v}$.

Algorithm 1. The user functions of gradient descent.

$\mathbf{c}_i$ is the centroid of cluster i, S_i is the set of records belonging to cluster i and $|S_i|$ is the number of elements in S_i. In one iteration, $\mathbf{c}_1, \mathbf{c}_2, \ldots, \mathbf{c}_l$ should be updated all together, which means that the clients should jointly calculate $(\sum_{\mathbf{x} \in S_1} \mathbf{x}, \sum_{\mathbf{x} \in S_2} \mathbf{x}, \ldots, \sum_{\mathbf{x} \in S_l} \mathbf{x}, |S_1|, |S_2|, \ldots, |S_l|)$.

As the change of a record belonging to cluster i may change $\sum_{\mathbf{x} \in S_i} \mathbf{x}$ and $|S_i|$ simultaneously, the servers in PEM should not only add noise to perturb the centroids, but also add noise to perturb the number of records of each cluster. Specifically, we modify the formula for centroid update as $\mathbf{c}_i = \frac{1}{|S_i| + \sum_{j=1}^{m} \mathbf{r}_{j,2}}(\sum_{\mathbf{x} \in S_i} \mathbf{x} + \sum_{j=1}^{m} \mathbf{r}_{j,1})$ where $\mathbf{r}_{j,1}$ is the noise for the sum of records and $\mathbf{r}_{j,2}$ is the noise for the number of records in each cluster. .

The records themselves and record-count obviously carry different amount of privacy information, and thus we add different levels of noises to the records dimensions and the count dimension. Using PEM's ϵ-spliting feature, we set the values of ϵ_1 and ϵ_2 as the users want, as long as their sum is ϵ. Denote L_1 to be the sensitivity of the sum of records $(\sum_{\mathbf{x} \in S_1} \mathbf{x}, \sum_{\mathbf{x} \in S_2} \mathbf{x}, \ldots, \sum_{\mathbf{x} \in S_l} \mathbf{x})$ and L_2 to be the sensitivity of the number of records in the clusters $(|S_1|, |S_2|, \ldots, |S_l|)$. It is easy to see that $L_1 = 2 \cdot L_{\mathbf{x}}$ and $L_2 = 2 \cdot 1$, where $L_{\mathbf{x}}$ is the maximum 1-norm of $\mathbf{x}$. There is a coefficient 2 here is because the change of an record may affect at most two clusters. Then we have $\mathbf{r}_{j,1} \sim \mathbf{Lap}^{ld}(TL_1/\epsilon_1)$ and $\mathbf{r}_{j,2} \sim \mathbf{Lap}^{l}(TL_2/\epsilon_2)$ where T is the number of iterations. Algorithm 2 shows the pseudo-code.

6.3 Apriori

Apriori is an algorithm for frequent itemset mining [17]. The *support* of a set of items (itemset) is the fraction of records containing the itemset respect to the database D. If the support of an itemset is above a preassigned minimum support, we call the itemset a *large itemset*. The target of frequent itemset mining is to find all the large itemsets. We denote I_k as the set of large itemsets of length k. To find large itemsets I_k, Apriori uses a function called *apriori-gen*, which takes I_{k-1} as an argument and generates candidates of k-itemsets denoted by I_k^*. Then Apriori calculates the count of each itemset in I_k^* and reserves the itemsets whose supports are above the minimum support. The set of reserved large k-itemsets from I_k^* is I_k.

Function *Init()*:

Initialize the iteration count T and the number of centroids l. Then initialize the rest parameters as follows: $\mathbf{c} = [\mathbf{c}_1, \mathbf{c}_2, \ldots, \mathbf{c}_l]$, $d_1 = \{1 \text{ to } ld\}$, $d_2 = \{ld + 1 \text{ to } ld + l\}$, $\epsilon = [d_1 : \epsilon_1, d_2 : \epsilon_2]$ and $L = [d_1 : 2L_{\mathbf{x}}, d_2 : 2]$.

Function *mapInClient*$(\mathbf{x}, \mathbf{c})$:

First we initialize the gradient $\mathbf{g} := [\mathbf{g}_1, \mathbf{g}_2, \cdots, \mathbf{g}_l, g_1, g_2, \cdots, g_l]$, where $\mathbf{g}_i = \mathbf{0}$ and $g_i = 0$ for each $i \in \{1 \text{ to } l\}$. Then we calculate $k = findCluster(\mathbf{x}, \mathbf{c})$ and set $\mathbf{g}_i = \mathbf{x}$ and $g_i = 1$. Return $\mathbf{g}$ as the result.

Function *reduceInAggregator*$(\mathbf{v})$:

Represent $\mathbf{v}$ as $[\mathbf{v}_1, \mathbf{v}_2, \cdots, \mathbf{v}_l, v_1, v_2, \cdots, v_l]$ and update the centroid $\mathbf{c} = [\mathbf{v}_1/v_1, \mathbf{v}_2/v_2, \ldots, \mathbf{v}_l/v_l]$.

Algorithm 2. The user functions of k-means.

As the change of a record will affect the count of itemsets, we should add noise to the count of the itemsets to preserve privacy. In PEM, in the situation where the dataset is distributed in multiple clients, *apriori-gen* can be done locally in each client given I_{k-1} and the count of each itemset in I_{k-1}. Filtering I_k^* to get I_k involves no raw data and thus can be done in the aggregator without privacy issue. We only need to add noise to the count of each itemset in I_k^*. Formally, we represent I_k^* as $[t_1, t_2, \ldots, t_{|I_k^*|}]$ where $t_i \in I_k^*$ is a candidate large itemset. In each iteration for generating I_k, $count(I_k^*, D) = [c_1, c_2, \ldots, c_{|I_k^*|}]$ is calculated, where c_i is the count of t_i respect to the dataset D. The sensitivity L of k-itemset counts is different for different k values. A record of length l contains at most $\binom{l}{k} = \frac{l!}{k!(l-k)!}$ itemsets of length k. The change of a record may affect the counts of at most $2 \cdot \binom{l}{k}$ itemsets. Then if we know the length l_m of the longest record beforehand, we can calculate the sensitivity of $count(I_k^*, D)$ as $L_k = \min(2 \cdot \binom{l_m}{k}, |I_k^*|)$. Algorithm 3 shows the user functions. We enable *dynamic L setting* and thus λ is modified in each iteration along with the sensitivity L_k.

Function *Init()*:

Set l_m as the length of the longest record, I_1 as the set of atomic items, and *minsup*. Initialize the iteration count T and the privacy parameter ϵ. Then set the sensitivity function $L(k, I_k^*)$ accordingly.

Function *mapInClient*$(\mathbf{x}, I_{k-1})$:

Set $I_k^* = apriori\text{-}gen(I_{k-1})$ and return $count(I_k^*, \mathbf{x})$.

Function *reduceInAggregator*$(\mathbf{v})$:

Here the parameter $\mathbf{v}$ is $count(I_k^*, \mathbf{x})$, i.e. the output of *mapInClient*. First we set $I_k^* = apriori\text{-}gen(I_{k-1})$ and I_k as the empty set $\{\}$. Then for each element $e \in I_k^*$, if $\mathbf{v}[e] > minsup$, put it in I_k. I_k is the set of large items of length k.

Algorithm 3. The user functions of Apriori.

7   Evaluation

7.1   Evaluation Setup

We empirically evaluate PEM in a cloud computing environment. We setup one aggregator and two servers ($m = 2$). Each of them runs on a virtual machine (VM) node. Each VM has 8 Xeon CPU cores, 16 GB RAM and 10GE ethernet. We emulate 100–1000 clients, each of which uses a sperate VM node with the same configuration. We use three open datasets from *UCI Machine Learning Repository* [1] for evaluation. We partition each dataset evenly onto the clients, emulating a horizontally partitioned dataset setting.

– The **Adult** dataset contains information of many people, including gender, age and salary. We clean the dataset and finally get 48,842 data records, each of which has 124 dimensions.
– The **Nursery** dataset is derived from a hierarchical decision model originally for nursery schools. There are 21,960 instances and 8 categorical attributes.
– The **Mushroom** dataset contains information of hypothetical samples corresponding to 23 species of gilled mushrooms. There are 8,124 instances with 22 attributes, each of which describes some shape information.

In our evaluation, we compare our algorithms with no-privacy versions. Meanwhile, many approaches add adequate noise on all clients, which means 100–1000 Laplace noises in our experiment setting. We call the approach *noise-only approach*, and we show the comparison with ours. Finally we show the performance overhead of PEM.

7.2   Performance of Algorithms

Distributed gradient descent. We use *logistic regression* as the example of gradient descent shown in Algorithm 1. With the **Adult** dataset, we construct a logistic regression model to predict whether each person has high income ($>$50K) or not ($\leq$50K). We preprocess the dataset using *one-hot encoding* and the sensitivity L is 28 here. Each accuracy number is obtained from a 10-fold validation with 1000 iterations. Using different values of ϵ, we compare the model prediction accuracy using different approaches including no privacy, noise-only (100 clients) and PEM approach. Using different numbers of clients, we compare the model prediction accuracy with $\epsilon = 1$. Figure 2(a) shows the comparison. As expected, the noise-only approach needs to add significant noise, reducing the prediction accuracy significantly. Using the severs to add noise significantly lowers the loss in model accuracy, even for small ϵ. In comparison, noise-only approach with too many clients will cause too much noise.

Distributed clustering. We use k-means for clustering as shown in Algorithm 2, with the **Nursery** dataset. There are 8 categorical attributes in the dataset. We preprocess using *one-hot encoding* and finally get 27 binary

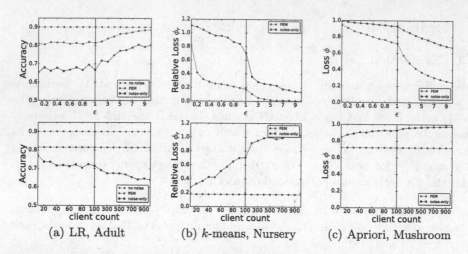

Fig. 2. Evaluation of algorithms using the according datasets with different values of ϵ (client count = 100) and different numbers of clients (fixing $\epsilon = 1$).

attributes. It is easy to see that $L_{\mathbf{x}} = 8$, and thus $L_1 = 16$. In this experiment, we partition the records into five clusters, i.e., $l = 5$. We split ϵ equally to ϵ_1 and ϵ_2, i.e. $\epsilon_1 = \epsilon_2 = \frac{1}{2}\epsilon$. We calculate the loss function ϕ as $\phi(D, \mathbf{c}) = \sum_{i=1}^{l} \sum_{\mathbf{x} \in S_i} \|\mathbf{x} - \mathbf{c}_i\|^2$. As our goal is to compare the loss of different approaches, we use the relative loss to evaluate the performance of different approaches, which is calculated as $\phi_r(D, \mathbf{c}) = \frac{\phi(D,\mathbf{c}) - \phi_0}{\phi_0}$ where ϕ_0 is the loss of the approach without noise. Each training has 50 iterations. Figure 2(b) shows the result.

Distributed frequent itemset mining. We perform distributed frequent itemset mining on the **Mushroom** dataset using Algorithm 3. Considering the itemsets and the counts of them, we define the loss function as $\phi(\mathbf{I}) = \frac{\sum_k \sum_{t \in I_k \cup I_{k0}} |I_k(t) - I_{k0}(t)|}{\sum_k \sum_{t \in I_k \cup I_{0k}} |I_k(t) + I_{k0}(t)|}$ where $\mathbf{I} = [I_2, I_3, \dots]$, I_{k0} is the set of large itemsets of length k generated without noise, and $I_k(t)$ is the (perturbed) count of t for $t \in I_k$. In this experiment, we consider the large itemsets of length no more than 4, which means $\mathbf{I} = [I_2, I_3, I_4]$. We set *minsup* to 0.01. Figure 2(c) shows the result.

7.3   Scalability

The main computation workload of PEM comes from two parts: (1) computation overhead: vector addition and noise generation; and (2) communication overhead. We show that the overhead is small in most of the cases. First, we increase the number of clients from 0 to 10000 and Fig. 3(a) shows the computation time for a scalar on different roles. The aggregator computation time is independent of the number of clients, as it only receives a vector from each server. Thus the

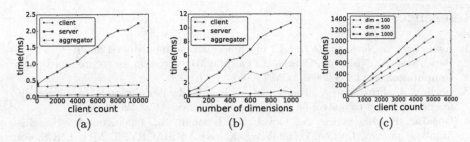

Fig. 3. (a) Computation overhead w.r.t. number of clients. (b) Computation overhead w.r.t. the dimensionality of feature vectors. (c) Overall overhead.

aggregator is unlikely to become the bottleneck. The server workload increases linearly with the number of clients. Fortunately, we can increase each logical server capacity by adding more nodes. Figure 3(b) shows that the computation time is linear to the dimensionality of the feature vector, as expected. Finally, we record the overall overhead of vector addition, including the computation and communication overhead. Figure 3(c) shows that, compared with the communication time, the computation (i.e. vector addition) time can be negligible.

8   Conclusion and Future Work

We present PEM as a practical tradeoff among privacy, utility and computation overhead. PEM is practical in that (1) it has simple assumptions: it only requires a few semi-honest servers and there is no restriction on the dataset itself; (2) it supports a large range of common applications; (3) PEM entails low computation overhead and is scalable to a large number of clients; and (4) all user-visible configuration parameters are intuitive and PEM automatically determines other internal parameters. Using algorithms on real datasets, we show that we can achieve the same level of privacy without the amount of accuracy degradation that previous systems suffer from. Our system also has low computation and communication cost, and thus is very practical.

There are lots of future directions along the lines of privacy. Firstly, we are extending our system to support more operations, such as handling vertically partitioned datasets. Secondly, we will provide the flexibility allowing the clients to choose different trust assumptions, so that the application programmers can choose their own tradeoffs. Last but not least, we will provide a permission system allowing different clients to see different levels of private content, like CryptDB [20] does, but on a much larger scale.

Acknowledgement. This research is supported in part by the National Natural Science Foundation of China (NSFC) grant 61532001, Tsinghua Initiative Research Program Grant 20151080475, MOE Online Education Research Center (Quantong Fund) grant 2017ZD203, and gift funds from Huawei and Ant Financial.

References

1. Asuncion, A., Newman, D.J.: UCI machine learning repository (2007)
2. Ben-David, A., Nisan, N., Pinkas, B.: FairplayMP: a system for secure multi-party computation. In: CCS 2008. ACM (2008)
3. Berlioz, A., Friedman, A., Kaafar, M.A., Boreli, R., Berkovsky, S.: Applying differential privacy to matrix factorization. In: RecSys. ACM (2015)
4. Bhaskar, R., Bhowmick, A., Goyal, V., Laxman, S., Thakurta, A.: Noiseless database privacy. In: Lee, D.H., Wang, X. (eds.) ASIACRYPT 2011. LNCS, vol. 7073, pp. 215–232. Springer, Heidelberg (2011). https://doi.org/10.1007/978-3-642-25385-0_12
5. Bogdanov, D., Laur, S., Willemson, J.: Sharemind: a framework for fast privacy-preserving computations. In: Jajodia, S., Lopez, J. (eds.) ESORICS 2008. LNCS, vol. 5283, pp. 192–206. Springer, Heidelberg (2008). https://doi.org/10.1007/978-3-540-88313-5_13
6. Burkhart, M., Strasser, M., Many, D., et al.: SEPIA: privacy-preserving aggregation of multi-domain network events and statistics. Network 1 (2010)
7. Chaudhuri, K., Sarwate, A.D., Sinha, K.: A near-optimal algorithm for differentially-private principal components. JMLR 14(1), 2905–2943 (2013)
8. Duan, Y.: Differential privacy for sum queries without external noise. In: ACM Conference on Information and Knowledge Management (CIKM) (2009)
9. Duan, Y., Canny, J., Zhan, J.: P4P: practical large-scale privacy-preserving distributed computation robust against malicious users. In: Proceedings of USENIX Security. USENIX Association (2010)
10. Dwork, C., McSherry, F., Nissim, K., Smith, A.: Calibrating noise to sensitivity in private data analysis. In: Halevi, S., Rabin, T. (eds.) TCC 2006. LNCS, vol. 3876, pp. 265–284. Springer, Heidelberg (2006). https://doi.org/10.1007/11681878_14
11. Dwork, C., Naor, M., Pitassi, T., Rothblum, G.N., Yekhanin, S.: Pan-private streaming algorithms. In: ICS (2010)
12. Dwork, C., Roth, A.: The algorithmic foundations of differential privacy. Found. Trends Theor. Comput. Sci. 9(34), 211–407 (2014)
13. Friedman, A., Sharfman, I., Keren, D., Schuster, A.: Privacy-preserving distributed stream monitoring. In: NDSS (2014)
14. Jia, Q., Guo, L., Jin, Z., Fang, Y.: Privacy-preserving data classification and similarity evaluation for distributed systems. In: ICDCS. IEEE (2016)
15. Li, N., Qardaji, W., Su, D.: On sampling, anonymization, and differential privacy or, k-anonymization meets differential privacy. In: Proceedings of the 7th ACM Symposium on Information, Computer and Communications Security. ACM (2012)
16. Machanavajjhala, A., Kifer, D., Gehrke, J., Venkitasubramaniam, M.: l-diversity: privacy beyond k-anonymity. ACM TKDD 1(1) (2007)
17. Margahny, M.H., Mitwaly, A.A.: Fast algorithm for mining association rules. In: Proceedings of the International Conferencs on Very Large Databases (VLDB), vol. 23, no. 3 (1994)
18. McSherry, F.D.: Privacy integrated queries: an extensible platform for privacy-preserving data analysis. In: Proceedings of SIGMOD. ACM (2009)
19. Pathak, M., Rane, S., Raj, B.: Multiparty differential privacy via aggregation of locally trained classifiers. In: NIPS (2010)
20. Popa, R.A., Redfield, C.M.S., Zeldovich, N., Balakrishnan, H.: CryptDB: protecting confidentiality with encrypted query processing. In: ACM SOSP (2011)

21. Rastogi, V., Nath, S.: Differentially private aggregation of distributed time-series with transformation and encryption. In: Proceedings of SIGMOD. ACM (2010)
22. Sarathy, R., Muralidhar, K.: Evaluating Laplace noise addition to satisfy differential privacy for numeric data. Trans. Data Priv. 4(1), 1–7 (2011)
23. Shamir, A.: How to share a secret. Commun. ACM 22(11), 612–613 (1979)
24. Shokri, R., Shmatikov, V.: Privacy-preserving deep learning. In: ACM Conference on Computer and Communications Security (2015)
25. Su, D., Cao, J., Li, N., Bertino, E., Jin, H.: Differentially private k-means clustering. In: Proceedings of CODASPY (2016)
26. Takabi, H., Koppikar, S., Zargar, S.T.: Differentially private distributed data analysis. In: 2016 IEEE 2nd International Conference on Collaboration and Internet Computing (CIC), pp. 212–218. IEEE (2016)
27. Xu, K., Yue, H., Guo, L., Guo, Y., Fang, Y.: Privacy-preserving machine learning algorithms for big data systems. In: Distributed Computing Systems (2015)
28. Yao, A.C.: Protocols for secure computations. In: SFCS 2008 23rd Annual Symposium on Foundations of Computer Science. IEEE (1982)
29. Zhang, N., Li, M., Lou, W.: Distributed data mining with differential privacy. In: 2011 IEEE International Conference on Communications (ICC), pp. 1–5. IEEE (2011)
30. Zinkevich, M.: Online convex programming and generalized infinitesimal gradient ascent. In: ICML (2003)

Probabilistic Models and Methods

Bayesian Heatmaps: Probabilistic Classification with Multiple Unreliable Information Sources

Edwin Simpson[1,2]([✉]), Steven Reece[2], and Stephen J. Roberts[2]

[1] Ubiquitous Knowledge Processing Lab, Department of Computer Science,
Technische Universität Darmstadt, Darmstadt, Germany
simpson@ukp.informatik.tu-darmstadt.de
[2] Department of Engineering Science, University of Oxford, Oxford, UK
{reece,sjrob}@robots.ox.ac.uk

Abstract. Unstructured data from diverse sources, such as social media and aerial imagery, can provide valuable up-to-date information for intelligent situation assessment. Mining these different information sources could bring major benefits to applications such as situation awareness in disaster zones and mapping the spread of diseases. Such applications depend on classifying the situation across a region of interest, which can be depicted as a spatial "heatmap". Annotating unstructured data using crowdsourcing or automated classifiers produces individual classifications at sparse locations that typically contain many errors. We propose a novel Bayesian approach that models the relevance, error rates and bias of each information source, enabling us to learn a spatial Gaussian Process classifier by aggregating data from multiple sources with varying reliability and relevance. Our method does not require gold-labelled data and can make predictions at any location in an area of interest given only sparse observations. We show empirically that our approach can handle noisy and biased data sources, and that simultaneously inferring reliability and transferring information between neighbouring reports leads to more accurate predictions. We demonstrate our method on two real-world problems from disaster response, showing how our approach reduces the amount of crowdsourced data required and can be used to generate valuable heatmap visualisations from SMS messages and satellite images.

1 Introduction

Social media enables members of the public to post real-time text messages, videos and photographs describing events taking place close to them. While many posts may be extraneous or misleading, social media nonetheless provides streams of up-to-date information across a wide area. For example, after the Haiti 2010 earthquake, Ushahidi gathered thousands of text messages that provided valuable first-hand information about the disaster situation [14]. An effective way to extract information from large unstructured datasets such as these is to employ crowds of non-expert annotators, as demonstrated by Galaxy Zoo [10]. Besides social media, crowdsourcing provides a means to obtain geo-tagged

© Springer International Publishing AG 2017
M. Ceci et al. (Eds.): ECML PKDD 2017, Part II, LNAI 10535, pp. 109–125, 2017.
https://doi.org/10.1007/978-3-319-71246-8_7

annotations from other unstructured data sources such as imagery from satellites or unmanned aerial vehicles (UAV).

In scenarios such as disaster response, we wish to infer the situation across a region of interest by combining annotations from multiple information sources. For example, we may wish to determine which areas are currently flooded, the level of damage to buildings in an earthquake zone, or the type of terrain in a specific area from a combination of SMS reports and satellite imagery. The situation across an area of interest can be visualised using a *heatmap* (e.g. Google Maps heatmap layer[1]), which overlays colours onto a map to indicate the intensity or probability of phenomena of interest. Probabilistic methods have been used to generate heatmaps from observations at sparse, point locations [1,8,9], using a Bayesian treatment of Poisson process models. However, these approaches model the rate of occurrence of events, so are not suitable for classification problems. Instead, a Gaussian process (GP) classifier can be used to model a class label that varies smoothly over space or time. This uses a latent function over input coordinates, which is mapped through a sigmoid function to obtain probabilities [16]. However, standard GP classifiers are unsuitable for heterogeneous, crowdsourced data since they do not account for the differing relevance, error rates and bias of individual information sources and annotators.

A key challenge in exploiting crowdsourced information is to account for its unreliability and combine it with trusted data as it becomes available, such as reports from experienced first responders in a disaster zone. For regression problems, differing levels of accuracy can be handled using sensor fusion approaches such as [12,25]. The approach of [25] uses heteroskedastic GPs to produce heatmaps that account for sensor accuracy through variance scaling. This method could be applied to spatial classification by mapping GPs through a softmax function. However, such an approach cannot handle label bias or accuracy that depends on the true class. Recently, [11], proposed learning a GP classifier from crowdsourced annotations, but their method uses a coin-flipping noise model that would suffer from the same drawbacks as adapting [25]. Furthermore they train the model using a maximum likelihood (ML) approach, which may incorrectly estimate reliability when data for some workers is insufficient [7,17,20].

For classification problems, each information source can be modelled by a confusion matrix [3], which quantifies the likelihood of observing a particular annotation from an information source given the true class label. This approach naturally accounts for bias toward a particular answer and varying accuracy depending on the true class, and has been shown to outperform techniques such as majority voting and weighted sums [7,17,20]. Recent extensions following the Bayesian treatment of [7] can further improve results: by identifying clusters of crowd workers with shared confusion matrices [13,23]; accounting for the time each worker takes to complete a task [24]; additionally modelling language features in text classification tasks [4,21]. However, these methods depend on

---

[1] https://developers.google.com/maps/documentation/javascript/examples/layer-hea tmap.

receiving multiple labels from different workers for the same data points, or, in the case of [4,21], on correlations between text features and target classes. None of the existing confusion matrix-based approaches can model the spatial distribution of each class, and therefore, when reports are sparsely distributed over an area of interest, they cannot compensate for the lack of data at each location.

In this paper, we propose a novel Bayesian approach to aggregating sparse, geo-tagged reports from sources of varying reliability, which combines independent Bayesian classifier combination (IBCC) [7] with a GP classifier to infer discrete state values across an area of interest. Our model, *HeatmapBCC*, assumes that states at neighbouring locations are correlated, allowing us to fuse neighbouring reports and interpolate between them to predict the state at locations with no reports. HeatmapBCC uses confusion matrices to model the error rates, relevance and bias of each information source, permitting the use of non-expert crowds providing heterogeneous annotations. The GP handles the uncertainty that arises from sparse spatial data in a principled Bayesian manner, allowing us to incorporate prior information, such as physical models of disaster events such as earthquakes, and visualise the resulting posterior distribution as a spatial heatmap. We derive a variational inference method that is able to learn the reliability model for each information source without the need for ground truth training data. This method learns full distributions over latent variables that can be used to prioritise locations for further data gathering using an active learning approach. The next section presents in detail the HeatmapBCC model, and provides details of our efficient approximate inference algorithm. The following section then provides an empirical evaluation of our method on both synthetic and real-world problems, showing that HeatmapBCC can outperform rival methods. We make our code publicly available at https://github.com/OxfordML/heatmap_expts.

2   The HeatmapBCC Model

Our goal is to classify locations of interest, e.g. to identify them as "flooded" or "not flooded". We can then choose locations in a grid over an area of interest and plot the classifications on a map as a spatial *heatmap*. The task is to infer a vector $t^* \in \{1, \ldots, J\}^{N^*}$ of target state values at N^* locations X^*, where J is the number of state values or classes. Each row x_i of matrix X^* is a coordinate vector that specifies a point on the map. We observe a matrix of potentially unreliable geo-tagged *reports*, $c \in \{1, \ldots, L\}^{N \times S}$, with L possible discrete values, from S different information sources at N training locations X.

HeatmapBCC assumes that each report label $c_i^{(s)}$, from information source s, at location x_i, is drawn from $c_i^{(s)} | t_i, \pi^{(s)} \sim \text{Categorical}(\pi_{t_i}^{(s)})$. The target state, t_i, selects the row, $\pi_{t_i}^{(s)}$, of a *confusion matrix* [3,20], $\pi^{(s)}$, which describes the errors and biases of s as a dependency between the report labels and the ground truth state, t_i. As per standard IBCC [7], the reports from each information source are conditionally independent of one another given target t_i, and each row of the confusion matrix is drawn from $\boldsymbol{\pi}_j^{(s)}|\boldsymbol{\alpha}_{0,j}^{(s)} \sim \text{Dirichlet}(\boldsymbol{\alpha}_{0,j}^{(s)})$. The hyperparameters $\boldsymbol{\alpha}_{0,j}^{(s)}$ encode the prior trust in s.

We assume that state t_i at location $\boldsymbol{x}_i$ is drawn from a categorical distribution, $t_i|\boldsymbol{\rho}_i \sim \text{Categorical}(\boldsymbol{\rho}_i)$, where $\rho_{i,j} = p(t_i = j|\boldsymbol{\rho}_i) \in [0,1]$ is the probability of state j at location $\boldsymbol{x}_i$. The generative process for state probabilities, $\boldsymbol{\rho}$, is as follows. First, draw latent functions for classes $j \in \{1,...,J\}$ from a Gaussian process prior: $f_j \sim \mathcal{GP}(m_j, k_{j,\theta}/\varsigma_j)$, where m_j is the prior mean function, k_j is the prior covariance function, $\boldsymbol{\theta}$ are hyperparameters of the covariance function, and ς_j is the inverse scale. Map latent function values $f_j(\boldsymbol{x}_i) \in \mathcal{R}$ to state probabilities: $\boldsymbol{\rho}_i = \sigma(f_1(\boldsymbol{x}_i),...,f_J(\boldsymbol{x}_i)) \in [0,1]^J$. Appropriate functions for σ include the logistic sigmoid and probit functions for binary classification, and softmax and multinomial probit for multi-class classification. We assume that ς_j is drawn from a conjugate gamma hyperprior, $\varsigma_j \sim \mathcal{G}(a_0, b_0)$, where a_0 is a shape parameter and b_0 is the inverse scale.

While the reports, $c_i^{(s)}$, are modelled in the same way as standard IBCC [7], HeatmapBCC introduces a location-specific state probability, $\boldsymbol{\rho}_i$, to replace the global class proportions, $\boldsymbol{\kappa}$, which IBCC [20] assumes are constant for all locations. Using a Gaussian process prior means the state probability varies reasonably smoothly between locations, thereby encoding correlations in the distribution over states at neighbouring locations. The covariance function is chosen to suit the scenario we wish to model and may be tailored to specific spatial phenomena (the geo-spatial impact of an earthquake, for example). The hyperparameters, $\boldsymbol{\theta}$, typically include a length-scale, l, which controls the smoothness of the function. Here, we assume a stationary covariance function of the form $k_{j,\theta}(\boldsymbol{x},\boldsymbol{x}') = k_j(|\boldsymbol{x}-\boldsymbol{x}'|, l)$, where k is a function of the distance between two points and the length-scale, l. The joint distribution for the complete model is:

$$p\left(\boldsymbol{c}, \boldsymbol{t}, \boldsymbol{f}_1,...,\boldsymbol{f}_J, \varsigma_1,...,\varsigma_J, \boldsymbol{\pi}^{(1)},...,\boldsymbol{\pi}^{(S)}|\boldsymbol{\mu}_1,...,\boldsymbol{\mu}_J, \boldsymbol{K}_1,...,\boldsymbol{K}_J, \boldsymbol{\alpha}_0^{(1)},...,\boldsymbol{\alpha}_0^{(S)}\right) =$$

$$\prod_{i=1}^{N}\left\{\rho_{i,t_i}\prod_{s=1}^{S}\pi_{t_i,c_i^{(s)}}^{(s)}\right\}\prod_{j=1}^{J}\left\{p\left(\boldsymbol{f}_j|\boldsymbol{\mu}_j, \boldsymbol{K}_j/\varsigma_j\right)p\left(\varsigma_j|a_0, b_0\right)\prod_{s=1}^{S}p\left(\boldsymbol{\pi}_j^{(s)}|\boldsymbol{\alpha}_{0,j}^{(s)}\right)\right\},$$

where $\boldsymbol{f}_j = [f_j(\boldsymbol{x}_1),...,f_j(\boldsymbol{x}_N)]$, $\boldsymbol{\mu}_j = [m_j(\boldsymbol{x}_1),...,m_j(\boldsymbol{x}_N)]$, and $\boldsymbol{K}_j \in \mathbb{R}^{N \times N}$ with elements $K_{j,n,n'} = k_{j,\theta}(\boldsymbol{x}_n, \boldsymbol{x}_{n'})$.

3   Variational Inference for HeatmapBCC

We use *variational Bayes (VB)* to efficiently approximate the posterior distribution over all latent variables, allowing us to handle streaming data reports online by restarting the VB algorithm from the previous estimate as new reports are received. To apply variational inference, we replace the exact posterior distribution with a variational approximation that factorises into separate latent variables and parameters:

$$p(\boldsymbol{t}, \boldsymbol{f}, \varsigma, \boldsymbol{\pi}^{(1)},...,\boldsymbol{\pi}^{(S)}|\boldsymbol{c}, \boldsymbol{\mu}, \boldsymbol{K}, \boldsymbol{\alpha}_0^{(1)},...,\boldsymbol{\alpha}_0^{(S)}) \approx q(\boldsymbol{t})\prod_{j=1}^{J}\left\{q(\boldsymbol{f}_j)q(\varsigma_j)\prod_{s=1}^{S}q\left(\boldsymbol{\pi}_j^{(s)}\right)\right\}.$$

We perform approximate inference by optimising the variational posterior using Algorithm 1. In the remainder of this section we define the variational factors $q()$, expectation terms, variational lower bound and prediction step required by the algorithm.

---

input : Hyperparameters $\boldsymbol{\alpha}_0^{(s)}$ $\forall s$, $\boldsymbol{\mu}_j$ $\forall j$, $\boldsymbol{K}$, a_0, b_0; observed report data $\boldsymbol{c}$
Initialise $q\left(\boldsymbol{f}_j\right)$ $\forall j$, $q\left(\boldsymbol{\pi}_j^{(s)}\right)$ $\forall j$ $\forall s$, and $q(\varsigma_j)\forall j$ randomly
while *variational lower bound not converged* **do**

> Calculate $\mathbb{E}\left[\log \rho\right]$ and $\mathbb{E}\left[\log \boldsymbol{\pi}^{(s)}\right]$, $\forall s$ given current factors $q\left(\boldsymbol{f}_j\right)$ and $q\left(\boldsymbol{\pi}_j^{(s)}\right)$
> Update $q(\boldsymbol{t})$ given $\mathbb{E}\left[\log \boldsymbol{\pi}^{(s)}\right]$, $\forall s$ and $\mathbb{E}\left[\log \rho\right]$
> Update $q\left(\boldsymbol{\pi}_j^{(s)}\right)$, $\forall j, \forall s$ given current estimate for $q(\boldsymbol{t})$
> Update $q\left(\boldsymbol{f}_j\right)$, $\forall j$ current estimates for $q(\boldsymbol{t})$ and $q(\varsigma_j), \forall j$
> Update $q(\varsigma_j), \forall j$ given current estimate for $q\left(\boldsymbol{f}_j\right)$

end
output: Use converged estimates to predict ρ^* and $\boldsymbol{t}^*$ at output points X^*

Algorithm 1. VB algorithm for HeatmapBCC

Variational Factor for Targets, $\boldsymbol{t}$:

$$\log q(\boldsymbol{t}) = \sum_{i=1}^{N}\left\{\mathbb{E}[\log \rho_{i,t_i}] + \sum_{s=1}^{S}\mathbb{E}\left[\log \pi_{t_i,c_i^{(s)}}^{(s)}\right]\right\} + \text{const.} \tag{1}$$

The variational factor $q(\boldsymbol{t})$ further factorises into individual data points, since the target value, t_i, at each input point, $\boldsymbol{x}_i$, is independent given the state probability vector $\boldsymbol{\rho}_i$, giving $r_{i,j} := q(t_i = j)$ where $q(t_i = j) = q(t_i = j, \boldsymbol{c}_i)/\sum_{\iota \in J} q(t_i = \iota, \boldsymbol{c}_i)$ and:

$$q(t_i = j, \boldsymbol{c}_i) = \exp\left(\mathbb{E}\left[\log \rho_{i,j}\right] + \sum_{s=1}^{S}\mathbb{E}\left[\log \pi_{j,c_i^{(s)}}^{(s)}\right]\right). \tag{2}$$

Missing reports in $\boldsymbol{c}$ can be handled simply by omitting the term $\mathbb{E}\left[\log \pi_{j,c_i^{(s)}}^{(s)}\right]$ for information sources, s, that have not provided a report $c_i^{(s)}$.

Variational Factor for Confusion Matrix Rows, $\boldsymbol{\pi}_j^{(s)}$:

$$\log q\left(\boldsymbol{\pi}_j^{(s)}\right) = \mathbb{E}_t\left[\log p\left(\boldsymbol{\pi}^{(s)}|\boldsymbol{t},\boldsymbol{c}\right)\right] = \sum_{l=1}^{L} N_{j,l}^{(s)}\log \pi_{j,l}^{(s)} + \log p\left(\boldsymbol{\pi}_j^{(s)}|\boldsymbol{\alpha}_{0,j}^{(s)}\right) + \text{const.},$$

where $N_{j,l}^{(s)} = \sum_{i=1}^{N} r_{i,j}\delta_{l,c_i^{(s)}}$ are pseudo-counts and δ is the Kronecker delta. Since we assumed a Dirichlet prior, the variational distribution is also a Dirichlet, $q(\pi_j^{(s)}) = \mathcal{D}(\pi_j^{(s)}|\alpha_j^{(s)})$, with parameters $\alpha_j^{(s)} = \alpha_{0,j}^{(s)} + N_j^{(s)}$, where $N_j^{(s)} = \left\{N_{j,l}^{(s)}|l \in [1,...,L]\right\}$. Using the digamma function, $\Psi()$, the expectation required for Eq. 2 is therefore:

$$\mathbb{E}\left[\log \pi_{j,l}^{(s)}\right] = \Psi\left(\alpha_{j,l}^{(s)}\right) - \Psi\left(\sum_{\iota=1}^{L} \alpha_{j,\iota}^{(s)}\right). \tag{3}$$

Variational Factor for Latent Function: The variational factor $q(\boldsymbol{f})$ factorises between target classes, since t_i at each point is independent given $\boldsymbol{\rho}$. Using the fact that $\mathbb{E}_{t_i}[\log \text{Categorical}([t_i = j]|\rho_{i,j})] = r_{i,j}\log \sigma(\boldsymbol{f})_{j,i}$, the factor for each class is:

$$\log q(\boldsymbol{f}_j) = \sum_{i=1}^{N} r_{i,j}\log \sigma(\boldsymbol{f})_{j,i} + \mathbb{E}_{\varsigma_j}\left[\log \mathcal{N}(\boldsymbol{f}_j|\boldsymbol{\mu}_j, \boldsymbol{K}_j/\varsigma_j)\right] + \text{const.} \tag{4}$$

This variational factor cannot be computed analytically, but can itself be approximated using a variational method based on the extended Kalman filter (EKF) [18,22] that is amenable to inclusion in our overall VB algorithm. Here, we present a multi-class variant of this method that applies ideas from [5]. We approximate the likelihood $p(t_i = j|\boldsymbol{\rho}_{i,j}) = \boldsymbol{\rho}_{i,j}$ with a Gaussian distribution, using $\mathbb{E}[\log \mathcal{N}([t_i = j]|\sigma(\boldsymbol{f})_{j,i}, v_{i,j})] = \log \mathcal{N}(r_{i,j}|\sigma(\boldsymbol{f})_{j,i}, v_{i,j})$ to replace Eq. 4 with the following:

$$\log q(\boldsymbol{f}_j) \approx \sum_{i=1}^{N} \log \mathcal{N}(r_{i,j}|\sigma(\boldsymbol{f})_{j,i}, v_{i,j}) + \mathbb{E}_{\varsigma_j}[\log \mathcal{N}(\boldsymbol{f}_j|\boldsymbol{\mu}_j, \boldsymbol{K}_j/\varsigma_j)] + \text{const}, \tag{5}$$

where $v_{i,j} = \rho_{i,j}(1 - \rho_{i,j})$ is the variance of the binary indicator variable $[t_i = j]$ given by the Bernoulli distribution. We approximate Eq. 5 by linearising $\sigma()$ using a Taylor series expansion to obtain a multivariate Gaussian distribution $q(\boldsymbol{f}_j) \approx \mathcal{N}\left(\boldsymbol{f}_j|\hat{\boldsymbol{f}}_j, \boldsymbol{\Sigma}_j\right)$. Consequently, we estimate $q\left(\boldsymbol{f}_j\right)$ using EKF-like equations [18,22]:

$$\hat{\boldsymbol{f}}_j = \boldsymbol{\mu}_j + \boldsymbol{W}\left(\boldsymbol{r}_{.,j} - \sigma(\hat{\boldsymbol{f}})_j + \boldsymbol{G}(\hat{\boldsymbol{f}}_j - \boldsymbol{\mu}_j)\right) \tag{6}$$

$$\boldsymbol{\Sigma}_j = \hat{\boldsymbol{K}}_j - \boldsymbol{W}\boldsymbol{G}_j\hat{\boldsymbol{K}}_j \tag{7}$$

where $\hat{\boldsymbol{K}}_j^{-1} = \boldsymbol{K}_j^{-1}\mathbb{E}[\varsigma_j]$ and $\boldsymbol{W} = \hat{\boldsymbol{K}}_j\boldsymbol{G}_j^T\left(\boldsymbol{G}_j\hat{\boldsymbol{K}}_j\boldsymbol{G}_j^T + \boldsymbol{Q}_j\right)^{-1}$ is the Kalman gain, $\boldsymbol{r}_{.,j} = [r_{1,j}, r_{N,j}]$ is the vector of probabilities of target state j computed using Eq. 2 for the input points, $\boldsymbol{G}_j \in \mathbb{R}^{N \times N}$ is the diagonal sigmoid Jacobian matrix and $\boldsymbol{Q}_j \in \mathbb{R}^{N \times N}$ is a diagonal observation noise variance matrix. The diagonal elements of $\boldsymbol{G}$ are $G_{j,i,i} = \sigma(\hat{\boldsymbol{f}}_{.,i})_j(1 - \sigma(\hat{\boldsymbol{f}}_{.,i})_j)$, where $\hat{\boldsymbol{f}} = \left[\hat{\boldsymbol{f}}_1, ..., \hat{\boldsymbol{f}}_J\right]$ is the matrix of mean values for all classes.

The diagonal elements of the noise covariance matrix are $Q_{j,i,i} = v_{i,j}$, which we approximate as follows. Since the observations are Bernoulli distributed with an uncertain parameter $\rho_{i,j}$, the conjugate prior over $\rho_{i,j}$ is a beta distribution with parameters $\sum_{j'=1}^{J} \nu_{0,j'}$ and $\nu_{0,j}$. This can be updated to a posterior Beta distribution $p(\rho_{i,j}|r_{i,j}, \boldsymbol{\nu}_0) = \mathcal{B}(\rho_{i,j}|\nu_{\neg j}, \nu_j)$, where $\nu_{\neg j} = \sum_{j'=1}^{J} \nu_{0,j'} - \nu_{0,j} + 1 - r_{i,j}$ and $\nu_j = \nu_{0,j} + r_{i,j}$. We now estimate the expected variance:

$$v_{i,j} \approx \hat{v}_{i,j} = \int \left(\rho_{i,j} - \rho_{i,j}^2\right) \mathcal{B}(\rho_{i,j}|\nu_{\neg j}, \nu_j) \, d\rho_{i,j} = \mathbb{E}[\rho_{i,j}] - \mathbb{E}\left[\rho_{i,j}^2\right] \quad (8)$$

$$\mathbb{E}[\rho_{i,j}] = \frac{\nu_j}{\nu_j + \nu_{\neg j}} \qquad \mathbb{E}\left[\rho_{i,j}^2\right] = \mathbb{E}[\rho_{i,j}]^2 + \frac{\nu_j \nu_{\neg j}}{(\nu_j + \nu_{\neg j})^2 (\nu_j + \nu_{\neg j} + 1)}. \quad (9)$$

We determine values for the prior beta parameters, $\nu_{0,j}$, by moment matching with the prior mean $\hat{\rho}_{i,j}$ and variance $u_{i,j}$ of $\rho_{i,j}$, found using numerical integration. According to Jensen's inequality, the convex function $\varphi(\boldsymbol{Q}) = \left(\boldsymbol{G}_j \boldsymbol{K}_j \boldsymbol{G}_j^T + \boldsymbol{Q}\right)^{-1}$ is a lower bound on $\mathbb{E}[\varphi(\boldsymbol{Q})] = \mathbb{E}\left[(\boldsymbol{G}_j \boldsymbol{K}_j \boldsymbol{G}_j^T + \boldsymbol{Q})^{-1}\right]$. Thus our approximation provides a tractable estimate of the expected value of $\boldsymbol{W}$.

The calculation of $\boldsymbol{G}_j$ requires evaluating the latent function at the input points $\hat{\boldsymbol{f}}_j$. Further, Eq. 6 requires $\boldsymbol{G}_j$ to approximate $\hat{\boldsymbol{f}}_j$, causing a circular dependency. Although we can fold our expressions for $\boldsymbol{G}_j$ and $\hat{\boldsymbol{f}}_j$ directly into the VB cycle and update each variable in turn, we found solving for $\boldsymbol{G}_j$ and $\hat{\boldsymbol{f}}_j$ each VB iteration facilitated faster inference. We use the following iterative procedure to estimate $\boldsymbol{G}_j$ and $\hat{\boldsymbol{f}}_j$:

1. Initialise $\sigma(\hat{\boldsymbol{f}}_{.,i}) \approx \mathbb{E}[\boldsymbol{\rho}_i]$ using Eq. 9.
2. Estimate $\boldsymbol{G}_j$ using the current estimate of $\sigma(\hat{f}_{j,i})$.
3. Update the mean $\hat{\boldsymbol{f}}_j$ using Eq. 6, inserting the current estimate of $\boldsymbol{G}$.
4. Repeat from step 2 until $\hat{\boldsymbol{f}}_j$ and $\boldsymbol{G}_j$ converge.

The latent means, $\hat{\boldsymbol{f}}$, are then used to estimate the terms $\log \rho_{i,j}$ for Eq. 2:

$$\mathbb{E}[\log \rho_{i,j}] = \hat{f}_{j,i} - \mathbb{E}\left[\log \sum_{j'=1}^{J} \exp(f_{j',i})\right]. \quad (10)$$

When inserted into Eq. 2, the second term in Eq. 10 cancels with the denominator, so need not be computed.

Variational Factor for Inverse Function Scale: The inverse covariance scale, ς_j, can also be inferred using VB by taking expectations with respect to $\boldsymbol{f}$:

$$\log q(\varsigma_j) = \mathbb{E}_\rho[\log p(\varsigma_j|\boldsymbol{f}_j)] = \mathbb{E}_{\boldsymbol{f}_j}[\log \mathcal{N}(\boldsymbol{f}_j|\mu_i, \boldsymbol{K}_j/\varsigma_j)] + \log p(\varsigma_j|a_0, b_0) + \text{const}$$

which is a gamma distribution with shape $a = a_0 + \frac{N}{2}$ and inverse scale $b = b_0 + \frac{1}{2}\mathrm{Tr}\left(K_j^{-1}\left(\Sigma_j + \hat{f}_j\hat{f}_j^T - 2\mu_{j,i}\hat{f}_j^T - \mu_{j,i}\mu_{j,i}^T\right)\right)$. We use these parameters to compute the expected latent model precision, $\mathbb{E}[\varsigma_j] = a/b$ in Eq. 7, and for the lower bound described in the next section we also require $\mathbb{E}_q[\log(\varsigma_j)] = \Psi(a) - \log(b)$.

Variational Lower Bound: Due to the approximations described above, we are unable to guarantee an increased variational lower bound for each cycle of the VB algorithm. We test for convergence of the variational approximation efficiently by comparing the variational lower bound $\mathcal{L}(q)$ on the model evidence calculated at successive iterations. The lower bound for HeatmapBCC is given by:

$$\mathcal{L}(q) = \mathbb{E}_q\left[\log p\left(c|t, \pi^{(1)}, ..., \pi^{(S)}\right)\right] + \mathbb{E}_q\left[\log \frac{p(t|\rho)}{q(t)}\right] + \sum_{j=1}^{J}\Bigg\{ \tag{11}$$

$$\mathbb{E}_q\left[\log \frac{p\left(f_j|\mu_j, K_j/\varsigma_j\right)}{q(f_j)}\right] + \mathbb{E}_q\left[\log \frac{p(\varsigma_j|a_0, b_0)}{q(\varsigma_j)}\right] + \sum_{s=1}^{S}\mathbb{E}_q\left[\log \frac{p\left(\pi_j^{(s)}|\alpha_{0,j}^{(s)}\right)}{q\left(\pi_j^{(s)}\right)}\right]\Bigg\}.$$

Predictions: Once the algorithm has converged, we predict target states, t^* and probabilities ρ^* at output points X^* by estimating their expected values. For a heatmap visualisation, X^* is a set of evenly-spaced points on a grid placed over the region of interest. We cannot compute the posterior distribution over ρ^* analytically due to the non-linear sigmoid function. We therefore estimate the expected values $\mathbb{E}[\rho_j^*]$ by sampling f_j^* from its posterior and mapping the samples through the sigmoid function. The multivariate Gaussian posterior of f_j^* has latent mean $\hat{f}^*$ and covariance Σ^*:

$$\hat{f}_j^* = \mu_j^* + W_j^*\left(r_j - \sigma(\hat{f}_j) + G(\hat{f}_j - \mu_j)\right) \tag{12}$$

$$\Sigma_j^* = \hat{K}_j^{**} - W_j^* G_j \hat{K}_j^*, \tag{13}$$

where μ_j^* is the prior mean at the output points, $\hat{K}_j^{**}$ is the covariance matrix of the output points, $\hat{K}_j^*$ is the covariance matrix between the input and the output points, and $W_j^* = \hat{K}_j^* G_j^T\left(G_j \hat{K}_j G_j^T + Q_j\right)^{-1}$ is the Kalman gain. The predictions for output states t^* are the expected probabilities $\mathbb{E}\left[t_{i,j}^*\right] = r_{i,j}^* \propto q(t_i = j, c)$ of each state j at each output point $x_i \in X^*$, computed using Eq. 2. In a multi-class setting, the predictions for each class could be plotted as separate heatmaps.

4   Experiments

We compare the efficacy of our approach with alternative methods on synthetic data and two real datasets. In the first real-world application we combine crowd-sourced annotations of images in the aftermath of a disaster, while in the second we aggregate crowdsourced labels assigned to geo-tagged text messages to predict emergencies in the aftermath of an Earthquake. All experiments are binary classification tasks where reports may be negative (recorded as $c_i^{(s)} = 1$) or positive ($c_i^{(s)} = 2$). In all experiments, we examine the effect of data sparsity using an *incremental train/test procedure*:

1. Train all methods on a random subset of reports (initially a small subset)
2. Predict states t^* at grid points in an area of interest. For HeatmapBCC, we use the predictions $\mathbb{E}[t_{i,j}^*]$ described in Sect. 3
3. Evaluate predictions using the area under the ROC curve (AUC) or cross entropy classification error
4. Increment subset of training labels at random and repeat from step 1.

Specific details vary in each experiment and are described below. We evaluate HeatmapBCC against the following alternatives: a Kernel density estimator (KDE) [15,19], which is a non-parametric technique that places a Gaussian kernel at each observation point, then normalises the sum of Gaussians over all observations; a GP classifier [18], which applies a Bayesian non-parametric approach but assumes reports are equally reliable; IBCC with VB [20], which performs no interpolation between spatial points, but is a state-of-the-art method for combining unreliable crowdsourced classifications; and an ad-hoc combination of IBCC and the GP classifier ($IBCC+GP$), in which the output classifications of IBCC are used as training labels for the GP classifier. This last method illustrates whether the single VB learning approach of HeatmapBCC is beneficial, for example, by transferring information between neighbouring data points when learning confusion matrices. For the first real dataset, we include additional baselines: SVM with radial basis function kernel; a K-nearest neighbours classifier with $n_{neighbours} = 5$ (NN); and majority voting (MV), which defaults to the most frequent class label (negative) in locations with no labels.

4.1 Synthetic Data

We ran three experiments with synthetic data to illustrate the behaviour of HeatmapBCC with different types of unreliable reporters. For each experiment, we generated 25 binary ground truth datasets as follows: obtain coordinates at all 1600 points in a 40×40 grid; draw latent function values f_x from a multivariate Gaussian distribution with zero mean and Matérn$\frac{3}{2}$ covariance with $l = 20$ and inverse scale 1.2; apply sigmoid function to obtain state probabilities, ρ_x; draw target values, t_x, at all locations.

Noisy reporters: The first experiment tests robustness to error-prone annotators. For each of the 25 ground truth datasets, we generated three crowds of 20 reporters. In each crowd, we varied the number of *reliable* reporters between 5, 10 and 15, while the remainder were *noisy* reporters with high random error rates. We simulated reliable reporters by drawing confusion matrices, $\boldsymbol{\pi}^{(s)}$, from beta distributions with parameter matrix set to $\alpha_{jj}^{(s)} = 10$ along the diagonals and 1

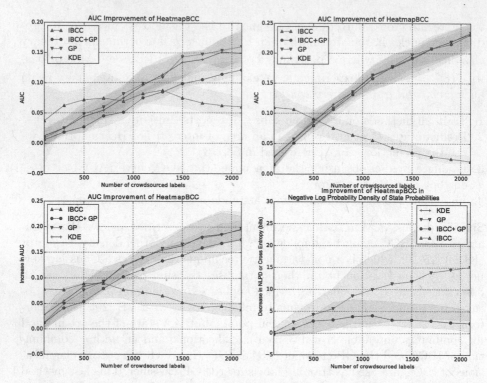

Fig. 1. Synthetic data, *noisy reporters*: median improvement of HeatmapBCC over alternatives over 25 datasets, against number of crowdsourced labels. Shaded areas show inter-quartile range. Top-left: AUC, 25% noisy reporters. Top-right: AUC, 50% noisy reporters. Bottom-left: AUC, 75% noisy reporters. Bottom-right: NLPD of state probabilities, ρ, with 50% noisy reporters.

elsewhere. For noisy workers, all parameters were set equally to $\alpha_{jl}^{(s)} = 5$. For each proportion of noisy reporters, we selected reporters and grid points at random, and generated 2400 reports by drawing binary labels from the confusion matrices $\pi^{(1)}, ..., \pi^{(20)}$. We ran the incremental train/test procedure for each crowd with each of the 25 ground truth datasets. For HeatmapBCC, GP and IBCC+GP the kernel hyperparameters were set as $l = 20$, $a_0 = 1$, and $b_0 = 1$. For HeatmapBCC, IBCC and IBCC+GP, we set confusion matrix hyperparameters to $\alpha_{j,j}^{(s)} = 2$ along the diagonals and $\alpha_{j,l}^{(s)} = 1$ elsewhere, assuming a weak tendency toward correct labels. For IBCC we also set $\nu_0 = [1, 1]$.

Figure 1 shows the median differences in AUC between HeatmapBCC and the alternative methods for *noisy reporters*. Plotting the difference between methods allows us to see consistent performance differences when AUC varies substantially between runs. More reliable workers increase the AUC improvement of HeatmapBCC. With all proportions of workers, the performance improvements are smaller with very small numbers of labels, except against IBCC, as none of the methods produce a confident model with very sparse data. As more labels are gathered, there are more locations with multiple reports, and IBCC is able to make good predictions at those points, thereby reducing the difference in AUC as the number of labels increases. However, for the other three methods, the difference in AUC continues to increase, as they improve more slowly as more labels are received. With more than 700 labels, using the GP to estimate the class labels directly is less effective than using IBCC classifications at points where we have received reports, hence the poorer performance of GP and IBCC+GP.

In Fig. 1 we also show the improvement in negative log probability density (NLPD) of state probabilities, ρ. We compare HeatmapBCC only against the methods that place a posterior distribution over their estimated state proba-bilities. As more labels are received, the IBCC+GP method begins to improve slightly, as it is begins to identify the noisy reporters in the crowd. The GP is much slower to improve due to the presence of these noisy labels.

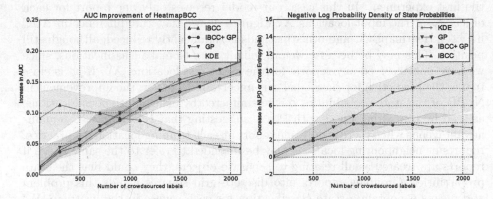

Fig. 2. Synthetic data, 50% *biased reporters*: median improvement of HeatmapBCC compared to alternatives over 25 datasets, against number of crowdsourced labels. Shaded areas showing the inter-quartile range. Left: AUC. Right: NLPD of state prob-abilities, ρ.

Biased reporters: The second experiment simulates the scenario where some reporters choose the negative class label overly frequently, e.g. because they fail to observe the positive state when it is present. We repeated the procedure used for noisy reporters but replaced the noisy reporters with *biased* reporters gener-ated using the parameter matrix $\alpha^{(s)} = \left[\begin{smallmatrix} 7 & 1 \\ 6 & 2 \end{smallmatrix}\right]$. We observe similar performance improvements to the first experiment with noisy reporters, as shown in Fig. 2, suggesting that HeatmapBCC is also better able to model biased reporters from sparse data than rival approaches. Figure 3 shows an example of the posterior distributions over t_x produced by each method when trained on 1500 random labels from a simulated crowd with 50% *biased reporters*. We can see that the ground truth appears most similar to the HeatmapBCC estimates, while IBCC is unable to perform any smoothing.

Continuous report locations: In the previous experiments we drew reports from discrete grid points so that multiple reporters produced noisy labels for the

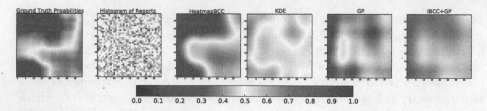

Fig. 3. Synthetic data, 50% *biased reporters*: posterior distributions. Histogram of reports shows the difference between positive and negative label frequencies at each grid square.

same target, t_x. The third experiment tests the behaviour of our model with reports drawn from continuous locations, with 50% noisy reporters drawn as in the first experiment. In this case, our model receives only one report for each object t_x at the input locations X. Figure 4 shows that the difference in AUC between HeatmapBCC and other methods is significantly reduced, although still positive. This may be because we are reliant on ρ to make classifications, since we have not observed any reports for the exact test locations X^*. If ρ_x is close to 0.5, the prediction for class label x is uncertain. However, the improvement in NLPD of the state probabilities ρ is less affected by using continuous locations, as seen by comparing Fig. 1 with Fig. 4, suggesting that HeatmapBCC remains advantageous when there is only one report at each training location. In practice, reports at neighbouring locations may be intended to refer to the same t_x, so if reports are treated as all relating to separate objects, they could bias the state probabilities. Grouping reports into discrete grid squares avoids this problem and means we obtain a state classification for each square in the heatmap. We therefore continue to use discrete grid locations in our real-world experiments.

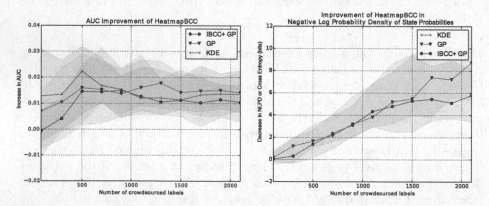

Fig. 4. Synthetic data, 50% *noisy reporters, continuous report locations*. Median improvement of HeatmapBCC compared to alternatives over 25 datasets, against number of crowdsourced labels. Shaded areas showing the inter-quartile range. Left: AUC. Right: NLPD of state probabilities, ρ.

4.2   Crowdsourced Labels of Satellite Images

We obtained a set of 5,477 crowdsourced labels from a trial run of the Zooniverse Planetary Response Network project[2]. In this application, volunteers labelled satellite images showing damage to Tacloban, Philippines, after Typhoon Haiyan/Yolanda. The volunteers' task was to mark features such as damaged buildings, blocked roads and floods. For this experiment, we first divided the area into a 132×92 grid. The goal was then to combine crowdsourced labels to classify grid squares according to whether they contain buildings with major damage or not. We treated cases where a user observed an image but did not mark any features as a set of multiple negative labels, one for each of the grid squares covered by the image. Our dataset contained 1,641 labels marking buildings with major structural damage, and 1,245 negative labels. Although this dataset does not contain ground truth annotations, it contains enough crowdsourced annotations that we can confidently determine labels for most of the region of interest using all data. The aim is to test whether our approach can replicate these results using only a subset of crowdsourced labels, thereby reducing the workload of the crowd by allowing for sparser annotations. We therefore defined gold-standard labels by running IBCC on the complete set of crowdsourced labels, and then extracting the IBCC posterior probabilities for 572 data points with ≥ 3 crowdsourced labels where the posterior of the most probable class ≥ 0.9. The IBCC hyperparameters were set to $\alpha_{0,j,j}^{(s)} = 2$ along the diagonals, $\alpha_{0,j,l}^{(s)} = 1$ elsewhere, and $\nu_0 = [100, 100]$.

We ran our incremental train/test procedure 20 times with initial subsets of 178 random labels. Each of these 20 repeats required approximately 45 min runtime on an Intel i7 desktop computer. The length-scales l for HeatmapBCC, GP and IBCC+GP were optimised at each iteration using maximum likelihood II by maximising the variational lower bound on the log likelihood (Eq. 11), as described in [16]. The inverse scale hyperparameters were set to $a_0 = 0.5$ and $b_0 = 5$, and the other hyperparameters were set as for gold label generation. We did not find a significant difference when varying diagonal confusion matrix values $\alpha_{j,j}^{(s)} = 2$ from 2 to 20.

In Fig. 5 (left) we can see how AUC varies as more labels are introduced, with HeatmapBCC, GP and IBCC+GP converging close to our gold-standard solution. HeatmapBCC performs best initially, potentially because it can learn a more suitable length-scale with less data than GP and IBCC+GP. SVM outperforms GP and IBCC+GP with 178 labels, but is outperformed when more labels are provided. Majority voting, nearest neighbour and IBCC produce much lower AUCs than the other approaches. The benefits of HeatmapBCC can be more clearly seen in Fig. 5 (right), which shows a substantial reduction in cross entropy classification error compared to alternative methods, indicating that HeatmapBCC produces better probability estimates.

---

[2] http://www.planetaryresponsenetwork.com/beta/.

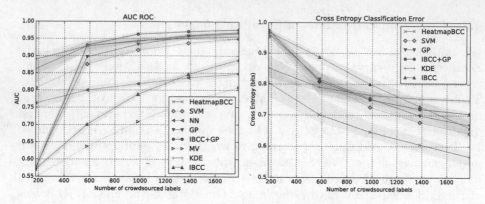

Fig. 5. Planetary response network, major structural damage data. Median values over 20 repeats against the number of randomly selected crowdsourced labels. Shaded areas show the inter-quartile range. Left: AUC. Right: cross entropy error.

4.3 Haiti Earthquake Text Messages

Here we aggregate text reports written by members of the public after the Haiti 2010 Earthquake. The dataset we use was collected and labelled by Ushahidi [14]. We have selected 2,723 geo-tagged reports that were sent mainly by SMS and were categorised by Ushahidi volunteers. The category labels describe the type of situation that is reported, such as "medical emergency" or "collapsed building". In this experiment, we aim to predict a binary class label, "emergency" or "no emergency" by combining all reports. We model each category as a different information source; if a category label is present for a particular message, we observe a value of 1 from that information source at the message's geo-location. This application differs from the satellite labelling task because many of the reports do not explicitly report emergencies and may be irrelevant. In the absence of ground truth data, we establish a gold-standard test set by training IBCC on all 2723 reports, placed into 675 discrete locations on a 100×100 grid. Each grid square has approximately 4 reports. We set IBCC hyper-parameters to $\alpha_{0,j,j}^{(s)} = 100$ along the diagonals, $\alpha_{0,j,l}^{(s)} = 1$ elsewhere, and $\nu_0 = [2000, 1000]$.

Since the Ushahidi data set contains only reports of emergencies, and does not contain reports stating that no emergency is taking place, we cannot learn the length-scale l from this data, and must rely on background knowledge. We therefore select another dataset from the Haiti 2010 Earthquake, which has gold standard labels, namely the building damage assessment provided by UNOSAT [2]. We expect this data to have a similar length-scale because the underlying cause of both the building damages and medical emergencies was an earthquake affecting built-up areas where people were present. We estimated l using maximum likelihood II optimisation, giving an optimal value of $l = 16$ grid squares. We then transferred this point estimate to the model of the Ushahidi data. Our experiment repeated the incremental train/test procedure 20 times with hyperparameters set to $a_0 = 1500$, $b_0 = 1500$, $\alpha_{0,j,j}^{(s)} = 100$ along the diagonals, $\alpha_{0,j,l}^{(s)} = 1$ elsewhere, and $\nu_0 = [2000, 1000]$.

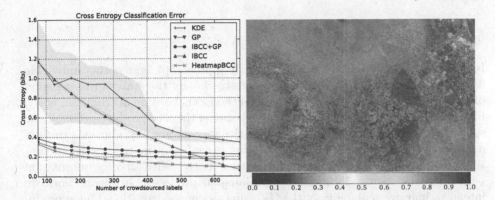

Fig. 6. Haiti text messages. Left: cross entropy error against the number of randomly selected crowdsourced labels. Lines show the median over 25 repeats, with shaded areas showing the inter-quartile range. Gold standard defined by running IBCC with 675 labels using a 100×100 grid. Right: heatmap of emergencies for part of Port-au-Prince after the 2010 Earthquake, showing high probability (dark orange) to low probability (blue). (Color figure online)

Figure 6 shows that HeatmapBCC is able to achieve low error rates when the reports are sparse. The IBCC and HeatmapBCC results do not quite converge due to the effect of interpolation performed by HeatmapBCC, which can still affect the results with several reports per grid square. The gold-standard predictions from IBCC also contain some uncertainty, so cross entropy does not reach zero, even with all labels. The GP alone is unable to determine the different reliability levels of each report type, so while it is able to interpolate between sparse reports, HeatmapBCC and IBCC detect the reliable data and produce different predictions when more labels are supplied. In summary, HeatmapBCC produces predictions with 439 labels (65%) that has an AUC within 0.1 of the gold standard predictions produced using all 675 labels, and reduces cross entropy to 0.1 bits with 400 labels (59%), showing that it is effective at predicting emergency states with reduced numbers of Ushahidi reports. Using an Intel i7 laptop, the HeatmapBCC inference over 675 labels required approximately one minute.

We use HeatmapBCC to visualise emergencies in Port-au-Prince, Haiti after the 2010 earthquake, by plotting the posterior class probabilities as the heatmap shown in Fig. 6. Our example shows how HeatmapBCC can combine reports from trusted sources with crowdsourced information. The blue area shows a negative report from a simulated first responder, with confusion matrix hyperparameters set to $\alpha_{0,j,j}^{(s)} = 450$ along the diagonals, so that the negative report was highly trusted and had a stronger effect than the many surrounding positive reports.

Uncertainty in the latent function f_j can be used to identify regions where information is lacking and further reconnaisance is necessary. Probabilistic heatmaps therefore offer a powerful tool for situation awareness and planning in disaster response.

5   Conclusions

In this paper we presented a novel Bayesian approach to aggregating unreliable discrete observations from different sources to classify the state across a region of space or time. We showed how this method can be used to combine noisy, biased and sparse reports and interpolate between them to produce probabilistic spatial heatmaps for applications such as situation awareness. Our experiments demonstrated the advantages of integrating a confusion matrix model to capture the unreliability of different information sources with sharing information between sparse report locations using Gaussian processes. In future work we intend to improve scalability of the GP using stochastic variational inference [6] and investigate clustering confusion matrices using a hierarchical prior, as per [13,23], which may improve the ability to learn confusion matrices when data for individual information sources is sparse.

Acknowledgments. We thank Brooke Simmons at Planetary Response Network for invaluable support and data. This work was funded by EPSRC ORCHID programme grant (EP/I011587/1).

References

1. Adams, R.P., Murray, I., MacKay, D.J.: Tractable nonparametric Bayesian inference in Poisson processes with Gaussian process intensities. In: Proceedings of the 26th Annual International Conference on Machine Learning, pp. 9–16. ACM (2009)
2. Corbane, C., Saito, K., Dell'Oro, L., Bjorgo, E., Gill, S.P., Emmanuel Piard, B., Huyck, C.K., Kemper, T., Lemoine, G., Spence, R.J., et al.: A comprehensive analysis of building damage in the 12 January 2010 MW7 Haiti earthquake using high-resolution satellite and aerial imagery. Photogramm. Eng. Remote Sens. **77**(10), 997–1009 (2011)
3. Dawid, A.P., Skene, A.M.: Maximum likelihood estimation of observer error-rates using the EM algorithm. J. Roy. Stat. Soc.: Ser. C (Appl. Stat.) **28**(1), 20–28 (1979)
4. Felt, P., Ringger, E.K., Seppi, K.D.: Semantic annotation aggregation with conditional crowdsourcing models and word embeddings. In: International Conference on Computational Linguistics, pp. 1787–1796 (2016)
5. Girolami, M., Rogers, S.: Variational Bayesian multinomial probit regression with Gaussian process priors. Neural Comput. **18**(8), 1790–1817 (2006)
6. Hensman, J., Matthews, A.G.d.G., Ghahramani, Z.: Scalable variational Gaussian process classification. In: International Conference on Artificial Intelligence and Statistics (2015)
7. Kim, H., Ghahramani, Z.: Bayesian classifier combination. Gatsby Computational Neuroscience Unit, Technical report GCNU-T, London, UK (2003)

8. Kom Samo, Y.L., Roberts, S.J.: Scalable nonparametric Bayesian inference on point processes with Gaussian processes. In: International Conference on Machine Learning, pp. 2227–2236 (2015)

9. Kottas, A., Sansó, B.: Bayesian mixture modeling for spatial poisson process intensities, with applications to extreme value analysis. J. Stat. Plan. Inference **137**(10), 3151–3163 (2007)

10. Lintott, C.J., Schawinski, K., Slosar, A., Land, K., Bamford, S., Thomas, D., Raddick, M.J., Nichol, R.C., Szalay, A., Andreescu, D., et al.: Galaxy Zoo: morphologies derived from visual inspection of galaxies from the sloan digital sky survey. Mon. Not. R. Astron. Soc. **389**(3), 1179–1189 (2008)

11. Long, C., Hua, G., Kapoor, A.: A joint Gaussian process model for active visual recognition with expertise estimation in crowdsourcing. Int. J. Comput. Vis. **116**(2), 136–160 (2016)

12. Meng, C., Jiang, W., Li, Y., Gao, J., Su, L., Ding, H., Cheng, Y.: Truth discovery on crowd sensing of correlated entities. In: 13th ACM Conference on Embedded Networked Sensor Systems, pp. 169–182. ACM (2015)

13. Moreno, P.G., Teh, Y.W., Perez-Cruz, F.: Bayesian nonparametric crowdsourcing. J. Mach. Learn. Res. **16**, 1607–1627 (2015)

14. Morrow, N., Mock, N., Papendieck, A., Kocmich, N.: Independent evaluation of the Ushahidi Haiti Project. Dev. Inf. Syst. Int. **8**, 2011 (2011)

15. Parzen, E.: On estimation of a probability density function and mode. Ann. Math. Stat. **33**(3), 1065–1076 (1962)

16. Rasmussen, C.E., Williams, C.K.I.: Gaussian Processes for Machine Learning, vol. 38, pp. 715–719. The MIT Press, Cambridge (2006)

17. Raykar, V.C., Yu, S.: Eliminating spammers and ranking annotators for crowdsourced labeling tasks. J. Mach. Learn. Res. **13**, 491–518 (2012)

18. Reece, S., Roberts, S., Nicholson, D., Lloyd, C.: Determining intent using hard/soft data and Gaussian process classifiers. In: 14th International Conference on Information Fusion, pp. 1–8. IEEE (2011)

19. Rosenblatt, M., et al.: Remarks on some nonparametric estimates of a density function. Ann. Math. Stat. **27**(3), 832–837 (1956)

20. Simpson, E., Roberts, S., Psorakis, I., Smith, A.: Dynamic Bayesian combination of multiple imperfect classifiers. In: Guy, T., Karny, M., Wolpert, D. (eds.) Decision Making and Imperfection, vol. 474, pp. 1–35. Springer, Heidelberg (2013). https://doi.org/10.1007/978-3-642-36406-8_1

21. Simpson, E.D., Venanzi, M., Reece, S., Kohli, P., Guiver, J., Roberts, S.J., Jennings, N.R.: Language understanding in the wild: combining crowdsourcing and machine learning. In: 24th International Conference on World Wide Web, pp. 992–1002 (2015)

22. Steinberg, D.M., Bonilla, E.V.: Extended and unscented Gaussian processes. In: Advances in Neural Information Processing Systems, pp. 1251–1259 (2014)

23. Venanzi, M., Guiver, J., Kazai, G., Kohli, P., Shokouhi, M.: Community-based Bayesian aggregation models for crowdsourcing. In: 23rd International Conference on World Wide Web, pp. 155–164 (2014)

24. Venanzi, M., Guiver, J., Kohli, P., Jennings, N.R.: Time-sensitive Bayesian information aggregation for crowdsourcing systems. J. Artif. Intell. Res. **56**, 517–545 (2016)

25. Venanzi, M., Rogers, A., Jennings, N.R.: Crowdsourcing spatial phenomena using trust-based heteroskedastic Gaussian processes. In: 1st AAAI Conference on Human Computation and Crowdsourcing (HCOMP) (2013)

Bayesian Inference for Least Squares Temporal Difference Regularization

Nikolaos Tziortziotis[1(✉)] and Christos Dimitrakakis[2,3]

[1] LIX, École Polytechnique, 91120 Palaiseau, France
ntziorzi@gmail.com
[2] University of Lille, 59650 Villeneuve-d'Ascq, France
christos.dimitrakakis@gmail.com
[3] SEAS, Harvard University, Cambridge, MA 02138, USA

Abstract. This paper proposes a fully Bayesian approach for Least-Squares Temporal Differences (LSTD), resulting in fully probabilistic inference of value functions that avoids the overfitting commonly experienced with classical LSTD when the number of features is larger than the number of samples. Sparse Bayesian learning provides an elegant solution through the introduction of a prior over value function parameters. This gives us the advantages of probabilistic predictions, a sparse model, and good generalisation capabilities, as irrelevant parameters are marginalised out. The algorithm efficiently approximates the posterior distribution through variational inference. We demonstrate the ability of the algorithm in avoiding overfitting experimentally.

1 Introduction

Value function estimation is an integral part of many reinforcement learning (RL) [29] algorithms (e.g., policy evaluation step of policy iteration) as it assesses the quality of a fixed control policy. This is straightforward in domains with a finite number of states. Large or infinite state spaces generally prohibit an explicit value function representation, but we can always represent the value function through a parameterized class of functions. In this paper we focus on the case of linear architectures where the values are approximated by a linear combination of a number of features. This approach is used by the Least-Squares Temporal Difference (LSTD) [6] algorithm, a temporal-difference algorithm that finds a linear approximation to the value function that minimizes the *mean squared projected bellman error* (MSPBE) [30].

The selection of features is critical for LSTD, as it determines the expressiveness of the value function representation. The richer the feature space is, the more likely that the value function space will contain a good approximation to the value function, but more data are needed [21]. This problem, already present in linear regression is only exacerbated in RL problems. Furthermore, using too many features makes use of the computed policies rather slow.

In linear regression, regularization is commonly used to control over-fitting, through a penalty term which discourages coefficients from reaching large values.

M. Ceci et al. (Eds.): ECML PKDD 2017, Part II, LNAI 10535, pp. 126–141, 2017.
https://doi.org/10.1007/978-3-319-71246-8_8

In regression problems, two of the most effective regularization approaches are ℓ_1 and ℓ_2-regularization [15] which involve adding a penalty term (ℓ_1 and ℓ_2 norms of the parameter vector, respectively) to the error function in order to discourage model's parameters from getting large values. In both schemes, a coefficient term λ, which typically must be selected in advance, governs the relative importance of the penalty term compared to the error function.

Bayesian reinforcement learning (BRL) (see [34] for an overview) is a framework for designing RL algorithms that models the reinforcement learning problem in a Bayesian decision theoretic manner. In *model-free* BRL, a probability distribution is maintained over the parameters of the value function, which quantifies our uncertainty over its parameters. One of the first such algorithm was Gaussian-process temporal-difference learning (GPTD) [9], which assumes that the unknown true values over the observed states are random variables generated by a Gaussian process. More specifically, GPTD incorporates a Gaussian prior over value functions and assumes a Gaussian noise model. Thus, the solution to the inference problem is given by the posterior distribution conditioned on the observed sequence of rewards. A sparse Bayesian extension of GPTD was proposed in [32,33], called RVMTD, where adopted a sparse kernelized Bayesian learning approach [31]. However, RVMTD minimizes the mean Bellman error instead of the MSPBE as in our case.

In this paper, we propose a Bayesian treatment of the LSTD algorithm, called BLSTD, that instead of seeking only a point estimate over the unknown value function parameters, actually considers the uncertainty on the value function. We adopt a fully probabilistic framework by introducing a stochastic variant of the standard Bellman operator as well as a prior distribution over the unknown model's parameters. To avoid overfitting, we further extend BLSTD algorithm with a sparse Bayesian learning approach [3,31], which we call VBLSTD. By using a tractable variational approach to automatically determine the model's complexity, we obviate the need to select a regularization parameter. We demonstrate the performance of the proposed algorithms on a number of domains, showing the ability of our model to avoid overfitting.

The remainder of the paper is organised as follows. Section 2 presents some preliminaries, review the LSTD algorithm and gives an overview of related work. Sections 3 introduce the Bayesian LSTD algorithm. In Sect. 4 we extend the Bayesian LSTD algorithm, presenting the VBLSTD algorithm that constitutes the main contribution of this paper. Our empirical analysis is presented in Sect. 5. We conclude the paper in Sect. 6 by discussing future directions.

2   Preliminaries and Related Work

A Markov Decision Process (MDP) is a tuple $\{S, A, P, r, \gamma\}$, where S is a set of states; A is a set of actions; $P(\cdot|s, a)$ is a transition probability kernel, defining the probability of next states in S for any state action pair $s \in S$ and $a \in A$; $r : S \to \mathbb{R}$ is a reward function and $\gamma \in [0, 1]$ is a constant discount factor. The policy $\pi : S \to A$ to be evaluated is a deterministic mapping from states to actions.

Value functions are of central interest in reinforcement learning. Briefly, value function V^π defines the expected discounted sum of rewards for the policy π, given that we start at state s: $V^\pi(s) \triangleq \mathbb{E}^\pi [\sum_{t=0}^\infty \gamma^t r(s_t)|s_0 = s]$, with $V^* \triangleq \sup_\pi V^\pi$. It is known [28] that the value function is the unique fixed-point of the Bellman operator T^π, i.e., $V^\pi = T^\pi V^\pi$, defined as:

$$(T^\pi V)(s) = r(s) + \gamma \int_{\mathcal{S}} V(s')dP(s'|s, \pi(s)), \tag{1}$$

or in a more compact form as $T^\pi V = \boldsymbol{r} + \gamma P^\pi V$, where V and $\boldsymbol{r}$ are vectors of size $|\mathcal{S}|$ that contains the state values and rewards, respectively. When the rewards and transition probabilities are known, the value function can be obtained analytically by solving the next linear system $V^\pi = (\boldsymbol{I} - \gamma P^\pi)^{-1}\boldsymbol{r}$.

In practice, however, the MDP is unknown, and we only have access to a set of n observations $\mathcal{D} = \{(s_i, r_i, s_i')\}_{i=1}^n$ generated by the policy we wish to evaluate, i.e., $s_i' \sim P(s_i, \pi(s_i))^1$. An additional difficulty is that when the state space is large (e.g., continuous) the value function cannot be represented exactly. It is then common to use some form of parametric value function approximation. In this paper we consider linear approximation architectures with parameters $\boldsymbol{\theta} \in \mathbb{R}^k$ over k features $\boldsymbol{\phi} : \mathcal{S} \to \mathbb{R}^k$, $\boldsymbol{\phi}(\cdot) = (\phi_1(\cdot), \ldots, \phi_k(\cdot))^\top$:

$$V_{\boldsymbol{\theta}}^\pi(s) = \boldsymbol{\phi}(s)^\top \boldsymbol{\theta} = \sum_{i=1}^k \phi_i(s)\theta_i.$$

Throughout the paper we denote by $\mathcal{F}$ the linear function space spanned by the features ϕ_i, i.e., $\mathcal{F} = \{f_{\boldsymbol{\theta}}|f_{\boldsymbol{\theta}}(\cdot) = \boldsymbol{\phi}(\cdot)^\top\boldsymbol{\theta}\}$. Roughly speaking, $\mathcal{F}$ contains all the value functions that can be represented by the features. Let us also introduce the projection operator Π onto $\mathcal{F}$, which takes any value function $\boldsymbol{u}$ and projects it to the nearest value function, such that $\Pi\boldsymbol{u} = V_{\boldsymbol{\theta}}^\pi$ where the corresponding parameters are the solution to the least-squares problem: $\boldsymbol{\theta} = \arg\min_{\boldsymbol{\theta}} \|V_{\boldsymbol{\theta}}^\pi - \boldsymbol{u}\|_D^2$ [30]. As the parameterization is linear, it is straightforward to show that the projection operator is linear and independent of the parameters $\boldsymbol{\theta}$ and given by $\Pi = \Phi C^{-1}\Phi^\top D$, where $\Phi \in \mathbb{R}^{|\mathcal{S}|\times k}$ is a matrix whose rows contain the feature vector $\boldsymbol{\phi}(s)^\top, \forall s \in \mathcal{S}$ and $C = \Phi^\top D\Phi$ is the Gram matrix.

2.1 Least Squares Temporal Difference

The least-squares temporal difference (LSTD) algorithm was introduced by Bradtke and Barto [6] and computes the fixed-point of the composed projection and Bellman operators: $V_{\boldsymbol{\theta}}^\pi = \Pi T^\pi V_{\boldsymbol{\theta}}^\pi$ (see Fig. 1). It can be seen as minimizing the *mean-square projected Bellman error* (MSPBE), i.e., the distance between $V_{\boldsymbol{\theta}}$ and its projected Bellman image onto $\mathcal{F}$:

$$\boldsymbol{\theta} = \arg\min_{\boldsymbol{\theta} \in \mathbb{R}^k} \|V_{\boldsymbol{\theta}}^\pi - \Pi T^\pi V_{\boldsymbol{\theta}}^\pi\|_D^2. \tag{2}$$

---

[1] With the starting state $s_0 \sim d(\cdot)$ sampled from some starting distribution d.
[2] The squared norm $\|\boldsymbol{u}\|_D^2 = \boldsymbol{u}^\top D\boldsymbol{u}$ is weighted by the non-negative diagonal matrix $D \in \mathbb{R}^{|\mathcal{S}|\times|\mathcal{S}|}$ with elements $d(s)$ on its diagonal.

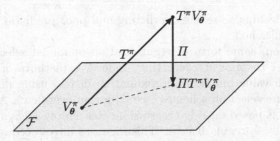

Fig. 1. A graphical representation of the LSTD problem. Here we can see the geometric relationship between the MSBE and the MSPBE. Figure adopted from [16].

As is shown in prior work [1], LSTD seen as solving the following nested optimization problem:

$$u^* = \arg\min_{u \in \mathbb{R}^k} \|\Phi u - T^\pi \Phi \theta\|_D^2, \qquad \theta = \arg\min_{\theta \in \mathbb{R}^k} \|\Phi \theta - \Phi u^*\|_D^2, \qquad (3)$$

where the first (projection) step finds the back-projection of $T^\pi V_\theta^\pi$ onto $\mathcal{F}$, and the second (fixed-point) step solves the fixed-point problem which minimizes the distance between V_θ^π and its projection.

As we discussed, usually the MDP model is unknown, or the full Φ matrices are too large to be formed, and so LSTD relies on sample-based estimates. Using a set $\mathcal{D}$ of samples from the MDP of interest, we can define $\tilde{\Phi} = [\phi(s_1)^\top; \ldots; \phi(s_n)^\top]$ and $\tilde{\Phi}' = [\phi(s_1')^\top; \ldots; \phi(s_n')^\top]$ to be the sampled feature matrices of successive transition states, and as $\tilde{R} = [r_i, \ldots, r_n]^\top$ the sampled reward vector. Given these samples, the sample-based LSTD solution is given by the empirical version of Eq. (3):

$$u^* = \tilde{C}^{-1}\tilde{\Phi}(\tilde{R} + \gamma\tilde{\Phi}'\theta),$$
$$\theta = (\tilde{\Phi}^\top(\tilde{\Phi} - \gamma\tilde{\Phi}'))^{-1}\tilde{\Phi}^\top\tilde{R} = A^{-1}b,$$

where we have defined

$$\tilde{C} \triangleq \tilde{\Phi}^\top\tilde{\Phi}, \quad A \triangleq \tilde{\Phi}^\top(\tilde{\Phi} - \gamma\tilde{\Phi}'), \quad \text{and} \quad b \triangleq \tilde{\Phi}^\top\tilde{R}.$$

As the number of samples n increases, the LSTD solution $\tilde{\Phi}\theta$ converges to the fixed-point of $\hat{\Pi}T^\pi$ [6,21,24]. We denote as $\hat{\Pi}$ the sample based feature space projector (empirical projection).

2.2 Review of Regularized LSTD Schemes

Despite the fact that LSTD offers an unbiased estimate of the value function [16], high-dimensional feature space create additional challenges. The larger the number of features is, the more samples required to estimate θ. In some cases, the number of features may even significantly outnumber the number of observed samples $n \gg k$, leading to severe overfitting and poor prediction as the matrix A will be ill-conditioned.

For this reason, some form of regularization or model selection should be adopted, in order to prevent overfitting. Indeed, a plethora of methods have been proposed in value function approximation in RL, using different regularization and feature selection schemes (see [7] for an overview). A common form of regularization is based on ridge regression: this simply adds a term λI to A, which is essentially ℓ_2-regularization. This idea was introduced and analysed by Farahmand et al. [10] for the $\ell_{2,2}$-LSTD algorithm, which uses an ℓ_2-penalty for both the projection and fixed-point steps. However, when the number of samples is much smaller than the number of features, ridge regression may fail, as it does not encourage sparsity.

On the other hand, as ℓ_1-penalties enforce sparsity, it is natural to consider those instead. The LASSO-TD variant[3] incorporates an ℓ_1-penalty in the projection step. LARS-TD [19] applies ℓ_1-regularization to the projection operator in the feature space $\mathcal{F}$, using a variant of LARS [8]. Finally, LC-TD [17] reformulates Lasso-TD as a linear complementary (LC) problem, allowing the usage of any efficient off-shelf solver. It should be emphasised that some of the solvers allow warm-starts, offering a significant computational advantage in the policy iteration context. In order for both LARS-TD and LC-TD to find a solution, matrix A is required to be a P-matrix[4]. The theoretical properties of the Lasso-TD problem were examined in [14], demonstrating that LARS-TD and LC-TD converge to the same solution. Particularly, it has been shown that Lasso-TD is guaranteed to have a unique fixed point. Additionally, Pires [27] suggests to solve the linear system of LSTD by including an ℓ_1-regularization term directly to it. This is a typical convex optimization problem where any standard solver can be used, being also applicable to off-policy learning.

Two closely related algorithms have been proposed in order to alleviate some of the limitations of Lasso-TD (e.g., P-matrix constraint), the ℓ_1-PBR (Projected Bellman residual) [11] and the $\ell_{2,1}$-LSTD [16]. Both of them place an ℓ_1-penalty term in the fix-point step, which actually penalizes the projected Bellman residual and yields a convex optimization problem. In contrast with ℓ_1-PBR, $\ell_{2,1}$-LSTD puts also an ℓ_2-penalty term on the operator problem. The Dantzig-LSTD algorithm, proposed by Geist et al. [12], integrates LSTD with the Dantzig selector, converting it into a standard linear program, that can be solved efficiently. Actually, it minimizes the sum of all parameters under the constraint that the linear system of LSTD is smaller than a predefined parameter λ in each dimension. An alternative Dantzig Selector temporal difference learning algorithm has been introduced recently by Liu et al. [22], called ODDS-TD. It is a two-stage algorithm that is also able to compute the optimal denoising matrix.

---

[3] Based on LASSO regression, which uses ℓ_1-regularization.

[4] P-matrix is a squared matrix with all of its principal minors positive (superset of the class of positive definite matrices).

3   Bayesian LSTD

In this section we present a Bayesian LSTD algorithm, called BLSTD. In our analysis, we model the fact that the transition distribution P is not known exactly by considering an *empirical Bellman operator*, given by the standard Bellman operator (1) plus additive white noise, $\epsilon \sim \mathcal{N}(0, \beta^{-1})$. For simplicity, we can assume that the noise term is state independent. Thus, the empirical Bellman operator can be written concisely as

$$\hat{T}^\pi V_\theta^\pi = r + \gamma P^\pi V_\theta^\pi + N, \quad N \sim \mathcal{N}(0, \beta^{-1}I).$$

In other words, our model says that $r + \gamma \hat{P}^\pi V_\theta^\pi$ is normally distributed with mean $r + \gamma P^\pi V_\theta^\pi$. We shall formulate a Bayesian regression model, that is based on a sample from this empirical Bellman operator.

As aforementioned, given the set of observations $\mathcal{D}$, LSTD seeks the value function parameters θ which are invariant with respect to the composed operator $\hat{\Pi}\hat{T}^\pi$:

$$V_\theta^\pi = \hat{\Pi}\hat{T}^\pi V_\theta^\pi \Leftrightarrow$$
$$\tilde{\Phi}^\top \tilde{R} = \tilde{\Phi}^\top (\tilde{\Phi} - \gamma \tilde{\Phi}')\theta + \tilde{\Phi}^\top N,$$

where we have rewritten the projection operators and approximate value function in terms of the feature matrix and parameter vectors. We can now reformulate this as the following linear regression model:

$$b = A\theta + \tilde{\Phi}^\top N.$$

The **likelihood function** for this model is given by:

$$p(b|\theta, \beta) = \mathcal{N}(b|A\theta, \beta^{-1}\tilde{C}).$$

Taking the logarithm of the likelihood, we have

$$\ln p(b|\theta, \beta) = \frac{k}{2}\ln(\beta) - \frac{1}{2}\ln(|\tilde{C}|) - \frac{k}{2}\ln(2\pi) - \frac{\beta}{2}E_\mathcal{D}(\theta), \tag{4}$$

where $E_\mathcal{D}$ corresponds to the MSPBE:

$$E_\mathcal{D}(\theta) = (b - A\theta)^\top \tilde{C}^{-1}(b - A\theta).$$

To complete our Bayesian model, we now introduce a **prior distribution** over the model parameters θ. Specifically, we consider a zero-mean isotropic Gaussian conjugate prior governed by a single precision parameter α,

$$p(\theta|\alpha) = \mathcal{N}(\theta|0, \alpha^{-1}I).$$

Thus, we model the *parametric* uncertainty [23], which arises if the true transition probabilities and expected rewards are not known and must be estimated.

Writing only the terms from the likelihood and prior depend on the model parameters, the log of the **posterior distribution** is given by

$$\ln p(\boldsymbol{\theta}|\mathcal{D}) \propto -\frac{\beta}{2}E_{\mathcal{D}}(\boldsymbol{\theta}) - \frac{\alpha}{2}\boldsymbol{\theta}^{\top}\boldsymbol{\theta}. \tag{5}$$

Taking the maximization of the posterior distribution with respect to $\boldsymbol{\theta}$ is equivalent to the minimization of the MSPBE with the addition of an ℓ_2-penalty ($\lambda = \alpha/\beta$). Thus, if hyperparameter α takes a large value, the total squared length of the parameter vector $\boldsymbol{\theta}$ is encouraged to be small. Completing the squares of Eq. (5),

$$\ln p(\boldsymbol{\theta}|\mathcal{D}) \propto -\frac{\beta}{2}(\boldsymbol{b} - A\boldsymbol{\theta})^{\top}\tilde{C}^{-1}(\boldsymbol{b} - A\boldsymbol{\theta}) - \frac{\alpha}{2}\boldsymbol{\theta}^{\top}\boldsymbol{\theta}$$

$$\propto -\frac{1}{2}\boldsymbol{\theta}^{\top}(a\boldsymbol{I} + \beta A^{\top}\tilde{C}^{-1}A)\boldsymbol{\theta} + \boldsymbol{\theta}^{\top}\beta A^{\top}\tilde{C}^{-1}\boldsymbol{b} + const$$

we get that the posterior distribution is also Gaussian,

$$p(\boldsymbol{\theta}|\mathcal{D}) = \mathcal{N}(\boldsymbol{\theta}|\boldsymbol{m}, S),$$

with the covariance and mean to be given as

$$S = (\alpha\boldsymbol{I} + \beta\underbrace{A^{\top}\tilde{C}^{-1}A}_{\Sigma})^{-1} \text{ and } \boldsymbol{m} = \beta SA^{\top}\tilde{C}^{-1}\boldsymbol{b},$$

respectively, where matrix $\Sigma \triangleq A^{\top}\tilde{C}^{-1}A$ is always positive definite. Hence, the predictive distribution of the value function over a new state s^* is estimated by averaging the output of all possible linear models w.r.t. the posterior distribution

$$p(V_{\boldsymbol{\theta}}^{\pi}(s^*)|s^*, \mathcal{D}) = \int_{\boldsymbol{\theta}} p(V_{\boldsymbol{\theta}}^{\pi}(s^*)|\boldsymbol{\theta}, s^*)dp(\boldsymbol{\theta}|\boldsymbol{b}, \alpha, \beta)$$

$$= \mathcal{N}(V_{\boldsymbol{\theta}}^{\pi}(s^*)|\phi(s^*)^{\top}\boldsymbol{m}, \phi(s^*)^{\top}S\phi(s^*)).$$

An online version of our model can also be derived easily, with the posterior distribution at any phase acting as the prior distribution for the subsequent transition [2] and by using the *matrix inversion lemma* for the covariance matrix.

Maximum likelihood. For illustrative purposes, consider also a maximum likelihood approach. Restricting respect to $\boldsymbol{\theta}$, we getting the gradient of the log likelihood (4):

$$-\frac{1}{2}\nabla_{\boldsymbol{\theta}}E_{\mathcal{D}}(\boldsymbol{\theta}) = -\beta A^{\top}\tilde{C}^{-1}(A\boldsymbol{\theta} - \boldsymbol{b}).$$

By setting the gradient equal to zero, we get the batch LSTD solution.

In conclusion, under our model, maximum a posteriori inference corresponds to ℓ_2-regularization, while maximum likelihood inference to standard LSTD. In the next section, we propose an extension of our model that also induces sparsity.

4   Variational Bayesian LSTD (VBLSTD)

We now extend our model through a hierarchical sparse Bayesian prior, and introduce a variational approach for inference. The hope is the resulting VBLSTD algorithm will be able to avoid the over-fitting problem through regularization. For the **prior distribution** over parameter vector $\boldsymbol{\theta}$, we use an approach similar to [31] where a sparse zero-mean Gaussian prior was considered. Specifically, our prior over the model's parameters $\boldsymbol{\theta}$ is given by:

$$p(\boldsymbol{\theta}|\boldsymbol{\alpha}) = \prod_{i=1}^{k} \mathcal{N}(\theta_i|0, \alpha_i^{-1}),$$

where $\boldsymbol{\alpha} = (\alpha_1, \ldots, \alpha_k)^{\top}$ are the parameters specifying our prior. Instead of selecting an arbitrary value for $\boldsymbol{\alpha}$, we select a **hyperprior** over $\boldsymbol{\alpha}$ of the form:

$$p(\boldsymbol{\alpha}) = \prod_{i=1}^{k} Gamma(\alpha_i|h_a, h_b),$$

where h_a, h_b are fixed parameters. The choice of the Gamma distribution for $\boldsymbol{\alpha}$ results in a marginal distribution $p(\boldsymbol{\theta})$ that is Student-t, which is known to enforce sparse representations. To complete the specification of our model, we define a Gamma hyperprior over the noise precision β:

$$p(\beta) = Gamma(\beta|h_c, h_d).$$

To get broad hyperpriors, we can set those parameters to some small value, e.g., $h_a = h_a = h_c = h_d = 10^{-6}$.

Bayesian inference requires the computation of the **posterior distribution** over all latent parameters $\mathcal{Z} = \{\boldsymbol{\theta}, \boldsymbol{\alpha}, \beta\}$ given the observations:

$$p(\boldsymbol{\theta}, \boldsymbol{\alpha}, \beta|\boldsymbol{b}) = \frac{p(\boldsymbol{b}|\boldsymbol{\theta}, \beta)p(\beta)p(\boldsymbol{\theta}|\boldsymbol{\alpha})p(\boldsymbol{\alpha})}{p(\boldsymbol{b})}.$$

As the direct computation of the marginal likelihood is analytically intractable, we resort to variational inference [3,18]. This introduces a variational approximation $\mathcal{Q}(\mathcal{Z})$ to the true distribution $p(\mathcal{Z}|\boldsymbol{b})$ over the latent variables, and the problem is defined as finding the approximation closest to the true posterior distribution in terms of KL divergence. The main insight in variational methods is the following identity,

$$\ln p(\boldsymbol{b}) = \mathcal{L}(\mathcal{Q}) + \mathrm{KL}(\mathcal{Q}\|p)$$

where we have defined,

$$\mathcal{L}(\mathcal{Q}) = \int \mathcal{Q}(\mathcal{Z}) \ln \left\{ \frac{p(\mathcal{Z}, \boldsymbol{b})}{\mathcal{Q}(\mathcal{Z})} \right\}, \tag{6}$$

$$\mathrm{KL}(\mathcal{Q}\|p) = -\int \mathcal{Q}(\mathcal{Z}) \ln \left\{ \frac{p(\mathcal{Z}|\boldsymbol{b})}{\mathcal{Q}(\mathcal{Z})} \right\}. \tag{7}$$

The KL$(\mathcal{Q}\|p)$ (7) represents the Kullback-Leibler divergence between the variational posterior distribution $\mathcal{Q}(\mathcal{Z})$ and the true posterior distribution $p(\mathcal{Z}|\boldsymbol{b})$ over the latent variables. As KL$(\mathcal{Q}\|p) \geq 0$, it follows that $\mathcal{L}(\mathcal{Q}) \leq \ln p(\boldsymbol{b})$, which means that $\mathcal{L}(\mathcal{Q})$ is a lower bound on $\ln p(\boldsymbol{b})$. Therefore, maximizing the evidence lower bound (ELBO, see [4] for an overview) $\mathcal{L}(\mathcal{Q})$ with respect to $\mathcal{Q}$ is equivalent to minimizing the KL$(\mathcal{Q}\|p)$, as the largest value of $\mathcal{L}(\mathcal{Q})$ will be achieved when the KL$(\mathcal{Q}\|p)$ becomes zero.

In our problem, we consider a **variational distribution** with a factorized Gaussian form over the latent variables (c.f. *mean field theory* [26]), such that $\mathcal{Q}(\mathcal{Z}) = \mathcal{Q}_\theta(\boldsymbol{\theta})\mathcal{Q}_\alpha(\boldsymbol{\alpha})\mathcal{Q}_\beta(\beta)$. Then the optimal distribution for each one of the factors can be written as:

$$\mathcal{Q}_\theta(\boldsymbol{\theta}) = \mathcal{N}(\boldsymbol{\theta}|\boldsymbol{m}, S) \tag{8}$$

$$\mathcal{Q}_\beta(\beta) = \mathcal{G}amma(\beta|\tilde{c}, \tilde{d}) \tag{9}$$

$$\mathcal{Q}_\alpha(\boldsymbol{\alpha}) = \prod_{i=1}^{k} \mathcal{G}amma(\alpha_i|\tilde{a}_i, \tilde{b}_i) \tag{10}$$

where,

$$S = (diag\,\mathbb{E}[\boldsymbol{\alpha}] + \mathbb{E}[\beta]\Sigma)^{-1}, \quad \boldsymbol{m} = \mathbb{E}[\beta]SA^\top\tilde{C}^{-1}\boldsymbol{b},$$

$$\tilde{a}_i = h_a + \frac{1}{2}, \quad \tilde{b}_i = h_b + \frac{1}{2}\,\mathbb{E}[\theta_i^2],$$

$$\tilde{c} = h_c + \frac{k}{2}, \quad \tilde{d} = h_d + \frac{1}{2}\|\boldsymbol{b} - A\boldsymbol{m}\|_{\tilde{C}}^2 + \frac{1}{2}\,\text{tr}(\Sigma S).$$

The required moments can be expressed as follows:

$$\mathbb{E}[\alpha_i] = \tilde{a}_i/\tilde{b}_i, \quad \mathbb{E}[\beta] = \tilde{c}/\tilde{d}, \text{ and } \mathbb{E}[\theta_i^2] = \boldsymbol{m}_i^2 + S_{ii}.$$

The variational posterior distributions given in Eqs. (8), (9) and (10) are then iteratively updated until convergence. As the evidence lower bound is convex with respect to each one of the factors, convergence is guaranteed.

Similarly to BLSTD, the **value function distribution** over a new state s^* can be approximated by averaging the output of all possible linear models w.r.t the variational posterior distribution $\mathcal{Q}_\theta(\boldsymbol{\theta})$

$$p(V_\theta^\pi(s^*)|s^*, \mathcal{D}) = \int_\theta p(V_\theta^\pi(s^*)|\boldsymbol{\theta}, s^*)d\mathcal{Q}_\theta(\boldsymbol{\theta})$$

$$= \mathcal{N}\left(V_\theta^\pi(s^*) \mid \boldsymbol{\phi}(s^*)^\top\boldsymbol{m}, \boldsymbol{\phi}(s^*)^\top S\boldsymbol{\phi}(s^*)\right).$$

This gives us not only a specific mean value function, but also effectively expresses our uncertainty about what the value function is through the covariance terms.

The lower bound is interesting to look at more closely, as it is the quantity that we maximizing. Furthermore, it can be used as a convergence criterion for the variational inference. If the difference between the lower bound on two successive iterations is lower than a threshold, we assume that our model converges. Algorithm 1 provides the pseudocode of the sparse Bayesian LSTD algorithm.

Remark 1. The lower bound can be written as

$$\mathcal{L}(\mathcal{Q}) = \frac{1}{2}\ln|S| - \frac{1}{2}|\tilde{C}| + \sum_{i=1}^{k}\{\ln\Gamma(\tilde{a}_i) - \tilde{a}_i\ln\tilde{b}_i\} + \ln\Gamma(\tilde{c}) - \tilde{c}\ln\tilde{d}$$

$$+ \frac{k}{2}(1 - \ln 2\pi) - k\ln\Gamma(h_a) + kh_a\ln h_b - \ln\Gamma(h_c) + h_c\ln h_d. \quad (11)$$

Proof. Decomposing Eq. (6) we obtain:

$$\mathcal{L}(\mathcal{Q}) = \mathbb{E}_{\theta,\beta}[\ln p(\boldsymbol{b}|\boldsymbol{\theta},\beta)] + \mathbb{E}_{\beta}[\ln p(\beta)] + \mathbb{E}_{\theta,\alpha}[\ln p(\boldsymbol{\theta}|\boldsymbol{\alpha})] + \mathbb{E}_{\alpha}[\ln p(\boldsymbol{\alpha})]$$
$$- \mathbb{E}_{\theta}[\ln \mathcal{Q}_{\theta}(\boldsymbol{\theta})] - \mathbb{E}_{\alpha}[\ln \mathcal{Q}_{\alpha}(\boldsymbol{\alpha})] - \mathbb{E}_{\beta}[\ln \mathcal{Q}_{\beta}(\beta)].$$

We now evaluate each term in turn.

$$\mathbb{E}_{\theta,\beta}[\ln p(\boldsymbol{b}|\boldsymbol{\theta},\beta)] = \frac{k}{2}(\psi(\tilde{c}) - \ln\tilde{d}) - \frac{k}{2}\ln 2\pi - \frac{1}{2}|\tilde{C}| - \frac{1}{2}\mathbb{E}[\beta]\{\|\boldsymbol{b} - A\boldsymbol{m}\|_{\tilde{C}}^2 + \text{tr}(\Sigma S)\}$$

$$\mathbb{E}_{\theta,\alpha}[\ln p(\boldsymbol{\theta}|\boldsymbol{\alpha})] = -\frac{k}{2}\ln 2\pi - \frac{1}{2}\sum_{i=1}^{k}(\psi(\tilde{a}_i) - \ln\tilde{b}_i) - \frac{1}{2}\sum_{i=1}^{k}\mathbb{E}[\alpha_i](m_i^2 + S_{ii})$$

$$\mathbb{E}_{\alpha}[\ln p(\boldsymbol{\alpha})] = -k\ln\Gamma(h_a) + kh_a\ln h_b + (h_a - 1)\sum_{i=1}^{k}(\psi(\tilde{a}_i) - \ln\tilde{b}_i) - h_b\sum_{i=1}^{k}\mathbb{E}[\alpha_i]$$

$$\mathbb{E}_{\beta}[\ln p(\beta)] = -\ln\Gamma(h_c) + h_c\ln h_d + (h_c - 1)(\psi(\tilde{c}) - \ln\tilde{d}) - h_d\mathbb{E}[\beta]$$

$$\mathbb{E}_{\theta}[\ln \mathcal{Q}_{\theta}(\boldsymbol{\theta})] = -\frac{1}{2}\ln|S| - \frac{k}{2}(1 + \ln 2\pi)$$

$$\mathbb{E}_{\alpha}[\ln \mathcal{Q}_{\alpha}(\boldsymbol{\alpha})] = \sum_{i=1}^{k}\{-\ln\Gamma(\tilde{a}_i) + \tilde{a}_i\ln\tilde{b}_i + (\tilde{a}_i - 1)(\psi(\tilde{a}_i) - \ln\tilde{b}_i) - \tilde{b}_i\mathbb{E}[\alpha_i]\}$$

$$\mathbb{E}_{\beta}[\ln \mathcal{Q}_{\beta}(\beta)] = -\ln\Gamma(\tilde{c}) + \tilde{c}\ln\tilde{d} + (\tilde{c} - 1)(\psi(\tilde{c}) - \ln\tilde{d}) - \tilde{d}\mathbb{E}[\beta].$$

Substituting back, we obtain the required result.   □

In the next section, we compare the Bayesian LSTD methods we derived with other state-of-the-art LSTD approaches for value function estimation.

5   Experiments

To analyze the performance of the proposed VBLSTD algorithm, we considered two discrete chain problems. Through our empirical analysis we examine both the convergence capabilities of VBLSTD on the *true* solution, as well as the ability of the VBLSTD algorithm in avoiding overfitting. In the first case, comparisons have been made with the vanilla LSTD algorithm, considering three different sizes of the Boyan's chain [5]. In the second case, comparisons have been conducted with the ℓ_2-LSTD (adding an ℓ_2-regularization factor to the projector operator), LarsTD [19] and OMPTD [25] algorithms. For that purpose, we considered the *corrupted chain problem* similar to [12,16,19].

In contrast to the VBLSTD algorithm, the performance of the three aforementioned algorithms are totally depended on the penalty parameter that should

---

Algorithm 1. VBLSTD($\mathcal{D}, \phi, \gamma$)

Initialization :

1      $\max_t = 1000$, $t = 0$;

2      $h_a = h_a = h_c = h_d = 10^{-6}$;

3 **begin**

4      $\boldsymbol{b} = \sum_{(s,r,s') \in \mathcal{D}} \phi(s) r$;

5      $\tilde{C} = \sum_{(s,r,s') \in \mathcal{D}} \phi(s) \phi(s)^\top$;

6      $A = \sum_{(s,r,s') \in \mathcal{D}} \phi(s) (\phi(s) - \gamma \phi(s'))^\top$;

7      $\Sigma = A^\top \tilde{C}^{-1} A$;

8      $\langle \beta \rangle = std(\boldsymbol{b})$;

9      $\langle \alpha_i \rangle = 0.01$, $\forall i = 1, \ldots, k$;

10   **repeat**

11          $t \leftarrow t + 1$;

12   1. Update $\mathcal{Q}_\theta$;

13            $S = (diag\, \mathbb{E}[\boldsymbol{\alpha}] + \mathbb{E}[\beta]\Sigma)^{-1}$;

14            $\boldsymbol{m} = \mathbb{E}[\beta] S A^\top \tilde{C}^{-1} \boldsymbol{b}$;

15            $\mathbb{E}[\theta_i^2] = \boldsymbol{m}_i^2 + S_{ii}$;

16   2. Update $\mathcal{Q}_\beta$;

17            $\tilde{c} = h_c + k/2$;

18            $\tilde{d} = h_d + \frac{1}{2}\|\boldsymbol{b} - A\boldsymbol{m}\|_{\tilde{C}}^2 + \frac{1}{2}\,\mathrm{tr}(\Sigma S)$;

19            $\mathbb{E}[\beta] = \tilde{c}/\tilde{d}$;

20   3.Update $\mathcal{Q}_\alpha$;

21   **for** $i = 1$ **to** k **do**

22            $\tilde{a}_i = h_a + 1/2$;

23            $\tilde{b}_i = h_b + \mathbb{E}[\theta_i^2]/2$;

24            $\mathbb{E}[\alpha_i] = \tilde{a}_i/\tilde{b}_i$;

25   **end**

26   4. Calculate bound $\mathcal{L}_t$, based on Eq.(11);

27   **until** *convergence or* $t > \max_t$;

28   **return** $\boldsymbol{m}$, S;

29 **end**

---

be defined explicitly in advance. Therefore, we have to answer the next question: which is the best value to set the regalurization factor? In the case of the ℓ_2-LSTD algorithm we adopted the same strategy with the one followed by Hoffman et al. [16]. Actually, we used a grid of 10 parameters logarithmically spaced between 10^{-6} and 10. In the case of the LarsTD and OMPTD algorithms, we computed the whole regularization path similar to [12] by setting the regularization factor equal to 10^{-7}. In all cases, the best prediction error has been reported.

In our experimental results, we illustrate the average root mean squared error with respect to the true value function, V^*. The optimal value function was computed explicitly since we examine discrete environments. It should be also noticed that for each run the algorithms were provided with the same rollouts of data. For each average, we also plot the 95% confidence interval for the accuracy

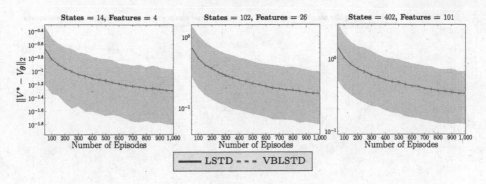

Fig. 2. Performance of policy evaluation on the Boyan's Chain for a fixed policy.

of the mean estimate with error bars. Additionally, we show the 90% percentile region of the runs, in order to indicate inter-run variability in performance.

5.1 Boyan's Chain

To demonstrate the ability of the VBLSTD algorithm to converge to the same solution with that of standard LSTD, we examine the Boyan's chain problem [5]. Actually, it is an N-state Markov chain with a single action. Each episode starts in state $N - 1$ and terminates when the (absorbing) state zero is reached. For each state, $s > 2$, we transit with equal probability in states $s - 1$ or $s - 2$, with reward -3. On the other hand, we deterministically transit from states 2 to 1 and state 1 to 0, where the received rewards are equal to -2 and 0, respectively. Similar to [13], three different problems sizes have been considered: $N = \{14, 102, 402\}$. The feature vectors that considered for the states' representation are exactly the same with those used by Geramifard et al. [13]. Figure 2 illustrates the performance of the VBLSTD and LSTD algorithms on the three different Boyan's chain problems, averaged over 1000 runs. In all these three problems, it is clear that the proposed VBLSTD algorithm converges to same solution with the one returned by the LSTD algorithm. It means that the VBLSTD algorithm discovers the global optimum solution (i.e., the solution that corresponds to the minimum MSPBE).

5.2 Corrupted Chain

In order to examine the sparsification properties of the VBLSTD algorithm, we consider the corrupted chain problem as in [12,16,19]. This is a 20-state, 2-actions MDP proposed in [20]. In this problem, the states are connected in a chain with the actions to indicate the direction (*left* or *right*), with the probability of success to be equal to 0.9. For instance, executing *left* action at state s, the system transitions to state $s - 1$ with probability 0.9 and to state $s + 1$ with probability 0.1. A reward of one is given only at the ends of the chain. Similar to [12,16,19], to represent the value function we will consider $k = 6 + \bar{s}$, 6 'relevant'

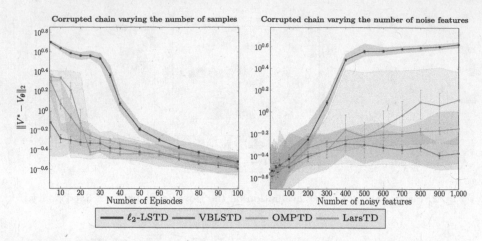

Fig. 3. Performance of policy evaluation on the Corrupted chain for a fixed policy. (Left) we consider $\bar{s} = 600$ 'irrelevant' features while varying the number of samples. The horizon of each episode is set equal to 20 steps. (Right) we use 400 transitions (20 rollouts of horizon 20) varying the number of 'irrelevant' features.

features (i.e., including a bias term and five RBF basis functions spaced evenly over the state space) and $\bar{s}$ additional 'irrelevant' (noise) features (containing random Gaussian noise, $\mathcal{N}(0,1)$). It should also be stressed that through our analysis we didn't perform any standarization over the feature matrices. Also, in the case of the VBLSTD, we keep the noise precision unchanged.

The results of our experiments are presented in Fig. 3, averaged over 30 runs. We report the prediction error between the estimated and the true value function on 1000 test points. The evaluated policy is the optimal one, which selects left action on the first 10 states and right action on the rest 10. The first (left) plot shows the results in the case where we have $\bar{s} = 600$ 'irrelevant' features while varying the number of samples (the horizon of each episode is equal to 20, started randomly on $\{1, \ldots, 20\}$). On the other hand, the second (right) plot depicts the results in the case where we sample 400 transitions (20 rollouts of horizon 20), varying the number of 'irrelevant' features. In both cases, it seems that the VBLSTD algorithm performs much better compared with the others three regularization schemes. The difference between them becomes more apparent as the number of irrelevant features increased. Additionally, it stems that the performance of VBLSTD is quite stable even when we select a large number of noise features. On the contrary, the OMPTD algorithm seems to become unstable when the number of noise features becomes large. Furthermore, we also note that the VBLSTD is not affected by overfitting when the number of features becomes greater than the number of samples. As it is expected, the performance of all algorithms is quite close even when large number of transitions are used for training or the number of noise features is quite small.

Finally, Fig. 4 illustrates the mean weights (solution), $\boldsymbol{\theta}$, for each one of the examined algorithms considering 600 irrelevant features. The number of training

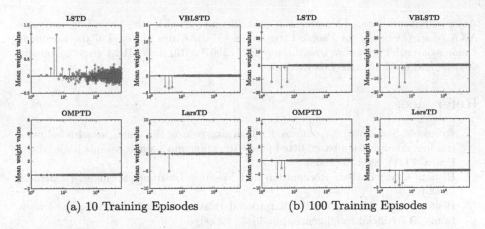

(a) 10 Training Episodes   (b) 100 Training Episodes

Fig. 4. The 606 mean weight values. The first weight is the bias term, the next 5 correspond to the relevant features (RBFs), the rest 600 correspond to the noise (irrelevant) features. The dichotomy between the 'relevant' and 'irrelevant' weights is apparent.

episodes that used for training on these two plots (Fig. 4) are 10 and 100, respectively. As we can easily verify, when the number of features exceeds the number of the training samples, we encounter the overfitting problem that leads to poor predictions. It is more apparent in the case of the LSTD algorithm. However, the VBLSTD algorithm achieves to avoid the overfitting problem, succeeding to identify the relevant features even in the case where the number of samples is much lower than that of the features. Last but not least, it should be highlighted that when the number of training samples is much higher than the number of features, the solutions of the LSTD and VBLSTD algorithms are quite similar.

6 Conclusion

In this paper we introduced a fully Bayesian framework for least-squares temporal difference learning algorithm, called BLSTD. This is achieved by adopting an explicit probabilistic model for the empirical Bellman operator and introducing a prior distribution over the unknown model's parameters. This gives us the advantage of not only having a point estimate over the unknown value function parameters, but also quantifying our uncertainty about the value function. We further extended this method to a sparse variational Bayes model, called VBLSTD. The main advantage of VBLSTD compared to other regularization schemes, is its ability to avoid over-fitting by determining the model's complexity in an automatic way. In practice, we verified that the VBLSTD algorithm solutions are at least as good as any other state-of-the-art algorithm, while being able to automatically ignore noisy features. We believe that this principled approach to policy evaluation can also lead to reinforcement learning algorithms with good exploration performance, something that we leave for future work.

Acknowledgements. This work was partially supported by the École Polytechnique AXA Chair (DaSciS), the People Programme (Marie Curie Actions) of the European Union's Seventh Framework Programme (FP7/2007–2013) under REA grant agreement 608743, and the Future of Life Institute.

References

1. Antos, A., Szepesvári, C., Munos, R.: Learning near-optimal policies with bellman-residual minimization based fitted policy iteration and a single sample path. Mach. Learn. **71**(1), 89–129 (2008)
2. Bishop, C.M.: Pattern Recognition and Machine Learning. Springer, Heidelberg (2006)
3. Bishop, C.M., Tipping, M.E.: Variational relevance vector machines. In: Uncertainty in Artificial Intelligence, pp. 46–53 (2000)
4. Blei, D.M., Kucukelbir, A., McAuliffe, J.D.: Variational inference: a review for statisticians. CoRR (2016)
5. Boyan, J.: Technical update: least-squares temporal difference learning. Mach. Learn. **49**(2), 233–246 (2002)
6. Bradtke, S., Barto, A.: Linear least-squares algorithms for temporal difference learning. Mach. Learn. **22**(1), 33–57 (1996)
7. Dann, C., Neumann, G., Peters, J.: Policy evaluation with temporal differences: a survey and comparison. J. Mach. Learn. Res. **15**, 809–883 (2014)
8. Efron, B., Hastie, T., Johnstone, L., Tibshirani, R.: Least angle regression. Ann. Stat. **32**, 407–499 (2004)
9. Engel, Y., Mannor, S., Meir, R.: Reinforcement learning with Gaussian process. In: International Conference on Machine Learning, pp. 201–208 (2005)
10. Farahmand, A.M., Ghavamzadeh, M., Szepesvári, C., Mannor, S.: Regularized policy iteration. Adv. Neural Inf. Process. Syst. **21**, 441–448 (2008)
11. Geist, M., Scherrer, B.: ℓ1-penalized projected Bellman residual. In: Sanner, S., Hutter, M. (eds.) EWRL 2011. LNCS (LNAI), vol. 7188, pp. 89–101. Springer, Heidelberg (2012). https://doi.org/10.1007/978-3-642-29946-9_12
12. Geist, M., Scherrer, B., Lazaric, A., Ghavamzadeh, M.: A Dantzig selector approach to temporal difference learning. In: International Conference on Machine Learning, pp. 1399–1406 (2012)
13. Geramifard, A., Bowling, M., Sutton, R.S.: Incremental least-square temporal difference learning. In: The Twenty-first National Conference on Artificial Intelligence (AAAI), pp. 356–361 (2006)
14. Ghavamzadeh, M., Lazaric, A., Munos, R., Hoffman, M.W.: Finite-sample analysis of Lasso-TD. In: International Conference on Machine Learning, pp. 1177–1184 (2011)
15. Hastie, T., Tibshirani, R., Friedman, J.: The Elements of Statistical Learning: Data Mining Inference and Prediction. Springer, Heidelberg (2009). https://doi.org/10.1007/978-0-387-84858-7
16. Hoffman, M.W., Lazaric, A., Ghavamzadeh, M., Munos, R.: Regularized least squares temporal difference learning with nested ℓ2 and ℓ1 penalization. In: Sanner, S., Hutter, M. (eds.) EWRL 2011. LNCS (LNAI), vol. 7188, pp. 102–114. Springer, Heidelberg (2012). https://doi.org/10.1007/978-3-642-29946-9_13
17. Johns, J., Painter-wakefield, C., Parr, R.: Linear complementarity for regularized policy evaluation and improvement. Adv. Neural Inf. Process. Syst. **23**, 1009–1017 (2010)

18. Jordan, M.I., Ghahramani, Z., Jaakkola, T.S., Saul, L.K.: An introduction to variational methods for graphical models. Mach. Learn. **37**(2), 183–233 (1999)
19. Kolter, J.Z., Ng, A.Y.: Regularization and feature selection in least-squares temporal difference learning. In: International Conference on Machine Learning, pp. 521–528 (2009)
20. Lagoudakis, M., Parr, R.: Least-squares policy iteration. J. Mach. Learn. Res. **4**, 1107–1149 (2003)
21. Lazaric, A., Ghavamzadeh, M., Munos, R.: Finite-sample analysis of LSTD. In: International Conference on Machine Learning, pp. 615–622 (2010)
22. Liu, B., Zhang, L., Liu, J.: Dantzig selector with an approximately optimal denoising matrix and its application in sparse reinforcement learning. In: Proceedings of the Thirty-Second Conference on Uncertainty in Artificial Intelligence, UAI (2016)
23. Mannor, S., Simester, D., Sun, P., Tsitsiklis, J.N.: Bias and variance in value function estimation. In: International Conference on Machine Learning (2004)
24. Nedić, A., Bertsekas, D.P.: Least squares policy evaluation algorithms with linear function approximation. Discrete Event Dyn. Syst. **13**(1), 79–110 (2003)
25. Painter-Wakefield, C., Parr, R.: Greedy algorithms for sparse reinforcement learning. In: International Conference on Machine Learning (2012)
26. Parisi, G.: Statistical field theory. In: Frontiers in Physics. Addison-Wesley, Boston (1988)
27. Pires, B.A.: Statistical analysis of l1-penalized linear estimation with applications (2011)
28. Puterman, M.L.: Markov Decision Processes: Discrete Stochastic Dynamic Programming. Wiley, New Jersey (2005)
29. Sutton, R.S., Barto, A.G.: Reinforcement Learning: An Introduction. MIT Press, Cambridge (1998)
30. Sutton, R., Maei, H., Precup, D., Bhatnagar, S., Silver, D., Szepesvári, C., Wiewiora, E.: Fast gradient-descent methods for temporal-difference learning with linear function approximation. In: International Conference on Machine Learning, pp. 993–1000 (2009)
31. Tipping, M.E.: Sparse Bayesian learning and the relevance vector machine. J. Mach. Learn. Res. **1**, 211–244 (2001)
32. Tziortziotis, N.: Machine Learning for Intelligent Agents. Ph.D. thesis, Department of Computer Science and Engineering, University of Ioannina, Greece (2015)
33. Tziortziotis, N., Blekas, K.: Value function approximation through sparse Bayesian modeling. In: Sanner, S., Hutter, M. (eds.) EWRL 2011. LNCS (LNAI), vol. 7188, pp. 128–139. Springer, Heidelberg (2012). https://doi.org/10.1007/978-3-642-29946-9_15
34. Vlassis, N., Ghavamzadeh, M., Mannor, S., Poupart, P.: Reinforcement learning. In: Wiering, M., Van Otterlo, M. (eds.) Bayesian Reinforcement Learning, vol. 12, pp. 359–386. Springer, Heidelberg (2012). https://doi.org/10.1007/978-3-642-27645-3_11

Discovery of Causal Models that Contain Latent Variables Through Bayesian Scoring of Independence Constraints

Fattaneh Jabbari[1]([✉]), Joseph Ramsey[2], Peter Spirtes[2], and Gregory Cooper[1]

[1] Intelligent Systems Program, University of Pittsburgh, Pittsburgh, PA, USA
{fattaneh.j,gfc}@pitt.edu
[2] Department of Philosophy, Carnegie Mellon University, Pittsburgh, PA, USA
{jdramsey,ps7z}@andrew.cmu.edu

Abstract. Discovering causal structure from observational data in the presence of latent variables remains an active research area. Constraint-based causal discovery algorithms are relatively efficient at discovering such causal models from data using independence tests. Typically, however, they derive and output only one such model. In contrast, Bayesian methods can generate and probabilistically score multiple models, outputting the most probable one; however, they are often computationally infeasible to apply when modeling latent variables. We introduce a hybrid method that derives a Bayesian probability that the set of independence tests associated with a given causal model are jointly correct. Using this constraint-based scoring method, we are able to score multiple causal models, which possibly contain latent variables, and output the most probable one. The structure-discovery performance of the proposed method is compared to an existing constraint-based method (RFCI) using data generated from several previously published Bayesian networks. The structural Hamming distances of the output models improved when using the proposed method compared to RFCI, especially for small sample sizes.

Keywords: Observational data · Latent (hidden) variable Constraint-based and Bayesian causal discovery · Posterior probability

1 Introduction

Much of science consists of discovering and modeling causal relationships [21, 29,33]. Causal knowledge provides insight into mechanisms acting currently and prediction of outcomes that will follow when actions are taken (e.g., the chance that a disease will be cured if a particular medication is taken).

There has been substantial progress in the past 25 years in developing computational methods to discover causal relationships from a combination of existing knowledge, experimental data, and observational data. Given the increasing amounts of data that are being collected in all fields of science, this line of

© Springer International Publishing AG 2017
M. Ceci et al. (Eds.): ECML PKDD 2017, Part II, LNAI 10535, pp. 142–157, 2017.
https://doi.org/10.1007/978-3-319-71246-8_9

research has significant potential to accelerate scientific causal discovery. Some of the most significant progress in causal discovery research has occurred using causal Bayesian networks (CBNs) [29, 33].

Considerable CBN research has focused on constraint-based and Bayesian approaches to learning CBNs, although other approaches are being actively developed and investigated [30]. A constraint-based approach uses tests of conditional independence; causal discovery occurs by finding patterns of conditional independence and dependence that are likely to be present only when particular causal relationships exist. A Bayesian approach to learning typically involves a heuristic search for CBNs that have relatively high posterior probabilities.

The constraint-based and the Bayesian approaches each have significant, but different, strengths and weaknesses. The constraint-based approach can model and discover causal models with hidden (latent) variables relatively efficiently (depending upon what the true causal structure is, which variables are measured, and how many and what kind of hidden confounders have not been measured). This capability is important because oftentimes there are hidden variables that cause measured variables to be statistically associated (confounded). If such confounded relationships are not revealed, erroneous causal discoveries may occur.

The constraint-based approaches do not, however, provide a meaningful summary score of the chance that a causal model is correct. Rather, a single model is derived and output, without quantification regarding how likely it is to be correct, relative to alternative models. In contrast, Bayesian methods can generate and probabilistically score multiple models, outputting the most probable one. By doing so, they may increase the chance of finding a model that is causally correct. They also can quantify the probability of the top scoring model relative to other models that are considered in the search. The top scoring model might be close, or alternatively far away, from other models, which could be helpful to know. The Bayesian scoring of causal models that contain hidden confounders is very expensive computationally, however. Consequently, the practical application of Bayesian methods is largely relegated to CBNs that do not contain hidden variables, which significantly decreases the general applicability of these methods for causal discovery. In addition, while constraint-based methods can incorporate domain beliefs known with certainty (e.g., that a gene X is regulated by gene Y), they cannot incorporate domain beliefs about what is likely but not certain (e.g., that there is a 0.8 chance that gene X is regulated by gene Z). In general, Bayesian methods can incorporate as prior probabilities domain beliefs about what is likely but not certain, which is a common situation.

The current paper investigates a hybrid approach that combines strengths of constraint-based and Bayesian methods. The hybrid method derives the probability that relevant constraints are true. Consider a causal model (or equivalence class of models) that entails a set of conditional independence constraints over the distribution of the measured variables. In the hybrid approach, the probability of the model being correct is equal to the probability that the constraints that uniquely characterize the model (or class of models) are correct. This hybrid method exhibits the computational efficiency of a constraint-based method combined with the Bayesian approaches ability to quantitatively compare alternative causal models according to their posterior probabilities and to incorporate non-certain background beliefs.

The remainder of this paper first provides relevant background in Sect. 2. Sections 3 and 4 then describe a method for the Bayesian scoring of constraints, how to combine it with a constraint-based learning method, and two techniques for evaluating the posterior probabilities of models that are output. Section 5 describes an evaluation of the method using data generated from existing CBNs.

2   Background

A causal Bayesian network (CBN) is a Bayesian network in which each arc is interpreted as a direct causal influence between a parent node (a cause) and a child node (an effect), relative to the other nodes in the network [29]. In this paper, we focus on the discovery of CBN structures because this task is generally the first and most crucial step in the causal discovery process. As shorthand, the term CBN will denote a CBN structure, unless specified otherwise. We also focus on learning CBNs from observational data, since this is among the most challenging causal learning tasks. General reviews of the topic are in [11,14,22].

2.1   Constraint-Based Learning of CBNs from Data

A constraint-based Bayesian network search algorithm searches for a set of Bayesian networks, all of which entail a particular set of conditional independence constraints, which we simply call *constraints*, that are judged to hold in a dataset of samples based on the results of tests applied to that data. It is usually not computationally or statistically feasible to actually test each possible constraint among the measure variables for more than a few dozen variables, so constraint-based algorithms typically select a sufficient subset of constraints to test. Generally, the subset of constraint tests that are performed within a sequence of such tests depends upon the results of previous tests.

Fast Causal Inference (FCI) [33] is a constraint-based causal discovery algorithm, which we discuss in more detail here because it serves as a good example of a constraint-based algorithm, and we use an adapted version of it, called Really Fast Causal Inference (RFCI) [10], in the research reported here. FCI takes as input observed sample data and optional deterministic background knowledge, and it outputs a graph, called a Partial Ancestral Graph (PAG). A PAG represents a Markov equivalence class of Bayesian networks (possibly with hidden variables) that entail the same constraints. A PAG model returned by FCI represents as much about the true causal graph as can be determined from the conditional independence relations among the observed variables [36]. In particular, under assumptions, the FCI algorithm has been shown to have correct output with probability 1.0 in the large sample limit, even if there are hidden confounders [36]. In addition, a modification of FCI can be implemented to run in polynomial time, if a maximum number of causes (parents) per node is specified [9].

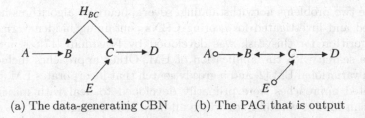

(a) The data-generating CBN   (b) The PAG that is output

Fig. 1. The PAG in (b) is learnable in the large sample limit from observational data generated by the causal model in (a), where H_{BC} is a hidden variable and the other variables are measured.

As an example, Fig. 1 shows in panel (b) the PAG that would be output by the FCI search if given a large enough sample of data from the data-generating CBN shown in panel (a), assuming the Markov and faithfulness[1] conditions hold [33]. In panel (b), the subgraph $B \leftrightarrow C$ represents that B and C are both caused by one or more hidden variables (i.e., they are confounded by a hidden variable). The subgraph $C \to D$ represents that C is a cause of D and that there are no hidden confounders of C and D. The subgraph $A \circ\!\!\to B$ represents that either A causes B, A and B are confounded by a hidden variable, or both. Another edge possibility, which does not appear in the example, is $X \circ\!\!-\!\!\circ Y$, which is compatible with the true causal model having X as cause of Y, Y as a cause of X, a hidden confounder of X and Y, or some acyclic combination of these three alternatives. The PAG in Fig. 1b indicates that not all the causal relationships in Fig. 1a can be learned from constraints on the data generated by that causal model, but some can be; in particular, Fig. 1b shows that is it possible to learn that B and C are both caused by a hidden variable(s) and that C causes D.

2.2 Bayesian Learning of CBNs from Data

Score-based methods derive a score for a CBN, given a dataset of samples and possibly prior knowledge or belief. Different types of scores have been developed and investigated, including the Minimum Description Length (MDL), Minimum Message Length (MML), and Bayesian scores [11,18]. There are two major problems when learning a CBN using Bayesian approaches:

- Problem 1 (model search): There is an infinite space of hidden-variable models, both in terms of parameters and hidden structure. Even when restrictions are assumed, the search space generally remains enormous in size, making it challenging to find the highest scoring CBNs.
- Problem 2 (model scoring): Scoring a given CBN with hidden variables is also challenging. In particular, marginalizing over the hidden variables greatly complicates Bayesian scoring in terms of accuracy and computational tractability.

---

[1] The faithfulness assumption states that if X and Y conditional on a set $\mathbf{Z}$ are d-connected in the structure of the data-generating CBN, then X and Y are dependent given $\mathbf{Z}$ in the probability distribution defined by the data-generating CBN.

These two problems notwithstanding, several heuristic algorithms have been developed and investigated for scoring CBNs containing hidden variables. An early algorithm for this task was developed by Friedman [17]; it interleaved structure search with the application of EM. Other approaches include those based on variational EM [2] and a greedy search that incorporates EM [5]. These and related approaches were primarily developed to deal with missing data, rather than hidden variables for which all data are missing.

Several Bayesian algorithms have been specifically developed to score CBNs with hidden variables, including methods that use a Laplace approximation [19], an approach that uses EM and a form of clustering [16], and a structural expectation propagation method [23]. However, these methods do not search over the space of all CBNs that include a given set of measured variables. Rather, they require that the user manually provides the proposed CBN models to be scored [19], they search a very restricted space of models, such as bipartite graphs [23] or trees of hidden structure [7,16], or they score ancestral relations between pairs of variables [28]. Thus, within a Bayesian framework, the automated discovery of CBNs that contain hidden variables remains an important open problem.

2.3   Hybrid Methods for Learning CBNs from Data

Researchers have also developed algorithms that combine constraint-based and Bayesian scoring approaches for learning CBNs [8,12,13,24–26,32,34,35]. However, these hybrid methods, except [8,24,26,34], do not include the possibility that the CBNs being modeled contain hidden variables. When a CBN can contain hidden variables, the Bayesian learning task is much more difficult.

In [8], a Bayesian method is proposed to score and rank order constraints; then, it uses those rank-ordered constraints as inputs to a constraint-based causal discovery method. However, it does not derive the posterior probability of a causal model from the probability of the constraints that characterize the model. The method in [26] models the possibility of hidden confounders but it does not provide any quantification of the output graph. In [34], a method is proposed to convert p-values to posterior probabilities of adjacencies and non-adjacencies in a graph; then, those probabilities are used to identify neighborhoods of the graph in which all relations have probabilities above a certain threshold. This method is, in fact, a post-processing step on the skeleton of the output network and not applicable to convert p-values to probabilities while running a constraint-based search method. It also does not provide a way of computing posterior probability of the whole output PAG. [24] introduces a logic-based method to reconstruct ancestral relations and score their marginal probabilities; it does not provide the probability of the output graph, however. In [24], authors mentioned that modeling the relationships among the constraints may be an improvement; in this paper, we propose an empirical way of modeling such relationships.

The research reported in [20] is the closest previous work of which we are aware to that introduced here. It describes how to score constraints on graphs by treating the constraints as independent of each other. The method is very expensive computationally, however, and is reported as working on up to only 7 measured variables. The method we introduce was feasibly applied to a dataset containing 70 variables and plausibly is practical for considerably larger datasets. Also, the method in [20], as described, is limited to deriving just the most probable graph, rather than deriving a set of graphs, as we do, which can be rank ordered, compared, and used to perform selective model averaging that derives (for example) distributions over edge types.

3   The Proposed Hybrid Method

This paper investigates a novel approach based on Bayesian scoring of constraints (BSC) that has major strengths of the constraint-based and Bayesian approaches. Namely, BSC uses a Bayesian method to score the constraints, rather than score the CBNs directly. The posterior probability of a CBN will be proportional to the posterior probability of the correctness of the constraints that characterize that CBN (or class of CBNs). The BSC approach, therefore, attenuates both problems of the Bayesian approach listed in Sect. 2.2:

- Problem 1 (model search): In the BSC approach, the search space is finite, not infinite as in the general Bayesian approach, because the number of possible constraints on a given set of measured variables is finite.
- Problem 2 (model scoring): In a constraint-based approach, the constraints are on measured variables only, as discussed in Sect. 2. Thus, when BSC uses a Bayesian approach to derive the probability of a set of constraints and thereby score a CBN, it needs only to consider measured variables. In contrast, a traditional Bayesian approach must marginalize over hidden variables, which is a difficult and computationally expensive operation.

3.1   Bayesian Scoring of Constraints (BSC)

This section describes how to score a constraint r_i. The term r_i denotes an arbitrary conditional independence of the form $(X_i \perp\!\!\!\perp Y_i | \mathbf{Z}_i)$ which is hypothesized to hold in the data-generating model that produced dataset D, where X_i and Y_i are variables of dataset D, and $\mathbf{Z}_i$ is a subset of variables not containing X_i or Y_i. Each r_i is called a conditional independence constraint, or constraint for short, where its value is either *true* or *false*. To score the posterior probability of a constraint r_i, we assume that the only parts of data D that influence belief about r_i are the data D_i, i.e. data about X_i, Y_i, and $\mathbf{Z}_i$. This is called the *data relevance assumption* which results in:

$$P(r_i|D) = P(r_i|D_i) .\tag{1}$$

Assuming uniform structure priors on constraints and applying Bayes rule result in the following equation:

$$P(r_i|D_i) = \frac{P(D_i|r_i)}{P(D_i|r_i) + P(D_i|\bar{r}_i)} .\tag{2}$$

Since we consider discrete variables in this paper, we can use the BDeu score in [18], which provides a closed-form solution for deriving marginal likelihoods, i.e. $P(D_i|r_i)$ and $P(D_i|\bar{r}_i)$, in Eq. (2). To derive a value for $P(D_i|r_i)$ (i.e., assuming X_i is independent of Y_i given $\mathbf{Z}_i$), we score the following BN structure, where $\mathbf{Z}_i$ is a set of parents nodes for X_i and Y_i:

$$X_i \leftarrow \mathbf{Z}_i \rightarrow Y_i$$

To compute $P(D_i|\bar{r}_i)$ (i.e., assuming X_i and Y_i are dependent given $\mathbf{Z}_i$) we score the following BN structure:

$$\mathbf{Z}_i \rightarrow X_i Y_i$$

where $X_i Y_i$ denotes a new node whose values are the Cartesian product of the values of X_i and Y_i. This is similar to scoring a DAG that consists of the following edges: $X_i \leftarrow \mathbf{Z}_i \rightarrow Y_i$ and $X_i \rightarrow Y_i$, which has been used previously [20]. In general, however, any Bayesian test of conditional independence can be used.

3.2 RFCI-BSC (RB)

This section describes an algorithm that combines constraint-based model search with the BSC method described in Sect. 3.1. As mentioned, RFCI [10] is a constraint-based algorithm for discovering the causal structure of the data-generating process in the presence of latent variables using Partial Ancestral Graphs (PAGs) as a representation, which encodes a Markov equivalence class of Bayesian networks (possibly with latent variables). RFCI has two stages. The first stage involves a selective search for the constraints among the measured variables, which is called *adjacency search*. The second stage involves determining the causal relationships among pairs of nodes that are directly connected according to the first stage; this stage is called the *orientation phase*.

We adapted RFCI to perform model search using BSC. We call this algorithm RFCI-BSC, or RB for short. During the first stage of search, when RFCI requests that an independence condition be tested, RB uses BSC to determine the probability p that independence holds. It then samples with probability p whether independence holds and returns that result to RFCI. To do so, it generates a random number U from Uniform$[0,1]$; if $U \leq p$ then it returns *true*, and otherwise, it returns *false*. Ultimately, RFCI will complete stage 1 in this manner, then stage 2, and finally return a PAG.

RB then repeats the procedure in the previous paragraph n times to generate up to n unique PAG models. Let each repetition be called a round. Since the set of constraints generated in each round is determined stochastically (i.e. sampling with probability p), these rounds will produce many different sets of constraints, and consequently, different PAGs. Algorithm 1 shows pseudo-code of the RB method that inputs dataset D and the number of rounds n. It then outputs a set of at most n PAGs and for each PAG, an associated set of constraints that were queried during the RFCI search. Note that RFCI* in this procedure denotes the RFCI search that uses BSC to evaluate each constraint, rather than using frequentist significant testing. The computational complexity of RB is $O(n)$

Algorithm 1. RB(D, n)

Input: dataset D, number of rounds n
Output: a set $\mathcal{G}$ containing PAG members G_j, a set $\mathbf{r}$ of constraints

1: Let $\mathcal{G}$ and $\mathbf{r}$ be empty sets
2: **for** $j = 1$ **to** n **do**
3:      $G_j, \mathbf{r_j} \leftarrow \mathrm{RFCI}^*(D)$ ▷ RFCI* uses BSC to evaluate each constraint
4:      $\mathcal{G} \leftarrow \mathcal{G} \cup G_j$
5:      $\mathbf{r} \leftarrow \mathbf{r} \cup \mathbf{r_j}$
6: **return** $\mathcal{G}$ and $\mathbf{r}$

---

times that of RFCI, since it calls RFCI n times. In the next section, we propose two methods to score each generated PAG model G_j.

4   Scoring a PAG Using RB

Let $\mathbf{r}$ be the union of all the independence conditions tested by RB over all rounds, which we will use to score each generated PAG model G_j. Based on the axioms of probability, we have the following equation:

$$P(G_j|D) = \sum_{\mathbf{r}} P(G_j|\mathbf{r}, D) \cdot P(\mathbf{r}|D) \ , \tag{3}$$

where the sum is over all possible value assignments to the constraints in set $\mathbf{r}$. Although Eq. (3) is valid, it does not provide a useful method for calculating $P(G_j|D)$. In this section, we propose a method to derive a way of computing $P(G_j|D)$ effectively.

Assume that the data only influence belief about a causal model via belief about the conditional independence constraints given by $\mathbf{r}$, i.e. $P(G_j|\mathbf{r}, D) = P(G_j|\mathbf{r})$, which is a standard assumption of constraint-based methods. Therefore, we can rewrite Eq. (3) as following:

$$P(G_j|D) = \sum_{\mathbf{r}} P(G_j|\mathbf{r}) \cdot P(\mathbf{r}|D) \ . \tag{4}$$

Although Eq. (4) is less general than a full Bayesian approach that integrates over CBN parameters, it is nonetheless more expressive than existing constraint-based methods that in essence assume that $P(\mathbf{r}|D) = 1$ for a set of constraints $\mathbf{r}$ that are derived using frequentist statistical tests.

Let $\mathbf{r}'_j$ denote the values of all the constraints in $\mathbf{r}$, according to the independencies implied by graph G_j as tested by RFCI. Since RFCI finds a set of sufficient independence conditions that distinguishes G_j from all other PAGs, so that $P(G_j|\mathbf{r} = \mathbf{r}'_j) = 1$ and $P(G_j|\mathbf{r} \neq \mathbf{r}'_j) = 0$, Eq. (4) becomes:

$$P(G_j|D) = \sum_{\mathbf{r}} P(G_j|\mathbf{r}) \cdot P(\mathbf{r}|D) = P(\mathbf{r} = \mathbf{r}'_j|D) \ . \tag{5}$$

Section 3.1 describes a method to compute the probability of one constraint given data, i.e. $P(r_i|D_i)$. Now, we need to extend it for a set of constraints, i.e. $P(\mathbf{r} = \mathbf{r}'_j|D)$ in Eq. (5). Applying the chain rule of probability, it becomes:

$$
P(\mathbf{r} = \mathbf{r}'_j|D) = P(r'_1, r'_2, r'_3, ..., r'_m|D) = \prod_{i=1}^{m} P(r'_i|r'_1, r'_2, ..., r'_{i-1}, D)
$$

$$
= \prod_{i=1}^{m} P(r'_i|r'_1, r'_2, ..., r'_{i-1}, D_i)(\text{assuming data relevance}) ,
$$

(6)

where r'_i denotes the value of ith constraint according its value given in $\mathbf{r}'_j$. Using Eq. (6), RB determines the most probable generated PAG and its posterior probability. For each pair of measured nodes, we can also use model averaging to estimate the probability distribution over each PAG edge type as follows: Since PAGs are being sampled (generated) according to their posterior distribution (under assumptions), the probability of edge E existing between nodes A and B is estimated as the fraction of the sampled PAGs that contain E between A and B. In the following subsections, we propose two methods to approximate the joint posterior probabilities of constraints.

4.1   BSC with Independence Assumption (BSC-I)

In the first method, we assume that constraints in set $\mathbf{r} = \{r_1, r_2, ..., r_m\}$, which is a set of all independence constraints obtained by running RB algorithm, are independent of each other. We call this approach BSC-I. Given this assumption and Eq. (6), BSC-I scores an output graph as follows:

$$
P(G_j|D) = P(\mathbf{r} = \mathbf{r}'_j|D) = \prod_{i=1}^{m} P(r'_i|D_i) ,
$$

(7)

where $P(r'_i|D_i)$ can be computed as described in Sect. 3.1.

4.2   BSC with Dependence Assumption (BSC-D)

In this scoring approach, we model the possibility that the constraints are dependent, which often happens. The relationships among the constraints can be complicated, and to our knowledge, they have not been modeled previously. In the remainder of this section, we introduce an empirical method to model the relationships among conditional constraints.

Similar to BSC-I, consider $\mathbf{r}$ as a set of all the independence constraints queried by the RB method. As we mentioned earlier, each constraint $r_i \in \mathbf{r}$ has the form $(X_i \perp\!\!\!\perp Y_i|\mathbf{Z}_i)$, where X_i and Y_i are variables of dataset D and $\mathbf{Z}_i$ is a subset of variables not containing X_i or Y_i. Each r_i can take two values, *true* (1) or *false* (0); therefore, it can be considered as a binary random variable. We build a dataset, D_r, of these binary random variables using bootstrap sampling [15]

Algorithm 2. EmpiricalDataCreation(D, n, **r**)

Input: dataset D, number of bootstraps n, and a set of constraints **r**
Output: empirical dataset D_r with n rows and $m = |\mathbf{r}|$ columns

1: Let $D_r[n, m]$ be a new 2-d array with n rows and m columns
2: **for** $b = 1$ **to** n **do**
3:     $sample_b \leftarrow$ Bootstrap(D)
4:     **for** $r_i \in \{r_1, r_2, \ldots, r_m\}$ **do**
5:         $p \leftarrow$ BSC(r_i, $sample_b$)
6:         **if** $p \geq 0.5$ **then**
7:             $D_r[b, i] \leftarrow 1$
8:         **else**
9:             $D_r[b, i] \leftarrow 0$
10: **return** $D_r[n, m]$

and the BSC method. To do so, we first bootstrap (re-sample with replacement) the data D; let $sample_b$ denote the resulting dataset. Then, for each constraint $r_i \in \mathbf{r}$, we compute the BSC score using $sample_b$ and set its value to 1 if its BSC score is more than or equal to 0.5, and 0 otherwise. We repeat this entire procedure n times to fill in n rows of empirical data for the constraints. Algorithm 2 provides pseudo-code of this procedure. It inputs the original dataset D, the number of bootstraps n, and a set of constraints **r**. It outputs an empirical dataset D_r with n rows and $m = |\mathbf{r}|$ columns. The Bootstrap(D) function in this procedure creates a bootstrap sample from D, and BSC(r_i, $sample_b$) computes the BSC score of constraint r_i using $sample_b$.

The empirical data D_r can then be used to learn the relations among the constraints **r**. We learn a Bayesian network because doing so can be done efficiently with thousands of variables, such networks are expressive in representing the joint relationships among the variables, and inference of the joint state of the variables (constraints in this application) can be derived efficiently. We use an optimized implementation of the Greedy Equivalence Search (GES) [6], which is called Fast GES (FGES) [31] to learn a Bayesian network structure, BN_r, that encodes the dependency relationships among the constraints **r**. We then apply a maximum *a posteriori* estimation method to learn the parameters of BN_r given D_r, which we denote as θ_r. Finally, we use BN_r and θ_r to factorize $P(\mathbf{r} = \mathbf{r}_j|D)$ and score the output PAG as follows:

$$P(G_j|D) = P(\mathbf{r} = \mathbf{r}'_j|D) = \prod_{i=1}^{m} P(r'_i|Pa(r_i), D) , \qquad (8)$$

where r'_i and $Pa(r_i)$ denotes the parents of variable r_i in $\mathbf{r}'_j$ and its parents in BN_r, respectively.

5   Evaluation

This section describes an evaluation of the RB method using each of the BSC-I and BSC-D scoring techniques, which we call RB-I and RB-D, respectively.

Algorithm 3. RB-I(D, n)

Input: dataset D, number of rounds n
Output: the most probable PAG $PAG\text{-}I$

1: Let $\mathcal{G}$ and $\mathbf{r}$ be empty sets
2: $\mathcal{G}, \mathbf{r} \leftarrow$ RB(D, n)
3: $PAG\text{-}I \leftarrow \arg \max_{G_i \in \mathcal{G}}$ BSC-I($G_i, \mathbf{r}, D$)
4: return $PAG\text{-}I$

Table 1. Information about the CBNs used in the simulation experiments.

Name	Alarm	Hailfinder	Hepar II
Domain	Medicine	Weather	Medicine
Number of nodes	37	56	70
Number of edges	46	66	123
Number of parameters	509	2656	1453
Average degree	2.49	2.36	3.51

Algorithm 3 provides pseudo-code of RB-I method, which inputs dataset D, the number of rounds n, and outputs the most probable PAG. It first runs the RB method (Algorithm 1) to get a set of PAGs $\mathcal{G}$ and constraints $\mathbf{r}$. It then computes the posterior probability of each PAG $G_i \in \mathcal{G}$ using BSC-I and returns the most probable PAG, which is denoted by $PAG\text{-}I$ in Algorithm 3. Note that RB-D would be exactly the same except for using BSC-D in line 3.

5.1   Experimental Methods

To perform an evaluation, we first simulated data from manually constructed, previously published CBNs, with some variables designated as being hidden. We then provided that data to each of RB-I and RB-D. We compared the most probable PAG output by each of these two methods to the PAG consistent with the data-generating CBN. In particular, we simulated data from the Alarm [3], Hailfinder [1], and Hepar II [27] CBNs, which we obtained from [4]. Table 1 shows some key characteristics of each CBN. Using these benchmarks is beneficial in multiple ways. They are more likely to represent real-world distributions. Also, we can evaluate the results using the true underlying causal model, which we know by construction; otherwise, it is rare to find known causal models on more than a few variables and associated real, observational data.

To evaluate the effect of sample size, we simulated 200 and 2000 cases randomly from each CBN, according to the encoded joint probability distribution. In each CBN, we randomly designated 0.0%, 10.0%, and 20.0% of the confounder nodes to be hidden, which means data about those nodes were not provided to the discovery algorithms. In applying the two versions of the RB algorithm, we sampled 100 PAG models, according to the method described in Sect. 3.2 (i.e.,

$n = 100$ in Algorithm 1). Also, we bootstrapped the data 500 times (i.e., $n = 500$ in Algorithm 2) to create the empirical data for BSC-D scoring. For each network, we repeated the analyses in this paragraph 10 times, each time randomly sampling a different dataset. We refer to one of these 10 repetitions as a run.

Let PAG-I and PAG-D denote the sampled models that had the highest posterior probability when using BSC-I (see Eq. (7)) and BSC-D (see Eq. (8)) scoring methods, respectively. Let PAG-CS denote the model returned by RFCI when using a chi-squared test of independence, which is the standard approach; we used $\alpha = 0.05$, which is a common alpha value used with RFCI. Let PAG-$True$ be the PAG that represents all the causal relationships that can be learned about a CBN in the large sample limit when assuming faithfulness and using independence tests that are applied to observational data on the measured variables in a CBN.

We compared the causal discovery performance of PAG-I, PAG-D, and PAG-CS using PAG-$True$ as the gold standard. For a given CBN (e.g., Alarm) we calculated the mean Structural Hamming Distance (SHD) between a given PAG G and PAG-$True$, which counts the number of different edge marks over all 10 runs. For example, if the output graph contains the edge $A \circ\!\!\rightarrow B$ while $B \rightarrow A$ exists in PAG-$True$, then edge-mark SHD of this edge is 2. Similarly, edge-mark SHD would be 1 if $A \rightarrow B$ is in the output PAG but $A \leftrightarrow B$ is in PAG-$True$. Clearly, any extra or missing edge would count as 2 in terms of edge-mark SHD. We also measured the number of extra and/or missing edges (regardless of edge type) between a given PAG G and PAG-$True$, which corresponds to the SHD between the skeletons (i.e., the adjacency graph) of the two PAGs. For instance, if one graph includes $A \circ\!\!-\!\!\circ B$ while there is no edge between these variables in the other one, then skeleton SHD would be 1. For each of the measurements, we calculated its mean and 95% confidence interval over the 10 runs.

5.2   Experimental Results

Figure 2 shows the experimental results. The diagrams on the left show the SHD between the skeletons of each PAG and PAG-$True$. The diagrams on the right-hand side represent the SHD of the edge marks between each output PAG and PAG-$True$. For each diagram, circles and squares represent the average results for datasets with 2000 and 200 cases, respectively. The vertical error bars in the diagrams represent the 95% confidence interval around the average values. Also, each column labeled as $H = 0.0$, 0.1, or 0.2 shows the proportion of hidden variables in each experiment. Figure 2a and 2b show that using the RB method always improves both performance measures for the Alarm network, especially for small sample sizes. Similar results were obtained on Hepar II network (Fig. 2e and 2f). For Hailfinder, we observed significant improvements on both the skeleton and edge-marks SHD when the sample size is 2000. The results show that the edge-mark SHD always improves when applying the RB method. We observed that BSC-I and BSC-D performed very similarly.

We found that using BSC-I and BSC-D may result in different probabilities for the generated PAGs; however, the ordering of the PAGs according to their

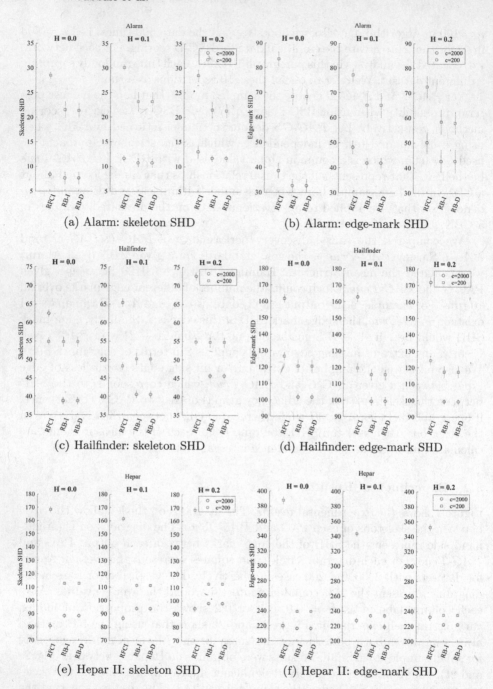

(a) Alarm: skeleton SHD   (b) Alarm: edge-mark SHD

(c) Hailfinder: skeleton SHD   (d) Hailfinder: edge-mark SHD

(e) Hepar II: skeleton SHD   (f) Hepar II: edge-mark SHD

Fig. 2. Skeleton and edge-mark SHD of output PAGs relative to the gold standards

posterior probabilities is almost always the same. We conjecture that performance of BSC-I is analogous to a naive Bayes classifier, which often performs classification well, even though it can be highly miscalibrated due to its universal assumption of conditional independence.

6 Discussion

This paper introduces a general approach for Bayesian scoring of constraints that is applied to learn CBNs which may contain hidden confounders. It allows the input of informative prior probabilities and the output of causal models that are quantified by their posterior probabilities. As a preliminary study, we implemented and experimentally evaluated two versions of the method called RB-I and RB-D. We compared these methods to a method that applies the RFCI algorithm using a chi-squared test.

For the edge-mark SHD, RB-I and RB-D had statistically significantly better results than RFCI for all three networks for any sample size and fraction of hidden variables. The skeleton SHD was better for most tested scenarios when using RB-I and RB-D, except for Hailfinder with $H = 0.1$ and 200 samples, and Hepar II with 2000 samples. Overall, the results indicate that RB tends to be more accurate than RFCI in predicting and orienting edges. Also, both RB-I and RB-D methods perform very similarly. We found out that posterior probabilities obtained by each of these methods are not equal but they result in the same most probable PAG. As the sample size increases, we expect the constraints to become independent of each other, but in our experiments, dependence did not matter for SHD, even with small sample sizes. Interestingly, this provides support that the simpler BSC-I method is sufficient for the purpose of finding the most probable PAG.

The RB method is a prototype that can be extended in numerous ways, including the following: (a) Develop more general tests of conditional independence to learn CBNs that contain continuous variables or a mixture of continuous and discrete variables; (b) Perform selective Bayesian model averaging of the edge probabilities as described in Sect. 4; (c) Incorporate informative prior probabilities on constraints. For example, one way to estimate the prior probability $P(r_i)$ for insertion into Eq. (2) is to use prior knowledge to define Maximal Ancestral Graph (MAG) edge probabilities for each pair of measured variables. Then, use those probabilities to stochastically generate a large set of graphs and retain those graphs that are MAGs. Finally, tally the frequency with which r_i holds in the set of MAGs as an estimate of $P(r_i)$.

The evaluation reported here can be extended in several ways, such as using additional manually constructed CBNs to generate data, evaluating a wider range of data sample sizes and fractions of hidden confounders, and applying additional algorithms as methods of comparison [8,17,23]. Despite its limitations, the current paper provides support that the Bayesian scoring of constraints is a promising hybrid approach for the problem of learning the most probable causal model that can include hidden confounders. The results suggest that further investigation of the approach is warranted.

Acknowledgments. Research reported in this publication was supported by grant U54HG008540 awarded by the National Human Genome Research Institute through funds provided by the trans-NIH Big Data to Knowledge initiative. The content is solely the responsibility of the authors and does not necessarily represent the official views of the National Institutes of Health.

References

1. Abramson, B., Brown, J., Edwards, W., Murphy, A., Winkler, R.L.: Hailfinder: a Bayesian system for forecasting severe weather. Int. J. Forecast. **12**(1), 57–71 (1996)
2. Beal, M.J., Ghahramani, Z.: The variational Bayesian EM algorithm for incomplete data: with application to scoring graphical model structures. In: Proceedings of the Seventh Valencia International Meeting, pp. 453–464 (2003)
3. Beinlich, I.A., Suermondt, H.J., Chavez, R.M., Cooper, G.F.: The ALARM monitoring system: a case study with two probabilistic inference techniques for belief networks. In: Hunter, J., Cookson, J., Wyatt, J. (eds.) AIME 89. LNMI, vol. 38, pp. 247–256. Springer, Heidelberg (1989). https://doi.org/10.1007/978-3-642-93437-7_28
4. Bayesian Network Repository. http://www.bnlearn.com/bnrepository/
5. Borchani, H., Ben Amor, N., Mellouli, K.: Learning Bayesian network equivalence classes from incomplete data. In: Todorovski, L., Lavrač, N., Jantke, K.P. (eds.) DS 2006. LNCS (LNAI), vol. 4265, pp. 291–295. Springer, Heidelberg (2006). https://doi.org/10.1007/11893318_29
6. Chickering, D.M.: Optimal structure identification with greedy search. J. Mach. Learn. Res. **3**, 507–554 (2002)
7. Choi, M.J., Tan, V.Y., Anandkumar, A., Willsky, A.S.: Learning latent tree graphical models. J. Mach. Learn. Res. **12**, 1771–1812 (2011)
8. Claassen, T., Heskes, T.: A Bayesian approach to constraint based causal inference. In: Proceedings of the Conference on Uncertainty in Artificial Intelligence, pp. 207–216 (2012)
9. Claassen, T., Mooij, J., Heskes, T.: Learning sparse causal models is not NP-hard. In: Proceedings of the Conference on Uncertainty in Artificial Intelligence (2013)
10. Colombo, D., Maathuis, M.H., Kalisch, M., Richardson, T.S.: Learning high-dimensional directed acyclic graphs with latent and selection variables. Ann. Stat. **40**(1), 294–321 (2012)
11. Daly, R., Shen, Q., Aitken, S.: Review: learning Bayesian networks: approaches and issues. Knowl. Eng. Rev. **26**(2), 99–157 (2011)
12. Dash, D., Druzdzel, M.J.: A hybrid anytime algorithm for the construction of causal models from sparse data. In: Proceedings of the Fifteenth Conference on Uncertainty in Artificial Intelligence, pp. 142–149 (1999)
13. De Campos, L.M., FernndezLuna, J.M., Puerta, J.M.: An iterated local search algorithm for learning Bayesian networks with restarts based on conditional independence tests. Int. J. Intell. Syst. **18**(2), 221–235 (2003)
14. Drton, M., Maathuis, M.H.: Structure learning in graphical modeling. Annu. Rev. Stat. Appl. **4**, 365–393 (2016)
15. Efron, B., Tibshirani, R.J.: An Introduction to the Bootstrap. CRC Press, Boca Raton (1994)
16. Elidan, G., Friedman, N.: Learning hidden variable networks: the information bottleneck approach. J. Mach. Learn. Res. **6**(Jan), 81–127 (2005)

17. Friedman, N.: The Bayesian structural EM algorithm. In: Proceedings of the Fourteenth Conference on Uncertainty in Artificial Intelligence, pp. 129–138 (1998)
18. Heckerman, D., Geiger, D., Chickering, D.M.: Learning Bayesian networks: the combination of knowledge and statistical data. Mach. Learn. **20**(3), 197–243 (1995)
19. Heckerman, D., Meek, C., Cooper, G.: A Bayesian approach to causal discovery. In: Glymour, C., Cooper, G.F. (eds.) Computation, Causation, and Discovery, pp. 141–165. MIT Press, Menlo Park, CA (1999)
20. Hyttinen, A., Eberhardt, F., Jrvisalo, M.: Constraint-based causal discovery: conflict resolution with answer set programming. In: Proceedings of the Conference on Uncertainty in Artificial Intelligence (UAI), pp. 340–349 (2014)
21. Illari, P.M., Russo, F., Williamson, J.: Causality in the Sciences. Oxford University Press, Oxford (2011)
22. Koski, T.J., Noble, J.: A review of Bayesian networks and structure learning. Math. Appl. **40**(1), 51–103 (2012)
23. Lazic, N., Bishop, C.M., Winn, J.M.: Structural Expectation Propagation (SEP): Bayesian structure learning for networks with latent variables. In: Proceedings of the Conference on Artificial Intelligence and Statistics (AISTATS), pp. 379–387 (2013)
24. Magliacane, S., Claassen, T., Mooij, J.M.: Ancestral causal inference. In: Advances in Neural Information Processing Systems, pp. 4466–4474 (2016)
25. Nandy, P., Hauser, A., Maathuis, M.H.: High-dimensional consistency in score-based and hybrid structure learning. arXiv preprint arXiv:1507.02608 (2015)
26. Ogarrio, J.M., Spirtes, P., Ramsey, J.: A hybrid causal search algorithm for latent variable models. In: Conference on Probabilistic Graphical Models, pp. 368–379 (2016)
27. Onisko, A.: Probabilistic causal models in medicine: application to diagnosis of liver disorders. Ph.D. dissertation, Institute of Biocybernetics and Biomedical Engineering, Polish Academy of Science, Warsaw (2003)
28. Parviainen, P., Koivisto, M.: Ancestor relations in the presence of unobserved variables. Mach. Learn. Knowl. Discov. Databases **6912**, 581–596 (2011)
29. Pearl, J.: Causality: Models, Reasoning, and Inference. Cambridge University Press, New York (2009)
30. Peters, J., Mooij, J., Janzing, D., Schlkopf, B.: Identifiability of causal graphs using functional models. In: Proceedings of the Conference on Uncertainty in Artificial Intelligence, pp. 589–598 (2012)
31. Ramsey, J.D.: Scaling up greedy equivalence search for continuous variables. CoRR, abs/1507.07749 (2015)
32. Singh, M., Valtorta, M.: Construction of claass network structures from data: a brief survey and an efficient algorithm. Int. J. Approx. Reason. **12**(2), 111–131 (1995)
33. Spirtes, P., Glymour, C.N., Scheines, R.: Causation, Prediction, and Search. MIT Press, Cambridge (2000)
34. Triantafillou, S., Tsamardinos, I., Roumpelaki, A.: Learning neighborhoods of high confidence in constraint-based causal discovery. In: van der Gaag, L.C., Feelders, A.J. (eds.) PGM 2014. LNCS (LNAI), vol. 8754, pp. 487–502. Springer, Cham (2014). https://doi.org/10.1007/978-3-319-11433-0_32
35. Tsamardinos, I., Brown, L.E., Aliferis, C.F.: The max-min hill-climbing Bayesian network structure learning algorithm. Mach. Learn. **65**(1), 31–78 (2006)
36. Zhang, J.: On the completeness of orientation rules for causal discovery in the presence of latent confounders and selection bias. Artif. Intell. **172**(16), 1873–1896 (2008)

Labeled DBN Learning with Community Structure Knowledge

E. Auclair, N. Peyrard$^{(\boxtimes)}$, and R. Sabbadin

MIAT, UR 875, Université de Toulouse, INRA,
31320 Toulouse, Castanet-Tolosan, France
nathalie.peyrard@inra.fr

Abstract. Learning interactions between dynamical processes is a widespread but difficult problem in ecological or human sciences. Unlike in other domains (bioinformatics, for example), data is often scarce, but expert knowledge is available. We consider the case where knowledge is about a limited number of interactions that drive the processes dynamics, and on a community structure in the interaction network. We propose an original framework, based on Dynamic Bayesian Networks with labeled-edge structure and parsimonious parameterization, and a Stochastic Block Model prior, to integrate this knowledge. Then we propose a restoration-estimation algorithm, based on 0-1 Linear Programing, that improves network learning when these two types of expert knowledge are available. The approach is illustrated on a problem of ecological interaction network learning.

Keywords: Labeled edge network learning
Dynamic Bayesian Network · Stochastic Block Model
0-1 linear programing · Trophic network

1 Introduction

Learning an interaction network between entities is a widespread problem in bioinformatics [29], ecology [19] or social sciences [17]. This problem is often formulated in the framework of Bayesian Networks (BN) [13]. When the state of variables changes through time, learning approaches based on Dynamic Bayesian Networks (DBN) have also been proposed [10]. Learning a DBN amounts to learning both its structure (i.e. the conditional independences between the variables) and its Transition Probability Tables (TPT). Several solution approaches to DBN learning exist. They generally extend the methods used for learning static BN (see [20] for a review). They often consist in defining a global score function on networks measuring their "goodness of fit" and in using search methods to find the DBN structure and TPT that jointly optimize the score function. While finding an optimal BN is NP-hard in general [4], it is not the case for DBN without synchronous edges where the global score function is decomposable into independent local scores (one per variable). This is because, without synchronous

© Springer International Publishing AG 2017
M. Ceci et al. (Eds.): ECML PKDD 2017, Part II, LNAI 10535, pp. 158–174, 2017.
https://doi.org/10.1007/978-3-319-71246-8_10

arcs, a DBN structure is acyclic, so there is no need to check a global acyclicity constraint on the learned network, as opposed to the BN case. Under this assumption, [6] has provided polynomial time algorithms for learning DBN structure in the case of minimum description length (MDL) and Bayesian Dirichlet equivalence (BDe) scores. [27] have extended these results to the Mutual Information Tests (MIT) score.

Even with the hypothesis of no synchronous edges, learning DBN structure remains difficult since in many problems where interactions are to be learned, observed data are scarce. On the other hand, expert knowledge is often available for such problems, that could be taken into account in the learning process. In this paper, we consider two different types of expert knowledge and show how to use this knowledge to improve DBN structure learning.

First, we consider information about the mechanisms driving the process dynamics (e.g. facilitation, competition, cooperation...). This may be useful in order to constrain some elements of the TPT. For instance, equality constraints between some elements of distinct TPTs have been studied by [22]. Here, we derive such equalities in the case of generalized'per contact' processes where the dynamics of a variable is the result of a limited number of interaction types. This enables us to define a parsimonious parameterization of the TPT from a labeled-edge structure of the DBN, using one label and one parameter per interaction type. Then, variables submitted to the same influences share the same TPT. This defines the general framework of DBN with labeled-edge structure and parsimonious parameterization of the TPT (Sect. 2.1). We will refer to them as Labeled DBN (L-DBN). We consider in particular the case of only two types of interactions: impulsion and inhibition.

The idea of labeling the edges of a BN to model the positive or negative influence of a variable on another has already been considered in the framework of qualitative BN [28]. However, such a labeled network is usually given as an input of a BN parameter learning problem, in order to constrain the learned CPT [8]. In this article we tackle the question of learning the labels together with the structure and the parameters.

The second type of information we consider is knowledge about the structure of the interaction network. Structural constraints can be imposed on the network to simplify the learning task by reducing the search space, independently of the physical meaning of the network. They can be local constraints on node degree or forbidden edges [3]. Global features have also been considered. For instance, in [24] an upper bound of the treewidth of the BN graph is introduced in the learning procedure. In [23], the authors introduce a prior on the partial ordering of the nodes and show how to learn a BN in a Bayesian framework. As opposed to these kind of constraints, we consider structural constraints linked to expert knowledge, and we formalize the introduction of knowledge about a community structure of the network, during L-DBN learning. Namely we assume that the nodes of the interaction network are grouped into communities. Social networks, as well as food webs, are naturally structured in communities of individuals defined by jobs, schools, etc., or by trophic levels... Knowing them provides some prior knowledge about the within and between communities interactions that can help the learning. We model this a priori as a Stochastic Block Model (SBM, [15]), in the spirit of [1] for static continuous data for learning Gaussian graphical models. In this paper, we extend SBM to multiple interaction types, in order to deal with the labeled edges of a L-DBN (Sect. 2.2).

In Sect. 3, we propose an iterative Restoration-Estimation (RE) algorithm for learning both the structure (edges and labels) and the parameters of a L-DBN model with SBM prior. In Sects. 4 and 5, we model and solve a problem of ecological interactions network learning by combining a L-DBN model related to causal independence BN models [14], the SBM prior and the RE algorithm. In Sect. 6, we evaluate, on synthetic ecological networks and on a real one, how the successive introductions of knowledge on interaction types and on network structure improve the quality of the restored network, compared to a learning approach based on a non-parameterized DBN.

2 Integrating Labeled Edges and Community Structure Knowledge in DBN

2.1 Labeled Dynamic Bayesian Networks

Let us consider a set $\{(X_1^t)_{t=1,...T}, \ldots, (X_n^t)_{t=1,...T}\}$ of n coupled random processes over horizon T. Then, denoting $X^t = \{X_1^t, \ldots, X_n^t\}$, a DBN allows to concisely represent the joint probability distribution $P(X^1, \ldots, X^T)$ under Markovian and stationarity assumptions, by exploiting conditional independence between the variables. These independences can be represented by a bipartite graph $\mathcal{G}_\rightarrow = (V, E)$ between two sets of vertices, both indexed by $\{1, \ldots, n\}$ and respectively representing the variables $\{X_1^t, \ldots, X_n^t\}$ and $\{X_1^{t+1}, \ldots, X_n^{t+1}\}$. In $\mathcal{G}_\rightarrow$, edges are directed from vertices at time t, to vertices at time $t+1$. The joint probability distribution writes $P(X^{t+1}|X^t) = \prod_{i=1}^n P(X_i^{t+1}|X_{Par(i,\mathcal{G}_\rightarrow)}^t)$, where $Par(i, \mathcal{G}_\rightarrow) = \{j, (j,i) \in E\}$.

The DBN framework enables a huge gain in space by representing individual tables $P_i(X_i^{t+1}|X_{Par(i,\mathcal{G}_\rightarrow)}^t)$ rather than directly the joint transition probability. However, when some domain-specific knowledge impose that some individual transition probabilities are identical, it is possible to save even more space. This will be the case, for instance, when a limited number L of interaction types between variables exists, and all interactions of same type have the same effect on a variable, regardless which parent variables are concerned. We can use these interaction types to define the TPT by a small number of parameters. This is the case for epidemic contact processes models [9], where there is only one interaction type (contamination) and the state of a variable X_i^t only depends on the number of infected parents (and not on the precise knowledge of which parents are infected). We generalize this idea with the L-DBN framework. To do so, we consider a labeled version of graph $\mathcal{G}_\rightarrow$, namely graph $\mathcal{LG}_\rightarrow = (V, E, \mathcal{L}, \lambda)$, where E is a set of edges, $\mathcal{L} = \{1, \ldots, L\}$ a set of edge labels (interaction types) and $\lambda : E \rightarrow \mathcal{L}$ a labeling function.

Definition 1. *A Labeled DBN is a DBN such that:*

- *In the graphical representation of the conditional independences of the global transition probability, each edge is labeled by a label $l \in \mathcal{L}$ (except edges from X_i^t to X_i^{t+1} if present). The set of parents of a vertex i connected through an edge with label l is denoted $Par^l(i, \mathcal{LG}_\rightarrow)$.*
- *Two parents in $Par^l(i, \mathcal{LG}_\rightarrow)$ are assumed indistinguishable in their influence on i, and each labeled influence applies independently. This means that the transition probability distribution of X_i^{t+1} only depends on the number of parents in each possible state for each label (and the state of X_i^t if the edge exists).*
- *Two individuals i and j, such that $card(Par^l(i, \mathcal{LG}_\rightarrow)) = card(Par^l(j, \mathcal{LG}_\rightarrow))$ for all $l \in \mathcal{L}$, have the same TPT.*
- *This transition probability distribution is defined as a function of a vector of parameters θ, of low dimension (one per label and, possibly, a further one to model transitions independent of the $\{Par^l(i, \mathcal{LG}_\rightarrow)\}_l$).*

Once the form of the parameterized transition function is given, the TPTs of a L-DBN can be modeled in a very concise way, by specifying only the labeled graph (sets $Par^l(i, \mathcal{LG}_\rightarrow)$) and the parameters vector θ. One advantage of using a parameterized representation of a L-DBN is that it can be learnt more efficiently from small data sets than a non-parameterized representation.

The L-DBN framework is very general. A family of L-BDN of interest is that of binary *per contact propagation* processes. In this case, X_i^t is a binary random variable: $X_i^t = 1$ is for presence and $X_i^t = 0$ represents absence. Two types of interactions are possible ($\mathcal{L} = \{+, -\}$): an *impulsion interaction* ($+$) from a variable X_i^t to a variable X_i^{t+1}, increases the probability of presence of the process i at $t + 1$; an *inhibition interaction* ($-$) from a variable X_j^t to a variable X_i^{t+1} decreases the probability of presence of the process i at $t + 1$ (as in qualitative BN, [28]). All edges of identical label have the same impact on the transition probabilities of the affected variables. We associate a parameter to each label: ρ^+ is the probability of success of an impulsion; ρ^- is the probability of success of an inhibition. We also assume that the success or failure of the influence of all parents are independent (as in a causal independence BN model, [14]). The L-DBN model when the parents only have an impact on the survival (e.g. for species in interactions) is as follows. First the probability of apparition at vertex i is independent from the state of the other variables. We model this by a parameter ε, interpretable as the probability of spontaneous apparition. Then, the probability of survival of a process i is the probability of success of at least one impulsion interaction and of failure of all inhibition interactions. Therefore, the survival of i is the result of independent coin flips. Let $N_{i,l}^t = |\{j \in Par^l(i, \mathcal{LG}_\rightarrow), X_j^t = 1\}|$, for $l \in \{+, -\}$, then,

$$P(X_i^{t+1} = 1 | X_i^t = 0) = \varepsilon$$
$$P(X_i^{t+1} = 1 | X_i^t = 1, N_{i,+}^t, N_{i,-}^t) = \left(1 - (1 - \rho^+)^{N_{i,+}^t}\right) \cdot (1 - \rho^-)^{N_{i,-}^t}.$$

Similarly, the L-DBN model when interactions have only an impact on apparition (e.g. for disease spread) is defined by

$$P(X_i^{t+1} = 1 | X_i^t = 1) = \varepsilon$$
$$P(X_i^{t+1} = 1 | X_i^t = 0, N_{i,+}^t, N_{i,-}^t) = \left(1 - (1 - \rho^+)^{N_{i,+}^t}\right) \cdot (1 - \rho^-)^{N_{i,-}^t}.$$

The family of per contact propagation processes also includes processes where
survival (or apparition) requires the success of all impulsion interactions and the
failure of one inhibition interaction, and processes defined by any other AND/OR
combination of independent events of impulsion and inhibition successes. In this
family, the TPT are defined by three parameters only: $\theta = \{\rho^+, \rho^-, \varepsilon\}$.

Figure 1 (left) shows the graphical representation of an example L-DBN struc-
ture $\mathcal{LG}_\rightarrow$ with $n = 4$. In this example, $Par^+(1, \mathcal{LG}_\rightarrow) = \emptyset$ and $Par^-(1, \mathcal{LG}_\rightarrow) =
\{2, 4\}$. Because the state of X_i^t determines whether the transition is a survival
or an apparition, we also add an edge from X_i^t to X_i^{t+1} (without label) not
associated to any parameter. This edge is known, it does not have to be learnt.
Figure 1 (right) shows an equivalent static representation of $\mathcal{LG}_\rightarrow$, where nodes
corresponding to variables X_i^t and X_i^{t+1} have been collapsed. This representation may have a natural meaning with respect to the represented process, as will
be the case with the ecological network case study we will describe in Sect. 4.
The meaning of the dashed boxes is related to this example (see Sect. 4.2).

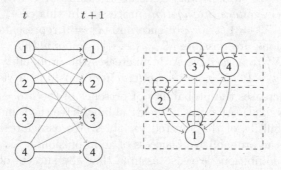

Fig. 1. The two graphical representations of the structure of a L-DBN with 4 variables
and 2 labels. Green and red edges represent respectively label '+' and '−' in the case
of ecological network. Black edges represent the unlabeled edges accounting for the
dependence of X_i^{t+1} on X_i^t. Left: $\mathcal{LG}_\rightarrow$, dynamics representation. Right: equivalent
static representation (dashed rectangles represent the blocks of the SBM) (Color figure
online).

2.2   Stochastic Block Models for L-DBN

In the above section, the labels and the parameterized TPT enable to encode
knowledge about the mechanisms underlying the dynamics of X^t, for a given
DBN structure. Now we present how to embed knowledge about the properties
of the structure $\mathcal{LG}_\rightarrow$ itself in the L-DBN model.

Let $\{G_{ij}^l\}_{1 \leq i,j \leq n, 1 \leq l \leq L}$ be a random binary vector representing the presence or absence of each type of edge from i to j: $G_{ij}^l = 1$ if $i \in Par^l(j, \mathcal{LG}_\rightarrow)$ and 0 otherwise. A realization of $\{G_{ij}^l\}_{1 \leq i,j \leq n, 1 \leq l \leq L}$ defines the labeled graph $\mathcal{LG}_\rightarrow$ of a L-DBN. Without prior information, the variables G_{ij}^l could be modeled as independent variables with uniform distribution. Instead, we assume that the vertices of the static representation of $\mathcal{LG}_\rightarrow$ are organized into B disjoints blocks, or communities, and block membership is indicated by a function $f : \{1, \ldots, n\} \rightarrow \{1, \ldots, B\}$. In the example of Fig. 1, there are three such blocks: $\{1\}, \{2\}, \{3, 4\}$. Then, we model the distribution of the $\{G_{ij}^l\}$s in the Stochastic Block Model (SBM) framework [15]. The SBM model makes only two assumptions: (1) the presence of an edge with label l from vertex i towards j (variable $\{G_{ij}^l\}$) is independent of the presence of an edge of the same label from vertex u towards v (variable $\{G_{uv}^l\}$), $\forall(i, j, u, v)$ and (2), the probability distribution of G_{ij}^l only depends on l, $f(i)$ and $f(j)$ (and not on the specific vertices i and j directly). Therefore, in the case of two labels $(L = 2)$ the joint distribution of the $\{G_{ij}^l\}_{1 \leq i,j \leq n, 1 \leq l \leq L}$ is fully determined by the two probabilities $P(G_{ij}^1 = 1 \mid f(i), f(j))$ and $P(G_{ij}^2 = 1 \mid f(i), f(j), G_{ij}^1)$. We will assume that these probabilities are parameterized by parameter ψ. Note that, unlike in most applications of SBM, we assume here that the block memberships are known while the edges are unknown and modeled by random variables G_{ij}^l. This is because our objective is to learn the network from the blocks and observations X_i^t, instead of learning the blocks from the network, as usual.

3   A Restoration-Estimation Procedure for L-DBN Structure Learning

L-DBN parameters and structure learning poses several difficulties. In the non-parameterized DBN learning case, when the structure of the DBN is known, analytic expressions for the estimators of the transition probabilities from counts on data are available [20]. An analytic expression of the solution for likelihood maximization is not available anymore in L-DBN since the tables are no longer independent. So we will have to rely on numerical solvers. Structure learning steps (for given model parameters) must also be handled differently: First, not only edges presence but also their labels must be learned. Then, usual score functions combine a term measuring how a network fits a dataset and a penalty term on the model complexity [26], to avoid over-fitting that occurs when increasing the number of edges in the learnt network. Such penalties are no longer relevant for a L-DBN, where the number of parameters is fixed and does not vary with $\mathcal{LG}_\rightarrow$: the model complexity does not increase with the number of edges. The BDe score [13] is not relevant either, due to the assumption of independence between parameters in the different tables, which does not hold in a L-DBN.

Therefore, we propose to maximize the non-penalized log-likelihood. Our restoration-estimation algorithm is an iterative algorithm which alternatively updates estimates of $\mathcal{LG}_\rightarrow^k$ and (θ^k, ψ^k) until a local maximum of $P(D, \mathcal{LG}_\rightarrow \mid \theta, \psi)$ is found, that is:

$$E \ step: \theta^{k+1}, \psi^{k+1} \leftarrow \arg\max_{\theta,\psi} \log P(D, \mathcal{LG}^k_\rightarrow | \theta, \psi),$$

$$R \ step: \mathcal{LG}^{k+1}_\rightarrow \leftarrow \arg\max_{\mathcal{LG}_\rightarrow} \log P(D, \mathcal{LG}_\rightarrow | \theta^{k+1}, \psi^{k+1}).$$

These two iterative steps can be rewritten as follows:

$$E: \theta^{k+1} \leftarrow \arg\max_{\theta} \log P(D|\mathcal{LG}^k_\rightarrow, \theta),$$

$$\psi^{k+1} \leftarrow \arg\max_{\psi} \log P(\mathcal{LG}^k_\rightarrow | \psi), \tag{1}$$

$$R: \mathcal{LG}^{k+1}_\rightarrow \leftarrow \arg\max_{\mathcal{LG}_\rightarrow}[\log P(D|\mathcal{LG}_\rightarrow, \theta^{k+1}) + \log P(\mathcal{LG}_\rightarrow | \psi^{k+1})]. \tag{2}$$

In the first step, given a fixed labeled graph $\mathcal{LG}^k_\rightarrow$, both the parameters θ^{k+1} of the L-DBN and the parameters ψ^{k+1} of the SBM are estimated by continuous optimization. In the second step, given fixed parameters values, the labeled graph $\mathcal{LG}^{k+1}_\rightarrow$ is updated by solving a 0-1 Integer Linear Program (ILP) [16]. In practice, since the log-likelihood is a decomposable score, it amounts to solving n 0/1 ILPs by defining for each vertex as many variables (i.e. an exponential number in k) as potential parents sets [5]. However, the structure of real problems often allows to decrease the number of introduced variables in the 0/1 ILP. In the next section, we illustrate this by instanciating the RE algorithm on a problem of ecological interaction network learning, in which the number of variables will only be quadratic in k.

4   Ecological Network Modeling

An ecological network describes interactions between species in a given environment. The learning problem is that of learning this network from time series of observations of the species. Interactions can be trophic (prey/predator), parasitic, competitive, ... They can model positive or negative influence on the species survival. It is therefore possible to label the edges of an ecological network with a '+' or a '−' label (absence of interaction is represented by an absence of edge). In practice, the main interactions between species are trophic interactions. They structure the community into trophic levels that are often known [21]. We now show how to take these labels and trophic levels into account to model species dynamics in the L-DBN framework with SBM prior.

4.1   L-DBN Species Transition Probabilities

We assume that the species dynamics are observed at regular time steps, and that occurrence observations are available: dataset D corresponds to the observation of the absence $(x_i^t = 0)$[1] or presence $(x_i^t = 1)$ of every species at time $t \in \{1, \ldots, T\}$. Information is also available on whether the observed area is

---

[1] Here and in the following, upper case letters are used for random variables, and lower case letters for a realization.

protected ($a^t = 1$) or not ($a^t = 0$). Labels of the edges in the associated graph $\mathcal{LG}_\rightarrow$ can take 2 values: '+' or '−'. An example of labeled ecological network with four species is shown in Fig. 1 (right). Then, the definition of the TPT is based on the following assumptions:

(a) a species survives if at least one positive influence succeeds and all negative fail;
(b) a species with empty $Par^+(i, \mathcal{LG}_\rightarrow)$ (for instance a species at the bottom of the trophic chain) cannot disappear if it is protected and all the species in $Par^-(i, \mathcal{LG}_\rightarrow)$ are absent;
(c) a species with non empty $Par^+(i, \mathcal{LG}_\rightarrow)$ cannot survive if all species are absent in $Par^+(i, \mathcal{LG}_\rightarrow)$;
(d) If $i \in Par^+(j, \mathcal{LG}_\rightarrow)$, then $i \notin Par^-(j, \mathcal{LG}_\rightarrow)$.

These assumptions form "hard" knowledge, which limits the set of possible ecological interaction networks, for a given observed dataset D. Then, the TPT of the L-DBN are defined from the vector of parameters $\theta = (\varepsilon, \rho^+, \rho^-, \mu)$, where ε is a probability of recolonization, ρ^+, ρ^-, are probabilities of success of positive and negative influences. $\mu \in [0, 1]$ is a penalty factor applied to recolonization and survival probabilities of species when the area is unprotected. We describe only the transition probabilities towards presence $P(X_i^{t+1} = 1 | X_i^t, a^t)$. All other transition probabilities are derived from these. Two situations are possible depending on whether species i is absent or present at time t:

(i) the probability for a species absent at t to colonize the observed area at $t+1$ is assumed fixed and independent of the presence of other species

$$P(X_i^{t+1} = 1 | X_i^t = 0, a^t) = \mu^{(1-a^t)}\varepsilon. \tag{3}$$

(ii) The probability for a species present at t to survive at $t+1$ is the probability of success of at least one positive influence (if needed) and the probability of failure of all negative influences, and these interaction events are independent. It is expressed as follows: if $Par^+(i, \mathcal{LG}_\rightarrow) = \emptyset$

$$P(X_i^{t+1} = 1 | X_i^t = 1, x_{Par^-(i,\mathcal{LG}_\rightarrow)\backslash i}^t, a^t) = \mu^{(1-a^t)}\left(1 - \rho^-\right)^{N_{i,-}^t}. \tag{4}$$

Else it is equal to $\mu^{(1-a^t)}\left(1 - \left(1 - \rho^+\right)^{N_{i,+}^t}\right)\left(1 - \rho^-\right)^{N_{i,-}^t}.$ (5)

When the area is unprotected ($a^t = 0$), the transition probabilities (3), (4) and (5) depend on parameter μ, to account for the loss in recolonization/survival probability.

4.2 SBM Model of the Prior on Ecological Links

The (known) trophic level of species i is denoted $TL(i)^2$. By convention, top predators have the largest trophic level, while basal species have trophic level 0.

---

[2] Trophic levels are represented in Fig. 1, right: $TL(1) = 0$, $TL(2) = 1$, $TL(3) = TL(4) = 2$.

Species feed on species in lower trophic levels. So, it is more likely that there is a '+' edge from i to j if $TL(j) > TL(i)$, assuming that most '+' edges model a trophic relation. We will assume here that all positive influences are prey-to-predator ones and that, furthermore, the closer the trophic levels, the more likely i is a prey of j. This a priori knowledge can be modeled by the following SBM, where the blocks are the trophic levels and the block membership function $f(i)$ is defined by $TL(i)$:

$$P\left(G_{ij}^+ = 1\right) = 0 \text{ if } TL(i) \geq TL(j) \text{ and } \frac{e^{\alpha \Delta_{ij}}}{1 + e^{\alpha \Delta_{ij}}} \text{ if } TL(i) < TL(j).$$

where $\Delta_{ij} = TL(i) - TL(j)$ and $\alpha > 0$.

Negative influences represent different phenomena (negative influence of the predator on its prey, but also parasitism, competition...). We consider a simple probability model for negative influences, only taking into account the relative position of trophic levels.

$$\text{If } TL(i) < TL(j), P\left(G_{ij}^- = 1 \mid G_{ij}^+ = 1\right) = 0 \text{ and } P\left(G_{ij}^- = 1 \mid G_{ij}^+ = 0\right) = \beta_2,$$
$$\text{If } TL(i) = TL(j), \qquad P\left(G_{ij}^- = 1\right) = \beta_2,$$
$$\text{If } TL(i) > TL(j), \qquad P\left(G_{ij}^- = 1\right) = \beta_1.$$

with $\beta_1 > \beta_2$, to represent the fact that predator-to-prey influences are the most frequent negative influences.

The vector $\psi = (\alpha, \beta_1, \beta_2)$ defines the prior on $\mathcal{LG}_\rightarrow$.

5   Ecological Network Learning Algorithm

In this section, we derive a version of the generic Restoration/Estimation algorithm of Sect. 3, specific to the L-DBN model of ecological network.

5.1   Expression of $\log P(D|\mathcal{LG}_\rightarrow, \theta)$

To express the data log-likelihood, we distinguish the basal species (nb species) that have non-empty $Par^+(i, \mathcal{LG}_\rightarrow)$, from the other ones (the basal species b, which have no prey). We also define the quantity $R_{i,C}^{t,d^+,d^-}$ equal to 1 if the species i is of class $C \in \{nb, b\}$ and at time t, $N_{i,+}^t = d^+$, $N_{i,-}^t = d^-$ and 0 otherwise. By convention, for a species of type b, we set $N_{i,+}^t = 0$. We also assume that the maximum overall number of incoming edges of any node i is fixed, equal to k. The log-likelihood of a dataset $D = \{x^1, \ldots, x^T\}$, for a given initial state x^0, can be computed as[3]:

$$\log P(D|\mathcal{LG}_\rightarrow, \theta) = \log P(x^1, \ldots, x^T \mid x^0, a, \theta, \mathcal{LG}_\rightarrow) = \sum_{i=1}^n score(i),$$

---

[3] $P_{\mathcal{LG}_\rightarrow, \theta}(x^0)$ will not be estimated.

where $score(i)$ is the contribution of species i to the log-likelihood:

$$score(i) = \sum_{t=0}^{T-1}(1 - x_i^t)\log(P_0^t(x_i^{t+1})) + \sum_{t=0}^{T-1} x_i^t \sum_{0 \leq d^+ + d^- \leq k} \log\left(P_{1,+}^{t,d^+,d^-}(x_i^{t+1})\right) R_{i,nb}^{t,d^+,d^-}$$

$$+ \sum_{t=0}^{T-1} x_i^t \sum_{d^-=0}^{k} \log\left(P_{1,b}^{t,0,d^-}(x_i^{t+1})\right) R_{i,b}^{t,0,d^-} \tag{6}$$

At time t, there is only one term among the three which is non-zero: either the one corresponding to the probability of transition from $x_i^t = 0$ to x_i^{t+1} ($P_0^t(x_i^{t+1})$) or from $x_i^t = 1$ to x_i^{t+1} for non-basal species ($P_{1,+}^{t,d^+,d^-}(x_i^{t+1})$) or from $x_i^t = 1$ to x_i^{t+1} for basal species ($P_{1,b}^{t,0,d^-}(x_i^{t+1})$). The probabilities in Eq. (6) are defined by Eqs. (7) and (8) :

$$\log\left(P_0^t(x_i^{t+1})\right) = x_i^{t+1} a^t \log \epsilon + (1 - x_i^{t+1}) a^t \log(1 - \epsilon)$$
$$+ x_i^{t+1}(1 - a^t)\log(\mu\epsilon) + (1 - x_i^{t+1})(1 - a^t)\log(1 - \mu\epsilon). \tag{7}$$

$$\log\left(P_{1,C}^{t,d^+,d^-}(x_i^{t+1})\right) = x_i^{t+1} a^t \log\left(P_{1\to1}^{1C}(d^+,d^-)\right) + (1 - x_i^{t+1}) a^t \log\left(P_{1\to0}^{1C}(d^+,d^-)\right)$$
$$+ x_i^{t+1}(1 - a^t)\log\left(P_{1\to1}^{0C}(d^+,d^-)\right)$$
$$+ (1 - x_i^{t+1})(1 - a^t)\log\left(P_{1\to0}^{0C}(d^+,d^-)\right), \tag{8}$$

where $P_{1\to x_i^{t+1}}^{a^tC}(d^+,d^-)$ is the probability to transition from $x_i^t = 1$ to x_i^{t+1} for species i of type C under action a^t, when it has d^+ favorable and d^- unfavorable species extant. Those probabilities are described in (4) and (5). Note that these expressions are linear functions of the variables $\{R_{i,C}^{t,d^+,d^-}\}$, given the data $\{x_i^t\}$, $\{a^t\}$ and parameters $(\varepsilon, \rho^+, \rho^-, \mu)$ of the model.

5.2   Restoration Step

Let us focus first on the graph update phase (2). If we ignore the SBM part for the moment, the maximization of the first term in (2) can be decomposed into n independent maximization problems (one per $score(i)$). Each maximization problem can be expressed as a 0-1 ILP by introducing auxiliary variables. The auxiliary variables and the linear constraints are provided in the appendix. The SBM term in expression (2) is also decomposable: $\log P(\mathcal{LG}_\to|\psi) = \sum_j score^{SBM}(j)$. The function $score^{SBM}$ writes (provided $\mathcal{LG}_\to$ only contains edges which are consistent with the SBM):

$$score^{SBM}(j) = \sum_{i,\Delta_{ij}=0} g_{ij}^- \log\beta_2 + (1 - g_{ij}^-)\log(1 - \beta_2)$$

$$+ \sum_{i,\Delta_{ij}<0} \alpha\Delta_{ij}g_{ij}^+ - \log(1 + \exp^{\alpha\Delta_{ij}}) + (1 - g_{ij}^+)(g_{ij}^- \log\beta_2 + (1 - g_{ij}^-)\log(1 - \beta_2))$$

$$+ \sum_{i,\Delta_{ij}>0} g_{ij}^- \log\beta_1 + (1 - g_{ij}^-)\log(1 - \beta_1).$$

This expression is not linear in the variables $\{g_{ij}^l\}$. We linearize it by adding an extra variable g_{ij}^{+-} equal to 1 if $g_{ij}^+ = 1$ and $g_{ij}^- = 1$ and 0 otherwise. So doing, the network optimization step (with or without SBM prior) can be performed by solving n independent 0-1 integer linear programs.

5.3   Parameters Estimation Step

Recall that in the parameters update phase (1), parameters vectors θ^{k+1} and ψ^{k+1} can be updated separately: The update of θ is performed using the interior point method for non-linear programming [2]. For β_1 and β_2 the solution of the update is analytic:

$$\beta_1^{k+1} = \frac{\sum_{(i,j),\Delta_{ij}>0} g_{ij}^-}{|\{(i,j),\Delta_{ij}>0\}|},$$

$$\beta_2^{k+2} = \frac{\sum_{(i,j),\Delta_{ij}\leq 0} g_{ij}^-(1-g_{ij}^+)}{\sum_{(i,j),\Delta_{ij}\leq 0}(1-g_{ij}^+)}.$$

The updated α is obtained as a (numerical) solution of the moment-matching equation:

$$\sum_{(i,j),\Delta_{ij}<0} \Delta_{ij} g_{ij}^+ = \sum_{(i,j),\Delta_{ij}<0} \Delta_{ij}\frac{\exp^{\alpha\Delta_{ij}}}{1+\exp^{\alpha\Delta_{ij}}}.$$

6   Experiments

We considered ecological network learning in situations where the sample size is small and we compared the behavior of 4 DBN learning methods corresponding to different levels of embedded a priori knowledge. First the Restoration-Estimation algorithm of Sect. 4 was applied to the L-DBN model of species dynamics (1) without additional knowledge (L-DBN-OK), (2) with a SBM prior (L-DBM-SBM), and (3) with 20% of variables G_{ij}^l known[4] and no SBM prior (L-DBM-20K). The restoration step was solved using the CPLEX solver. We also applied MIT [27] which optimizes a mutual information test score and works with a full (non-parameterized) representation of the TPT. For comparison purposes, we have enriched MIT with an edge-labeling method using the notion of qualitative influence from [28] In qualitative influence, positive and negative influences of a binary variable Y on a binary variable X is defined as follows:

$$Y \xrightarrow{+} X \ iff \ P(X=1|Y=1,Z) \geq P(X=1|Y=0,Z), \forall Z,$$

$$Y \xrightarrow{-} X \ iff \ P(X=0|Y=1,Z) \geq P(X=0|Y=0,Z), \forall Z,$$

where Z is the set of other variables influencing X. Replacing probabilities with data counts, we used these definitions to (partially) label the structure learned

---

[4] Known variables were selected uniformly at random.

by MIT (links between variables for which counts do not satisfy any of the above conditions remain unlabeled).

Synthetic networks. We have generated ten synthetic networks of 20 species according to a SBM model with $\alpha = 1/\sqrt{20}, \beta_1 = \alpha/2, \beta_2 = \beta_1/2$. For each of these networks, we have generated 10 data sets, a data set corresponding to a simulated trajectory of length 30 of the species dynamics, with no protection action the first 12 years and protection after. Values of the L-DBN parameters $(\epsilon, \mu, \rho^+, \rho^-)$ were all set to 0.8. The RE algorithm was applied to each data set. So we obtained 10 restored graphs for a single synthetic one. We ordered learnt edges by their decreasing occurrence frequency in these 10 restored graphs, and defined the aggregated graph of size x as the restored graph composed of the x first edges in this ordering. Figure 2 shows the joint evolution of the precision and recall of '+' and '−' edges when the number of edges in the aggregated graph changes. Results for MIT are not reported because precision and recall were close to zero, showing the difficulty to learn both a DBN structure and its TPT in a non parameterized model, when data are scarce.

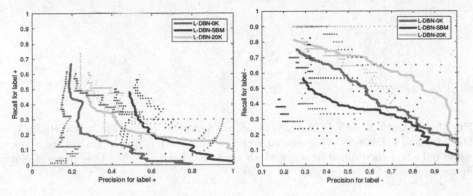

Fig. 2. Synthetic networks. Precision and recall for '+' (left) and '−'(right) edges, for the L-DBN-0K, L-DBN-SBM and L-DBN-20K learning methods. Plain lines are mean values: one dot in these lines is obtained by averaging results for a given value of edges in the aggregated graph, over the 10 data sets. Dotted lines are worse and best cases among the 10 data sets.

We observed that when incorporating a SBM prior in the learning procedure of a L-DBN, fewer edges are learnt. Let us denote by x_{SBM} the maximum number of edges in the aggregated graph built from the 10 L-DBN-SBM restored graphs. When comparing the aggregated graphs with x_{SBM} edges for the different methods, we observed that the one provided by L-DBN-SBM leads to the best precision and recall for '+' edges, and to the best precision and recall when the two labels are not distinguished. Here the SBM knowledge was more helpful than the knowledge of 20% of the edges in the learning process. However, the L-DBN-SBM method was less efficient for learning the '−' edges. This is not

surprising since the prior knowledge embedded in our SBM model is stronger for
'+' edges than for '−' edges (it depends on the TL differences).

Real ecological network. We applied the MIT, L-DBN-0K and L-DBN-SBM
learning methods on data generated with a L-DBN with the same parameter
values as above, but for the real ecological network structure of the Alaskan
food web [7]. This network is composed of 13 species, that can be grouped into
5 trophic levels, and contains 21 '+' edges and no '−' edges (see Fig. 3 top left).
Here also 10 data sets were used to build an aggregated graph. The precision
and recall reached for the aggregated graph composed of all edges learnt at
least once were respectively (0.47, 0.33), (0.26, 0.86) and (0.49, 0.81) for MIT,
L-DBN-0K and L-DBN-SBM. L-DBN-0K and L-DBN-SBM both learn fewer
'−' edges than '+' edges. However, the L-DBN-SBM algorithm provided more
parsimonious graphs (35 edges instead of 85 for L-DBN-0K). Figure 3 (bottom
left and right) illustrate the gain in integrating the SBM prior: for instance,
without SBM knowledge, the information that species do not feed on the same
trophic level can not be recovered from the data alone.

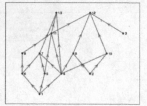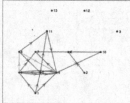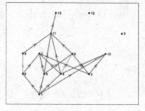

Fig. 3. Alaskan food web. Left: real network, with only '+' edges, Middle and Right: L-
DBN-0K and L-DBN-SBM aggregated graphs with 21 edges. Blue edges are '+' edges,
while red edges correspond to edges which are learnt both as '+' and '−' edges (Color
figure online).

7   Conclusion

We proposed an approach to improve learning of a Dynamic Bayesian Network
(DBN) structure *(without synchronous edges)* when data are scarce. The approach combines the definition of a family of parameterized DBN with labeled
edges, an a priori Stochastic Block Model (SBM) on the DBN structure and
a Restoration-Estimation (RE) learning algorithm. To define a parsimonious
parameterization we make the assumption of identical transition probabilities
tables for all variables submitted to the same number of each possible type
of influence. This is a restrictive but necessary assumption in situations where
there is not enough data to learn more complex models. The proposed modeling
framework enables us to take into account expert knowledge to help the learning. Our experiments show that by limiting the number of parameters describing the DBN, and by introducing community structure knowledge via SBM,

we can improve learning quality compared to a method based on a full non-parameterized representation of the DBN.

The RE algorithm is a greedy iterative two-steps algorithm. It includes a structure improvement step modeled as n 0-1 integer linear programs, one per variable of the DBN. This procedure is generic since the log-likelihood for a DBN can always be decomposed as a linear function of variables describing the graph structure, as in [5], and as soon as additional constraints on these variables are linear, ILP can be applied. Still, for a specific L-DBN, it is worth deriving a specific ILP model, which will require fewer variables, as we examplified on the problem of learning an ecological interaction network from temporal data of presence/absence of species.

In the ecological network application, the L-DBN transition function is merely an extension of a generic contact process model [12] to more than one influence type. It can also be seen as a DBN with a causal independence model [14] for each transition probabilities table, where each parent's influence is either positive or negative as in a qualitative BN model [18,28]. The proposed model may seem simple compared to the complexity on an ecological network. For instance, we assume identical strengths of positive and negative influences for all species and stationarity of the interaction network structure. The model could be straightforwardly extended to more than two labels, in order to relax the first assumption. Stationarity is more critical and cannot be relaxed without modifying the learning algorithm. Still, propagation by contact models are encountered in several other domains such as fire propagation, health management (disease propagation, [25]), social networks (rumor propagation, [11]), computer science (network security). Therefore (probably with an adaptation of the SBM prior model) the L-DBN model for ecological network and the associated RE algorithm could be useful for learning interaction networks in a wide range of applications.

Appendix: Maximization of the Data Log-likelihood as a 0-1 ILP

We describe how the problem of maximizing the log-likelihood $\log P(D|\mathcal{LG}_\rightarrow, \theta^{t+1})$ over the variables g_{ij}^l defining a DBN structure can be expressed as n independent 0-1 ILP. The expression of the likelihood is an expression of the variables $\left\{ R_{i,C}^{t,d^+,d^-} \right\}$ which are themselves functions of the variables g_{ij}^l, as well as of the observed data D. In the following we show how to define the binary variables $\left\{ R_{i,C}^{t,d^+,d^-} \right\}$ from linear constraints involving the binary variables $\{g_{ij}^l\}$, the data D and some other auxiliary binary variables. So doing, we will have defined one 0-1 integer linear program per species, which can be solved by classical solvers. The 0-1 integer linear program for a species i describes the vertices pointing to i that maximize the quantity $score(i)$. The following auxiliary binary variables are defined for a particular species i.

- *Non-basal species*: $\left\{h_i^{nb}\right\}$. $h_i^{nb} = 1$ iff $Par^+(i, \mathcal{LG}_\rightarrow) \neq \emptyset$. These variables are defined for all $i \in \{1,..,n\}$ by the constraints described in (9).

- *Lower bound on the number of extant parents*: $\left\{M_{i,l}^{t,d}\right\}$. $M_{i,l}^{t,d} = 1$ iff $N_{i,l}^{t} \geq d$ (the species i has at least d parents of label type l extant at time t). k is the maximum allowed number of parents of any label. These variables are defined for all $i \in \{1,..,n\}, t \in \{1,..,T\}, d \in \{0,..,k\}, l \in \{+,-\}$ by the constraints in (10).

- *Upper bound on the number of extant parents*: $\left\{\nu_{i,l}^{t,d}\right\}$. $\nu_{i,l}^{t,d} = 1$ iff $N_{i,l}^{t} \leq d$ (the species i has at most d parents of label type l extant at time t). These variables are defined for all $i \in \{1,..,n\}, t \in \{1,..,T\}, d \in \{0,..,k\}, l \in \{+,-\}$ by the constraints in (11).

- *Number of extant parents*: $\left\{\Lambda_{i,l}^{t,d}\right\}$. $\Lambda_{i,l}^{t,d} = 1$ iff $N_{i,l}^{t} = d$ (species i has exactly d parents of label type l extant at time t). These variables are defined for all $i \in \{1,..,n\}, t \in \{1,..,T\}, d \in \{0,..,k\}, l \in \{+,-\}$ by the set of constraints in (12).

Now, we are ready to write the linear constraints defining the binary variables $R_{i,C}^{t,d^+,d^-}$. Recall that $R_{i,C}^{t,d^+,d^-} = 1$ if and only if the species i is of type $C \in \{nb, b\}$ and has exactly d^+ parents of type $+$ and d^- parents of type $-$ extant at time t. Thus, $R_{i,nb}^{t,d^+,d^-} = 1$ iff $h_i^{nb} = \Lambda_{i,+}^{t,d^+} = \Lambda_{i,-}^{t,d^-} = 1$. $R_{i,b}^{t,d^+,d^-} = 1$ iff $h_i^{nb} = 0$, $\Lambda_{i,+}^{t,d^+} = \Lambda_{i,-}^{t,d^-} = 1$.

Variables $\left\{R_{i,nb}^{t,d^+,d^-}\right\}$ are defined by the set of constraints (13–14):

$\forall i,j \in \{1,...,n\}$,

$$h_i^{nb} \leq \sum_{j=1}^{n} g_{ji}^{+} \tag{9}$$

$\forall i \in \{1,..,n\}, t \in \{1,..,T\}, d \in \{0,..,k\}, l \in \{+,-\}$,

$$
\begin{aligned}
M_{i,l}^{t,d} \cdot (d+1) - \sum_{j=1}^{n}\left(g_{ji}^{l} \cdot x_j^t\right) &\leq 1 \\
M_{i,l}^{t,d} \cdot (k+1-d) - \sum_{j=1}^{n}\left(g_{ji}^{l} \cdot x_j^t\right) &> -d
\end{aligned}
\tag{10}
$$

$$
\begin{aligned}
\nu_{i,l}^{t,d} \cdot (k+1-d) + \sum_{j=1}^{n}\left(g_{ji}^{l} \cdot x_j^t\right) &\leq k+1 \\
\nu_{i,l}^{t,d} \cdot (d+1) + \sum_{j=1}^{n}\left(g_{ji}^{l} \cdot x_j^t\right) &> d
\end{aligned}
\tag{11}
$$

$$\Lambda_{i,l}^{t,d} - M_{i,l}^{t,d} \leq 0 \;\; ; \;\; \Lambda_{i,l}^{t,d} - \nu_{i,l}^{t,d} \leq 0 \;\; ; \;\; \Lambda_{i,l}^{t,d} - M_{i,l}^{t,d} - \nu_{i,l}^{t,d} \geq -1 \tag{12}$$

$\forall i \in \{1,..,n\}, t \in \{1,..,T\}, d^+ \in \{0,..,k\}, d^- \in \{0,..,k-d^+\}$,

$$R_{i,b}^{t,d^-} \leq \Lambda_{i,l}^{t,d^-} \;\; ; \;\; R_{i,b}^{t,d^-} \leq 1 - h_i^{nb} \;\; ; \;\; R_{i,b}^{t,d^-} \geq -h_i^{nb} + \Lambda_{i,-}^{t,d^-} \tag{13}$$

$$
\begin{aligned}
R_{i,nb}^{t,d^+,d^-} &\leq \Lambda_{i,l}^{t,d^+} \;\; ; \;\; R_{i,nb}^{t,d^+,d^-} \leq \Lambda_{i,l}^{t,d^-} \;\; ; \;\; R_{i,nb}^{t,d^+,d^-} \leq h_i^{nb} \\
R_{i,nb}^{t,d^+,d^-} &\geq h_i^{nb} + \Lambda_{i,+}^{t,d^+} + \Lambda_{i,-}^{t,d^-} - 2
\end{aligned}
\tag{14}
$$

At this stage, all the variables needed for the calculation of $score(i)$, defined in (6), have been introduced. We incorporate further constraints in the ILP formulation to model hard expert knowledge about the ecological network. The species i has at most k parents (constraint (15)). There exists at most one edge from j to i (constraint (16)). If species i is non-basal, it will become extinct at time $t + 1$ if it has no prey at time t (constraint (17)). If the species is basal, it will remain extant at time $t + 1$ if it is extant at time t and has no negative influence (constraint (18)).

$$\sum_{j=1}^{n} \sum_{l \in \{+,-\}} g_{ij}^l \leq k, \forall i = 1, \ldots, n \tag{15}$$

$$g_{ji}^+ + g_{ji}^- \leq 1, \forall i = 1, \ldots, n; j = 1, \ldots, n \tag{16}$$

$$R_{i,nb}^{t,d^+=0,d^-} \cdot x_i^{t+1} = 0, \forall i, t, d^- \tag{17}$$

$$R_{i,b}^{t,d^+,d^-=0} \cdot x_i^t \cdot (1 - x_i^{t+1}) = 0, \forall i, t, d^+. \tag{18}$$

The problem of finding the ecological network structure which optimizes the log-likelihood is now modeled as a set of n 0-1 integer linear programs, whose variables are all the $\{g_{ij}^l, M_{i,l}^{t,d}, \nu_{i,l}^{t,d}, \Lambda_{i,l}^{t,d}, R_{i,V}^{t,d^+,d^-}\}$ with constraints (9–18).

Note that the total number of variables and constraints of the 0-1 linear program for a species is linear in n and quadratic in k. Thus, the complexity of the graph update phase is "simply" exponential in n.

References

1. Ambroise, C., Chiquet, J., Matias, C.: Inferring sparse Gaussian graphical models with latent structure. Electron. J. Stat. **3**, 205–238 (2009)
2. Byrd, R.H., Hribar, M.E., Nocedal, J.: An interior point algorithm for large-scale nonlinear programming. SIAM J. Optim. **9**(4), 877–900 (1999)
3. de Campos, C.P., Ji, Q.: Efficient structure learning of Bayesian networks using constraints. J. Mach. Learn. Res. **12**, 663–689 (2011)
4. Chickering, D.M.: Learning Bayesian networks is NP-complete. In: Learning from data, pp. 121–130 (1996)
5. Cussens, J.: Bayesian network learning with cutting planes. In: UAI 2011, Proceedings of the Twenty-Seventh Conference on Uncertainty in Artificial Intelligence (2011)
6. Dojer, N.: Learning Bayesian networks does not have to be NP-hard. In: Královič, R., Urzyczyn, P. (eds.) MFCS 2006. LNCS, vol. 4162, pp. 305–314. Springer, Heidelberg (2006). https://doi.org/10.1007/11821069_27
7. Estes, J.A., Doak, D.F., Springer, A.M., Williams, T.M.: Causes and consequences of marine mammal population declines in southwest Alaska: a food-web perspective. Philos. Trans. R. Soc. Lond. B: Biol. Sci. **364**, 1647–1658 (2009)

8. Feelders, A., van der Gaag, L.: Learning Bayesian network parameters under order constraints. Int. J. Approx. Reason. **42**, 37–53 (2006)
9. Franc, A.: Metapopulation dynamics as a contact process on a graph. Ecol. Complex. **1**(1), 49–63 (2004)
10. Ghahramani, Z.: Learning dynamic Bayesian networks. In: Giles, C.L., Gori, M. (eds.) NN 1997. LNCS, vol. 1387, pp. 168–197. Springer, Heidelberg (1998). https://doi.org/10.1007/BFb0053999
11. Gomez Rodriguez, M., Leskovec, J., Krause, A.: Inferring networks of diffusion and influence. In: Proceedings of the 16th ACM SIGKDD International Conference on Knowledge Discovery and Data Mining, pp. 1019–1028. ACM (2010)
12. Harris, T.E.: Contact interactions on a lattice. Ann. Probab. **2**, 969–988 (1974)
13. Heckerman, D., Geiger, D., Chickering, D.M.: Learning Bayesian networks: the combination of knowledge and statistical data. Mach. Learn. **20**(3), 197–243 (1995)
14. Heckerman, D., Breese, J.S.: Causal independence for probability assessment and inference using Bayesian networks. IEEE Trans. Syst. Man Cybern. Part A **26**(6), 826–831 (1996)
15. Holland, P., Laskey, K., Leinhardt, S.: Stochastic blockmodels: first steps. Soc. Netw. **5**(2), 109–137 (1983)
16. Korhonen, J.H., Parviainen, P.: Tractable Bayesian network structure learning with bounded vertex cover number. In: Advances in Neural Information Processing Systems, vol. 28 (2015)
17. Liben-Nowell, D., Kleinberg, J.: The link-prediction problem for social networks. J. Am. Soc. Inf. Sci. Technol. **58**(7), 1019–1031 (2007)
18. Lucas, P.J.F.: Bayesian network modelling through qualitative patterns. Artif. Intell. **163**(2), 233–263 (2005)
19. Milns, I., Beale, C.M., Smith, V.A.: Revealing ecological networks using Bayesian network inference algorithms. Ecology **91**(7), 1892–1899 (2010)
20. Murphy, K.: Dynamic Bayesian Networks: representation, inference and learning. Ph.D. thesis, University of California (2002)
21. Newman, M.E.J., Clauset, A.: Structure and inference in annotated networks. Nat. Commun., 7 (2016)
22. Niculescu, R.S., Mitchell, T.M., Rao, R.B.: Bayesian network learning with parameter constraints. J. Mach. Learn. Res. **7**(July), 1357–1383 (2006)
23. Oyen, D., Anderson, B., Anderson-Cook, C.: Bayesian networks with prior knowledge for malware phylogenetics. In: AAAI-2016 Workshop on Artificial Intelligence and Cybersecurity (2016)
24. Parviainen, P., Farahani, H.S., Lagergren, J.: Learning bounded tree-width Bayesian networks using integer linear programming. In: Seventeenth International Conference on Articial Intelligence and Statistics (AISTAT 2014) (2014)
25. Salathé, M., Jones, J.H.: Dynamics and control of diseases in networks with community structure. PLOS Comput. Biol. **6**(4), e1000736 (2010)
26. Schwarz, G.: Estimating the dimension of a model. Ann. Stat. **6**(2), 461–464 (1978)
27. Vinh, N.X., Chetty, M., Coppel, R., Wangikar, P.P.: Polynomial time algorithm for learning globally optimal dynamic Bayesian network. In: Lu, B.-L., Zhang, L., Kwok, J. (eds.) ICONIP 2011. LNCS, vol. 7064, pp. 719–729. Springer, Heidelberg (2011). https://doi.org/10.1007/978-3-642-24965-5_81
28. Wellman, M.P.: Fundamental concepts of qualitative probabilistic networks. Artif. Intell. **44**, 257–303 (1990)
29. Yu, J., Smith, V.A., Wang, P.P., Hartemink, A.J., Jarvis, E.D.: Advances to Bayesian network inference for generating causal networks from observational biological data. Bioinformatics **20**(18), 3594–3603 (2004)

Multi-view Generative Adversarial Networks

Mickaël Chen[✉] and Ludovic Denoyer

Sorbonne Universités, UPMC Univ Paris 06, UMR 7606, LIP6, 75005 Paris, France
{mickael.chen,ludovic.denoyer}@lip6.fr

Abstract. Learning over multi-view data is a challenging problem with strong practical applications. Most related studies focus on the classification point of view and assume that all the views are available at any time. We consider an extension of this framework in two directions. First, based on the BiGAN model, the Multi-view BiGAN (MV-BiGAN) is able to perform density estimation from multi-view inputs. Second, it can deal with missing views and is able to update its prediction when additional views are provided. We illustrate these properties on a set of experiments over different datasets.

1 Introduction

Many concrete applications involve multiple sources of information generating different views on the same object [4]. If we consider human activities for example, GPS values from a mobile phone, navigation traces over the Internet, or even photos published on social networks are different views on a particular user. In multimedia applications, views can correspond to different modalities [2] such as sounds, images, videos, sequences of previous frames, etc.

The problem of multi-view machine learning has been extensively studied during the last decade, mainly from the classification point of view. In that case, one wants to predict an output y based on multiple views acquired on an unknown object x. Different strategies have been explored but a general common idea is based on the (early or late) fusion of the different views at a particular level of a deep architecture [17,24,29].

The existing literature mainly explores problems where outputs are chosen in a discrete set (e.g. categorization), and where all the views are available. An extension of this problem is to consider the density estimation problem where one wants to estimate the conditional probabilities of the outputs given the available views. As noted by [15], minimizing classical prediction losses (e.g. Mean square error) will not capture the different output distribution modalities.

In this article, we propose a new model able to estimate a distribution over the possible outputs given any subset of views on a particular input. This model is based on the (Bidirectional) *Generative Adversarial Networks* (BiGAN) formalism. More precisely, we bring two main contributions: first, we propose the CV-BiGAN (*Conditional Views BiGAN* – Sect. 3) architecture that allows one to model a conditional distribution $P(y|.)$ in an original way. Second, on top of this architecture, we build the Multi-view BiGANs (MV-BiGAN – Sect. 4) which

M. Ceci et al. (Eds.): ECML PKDD 2017, Part II, LNAI 10535, pp. 175–188, 2017.
https://doi.org/10.1007/978-3-319-71246-8_11

is able to both **predict when only one or few views are available**, and to **update its prediction if new views are added**. We evaluate this model on different multi-views problems and different datasets (Sect. 5). The related work is provided in Sect. 6 and we propose some future research directions in Sect. 7.

2 Background and General Idea

2.1 Notations and Task

Let us denote $\mathcal{X}$ the space of objects on which different views will be acquired. Each possible input $x \in \mathcal{X}$ is associated to a target prediction $y \in \mathbb{R}^n$. A classical machine learning problem is to estimate $P(y|x)$ based on the training set. But we consider instead a multi-view problem in which different views on x are available, x being unknown. Let us denote V the number of possible views and $\tilde{x}_k$ the k-th view over x. The description space for view k is $\mathbb{R}^{n_k}$ where n_k is the number of features in view k. Moreover, we consider that some of the V views can be missing. The subset of available views for input x^i will be represented by an index vector $s^i \in \mathcal{S} = \{0,1\}^V$ so that $s_k^i = 1$ if view k is available and $s_k^i = 0$ elsewhere. Note that all the V views will not be available for each input x, and the prediction model must be able to predict an output given any subset of views $s \in \{0;1\}^V$.

In this configuration, our objective is to estimate the distributions $p(y|v(s,x))$ where $v(s,x)$ is the set of views $\tilde{x}_k$ so that $s_k = 1$. This distribution p will be estimated based on a training set $\mathcal{D}$ of N training examples. Each example is composed of a subset of views $s^i, v(s^i, x^i)$ associated to an output y^i, so that $\mathcal{D} = \{(y^1, s^1, v(s^1, x^1)), ..., (y^N, s^N, v(s^N, x^N))\}$ where s^i is the index vector of views available for x^i. Note that x^i is not directly known in the training set but only observed through its associated views.

2.2 Bidirectional Generative Adversarial Nets (BiGAN)

We quickly remind the principle of BiGANs since our model is an extension of this technique. Generative Adversarial Networks (GAN) have been introduced by [10] and have demonstrated their ability to model complex distributions. They have been used to produce compelling natural images from a simple latent distribution [6,19]. Exploring the latent space has uncovered interesting, meaningful patterns in the resulting outputs. However, GANs lack the ability to retrieve a latent representation given an output, missing out an opportunity to exploit the learned manifold. Bidirectional Generative Adversarial Networks (BiGANs) have been proposed by [7,8], independently, to fill that gap. BiGANs simultaneously learn both an encoder function E that models the encoding process $P_E(z|y)$ from the space $\mathbb{R}^n$ to a latent space $\mathbb{R}^Z$, and a generator function G that models the mapping distribution $P_G(y|z)$ of any latent point $z \in \mathbb{R}^Z$ to a possible object $y \in \mathbb{R}^n$. From both the encoder distribution and the generator distribution, we can model two joint distributions, respectively denoted $P_E(y,z)$ and $P_G(y,z)$:

$$P_G(y, z) = P(z)P_G(y|z)$$
$$P_E(y, z) = P(y)P_E(z|y) \tag{1}$$

assuming that $P(z) = \mathcal{N}(0, 1)$ and $P(y)$ can be estimated over the training set by a uniform sampling. The BiGAN framework also introduces a discriminator network D_1 whose task is to determine whether a pair (y, z) is sampled from $p_G(y, z)$ or from $p_E(y, z)$, while E and G are trained to fool D_1, resulting in the following learning problem:

$$\min_{G,E} \max_{D_1} \mathbb{E}_{y \sim P(y), z \sim P_E(z|y)} \left[\log D_1(y, z) \right]$$
$$+ \mathbb{E}_{z \sim P(z), y \sim P_G(y|z)} \left[1 - \log D_1(y, z) \right] \tag{2}$$

It can be shown, by following the same steps as in [10], that the optimization problem described in Eq. 2 minimizes the Jensen-Shanon divergence between $P_E(y, z)$ and $P_G(y, z)$, allowing the model to learn both a decoder and a generator over a training set that will model the joint distribution of (y, z) pairs. As proposed by [8], we consider in the following that $P_G(y|z)$ is modeled by a deterministic non-linear model G so that $G(z) - y$, and P_E as a diagonal Gaussian distribution $E(z) = (\mu(y), \sigma(y))$. G, μ and σ are estimated by using gradient-based descent techniques.

2.3   General Idea

We propose a model based on the Generative Adversarial Networks paradigm adapted to the multi-view prediction problem. Our model is based on two different principles:

Conditional Views BiGANs (CV-BiGAN): First, since one wants to model an output distribution based on observations, our first contribution is to propose an adaptation of BiGANs to model conditional probabilities, resulting in a model able to learn $P(y|\tilde{x})$ where $\tilde{x}$ can be either a single view or an aggregation of multiple views. If conditional GANs have already been proposed in the literature (see Sect. 6) they are not adapted to our problem which require explicit mappings between input space to latent space, and from latent space to output space.

Multi-View BiGANs (MV-BiGAN): On top of the CV-BiGAN model, we build a multi-view model able to estimate the distribution of possible outputs based on any subset of views $v(s, x)$. If a natural way to extend the Conditional BiGANS for handling multi-view is to define a mapping function which map the set of views to a representation space (see Sect. 4.1) the resulting model has shown undesirable behaviors (see Sect. 5.1). Therefore, we propose to constrain the model based on the idea that adding one more view to any subset of views must decrease the uncertainty on the output distribution i.e. the more views are provided, the less variance the output distribution has. This behavior is encouraged by using a Kullback-Leibler divergence (KL) regularization (see Sect. 4.2).

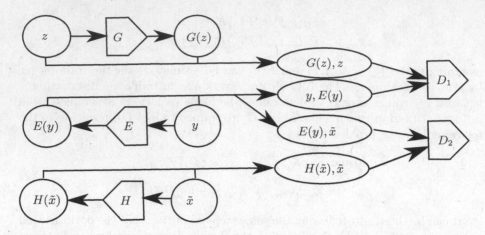

Fig. 1. The CV-BiGAN architecture. The two top levels correspond to the BiGAN model, while the third level is added to model the distribution over the latent space given the input of the CV-BiGAN. The discriminator D_2 is used to constraint $P(z|y)$ and $P(z|\tilde{x})$ to be as close as possible.

3   The Conditional BiGAN Model (CV-BiGAN)

Our first objective is to extend the BiGAN formalism to handle an input space (e.g. a single observed view) in addition to the output space $\mathbb{R}^n$. We will denote $\tilde{x}$ the observation and y the output to predict. In other words, we wish to capture the conditional probability $P(y|\tilde{x})$ from a given training dataset. Assuming one possesses a bidirectional mapping between the input space and an associated representation space, i.e. $P_E(z|y)$ and $P_G(y|z)$, one can equivalently capture $P(z|\tilde{x})$. The CV-BiGAN model keeps the encoder E and generator G defined previously but also includes an additional encoder function denoted H which goal is to map a value $\tilde{x}$ to the latent space $\mathbb{R}^Z$. Applying H on any value of $\tilde{x}$ results in a distribution $P_H(z|\tilde{x}) = \mathcal{N}(\mu_H(\tilde{x}), \sigma_H(\tilde{x}))$ so that a value of z can be sampled from this distribution. This would then allow one to recover a distribution $P(y|\tilde{x})$.

Given a pair $(\tilde{x}, y)$, we wish a latent representation z sampled from $P_H(z|\tilde{x})$ to be similar to one from $P_E(z|y)$. As our goal here is to learn $P(z|\tilde{x})$, we define two joint distributions between $\tilde{x}$ and z:

$$P_H(\tilde{x}, z) = P_H(z|\tilde{x})P(\tilde{x})$$
$$P_E(\tilde{x}, z) = \sum_y P_E(z|y)P(\tilde{x}, y) \tag{3}$$

Minimizing the Jensen-Shanon divergence between these two distributions is equivalent to solving the following adversarial problem:

$$\min_{E,H} \max_{D_2} \mathbb{E}_{\tilde{x},y \sim p(\tilde{x},y), z \sim p_E(z|y)} \left[\log D_2(\tilde{x}, z) \right]$$
$$+ \mathbb{E}_{\tilde{x},y \sim p(\tilde{x},y), z \sim p_H(z|x)} \left[1 - \log D_2(\tilde{x}, z) \right] \tag{4}$$

Note that when applying stochastic gradient-based descent techniques over this objective function, the probability $P(\tilde{x}, y)$ is approximated by sampling uniformly from the training set. We can sample from $P_H(\tilde{x}, z)$ and $P_E(\tilde{x}, z)$ by forwarding the pair $(\tilde{x}, y)$ into the corresponding network.

By merging the two objective functions defined in Eqs. 2 and 4, the final learning problem for our Conditionnal BiGANs is defined as:

$$\min_{G,E,H} \max_{D_1,D_2} \mathbb{E}_{\tilde{x},y \sim P(\tilde{x},y), z \sim P_E(z|y)} \left[\log D_1(y, z) \right]$$
$$+ \mathbb{E}_{z \sim P(z), y \sim P_G(y|z)} \left[1 - \log D_1(y, z) \right]$$
$$+ \mathbb{E}_{\tilde{x},y \sim P(\tilde{x},y), z \sim p_E(z|y)} \left[\log D_2(\tilde{x}, z) \right] \tag{5}$$
$$+ \mathbb{E}_{\tilde{x},y \sim P(\tilde{x},y), z \sim P_H(z|\tilde{x})} \left[1 - \log D_2(\tilde{x}, z) \right]$$

The general idea of CV-BiGAN is illustrated in Fig. 1.

4   Multi-View BiGAN

4.1   Aggregating Multi-views for CV-BiGAN

We now consider the problem of computing an output distribution conditioned by multiple different views. In that case, we can use the CV-BiGAN Model (or other conditional approaches) conjointly with a model able to aggregate the different views where A is the size of the aggregation space. Instead of considering the input $\tilde{x}$, we define an aggregation model Ψ. $\Psi(v(s, x))$ will be the representation of the aggregation of all the available views $\tilde{x}_k$[1]:

$$\Psi(v(s,x)) = \sum_{k=1}^{V} s_k \phi_k(\tilde{x}^k) \tag{6}$$

where ϕ_k is a function that will be learned that maps a particular view in $\mathbb{R}^{n_k}$ to the aggregation space $\mathbb{R}^A$. By replacing $\tilde{x}$ in Eq. 5, one can then simultaneously learn the functions ϕ_k and the distributions P_H, P_E and P_D, resulting in a multi-view model able to deal with any subset of views (Fig. 2).

4.2   Uncertainty Reduction Assumption

However, the previous idea suffers from a very high instability when learning, as it is usually noted with complex GANs architectures (see Sect. 5). In order to stabilize our model, we propose to add a regularization based on the idea that adding new views to an existing subset of views should reduce the uncertainty over the outputs. Indeed, under the assumption that views are consistent one another, adding a new view should allow to refine the predictions and reduce the variance of the distribution of the outputs.

---

[1]   Note that other aggregation scheme can be used like recurrent neural networks for example.

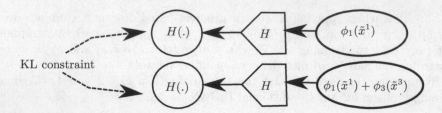

Fig. 2. The MV-BiGAN additional components. In this example, we consider a case where only $\tilde{x}^1$ is available (top level) and a second case where both $\tilde{x}^1$ and $\tilde{x}^3$ are available. The distribution $P(z|\tilde{x}^1, \tilde{x}^3)$ is encouraged to be "included" in $P(z|\tilde{x}^1)$ by the KL constraint. The aggregation of the views is made by the ϕ_k functions that are learned conjointly with the rest of the model.

Let us consider an object x and two index vectors s and s' such that $v(x,s) \subset v(x,s')$ i.e. $\forall k, s'_k \geq s_k$. Then, intuitively, $P(x|v(x,s'))$ should be "included" in $P(x|v(x,s))$. In the CV-GAN model, since $P(y|z)$ is deterministic, this can be enforced at a latent level by minimizing $KL(P(z|v(x,s'))||P(z|v(x,s)))$. By assuming those two distributions are diagonal gaussian distributions (i.e. $P(z|v(x,s')) = \mathcal{N}(\mu_1, \Sigma_1)$ and $P(z|v(x,s) = \mathcal{N}(\mu_2, \Sigma_2)$ where Σ_k are diagonal matrices with diagonal elements $\sigma_{k(i)}$), the KL divergence can be computed as in Eq. 7 and differentiated.

$$KL(P(z|v(x,s'))||P(z|v(x,s))) =$$
$$\frac{1}{2}\sum_{i=1}^{Z}\left(-1 - \log\left(\frac{\sigma_{1(i)}^2}{\sigma_{2(i)}^2}\right) + \frac{\sigma_{1(i)}^2}{\sigma_{2(i)}^2} + \frac{(\mu_{1(i)} - \mu_{2(i)})^2}{\sigma_{2(i)}^2}\right) \quad (7)$$

Note that this divergence is written on the estimation made by the function H and will act as a regularization over the latent conditional distribution.

The final objective function of the MV-BiGAN can be written as:

$$\begin{aligned}
\min_{G,E,H}\max_{D_1,D_2} & \; \mathbb{E}_{s,x,y\sim P(s,x,y),z\sim P_E(z|y)}\left[\log D_1(y,z)\right] \\
& + \mathbb{E}_{z\sim P(z),y\sim P_G(y|z)}\left[1 - \log D_1(y,z)\right] \\
& + \mathbb{E}_{s,x,y\sim P(s,x,y),z\sim P_E(z|y)}\left[\log D_2(v(x,s),z)\right] \\
& + \mathbb{E}_{s,x,y\sim P(s,x,y),z\sim P_H(z|v(x,s))}\left[1 - \log D_2(v(x,s),z)\right] \\
& + \lambda\mathbb{E}_{x\sim P(x)}\sum_{\substack{s,s'\in\mathcal{S}_x \\ \forall k, s'_k \geq s_k}}KL(H(v(x,s'))||H(v(x,s)))
\end{aligned} \quad (8)$$

where λ controls the strength of the regularization. Note that aggregation models Ψ are included into H and D_2 and can be optimized conjointly in this objective function.

4.3   Learning the MV-BiGAN

The different functions E, G, H, D_1 and D_2 are implemented as parametric neural networks and trained by mini-batch stochastic gradient descent (see Sect. 5.4 for more details concerning the architectures).We first update the discriminators networks D_1 and D_2, then we update the generator and encoders G, E and H with gradient steps in the opposite direction.

As with most other implementation of GAN-based models, we find that using an alternative objective proposed by [10] for E, G and H instead leads to more stable training. The new objective consist of swapping the labels for the discriminators instead of reversing the gradient. We also find that we can update all the modules in one pass instead of taking alternate gradient steps while obtaining similar results.

Note that the MV-BiGAN model is trained based on datasets where all the V views are available for each data point. In order to generate examples where only subsets of views are available, the ideal procedure would be to consider all the possible subsets of views. Due to the number of data points that would be generated by such a procedure, we build random sequences of incremental sets of views and enforce the KL regularization over successive sets.

5   Experiments

We evaluate our model on three different types of experiments, and on two differents datasets. The first dataset we experiment on is the MNIST dataset of handwritten digits. The second dataset is the CelebA [14] dataset composed of both images of faces and corresponding attributes. The MNIST dataset is used to illustrate the ability of the MV-BiGAN to handle different subset of views, and to update its prediction when integrating new incoming views. The CelebA dataset is used to demonstrate the ability of MV-BiGAN to deal with different types (heterogeneous) of views.

5.1   MNIST, 4 Views

We consider the problem where 4 different views can be available, each view corresponding to a particular quarter of the final image to predict – each view is a vector of $\mathbb{R}^{(14 \times 14)}$. The MV-BiGAN is used here to recover the original image. The model is trained on the MNIST training digits, and results are provided on the MNIST testing dataset.

Figure 3 illustrates the results obtained for some digits. In this figure, the first column displays the input (the subset of views), while the other columns shows predicted outputs sampled by the MV-BiGAN. An additional view is added between each row. This experiment shows that when new views are added, the diversity in the predicted outputs decreases due to the KL-contraint introduced in the model, which is the desired behavior i.e. more information implied less variance. When removing the KL constraint (Fig. 4), the diversity still remains important, even if many views are provided to the model. This show the importance of the KL regularization term in the MV-BiGAN objective.

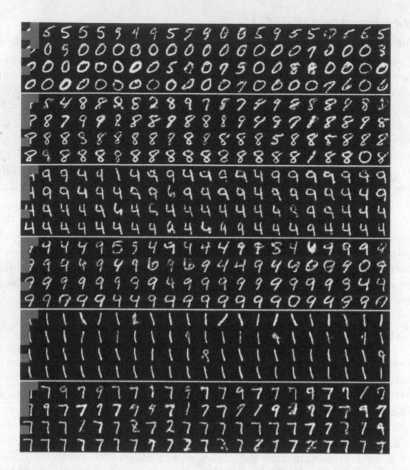

Fig. 3. Results of the MV-BiGAN on sequences of 4 different views. The first column corresponds to the provided views, while the other columns correspond to outputs sampled by the MV-BiGAN.

Fig. 4. Comparaison between MV-BiGAN with (top) and without (bottom) KL-constraint.

5.2 MNIST, Sequence of Incoming Views

We made another set of experiments where the views correspond to images with missing values (missing values are replaced by 0.5). This can be viewed as a data imputation problem – Fig. 5. Here also, the behavior of the MV-BiGAN exhibits interesting properties: the model is able to predict the desired output as long as enough information has been provided. When only non-informative views are provided, the model produces digits with a high diversity, the diversity decreasing when new information is added.

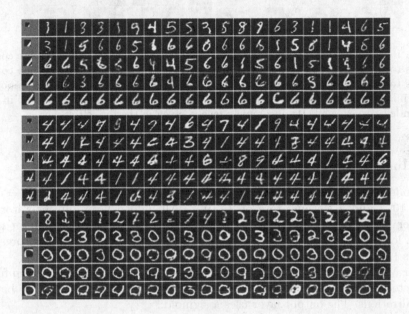

Fig. 5. MV-BiGAN with sequences of incoming views. Here, each view is a 28 × 28 matrix (values are between 0 and 1 with missing values replaced by 0.5).

5.3 CelebA, Integrating Heterogeneous Information

At last, the third experiment aims at measuring the ability of MV-BiGAN to handle heterogeneous inputs. We consider two views: (i) the attribute vector containing information about the person in the picture (hair color, sex, ...), and (ii) a incomplete face. Figure 6 illustrates the results obtained on two faces. The first line corresponds to the faces generated based on the attribute vector. One can see that the attribute information has been captured by the model: for example, the sex of the generated face is constant (only women) showing that MV-BiGan has captured this information from the attribute vector. The second line corresponds to the faces generated when using the incomplete face as an input. One can also see that the generated outputs are "compatible" with the incomplete information provided to the model. But the attribute are not considered (for example, women and men are generated). At last, the third line corresponds to images generated based on the two partial views (attributes and incomplete face) which are close to the ground-truth image (bottom left). Note that, in this set of experiments, the convergence of the MV-BiGAN was quite difficult to obtain, and the quality of the generated faces is still not satisfying.

Fig. 6. Results obtained on the CelebA dataset for two examples. The first line corresponds to the images generated based on the attribute vector, the second line corresponds to images generated based on the incomplete face, the third line corresponds to the images generated based on the two views. The groundthruth face is given in the bottom-left corner, while the incomplete face is given in the top-left corner.

5.4   Implementation Details

All models are optimized using Adam with standard hyperparameters $\beta_1 = 0.5$, $\beta_2 = 10^{-3}$ and a learning rate of $2 \cdot 10^{-5}$. All hidden layers in generator or encoder networks are followed by a rectified linear unit. In discriminator networks, leaky rectified linear units of slope 0.2 are used instead. Latent representations $(\mu, \log(\sigma^2))$ are of size 2×128.

For **MNIST experiments**, the generator function G has three hidden fully connected layers. The second and the third hidden layers are followed by batch normalizations. The output layer uses a sigmoid.

The aggregation model Ψ is a sum of mapping functions ϕ_k. Each ϕ_k is a simple linear transformation. The encoding functions E and H are both neural networks that include an aggregation network followed by two fully connected layers. A batch normalization is added after the second layer. They output a pair of vectors $(\mu, \log(\sigma^2))$. The output layers has a tanh for μ and a negative exponential linear unit for $\log \sigma^2$.

The discriminator D_1 has three fully connected layers with batch normalization at the third layer. A sigmoid is applied to the outputs. The vector z is concatenated to the representation at the second layer.

The discriminator D_2 is similar to E and H except it uses a sigmoid at the output level. z is concatenated directly to the aggregation vector $\Psi(v(x, s))$.

All hidden layers and the aggregation space are of size 1500. λ is set to $1 \cdot 10^{-5}$. Minibatch size is set to 128. The models have been trained for 300 epochs.

For **CelebA experiments**, the generator function G is a network of transposed convolution layers described in Table 1.

The mapping functions ϕ_k for images are convolution networks (Table 1). For attribute vectors, they are linear transformations. E and H are neural networks

Table 1. Convolution architectures used in our experiments on the CelebA dataset. The top part is used for encoding images into the aggregation space. The bottom part is used in G to generate images from a vector z.

Operation	Kernel	Strides	Padding	Feature maps	BN	Nonlinearity
Convolution	4×4	2×2	1×1	64	$\times$	Leaky ReLU
Convolution	4×4	2×2	1×1	128	$\checkmark$	Leaky ReLU
Convolution	4×4	2×2	1×1	256	$\checkmark$	Leaky ReLU
Convolution	4×4	2×2	1×1	512	$\checkmark$	Leaky ReLU
Convolution	4×4	1×1		Output size	$\times$	Linear
Transposed convolution	4×4	1×1		512	$\checkmark$	ReLU
Transposed convolution	4×4	2×2	1×1	256	$\checkmark$	ReLU
Transposed convolution	4×4	2×2	1×1	128	$\checkmark$	ReLU
Transposed convolution	4×4	2×2	1×1	64	$\checkmark$	ReLU
Transposed convolution	4×4	2×2	1×1	3	$\times$	Tanh

with one hidden layer on top of the aggregation model. The hidden layer is followed by a batch normalization. The output layer is the same as in the MNIST experiments. The discriminator D_1 is a transposed convolution network followed by a hidden fully connected layer before the output layer. z is concatenated at the hidden fully connected level. As in the MNIST experiments, the discriminator D_2 is similar to E and H, and z is concatenated directly to the aggregation vector $\Psi(v(x,s))$. Aggregation space is of size 1000. λ is set to $1 \cdot 10^{-3}$, and mini-batch size is 16. The model has been trained for 15 epochs.

6   Related Work

Multi-view and Representation Learning: Many application fields naturally deal with multi-view data with true advantages. For example, in the multimedia domain, dealing with a bunch of views is usual [2]: text, audio, images (different framings from videos) are starting points of these views. Besides, multimedia learning tasks from multi-views led to a large amount of fusion-based ad-hoc approaches and experimental results. The success of multi-view supervised learning approaches in the multimedia community seems to rely on the ability of the systems to deal with the complementary of the information carried by each modality. Comparable studies are of importance in many domains, such as bioinformatics [23], speech recognition [1,13], signal-based multimodal integration [31], gesture recognition [30], etc.

Moreover, multi-view learning has been theoretically studied mainly under the semi-supervised setting, but only with two facing views [5,11,25,26]. In parallel, ensemble-based learning approaches have been theoretically studied, in the supervised setting: many interesting results should concern multi-view learning, as long as the ensemble is built upon many views [20,32]. From the representation learning point of view, recent models are based on the incorporation of some "fusion" layers in the deep neural network architecture as in [17] or [24] for example. Some other interesting models include the multiview perceptron [33].

Estimating Complex Distributions: While deep learning has shown great results in many classification task for a decade, training deep generative models still remains a challenge. Deep Boltzmann Machines [21] are un-directed graphical models organized in a succession of layers of hidden variables. In a multi-view setting, they are able to deal with missing views and have been used to capture the joint distribution in bi-modal text and image data [22,24]. Another trend started with denoising autoencoder [28], which aims to reconstruct a data from a noisy input have been proved to possess some desirable properties for data generation [3]. The model have been generalized under the name Generative Stochastic Networks by replacing the noise function C with a mapping to a latent space [27]. Pulling away from the mixing problems encountered in previous approaches, Variational Autoencoders [12] attempts to map the input distribution to a latent distribution which is easy to sample from. The model is trained by optimizing a variational bound on the likelihood, using stochastic gradient descent methods. The Kullback-Leibler regularizer on the latent Gaussian representations used in our model is reminiscent of the one introduced in the variational lower bound used by the VAE.

The BiGAN model [7,8] that serves as a basis for our work is an extension of the Generative Adversarial Nets [10]. A GAN extension that captures conditional probabilities (CGAN) has been proposed in [16]. However, as noted by [15,18], they display very unstable behavior. More specifically, CGAN have been able to generate image of faces conditioned on an attribute vector [9], but fail to model image distribution conditioned on a part of the image or on previous frames. In both CGAN and CVBiGAN, the generation process uses random noise to be able to generate a diversity of outputs from the same input. However, in a CGAN, the generator concatenate an independent random vector to the input while CV-BiGAN learns a stochastic latent representation of the input. Also, some of the difficulties of CGAN in handling images as both inputs $\tilde{x}$ and outputs $\tilde{y}$ stem from the fact that CGAN's discriminator directly compares $\tilde{x}$ and y. In CV-BiGAN, neither discriminators has access to both $\tilde{x}$ and y but only to a latent representation z and either $\tilde{x}$ or y.

7   Conclusion and Perspectives

We have proposed the CV-BiGAN model for estimating conditional densities, and its extension MV-BiGAN to handle multi-view inputs. The MV-BiGAN model is able to both handle subsets of views, but also to update its prediction when new views are added. It is based on the idea that the uncertainty of the prediction must decrease when additional information is provided, this idea being handled through a KL constraint in the latent space. This work opens different research directions. The first one concerns the architecture of the model itself since the convergence of MV-BiGAN is still difficult to obtain and has a particularly high training cost. Another direction would be to see if this family of model could be used on data streams for anytime prediction.

Acknowledgments. This work was supported by the French project LIVES ANR-15-CE23-0026-03.

References

1. Arora, R., Livescu, K.: Kernel CCA for multi-view acoustic feature learning using articulatory measurements. In: MLSP (2012)
2. Atrey, P.K., Hossain, M.A., El Saddik, A., Kankanhalli, M.S.: Multimodal fusion for multimedia analysis: a survey. Multimed. Syst. **16**(6), 345–379 (2010)
3. Bengio, Y., Yao, L., Alain, G., Vincent, P.: Generalized denoising auto-encoders as generative models. In: Advances in Neural Information Processing Systems, pp. 899–907 (2013)
4. Cesa-Bianchi, N., Hardoon, D.R., Leen, G.: Guest editorial: learning from multiple sources. Mach. Learn. **79**(1), 1–3 (2010)
5. Chapelle, O., Schölkopf, B., Zien, A.: Semi-supervised Learning. MIT Press, Cambridge (2006)
6. Denton, E.L., Chintala, S., Fergus, R., et al.: Deep generative image models using a Laplacian pyramid of adversarial networks. In: Advances in Neural Information Processing Systems, pp. 1486–1494 (2015)
7. Donahue, J., Krähenbühl, P., Darrell, T.: Adversarial feature learning. In: Proceedings of the 5th International Conference on Learning Representations (ICLR), no. 2017 (2016, in press)
8. Dumoulin, V., Belghazi, I., Poole, B., Lamb, A., Arjovsky, M., Mastropietro, O., Courville, A.: Adversarially learned inference. In: Proceedings of the 5th International Conference on Learning Representations (ICLR), no. 2017 (2016, in press)
9. Gauthier, J.: Conditional generative adversarial nets for convolutional face generation. Class Project for Stanford CS231N: Convolutional Neural Networks for Visual Recognition, Winter semester 2014 (2014)
10. Goodfellow, I., Pouget-Abadie, J., Mirza, M., Xu, B., Warde-Farley, D., Ozair, S., Courville, A., Bengio, Y.: Generative adversarial nets. In: Advances in Neural Information Processing Systems, pp. 2672–2680 (2014)
11. Johnson, R., Zhang, T.: Semi-supervised learning with multi-view embedding: theory and application with convolutional neural networks. CoRR abs/1504.012555v1 (2015)
12. Kingma, D.P., Welling, M.: Auto-encoding variational Bayes. In: Proceedings of the 2nd International Conference on Learning Representations (ICLR), no. 2014 (2013)
13. Koço, S., Capponi, C., Béchet, F.: Applying multiview learning algorithms to human-human conversation classification. In: INTERSPEECH (2012)
14. Liu, Z., Luo, P., Wang, X., Tang, X.: Deep learning face attributes in the wild. In: Proceedings of International Conference on Computer Vision (ICCV) (2015)
15. Mathieu, M., Couprie, C., LeCun, Y.: Deep multi-scale video prediction beyond mean square error. arXiv preprint arXiv:1511.05440 (2015)
16. Mirza, M., Osindero, S.: Conditional generative adversarial nets. arXiv preprint arXiv:1411.1784 (2014)

17. Ngiam, J., Khosla, A., Kim, M., Nam, J., Lee, H., Ng, A.Y.: Multimodal deep learning. In: Proceedings of the 28th International Conference on Machine Learning (ICML) (2011)

18. Pathak, D., Krahenbuhl, P., Donahue, J., Darrell, T., Efros, A.A.: Context encoders: feature learning by inpainting. In: Proceedings of the IEEE Conference on Computer Vision and Pattern Recognition, pp. 2536–2544 (2016)

19. Radford, A., Metz, L., Chintala, S.: Unsupervised representation learning with deep convolutional generative adversarial networks. arXiv preprint arXiv:1511.06434 (2015)

20. Rokach, L.: Ensemble-based classifiers. Artif. Intell. Rev. **33**(1–2), 1–39 (2010)

21. Salakhutdinov, R., Hinton, G.E.: Deep Boltzmann machines. In: AISTATS, vol. 1, no. 3 (2009)

22. Sohn, K., Shang, W., Lee, H.: Improved multimodal deep learning with variation of information. In: Advances in Neural Information Processing Systems, pp. 2141–2149 (2014)

23. Sokolov, A., Ben-Hur, A.: Multi-view prediction of protein function. In: ACM-BCB, pp. 135–142 (2011)

24. Srivastava, N., Salakhutdinov, R.R.: Multimodal learning with deep Boltzmann machines. In: Advances in Neural Information Processing Systems, pp. 2222–2230 (2012)

25. Sun, S.: A survey of multi-view machine learning. Neural Comput. Appl. **23**(7–8), 2031–2038 (2013)

26. Sun, S., Taylor, J.S.: PAC-Bayes analysis of multi-view learning. CoRR abs/1406.5614 (2014)

27. Thibodeau-Laufer, E., Alain, G., Yosinski, J.: Deep generative stochastic networks trainable by backprop (2014)

28. Vincent, P., Larochelle, H., Bengio, Y., Manzagol, P.A.: Extracting and composing robust features with denoising autoencoders. In: Proceedings of the 25th international conference on Machine learning, pp. 1096–1103. ACM (2008)

29. Wang, W., Arora, R., Livescu, K., Bilmes, J.: On deep multi-view representation learning. In: Proceedings of the 32st International Conference on Machine Learning (ICML 2015), pp. 1083–1092 (2015)

30. Wu, J., Cheng, J., Zhao, C., Lu, H.: Fusing multi-modal features for gesture recognition. In: ICMI, pp. 453–460 (2013)

31. Wu, L., Oviatt, S., Cohen, P.: Multimodal integration: a statistical view. MM **1**(4), 334–341 (1999)

32. Zhang, J., Zhang, D.: A novel ensemble construction method for multi-view data using random cross-view correlation between within-class examples. Pattern Recogn. **44**(6), 1162–1171 (2011)

33. Zhu, Z., Luo, P., Wang, X., Tang, X.: Multi-view perceptron: a deep model for learning face identity and view representations. In: Advances in Neural Information Processing Systems, pp. 217–225 (2014)

Online Sparse Collapsed Hybrid Variational-Gibbs Algorithm for Hierarchical Dirichlet Process Topic Models

Sophie Burkhardt$^{(\boxtimes)}$ and Stefan Kramer

Johannes Gutenberg-Universität, Mainz, Germany
{burkhardt,kramer}@informatik.uni-mainz.de

Abstract. Topic models for text analysis are most commonly trained using either Gibbs sampling or variational Bayes. Recently, hybrid variational-Gibbs algorithms have been found to combine the best of both worlds. Variational algorithms are fast to converge and more efficient for inference on new documents. Gibbs sampling enables sparse updates since each token is only associated with one topic instead of a distribution over all topics. Additionally, Gibbs sampling is unbiased. Although Gibbs sampling takes longer to converge, it is guaranteed to arrive at the true posterior after infinitely many iterations. By combining the two methods it is possible to reduce the bias of variational methods while simultaneously speeding up variational updates. This idea has previously been applied to standard latent Dirichlet allocation (LDA). We propose a new sampling method that enables the application of the idea to the nonparametric version of LDA, hierarchical Dirichlet process topic models. Our fast sampling method leads to a significant speedup of variational updates as compared to other sampling methods. Experiments show that training of our topic model converges to a better log-likelihood than previously existing variational methods and converges faster than Gibbs sampling in the batch setting.

1 Introduction

Topic models based on latent Dirichlet allocation (LDA) are a common tool for analyzing large collections of text. They are used for extracting common themes and provide a probabilistic clustering of documents. The topics, usually represented by word clouds of frequent words, can be interpreted and used to understand the content of a text corpus. Since scalability is an important factor when modeling large datasets, various online algorithms have been developed to handle streams of documents. A generalization of LDA that allows for asymmetric topic priors and an unbounded number of topics is provided by nonparametric topic models. These are based on hierarchical Dirichlet processes (HDPs) [16].

Electronic supplementary material The online version of this chapter (https://doi.org/10.1007/978-3-319-71246-8_12) contains supplementary material, which is available to authorized users.

© Springer International Publishing AG 2017
M. Ceci et al. (Eds.): ECML PKDD 2017, Part II, LNAI 10535, pp. 189–204, 2017.
https://doi.org/10.1007/978-3-319-71246-8_12

The two main algorithms for training LDA topic models are Gibbs sampling and variational Bayes. While Gibbs sampling is an unbiased method, it takes very long on large datasets to converge. To make Gibbs sampling online, one has to use particle samplers which are rather inefficient. Variational Bayes on the other hand may be combined with stochastic gradients to be trained online.

Hybrid methods have gained popularity in recent years, especially in deep neural networks where black box variational inference is a very efficient training algorithm [11]. Sampling can be used to approximate the gradient in variational Bayes which leads to a second source of stochasticity (in addition to random choices of data subsets). In previous work this was applied to parametric topic models [9] and a very efficient variant of this algorithm was recently proposed that takes advantage of the sparsity in topic distributions during sampling [14]. Hybrid algorithms have the combined advantages of a reduced bias of the variational method and a faster convergence as compared to pure Gibbs sampling, as well as the possibility of online training through stochastic gradient estimation.

We will first introduce topic models and the two main training algorithms, Gibbs sampling and variational Bayes. Following this, we briefly introduce nonparametric topic models and present an efficient sampling method. We then show how this sampling method can be used to construct a hybrid Variational-Gibbs method. To further speed up our algorithm, we propose a more efficient sampling algorithm. Our experiments show that our method converges better than the purely variational topic modeling method as well as the Gibbs sampler. Supplementary material for this paper is available at https://www.datamining. informatik.uni-mainz.de/files/2015/06/supplement.pdf.

2 Background

In this section, we will first provide the necessary background in latent Dirichlet allocation (LDA). Second, we will introduce the nonparametric version, hierarchical Dirichlet processes (HDP).

2.1 Latent Dirichlet Allocation

Latent Dirichlet allocation (LDA) is a generative model of document collections where each document is modeled as a mixture of latent topics. Each topic $k \in 1, \ldots, K$ is represented by a multinomial distribution ϕ_k over words that is assumed to be drawn from a Dirichlet distribution $Dir(\beta)$. The dth document is generated by drawing a distribution over topics from a Dirichlet $\theta_d \sim Dir(\alpha)$, and for the nth word token in the document, first drawing a topic indicator $z_{dn} \sim \theta_d$ and finally drawing a word $w_{dn} \sim \phi_{z_{dn}}$.

To learn a model over an observed document collection W, we need to estimate the posterior distribution over the latent variables z, θ, and ϕ.

$$p(\phi, \theta, z|W, \alpha, \beta) = \prod_{k=1}^{K} p(\phi_k|\beta) \prod_{d=1}^{D} p(\theta_d|\alpha) \prod_{n=1}^{N_d} p(z_{dn}|\theta_d)p(w_{dn}|\phi_{z_{dn}}) \quad (1)$$

The most popular training algorithms are variational Bayesian inference and Gibbs sampling and will briefly be introduced in the following sections.

2.2 Variational Bayesian Inference

In variational Bayesian inference a variational distribution is introduced to approximate the posterior by minimizing the KL divergence between the variational distribution and the true posterior. Usually, a fully factorized variational distribution is chosen:

$$q(\phi, \theta, z | \tilde{\beta}, \tilde{\alpha}, \tilde{\theta}) = \prod_k^K q(\phi_k | \tilde{\beta}_k) \prod_d^D q(\theta_d | \tilde{\alpha}_d) \prod_n^{N_d} q(z_{dn} | \tilde{\theta}_{dn}) \qquad (2)$$

The evidence lower bound (ELBO) that is to be maximized is given as follows:

$$\log p(W) \geq \mathcal{L}(\tilde{\beta}, \tilde{\alpha}, \tilde{\theta}) \triangleq \mathbb{E}_q[\log p(\phi, \theta, z, W)] + \mathcal{H}(q(\phi, \theta, z)), \qquad (3)$$

where $\mathcal{H}$ denotes the entropy.

By calculating the gradient of the ELBO with respect to the variational parameters, the parameters can be updated until convergence. The local/document-level update equations for collapsed variational Bayes (CVB0 [1,6]), holding global variational parameter $\tilde{\beta}$ fixed, are:

$$\tilde{\alpha}_{dk} = \alpha + \sum_{n=1}^{N_d} \tilde{\theta}_{dnk} \qquad (4)$$

$$\tilde{\theta}_{dnk} \propto \frac{\tilde{\beta}_{wk} + \beta}{\sum_v (\tilde{\beta}_{vk} + \beta_v)} (\tilde{\alpha}_{dk} + \alpha), \qquad (5)$$

where α and β are hyperparameters and N_d is the number of words in document d. Based on the local variational parameters $\tilde{\theta}$, the global parameter $\tilde{\beta}$ can be updated as follows:

$$\tilde{\beta}_{vk} = \beta + \sum_{d=1}^{D} \sum_{n=1}^{N_d} \tilde{\theta}_{dnk} \mathbb{1}[w_{dn} = v], \qquad (6)$$

where D is the number of documents. $\mathbb{1}[w_{dn} = v]$ is one if word $w_{dn} = v$ and zero otherwise.

2.3 Gibbs Sampling

Gibbs sampling does not have to resort to a factorized variational distribution, which is why the method is unbiased. Through integrating out the latent variables ϕ and θ, the model can be efficiently trained. Convergence is slower for Gibbs sampling since updates only involve a sampled topic instead of the full distribution over topics as in variational Bayes.

The conditional probabilities for training an LDA topic model are [7]:

$$p(z_i = k | z_{-i}, d, w) \propto \frac{n_{wk} + \beta}{\sum_v n_{vk} + \beta_v} (n_{dk} + \alpha), \qquad (7)$$

where n_{wk} and n_{dk} are the respective counts of topics k with words w or in documents d and α and β are hyperparameters as before. z_{-i} are all topic indicators except the one for token i.

2.4   Hierarchical Dirichlet Processes

For hierarchical Dirichlet process (HDP) topic models [16], the multinomial distribution θ from LDA is drawn from an HDP instead of a Dirichlet distribution: $\theta \sim DP(G_0, b_1), G_0 \sim DP(H, b_0)$. The base distribution G_0 of the first Dirichlet process (DP) is again drawn from a DP with base distribution H. This is why it is called a *hierarchical* DP.

A DP is a prior for a multinomial with a potentially unbounded number of topics. Drawing different multinomials from a DP results in multinomials of different sizes. Because the prior is hierarchical, there is a local topic distribution θ for each document and a global topic distribution G_0 which is shared among all documents. The advantage of this global topic distribution is that it allows topics of widely varying frequencies whereas in standard LDA with a symmetric prior α, all topics are expected to have the same frequency. The asymmetric prior of HDP usually leads to a better representation and higher log-likelihood of the dataset [16].

2.5   Sampling for HDP

Sampling methods for HDPs are mostly based on the Chinese restaurant process metaphor. Each word token is assumed to be a customer entering a restaurant, and sitting down at a certain table where a specific dish is served. Each table is associated with one dish which corresponds to a topic in a topic model. The probability for a customer to sit down at a certain table is proportional to the number of customers already sitting at that table. With a certain probability α, the customer sits down at a new table. In this case a topic is sampled from the base distribution. For an HDP topic model, each document corresponds to a restaurant. The topics in each document-restaurant are drawn from a global restaurant. Because all documents share the same global restaurant, the topics are shared. If a new table is added to a document restaurant, a pseudo customer enters the global restaurant. If a new table is opened in the global restaurant, a new topic is added to the topic model.

In terms of the statistics that need to be kept, in the basic version we need to store for each word not only the sampled topic, but also the table it is associated with. Also, we need to store the corresponding topic for each table.

Three basic sampling methods were introduced in Teh *et al.* [16], two methods are directly based on the Chinese restaurant representation, the third is the direct assignment sampler. The first two methods sample a table for each word and a topic for each table. This can be very slow and requires to store separate counts for each table. The direct assignment sampler does not sample an individual table but assigns a topic to each word token directly, and instead of keeping the statistics for each table separately, it simply samples the number of tables that are associated with a certain topic. While this sampler has improved convergence over the other sampling methods, it needs to sample $s_k \in \{1, \ldots, N_k\}$, the number of tables for topic k which can be very inefficient when the number of customers per topic N_k is very large. A further improved version was therefore introduced by Chen *et al.* [5]. In this version another auxiliary variable u is introduced which is sampled for each customer and determines whether or not the customer sits down at a new table or an existing one. This is then used to update the table count s_k. u itself does not have to be kept in memory but can be sampled when needed. This way, the memory requirements are similar to Teh's auxiliary variable sampler, the sampling process itself is more efficient, and convergence is improved. For this reason we base our sampler on the sampling method by Chen *et al.*. We will only give the sampling equations here and refer to the original publication for more details.

$P(z_i = k|rest)$, the conditional probability of assigning topic k to token i in document j, is given as follows:

If the topic is new for the root restaurant (table indicator is zero):

$$P(z_i = k_{new}, u_i = 0|rest) \propto \frac{b_0 b_1}{(b_0 + M.)(b_1 + N_j)} \frac{N_{wk} + \beta}{\sum_{w'} (N_{w'k} + \beta)} \quad (8)$$

If the topic is new for the base restaurant (e.g. a document), but not for the root restaurant (table indicator is one):

$$P(z_i = k, u_i = 1|rest) \propto \frac{b_1 * M_k^2}{(M_k + 1)(b_0 + M.)(b_1 + N_j)} \frac{N_{wk} + \beta}{\sum_{w'} (N_{w'k} + \beta)} \quad (9)$$

If the topic exists at the base restaurant and a new table is opened (table indicator is one):

$$P(z_i = k, u_i = 1|rest) \propto \frac{b_1}{b_1 + N_j} \frac{S_{m_{jk}+1}^{n_{jk}+1}}{S_{m_{jk}}^{n_{jk}}} \frac{m_{jk} + 1}{n_{jk} + 1} \frac{M_k^2}{(b_0 + M.)(M_k + 1)}$$
$$\frac{N_{wk} + \beta}{\sum_{w'} (N_{w'k} + \beta)} \quad (10)$$

If the topic exists at the base restaurant and an old table is chosen (table indicator is two):

$$P(z_i = k, u_i = 2|rest) \propto \frac{S_{m_{jk}}^{n_{jk}+1}}{S_{m_{jk}}^{n_{jk}}} \frac{n_{jk} - m_{jk} + 1}{(n_{jk} + 1)(b_1 + N_j)} \frac{N_{wk} + \beta}{\sum_{w'} (N_{w'k} + \beta)} \quad (11)$$

In the above equations, b_0 and b_1 are hyperparameters, M_k is the overall number of tables for topic k, N_{wk} is the overall number of customers for topic k and word w, N_j is the overall number of customers for restaurant j, n_{jk} is the number of customers in restaurant j for topic k, and m_{jk} is the number of tables in restaurant j for topic k. S_m^n are generalized Stirling numbers that can be efficiently precomputed and retrieved in $O(1)$ [2].

3   Proposed Method – Hybrid Variational-Gibbs

We will now describe how the sampling method above can be used to construct a hybrid Variational-Gibbs training algorithm for the HDP. Our algorithm is online since it is based on stochastic gradient ascent [6,9,10]. This means our model can be continuously updated with new batches of data.

Taking up Sect. 2.2 on variational inference and following Hoffman *et al.* [9], the natural gradient of the ELBO with respect to $\tilde{\beta}$ is defined as

$$\mathbb{E}_q[N_{dkw}] + \frac{1}{D}(\beta - \tilde{\beta}_{kw}) \tag{12}$$

To evaluate the expectation in this equation we would need to evaluate all possible topic configurations for each document. For using stochastic gradient ascent however, an approximation is sufficient. This is where Gibbs sampling comes in. By taking samples from the distribution q^* we can approximate the expectation in the above equation.

$$q^*(z_{di} = k|z_{-i}) \propto \exp\{\mathbb{E}_{q(\neg z_d)} \log(p(z_d|b_1, G_0)p(w_d|z_d, \phi))\}, \tag{13}$$

where $\neg z_d$ denotes all topic indicators z except the ones for document d. This distribution is difficult to normalize since we would have to consider all possible topic configurations z_d. However, we can easily sample from it and estimate the variational Dirichlet parameters as follows [9]:

$$\tilde{\beta}_{kv} = \beta + \sum_d \sum_i \mathbb{E}_q[\mathbb{1}[z_{di} = k]\mathbb{1}[w_{di} = v]], \tag{14}$$

where the expectation is approximated by the samples from q^*.

In contrast to Hoffman *et al.*, we have an additional variational distribution over the topics G_0. This is the global topic prior. The global variational distribution for G_0 and the mixture components ϕ is

$$q(G_0, \phi|\tilde{\gamma}, \tilde{\beta}) = \prod_k q(G_{0_k}|\tilde{\gamma})q(\phi_k|\tilde{\beta}_k), \tag{15}$$

where $\tilde{\gamma}$ and $\tilde{\beta}$ are Dirichlet parameters.

The variational Dirichlet parameter for the global topic distribution is analogously estimated as follows:

$$\tilde{\gamma}_k = \gamma + \sum_d \sum_i \mathbb{E}_q[\mathbb{1}[z_{di} = k]\mathbb{1}[u_{di} = 1|u_{di} = 0]], \tag{16}$$

where γ is a hyperparameter, and $\mathbb{1}[u = 1|u = 0]$ is one if the table indicator u is either zero or one, which means that a new table is being opened, and otherwise zero.

The expectations in Eqs. 14 and 16 can be estimated by sampling from q^* which is given by the following set of equations (compare to Eqs. 8–11, differences are highlighted in bold):

If the topic is new for the root restaurant (table indicator is zero):

$$q^*(z_{di} = k, u = 0|z_{-i}) \propto \frac{b_0 b_1}{(b_0 + \sum_{k'} \tilde{\gamma}_{k'})(b_1 + N_j)} \exp\left(\mathbb{E}[\log\phi_{wk}]\right) \tag{17}$$

If the topic is new for the base restaurant (e.g. a document), but not for the root restaurant (table indicator is one):

$$q^\star(z_{di} = k, u = 1|z_{-i}) \propto \frac{b_1 * \tilde{\gamma}_k^2}{(\tilde{\gamma}_k + 1)\left(\sum_{k'} \tilde{\gamma}_{k'} + b_0\right)(b_1 + N_j)} \exp\left(\mathbb{E}[\log\phi_{wk}]\right) \tag{18}$$

If the topic exists at the base restaurant and a new table is opened ($u = 1$):

$$q^\star(z_{di} = k, u = 1|z_{-i}) \propto \frac{b_1}{b_1 + N_j} \frac{S_{m_{jk}+1}^{n_{jk}+1}}{S_{m_{jk}}^{n_{jk}}} \frac{m_{jk} + 1}{n_{jk} + 1}$$
$$\frac{\tilde{\gamma}_k^2}{(\sum_{k'} \tilde{\gamma}_{k'} + b_0)(\tilde{\gamma}_k + 1)} \exp\left(\mathbb{E}[\log\phi_{wk}]\right) \tag{19}$$

If the topic exists at the base restaurant and an old table is chosen ($u = 2$):

$$q^\star(z_{di} = k, u = 2|z_{-i}) \propto \frac{S_{m_{jk}}^{n_{jk}+1}}{S_{m_{jk}}^{n_{jk}}} \frac{n_{jk} - m_{jk} + 1}{n_{jk} + 1} \exp\left(\mathbb{E}[\log\phi_{wk}]\right) \tag{20}$$

In the above equations, the number of tables M_k (Eqs. 8 and 9) is substituted by the global variational parameter $\tilde{\gamma}$. $\exp(\mathbb{E}[\log\phi_{wk}])$ is expensive to compute, since $\log(\phi_{wk}) = \psi(\tilde{\beta}_{wk}) - \psi(\sum_w \tilde{\beta}_{wk})$, where $\psi(\cdot)$ is the digamma function. Following Wang and Blei [18] and Li *et al.* [14] we use $\frac{\beta_{wk}+\beta}{\sum_{w'} (\beta_{w'k}+\beta)}$ instead. The remaining variables are the local counts equivalent to the counts in Eqs. 8 to 11.

Updating variational parameters $\tilde{\beta}$ and $\tilde{\gamma}$ for one minibatch M is done as follows, where the counts for one minibatch are scaled by $\frac{|D|}{B|M|}$ for B burn-in iterations, to arrive at the expectation for the whole corpus and ρ is a parameter between zero and one.

$$\tilde{\beta}_{kw} = (1 - \rho_t)\tilde{\beta}_{kw} + \rho_t \left(\beta + \frac{|D|}{S|M|} \sum_{d \in M} N_{dkw}\right) \tag{21}$$

$$\tilde{\gamma}_k = (1 - \rho_t)\tilde{\gamma}_k + \rho_t \left(\gamma + \frac{|D|}{S|M|} \sum_{d \in M} \sum_{n \in d} \mathbb{1}[z_{dn} = k]\mathbb{1}[u_{dn} = 1|u_{dn} = 0]\right) \tag{22}$$

Summing up this section, we make use of the table indicators from Chen *et al.*'s sampling method for the HDP to be able to approximate the global topic distribution from minibatch samples. This yields an online algorithm for the HDP topic model.

3.1   Doubly Sparse Sampling for HDP

Having introduced our hybrid variational algorithm based on the table indicator sampling scheme, we will now introduce a doubly sparse sampling method for the nonparametric topic model [4]. This is similar to Li *et al.*'s [14] method for the parametric topic model and would not be possible for the direct assignment

---

Algorithm 1. Train Topic Model

Input: Dataset D

1 **repeat**
2 | $M \leftarrow$ get minibatch from D
3 | compute q^e (Eqs. 23,17,18) and $Q = \sum q^e$
4 | $A_w \leftarrow$ computeAliasTable$(\frac{q_w^e}{Q_w})$ for each word w (see supplement)
5 | **for** *document* $d \in M$ **do**
6 | | $z_d \leftarrow$ initialize randomly
7 | | **for** *iteration* $i = 1, \ldots, S + B$ **do**
8 | | | **for** *token* $n = 1, \ldots, N_d$ **do**
9 | | | | $z_{dn}, u_{dn} \leftarrow$ Sample(A, w_{dn}) (Algorithm 2)
10 | | | **if** $i > B$ **then**
11 | | | | Save sample
12 | update $\tilde{\beta}$ and $\tilde{\gamma}$ (Eqs. 21 and 22)
13 **until** *convergence*

Algorithm 2. Sample(A, w)

---

1 compute $\tilde{p}, \tilde{P} = \sum \tilde{p}, \Delta$ (Eqs. 24,25), $i = -1$, $u \leftarrow 1$
2 sample $r \sim$ Uniform$(0, \tilde{P} + \tilde{Q})$
3 **if** $r < \tilde{P}$ **then**
4 | **while** $r > 0$ **do**
5 | | $i \leftarrow i + 1$, $t \leftarrow i/2$, $u \leftarrow 2 - (i \mod 2)$, $r \leftarrow r - \tilde{p}_{j,w}(t, u)$
6 **else**
7 | **repeat**
8 | | $t \leftarrow$ sample from Alias A_w
9 | **until** t *is new in document*
10 **return** u, t

---

sampler. Hereby, we take advantage of the fact that the number of topics that occur in one document K_d is usually much lower than the total number of topics K. Furthermore, we improve this sampling method to make it more memory efficient.

Making Table Indicator Sampling Sparse. To obtain sparsity, the topic distribution can be divided into three parts according to the table indicators [4]:

$$q^\star(z = k|z_{-i}) = q^\star(z = k, u = 0|z_{-i}) +$$
$$q^\star(z = k, u = 1|z_{-i}) + q^\star(z = k, u = 2|z_{-i}).$$

The last part is sparse since it is only nonzero for the topics that occur with the document. Therefore, we can use alias-sampling to save the distribution for the dense part (The algorithm is provided in the supplementary material.) and subsequently draw samples from it in $O(1)$, whereas the sparse part can be computed in $O(K_j)$ since it is only necessary to iterate over those topics that occur within the jth document.

Formally, we rewrite the topic distribution $q^\star(k)$ over topics k as a combination of a dense distribution q_{jw} and a sparse distribution p_{jw}, where w is a word, and j is a document-restaurant. The normalization terms are given by $P_{jw} = \sum_k p_{jw}(k)$ and $Q_{jw} = \sum_k q_{jw}(k)$. The resulting distribution is given by:

$$q^\star(k) := \frac{p_{jw}(k) + q_{jw}(k)}{P_{jw} + Q_{jw}}$$

We define the stale distribution q_{jw} as the distribution over all topics and a table indicator of 0 or 1:

$$q_{jw}(k) := q^\star(z = k, u = 0|rest) + q^\star(z = k, u = 1|rest) \tag{23}$$

The fresh distribution p_{jw} is defined as the distribution over all topics that exist in restaurant j and a table indicator of 2:

$$p_{jw}(k) := q^\star(z = k, u = 2|rest)$$

Improving the Sparse Sampler. As can be inferred from the above equations and Eqs. 18 and 19, q_{jw} depends on document j. If topic k exists in document j, the probability is given by Eq. 19, otherwise by Eq. 18. This means we would have to save topic distributions for every single document. It would be more appealing to have one topic distribution that represents the global topic distribution and can be used for all documents. The solution we propose to improve the method described above is as follows:

We simply assume for each topic that it does not exist in the document and save the resulting distribution q_w^e for an empty pseudo document e. This is equivalent to replacing Eq. 19 with Eq. 18. Now we only need to add a subsequent rejection step for the case where we sample a topic from this distribution that exists in the current document. If this happens, we simply discard it and draw a new sample. Since each sample is drawn in $O(1)$ this does not significantly increase the complexity as long as the basic assumption holds that $K_d \ll K$. The sparse distribution is now over table indicators one and two instead of just two, to account for the case where a new table is opened for an existing topic.

$$\tilde{p}_{jw}(k, u') := q^\star(z = k, u = u'|rest)\mathbb{1}[n_{jk} > 0], \tag{24}$$

where $\mathbb{1}[n_{jk} > 0]$ is one if the number of tokens in document-restaurant j associated with topic k is at least one and zero otherwise. Accordingly, the normalization sum is $\tilde{P}_{jw} = \sum_k \sum_u \tilde{p}_{jw}(k, u)$.

We need to subtract an amount Δ_j from the normalization sum Q_w which is different for each document j and accounts for the topics that are present in document j and would be rejected if drawn from distribution q. We call it the discard mass Δ and it is defined as follows:

$$\Delta_j := \sum q_k^e \mathbb{1}[n_{jk} > 0] \tag{25}$$

Since Δ_j can be computed in $O(K_j)$ time, it does not add to the overall computational complexity. Following this, the normalization sum is now given by $\tilde{Q}_{jw} = Q_w - \Delta_j$, where $Q_w = \sum q_w^e$.

Note that since the distribution does not change during the gradient estimation for one minibatch, i.e. we sample from the exact distribution, we do not need to add a Metropolis-Hastings acceptance step as in Li et al. [13].

The whole algorithm is summed up in Algorithm 1. For each minibatch, the dense distributions q^e are computed for each word that occurs in the minibatch. Alias tables are computed for these distributions (see the supplementary material) to be able to sample from them in $O(1)$. For each document d in the minibatch, the topics z_d are sampled using Algorithm 2 and stored after a burn-in period of B iterations. The stored samples are finally used to update the global variational distributions.

4   Experiments

4.1   Algorithms

We compare three different topic models:

1. Our method OSCHVB-HDP (Online Sparse Collapsed Hybrid Variational Bayes): The implementation is done in Java.
2. Wang et al.'s stochastic mean-field variational HDP (SMF-HDP) [19]: The python implementation provided by the author was used. Unfortunately this does not allow doing a fair runtime comparison. We generally observed our method to be faster, however, this could be due to the different programming languages.
3. The Gibbs sampling algorithm by Chen et al. [5]: Our own Java implementation was used to make the runtime comparable to our hybrid algorithm; the code only differs in the implementation of the sampling step.

4.2   Parameter Settings

The Dirichlet parameter for the topic-word distributions β is always set to 0.01, a standard default value.

The number of sampling iterations S and the number of burn-in iterations B are parameters in the hybrid algorithm. In accordance with the existing literature, we chose $S = B = 5$. The same parameters were used during evaluation.

We used the same update parameter $\rho_t = \frac{1}{(1+n)^{0.6}}$ for all algorithms, where n is the number of batches that have been processed up to time t.

For the mean-field variational algorithm by Wang et al. [19] we used the public python implementation provided by the authors. As parameter settings we chose the default settings that were also used by Wang and Blei [18], the second level truncation was set to 20. A batch size of 100 was chosen as our default batch size. For our nonparametric method we used the same hyperparameters $b_1 = b_0 = 1.0$.

4.3   Datasets

We used four publicly available datasets (Table 1) and preprocessed them as follows:

1. `BioASQ`:
   This dataset consists of paper abstracts from the PubMed database. It was made available for the BioASQ competition, a large-scale semantic indexing challenge [17]. We separated the 500,000 most recent documents plus 10,000 documents as a separate test dataset. After stopword removal we kept the 20,000 most frequent features.
2. `Enron`:
   The `Enron` dataset consists of ca. 500,000 emails and is available at https://www.cs.cmu.edu/~./enron/[12]. We removed the header, tokenized the emails, removed stopwords and kept the 20,000 most frequent features. We randomly separated 10,000 documents as a testing dataset.
3. `NIPS`:
   This dataset is available in a preprocessed format from the UCI Machine Learning Repository [15]. It has 5,812 documents and 11,463 features and is the second smallest dataset that we used. It consists of NIPS conference papers published between 1987 and 2015. In comparison to the other datasets, the individual documents are large. 1000 documents were separated as a test-set.
4. `KOS`:
   The 3430 blog entries of this dataset were originally extracted from http://www.dailykos.com/, the dataset is available in the UCI Machine Learning Repository https://archive.ics.uci.edu/ml/datasets/Bag+of+Words. The number of features is 6906.

Table 1. Statistics for the datasets used in our experiments, $|D|$ train and test: the number of documents in the train- and testset, respectively, $|V|$: size of vocabulary

Dataset	$|D|$ train	$|D|$ test	$|V|$
Enron	507,401	10,000	20,000
BioASQ	500,000	10,000	20,000
NIPS	4,811	1,000	11,463
KOS	2930	500	6906

4.4   Evaluation

We evaluated the models on the per-word-log-likelihood according to Heinrich [8]:

$$\frac{\log p(w|M)}{|n_d|} = \left(\sum_{w \in d} n_{dw} \log \left(\sum_{k=1}^{K} \phi_{kw}\theta_{kw} \right) \right) / |n_d|, \tag{26}$$

where n_{dw} is the number of times word w occurs in document d, K is the overall number of topics, and ϕ are the model parameters. θ are the document specific parameters that need to be estimated using the model. In our case, we run the sampler with a fixed point estimate of parameters ϕ and estimate θ analogously to the training procedure with S samples that are saved after B burn-in iterations.

4.5   Experimental Results

Mean-Field vs. Hybrid Approach. First of all we compare our hybrid approach to Wang et al.'s SMF-HDP [19] on the three largest datasets. We notice that the performance of SMF-HDP heavily depends on the batch size. Small batch sizes lead to a much worse performance (see Fig. 1). The same observation was made by Wang and Blei [18]. SMF-HDP starts with the maximum topic number and then reduces the number of topics. In our experiments, often only a handful of topics remained for small batch sizes. Wang and Blei hypothesized that this is due to the algorithm being strongly dependent on the initialization and not being able to add topics occurring in later batches that had not been present from the start. This behavior is problematic, especially in the online setting where it is not guaranteed that the first batch contains all the topics. Our method is more robust and better suited to settings where small batch sizes are a requirement.

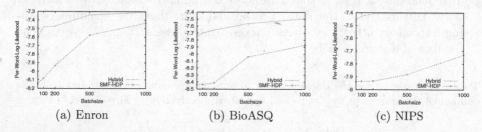

(a) Enron   (b) BioASQ   (c) NIPS

Fig. 1. Effect of batch size on the log-likelihood after 1000 updates. The truncation was set to 100 topics.

Second, we compare the two algorithm for different settings of the truncation for the number of topics (50, 100, 200, 500, 1000). The results are shown in Fig. 2. While the truncation does not seem to influence the performance of SMF-HDP for small batch sizes, our method has an improved performance for higher topic numbers on all three datasets. We see therefore, that it is not necessary to start with one topic and add more topics subsequently as was suggested by Wang and Blei [18]. Agreeing with the observations in previous work [3], we find that it is beneficial to start out with the maximum number of topics and subsequently reduce it. Overall, our method has a higher log-likelihood for all truncation settings.

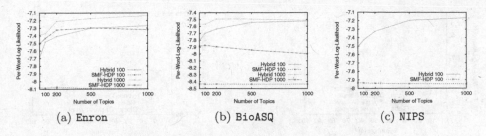

(a) Enron   (b) BioASQ   (c) NIPS

Fig. 2. Effect of truncation on the log-likelihood after 1000 updates. The batch size was set to 100 and 1000 documents as mentioned in the plot labels. For Enron we only used a batch size of 100 since larger batch sizes lead to a substantial increase in runtime. Our method has a higher log-likelihood for all settings.

Gibbs Sampling vs. Hybrid Approach. Since training and evaluation of the Gibbs sampler take too long on the large datasets, we also included the smaller KOS dataset in our experiments. We compare the convergence of the Gibbs sampler to the convergence of our method by measuring the testset per-word-log-likelihood after each iteration over the full dataset for the Gibbs sampler, and evaluating our method after each batch. Figure 3a shows the performance of four different methods trained with a truncation to 100 topics. We can see that the two hybrid methods converge much faster initially than the Gibbs samplers. Comparing the sparse and the original sampler, the sparse sampler is worse in the beginning since it does not sample from the true distribution, but manages to catch up and even surpass the original sampler due to its faster sampling[1].

Figure 3b shows the performance for a truncation to 1000 topics. We can see that here the difference in log-likelihood between the sparse and the original sampling method is much bigger. This is because the number of topics per document K_d does not grow that much when the number of total topics is increased. Therefore, while for small topic numbers the differences might be negligible in practice, for higher topic numbers, our sparse sampling method is preferable. This performance improvement in the experiments shows that sparseness is actually achieved and the assumption $K_d \ll K$ is justified in practice.

For the NIPS dataset, the difference between 100 and 1000 topics is even bigger (Fig. 3c and d). NIPS has very long documents which means that it has more topics per document on average. With only 100 topics, it is possible that almost all topics are present in the document. Therefore, the original sampler is faster than the sparse sampler for 100 topics, but not for 1000 topics, where it is the other way around. The Gibbs samplers have barely even started to converge in the first 1000 s where the convergence for the hybrid methods is far ahead.

---

[1] Note that our implementation of the hybrid method uses the sparse sampling method, but does not use the sparse updating introduced by Hoffman *et al.* [9]. Therefore, a further speedup is possible.

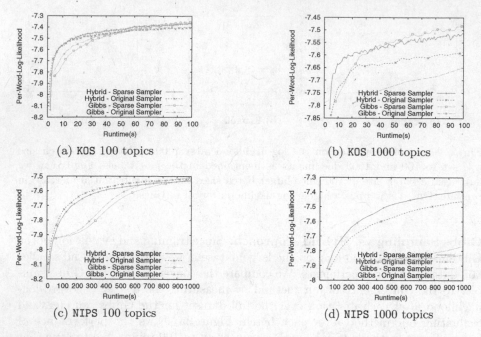

(a) KOS 100 topics   (b) KOS 1000 topics

(c) NIPS 100 topics   (d) NIPS 1000 topics

Fig. 3. Comparison of runtime with 100 and 1000 topics, respectively. The hybrid methods were trained with a batch size of 100 documents. Performance was evaluated on a separate testset. Our hybrid method with the sparse sampling algorithm converges faster than the other methods.

5 Related Work

A hybrid Variational-Gibbs method for parametric LDA topic models was introduced by Hoffman *et al.* [9]. In this work, a sparse update scheme was proposed that allowed to do variational updates for only the topic-word-combinations that were actually sampled. Experiments by Hoffman *et al.* showed that the method is faster especially for large topic numbers. Additionally the convergence is improved as compared to other variational methods since the variational distribution considered is not completely factorized but considers each document as a unity.

A doubly sparse method was build on top of the sparse hybrid model by Li *et al.* [14]. They used a fast Gibbs sampling method to further speed up sampling for parametric LDA. Thereby they exploited sparseness in the variational updates as well as the document-topic distributions. We do the same, only for nonparametric topic models.

Another extension of the original work by Hoffman *et al.* was proposed by Wang and Blei [18] who developed a similar method for the nonparametric HDP. Unfortunately, we were not able to compare to this method since the code is not publicly available. The main contribution of this work is the development of a truncation-free variational method that allows the number of topics to grow. This is made possible by the sampling step which does not depend on a truncation as in pure variational methods. In contrast to our work, their method builds on Teh's direct assignment sampler [16], whereas our method relies on the more advanced table indicator sampler proposed by Chen *et al.* [5]. Also, our method starts out with the maximum number of topics, and subsequently removes topics (by letting their expected counts approach zero over time). This was found to be beneficial in previous work [3]. Finally, their method is not sparse which is due to the direct assignment sampling scheme which cannot be made sparse as easily as the table indicator sampling scheme.

6   Conclusion

To conclude, we introduced a hybrid sparse Variational-Gibbs nonparametric topic model that can be trained online on large-scale or streaming datasets. Experiments on three large-scale test datasets as well as one smaller dataset were conducted. We found our method to be superior to the purely variational Bayes mean field approach in per-word log-likelihood. Additionally, it is more robust to different settings of the batch size. Compared to the pure Gibbs sampler it converges faster with improved log-likelihood. In the future, we would like to apply our method to hierarchical topic models with more levels.

References

1. Asuncion, A., Welling, M., Smyth, P., Teh, Y.W.: On smoothing and inference for topic models. In: Proceedings of the 25th Conference on Uncertainty in Artificial Intelligence, UAI 2009, pp. 27–34. AUAI Press, Arlington (2009)
2. Buntine, W., Hutter, M.: A Bayesian view of the Poisson-Dirichlet process (2010). arXiv preprint arXiv:1007.0296
3. Buntine, W.L., Mishra, S.: Experiments with non-parametric topic models. In: Proceedings of the 20th ACM SIGKDD International Conference on Knowledge Discovery and Data Mining, KDD 2014, pp. 881–890. ACM, New York (2014)
4. Burkhardt, S., Kramer, S.: Multi-label classification using stacked hierarchical Dirichlet processes with reduced sampling complexity. In: ICBK 2017 - International Conference on Big Knowledge. Hefei, China (2017, to appear)
5. Chen, C., Du, L., Buntine, W.: Sampling table configurations for the hierarchical Poisson-Dirichlet process. In: Gunopulos, D., Hofmann, T., Malerba, D., Vazirgiannis, M. (eds.) ECML PKDD 2011. LNCS (LNAI), vol. 6911, pp. 296–311. Springer, Heidelberg (2011). https://doi.org/10.1007/978-3-642-23780-5_29
6. Foulds, J., Boyles, L., DuBois, C., Smyth, P., Welling, M.: Stochastic collapsed variational Bayesian inference for latent Dirichlet allocation. In: Proceedings of the 19th ACM SIGKDD International Conference on Knowledge Discovery and Data Mining, KDD 2013, pp. 446–454. ACM, New York (2013)
7. Griffiths, T.L., Steyvers, M.: Finding scientific topics. In: Proceedings of the National Academy of Sciences of the United States of America. vol. 101, pp. 5228–5235. National Acad. Sciences (2004)
8. Heinrich, G.: Parameter estimation for text analysis. Technical report, Fraunhofer IGD (2004)

9. Hoffman, M., Blei, D.M., Mimno, D.M.: Sparse stochastic inference for latent Dirichlet allocation. In: Proceedings of the 29th International Conference on Machine Learning (ICML2012), pp. 1599–1606 (2012)

10. Hoffman, M.D., Blei, D.M., Bach, F.R.: Online learning for latent Dirichlet allocation. In: Lafferty, J.D., Williams, C.K.I., Shawe-Taylor, J., Zemel, R.S., Culotta, A. (eds.) Advances in Neural Information Processing Systems 23, pp. 856–864. Curran Associates, Inc., New York (2010)

11. Kingma, D.P., Welling, M.: Auto-encoding variational bayes. CoRR abs/1312.6114 (2013)

12. Klimt, B., Yang, Y.: Introducing the Enron corpus. In: CEAS 2004 - First Conference on Email and Anti-Spam, 30–31 July 2004, Mountain View, California, USA (2004)

13. Li, A.Q., Ahmed, A., Ravi, S., Smola, A.J.: Reducing the sampling complexity of topic models. In: Proceedings of the 20th ACM SIGKDD International Conference on Knowledge Discovery and Data Mining, KDD 2014, pp. 891–900. ACM, New York (2014)

14. Li, X., OuYang, J., Zhou, X.: Sparse hybrid variational-Gibbs algorithm for latent Dirichlet allocation. In: Proceedings of the 2016 SIAM International Conference on Data Mining, Miami, Florida, USA, 5–7 May 2016, pp. 729–737 (2016)

15. Perrone, V., Jenkins, P.A., Spano, D., Teh, Y.W.: Poisson random fields for dynamic feature models (2016). arXiv e-prints: arXiv:1611.07460

16. Teh, Y.W., Jordan, M.I., Beal, M.J., Blei, D.M.: Hierarchical Dirichlet processes. J. Am. Stat. Assoc. **101**, 1566–1581 (2004)

17. Tsatsaronis, G., Balikas, G., Malakasiotis, P., Partalas, I., Zschunke, M., Alvers, M.R., Weissenborn, D., Krithara, A., Petridis, S., Polychronopoulos, D., Almirantis, Y., Pavlopoulos, J., Baskiotis, N., Gallinari, P., Artieres, T., Ngonga, A., Heino, N., Gaussier, E., Barrio-Alvers, L., Schroeder, M., Androutsopoulos, I., Paliouras, G.: An overview of the bioasq large-scale biomedical semantic indexing and question answering competition. BMC Bioinform. **16**, 138 (2015)

18. Wang, C., Blei, D.M.: Truncation-free online variational inference for Bayesian nonparametric models. In: Pereira, F., Burges, C.J.C., Bottou, L., Weinberger, K.Q. (eds.) Advances in Neural Information Processing Systems, vol. 25, pp. 413–421. Curran Associates, Inc., New York (2012)

19. Wang, C., Paisley, J.W., Blei, D.M.: Online variational inference for the hierarchical Dirichlet process. In: AISTATS, vol. 15, pp. 752–760. PMLR, Fort Lauderdale (2011)

PAC-Bayesian Analysis for a Two-Step Hierarchical Multiview Learning Approach

Anil Goyal[1,2(✉)], Emilie Morvant[1],
Pascal Germain[3,4], and Massih-Reza Amini[2]

[1] Univ Lyon, UJM-Saint-Etienne, CNRS, Institut d'Optique Graduate School,
Laboratoire Hubert Curien UMR 5516, 42023 Saint-Etienne, France
{anil.goyal,emilie.morvant}@univ-st-etienne.fr
[2] Univ. Grenoble Alps, Laboratoire d'Informatique de Grenoble, AMA,
Centre Equation 4, BP 53, 38041 Grenoble Cedex 9, France
massih-reza.amini@imag.fr
[3] Département d'informatique de l'ENS École Normale Supérieure, CNRS,
PSL Research University, 75005 Paris, France
pascal.germain@inria.fr
[4] INRIA, Paris, France

Abstract. We study a two-level multiview learning with more than two views under the PAC-Bayesian framework. This approach, sometimes referred as late fusion, consists in learning sequentially multiple view-specific classifiers at the first level, and then combining these view-specific classifiers at the second level. Our main theoretical result is a generalization bound on the risk of the majority vote which exhibits a term of diversity in the predictions of the view-specific classifiers. From this result it comes out that controlling the trade-off between diversity and accuracy is a key element for multiview learning, which complements other results in multiview learning. Finally, we experiment our principle on multiview datasets extracted from the Reuters RCV1/RCV2 collection.

Keywords: PAC-Bayesian theory · Multiview learning

1 Introduction

With the ever-increasing observations produced by more than one source, multiview learning has been expanding over the past decade, spurred by the seminal work of Blum and Mitchell [4] on co-training. Most of the existing methods try to combine multimodal information, either by directly merging the views or by combining models learned from the different views[1] [28], in order to produce a model more reliable for the considered task. Our goal is to propose a theoretically

---

[1] The fusion of descriptions, *resp.* of models, is sometimes called Early Fusion, *resp.* Late Fusion.

© Springer International Publishing AG 2017
M. Ceci et al. (Eds.): ECML PKDD 2017, Part II, LNAI 10535, pp. 205–221, 2017.
https://doi.org/10.1007/978-3-319-71246-8_13

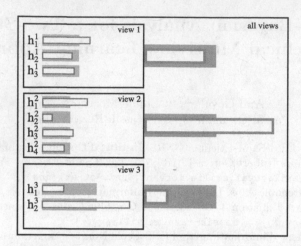

Fig. 1. Example of the multiview distributions hierarchy with 3 views. For all views $v \in \{1,2,3\}$, we have a set of voters $\mathcal{H}_v = \{h_1^v, \ldots, h_{n_v}^v\}$ on which we consider prior P_v view-specific distribution (in blue), and we consider a hyper-prior π distribution (in green) over the set of 3 views. The objective is to learn a posterior Q_v (in red) view-specific distributions and a hyper-posterior ρ distribution (in orange) leading to a good model. The length of a rectangle represents the weight (or probability) assigned to a voter or a view. (Color figure online)

grounded criteria to "correctly" combine the views. With this in mind we propose to study multiview learning through the PAC-Bayesian framework (introduced in [21]) that allows to derive generalization bounds for models that are expressed as a combination over a set of voters. When faced with learning from one view, the PAC-Bayesian theory assumes a prior distribution over the voters involved in the combination, and aims at learning—from the learning sample—a posterior distribution that leads to a well-performing combination expressed as a weighted majority vote. In this paper we extend the PAC-Bayesian theory to multiview with more than two views. Concretely, given a set of view-specific classifiers, we define a hierarchy of posterior and prior distributions over the views, such that (i) for each view v, we consider prior P_v and posterior Q_v distributions over each view-specific voters' set, and (ii) a prior π and a posterior ρ distribution over the set of views (see Fig. 1), respectively called hyper-prior and hyper-posterior[2]. In this way, our proposed approach encompasses the one of Amini et al. [1] that considered uniform distribution to combine the view-specific classifiers' predictions. Moreover, compared to the PAC-Bayesian work of Sun et al. [29], we are interested here to the more general and natural case of multiview learning with more than two views. Note also that Lecué and Rigollet [18]

---

[2] Our notion of hyper-prior and hyper-posterior distributions is different than the one proposed for lifelong learning [25], where they basically consider hyper-prior and hyper-posterior over the set of possible priors: The prior distribution P over the voters' set is viewed as a random variable.

proposed a non-PAC-Bayesian theoretical analysis of a combination of voters (called Q-Aggregation) that is able to take into account a prior and a posterior distribution but in a single-view setting.

Our theoretical study also includes a notion of disagreement between all the voters, allowing to take into account a notion of diversity between them which is known as a key element in multiview learning [1, 6, 13, 20]. Finally, we empirically evaluate a two-level learning approach on the Reuters RCV1/RCV2 corpus to show that our analysis is sound.

In the next section, we recall the general PAC-Bayesian setup, and present PAC-Bayesian *expectation bounds*—while most of the usual PAC-Bayesian bounds are *probabilistic bounds*. In Sect. 3, we then discuss the problem of multiview learning, adapting the PAC-Bayesian expectation bounds to the specificity of the two-level multiview approach. In Sect. 4, we discuss the relation between our analysis and previous works. Before concluding in Sect. 6, we present experimental results obtained on a collection of the Reuters RCV1/RCV2 corpus in Sect. 5.

2   The Single-View PAC-Bayesian Theorem

In this section, we state a *new* general mono-view PAC-Bayesian theorem, inspired by the work of Germain et al. [10], that we extend to multiview learning in Sect. 3.

2.1   Notations and Setting

We consider binary classification tasks on data drawn from a fixed yet unknown distribution $\mathcal{D}$ over $\mathcal{X} \times \mathcal{Y}$, where $\mathcal{X} \subseteq \mathbb{R}^d$ is a d-dimensional input space and $\mathcal{Y} = \{-1, +1\}$ the label/output set. A learning algorithm is provided with a training sample of m examples denoted by $S = \{(x_i, y_i)\}_{i=1}^m \in (\mathcal{X} \times \mathcal{Y})^m$, that is assumed to be independently and identically distributed (*i.i.d.*) according to $\mathcal{D}$. The notation $\mathcal{D}^m$ stands for the distribution of such a m-sample, and $\mathcal{D}_\mathcal{X}$ for the marginal distribution on $\mathcal{X}$. We consider a set $\mathcal{H}$ of classifiers or voters such that $\forall h \in \mathcal{H}$, $h : \mathcal{X} \to \mathcal{Y}$. In addition, PAC-Bayesian approach requires a prior distribution P over $\mathcal{H}$ that models *a priori* belief on the voters from $\mathcal{H}$ before the observation of the learning sample S. Given $S \sim \mathcal{D}^m$, t a posterior distribution Q over $\mathcal{H}$ leading to an accurate Q-weighted majority vote $B_Q(x)$ defined as

$$B_Q(x) = \text{sign}\left[\underset{h \sim Q}{\mathbb{E}} \, h(x)\right].$$

In other words, one wants to learn Q over $\mathcal{H}$ such that it minimizes the true risk $R_\mathcal{D}(B_Q)$ of $B_Q(x)$:

$$R_\mathcal{D}(B_Q) = \underset{(x,y) \sim \mathcal{D}}{\mathbb{E}} \mathbb{1}_{[B_Q(x) \neq y]},$$

where $\mathbb{1}_{[\pi]} = 1$ if predicate π holds, and 0 otherwise. However, a PAC-Bayesian generalization bound does not directly focus on the risk of the deterministic Q-weighted majority vote B_Q. Instead, it upper-bounds the risk of the stochastic Gibbs classifier G_Q, which predicts the label of an example x by drawing h from $\mathcal{H}$ according to the posterior distribution Q and predicts $h(x)$. Therefore, the true risk $R_D(G_Q)$ of the Gibbs classifier on a data distribution $\mathcal{D}$, and its empirical risk $R_S(G_Q)$ estimated on a sample $S \sim \mathcal{D}^m$ are respectively given by

$$R_D(G_Q) = \underset{(x,y)\sim\mathcal{D}}{\mathbb{E}} \underset{h\sim Q}{\mathbb{E}} \mathbb{1}_{[h(x)\neq y]}, \text{ and } R_S(G_Q) = \frac{1}{m}\sum_{i=1}^{m} \underset{h\sim Q}{\mathbb{E}} \mathbb{1}_{[h(x_i)\neq y_i]}.$$

The above Gibbs classifier is closely related to the Q-weighted majority vote B_Q. Indeed, if B_Q misclassifies $x \in \mathcal{X}$, then at least half of the classifiers (under measure Q) make an error on x. Therefore, we have

$$R_D(B_Q) \leq 2R_D(G_Q). \tag{1}$$

Thus, an upper bound on $R_D(G_Q)$ gives rise to an upper bound on $R_D(B_Q)$. Other tighter relations exist [10,14,16], such as the so-called C-Bound [14] that involves the *expected disagreement* $d_D(Q)$ between all the pair of voters, and that can be expressed as follows (when $R_D(G_Q) \leq \frac{1}{2}$):

$$R_D(B_Q) \leq 1 - \frac{(1 - 2R_D(G_Q))^2}{1 - 2d_D(Q)}, \text{ where } d_D(Q) = \underset{x\sim\mathcal{D}_\mathcal{X}}{\mathbb{E}} \underset{(h,h')\sim Q^2}{\mathbb{E}} \mathbb{1}_{[h(x)\neq h'(x)]}. \tag{2}$$

Moreover, Germain et al. [10] have shown that the Gibbs classifier's risk can be rewritten in terms of $d_D(Q)$ and *expected joint error* $e_D(Q)$ between all the pair of voters as

$$R_D(G_Q) = \frac{1}{2}d_D(Q) + e_D(Q), \tag{3}$$

$$\text{where } e_D(Q) = \underset{(x,y)\sim\mathcal{D}}{\mathbb{E}} \underset{(h,h')\sim Q^2}{\mathbb{E}} \mathbb{1}_{[h(x)\neq y]} \mathbb{1}_{[h'(x)\neq y]}.$$

It is worth noting that from multiview learning standpoint where the notion of diversity among voters is known to be important [1,2,13,20,29], Eqs. (2) and (3) directly capture the trade-off between diversity and accuracy. Indeed, $d_D(Q)$ involves the diversity between voters [23], while $e_D(Q)$ takes into account the errors. Note that the principle of controlling the trade-off between diversity and accuracy through the C-bound of Eq. (2) has been exploited by Laviolette et al. [17] and Roy et al. [26] to derive well-performing PAC-Bayesian algorithms that aims at minimizing it. For our experiments in Sect. 5, we make use of CqBoost [26]—one of these algorithms—for multiview learning.

Last but not least, PAC-Bayesian generalization bounds take into account the given prior distribution P on $\mathcal{H}$ through the Kullback-Leibler divergence between the learned posterior distribution Q and P:

$$\mathrm{KL}(Q\|P) = \mathop{\mathbb{E}}_{h \sim Q} \ln \frac{Q(h)}{P(h)}.$$

2.2   A New PAC-Bayesian Theorem as an Expected Risk Bound

In the following we introduce a new variation of the general PAC-Bayesian theorem of Germain et al. [9,10]; it takes the form of an upper bound on the "deviation" between the true risk $R_{\mathcal{D}}(G_Q)$ and empirical risk $R_S(G_Q)$ of the Gibbs classifier, according to a convex function $D{:}[0,1] \times [0,1] \to \mathbb{R}$. While most of the PAC-Bayesian bounds are probabilistic bounds, we state here an *expected risk bound*. More specifically, Theorem 1 below is a tool to upper-bound $\mathbb{E}_{S \sim \mathcal{D}^m} R_{\mathcal{D}}(G_{Q_S})$—where Q_S is the posterior distribution outputted by a given learning algorithm after observing the learning sample S—while PAC-Bayes usually bounds $R_{\mathcal{D}}(G_Q)$ uniformly for all distribution Q, but with high probability over the draw of $S \sim \mathcal{D}^m$. Since by definition posterior distributions are data dependent, this different point of view on PAC-Bayesian analysis has the advantage to involve an expectation over all the possible learning samples (of a given size) in bounds itself.

Theorem 1. *For any distribution $\mathcal{D}$ on $\mathcal{X} \times \mathcal{Y}$, for any set of voters $\mathcal{H}$, for any prior distribution P on $\mathcal{H}$, for any convex function $D : [0,1] \times [0,1] \to \mathbb{R}$, we have*

$$D \left(\mathop{\mathbb{E}}_{S \sim \mathcal{D}^m} R_S(G_{Q_S}), \mathop{\mathbb{E}}_{S \sim \mathcal{D}^m} R_{\mathcal{D}}(G_{Q_S}) \right)$$
$$\leq \frac{1}{m} \left[\mathop{\mathbb{E}}_{S \sim \mathcal{D}^m} \mathrm{KL}(Q_S\|P) + \ln \left(\mathop{\mathbb{E}}_{S \sim \mathcal{D}^m} \mathop{\mathbb{E}}_{h \sim P} e^{m\, D(R_S(h), R_{\mathcal{D}}(h))} \right) \right],$$

where $R_{\mathcal{D}}(h)$ and $R_S(h)$ are respectively the true and the empirical risks of individual voters.

Similarly to Germain et al. [9,10], by selecting a well-suited deviation function D and by upper-bounding $\mathbb{E}_S \mathbb{E}_h e^{m\, D(R_S(h), R_{\mathcal{D}}(h))}$, we can prove the *expected bound* counterparts of the classical PAC-Bayesian theorems of Catoni [5], McAllester [21], Seeger [27]. The proof presented below borrows the straightforward proof technique of Bégin et al. [3]. Interestingly, this approach highlights that the expectation bounds are obtained simply by replacing the *Markov inequality* by the *Jensen inequality* (respectively Theorems 5 and 6, in Appendix).

Proof of Theorem 1. The last three inequalities below are obtained by applying Jensen's inequality on the convex function D, the change of measure inequality [as stated by [3], Lemma 3], and Jensen's inequality on the concave function ln.

$$mD\left(\underset{S\sim\mathcal{D}^m}{\mathbb{E}} R_S(G_{Q_S}), \underset{S\sim\mathcal{D}^m}{\mathbb{E}} R_{\mathcal{D}}(G_{Q_S})\right)$$

$$= mD\left(\underset{S\sim\mathcal{D}^m}{\mathbb{E}} \underset{h\sim Q_S}{\mathbb{E}} R_S(h), \underset{S\sim\mathcal{D}^m}{\mathbb{E}} \underset{h\sim Q_S}{\mathbb{E}} R_{\mathcal{D}}(h)\right)$$

$$\leq \underset{S\sim\mathcal{D}^m}{\mathbb{E}} \underset{h\sim Q_S}{\mathbb{E}} mD\left(R_S(h), R_{\mathcal{D}}(h)\right)$$

$$\leq \underset{S\sim\mathcal{D}^m}{\mathbb{E}} \left[\mathrm{KL}(Q_S\|P) + \ln\left(\underset{h\sim P}{\mathbb{E}} e^{m\,D(R_S(h),R_{\mathcal{D}}(h))}\right)\right]$$

$$\leq \underset{S\sim\mathcal{D}^m}{\mathbb{E}} \mathrm{KL}(Q_S\|P) + \ln\left(\underset{S\sim\mathcal{D}^m}{\mathbb{E}} \underset{h\sim P}{\mathbb{E}} e^{m\,D(R_S(h),R_{\mathcal{D}}(h))}\right).$$

$\square$

Since the C-bound of Eq. (2) involves the expected disagreement $d_{\mathcal{D}}(Q)$, we also derive below the expected bound that upper-bounds the deviation between $\mathbb{E}_{S\sim\mathcal{D}^m} d_S(Q_S)$ and $\mathbb{E}_{S\sim\mathcal{D}^m} d_{\mathcal{D}}(Q_S)$ under a convex function D. Theorem 2 can be seen as the *expectation* version of probabilistic bounds over $d_S(Q_S)$ proposed by Germain et al. [10], Lacasse et al. [14].

Theorem 2. *For any distribution $\mathcal{D}$ on $\mathcal{X}\times\mathcal{Y}$, for any set of voters $\mathcal{H}$, for any prior distribution P on $\mathcal{H}$, for any convex function $D : [0,1]\times[0,1] \to \mathbb{R}$, we have*

$$D\left(\underset{S\sim\mathcal{D}^m}{\mathbb{E}} d_S(Q_S), \underset{S\sim\mathcal{D}^m}{\mathbb{E}} d_{\mathcal{D}}(Q_S)\right)$$

$$\leq \frac{2}{m}\left[\underset{S\sim\mathcal{D}^m}{\mathbb{E}} \mathrm{KL}(Q_S\|P) + \ln\sqrt{\underset{S\sim\mathcal{D}^m}{\mathbb{E}} \underset{(h,h')\sim P^2}{\mathbb{E}} e^{mD(d_S(h,h'),d_{\mathcal{D}}(h,h'))}}\right],$$

where $d_{\mathcal{D}}(h,h') = \mathbb{E}_{x\sim\mathcal{D}_{\mathcal{X}}} \mathbb{1}_{[h(x)\neq h'(x)]}$ is the disagreement of voters h and h' on the distribution $\mathcal{D}$, and $d_S(h,h')$ is its empirical counterpart.

Proof. First, we apply the exact same steps as in the proof of Theorem 1:

$$mD\left(\underset{S\sim\mathcal{D}^m}{\mathbb{E}} d_S(Q_S), \underset{S\sim\mathcal{D}^m}{\mathbb{E}} d_{\mathcal{D}}(Q_S)\right)$$

$$= mD\left(\underset{S\sim\mathcal{D}^m}{\mathbb{E}} \underset{(h,h')\sim Q_S^2}{\mathbb{E}} d_S(h,h'), \underset{S\sim\mathcal{D}^m}{\mathbb{E}} \underset{(h,h')\sim Q_S^2}{\mathbb{E}} d_{\mathcal{D}}(h,h')\right)$$

$$\vdots$$

$$\leq \underset{S\sim\mathcal{D}^m}{\mathbb{E}} \mathrm{KL}(Q_S^2\|P^2) + \ln \underset{S\sim\mathcal{D}^m}{\mathbb{E}} \underset{(h,h')\sim P^2}{\mathbb{E}} e^{mD(d_S(h,h'),d_{\mathcal{D}}(h,h'))}.$$

Then, we use the fact that $\mathrm{KL}(Q_S^2\|P^2) = 2\,\mathrm{KL}(Q_S\|P)$ [see [10], Theorem 25]. $\square$

In the following we provide an extension of this PAC-Bayesian framework to multiview learning with more than two views.

3   Multiview PAC-Bayesian Approach

3.1   Notations and Setting

We consider binary classification problems where the multiview observations $\mathbf{x} = (x^1, \ldots, x^V)$ belong to a multiview input set $\mathcal{X} = \mathcal{X}_1 \times \ldots \times \mathcal{X}_V$, where $V \geq 2$ is the number of views of not-necessarily the same dimension. We denote $\mathcal{V}$ the set of the V views. In binary classification, we assume that examples are pairs $(\mathbf{x}, y)$, with $y \in \mathcal{Y} = \{-1, +1\}$, drawn according to an unknown distribution $\mathcal{D}$ over $\mathcal{X} \times \mathcal{Y}$. To model the two-level multiview approach, we follow the next setting. For each view $v \in \mathcal{V}$, we consider a view-specific set $\mathcal{H}_v$ of voters $h : \mathcal{X}_v \to \mathcal{Y}$, and a prior distribution P_v on $\mathcal{H}_v$. Given a *hyper-prior* distribution π over the views $\mathcal{V}$, and a multiview learning sample $S = \{(\mathbf{x}_i, y_i)\}_{i=1}^m \sim (\mathcal{D})^m$, our PAC-Bayesian learner objective is twofold: *(i)* finding a posterior distribution Q_v over $\mathcal{H}_v$ for all views $v \in \mathcal{V}$; *(ii)* finding a *hyper-posterior* distribution ρ on the set of views $\mathcal{V}$. This hierarchy of distributions is illustrated by Fig. 1. The learned distributions express a multiview weighted majority vote B_ρ^{MV} defined as

$$B_\rho^{\mathrm{MV}}(\mathbf{x}) = \mathrm{sign} \left[\mathop{\mathbb{E}}_{v \sim \rho} \mathop{\mathbb{E}}_{h \sim Q_v} h(x^v) \right].$$

Thus, the learner aims at constructing the posterior and hyper-posterior distributions that minimize the true risk $R_\mathcal{D}(B_\rho^{\mathrm{MV}})$ of the multiview weighted majority vote:

$$R_\mathcal{D}(B_\rho^{\mathrm{MV}}) = \mathop{\mathbb{E}}_{(\mathbf{x}, y) \sim \mathcal{D}} \mathbb{1}_{[B_\rho^{\mathrm{MV}}(\mathbf{x}) \neq y]}.$$

As pointed out in Sect. 2, the PAC-Bayesian approach deals with the risk of the stochastic Gibbs classifier G_ρ^{MV} defined as follows in our multiview setting, and that can be rewritten in terms of *expected disagreement* $d_\mathcal{D}^{\mathrm{MV}}(\rho)$ and *expected joint error* $e_\mathcal{D}^{\mathrm{MV}}(\rho)$:

$$\begin{aligned}
R_\mathcal{D}(G_\rho^{\mathrm{MV}}) &= \mathop{\mathbb{E}}_{(\mathbf{x}, y) \sim \mathcal{D}} \mathop{\mathbb{E}}_{v \sim \rho} \mathop{\mathbb{E}}_{h \sim Q_v} \mathbb{1}_{[h(x^v) \neq y]} \\
&= \tfrac{1}{2} d_\mathcal{D}^{\mathrm{MV}}(\rho) + e_\mathcal{D}^{\mathrm{MV}}(\rho),
\end{aligned} \tag{4}$$

$$\text{where } d_\mathcal{D}^{\mathrm{MV}}(\rho) = \mathop{\mathbb{E}}_{\mathbf{x} \sim \mathcal{D}_\mathcal{X}} \mathop{\mathbb{E}}_{v \sim \rho} \mathop{\mathbb{E}}_{v' \sim \rho} \mathop{\mathbb{E}}_{h \sim Q_v} \mathop{\mathbb{E}}_{h' \sim Q_{v'}} \mathbb{1}_{[h(x^v) \neq h'(x^{v'})]},$$

$$\text{and } e_\mathcal{D}^{\mathrm{MV}}(\rho) = \mathop{\mathbb{E}}_{(\mathbf{x}, y) \sim \mathcal{D}} \mathop{\mathbb{E}}_{v \sim \rho} \mathop{\mathbb{E}}_{v' \sim \rho} \mathop{\mathbb{E}}_{h \sim Q_v} \mathop{\mathbb{E}}_{h' \sim Q_{v'}} \mathbb{1}_{[h(x^v) \neq y]} \mathbb{1}_{[h'(x^{v'}) \neq y]}.$$

Obviously, the empirical counterpart of the Gibbs classifier's risk $R_\mathcal{D}(G_\rho^{\mathrm{MV}})$ is

$$\begin{aligned}
R_S(G_\rho^{\mathrm{MV}}) &= \frac{1}{m} \sum_{i=1}^m \mathop{\mathbb{E}}_{v \sim \rho} \mathop{\mathbb{E}}_{h \sim Q_v} \mathbb{1}_{[h(x_i^v) \neq y_i]} \\
&= \frac{1}{2} d_S^{\mathrm{MV}}(\rho) + e_S^{\mathrm{MV}}(\rho)
\end{aligned}$$

where $d_S^{\mathrm{MV}}(\rho)$ and $e_S^{\mathrm{MV}}(\rho)$ are respectively the empirical estimations of $d_{\mathcal{D}}^{\mathrm{MV}}(\rho)$ and $e_{\mathcal{D}}^{\mathrm{MV}}(\rho)$ on the learning sample S. As in the single-view PAC-Bayesian setting, the multiview weighted majority vote B_ρ^{MV} is closely related to the stochastic multiview Gibbs classifier G_ρ^{MV}, and a generalization bound for G_ρ^{MV} gives rise to a generalization bound for B_ρ^{MV}. Indeed, it is easy to show that $R_{\mathcal{D}}(B_\rho^{\mathrm{MV}}) \leq 2R_{\mathcal{D}}(G_\rho^{\mathrm{MV}})$, meaning that an upper bound over $R_{\mathcal{D}}(G_\rho^{\mathrm{MV}})$ gives an upper bound for the majority vote. Moreover the C-Bound of Eq. (2) can be extended to our multiview setting by Lemma 1 below. Equation (5) is a straightforward generalization of the single-view C-bound of Eq. (2). Afterward, Eq. (6) is obtained by rewriting $R_{\mathcal{D}}(G_\rho^{\mathrm{MV}})$ as the ρ-average of the risk associated to each view, and lower-bounding $d_{\mathcal{D}}^{\mathrm{MV}}(\rho)$ by the ρ-average of the disagreement associated to each view.

Lemma 1. *Let $V \geq 2$ be the number of views. For all posterior $\{Q_v\}_{v=1}^V$ and hyper-posterior ρ distribution, if $R_{\mathcal{D}}(G_\rho^{\mathrm{MV}}) < \frac{1}{2}$, then we have*

$$R_{\mathcal{D}}(B_\rho^{\mathrm{MV}}) \leq 1 - \frac{\left(1 - 2R_{\mathcal{D}}(G_\rho^{\mathrm{MV}})\right)^2}{1 - 2d_{\mathcal{D}}^{\mathrm{MV}}(\rho)} \tag{5}$$

$$\leq 1 - \frac{\left(1 - 2\mathbb{E}_{v \sim \rho} R_{\mathcal{D}}(G_{Q_v})\right)^2}{1 - 2\mathbb{E}_{v \sim \rho} d_{\mathcal{D}}(Q_v)}. \tag{6}$$

Proof. Equation (5) follows from the Cantelli-Chebyshev's inequality (Theorem 7, in Appendix). To prove Eq. (6), we first notice that in the binary setting where $y \in \{-1, 1\}$ and $h : \mathcal{X} \to \{-1, 1\}$, we have $\mathbb{1}_{[h(x^v) \neq y]} = \frac{1}{2}(1 - y\,h(x^v))$, and

$$R_{\mathcal{D}}(G_\rho^{\mathrm{MV}}) = \mathop{\mathbb{E}}_{(\mathbf{x},y) \sim \mathcal{D}} \mathop{\mathbb{E}}_{v \sim \rho} \mathop{\mathbb{E}}_{h \sim Q_v} \mathbb{1}_{[h(x^v) \neq y]}$$

$$= \frac{1}{2}\left(1 - \mathop{\mathbb{E}}_{(\mathbf{x},y) \sim \mathcal{D}} \mathop{\mathbb{E}}_{v \sim \rho} \mathop{\mathbb{E}}_{h \sim Q_v} y\,h(x^v)\right)$$

$$= \mathop{\mathbb{E}}_{v \sim \rho} R_{\mathcal{D}}(G_{Q^v}).$$

Moreover, we have

$$d_{\mathcal{D}}^{\mathrm{MV}}(\rho) = \mathop{\mathbb{E}}_{\mathbf{x} \sim \mathcal{D}_{\mathcal{X}}} \mathop{\mathbb{E}}_{v \sim \rho} \mathop{\mathbb{E}}_{v' \sim \rho} \mathop{\mathbb{E}}_{h \sim Q_v} \mathop{\mathbb{E}}_{h' \sim Q_{v'}} \mathbb{1}_{[h(x^v) \neq h'(x^{v'})]}$$

$$= \frac{1}{2}\left(1 - \mathop{\mathbb{E}}_{\mathbf{x} \sim \mathcal{D}_{\mathcal{X}}} \mathop{\mathbb{E}}_{v \sim \rho} \mathop{\mathbb{E}}_{v' \sim \rho} \mathop{\mathbb{E}}_{h \sim Q_v} \mathop{\mathbb{E}}_{h \sim Q_{v'}} h(x^v) \times h'(x^{v'})\right)$$

$$= \frac{1}{2}\left(1 - \mathop{\mathbb{E}}_{\mathbf{x} \sim \mathcal{D}_{\mathcal{X}}} \left[\mathop{\mathbb{E}}_{v \sim \rho} \mathop{\mathbb{E}}_{h \sim Q_v} h(x^v)\right]^2\right).$$

From Jensen's inequality (Theorem 6, in Appendix) it comes

$$d_{\mathcal{D}}^{\mathrm{MV}}(\rho) \geq \frac{1}{2}\left(1 - \mathop{\mathbb{E}}_{\mathbf{x}\sim\mathcal{D}_{\mathcal{X}}} \mathop{\mathbb{E}}_{v\sim\rho}\left[\mathop{\mathbb{E}}_{h\sim Q_v} h(x^v)\right]^2\right)$$

$$= \mathop{\mathbb{E}}_{v\sim\rho}\left[\frac{1}{2}\left(1 - \mathop{\mathbb{E}}_{\mathbf{x}\sim\mathcal{D}_{\mathcal{X}}}\left[\mathop{\mathbb{E}}_{h\sim Q_v} h(x^v)\right]^2\right)\right]$$

$$= \mathop{\mathbb{E}}_{v\sim\rho} d_{\mathcal{D}}(Q_v).$$

By replacing $R_{\mathcal{D}}(G_\rho^{\mathrm{MV}})$ and $d_{\mathcal{D}}^{\mathrm{MV}}(\rho)$ in Eq. (5), we obtain

$$1 - \frac{\left(1 - 2R_{\mathcal{D}}(G_\rho^{\mathrm{MV}})\right)^2}{1 - 2d_{\mathcal{D}}^{\mathrm{MV}}(\rho)} \leq 1 - \frac{\left(1 - 2\mathbb{E}_{v\sim\rho}R_{\mathcal{D}}(G_{Q_v})\right)^2}{1 - 2\mathbb{E}_{v\sim\rho}d_{\mathcal{D}}(Q_v)}.$$

$\square$

Similarly than for the mono-view setting, Eqs. (4) and (5) suggest that a good trade-off between the risk of the Gibbs classifier G_ρ^{MV} and the disagreement $d_{\mathcal{D}}^{\mathrm{MV}}(\rho)$ between pairs of voters will lead to a well-performing majority vote. Equation (6) exhibits the role of diversity among the views thanks to the disagreement's expectation over the views $\mathbb{E}_{v\sim\rho}d_{\mathcal{D}}(Q_v)$.

3.2 General Multiview PAC-Bayesian Theorems

Now we state our general PAC-Bayesian theorem suitable for the above multiview learning setting with a two-level hierarchy of distributions over views (or voters). A key step in PAC-Bayesian proofs is the use of a *change of measure inequality* [22], based on the Donsker-Varadhan inequality [8]. Lemma 2 below extends this tool to our multiview setting.

Lemma 2. *For any set of priors $\{P_v\}_{v=1}^V$ and any set of posteriors $\{Q_v\}_{v=1}^V$, for any hyper-prior distribution π on views $\mathcal{V}$ and hyper-posterior distribution ρ on $\mathcal{V}$, and for any measurable function $\phi : \mathcal{H}_v \to \mathbb{R}$, we have*

$$\mathop{\mathbb{E}}_{v\sim\rho} \mathop{\mathbb{E}}_{h\sim Q_v} \phi(h) \leq \mathop{\mathbb{E}}_{v\sim\rho} \mathrm{KL}(Q_v\|P_v) + \mathrm{KL}(\rho\|\pi) + \ln\left(\mathop{\mathbb{E}}_{v\sim\pi} \mathop{\mathbb{E}}_{h\sim P_v} e^{\phi(h)}\right).$$

Proof. We have

$$\mathop{\mathbb{E}}_{v\sim\rho} \mathop{\mathbb{E}}_{h\sim Q_v} \phi(h) = \mathop{\mathbb{E}}_{v\sim\rho} \mathop{\mathbb{E}}_{h\sim Q_v} \ln e^{\phi(h)}$$

$$= \mathop{\mathbb{E}}_{v\sim\rho} \mathop{\mathbb{E}}_{h\sim Q_v} \ln\left(\frac{Q_v(h)}{P_v(h)}\frac{P_v(h)}{Q_v(h)}e^{\phi(h)}\right)$$

$$= \mathop{\mathbb{E}}_{v\sim\rho}\left[\mathop{\mathbb{E}}_{h\sim Q_v} \ln\left(\frac{Q_v(h)}{P_v(h)}\right) + \mathop{\mathbb{E}}_{h\sim Q_v} \ln\left(\frac{P_v(h)}{Q_v(h)}e^{\phi(h)}\right)\right].$$

According to the Kullback-Leibler definition, we have

$$\mathop{\mathbb{E}}_{v\sim\rho} \mathop{\mathbb{E}}_{h\sim Q_v} \phi(h) = \mathop{\mathbb{E}}_{v\sim\rho}\left[\mathrm{KL}(Q_v\|P_v) + \mathop{\mathbb{E}}_{h\sim Q_v} \ln\left(\frac{P_v(h)}{Q_v(h)}e^{\phi(h)}\right)\right].$$

By applying Jensen's inequality (Theorem 6, in Appendix) on the concave function ln, we have

$$
\mathop{\mathbb{E}}_{v\sim\rho}\mathop{\mathbb{E}}_{h\sim Q_v}\phi(h) \le \mathop{\mathbb{E}}_{v\sim\rho}\left[\mathrm{KL}(Q_v\|P_v) + \ln\left(\mathop{\mathbb{E}}_{h\sim P_v}e^{\phi(h)}\right)\right]
$$

$$
= \mathop{\mathbb{E}}_{v\sim\rho}\mathrm{KL}(Q_v\|P_v) + \mathop{\mathbb{E}}_{v\sim\rho}\ln\left(\frac{\rho(v)}{\pi(v)}\frac{\pi(v)}{\rho(v)}\mathop{\mathbb{E}}_{h\sim P_v}e^{\phi(h)}\right)
$$

$$
= \mathop{\mathbb{E}}_{v\sim\rho}\mathrm{KL}(Q_v\|P_v) + \mathrm{KL}(\rho\|\pi) + \mathop{\mathbb{E}}_{v\sim\rho}\ln\left(\frac{\pi(v)}{\rho(v)}\mathop{\mathbb{E}}_{h\sim P_v}e^{\phi(h)}\right).
$$

Finally, we apply again the Jensen inequality (Theorem 6) on ln to obtain the lemma.   □

Based on Lemma 2, the following theorem can be seen as a generalization of Theorem 1 to multiview. Note that we still rely on a general convex function $D : [0,1] \times [0,1] \to \mathbb{R}$, that measures the "deviation" between the empirical disagreement/joint error and the true risk of the Gibbs classifier.

Theorem 3. *Let $V \ge 2$ be the number of views. For any distribution $\mathcal{D}$ on $\mathcal{X}\times\mathcal{Y}$, for any set of prior distributions $\{P_v\}_{v=1}^V$, for any hyper-prior distribution π over $\mathcal{V}$, for any convex function $D : [0,1]\times[0,1]\to\mathbb{R}$, we have*

$$
D\Big(\tfrac{1}{2}\mathop{\mathbb{E}}_{S\sim\mathcal{D}^m}d_S^{\mathrm{MV}}(\rho_S) + \mathop{\mathbb{E}}_{S\sim\mathcal{D}^m}e_S^{\mathrm{MV}}(\rho_S), \mathop{\mathbb{E}}_{S\sim\mathcal{D}^m}R_{\mathcal{D}}(G_{\rho_S}^{\mathrm{MV}})\Big) \le \frac{1}{m}\Bigg[\mathop{\mathbb{E}}_{S\sim\mathcal{D}^m}\mathop{\mathbb{E}}_{v\sim\rho_S}\mathrm{KL}(Q_{v,S}\|P_v)
$$

$$
+ \mathop{\mathbb{E}}_{S\sim\mathcal{D}^m}\mathrm{KL}(\rho_S\|\pi) + \ln\left(\mathop{\mathbb{E}}_{S\sim\mathcal{D}^m}\mathop{\mathbb{E}}_{v\sim\pi}\mathop{\mathbb{E}}_{h\sim P_v}e^{mD(R_S(h),R_{\mathcal{D}}(h))}\right)\Bigg].
$$

Proof. We follow the same steps as in Theorem 1 proof.

$$
mD\Big(\mathop{\mathbb{E}}_{S\sim\mathcal{D}^m}R_S(G_{\rho_S}^{\mathrm{MV}}), \mathop{\mathbb{E}}_{S\sim\mathcal{D}^m}R_{\mathcal{D}}(G_{\rho_S}^{\mathrm{MV}})\Big)
$$

$$
= mD\Big(\mathop{\mathbb{E}}_{S\sim\mathcal{D}^m}\mathop{\mathbb{E}}_{v\sim\rho_S}\mathop{\mathbb{E}}_{h\sim Q_{v,S}}R_S(h), \mathop{\mathbb{E}}_{S\sim\mathcal{D}^m}\mathop{\mathbb{E}}_{v\sim\rho_S}\mathop{\mathbb{E}}_{h\sim Q_{v,S}}R_{\mathcal{D}}(h)\Big)
$$

$$
\le \mathop{\mathbb{E}}_{S\sim\mathcal{D}^m}\mathop{\mathbb{E}}_{v\sim\rho_S}\mathop{\mathbb{E}}_{h\sim Q_{v,S}}mD\left(R_S(h), R_{\mathcal{D}}(h)\right)
$$

$$
\le \mathop{\mathbb{E}}_{S\sim\mathcal{D}^m}\left[\mathop{\mathbb{E}}_{v\sim\rho_S}\mathrm{KL}(Q_{v,S}\|P_v) + \mathrm{KL}(\rho_S\|\pi) + \ln\left(\mathop{\mathbb{E}}_{v\sim\pi}\mathop{\mathbb{E}}_{h\sim P_v}e^{mD(R_S(h),R_{\mathcal{D}}(h))}\right)\right],
$$

where the last inequality is obtained using Lemma 2. After distributing the expectation of $S \sim \mathcal{D}^m$, the final statement follows from Jensen's inequality (Theorem 6)

$$
\mathop{\mathbb{E}}_{S\sim\mathcal{D}^m}\ln\left(\mathop{\mathbb{E}}_{v\sim\pi}\mathop{\mathbb{E}}_{h\sim P_v}e^{mD(R_S(h),R_{\mathcal{D}}(h))}\right) \le \ln\left(\mathop{\mathbb{E}}_{S\sim\mathcal{D}^m}\mathop{\mathbb{E}}_{v\sim\pi}\mathop{\mathbb{E}}_{h\sim P_v}e^{mD(R_S(h),R_{\mathcal{D}}(h))}\right),
$$

and from Eq. (3): $R_S(G_{\rho_S}^{\mathrm{MV}}) = \tfrac{1}{2}d_S^{\mathrm{MV}}(\rho_S) + e_S^{\mathrm{MV}}(\rho_S)$.   □

It is interesting to compare this generalization bound to Theorem 1. The main difference relies on the introduction of view-specific prior and posterior distributions, which mainly leads to an additional term $\mathbb{E}_{v\sim\rho}\mathrm{KL}(Q_v\|P_v)$, expressed as the expectation of the view-specific Kullback-Leibler divergence term over the views $\mathcal{V}$ according to the hyper-posterior distribution ρ. We also introduce the empirical disagreement allowing us to directly highlight the presence of the diversity between voters and between views. As Theorems 1 and 3 provides a tool to derive PAC-Bayesian generalization bounds for a multiview supervised learning setting. Indeed, by making use of the same trick as Germain et al. [9,10], the generalization bounds can be derived from Theorem 3 by choosing a suitable convex function D and upper-bounding $\mathbb{E}_S \mathbb{E}_v \mathbb{E}_h e^{m\,D(R_S(h),R_{\mathcal{D}}(h))}$. We provide an example of such a specialization in Sect. 3.3, by following McAllester's [21] point of view. Note that we provide the specialization to the two other classical PAC-Bayesian approaches of Catoni [5], Langford [15], Seeger [27] in our research report Goyal et al. [11, Sect. 3.3.].

Following the same approach, we can obtain a mutiview bound for the expected disagreement.

Theorem 4. *Let $V \geq 2$ be the number of views. For any distribution $\mathcal{D}$ on $\mathcal{X} \times \mathcal{Y}$, for any set of prior distributions $\{P_v\}_{v=1}^{V}$, for any hyper-prior distribution π over $\mathcal{V}$, for any convex function $D : [0,1] \times [0,1] \to \mathbb{R}$, we have*

$$D\Big(\underset{S\sim\mathcal{D}^m}{\mathbb{E}}\, d_S^{\mathrm{MV}}(\rho_S),\ \underset{S\sim\mathcal{D}^m}{\mathbb{E}}\, d_{\mathcal{D}}^{\mathrm{MV}}(\rho_S) \Big) \leq \frac{2}{m}\Bigg[\underset{S\sim\mathcal{D}^m}{\mathbb{E}}\ \underset{v\sim\rho_S}{\mathbb{E}}\, \mathrm{KL}(Q_{v,S}\|P_v) + \underset{S\sim\mathcal{D}^m}{\mathbb{E}}\, \mathrm{KL}(\rho_S\|\pi)$$
$$+ \ln\sqrt{ \underset{S\sim\mathcal{D}^m}{\mathbb{E}}\ \underset{(h,h')\sim P^2}{\mathbb{E}}\, e^{m D(d_S(h,h'),d_{\mathcal{D}}(h,h'))} }\ \Bigg].$$

Proof. The result is obtained straightforwardly by following the proof steps of Theorem 3, using the disagreement instead of the Gibbs risk. Then, similarly at what we have done to obtain Theorem 2, we substitute $\mathrm{KL}(Q_{v,S}^2\|P_v^2)$ by $2\,\mathrm{KL}(Q_{v,S}\|P_v)$, and $\mathrm{KL}(\rho_S^2\|\pi^2)$ by $2\,\mathrm{KL}(\rho_S\|\pi)$. □

3.3 Specialization of Our Theorem to the McAllester's Approach

We derive here the specialization of our multiview PAC-Bayesian theorem to the McAllester [22]'s point of view. To do so, we follow the same principle as Germain et al. [9,10] to obtain Corollary 1.

Corollary 1. *Let $V \geq 2$ be the number of views. For any distribution $\mathcal{D}$ on $\mathcal{X} \times \mathcal{Y}$, for any set of prior distributions $\{P_v\}_{v=1}^{V}$, for any hyper-prior distribution π over $\mathcal{V}$, we have*

$$\underset{S\sim\mathcal{D}^m}{\mathbb{E}}\, R_{\mathcal{D}}(G_{\rho_S}^{\mathrm{MV}}) \leq \frac{1}{2} \underset{S\sim\mathcal{D}^m}{\mathbb{E}}\, d_S^{\mathrm{MV}}(\rho_S) + \underset{S\sim\mathcal{D}^m}{\mathbb{E}}\, e_S^{\mathrm{MV}}(\rho_S) +$$
$$\sqrt{ \frac{ \underset{S\sim\mathcal{D}^m}{\mathbb{E}}\ \underset{v\sim\rho_S}{\mathbb{E}}\, \mathrm{KL}(Q_{v,S}\|P_v) + \underset{S\sim\mathcal{D}^m}{\mathbb{E}}\, \mathrm{KL}(\rho_S\|\pi) + \ln\frac{2\sqrt{m}}{\delta} }{ 2m } },$$

Proof. To prove the above result, we apply Theorem 3 with $D(a,b) = 2(a-b)^2$, and we upper-bound $\underset{S\sim\mathcal{D}^m}{\mathbb{E}}\ \underset{v\sim\pi}{\mathbb{E}}\ \underset{h\sim P_v}{\mathbb{E}}\, e^{m\,D(R_S(h),R_{\mathcal{D}}(h))}$. According to Pinsker's inequality, we have $D(a,b) \leq \mathrm{kl}(a,b) = a \ln \frac{a}{b} + (1-a) \ln \frac{1-a}{1-b}$. Then by considering $R_S(h)$ as a random variable which follows a binomial distribution of m trials with a probability of success $R(h)$, we obtain

$$
\mathop{\mathbb{E}}_{S \sim \mathcal{D}^m} \mathop{\mathbb{E}}_{v \sim \pi} \mathop{\mathbb{E}}_{h \sim P_v} e^{m\, D(R_S(h), R_\mathcal{D}(h))} \leq \mathop{\mathbb{E}}_{S \sim \mathcal{D}^m} \mathop{\mathbb{E}}_{v \sim \pi} \mathop{\mathbb{E}}_{h \sim P_v} e^{m\, \mathrm{kl}(R_S(h), R_\mathcal{D}(h))}
$$

$$
= \mathop{\mathbb{E}}_{v \sim \pi} \mathop{\mathbb{E}}_{h \sim P_v} \mathop{\mathbb{E}}_{S \sim \mathcal{D}^m} \left[\frac{R_S(h)}{R_\mathcal{D}(h)}\right]^{mR_S(h)} \left[\frac{1-R_S(h)}{1-R_\mathcal{D}(h)}\right]^{m(1-R_S(h))}
$$

$$
= \mathop{\mathbb{E}}_{v \sim \pi} \mathop{\mathbb{E}}_{h \sim P_v} \sum_{k=0}^{m} \mathop{\mathrm{Pr}}_{S \sim \mathcal{D}^m}\left[R_S(h) = \tfrac{k}{m}\right] \left[\frac{k/m}{R_\mathcal{D}(h)}\right]^{k} \left[\frac{1-k/m}{1-R_\mathcal{D}(h)}\right]^{m-k}
$$

$$
= \sum_{k=0}^{m} \binom{m}{k} \left[\frac{k}{m}\right]^{k} \left[1 - \frac{k}{m}\right]^{m-k}
$$

$$
\leq 2\sqrt{m}.
$$

$\square$

4   Discussion on Related Work

In this section, we discuss two related theoretical studies of multiview learning related to the notion of Gibbs classifier.

Amini et al. [1] proposed a Rademacher analysis of the risk of the stochastic Gibbs classifier over the view-specific models (for more than two views) where the distribution over the views is restricted to the uniform distribution. In their work, each view-specific model is found by minimizing the empirical risk: $h_v^* = \mathop{\mathrm{argmin}}_{h \in \mathcal{H}_v} \frac{1}{m} \sum_{(\mathbf{x},y) \in S} \mathbb{1}_{[h(x^v) \neq y]}$. The prediction for a multiview example $\mathbf{x}$ is then based over the stochastic Gibbs classifier defined according to the uniform distribution, i.e., $\forall v \in V$, $\rho(v) = \frac{1}{V}$. The risk of the multiview classifier Gibbs is hence given by

$$
R_\mathcal{D}(G_{\rho=1/V}^{\mathrm{MV}}) = \mathop{\mathbb{E}}_{(\mathbf{x},y) \sim \mathcal{D}} \frac{1}{V} \sum_{v=1}^{V} \mathbb{1}_{[h_v^*(x^v) \neq y]}.
$$

Moreover, Sun et al. [29] proposed a PAC-Bayesian analysis for multiview learning over the concatenation of the views, where the number of views is set to two, and deduced a SVM-like learning algorithm from this framework. The key idea of their approach is to define a prior distribution that promotes similar classification among the two views, and the notion of diversity among the views is handled by a different strategy than ours. We believe that the two approaches are complementary, as in the general case of more than two views that we consider in our work, we can also use a similar informative prior as the one proposed by Sun et al. [29] for learning.

Table 1. Accuracy and $F1$-score averages for all the classes over 20 random sets. Note that the results are obtained for different sizes m of the learning sample and are averaged over the six *one-vs-all* classification problems. Along the columns, best results are in bold. $\downarrow$ indicates statistically significantly worse performance than the best result, according to Wilcoxon rank sum test ($p < 0.02$) [19].

Strategy	$m = 150$		$m = 200$		$m = 250$		$m = 300$	
	Accuracy	F_1	Accuracy	F_1	Accuracy	F_1	Accuracy	F_1
Mono_v	.8516±.0031↓	.1863±.0299↓	.8424±.0272↓	.3056±.0233↓	.8691±.0017↓	.3352±.0164↓	.8770±.0018↓	.4103±.0158↓
$\text{Concat}_{\text{SVM}}$	.8507±.0051↓	.1577±.0403↓	.8615±.0018↓	.2505±.0182↓	.8674±.0026↓	.3006±.0267↓	.8746±.0022↓	.3647±.0258↓
Aggreg_{P}	.8521±.0041↓	.1810±.0305↓	.8420±.0385↓	.2852±.0339↓	.8676±.0023↓	.3027±.0234↓	.8774±.0021↓	.3945±.0185↓
Aggreg_{L}	.8507±.0043↓	.1653±.0336↓	.8477±.0377↓	.2806±.0244↓	.8682±.0022↓	.3116±.0210↓	.8773±.0024↓	.3943±.0204↓
$\text{Fusion}^{\text{all}}_{\text{SVM}}$	.8568±.0087↓	.3899±.0789↓	.8527±.0406↓	.5027±.0780	.8490±.0716↓	**.5399±.0585**	.8422±.0526↓	**.5779±.0422**
$\text{Fusion}^{\text{all}}_{\text{Cq}}$	**.8692±.0059**	**.4298±.0570**	**.8768±.0082**	**.5066±.0402**	**.8846 ±.0047**	.5365±.0371	**.8881± .0060**	.5705±.0286

5 Experiments

In this section, we present experiments to highlight the usefulness of our theoretical analysis by following a two-level hierarchy strategy. To do so, we learn a multiview model in two stages by following a classifier late fusion approach [28] (sometimes referred as stacking [30]). Concretely, we first learn view-specific classifiers for each view at the base level of the hierarchy. Each view-specific classifier is expressed as a majority vote of kernel functions. Then, we learn weighted combination based on predictions of view-specific classifiers. It is worth noting that this is the procedure followed by Morvant et al. [23] in a PAC-Bayesian fashion, but without any theoretical justifications and in a ranking setting.

We consider a publicly available multilingual multiview text categorization corpus extracted from the Reuters RCV1/RCV2 corpus [1][3], which contains more than 110,000 documents from five different languages (English, German, French, Italian, Spanish) distributed over six classes. To transform the dataset into a binary classification task, we consider six *one-versus-all* classification problems: For each class, we learn a multiview binary classification model by considering all documents from that class as positive examples and all others as negative examples. We then split the dataset into training and testing sets: we reserve a test sample containing 30% of total documents. In order to highlight the benefits of the information brought by multiple views, we train the models with small learning sets by randomly choosing the learning sample S from the remaining set of the documents; the number of learning examples m considered are: 150, 200, 250 and 300. For each fusion-based approach, we split the learning sample S into two parts: S_1 for learning the view-specific classifier at the first level and S_2 for learning the final multiview model at the second level; such that $|S_1| = \frac{3}{5}m$ and $|S_2| = \frac{2}{5}m$ (with $m = |S|$). In addition, the reported results are averaged on 20 runs of experiments, each run being done with a new random learning sample.

---

[3] https://archive.ics.uci.edu/ml/datasets/Reuters+RCV1+RCV2+Multilingual,
    +Multiview+Text+Categorization+Test+collection.

Since the classes are highly unbalanced, we report in Table 1 the accuracy along with the $F1$-measure, which is the harmonic average of precision and recall, computed on the test sample.

To assess that multiview learning with late fusion makes sense for our task, we consider as baselines the four following one-step learning algorithms (provided with the learning sample S). First, we learn a view-specific model on each view and report, as $Mono_v$, their average performance. We also follow an early fusion procedure, referred as $Concat_{SVM}$, consisting of learning one single model using SVM [7] over the simple concatenation of the features of five views. Moreover, we look at two simple voters' combinations, respectively denoted by $Aggreg_P$ and $Aggreg_L$, for which the weights associated with each view follow the uniform distribution. Concretely, $Aggreg_P$, respectively $Aggreg_L$, combines the real-valued prediction, respectively the labels, returned by the view-specific classifiers. In other words, we have

$$\texttt{Aggreg}_\texttt{P}(\mathbf{x}) = \tfrac{1}{5}\sum_{v=1}^{5} h^v(x^v), \text{and } \texttt{Aggreg}_\texttt{L}(\mathbf{x}) = \tfrac{1}{5}\sum_{v=1}^{5} \text{sign}\left[h^v(x^v)\right],$$

with $h^v(x^v)$ the real-valued prediction of the view-specific classifier learned on view v.

We compare the above one-step methods to the two following late fusion approaches that only differ at the second level. Concretely, at the first level we construct from S_1 different view-specific majority vote expressed as linear SVM models[4] with different hyperparameter C values (12 values between 10^{-8} and 10^3): We do not perform cross-validation at the first level. This has the advantage to (i) lighten the first level learning process, since we do not need to validate models, and (ii) to potentially increase the expressivity of the final model.

At the second level, as it is often done for late fusion, we learn from S_2 the final weighted combination over the view specific voters using a RBF kernel. The methods referred as $Fusion_{SVM}^{all}$, respectively $Fusion_{Cq}^{all}$, make use of SVM, respectively the PAC-Bayesian algorithm CqBoost [26]. Note that, as recalled in Sect. 2, CqBoost is an algorithm that tends to minimize the C-Bound of Eq. (2): it directly captures a trade-off between accuracy and disagreement.

We follow a 5-fold cross-validation procedure for selecting the hyperparameters of each learning algorithm. For $Mono_v$, $Concat_{SVM}$, $Aggreg_P$ and $Aggreg_L$ the hyperparameter C is chosen over a set of 12 values between 10^{-8} and 10^3. For $Fusion_{SVM}^{all}$ and $Fusion_{Cq}^{all}$ the hyperparameter γ of the RBF kernel is chosen over 9 values between 10^{-6} and 10^2. For $Fusion_{SVM}^{all}$, the hyperparameter C is chosen over a set of 12 values between 10^{-8} and 10^3. For $Fusion_{Cq}^{all}$, the hyperparameter μ is chosen over a set of 8 values between 10^{-8} and 10^{-1}. Note that we made use of the *scikit-learn* [24] implementation for learning our SVM models.

First of all, from Table 1, the two-step approaches provide the best results on average. Secondly, according to a Wilcoxon rank sum test [19] with $p < 0.02$, the PAC-Bayesian late fusion based approach $Fusion_{Cq}^{all}$ is significantly the best method—in terms of accuracy, and except for the smallest learning sample size

---

[4] We use linear SVM model as it is usually done for text classification tasks [*e.g.*, 12].

($m = 150$), Fusion$_{Cq}^{all}$ and Fusion$_{SVM}^{all}$ produce models with similar $F1$-measure. We can also remark that Fusion$_{Cq}^{all}$ is more "stable" than Fusion$_{SVM}^{all}$ according to the standard deviation values. These results confirm the potential of using PAC-Bayesian approaches for multiview learning where we can control a trade-off between accuracy and diversity among voters.

6   Conclusion and Future Work

In this paper, we proposed a first PAC-Bayesian analysis of weighted majority vote classifiers for multiview learning when observations are described by more than two views. Our analysis is based on a hierarchy of distributions, *i.e.* weights, over the views and voters: *(i)* for each view v a posterior and prior distributions over the view-specific voter's set, and *(ii)* a hyper-posterior and hyper-prior distribution over the set of views. We derived a general PAC-Bayesian theorem tailored for this setting, that can be specialized to any convex function to compare the empirical and true risks of the stochastic Gibbs classifier associated with the weighted majority vote. We also presented a similar theorem for the expected disagreement, a notion that turns out to be crucial in multiview learning. Moreover, while usual PAC-Bayesian analyses are expressed as probabilistic bounds over the random choice of the learning sample, we presented here bounds in expectation over the data, which is very interesting from a PAC-Bayesian standpoint where the posterior distribution is data dependent. According to the distributions' hierarchy, we evaluated a simple two-step learning algorithm (based on late fusion) on a multiview benchmark. We compared the accuracies while using SVM and the PAC-Bayesian algorithm CqBoost for weighting the view-specific classifiers. The latter revealed itself as a better strategy, as it deals nicely with accuracy and the disagreement trade-off promoted by our PAC-Bayesian analysis of the multiview hierarchical approach.

We believe that our theoretical and empirical results are a first step toward the goal of theoretically understanding the multiview learning issue through the PAC-Bayesian point of view, and toward the objective of deriving new multiview learning algorithms. It gives rise to exciting perspectives. Among them, we would like to specialize our result to linear classifiers for which PAC-Bayesian approaches are known to lead to tight bounds and efficient learning algorithms [9]. This clearly opens the door to derive theoretically founded algorithms for multiview learning. Another possible algorithmic direction is to take into account a second statistical moment information to link it explicitly to important properties between views, such as diversity or agreement [1,13]. A first direction is to deal with our multiview PAC-Bayesian C-Bound of Lemma 1— that already takes into account such a notion of diversity [23]—in order to derive an algorithm as done in a mono-view setting by Laviolette et al. [17,26]. Another perspective is to extend our bounds to diversity-dependent priors, similarly to the approach used by Sun et al. [29], but for more than two views. This would allow to additionally consider an *a priori* knowledge on the diversity. Moreover, we would like to explore the *semi-supervised* multiview learning where one has access to unlabeled data $S_u = \{\mathbf{x}_j\}_{j=1}^{m_u}$ along with labeled data $S_l = \{(\mathbf{x}_i, y_i)\}_{i=1}^{m_l}$ during training. Indeed, an interesting behaviour of our theorem is that it can be easily extended to this situation: the bound will be a concatenation of a bound over $\frac{1}{2}d_{S_u}^{\mathrm{MV}}(\rho)$ (depending on m_u) and a bound over $e_{S_l}^{\mathrm{MV}}(\rho)$ (depending on m_s). The main difference with the supervised bound is that the Kullback-Leibler divergence will be multiplied by a factor 2.

Acknowledgments. This work was partially funded by the French ANR project LIVES ANR-15-CE23-0026-03, the "Région Rhône-Alpes", and the CIFAR program in Learning in Machines & Brains.

Appendix—Mathematical Tools

Theorem 5 (Markov's ineq.). *For any random variable X s.t. $\mathbb{E}(|X|) = \mu$, for any $a > 0$, we have $\mathbb{P}(|X| \geq a) \leq \dfrac{\mu}{a}$.*

Theorem 6 (Jensen's ineq.). *For any random variable X, for any concave function g, we have $g(\mathbb{E}[X]) \geq \mathbb{E}[g(X)]$.*

Theorem 7 (Cantelli-Chebyshev ineq.). *For any random variable X s.t. $\mathbb{E}(X) = \mu$ and $\mathbf{Var}(X) = \sigma^2$, and for any $a > 0$, we have $\mathbb{P}(X - \mu \geq a) \leq \dfrac{\sigma^2}{\sigma^2 + a^2}$.*

References

1. Amini, M.-R., Usunier, N., Goutte, C.: Learning from multiple partially observed views - an application to multilingual text categorization. In: NIPS, pp. 28–36 (2009)
2. Atrey, P.K., Hossain, M.A., El-Saddik, A., Kankanhalli, M.S.: Multimodal fusion for multimedia analysis: a survey. Multimedia Syst. **16**(6), 345–379 (2010)
3. Bégin, L., Germain, P., Laviolette, F., Roy, J.-F.: PAC-Bayesian bounds based on the Rényi divergence. In: AISTATS, pp. 435–444 (2016)
4. Blum, A., Mitchell, T.M.: Combining Labeled and Unlabeled Data with Co-training. In: COLT, pp. 92–100 (1998)
5. Catoni, O.: PAC-Bayesian Supervised Classification: The Thermodynamics of Statistical Learning, vol. 56. Institute of Mathematical Statistic, Shaker Heights (2007)
6. Chapelle, O., Schlkopf, B., Zien, A.: Semi-Supervised Learning, 1st edn. The MIT Press, Cambridge (2010). ISBN 0262514125, 9780262514125
7. Cortes, C., Vapnik, V.: Support-vector networks. Mach. Learn. **20**(3), 273–297 (1995)
8. Donsker, M.D., Varadhan, S.S.: Asymptotic evaluation of certain markov process expectations for large time, I. Commun. Pure Appl. Math. **28**(1), 1–47 (1975)
9. Germain, P., Lacasse, A., Laviolette, F., Marchand, M.: PAC-Bayesian learning of linear classifiers. In: ICML, pp. 353–360 (2009)
10. Germain, P., Lacasse, A., Laviolette, F., Marchand, M., Roy, J.: Risk bounds for the majority vote: from a PAC-Bayesian analysis to a learning algorithm. JMLR **16**, 787–860 (2015)

11. Goyal, A., Morvant, E., Germain, P., Amini, M.-R.: PAC-Bayesian analysis for a two-step hierarchical multiview learning approach. arXiv preprint arXiv:1606.07240 (2016)
12. Joachims, T.: Text categorization with support vector machines: learning with many relevant features. In: Nédellec, C., Rouveirol, C. (eds.) ECML 1998. LNCS, vol. 1398, pp. 137–142. Springer, Heidelberg (1998). https://doi.org/10.1007/BFb0026683. ISBN 3-540-64417-2
13. Kuncheva, L.I.: Combining Pattern Classifiers: Methods and Algorithms. Wiley-Interscience, Hoboken (2004). ISBN 0471210781
14. Lacasse, A., Laviolette, F., Marchand, M., Germain, P., Usunier, N.: PAC-Bayes bounds for the risk of the majority vote and the variance of the Gibbs classifier. In: NIPS, pp. 769–776 (2006)
15. Langford, J.: Tutorial on practical prediction theory for classification. JMLR **6**, 273–306 (2005)
16. Langford, J., Shawe-Taylor, J.: PAC-Bayes & margins. In: NIPS, pp. 423–430. MIT Press (2002)
17. Laviolette, F., Marchand, M., Roy, J.-F.: From PAC-Bayes bounds to quadratic programs for majority votes. In: ICML (2011)
18. Lecué, G., Rigollet, P.: Optimal learning with Q-aggregation. Ann. Statist. **42**(1), 211–224 (2014). https://doi.org/10.1214/13-AOS1190
19. Lehmann, E.: Nonparametric Statistical Methods Based on Ranks. McGraw-Hill, New York (1975)
20. Maillard, O.-A., Vayatis, N.: Complexity versus agreement for many views. In: Gavaldà, R., Lugosi, G., Zeugmann, T., Zilles, S. (eds.) ALT 2009. LNCS (LNAI), vol. 5809, pp. 232–246. Springer, Heidelberg (2009). https://doi.org/10.1007/978-3-642-04414-4_21
21. McAllester, D.A.: Some PAC-Bayesian theorems. Mach. Learn. **37**, 355–363 (1999)
22. McAllester, D.A.: PAC-Bayesian stochastic model selection. Mach. Learn. **51**, 5–21 (2003)
23. Morvant, E., Habrard, A., Ayache, S.: Majority vote of diverse classifiers for late fusion. In: Fränti, P., Brown, G., Loog, M., Escolano, F., Pelillo, M. (eds.) S+SSPR 2014. LNCS, vol. 8621, pp. 153–162. Springer, Heidelberg (2014). https://doi.org/10.1007/978-3-662-44415-3_16
24. Pedregosa, F., Varoquaux, G., Gramfort, A., Michel, V., Thirion, B., Grisel, O., Blondel, M., Prettenhofer, P., Weiss, R., Dubourg, V., Vanderplas, J., Passos, A., Cournapeau, D., Brucher, M., Perrot, M., Duchesnay, E.: Scikit-learn: machine learning in Python. J. Mach. Learn. Res. **12**, 2825–2830 (2011)
25. Pentina, A., Lampert, C.H.: A PAC-Bayesian bound for lifelong learning. In: ICML, pp. 991–999 (2014)
26. Roy, J.-F., Marchand, M., Laviolette, F.: A column generation bound minimization approach with PAC-Bayesian generalization guarantees. In: Proceedings of the 19th International Conference on Artificial Intelligence and Statistics, pp. 1241–1249 (2016)
27. Seeger, M.W.: PAC-Bayesian generalisation error bounds for gaussian process classification. JMLR **3**, 233–269 (2002)
28. Snoek, C., Worring, M., Smeulders, A.W.M.: Early versus late fusion in semantic video analysis. In: ACM Multimedia, pp. 399–402 (2005)
29. Sun, S., Shawe-Taylor, J., Mao, L.: PAC-Bayes analysis of multi-view learning. CoRR, abs/1406.5614 (2016)
30. Wolpert, D.H.: Stacked generalization. Neural Netw. **5**(2), 241–259 (1992)

Partial Device Fingerprints

Michael Ciere[⊠], Carlos Gañán, and Michel van Eeten

Delft University of Technology, Delft, The Netherlands
m.ciere@tudelft.nl

Abstract. In computing, remote devices may be identified by means of *device fingerprinting*, which works by collecting a myriad of client-side attributes such as the device's browser and operating system version, installed plugins, screen resolution, hardware artifacts, Wi-Fi settings, and anything else available to the server, and then merging these attributes into uniquely identifying fingerprints. This technique is used in practice to present personalized content to repeat website visitors, detect fraudulent users, and stop masquerading attacks on local networks. However, device fingerprints are seldom uniquely identifying. They are better viewed as *partial device fingerprints*, which do have some discriminatory power but not enough to uniquely identify users. How can we infer from partial fingerprints whether different observations belong to the same device? We present a mathematical formulation of this problem that enables probabilistic inference of the correspondence of observations. We set out to estimate a correspondence probability for every pair of observations that reflects the plausibility that they are made by the same user. By extending probabilistic data association techniques previously used in object tracking, traffic surveillance and citation matching, we develop a general-purpose probabilistic method for estimating correspondence probabilities with partial fingerprints. Our approach exploits the natural variation in fingerprints and allows for use of situation-specific knowledge through the specification of a generative probability model. Experiments with a real-world dataset show that our approach gives calibrated correspondence probabilities. Moreover, we demonstrate that improved results can be obtained by combining device fingerprints with behavioral models.

1 Introduction

In networking, remote computers may be partially identified from information they disclose about themselves. This is called device fingerprinting. In the most general terms, device fingerprinting refers to any active or passive collection of meta-data for the purpose of host identification. These meta-data can for instance be browser user-agents, hard drive serial numbers, hardware artifacts such as clock skew, or implementation and configuration details of various protocols [4,10,14,19,21,22]. Device fingerprints can be used both to distinguish hosts — for instance as a security mechanism against masquerading attacks on local networks — and to identify or track them, for instance to present personalized advertisements or to prevent fraud by recognizing blacklisted devices.

© Springer International Publishing AG 2017
M. Ceci et al. (Eds.): ECML PKDD 2017, Part II, LNAI 10535, pp. 222–237, 2017.
https://doi.org/10.1007/978-3-319-71246-8_14

Ideally, the device fingerprints are both highly diverse and stable over time, like human fingerprints, so that they allow us to safely conclude whether or not two hosts are one and the same, even when considerable time has passed between observing them. In practice, however, device fingerprints are typically not completely unique across devices, which means they sometimes fail to distinguish hosts, or they are not stable, diminishing our ability to identify previously seen hosts over time. In some cases there is, in fact, a trade-off between uniqueness and stability, in that fingerprints can be made more unique by including unstable attributes.

Various scholars have addressed a lack of stability with heuristic methods that merge fingerprints based on some measure of similarity [10,28]. However, little has been said about what to do when fingerprints are not completely unique. In this paper we consider *partial device fingerprints*. To be precise, the kind of fingerprint we are interested in is that which is not perfectly unique — multiple devices may have the same fingerprint — but it is stable and free of noise, at least in the time frame in which they are used, so we can distinguish devices with different partial fingerprints with absolute certainty. The challenge is to determine whether devices that show up with the same partial fingerprint are in fact one and the same.

If fingerprints are almost unique, we may be satisfied to map every fingerprint to a single device and accept the small number of false positive matches. As the diversity of fingerprints decreases, however, this may result in too many devices getting clumped together. A particularly unfortunate scenario is when the majority of fingerprints is unique, but a small number of fingerprints is shared by a disproportionate group of devices. In that case the discriminatory power of the unique fingerprints is diluted by the common ones.

So far, no-one has investigated automated methods for identifying devices from partial fingerprints. Our contribution is the development of a general-purpose probabilistic method that allows one to calculate for every pair of observations (o_1, o_2) with matching device fingerprints a *correspondence probability* $P(\text{device}(o_1) = \text{device}(o_2))$, which reflects the plausibility that the observations originate from the same device. These correspondence probabilities have a self-contained measure of uncertainty, which may be large or small depending on the prevalence of a fingerprint. The strength of our method lies in the ability to use partial device fingerprints in combination with user modeling, which we demonstrate on a real-world dataset. Our mathematical formulation of the partial fingerprint problem reveals similarities to *data association* problems studied in Artificial Intelligence, and we draw upon methods previously developed for those problems. Our main developments in this regard are a new approach to dealing with an unknown number of entities and a way to exploit the variation in fingerprint prevalence.

2   Problem Setting

The problem we are interested in can be formulated as follows. Let $O = \{o_1, o_2, ...\}$ be a set of observations, each one tagged with a partial fingerprint

$f_1, f_2, ...$. The observations can take any form. Our only assumption is that they are exchangeable, which means O is invariant to permutations of the labeling. Let the unobserved *assignment* ω be a partition of $\{1, ..., |O|\}$ such that each subset in the partition contains the indices of observations made by a single device. The assignment defines an equivalence relation $o_a \sim o_b$ for observation pairs (o_a, o_b) belonging to the same device. We assume that fingerprints are *stable*; that is, if $f_a \neq f_b$ then $o_a \nsim o_b$. The converse is not true: $f_a = f_b$ does not imply $o_a \sim o_b$, as different devices may have the same fingerprint.

Let the *complete-data* $Y = \{O, \omega\}$ be the union of the observed data O and the unobserved assignment ω. We only observe Y completely if the fingerprints are unique, because then we can infer ω by putting each fingerprint in its own equivalence class. If however we have partial fingerprints, then multiple assignments $\omega \in \Omega$ are plausible. Hence, our observations O can be seen as incomplete-data, and the assignment ω as missing data.

Analogous problems have been studied in the Artificial Intelligence field where they are referred to as *data association* problems [2]. A general formulation was given in [13]. Applications include object tracking, robotic map-building, surveillance, and citation matching [3,6,15,18,25,27]. These problems have in common that observed objects or entities are to some degree indistinguishable, and the challenge is to determine whether two identical-looking objects are in fact one and the same.

In all data association problems, uncertainty about the correct assignment ω is unavoidable. This limits the usefulness of heuristic imputation of correspondences. Several AI scholars have applied probabilistic methods that capture the uncertainty due to measurement error and unpredictable trajectories in a generative probability model (e.g., [7,17,25]). This model can be designed to include prior information and assumptions about the appearance and behavior of entities.

The complete-data $Y = \{O, \omega\}$ can be written as a set of observation sequences $\{T_1, T_2, ..., T_N\}$, $N = |\omega|$, each one belonging to a unique device. Borrowing a term from the object tracking literature, we call such sequences *trajectories*. A probabilistic model $P(Y) = P(T_1, ..., T_N)$ can be designed to capture patterns and regularities in these trajectories. Using Bayes' law, this model then implies a conditional probability distribution over assignments given the observed data

$$P(\omega|O) = \frac{P(O, \omega)}{P(O)} \propto P(Y)$$

where we ignore the normalization constant $P(O)$ since it is constrained to make the probabilities sum to one. A correspondence probability $P(o_a \sim o_b)$ can then be estimated by summing over the unobserved assignment:

$$P(o_a \sim o_b) = \sum_{\omega \in \Omega:\, o_a \sim o_b} P(\omega|O) \propto \sum_{\omega \in \Omega:\, o_a \sim o_b} P(\omega, O). \tag{1}$$

In the following section we explain how such a generative probability model can be set up, and in Sect. 4 we explain how correspondence probabilities can be computed according to (1).

3   Designing a Complete-Data Probability Model

We consider probability models that consist of three independent components: a prior distribution on the number of users, a prior distribution on the fingerprint counts, and a trajectory model. The parameters of these models can often be estimated from other data or prior information; for instance, if partial fingerprints are used to track website visitors who have cookies turned off, the complete-data model can be fitted to the data from users who have cookies turned on. Alternatively, model parameters can be estimated from the incomplete-data using a stochastic EM scheme that we will explain in Sect. 4.2.

3.1   Dealing with an Unknown Population Size

The number of devices N observed in O is often unknown, and moreover it may vary over time as new observations come in. A similar situation was encountered in [16], who suggested estimating N by doing a grid search to optimize a pre-specified criterion. In the partial fingerprint context this approach is not attractive, as taking N to be a fixed parameter does not lead to a generative model. Furthermore, if one is looking to simultaneously estimate the size of multiple subpopulations, a grid search is not workable because of the curse of dimensionality.

An idea borrowed from ecology is to introduce an artificial supercommunity of size $S > N$ from which devices are randomly selected (e.g. [9]). To be precise, we fix S at some value that we know is larger than N, and for each device $i = 1, ..., S$ we introduce a random variable $z_i \in \{0, 1\}$ that determines whether device i is *available*. If a device has at least one observation in O, it is by definition available; otherwise it is either unavailable or available and unobserved. Hence, the size of the observed population is $N = \sum_{i=1}^{S} z_i$, which is a random variable.

One advantage of this approach is that we may introduce user-level parameters ν_i for $i = 1, ..., S$, which is of fixed size even though N is unknown. A second advantage is that it is straightforward to use prior information on N. In the simplest case we may set $N \sim U(1, S)$, but another option is to use $N \sim \text{Binomial}(S, \pi)$ with some informative prior on π that reflects prior knowledge. In addition, if we want the model to work for varying observation windows, we may choose to model the arrival and departure of users. This allows for online deployment of the model with a continuously expanding observation window.

3.2   Modeling Variation in Fingerprint Prevalence

The overall prevalence of a partial fingerprint can be used to inform correspondence probabilities. The intuition is that if we have never seen fingerprint f before and then observe it twice, it seems likely that these observations correspond to the same device; if on the other hand this fingerprint is known to occur frequently, we are less certain of this correspondence.

The simplest way to incorporate this idea in a probability model is to assume that fingerprints are drawn independently from a categorical distribution

$f \sim \text{Categorical}(\mathbf{p})$ with $\mathbf{p} = (p_1, ..., p_K)$, $\sum_{f=1}^{K} p_f = 1$. Then for N users, the probability of generating fingerprint f_1 for the first user, f_2 for the second user, and so on, is

$$P(\mathbf{f}) = p_{f_1} p_{f_2} \cdots p_{f_N} = \prod_{f=1}^{K} p_f^{n_f}, \tag{2}$$

where n_f is the number of users with fingerprint f.

Now imagine two complete-data realizations Y_1 and Y_2, which can both be obtained by adding an observation o_{new} with a fingerprint f to Y_0, with one difference: in Y_1, the new observation is assigned to a device that exists in Y_0, while in Y_2, it is assigned to a new device. If Y_1 has N devices with n_f occurrences of fingerprint f, then Y_2 has $N+1$ devices with $n_f + 1$ occurrences of fingerprint f. Hence, in the probability ratio $\frac{P(Y_1)}{P(Y_2)}$ all terms in (2) cancel out except for a $\frac{1}{p_f}$. This term is inversely proportional to p_f, the population frequency of fingerprint f, and consequently the hypothesis that o_{new} is made by a device previously seen with the same fingerprint is more plausible if the fingerprint is known to occur rarely.

In practice we do not know the true fingerprint population proportions $\mathbf{p}$. If we put a Dirichlet(α) prior on $\mathbf{p}$ for some $\alpha > 0$ and then integrate out the uncertainty in the fingerprint probabilities $\mathbf{p}$, the fingerprint counts follow a compound Dirichlet-Categorical distribution, which we can derive as

$$P(\mathbf{f}|\alpha) = \int_{\mathbf{p}} P(\mathbf{f}|\mathbf{p}) P(\mathbf{p}|\alpha) \, d\mathbf{p}$$
$$= \frac{\Gamma(A)}{\Gamma(\alpha)^K} \frac{\prod_{f=1}^{K} \Gamma(n_f + \alpha)}{\Gamma(N+A)}.$$

Now the probability ratio $\frac{P(Y_1)}{P(Y_2)}$ contains a term

$$\frac{N+A}{n_f + \alpha}. \tag{3}$$

This is inversely proportional to the smoothed fingerprint count $n_f + \alpha$, which is in line with our intuition: if we have seen few occurrences of f, we are inclined to believe that o_{new} belongs to a previously seen device.

Thus far we have assumed that the number of distinct fingerprints K that could ever occur is a finite number known to us. We may circumvent this by specifying a fingerprint distribution with an infinite number of fingerprints. If we fix A and write $\alpha = A/K$, then as K goes to infinity the term (3) goes to $(N+A)/n_f$. The corresponding generative process is known as the *Chinese restaurant process* (CRP) [1,26]. Ignoring the order of users within groups, it has probability mass function

$$P(\mathbf{f}) = \frac{\Gamma(A) \, A^{|F|}}{\Gamma(A+N)} \prod_{f \in F} \Gamma(n_f) \tag{4}$$

where F is the set of observed fingerprints. The same distribution was introduced in the context of population genetics by Ewens [11].

A simple generalization of this distribution uses a discount parameter $0 < d < 1$ with

$$P(\mathbf{f}) = \frac{\Gamma(A)}{\Gamma(A+N)} \frac{d^{|F|}\Gamma(A/d + |F|)}{A/d} \prod_{f \in F} \frac{\Gamma(n_f - d)}{\Gamma(1-d)}. \tag{5}$$

This is sometimes called the two-parameter Poisson-Dirichlet distribution, and it is regularly used in similar situations such as the modeling of word frequencies.

3.3   Modeling User Appearance and Behavior

The last component of the generative probability model is a trajectory model $P(T)$. We assume that trajectories are independent, i.e. $P(T_1, ..., T_N) = \prod_{i=1}^{N} P(T_i)$, although the $P(T_i)$ may depend on shared hyperparameters.

We propose a simple trajectory model that can be used in all partial fingerprint situations. Assume that the number of observations in a trajectory follows a distribution $P(n = |T_i|)$ and that given $|T_i|$, the times at which these observations occur are drawn uniformly from the observation window. Then the probability of a trajectory T_i is

$$P(T_i) = P(|T_i|) \cdot |T_i|!$$

where the factorial term arises from the fact that the observations in a trajectory are always ordered in time.

More advanced trajectory models use domain-specific assumptions about the *coherence* of observation sequences. This can be achieved by modeling a sequence $T_i = o_1^i, o_2^i, ..., o_{|T_i|}^i$ as a Markov process

$$P(T_i) = p(o_1^i) \prod_{j=2}^{|T_i|} p(o_j^i | o_{j-1}^i)$$

where the transition probability $p(o_j^i | o_{j-1}^i)$ may use any information available at the observation-level.

If devices are observed repeatedly, there may be value in modeling the behavioral characteristics of users. For instance, a user may have a tendency to appear at a specific time of day. If this user's fingerprint appears at a non-typical time, we may be inclined to say this is a different user with the same fingerprint.

Such patterns can be used to inform correspondence probabilities by introducing user-level parameters ν_i in a trajectory model $P(T_i) = P(T|\nu_i)$. Alternatively, one may treat the user-level parameters as nuisance parameters and integrate them out of the likelihood. The probability of a trajectory T then changes from $P(T|\nu)$ to $P(T|\phi) = \int_\nu P(T|\nu)P(\nu|\phi)\,d\nu$ where $P(\nu|\phi)$ is a hierarchical distribution. For convenient modeling choices this integral may have a closed-form solution.

3.4 Putting the Components Together

If we put these three model components together, we get a generative complete-data probability model

$$P(Y) = p(N) \cdot N! \cdot \frac{\Gamma(A)\, A^{|F|}}{\Gamma(A+N)} \prod_{f \in F} \Gamma(n_f) \cdot \prod_{i=1}^{N} P(T_i). \tag{6}$$

The $N!$ term is needed to account for the fact that there is no natural order of users in our complete-data formulation.

4 Inference

4.1 Calculating Correspondence Probabilities from a Complete-Data Probability Model

The calculation of correspondence probabilities from a complete-data model as per (1) involves a summation over all possible assignments $\omega \in \Omega$. In general, to estimate any function of the complete-data $f(Y)$ conditional on the observations O and the probability model $P(Y)$ we must average over the uncertainty in ω. This is cumbersome. If there are fingerprints with more than a few observations, then a sum over $\omega \in \Omega$ is intractable, as the number of possible assignments $\omega \in \Omega$ grows combinatorially in the number of observations. A solution used extensively in the data association literature is to draw a sample $\omega^{(1)}, ..., \omega^{(M)}$ from $P(\omega|O) \propto P(Y)$ using a Markov-Chain Monte Carlo (MCMC) approach [12].

MCMC methods work by starting with a random guess ω_0 and then generating a Markov chain $\omega_0, \omega_1, \omega_2, ...$ using a transition kernel density $P(\omega'|\omega)$. For a suitably chosen transition density, the sequence $\omega_1, \omega_2, ...$ has the desired distribution as its stationary distribution. One way to achieve this is offered by the Metropolis-Hastings algorithm, which works as follows. First, the assignment is initialized to some value $\omega^{(0)} \in \Omega$. Then, in every iteration $t = 1, ..., T$, a candidate transition $\omega \to \omega'$ is randomly drawn from a proposal distribution $q(\omega'|\omega)$ and accepted with probability

$$A(\omega'|\omega) = \min\left(1, \frac{P(\omega'|O)\, q(\omega|\omega')}{P(\omega|O)\, q(\omega'|\omega)}\right). \tag{7}$$

If the proposal is accepted we set $\omega^{(t+1)} = \omega'$, and otherwise $\omega^{(t+1)} = \omega^{(t)}$. If the proposal distribution is reversible, that is, $q(\omega|\omega') > 0$ if and only if $q(\omega'|\omega) > 0$, then the resulting Markov Chain is ergodic with stationary distribution $P(\omega|O)$, as desired.

The idea is to construct a proposal distribution that proposes small changes to the assignment, by at most a few correspondences. Figure 1 shows four types of simple reversible transitions. It can easily be verified that every possible assignment can eventually be reached by making only such transitions. Various proposal mechanisms using some combination of the above transitions have been suggested in the data association literature [18, 23–25].

The computational efficiency of a Metropolis-Hastings algorithm designed in this manner lies in the fact that the acceptance probability (7) is always simple to compute, because most terms cancel out. The first half of the probability ratio in (7) can be written as

$$\frac{P(\omega'|O)}{P(\omega|O)} = \frac{P(\omega',O)}{P(\omega,O)} = \frac{P(Y')}{P(Y)}.$$

Plugging in the three-component model (6) yields

$$\frac{P(Y')}{P(Y)} = \frac{p(N') \cdot N'! \cdot \frac{\Gamma(A) A^{|F|}}{\Gamma(A+N')} \prod_{f \in F} \Gamma(n'_f) \cdot \prod_{i=1}^{N} P(T'_i)}{p(N) \cdot N! \cdot \frac{\Gamma(A) A^{|F|}}{\Gamma(A+N)} \prod_{f \in F} \Gamma(n_f) \cdot \prod_{i=1}^{N} P(T_i)}. \tag{8}$$

When the proposal is a simple transition shown in Fig. 1, Y and Y' differ by at most three trajectories T_i and only one fingerprint count n_f, which means almost all terms in (8) cancel out and a simple expression remains.

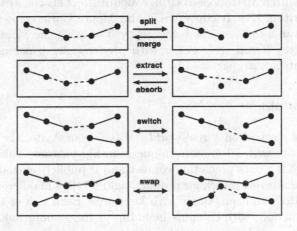

Fig. 1. Visualization of Markov transitions used in MCMC sampling from the conditional distribution of assignments

4.2 Parameter Estimation

We may estimate the parameter vector $\boldsymbol{\theta}$ of the probability model by maximizing the marginal likelihood, obtained by summing out the unobserved assignment ω

$$L(\boldsymbol{\theta}|O) = P(O|\boldsymbol{\theta}) = \sum_{\omega \in \Omega} P(O,\omega|\boldsymbol{\theta}).$$

The summation is intractable. A solution proposed by [25] is to use a stochastic Expectation-Maximization (EM) scheme. The EM algorithm works by first setting the parameter vector to an initial guess θ_0 and then alternating the following two steps [8, 20]:

E: Compute the posterior distribution over the missing data ω given the observed data O and the current parameter guess:

$$\tilde{P}^{(t)}(\omega) = P(\omega|O, \boldsymbol{\theta}^{(t-1)}).$$

M: Update the parameters to $\theta^{(t)}$ by maximizing the expectation of the complete data log-likelihood with respect to $\tilde{P}^{(t)}$:

$$\theta^{(t)} = \arg \max_{\theta} \mathbf{E}_{\tilde{P}^{(t)}}[\log P(\omega, O|\boldsymbol{\theta})]$$

$$= \arg \max_{\theta} \sum_{\omega \in \Omega} \left[\tilde{P}^{(t)}(\omega) \cdot \log P(O, \omega|\boldsymbol{\theta}) \right].$$

If the complete-data log-likelihood is an exponential family, the maximization in the M-step depends only on a fixed number of sufficient statistics of $\tilde{P}^{(t)}(\omega)$. These sufficient statistics may be approximated by drawing a sample from $P(\omega|O, \boldsymbol{\theta}^{(t-1)})$ with a Metropolis-Hastings algorithm. This constitutes a Monte-Carlo EM algorithm [29]. It converges if the Monte Carlo error is kept small enough, which can most easily be achieved by increasing the Monte Carlo sample size with every iteration — either with predetermined increments or with data-driven strategies as suggested by [5].

5   Experiments

We validate our method on a real-world dataset from Avito, the largest Russian classifieds site with 70 million unique monthly visitors. This dataset was previously used for a data prediction contest and is publicly available[1]. Users of Avito browse the site and search for products and services in different categories, filtered by location, and sometimes with keywords. Each row of the dataset is a single search action, with columns including a timestamp, indicators of the category, location, and keywords used in the search, as well as anonymous identifiers of the device model, browser version, and browser User-Agent of the client device. In addition, for every search a user identifier is given, obtained from cookies stored on their machine. The dataset contains a sample of 4.3 million users and spans from April 25th, 2015, to May 20th, 2015. In this period these users made 112 million searches.

For every user we constructed a device fingerprint from the user's device type and model (e.g. Samsung GT-i9500, iPhone, etc.) and browser family (Chrome, Safari, etc.). A total of 9252 distinct fingerprints were found, among 4.3 million unique users. The distribution of fingerprints across users, shown in Fig. 2, is highly skewed: the ten most common fingerprints account for 92% of users.

While these device fingerprints are far from uniquely identifying, we can still use them in combination with other information to discover correspondences

---

[1] https://www.kaggle.com/c/avito-context-ad-clicks/data.

between searches. Our goal in this section is to estimate correspondence probabilities for searches by using these partial fingerprints in combination with three models: (1) a model that only considers the number of searches made by each user, (2) the same model, but with one extra parameter left to be estimated from the data with MCMC-EM, and (3) a model that aims to exploit consistency in the sequence of searches made by a user.

All experiments are run on a test set consisting of all the searches made by a sample of 5000 users. These users made a total of 124331 searches with 218 unique fingerprints. Among those fingerprints, 55 were unique to a single device in the test dataset; the most common fingerprint was observed with 1878 different devices. The number of searches made by users follows a heavy-tailed distribution with a median of 5 and a maximum of almost 2000 searches.

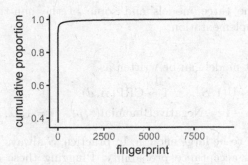

Fig. 2. Distribution of partial fingerprints

5.1 Evaluation Metrics

We evaluate the performance of our method with the three models based on the estimated pairwise correspondence probabilities $P(o_i \sim o_j), i \neq j, f_i = f_j$. Our first performance metric is the Brier Score, calculated as

$$BS = \frac{1}{\#\text{pairs}} \sum_{i \neq j, \ f_i = f_j} (\hat{p}_{ij} - y_{ij})^2$$

where $y_{ij} = 1$ if $o_i \sim o_j$ and $y_{ij} = 0$ if $o_i \not\sim o_j$, $\hat{p}_{ij}$ is the estimated correspondence probability for o_i and o_j, and the summation runs over all observation pairs that have the same partial fingerprint. Since there are too many such pairs to enumerate (namely $N(N-1)/2$), we approximate the sum using a sample of 5000 observation pairs. We draw this sample in two different ways: (1) by drawing observations pairs u.a.r. from all possible pairs, and (2) by first drawing one observation u.a.r. and then another with the same fingerprint. Since the number of possible pairs for a fingerprint with n observations is $n(n-1)/2$ rather than n, the first sample will contain many more pairs from the most common fingerprints than the second sample.

The second performance metric is the Logarithmic Loss

$$LL = \frac{-1}{\#\text{pairs}} \sum_{i \neq j, \, f_i = f_j} (y_{ij} \log(p_{ij}) + (1 - y_{ij}) \log(1 - \hat{p}_{ij})) .$$

The summation is approximated using the same two samples of 5000 pairs. From an information-theoretic perspective, this logarithmic loss is essentially minus the expected cross-entropy between the estimated correspondence probabilities and the true correspondences. A higher log-loss indicates a higher expected surprisal.

5.2 Models and Implementation

We now outline the three models and some of the non-trivial calculations involved in their implementation.

Model 1. The first model can be written as

$$N \sim \text{U}(1, S), \quad \mathbf{f} \sim \text{CRP}(A, d),$$
$$|T_i| - 1 \sim \text{NegativeBinomial}(r, p), \quad i = 1, ..., N \tag{9}$$

where S is assumed to be large enough. In practice, S always cancels out of the Metropolis-Hastings acceptance probability. Plugging these distributions into the complete-data likelihood (6) gives

$$P(Y|M_1, A, r, p) = \frac{1}{S - 1} N! \times \frac{\Gamma(A) \, A^{|F|}}{\Gamma(A + N)} \prod_{f \in F} \Gamma(n_f)$$

$$\times \prod_{i=1}^{N} \left(\binom{|T_i| + r - 2}{|T_i| - 1} (1 - p)^r \, p^{|T_i| - 1} \, |T_i|! \right) .$$

The parameters A, r and p are estimated from the rows in the data set that are not in the test sample. Using maximum-likelihood, we found the values $\hat{r} = 0.26$ and $\hat{p} = 0.012$. The fingerprint distribution parameters A and d were found with the following procedure. First, we estimated the empirical relationship between the number of users N and the number of unique fingerprints $|F|$ by subsampling N users for a range of N values and each time counting the number of unique fingerprints. Then we fitted the distribution $\mathbf{f} \sim \text{CRP}(A, d)$ by matching the expected number of fingerprints $E(|F|; A, d)$ to the empirical estimates using least squares.

Model 2. The second model is similar to the first model, but instead of a uniform prior on the total number of observed users N a binomial prior is used, i.e. $N \sim \text{B}(S, \rho)$ where $S \gg N$ is fixed at some large value. We estimate ρ from the incomplete-data O using the MCMC-EM algorithm described in Sect. 4.2.

In the M-step we set

$$\rho^{(t)} = \arg\max_{\boldsymbol{\theta}} \mathbf{E}_{\tilde{P}(t)}[\log P(\omega, O|\boldsymbol{\theta})]$$

$$\approx \arg\max \sum_{m=1}^{M} \left[\log P(O, \omega^{(m)}|\rho)\right]$$

$$= \frac{1}{M} \sum_{m=1}^{M} \frac{N^{(m)}}{S}$$

where $\omega^{(1)}, ..., \omega^{(M)}$ are MCMC samples drawn in the E-step. The advantage of this model is that prior information on ρ can be used to get better results and faster convergence of the MCMC algorithm.

Model 3. The third model is an extended version of Model 1. This model is designed to exploit coherence in the searches that users make. Recall that the data set includes for every observation an identifier for the location used in the search; users tend to repeatedly use a small number of locations in all their searches. Let z_j^i be the location used in search o_j^i by user i. We assume, for a trajectory T_i, that the associated sequence of searches $\mathbf{z}^i = (z_1^i, z_2^i, ..., z_{|T_i|}^i)$ follows a Dirichlet-Categorical distribution

$$P(\mathbf{z}^i|\boldsymbol{\alpha}) = \frac{\Gamma(\sum_{k=1}^{K}\alpha_k)}{\Gamma(|T_i| + \sum_{k=1}^{K}\alpha_k)} \frac{\prod_{k=1}^{K}\Gamma(n_{ik} + \alpha_k)}{\prod_{k=1}^{K}\Gamma(\alpha_k)}$$

where $n_{ik} = \sum_{j=1}^{|T_i|} I[z_j^i = k]$. A draw from this distribution can be generated by first drawing a probability vector $\mathbf{p}$ from a Dirichlet distribution with parameter vector $\boldsymbol{\alpha}$ and then drawing $|T_i|$ observations from a categorical distribution with probability vector $\mathbf{p}$. The conditional probability distribution of the location z_j^i used in the j'th search by user i given this user's previous searches equals

$$P(z_j^i = k|z_1^i, ..., z_{j-1}^i) = \frac{\alpha_k + n_{ik}^{j-1}}{\sum_{k=1}^{K} \alpha_k + (j-1)}$$

where n_{ik}^{j-1} is the number of previous searches made with the same location. Since the conditional probability of location k increases with the number of times it has been used before, this model incorporates the idea that users tend to keep using the same locations.

As before, the parameter vector $\boldsymbol{\alpha}$ was estimated a priori from the data excluding the test sample.

Implementation Details. For all models, data preprocessing was done in R, and core computations were implemented in C++ functions which were called from R using the Rcpp module. Code is available online[2].

---

[2] https://github.com/michaelciere/partial-fingerprints.

For the third model, to get satisfactory acceptance rates an additional type of Metropolis transition was used that extracts or absorbs a group observations with the same LocationID from a user's trajectory in one go.

For all three models a total of 250×10^6 MCMC iterations was used, starting from a randomly initialized assignment $\omega^{(0)}$. For the second model, these iterations were spread out over 250 MCMC-EM iterations. Convergence was confirmed by manual inspection. The performance metrics were then computed with another 50×10^6 samples. Running on a machine with a 2.0 GHz Intel Xeon CPU and plenty of memory (the algorithm needs about 5 GB with this dataset), the MCMC algorithm took about ten seconds per 10^6 samples.

5.3   Results

Table 1 shows the performance of the three models. As a benchmark this table also shows the scores obtained by naively assuming that observations with matching fingerprints are always made by the same user.

Table 1. Calibration of the correspondence probabilities estimated on the test dataset using the three models and a naive benchmark model which assumes observations with matching fingerprints correspond to one device with probability one. Lower scores mean better calibration. The metrics on the first and third row are computed on a sample of observations pairs randomly drawn from the set of all possible pairs with matching fingerprints. The metrics on the second and fourth row are computed on a stratified sample in which the number of pairs with fingerprint f is proportional to n_f.

	Naive	Model 1	Model 2	Model 3
Brier score	0.925	0.0390	0.0366	0.0224
Brier score (weighted)	0.945	0.0313	0.0309	0.0150
Log-loss	12.8	0.249	0.222	0.117
Log-loss (weighted)	13.1	0.216	0.222	0.0919

All three models give calibrated correspondence estimates. This can also be seen in the calibration plots in Fig. 3. Results for model 2, in which an extra model parameter was estimated from the incomplete-data, are comparable with the results from model 1. The correspondence probability estimates obtained by model 3, which attempts to capture coherence in the locations by which users filter their searches, are markedly better than those for the first two models. This improvement is larger for the weighted scores, which can be explained by the fact that the most common fingerprints, which have a smaller influence on the weighted scores, are shared by so many users that estimating correspondences remains difficult even when locations are taken into account. In addition, for these common fingerprints MCMC sampling is difficult and many iterations are needed for the correspondence probability estimates to converge.

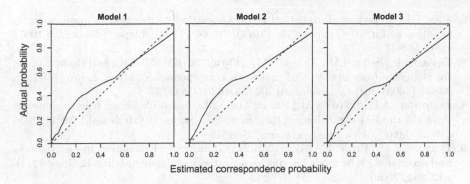

Fig. 3. LOWESS-smoothed calibration curves of estimated correspondence probabilities.

6 Conclusions

This paper has introduced a general-purpose probabilistic method for identifying devices from non-unique device fingerprints. Our experiments have shown that this method produces calibrated correspondence probability estimates. Especially promising are the results obtained by combining partial device fingerprints with user modeling. Many challenges remain. For instance, we have assumed in this work that the device fingerprints are stable, meaning they are measured without noise. When fingerprints are noisy, this noise has to be modeled as well, which introduces additional complications. Furthermore, online deployment of this method would require one to model the arrival and departure of devices. Resolving these issues opens the door to new uses of device fingerprinting in fraud detection, content personalization, and automated authentication.

References

1. Antoniak, C.E.: Mixtures of dirichlet processes with applications to Bayesian nonparametric problems. Ann. Stat. **2**(6), 1152–1174 (1974). http://projecteuclid.org/euclid.aos/1176342871
2. Bar-Shalom, Y., Fortmann, T.E.: Tracking and Data Association. Academic Press, New York (1988)
3. Blackman, S.S.: Multiple-target Tracking with Radar Applications. Artech House, Norwood (1986)
4. Brik, V., Banerjee, S., Gruteser, M., Oh, S.: Wireless device identification with radiometric signatures. In: Proceedings of the 14th ACM International Conference on Mobile Computing and Networking, MobiCom 2008, pp. 116–127. ACM, New York (2008). http://doi.acm.org/10.1145/1409944.1409959
5. Caffo, B.S., Jank, W., Jones, G.L.: Ascent-based Monte Carlo expectation maximization. J. Royal Stat. Soc.: Series B (Stat. Methodol.) **67**(2), 235–251 (2005). http://onlinelibrary.wiley.com/doi/10.1111/j.1467-9868.2005.00499.x/abstract

6. Cox, I.J.: A review of statistical data association techniques for motion correspondence. Int. J. Comput. Vis. **10**(1), 53–66 (1993). https://doi.org/10.1007/BF01440847

7. Dellaert, F., Seitz, S.M., Thorpe, C.E., Thrun, S.: EM, MCMC, and chain flipping for structure from motion with unknown correspondence. Mach. Learn. **50**(1–2), 45–71 (2003). https://doi.org/10.1023/A:1020245811187

8. Dempster, A.P., Laird, N.M., Rubin, D.B.: Maximum likelihood from incomplete data via the EM algorithm. J. Roy. Stat. Soc. Series B (Methodol.) **39**(1), 1–38 (1977). http://www.jstor.org/stable/2984875

9. Dorazio, R.M., Royle, J.A., Sderstrm, B., Glimskr, A.: Estimating species richness and accumulation by modeling species occurrence and detectability. Ecology **87**(4), 842–854 (2006)

10. Eckersley, P.: How unique is your web browser? In: Atallah, M.J., Hopper, N.J. (eds.) PETS 2010. LNCS, vol. 6205, pp. 1–18. Springer, Heidelberg (2010). https://doi.org/10.1007/978-3-642-14527-8_1

11. Ewens, W.J.: The sampling theory of selectively neutral alleles. Theor. Popul. Biol. **3**(1), 87–112 (1972). http://www.sciencedirect.com/science/article/pii/0040580972900354

12. Gilks, W.R.: Markov chain Monte Carlo. In: Encyclopedia of Biostatistics. Wiley, Hoboken (2005). http://onlinelibrary.wiley.com/doi/10.1002/0470011815.b2a14021/abstract

13. Huang, T., Russell, S.: Object identification in a Bayesian context. In: Proceedings of the Fifteenth International Joint Conference on Artifical Intelligence, IJCAI 1997, vol. 2, pp. 1276–1282. Morgan Kaufmann Publishers Inc., San Francisco, CA, USA (1997). http://dl.acm.org/citation.cfm?id=1622270.1622340

14. Kohno, T., Broido, A., Claffy, K.C.: Remote physical device fingerprinting. IEEE Trans. Dependable Secure Comput. **2**(2), 93–108 (2005)

15. Luo, J., Pattipati, K.R., Willett, P.K., Hasegawa, F.: Near-optimal multiuser detection in synchronous CDMA using probabilistic data association. IEEE Commun. Lett. **5**(9), 361–363 (2001)

16. Marinakis, D., Dudek, G.: Occam's Razor applied to network topology inference. IEEE Trans. Robot. **24**(2), 293–306 (2008)

17. Marinakis, D., Dudek, G., Fleet, D.J.: Learning sensor network topology through Monte Carlo expectation maximization. In: Proceedings of the 2005 IEEE International Conference on Robotics and Automation, pp. 4581–4587, April 2005

18. Marinakis, D., Dudek, G.: Topological mapping through distributed, passive sensors. In: Proceedings of the 20th International Joint Conference on Artifical Intelligence, IJCAI 2007, pp. 2147–2152. Morgan Kaufmann Publishers Inc., San Francisco, CA, USA (2007). http://dl.acm.org/citation.cfm?id=1625275.1625622

19. Mowery, K., Bogenreif, D., Yilek, S., Shacham, H.: Fingerprinting information in Javascript implementations. ResearchGate, May 2012

20. Neal, R.M., Hinton, G.E.: Learning in Graphical Models, pp. 355–368. MIT Press, Cambridge (1999). http://dl.acm.org/citation.cfm?id=308574.308679

21. Neumann, C., Heen, O., Onno, S.: An empirical study of passive 802.11 device fingerprinting. In: 2012 32nd International Conference on Distributed Computing Systems Workshops, pp. 593–602, June 2012

22. Nikiforakis, N., Kapravelos, A., Joosen, W., Kruegel, C., Piessens, F., Vigna, G.: Cookieless Monster: exploring the ecosystem of web-based device fingerprinting. In: 2013 IEEE Symposium on Security and Privacy (SP), pp. 541–555, May 2013

23. Oh, S., Russell, S., Sastry, S.: Markov chain Monte Carlo data association for multi-target tracking. IEEE Trans. Autom. Control **54**(3), 481–497 (2009)

24. Oh, S., Russell, S., Sastry, S.: Markov chain Monte Carlo data association for general multiple-target tracking problems. In: 43rd IEEE Conference on Decision and Control, 2004. CDC, vol. 1, pp. 735–742, December 2004

25. Pasula, H., Russell, S.J., Ostland, M., Ritov, Y.: Tracking many objects with many sensors. In: Proceedings of the Sixteenth International Joint Conference on Artificial Intelligence, pp. 1160–1171. IJCAI 1999. Morgan Kaufmann Publishers Inc., San Francisco (1999). http://dl.acm.org/citation.cfm?id=646307.687451

26. Pitman, J.: Combinatorial Stochastic Processes. Lecture Notes in Mathematics, vol. 1875. Springer, Heidelberg (2006). https://doi.org/10.1007/b11601500

27. Sittler, R.W.: An optimal data association problem in surveillance theory. IEEE Trans. Mil. Electron. **8**(2), 125–139 (1964)

28. Spooren, J., Preuveneers, D., Joosen, W.: Mobile device fingerprinting considered harmful for risk-based authentication. In: Proceedings of the Eighth European Workshop on System Security, EuroSec 2015, pp. 6:1–6:6. ACM, New York (2015). http://doi.acm.org/10.1145/2751323.2751329

29. Wei, G.C.G., Tanner, M.A.: A Monte Carlo implementation of the EM algorithm and the poor man's data augmentation algorithms. J. Am. Stat. Assoc. **85**(411), 699–704 (1990). https://doi.org/10.1080/01621459.1990.10474930

Robust Multi-view Topic Modeling
by Incorporating Detecting Anomalies

Guoxi Zhang[1(✉)], Tomoharu Iwata[2], and Hisashi Kashima[1]

[1] Graduate School of Informatics, Kyoto University, Kyoto, Japan
guoxi@ml.ist.i.kyoto-u.ac.jp, kashima@i.kyoto-u.ac.jp
[2] NTT Communication Science Laboratories, Kyoto, Japan
iwata.tomoharu@lab.ntt.co.jp

Abstract. Multi-view text data consist of texts from different sources. For instance, multilingual Wikipedia corpora contain articles in different languages which are created by different group of users. Because multi-view text data are often created in distributed fashion, information from different sources may not be consistent. Such inconsistency introduce noise to analysis of such kind of data. In this paper, we propose a probabilistic topic model for multi-view data, which is robust against noise. The proposed model can also be used for detecting anomalies. In our experiments on Wikipedia data sets, the proposed model is more robust than existing multi-view topic models in terms of held-out perplexity.

1 Introduction

Multi-view text data consist of texts from different information sources. A view of an instance refers to a part that is from some information source. For example, in a English-Japanese bilingual corpora, an document has two views: English article and Japanese article. Multi-view text data are considered as comparable if views of a document are description of the same target. Multi-view topic modeling is the task of extracting aligned topics from comparable multi-view text data, which are tuples of semantically similar topics of different views. Aligned topics facilitate construction of bilingual lexicon of semantically related words, which can be useful in cross-lingual information retrieval [16]. Aligned topics can also be used to transfer knowledge from one language to another in cross-lingual document classification [5]. Moreover, on data consist of texts and social annotation, complementary information in tags can be utilized to improve performance of clustering tasks [14] using aligned topics.

In a multi-view topic models, a view of a document is modeled as a mixture of topics, which are Categorical distributions over words. The mixture weights, which are often called topic proportions, can be considered as low-dimensional representation of documents. It is often assumed in existing multi-view topic models that different views of the same document are semantically consistent. Under this assumption, topic proportions are shared across all views of a document [11]. However, for multilingual corpora that are managed in distributed

M. Ceci et al. (Eds.): ECML PKDD 2017, Part II, LNAI 10535, pp. 238–250, 2017.
https://doi.org/10.1007/978-3-319-71246-8_15

fashion, this assumption does not necessarily hold. For example, since articles in different Wikipedia languages are usually managed by different communities, they often differ in details. Figure 6 shows an example for this. This bilingual document contains Japanese article and Finnish article about Orne, a province in France. Compared to Japanese article, Finnish article contains more information about history of Orne, so it should have larger weights for topics that related to history.

Documents that have inconsistent weights can be regarded as multi-view anomalies [8]. Although inconsistency in content should incur difference in topic proportions, existing models are not capable of depicting it while learning low-dimensional representation of documents. In this paper we propose a multi-view topic model which models data and detects anomalous instances simultaneously. Appropriate number of topic proportions variable are inferred for anomalous instances, to model the inconsistent views. As a result, the proposed model is more robust to multi-view anomalies, and also applicable for the multi-view anomaly detection task. The proposed model is beneficial in at least two applications. In large enterprise with global business, managing information consistency in multilingual documents is an important but expensive task [6]. Cost of management can be reduced if anomalous documents are detected automatically. In cross-cultural analysis [9], documents with inconsistent views are used to analyze cultural difference. We can reduce cost of obtaining samples by using the proposed model to identify anomalous documents from large datasets automatically.

In the proposed model, documents that contain inconsistent views are regarded as anomalies, and such views have distinct topic proportions. Views of a non-anomalous documents share the same topic proportions variable. We use Dirichlet process as the prior for topic proportions variables to infer the appropriate number of topic proportions variable for each document. Based on collapsed Gibbs sampling, we derive efficient inference procedures for the proposed model. To our knowledge, this is the first model that addresses the problem of multi-view anomaly detection in the literature of topic modeling. Performance of the proposed model is examined on ten bilingual Wikipedia corpora. It is demonstrated that the proposed model is more robust than existing multi-view topic models, in terms of held-out perplexity. In addition, compared to existing multi-view anomaly detection methods the proposed model is more efficient and has higher anomaly detection performance on multi-view text data.

The rest of this paper is organized as the following. Section 2 includes related work on topic modeling and multi-view anomaly detection. The proposed model and its inference method are presented in Sect. 3. Section 4 contains evaluation of models' generalization ability in terms of held-out perplexity on Wikipedia corpora. Section 5 contains evaluation of multi-view anomaly detection. In Sect. 6, examples of aligned topics and multi-view anomalies in a Wikipedia corpus are presented. We conclude this paper in Sect. 7.

2   Related Work

Topic models, such as Latent Dirichlet Allocation (LDA) [4], are analysis tool for discrete data. Polylingual Topic Model (PLTM) [11] is an extension of LDA to comparable multi-view setting, and is demonstrated to be useful in various applications, such as cross-cultural understanding, cross-lingual semantic similarity calculation and cross-lingual document classification [16]. Based on the fact that views of a document are information about the same target from different perspective, topics of different views are aligned by sharing topic proportions variable among all views of a document in PLTM. While information in different views are utilized jointly, this model assumption is not valid for data that contain multi-view anomalies. Correspondence LDA [3] and symmetric correspondence LDA [7] are another kinds of multi-view topic models, which extract direct correspondence between topics of different views. However, in these models distinct topic proportions variables are inferred for views of the same document, in regardless of view consistency. Hence they are not applicable in detecting multi-view anomalies and obtaining low-dimensional representation of multi-view documents. Moreover, in existing models topics are to be aligned without considering view consistency, so on noisy data that contains a lot of multi-view anomalies their performance may degenerate.

Various methods can be applied to the task of multi-view anomaly detection. In probabilistic canonical correlation analysis (PCCA) [2] a shared latent vector among all views and its projection matrices for each view are estimated. The reconstruction error is considered as anomaly score, based on the idea that high reconstruction error indicates views are inconsistent. In [10] the authors propose a robust version of PCCA by detecting multi-view anomalies during estimating parameters. Nevertheless, this model assumes Gaussian error so it may not be suitable for textual data. Moreover, textual data have high-dimensional features, which leads to efficiency issue when applying that method.

3   Proposed Model

3.1   Generative Process

Suppose there are D documents, and each of them contains L views. In the proposed model, views of a document are grouped into clusters. The proposed model assumes that each document can have a countably infinite number of clusters. A topic proportion vector θ_{dy} is generated for each cluster y in document d, and it is then used to generate words in each view of y. As a result, views in the same cluster share the same topic proportions vector, and views belong to different cluster have distinct topic proportions vectors. Consequently, multi-view anomalies are identified by the number of clusters they have. A document is a normal document if it has only one cluster, and is an anomaly if it has more than one cluster.

Specifically, we use Stick-breaking process [15] to generate clusters and cluster assignments of views. The probability that a view belongs to some cluster is related to the proportions of its words' topic assignments. In anomalous documents, such proportions are different in different views, causing its views to be assigned to different clusters. Meanwhile, in normal documents such proportions of views are similar, so views are assigned to the same cluster.

The generative process of the proposed model is described as the following, and the graphical model representation is shown in Fig. 1.

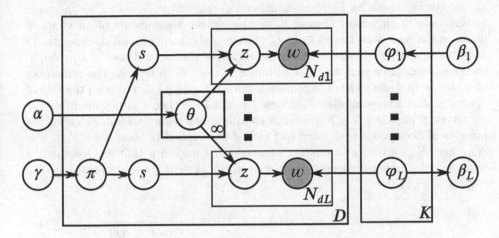

Fig. 1. Graphical model representation of the proposed model.

For each $\ell = 1, 2, \ldots, L$ and $k = 1, 2, \ldots, K$, generate a topic $\phi_{\ell k} \in R^{V_\ell}$ with a symmetric prior $\beta_\ell e \in R^K$, where $\beta_\ell \in R$ and $e \in R^K$ is all-ones vector. V_ℓ is the number of unique words in view ℓ.

$$\phi_{\ell k} \sim \text{Dirichlet}(\beta_\ell). \tag{1}$$

For each document d, generate mixture weights π_d by the stick-breaking process with concentration parameter γ, which generates mixture wights of the Dirichlet process.

$$\pi_d \sim \text{Stick}(\gamma). \tag{2}$$

For each view ℓ of the document d, generate cluster assignments $s_{d\ell}$ from π_d:

$$s_{d\ell} \sim \text{Category}(\pi_d). \tag{3}$$

Then generate topic proportions θ_{dy} for cluster y of d using asymmetric prior $\alpha \in R^K$:

$$\theta_{dy} \sim \text{Dirichlet}(\alpha). \tag{4}$$

Finally, generate topic assignment $z_{d\ell n}$ of nth word in view ℓ of d, and the corresponding word $w_{d\ell n}$, for $n = 1, \ldots, N_{d\ell}$, where $N_{d\ell}$ is the number of words in view ℓ of document d.

$$z_{d\ell n} \sim \mathrm{Dirichlet}(\theta_{ds_{d\ell}}), \tag{5}$$

$$w_{d\ell n} \sim \mathrm{Category}(\phi_{\ell z_{d\ell n}}). \tag{6}$$

3.2   Inference

Collapsed Gibbs Sampling. In the following inference procedure, θ, π and ϕ are marginalized out by Dirichlet-multinomial conjugacy. Denote Z as the topic assignments of all words. Denote S_d as the cluster assignments of all views in document d and S as the identity of cluster assignments in all documents. In order to simplify expression, denote subscript $d\ell n$ as J, and use $\backslash J$ to refer to the remaining after removing $z_{d\ell n}$. Similarly, use $\backslash d\ell$ to refer to the remaining of a cluster in d after view ℓ is removed. For example, $y \backslash d\ell$ refers to the rest of cluster y after removing view ℓ. If view ℓ is not in y, then y is not modified.

Given S and $Z_{\backslash J}$, Eq. 7 is used to sample a new value for $z_{d\ell n}$. Denote the number of occurrence that word t in view ℓ is assigned to topic k as $N_{\ell kt}$. Use $N_{d\ell k}$ and N_{dyk} to refer to number of words in ℓ and in y that are assigned to topic k in document d. Denote number of words in view ℓ^{th} that are assigned to topic k as $N_{\ell k}$.

$$P(z_{d\ell n} = k \mid Z_{\backslash J}, S) \propto (N_{ds_{d\ell}k \backslash J} + \alpha_k)\frac{N_{\ell k w_{d\ell n} \backslash J} + \beta_\ell}{N_{\ell k \backslash J} + \beta_\ell V_\ell}. \tag{7}$$

For each document d, given Z and $S_{d\backslash d\ell}$, Eq. 8 is used for sampling a new value for $s_{d\ell}$. ℓ, a view of document d, could be assigned to an existing cluster y or a new cluster $\tilde{y}$. Denote number of words y contains as N_{dy} and number of words in y that are assigned to topic k as N_{dyk}. Denote number of views in cluster y of document d as L_{dy}. $\bar{\alpha} = \sum_{k=1}^{K} \alpha_k$. $\Gamma(\cdot)$ refers to the gamma function.

$$\begin{aligned}
&P(s_{d\ell} = y \mid Z, S_{d\backslash d\ell}) \propto L_{dy\backslash d\ell} \\
&\times \left[\prod_{k:N_{d\ell k}>0} \frac{\Gamma(N_{dyk\backslash d\ell} + N_{d\ell k} + \alpha_k)}{\Gamma(N_{dyk\backslash d\ell} + \alpha_k)} \right] \frac{\Gamma(N_{dy\backslash d\ell} + \bar{\alpha})}{\Gamma(N_{dy} + \bar{\alpha})}, \\
&P(s_{d\ell} = \tilde{y} \mid Z, S_{d\backslash d\ell}) \propto \gamma \\
&\times \left[\prod_{k:N_{d\ell k}>0} \frac{\Gamma(N_{d\ell k} + \alpha_k)}{\Gamma(\alpha_k)} \right] \frac{\Gamma(\bar{\alpha})}{\Gamma(N_{d\ell} + \bar{\alpha})}.
\end{aligned} \tag{8}$$

Hyper-parameter Estimation. Hyper parameters α and β smooth word counts in inference. They can be either set to some small values or optimized by placing Gamma priors on them and then using fixed-point iteration method [12]. As demonstrated in [1], the later approach reduce performance difference that is resulted from learning algorithm. Thus we optimize these hyper parameters using the approached introduced in [12], as Eq. 9. Y_d denotes the set of clusters in document d. $\Psi(\cdot)$ refers to the digamma function.

$$\alpha_k^{new} = \alpha_k \frac{\sum_{d=1}^{D}(\sum_{y \in Y_d} \Psi(N_{dyk} + \alpha_k) - |Y_d|\Psi(\alpha_k))}{\sum_{d=1}^{D}(\sum_{y \in Y_d} \Psi(N_{dy} + \bar{\alpha}) - N_{dy}|Y_d|\Psi(\bar{\alpha}))}$$

$$\beta_\ell^{new} = \beta_\ell \frac{\sum_{k=1}^{K}\sum_{t=1}^{V_\ell} \Psi(N_{\ell kt} + \beta_\ell) - KV_\ell\Psi(\beta_\ell)}{V_\ell \sum_{k=1}^{K} \Psi(N_{\ell k} + V_\ell\beta_\ell) - KV_\ell\Psi(\beta_\ell)}$$

$$(9)$$

Estimation of Θ and Φ. After iteratively sampling and updating hyper-parameters, point estimates for Θ and Φ are made:

$$\theta_{yk} = \frac{N_{dyk} + \alpha_k}{N_{ds} + \bar{\alpha}},$$

$$\phi_{\ell kt} = \frac{\beta_\ell + N_{\ell kt}}{N_{\ell k} + V_\ell\beta_\ell}.$$

$$(10)$$

Anomaly Score. Because view consistency is modeled stochastically using the Dirichlet process, we use the probability that a document has more than one clusters as anomaly score. High value indicates views in a document tend to diverge, so probably it is a multi-view anomaly. As shown Eq. 11, such anomaly score is estimated with samples of S generated using the Gibbs sampler Eq. 8. T refers the total number of iterations in model training. $\left|Y_d^{(t)}\right|$ is the number of clusters in document d in iteration t. $I(\cdot)$ is the indicator function. In experiments we use sufficiently large T to ensure $score_d$ converges.

$$score_d = \frac{1}{T} \sum_{t=1}^{T} I(\left|Y_d^{(t)}\right| > 1),$$

$$(11)$$

4   Held-Out Perplexity Evaluation

4.1   Dataset

We collect 34024 articles in Japanese, German, French, Italian, English, and Finnish from Wikipedia. This data is preprocessed by removing general stop words and corpus stop words, which are words with frequency larger than 3402. We also remove words with frequency lower than 100 to reduce the size of vocabulary. After preprocessing, the vocabularies of each language are 12148, 17375, 12813, 16291, 22500, and 7910. From this corpus we select ten bilingual corpora for experiments. They are Japanese - Finnish, Japanese - German, Japanese - French, Japanese - Italian, Japanese - English, English - Germany, English - Finnish, English - Japanese, English - French and English - Italian. We filter out article pairs whose both views are shorter than five words. The numbers of documents in these ten bilingual corpora are 33652, 33668, 33658, 33653, 33854, 33829, 33813, 33854, 33822, and 33814. From each bilingual corpus ten datasets are randomly sampled for experiments, each of them contains 5000 documents.

To quantitively examine models' performance when multi-view anomalies are present, view-swapping is performed to generate multi-view anomalies, as used in [8, 10]. Specifically, $10\%, 20\%, 30\%, 40\%, 50\%$ of documents in each dataset are randomly selected as anomalies, and their views are swapped . As a result, these datasets contain multi-view anomalies with ratio $10\%, ..., 50\%$. Because data of each view are not modified, these datasets can be used to investigate model's performance against multi-view anomalies.

4.2   Settings

Perplexity of held-out corpus is selected as an evaluation metric. Low perplexity indicates good generalization ability. The proposed model is compared with PLTM and CorrLDA to examine the effect of anomaly detection in multi-view topic modeling.

Perplexity is calculated using Eq. 12. As perplexity of CorrLDA depends on choice of pivot view, we report the average of for different choice of pivot view. The held-out corpus is constructed by randomly selecting 20% of documents and then randomly selecting half of their words in each view. Denote the set of index of documents chosen as D^{test}. Denote the set of words chosen in document d as w_d^{test}. Denote the total number of words chosen as N^{test}.

$$
\text{perplexity} = \exp\left(-\frac{\sum\limits_{d \in D^{\text{test}}} \sum\limits_{\ell=1}^{L} \sum\limits_{t \in w_{d\ell}^{\text{test}}} \ln(\sum\limits_{k=1}^{K} \theta_{ds_{d\ell}k} \phi_{\ell k t})}{N^{\text{test}}} \right) \tag{12}
$$

In all experiments, initial value of α_k, β and γ are set to 0.05. Gibbs sampling is executed for 1000 iterations. The proposed model is initialized by using single cluster for every documents in the first 256 iterations. After that parameters are learned using procedures described in Sect. 3.2.

4.3   Results

Figure 2 shows the average of held-out perplexities and their standard errors on Japanese - Finnish dataset containing 30% multi-view anomalies. Number of topic K varies from 100 to 700. With the same K, the proposed model always achieves the lowest perplexity. As perplexities stop decreasing after $K \geq 700$, further increasing number of topics provides no improvement generalization ability. Thus when multi-view anomalies exist, the proposed model outperforms all alternative methods in irrespective of number of topics.

Figure 3 shows average held-out perplexities and their standard errors on the Japanese - Finnish corpus. Anomaly ratio varies from 0 to 0.5. It is shown that

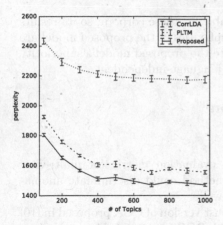

Fig. 2. Average held-out perplexities and their standard errors on Japanese - Finnish dataset with 30% multi-view anomalies

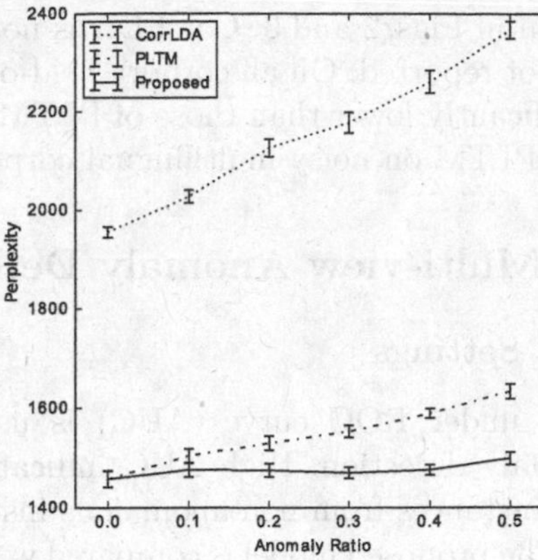

Fig. 3. Average held-out perplexities and their standard error on Japanese - Finnish dataset when anomaly ratio varies. K = 700.

as the anomaly ratio increases, perplexities of CorrLDA and PLTM increase significantly. Because view-swapping does not modify content of each view, this performance degeneration could only result from inconsistency among views. Meanwhile, perplexity of the proposed model increases very slowly when the anomaly ratio increases. Note that in Fig. 2, the proposed model has the lowest perplexity in regardless of K. We conclude that the proposed model has the best generative ability when multi-view anomalies are present on this bilingual dataset.

Table 1. Average held-out perplexities and their standard errors on 10 bilingual corpora

	English - Germany	English - Finnish	English - French	English - Italian
Proposed	**2828.3±16.8**	**2505.9±11.4**	**2529.3±12.3**	**2664.6±13.9**
PLTM	3036.9±18.1	2602.1±18.7	2664.9±17.93	2841.5±25.3

	English - Japanese	Japanese - Germany	Japanese - Finnish	Japanese - English
Proposed	**2215.3±11.4**	**2130.3± 16.8**	**1470.3±11.5**	**2226.0±13.4**
PLTM	2399.7±18.0	2306.4±13.9	1553.3± 12.0	2368.3±12.5

	Japanese - French	Japanese - Italian
Proposed	**1728.4±11.2**	**1940.9±15.4**
PLTM	1866.9±15.2	2130.9±14.2

Datasets contain 30% multi-view anomalies. $K = 700$.

Table 1 shows average held-out perplexities and their standard errors on all the ten bilingual corpora with 30% multi-view anomalies for K equals to 700. As shown in Figs. 2 and 3, CorrLDA is not suitable for these corpora, so its results are not reported. On all corpora held-out perplexities of the proposed model are significantly lower than those of PLTM. Hence the proposed model's superiority over PLTM on noisy multilingual corpora is language-independent.

5 Multi-view Anomaly Detection

5.1 Settings

Area under ROC curve (AUC) is used as evaluation metric for multi-view anomaly detection. High AUC indicates a method could discriminate anomalous instances from non-anomalous instance well.

The proposed model is compared with robust version of CCA proposed in [10] (RCCA), one-class SVM (OCSVM) and PLTM. RCCA is included in comparison because it also uses Dirichlet process and is reported to be effective on continuous data. OCSVM is a representative method for single-view anomaly detection. It is included into experiments to investigate whether methods for single view anomaly detection are also applicable for detecting multi-view anomalies. In experiments OCSVM implementation in scikit-learn package [13] with radial-basis function kernel is used. It is applied into multilingual setting by using bag-of-word representation and appending one view at the end of another.

We also report results of classification by using PLTM's perplexity of training data as anomaly score. Inasmuch as model assumption of PLTM is not valid on anomalous documents, perplexities of such documents are higher than non-anomalous documents. Because the proposed model reduces to PLTM if cluster assignments of views are fixed to be the same, comparison between the proposed model and this method demonstrates the efficacy of using Dirichlet process to model view consistency.

5.2 Dataset

As RCCA does not scale well on high dimensional textual data, we have to carry out comparison experiments with data of smaller size. On Japanese - Finnish dataset, sizes of vocabulary are reduced 100 by removing low-frequency words. Documents that have view shorter than 50 words are removed. From the remaining ten datasets are sampled, each of them contains 100 documents.

5.3 Results

Figure 4 shows AUC of multi-view anomaly detection when anomaly ratio varies. AUC of RCCA and OCSVM are around 0.5 in all cases, which means they barely discriminate anomalies from non-anomalies. AUC of PLTM is around 0.6, and that of the proposed method is around 0.7. Thus the proposed model outperforms all alternative methods.

Figure 5 shows AUC on dataset containing 20% anomalies with various number of topics K. K correspond to dimension of latent space in RCCA. It is

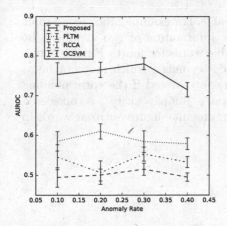

Fig. 4. AUC and their standard errors. Dimension of latent spaces are set to 8 for RCCA. $K = 8$ is used for the proposed model and PLTM

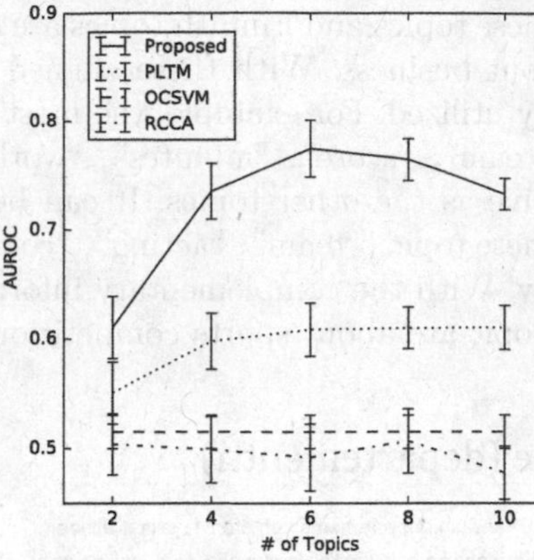

Fig. 5. AUC and their standard errors with anomaly rate equals to 20%.

shown that $K = 4$ is enough for the proposed model and PLTM to achieve their best performance, and the proposed method outperform all comparing methods. Meanwhile, AUC of RCCA is around 0.5 for all cases, which means increasing dimension of latent space cannot improve performance of anomaly detection.

6   Examples of Aligned Topics and Anomalies

In previous sections we demonstrate the proposed model's efficacy in modeling multi-view text data with manually created multi-view anomalies and detecting such anomalies. In this section we present topics extracted by the proposed model and example of multi-view anomalies detected from the original data.

Table 2. An example of aligned Finnish(fi) and Japanese(ja) topics.

fi	final, fantasy, Nintendo, iv, crystal
ja	magic, final, fantasy, character, combat
fi	cooperate, business, production, economics, formula
ja	capital, company, market, analysis, cost
fi	married, spatula, marry, wife, son
ja	marriage, girlfriend, father, mother, daughter
fi	team, score, minutes, world, seconds
ja	team, acting, competition, jump, skate

Examples of most probable words of aligned topics extracted from original Japanese - Finnish corpus are presented in Table 2. Relatedness between

Japanese topics and Finnish topics are observable. For example, the second topic is about business. With these aligned topics, information of two views can be jointly utilized. For example, the most probable words for fourth Finnish topics are "team", "score", "minutes", "world" and "seconds", which may not be as cohesive as the other topics. It can be better interpreted if the corresponding Japanese topic ("team", "acting", "competition", "jump", "skate") is considered jointly. With the complementary information, one may figure out that words in this topic are about sports competitions.

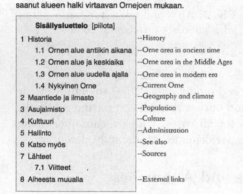
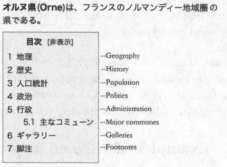

Fig. 6. Article for Orne in Finnish (left) and its counterpart in Japanese (right).

In addition, an example of multi-view anomaly detected from original Japanese - Finnish corpus is shown in Fig. 6. Screenshots are captured in Feb 11th, 2017. These two articles are about Orne, a province in France. While they contain common sections, Finnish and Japanese articles differ significantly in history section. For applications in which inconsistency is detrimental, we can use the proposed model to detect and process documents like this automatically.

7   Conclusion

Since multi-view text data are often managed in distributed fashion, they may contain multi-view anomalies and pose challenge on topic modeling. In this paper a probabilistic topic model is proposed for multi-view topic modeling, which is capable of modeling joint distribution of views and detecting anomalies simultaneously. In our experiments on ten bilingual Wikipedia corpora, it is demonstrated that the proposed model is more robust than existing multi-view topic models against multi-view anomalies. In addition, from comparison with other multi-view anomaly detection methods it is shown that the proposed model is more effective on textual data. Future work of the proposed model includes applying to multi-modal text data.

References

1. Asuncion, A., Welling, M., Smyth, P., Teh, Y.W.: On smoothing and inference for topic models. In: Proceedings of the Twenty-Fifth Conference on Uncertainty in Artificial Intelligence, pp. 27–34. AUAI Press (2009)
2. Bach, F.R., Jordan, M.I.: A probabilistic interpretation of canonical correlation analysis. Technical report 688, Department of Statistics, University of California, Berkeley (2005)
3. Blei, D.M., Jordan, M.I.: Modeling annotated data. In: Proceedings of the 26th Annual International ACM SIGIR Conference on Research and Development in Informaion Retrieval, pp. 127–134. ACM (2003)
4. Blei, D.M., Ng, A.Y., Jordan, M.I.: Latent dirichlet allocation. J. Mach. Learn. Res. 3(Jan), 993–1022 (2003)
5. De Smet, W., Tang, J., Moens, M.-F.: Knowledge transfer across multilingual corpora via latent topics. In: Huang, J.Z., Cao, L., Srivastava, J. (eds.) PAKDD 2011. LNCS (LNAI), vol. 6634, pp. 549–560. Springer, Heidelberg (2011). https://doi.org/10.1007/978-3-642-20841-6_45
6. Duh, K., Yeung, C.-M.A., Iwata, T., Nagata, M.: Managing information disparity in multilingual document collections. ACM Trans. Speech Lang. Process. (TSLP) 10(1), 1 (2013)
7. Fukumasu, K., Eguchi, K., Xing, E.P.: Symmetric correspondence topic models for multilingual text analysis. In: Advances in Neural Information Processing Systems 25, pp. 1286–1294. Curran Associates Inc. (2012)
8. Gao, J., Fan, W., Turaga, D., Parthasarathy, S., Han, J.: A spectral framework for detecting inconsistency across multi-source object relationships. In: 2011 IEEE 11th International Conference on Data Mining (ICDM), pp. 1050–1055. IEEE (2011)
9. Hara, N., Shachaf, P., Hew, K.F.: Cross-cultural analysis of the Wikipedia community. J. Am. Soc. Inform. Sci. Technol. 61(10), 2097–2108 (2010)
10. Iwata, T., Yamada, M.: Multi-view anomaly detection via robust probabilistic latent variable models. In: Lee, D.D., Sugiyama, M., Luxburg, U.V., Guyon, I., Garnett, R. (eds.) Advances in Neural Information Processing Systems 29, pp. 1136–1144. Curran Associates Inc., New York (2016)
11. Mimno, D., Wallach, H.M., Naradowsky, J., Smith, D.A., McCallum, A.: Polylingual topic models. In: Proceedings of the 2009 Conference on Empirical Methods in Natural Language Processing: Volume 2-Volume 2, pp. 880–889. Association for Computational Linguistics (2009)
12. Minka, T.P.: Estimating a Dirichlet distribution (2000). https://tminka.github.io/papers/dirichlet/. Accessed 10 January 2017
13. Pedregosa, F., Varoquaux, G., Gramfort, A., Michel, V., Thirion, B., Grisel, O., Blondel, M., Prettenhofer, P., Weiss, R., Dubourg, V., Vanderplas, J., Passos, A., Cournapeau, D., Brucher, M., Perrot, M., Duchesnay, E.: Scikit-learn: machine learning in Python. J. Mach. Learn. Res. 12, 2825–2830 (2011)
14. Ramage, D., Heymann, P., Manning, C.D., Garcia-Molina, H.: Clustering the tagged web. In: Proceedings of the Second ACM International Conference on Web Search and Data Mining, pp. 54–63. ACM (2009)

15. Sethuraman, J.: A constructive definition of dirichlet priors. Stat. Sin. 639–650 (1994)
16. Vulić, I., De Smet, W., Tang, J., Moens, M.-F.: Probabilistic topic modeling in multilingual settings: an overview of its methodology and applications. Inf. Process. Manag. **51**(1), 111–147 (2015)

Recommendation

A Regularization Method with Inference of Trust and Distrust in Recommender Systems

Dimitrios Rafailidis[1(✉)] and Fabio Crestani[2]

[1] Department of Computer Science, University of Mons, Mons, Belgium
dimitrios.rafailidis@umons.ac.be
[2] Faculty of Informatics, Università della Svizzera italiana, Lugano, Switzerland
fabio.crestani@usi.ch

Abstract. In this study we investigate the recommendation problem with trust and distrust relationships to overcome the sparsity of users' preferences, accounting for the fact that users trust the recommendations of their friends, and they do not accept the recommendations of their foes. In addition, not only users' preferences are sparse, but also users' social relationships. So, we first propose an inference step with multiple random walks to predict the implicit-missing trust relationships that users might have in recommender systems, while considering users' explicit trust and distrust relationships during the inference. We introduce a regularization method and design an objective function with a social regularization term to weigh the influence of friends' trust and foes' distrust degrees on users' preferences. We formulate the objective function of our regularization method as a minimization problem with respect to the users' and items' latent features and then we solve our recommendation problem via gradient descent. Our experiments confirm that our approach preserves relatively high recommendation accuracy in the presence of sparsity in both the users' preferences and social relationships, significantly outperforming several state-of-the-art methods.

Keywords: Recommender systems · Collaborative filtering
Social relationships · Regularization

1 Introduction

Recommender systems have widely followed the collaborative filtering strategy, where similar-minded users tend to get similar recommendations [8]. In real-world scenarios, the main limitation of collaborative filtering strategy is the data sparsity of users' preferences, significantly degrading the recommendation accuracy. To overcome the data sparsity problem and generate trust-based recommendations, several models exploit the selections of trust friends [4,12]. In trust-based recommendations, models consider that people tend to rely on recommendations from their friends [6,10]. In addition, in online networks users may establish both trust and distrust relationships, while the vast majority of

© Springer International Publishing AG 2017
M. Ceci et al. (Eds.): ECML PKDD 2017, Part II, LNAI 10535, pp. 253–268, 2017.
https://doi.org/10.1007/978-3-319-71246-8_16

users have unknown relationships. Epinions[1], an e-commerce site for reviewing and rating products, allows users to evaluate others based on the quality of their reviews, and form trust and distrust relations with them. In Slashdot[2] users post news and comments, and tag other users as friends or foes. The analyses in related studies conclude that users might accept recommendations from their trusted friends, but will certainly exclude recommendations from their distrusted foes [20, 21, 23]. In this respect, more recently a few attempts have been made to exploit both trust and distrust relationships at the same time in recommender systems [1, 2, 15, 19]. Following the collaborative filtering strategy, these models assume that the latent features of trust/distrust users should be as close/far as possible. However, such models exploit trust and distrust relationships in an *explicit* manner, that is users' direct friends or foes, and do not *infer implicit (indirect)* social relationships that users establish in recommender systems. Due to the presence of sparsity in social relationships, models that use trust and distrust relationships exploit a limited number of social relationships, thus not performing well as we will show in our experiments in Sect. 5.

Inferring social relationships of trust and distrust users is a challenging task [20]. Given explicit social relationships, the goal is to infer the indirect relationships of trust and distrust users. Trust relationships show strong transitivity, which means that inferring trust relationships can be computed in a network of trust users, mainly because if two users a and b are friends and a third user c is friend with a, then user c might be a friend of b as well. However, recent studies showed that distrust is certainly not transitive [3, 21, 23]. Therefore, distrust cannot be considered as the negative of trust when inferring users' distrust relationships. Accounting for the transitivity of trust relationships, a few prediction models have been proposed to infer the implicit trust relationships, while exploiting explicit distrust relationships in their predictions [7, 20]. Nonetheless, these models are designed to predict missing-trust relationships and not to generate recommendations. Therefore, a pressing challenge resides on how to infer trust relationships of users with their distrust relationships to boost the recommendation accuracy.

1.1 Contribution

To overcome the shortcomings of existing methods our contributions are summarized as follows, first we infer implicit trust relationships with multiple random walks while considering the users' explicit distrust relationships during the inference. By significantly enhancing users' relationships and reducing the number of unknown relationships, we formulate a social regularization term to weigh users' trust and distrust degrees and capture the correlations of users' preferences with those of their friends and foes. We introduce a regularization method and design an objective function as a minimization problem with respect to the latent features, and solve our recommendation problem via gradient descent.

---

[1] http://www.epinions.com/.

[2] https://slashdot.org/.

Our experiments show that the proposed approach is superior to all the competitors at different levels of sparsity in users' preferences, as well as in users' social relationships.

The remainder of the paper is organized as follows, Sect. 2 reviews the related work and in Sect. 3 we formally define our problem. Section 4 details the proposed model, Sect. 5 presents the experimental results and Sect. 6 concludes the study.

2   Related Work

In trust-based recommender systems, many different strategies have been introduced to exploit the selections of trust users. Jamali and Ester [6] extend the probabilistic matrix factorization of [12] by weighting the user latent factors based on their trust relationships. In [10], a trust-based ensemble method is presented to combine matrix factorization with a social-based neighborhood model. In [5], a random walk model is introduced to incorporate a trust-based approach into a neighborhood-based collaborative filtering strategy. In this study, authors run random walks on the trust network, formed by the trust relationships, and then perform a probabilistic item selection strategy to generate recommendations. The random walk performs the search in the trust network and the item selection part considers ratings on similar items to avoid going too deep in the network. However, TrustWalker does not infer implicit (indirect) trust relationships, as it performs random walks to generate recommendations using a neighborhood-based model. Guo et al. [4] extend SVD++ [8], to learn both the user preferences and the social influence of her friends. However, all the above studies ignore users' distrust relationships when generating recommendations, an important factor to boost the recommendation accuracy [23]. Ma et al. [11] present a social regularization method by sharing a common user-item matrix, factorized by ratings and social relationships. This work introduces a trust-based as well as a distrust-based model to exploit trust and distrust relationships in each model separately. The goal of the trust-based model is to minimize the distances of latent features between trust users, while the distrust-based model tries to maximize the latent features' distances between distrust users.

Recently, a few attempts have been made to exploit both trust and distrust relationships at the same time in recommender systems. Forsati et al. [2] incorporate trust and distrust relationships into a matrix factorization framework, by formulating a hinge loss function. This method assumes that the trust/distrust relationships between users are considered as similarity/dissimilarity in their preferences, and then the latent features are computed in a manner such that the latent features of foes who are distrusted by a certain user have a guaranteed minimum dissimilarity gap from the maximum dissimilarity of friends who are trusted by this user. This means that when the user agrees on an item with one of her trusted friends, she will probably disagree on the same item with her distrusted foes, assuming a minimum predefined margin. In [1], a recommendation strategy is introduced to rank the latent features of users, based on the users' trust and distrust relationships. This method also considers the neutral relationships (of users who have no relation to a certain user), aiming to rank the user latent features after the trust relationships and before the distrust ones. In [15] a signed graph is constructed, considering positive and negative weights for the trust and distrust relationships, respectively. Then, a spectral clustering approach is presented to generate clusters in the signed graph. The clusters are extracted on condition that users with positive connections should lie close, while users with negative ones should lie far. Following a joint non-negative matrix factorization framework the final recommendations are generated, by co-factorizing the user-item and user-cluster associations. Tang et al. [19] form a signed graph with trust and distrust relationships and capture local and global information from the graph. Local information reveals the correlations among a user and her friends/foes, and the global information reveals the reputation of the user in the whole social network, as users tend to trust users with high global reputation. Then, they exploit both local and global information in a matrix factorization technique to produce recommendations.

3 Problem Formulation

Let n and m be the numbers of users and items. Given a rating matrix $\mathbf{R} \in \mathbb{R}^{n \times m}$, each entry $\mathbf{R}_{iq}$ corresponds to the rating that user i has assigned to item q, with $i = 1 \ldots n$ and $q = 1 \ldots m$. If a user i has not rated an item q, then we set $\mathbf{R}_{iq} = 0$. In our approach we follow the collaborative filtering strategy of matrix factorization. This means that by factorizing the matrix $\mathbf{R}$, the recommendations are in its low-rank approximation $\hat{\mathbf{R}} \in \mathbb{R}^{n \times m}$. Matrix $\hat{\mathbf{R}}$ can be calculated as $\hat{\mathbf{R}} = \mathbf{U}\mathbf{V}^{\top}$, where $\mathbf{U} \in \mathbb{R}^{n \times d}$ is the user factor matrix, and $\mathbf{V} \in \mathbb{R}^{m \times d}$ is the item factor matrix with d the number of latent dimensions. To consider users' trust and distrust relationships, we also form a graph $\mathcal{G}$ with $n = |\mathcal{V}|$ nodes and $i, j \in \mathcal{V}$. Two nodes are connected with edges in the form $(i, j) \in \mathcal{E}$. The edges are considered weighted, and in our setting we consider positive and negative weights to express trust and distrust relationships, respectively. Both positive and negative weights are stored in an adjacency matrix $\mathbf{A} \in \mathbb{R}^{n \times n}$. In our approach we generate two different graphs, a graph G_+ which contains only the positive edges and a second graph G_- with the negative ones. Given $\mathcal{E} \equiv \mathcal{E}_+ \cup \mathcal{E}_-$, we compute two different adjacency matrices $\mathbf{A}_+ \in \mathbb{R}^{n \times n}$ and $\mathbf{A}_- \in \mathbb{R}^{n \times n}$, corresponding to the weights of the positive $(i, j)_+ \in \mathcal{E}_+$ and negative $(i, j)_- \in \mathcal{E}_-$ edges/relationships. In addition, $\forall (i, j)_- \in \mathcal{E}_-$ we set $(\mathbf{A}_-)_{ij} = |\mathbf{A}_{ij}|$, storing the absolute values of the negative weights. With this setting, the goal of the proposed approach is formally defined as follows:

Definition 1 (Problem). *"Given (i) the rating matrix $\mathbf{R}$ and (ii) the adjacency matrices $\mathbf{A}_+$ and $\mathbf{A}_-$ with the trust and distrust relationships, the goal of the proposed approach is first to infer the implicit trust relationships based on users' explicit trust and distrust relationships, and then compute the low-rank approximation $\hat{\mathbf{R}} = \mathbf{U}\mathbf{V}^{\top}$ to generate the final recommendations."*

4   Proposed Approach

4.1   Social Inference via Multiple Random Walks

To infer the implicit trust relationships, we perform random walks on the n nodes in graph $\mathcal{G}_+$, by taking into account the distrust relationships in graph $\mathcal{G}_-$ during the inference. In particular, the proposed approach runs multiple random walks on the graph $\mathcal{G}_+$ with the trust relationships and then filters out the inferred trust relationships by considering the distrust relationships in graph $\mathcal{G}_-$. The main reason that we avoid to perform random walks on graph $\mathcal{G}_-$ is that *distrust is not transitive*, as opposed to trust [7,20,21,23]. Next, we present the case of performing a single random walk on graph $\mathcal{G}_+$ and we show how to perform multiple random walks to better infer the implicit trust relationships.

Single Random Walk. Given a source node *sou* and a target node *tar* in graph $\mathcal{G}_+$, with $(sou, tar) \notin \mathcal{E}_+$, the goal is to start a random walk from *sou* to reach *tar* to infer their trust relationships, denoted as $(\overline{\mathbf{A}}_+)_{sou,tar}$. We assume that the walk moves from one node to a neighbourhood node at each step, and at time t has moved to node i. The walk chooses whether to move to another node with probability ξ_t or terminate the walk with probability $1 - \xi_t$. In the case of terminating the walk, the value $(\overline{\mathbf{A}}_+)_{sou,tar}$ is returned only if edge $(i, tar) \in \mathcal{E}_+$, and 0 otherwise. The *transition probability* of moving from a current node i to another node j is calculated as follows:

$$p_+(j|i) = (\mathbf{A}_+)_{ij}/d_i \tag{1}$$

where $d_i = \sum_j (\mathbf{A}_+)_{ij}$ is the degree of i. The *transition matrix* $\mathbf{T}_+ \in \mathbb{R}^{n \times n}$ of a random walk is given by

$$\mathbf{T}_+ = \mathbf{D}_+^{-1} \mathbf{A}_+ \tag{2}$$

where $(\mathbf{D}_+)_{ii} = d_i$ is the degree diagonal matrix, with $\mathbf{D}_+ \in \mathbb{R}^{n \times n}$. A vector $\mathbf{p}_+^{(t)} \in \mathbb{R}^n$ represents the visiting distribution over all n nodes at a certain time t. With these settings, if the walk continues at the next time $t+1$, the distribution vector will be updated as follows:

$$\mathbf{p}_+^{(t+1)} = \mathbf{p}_+^{(t)} \times \mathbf{T}_+ \tag{3}$$

In the case of isolated users, random walks are not performed, as these users have not expressed their social preferences.

Multiple Random Walks. Instead of performing a single walk, we run multiple random walks from a source node in graph $\mathcal{G}_+$ to better infer the implicit trust relationships. The main reason that we can achieve better inference is that *multiple random walks start from the source user sou to seek more alternatives for the implicit (indirect) relationship to the target user tar*. Given the graph $\mathcal{G}_+$, we define s as the total length of a single walk for which we recursively update the distribution vector $\mathbf{p}^{(s)}$. For a target node tar we consider all its in-linked edges, denoted by $(\mathbf{A}_+)_{*tar}$, that is the tar-th column vector of $\mathbf{A}_+$. With these settings, the returned value for a random walk terminated at time s is:

$$(\overline{\mathbf{A}}_+)_{sou,tar}|s = \mathbf{p}^{(s)}(\mathbf{A}_+)_{*tar} \tag{4}$$

Theoretically, we can perform random walks with infinite lengths from the source node. Aggregating the multiple random walks from the source node we have:

$$(\overline{\mathbf{A}}_+)_{sou,tar} = \sum_{t=1}^{\infty} \omega_+(t)\mathbf{p}_+^{(0)}\mathbf{T}_+^t(\mathbf{A}_+)_{*tar} \tag{5}$$

where $\mathbf{p}_+^{(0)}$ is the starting distribution of a walk on $\mathcal{G}_+$ and $\omega_+(t)$ expresses the probability that a random walk will terminate at a certain time t:

$$\omega_+(t) = p_+(s = t|\xi) = \xi_t \prod_{i=1}^{t-1} (1 - \xi_i) \tag{6}$$

Therefore, the adjacency matrix $\overline{\mathbf{A}}_+$ with the inferred trust relationships is calculated as follows:

$$\overline{\mathbf{A}}_+ = \sum_{t=1}^{\infty} \omega_+(t)\mathbf{T}_+^t\mathbf{A}_+ \tag{7}$$

In our implementation, we avoid long (infinite) walks on the graph, following the idea of the "six degrees of separation", that is most nodes can be reached with a six step walk length [13]. This means that if a walk has reached more than six steps, then the walk is terminated. In practice, we observed that random walks do not reach more than 4 steps in our experiments with $\xi_t = 0.85$, equal to the dampening factor of PageRank [14]. Next, the inferred trust relationships are stored in matrix $\overline{\mathbf{A}}_+$.

When performing multiple random walks on graph $\mathcal{G}_+$, the distrust relationships in graph $\mathcal{G}_-$ are ignored. Consequently, an inferred trust relationship between a source user sou and a target user tar in $\overline{\mathbf{A}}_+$, might have a distrust relationship between sou and tar in graph $\mathcal{G}_-$. Since users do not accept the recommendations of distrust users [2,7], we recompute matrix $\overline{\mathbf{A}}_+$ by setting $\overline{\mathbf{A}}_+ \leftarrow 0$, if $(\overline{\mathbf{A}}_+)_{ij} > 0 \wedge (\mathbf{A}_-)_{ij} > 0, \forall i, j = 1 \dots n$. Finally, the filtered trust relationships and their positive weights are stored into the initial adjacency matrix with the trust relationships, by setting $\mathbf{A}_+ \leftarrow \overline{\mathbf{A}}_+$.

4.2 Regularization Method

To generate the recommendations in our regularization method, we have to compute the low rank approximation $\hat{\mathbf{R}}$ based on the inferred trust relationships in $\mathbf{A}_+$ and the distrust relationships in $\mathbf{A}_-$. We first capture the user-based similarities using the rating matrix $\mathbf{R}$. If users i and j have interacted with at least a common item q, then users i and j are connected based on their preferences [22]. The preferences' connections/similarities are stored in a similarity matrix $\mathbf{S} \in \mathbb{R}^{n \times n}$, whose ij-th entries are calculated as follows:

$$\mathbf{S}_{ij} = \begin{cases} \dfrac{\sum\limits_{q=1}^{m} \mathbf{R}_{iq}\mathbf{R}_{jq}}{\sqrt{\sum\limits_{q=1}^{m} \mathbf{R}_{iq}^2}\sqrt{\sum\limits_{q=1}^{m} \mathbf{R}_{jq}^2}} & , \quad \text{if users } i \text{ and } j \text{ are connected} \\ 0 & , \quad \text{otherwise} \end{cases} \tag{8}$$

with $i, j = 1, \ldots, n$.

Next, we form the neighbourhoods $\mathcal{N}_+^i$ and $\mathcal{N}_-^i$ based on the adjacency matrices $\mathbf{A}_+$ and $\mathbf{A}_-$, respectively, where $\mathbf{A}_+$ contains the explicit as well as the inferred implicit trust relationships. Given the latent vector $\mathbf{U}_i$ of user i and the latent vector $\mathbf{U}_j$ of her friend j, with $j \in \mathcal{N}_+^i$, their distance $||\mathbf{U}_i - \mathbf{U}_j||_2^2$ should be as close as possible, weighted by the trust degree $(\mathbf{A}_+)_{ij}$ and the preference similarity $\mathbf{S}_{ij}$. The reason that we consider the similarity between trust users i and j is that trust friends do not necessarily have similar preferences [4]. The higher the similarity between two friends the most likely would be that they have similar preferences. Thus, we weigh the influence of trust friends i and j as follows:

$$\sum_{j \in \mathcal{N}_+^i} (\mathbf{A}_+)_{ij}\mathbf{S}_{ij}||\mathbf{U}_i - \mathbf{U}_j||_2^2 \tag{9}$$

Accordingly, for a distrusted user $k \in \mathcal{N}_-^i$, we have to penalize their distance $||\mathbf{U}_i - \mathbf{U}_k||_2^2$, weighted by the distrust degree $(\mathbf{A}_-)_{ik}$ and their preference dissimilarity $1 - \mathbf{S}_{ik}$ as follows:

$$-\sum_{k \in \mathcal{N}_-^i} (\mathbf{A}_-)_{ik}(1 - \mathbf{S}_{ik})||\mathbf{U}_i - \mathbf{U}_k||_2^2 \tag{10}$$

By aggregating Eqs. (9) and (10), we have the following regularization term with respect to a latent vector $\mathbf{U}_i$, to measure the weighted influence of trust and distrust relationships:

$$\Psi(\mathbf{U}_i) = \sum_{j \in \mathcal{N}_+^i} (\mathbf{A}_+)_{ij}\mathbf{S}_{ij}||\mathbf{U}_i - \mathbf{U}_j||_2^2 - \sum_{k \in \mathcal{N}_-^i} (\mathbf{A}_-)_{ik}(1 - \mathbf{S}_{ik})||\mathbf{U}_i - \mathbf{U}_k||_2^2 \tag{11}$$

To compute the latent matrices $\mathbf{U}$ and $\mathbf{V}$ we formulate the following objective function $\mathfrak{L}$ as a minimization problem:

$$\min_{\mathbf{U},\mathbf{V}} \mathfrak{L}(\mathbf{U}, \mathbf{V}) = ||\mathbf{R} - \mathbf{U}\mathbf{V}^\top||_F^2 + \lambda(||\mathbf{U}||_F^2 + ||\mathbf{V}||_F^2) + \alpha \sum_{i=1}^{n} \Psi(\mathbf{U}_i) \tag{12}$$

where the second and third terms are the regularization term to avoid model overfitting and the social regularizer of Eq. (11), respectively. Parameters λ and α control the influences of the respective regularization terms.

To optimize Eq. (12) we use a gradient-based strategy. As the term $\Psi(\mathbf{U}_i)$ in Eq. (11) may become negative[3], making the objective function $\mathcal{L}$ non-convex, we define an auxiliary matrix $\mathbf{H} \in \mathbb{R}^{n \times d}$. During the optimization of the objective function $\mathcal{L}$ for the iteration $iter$ and $i = 1 \ldots : n$ we set $\mathbf{H}_i^{iter} = 1$ if $\Psi(\mathbf{U}_i) > 0$, and 0 otherwise. Thus, at the iteration $iter$ we have the following objective function:

$$\min_{\mathbf{U},\mathbf{V}} \mathcal{L}(\mathbf{U}, \mathbf{V}) = ||\mathbf{R} - \mathbf{U}\mathbf{V}^\top||_F^2 + \lambda(||\mathbf{U}||_F^2 + ||\mathbf{V}||_F^2)$$

$$+ \alpha \sum_{i=1}^n \mathbf{H}_i^{iter} \left[\sum_{j \in \mathcal{N}_+^i} (\mathbf{A}_+)_{ij} \mathbf{S}_{ij} ||\mathbf{U}_i - \mathbf{U}_j||_2^2 - \sum_{k \in \mathcal{N}_-^i} (\mathbf{A}_-)_{ik}(1 - \mathbf{S}_{ik})||\mathbf{U}_i - \mathbf{U}_k||_2^2 \right]$$

$$(13)$$

As $\mathcal{L}$ becomes convex with the auxiliary matrix $\mathbf{H}^{iter}$, we follow a gradient-based optimization strategy to calculate matrices/variables $\mathbf{U}$ and $\mathbf{V}$. Given a learning parameter η we have the following update rules:

$$\mathbf{U}_i^{iter+1} \leftarrow \mathbf{U}_i^{iter} - \eta \frac{\partial \mathcal{L}}{\partial \mathbf{U}_i^{iter}}, \quad i = 1 \ldots n$$

$$\mathbf{V}_q^{iter+1} \leftarrow \mathbf{V}_q^{iter} - \eta \frac{\partial \mathcal{L}}{\partial \mathbf{V}_q^{iter}}, \quad q = 1 \ldots m \qquad (14)$$

where the respective gradients at iteration $iter$ are calculated as follows:

$$\frac{\partial \mathcal{L}}{\partial \mathbf{U}_i^{iter}} = 2 \sum_{q=1}^m \mathbb{I}_{iq}^R (\mathbf{U}_i^\top \mathbf{V}_q - \mathbf{R}_{iq})\mathbf{V}_q + 2\lambda\mathbf{U}_i$$

$$+ 2\alpha\mathbf{H}_i^{iter} \sum_{j \in \mathcal{N}_+^i} (\mathbf{A}_+)_{ij}\mathbf{S}_{ij}(\mathbf{U}_i - \mathbf{U}_j) \qquad (15)$$

$$- 2\alpha\mathbf{H}_i^{iter} \sum_{k \in \mathcal{N}_-^i} (\mathbf{A}_-)_{ik}(1 - \mathbf{S}_{ik})(\mathbf{U}_i - \mathbf{U}_k)$$

$$\frac{\partial \mathcal{L}}{\partial \mathbf{V}_q^{iter}} = 2 \sum_{i=1}^n \mathbb{I}_{iq}^R (\mathbf{U}_i^\top \mathbf{V}_q - \mathbf{R}_{iq})\mathbf{U}_i + 2\lambda\mathbf{V}_q \qquad (16)$$

where $\mathbb{I}_{iq}^R \in \{0,1\}^{n \times m}$ is an indicator matrix, with $\mathbb{I}_{iq}^R = 1$ if $\mathbf{R}_{iq} > 0$ and 0 otherwise. Based on the gradients in Eqs. (15) and (16) we can solve the minimization problem of Eq. (13) using the update rules in Eq. (14). Having computed matrices $\mathbf{U}$ and $\mathbf{V}$, we reconstruct the initial rating matrix $\mathbf{R}$ by computing $\hat{\mathbf{R}} = \mathbf{U}\mathbf{V}^\top$ to generate the final recommendations.

---

[3] If $\sum_{k \in \mathcal{N}_-^i} (\mathbf{A}_-)_{ik}(1 - \mathbf{S}_{ik})||\mathbf{U}_i - \mathbf{U}_k||_2^2 > \sum_{j \in \mathcal{N}_+^i} (\mathbf{A}_+)_{ij}\mathbf{S}_{ij}||\mathbf{U}_i - \mathbf{U}_j||_2^2$.

5   Experimental Evaluation

5.1   Experimental Setup

In our experiments, we use a real-world dataset from Epinions[4] [3]. For comparison reasons, we conduct our experiments on a down-sampled dataset, at the same scale as in [2,15]. The down-sampled dataset contains $n = 119,867$ users, $m = 676,436$ product-items and 12,328,927 ratings, including 452,123 trust and 92,417 distrust social relationships. The reason for selecting the dataset is that it is among the most challenging datasets in the relevant literature, as it contains many users and items with high sparsity[5] in users' preferences, as well as it includes users' trust and distrust relationships.

To evaluate the performance of the proposed method, we randomly select a percentage of ratings (i, q) as a test set $\mathcal{T}$, while the remaining ratings are used to train our model. Following relevant studies [2,11], we evaluate the performance of our model in terms of Mean Absolute Error (MAE) and Root Mean Squared Error (RMSE), which are formally defined as follows

$$\text{MAE} = \frac{\sum_{(i,q)\in\mathcal{T}} |\mathbf{R}_{iq} \cdot \hat{\mathbf{R}}_{iq}|}{|\mathcal{T}|}$$

$$\text{RMSE} = \sqrt{\frac{\sum_{(i,q)\in\mathcal{T}} \left(\mathbf{R}_{iq} - \hat{\mathbf{R}}_{iq}\right)^2}{|\mathcal{T}|}}$$

where $\hat{\mathbf{R}}$ is the prediction of our model e.g., the low rank approximation matrix, and $\mathbf{R}$ is the rating matrix with the ratings of the test set $\mathcal{T}$. The difference between the two evaluation metrics is that RMSE emphasizes more in larger prediction errors than MAE. We repeated our experiments five times and we averaged our results over the five runs.

5.2   Compared Methods

In our experiments, we use the following baseline methods:

- **NMF** [9]: a baseline non-negative matrix factorization method, which does not consider neither trust or distrust relationships.
- **MF-distrust** [11]: a matrix factorization strategy that incorporates the distrust information, trying to maximize the user latent features of users who are connected with a explicit distrust social relationship. This strategy does not use trust relationships.
- **TrustWalker** [5]: a random walk model that exploits explicit trust relationships in a neighborhood-based collaborative filtering strategy. This model ignores users' distrust relationships.

---

[4] http://www.trustlet.org/epinions.html.
[5] It approximately includes 0.02% of all entries in the rating matrix $\mathbf{R}$.

- **TrustSVD** [4]: a model that extends SVD++ [8] to learn both the user preferences and the social influence of her friends. This method does not exploit distrust relationships as well.

We also compare the proposed approach with the following competitive strategies that exploit both trust and distrust relationships:

- **MF-TD** [2]: a method that performs matrix factorization such that the latent features of foes who are distrusted by a certain user have a guaranteed minimum dissimilarity gap from the worst dissimilarity of friends who are trusted by this user.
- **JNMF-SG** [15]: a method that co-factorizes user-item and user-cluster associations, by partitioning users into clusters with a spectral clustering approach based on users' trust and distrust explicit relationships.
- **RecSSN** [19]: a recommendation method in social signed networks, that considers trust and distrust explicit relationships when generating recommendations. RecSSN captures both local and global information from the signed graph and then exploits both types of information in a matrix factorization scheme.
- **RF**: a variant of the proposed method, where we avoid the inference step of Sect. 4.1 and use only the trust and distrust explicit (direct) relationships. This variant is used to show the importance of the social inference step in our regularization method, when implicit trust relationships are missing.
- **MRW-RF**: the proposed method that infers social relationships in our regularization method to exploit explicit and implicit relationships.

The parameters of the examined methods have been determined via cross-validation and in our experiments we report the best results. The parameter analysis of the proposed method is further studied in Sect. 5.5.

5.3   Comparison with State-of-the-Art

In Fig. 1 we evaluate the examined models in terms of MAE and RMSE. To show the negative effect of sparsity we train the models with different percentages of ratings. In this set of experiments we use all social explicit relationships. We observe that reducing the training set sizes degrades the recommendation accuracy, indicated by larger errors of MAE and RMSE in all models. Compared to baseline NMF, the trust-based models TrustWalker and TrustSVD are less affected by the presence of sparsity in the reduced training sets, as they exploit users' trust explicit relationships. However, both TrustWalker and TrustSVD ignore users' distrust relationships, which explains their limited performance. Also, MF-distrust can reduce the sparsity problem using the distrust relationships. The main reason that MF-distrust performs lower than TrustWalker and TrustSVD is that there are less users' distrust relationships than trust ones in the Epinions dataset (Sect. 5.1). This complies with several studies reporting that users tend to establish less distrust relationships than trust ones [19,20].

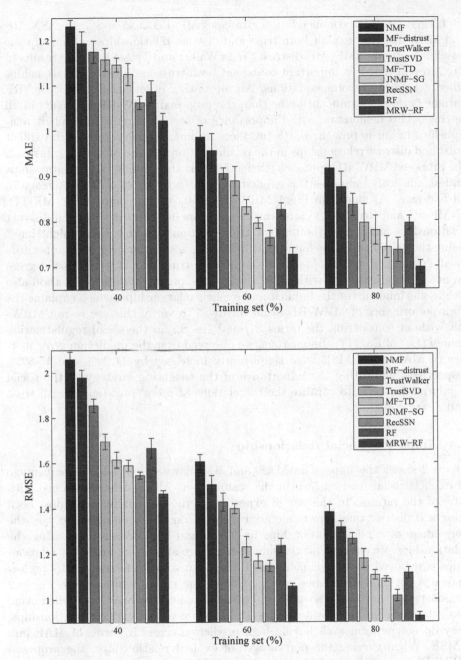

Fig. 1. Effect on MAE and RMSE when varying the percentage of ratings in the training set, using all explicit relationships.

Clearly, all the recommendation strategies MF-TD, JNMF-SG, RecSSN, RF and MRW-RF that exploit both trust and distrust relationships outperform the baseline methods NMF, MF-distrust, TrustWalker and TrustSVD. The results in Fig. 1 reveal that it is important to exploit both trust and distrust relationships when generating recommendations. An interesting observation is that our RF variant performs significant lower than the proposed MRW-RF approach in all settings. This demonstrates the importance of social inference of implicit relationships in our approach, as RF produces recommendations only with explicit trust and distrust relationships in the regularization method of Sect. 4.2. Instead, the proposed MRW-RF approach efficiently infers the implicit trust relationships and significantly enhances the explicit ones by a factor of 1.72 on average in all five runs. As shown in Fig. 1 MRW-RF beats all its competitors, MF-TD, JNMF-SG and RecSSN, because the competitors use explicit trust and distrust relationships and ignore the implicit trust relationships that users might have. Using the paired t-test, we found that MRW-RF is superior over the competitors in all runs, for $p < 0.01$. Moreover, as friends' trust and foes' distrust degrees do not necessarily match with users' preferences, our regularization method also weighs the influence of the implicit and explicit relationships, which explains the high performance of MRW-RF in all settings. To verify this, we re-ran MRW-RF without considering the terms $\mathbf{S}_{ij}$ and $1 - \mathbf{S}_{ik}$ in the social regularization term $\Psi(\mathbf{U}_i)$ of Eq. (11). In our runs, we observed that the prediction error metrics of MAE and RMSE were significantly increased by 11.29% and 15.62%, respectively, indicating the importance of the weighting strategy in the social regularization term to capture the correlations of users' preferences with trust and distrust degrees.

5.4   Impact of Social Relationships

Figure 2 shows the impact on MAE and RMSE when varying the percentage of explicit social relationships in the training set. All models are trained with 50% of the ratings. In this set of experiments the explicit relationships, both trust and distrust ones, are randomly removed from the training set to vary the percentage of explicit relationships in 40, 60 and 80%. When downsizing the relationships, we ensure that the same percentage of trust and distrust relationships is removed. The baseline method NMF is used for reference, as its performance is not influenced when varying the percentages of relationships. Figure 2 demonstrates that all the social-based models are negatively affected when using less relationships. Since state-of-the-art models use only explicit relationships, they do not perform well, having high prediction errors in terms of MAE and RMSE. When varying the percentages of explicit relationships, the proposed MRW-RF approach preserves the recommendation accuracy relatively high, as MRW-RF correctly infers the implicit relationships with the inference step of Sect. 4.1. In the case of 40, 60 and 80% relationships, our inference step significantly enhances the reduced relationships, by 3.28, 2.64 and 2.11, respectively. Consequently, MRW-RF is less affected in the presence of sparsity in social relationships. While state-of-the-art methods have limited performance as there are

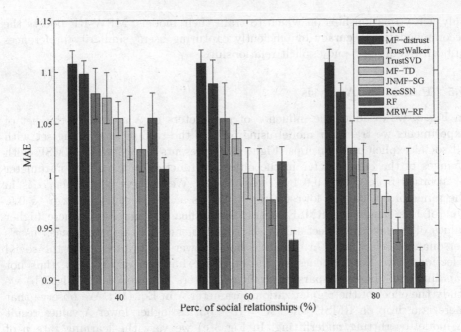

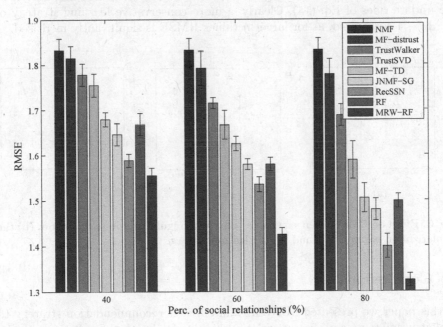

Fig. 2. Effect on MAE and RMSE when varying the percentage of explicit relationships and using 50% of the ratings as training set.

only a few relationships on which to train their models, MRW-RF boosts the recommendation accuracy by efficiently capturing users' similarity preferences in both the implicit and explicit relationships.

5.5   Parameter Analysis

In Fig. 3, we evaluate the influence of parameters α, λ and η. In this set of experiments we train our model using 50% of the ratings as training set with all social explicit relationships. Figure 3(a) presents the effect on RMSE with changes of the α parameter in the objective function of Eq. (13). Parameter α is varied from 0.1 to 0.9 by a step of 0.1. We observe that when α is in the range of 0.4–0.6 the lowest error RMSE is achieved, thus we fix α to 0.5. Out of this range, the RMSE metric is significantly increased, where higher values of α make the selections of the social friends/foes dominate each user's personalized selections. On the other hand, lower values of α make the social selections have less influence on the objective function of Eq. (13), thus not handling well the data sparsity problem in users' preferences. In Fig. 3(b) we study the effect of the regularization parameter λ in Eq. (13). We observe that there is a drop on RMSE when $\lambda = 1e - 03$, as higher/lower λ values result in model overfitting/underfitting. In Fig. 3(c) we vary the learning rate η of the update rules of Eq. (14). Clearly, a more conservative learning strategy of $\eta = 1e - 4$ is required, as for larger η values RMSE is significantly increased.

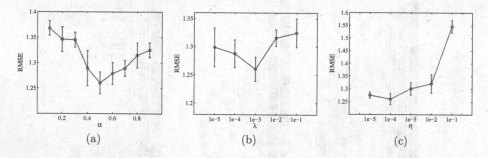

Fig. 3. Effect on RMSE when varying (a) the social regularization parameter α, (b) the regularization parameter λ and (c) the learning rate η.

6   Conclusions

In this paper we presented MRW-RF, an efficient recommendation strategy to exploit users' trust and distrust relationships and solve the sparsity in both the users' preferences and social relationships. The two key factors of the proposed approach are (i) the correct inference of the missing-implicit trust relationships while considering users' distrust relationships during the inference step, and (ii) the capture of the influence of friends and trust degrees on users' preferences in our regularization method. In the proposed inference step, MRW-RF efficiently computes the trust degrees of implicit relationships and significantly enhances the explicit ones. In doing so, the proposed approach handles the sparsity in users' relationships, which plays a crucial role on boosting the recommendation accuracy when a part of the social relationships is not available. Then, in our regularization method we weigh the influence of the inferred/implicit and explicit social relationships by taking into account the users' preferences of friends and foes. Hence, our proposed MRW-RF method achieves high recommendation accuracy by exploiting the selection of the inferred and explicit friends, as well as the selection of explicit foes. Our experiments demonstrate the superiority of MRW-RF over several baselines when varying the sparsity in users' preferences and in social relationships. By enhancing users' social relationships in the inference step and efficiently incorporating the trust and distrust degrees into our regularization method, our approach significantly outperforms competitors in all settings. As future work we plan to investigate the performance of social inference of evolving trust and distrust relationships, to capture users' preference dynamics, a challenging task for recommender systems [16–18].

Acknowledgments. Dimitrios Rafailidis was supported by the COMPLEXYS and INFORTECH Research Institutes of University of Mons.

References

1. Forsati, R., Barjasteh, I., Masrour, F., Esfahanian, A., Radha, H.: Pushtrust: an efficient recommendation algorithm by leveraging trust and distrust relations. In: Proceedings of the 9th ACM Conference on Recommender Systems, Vienna, Austria, pp. 51–58 (2015)
2. Forsati, R., Mahdavi, M., Shamsfard, M., Sarwat, M.: Matrix factorization with explicit trust and distrust side information for improved social recommendation. ACM Trans. Inf. Syst. **32**(4), 17:1–17:38 (2014)
3. Guha, R.V., Kumar, R., Raghavan, P., Tomkins, A.: Propagation of trust and distrust. In: Proceedings of the 13th ACM International Conference on World Wide Web, New York, NY, USA, pp. 403–412 (2004)
4. Guo, G., Zhang, J., Yorke-Smith, N.: TrustSVD: Collaborative filtering with both the explicit and implicit influence of user trust and of item ratings. In: Proceedings of the 29th AAAI Conference on Artificial Intelligence, Austin, Texas, USA, pp. 123–129 (2015)
5. Jamali, M., Ester, M.: TrustWalker: a random walk model for combining trust-based and item-based recommendation. In: Proceedings of the 15th ACM SIGKDD International Conference on Knowledge Discovery and Data Mining, Paris, France, pp. 397–406 (2009)
6. Jamali, M., Ester, M.: A matrix factorization technique with trust propagation for recommendation in social networks. In: Proceedings of the 2010 ACM Conference on Recommender Systems, Barcelona, Spain, pp. 135–142 (2010)
7. Jang, M., Faloutsos, C., Kim, S., Kang, U., Ha, J.: PIN-TRUST: fast trust propagation exploiting positive, implicit, and negative information. In: Proceedings of the 25th ACM International on Conference on Information and Knowledge Management, Indianapolis, IN, USA, pp. 629–638 (2016)

8. Koren, Y.: Factorization meets the neighborhood: a multifaceted collaborative filtering model. In: Proceedings of the 14th ACM SIGKDD International Conference on Knowledge Discovery and Data Mining, Las Vegas, Nevada, USA, pp. 426–434 (2008)

9. Lee, D.D., Seung, H.S.: Algorithms for non-negative matrix factorization. In: Advances in Neural Information Processing Systems, Denver, CO, USA, pp. 556–562 (2000)

10. Ma, H., King, I., Lyu, M.R.: Learning to recommend with social trust ensemble. In: Proceedings of the 32nd Annual International ACM SIGIR Conference on Research and Development in Information Retrieval, Boston, MA, USA, pp. 203–210 (2009)

11. Ma, H., Lyu, M.R., King, I.: Learning to recommend with trust and distrust relationships. In: Proceedings of the 2009 ACM Conference on Recommender Systems, New York, NY, USA, pp. 189–196 (2009)

12. Ma, H., Yang, H., Lyu, M.R., King, I.: Sorec: social recommendation using probabilistic matrix factorization. In: Proceedings of the 17th ACM Conference on Information and Knowledge Management, Napa Valley, California, USA, pp. 931–940 (2008)

13. Milgram, S.: The small world problem. Psychol. Today **2**, 60–67 (1967)

14. Page, L., Brin, S., Motwani, R., Winograd, T.: The pageRank citation ranking: bringing order to the web. Technical report, Stanford Digital Libraries SIDL-WP-1999-0120 (1999)

15. Rafailidis, D.: Modeling trust and distrust information in recommender systems via joint matrix factorization with signed graphs. In: Proceedings of the 31st Annual ACM Symposium on Applied Computing, Pisa, Italy, pp. 1060–1065 (2016)

16. Rafailidis, D., Kefalas, P., Manolopoulos, Y.: Preference dynamics with multimodal user-item interactions in social media recommendation. Expert Syst. Appl. **74**, 11–18 (2017)

17. Rafailidis, D., Nanopoulos, A.: Modeling the dynamics of user preferences in coupled tensor factorization. In: Proceedings of the 8th ACM Conference on Recommender Systems, Foster City, Silicon Valley, CA, USA, pp. 321–324 (2014)

18. Rafailidis, D., Nanopoulos, A.: Modeling users preference dynamics and side information in recommender systems. IEEE Trans. Syst. Man Cybern. Syst. **46**(6), 782–792 (2016)

19. Tang, J., Aggarwal, C.C., Liu, H.: Recommendations in signed social networks. In: Proceedings of the 25th ACM International Conference on World Wide Web, Montreal, Canada, pp. 31–40 (2016)

20. Tang, J., Chang, Y., Aggarwal, C., Liu, H.: A survey of signed network mining in social media. ACM Comput. Surv. **49**(3), 42:1–42:37 (2016)

21. Tang, J., Hu, X., Chang, Y., Liu, H.: Predictability of distrust with interaction data. In: Proceedings of the 23rd ACM International Conference on Conference on Information and Knowledge Management, Shanghai, China, pp. 181–190 (2014)

22. Tang, J., Hu, X., Liu, H.: Social recommendation: a review. Soc. Netw. Anal. Min. **3**(4), 1113–1133 (2013)

23. Victor, P., Cornelis, C., Cock, M.D., Teredesai, A.: Trust-and distrust-based recommendations for controversial reviews. IEEE Intell. Syst. **26**(1), 48–55 (2011)

A Unified Contextual Bandit Framework for Long- and Short-Term Recommendations

M. Tavakol[1,2]($\boxtimes$) and U. Brefeld[1]

[1] Leuphana Universität Lüneburg, Lüneburg, Germany
{tavakol,brefeld}@leuphana.de
[2] Technische Universität Darmstadt, Darmstadt, Germany

Abstract. We present a unified contextual bandit framework for recommendation problems that is able to capture long- and short-term interests of users. The model is devised in dual space and the derivation is consequentially carried out using Fenchel-Legrende conjugates and thus leverages to a wide range of tasks and settings. We detail two instantiations for regression and classification scenarios and obtain well-known algorithms for these special cases. The resulting general and unified framework allows for quickly adapting contextual bandits to different applications at-hand. The empirical study demonstrates that the proposed long- and short-term framework outperforms both, short-term and long-term models on data. Moreover, a tweak of the combined model proves beneficial in cold start problems.

Keywords: Recommendation · Contextual bandits
Dual optimization · Personalization

1 Introduction

Recommender systems are designed to serve user needs. While some needs arise on short notice due to weather changes, news articles, or advertisements, others manifest over a long time span and express general interest in, for example, cars, stock markets, or garments in favored colors. User needs are therefore driven by individual *long-term* and collective *short-term* interests where the latter is highly influenced by the zeitgeist and common trends.

Traditional recommender systems, however, focus on only one aspect of recommendation, that is either on a personalized long-term, or an ad-hoc short-term approach. Collaborative filtering-based methods [6,8], for example, aim to consider long-term preferences of users, while others aim topics of user sessions and focus on short-term interests [2,13,15]. In general, context-aware approaches [9], and their kernelized variants [4,14], may be leveraged to meet both aspects. On the other hand, some recent works focus on context-aware bandits for personalization purposes. Collaborative contextual bandits are introduced in [16] where the context and payoffs are shared among the neighboring users to reduce learning complexity and overall regret. In addition, contextual bandits are used to

© Springer International Publishing AG 2017
M. Ceci et al. (Eds.): ECML PKDD 2017, Part II, LNAI 10535, pp. 269–284, 2017.
https://doi.org/10.1007/978-3-319-71246-8_17

learn the latent structure of users in probabilistic settings to cope with cold-start scenarios [12,17]. Nevertheless, these methods are usually tailored to solve very specific recommendation tasks and may not be applicable to different scenarios. Therefore, a more flexible and comprehensive approach is required to cope with diverse facets of recommendation.

In this paper, we present a unified contextual bandit framework to capture long- and short-term interests of users. The underlying model consists of a contextual (the short-term) and an individual user-based (the long-term) part to determine the expected reward,

$$\mathbb{E}[r_{t,a_i}|u_j] = \underbrace{\boldsymbol{\theta}_i^\top \boldsymbol{x}_t}_{Short-term} + \underbrace{\boldsymbol{\beta}_j^\top \boldsymbol{z}_{a_i}}_{Long-term} + b_i.$$

In the above composition, the expected reward is computed from two distinct parts. The first term models the short-term behavior for a given context $\boldsymbol{x}_t$ at time t. The context determines the recent trend or the topical interest of the current session. In the short-term part, the outcome of choosing each arm a_i for the given context $\boldsymbol{x}_t$ is specified linearly and by its weight vector, $\boldsymbol{\theta}_i$.

The long-term model, on the other hand, allows to capture individual interests for user u_j across item features, $\boldsymbol{z}_{a_i}$ (describing item a_i). We propose to connect the short-term and long-term recommendation in one unified model. Note that b_i acts as constant term in the linear model for each arm. The optimization is performed simultaneously for all the arms so that the short-term part serves as a joint popularity-based predictor while the long-term part acts as an individual offset. All derivations are carried out in the dual space using Fenchel-Legendre conjugates of the loss functions which renders our approach as a framework for a wide range of loss functions. We obtain LinUCB [9] and LogUCB [10] as special cases for regression and classification scenarios, respectively.

The next section derives a generalized recommendation model in dual space which is followed by its instantiations for regression and classification scenarios. Section 3 contains our main contribution and presents the combination of long-term and short-term recommender systems within the unified framework with potential optimization methods. Additionally, possible extensions for our proposed approach is discussed in Sect. 4. We present empirical studies in Sects. 5 and 6 concludes.

2   Linear Bandits in Dual Space

In this paper, we focus on sequential recommender systems for m users, $U = \{u_1, u_2, ..., u_m\}$, and n items, $A = \{a_1, a_2, ..., a_n\}$. Every item a_i is characterized by a set of attributes given by a feature vector $\boldsymbol{z}_{a_i} \in \mathbb{R}^k$. At each time step t, the goal of the system is to recommend items for the actual context of the ongoing session, which is described by a feature vector $\boldsymbol{x}_t \in \mathbb{R}^d$. In the following, we show how to derive the general optimization framework for linear bandits in dual space considering short-term information.

2.1   General Optimization

Assume that the learning procedure for every item (arm) consists of T_i trials, and for every context $\boldsymbol{x}_t$ the reward r_t is obtained. Therefore, $\{(\boldsymbol{x}_t, r_t)\}_{t=1}^{T_i}$ is the set of T_i samples and their corresponding rewards. The reward corresponds to the user feedback w.r.t. the recommended items; its domain depends on the application at-hand; e.g., $r_t \in \{1, 0\}$ for click/no click. We deploy a contextual bandit framework with linear payoff function for arm a_i,

$$h_{\boldsymbol{\theta}_i, b_i}^{(i)}(\boldsymbol{x}_t) = \boldsymbol{\theta}_i^\top \boldsymbol{x}_t + b_i,$$

where hypothesis h predicts the expected payoff for the i-th arm, $\mathbb{E}[r_{t,a_i}]$, and $\boldsymbol{\theta}$ contains the model parameters. The bandit framework learns every hypothesis $h^{(i)}$ independently of the other arms. We therefore discard the index i in the remainder of this section for ease of notation and address the problem for a single arm.

Given an arbitrary loss function $V(\cdot, r_t)$, and using l_2 norm regularizer, the optimization problem can be stated as

$$\inf_{\boldsymbol{\theta}, b}\quad \frac{1}{T}\sum_{t=1}^{T} V(\boldsymbol{\theta}^\top \boldsymbol{x}_t + b, r_t) + \frac{\lambda}{2}\|\boldsymbol{\theta}\|^2.$$

We rewrite the objective by incorporating y_t as shorthand for the predicted payoff. Using $C = \frac{1}{\lambda T}$ gives

$$\inf_{\boldsymbol{\theta}, b, y}\quad C\sum_{t=1}^{T} V(y_t, r_t) + \frac{1}{2}\|\boldsymbol{\theta}\|^2 \quad s.t. \quad \forall t : \boldsymbol{\theta}^\top \boldsymbol{x}_t + b = y_t.$$

The equivalent unconstrained problem is derived by incorporating Lagrange multipliers, $\boldsymbol{\alpha} \in \mathbb{R}^T$,

$$\sup_{\boldsymbol{\alpha}}\ \inf_{\boldsymbol{\theta}, y, b}\quad C\sum_{t=1}^{T} V(y_t, r_t) + \frac{1}{2}\|\boldsymbol{\theta}\|^2 - \sum_{t=1}^{T} \alpha_t(\boldsymbol{\theta}^\top \boldsymbol{x}_t + b - y_t).$$

Setting the partial derivatives w.r.t. b and $\boldsymbol{\theta}$ to zero, leads to the following condition

$$\mathbf{1}^\top \boldsymbol{\alpha} = 0 \quad \text{and} \quad \boldsymbol{\theta} = \sum_{t=1}^{T} \alpha_t \boldsymbol{x}_t = X^\top \boldsymbol{\alpha},$$

where $X \in \mathbb{R}^{T \times d}$ is the design matrix given by the training data. Substituting the optimality conditions into the optimization function yields

$$\sup_{\boldsymbol{\alpha}, \mathbf{1}^\top \boldsymbol{\alpha} = 0}\ \inf_{y}\quad C\sum_{t=1}^{T}\left(V(y_t, r_t) + \frac{1}{C}\alpha_t y_t\right) - \frac{1}{2}\boldsymbol{\alpha}^\top X X^\top \boldsymbol{\alpha}.$$

Moreover, we move the infimum inside the summation as it solely depends on the first term. Using $\inf_w f(w) = -\sup_w -f(w)$, we obtain

$$\sup_{\alpha, 1^\top \alpha = 0} \quad -C \sum_{t=1}^{T} \sup_{y_t} \left(-\frac{\alpha_t}{C} y_t - V(y_t, r_t) \right) - \frac{1}{2} \alpha^\top X X^\top \alpha.$$

Recall that the Fenchel-Legendre conjugate of a function g is defined as $g^*(u) = \sup_x u^\top x - g(x)$ [3]. Thus, the dual loss is given by

$$V^*\left(-\frac{\alpha_t}{C}, r_t \right) = \sup_{y_t} -\frac{\alpha_t}{C} y_t - V(y_t, r_t).$$

(for a comprehensive list of dual losses see [11]). The generalized optimization problem in dual space is therefore reduces to

$$\sup_{\alpha, 1^\top \alpha = 0} \quad -C \sum_{t=1}^{T} V^*\left(-\frac{\alpha_t}{C}, r_t \right) - \frac{1}{2} \alpha^\top X X^\top \alpha. \tag{1}$$

2.2 Upper Confidence Bound

The challenge in bandit-based approaches is to balance exploration and exploitation to minimize the regret. Auer [1] demonstrates that confidence bounds provide useful means to balance the two oppositional strategies. The idea is to use the predicted reward together with its confidence interval to reflect the uncertainty of the model given the actual context. Thus, gathering enough information to reduce the uncertainty in a multi-armed bandit is as important as maximizing the reward.

In our contextual bandit, the expected payoff is approximated by a linear model with an arbitrary loss function where a general optimization approach is used to estimate the parameters. The uncertainty U of the obtained value for each arm is therefore proportional to the standard deviation σ of the expected payoff, $U = c\sigma$, where the variance σ^2 is estimated from training points in neighbouring contexts as well as the model parameters. The uncertainty is added as an upper bound to the prediction to produce a confidence bound for selection strategy across the arms. The computation of the confidence bound depends on the choice of the loss function. We illustrate the obtained bounds for two special cases in the remainder.

2.3 Instantiations

In the following parts, we demonstrate two well-known optimization problems which can be recovered from Eq. (1) by substituting the corresponding loss functions. The instantiations illustrate how a general platform simplifies comparing and analyzing various loss functions in different situations.

Squared Loss. The first instantiation deals with regression scenarios for real-valued payoffs, $r_t \in \mathbb{R}$. The squared loss function and its dual are given by

$$V(y_t, r_t) = \frac{1}{2}(y_t - r_t)^2 \quad \text{and} \quad V^*(s_t, r_t) = \frac{1}{2}s_t^2 + s_t r_t,$$

where the latter can be rewritten as

$$V^*\Big(-\frac{\alpha_t}{C}, r_t\Big) = \frac{1}{2C^2}\alpha_t^2 - \frac{1}{C}\alpha_t r_t.$$

Incorporating the conjugate loss function into Eq. (1) gives

$$\max_{\alpha, \mathbf{1}^\top \alpha = 0} \quad -\frac{1}{2C}\alpha^\top \alpha + \alpha^\top r - \frac{1}{2}\alpha^\top XX^\top \alpha, \tag{2}$$

where the supremum becomes a maximum as the loss function is continuous. The equivalent problem in the primal space corresponds to ridge regression where parameters are determined by optimizing the regularized sum of squared errors,

$$\min_{\theta, b} \quad \frac{1}{T}\sum_{t=1}^{T}\frac{1}{2}(\theta^\top x_t + b - r_t)^2 + \frac{\lambda}{2}\theta^\top \theta.$$

To obtain θ, we set its gradient to 0 which yields $\theta = -\frac{1}{\lambda T}\sum_{t=1}^{T}(\theta^\top x_t + b - r_t)x_t$. The relation $\alpha_t = -\frac{1}{\lambda T}(\theta^\top x_t + b - r_t)$ holds and we have

$$\theta = \sum_{t=1}^{T}\alpha_t x_t = X^\top \alpha.$$

For the threshold parameter b, we obtain the equation $\frac{1}{T}\sum_{t=1}^{T}(\theta^\top x_t + b - r_t) = 0$, and thus arrive at the optimality conditions

$$-\lambda \sum_{t=1}^{T}\alpha_t = 0 \quad \Rightarrow \quad \mathbf{1}^\top \alpha = 0.$$

Expanding the terms in the summation and substituting the optimality conditions leads to the optimization problem

$$\min_{\alpha, \mathbf{1}^\top \alpha = 0} \quad C\Big(\frac{1}{2}\alpha^\top XX^\top XX^\top \alpha - r^\top XX^\top \alpha\Big) + \frac{1}{2}\alpha^\top XX^\top \alpha,$$

where $C = \frac{1}{\lambda T}$. By removing $XX^\top$ from all the terms and converting the minimization into a maximization, we have

$$\max_{\alpha, \mathbf{1}^\top \alpha = 0} \quad -\frac{1}{2}\alpha^\top XX^\top \alpha + r^\top \alpha - \frac{1}{2C}\alpha^\top \alpha,$$

which precisely recovers Eq. (2). The confidence bound for the linear bandit with square loss is given by (cmp. also [9])

$$U = c\sqrt{x_t^\top (X^\top X + \lambda I)^{-1} x_t}.$$

Logistic Loss. In this section, we derive the optimization problem for the logistic loss which is defined as

$$V(y_t, r_t) = \log(1 + \exp(-y_t r_t)).$$

The conjugate of loss function is given by

$$V^*(-\frac{\alpha_t}{r_t}, r_t) = (1 - \frac{\alpha_t}{Cr_t})\log(1 - \frac{\alpha_t}{Cr_t}) + \frac{\alpha_t}{Cr_t}\log(\frac{\alpha_t}{Cr_t}),$$

and incorporating the latter into Eq. (1) leads to Eq. (3)

$$\max_{\alpha, 1^\top \alpha = 0} -C\sum_{t=1}^{T}[(1 - \frac{\alpha_t}{Cr_t})\log(1 - \frac{\alpha_t}{Cr_t}) + \frac{\alpha_t}{Cr_t}\log(\frac{\alpha_t}{Cr_t})]$$
$$-\frac{1}{2}\alpha^\top XX^\top \alpha. \tag{3}$$

The analogous problem in primal space is known as a logistic regression [7] and gives

$$\min_{\hat{\alpha}} \quad \frac{1}{2}\left\|\sum_{t=1}^{T}\hat{\alpha}_t r_t x_t\right\|^2 + C\sum_{t=1}^{T}G(\frac{\hat{\alpha}_t}{C}), \quad s.t. \quad \sum_{t=1}^{T}\hat{\alpha}_t r_t = 0,$$

where $G(\delta) = \delta \log \delta + (1 - \delta)\log(1 - \delta)$. Setting $\alpha_t = \hat{\alpha}_t r_t$, and converting the minimization into a maximization recovers Eq. (3).

The covariance of the parameters for the logistic regression problem is given by $\Sigma = X^T V X$, where V is diagonal matrix of $\pi(1 - \pi)$, and π is computed by the sigmoid function ρ, i.e., $\pi = \rho(X^\top \theta)$. Consequentially, the lower and upper confidence bounds are given by

$$U_{lo} = \rho(\hat{r}_t - c\sqrt{x_t^\top \Sigma^{-1} x_t}), \quad U_{up} = \rho(\hat{r}_t + c\sqrt{x_t^\top \Sigma^{-1} x_t}),$$

respectively [5]. The confidence bound for the contextual bandit is therefore $U = U_{up} - U_{lo}$. Mahajan et al. [10] introduce a variance approximation technique to obtain the confidence bound for logistic loss for probit functions.

3   A Unified Contextual Bandit

In our setting, personalized and user specific information cannot simply be incorporated into the bandit by another type of context. Instead, we suggest to incorporate a long-term model into the short-term approach of the previous section. Therefore, we are able to model the behavior of users for the recommendation process. The long-term part captures the interests of user u_t for every arm a_i. We thus assume a separate set of parameters for the personalized part of the model, given by $\beta_j \in \{\beta_1, ..., \beta_m\}$, where $\beta_j \in \mathbb{R}^k$. The long-term preferences of users are also modelled by a linear relationship $\beta_j^\top z_{a_i}$. For user $u_t \equiv u_j$, the joint long- and short-term model is

$$h^{(i)}_{\theta_i, \beta_t, b_i}(x_t, z_{a_i}) = \theta_i^\top x_t + \beta_t^\top z_{a_i} + b_i.$$

3.1   The Objective Function

As in Sect. 2, all the parameters of the short-term model are still independent from every other item as well as the user parameters among themselves. However, user parameters $\{\beta_1, ..., \beta_m\}$ are shared across the arms and that makes the objective function to be connected for all the arms and users. Hence, the general optimization problem with arbitrary loss function, $V(\cdot, r_t)$ becomes

$$\inf_{\substack{\theta_1, ..., \theta_n \\ \beta_1, ..., \beta_m \\ b}} \frac{1}{T} \sum_{t=1}^{T} V(\theta_t^\top x_t + \beta_t^\top z_t + b_t, r_t) + \frac{\lambda}{2} \sum_i \|\theta_i\|^2 + \frac{\hat{\mu}}{2} \sum_j \|\beta_j\|^2$$

where λ and $\hat{\mu}$ are the regularization parameters for the item and user weights, respectively. Let $C = \frac{1}{\lambda T}$, $\mu = \frac{\hat{\mu}}{\lambda}$, and $y = (\ldots, y_t, \ldots)^\top$, we have

$$\inf_{\substack{\theta_1, ..., \theta_n \\ \beta_1, ..., \beta_m \\ b, y}} C \sum_{t=1}^{T} V(y_t, r_t) + \frac{1}{2} \sum_i \|\theta_i\|^2 + \frac{\mu}{2} \sum_j \|\beta_j\|^2$$

$$s.t. \quad \forall t: \quad \theta_t^\top x_t + \beta_t^\top z_t + b_t = y_t,$$

which results in the Lagrange function

$$\sup_\alpha \inf_{\substack{\theta_1, ..., \theta_n \\ \beta_1, ..., \beta_m \\ b, y}} C \sum_{t=1}^{T} V(y_t, r_t) + \frac{1}{2} \sum_i \|\theta_i\|^2 + \frac{\mu}{2} \sum_j \|\beta_j\|^2$$

$$- \sum_{t=1}^{T} \alpha_t (\theta_t^\top x_t + \beta_t^\top z_t + b_t - y_t).$$

Note that $\{\theta_t, z_t\} \in \{\{\theta_1, z_{a_1}\}, \ldots, \{\theta_n, z_{a_n}\}\}$, $\beta_t \in \{\beta_1, \ldots, \beta_m\}$, and $b_t \in \{b_1, \ldots, b_n\}$. The derivatives with respect to θ_i generate

$$\theta_i = \sum_{\substack{t \\ \theta_t = \theta_i}} \alpha_t x_t = \sum_t \delta_{it} \alpha_t x_t = (X \circ \delta_i)^\top \alpha.$$

In the above equation, $\delta_i \in \mathbb{R}^T$ is a binary vector which is 1 when $\theta_t = \theta_i$, and zero otherwise. $X \in \mathbb{R}^{T \times d}$ is the design matrix of input vectors, and $\circ$ is element-wise product (each element in the vector multiplies by a row in the matrix). We compute the derivations for β_j,

$$\beta_j = \frac{1}{\mu} \sum_{\substack{t \\ \beta_t = \beta_j}} \alpha_t z_t = \frac{1}{\mu} \sum_t \phi_{jt} \alpha_t z_t = \frac{1}{\mu} (Z \circ \phi_j)^\top \alpha,$$

where again $\phi_j \in \mathbb{R}^T$ is the indicator vector for the corresponding user and Z is the design matrix for the items features. Additionally, the derivatives w.r.t. b_i gives

$$\forall i, \quad \sum_{t:b_t=b_i} \alpha_t = 0 \quad \rightarrow \mathbf{1}^\top \alpha = 0.$$

Substituting the obtained conditions in the original problem leads to

$$\sup_{\alpha, \mathbf{1}^\top \alpha=0} \quad \inf_{y} \quad C\sum_{t=1}^{T}[V(y_t, r_t) + \frac{1}{C}\alpha_t y_t]$$
$$-\frac{1}{2}\sum_{i}\alpha^\top(X \circ \delta_i)(X \circ \delta_i)^\top \alpha - \frac{1}{2\mu}\sum_{j}\alpha^\top(Z \circ \phi_j)(Z \circ \phi_j)^\top \alpha,$$

which can be written as

$$\sup_{\alpha, \mathbf{1}^\top \alpha=0} \quad -C\sum_{t=1}^{T}\sup_{y_t}(-\frac{\alpha_t}{C}y_t - V(y_t, r_t))$$
$$-\frac{1}{2}\sum_{i}\alpha^\top(X \circ \delta_i)(X \circ \delta_i)^\top \alpha - \frac{1}{2\mu}\sum_{j}\alpha^\top(Z \circ \phi_j)(Z \circ \phi_j)^\top \alpha.$$

Finally, by converting the first term to the conjugate of the loss function using Fenchel-Legendre conjugates, we obtain

$$\sup_{\alpha, \mathbf{1}^\top \alpha=0} \quad -C\sum_{t=1}^{T}V^*(-\frac{\alpha_t}{C}, r_t) - \frac{1}{2}\sum_{i}\alpha^\top(X \circ \delta_i)(X \circ \delta_i)^\top \alpha$$
$$-\frac{1}{2\mu}\sum_{j}\alpha^\top(Z \circ \phi_j)(Z \circ \phi_j)^\top \alpha. \tag{4}$$

Equation (4) constitutes a generalized optimization problem for contextual bandits with arbitrary loss function. It contains the short-term model in Eq. (1) as a special case when no personal long-term interests need to be captured.

3.2 Optimization

Equation (4) can be optimized with various optimization methods depending on the loss function as well as standard techniques such as gradient-based approaches. For real-time applications and online scenarios, model updates can be performed using (mini-) batches at regular intervals as well, for efficiency. The objective function needs to be maximized w.r.t. the dual parameters α and is given by

$$\sup_{\alpha, \mathbf{1}^\top \alpha=0} \quad -C\mathbb{I}^\top V^*(-\frac{\alpha}{C}, r) - \frac{1}{2}\sum_{i}\alpha^\top(X \circ \delta_i)(X \circ \delta_i)^\top \alpha$$
$$-\frac{1}{2\mu}\sum_{j}\alpha^\top(Z \circ \phi_j)(Z \circ \phi_j)^\top \alpha.$$

The gradient w.r.t. $\boldsymbol{\alpha}$ is obtained by computing the derivatives

$$-C\frac{\partial V^*(-\frac{\alpha}{C}, r)}{\partial \alpha} - [\sum_i (X \circ \delta_i)(X \circ \delta_i)^\top]\alpha - \frac{1}{\mu}[\sum_j (Z \circ \phi_j)(Z \circ \phi_j)^\top]\alpha - \gamma\mathbb{I} = 0.$$

The actual form of the gradient depends on the dual loss V^* and further derivations are omitted accordingly. Note that instantiations often give rise to more sophisticated and efficient optimization techniques than the general form in Eq. (4) allows, see also Sect. 2.3. Nevertheless, the sketched gradient-based approach will always work in case a general optimizer is needed, e.g., in cases where several loss functions should be tried out. Once the optimal parameters, $\boldsymbol{\alpha}^{opt}$, have been found, they can be used to compute the primal parameters

$$\theta_i = (X \circ \delta_i)^\top \alpha^{opt}, \quad \beta_j = \frac{1}{\mu}(Z \circ \phi_j)^\top \alpha^{opt}.$$

Alternatively, kernels $K_X = \phi_X(X, X)$ and $K_Z = \phi_Z(Z, Z)$ could be deployed in the dual representation to allow for non-linear transformations and convolutions of the feature space.

Once the required parameters are found, the payoff estimates are used together with the respective confidence interval U of the arm to choose the arm with the maximum upper confidence value according to

$$a_t = \arg\max_{a_i \in A} \quad \theta_i^\top x_t + \beta_t^\top z_{a_i} + b_i + U_{i,t}.$$

Learning with Squared Loss. In this section, we present the optimization algorithm for a special case of unified contextual bandit framework with squared loss. As it is mentioned in Sect. 2.3, the conjugate of squared loss is given by

$$V^*(-\frac{\alpha_t}{C}, r_t) = \frac{1}{2C^2}\alpha_t^2 - \frac{1}{C}\alpha_t r_t,$$

which leads to the following objective

$$\max_{\alpha, \mathbf{1}^\top \alpha = 0} \quad -\frac{1}{2C}\alpha^\top\alpha + r^\top\alpha - \frac{1}{2}\sum_i \alpha^\top(X \circ \delta_i)(X \circ \delta_i)^\top\alpha$$

$$-\frac{1}{2\mu}\sum_j \alpha^\top(Z \circ \phi_j)(Z \circ \phi_j)^\top\alpha.$$

The summation $\sum_i (X \circ \delta_i)(X \circ \delta_i)^\top$ is equivalent to $(\sum_i \delta_i \otimes \delta_i^\top) \circ XX^\top$, where $\otimes$ stands for the vector outer product. Considering the same equivalency for the last term as well, we rewrite the equation as follows

$$\max_{\alpha, \mathbf{1}^\top \alpha = 0} \quad -\frac{1}{2C}\alpha^\top\alpha + r^\top\alpha$$

$$-\frac{1}{2}\alpha^\top[(\sum_i \delta_i \otimes \delta_i^\top) \circ XX^\top + \frac{1}{\mu}(\sum_i \phi_i \otimes \phi_i^\top) \circ ZZ^\top]\alpha.$$

By using *min* instead of *max*, setting $P = \frac{1}{C}\mathbb{I} + (\sum_i \boldsymbol{\delta}_i \otimes \boldsymbol{\delta}_i^\top) \circ XX^\top + \frac{1}{\mu}(\sum_i \boldsymbol{\phi}_i \otimes \boldsymbol{\phi}_i^\top) \circ ZZ^\top$, and $\mathbf{q} = -\mathbf{r}$, the problem becomes a standard quadratic optimization with a constraint,

$$\min_{\boldsymbol{\alpha}, \mathbf{1}^\top \boldsymbol{\alpha} = 0} \frac{1}{2}\boldsymbol{\alpha}^\top P\boldsymbol{\alpha} + \mathbf{q}^\top \boldsymbol{\alpha}. \tag{5}$$

Algorithm 1 summarizes the procedure of optimizing for the squared loss. In each iteration, the algorithm computes the UCB value of all arms for the observed user, and in line 14 chooses the arm with the highest value. The required parameters for the quadratic optimization are updated from line 15 to 22 which leads to optimizing $\boldsymbol{\alpha}$. The obtained vector is used to update the model parameters. Note that the objective function is optimized for all the parameters, therefore, it affects them all and not just one user and one item. In this algorithm, we assume that the covariance matrices of item and user parameters are independent from each other. Hence, we discard the correlation between them and obtain the variance by summing them as $\mathbf{z}_a^\top A_{u_t}^{-1}\mathbf{z}_a + \mathbf{x}_t^\top A_a^{-1}\mathbf{x}_t$ (line 11) in order to compute the confidence bound.

Algorithm 1. Short- and long-term regression UCB

1: Inputs: c, C, and μ
2: Initialize $X \leftarrow \emptyset_{0 \times d}$, $\quad Z \leftarrow \emptyset_{0 \times k}$, $\quad \mathbf{r} \leftarrow \emptyset$
3: **for** $t = 1, 2, ..., T$ **do**
4: $\quad$ **if** u_t is new **then** $\qquad\qquad$ (Observe the user u_t and context $\mathbf{x}_t \in \mathbb{R}^{d \times 1}$)
5: $\qquad$ $A_{u_t} \leftarrow \mathbb{I}_k \cdot \mu$, $\quad \boldsymbol{\beta}_{u_t} \leftarrow \mathbf{0}_{k \times 1}$, $\quad \boldsymbol{\phi}_{u_t} \leftarrow \mathbf{0}_{t \times 1}$
6: $\quad$ **end if**
7: $\quad$ **for all** $a \in A_t$ **do**
8: $\qquad$ **if** a is new **then** $\qquad$ (Observe the features of arm $\mathbf{z}_a \in \mathbb{R}^{k \times 1}$)
9: $\qquad\quad$ $A_a \leftarrow \mathbb{I}_d$, $\quad \boldsymbol{\theta}_a \leftarrow \mathbf{0}_{d \times 1}$, $\quad \boldsymbol{\delta}_a \leftarrow \mathbf{0}_{t \times 1}$
10: $\qquad$ **end if**
11: $\qquad$ $s_{t,a} = \mathbf{z}_a^\top A_{u_t}^{-1}\mathbf{z}_a + \mathbf{x}_t^\top A_a^{-1}\mathbf{x}_t$
12: $\qquad$ $p_{t,a} = \boldsymbol{\theta}_a^\top \mathbf{x}_t + \boldsymbol{\beta}_{u_t}^\top \mathbf{z}_a + c\sqrt{s_{t,a}}$
13: $\quad$ **end for**
14: $\quad$ Choose arm $a_t = \arg\max_a p_{t,a}$ with tie broken randomly, and observe payoff r_t
15: $\quad$ $A_{a_t} = A_{a_t} + \mathbf{x}_t\mathbf{x}_t^\top$
16: $\quad$ $A_{u_t} = A_{u_t} + \mathbf{z}_{a_t}\mathbf{z}_{a_t}^\top$
17: $\quad$ $X \leftarrow [X; \mathbf{x}_t^\top]$ $\qquad\qquad$ (Append vertically)
18: $\quad$ $Z \leftarrow [Z; \mathbf{z}_{a_t}^\top]$ $\qquad\qquad$ (Append vertically)
19: $\quad$ $\mathbf{r} \leftarrow [\mathbf{r}, r_t]$
20: $\quad$ **for all** $a \in A_t$ and $u \in U_t$ **do**
21: $\qquad$ Update $\boldsymbol{\delta}_a$ and $\boldsymbol{\phi}_u$
22: $\quad$ **end for**
23: $\quad$ **for all** $a \in A_t$ and $u \in U_t$ **do** (Obtain $\boldsymbol{\alpha}$ by optimizing Eq. 5)
24: $\qquad$ $\boldsymbol{\theta}_a = (X \circ \boldsymbol{\delta}_a)^\top \boldsymbol{\alpha}$
25: $\qquad$ $\boldsymbol{\beta}_u = (Z \circ \boldsymbol{\phi}_u)^\top \boldsymbol{\alpha}$
26: $\quad$ **end for**
27: **end for**

Learning with Logistic Loss. Another special case of our unified framework is to apply the logistic loss for the optimization process. As we introduced in Sect. 2.3, the conjugate of logistic loss is as follows

$$V^*(-\frac{\alpha_t}{r_t}, r_t) = (1 - \frac{\alpha_t}{Cr_t}) \log(1 - \frac{\alpha_t}{Cr_t}) + \frac{\alpha_t}{Cr_t} \log(\frac{\alpha_t}{Cr_t}).$$

Employing the above conjugate into the Eq. (4) leads to

$$\min_{\alpha, 1^\top \alpha = 0} \quad C \sum_{t=1}^{T} [(1 - \frac{\alpha_t}{Cr_t}) \log(1 - \frac{\alpha_t}{Cr_t}) + \frac{\alpha_t}{Cr_t} \log(\frac{\alpha_t}{Cr_t})]$$
$$+ \frac{1}{2}\alpha^\top [(\sum_i \delta_i \otimes \delta_i^\top) \circ XX^\top + \frac{1}{\mu}(\sum_i \phi_i \otimes \phi_i^\top) \circ ZZ^\top]\alpha.$$

The procedure for learning the model is similar to Algorithm 1 in the previous section. Nevertheless, the objective function in line 23 needs to be optimized differently, and also computing $s_{t,a}$ in line 11. In the latter, the covariance matrix is computed for both set of parameters, $\Sigma_a = X^T V_a X$ and $\Sigma_{ut} = Z^T V_{ut} Z$, respectively. Therefore, $x_t^\top \Sigma_a^{-1} x_t + z_t^\top \Sigma_{ut}^{-1} z_t$ is used as the variance in computing the lower and upper confidence bounds (see Sect. 2.3). Note that gradient based methods are still applicable in the optimization part.

4 Discussion

In the following, we discuss some potential alternatives of our proposed approach which are suitable for particular circumstances.

4.1 Complexity of the Model

The presented unified model in Sect. 3 combines the contextual item model with the user interest in one framework. The model is therefore more than the vanilla bandit-based approaches that only model one of those. However, the model contains many parameters and the optimization part becomes more and more complex as the system size (both the number of items and users) grows. We propose to simplify the approach in two different directions; relaxing the item model or discarding the personalized term. Hence, we introduce four simplified cases of the combined approach as follows.

1. **Short-Term:** To model the payoff function only for the items, no personalization (aka. LinUCB [9]): $\mathbb{E}[r_{t,a_i}] = \theta_i^\top x_t$.
2. **Short-Term+Average:** Considering an average term for all the items, no personalization (resembling HybridUCB [9]): $\mathbb{E}[r_{t,a_i}] = \theta_i^\top x_t + \beta^\top z_{a_i}$.
3. **Long-Term:** Only personalized model: $\mathbb{E}[r_{t,a_i}|u_j] = \beta_j^\top z_{a_i}$.
4. **Long-Term+Average:** Incorporating the average term into the personalized model: $\mathbb{E}[r_{t,a_i}|u_j] = \beta_j^\top z_{a_i} + \theta^\top z_{a_i}$.

These cases are easily derivable from equations in Sect. 3. Note that the average part in case 2 and 4 depicts the item popularity in the recommender systems. We further examine the benefits of average models in Sect. 5.

4.2 Preference Based Bandits

One natural extension of our approach is to characterize the model in the preference-based setting. There are many systems with no available quantitative feedback, whereas the feedback is provided in terms of pairwise comparison between items. In such cases, the preferences are used in the learning process and the rankings are predicted directly from the model. In this section, we discuss how to phrase our bandit framework in a preference-based context.

We consider the contextual bandit problem in a way that the context is specified by the features of items to recommend. The model is thus defined by a single bandit which learns the preferences between items for all the users. Assume that $\mathbf{z}_i$ and $\mathbf{z}_k$ are the features of items a_i and a_k, respectively, and we assign $\mathbf{z}_{i \succ k} := \mathbf{z}_i - \mathbf{z}_k$ to show the preference of item a_i over a_k. The payoff is therefore determined as a linear model of the preference,

$$\mathbb{E}[r_{t,i \succ k} | u_t = u_j] = \boldsymbol{\theta}^\top \mathbf{z}_{i \succ k} + \boldsymbol{\beta}_t^\top \mathbf{z}_{i \succ k},$$

where $\boldsymbol{\theta}$ is the weight vector for the average model, while $\boldsymbol{\beta}_t = \boldsymbol{\beta}_j$ is the individual parameter for user j which acts as a personal offset. The above equation is theoretically analogous to the case number 4 in the previous section.

5 Empirical Study

The purpose of this section is to evaluate the performance of our combined contextual bandit approach compare to either short-term or long-term models. We use the squared loss in our experiments as in Algorithm 1. The quality of recommendation is measured via normalized average rank. For every test instance, a ranking of all items is inferred by the model. The position of the actually clicked item in the ranking is then normalized (divided by the total number of items) and averaged over all test samples. The empirical study illustrates that adding a long-term model describing the user preferences improves the short-term recommendation. Additionally, we show that the simplified average models are beneficial in cold start scenarios.

5.1 Data

The experiments are conducted on a real-world dataset from Zalando, a large European online fashion retailer, with anonymized click history of various users. The data is collected over time and bucketized into consecutive sessions. Each user interacts with the system in different sessions, and each session contains a sequence of products views. Products are described with some categorical attributes, such as *category*, *brand*, *color*, *gender*, *price level*, and *action*. We apply a one-hot encoding of the categorical features and enrich the representation by three additional features: the item popularity for each item, and "sale to view" as well as "view to action" ratios per user. The augmented dataset encompasses users with at least 5 sessions, where all sessions with more than one click are considered a valid session.

5.2   Overall Performance

In the first experiment, we examine how the combined approach performs on data sets of different sizes compared to the long- and short-term models and a matrix factorization baseline [8]. The parameters of the latter are optimized by model selection (200 factors, regularization constant 0.1). We thus generate several subsets of data by randomly sampling different numbers of users to obtain sets with about 1k user transactions to 15k. We split each set into training and test sets by reserving 80% of sessions for the former and assigning the rest to the test set. Note that there is no new user or new item in the test data.

The context in our setup is the feature vector of the previously viewed product. Therefore, the first click of each session is discarded and kept as the context for the next click. The reward value for each action is either 1 for the correct arm or -1 otherwise. We consider a fixed $c = 2.36$, and set regularization parameters $\lambda = \mu = 1$ for simplicity. Figure 1 depicts the results for our approach as well as the long- and short-term models averaged over several runs.

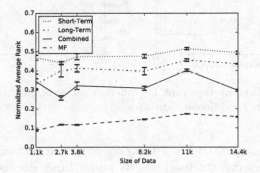

Fig. 1. Normalized average rank for different data sizes.

The figure shows that the combined approach outperforms both the long- and short-term methods in terms of average rank (lower is better). The short-term approach performs worse than the other two, since the data is obtained by sampling users, and there are many more items than users. However, the size of data does not change the behavior of the tested methods significantly apart from the combined model that improves performance with increasing data sizes; an indicator for the necessity of experiments at even larger scales. The matrix factorization baseline performs best when all users and items are known.

5.3   Cold Start

One of the main contributions of our proposed approach in the contextual setting is the ability to generalize over different items for individual users. This advantage suits well in cold start situations where content is highly dynamic and item and user sets change frequently. First, we demonstrate the behavior of the combined approach when new users and items appear in the test set.

We create a subset of data from all the sessions of 100 randomly selected users. The data contains 1,295 sessions which gives an average of 13 sessions per user, and about 8,000 products. We split the data into training and test sets with different ratios for the percentage of new items and new users in the test data. To this purpose, we leave $d\%$ of the users and $e\%$ of the products to appear only in the test set such that it realizes a ratio of $\frac{e}{d}$ for the new items over new users. We train our combined approach as well as both long- and short-term models, where the context and reward setup is as in the previous section. Figure 2 (left) shows the behavior of different approaches; results are averaged over multiple runs.

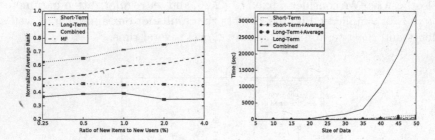

Fig. 2. Left: normalized average rank for different ratios of new items to new users. Right: execution time for different data sizes.

The first impression from Fig. 2 (left) suggests that although the combined method still outperforms the baselines, its performance declines a bit near the ratio of 1 when many new users and items are available. Unsurprisingly, the short-term model performs better for scenarios with only a few new items. This holds vice versa also for the long-term model that performs better for scenarios with almost constant sets of users. The performance of matrix factorization degrades significantly in the new setting which confirms the robustness of our combined method in real scenarios. However, the robustness comes at the cost of run-time: the combined approach is computationally expensive because of the involved convex optimization. The run-time analysis in Fig. 2 (right) displays the exponential growth in execution time in comparison to the other approaches discussed in Sect. 4.1.

In this section, we focus on the evaluation of adding average models to the long- and short-term approaches. We conduct the experiments on a medium sized dataset to evaluate their performance. The dataset in this experiment contains all transactions of 500 random users. We split the data by modifying the percentage of new users and new items in the test set and analyze two cases. Figure 3 shows how adding the average term significantly improves the performance of both long-term and short-term models, respectively. As in the previous experiment, in Fig. 3a, the performance of the long-term model decreases for increasing numbers of new users. By contrast, extending the long-term model by an average model remedies this effect and the extended model is able to cope with the challenging scenario and even improves performance. Similar behavior is shown in Fig. 3b where short-term model, augmented by an average model, eliminates the shortcomings of the short-term model in dealing with new items. By contrast, collaborative filtering fails to catch up and performs poorly in both scenarios. As a result, maintaining additional average models is an effective and efficient means in cold start situations. The experiments however also show that there is no one model that rules them all; instead, the model of choice depends clearly on the intrinsic dynamics of the applications.

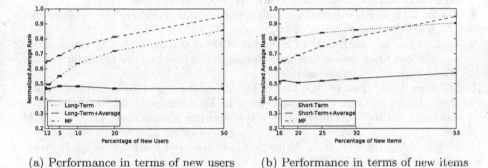

(a) Performance in terms of new users   (b) Performance in terms of new items

Fig. 3. Normalized average rank for the data with new items and users.

6   Conclusion

In this paper, we presented a unified model for short-term and long-term recommendation in a multi-armed bandit framework. The model incorporated the information from the actual context as well as the long-term preferences of the users into a single contextual bandit. We transformed the optimization problem of our bandits into the dual space considering a linear payoff model for the arms.

Addressing the problem in dual space led to a generalized optimization problem where arbitrary loss functions could be used to reshape the payoff function according to the application at-hand. As a result, applying contextual bandits for long- and short-term recommendations is considerably simplified. The experiments show that adding an average model to short- and long-term models leads to robust methods that clearly outperform their vanilla peers in terms of normalized average rank.

References

1. Auer, P.: Using confidence bounds for exploitation-exploration trade-offs. J. Mach. Learn. Res. **3**, 397–422 (2003)

2. Barbieri, N., Manco, G., Ritacco, E., Carnuccio, M., Bevacqua, A.: Probabilistic topic models for sequence data. Mach. Learn. **93**(1), 5–29 (2013)
3. Boyd, S., Vandenberghe, L.: Convex Optimization. Cambridge University Press, Cambridge (2004)
4. Deshmukh, A.A., Dogan, U., Scott, C.: Multi-task learning for contextual bandits. arXiv preprint arXiv:1705.08618 (2017)
5. Dybowski, R., Roberts, S.: Confidence intervals and prediction intervals for feed-forward neural networks. Clin. Appl. Artif. Neural Netw. 298–326 (2001)
6. Hu, Y., Koren, Y., Volinsky, C.: Collaborative filtering for implicit feedback datasets. In: Proceedings of the 8th IEEE International Conference on Data Mining, pp. 263–272. IEEE (2008)
7. Keerthi, S.S., Duan, K.B., Shevade, S.K., Poo, A.N.: A fast dual algorithm for kernel logistic regression. Mach. Learn. **61**(1–3), 151–165 (2005)
8. Koren, Y., Bell, R., Volinsky, C.: Matrix factorization techniques for recommender systems. Computer **8**, 30–37 (2009)
9. Li, L., Chu, W., Langford, J., Schapire, R.E.: A contextual-bandit approach to personalized news article recommendation. In: Proceedings of the International World Wide Web Conference (2010)
10. Mahajan, D.K., Rastogi, R., Tiwari, C., Mitra, A.: LogUCB: an explore-exploit algorithm for comments recommendation. In: Proceedings of the 21st ACM International Conference on Information and Knowledge Management, pp. 6–15. ACM (2012)
11. Rifkin, R.M., Lippert, R.A.: Value regularization and fenchel duality. J. Mach. Learn. Res. **8**, 441–479 (2007)
12. Tang, L., Jiang, Y., Li, L., Zeng, C., Li, T.: Personalized recommendation via parameter-free contextual bandits. In: Proceedings of the 38th International ACM SIGIR Conference on Research and Development in Information Retrieval, pp. 323–332. ACM (2015)
13. Tavakol, M., Brefeld, U.: Factored MDPs for detecting topics of user sessions. In Proceedings of the 8th ACM Conference on Recommender Systems, pp. 33–40. ACM (2014)
14. Valko, M., Korda, N., Munos, R., Flaounas, I., Cristianini, N.: Finite-time analysis of kernelised contextual bandits. arXiv preprint arXiv:1309.6869 (2013)
15. Wang, C., Blei, D.M.: Collaborative topic modeling for recommending scientific articles. In: Proceedings of the 17th ACM SIGKDD International Conference on Knowledge Discovery and Data Mining, pp. 448–456. ACM (2011)
16. Wu, Q., Wang, H., Gu, Q., Wang, H.: Contextual bandits in a collaborative environment. In: Proceedings of the 39th International ACM SIGIR conference on Research and Development in Information Retrieval, pp. 529–538. ACM (2016)
17. Zhou, L., Brunskill, E.: Latent contextual bandits and their application to personalized recommendations for new users. arXiv preprint arXiv:1604.06743 (2016)

Perceiving the Next Choice with Comprehensive Transaction Embeddings for Online Recommendation

Shoujin Wang[✉], Liang Hu, and Longbing Cao

Advanced Analytics Institute, University of Technology Sydney,
Sydney, NSW 2007, Australia
{Shoujin.Wang,Liang.Hu-2}@student.uts.edu.au, Longbing.Cao@uts.edu.au

Abstract. To predict customer's next choice in the context of what he/she has bought in a session is interesting and critical in the transaction domain especially for online shopping. Precise prediction leads to high quality recommendations and thus high benefit. Such kind of recommendation is usually formalized as transaction-based recommender systems (TBRS). Existing TBRS either tend to recommend popular items while ignore infrequent and newly-released ones (e.g., pattern-based RS) or assume a rigid order between items within a transaction (e.g., Markov Chain-based RS) which does not satisfy real-world cases in most time. In this paper, we propose a neural network-based comprehensive transaction embedding model (NTEM) which can effectively perceive the next choice in a transaction context. Specifically, we learn these comprehensive embeddings of both items and their features from relaxed ordered transactions. The relevance between items revealed by the transactions is encoded into such embeddings. With rich information embedded, such embeddings are powerful to predict the next choices given those already bought items. NTEM is a shallow wide-in-wide-out network, which is more efficient than deep networks considering large numbers of items and transactions. Experimental results on real-world datasets show that NTEM outperforms three typical TBRS models FPMC, PRME and GRU4Rec in terms of recommendation accuracy and novelty. Our implementation is available at https://github.com/shoujin88/NTEM-model.

1 Introduction

1.1 Target Problem and Motivation

Nowadays, recommender systems (RS) play an important role in real-world business especially in the e-commerce domain. For example, the RS behind thousands of websites (e.g., Amazon) provide magic power to help end users to discover and make choices from a huge number of items. As a result, RS can not only improve customers' shopping experience but also increase the business profits. Although lots of work has been done to produce high quality recommendations, some issues are still challenging and need more efforts. One of them is to enable

© Springer International Publishing AG 2017
M. Ceci et al. (Eds.): ECML PKDD 2017, Part II, LNAI 10535, pp. 285–302, 2017.
https://doi.org/10.1007/978-3-319-71246-8_18

RS to dynamically perceive customers' next choice on thousands of candidate items online according to what they have just put in the shopping carts. In this paper, we call the next item to choose as the target item while those items having been added to cart are treated as its context. The challenges come from two sides: on one hand, the context is dynamic along with the shopping transaction; on the other hand, the RS needs to keep updating the recommendations when the context was changed. For instance, suppose a customer Robin starts an online shopping transaction on Amazon: first, he bought a cellphone and then a protective film may be his next choice, so good RS should be capable of perceiving the protective film according to the context of cellphone. When Robin has bought the cellphone and protective film, his next choice is probably an earphone rather than buying another type of protective film again. Therefore, the recommended item should be changed from the protective film to the earphone accordingly. This dynamic recommendation process keeps updating until the transaction is finished.

This kind of recommender systems are also called transaction-based RS (TBRS) [11] as they work on transactional data, which are different from the rating-based RS (RBRS) [2] working on rating data. Although existing TBRS can recommend next items given the context, most of them treat the context as static rather than dynamic and can only work on static transactional data, an example is the pattern-based RS [23]. Moreover, it is quite hard for them to keep updating recommendations due to the long computational time, like deep network-based RS [10, 21]. As a result, they do not work efficiently for online recommendations. RBRS has been well studied but it cannot tackle our problems here due to the lack of consideration of context, existing TBRS cannot perform well either for online recommendations due to the aforementioned reasons. In this paper, we focus on TBRS and target at handling the dynamic context and producing online recommendations.

Due to the lack of effective approach to model the context of a transaction event, most current TBRS cannot capture the intra-session relevance over items perfectly and thus cannot produce high quality recommendations. They tend to recommend those popular and long-released items while ignore those less popular or newly-released ones. In practice, customers may not always need popular or similar items to form immutable shopping behaviors, instead, they want to explore something novel or unpopular. For example, when the first smart-phone *iPhone* appeared in 2010, most customers prefer to it rather than *Nokia*, a popular function-phone (not smart-phone) brand lasted for more than a decade at that time. Therefore, more sensible RS which can bring surprising experience to users by recommending novel but relevant items is increasingly important. In this paper, novel items refer to unpopular or newly-released items while "relevant" infers the recommended items are strongly relevant to the context. It's clear that an effective approach to model the dynamic context in real-time is necessary to produce high quality online recommendations.

Content-based filtering (CBF) [14] and collaborative filtering (CF) [12] are two approaches that are most commonly used in RS. They are not applicable to TBRS directly though they perform well in RBRS. This is because these methods are actually designed to work on rating matrix in RBRS, which is quite different from the shopping-basket data in TBRS, thus it is hard for them to capture the relevance between items embedded in transactional data.

Pattern-based recommendation [23] is a straightforward solution to transaction-based recommendation issues. It first captures associations between items and then recommends items associated to the context items. Although simple and effective sometimes, patterns are extracted from those frequent items due to the "support" measure, whereas those infrequent ones are missed. As a result, the discovered patterns can neither capture the relevance between all items nor achieve the goal in this work. In addition, the sequential pattern mining assumes a rigid order on transaction data [9]. For instance, it makes no difference whether milk or bread is put into the cart firstly. Markov Chain (MC) [18] is another straightforward way to model sequential data and thus can be used for TBRS on sequence data. However, MC can only capture the transition from one item to another rather than from a context item set to an item. Recently, matrix factorization (MF) [5] is used to factorize the transition probability from current item to the next one to the latent factors of each item and user. Similar to sequential patterns, both MC and MF were originally designed for time-series data with rigid natural order, which limit their applications in TBRS, where the order between items within a transaction usually makes no difference. Moreover, they cannot handle those novel items well as the transitions between them and other items tend to be weak due to their low frequencies.

The above illustrations reveal the difficulty of modeling the context especially for dynamic context. In practice, the next choice is not only affected by one item or part of items in front of the target item, but by all items bought in the transaction event (i.e., the whole context). It is important to model the whole context and learn the relevance between the whole context and the target item. Furthermore, the intra-transaction relevance over items are greatly driven by their intrinsic nature, in another words, there are complex coupling relations [4] between item features and item relevance. Such relations are particularly critical for those novel items with quite a few transaction records. For example, milk and bread are always bought together probably due to their different but closely related categories "drink" and "food". This indicates that not only the context items but also the features of these items can affect the choice of the target item. To capture the indicators on the next choice as more as possible, we propose a neural-network-based comprehensive transaction embedding model (NTEM) to learn the embeddings of both items and their features when modeling the relevance between the context and the known choice. The model is comprehensive for several reasons: it models the relations between the target item and the whole context rather than part of it. It learns the embedding of the two important aspects (e.g., items and their features) of a transaction at the same time. During model training, the coupling relations between item relevance and item features are learnt and encoded into feature embeddings, which is useful for novel item discovery and recommendation. Though comprehensive, the embedding model has a shallow and wide network structure containing only one hidden layer, which guarantees its efficiency to find the best next choices over thousands of candidate items when the given context changes over time. This is suitable for online recommendation.

1.2 Our Design and Main Contributions

Inspired by the great success of modern word embedding models, such as Word2Vec [15], in natural language processing (NLP) domain, we propose a shallow and wide network-based transaction embedding model (NTEM) to learn the relevance between different items efficiently on a large number of items with its wide-in-wide-out structure. Such relevance is learnt by capturing both the explicit relevance between items from the shopping-basket data and the implicit relevance from item features together with the coupling relations between them and the item relevance. It is noted that a deep structure is not efficient for online recommendation due to the long computational time needed to deal with thousands of items in real time. The Word2Vec cannot be directly applied to RS for two reasons: on one hand, it lacks of necessary element to incorporate the item features. On the other hand, the words in NLP often have a strict sequence, which is different from our case.

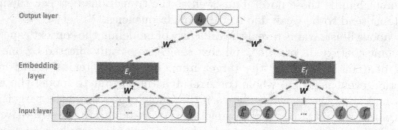

Fig. 1. The NTEM architecture, which learns item embeddings and feature embeddings for target item prediction based on contextual items and their features

Our model, NTEM has a three-layer network structure consisting of input layer, embedding layer and output layer as shown in Fig. 1. The input layer contains double wide-in data vectors, the contextual itemset is collected from one while their corresponding features are acquired from the other. The embedding layer learns the item embeddings and feature embeddings respectively. The target item is then predicted by the output layer taking the embeddings of contextual items and features as the input. The NTEM learns the relevance between items with comprehensive transaction embeddings using a concise network structure. The main contributions of this work are summarized below.

(1) We model the whole context using a comprehensive network-based transaction embedding model for the next choice prediction.

(2) A TBRS model is proposed, which does not require the strict order over items within one transaction. This is more consistent with the real-world case.

(3) We incorporate item features into the model and encode the feature-item relevance coupling relations into feature embeddings, which makes our model also work well on cold start cases.

(4) We propose a shallow and wide network, which recommends the next item efficiently on large number of items under dynamic context.

2 Related Work

Existing transaction-based recommender system (TBRS) can be roughly divided into pattern mining-based and model-based ones, we will briefly review the literature on these two.

Pattern mining-based approaches are an intuitive solution to TBRS and widely used in real-world business. [13] proposed to adjust the minimum support according to the various frequencies of users and items. The mined association rules between users as well as between items are then used for making recommendations. [1] introduced relation rule mining to discover relations between different objects and such relations are then used for recommendations. Considering the order between items, some sequential pattern mining (SPM) based recommendation methods are proposed. [23] introduced a personalized sequential pattern mining-based recommendation framework. Using a novel Competence Score measure, the proposed framework effectively learns user-specific sequence importance knowledge for accurate personalized recommendation. Although simple and effective, such kind of approaches may lose information of those infrequent items due to the "minimum support" constraint. In addition, they can not be applied to online recommendation directly as the dynamic context may contain arbitrary items, it probably fail to match any mined pattern.

Overall, there are mainly two kinds of model-based approaches for TBRS: Markov Chain (MC) based ones and matrix-factorization (MF) based ones. [18] used Markov Chain to estimate the transition probability from current item to the next item and thus make prediction based on this probability. [22] proposed Personalized Markov Embedding (PME) to map the users and songs to an Euclidean space by modeling the sequential singing behaviours, the prediction and recommendation are conducted on the base of the embeddings. Although sequential behaviour prediction based on Markov Chain is effective for capturing the transition preferences of certain users and thus make good recommendations, it is essentially based on item order within a transaction, which is not always available in real-world business and such method only captures the first-order dependency between items. Recently, MF-based approaches are also developed for TBRS. [18] combined MF and MC for next-item recommendation, the latent user and item representations from the {user, item, last item} triplets are learned, and then the inner product of these latent vectors is used to perform next-item recommendation. Following factorization machine [17], a pairwise factorization model is proposed in [5] to learn the latent vectors of user, last item and item for next-song recommendation. Similar to MC-based model, MF-based ones depend strongly on the rigid order between items. The working mechanism limits it to model the relations between one item and another rather than between a contextual itemset and the target item. Furthermore, MF-based methods can not work well on sparse dataset, while the data on novel items usually has a large chance to be sparse due to their low frequencies.

3   Transaction Embeddings for Online Recommendation

In this section, we start with the problem formulation, then we talk about the proposed NTEM model including the network architecture in the model and the model construction, finally we illustrate how to train the model and how to make prediction and thus produce recommendations for online shopping using the trained model.

3.1   Problem Formalization

Let $T = \{t_1, t_2...t_{|T|}\}$ be a set of transactions, each transaction $t = \{i_1, i_2...i_{|t|}\}$ contains a set of items, where $|T|$ denotes the number of elements in set T. All the items occurred in all transactions constitute the whole item set $I = \{i_1, i_2...i_{|I|}\}$. Let $F = \{f^1, f^2...f^{|F|}\}$ be a set a features which describe the items from I. Each item i is described by a set of feature values $F_i = \{f_i^1, f_i^2...f_i^{|F|}\}$. Note that the items in one transaction t may not have a rigid order, which is consistent with the real-world cases. Given the set of context itemset $\mathbf{c}$, our NTEM is constructed and trained as a probabilistic classifier that learns to predict a conditional probability distribution $P(i_s|\mathbf{c})$, where $\mathbf{c} \subseteq t \backslash i_s$ is the context from a transaction $t \in T$ w.r.t. the target item i_s. This is similar to the bag of word (BOW) model in natural language processing, which trains a classifier to learn a conditional probability distribution $P(w_j|w_I)$, where w_I is the context consisting of several words of the target word w_j [19]. Similar to BOW, for each target item $i_s \in t$, the transaction context is $\mathbf{c} = t \backslash i_s$, namely all the items except the target one in the transaction are picked up as the context. Totally $|t|$ training instances are built for each transaction t by picking up one item as the target one each time.

Since we want to capture more information from the context for prediction, the features of items are added to the model as part of context, which result in $\mathbf{c}_f = < \mathbf{c}, F_c >$, where $F_c = \{F_i | i \in \mathbf{c}\}$ is the corresponding features of the items from $\mathbf{c}$. Thus our NTEM model is refined to predict the conditional probability distribution $P(i_s|\mathbf{c}, F_c)$ when the transaction-feature context $< \mathbf{c}, F_c >$ is given. We call $\mathbf{c}$ and F_c as transaction context and feature context respectively in this paper. Thus, the TBRS is reduced to ranking all candidate items in terms of their conditional probability over the given transaction-feature context. Note that in the prediction stage, the conditional probability is computed based on the embeddings of item set $\mathbf{c}$ and its corresponding features F_c learned in the training stage.

Particularly, the incorporation of features contributes greatly to the recommendation of novel items. Due to the low frequencies of novel items in training set, the embeddings of these items may not be learned well during the model training process and it leads to poor prediction. Thanks to the feature embeddings synchronously learned with the item embeddings, the intra-transaction item relevance can be partly encoded into feature value embeddings of novel items. In addition, part of the feature values of novel items may already be embedded when encoding those of frequent items as some feature values may be shared between frequent items and infrequent ones.

3.2   Neural-Network-Based Transaction Embedding Model (NTEM)

In this section, we mainly talk about the details of constructing NTEM and learning its parameters.

Giving a context $< \mathbf{c}, F_c >$ to the input layer, the input units in the bottom left corner of Fig. 1 constitute a one-hot encoding vector where only the units at position i_j ($i_j \in \mathbf{c}$) is set to 1 and all other units are set to 0. For each $i \in \mathbf{c}$, we encode it in the same way as i_j. Note that items may have both numerical and categorical features in real-world business. In this work, we only consider those categorical features. For a value from a categorical feature f ($f \in F$) with m different values, we transform it to a $1 \times m$ vector using one-hot encoding. Suppose $V = \sum n_k$ is the total number of distinct values of all features, where n_k is the number of distinct values in feature f^k. For the features of a given item, a $1 \times V$ vector is achieved by doing the transformation of all feature values first and then concatenate all the transformed vectors together. Thus the input layer for each training example consists of $|\mathbf{c}|$ item vector with length $|I|$ and $|\mathbf{c}|$ item feature vectors with length V.

In the input layer, items and item features are represented by sparse one-hot item and feature vectors. In NTEM, we create an embedding mechanism to map these vectors to an informative and lower-dimensional vector representation in the embedding layer, where a K-dimension vector $\mathbf{E}_i \in [0,1]^K$ is used to represent the item embedding. The transaction context weight matrix $\mathbf{W}^t \in \mathbb{R}^{K \times |I|}$ is used to fully connect between input-layer and embedding-layer. Where the ith column $\mathbf{W}^t_{:,i}$ encodes the one-hot vector of item i to the embedding $\mathbf{E}_i$ using the commonly used logistic function $\sigma(\cdot)$.

$$\mathbf{E}_i = \sigma(\mathbf{W}^t_{:,i}) \tag{1}$$

To make the training and prediction more stable, here we use the nonlinear embeddings as they are bounded in $[0, 1]$ compared to the linear embeddings. Furthermore, the nonlinear embeddings are more expressive than linear one though they may involve a little more computation cost. After embedding all items in $\mathbf{c}$, we can obtain the embedding $\mathbf{E}_c \in [0,1]^L$ of transaction context $\mathbf{c}$ by combining all embeddings of items in such context. As illustrated in the following equation, the transaction context embedding is built as a combination of $\mathbf{E}_i, i \in \mathbf{c}$.

$$\mathbf{E}_c = \sum_{i \in c} \omega_i \mathbf{E}_i \tag{2}$$

where $\sum_{i \in c} \omega_i = 1$. The combination weight ω_i for each item i in context $\mathbf{c}$ can be assigned to different values according to specific applications. For instance, in sequential data, the weights decay with time span to the target item. As illustrated in the introduction part, we treat the items within a transaction as unordered, so uniform weights are used in this paper, i.e. the items in context are equally important for the prediction of the target item.

Similarly, we use the feature context weight matrix $\mathbf{W}^f \in \mathbb{R}^{L \times V}$ to encode the one-hot item feature vector F_i of item i to the embedding $\mathbf{E}_{F_i} \in [0,1]^L$.

$$\mathbf{E}_{F_i} = \sigma(\mathbf{W}^f_{:,F_i}) \tag{3}$$

Similar to transaction context, we combine the embeddings of features of all items from $\mathbf{c}$ to construct the whole feature embedding $\mathbf{E}_{F_c}$ as below.

$$\mathbf{E}_{F_c} = \sum_{i \in \mathbf{c}} \omega_{F_i} \mathbf{E}_{F_i} \tag{4}$$

where $\sum_{i \in c} \omega_{F_i} = 1$. Uniform weights are assigned to the feature embedding of each item for the same reason as ω_i illustrated above.

The output weight matrices $\mathbf{W}^o \in \mathbb{R}^{|I| \times K}$ and $\mathbf{W}^p \in \mathbb{R}^{|I| \times L}$ is used to fully connect the embedding-layer and output-layer as depicted in the top of Fig. 1. With the embeddings of given contextual itemset $\mathbf{c}$ and its features F_c, plus the weight metrics $\mathbf{W}^o$ and $\mathbf{W}^p$, the score S_{i_s} of a target item i_s w.r.t. the given context $< \mathbf{c}, F_c >$ is computed as:

$$S_{i_s}(\mathbf{c}, F_c) = \mathbf{W}^o_{s,:} \mathbf{E_c} + \mathbf{W}^p_{s,:} \mathbf{E}_{F_c} \tag{5}$$

where $W^o_{s,:}$ denotes the sth row of W^o. This score quantifies the relevance of the target item i_s w.r.t. the given context $< \mathbf{c}, F_c >$. As a result, the conditional probability distribution $P_\Theta(i_s | \mathbf{c}, F_c)$ can be defined in terms of softmax function, which is commonly used in neural network or regression model.

$$P_\Theta(i_s | \mathbf{c}, F_c) = \frac{exp(S_{i_s}(\mathbf{c}, F_c))}{Z(\mathbf{c}, F_c)} \tag{6}$$

where $Z(\mathbf{c}, F_c) = \sum_{i \in I} exp(S_i(\mathbf{c}, F_c))$ is the normalization constant and $\Theta = \{\mathbf{W}^t, \mathbf{W}^f, \mathbf{W}^o, \mathbf{W}^p\}$ is the model parameters. Thus a probabilistic classifier modeled by our NTEM is obtained.

3.3 Learning and Prediction

In the above subsection, we have built a probabilistic classifier based on the transaction and item feature information data $b = < \mathbf{g}, i_g >$, where $\mathbf{g} = < \mathbf{c}, F_c >$ is the input data, namely the transaction-feature context, and i_g is the corresponding observed output, namely an item bought together with the given transaction context $\mathbf{c}$. Given a training dataset $D = \{< \mathbf{g}, i_g >\}$, the joint probability distribution over it is obtained:

$$P_\Theta(D) \propto \prod_{b \in D} P_\Theta(i_g | \mathbf{c}, F_c) \qquad (7)$$

As a result, the model parameters Θ can be learned by maximizing the conditional log-likelihood:

$$L_\Theta = \sum_{b \in D} log P_\Theta(i_g | \mathbf{c}, F_c) = \sum_{b \in D} S_{i_g}(\mathbf{c}, F_c) - log Z(\mathbf{c}, F_c) \qquad (8)$$

Evaluating L_Θ and evaluating the corresponding log-likelihood gradient involve the normalization term $Z(\mathbf{c}, F_c)$, which needs to sum $exp(S_{i_g}(\mathbf{c}, F_c))$ over the whole itemset for each training instance. That is to say, training this model take $|I| \times |D|$ times of computation to get the normalization constant for each iteration, which makes the training process intractable. To tackle this problem, we adopt a sub-sampling approach, namely noise-constrictive estimation (NCE) [8] to deal with the normalization calculation of softmax function in the training process. We sample 50 negative items each time in the experiment.

All the parameters Θ are learned by back propagation. Algorithm 1 summarizes the learning process briefly. In Algorithm 1, $\odot$ denotes element-wise product operation, i_o is the output item which includes both positive example and noise examples. $i_j \in \mathbf{c}$ is the input item from the context and ω_{i_j} is the corresponding combination weight used in Eq. 2. Similarly, F_{i_j} is the corresponding input feature of item i_j and $\omega_{F_{i_j}}$ is its corresponding combination weight used in Eq. 4.

Algorithm 1. NTEM Parameter Learning Using Gradient Descent

1: $l \leftarrow 0$
2: **while** not converged **do**
3:    Compute $w_{i_o}^o$-gradient (Eq. 1): $g_{w_{i_o}^o} \leftarrow \mathbf{E}_c^\top$
4:    Compute $w_{i_o}^p$-gradient (Eq. 3): $g_{w_{i_o}^p} \leftarrow \mathbf{E}_{F_c}^\top$
5:    Compute $w_{:,i_j}^t$-gradient (Eq. 5):
     $g_{w_{:,i_j}^t} \leftarrow \omega_{i_j} \mathbf{W}_{i_o,:}^{o\top} \odot \mathbf{E}_i \odot (1 - \mathbf{E}_i)$
6:    Compute $w_{:,F_{i_j}}^f$-gradient (Eq. 5):
     $g_{w_{:,F_{i_j}}^f} \leftarrow \omega_{F_{i_j}} \mathbf{W}_{F_{i_o},:}^{p\top} \odot \mathbf{E}_{F_i} \odot (1 - \mathbf{E}_{F_i})$
7:    Perform SGD-updates for $w_{i_o}^o$, $w_{i_o}^p$, $w_{:,i_j}^t$ and $w_{:,i_j}^t$:
     $w_{i_o}^o \leftarrow w_{i_o}^o + S_{i_o}^l(g) g_{w_{i_o}^o}$, $\quad w_{i_o}^p \leftarrow w_{i_o}^p + S_{i_o}^l(g) g_{w_{i_o}^p}$, $\quad w_{:,i_j}^t \leftarrow w_{:,i_j}^t + S_{i_o}^l(g) g_{w_{:,i_j}^t}$, $w_{:,F_{i_j}}^f \leftarrow w_{:,F_{i_j}}^f + S_{i_o}^l(g) g_{w_{:,F_{i_j}}^f}$
8:    $l \leftarrow l + 1$
9: **end while**

Due to the large computation cost and the strength in matrix calculation of GPUs, a GPU-based adaptive stochastic gradient descent (SGD) optimizer is designed to speed up the training process.

4 Experiments and Evaluation

4.1 Experimental Setup

Data Preparation. We evaluate our method on two real-world transaction data sets: IJCAI-15[1] and Tafang[2]. First, a shopping-basket-based transaction table and an item information table are extracted from each of the original data. The transaction table contains multiple transactions and each transaction consists of multiple items. Note that those transactions containing only one item are removed as they can not fit our model. This is because, we use at least one item as context for constructing the embeddings. The item information table contains the feature values of each item occurred in the transaction table, note that we only focus on those categorical features in this paper. Secondly, the transaction table is splitted into training and testing set. Specifically, we randomly choose 20% from the transactions happened in last 30 days as the testing set, while others are used for training. The item information table is used in both training and testing processes. Finally, to test the performance of our proposed model under different cold-start levels, part of the transactions in training set are removed following certain rules. To be specific, we construct 4 different training sets with a drop rate of $0, 40\%, 80\%, 95\%$ respectively. Taking the one with drop rate of 40% as an example, for each target item selected in the testing set, 40% of all the transactions containing it in the training set are dropped. The characteristics of the datasets are shown in Table 1.

Table 1. Statistics of experimental datasets

Statistics	IJCAI-15	Tafang
#Transactions	144,936	19,538
#Items	27,863	5,263
#Features	3	1
Avg. transaction length	2.91	7.41
#Training transactions	141,840	18,840
#Training instances	412,679	141,768
#Testing transactions	3,096	698
#Testing instances	9,030	3,150

---

[1] https://tianchi.aliyun.com/datalab/dataSet.htm?id=1.

[2] http://stackoverflow.com/questions/25014904/download-link-for-ta-feng-grocery-dataset.

During the training, the transactions in the training set together with the corresponding item features are imported into the model in batches to learn embeddings of each item and each feature value. In the testing process, the learned embeddings are used to predict the target item given the $(n-1)$ ones. The real target item in the testing set is used as the ground truth. We calculate accuracy measures Recall@K [24] and MRR [5] by comparing the predicted results and the ground truth. We also calculate the recommendation novelty measures: global novelty and local novelty by comparing the recommendation list and the whole item population and the given context item set respectively. Finally, the performance of our proposed method is evaluated by comparing it and other related ones in terms of recommendation accuracy and novelty.

Evaluated Methods. In the experiments, we compare our proposed NTEM with the following commonly used baselines:

- *FPMC:* A model that combines matrix factorization and Markov Chains for next-basket recommendation. The model factorize the personal transition matrix between items with a pairwise interaction model [18].
- *PRME:* A personalized ranking metric embedding method (PRME) to model personalized check-in sequences. The learned PRME is then used to recommend next new point of interest (POI) of users [6].
- *GRU4Rec:* A RNN-based approach for session-based recommendations by modeling the whole session using a modified RNN [10].

4.2   Performance Evaluation

Except for the traditional recommendation accuracy, we also evaluate the recommendation novelty to test the capability of the proposed method to recommend novel items.

Accuracy Evaluation. We use the following widely used accuracy metrics for transaction-based recommendation to evaluate all the comparison approaches.The result of each approach is given in Table 2.

- *REC@K:* It measures the recall of the top-K ranked items in the recommendation list over all the testing instances [24]. Recall that in the real-world case most customers are only interested in the items recommended on the first one or two pages, so here we choose $K \in \{10, 50\}$. In practice, it's a big challenge to find the extract one true item from thousands of items.
- *MRR:* It measures the mean reciprocal rank of the predictive positions of the true target item on all the testing instances [5].

Table 2 demonstrates the results of REC@10, REC@50 and MRR over the testing sets of different cold-start levels. The number of factor is set to 10 for training FPMC as the best performance is achieved in such setting. The performance of FPMC on both data sets is quite poor, even in the warm start situation

Table 2. Accuracy comparisons between different recommendation models

Scenario	Model	IJCAI-15			Tafang		
		REC@10	REC@50	MRR	REC@10	REC@50	MRR
drop 0	FPMC	0.0016	0.0025	0.0031	0.0189	0.0216	0.0089
	PRME	0.0555	0.0612	0.0405	0.0212	0.0305	0.0102
	GRU4Rec	0.1182	0.1566	0.0965	0.0428	0.0887	0.0221
	NTEM	0.2026	0.3224	0.1125	0.0689	0.1716	0.0231
drop 40%	FPMC	0.0012	0.0021	0.0026	0.0008	0.0010	0.0058
	PRME	0.0327	0.0411	0.0312	0.0102	0.0205	0.0095
	GRU4Rec	0.1108	0.1356	0.0868	0.0330	0.0659	0.0196
	NTEM	0.1928	0.2794	0.1117	0.0575	0.1049	0.0377
drop 80%	FPMC	0.0009	0.0017	0.0021	0.0005	0.0008	0.0020
	PRME	0.0212	0.0287	0.0215	0.0084	0.0125	0.0056
	GRU4Rec	0.0493	0.0611	0.0398	0.0110	0.0244	0.0054
	NTEM	0.1098	0.1450	0.0686	0.0254	0.0494	0.0072
drop 95%	FPMC	0.0003	0.0008	0.0012	0.0002	0.0004	0.0008
	PRME	0.0089	0.0113	0.0105	0.0071	0.0096	0.0043
	GRU4Rec	0.0233	0.0337	0.0173	0.0101	0.0172	0.0042
	NTEM	0.0318	0.0639	0.0173	0.0215	0.0305	0.0068

(e.g., REC@50=0.0025 when drop rate=0 on IJCAI-15). The main reason is that both data sets are extremely sparse and thus a very large but quite sparse matrix is constructed to train the MF model for each data set. For instance, in IJCAI-15 data set, each user only has an average of 3.6 transactions and each transaction only contains an average of 2.91 items of over 27,000 ones. This leads to that each row of the constructed matrix contains less than two items (cf. the avg. transaction length in Table 1, note that one out of the items need to be taken out as the output) and all other entries are empty. By calculation, we found the non-empty entries in the constructed matrix of IJCAI-15 account for less than 0.01%. We set the embedding dimensions to 60 as suggested in [6] when training PRME model. The performance of PRME is a little better than FPMC, but it's still poor especially in those cold start cases. This is because PRME is a fist-order MC model, which learns the transition probability over successive item instead of the whole context. Namely, this model only predict the target item based on its previous one while ignore all those bought before in the same transaction, which lost some information. Furthermore, in the real-word, the choice of items does not follow a rigid sequence assumed by such kind of models. Benefiting from the deep structure, GRU4Rec achieves much better performance compared with FPMC and PRME models. Especially, when we drop no more than 40% transactions, the REC@10 is above 10% on IJCAI-15, which can lead to a relative accurate recommendation in real-world business.

The batch size is empirically set to 200 and number of hidden units for the item and feature value embeddings are set to 50 and 20 respectively. We run 60 epochs to train our NTEM model. It achieves much better performance than GRU4Rec, where the REC@10 and REC@50 exceed 20% and 30% respectively when drop nothing on IJCAI-15. In the extremely cold start case (drop 95%), it also achieves obvious better performance than other methods on both data sets. The REC@10 and REC@50 of our model exceed 3% and 6% respectively, compared to around 2% and 3% of GRU4Rec on IJCAI-15. The higher MRR of our model also shows that we can accurately put the customers' desired items in the front of the recommendation list. The reason is multifaceted. First of all, we use a complete context to predict one target item, and we donot assume a rigid order between the items within one transaction. This makes the input more informative and consistent with the reality compared with those models which only utilize part of the context and assume a rigid order between items. Second, more information is used by our model by incarnating the item features into it. More importantly, the coupling relations between item features and its transactions are learned and encoded into item and its feature embeddings in the back-forward propagation training mechanism. Thanks to the features and the coupling relations, additional information is provided to help with the item prediction and recommendation, especially for those novel items which lack of sufficient transaction information. What's more important, our model has a very concise structure which is easy to train. This shallow structure is efficient enough to retrain and recompute the score for ranking of all candidate items in an incremental dataset for online recommendations. However, GRU4Rec is a deep RNN consisting of GRU layers, which makes it more time consuming when new transaction records are added into the dataset.

Novelty Evaluation. Except for accuracy, novelty is another important quality which should be considered in real-world recommendation scenarios [20]. Recall that in this paper we also try to recommend those infrequent or unpopular items, namely novel items, so the novelty is particularly important to measure the recommendation quality of our model. Specifically, we evaluate the recommendation novelty from both global perspective and local perspective by using different metrics.

- *Global novelty:* intuitively, the novelty of an item from the global perspective can be defined as the opposite of item popularity w.r.t the whole population. The item is novel if it occurs in few transactions, i.e., the item is far in the long tail of the popularity distribution [16]. Inspired by the inverse user frequency (IUF) proposed in [3,7], we define the concept of inverse transaction frequency (ITF) as $ITF = -log_2|T_i|/|T|$, where $T_i = \{t \in T | i \in t\}$ denotes the set of transactions containing item i. Similar to the novelty metric MIUF defined in [16], here we use the average of the ITF (MITF) of the recommended items to measure the global novelty of a recommendation:

$$MITF = -\frac{1}{|R|} \sum_{i \in R} log_2 \frac{|T_i|}{|T|} \qquad (9)$$

where R is the set of recommended items in one recommendation and T is the whole set of transactions. For all the N recommendations on a certain testing dataset, the global recommendation novelty is defined as the mean of $MITF$ of each recommendation:

$$M^2ITF = \frac{1}{N} \sum MITF \qquad (10)$$

- *Local novelty:* different from global novelty, local novelty refers to the difference of recommended items w.r.t the previous experience of the users. Generally, if the recommended items are more different from the already bought items, the more novel these items are. In our model, the already bought items correspond to the context **c** used for recommendation R. Given a recommendation list R, the more items in R having been seen in the context **c**, the less novelty the recommendation is. Based on this observation, the context-aware novelty (CAN) of recommendation R can be defined as:

$$CAN = 1 - \frac{|R \cap \mathbf{c}|}{|R|} \qquad (11)$$

Similarly, for all the N recommendations on a certain testing set, the local recommendation novelty is defined as the mean of CAN ($MCAN$) of all recommendations:

$$MCAN = \frac{1}{N} \sum CAN \qquad (12)$$

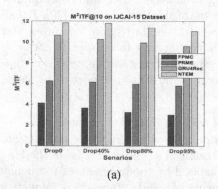
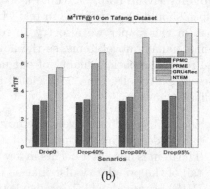

(a)   (b)

Fig. 2. M^2ITF comparisons on top-10 items, NTEM achieves higher global novelty than other approaches

Figures 2 and 3 show the results of global novelty and local novelty comparisons of top-10 recommendations over testing sets using the aforementioned

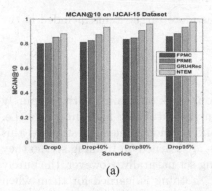
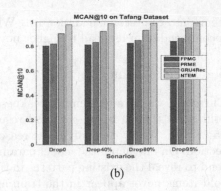

(a)   (b)

Fig. 3. $MCAN$ comparisons on top-10 items, NTEM achieves higher local novelty than other approaches

measures $M^2ITF@10$ and $MCAN@10$ respectively on both data sets. Overall, our proposed NTEM achieves both higher global novelty and local novelty compared to other approaches. FPMC is not trained well on such sparse dataset and can not work well, so its recommendations are not precise and have a lot of randomness. Recall the fact that most of the candidate items in the dataset are frequent and not novel. Accordingly, we get low M^2ITF and $MCAN$ on FPMC. PRME is a first-order Markov Chain model in which a sequential hypothesis is forced between the items within a transaction and it predicts the target item by only utilizing the one before it in the transaction. This is not always the case in the real world and it leads to information loss by ignoring other items in the transaction, so PRME also gets low novelties here. GRU units in GRU4Rec can accumulate the effect of all the sequential items in a transaction, it makes use of all other items in the transaction to predict the target one. Taking the advantage of the deep structure, it achieves relative high novelties. As a result, GRU4Rec is a relative good TBRS to make novel recommendations.

Our proposed NTEM do not have the sparse issue as FPMC due to its totally different work mechanism from matrix factorization used in FPMC, so it works well even on a sparse data set as shown in our experiment. Furthermore, NTEM does not assume a sequence between the items within a transaction, which is closer to the actual situation. More importantly, we make full use of all the other items in a transaction to predict the target one, which makes the prediction more solid by utilizing richer information. For the cold start issue, we incorporate the item features into the model and take the advantage of the information flow on feature values for prediction when the available transaction information is limited. So NTEM is more easily to provide novel recommendations and to achieve higher global and local novelties.

In practice, both the global and local novelties are somehow related to the recommendation accuracy here, specifically, some kind of positive relations exist between them. In the cold start scenarios, as all the target items are infrequent or novel, higher recommendation accuracy means more chance to get the target items into the recommendation list. Whereas more target items included in the recommendation list lead to higher novelty.

5 Discussions

Currently, we have to admit a weakness of our work on dealing with the purely cold items though it can get good results in the extremely high drop rate (i.e. 95%) case. We try to analyze the reasons and provide some suggestions in this section. On one hand, the model parameters are learned on training set, so they tend to reveal the existing patterns in training set naturally. However, the purely cold items never appear in the training set, nothing is learned for them when training. On the other hand, for those purely cold items, only their features are used for predictions. This requires their feature values have occurred frequently in the training set and thus good embeddings can be learned. In addition, the item features also affect the prediction of new items greatly. In general, high-dimensional features have stronger capabilities to deliver information than low-dimensional ones. Unfortunately, the shopping-basket datasets in transaction domain usually have diverse feature values and low dimensions. For instance, the IJCAI-15 dataset only has three features: category, brand and seller. The average frequency of feature values in brand and seller is below five, which is quite infrequent. This leads to the feature values of new items in testing set occur too few times or even never occur, which leads to learning poor embeddings of them and thus poor information delivering capacity. Some potential solutions to improving our model for dealing with purely cold items may be: (1) finding more suitable datasets with high-dimensional features and intensive feature values, like audio data, image data or text data; (2) only using those features having already occurred in training set.

6 Conclusions

Perceiving the next choice in a dynamic context for onling recommendation is de-manding but challenging. In this paper, we propose NTEM to build a more precise and efficient TBRS, it also shows great potential to improve the recommendation novelty. The empirical evaluation on real-world e-commerce datasets shows the com-prehensive superiority of our methods over other state-of-the-art ones. In the future, we will extend our model to other domains, like relational learning on social networks.

Acknowledgment. This work is partially sponsored by the China Scholarship Council (CSC Student ID 201406890033).

References

1. Adda, M., Missaoui, R., Valtchev, P., Djeraba, C.: Recommendation strategy based on relationrule mining. In: IJCAI Workshop on Intelligent Techniques for Web Personalization(ITWP 2005), pp. 33–40 (2005)

2. Adomavicius, G., Tuzhilin, A.: Context-aware recommender systems. In: Ricci, F., Rokach, L., Shapira, B. (eds.) Recommender Systems Handbook, pp. 191–226. Springer, Boston, MA (2015). https://doi.org/10.1007/978-1-4899-7637-6_6
3. Breese, J.S., Heckerman, D., Kadie, C.: Empirical analysis of predictive algorithms for collaborative filtering. In: Proceedings of the Fourteenth Conference on Uncertainty in Artificial Intelligence, Morgan Kaufmann Publishers Inc., pp. 43–52 (1998)
4. Cao, L.: Coupling learning of complex interactions. Inf. Process. Manag. **51**(2), 167–186 (2015)
5. Chou, S.Y., Yang, Y.H., Jang, J.S.R., Lin, Y.C.: Addressing cold start for next-song recommendation. In: Proceedings of the 10th ACM Conference on Recommender Systems, pp. 115–118. ACM (2016)
6. Feng, S., Li, X., Zeng, Y., Cong, G., Chee, Y.M., Yuan, Q.: Personalized ranking metric embedding for next new POI recommendation. In: IJCAI, pp. 2069–2075 (2015)
7. Ricci, F., Rokach, L., Shapira, B.: Recommender Systems Handbook, 2nd edn. Springer, Heidelberg (2015). https://doi.org/10.1007/978-0-387-85820-3
8. Gutmann, M.U., Hyvärinen, A.: Noise-contrastive estimation of unnormalized statistical models, with applications to natural image statistics. J. Mach. Learn. Res. **13**, 307–361 (2012)
9. Han, J., Pei, J., Mortazavi-Asl, B., Chen, Q., Dayal, U., Hsu, M.C.: FreeSpan: frequent pattern-projected sequential pattern mining. In: Proceedings of the Sixth ACM SIGKDD International Conference on Knowledge Discovery and Data Mining, pp. 355–359. ACM (2000)
10. Hidasi, B., Karatzoglou, A., Baltrunas, L., Tikk, D.: Session-based recommendations with recurrent neural networks. arXiv preprint arXiv:1511.06939 (2015)
11. Huang, Z., Zeng, D.D.: Why does collaborative filtering work? transaction-based recommendation model validation and selection by analyzing bipartite random graphs. INFORMS J. Comput. **23**(1), 138–152 (2011)
12. Koren, Y.: Factorization meets the neighborhood: a multifaceted collaborative filtering model. In: Proceedings of the 14th ACM SIGKDD International Conference on Knowledge Discovery and Data Mining, pp. 426–434. ACM (2008)
13. Lin, W., Alvarez, S.A., Ruiz, C.: Efficient adaptive-support association rule mining for recommender systems. Data Min. Knowl. Discov. **6**(1), 83–105 (2002)
14. Melville, P., Mooney, R.J., Nagarajan, R.: Content-boosted collaborative filtering for improved recommendations. In: Aaai/iaai, pp. 187–192 (2002)
15. Mikolov, T., Sutskever, I., Chen, K., Corrado, G.S., Dean, J.: Distributed representations of words and phrases and their compositionality. In: Advances in neural information processing systems, pp. 3111–3119 (2013)
16. Park, Y.J., Tuzhilin, A.: The long tail of recommender systems and how to leverage it. In: Proceedings of the 2008 ACM Conference on Recommender Systems, pp. 11–18. ACM (2008)
17. Rendle, S.: Factorization machines with libfm. ACM Trans. Intell. Syst. Technol. (TIST) **3**(3), 57 (2012)
18. Rendle, S., Freudenthaler, C., Schmidt-Thieme, L.: Factorizing personalized markov chains for next-basket recommendation. In: Proceedings of the 19th International Conference on World Wide Web, pp. 811–820. ACM (2010)
19. Rong, X.: word2vec parameter learning explained. arXiv preprint arXiv:1411.2738 (2014)

20. Vargas, S., Castells, P.: Rank and relevance in novelty and diversity metrics for recommender systems. In: Proceedings of the Fifth ACM Conference on Recommender Systems, pp. 109–116. ACM (2011)
21. Wang, S., Liu, W., Wu, J., Cao, L., Meng, Q., Kennedy, P.J.: Training deep neural networks on imbalanced data sets. In: 2016 International Joint Conference on Neural Networks (IJCNN), pp. 4368–4374. IEEE (2016)
22. Wu, X., Liu, Q., Chen, E., He, L., Lv, J., Cao, C., Hu, G.: Personalized next-song recommendation in online karaokes. In: Proceedings of the 7th ACM Conference on Recommender Systems, pp. 137–140. ACM (2013)
23. Yap, G.-E., Li, X.-L., Yu, P.S.: Effective next-items recommendation via personalized sequential pattern mining. In: Lee, S., Peng, Z., Zhou, X., Moon, Y.-S., Unland, R., Yoo, J. (eds.) DASFAA 2012. LNCS, vol. 7239, pp. 48–64. Springer, Heidelberg (2012). https://doi.org/10.1007/978-3-642-29035-0_4
24. Yuan, Q., Cong, G., Ma, Z., Sun, A., Thalmann, N.M.: Time-aware point-of-interest recommendation. In: Proceedings of the 36th International ACM SIGIR Conference on Research and Development in Information Retrieval, pp. 363–372. ACM (2013)

Regression

Adaptive Skip-Train Structured Regression for Temporal Networks

Martin Pavlovski[1,2], Fang Zhou[1], Ivan Stojkovic[1,3], Ljupco Kocarev[2],
and Zoran Obradovic[1(✉)]

[1] Temple University, Philadelphia, PA 19122, USA
zoran.obradovic@temple.edu
[2] Macedonian Academy of Sciences and Arts, Skopje 1000, Republic of Macedonia
[3] University of Belgrade, 11120 Belgrade, Serbia

Abstract. A broad range of high impact applications involve learning a predictive model in a temporal network environment. In weather forecasting, predicting effectiveness of treatments, outcomes in healthcare and in many other domains, networks are often large, while intervals between consecutive time moments are brief. Therefore, models are required to forecast in a more scalable and efficient way, without compromising accuracy. The Gaussian Conditional Random Field (GCRF) is a widely used graphical model for performing structured regression on networks. However, GCRF is not applicable to large networks and it cannot capture different network substructures (communities) since it considers the entire network while learning. In this study, we present a novel model, Adaptive Skip-Train Structured Ensemble (AST-SE), which is a sampling-based structured regression ensemble for prediction on top of temporal networks. AST-SE takes advantage of the scheme of ensemble methods to allow multiple GCRFs to learn from several subnetworks. The proposed model is able to automatically skip the entire training or some phases of the training process. The prediction accuracy and efficiency of AST-SE were assessed and compared against alternatives on synthetic temporal networks and the H3N2 Virus Influenza network. The obtained results provide evidence that (1) AST-SE is ~140 times faster than GCRF as it skips retraining quite frequently; (2) It still captures the original network structure more accurately than GCRF while operating solely on partial views of the network; (3) It outperforms both unweighted and weighted GCRF ensembles which also operate on subnetworks but require retraining at each timestep. Code and data related to this chapter are available at: https://doi.org/10.6084/m9.figshare.5444500.

1 Introduction

A variety of real-world prediction problems involve temporal network analysis to forecast future events. In particular, structured regression models are widely used for severe weather forecasting by learning past weather-related measurements while considering the network structure among measurement stations [13]. These models are also applied for predicting future hospital admissions based on past

© Springer International Publishing AG 2017
M. Ceci et al. (Eds.): ECML PKDD 2017, Part II, LNAI 10535, pp. 305–321, 2017.
https://doi.org/10.1007/978-3-319-71246-8_19

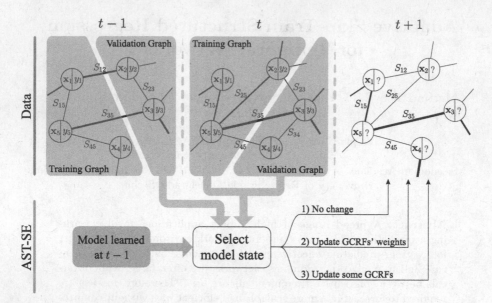

Fig. 1. Skip-training between consecutive timesteps.

admissions and couplings between hospitals [13]; predicting disease occurrence, knowing which diseases co-occur and how frequently each occurred in the past.

Forecasting in temporal networks is commonly approached by employing a single structured model to learn the relationship between the response variables and the explanatory variables, along with the correlations between nodes, from multiple past timesteps, in order to predict the response for each node in one or multiple upcoming timesteps [6,15]. However, issues arise when the time for prediction is limited, and the large size of networks increases the computational and space complexity when learning from multiple previous timesteps. To overcome these limitations, one can train a simple unstructured learner at the current timestep in order to perform a one-step ahead prediction. Although unstructured learners can be rapidly trained, they do not always obtain accurate predictions since they are not capable of capturing between-node correlations. Structured learners such as the Gaussian Conditional Random Fields (GCRFs) may be more accurate, but they require more time for retraining at each timestep. Moreover, they consider the whole network structure while learning, without taking advantage of useful substructures within the network. Taking these issues into account, while noticing that the data distribution does not change frequently in most temporal settings, we propose a new model that can automatically decide whether to skip the majority of unnecessary computation, which comes from retraining at each timestep, and make predictions in a more timely and accurate manner.

Inspired by this insight, we propose a multi-state model, Adaptive Skip-Train Structured Ensemble (AST-SE). The AST-SE is outlined in Fig. 1. First, in order to achieve greater predictive performance, AST-SE incorporates multiple GCRFs into a single composite structured ensemble. Then, to capture the hidden network substructures, GCRFs are trained simultaneously on subnetworks, generated by subsampling which decreases complexity and increases scalability. In addition, at each timestep, AST-SE automatically determines whether it should (1) simply skip training if the model learned from the previous data obtains comparable accuracy on the present validation data (*State 1*); (2) assign new weights to its components based on the present training data (*State 2*); or (3) retrain some of its GCRF components (*State 3*). Its accuracy and efficiency were compared to ensemble and non-ensemble based alternatives on both synthetic temporal networks and on a real-world application. AST-SE has shown to outperform its competitors, while learning in a more efficient, scalable, and potentially more accurate manner.

The main characteristics of AST-SE are summarized in the following:

1. **Efficiency:** AST-SE is ~140 and ~4.5 times faster than GCRF and ensemble-based alternatives, respectively, in case when its components are run in parallel on the H3N2 Virus Influenza network.
2. **Scalability:** AST-SE focuses only on partial views of a network, and therefore it is scalable as the network size expands.
3. **Accuracy:** While being fast and scalable with vast network sizes, AST-SE also obtains a ~34–41% decrease in mean squared error on average, when compared against alternatives on the H3N2 Virus Influenza network.

2   Related Work

Structured Regression. The Gaussian Conditional Random Field model is a popular graphical model for structured regression. It was originally proposed in [9] for regression in remote sensing applications. The framework is further extended for *temporal prediction tasks*, along with a method for uncertainty propagation tracking [6]. Models for spatio-temporal prediction are also adapted for semi-supervised learning, and can handle missing information in targets [12], and attributes [11]. The extension of the framework for directed graphs is proposed in [16].

Some approaches [13] further extended the model's expressiveness, which allowed *more accurate predictions*. Finally, *scalabillity and computational efficiency* have also been addressed. One fast approximation for structured regression utilizes graph compression to reduce the computational burden in large graphs [18]. However, due to approximation, those approaches lose information, which leads to the implementation of a fast model with exact inference in [5].

Ensemble Methods. Ensemble learning has been thoroughly researched over the past two decades. The main idea of ensemble methods is to improve the predictive performance of a single learner by generating multiple versions of it and learning each on a different data subset. Predictions made by these learners are then combined according to a certain aggregation scheme. Various methods for ensemble generation have been developed using different data replication techniques and learner aggregation schemes, with Bagging [1,2] and Boosting [4] being among the most popular. Although, recent ensemble approaches [7,10] have been focused on regression, they do not consider the structural component while learning. To the best of our knowledge, the power of ensemble learning has not been exploited when dealing with structured data in a temporal setting. For more details on ensembles in general, refer to [3].

3 Preliminaries

3.1 Problem Statement

Assume that a network of N nodes changes over time. At timestep t, the network is represented by the weighted attributed graph $G^{(t)} = (V^{(t)}, E^{(t)}, \mathbf{X}^{(t)}, \mathbf{y}^{(t)})$ comprised of a set $V^{(t)}$ of N nodes, a set of edges $E^{(t)} = \{(v_i^{(t)}, v_j^{(t)}) | S_{ij}^{(t)} > 0\} \subseteq V^{(t)} \times V^{(t)}$, N D-dimensional input vectors (attributes) organized in a matrix $\mathbf{X}^{(t)} = [\mathbf{x}_1^{(t)}, \ldots, \mathbf{x}_N^{(t)}]^\top$, and an output (target) vector $\mathbf{y}^{(t)}$. A node v_i is associated with its attribute vector $\mathbf{x}_i^{(t)}$ and a corresponding output (target) value $y_i^{(t)}$, for each $i = 1, \ldots, N$, while an edge $(v_i^{(t)}, v_j^{(t)})$ connects nodes $v_i^{(t)}$ and $v_j^{(t)}$ only if the element $S_{ij}^{(t)}$ in the similarity matrix $\mathbf{S}^{(t)}$ is positive. In this temporal formulation, the objective is to predict the outputs $\mathbf{y}^{(t+1)}$ for all nodes in the next timestep, given the unobserved graph $G^{(t+1)} = (V^{(t+1)}, E^{(t+1)}, \mathbf{X}^{(t+1)})$.

In this work, we consider networks with a fixed number of nodes among all timesteps. In a more general case, N can change over time, and our model can also be directly applied to such case.

3.2 Gaussian Conditional Random Fields

Continuous Conditional Random Fields (CCRFs) [8] address the above-described problem by modeling the conditional distribution of $\mathbf{y}^{(t)}$, given $\mathbf{X}^{(t)}$, as

$$P(\mathbf{y}^{(t)}|\mathbf{X}^{(t)}) = \frac{1}{Z(\mathbf{X}^{(t)}, \alpha^{(t)}, \beta^{(t)})} \exp\left\{\sum_{i=1}^{N} A(\alpha^{(t)}, y_i^{(t)}, \mathbf{X}^{(t)}) + \sum_{i \sim j} I(\beta^{(t)}, y_i^{(t)}, y_j^{(t)})\right\}, \tag{1}$$

where the interaction potential function $A(\alpha^{(t)}, y_i^{(t)}, \mathbf{X}^{(t)})$ models the relationship between $y_i^{(t)}$ and all attribute vectors in $\mathbf{X}^{(t)}$, while the pairwise interactions between $y_i^{(t)}$ and $y_j^{(t)}$ are captured by an interaction potential function $I(\beta^{(t)}, y_i^{(t)}, y_j^{(t)})$, for all $i, j = 1, \ldots, N$. Integrating the entire term in the exponent over $\mathbf{y}$ gives the value of the normalization constant $Z(\mathbf{X}^{(t)}, \alpha^{(t)}, \beta^{(t)})$. Typically, both functions are defined by combining the parameters $\alpha^{(t)}$ and $\beta^{(t)}$ with a feature function $f(y_i^{(t)}, \mathbf{X}^{(t)})$ and a pairwise interaction function $g(y_i^{(t)}, y_j^{(t)})$, respectively. Defining $f(y_i^{(t)}, \mathbf{X}^{(t)})$ and $g(y_i^{(t)}, y_j^{(t)})$ as quadratic functions

$$f(y_i^{(t)}, \mathbf{X}^{(t)}) = -\left(y_i^{(t)} - R_i(\mathbf{X}^{(t)})\right)^2; \quad g(y_i^{(t)}, y_j^{(t)}) = -S_{ij}^{(t)}\left(y_i^{(t)} - y_j^{(t)}\right)^2, \tag{2}$$

yields the following expression for the conditional probability

$$P(\mathbf{y}^{(t)}|\mathbf{X}^{(t)}) = \frac{1}{Z(\mathbf{X}^{(t)}, \alpha^{(t)}, \beta^{(t)})} \exp\left\{ -\alpha^{(t)} \sum_{i=1}^{N} \left(y_i^{(t)} - R_i(\mathbf{X}^{(t)}) \right)^2 \right.$$
$$\left. -\beta^{(t)} \sum_{i \sim j} S_{ij}^{(t)} \left(y_i^{(t)} - y_j^{(t)} \right)^2 \right\}, \tag{3}$$

where $R_i(\mathbf{X}^{(t)})$ denotes the prediction for the i-th node, made by the k-th unstructured predictor at timestep t. The first term in the exponent controls the relevance of each unstructured predictor. The second term models the dependencies among the output values by considering a symmetric similarity matrix $\mathbf{S}^{(t)} = [S_{ij}^{(t)}]_{N \times N}$, thus defining an undirected weighted graph. The larger the weight of an edge $(v_i^{(t)}, v_j^{(t)})$, the more similar $y_i^{(l)}$ and $y_j^{(t)}$ are. Of course, a weight of zero indicates no connection between a pair of nodes.

Since the exponent in Eq. (3) is composed of quadratic functions of $\mathbf{y}^{(t)}$, the conditional probability distribution can be transposed directly onto a multivariate Gaussian distribution,

$$P(\mathbf{y}^{(t)}|\mathbf{X}^{(t)}) = \frac{1}{(2\pi)^{N/2}|\mathbf{\Sigma}^{(t)}|^{1/2}} \exp\left\{ -\frac{1}{2}(\mathbf{y}^{(t)} - \boldsymbol{\mu}^{(t)})^\top \mathbf{\Sigma}^{(t)^{-1}}(\mathbf{y}^{(t)} - \boldsymbol{\mu}^{(t)}) \right\}. \tag{4}$$

Therefore the resulting model is referred to as Gaussian CRF (GCRF). Setting (3) and (4) equal to each other results in the precision matrix

$$\mathbf{Q}^{(t)} = \alpha^{(t)}\mathbf{I} + \beta^{(t)}\mathbf{L}^{(t)}, \tag{5}$$

where $\mathbf{I}$ is an identity matrix, and $\mathbf{L}^{(t)}$ is the Laplacian matrix of $\mathbf{S}^{(t)}$. The precision matrix $\mathbf{Q}^{(t)}$, being the first canonical parameter of the Gaussian distribution, can be used to directly calculate $\mathbf{\Sigma}^{(t)} = \frac{1}{2}\mathbf{Q}^{(t)^{-1}}$. The second canonical parameter is simply a weighted combination of all unstructured predictions $\mathbf{R}^{(t)} = [R_1(\mathbf{X}^{(t)}), \dots, R_N(\mathbf{X}^{(t)})]^\top$, i.e. $\mathbf{b}^{(t)} = 2\mathbf{R}^{(t)}\alpha^{(t)}$. Finally, learning a GCRF model at timestep t comes down to determining the optimal parameters that maximize the conditional log-likelihood

$$[\alpha^{(t)}, \beta^{(t)}]^\top = \arg\max_{\alpha, \beta} L(\alpha, \beta) = \arg\max_{\alpha, \beta} \log\left(P(\mathbf{y}^{(t)}|\mathbf{X}^{(t)}; \alpha, \beta) \right), \tag{6}$$

such that $\alpha, \beta > 0$ is satisfied to guarantee the positive semi-definiteness of $\mathbf{Q}^{(t)}$. Upon learning, predictions for the nodes in the next timestep are simply made by using the canonical parameters to directly calculate the distribution's expected value, that is,

$$\boldsymbol{\mu}^{(t+1)} = \left(\alpha^{(t)}\mathbf{I} + \beta^{(t)}\mathbf{L}^{(t+1)} \right)^{-1} \mathbf{R}^{(t+1)}\alpha^{(t)}. \tag{7}$$

Note that, in the general case, multiple unstructured predictors and different similarity matrices can be used by the GCRF.

4   Methodology

In this section we provide a detailed description of the proposed model, called *Adaptive Skip-Train Structured Ensemble* (AST-SE). First, we briefly introduce the major component, which applies ensemble learning in the structured regression realm. Thereafter, we explain how AST-SE can skip-train on top of temporal networks. As for the computational complexity of AST-SE, it is discussed in Sect. 5.2, along with the complexities of its competitors.

4.1   Generating GCRF Ensembles by Network Subsampling

In the traditional GCRF, the relationship between the influence of unstructured predictors and the influence of the dependencies among the outputs is modeled through a single pair of α and β. However, in the real-world datasets, a single pair of α and β cannot fully capture such relationships over the whole network. One straightforward solution is to model relationships for each node and each link, which increases the complexity of the model [13]. Therefore, in our proposed model, AST-SE, multiple graphical models are employed in order to learn different relationships using network sub-structures. The model takes advantage of the scheme of ensemble methods to incorporate multiple GCRFs to learn from several replicas of the available data by utilizing sampling techniques.

Subbagging [1] (a variation of bagging that considers subsampling, i.e. sampling at random, but without replacement, to generate multiple training subsets) is one of the most popular sampling-based ensemble methods, and it has shown to reduce variance and improve stability, as well as to aid overfitting avoidance. Since we are dealing with networked data, subbagging is applied in AST-SE as it is easily scalable to large networks and it is more suitable to sample networks without replacement, so that nodes and edges are not duplicated within a single subnetwork, but can be shared among multiple subnetworks. Henceforth, by sampling multiple subnetworks and aggregating the knowledge gathered from different graphical models that operate among these subnetworks, AST-SE learns hidden substructures within the original network.

Now, let $\phi : (\mathbb{N}^N, \mathbb{R}^{N \times N}, \mathbb{R}^{N \times D}) \mapsto \mathbb{R}^N$ denote the outcome of a GCRF model, that maps a graph G to a vector μ containing the predictions for all nodes in G. In order to generate a graphical ensemble model, the graph at the current timestep $G^{(t)}$ is subsampled M times, thus resulting in M subgraphs $G_1^{(t)}, \ldots, G_M^{(t)}$ such that $N_m = |V_m^{(t)}| = \eta N$, where $\eta \in [0, 1]$, for each $m = 1, \ldots, M$. Thereafter, a single GCRF model $\phi_m^{(t)}$ is trained on each subgraph $G_m^{(t)}$.

One direct way to predict the outputs for all nodes at the next timestep is to aggregate the predictions made by all M GCRFs,

$$\Phi^{(t)}(G^{(t+1)}) = \frac{1}{M} \sum_{m=1}^{M} \phi_m^{(t)}(G^{(t+1)}). \tag{8}$$

However, not all sampled $G_m^{(t)}$ match some of the representative subgraphs. Therefore, one convenient way to overcome this issue is to assign a weight to each GCRF within the ensemble. This way, GCRFs trained on more representative samples of $G^{(t)}$ should obtain more accurate predictions for all nodes in $G^{(t)}$, thus gaining higher weights. The overall performance of this ensemble model is evaluated by minimizing the regularized quadratic loss function,

$$\ell(\{\phi_m^{(t)}\}, \boldsymbol{\omega}, G^{(t)}) = \frac{1}{N} \sum_{i=1}^{N} \left(y_i^{(t)} - \sum_{m=1}^{M} \omega_m \phi_m^{(t)}(G^{(t)}) \right)^2 + \lambda \sum_{m=1}^{M} |\omega_m|, \qquad (9)$$

where $\lambda \geq 0$ is a regularization parameter, and the weights are obtained by

$$\boldsymbol{\omega}^{(t)} = \arg\min_{\boldsymbol{\omega}} \ell(\{\phi_m^{(t)}\}, \boldsymbol{\omega}, G^{(t)}), \quad \text{s.t. } 0 < \omega_m \leq 1, \sum_{m=1}^{M} \omega_m = 1. \qquad (10)$$

Once the weights are learned, predictions for $G^{(t+1)}$ made by all GCRFs in the model sequence $\{\phi_m^{(t)}\}$ are combined in the weighted mixture,

$$\Phi^{(t)}(G^{(t+1)}) = \sum_{m-1}^{M} \omega_m^{(t)} \phi_m^{(t)}(G^{(t+1)}). \qquad (11)$$

4.2   Adaptive Skip-Training in a Temporal Environment

In order to predict the outputs for all nodes at timestep $t+1$, one can train a single GCRF or even a GCRF ensemble model (described in Sect. 4.1) at timestep t. However, repetitive retraining at each timestep can be often redundant because data distributions are similar in consecutive timesteps, and sometimes even infeasible. For instance, in a case when the number of nodes is large and both learning and inference must be attained within small time intervals between consecutive timesteps.

To overcome this issue, we propose a multi-state model that tends to learn over time in a more adaptive, pragmatic and efficient manner. Such a model can be adaptive to an extent where it is able to detect and learn changes in a network once it is necessary while maintaining accuracy. Changing through 3 different states as time passes, AST-SE adapts accordingly. *State 1* suggests that the model trained using the previous data is sufficient for prediction on the present data, i.e. the previously learned model obtains comparable accuracy on the present data. On the other hand, when in *State 2*, the model needs to slightly change by updating the weights of its GCRF components based on the present data. Lastly, *State 3* adapts to the present data by updating only some of the GCRF components.

The network at timestep t is split into two parts. One is for training, $G_{train}^{(t)}$, and the other is for validation, $G_{val}^{(t)}$. Initially $(t = 0)$, there is no previous data and therefore AST-SE is trained as a weighted structured ensemble using $G_{train}^{(0)}$.

When $t > 0$, several criteria are examined to determine which state should be selected and adaptive (skip-)training is performed accordingly through the following procedure:

Phase I. First, the model's state is initialized to *State 1* assuming that data distribution in the present timestep is similar to the previous one. Efficiency is maximized by relying solely on knowledge gathered in the past. This way, neither retraining nor weight updating is needed, meaning that both the GCRF components $\{\phi_m^{(t-1)}\}$ and their corresponding weights $\omega^{(t-1)}$ from the previous timestep are combined to predict outcomes at timestep $t + 1$, i.e.

$$\Phi_1^{(t)}(G^{(t+1)}) = \sum_{m=1}^{M} \omega_m^{(t-1)} \phi_m^{(t-1)}(G^{(t+1)}). \tag{12}$$

State 1 is selected if the previously learned model can obtain similar accuracy on the present data. This occurs when the loss obtained on the current data using the previous model, $\ell_1^{(t)} = \ell(\{\phi_m^{(t-1)}\}, \omega^{(t-1)}, G_{val}^{(t)})$, is not larger than the loss obtained on the previous data $\ell_0^{(t)} = \ell(\{\phi_m^{(t-1)}\}, \omega^{(t-1)}, G_{val}^{(t-1)})$. Once the condition is satisfied, the procedure selects $\Phi_1^{(t)}$ to perform prediction for the next timestep.

Phase II. However, relying entirely on past knowledge may cause predictive performance to deteriorate, especially when the data distribution slightly changes between consecutive timesteps. A fast way to retrain AST-SE is to update the weights of the GCRF components obtained at $t - 1$ using the current training graph $G_{train}^{(t)}$:

$$\Phi_2^{(t)}(G^{(t+1)}) = \sum_{m=1}^{M} \omega_m^{(t)} \phi_m^{(t-1)}(G^{(t+1)}), \tag{13}$$

where $\omega^{(t)} = \arg\min_{\omega} \ell(\{\phi_m^{(t-1)}\}, \omega, G_{train}^{(t)})$. This compels AST-SE to adapt to current data, while avoiding to retrain its GCRF components. The performance of $\Phi_2^{(t)}$ is assessed on the present validation graph using Eq. (9) to calculate the loss $\ell_2^{(t)} = \ell(\{\phi_m^{(t-1)}\}, \omega^{(t)}, G_{val}^{(t)})$. If this loss is smaller than or equal to $\ell_0^{(t)}$, then the procedure selects $\Phi_2^{(t)}$ and prediction is performed for the next timestep.

Phase III. Once this phase is reached, retraining must be performed in order to obtain a lower loss. However, AST-SE still tends to skip training when possible. AST-SE automatically selects models to retrain based on the largest increase in ascending order of weights. Therefore, instead of retraining all GCRF components, a model selection is performed by sorting their weights obtained at $t - 1$. The sorted weight sequence $\omega_{s_1}^{(t-1)} \le \omega_{s_2}^{(t-1)} \le \cdots \le \omega_{s_M}^{(t-1)}$ is then used to determine a threshold value M^* for model selection,

$$M^* = \arg\max_{m \in [2,M]} \left(\omega_{s_{m-1}}^{(t-1)} - \omega_{s_m}^{(t-1)} \right), \tag{14}$$

Algorithm 1. Adaptive Skip-Train Structured Ensemble (at timestep t)

Input:

    GCRF components from previous timestep $\{\phi_m^{(t-1)}\}$, along with their weights $\omega^{(t-1)}$

    Training graph $G_{train}^{(t-1)}$ and validation graph $G_{val}^{(t-1)}$

    Attributed graph $G^{(t)} = (V^{(t)}, E^{(t)}, \mathbf{X}^{(t)}, \mathbf{y}^{(t)})$

Procedure:

    $(G_{train}^{(t)}, G_{val}^{(t)}) \leftarrow \mathbf{Split}(G^{(t)})$

    $\Phi_1^{(t)} \leftarrow \sum_{m=1}^{M} \omega_m^{(t-1)} \phi_m^{(t-1)}$ $\triangleright$ Eq. (12)

    $state^{(t)} \leftarrow 1$ $\triangleright$ Initialize model's state to 1

    if $t > 0$ **then**,

Phase I

    $\ell_0^{(t)} \leftarrow \ell(\{\phi_m^{(t-1)}\}, \omega^{(t-1)}, G_{val}^{(t-1)})$

    $\ell_1^{(t)} \leftarrow \ell(\{\phi_m^{(t-1)}\}, \omega^{(t-1)}, G_{val}^{(t)})$

    if $\ell_1^{(t)} \leq \ell_0^{(t)}$ **then**

        $state^{(t)} \leftarrow 1$ $\triangleright$ Remain in *State 1*

    else

Phase II

    $\omega^{(t)} \leftarrow \arg\min_\omega \ell(\{\phi_m^{(t-1)}\}, \omega, G_{train}^{(t)})$

    $\Phi_2^{(t)} \leftarrow \sum_{m=1}^{M} \omega_m^{(t)} \phi_m^{(t-1)}$ $\triangleright$ Eq. (13)

    $\ell_2^{(t)} \leftarrow \ell(\{\phi_m^{(t-1)}\}, \omega^{(t)}, G_{val}^{(t)})$

    if $\ell_2^{(t)} \leq \ell_0^{(t)}$ **then**

        $state^{(t)} \leftarrow 2$ $\triangleright$ Set model's state to 2

    else

Phase III

    $[\omega_{s_1}^{(t-1)}, \omega_{s_2}^{(t-1)}, \ldots, \omega_{s_M}^{(t-1)}] \leftarrow \mathbf{Sort}(\omega^{(t-1)})$

    $M^* \leftarrow \arg\max_{m \in [2,M]} \left(\omega_{s_{m-1}}^{(t-1)} - \omega_{s_m}^{(t-1)} \right)$

    $\{\phi_m^{(t)'}\} \leftarrow \{\phi_m^{(t)}\}_{m=1}^{M^*-1} \cup \{\phi_{s_m}^{(t-1)}\}_{m=M^*}^{M}$

    $\omega^{(t)} \leftarrow \arg\min_\omega \ell(\{\phi_m^{(t)'}\}, \omega, G_{train}^{(t)})$

    $\Phi_3^{(t)} \leftarrow \sum_{m=1}^{M} \omega_m^{(t)} \phi_m^{(t)'}$ $\triangleright$ Eq. (15)

    $\ell_3^{(t)} \leftarrow \ell(\{\phi_m^{(t)'}\}, \omega^{(t)}, G_{val}^{(t)})$

    if $\ell_3^{(t)} \leq \ell_0^{(t)}$ **then**

        $state^{(t)} \leftarrow 3$ $\triangleright$ Set model's state to 3

    else

        $state^{(t)} \leftarrow \arg\min_{p=1,2,3} \ell_p$ $\triangleright$ Choose the min-loss state (Eq. (16))

        end if

      end if

    end if

  end if

Output:

    Return $\Phi_{state^{(t)}}^{(t)}$

---

thus pruning those GCRFs whose weights preceded the largest weight increase in the sorted sequence. Upon removal, exactly $M^* - 1$ new GCRF components are trained on $G_{train}^{(t)}$ and added to the ensemble. In addition, as in Phase II, new weights $\boldsymbol{\omega}^{(t)}$ are obtained from the present training graph $G_{train}^{(t)}$,

$$\Phi_3^{(t)}(G^{(t+1)}) = \sum_{m=1}^{M^*-1} \omega_m^{(t)} \phi_m^{(t)}(G^{(t+1)}) + \sum_{m=M^*}^{M} \omega_m^{(t)} \phi_{s_m}^{(t-1)}(G^{(t+1)}) \quad (15)$$

Accordingly, if the loss of this fused ensemble $\ell_3^{(t)} = \ell(\{\phi_m^{(t)}\}_{m=1}^{M^*-1} \cup \{\phi_{s_m}^{(t-1)}\}_{m=M^*}^{M}, \boldsymbol{\omega}^{(t)}, G_{val}^{(t)})$ is lower than or equal to $\ell_0^{(t)}$, then $\Phi_3^{(t)}$ is considered as the final AST-SE choice at timestep t. Otherwise, the state of the final AST-SE outcome $\Phi_{p^*}^{(t)}$ is chosen as the state in which the minimum loss was obtained, i.e.

$$p^* = \arg\min_{p=1,2,3} \ell_p. \quad (16)$$

The above-described procedure is repeated at each timestep $t = 1, \ldots, T$. Its algorithmic description is presented in Algorithm 1.

5   Experimental Evaluation

5.1   Experimental Setup

In order to inspect the predictive ability of AST-SE and its competitors, experiments were performed to analyze their predictive performance on: (1) synthetically generated temporal networks, and (2) gene expression network [17] - a real-world temporal network. In each experiment, given a training graph, $M = 30$ GCRF models were used by the ensemble approaches, while $\eta = 30\%$ of the nodes in the original training graph were sampled to construct the subgraph for each GCRF. At each timestep t, the training graph $G_{train}^{(t)}$ for AST-SE was constructed by sampling 80% of the nodes in $G^{(t)}$, along with the existing edges between them, while the rest were used for validation. For the alternatives, the whole graph $G^{(t)}$ was used for training.

Mean squared error (MSE) was calculated for all models when they were tested on the network at timestep $t + 1$. In addition, to assess efficiency, the execution time of all models was measured. Since the components within the ensemble-based models are decoupled in time, the execution time for each of these models was measured as their components are run in parallel. Here, we report both the average MSEs and average execution times, along with the corresponding 90% confidence intervals. All experiments were run on Windows with 64 GB memory and 3.4 GHz CPU. The code was written in MATLAB and is publicly available at https://github.com/martinpavlovski/AST-SE.

5.2   Baselines

AST-SE was compared against multiple alternatives including both standard and ensemble-based models. Each one is briefly described in the following:

- *LR:* An L1-regularized linear regression. LR was employed as an unstructured predictor for each of the following models in order to achieve efficiency.
- *GCRF:* Standard GCRF [9] model that enables the chosen unstructured predictor to learn the network structure.
- *SE:* Structured ensemble composed of multiple GCRF models. Predictions for the next timestep are made according to Eq. (8).
- *WSE:* Weighted structured ensemble that combines the predictions of multiple GCRFs in a weighted mixture (refer to Eq. (11)) in order to predict the nodes' outputs in the next timestep.

The computational complexities of all models listed above are presented in Table 1. The computational complexity of all structured models is calculated in case learning is attained according to the original GCRF optimization procedure [9] which takes $\mathcal{O}(IN^3)$. However, the standard GCRF can be replaced by a faster variant called Unimodal GCRF (UmGCRF)[5]. In such case, the computational complexities of all structured models (GCRF, SE, WSE, and AST-SE) will decrease proportionally.

Table 1. Computational complexity of all models in terms of $\mathcal{O}$ notation. $\delta = |V_{train}^{(t)}|/N$, η is the fraction of the graph used for training, and M is the number of GCRF components within an ensemble, while I and I' denote the number of gradient ascent iterations needed to learn a GCRF and the number of optimization iterations needed to obtain weights for multiple GCRF components, respectively. Note that the computational complexity of each ensemble-based model is calculated as its components are run in parallel.

Model		Complexity
LR		$\mathcal{O}(d^3 + d^2 N)$
GCRF		$\mathcal{O}(d^3 + d^2 N + IN^3)$
SE		$\mathcal{O}(d^3 + d^2(\eta N) + I(\eta N)^3)$
WSE		$\mathcal{O}(d^3 + d^2(\eta N) + I(\eta N)^3 + N^3 + I'(NM))$
AST-SE	State 1	$\mathcal{O}(((1-\delta)N)^3)$
	State 2	$\mathcal{O}(((1-\delta)N)^3 + (\delta N)^3 + I'(\delta N M))$
	State 3	$\mathcal{O}(((1-\delta)N)^3 + (\delta N)^3 + I'(\delta N M) + d^3 + d^2(\delta \eta N) + I(\delta \eta N)^3 + I'(\delta \eta N M))$

5.3   Experiments on Synthetic Temporal Networks

First, we briefly describe the synthetic temporal networks used in the experiments, their node attributes and edge weights. The structures of the networks were generated using an Erdős-Rényi random graph model with $N = 10,000$ nodes, while an $N \times D$ attribute matrix $\mathbf{X}$ was generated for the node attributes, such that each attribute x_{id} is normally distributed according to $\mathcal{N}(0,1)$. Then, assuming that the attributes have linear relationship with the final outputs, we randomly generated parameters $\boldsymbol{\theta}$ and used them to get an artificial output of an unstructured predictor. That is,

$$R_i = \theta_0 + \theta_1 x_{i1} + \cdots + \theta_D x_{iD} + \epsilon_i, \quad \forall i = 1, \ldots, N, \tag{17}$$

where $\theta_0, \theta_1, \ldots, \theta_D (D = 5)$ and ϵ_i were randomly sampled from $\mathcal{U}(-1,1)$ and $\mathcal{N}(0,1/3)$, respectively. A weight was assigned to each edge as $S_{ij} = e^{-|R_i - R_j|}$. Then, noise sampled from $\mathcal{N}(0,2/3)$ was added to $\mathbf{R} = [R_1, \ldots, R_N]^\top$, thus yielding $\hat{\mathbf{R}}$.

Temporal networks were constructed assuming that there are 5 different substructures (communities) in each network, meaning that the influence of α and β is different among communities. The similarity matrix $\mathbf{S}$ was divided into 5 disjoint submatrices $\mathbf{S}_1, \ldots, \mathbf{S}_5$. By utilizing GCRF in a generative manner, the Laplacians of these submatrices, along with their own α and β, and the noisy predictions $\hat{\mathbf{R}}_m$, were used to generate the nodes' outputs for each subgraph $\mathbf{y}_m = (\alpha_m \mathbf{I} + \beta_m \mathbf{L}_m)^{-1} \hat{\mathbf{R}}_m \alpha_m$, such that the values of α_m and β_m were set in advance. Finally, all $\mathbf{y}_m$ and $\mathbf{S}_m$ were combined accordingly in a single $\mathbf{y}$ and $\mathbf{S}$, respectively.

The above-described procedure was repeated T times in order to generate $\mathbf{X}$, $\mathbf{y}$ and $\mathbf{S}$ for T timesteps using a different set of α and β parameters at each timestep. α and β were set according to two scenarios. In the first scenario, we consider only one data distribution change at timestep 6. In timesteps 1 to 5 distributions are similar and also at timesteps 6 to 10. In the second scenario, the distribution is changed more frequently among timesteps. The GCRF parameter values in case of both scenarios are summarized in Table 2.

Table 2. GCRF parameter values used to generate the synthetic data. Note that column 3 and 4 contain the values of α and β, or the intervals from which their values were uniformly sampled.

Scenario	Timestep	GCRF parameters	
		α	β
#1	$t = 1, \ldots, 5$	1	$[2, 5]$
	$t = 6, \ldots, 10$	$[2, 5]$	1
#2	$t = 1, 2$	1	$[2, 5]$
	$t = 3, 4$	1	$[6, 9]$
	$t = 5$	1	1
	$t = 6, 7$	$[2, 5]$	1
	$t = 8, 9$	$[6, 9]$	1
	$t = 10$	1	1

Upon generation, the performance of AST-SE and its alternatives were evaluated under each synthetic scenario. The obtained results are reported in Table 3. They provide evidence that AST-SE outperforms all of its competitors in terms of accuracy, and it is the second fastest among them.

Table 3. Synthetic scenarios #1 and #2 - Testing MSE and execution time (in seconds), averaged over all timesteps.

Model	Scenario #1		Scenario #2	
	MSE	Execution time	MSE	Execution time
LR	0.29 ± 0.0077	**0.01 ± 0.0110**	0.28 ± 0.0094	**0.01 ± 0.0117**
GCRF	0.25 ± 0.0083	5014.65 ± 360.4290	0.26 ± 0.0089	5254.84 ± 259.3738
SE	0.23 ± 0.0083	188.30 ± 9.0330	0.23 ± 0.0084	193.30 ± 11.1379
WSE	0.21 ± 0.0089	207.52 ± 3.8419	0.21 ± 0.0047	217.14 ± 7.1168
AST-SE	**0.19 ± 0.0051**	15.38 ± 28.0564	**0.20 ± 0.0066**	70.16 ± 34.4926

Accuracy: The results show that, under Scenario #1 AST-SE outperforms all of its competitors by a significant margin. Moreover, its average MSE has the tightest confidence interval. As for Scenario #2, although the data generated according to this scenario changes quite frequently, AST-SE still manages to obtain the lowest MSE among alternatives, while being the second most stable model with respect to its confidence interval. The models' accuracy was further analyzed by observing their MSEs obtained at each individual timestep. Figure 2 shows that under both scenarios AST-SE manifests consistent accuracy by maintaining the lowest MSE at almost all timesteps.

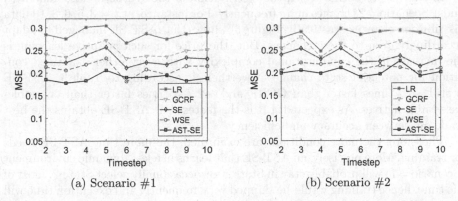

(a) Scenario #1   (b) Scenario #2

Fig. 2. Testing MSEs over time. Note that the x-axis starts from 2 since testing starts after all models are trained at timestep 1.

Efficiency: In order to examine the efficiency of AST-SE, its states were observed over time and are illustrated in Fig. 3. The frequency of fluctuations in the data

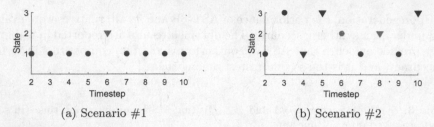

(a) Scenario #1   (b) Scenario #2

Fig. 3. AST-SE states selected over time (State 1 - blue, State 2 - red, State 3 - black). Circles depict states that were selected directly, while triangles depict states that were indirectly selected, i.e. they were selected in case all phases were passed and they obtained the minimum loss. (Color figure online)

distribution is different for each scenario, while the ability of ASE-SE to react to such fluctuations accordingly is evident in both cases. First, according to Fig. 3a, AST-SE stays in State 1 almost all of the time. A change in its state occurs exactly at timestep 6, at which the structure of the data generated according to $\alpha = 1, \beta \in [2, 5]$ suddenly changes to a structure that holds $\alpha \in [2, 5], \beta = 1$ (see Table 2). More precisely, the model's state changes to State 2 in which it needs to update the weights of its GCRF components in order to adapt to the new data distribution. At the next timestep, the model returns to State 1 since there are no drastic changes in the values of α and β and stays in this state till the last timestep. Overall, by changing states only at two timesteps (6 and 7) under Scenario #1, AST-SE is ~320 times faster than GCRF and ~12–13 times faster than SE and WSE when all GCRF components within SE, WSE and AST-SE are run in parallel. The only model faster than AST-SE is LR, but LR obtains much higher MSE than ASE-SE.

In contrast to Scenario #1, the distribution of the synthetic data generated under Scenario #2 changes quite frequently. For instance, at $t = 3, 5, 6, 8, 10$ data distribution changes drastically. Figure 3b shows that AST-SE managed to adapt accordingly even to these changes. But, the price for such adaptive learning is the increase in the computational complexity with every other examined condition for potential state change. Nevertheless, according to Table 3, AST-SE is still ~75 times faster than GCRF and ~2.7–3 times faster than ensemble-based alternatives. As expected, LR is the fastest but AST-SE obtains the best trade-off between accuracy and efficiency.

Therefore, the more conditions are examined, the more time AST-SE needs for training. More precisely, an AST-SE that learns in a less dynamic environment (Scenario #1) will probably stay in State 1, or occasionally select State 2, most of the time, hence training would be skipped very frequently and execution time will be reduced. On the contrary, learning in a highly dynamic environment in which the distribution of the data changes all the time (Scenario #2) will impose AST-SE to change between states quite often. According to all previously presented results, it can be inferred that AST-SE may handle both scenarios, but it certainly works much better for problems in which there are no drastic changes over time. Many real-world problems are characterized by such properties.

5.4   Performance on a Real-World Application

The performance of AST-SE against its competitors was also examined on a real-world data, the H3N2 Influenza Virus dataset. This dataset contains temporally collected gene expression measurements (12,032 genes) of a human subject infected with the H3N2 virus [17]. Blood samples were collected on multiple occasions (16 time points drawn approximately once every eight hours) during the five-day period, after the virus was inoculated in the subject. The task is to predict the expression value for all genes (nodes) at the next timestep using the previous 3 timesteps as features. The similarity structure among the genes was constructed by estimating the sparse inverse covariance matrix from the expression data, using the algorithm proposed in [14].

The results of all models are summarized in Table 4. Clearly, AST-SE outperforms all baselines in terms of accuracy. It obtains the lowest average MSE and seems to be the most stable, as it has the tightest confidence interval for its average MSE. As to execution time, LR is by far the fastest approach. However, it obtains a high MSE. Furthermore, although AST-SE is only the second fastest among all models, it is approximately ~34–41% more accurate compared to all of them including LR, thus providing the best trade-off between accuracy and efficiency. In other words, notwithstanding LR, which is an unstructured predictor, AST-SE is the fastest among the structured approaches. More precisely, when its components are run in parallel, in conducted experiments it was approximately 4.5 times faster than SE and WSE, while being ~140 times faster (on average) than a standard GCRF. One reason for this is that, over the whole time span, *State 1* was more frequently chosen than *State 2* and *State 3* (see Fig. 4). Another reason is that the ultimate scenario of passing through all 3 states to choose the best one (that is, indirectly selected) is the case in only 4 timesteps out of 12. What is most surprising, is that a model that skips the entire or some parts of the training process so frequently while operating solely on partial views can still capture the original network structure more accurately and in a more efficient manner than a graphical model that takes the whole network structure into account. According to this, the performance of AST-SE on both synthetic and gene expression data are consistent.

Table 4. Testing MSE and execution time (in seconds), averaged over all timesteps.

Model	MSE	Execution time
LR	0.38 ± 0.19	**0.10 ± 0.03**
GCRF	0.39 ± 0.21	9082.71 ± 1898.43
SE	0.39 ± 0.21	297.29 ± 19.42
WSE	0.35 ± 0.19	309.32 ± 19.44
AST-SE	**0.23 ± 0.07**	64.00 ± 45.73

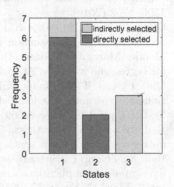

Fig. 4. Selected AST-SE states.

6   Conclusion

In this study, we introduced AST-SE, a novel ensemble-based model for structured regression on temporal networks. This model extends the concept of ensemble learning in temporal environments by employing multiple GCRF models to capture different network substructures and combining them into a single composite ensemble in order to achieve greater predictive power. Changing between states, at each timestep, AST-SE is able to automatically detect changes occurring over time in the data distribution, and to adapt accordingly by partially, or even completely skipping the retraining process. According to the experimental results on both synthetic and real-world data, AST-SE achieves a significant reduction in execution time, while maintaining sufficient accuracy. Nevertheless, our future plans are directed towards developing even more intelligent and advanced methodologies for detecting changes in temporal data distributions.

Acknowledgments. This research was supported in part by DARPA grant No. FA9550-12-1-0406 negotiated by AFOSR, the National Science Foundation grants NSF-SES-1447670, NSF-IIS-1636772, Temple University Data Science Targeted Funding Program, NSF grant CNS-1625061, Pennsylvania Department of Health CURE grant and ONR/ONR Global (grant No. N62909-16-1-2222).

References

1. Andonova, S., Elisseeff, A., Evgeniou, T., Pontil, M.: A simple algorithm for learning stable machines. In: ECAI, pp. 513–517. IOS Press (2002)
2. Breiman, L.: Bagging predictors. Mach. Learn. **24**(2), 123–140 (1996)
3. Dietterich, T.G.: Ensemble methods in machine learning. In: Kittler, J., Roli, F. (eds.) MCS 2000. LNCS, vol. 1857, pp. 1–15. Springer, Heidelberg (2000). https://doi.org/10.1007/3-540-45014-9_1
4. Freund, Y., Schapire, R.E.: A desicion-theoretic generalization of on-line learning and an application to boosting. In: Vitányi, P. (ed.) EuroCOLT 1995. LNCS, vol. 904, pp. 23–37. Springer, Heidelberg (1995). https://doi.org/10.1007/3-540-59119-2_166
5. Glass, J., Ghalwash, M.F., Vukicevic, M., Obradovic, Z.: Extending the modelling capacity of Gaussian conditional random fields while learning faster. In: AAAI, pp. 1596–1602 (2016)
6. Gligorijevic, D., Stojanovic, J., Obradovic, Z.: Uncertainty propagation in long-term structured regression on evolving networks. In: AAAI (2016)
7. Mendes-Moreira, J., Soares, C., Jorge, A.M., Sousa, J.F.D.: Ensemble approaches for regression: a survey. ACM Comput. Surv. (CSUR) **45**(1), 10 (2012)
8. Qin, T., Liu, T.Y., Zhang, X.D., Wang, D.S., Li, H.: Global ranking using continuous conditional random fields. In: Advances in Neural Information Processing Systems, pp. 1281–1288 (2009)
9. Radosavljevic, V., Vucetic, S., Obradovic, Z.: Continuous conditional random fields for regression in remote sensing. In: ECAI (2010)
10. Ren, Y., Zhang, L., Suganthan, P.N.: Ensemble classification and regression-recent developments, applications and future directions [review article]. IEEE Comput. Intell. Mag. **11**(1), 41–53 (2016)

11. Stojanovic, J., Gligorijevic, D., Obradovic, Z.: Modeling customer engagement from partial observations. In: CIKM, pp. 1403–1412 (2016)
12. Stojanovic, J., Jovanovic, M., Gligorijevic, D., Obradovic, Z.: Semi-supervised learning for structured regression on partially observed attributed graphs. In: SDM (2015)
13. Stojkovic, I., Jelisavcic, V., Milutinovic, V., Obradovic, Z.: Distance based modeling of interactions in structured regression. In: IJCAI (2016)
14. Stojkovic, I., Jelisavcic, V., Milutinovic, V., Obradovic, Z.: Fast sparse gaussian markov random fields learning based on cholesky factorization. In: IJCAI (2017)
15. Stojkovic, I., Obradovic, Z.: Predicting sepsis biomarker progression under therapy. In: IEEE CBMS (2017)
16. Vujicic, T., Glass, J., Zhou, F., Obradovic, Z.: Gaussian conditional random fields extended for directed graphs. Mach. Learn. **106**, 1–18 (2017)
17. Zaas, A.K., Chen, M., Varkey, J., Veldman, T., et al.: Gene expression signatures diagnose influenza and other symptomatic respiratory viral infections in humans. Cell Host & Microbe **6**(3), 207–217 (2009)
18. Zhou, F., Ghalwash, M., Obradovic, Z.: A fast structured regression for large networks. In: 2016 IEEE International Conference on Big Data, pp. 106–115 (2016)

ALADIN: A New Approach for Drug–Target Interaction Prediction

Krisztian Buza[1(✉)] and Ladislav Peska[2]

[1] Knowledge Discovery and Machine Learning,
Rheinische Friedrich-Wilhelms-Universität Bonn, Bonn, Germany
buza@iai.uni-bonn.de
[2] Faculty of Mathematics and Physics, Charles University, Prague, Czech Republic
peska@ksi.mff.cuni.cz
https://www.kdml.iai.uni-bonn.de/

Abstract. Due to its pharmaceutical applications, one of the most prominent machine learning challenges in bioinformatics is the prediction of drug–target interactions. State-of-the-art approaches are based on various techniques, such as matrix factorization, restricted Boltzmann machines, network-based inference and bipartite local models (BLM). In this paper, we extend BLM by the incorporation of a hubness-aware regression technique coupled with an enhanced representation of drugs and targets in a multi-modal similarity space. Additionally, we propose to build a projection-based ensemble. Our *Advanced Local Drug-Target Interaction Prediction* technique (ALADIN) is evaluated on publicly available real-world drug–target interaction datasets. The results show that our approach statistically significantly outperforms BLM-NII, a recent version of BLM, as well as NetLapRLS and WNN-GIP.
Code related to this chapter is available at:
https://github.com/lpeska/ALADIN
Data related to this chapter are available at:
https://zenodo.org/record/556337#.WPiAzIVOIdV
Supplementary material is available at:
http://www.biointelligence.hu/dti/

Keywords: Drug–target interaction prediction
Bipartite local models · ALADIN

1 Introduction

Prediction of drug–target interactions is one of the most prominent machine learning applications in the pharmaceutical industry, the importance of which is underlined by the fact that both time and expenditure related to drug development are enormous: on average, it costs $\approx\$1.8$ billion and takes more than 10 years to bring a new drug to the market [17]. Drug–target interaction prediction (DTI) techniques promise to reduce the aforementioned cost and time, and to support drug repositioning [40], i.e., the use of an existing medicine to treat a disease that has not been treated with that drug yet.

© Springer International Publishing AG 2017
M. Ceci et al. (Eds.): ECML PKDD 2017, Part II, LNAI 10535, pp. 322–337, 2017.
https://doi.org/10.1007/978-3-319-71246-8_20

Computational methods for DTI include approaches based on molecular docking simulations [9,15] and ligand chemistry [21,25]. Furthermore, text mining techniques have been proposed to identify biomedical entities and relations between them [7,13,28,42]. However, a serious limitation of docking-based approaches is that they require information about the three-dimensional structure of candidate drugs and targets which is often not available, especially for G-protein coupled receptors (GPCRs) and ion channels. Additionally, the performance of ligand-based approaches is known to decrease if only few ligands are known. Therefore, machine learning techniques have been proposed for DTI [11,19,39]. Recent approaches are based on matrix factorization [5,14,41], support vector regression [34,35], restricted Boltzmann machines [37], network-based inference [8,10], decision lists [30] and bipartite local models (BLM) [4] with semi-supervised prediction [38], improved kernels [22] and the incorporation of neighbor-based interaction-profile inferring [23].

Real-world datasets in biology, chemistry and medicine [1], including drug–target interaction networks, have been shown to contain hubs, i.e., vertices that are connected to surprisingly many other vertices. For example, in the *Enzyme* dataset (described in Sect. 5.1), the vast majority of targets have less then 5 interactions, while some of the targets are very popular: each of 30 most popular targets interacts with 20 drugs at least. Despite such observations, none of the aforementioned variants of BLM took the presence of hubs into account. Furthermore, the presence of hubs has been observed in nearest neighbor graphs [29], which lead to the development of hubness-aware classifiers [33] and regression techniques [6]. Although hubness-aware techniques are among the most promising recent machine learning approaches, their potential to enhance drug–target interaction prediction methods has not been exploited yet.

In this paper, we extend BLM by the incorporation of a hubness-aware regression approach. Additionally, we propose an enhanced representation of drugs and targets in a multi-modal similarity space and build a projection-based ensemble. We call the resulting approach $\underline{A}dvanced$ $\underline{Loc}al$ $\underline{D}rug$-$Target$ $\underline{I}nteraction$ $Prediction$, or ALADIN for short. In order to assist reproducibility of our work, we perform experiments on publicly available real-world drug–target interaction datasets. The results show that our approach outperforms BLM-NII [23], a recent version of BLM, and two other drug–target prediction techniques.

The rest of this paper is organized as follows: in Sect. 2, we define the drug–target interaction prediction problem, this is followed by the review of BLM and hubness-aware regression in Sect. 3. We describe our approach, ALADIN, in Sect. 4 and present the results of experimental evaluation in Sect. 5. Finally, we conclude in Sect. 6.

2   Basic Notation and Problem Formulation

First, we define the Drug–Target Interaction Prediction problem. We are given a set $\mathcal{D} = \{d_1, \ldots, d_n\}$ of n drugs, a set $\mathcal{T} = \{t_1, \ldots, t_m\}$ of m pharmaceutical targets, an $n \times n$ drug similarity matrix $\mathcal{S}^D$, an $m \times m$ target similarity matrix

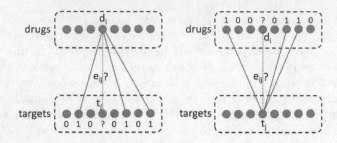

Fig. 1. Two independent predictions of Bipartite Local Models.

$\mathcal{S}^T$ and an $n \times m$ interaction matrix $\mathcal{M}$. Each entry $s_{i,j}^D$ of $\mathcal{S}^D$ (and $s_{i,j}^T$ of $\mathcal{S}^T$, resp.) describe the similarity between drugs d_i and d_j (targets t_i and t_j). Each entry $m_{i,j}$ of $\mathcal{M}$ denotes if drug d_i and target t_j are known to interact:

$$m_{i,j} = \begin{cases} 1 & \text{if there is a known interaction between } d_i \text{ and } t_j \\ 0 & \text{otherwise.} \end{cases}$$

This formulation is in accordance with the usual setting in which only positive information is available: in case if $m_{i,j} = 0$, the corresponding drug d_i and target t_j *may* or *may not* interact, therefore, we call $u_{i,j} = (d_i, t_j)$ an *unknown* pair. The task is to predict the likelihood of interaction for each unknown pair.

At the first glance, the above DTI problem seem to be similar to the problems considered in the recommender systems community. Note, however, that most recommender techniques consider only the interactions ("ratings") because even a few ratings are thought to be more informative than metadata, such as users' similarity based on their demographic information [27]. In contrast, drug–drug and target–target similarities play an essential role in DTI.

3   Background

In this section, we review the BLM approach and hubness-aware error correction for nearest neighbor regression.

3.1   Bipartite Local Model

BLM considers DTI as a link prediction problem in bipartite graphs [4]. The vertices in one of the vertex classes correspond to drugs, whereas the vertices in the other vertex class correspond to targets. There is an edge between drug d_i and target t_j if and only if $m_{i,j} = 1$.

The likelihood of unknown interactions is predicted as follows: we consider an unknown pair $u_{i,j} = (d_i, t_j)$ and calculate the likelihood of interaction as the aggregate of two independent predictions.

The first prediction (Fig. 1, left panel) is based on the relations between d_i and the targets. Each target t_k (except t_j) is labeled as "1" or "0" depending on

$m_{i,k}$. Then a model is trained to distinguish "1"-labeled and "0"-labeled targets. Subsequently, this model is applied to predict the likelihood of interaction for the unknown pair $u_{i,j}$. This first prediction is denoted by $\hat{y}'_{i,j}$.

The second prediction, $\hat{y}''_{i,j}$, is obtained in a similar fashion, but instead of considering the interactions of drug d_i and labeling the targets, the interactions of target t_j are considered and drugs are labeled (Fig. 1, right panel). The models that make the first and second predictions are called *local models*.

In order to obtain the final prediction of the BLM, we average the predictions of the aforementioned local models:

$$\hat{y}_{i,j} = \frac{\hat{y}'_{i,j} + \hat{y}''_{i,j}}{2} \tag{1}$$

Note that instead of averaging, other aggregation functions, such as minimum or maximum are possible as well.

BLM is a generic framework in which various regressors or classifiers can be used as local models. Bleakley and Yamanishi [4] used support vector machines with a domain-specific kernel. In contrast, we propose to use a hubness-aware regression technique, ECkNN, which is described next.

3.2 ECkNN: k-Nearest Neighbor Regression with Error Correction

In the last decades, various regression schemes have been introduced, such as linear and polynomial regression, support vector regression, neural networks, etc. One of the most popular regression techniques is based on k-nearest neighbors: when predicting the numeric label on an instance x with k-nearest neighbor regression, the k-nearest neighbors of x (i.e., k instances that are most similar to x) are determined and the average of their labels is calculated as the predicted label of x. In our case, instances may either correspond to drugs or targets, depending on whether the first or the second BLM-prediction is calculated.

While being intuitive and simple to implement, k-nearest neighbor regression is well-understood from the point of view of theory as well, see e.g. [3], and the references therein for an overview of the most important theoretical results. The theoretical results are also justified by empirical studies: for example, in their recent paper, Stensbo-Smidt et al. found that nearest neighbor regression outperforms model-based prediction of star formation rates [31], while Hu et al. showed that a model based on k-nearest neighbor regression is able to estimate the capacity of lithium-ion batteries [18].

Despite all of the aforementioned advantages of k-nearest neighbor regression, one of its recently explored shortcomings is its suboptimal performance in the presence of bad hubs. Intuitively, bad hubs are instances that appear as nearest neighbors of many other instances, but have substantially different labels from those instances. The presence of bad hubs has been shown to be related to the intrinsic dimensionality of the data. This means, roughly speaking, that bad hubs are expected in complex data, such as drug–target interaction data. For a more detailed discussion, we refer to [6].

In order to alleviate the detrimental effect of bad hubs, in [6] we proposed an error correction technique which is reviewed next. We define the *corrected label* $y_c(x)$ of a training instance x as

$$y_c(x) = \begin{cases} \frac{1}{|\mathcal{R}_x|} \sum\limits_{x_i \in \mathcal{R}_x} y(x_i) & \text{if } |\mathcal{R}_x| \geq 1 \\ y(x), & \text{otherwise} \end{cases}, \tag{2}$$

where $y(x_i)$ denotes the original (i.e., uncorrected) label of instance x_i, and $\mathcal{R}_x$ is the set of "reverse neighbors", i.e. the set of training instances that have x as one of their k-nearest neighbors:

$$\mathcal{R}_x = \{\forall x_i | x \in \mathcal{N}(x_i)\} \tag{3}$$

where $\mathcal{N}(x_i)$ denotes the set of k-nearest neighbors of x_i.

In order to make predictions, k-nearest neighbor regression with error correction (ECkNN) uses the corrected labels. Given a "new" (unlabeled) instance x', its predicted label $\hat{y}(x')$ is calculated as follows:

$$\hat{y}(x') = \frac{1}{k} \sum\limits_{x_i \in \mathcal{N}(x')} y_c(x_i). \tag{4}$$

4   Our Approach

Next, we present ALADIN, our Advanced Local Drug-Target Interaction Prediction approach. Following subsections describe the components of ALADIN.

4.1   Similarity-Based Representation

The given drug–drug similarities allow us to represent drugs in the similarity space: in particular, drug d_i is represented by the vector $(s_{i,1}^D, \ldots s_{i,n}^D)$. Given the target similarity matrices, targets may be represented in an analogous way, i.e., using their similarities to all the targets.

Additionally to the given drug–drug and target–target similarities, we propose to compute drug–drug and target–target similarities based on the known interactions (i.e., interactions in the training set). In particular, using the interaction matrix, we calculate the Jaccard-similarity between drugs as well as between targets. Thus the enhanced similarity-based representation of a drug (or target, respectively) consists of its chemical (genetic) similarity to all the drugs (targets) and its interaction-based similarity to all the drugs (targets). This is illustrated in Fig. 2.

4.2   Projection-Based Ensemble

We propose to build a projection-based ensemble of BLMs as follows. Given the enhanced similarity-based representation of drugs and targets, we select a

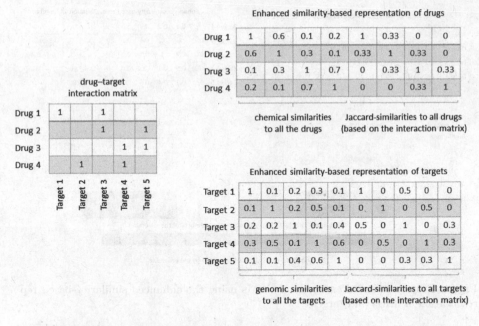

Fig. 2. Illustration of enhanced similarity-based representation of drugs and targets

random subset of features and use only the selected features when training the local models (ECkNN) and making predictions. Denoting the size of the set of selected features by F_D and F_T (for drugs and targets, respectively), the above procedure first projects drugs into F_D-dimensional, and targets into an F_T-dimensional subspace. Subsequently, these lower dimensional representations are used with the prediction models.

The above process of random selection of features and making predictions using the resulting lower-dimensional representation is repeated N-times. This results in an ensemble of N prediction models. As each member of the ensemble is constructed in the same way, their expected prediction accuracies will be similar, therefore, we propose to average the predictions of the members of the ensemble. Thus the final output of the ensemble is:

$$\hat{y}_{i,j} = \frac{1}{N} \sum_{l=1}^{N} \hat{y}_{i,j}^{(l)} \tag{5}$$

where $\hat{y}_{i,j}^{(l)}$ is the prediction of the l-th BLM for the unknown pair $u_{i,j}$.

The projection-based ensemble is illustrated in Fig. 3 for $N = 2$ base prediction models with $F_D = F_T = 3$ features selected from the enhanced similarity-based representation.

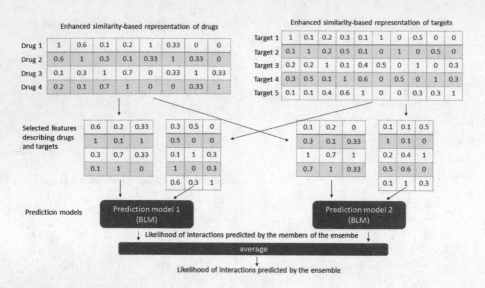

Fig. 3. Projection-based ensemble of BLMs using the enhanced similarity-based representation of drugs and targets.

4.3 Prediction for New Drugs and Targets

One of the shortcomings of the BLM approach is that it does not handle the case of new drugs/targets. With new drug (target, resp.), we mean a drug d (target t) that does not have any known interaction in the training data. In such cases, BLM labels all targets (drugs) as "0", consequently, no reasonable local model can be learned. In order to alleviate this problem, we use the weighted profile [39] approach to obtain predictions for new drugs/targets.

Given a new drug d_i, and a target t_j, we predict the likelihood of the interaction between d_i and t_j as follows:

$$\hat{y}'_{i,j} = \frac{\sum\limits_{d_k \in \mathcal{D} \setminus \{d_i\}} m_{k,j} \mathcal{S}^D_{i,k}}{\sum\limits_{d_k \in \mathcal{D} \setminus \{d_i\}} \mathcal{S}^D_{i,k}}. \tag{6}$$

The intuition behind Eq. (6) is that similar drugs are likely to behave similarly in terms of their interaction with a given target. Therefore, drugs are weighed according to their similarity to the new drug d_i and we calculate the weighted average of the known interactions of other drugs with the same target.

The case of new targets is analogous. Given a new target t_j and a drug d_i, the weighted profile approach can be used to calculate the prediction for the likelihood of the interaction between d_i and t_j as follows:

$$\hat{y}''_{i,j} = \frac{\sum\limits_{t_k \in \mathcal{T} \setminus \{t_j\}} m_{i,k} \mathcal{S}^T_{j,k}}{\sum\limits_{t_k \in \mathcal{T} \setminus \{t_j\}} \mathcal{S}^T_{j,k}}. \tag{7}$$

Algorithm 1. Advanced Local Drug-Target Interaction Prediction (ALADIN)

Require: Drug–Target interaction matrix I, Drug–drug similarity matrix S^D, Target–target similarity matrix S^T, number of nearest neighbors k, ensemble size N, number of selected features F_D, F_T

Ensure: Likelihood of drug–target interactions

1: $D \leftarrow$ enhanced similarity-based representations of drugs
2: $T \leftarrow$ enhanced similarity-based representations of targets
3: **for** $l = 1 \ldots N$ **do**
4:     $D' \leftarrow$ random subset of D with F_D features
5:     $T' \leftarrow$ random subset of T with F_T features
6:     Predict interaction scores with BLM using ECkNN as local model and
      D' and T' as the representation of drugs and targets.
      (Use the weighted profile approach instead of BLM in case of new
      drugs/targets.)
7: **end for**
8: Average the predictions made in each execution of the loop

---

Although the weighted profile approach is more general than BLM, in the sense that it can be used for new drugs/targets as well, the predictions of the weighted profile approach are less accurate than the predictions of BLM. Therefore, we use the weighted profile approach instead of BLM *only* in case of *new* drugs and targets. We summarize the proposed approach in Algorithm 1.

5 Experimental Evaluation

In order to assist reproducibility of our work, we evaluated our approach on publicly available real-word drug–target interaction data. Next we describe the data and the experimental protocol in detail. This is followed by the discussion of our experimental results.[1]

5.1 Experimental Settings

Datasets. We performed experiments on five drug–target interaction datasets (Table 1), namely Enzyme, Ion Channel, G-protein coupled receptors (GPCR), Nuclear Receptors (NR), and Kinase.[2] These datasets have been used in various studies previously, see e.g. [4,12,14,24,38,39].

The first four datasets contain binary interaction matrices between drugs and targets, each entry of which indicates whether the interaction between the corresponding drug and target is known. In contrast, Kinase contains continuous values of binding affinity for all drug–target pairs of the dataset. In order to produce a binary interaction matrix, we used the same cutoff threshold as Pahikkala et al. [24].

---

[1] See http://www.biointelligence.hu/dti for further results.
[2] The datasets are available at https://zenodo.org/record/556337#.WPiAzIVOIdV.

Table 1. Number of drugs, targets and interactions in the datasets used in our study.

Dataset	# Drugs	# Targets	# Interactions
Enzyme	445	664	2926
Ion channels	210	204	1476
G-protein coupled receptors (GPCR)	223	95	635
Nuclear Receptors (NR)	54	26	90
Kinase	68	442	1527

Additionally, each dataset contains a drug–drug similarity matrix and a target–target similarity matrix. In case of the Enzyme, Ion Channel, GPCR and NR datasets, chemical structure similarities between drugs were computed using the SIMCOMP algorithm [16], while the Kinase dataset contains 2D Tanimoto coefficients. Similarities between targets were determined by the Smith-Waterman algorithm, see [12, 39] for details.

Evaluation Protocol. Although leave-one-out cross-validation is popular in the DTI literature [4, 22, 23], in their recent study, Pahikkala et al. [24] argue that it may lead to overoptimistic results. Thus, we performed experiments according to the interaction-based 5×5-fold cross-validation protocol (in each round of the cross-validation, the test set contains one fifth of all the drug–target pairs).

Evaluation Metrics. We evaluated the predictions both in terms of Area Under ROC Curve (AUC) and Area Under Precision-Recall Curve (AUPR). AUC and AUPR values were calculated in each round of the cross-validation. We report averaged values. Additionally, we performed paired t-test at significance level $p = 0.01$ in order to judge if the observed differences are statistically significant.

Baselines. We compared our approach, ALADIN, with other drug–target interaction prediction techniques, such as BLM-NII, NetLapRLS and WNN-GIP. BLM-NII is a recent version of BLM that extends BLM with "neighbor-based interaction-profile inferring" [23]. NetLapRLS stands for "net Laplacian regularized least squares" [38], while WNN-GIP is a combination of weighted nearest neighbor and Gaussian interaction profile kernels [36].

Parameter Settings. We set the number of base prediction models (N) to 25 for ALADIN.[3] Other hyperparameters of ALADIN, whenever not indicated

---

[3] In our initial experiments, we observed that increasing the number of base models results in asymptotically increasing performance. For example, we obtained AUPR of 0.835, 0.867 and 0.871 with 5, 25 and 100 base models on the Ion Channel dataset. We made similar observations on the other datasets both in terms of AUC and AUPR.

otherwise, were learned via grid-search in internal 5-fold cross-validation on the training data. In particular: the number of nearest neighbors for the local model, ECkNN, and the number of selected features, were chosen from {3, 5, 7} and {10, 20, 50} respectively.

Hyperparameters of the baselines were learned similarly. In particular: for BLM-NII, the `max` function was used to generate final predictions and the weight α for the combination of structural and collaborative similarities was chosen from {0.0, 0.1, ..., 1.0}. In WNN-GIP, the decay hyperparameter T was chosen from {0.1, 0.2, ..., 1.0} and the weight α for combination of structural and collaborative similarities was chosen from {0.0, 0.1, ..., 1.0}. The hyperparameters[4] of NetLapRLS, were chosen from {$10^{-6}, 10^{-5}, ..., 10^{2}$}.

Implementation. We implemented our approach, ALADIN, in Python.[5] We used the ECkNN implementation from the publicly available PyHubs library[6] and methods from the NumPy machine learning library for the calculation of AUC and AUPR. We used implementations of NetLapRLS, BLM-NII and WNN-GIP from the publicly available PyDTI software library.[7]

5.2 Experimental Results

Our results are shown in Figs. 4 and 5. The symbols $+/-$ denote if the differences between the best-performing approach and other methods are statistically significant ($+$) or not ($-$).

As one can see, our approach, ALADIN outperformed its competitors, Net-LapRLS, BLM-NII and WNN-GIP, on the Enzyme, Ion Channel, GPRC and Kinase datasets both in terms of AUC and AUPR. In the vast majority of the cases, the difference is statistically significant. In case of the NR dataset, the difference between ALADIN, BLM-NII and WNN-GIP is not significant. Note, however, that NR is an exceptionally small dataset, therefore, the results obtained on NR are likely to be less stable compared to other datasets.

Additionally, we examined the contribution of hubness-aware error correction: in particular, we run ALADIN with simple kNN regression instead of ECkNN. We found that *ALADIN with ECkNN* systematically outperformed *ALADIN with kNN* on all the examined datasets. The difference was statistically significant in most of the cases. In terms of AUC, we observed the largest difference on the Kinase dataset (0.93 versus 0.90), whereas in terms of AUPR, the largest difference was observed on the Enzyme dataset (0.83 versus 0.73). These results indicate that error correction is essential for accurate predictions.[8]

Therefore, using $N = 25$ base models seems to be a fair compromise between runtime and prediction quality.

[4] $\beta = \beta_{drug} = \beta_{target}$ and $\gamma = \gamma_{drug} = \gamma_{target}$.

[5] See https://github.com/lpeska/ALADIN for our codes.

[6] https://sourceforge.net/projects/pyhubs/.

[7] https://github.com/stephenliu0423/PyDTI.

[8] These results are in accordance with our further observations: considering the input data of the local models, the skewness of the distribution of bad k-nearest neighbor

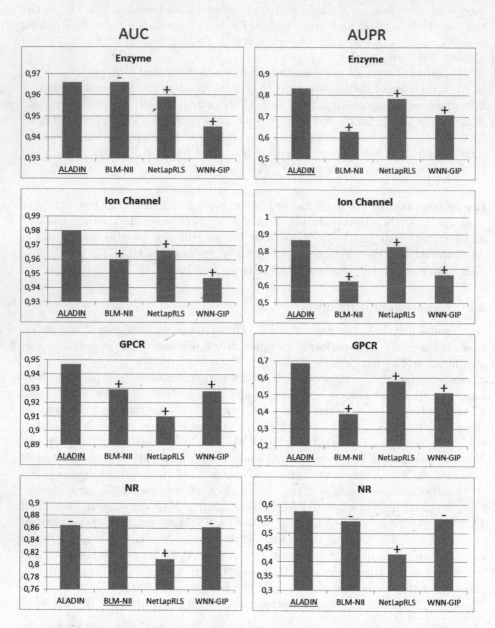

Fig. 4. Experimental results: the performance of ALADIN and its competitors in terms of AUC (left) and AUPR (right) on the Enzyme, Ion Channel, GPCR and NR datasets. The best-performing method is underlined. The symbols $+/-$ denote if the differences between the best-performing approach and other methods are statistically significant ($+$) or not ($-$).

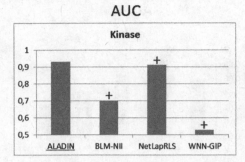
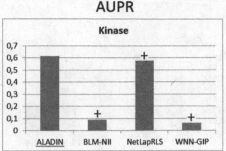

Fig. 5. Experimental results: the performance of ALADIN and its competitors in terms of AUC (left) and AUPR (right) on the Kinase dataset. The best-performing method is <u>underlined</u>. The symbols $+/-$ denote if the differences between the best-performing approach and other methods are statistically significant $(+)$ or not $(-)$.

Furthermore, we examined how ALADIN's performance depend on k, the number of nearest neighbors in ECkNN. As one can see in Fig. 6, high performance is maintained for various k values and $k = 3$ seems to result in good results both in terms of AUC and AUPR.

5.3 Application for the Prediction of New Interactions

Next, we illustrate that, besides achieving high accuracy in terms of AUC and AUPR, the predictions of ALADIN may be relevant for pharmaceutical applications as well. We begin this discussion by noting that the drug–target interactions contained in the Enzyme, Ion Channel, GPCR and NR datasets were extracted from the Kyoto Encyclopedia of Genes and Genomes[9] (KEGG) *several years ago* and, in order to allow for comparison of prediction techniques, they have been kept unchanged. However, in the mean time, additional drug–target interactions have been validated chemically and the results have been uploaded to databases, such as KEGG, DrugBank[10] or Matador[11].

Therefore, in order to demonstrate that our approach is able to predict new interactions, we trained ALADIN and its competitors, BLM-NII, NetLapRLS and WNN-GIP using all the interactions of the original datasets, and ranked the non-interacting drug–target pairs of the original datasets according to their predicted interaction scores. For simplicity, we use the term *predicted new interactions* for the top-ranked 20 drug–target pairs. We say that a predicted new interaction is *validated* if it is included in the current version of KEGG, Drug-Bank or Matador.

---

occurrences (with $k = 3$), which is often used to quantify the presence of bad hubs [33], is remarkably high, between 1.61 and 11.13.

[9] http://www.kegg.jp/.

[10] https://www.drugbank.ca/.

[11] http://matador.embl.de/.

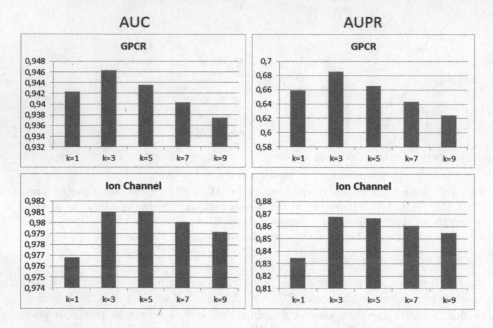

Fig. 6. ALADIN's performance in case of various k values in ECkNN.

In terms of the number of validated interactions, ALADIN had the best over-all performance. For example, on the Ion Channel and NR datasets, ALADIN was able to predict 12 and 8 validated interactions, whereas none of its competi-tors was able to predict more than 6 validated interactions on these datasets.

Most notably, numerous validated interactions were only predicted by our approach, for example, on the Enzyme dataset, the interactions between *Ibupro-fen* (D00126) and *arachidonate 15-lipoxygenase* (hsa:246) and its second type (hsa:247); as well as the interaction between *Phentermine* (D05458) and *mono-amine oxidase A* (hsa:4128); and the interaction between *Dyphylline* (D00691) and *phosphodiesterase 7A* (hsa:5150). On the GPCR dataset, only ALADIN was able to predict the validated interaction between *Theophylline sodium acetate* (D01712) and *adenosine A2b receptor* (hsa:136), as well as the interaction between *Loxapine* (D02340) and *dopamine receptor D1* (hsa:1812).

6   Conclusions and Outlook

In this paper, we considered the drug–target interaction prediction problem which has important applications in understanding the mechanisms of how drugs effect, drug repositioning and prediction of adverse effects. We proposed an extension of BLM, one of the most prominent DTI models. In particular, we proposed the ALADIN approach which represents drugs and targets in a multi-modal similarity space, uses ECkNN, a hubness-aware regression approach as local model in BLM and builds a projection-based ensemble.

We performed experiments on widely-used publicly-available datasets, the results of which show that our approach is superior to BLM-NII, NetLapRLS and WNN-GIP. We also demonstrated that our approach is able to predict chemically validated new drug–target interactions.

While DTI is an essential task, we point out that ALADIN may be adapted for the prediction of interactions between other biomedical entities, such as protein–RNA interactions [32] or protein–protein interactions [2].

Furthermore, we believe that our approach may motivate new recommender systems techniques as well. Although it was shown that only a few ratings per user may be more relevant than content-based metadata [27], we argue that the continuous flow of new users causes ongoing cold start problem [20,26] in many cases, such as small e-commerce enterprises. This indicates that hybrid prediction models incorporating both relevance feedback and metadata may be desirable. Methods like ALADIN can be applied in such domains, e.g., as a part of an alternating hybrid approach, where users with sufficient feedback receive purely collaborative recommendations.

Acknowledgment. Ladislav Peska was supported by the Charles University grant P46.

References

1. Barabási, A.L., Gulbahce, N., Loscalzo, J.: Network medicine: a network-based approach to human disease. Nat. Rev. Genet. **12**(1), 56–68 (2011)
2. Besemann, C., Denton, A., Yekkirala, A.: Differential association rule mining for the study of protein-protein interaction networks. In: 4th International Conference on Data Mining in Bioinformatics, pp. 72–80. Springer, Heidelberg (2004). https://dl.acm.org/citation.cfm?id=3000590
3. Biau, G., Cérou, F., Guyader, A.: On the rate of convergence of the bagged nearest neighbor estimate. J. Mach. Learn. Res. **11**, 687–712 (2010)
4. Bleakley, K., Yamanishi, Y.: Supervised prediction of drug-target interactions using bipartite local models. Bioinformatics **25**(18), 2397–2403 (2009)
5. Bolgar, B., Antal, P.: Bayesian matrix factorization with non-random missing data using informative Gaussian process priors and soft evidences. J. Mach. Learn. Res. **52**, 25–36 (2016)
6. Buza, K., Nanopoulos, A., Nagy, G.: Nearest neighbor regression in the presence of bad hubs. Knowl.-Based Syst. **86**, 250–260 (2015)
7. Cellier, P., Charnois, T., Plantevit, M.: Sequential patterns to discover and characterise biological relations. In: Gelbukh, A. (ed.) CICLing 2010. LNCS, vol. 6008, pp. 537–548. Springer, Heidelberg (2010). https://doi.org/10.1007/978-3-642-12116-6_46
8. Chen, X., Liu, M.X., Yan, G.Y.: Drug-target interaction prediction by random walk on the heterogeneous network. Mol. BioSyst. **8**(7), 1970–1978 (2012)
9. Cheng, A.C., Coleman, R.G., Smyth, K.T., Cao, Q., Soulard, P., Caffrey, D.R., Salzberg, A.C., Huang, E.S.: Structure-based maximal affinity model predicts small-molecule druggability. Nat. Biotechnol. **25**(1), 71–75 (2007)

10. Cheng, F., Liu, C., Jiang, J., Lu, W., Li, W., Liu, G., Zhou, W., Huang, J., Tang, Y.: Prediction of drug-target interactions and drug repositioning via network-based inference. PLoS Comput. Biol. **8**(5), e1002503 (2012)

11. Davis, J., Santos Costa, V., Ray, S., Page, D.: An integrated approach to feature invention and model construction for drug activity prediction. In: Proceedings of the 24th International Conference on Machine Learning, pp. 217–224 (2007)

12. Davis, M.I., Hunt, J.P., Herrgard, S., Ciceri, P., Wodicka, L.M., Pallares, G., Hocker, M., Treiber, D.K., Zarrinkar, P.P.: Comprehensive analysis of kinase inhibitor selectivity. Nat. Biotechnol. **29**(11), 1046–1051 (2011)

13. Fayruzov, T., De Cock, M., Cornelis, C., Hoste, V.: Linguistic feature analysis for protein interaction extraction. BMC Bioinform. **10**(1), 374 (2009)

14. Gönen, M.: Predicting drug-target interactions from chemical and genomic kernels using Bayesian matrix factorization. Bioinformatics **28**(18), 2304–2310 (2012)

15. Halperin, I., Ma, B., Wolfson, H., Nussinov, R.: Principles of docking: an overview of search algorithms and a guide to scoring functions. Proteins: Struct. Func. Bioinform. **47**(4), 409–443 (2002)

16. Hattori, M., Okuno, Y., Goto, S., Kanehisa, M.: Development of a chemical structure comparison method for integrated analysis of chemical and genomic information in the metabolic pathways. J. Am. Chem. Soc. **125**(39), 11853–11865 (2003)

17. Morgan, S., Grootendorst, P., Lexchin, J., Cunningham, C., Greyson, D.: The cost of drug development: a systematic review. Health Policy **100**(1), 4–17 (2011)

18. Hu, C., Jain, G., Zhang, P., Schmidt, C., Gomadam, P., Gorka, T.: Data-driven method based on particle swarm optimization and k-nearest neighbor regression for estimating capacity of lithium-ion battery. Appl. Energy **129**, 49–55 (2014)

19. Jamali, A.A., Ferdousi, R., Razzaghi, S., Li, J., Safdari, R., Ebrahimie, E.: Drugminer: comparative analysis of machine learning algorithms for prediction of potential druggable proteins. Drug Discov. Today **21**(5), 718–724 (2016)

20. Kaminskas, M., Bridge, D., Foping, F., Roche, D.: Product-seeded and basket-seeded recommendations for small-scale retailers. J. Data Semant. **6**, 1–12 (2016). https://link.springer.com/article/10.1007/s13740-016-0058-3

21. Keiser, M.J., Roth, B.L., Armbruster, B.N., Ernsberger, P., Irwin, J.J., Shoichet, B.K.: Relating protein pharmacology by ligand chemistry. Nat. Biotechnol. **25**(2), 197–206 (2007)

22. van Laarhoven, T., Nabuurs, S.B., Marchiori, E.: Gaussian interaction profile kernels for predicting drug-target interaction. Bioinformatics **27**(21), 3036–3043 (2011)

23. Mei, J.P., Kwoh, C.K., Yang, P., Li, X.L., Zheng, J.: Drug-target interaction prediction by learning from local information and neighbors. Bioinformatics **29**(2), 238–245 (2013)

24. Pahikkala, T., Airola, A., Pietilä, S., Shakyawar, S., Szwajda, A., Tang, J., Aittokallio, T.: Toward more realistic drug-target interaction predictions. Briefings Bioinform. **16**(2), 325–337 (2015)

25. Pérot, S., Regad, L., Reynès, C., Spérandio, O., Miteva, M.A., Villoutreix, B.O., Camproux, A.C.: Insights into an original pocket-ligand pair classification: a promising tool for ligand profile prediction. PloS One **8**(6), e63730 (2013)

26. Peska, L., Vojtas, P.: Recommending for disloyal customers with low consumption rate. In: Geffert, V., Preneel, B., Rovan, B., Štuller, J., Tjoa, A.M. (eds.) SOFSEM 2014. LNCS, vol. 8327, pp. 455–465. Springer, Cham (2014). https://doi.org/10.1007/978-3-319-04298-5_40

27. Piląszy, I., Tikk, D.: Recommending new movies: even a few ratings are more valuable than metadata. In: 3rd ACM Conference on Recommender Systems, pp. 93–100 (2009)
28. Plantevit, M., Charnois, T., Klema, J., Rigotti, C., Crémilleux, B.: Combining sequence and itemset mining to discover named entities in biomedical texts: a new type of pattern. Int. J. Data Min. Model. Manag. 1(2), 119–148 (2009)
29. Radovanović, M., Nanopoulos, A., Ivanović, M.: Hubs in space: popular nearest neighbors in high-dimensional data. J. Mach. Learn. Res. 11, 2487–2531 (2010)
30. Sönströd, C., Johansson, U., Norinder, U., Boström, H.: Comprehensible models for predicting molecular interaction with heart-regulating genes. In: 7th IEEE International Conference on Machine Learning and Applications, pp. 559–564 (2008)
31. Stensbo-Smidt, K., Igel, C., Zirm, A., Pedersen, K.S.: Nearest neighbour regression outperforms model-based prediction of specific star formation rate. In: IEEE International Conference on Big Data, pp. 141–144 (2013)
32. Stražar, M., Žitnik, M., Zupan, B., Ule, J., Curk, T.: Orthogonal matrix factorization enables integrative analysis of multiple RNA binding proteins. Bioinformatics 32(10), 1527–1535 (2016)
33. Tomašev, N., Buza, K., Marussy, K., Kis, P.B.: Hubness-aware classification, instance selection and feature construction: survey and extensions to time-series. In: Stańczyk, U., Jain, L.C. (eds.) Feature Selection for Data and Pattern Recognition. SCI, vol. 584, pp. 231–262. Springer, Heidelberg (2015). https://doi.org/10.1007/978-3-662-45620-0_11
34. Ullrich, K., Kamp, M., Gärtner, T., Vogt, M., Wrobel, S.: Ligand-based virtual screening with co-regularised support vector regression. In: 16th IEEE International Conference on Data Mining Workshops, pp. 261–268 (2016)
35. Ullrich, K., Mack, J., Welke, P.: Ligand affinity prediction with multi-pattern kernels. In: Calders, T., Ceci, M., Malerba, D. (eds.) DS 2016. LNCS (LNAI), vol. 9956, pp. 474–489. Springer, Cham (2016). https://doi.org/10.1007/978-3-319-46307-0_30
36. van Laarhoven, T., Marchiori, E.: Predicting drug-target interactions for new drug compounds using a weighted nearest neighbor profile. PloS One 8(6), e66952 (2013)
37. Wang, Y., Zeng, J.: Predicting drug-target interactions using restricted Boltzmann machines. Bioinformatics 29(13), i126–i134 (2013)
38. Xia, Z., Wu, L.Y., Zhou, X., Wong, S.T.: Semi-supervised drug-protein interaction prediction from heterogeneous biological spaces. BMC Syst. Biol. 4(Suppl 2), S6 (2010)
39. Yamanishi, Y., Araki, M., Gutteridge, A., Honda, W., Kanehisa, M.: Prediction of drug-target interaction networks from the integration of chemical and genomic spaces. Bioinformatics 24(13), i232–i240 (2008)
40. Zhang, P., Agarwal, P., Obradovic, Z.: Computational drug repositioning by ranking and integrating multiple data sources. In: Blockeel, H., Kersting, K., Nijssen, S., Železný, F. (eds.) ECML PKDD 2013. LNCS (LNAI), vol. 8190, pp. 579–594. Springer, Heidelberg (2013). https://doi.org/10.1007/978-3-642-40994-3_37
41. Zheng, X., Ding, H., Mamitsuka, H., Zhu, S.: Collaborative matrix factorization with multiple similarities for predicting drug-target interactions. In: 19th ACM SIGKDD International Conference on Knowledge Discovery and Data Mining, pp. 1025–1033 (2013)
42. Zhu, S., Okuno, Y., Tsujimoto, G., Mamitsuka, H.: A probabilistic model for mining implicit chemical compound-gene relations from literature. Bioinformatics 21(Suppl. 2), ii245–ii251 (2005)

Co-Regularised Support Vector Regression

Katrin Ullrich[1(✉)], Michael Kamp[1,2], Thomas Gärtner[3], Martin Vogt[1,4], and Stefan Wrobel[1,2]

[1] University of Bonn, Bonn, Germany
ullrich@iai.uni-bonn.de, {michael.kamp,stefan.wrobel}@iais.fraunhofer.de,
martin.vogt@bit.uni-bonn.de
[2] Fraunhofer IAIS, Sankt Augustin, Germany
[3] University of Nottingham, Nottingham, UK
thomas.gaertner@nottingham.ac.uk
[4] B-IT, LIMES Program Unit, Bonn, Germany

Abstract. We consider a semi-supervised learning scenario for regression, where only few labelled examples, many unlabelled instances and different data representations (multiple views) are available. For this setting, we extend support vector regression with a co-regularisation term and obtain co-regularised support vector regression (CoSVR). In addition to labelled data, co-regularisation includes information from unlabelled examples by ensuring that models trained on different views make similar predictions. Ligand affinity prediction is an important real-world problem that fits into this scenario. The characterisation of the strength of protein-ligand bonds is a crucial step in the process of drug discovery and design. We introduce variants of the base CoSVR algorithm and discuss their theoretical and computational properties. For the CoSVR function class we provide a theoretical bound on the Rademacher complexity. Finally, we demonstrate the usefulness of CoSVR for the affinity prediction task and evaluate its performance empirically on different protein-ligand datasets. We show that CoSVR outperforms co-regularised least squares regression as well as existing state-of-the-art approaches for affinity prediction. Code and data related to this chapter are available at: https://doi.org/10.6084/m9.figshare.5427241.

Keywords: Regression · Kernel methods · Semi-supervised learning
Multiple views · Co-regularisation · Rademacher complexity
Ligand affinity prediction

1 Introduction

We investigate an algorithm from the intersection field of semi-supervised and multi-view learning. In semi-supervised learning the lack of a satisfactory number of labelled examples is compensated by the usage of many unlabelled instances from the respective feature space. Multi-view regression algorithms utilise different data representations to train models for a real-valued quantity. Ligand affinity prediction is an important learning task from chemoinformatics since

© Springer International Publishing AG 2017
M. Ceci et al. (Eds.): ECML PKDD 2017, Part II, LNAI 10535, pp. 338–354, 2017.
https://doi.org/10.1007/978-3-319-71246-8_21

many drugs act as protein ligands. It can be assigned to this learning scenario in a very natural way. The aim of affinity prediction is the determination of binding affinities for small molecular compounds—the ligands—with respect to a bigger protein using computational methods. Besides a few labelled protein-ligand pairs, millions of small compounds are gathered in molecular databases as ligand candidates. Many different data representations—the so-called molecular fingerprints or views—exist that can be used for learning. Affinity prediction and other applications suffer from little label information and the need to choose the most appropriate view for learning. To overcome these difficulties, we propose to apply an approach called co-regularised support vector regression. We are the first to investigate support vector regression with co-regularisation, i.e., a term penalising the deviation of predictions on unlabelled instances. We investigate two loss functions for the co-regularisation. In addition to variants of our multi-view algorithm with a reduced number of optimisation variables, we also derive a transformation into a single-view method. Furthermore, we prove upper bounds for the Rademacher complexity, which is important to restrict the capacity of the considered function class to fit random data. We will show that our proposed algorithm outperforms affinity prediction baselines.

The strength of a protein-compound binding interaction is characterised by the real-valued *binding affinity*. If it exceeds a certain limit, the small compound is called a *ligand* of the protein. Ligand-based classification models can be trained to distinguish between ligands and non-ligands of the considered protein (e.g., with support vector machines [6]). Since framing the biological reality in a classification setting represents a severe simplification of the biological reality, we want to predict the strength of binding using regression techniques from machine learning. Both classification and regression methods are also known under the name of *ligand-based virtual screening*. (In the context of regression, we will use the name *ligands* for all considered compounds.) Various approaches like *neural networks* [7] have been applied. However, *support vector regression* (SVR) is the state-of-the-art method for affinity prediction studies (e.g., [12]).

As mentioned above, in the context of affinity prediction one is typically faced with the following practical scenario: for a given protein, only few ligands with experimentally identified affinity values are available. In contrast, the number of synthesizable compounds gathered in molecular databases (such as ZINC, BindingDB, ChEMBL[1]) is huge which can be used as unlabelled instances for learning. Furthermore, different free or commercial vectorial representations or *molecular fingerprints* for compounds exist. Originally, each fingerprint was designed towards a certain learning purpose and, therefore, comprises a characteristic collection of physico-chemical or structural molecular features [1], for example, predefined key properties (Maccs fingerprint) or listed subgraph patterns (ECFP fingerprints).

The canonical way to deal with multiple fingerprints for virtual screening would be to extensively test and compare different fingerprints [6] or perform time-consuming preprocessing feature selection and recombination steps [8].

---

[1] zinc.docking.org, www.bindingdb.org, www.ebi.ac.uk/chembl.

Other attempts to utilise multiple views for one prediction task can be found in the literature. For example, Ullrich et al. [13] apply multiple kernel learning. However, none of these approaches include unlabelled compounds in the affinity prediction task. The semi-supervised *co-regularised least squares regression* (CoRLSR) algorithm of Brefeld et al. [4] has been shown to outperform single-view *regularised least squares regression* (RLSR) for UCI datasets[2]. Usually, SVR shows very good predictive results having a lower generalisation error compared to RLSR. Aside from that, SVR represents the state-of-the-art in affinity prediction (see above). For this reason, we define *co-regularised support vector regression* (CoSVR) as an ε-insensitive version of co-regularisation. In general, CoSVR—just like CoRLSR—can be applied on every regression task with multiple views on data as well as labelled and unlabelled examples. However, learning scenarios with high-dimensional sparse data representations and very few labelled examples—like the one for affinity prediction—could benefit from approaches using co-regularisation. In this case, unlabelled examples can contain information that could not be extracted from a few labelled examples because of the high dimension and sparsity of the data representation.

A view on data is a representation of its objects, e.g., with a particular choice of features in $\mathbb{R}^d$. We will see that feature mappings are closely related to the concept of *kernel functions*, for which reason we introduce CoSVR theoretically in the general framework of kernel methods. Within the research field of *multi-view learning*, CoSVR and CoRLSR can be assigned to the group of co-training style [16] approaches that simultaneously learn multiple predictors, each related to a view. Co-training style approaches enforce similar outcomes of multiple predictor functions for unlabelled examples, measured with respect to some loss function. In the case of co-regularisation for regression the empirical risks of multiple predictors (*labelled error*) plus an error term for unlabelled examples (*unlabelled error, co-regularisation*) are minimised.

The idea for mutual influence of multiple predictors appeared in the paper of Blum and Mitchell [2] on classification with co-training. Wang et al. [14] combined the technique of co-training with SVR with a technique different from co-regularisation. Analogous to CoSVR, CoRLSR is a semi-supervised and multi-view version of RLSR that requires the solution of a large system of equations [4]. A co-regularised version for support vector machine classification SVM-2K already appeared in the paper of Farquhar et al. [5], where the authors define a co-regularisation term via the ε-insensitive loss on labelled examples. It was shown by Sindhwani and Rosenberg [11] that co-regularised approaches applying the squared loss function for the unlabelled error can be transformed into a standard SVR optimisation with a particular fusion kernel. A bound on the empirical Rademacher complexity for co-regularised algorithms with Lipschitz continuous loss function for the labelled error and squared loss function for the unlabelled error was proven by Rosenberg and Bartlett [9].

A preliminary version of this paper was published at the *Data Mining in Biomedical Informatics and Healthcare* workshop held at ICDM 2016. There, we

---

[2] UCI machine learning repository, http://archive.ics.uci.edu/ml.

considered only the CoSVR special case ε-CoSVR and its variants with reduced numbers of variables (for the definitions consult Definitions 1 –3 below) focusing the application of ligand affinity prediction. The ℓ_2-CoSVR case (see below) with its theoretical properties (Lemmas $1(ii)$ - $3(ii)$, $6(ii)$) and practical evaluation, as well as the faster Σ-CoSVR (Sect. 3.3) variant are novel contributions in the present paper.

In the following section, we will present a short summary of kernels and multiple views, as well as important notation. We define CoSVR and variants of the base algorithm in Sect. 3. In particular, a Rademacher bound for CoSVR will be proven in Sect. 3.5. Subsequently, we provide a practical evaluation of CoSVR for ligand affinity prediction in Sect. 4 and conclude with a brief discussion in Sect. 5.

2   Kernels and Multiple Views

We consider an arbitrary instance space $\mathcal{X}$ and the real numbers as label space $\mathcal{Y}$. We want to learn a function f that predicts a real-valued characteristic of the elements of $\mathcal{X}$. Suppose for training purposes we have sets $X = \{x_1, \ldots, x_n\} \subset \mathcal{X}$ of labelled and $Z = \{z_1, \ldots, z_m\} \subset \mathcal{X}$ of unlabelled instances at our disposal, where typically $m \gg n$ holds true. With $\{y_1, \ldots, y_n\} \subset \mathcal{Y}$ we denote the respective labels of X. Furthermore, assume the data instances can be represented in M different ways. More formally, for $v \in \{1, \ldots, M\}$ there are functions $\Phi_v : \mathcal{X} \to \mathcal{H}_v$, where $\mathcal{H}_v$ is an appropriate inner product space. Given an instance $x \in \mathcal{X}$, we say that $\Phi_v(x)$ is the v-th view of x. If $\mathcal{H}_v$ equals $\mathbb{R}^d$ for some finite dimension d, the intuitive names $(v$-th$)$ feature mapping and feature space are used for Φ_v and $\mathcal{H}_v$, respectively. If in the more general case $\mathcal{H}_v$ is a Hilbert space, d can even be infinite (see below). For view v the predictor function $f_v : \mathcal{X} \to \mathbb{R}$ is denoted with (single) view predictor. View predictors can be learned independently for each view utilising an appropriate regression algorithm like SVR or RLSR. As a special case we consider concatenated predictors f_v in Sect. 4 where the corresponding view v results from a concatenation of finite dimensional feature representations $\Phi_1, \ldots, \Phi_M$. Having different views on the data, an alternative is to learn M predictors $f_v : \mathcal{X} \to \mathbb{R}$ simultaneously that depend on each other, satisfying an optimisation criterion involving all views at once. Such a criterion could be the minimisation of the labelled error in line with co-regularisation which will be specified in the following subsection. The final predictor f will then be the average of the predictors f_v.

A function $k : \mathcal{X} \times \mathcal{X} \to \mathbb{R}$ is said to be a kernel if it is symmetric and positive semi-definite. Indeed, for every kernel k there is a feature mapping $\Phi : \mathcal{X} \to \mathcal{H}$ such that $\mathcal{H}$ is a reproducing kernel Hilbert space (RKHS) and $k(x_1, x_2) = \langle \Phi(x_1), \Phi(x_2) \rangle_{\mathcal{H}}$ holds true for all $x_1, x_2 \in \mathcal{X}$ (Mercer's theorem). Thus, the function k is the corresponding reproducing kernel of $\mathcal{H}$, and for $x \in \mathcal{X}$ the mappings $\langle \Phi(x), \Phi(\cdot) \rangle = k(x, \cdot)$ are functions defined on $\mathcal{X}$. Choosing RKHSs $\mathcal{H}_v$ of multiple kernels k_v as candidate spaces for the predictors f_v, the representer theorem of Schölkopf et al. [10] allows for a parameterisation of the optimisation problems for co-regularisation presented below. A straightforward modification of the representer theorem's proof leads to a representation of the predictors f_v as finite kernel expansion

$$f_v(\cdot) = \sum_{i=1}^{n} \pi_{vi} k_v(x_i, \cdot) + \sum_{j=1}^{m} \pi_{v(j+n)} k_v(z_j, \cdot) \tag{1}$$

with linear coefficients $\pi_v \in \mathbb{R}^{n+m}$, centered at labelled and unlabelled instances $x_i \in X$ and $z_j \in Z$, respectively.

The kernel matrices $K_v = \{k_v(x_i, x_j)\}_{i,j=1}^{n+m}$ are the *Gram matrices* of the v-th view kernel k_v over labelled and unlabelled examples and have decompositions into an upper and a lower part $L_v \in \mathbb{R}^{n \times (n+m)}$ and $U_v \in \mathbb{R}^{m \times (n+m)}$, respectively. We will consider the submatrices $k(Z, x) := (k(z_1, x), \ldots, k(z_m, x))^T$ and $k(Z, Z) := \{k(z_j, z_{j'})\}_{j,j'=1}^{m}$ of a Gram matrix with kernel k. If $\mathcal{H}_1$ and $\mathcal{H}_2$ are RKHSs then their sum space $\mathcal{H}_\Sigma$ is defined as $\mathcal{H}_\Sigma := \{f : f = f_1 + f_2, f_1 \in \mathcal{H}_1, f_2 \in \mathcal{H}_2\}$. With $Y = (y_1, \ldots, y_n)^T \in \mathbb{R}^n$ we denote the vector of labels. We will abbreviate $v \in \{1, \ldots, M\}$ with $v \in [\![M]\!]$. And finally, we will utilise the squared loss $\ell_2(y, y') = \|y - y'\|^2$ and the ε-insensitive loss $\ell_\varepsilon(y, y') = \max\{0, |y - y'| - \varepsilon\}$, $y, y' \in \mathcal{Y}$.

3   The CoSVR Algorithm: Variants and Properties

3.1   Base CoSVR

In order to solve a regression task in the presence of multiple views $v = 1, \ldots, M$, the approach of *co-regularisation* is to jointly minimise two error terms involving M predictor functions $f_1, \ldots, f_M$. Firstly, every view predictor f_v is intended to have a small training error with respect to labelled examples. Secondly, the difference between pairwise view predictions over unlabelled examples should preferably be small. We introduce *co-regularised support vector regression* (CoSVR) as an ε-insensitive loss realisation of the co-regularisation principle.

Definition 1. *For $v \in \{1, \ldots, M\}$ let $\mathcal{H}_v$ be RKHSs. The co-regularised empirical risk minimisation*

$$\min_{f_v \in \mathcal{H}_v} \sum_{v=1}^{M} \left(\frac{\nu_v}{2} \|f_v\|^2 + \sum_{i=1}^{n} \ell^L(y_i, f_v(x_i)) \right) \tag{2}$$

$$+ \lambda \sum_{u,v=1}^{M} \sum_{j=1}^{m} \ell^U(f_u(z_j), f_v(z_j)),$$

where $\nu_v, \lambda \geq 0$ is called co-regularised support vector regression (CoSVR) if $\ell^L = \ell_{\varepsilon^L}$, $\varepsilon^L \geq 0$, and ℓ^U is an arbitrary loss function for regression. Furthermore, we define ε-CoSVR to be the special case where $\ell^U = \ell_{\varepsilon^U}$, $\varepsilon^U \geq 0$, as well as ℓ_2-CoSVR to satisfy $\ell^U = \ell_2$.

The minimum in (2) is taken over all f_v, $v = 1, \ldots, M$. For reasons of simplification we will abbreviate $\min_{f_1 \in \mathcal{H}_1, \ldots, f_M \in \mathcal{H}_M}$ with $\min_{f_v \in \mathcal{H}_v}$. Note that the loss function parameters ε^L and ε^U can have different values. The parameters ν_v and λ are trade-off parameters between empirical risk and co-regularisation term. The added norm terms $\|f_v\|$ prevent overfitting. We will also refer to the empirical risk term with loss function ℓ^L as *labelled error* and to the co-regularisation term with ℓ^U as *unlabelled error*. In the case of $\ell^L = \ell^U = \ell_2$, the optimisation in (2) is known as *co-regularised least squares regression* (CoRLSR). Brefeld et al. [4] found a closed form solution for CoRLSR as linear system of equations in $M(n + m)$ variables. In the following, we present a solution for ε-CoSVR and ℓ_2-CoSVR.

Lemma 1. *Let $\nu_v, \lambda, \varepsilon^L, \varepsilon^U \geq 0$. We use the notation introduced above. In particular, $\pi_v \in \mathbb{R}^{n+m}$ denote the kernel expansion coefficients of the single view predictors f_v from (1), whereas $\alpha_v, \hat{\alpha}_v \in \mathbb{R}^n$ and $\gamma_{uv} \in \mathbb{R}^m$ are dual variables.*

(i) The dual optimisation problem of ε-CoSVR equals

$$\max_{\alpha_v, \hat{\alpha}_v \in \mathbb{R}^n, \gamma_{uv} \in \mathbb{R}^m} \sum_{v=1}^{M} \left(-\frac{1}{2\nu_v} \begin{pmatrix} \alpha \\ \gamma \end{pmatrix}_v^T K_v \begin{pmatrix} \alpha \\ \gamma \end{pmatrix}_v + (\alpha_v - \hat{\alpha}_v)^T Y \right.$$

$$\left. -(\alpha_v + \hat{\alpha}_v)^T \varepsilon^L \mathbf{1}_n - \sum_{u=1}^{M} \gamma_{uv}^T \varepsilon^U \mathbf{1}_m \right)$$

$$s.t. \left\{ \begin{array}{l} \mathbf{0}_n \leq \alpha_v, \hat{\alpha}_v \leq \mathbf{1}_n \\ \mathbf{0}_m \leq \gamma_{uv} \leq \lambda \mathbf{1}_m \end{array} \right\}_{v \in [\![M]\!], \, (u,v) \in [\![M]\!]^2},$$

where $\pi_v^T = \frac{1}{\nu_v}(\alpha \mid \gamma)_v^T$ and $(\alpha \mid \gamma)_v^T = (\alpha_v - \hat{\alpha}_v \mid \sum_{u=1}^{M}(\gamma_{uv} - \gamma_{vu}))^T$.

(ii) The dual optimisation problem of ℓ_2-CoSVR is

$$\max_{\alpha_v, \hat{\alpha}_v \in \mathbb{R}^n, \gamma_{uv} \in \mathbb{R}^m} \sum_{v=1}^{M} \left(-\frac{1}{2\nu_v} \begin{pmatrix} \alpha \\ \gamma \end{pmatrix}_v^T K_v \begin{pmatrix} \alpha \\ \gamma \end{pmatrix}_v + (\alpha_v - \hat{\alpha}_v)^T Y \right.$$

$$\left. -\varepsilon^L (\alpha_v + \hat{\alpha}_v)^T \mathbf{1}_n - \frac{1}{4\lambda} \sum_{u=1}^{M} \gamma_{uv}^T \gamma_{uv} \right)$$

$$s.t. \left\{ \begin{array}{l} \mathbf{0}_n \leq \alpha_v, \hat{\alpha}_v \leq \mathbf{1}_n \\ \gamma_{uv} = \frac{2\lambda}{\nu_u} U_u \begin{pmatrix} \alpha \\ \gamma \end{pmatrix}_u - \frac{2\lambda}{\nu_v} U_v \begin{pmatrix} \alpha \\ \gamma \end{pmatrix}_v \end{array} \right\}_{v \in [\![M]\!], \, (u,v) \in [\![M]\!]^2},$$

where $\pi_v^T = \frac{1}{\nu_v}(\alpha \mid \gamma)_v^T$ and $(\alpha \mid \gamma)_v^T = (\alpha_v - \hat{\alpha}_v \mid \sum_{u=1}^{M}(\gamma_{uv} - \gamma_{vu}))^T$.

Remark 1. The proofs of Lemma 1 as well as Lemmas 2 and 3 below use standard techniques from *Lagrangian dualisation* (e.g., [3]). They can be found in our CoSVR repository (see footnote 3).

We choose the concatenated vector representation $(\alpha \mid \gamma)_v^T \in \mathbb{R}^{n+m}$ in order to show the correspondence between the two problems ε-CoSVR and ℓ_2-CoSVR and further CoSVR variants below. Additionally, the similarities with and differences to the original SVR dual problem are obvious. We will refer to the optimisation in Lemma 1 as the base CoSVR algorithms.

3.2   Reduction of Variable Numbers

The dual problems in Lemma 1 are quadratic programs. Both depend on $2Mn + M^2m$ variables, where $m \gg n$. If the number of views M and the number of unlabelled examples m are large, the base CoSVR algorithm might cause problems with respect to runtime because of the large number of resulting variables. In order to reduce this number, we define modified versions of base CoSVR. We denote the variant with a modification in the labelled error with $CoSVR^{mod}$ and in the unlabelled error with $CoSVR_{mod}$.

Modification of the Empirical Risk. In base CoSVR the empirical risk is meant to be small for each single view predictor individually using examples and their corresponding labels. In the $CoSVR^{mod}$ variant the average prediction, i.e., the final predictor, is applied to define the labelled error term.

Definition 2. *The co-regularised support vector regression problem with modified constraints for the labelled examples ($CoSVR^{mod}$) is defined as*

$$\min_{f_v \in \mathcal{H}_v} \sum_{v=1}^{M} \frac{\nu_v}{2} \|f_v\|^2 + \sum_{i=1}^{n} \ell_{\varepsilon^L}(y_i, f^{\mathrm{avg}}(x_i))$$

$$+ \lambda \sum_{u,v=1}^{M} \sum_{j=1}^{m} \ell^U(f_u(z_j), f_v(z_j)),$$

where $f^{\mathrm{avg}} := \frac{1}{M} \sum_{v=1}^{M} f_v$ is the average of all single view predictors. We denote the case $\ell^U = \ell_{\varepsilon^U}$, $\varepsilon^U \geq 0$, with ε-$CoSVR^{mod}$ and the case $\ell^U = \ell_2$ with ℓ_2-$CoSVR^{mod}$.

In the following lemma we present solutions for ε-$CoSVR^{mod}$ and ℓ_2-$CoSVR^{mod}$.

Lemma 2. *Let $\nu_v, \lambda, \varepsilon^L, \varepsilon^U \geq 0$. We utilise dual variables $\alpha, \hat{\alpha} \in \mathbb{R}^n$ and $\gamma_{uv} \in \mathbb{R}^m$.*

(i) The ε-$CoSVR^{mod}$ dual optimisation problem can be written as

$$\max_{\alpha, \hat{\alpha} \in \mathbb{R}^n, \ \gamma_{uv} \in \mathbb{R}^m} \sum_{v=1}^{M} \left(-\frac{1}{2\nu_v} \binom{\alpha}{\gamma}_v^T K_v \binom{\alpha}{\gamma}_v + (\alpha - \hat{\alpha})^T Y \right.$$

$$\left. -(\alpha + \hat{\alpha})^T \varepsilon^L \mathbf{1}_n - \sum_{u=1}^{M} \gamma_{uv}^T \varepsilon^U \mathbf{1}_m \right)$$

$$s.t. \left\{ \begin{matrix} \mathbf{0}_n \leq \alpha, \hat{\alpha} \leq \mathbf{1}_n \\ \mathbf{0}_m \leq \gamma_{uv} \leq \lambda \mathbf{1}_m \end{matrix} \right\}_{v \in [\![M]\!]},$$

where $\pi_v^T = \frac{1}{\nu_v}(\alpha \mid \gamma)_v^T$ and $(\alpha \mid \gamma)_v^T = (\frac{1}{M}(\alpha - \hat{\alpha}) \mid \sum_{u=1}^{M}(\gamma_{uv} - \gamma_{vu}))^T$.

(ii) The ℓ_2-CoSVR^{mod} dual optimisation problem equals

$$\max_{\alpha, \hat{\alpha} \in \mathbb{R}^n, \ \gamma_{uv} \in \mathbb{R}^m} \sum_{v=1}^{M} \left(-\frac{1}{2\nu_v} \binom{\alpha}{\gamma}_v^T K_v \binom{\alpha}{\gamma}_v + (\alpha_v - \hat{\alpha}_v)^T Y \right.$$

$$\left. -(\alpha_v + \hat{\alpha}_v)^T \varepsilon^L \mathbf{1}_n - \frac{1}{4\lambda} \sum_{u=1}^{M} \gamma_{uv}^T \gamma_{uv} \right)$$

$$s.t. \left\{ \begin{array}{l} \mathbf{0}_n \leq \alpha, \hat{\alpha} \leq \mathbf{1}_n \\ \gamma_{uv} = \frac{2\lambda}{\nu_u} U_u \binom{\alpha}{\gamma}_u - \frac{2\lambda}{\nu_v} U_v \binom{\alpha}{\gamma}_v \end{array} \right\}_{v \in [\![M]\!]},$$

where $\pi_v^T = \frac{1}{\nu_v}(\alpha \mid \gamma)_v^T$ and $(\alpha \mid \gamma)_v^T = (\frac{1}{M}(\alpha - \hat{\alpha}) \mid \sum_{u=1}^{M}(\gamma_{uv} - \gamma_{vu}))^T$.

We can also reduce the number of variables more effectively using modified constraints for the co-regularisation term. Whereas the CoSVR^{mod} algorithm is rather important from a theoretical perspective (see Sect. 3.3), the variant presented in the next section is very beneficial from a practical perspective if the number of views M is large.

Modification of the Co-regularisation. The unlabelled error term of base CoSVR bounds the pairwise distances of view predictions, whereas now in CoSVR$_{mod}$ only the disagreement between predictions of each view and the average prediction of the residual views will be taken into account.

Definition 3. *We consider RKHSs $\mathcal{H}_1, \ldots, \mathcal{H}_M$ as well as constants $\varepsilon^L, \varepsilon^U, \nu_v, \lambda \geq 0$. The co-regularised support vector regression problem with modified constraints for the unlabelled examples (CoSVR$_{mod}$) is defined as*

$$\min_{f_v \in \mathcal{H}_v} \sum_{v=1}^{M} \left(\frac{\nu_v}{2} \|f_v\|^2 + \sum_{i=1}^{n} \ell_{\varepsilon^L}(y_i, f_v(x_i)) \right) \tag{3}$$

$$+ \lambda \sum_{v=1}^{M} \sum_{j=1}^{m} \ell^U \left(f_v^{\mathrm{avg}}(z_j), \ f_v(z_j) \right),$$

where now $f_v^{\mathrm{avg}} := \frac{1}{M-1} \sum_{u=1}^{M, u \neq v} f_u$ is the average of view predictors besides view v. We denote the case $\ell^U = \ell_{\varepsilon^U}$, $\varepsilon^U \geq 0$, with ε-CoSVR$_{mod}$ and the case $\ell^U = \ell_2$ with ℓ_2-CoSVR$_{mod}$.

Again we present solutions for ε-CoSVR$_{mod}$ and ℓ_2-CoSVR$_{mod}$.

Lemma 3. *Let $\nu_v, \lambda, \varepsilon^L, \varepsilon^U \geq 0$. We utilise dual variables $\alpha_v, \hat{\alpha}_v \in \mathbb{R}^n$ and $\gamma_v, \hat{\gamma}_v \in \mathbb{R}^m$, as well as $\gamma_v^{avg} := \frac{1}{M-1} \sum_{u=1}^{M, u \neq v} \gamma_u$ and $\hat{\gamma}_v^{avg} := \frac{1}{M-1} \sum_{u=1}^{M, u \neq v} \hat{\gamma}_u$ analogous to the residual view predictor average.*

(i) The ε-CoSVR$_{mod}$ dual optimisation problem can be written as

$$\max_{\alpha_v,\hat{\alpha}_v\in\mathbb{R}^n,\ \gamma_v,\hat{\gamma}_v\in\mathbb{R}^m} \sum_{v=1}^{M}\left(-\frac{1}{2\nu_v}\binom{\alpha}{\gamma}_v^T K_v \binom{\alpha}{\gamma}_v + (\alpha-\hat{\alpha})^T Y\right.$$

$$\left. -(\alpha_v+\hat{\alpha}_v)^T \varepsilon^L \mathbf{1}_n - (\gamma_v+\hat{\gamma}_v)\varepsilon^U \mathbf{1}_m\right)$$

$$s.t. \left\{\begin{array}{l} \mathbf{0}_n \le \alpha_v,\hat{\alpha}_v \le \mathbf{1}_n \\ \mathbf{0}_m \le \gamma_v,\hat{\gamma}_v \le \lambda\mathbf{1}_m \end{array}\right\}_{v\in[\![M]\!]},$$

where $\pi_v^T = \frac{1}{\nu_v}(\alpha\mid\gamma)_v^T$ and $(\alpha\mid\gamma)_v^T = (\alpha_v-\hat{\alpha}_v \mid (\gamma_v-\gamma_v^{avg})-(\hat{\gamma}_v-\hat{\gamma}_v^{avg}))^T$.

(ii) The ℓ_2-CoSVR$_{mod}$ dual optimisation problem equals

$$\max_{\alpha_v,\hat{\alpha}_v\in\mathbb{R}^n,\ \gamma_v\in\mathbb{R}^m} \sum_{v=1}^{M}\left(-\frac{1}{2\nu_v}\binom{\alpha}{\gamma}_v^T K_v \binom{\alpha}{\gamma}_v + (\alpha_v-\hat{\alpha}_v)^T Y\right.$$

$$\left. -(\alpha_v+\hat{\alpha}_v)^T \varepsilon^L \mathbf{1}_n - \frac{1}{4\lambda}\sum_{u=1}^{M}\gamma_v^T\gamma_v\right)$$

$$s.t. \left\{\begin{array}{l} \mathbf{0}_n \le \alpha_v,\hat{\alpha}_v \le \mathbf{1}_n \\ \gamma_v = \frac{1}{M-1}\sum_{u=1}^{M,u\ne v}\frac{2\lambda}{\nu_u}U_u\binom{\alpha}{\gamma}_u - \frac{2\lambda}{\nu_v}U_v\binom{\alpha}{\gamma}_v \end{array}\right\}_{v\in[\![M]\!]},$$

where $\pi_v^T = \frac{1}{\nu_v}(\alpha\mid\gamma)_v^T$ and $(\alpha\mid\gamma)_v^T = (\alpha_v-\hat{\alpha}_v \mid \gamma_v-\gamma_v^{avg})^T$.

Remark 2. If we combine the modifications in the labelled and unlabelled error term we canonically obtain the variants ε-CoSVR$_{mod}^{mod}$ and ℓ_2-CoSVR$_{mod}^{mod}$.

In the base CoSVR versions the semi-supervision is realised with proximity constraints on pairs of view predictions. We show in the following lemma that the constraints of the closeness of one view prediction to the average of the residual predictions implies a closeness of every pair of predictions.

Lemma 4. *Up to constants, the unlabelled error bound of CoSVR$_{mod}$ is also an upper bound of the unlabelled error of base CoSVR.*

Proof. We consider the settings of Lemmas $1(i)$ and $3(i)$. For part (ii) the proof is equivalent with $\varepsilon^U = 0$. In the case of $M = 2$, modified and base algorithm fall together which shows the claim. Now let $M > 2$. Because of the definition of the ε-insensitive loss we know that $|f_v(z_j)-f_v^{avg}(z_j)| \le \varepsilon^U +c_{vj}$, where $c_{vj} \ge 0$ is the actual loss value for fixed v and j. We denote $c_j := \max_{v\in\{1,\ldots,M\}}\{c_{1j},\ldots,c_{Mj}\}$ and, hence, $|f_v(z_j)-f_v^{avg}(z_j)| \le \varepsilon^U +c_j$ for all $v \in \{1,\ldots,M\}$. Now we conclude for $j \in \{1,\ldots,m\}$ and $(u,v) \in \{1,\ldots,M\}^2$

$$|f_u(z_j) - f_v(z_j)|$$
$$\le |f_u(z_j)-f_u^{avg}(z_j)| + |f_u^{avg}(z_j)-f_v^{avg}(z_j)| + |f_v^{avg}(z_j)-f_v(z_j)|$$
$$\le \varepsilon^U +c_j + \frac{1}{M-1}|f_v(z_j)-f_u(z_j)| + \varepsilon^U +c_j,$$

and therefore, $|f_u(z_j) - f_v(z_j)| \leq \frac{2(M-1)}{M-2}(\varepsilon^U + c_j)$. As a consequence we deduce from $\sum_{v=1}^{M} \sum_{j=1}^{m} \ell_{\varepsilon^U}(f_v^{\text{avg}}(z_j), f_v(z_j)) \leq M \sum_{j=1}^{m} c_j =: B$ that also the labelled error of CoSVR can be bounded $\sum_{u,v=1}^{M} \sum_{j=1}^{m} \ell_{\tilde{\varepsilon}}(f_u(z_j), f_v(z_j)) \leq \tilde{B}$ for $\tilde{\varepsilon} = \frac{2(M-1)}{M-2}\varepsilon^U$ and $\tilde{B} = \frac{2M(M-1)}{(M-2)}B$, which finishes the proof. $\qquad\square$

3.3 Σ-CoSVR

Sindhwani and Rosenberg [11] showed that under certain conditions co-regularisation approaches of two views exhibit a very useful property. If $\ell^U = \ell_2$ and the labelled loss is calculated utilising an arbitrary loss function for the average predictor f^{avg}, the resulting multi-view approach is equivalent with a single-view approach of a fused kernel. We use the notion from Sect. 2.

Definition 4. *Let* $\lambda, \nu_1, \nu_2, \varepsilon^L \geq 0$ *be parameters and the Gram submatrices* $k(Z, x)$ *and* $k(Z, Z)$ *be defined as in Sect. 2. We consider a merged kernel* k_Σ *from two view kernels* k_1 *and* k_2

$$k_\Sigma(x, x') := k^\oplus(x, x') - k^\ominus(Z, x)^T \left(\tfrac{1}{\lambda}I_m + k^\oplus(Z, Z)\right)^{-1} k^\ominus(Z, x'), \qquad (4)$$

for $x, x' \in \mathcal{X}$, *where* $k^\oplus := \frac{1}{\nu_1}k_1 + \frac{1}{\nu_2}k_2$ *and* $k^\ominus := \frac{1}{\nu_1}k_1 - \frac{1}{\nu_2}k_2$. *We denote the SVR optimisation*

$$\operatorname*{argmin}_{f \in \mathcal{H}_\Sigma} \|f\|^2 + \sum_{i=1}^{n} \ell_{\varepsilon^L}(y_i, \tfrac{1}{2}f(x_i)), \qquad (5)$$

Σ-*co-regularised support vector regression* (Σ-*CoSVR*), *where* $\mathcal{H}_\Sigma$ *is the RKHS of* k_Σ.

Please notice that for each pair (x, x') the value of $k_\Sigma(x, x')$ is calculated in (4) with k_1 and k_2 including not only x and x' but also unlabelled examples $z_1, \ldots, z_m$. Hence, the optimisation problem in (5) is a standard SVR with additional information about unlabelled examples incorporated in the RKHS $\mathcal{H}_\Sigma$.

Lemma 5. *The algorithms* ℓ_2-*CoSVR*^{mod} *and* Σ-*CoSVR are equivalent and* $\mathcal{H}_\Sigma$ *is the sum space* $\mathcal{H}_\Sigma = \{f : \mathcal{X} \to \mathbb{R} \mid f = f_1 + f_2, f_1 \in \mathcal{H}_1, f_2 \in \mathcal{H}_2\}$.

Proof. The proof is an application of Theorem 2.2. of Sindhwani and Rosenberg [11] for the loss function V being equal to the ε-insensitive loss with $\varepsilon = \varepsilon^L$, the parameter of the labelled error of ℓ_2-CoSVR^{mod}. $\qquad\square$

As Σ-CoSVR can be solved as a standard SVR algorithm we obtained a much faster co-regularisation approach. The information of the two views and the unlabelled examples are included in the candidate space $\mathcal{H}_\Sigma$ and associated kernel k_Σ.

3.4 Complexity

The CoSVR variants and CoRLSR mainly differ in the number of applied loss functions and the strictness of constraints. This results in different numbers of variables and constraints in total, as well as potentially non-zero variables (referred to as *sparsity*, compare Table 1). All presented problems are convex QPs with positive semi-definite matrices in the quadratic terms. As the number m of unlabelled instances in real-world problems is much greater than n, the runtime of a QP-solver is dominated by the respective second summand in the constraints column of Table 1. Because of the ε-insensitive loss the number of actual non-zero variables in the learned model will be even smaller for the CoSVR variants than the numbers reported in the sparsity column of Table 1. In particular, for the modified variants this will allow for a more efficient model storage compared to CoRLSR. Indeed, according to the *Karush-Kuhn-Tucker conditions* (e.g., [3]), only for active inequality constraints the corresponding dual γ-variables can be non-zero. In this sense the respective unlabelled $z_j \in Z$ are *unlabelled support vectors*. This consideration is also valid for the α-variables and support vectors $x_i \in X$ as we use the ε-insensitive loss for the labelled error in all CoSVR versions. And finally, in the two-view case with $M = 2$ the modified version with respect to the unlabelled error term and the base version coincide.

Table 1. Number of variables, constraints, and potential non-zero variables for different CoSVR versions and CoRLSR. The respective CoSVR^{mod} variant is included by cancelling the $\{M\}$-factor.

Algorithm	Variables	Constraints	Sparsity
ε-CoSVR	$2\{M\}n + M^2m$	$4\{M\}n + 2M^2m$	$\{M\}n + \frac{1}{2}(M^2 - M)m$
ℓ_2-CoSVR	$2\{M\}n + M^2m$	$4\{M\}n + M^2m$	$\{M\}n + M^2m$
ε-CoSVR$_{mod}$	$2\{M\}n + 2Mm$	$4\{M\}n + 4Mm$	$\{M\}n + Mm$
ℓ_2-CoSVR$_{mod}$	$2\{M\}n + Mm$	$4\{M\}n + Mm$	$\{M\}n + Mm$
Σ-CoSVR	$2n$	$4n$	n
CoRLSR	$Mn + Mm$	0	$Mn + Mm$

3.5 A Rademacher Bound for CoSVR

Similarly to the result of Rosenberg and Bartlett [9] we want to prove a bound on the *empirical Rademacher complexity* $\hat{\mathcal{R}}_n$ of CoSVR in the case of $M = 2$. Note that, despite the proof holding for the special case of $M = 2$, the CoSVR method in general is applicable to arbitrary numbers of views. The empirical Rademacher complexity is a data-dependent measure for the capacity of a function class $\mathcal{H}$ to fit random data and is defined as

$$\hat{\mathcal{R}}_n(\mathcal{H}) = \mathbb{E}^\sigma \left[\sup_{f \in \mathcal{H}} \left| \frac{2}{n} \sum_{i=1}^n \sigma_i f(x_i) \right| : \{x_1, \ldots, x_n\} = X \right].$$

The random data are represented via *Rademacher random variables* $\sigma = (\sigma_1, \ldots, \sigma_n)^T$. We consider ε-CoSVR and ℓ_2-CoSVR and define bounded versions $\mathcal{H}_\Sigma^\varepsilon$ and $\mathcal{H}_\Sigma^2$ of the sum space $\mathcal{H}_\Sigma$ from Sect. 2 for the corresponding versions. Obviously, a pair $(\pi_1, \pi_2) \in \mathbb{R}^{(n+m) \times (n+m)}$ of kernel expansion coefficients (see (1)) represents an element of $\mathcal{H}_\Sigma$. For ε-CoSVR and ℓ_2-CoSVR we set

$$\mathcal{H}_\Sigma^\varepsilon := \{ (\pi_1, \pi_2) \in \mathcal{H}_\Sigma : -\mu 1_{n+m} \le \pi_1, \pi_2 \le \mu 1_{n+m} \}, \quad \text{and} \quad (6)$$

$$\mathcal{H}_\Sigma^2 := \{ (\pi_1, \pi_2) \in \mathcal{H}_\Sigma : \nu_1 \pi_1^T K_1 \pi_1 + \nu_2 \pi_2^T K_2 \pi_2$$
$$+ \lambda (U_1 \pi_1 - U_2 \pi_2)^T (U_1 \pi_1 - U_2 \pi_2) \le 1 \}, \quad (7)$$

respectively. In (6) μ is an appropriate constant according to Lemmas 1 and 2. The definition in (7) follows the reasoning of Rosenberg and Bartlett [9]. Now we derive a bound on the empirical Rademacher complexity of $\mathcal{H}_\Sigma^\varepsilon$ and $\mathcal{H}_\Sigma^2$, respectively. We point out that the subsequent proof is also valid for the modified versions with respect to the empirical risk. For two views the base and modified versions with respect to the co-regularisation fall together anyway. For reasons of simplicity, in the following lemma and proof we omit ^{mod} and $_{mod}$ for the CoSVR variants. Furthermore, we will apply the infinity vector norm $\|v\|_\infty$ and row sum matrix norm $\|L\|_\infty$ (consult, e.g., Werner [15]).

Lemma 6. *Let $\mathcal{H}_\Sigma^\varepsilon$ and $\mathcal{H}_\Sigma^2$ be the function spaces in (6) and (7) and, without loss of generality, let $\mathcal{Y} = [-1, 1]$.*

(i) The empirical Rademacher complexity of ε-CoSVR can be bounded via

$$\hat{\mathcal{R}}_n(\mathcal{H}_\Sigma^\varepsilon) \le \frac{2s}{n} \mu (\|L_1\|_\infty + \|L_2\|_\infty),$$

where μ is a constant dependent on the regularisation parameters and s is the number of potentially non-zero variables in the kernel expansion vector $\pi \in \mathcal{H}_\Sigma^\varepsilon$.

(ii) The empirical Rademacher complexity of ℓ_2-CoSVR has a bound

$$\hat{\mathcal{R}}_n(\mathcal{H}_\Sigma^2) \le \frac{2}{n} \sqrt{tr_n(K_\Sigma)},$$

where $tr_n(K_\Sigma) := \sum_{i=1}^n k_\Sigma(x_i, x_i)$ with the sum kernel k_Σ from (4).

Our proof applies Theorems 2 and 3 of Rosenberg and Bartlett [9].

Proof. At first, using Theorem 2 of Rosenberg and Bartlett [9], we investigate the general usefulness of the empirical Rademacher complexity $\hat{\mathcal{R}}_n$ of $\mathcal{H}_\Sigma^{loss}$ in the CoSVR scenario. The function space $\mathcal{H}_\Sigma^{loss}$ can be either $\mathcal{H}_\Sigma^\varepsilon$ or $\mathcal{H}_\Sigma^2$. Theorem 2 requires two preconditions. First, we notice that the ε-insensitive loss function utilising the average predictor $\ell^L(y, f(x)) = \max\{0, |y - (f_1(x) + f_2(x))/2| - \varepsilon^L\}$

maps into $[0, 1]$ because of the boundedness of $\mathcal{Y}$. Second, it is easy to show that ℓ^L is *Lipschitz continuous*, i.e. $|\ell^L(y, y') - \ell^L(y, y'')|/|y' - y''| \leq C$, for some constant $C > 0$. With similar arguments one can show that the ε-insensitive loss function of base CoSVR is Lipschitz continuous as well. According to Theorem 2 of Rosenberg and Bartlett [9], the expected loss $\mathbb{E}_{(X,Y) \sim \mathcal{D}} \, \ell^L(f(X), Y)$ can then be bounded by means of the empirical risk and the empirical Rademacher complexity

$$\mathbb{E}_{\mathcal{D}} \, \ell^L(f(X), Y) \leq \frac{1}{n} \sum_{i=1}^{n} l^L(f(x_i), y_i) + 2C\hat{\mathcal{R}}_n(\mathcal{H}_{\Sigma}^{loss}) + \frac{2 + 3\sqrt{\ln(2/\delta)/2}}{\sqrt{n}}$$

for every $f \in \mathcal{H}_{\Sigma}^{loss}$ with probability at least $1 - \delta$. Now we continue with the cases (i) and (ii) separately.

(i) We can reformulate the empirical Rademacher complexity

$$\hat{\mathcal{R}}_n(\mathcal{H}_{\Sigma}^{\varepsilon}) = \frac{2}{n} \mathbb{E}^{\sigma} \left[\sup_{(\pi_1 \mid \pi_2)^T \in \mathcal{K}} \left| \sigma^T (L_1 \pi_1 + L_2 \pi_2) \right| \right],$$

where $\mathcal{K} := \{(\pi_1 \mid \pi_2)^T \in \mathbb{R}^{2(n+m)} : -\mu \mathbf{1}_{n+m} \leq \pi_1, \pi_2 \leq \mu \mathbf{1}_{n+m}\}$. The kernel expansion π of ε-CoSVR optimisation is bounded because of the box constraints in the respective dual problems. Therefore, π lies in the ℓ_1-ball of dimension s scaled with $s\mu$, i.e., $\pi \in s\mu \cdot B_1$. The dimension s is the sparsity of π, and thus, the number of expansion variables π_{vj} different from zero. From the dual optimisation problem we know that $s \ll 2(n+m)$. It is a fact that $\sup_{\pi \in s\mu \cdot B_1} |\langle v, \pi \rangle| = s\mu \|v\|_{\infty}$ (see Theorems II.2.3 and II.2.4 in Werner [15]). Let $L \in \mathbb{R}^{n \times 2(n+m)}$ be the concatenated matrix $L = (L_1 \mid L_2)$, where L_1 and L_2 are the upper parts of the Gram matrices K_1 and K_2 according to Sect. 2. From the definition we see that $v = \sigma^T L$ and, hence,

$$s\mu \|v\|_{\infty} = s\mu \|\sigma^T L\|_{\infty} \leq s\mu \|\sigma\|_{\infty} \|L\|_{\infty} \leq s\mu \|L\|_{\infty}$$

$$= s\mu \max_{i=1,\ldots,n} \sum_{j=1}^{n+m} \sum_{v=1,2} |k_v(x_i, x_j)|.$$

Finally, we obtain the desired upper bound for the empirical Rademacher complexity of ε-CoSVR

$$\hat{\mathcal{R}}_n(\mathcal{H}_{\Sigma}^{\varepsilon}) \leq \frac{2}{n} \mathbb{E}^{\sigma} s\mu \|L\|_{\infty} \leq \frac{2s}{n} \mu(\|L_1\|_{\infty} + \|L_2\|_{\infty}).$$

(ii) Having the Lipschitz continuity of the ε-insensitive loss ℓ^L, the claim is a direct consequence of Theorem 3 in the work of Rosenberg and Bartlett [9], which finishes the proof.   □

4 Empirical Evaluation

In this section we evaluate the performance of the CoSVR variants for predicting the affinity values of small compounds against target proteins.

Our experiments are performed on 24 datasets consisting of ligands and their affinity to one particular human protein per dataset, gathered from BindingDB. Every ligand is a single molecule in the sense of a connected graph and all ligands are available in the standard molecular fingerprint formats ECFP4, GpiDAPH3, and Maccs. All three formats are binary and high-dimensional. An implementation of the proposed methods and baselines, together with the datasets and experiment descriptions are available as open source[3].

We compare the CoSVR variants ε-CoSVR, ℓ_2-CoSVR, and Σ-CoSVR against CoRLSR, as well as SVR with a single-view (SVR([fingerprint name])) in terms of root mean squared error (RMSE) using the linear kernel. We take the two-view setting in our experiments as we want to include Σ-CoSVR results in the evaluation. Another natural baseline is to apply SVR to a new view that is created by concatenating the features of all views (SVR(concat)). We also compare the CoSVR variants against an oracle that chooses the best SVR for each view and each dataset (SVR(best)) by taking the result with the best performance in hindsight.

We consider affinity prediction as semi-supervised learning with many unlabelled data instances. Therefore, we split each labelled dataset into a labelled (30% of the examples) and an unlabelled part (the remaining 70%). For the co-regularised algorithms, both the labelled and unlabelled part are employed for training, i.e., in addition to labelled examples they have access to the entire set of unlabelled instances without labels. Of course, the SVR baselines only consider the labelled examples for training. For all algorithms the unlabelled part is used for testing. The RMSE is measured using 5-fold cross-validation. The parameters for each approach on each dataset are optimised using grid search with 5-fold cross-validation on a sample of the training set.

In Fig. 1 we present the results of the CoSVR variants compared to CoRLSR 1(a), SVR(concat) 1(b), and SVR(best) 1(c) for all datasets using the fingerprints GpiDAPH3 and ECFP4. Figure 1(a), (b), indicate that all CoSVR variants outperform CoRLSR and SVR(concat) on the majority of datasets. Figure 1(c) indicates that SVR(best) performs better than the other baselines but is still outperformed by ε-CoSVR and ℓ_2-CoSVR. Σ-CoSVR performs similar to SVR(best).

The indications in Fig. 1 are substantiated by a *Wilcoxon signed-rank test* on the results (presented in Table 2). In this table, we report the test statistics (Z and p-value). Results in which a CoSVR variant statistically significantly outperforms the baselines (for a significance level $p < 0.05$) are marked in bold. The test confirms that all CoSVR variants perform statistically significantly better than CoRLSR and SVR(concat). Moreover, ε-CoSVR and ℓ_2-CoSVR statistically significantly outperform an SVR trained on each individual view as well as taking the best single-view SVR in hindsight. Although Σ-CoSVR performs slightly better than SVR(best), the advantage is not statistically significant.

---

[3] CoSVR open source repository, https://bitbucket.org/Michael_Kamp/cosvr.

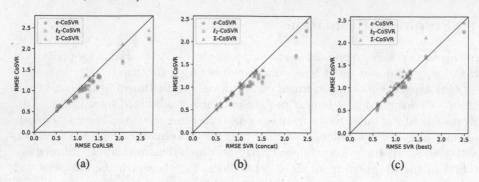

Fig. 1. Comparison of ε-CoSVR, ℓ_2-CoSVR, and Σ-CoSVR with the baselines CoRLSR, SVR(concat), and SVR(best) on 24 datasets using the fingerprints Gpi-DAPH3 and ECFP4 in terms of RMSEs. Each point represents the RMSEs of the two methods compared on one dataset.

Table 2. Comparing RMSEs using Wilcoxon signed-rank test (hypothesis test on whether CoSVR has significantly smaller RMSEs than the baselines).

baseline	ε-CoSVR		ℓ_2-CoSVR		Σ-CoSVR	
	Z	p-value	Z	p-value	Z	p-value
CoRLSR	8.0	< 0.00005	13.0	< 0.00009	70.0	< 0.02226
SVR(GpiDAPH3)	1.0	< 0.00002	1.0	< 0.00002	1.0	< 0.00002
SVR(ECFP4)	22.5	< 0.00027	44.0	< 0.00738	94.0	< 0.1096
SVR(concat)	3.0	< 0.00003	24.0	< 0.00032	79.5	< 0.04397
SVR(best)	27.0	< 0.00044	56.0	< 0.02208	88.0	< 0.07649

In Table 3 we report the average RMSEs of all CoSVR variants, CoRLSR and the single-view baselines for all combinations of the fingerprints Maccs, Gpi-DAPH3, and ECFP4. In terms of average RMSE, ε-CoSVR and ℓ_2-CoSVR out-perform the other approaches for the view combination Maccs and GpiDAPH3, as well as GpiDAPH3 and ECFP4. For the views Maccs and ECFP4, these CoSVR variants have lower average RMSE than CoRLSR and the single-view SVRs. However, for this view combination, the SVR(best) baseline outperforms CoSVR. Note that SVR(best) is only a hypothetical baseline, since the best view varies between datasets and is thus unknown in advance. The Σ-CoSVR per-forms on average similar to CoRLSR and the SVR(concat) baseline and slightly worse than SVR(best). To avoid confusion about the different performances of Σ-CoSVR and ℓ_2-CoSVR, we want to point out that Σ-CoSVR equals ℓ_2-CoSVR^{mod} (see Lemma 5) and not ℓ_2-CoSVR (equivalent with ℓ_2-CoSVR$_{mod}$ for $M = 2$) which we use for our experiments.

The advantage in learning performance of ε-CoSVR and ℓ_2-CoSVR comes along with the cost of a higher runtime as shown in Fig. 2. In concordance with the theory, Σ-CoSVR equalises the runtime disadvantage with a runtime similar to the single-view methods.

In conclusion, co-regularised support vector regression techniques are able to exploit the information from unlabelled examples with multiple sparse views in the practical setting of ligand affinity prediction. They perform better than the state-of-the-art single-view approaches [12], as well as a concatenation of features from multiple views. In particular, ε-CoSVR and ℓ_2-CoSVR outperform the multi-view approach CoRLSR [4] and SVR on all view combinations. ℓ_2-CoSVR outperforms SVR(concat) on all, ε-CoSVR on 2 out of 3 view combinations. Moreover, both variants outperform SVR(best) on 2 out of 3 view combinations.

Table 3. Average RMSEs for all combinations of the fingerprints Maccs, GpiDAPH3, and ECFP4

Method	View combinations		
	Maccs, ECFP4	Maccs, GpiDAPH3	GpiDAPH3, ECFP4
ε-CoSVR	1.035	**1.016**	1.049
ℓ_2-CoSVR	1.007	1.019	1.062
Σ-CoSVR	1.116	1.114	1.151
CoRLSR	1.06	1.073	1.199
SVR(view1)	1.04	1.041	1.355
SVR(view2)	1.094	1.37	1.106
SVR(concat)	1.011	1.12	1.194
SVR(best)	**0.966**	1.027	1.104

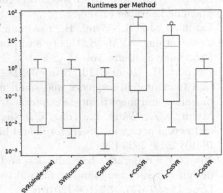

Fig. 2. Runtimes of the CoSVR variants, CoRLSR, and single-view SVRs on 24 ligand datasets and all view combinations (runtime in log-scale).

5   Conclusion

We proposed CoSVR as a semi-supervised multi-view regression method that copes with the practical challenges of few labelled data instances and multiple adequate views on data. Additionally, we presented CoSVR variants with considerably reduced numbers of variables and a version with substantially decreased runtime. Furthermore, we proved upper bounds on the Rademacher complexity for CoSVR. In the experimental part, we applied CoSVR successfully to the problem of ligand affinity prediction. The variants ε-CoSVR and ℓ_2-CoSVR empirically outperformed the state-of-the-art approaches in ligand-based virtual screening. However, this performance came at the cost of solving a more complex optimisation problem resulting in a higher runtime than single-view approaches. The variant Σ-CoSVR still outperformed most state-of-the-art approaches with the runtime of a single-view approach.

References

1. Bender, A., Jenkins, J.L., Scheiber, J., Sukuru, S.C.K., Glick, M., Davies, J.W.: How similar are similarity searching methods? A principal component analysis of molecular descriptor space. J. Chem. Inf. Model **49**(1), 108–119 (2009)
2. Blum, A., Mitchell, T.: Combining labeled and unlabeled data with co-training. In: Proceedings of the 11th Annual Conference on Learning Theory (1998)
3. Boyd, S., Vandenberghe, L.: Convex Optimization. Cambridge University Press, Cambridge (2004)
4. Brefeld, U., Gärtner, T., Scheffer, T., Wrobel, S.: Efficient co-regularised least squares regression. In: Proceedings of the 23rd International Conference on Machine Learning (2006)
5. Farquhar, J.D.R., Meng, H., Szedmak, S., Hardoon, D., Shawe-Taylor, J.: Two view learning: SVM-2K, theory and practice. In: Advances in Neural Information Processing Systems, vol. 18 (2006)
6. Geppert, H., Humrich, J., Stumpfe, D., Gärtner, T., Bajorath, J.: Ligand prediction from protein sequence and small molecule information using support vector machines and fingerprint descriptors. J. Chem. Inf. Model **49**(4), 767–779 (2009)
7. Myint, K.Z., Wang, L., Tong, Q., Xie, X.Q.: Molecular fingerprint-based artificial neural networks QSAR for ligand biological activity predictions. Mol. Pharm. **9**(10), 2912–2923 (2012)
8. Nisius, B., Bajorath, J.: Reduction and recombination of fingerprints of different design increase compound recall and the structural diversity of hits. Chem. Biol. Drug Des. **75**(2), 152–160 (2010)
9. Rosenberg, D.S., Bartlett, P.L.: The Rademacher complexity of co-regularized kernel classes. In: Proceedings of the 11th International Conference on Artificial Intelligence and Statistics (2007)
10. Schölkopf, B., Herbrich, R., Smola, A.J.: A generalized representer theorem. In: Helmbold, D., Williamson, B. (eds.) COLT 2001. LNCS (LNAI), vol. 2111, pp. 416–426. Springer, Heidelberg (2001). https://doi.org/10.1007/3-540-44581-1_27
11. Sindhwani, V., Rosenberg, D.S.: An RKHS for multi-view learning and manifold co-regularization. In: Proceedings of the 25th International Conference on Machine Learning (2008)
12. Sugaya, N.: Ligand efficiency-based support vector regression models for predicting bioactivities of ligands to drug target proteins. J. Chem. Inf. Model **54**(10), 2751–2763 (2014)
13. Ullrich, K., Mack, J., Welke, P.: Ligand affinity prediction with multi-pattern kernels. In: Calders, T., Ceci, M., Malerba, D. (eds.) DS 2016. LNCS (LNAI), vol. 9956, pp. 474–489. Springer, Cham (2016). https://doi.org/10.1007/978-3-319-46307-0_30
14. Wang, X., Ma, L., Wang, X.: Apply semi-supervised support vector regression for remote sensing water quality retrieving. In: IEEE International Geoscience and Remote Sensing Symposium (2010)
15. Werner, D.: Funktionalanalysis. Springer, Heidelberg (1995). https://doi.org/10.1007/978-3-642-21017-4_2
16. Xu, C., Tao, D., Xu, C.: A Survey on Multi-view Learning. arXiv (2013)

Online Regression with Controlled Label Noise Rate

Edward Moroshko$^{(\boxtimes)}$ and Koby Crammer

Department of Electrical Engineering, The Technion, Haifa, Israel
edward.moroshko@gmail.com, koby@ee.technion.ac.il

Abstract. Many online regression (and adaptive filtering) algorithms
are linear, use additive update and designed for the noise-free setting.
We consider the practical setting where the algorithm's feedback is noisy,
rather than a clean label. We propose a new family of algorithms which
modifies the learning rate based on the noise-variance of the feedback
(labels), by shrinking both inputs and feedbacks, based on the amount
of noise per input instance. We consider both settings, where the noise
is either given or estimated. Empirical study with both synthetic and
real-world speech data shows that our algorithms improve the overall
performance of the regressor, even when there is no additional explicit
information (i.e. amount of noise). We also consider a more general setting where an algorithm can sample more than single (noisy) label, yet
there is a total (or average) budget for the feedback. We propose a few
strategies how to effectively spend the given budget, which are based on
noise-variance estimation and our shrinkage rule. We show empirically
that our approach outperforms other naive approaches.

Keywords: Online learning · Regression · Adaptive filtering
Label-Noise

1 Introduction

Many online regression algorithms (aka adaptive filters in the signal processing
community [19]) are designed for a noise-free setting, or designed for the general
case where there is some noise, yet only global properties of it are known, and no
additional (or local) per-input knowledge is known or taken into consideration.
However, there are practical applications where it is possible to estimate the noise
for each input example, or the noise properties may be known. For example, in
adaptive filtering systems designed for acoustic echo cancellation, the label noise
variance can be estimated during silences [18]. In channel equalization systems
that used in digital receivers, the noise variance per-input instance may be given
as side information to the equalizer from other channel. In other cases where
privacy is an issue, data may be "sanitized", which corresponds to perturbing
data items with some noise [6].

We propose new online regression algorithms for the case that the variance
of the noise per input-instance is known. Our algorithms are similar to the

© Springer International Publishing AG 2017
M. Ceci et al. (Eds.): ECML PKDD 2017, Part II, LNAI 10535, pp. 355–369, 2017.
https://doi.org/10.1007/978-3-319-71246-8_22

Widrow-Hoff algorithm [24] and are derived from an optimization of an objective capturing the change of a model and loss (or cost) on the current input [13]. We also derive a few simple strategies to estimate the noise variance, and yield algorithms that do not assume any knowledge about the statistical properties of the noise. All of our algorithms modify the learning rate based on the noise (directly or indirectly). We evaluate our algorithms in two synthetic settings and also with a real-world speech filtering task, and show that our algorithms perform well, especially when the noise is not stationary. We then extend the setting to the case when a learning algorithm is allowed to measure (or query) more than single (noisy) feedback label, and propose a label sampling rule that outperforms other possible approaches.

Most of previous work on label noise consider the *classification* setting, see two previous works [14,16] and the references therein. Cesa-Bianchi et al. [5] considered online learning with noisy data and square loss. In their setting the features and the labels are noisy, and they analyzed the effect of noise on the stochastic gradient descent (SGD) algorithm. Their regret bound scales with the amount of noise. In addition, for the case where the feature's noise covariance is known they suggested an adaptation of the SGD algorithm. We focus on *label* noise and propose another modification. Note that in the setting of Cesa-Bianchi et al. [5] there is a global known bound on the noise variance for all examples, however we assume that per-example noise variance is known (or estimated).

The rest of the paper is organized as follows. In Sect. 2 we formally introduce the setting of the problem. Then, in Sect. 3 we derive a scaling rule to deal with label noise. In Sect. 4 we perform extensive empirical evaluation that demonstrate the usefulness of our approach, even when the noise variance is unknown. We summarize in Sect. 5 and outline some future research.

2   Problem Setting

We consider online linear regression with noisy feedback (or measurement). On each round t, a learning algorithm observes an input instance $x_t \in \mathbb{R}^d$ and outputs a prediction $\hat{y}_t \in \mathbb{R}$. After an algorithm outputs a prediction, it has access to a *noisy* unbiased version $\tilde{y}_t$ of the true label $y_t \in \mathbb{R}$, i.e. $\mathbb{E}\tilde{y}_t = y_t$ where the noise's variance is finite, $\mathbb{E}(\tilde{y}_t - y_t)^2 \doteq v_t$, We first assume that v_t is known to the learning algorithm. Later, in Sect. 4 we consider also the case where v_t is unknown. After observing the noisy label $\tilde{y}_t$ an algorithm may update its prediction rule, and then proceeds to the next round.

The loss of an algorithm on round t is measured by the expected square loss $\ell_t(alg) = \mathbb{E}(\hat{y}_t - y_t)^2$ where the expectation is with respect to labels noise distribution and any randomization in an algorithm. The total loss after T rounds is $L_T(alg) = \sum_{t=1}^{T} \ell_t(alg)$. Note that an algorithm is evaluated against the true labels y_t and not the noisy labels $\tilde{y}_t$.

3   Online Regression with Label Noise

A standard paradigm for deriving learning algorithms is minimizing a loss function under a regularization constraint. Equivalently, one can minimize an unconstrained objective which is a sum of the loss and a regularization term. In online learning, a common practice (e.g. [9,13]) to derive an update rule for the learned model parameters $\mathcal{M}$ at time step t is to minimize the function,

$$C\left(\mathcal{M}\right) = dis\left(\mathcal{M}, \mathcal{M}_{t-1}\right) + \eta\ell\left(\mathcal{M}, (x_t, y_t)\right), \tag{1}$$

where $\eta > 0$ is a tradeoff paramater between being correct on the current input and being conservative, i.e. not being far from the previous model [13]. Here, $dis\left(\mathcal{M}, \mathcal{M}_{t-1}\right)$ is a dissimilarity (distance) measure between a new model $\mathcal{M}$ and the previously learned model $\mathcal{M}_{t-1}$, and $\ell\left(\mathcal{M}, (x_t, y_t)\right)$ is the loss of the new model $\mathcal{M}$ on the example (x_t, y_t). First-order algorithms [4,8,13], such as the Widrow-Hoff algorithm [24] (known also as LMS or its variant NLMS [19]), often employ a weight vector $\mathcal{M} = \{w\}$, while second-order algorithms [9–11,15,20,23] model also second-order covariance-like information, maintaining both a vector w and a PSD matrix Σ, i.e. $\mathcal{M} = \{w, \Sigma\}$. Kivinen and Warmuth [13] derived first-order algorithms for regression by minimizing (1) with the squared loss. For the case when dis is the euclidean distance, minimizing (1) results in a simple additive update rule for w. For the case when dis is the relative entropy function, minimizing (1) results in the exponentiated gradient (EG) algorithm. This paradigm was later applied to classification, where the hinge loss was used rather than the square loss (e.g. first-order passive-aggressive algorithms [8] and second-order confidence weighted algorithms [9]).

We describe a framework for online regression with label noise. In order to deal with label noise we shrink each noisy example $(x_t, \tilde{y}_t)$ by a factor $\sqrt{\alpha_t}$, where $\alpha_t \in [0, 1]$. (The motivation for using a square-root will be clear shortly.) Intuitively, when the noise variance v_t is large for an example (very noisy label), we like α_t to be small so that this example will have a small affect on the learned model $\mathcal{M}$. Our main goal is to derive rules for $\alpha_t = \alpha_t(v_t)$. As one cannot directly minimize (1) (because the clean y_t is unknown), we approximate the objective (1) to be

$$\widetilde{C}(\mathcal{M}; \alpha_t) = d\left(\mathcal{M}, \mathcal{M}_{t-1}\right) + \eta\ell\left(\mathcal{M}, (\sqrt{\alpha_t}x_t, \sqrt{\alpha_t}\tilde{y}_t)\right), \tag{2}$$

where we replaced x_t and y_t in (1) with $\sqrt{\alpha_t}x_t$ and $\sqrt{\alpha_t}\tilde{y}_t$ respectively.

Our algorithm sets the weight α_t to be the solution of the following problem,

$$\alpha_t = \arg\min_\alpha \mathbb{E}\left[\min_\mathcal{M} \widetilde{C}\left(\mathcal{M}; \alpha\right)\right],$$

where the expectation is with respect to the labels noise distribution. That is, α_t is chosen to minimize the average (over noise) optimal (over models, per noise instance) objective over all instances of noise. In other words, we seek for the single (or shared) value α_t that performs well on average.

Let us now develop the above equation. We start with some notation, let

$$\mathcal{M}_t(\alpha) = \arg\min_{\mathcal{M}} \widetilde{C}(\mathcal{M}; \alpha)$$

to be the updated model at round t. We substitute $\mathcal{M}_t$ in (1),

$$C(\mathcal{M}_t(\alpha); \alpha) = d(\mathcal{M}_t(\alpha), \mathcal{M}_{t-1}) + \eta\ell(\mathcal{M}_t(\alpha), (x_t, y_t)).$$

Next we optimize for α_t,

$$\alpha_t = \arg\min_{\alpha} \mathbb{E}[C(\mathcal{M}_t; \alpha)],$$

where the expectation is with respect to the labels noise distribution.

In this work we adapt the additive update rule from [13] to the noisy setting. The algorithm maintains a weight vector $w \in \mathbb{R}^d$. Given a new example (x_t, y_t) the algorithm predicts $\hat{y}_t = w_{t-1}^\top x_t$ and then sets the weight vector to be the minimizer of the following objective,

$$C(w) = \frac{1}{2}\|w - w_{t-1}\|^2 + \frac{1}{2r}\left(y_t - w^\top x_t\right)^2, \qquad (3)$$

where $r > 0$ is a tradeoff parameter. Note that for the special case $r \to 0$ we have that (3) becomes $\min_w \|w - w_{t-1}\|^2$ s.t. $w^\top x_t = y_t$, which recovers the normalized least mean squares (NLMS) algorithm [2]. Substituting $\sqrt{\alpha_t}x_t$ and $\sqrt{\alpha_t}\tilde{y}_t$ instead of x_t and y_t in (3) we get

$$\widetilde{C}(w; \alpha_t) = \frac{1}{2}\|w - w_{t-1}\|^2 + \frac{\alpha_t}{2r}\left(\tilde{y}_t - w^\top x_t\right)^2.$$

By setting $\nabla_w \widetilde{C}(w; \alpha_t) = 0$ the update rule for w becomes

$$\begin{aligned}
w_t &= w_{t-1} + \frac{\left(\sqrt{\alpha_t}\tilde{y}_t - w_{t-1}^\top\sqrt{\alpha_t}x_t\right)\sqrt{\alpha_t}x_t}{r + \left\|\sqrt{\alpha_t}x_t\right\|^2} \\
&= w_{t-1} + \frac{\left(\tilde{y}_t - w_{t-1}^\top x_t\right)x_t}{r_t + \|x_t\|^2},
\end{aligned} \qquad (4)$$

where $r_t \doteq \frac{r}{\alpha_t}$. Next, we substitute (4) in (3),

$$\begin{aligned}
C(w_t) =\ &\frac{1}{2}\left\|\frac{\left(\tilde{y}_t - w_{t-1}^\top x_t\right)x_t}{r_t + \|x_t\|^2}\right\|^2 + \frac{1}{2r}\left(y_t - w_{t-1}^\top x_t - \frac{\left(\tilde{y}_t - w_{t-1}^\top x_t\right)\|x_t\|^2}{r_t + \|x_t\|^2}\right)^2 \\
=\ &\frac{1}{2r}\frac{\left(\tilde{y}_t - w_{t-1}^\top x_t\right)^2\|x_t\|^2\left(r + \|x_t\|^2\right)}{\left(r_t + \|x_t\|^2\right)^2} + \frac{1}{2r}\left(y_t - w_{t-1}^\top x_t\right)^2 \\
&- \frac{1}{r}\frac{\left(y_t - w_{t-1}^\top x_t\right)\left(\tilde{y}_t - w_{t-1}^\top x_t\right)\|x_t\|^2}{r_t + \|x_t\|^2}.
\end{aligned}$$

Taking expectation and using $\mathbb{E}\left[\tilde{y}_t\right] = y_t$ and $\mathbb{E}\left[\left(\tilde{y}_t - y_t\right)^2\right] = v_t$, we get

$$\mathbb{E}\left[C\left(w_t\right)\right] = \frac{1}{2r} \frac{\left(\left(y_t - w_{t-1}^\top x_t\right)^2 + v_t\right) \|x_t\|^2 \left(r + \|x_t\|^2\right)}{\left(r_t + \|x_t\|^2\right)^2}$$
$$+ \frac{1}{2r}\left(y_t - w_{t-1}^\top x_t\right)^2 - \frac{1}{r}\frac{\left(y_t - w_{t-1}^\top x_t\right)^2 \|x_t\|^2}{r_t + \|x_t\|^2}.$$

From $\frac{d}{dr_t}\left(\mathbb{E}\left[C\left(w_t\right)\right]\right) = 0$ we get the optimal value of r_t,

$$r_t = r + \frac{r + \|x_t\|^2}{\left(y_t - w_{t-1}^\top x_t\right)^2} v_t,$$

and

$$\alpha_t = \frac{r}{r_t} = \frac{1}{1 + \frac{r + \|x_t\|^2}{r\left(y_t - w_{t-1}^\top x_t\right)^2} v_t}. \tag{5}$$

By substituting the optimal α_t back in (4) we have that

$$w_t = w_{t-1} + \left(\frac{1}{1 + \frac{v_t}{(y_t - w_{t-1}^\top x_t)^2}}\right) \frac{(\tilde{y}_t - w_{t-1}^\top x_t) x_t}{r + \|x_t\|^2}.$$

We observe from the above update-rule that the learning rate is adapted according to the ratio of the noise variance and the algorithm's loss (with the clean label) on the current example. When the noise variance is large, compared to the loss suffered by the algorithm (with the clean label), the learning rate is small, as the uncertainty (due to label noise) of the update is large compared to the need to update (large loss).

However, the value of α_t in (5) depends on the clean label y_t which is unknown. Instead, in the experiments below we consider a few variants to replace y_t in (5). We also consider a more general form, which tries to imitate the dependence in (5) of α_t on v_t,

$$\alpha_t = \frac{1}{1 + \beta v_t}, \tag{6}$$

for a parameter $\beta \geq 0$. The parameter β is tuned on some grid (see below). A summary of the algorithm, which we call ORS (Online Regression with Scaling), appears in Fig. 1.

Remark 1. *An alternative way to derive the formula (5) is the following. Consider minimization of $C(w)$ in (3), which yields the update rule*

$$w_t = w_{t-1} + \frac{\left(y_t - w_{t-1}^\top x_t\right) x_t}{r + \|x_t\|^2}.$$

Parameter: $r > 0, \beta \geq 0$
Initialize: Set $w_0 = 0 \in \mathbb{R}^d$
For $t = 1, \ldots, T$ do
  − Receive an instance $x_t \in \mathbb{R}^d$
  − Output prediction $\hat{y}_t = w_{t-1}^\top x_t$
  − Receive a noisy label $\tilde{y}_t \in \mathbb{R}$
  − Set scaling factor α_t according to (5) or (6)
  − Update

$$w_t \leftarrow w_{t-1} + \frac{\left(\tilde{y}_t - w_{t-1}^\top x_t\right) x_t}{\frac{r}{\alpha_t} + \|x_t\|^2}$$

Fig. 1. ORS - Online Regression with Scaling.

Substitute $\sqrt{\alpha_t} x_t$ *and* $\sqrt{\alpha_t} \tilde{y}_t$ *instead of* x_t *and* y_t *(and using* $\tilde{w}_t$ *to denote the new vector based on the noisy label) we get,*

$$\tilde{w}_t = w_{t-1} + \frac{\left(\tilde{y}_t - w_{t-1}^\top x_t\right) x_t}{\frac{r}{\alpha_t} + \|x_t\|^2}.$$

We minimize the mean-square-error,

$$\alpha_t = \arg\min_\alpha \mathbb{E} \|\tilde{w}_t - w_t\|^2$$

$$= \arg\min_\alpha \mathbb{E}\left[\left(\frac{\tilde{y}_t - w_{t-1}^\top x_t}{\frac{r}{\alpha} + \|x_t\|^2} - \frac{y_t - w_{t-1}^\top x_t}{r + \|x_t\|^2} \right)^2 \|x_t\|^2 \right]$$

$$= \arg\min_\alpha \mathbb{E}\left[\left(\frac{\tilde{y}_t - w_{t-1}^\top x_t}{\frac{r}{\alpha} + \|x_t\|^2} \right)^2 - 2 \frac{\left(\tilde{y}_t - w_{t-1}^\top x_t\right)\left(y_t - w_{t-1}^\top x_t\right)}{\left(\frac{r}{\alpha} + \|x_t\|^2\right)\left(r + \|x_t\|^2\right)} \right.$$

$$\left. + \left(\frac{y_t - w_{t-1}^\top x_t}{r + \|x_t\|^2} \right)^2 \right]$$

$$= \arg\min_\alpha \left(\frac{\left(y_t - w_{t-1}^\top x_t\right)^2 + v_t}{\left(\frac{r}{\alpha} + \|x_t\|^2\right)^2} - 2 \frac{\left(y_t - w_{t-1}^\top x_t\right)^2}{\left(\frac{r}{\alpha} + \|x_t\|^2\right)\left(r + \|x_t\|^2\right)} \right).$$

Setting the derivative of the objective to 0 we get

$$\frac{2r}{\alpha^2} \frac{\left(y_t - w_{t-1}^\top x_t\right)^2 + v_t}{\left(\frac{r}{\alpha} + \|x_t\|^2\right)^3} - \frac{2r\left(y_t - w_{t-1}^\top x_t\right)^2}{\alpha^2 \left(\frac{r}{\alpha} + \|x_t\|^2\right)^2 \left(r + \|x_t\|^2\right)} = 0$$

$$\Rightarrow \frac{\left(y_t - w_{t-1}^\top x_t\right)^2 + v_t}{\frac{r}{\alpha} + \|x_t\|^2} = \frac{\left(y_t - w_{t-1}^\top x_t\right)^2}{r + \|x_t\|^2}.$$

Solving for α *we get (5).*

3.1   Discussion

There is a line of research in adaptive filtering theory that discusses NLMS algorithms with *variable* step size (VSS) [7,12,17,18,22,25]. In all these works the update rule is *assumed* beforehand to be of the additive form $w_t = w_{t-1} + \mu_t x_t e_t$ where $e_t = \tilde{y}_t - w_{t-1}^\top x_t$ is the error signal, and the only question is how to choose the step size to control the tradeoff between fast convergence rate and low steady-state misalignment. A common approach is to require $y_t = w_t^\top x_t$, which leads to $\mu_t = \left(1 - \frac{\varepsilon_t}{e_t}\right) / \|x_t\|^2$ where $\varepsilon_t = \tilde{y}_t - y_t$ is the instantaneous noise. As this result is impractical from many points of view (ε_t is unknown, μ_t could become negative), a few modifications were advised. One approach [12] suggested to employ a bound on the instantaneous noise, and also limit μ_t below by 0. Another approach [1], is to use $\mu_t = \left(1 - \frac{\sqrt{v_t}}{e_t}\right) / \|x_t\|^2$ where it is assumed that the noise variance v_t is known. In case the noise variance is unknown some heuristics have been proposed how to estimate it in some specific adaptive filtering applications (e.g in echo cancellation, it can be estimated during silences [18]). Valin and Collins [21] proposed another modification with the INLMS algorithm, which adapts the learning rate to achieve fast convergence.

In contrary to this previous work we do not assume any type of the update rule. We solve an optimization problem, which can also be generalized to other loss functions and dissimilarity measures. For the case of square loss and euclidean distance we derive an update rule which turns out to be NLMS-like with step size $\mu_t = 1 / \left(\frac{r}{\alpha_t} + \|x_t\|^2\right)$. Here the regularization parameter r is modulated by the label noise variance according to (5).

4   Experimental Study

We evaluated the update rule (4) with adaptive scaling (5) and (6). In Sect. 4.1 we consider the case of a single noisy label, and compare a few variants in case the noise variance is known or unknown. Then, in Sect. 4.2 we assume that $k > 1$ noisy labels per example can be sampled, and propose a label sampling rule that uses the instantaneous margin, which together with scaling outperforms other baselines.

4.1   Single Feedback Label

We use two synthetic and one real-world datasets, and compare the mean square error during the algorithm's run-time. We distinguish between two settings:

Known variance: in this case we assume that the label noise variance is known, so that we can employ it in (5). We consider the following variants:

- 'Clean feedback' - there is no label noise. This is the *only* variant without label noise and is used for reference.

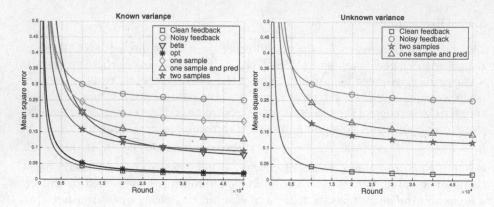

Fig. 2. Synthetic dataset. Left - known variance, right - unknown variance

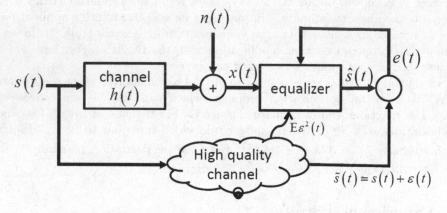

Fig. 3. Communication scheme with noisy equalizer feedback.

- 'Noisy feedback' - the algorithm does not apply scaling for examples ($\alpha_t = 1$).
- 'beta' - (6) is used as a scaling rule.
- 'opt' - (5) is used as a scaling rule, when we assume that y_t is *known*. Note that the update (4) is still performed with a noisy label $\tilde{y}_t$.
- 'one sample' - (5) is used, approximating y_t with $\tilde{y}_t$.
- 'one sample and pred' - (5) is used, approximating y_t with the average of the label $\tilde{y}_t$ and the prediction $\hat{y}_t$.
- 'two samples' - in this variant we assume that the algorithm has access to 2 samples of the label. In this case we employ the average of the labels as an approximation to y_t in (5) and $\tilde{y}_t$ in (4). The motivation for this variant will become clear when we discuss the unknown variance algorithms now.

Unknown variance: in this case we assume that the label's noise variance is unknown, and to employ (5) the algorithm should estimate the variance in some way. We consider the following variants:

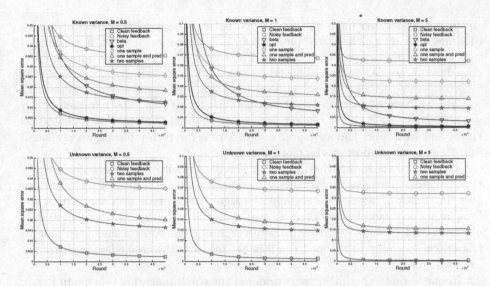

Fig. 4. Communication scheme evaluation. Top - known variance for different noise levels, bottom - unknown variance for different noise levels

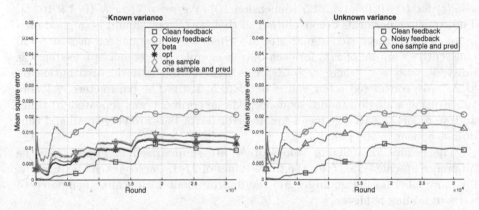

Fig. 5. Adaptive speech filter evaluation with known variance (left) and unknown variance (right).

- 'Clean feedback' - there is no label noise. This is the *only* variant without label noise and used for reference.
- 'Noisy feedback' - the algorithm does not apply scaling of examples ($\alpha_t = 1$).
- 'one sample and pred' - (5) is used, when we employ $\tilde{y}_t$ and the prediction $\hat{y}_t$ to approximate $y_t = (\tilde{y}_t + \hat{y}_t)/2$ and $v_t = (\tilde{y}_t - \hat{y}_t)^2/4$.
- 'two samples' - in this variant we assume that the algorithm has access to 2 noisy samples of the label - $\tilde{y}_{t,1}$ and $\tilde{y}_{t,2}$. We approximate $y_t = (\tilde{y}_{t,1} + \tilde{y}_{t,2})/2$ and $v_t = (\tilde{y}_{t,1} - \tilde{y}_{t,2})^2/4$ in (5). Also we employ $(\tilde{y}_{t,1} + \tilde{y}_{t,2})/2$ in (4).

In our first experiment we used a synthetic dataset with $50,000$ examples of dimension $d = 20$. The inputs $x_t \in \mathbb{R}^{20}$ were drawn from a zero-mean unit-covariance Gaussian distribution. The target $u \in \mathbb{R}^{20}$ is a zero-mean unit-covariance Gaussian vector. The labels were set according to $y_t = x_t^\top u + n_t$ where $n_t \sim \mathcal{N}(0, 0.01)$. We introduced label noise $\tilde{y}_t = y_t + \epsilon_t$ where $v_t = \mathbb{E}\epsilon_t^2$ was chosen from uniform distribution on $[0, M]$ for $M = 5$. The parameters r, β were tuned using a single random sequence. The experiment was repeated 20 times and we report the average mean-square-error (error bars are very small and thus do not appear in the plots).

The results are summarized in Fig. 2. For the known variance case we observe that (6) outperforms other variants. However, it requires tuning of the β parameter. In addition, taking into account the algorithm's prediction can help in the long run ('one sample and pred' curve). For the unknown variance case it is clearly better to employ (5) with the prediction ('one sample and pred' curve) than doing nothing. In the long run it is not far from the case when two samples are available.

In our second experiment we considered the communication scheme in Fig. 3. The $50,000$ input signal samples $s(t)$ were drawn from a zero-mean unit-variance Gaussian distribution. The channel $h(t)$ is a typical band-limited channel $h(t) = sinc(t)$ for $t = 0, 0.5, 1, ..., 4.5$ (dimension 10). We set $n(t) \sim \mathcal{N}(0, 4 \times 10^{-4})$. The equalizer performs deconvolution so that its output should be a good estimate of the transmitted signal $s(t)$. To this end, the equalizer uses $d = 20$ consecutive samples of $x(t)$ for each output sample. The equalizer's feedback is noisy, $\tilde{s}_t = s_t + \epsilon_t$ where $v_t = \mathbb{E}\epsilon_t^2$ was chosen from uniform distribution on $[0, M]$. We considered a few values of M (0.5, 1, 5). The parameters r, β were tuned using a single random sequence. The experiment was repeated 20 times and we report the average mean-square-error (error bars are very small and thus do not appear in the plots).

The results for different values of M are summarized in Fig. 4, note the difference in scale (y-axis) for the three values of M, i.e. across each line. Clearly, for more label noise (i.e. larger M) the improvement with scaling compared to without scaling is bigger.

In our third experiment we considered training of adaptive speech filter. The general scheme is similar to Fig. 3. A clean speech signal was passed through a typical room impulse response $h(t)$ of length 1024. We set $n(t) \sim \mathcal{N}(0, 4 \times 10^{-4})$. The purpose is to train adaptive filter that will recover the original speech signal. To this end, the filter uses $d = 1024$ consecutive samples of $x(t)$ for each output sample. We assume that the feedback of the filter was recorded with hammer noise in the background. The parameters r, β were tuned on 20% of the signal. The results are summarized in Fig. 5. Again, scaling can improve the performance of the adaptive filter. We emphasis, this approach does not have any additional information beyond the noisy label, and yet it works much better than just using the noisy label.

4.2   From 1 to k Labels

In this part we assume that the algorithm has access to a single feedback label or more each round. Specifically, there is a budget parameter B which defines the average number of labels per round. In other words, for n rounds the total number of labels is $B \times n$. We assume that each round the algorithm has access to a single noisy label or $k = \lceil B \rceil$ noisy labels. Note that for $k > 1$ the noise variance is reduced by a factor of k, which allows a more accurate update.

We employ the first synthetic dataset from Sect. 4.1, with $n = 10000$ examples. In addition to the uniform noise variance distribution (as described in Sect. 4.1) we consider also the cases of linearly increasing noise variance (from 0 to 5) and linearly decreasing noise variance (from 5 to 0).

We consider the following strategies for spending the given labels budget B:

- 'const' - in this case, in each round, k labels are sampled with probability $p = (B-1)/(k-1)$.
- 'begin' - in this case, k labels are sampled at the beginning of the algorithm's runtime. For total n rounds, k labels are sampled for the first $n \times p$ rounds.
- 'auto' - in this case, in each round 1 label $\tilde{y}_t$ is sampled and then we use the prediction $\hat{y}_t$ to decide whether to sample $k-1$ more labels. Our decision rule is probabilistic, and we decide to stay with 1 label with probability $\frac{a}{a+(\tilde{y}_t - \hat{y}_t)^2}$ for a positive parameter a. The parameter a is tuned manually so that the total number of labels is (close to) $B \times n$. A disadvantage of this strategy is that given some budget B it is not clear how to set a to spend the budget. A similar issue has been reported in randomized selective sampling algorithms [3]. To solve it, we propose the next strategy.
- 'auto self tuned' - this strategy is similar to 'auto', yet the algorithm receives as input the given budget B and tunes on the fly the parameter a as following: we start by setting $a = 1$, then each block of 100 rounds we check the instantaneous labels budget until the current round. If it is larger than the given budget, we try to decrease it by setting $a \leftarrow 2a$. However, if the instantaneous budget is smaller than the given budget we try to increase it by setting $a \leftarrow a/2$. In order to make sure that we do not sample more labels than the given budget, we take a safety margin and actually guide the algorithm to sample 5% less than the given budget. We call it limited budget, which equals $0.95B$.

We run the update (4) with scaling (5). For rounds where we have only a single label, we employ the prediction and update as described above for the 'one sample and pred' case for unknown variance. For rounds where we have k labels, we employ the sample mean (in (4) and (5)) and the sample variance (in (5)). In addition, we run a version of the algorithm (4) without scaling (i.e. $\alpha_t = 1$), which corresponds to "no scaling" lines in the plots. In this case the update (4) is performed with the single label or the mean for k labels.

The results are reported in Fig. 6, where the error bars correspond to the 95% confidence interval over 20 runs. Clearly, for all 3 noise models the scaling approach outperforms the 'no scaling' approach. Also, the 'auto' sampling strategy

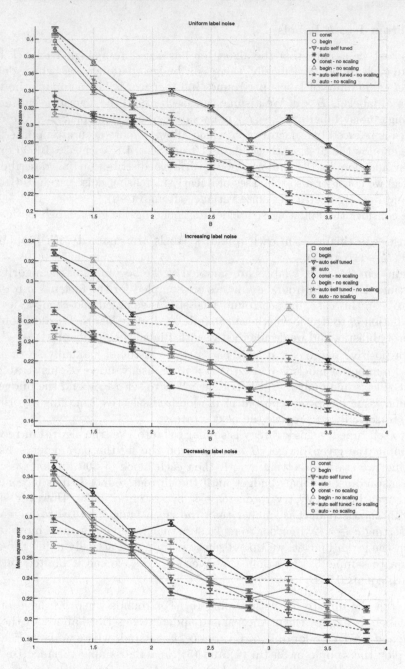

Fig. 6. Synthetic dataset with variable labels budget.

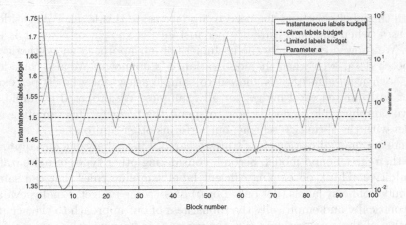

Fig. 7. Evaluation of the self-tuned scheme.

usually outperform 'begin' and 'const'. We note also that the self tuned version ('auto self tuned'), which does not require any manual tuning and samples 5% less labels on average, works quite well and not too much worse than 'auto'.

For decreasing noise it is beneficial to sample k labels at the beginning. This is expected as it allows learning with effectively smaller noise at the beginning, brining the algorithm closer (compared to using only a single label) to a good model. Then the algorithm can employ the predictions together with only a single noisy label.

From the uniform noise plot, we observe that for $B > 2$ 'auto' outperform 'begin' (with scaling). This is expected, as for larger B (i.e. more labels) the algorithm makes more accurate predictions, which in turn leads to a better estimation of the margin in the 'auto' sampling rule. Also note that for the 'no scaling' case, 'begin' performs like 'const'. This is because the 'no scaling' algorithm does not use the predictions to improve its performance. From the increasing noise plot, we observe that the 'begin' strategy, which spends the labels budget at the beginning, does not perform well. This is expected as we need more labels where the noise variance is large.

Finally, in Fig. 7 we demonstrate how the instantaneous budget and the parameter a change during a run of the 'auto self tuned' version (for a given budget $B = 1.5$). We note that the instantaneous budget converges close to the limited budget, and any case we do not sample more labels than the given budget.

5   Summary and Future Directions

We proposed a shrinkage scheme for online regression (adaptive filtering) with noisy feedback. Our algorithms are theoretically motivated by minimizing the average (over noise) optimal objective which captures the change of a model and loss. Our algorithms adapt the learning rate by taking in account the tradeoff between the label noise and the need to update using this label. We considered both cases, when the noise-variance per instance is known, or when it is estimated. We showed empirically the usefulness of our approach on synthetic and real speech data, even when the noise variance is unknown. We also considered the case when an algorithm is allowed to sample a few noisy labels per example, yet with limited total budget for all examples. We proposed a simple and effective sampling rule to decide how to spend the budget. This sampling rule, together with the shrinkage scheme, outperforms other approaches.

There are many directions for future research. First, analyzing our approach, e.g. in the regret bound model. Second, considering other types of noise, rather than additive. Third, developing efficient label sampling rules that can sample any number of noisy labels (not only 1 or k) given some budget. Finally, evaluating theoretically and empirically the robustness of our approach to the accurate estimation of the noise variance (i.e. the variance is known with some error).

References

1. Benesty, J., Rey, H., Vega, L.R., Tressens, S.: A nonparametric VSS NLMS algorithm. IEEE Sig. Process. Lett. **13**(10), 581–584 (2006)
2. Bershad, N.J.: Analysis of the normalized LMS algorithm with Gaussian inputs. IEEE Trans. Acoust. Speech Sig. Process. **34**(4), 793–806 (1986)
3. Cesa-Bianchi, N., Gentile, C., Zaniboni, L.: Worst-case analysis of selective sampling for linear classification. J. Mach. Learn. Res. **7**, 1205–1230 (2006)
4. Cesa-Bianchi, N., Long, P.M., Warmuth, M.K.: Worst case quadratic loss bounds for on-line prediction of linear functions by gradient descent. Technical Report IR-418, University of California, Santa Cruz, CA, USA (1993)
5. Cesa-Bianchi, N., Shalev-Shwartz, S., Shamir, O.: Online learning of noisy data. IEEE Trans. Inf. Theory **57**(12), 7907–7931 (2011)
6. Chawla, S., Dwork, C., McSherry, F., Smith, A.D., Wee, H.: Toward privacy in public databases. In: Proceedings of Theory of Cryptography, Second Theory of Cryptography Conference, TCC 2005, Cambridge, MA, USA, 10–12 February 2005, pp. 363–385 (2005)
7. Ciochină, S., Paleologu, C., Benesty, J.: An optimized NLMS algorithm for system identification. Sig. Process. **118**(C), 115–121 (2016)
8. Crammer, K., Dekel, O., Keshet, J., Shalev-Shwartz, S., Singer, Y.: Online passive-aggressive algorithms. JMLR **7**, 551–585 (2006)
9. Crammer, K., Kulesza, A., Dredze, M.: Adaptive regularization of weighted vectors. In: Advances in Neural Information Processing Systems, vol. 23 (2009)
10. Dredze, M., Crammer, K., Pereira, F.: Confidence-weighted linear classification. In: International Conference on Machine Learning (2008)
11. Forster, J.: On relative loss bounds in generalized linear regression. In: Ciobanu, G., Păun, G. (eds.) FCT 1999. LNCS, vol. 1684, pp. 269–280. Springer, Heidelberg (1999). https://doi.org/10.1007/3-540-48321-7_22
12. Gollamudi, S., Nagaraj, S., Kapoor, S., Huang, Y.F.: Set-membership filtering and a set-membership normalized LMS algorithm with an adaptive step size. IEEE Sig. Process. Lett. **5**(5), 111–114 (1998)
13. Kivinen, J., Warmuth, M.K.: Exponential gradient versus gradient descent for linear predictors. Inf. Comput. **132**, 132–163 (1997)

14. Liu, T., Tao, D.: Classification with noisy labels by importance reweighting. IEEE Trans. Pattern Anal. Mach. Intell. **38**(3), 447–461 (2016)
15. Moroshko, E., Crammer, K.: Weighted last-step min-max algorithm with improved sub-logarithmic regret. Theor. Comput. Sci. **558**, 107–124 (2014)
16. Natarajan, N., Dhillon, I.S., Ravikumar, P.K., Tewari, A.: Learning with noisy labels. Adv. Neural Inf. Process. Syst. **26**, 1196–1204 (2013)
17. Paleologu, C., Ciochină, S., Benesty, J., Grant, S.L.: An overview on optimized NLMS algorithms for acoustic echo cancellation. EURASIP J. Appl. Sig. Process. **2015**, 97 (2015)
18. Paleologu, C., Ciochina, S., Benesty, J.: Variable step-size NLMS algorithm for under-modeling acoustic echo cancellation. IEEE Sig. Process. Lett. **15**, 5–8 (2008)
19. Sayed, A.H.: Adaptive Filters. Wiley, New York (2008)
20. Vaits, N., Crammer, K.: Re-adapting the regularization of weights for non-stationary regression. In: The 22nd International Conference on Algorithmic Learning Theory, ALT 2011 (2011)
21. Valin, J., Collings, I.B.: Interference-normalized least mean square algorithm. IEEE Sig. Process. Lett. **14**(12), 988–991 (2007)
22. Vega, L., Rey, H., Benesty, J., Tressens, S.: A new robust variable step-size NLMS algorithm. Trans. Sig. Proc. **56**(5), 1878–1893 (2008)
23. Vovk, V.: Competitive on-line statistics. Int. Stat. Rev. **69**, 213–248 (2001)
24. Widrow, B., Hoff Jr., M.E.: Adaptive switching circuits (1960)
25. Yu, Y., Zhao, H.: A novel variable step size NLMS algorithm based on the power estimate of the system noise. CoRR abs/1504.05323 (2015)

Reinforcement Learning

Generalized Inverse Reinforcement Learning with Linearly Solvable MDP

Masahiro Kohjima[✉], Tatsushi Matsubayashi, and Hiroshi Sawada

NTT Service Evolution Laboratories, NTT Corporation, Kanagawa, Japan
{kohjima.masahiro,matsubayashi.tatsushi,sawada.hiroshi}@lab.ntt.co.jp

Abstract. In this paper, we consider a generalized variant of *inverse reinforcement learning* (IRL) that estimates both a cost (negative reward) function and a transition probability from observed optimal behavior. In theoretical studies of standard IRL, which estimates only the cost function, it is well known that IRL involves a non-identifiable problem, i.e., the cost function cannot be determined uniquely. This problem has been solved by using a new class of Markov decision process (MDP) called a linearly solvable MDP (LMDP). In this paper, we investigate whether a non-identifiable problem occurs in the generalized variant of IRL (gIRL) using the framework of LMDP and construct a new gIRL method. The contributions of this study are summarized as follows: (i) We point out that gIRL with LMDP suffers from a non-identifiable problem. (ii) We propose a Bayesian method to escape the non-identifiable problem. (iii) We validate the proposed method by performing an experiment on synthetic data and real car probe data.

Keywords: Inverse reinforcement learning · Linearly solvable MDP
Bayesian method

1 Introduction

Inverse reinforcement learning (IRL) is a method that estimates the cost (negative reward) function of a certain class of Markov decision process (MDP) from an agent's optimal behavior. Since designing a truly effective cost function is regarded as a difficult problem in various applications of *reinforcement learning* (RL) including robot control tasks, IRL attracted the attention of robotics researchers from an early stage [1]. Its application area is now spreading and the effectiveness of IRL has been reported for taxi driver destination prediction [2], preferred route estimation after a natural disaster [3], and natural language processing [4]. These studies show that IRL can estimate the cost functions of entities, such as people and animals, whose internal structure is unobservable and whose preferences remain vague.

In this paper, we consider a generalized variant of IRL that simultaneously estimates both the cost function and transition probability. Since this problem is a generalization of existing IRL methods that estimate only the cost function,

M. Ceci et al. (Eds.): ECML PKDD 2017, Part II, LNAI 10535, pp. 373–388, 2017.
https://doi.org/10.1007/978-3-319-71246-8_23

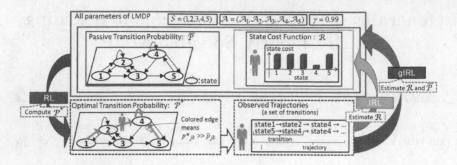

Fig. 1. Input and output of RL, IRL and gIRL with linearly solvable MDP.

we call it *generalized IRL* (gIRL). Figure 1 shows the input and output of RL, IRL and gIRL. Specifically, as shown in the figure, we tackle the gIRL using the framework of the linearly solvable MDP (LMDP) [5].

LMDP has been proposed as a new class of MDP where a forward problem (RL) is more easily solved than with standard MDP [5]. Dvijotham showed that IRL with the LMDP has a unique solution [6], i.e., the cost function generating agent behavior is uniquely identified. It is regarded as important result in IRL. The first IRL paper [7] proved the existence of a non-identifiable problem with standard MDP, and therefore the cost function is not unique and that a cost function with entirely zero values is always one of the solutions. Until Dvijotham's paper was published it had remained an open issue as to whether it was possible to avoid a non-identifiable problem.[1] However, a gIRL with the LMDP has not yet been studied. Since the number of transition probability parameters of the LMDP is smaller than that of a standard MDP, the use of the LMDP is suitable for gIRL.

The study most closely related to ours is the work reported by Makino and Takeuchi [8], which considers gIRL on the partially observable MDP for constructing efficient apprenticeship learning methods. It is experimentally confirmed that gIRL formulation contributes to the realization of a more effective policy [8]. However, the theoretical aspect of gIRL remained unknown and there is no gIRL method for the LMDP.

In this paper, we provide a theoretical analysis and a new method for gIRL. Beginning with an investigation as to whether a non-identifiable problem occurs in gIRL with LMDP, we establish a new formulation of gIRL and new gIRL methods. We apply the proposed method with both synthetic data and real car probe data collected in Yokohama City, Japan. The contributions of this paper can be summarized as follows:

---

[1] Although Ziebart et al. [2] also solve the non-identifiable problem by using the maximum entropy principle, Dvijotham and Todorov show that Ziebart's formulation is equivalent to the special case of an inverse problem of LMDP [6].

- We point out that generalized IRL using the framework of LMDP involves a non-identifiable problem; the cost function and transition probability cannot be uniquely estimated. This is because we cannot distinguish between the effect of the cost function and that of the transition probability on the observed transitions.
- To avoid the non-identifiable problem, we adopt a Bayesian approach with hyperparameters, which is also used approach for IRL with a standard MDP [8–11]. We also extend the LMDP to a multiple intention setting [11, 12] and use it to formulate generalized IRL. This enables us to apply the proposed method to many practical problems such as traffic data analysis. Our new Bayesian gIRL method with the extended LMDP can estimate the cost functions, the transition probability, and the hyperparameters.
- We confirm the effectiveness of the proposed method by performing numerical experiments using both synthetic data and real car probe data. The result of our car probe data experiment shows that the proposed method can estimate the LMDP parameters, which reflect car drivers' behavior.

The rest of this paper is organized as follows. In Sect. 2, we introduce the LMDP. The non-identifiable problem of the LMDP is illustrated in Sect. 3. Section 4 presents the extended LMDP and Sect. 5 introduces the proposed gIRL method. Section 6 is devoted to the experimental evaluation and Sect. 7 concludes the paper.

2   Linearly Solvable MDP (LMDP)

In this section, a basic property of the LMDP [5] is introduced. Although the definition of the LMDP is similar to that of the MDP, its difference is critical in creating solutions to the forward and inverse problems. Note that this work focuses on an "infinite horizon discounted cost" case [13]; however, its application to other settings is straightforward.

Definition of LMDP: The LMDP is defined by the quintuplet $\{\mathcal{S}, \mathcal{A}, \bar{\mathcal{P}}, \mathcal{R}, \gamma\}$, where $\mathcal{S} = \{1, 2, \cdots, S\}$ is a finite set of *states* and S is the number of states. $\mathcal{A} = \{\mathcal{A}_1, \mathcal{A}_2, \cdots, \mathcal{A}_S\}$ is a set of admissible actions at each states. $\bar{\mathcal{P}} = \{\bar{p}_{jk}\}_{j,k=1}^{S}$ indicates *passive transition probabilities*, each element of which defines the transition probability from state j to state k when an action is *not* executed. $\mathcal{R} : \mathcal{S} \to \mathbb{R}$ is a *state cost function* (*negative reward function*) and we denote the state cost at state j as r_j. $\gamma \in [0, 1)$ is a *discount factor*.

In the LMDP, *action* $\boldsymbol{a}$ is a continuous valued $\mathbb{R}^S$ dimensional vector and the *action transition probability* from state j to state k when action $\boldsymbol{a}_j = \{a_{jk}\}_{k=1}^{S}$ is executed is defined by

$$p_{jk}(\boldsymbol{a}_j) = \bar{p}_{jk} \exp(a_{jk}). \tag{1}$$

Note that any action executed at state j, $\boldsymbol{a}_j$, must belong to a set of admissible actions, $\mathcal{A}_j$, which is defined as

$$\mathcal{A}_j = \Big\{ \boldsymbol{a}_j \in \mathbb{R}^S \big| \sum_k p_{jk}(\boldsymbol{a}_j) = 1; \bar{p}_{jk} = 0 \to a_{jk} = 0 \Big\}, \tag{2}$$

so that the sum of the probabilities equals one. Therefore, the transition probability itself can be controlled by an action. To execute a certain action, it is necessary to pay the action cost defined by *action cost function*. The action cost when action a_j is executed in state j is defined as

$$q_j(a_j) = KL(p_j(a_j)\|p_j(0)), \tag{3}$$

where $KL(\cdot\|\cdot)$ is the Kullback-Leibler divergence and $p_j(a) = \{p_{jk}(a)\}_{k=1}^S$. Thus, the action cost increases as $p_{jk}(a)$ deviates further from a passive transition $\bar{p}_{jk}$. Note that when the action is a zero vector, $a = 0$, $p_{jk}(0)$ equals the passive transition probability $\bar{p}_{jk}$ and the action cost $q_j(0) = 0$. Intuitively, the LMDP is a class of MDP in which the transition probability itself can be controlled by the payment of the action cost. Unlike standard MDP where the transition probability is defined separately for each action, the passive transition probability substantially determines the transition probability of all the actions. Thus, we consider that it is suitable to use the LMDP for gIRL.

Let $\pi = \{a_j\}_{j=1}^S$ be a policy whose element a_j indicates the action executed in state j. The *value function* of policy π, $v^\pi = \{v_j^\pi\}_{j=1}^S$, is defined such that element v_j^π indicates the expected sum of the future cost from state j when following policy π,

$$v_j^\pi = \lim_{T\to\infty} \mathbb{E}_{d^T}\left[\sum_{t=1}^T \gamma^{t-1}\{r_{s_t} + q_{s_t}(a_{s_t})\}\Big|s_1 = j\right]. \tag{4}$$

Here $\mathbb{E}_{d^T}$ denotes the expectation over trajectory $d^T = \{s_t\}_{t=1}^T$, the transitions from $t = 1$ to T where s_t denotes the visit state at time t, which follow probability $P(d^T|\bar{\mathcal{P}}, \pi) = p_{s_1}^{ini}\prod_{t=1}^{T-1} p_{s_t s_{t+1}}(a_{s_t})$. p^{ini} is the initial state distribution.

Forward Problem with LMDP: The forward problem with the LMDP is to obtain the optimal policy $\pi^* = \{a_j^*\}_{j=1}^S$ that minimizes the expected sum of the future cost. The optimal action in state j is given by

$$a_j^* = \arg\min_{a_j \in \mathcal{A}_j}\left\{r_j + q_j(a_j) + \gamma\sum_{k=1}^S p_{jk}(a_j)v_k\right\} = -\gamma v_j - \log\left(\sum_{k=1}^S \bar{p}_{jk}\exp(-\gamma v_k)\right), \tag{5}$$

where $v = \{v_j\}_{j=1}^S$ is the optimal value function $v_j = \min_\pi v_j^\pi$ that can be computed by solving the optimal equation [5]. Inserting Eq. (5) into Eq. (1), *optimal transition probability*, the action transition probability when the optimal action being executed is written as

$$p_{jk}^* = p_{jk}(a_j^*) = \frac{\bar{p}_{jk}\exp(-\gamma v_k)}{\sum_\ell \bar{p}_{j\ell}\exp(-\gamma v_\ell)}. \tag{6}$$

We emphasize that the above form of optimal transition probability is a direct consequence of the LMDP unlike Bayesian IRL, which uses the value function as a potential function [9].

3   Generalized IRL and the Non-identifiable Problem

3.1   Generalized Inverse Reinforcement Learning

This section illustrates the non-identifiable problem of generalized IRL (gIRL) with the LMDP. The purpose of gIRL is to estimate the state cost function and passive transition probability of the LMDP from a transition log that follows the optimal transition probability Eq. (6). Figure 1 illustrates the forward and inverse problems of the LMDP.

A key motivation for gIRL can be explained as follows. Let us consider a case where the cost function must be estimated only from the past movements of a person who is interested in a certain place in a city. In this case, the passive transition probability between places, which can be interpreted as the transition probability of a person who has a uniform state cost function (same degree of interest in each place), is of course unknown and cannot be observed. That is, gIRL is useful for estimating a state cost function when only a set of past movements is available, which is a common setting in various machine learning problems.

To determine whether the state cost function and passive transition probability can be uniquely estimated or not, we consider a case where the amount of available data is sufficiently large. In this case, the optimal transition probability itself can be observed. Therefore, we need to seek the corresponding relation between the optimal transition probability and a pair consisting of a state cost function and a passive transition probability.

3.2   Toy Example of Non-identifiable Problem

Figure 2 shows a toy example in which LMDPs with different passive transitions and value functions provide equivalent optimal transition probabilities. In the left dotted box in Fig. 2, the passive transition from state-1 to state-1, $\bar{p}_{11}$, is p and the value function of state-1, v_1, is v. Similarly, in the right dotted box, they are $p' = p/(K - pK + p)$ and $v' = -\log(K)/\gamma + v$. We can easily confirm that the optimal transition probabilities for both LMDPs are equivalent for arbitrary constant K. This means that we cannot identify the passive transition probability and value function simultaneously from the optimal transition probability. By considering the transition probability and value function to be parameters and the optimal transition probability to be a probabilistic model, this is a case where the model is *non-identifiable*[2]. Note that the above examples are compatible with the claim made by Dvijotham and Todorov [6]. Since they consider a case where the passive transition probability is known, the value function can be uniquely estimated.

Non-identifiability implies the impossibility of estimating the state cost function uniquely. Let us consider the setting of the constant value $K = \exp(\gamma v)$ in

---

[2] The probabilistic model P_θ is called *identifiable* in statistics if parameter $\theta_1 \neq \theta_2$, then distributions P_{θ_1} and P_{θ_2} are different [14]. A model which is not *identifiable* is called *non-identifiable*.

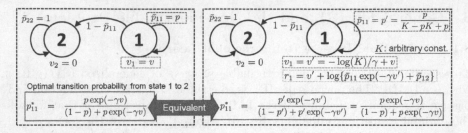

Fig. 2. An example that indicates that LMDPs with different passive transitions and value functions provides an equivalent optimal-transition probability. This implies that the passive transition probability and the value function cannot be uniquely estimated even if the optimal transition probability itself is observed. We call this problem the non-identifiable problem of gIRL with LMDP.

the previous toy example. In this case, the value and the state cost of state 1 become 0, $v_1 = r_1 = 0$, and the passive transition equals the optimal transition, $\bar{p}_{11} = p_{11}^*$. This means that the optimal transition probability of the left LMDP can be reproduced by the right LMDP with a state cost function whose values are all zero. This fact immediately leads to the following theorem:

Theorem 1 (Non-identifiability of gIRL with LMDP). *Let $\mathcal{S}$ and γ be a set of states and a discount factor. Then, the mapping from a pair consisting of passive transition probability $\bar{\mathcal{P}} \in [0,1]^{S \times S}$ and state cost function $\mathcal{R} \in \mathbb{R}^S$ to the optimal transition probability of the LMDP $(\mathcal{S}, \mathcal{A}, \bar{\mathcal{P}}, \mathcal{R}, \gamma)$ is not one-to-one.*

Proof. *When $\mathcal{R} = \mathbf{0}$, the passive transition probability and optimal transition probability are identical. Then, for any LMDP, there exists an LMDP that has an all zero state cost function and a passive transition probability that is identical to the optimal transition probability of a given LMDP.* □

This is an obviously unacceptable result because the transition probability and cost function have different roles in RL; cost is a target of the agent to be minimized, the transition probability determines the possible movements of the agent. Their two roles should not be mixed.

It is well-known for IRL with standard MDP that the cost function with entirely zero values is always one of the solutions [7]. Therefore, our observation indicates that the generalized IRL problem with the LMDP also raises similar theoretical concerns.

Remark 1. Note that we do not view this as a problem that the optimal transition is consistent by transformation $v_i' = v_i + c$ for all states i using a common constant value c while the passive transition probability remains fixed; this is because the magnitude relation of the value function holds. This type of degree of freedom can be removed by, for example, setting the value function of a certain state at zero. The problem tackled in this paper is the non-identifiability of the value function and the passive transition probability.

3.3 Approach for Non-identifiable Problem

We confirmed above that gIRL with LMDP suffers from non-identifiability. This subsection introduces an idea that can avoid this problem. A promising approach is to introduce hyperparameters. This approach is the same as that used by Ng and Russel for IRL with MDP to avoid non-identifiability [7]. They introduce hyperparameters to make the cost function sparse, i.e., the cost becomes zero in many states. However, as stated in their paper, a remaining problem was that the estimated result strongly depends on the manual setting of the hyperparameters. Therefore, we construct a gIRL method with a Bayesian framework that can estimate hyperparameters. By automatically estimating the hyperparameters, their dependency is weakened. The Bayesian approach is promising since its effectiveness has already been confirmed for standard IRL with an MDP [9–11]. We also introduce a new gIRL formulation for a later experiment.

Our new formulation considers a collection of LMDPs that share state and passive transition probabilities. The setting is referred to as multiple intention or multitask IRL in the literature [11,12]. The following example explains the motivation behind using this new formulation. Again, let us consider a case where the state cost function of a certain person needs to be extracted. If the trajectories of several people are available, the task seems obvious when we consider that only the cost function alone depends on each person and the passive transition probability is not person dependent. Thus, our new gIRL is formulated as the problem of estimating everybody's cost functions and common passive transition probabilities from observed trajectories. Thanks to this formulation, hyperparameters for the cost function are defined as common parameters among all people; this may contribute to performance improvement similar to that described in [11]. The next two sections present a rigorous formulation and an estimation algorithm.

4   Shared-Parameter LMDPs

In this section, we re-formulate gIRL. We consider a collection of LMDPs that share states $\mathcal{S}$, passive transition probability $\bar{\mathcal{P}}$, and discount factor γ. Each LMDP has its own state cost function, $\mathcal{R}_i$, where i is the index of the LMDP. We call this collection of LMDPs, *shared-parameter LMDPs* (SP-LMDPs). We formulate gIRL as an inverse learning problem to estimate the passive transition probability and all the state cost functions in the SP-LMDPs. Figure 3 shows all the parameters of the SP-LMDPs. gIRL with the SP-LMDPs is a natural extension of IRL because gIRL with SP-LMDPs can be seen as a setting in which multiple state cost functions are estimated. Each cost function may be the cost function of a different person, animal and so on. We emphasize that the setting at which multiple cost functions are defined on a *standard* MDP has already been studied [11,12] but it has not been studied for an LMDP.

We provide a formal definition of SP-LMDPs as follows. SP-LMDPs are defined by the quintuplet $\{\mathcal{S}, \mathcal{A}, \bar{\mathcal{P}}, \mathcal{R}, \gamma\}$. The definitions of $\mathcal{S}$, $\mathcal{A}$, $\mathcal{P}$ and γ are equivalent to those for an LMDP while that of $\mathcal{R}$ is different. $\mathcal{R} = (\mathcal{R}_1, \cdots, \mathcal{R}_I)$

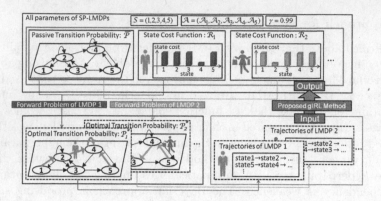

Fig. 3. Forward and inverse problems of SP-LMDPs. We call this inverse problem, which estimates all the state cost functions and passive transition probabilities of SP-LMDPs, gIRL with SP-LMDPs.

is a set of *state cost functions* and $\mathcal{R}_i = \{r_{ij}\}_{j=1}^S$. I is the number of functions in the set. From this definition, we can construct I LMDP; the i-th LMDP is defined as $\{\mathcal{S}, \mathcal{A}, \bar{\mathcal{P}}, \mathcal{R}_i, \gamma\}$. Note that SP-LMDPs with $I = 1$ reduce to an LMDP.

Since the forward problem of the i-th LMDP can be solved independently following the method explained in Sect. 2, SP-LMDPs pose no difficulty in solving the forward problem. Let us define $V = \{v_i\}_{i=1}^I$. $v_i = \{v_{ij}\}_{j=1}^S$ is the optimal value function of the i-th LMDP. This optimal value function satisfies the following optimal equation.

$$v_{ij} = \min_{a_{ij} \in \mathcal{A}_j} \left\{ r_{ij} + q_j(a_{ij}) + \gamma \sum_k p_{jk}(a_{ij}) v_{ik} \right\} = r_{ij} - \log\left(\sum_k \bar{p}_{jk} \exp(-\gamma v_{ik}) \right).$$

(7)

Then, the optimal transition probability from state j to state k is, for the i-th LMDP, given by

$$p_{ijk}^* = \frac{\bar{p}_{jk} \exp(-\gamma v_{ik})}{\sum_\ell \bar{p}_{j\ell} \exp(-\gamma v_{i\ell})}.$$

(8)

The above optimal transition probability shows that the agent executing the optimal policy tends to move adjacent states whose value functions are small.

5   Proposed Generalized IRL Method

5.1   Bayesian Modeling

This subsection details the proposed gIRL method, which can estimate both the state cost functions and passive transition probabilities with SP-LMDPs from observed transitions. We denote the transition logs of the i-th LMDP as $\mathcal{D}_i$ and the number of observed transitions from state j to state k in the i-th

LMDP as n_{ijk}. We also denote all the transition logs as $\mathcal{D} = \{\mathcal{D}_i\}_{i=1}^I$. Our gIRL method is naturally derived by considering that each transition is generated by the probability defined in Eq. (8) which has parameters $V, \bar{\mathcal{P}}$. In this section, we re-parametrize $\bar{p}_{jk}$ as $w_{jk} = -\log \bar{p}_{jk}$. We define $W = \{w_j\}_{j=1}^S$ and $w_j = \{w_{jk}\}_{k=1}^S$. Then, the probability that transition log $\mathcal{D}$ is generated given parameter V, W can be written as

$$P(\mathcal{D}|V, W) = \prod_i \prod_{j,k \in \mathcal{S}} \left(\frac{\exp(-w_{jk} - \gamma v_{ik})}{\sum_\ell \exp(-w_{j\ell} - \gamma v_{i\ell})} \right)^{n_{ijk}}. \tag{9}$$

We can avoid the ill-posedness of gIRL, and also obtain the full parameter-estimation procedure by adopting a Bayesian approach. We define a Gaussian prior distribution on v_i and w_j for all i, j given by

$$P(V|\alpha) = \prod_{i,j=1}^{I,S} \mathcal{N}(v_{ij}|0, \frac{1}{\alpha}), \ P(W|\beta) = \prod_{j=1}^S \prod_{k \in \Omega_j^{\mathrm{fr}}} \mathcal{N}(w_{jk}|0, \frac{1}{\beta}). \tag{10}$$

Note that Ω_j^{fr} denotes a set of reachable states "from" state j by a one step transition.[3] We also used a conjugate gamma prior on the hyper-parameters, similar to [15]:

$$P(\alpha) = \mathcal{G}(\alpha|a_0, b_0) = \frac{u_0^{b_0}}{\Gamma(a_0)} \alpha^{a_0-1} e^{-b_0\alpha}, \ P(\beta) = \mathcal{G}(\beta|a_0, b_0) = \frac{a_0^{b_0}}{\Gamma(a_0)} \beta^{a_0-1} e^{-b_0\beta}. \tag{11}$$

We set $a_0 = 10^{-1}$ and $b_0 = 10^{-2}$ in an experiment described later. Summarizing the above, we denote the joint distribution of all the parameters and the set of trajectories as

$$P(\mathcal{D}, V, W, \alpha, \beta) = P(\mathcal{D}|V, W) \underbrace{P(V|\alpha)P(W|\beta)P(\alpha)P(\beta)}_{P(V, W, \alpha, \beta)}. \tag{12}$$

Figure 4(a) shows a graphical model representation. The posterior distribution of parameters is given by

$$P(V, W, \alpha, \beta|\mathcal{D}) = P(\mathcal{D}, V, W, \alpha, \beta)/P(\mathcal{D}), \tag{13}$$

where $P(\mathcal{D})$ is the marginal likelihood $P(\mathcal{D}) = \int P(\mathcal{D}, V, W, \alpha, \beta) dV dW d\alpha d\beta$. Since the exact computation of the marginal likelihood is infeasible, we adopt the variational Bayesian (VB) approach [16] to obtain the posterior distribution.

5.2   Variational Bayes

The VB algorithm is designed to obtain the variational distributions that approximate the posterior distribution. The variational distribution $q(V, W, \alpha, \beta)$ is estimated by minimizing functional $\tilde{\mathcal{F}}[q, \eta, \xi]$, which is defined by

---

[3] If such adjacency information is not available, consider Ω_j^{fr} as a set of all states $\mathcal{S}$.

$$\tilde{\mathcal{F}}[q, \boldsymbol{\eta}, \boldsymbol{\xi}] := \mathbb{E}_q \left[\log \frac{q(\boldsymbol{V}, \boldsymbol{W}, \alpha, \beta)}{h(\boldsymbol{V}, \boldsymbol{W}, \boldsymbol{\eta}, \boldsymbol{\xi}) P(\boldsymbol{V}, \boldsymbol{W}, \alpha, \beta)} \right] \tag{14}$$

under the constraint that the parameters are independent: $q(\boldsymbol{V}, \boldsymbol{W}, \alpha, \beta) = q(\boldsymbol{V})q(\boldsymbol{W})q(\alpha)q(\beta)$. Note that $h(\boldsymbol{V}, \boldsymbol{W}, \boldsymbol{\eta}, \boldsymbol{\xi})$ is a lower bound of the likelihood function (Eq. (9)), i.e., $h(\boldsymbol{V}, \boldsymbol{W}, \boldsymbol{\eta}, \boldsymbol{\xi}) \leq P(\mathcal{D}|\boldsymbol{V}, \boldsymbol{W})$ for all $\boldsymbol{V}, \boldsymbol{W}$. $\boldsymbol{\eta}$ and $\boldsymbol{\xi}$ are auxiliary variables. The functional $\tilde{\mathcal{F}}[q, \boldsymbol{\eta}, \boldsymbol{\xi}]$ is an upper bound of the negative log marginal likelihood $-\log P(\mathcal{D})$. By minimizing $\tilde{\mathcal{F}}[q, \boldsymbol{\eta}, \boldsymbol{\xi}]$, we can indirectly minimize the Kullback-Leibler (KL) divergence between the variational distributions and posterior distribution.

Figure 4(b) makes it easier to understand our optimization scheme. We define functional $\bar{\mathcal{F}}$ as follows:

$$\bar{\mathcal{F}}[q] := \mathbb{E}_q \left[\log \frac{q(\boldsymbol{V})q(\boldsymbol{W})q(\alpha)q(\beta)}{P(\mathcal{D}, \boldsymbol{V}, \boldsymbol{W}, \alpha, \beta)} \right]. \tag{15}$$

This is also an upper bound of the negative log marginal likelihood, and its difference is given by the KL divergence between variational distributions and the posterior distribution (See the green box in Fig. 4(b)). $\tilde{\mathcal{F}}[q, \boldsymbol{\eta}, \boldsymbol{\xi}]$ is always greater than $\bar{\mathcal{F}}[q]$, and its difference is given by the average log ratio of function h and the likelihood function (See the blue box). Since log marginal likelihood does not depend on variational distributions, minimizing $\tilde{\mathcal{F}}$ w.r.t. variational distribution q corresponds to minimizing the sum of the KL divergence and the average log bound-likelihood ratio. Minimizing $\tilde{\mathcal{F}}$ w.r.t. auxiliary variables $\boldsymbol{\eta}$ and $\boldsymbol{\xi}$ corresponds to minimizing the average log bound-likelihood ratio. Iterating this procedure yields a variational distribution.

Remark 2. For probabilistic models belonging to an exponential family, the VB algorithm is derived by using $\bar{\mathcal{F}}[q]$ as the objective functional. However, since the softmax function in Eq. (9) breaks the conjugate-exponential structure in our model, we make use of upper bound function h. The use of a bound function in VB can be found in logistic regression [17], mixture of experts [15] and the correlated topic model [18].

There are various choices for function h since several bounds of the softmax function have been derived [18–21]. From here, we use the following definition of function h, which is a quadratic form with respect to v_{ij}, w_{jk}, by using the bound described by Bouchard [20]. This choice yields an analytical update equation that is easy to implement.

$$\log h(\boldsymbol{V}, \boldsymbol{W}, \boldsymbol{\eta}, \boldsymbol{\xi}) = \sum_{j,k} - n_{.jk} w_{jk} + \sum_{ij} - n_{i \cdot j} \gamma v_{ij}$$
$$- \sum_{ij} n_{ij \cdot} \left\{ \eta_{ij} + \sum_{\ell} f(-w_{j\ell} - \gamma v_{i\ell}, \eta_{ij}, \xi_{ij\ell}) \right\}, \tag{16}$$

$$f(x_\ell, \eta, \xi_\ell) = \log(1 + e^{\xi_\ell}) + (x_\ell - \eta - \xi_\ell)/2 + \lambda(\xi_\ell)\{(x_\ell - \eta)^2 - \xi_\ell^2\}, \tag{17}$$

where $\lambda(\xi_\ell) = \frac{1}{2\xi_\ell}(\sigma(\xi_\ell) - 1/2)$ and $\sigma(\cdot)$ is a sigmoid function. The dot index means that the corresponding index is summed out: $n_{.jk} = \sum_i n_{ijk}$, $n_{i \cdot k} = \sum_j n_{ijk}$, $n_{ij \cdot} = \sum_k n_{ijk}$. We can easily confirm that this h is a lower bound of likelihood $P(\mathcal{D}|\boldsymbol{V}, \boldsymbol{W})$ by the following theorem.

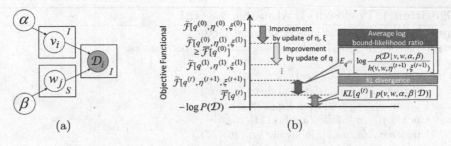

(a)   (b)

Fig. 4. (a) Graphical model. Shaded nodes indicate observed variables. Dependency on a_0 and b_0 is omitted for clarity. (b) Optimization scheme of the proposed algorithm. (Color figure online)

Theorem 2 (Bouchard) [20]. *For any $x \subset \mathbb{R}^L$, any $\eta \in \mathbb{R}$ and any $\xi \in [0, \infty)^L$, the following inequality holds:* $\log \left(\sum_{\ell=1}^{L} e^{x_\ell} \right) \leq \eta + \sum_{\ell=1}^{L} f(x_\ell, \eta, \xi_\ell)$.

We construct an algorithm that iteratively updates the variational distribution q and auxiliary variables $\boldsymbol{\xi}, \boldsymbol{\eta}$. Algorithm 1 summarizes the parameter estimation procedure. Parameter update is explained as follows.

Update of Variational Distribution q: With the variational method, the optimal variational distribution must satisfy the following optimal equation:

$$q(\boldsymbol{V}) \propto \exp \left(\mathbb{E}_{q(\boldsymbol{W})q(\alpha)} \left[\log h(\boldsymbol{V}, \boldsymbol{W}, \boldsymbol{\eta}, \boldsymbol{\xi}) p(\boldsymbol{V}|\alpha) \right] \right), \tag{18}$$

$$q(\boldsymbol{W}) \propto \exp \left(\mathbb{E}_{q(\boldsymbol{V})q(\beta)} \left[\log h(\boldsymbol{V}, \boldsymbol{W}, \boldsymbol{\eta}, \boldsymbol{\xi}) p(\boldsymbol{W}|\beta) \right] \right), \tag{19}$$

$$q(\alpha) \propto \exp \left(\mathbb{E}_{q(\boldsymbol{V})} \left[\log p(\boldsymbol{V}|\alpha) p(\alpha) \right] \right), \tag{20}$$

$$q(\beta) \propto \exp \left(\mathbb{E}_{q(\boldsymbol{W})} \left[\log p(\boldsymbol{W}|\beta) p(\beta) \right] \right). \tag{21}$$

The above distributions are given by elementwise Gaussian distribution $q(v_{ij}) = \mathcal{N}(v_{ij}|\mu_{ij}^v, (\sigma_{ij}^v)^2)$, $q(w_{jk}) = \mathcal{N}(w_{jk}|\mu_{jk}^w, (\sigma_{jk}^w)^2)$ and gamma distributions $q(\alpha) = \mathcal{G}(\alpha|a_\alpha, b_\alpha)$, $q(\beta) = \mathcal{G}(\beta|a_\beta, b_\beta)$, where $\mu_{ij}^v, \sigma_{ij}^v, \mu_{jk}^w, \sigma_{jk}^w, a_\alpha, b_\alpha, a_\beta, b_\beta$ are variational parameters.

$$\mu_{ij}^v = \left[-n_{i \cdot j} + \sum_{k \in \Omega_j^{\text{to}}} \left\{ \frac{n_{ik \cdot}}{2} - 2n_{ik \cdot} \lambda(\xi_{ikj})(\bar{w}_{kj} + \eta_{ik}) \right\} \right] \gamma (\sigma_{ij}^v)^2, \tag{22}$$

$$\sigma_{ij}^v = \left\{ \bar{\alpha} + \sum_{k \in \Omega_j^{\text{to}}} 2n_{ik \cdot} \lambda(\xi_{ikj}) \gamma^2 \right\}^{-\frac{1}{2}}, \tag{23}$$

$$\mu_{jk}^w = \left[-n_{\cdot jk} + \frac{n_{\cdot j \cdot}}{2} + \sum_i 2n_{ij \cdot} \lambda(\xi_{ijk})(-\gamma \bar{v}_{ik} - \eta_{ij}) \right] (\sigma_{jk}^w)^2, \tag{24}$$

$$\sigma_{jk}^w = \left\{ \bar{\beta} + \sum_i 2n_{ij \cdot} \lambda(\xi_{ijk}) \right\}^{-\frac{1}{2}}, \tag{25}$$

$$a_\alpha = a_0 + \frac{IS}{2}, \ b_\alpha = b_0 + \frac{1}{2} \sum_{ij} \mathbb{E}_{q(\boldsymbol{V})}[v_{ij}^2]. \tag{26}$$

$$a_\beta = a_0 + \frac{\sum_j |\Omega_j^{\text{fr}}|}{2}, \ b_\beta = b_0 + \frac{1}{2} \sum_j \sum_{k \in \Omega_j^{\text{fr}}} \mathbb{E}_{q(\boldsymbol{W})}[w_{jk}^2]. \tag{27}$$

Algorithm 1. Proposed VB Algorithm for gIRL

---

input $\mathcal{D}$: observed transitions, γ: discount factor
output $\mu_{ij}^v, \sigma_{ij}^v, \mu_{jk}^w, \sigma_{jk}^w, a_\alpha, b_\alpha, a_\beta, b_\beta$: variational parameters.
1: Initialization.
2: **repeat**
3:    //parameters for variational distribution q
4:    Update $\mu_{ij}^v, \sigma_{ij}^v$ following Eq.(22)(23).
5:    Update $\mu_{jk}^w, \sigma_{jk}^w$ following Eq.(24)(25).
6:    Update $a_\alpha, b_\alpha, a_\beta, b_\beta$ following Eq.(26)(27).
7:    //auxiliary variables $\boldsymbol{\xi}, \boldsymbol{\eta}$
8:    Update $\xi_{ij\ell}, \eta_{ij}$ following Eq.(29)(30).
9: **until** converge

Note that Ω_j^{to} denotes a set of states that can reach "to" state j by a one-step transition and some statistics are given by the following equations: $\bar{v}_{ij} = \mu_{ij}^v$, $\bar{w}_{jk} = \mu_{jk}^w$, $\bar{\alpha} = a_\alpha/b_\alpha$, $\bar{\beta} = a_\beta/b_\beta$, $\mathbb{E}_{q(V)}[v_{ij}^2] = (\sigma_{ij}^v)^2 + (\mu_{ij}^v)^2$, $\mathbb{E}_{q(W)}[w_{jk}^2] = (\sigma_{jk}^w)^2 + (\mu_{jk}^w)^2$.

The proposed algorithm works by iteratively updating the variational parameters. Note that the objective functional is monotonically decreased by the updates and thus converges to a local minimum.

Update of Auxiliary Parameter $\boldsymbol{\eta}, \boldsymbol{\xi}$: Since only the term $\log h(\boldsymbol{V}, \boldsymbol{W}, \boldsymbol{\eta}, \boldsymbol{\xi})$ depends on $\boldsymbol{\xi}$ in the objective functional Eq. (14), at the optimal point, its partial derivative must satisfy

$$\frac{\partial}{\partial \xi_{ij\ell}} \mathbb{E}_{q(V)q(W)q(\alpha)q(\beta)} \Big[-\log h(\boldsymbol{V}, \boldsymbol{W}, \boldsymbol{\eta}, \boldsymbol{\xi}) \Big] = 0$$
$$\Leftrightarrow (\sigma_{j\ell}^w)^2 + \gamma^2 (\sigma_{i\ell}^v)^2 + (-\bar{w}_{j\ell} - \gamma \bar{v}_{i\ell} - \eta_{ij})^2 - \xi_{ij\ell}^2 = 0. \tag{28}$$

Therefore, we develop the following update rule:

$$(\xi_{ij\ell}^{\mathrm{new}})^2 \leftarrow (\sigma_{j\ell}^w)^2 + \gamma^2 (\sigma_{i\ell}^v)^2 + (-\bar{w}_{j\ell} - \gamma \bar{v}_{i\ell} - \eta_{ij})^2. \tag{29}$$

Similarly, the update rule for η is given by

$$\eta_{ij}^{\mathrm{new}} \leftarrow \Big\{ \frac{1}{2} \big(\frac{|\Omega_j^{\mathrm{fr}}|}{2} - 1 \big) + \sum_{\ell \in \Omega_j^{\mathrm{fr}}} \lambda(\xi_{ij\ell})(-\bar{w}_{j\ell} - \gamma \bar{v}_{i\ell}) \Big\} \Big/ \Big\{ \sum_{\ell \in \Omega_j^{\mathrm{fr}}} \lambda(\xi_{ij\ell}) \Big\}. \tag{30}$$

In the process of parameter estimation, state cost function $\mathcal{R}$ need not be considered. However, if necessary, using optimal Eq. (7), the estimated function $\hat{\mathcal{R}} = \{\hat{r}_{ij}\}$ is obtained as $\hat{r}_{ij} = \bar{v}_{ij} + \log\big(\sum_k \exp(-\bar{w}_{jk} - \gamma \bar{v}_{ik})\big)$ after parameter estimation. Then, the estimated value function is the optimal value function of the LMDP with the above estimated state cost function.

6 Numerical Experiment

6.1 Experimental Settings

This section confirms the validity of the proposed method. We conduct a numerical experiment to determine (i) convergence property, (ii) predictive performance and (iii) parameter visualization.

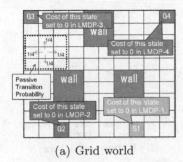

(a) Grid world   (b) Yokohama

Fig. 5. Settings for (a) grid world where uniform passive transition probability and four types of state cost functions are set and (b) yokohama using real car probe data

Data Description: We prepare two experiment settings: grid-world and yokohama. In the grid-world experiment, we set the passive transition probability of each state (vertical and horizontal) at a uniform probability (if walls or obstacles exist, self-transition is to be considered) and prepared four different types of state cost functions: $\mathcal{R}_1, \mathcal{R}_2, \mathcal{R}_3$ and $\mathcal{R}_4$. The cost of each function is set at 0 just for the corresponding goal state shown in Fig. 5(a) and at 1 for the other states. By computing the true optimal transition probability of each LMDP, we generate training and test data in an *iid* manner in each state. In the yokohama experiment, we use real car probe data provided by NAVITIME JAPAN Co, Ltd. This dataset is a collection of GPS trajectories of users who used a car navigation application on smartphones in Kanagawa Prefecture, Japan. In particular, we used the trajectories for the Minato-Mirai-21 district in Yokohama. We use the log data recorded during the holiday period from 2015.4.13 to 2015.5.1 (5 days in total) as training data and the log data of 2015.5.2 as test data. By applying a landmark graph construction algorithm [22], we construct the abstract street network as shown in Fig. 5(b). We convert the GPS into transition data between the nodes (states) of this graph. We treat the logs of 10:00–12:59, 14:00–16:59, 17:00–19:59 as the logs of LMDP1, 2 and 3, respectively.

Predictive Performance Measurement: To evaluate the predictive performance, we use the negative test log likelihood. A lower value indicates that the method extracts the parameter that reflects the agent's behavior more precisely. The negative test log likelihood is defined as $(1/\mathcal{T}) \sum_{i=1}^{I} \sum_{j,k \in \mathcal{S}} -n_{ijk}^{\text{test}} \log \hat{p}_{ijk}^*$, where $\mathcal{T}$ is the number of test datasets and n_{ijk}^{test} indicates the number of transitions from state j to state k in the i-th LMDP. $\hat{p}_{ijk}^*$ is computed by substituting $\bar{v}_{ij}$ and $\bar{w}_{kj}$ into Eq. (8). We compare the proposed method with Random and Dvijotham's method [6]. Since Dvijotham's method can estimate only the cost function, we set the passive transition probability at a uniform probability. Moreover, to investigate the effect of passive transition probability estimation, we also make a comparison with the proposed method, which does not learn the passive transition probability (fixed at a random initial value).

6.2   Results

Convergence Behavior: Figure 6 shows the convergence behavior of the objective function and hyperparameters. We can confirm that they both converge to certain values by iterating the update process. This shows that the proposed method can estimate

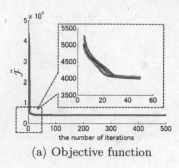

(a) Objective function

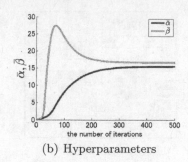

(b) Hyperparameters

Fig. 6. Convergence behavior of (a) objective function and (b) hyperparameters in grid world experiment with $n_{ij.} = 5$. (a) shows the result of 10 random initialization settings and (b) shows one of the paths from the initial point.

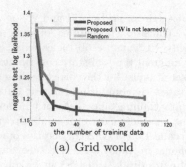

(a) Grid world

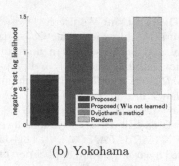

(b) Yokohama

Fig. 7. Comparison of predictive performance of (a) grid world and of (b) yokohama experiment. Lower values are better.

hyperparameters. In terms of convergence speed, Fig. 6(a) shows that the objective function basically converges within 50 iterations. In contrast, Fig. 6(b) shows that more than 200 iterations are needed for hyperparameter convergence. These results imply that a relatively longer running time is required in order to learn the hyperparameters.

Predictive Performance: Figure 7(a) shows the predictive performance in the grid world experiment. In comparison with the proposed method without learning the passive transition probability, the proposed method shows better predictive performance. This result shows that estimating the passive transition probability contributes to better performance. Figure 7(b) shows the predictive performance in the yokohama experiment. Dvijotham's method is competitive with the proposed method that does not learn the passive transition probability but the proposed method outperforms them. This also confirms the effectiveness of the proposed method.

Parameter Visualization: Figure 8(a) shows the estimated parameters in the grid world experiment for various numbers of observed transitions $n_{ij.}$ (visualization of LMDP-3 and 4 is omitted due to lack of space). We can confirm that as the number of observed transitions increases, the estimated parameters more closely approach the true parameters. Figure 8(b) shows the estimated parameters of the yokohama experiment[4].

---

[4] This figure is drawn by QGIS using the data interpolation plugin.

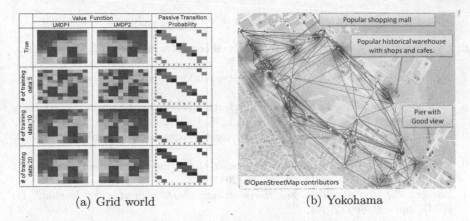

(a) Grid world   (b) Yokohama

Fig. 8. (a) True and estimated value functions of LMDP 1,2 and passive transition probabilities for states 1–10 of grid world for various numbers of observed transitions $n_{ij} = 5, 10, 20$. (b) Estimated value function of LMDP1 of yokohama. Value functions are visualized by a heat map with colors ranging from red to blue. (Color figure online)

Although we are unable to know the true parameters behind the real car probe data, we observe that the state near attractive locations has a lower value function value. We can predict that the agent (car driver) tends to move to the locations. This result implies that the proposed method estimates parameters that reflect car drivers' behavior.

7   Conclusion and Future Work

In this paper, we tackled the gIRL problem to estimate both the state cost function and transition probability from the observed optimal behavior of agents. We showed that gIRL with an LMDP suffers from a non-identifiable problem and, in response, we proposed a variational Bayesian gIRL algorithm with SP-LMDPs. The result of our experiment shows the effectiveness of the proposed method. Since the application area of our method is not limited to traffic data, we plan a further investigation into practical applications. We also consider that analyzing the theoretical performance constitutes important future research.

References

1. Abbeel, P., Coates, A., Quigley, M., Ng, A.Y.: An application of reinforcement learning to aerobatic helicopter flight. In: NIPS, pp. 1–8 (2007)
2. Ziebart, B.D., Maas, A.L., Bagnell, J.A., Dey, A.K.: Maximum entropy inverse reinforcement learning. In: AAAI, pp. 1433–1438 (2008)
3. Song, X., Zhang, Q., Sekimoto, Y., Shibasaki, R.: Intelligent system for urban emergency management during large-scale disaster. In: AAAI, pp. 458–464 (2014)
4. Neu, G., Szepesvári, C.: Training parsers by inverse reinforcement learning. Mach. Learn. **77**(2–3), 303–337 (2009)

5. Todorov, E.: Linearly-solvable Markov decision problems. In: NIPS, pp. 1369–1376 (2006)
6. Dvijotham, K., Todorov, E.: Inverse optimal control with linearly-solvable MDPs. In: ICML, pp. 335–342 (2010)
7. Ng, A.Y., Russell, S.: Algorithms for inverse reinforcement learning. In: ICML, pp. 663–670 (2000)
8. Makino, T., Takeuchi, J.: Apprenticeship learning for model parameters of partially observable environments. In: ICML, pp. 1495–1502 (2012)
9. Ramachandran, D., Amir, E.: Bayesian inverse reinforcement learning. In: IJCAI, pp. 2586–2591 (2007)
10. Rothkopf, C.A., Dimitrakakis, C.: Preference elicitation and inverse reinforcement learning. In: ECML PKDD, pp. 34–48 (2011)
11. Lazaric, A., Ghavamzadeh, M.: Bayesian multi-task reinforcement learning. In: ICML, pp. 599–606 (2010)
12. Babes, M., Marivate, V., Subramanian, K., Littman, M.L.: Apprenticeship learning about multiple intentions. In: ICML, pp. 897–904 (2011)
13. Puterman, M.L.: Markov Decision Processes: Discrete Stochastic Dynamic Programming. Wiley Series in Probability and Statistics. Wiley, Hoboken (2005)
14. Van der Vaart, A.W.: Asymptotic Statistics. Cambridge University Press, Cambridge (2000)
15. Bishop, C.M., Svenskn, M.: Bayesian hierarchical mixtures of experts. In: UAI, pp. 57–64 (2002)
16. Jordan, M.I., Ghahramani, Z., Jaakkola, T.S., Saul, L.K.: An introduction to variational methods for graphical models. Mach. Learn. **37**(2), 183–233 (1999)
17. Jaakkola, T., Jordan, M.I.: A variational approach to Bayesian logistic regression models and their extensions. In: AISTATS (1997)
18. Blei, D.M., Lafferty, J.D.: A correlated topic model of science. Ann. Appl. Stat. **1**, 17–35 (2007)
19. Böhning, D.: Multinomial logistic regression algorithm. Ann. Inst. Stat. Math. **44**(1), 197–200 (1992)
20. Bouchard, G.: Efficient bounds for the softmax function and applications to approximate inference in hybrid models. In: NIPS 2007 Workshop for Approximate Bayesian Inference in Continuous/Hybrid Systems (2007)
21. Jebara, T., Choromanska, A.: Majorization for CRFs and latent likelihoods. In: NIPS, pp. 557–565 (2012)
22. Yuan, J., Zheng, Y., Zhang, C., Xie, W., Xie, X., Sun, G., Huang, Y.: T-drive: driving directions based on taxi trajectories. In: SIGSPATIAL, pp. 99–108 (2010)

Max K-Armed Bandit: On the ExtremeHunter Algorithm and Beyond

Mastane Achab[1($\boxtimes$)], Stephan Clémençon[1], Aurélien Garivier[2],
Anne Sabourin[1], and Claire Vernade[1]

[1] LTCI, Télécom ParisTech, Université Paris-Saclay, Paris, France
mastane.achab@telecom-paristech.fr
[2] IMT, Université de Toulouse, Toulouse, France

Abstract. This paper is devoted to the study of the *max K-armed bandit problem*, which consists in sequentially allocating resources in order to detect extreme values. Our contribution is twofold. We first significantly refine the analysis of the EXTREMEHUNTER algorithm carried out in Carpentier and Valko (2014), and next propose an alternative approach, showing that, remarkably, Extreme Bandits can be reduced to a classical version of the bandit problem to a certain extent. Beyond the formal analysis, these two approaches are compared through numerical experiments.

1 Introduction

In a classical multi-armed bandit (MAB in abbreviated form) problem, the objective is to find a strategy/policy in order to sequentially explore and exploit K sources of gain, referred to as *arms*, so as to maximize the expected cumulative gain. Each arm $k \in \{1, \ldots, K\}$ is characterized by an unknown probability distribution ν_k. At each round $t \geq 1$, a strategy π picks an arm $I_t = \pi((I_1, X_{I_1,1}), \ldots, (I_{t-1}, X_{I_{t-1},t-1}))$ and receives a random reward $X_{I_t,t}$ sampled from distribution ν_{I_t}. Whereas usual strategies aim at finding and exploiting the arm with highest expectation, the quantity of interest in many applications such as medicine, insurance or finance may not be the sum of the rewards, but rather the *extreme* observations (even if it might mean replacing loss minimization by gain maximization in the formulation of the practical problem). In such situations, classical bandit algorithms can be significantly sub-optimal: the "best" arm should not be defined as that with highest expectation, but as that producing the maximal values. This setting, referred to as *extreme bandits* in Carpentier and Valko (2014), was originally introduced by Cicirello and Smith (2005) by the name of *max K-armed bandit problem*. In this framework, the goal pursued is to obtain the highest possible reward during the first $n \geq 1$ steps. For a given arm k, we denote by

$$G_n^{(k)} = \max_{1 \leq t \leq n} X_{k,t}$$

© Springer International Publishing AG 2017
M. Ceci et al. (Eds.): ECML PKDD 2017, Part II, LNAI 10535, pp. 389–404, 2017.
https://doi.org/10.1007/978-3-319-71246-8_24

the maximal value taken until round $n \geq 1$ and assume that, in expectation, there is a unique optimal arm

$$k^* = \arg \max_{1 \leq k \leq K} \mathbb{E}[G_n^{(k)}].$$

The expected *regret* of a strategy π is here defined as

$$\mathbb{E}[R_n] = \mathbb{E}[G_n^{(k^*)}] - \mathbb{E}[G_n^{(\pi)}], \tag{1}$$

where $G_n^{(\pi)} = \max_{1 \leq t \leq n} X_{I_t, t}$ is the maximal value observed when implementing strategy π. When the supports of the reward distributions (*i.e.* the ν_k's) are bounded, no-regret is expected provided that every arm can be sufficiently explored, refer to Nishihara et al. (2016) (see also David and Shimkin (2016) for a PAC approach). If infinitely many arms are possibly involved in the learning strategy, the challenge is then to explore and exploit optimally the unknown reservoir of arms, see Carpentier and Valko (2015). When the rewards are unbounded in contrast, the situation is quite different: the best arm is that for which the maximum $G_n^{(k)}$ tends to infinity faster than the others. In Nishihara et al. (2016), it is shown that, for unbounded distributions, no policy can achieve no-regret without restrictive assumptions on the distributions. In accordance with the literature, we focus on a classical framework in extreme value analysis. Namely, we assume that the reward distributions are *heavy-tailed*. Such Pareto-like laws are widely used to model extremes in many applications, where a conservative approach to risk assessment might be relevant (*e.g.* finance, environmental risks). Like in Carpentier and Valko (2014), rewards are assumed to be distributed as second order Pareto laws in the present article. For the sake of completeness, we recall that a probability law with cdf $F(x)$ belongs to the (α, β, C, C')-second order Pareto family if, for every $x \geq 0$,

$$|1 - Cx^{-\alpha} - F(x)| \leq C'x^{-\alpha(1+\beta)}, \tag{2}$$

where α, β, C and C' are strictly positive constants, see *e.g.* Resnick (2007). In this context, Carpentier and Valko (2014) have proposed the EXTREMEHUNTER algorithm to solve the *extreme bandit* problem and provided a regret analysis.

The contribution of this paper is twofold. First, the regret analysis of the EXTREMEHUNTER algorithm is significantly improved, in a nearly optimal fashion. This essentially relies on a new technical result of independent interest (see Theorem 1 below), which provides a bound for the difference between the expectation of the maximum among independent realizations $X_1, \ldots, X_T$ of a (α, β, C, C')-second order Pareto distribution, $\mathbb{E}[\max_{1 \leq i \leq T} X_i]$ namely, and its rough approximation $(TC)^{1/\alpha} \Gamma(1 - 1/\alpha)$. As a by-product, we propose a more simple EXPLORE-THEN-COMMIT strategy that offers the same theoretical guarantees as EXTREMEHUNTER. Second, we explain how extreme bandit can be reduced to a classical bandit problem to a certain extent. We show that a regret-minimizing strategy such as ROBUST-UCB (see Bubeck et al. (2013)), applied on correctly left-censored rewards, may also reach a very good performance. This claim is supported by theoretical guarantees on the number of pulls of the best arm k^* and by numerical experiments both at the same time. From a practical angle, the main drawback of this alternative approach consists in the fact that its implementation requires some knowledge of the complexity of the problem (*i.e.* of the gap between the first-order Pareto coefficients of the first and second arms). In regard to its theoretical analysis, efficiency is proved for large horizons only.

This paper is organized as follows. Section 2 presents the technical result mentioned above, which next permits to carry out a refined regret analysis of the ExtremeHunter algorithm in Sect. 3. In Sect. 4, the regret bound thus obtained is proved to be nearly optimal: precisely, we establish a lower bound under the assumption that the distributions are close enough to Pareto distributions showing the regret bound is sharp in this situation. In Sect. 5, reduction of the extreme bandit problem to a classical bandit problem is explained at length, and an algorithm resulting from this original view is then described. Finally, we provide a preliminary numerical study that permits to compare the two approaches from an experimental perspective. Due to space limitations, certain technical proofs are deferred to the Supplementary Material[1].

2 Second-Order Pareto Distributions: Approximation of the Expected Maximum Among i.i.d. Realizations

In the extreme bandit problem, the key to controlling the behavior of explore-exploit strategies is to approximate the expected payoff of a fixed arm $k \in \{1, \ldots, K\}$. The main result of this section, stated in Theorem 1, provides such control: it significantly improves upon the result originaly obtained by Carpentier and Valko (2014) (see Theorem 1 therein). As shall be next shown in Sect. 3, this refinement has substantial consequences on the regret bound.

In Carpentier and Valko (2014), the distance between the expected maximum of independent realizations of a (α, β, C, C')-second order Pareto and the corresponding expectation of a Fréchet distribution $(TC)^{1/\alpha}\Gamma(1 - 1/\alpha)$ is controlled as follows:

$$\left| \mathbb{E}\left[\max_{1 \le i \le T} X_i \right] - (TC)^{1/\alpha}\Gamma(1 - 1/\alpha) \right| \le \frac{4D_2 C^{1/\alpha}}{T^{1-1/\alpha}} + \frac{2C'D_{\beta+1}}{C^{\beta+1-1/\alpha}T^{\beta-1/\alpha}}$$
$$+ (2C'T)^{\frac{1}{(1+\beta)\alpha}}.$$

Notice that the leading term of this bound is $(2C'T)^{1/((1+\beta)\alpha)}$ as $T \to +\infty$. Below, we state a sharper result where, remarkably, this (exploding) term disappears, the contribution of the related component in the approximation error decomposition being proved as (asymptotically) negligible in contrast.

Theorem 1 (Fréchet approximation bound). *If $X_1, \ldots, X_T$ are i.i.d. r.v.'s drawn from a (α, β, C, C')-second order Pareto distribution with $\alpha > 1$*

---

[1] See the full-length paper: http://arxiv.org/abs/1707.08820.

and $T \geq Q_1$, where Q_1 is the constant depending only on α, β, C and C' given in Eq. (3) below, then,

$$\left| \mathbb{E}\left[\max_{1 \leq i \leq T} X_i \right] - (TC)^{1/\alpha} \Gamma(1 - 1/\alpha) \right|$$

$$\leq \frac{4D_2 C^{1/\alpha}}{T^{1-1/\alpha}} + \frac{2C'D_{\beta+1}}{C^{\beta+1-1/\alpha}T^{\beta-1/\alpha}} + 2(2C'T)^{\frac{1}{(1+\beta)\alpha}} e^{-HT^{\frac{\beta}{\beta+1}}}$$

$$= \underset{T \to \infty}{o} (T^{1/\alpha}),$$

where $H = C(2C')^{1/(\alpha(1+\beta))}/2$. In particular, if $\beta \geq 1$, we have:

$$\left| \mathbb{E}\left[\max_{1 \leq i \leq T} X_i \right] - (TC)^{1/\alpha} \Gamma(1 - 1/\alpha) \right| = o(1) \text{ as } T \to +\infty.$$

We emphasize that the bound above shows that the distance of $\mathbb{E}[\max_{1 \leq i \leq T} X_i]$ to the Fréchet mean $(TC)^{1/\alpha}\Gamma(1 - \frac{1}{\alpha})$ actually vanishes as $T \to \infty$ as soon as $\beta \geq 1$, a property that shall be useful in Sect. 3 to study the behavior of learning algorithms in the extreme bandit setting.

Proof. Assume that $T \geq Q_1$, where

$$Q_1 = \frac{1}{2C'} \max \left\{ (2C'/C)^{(1+\beta)/\beta}, (8C)^{1+\beta} \right\}. \tag{3}$$

As in the proof of Theorem 1 in Carpentier and Valko (2014), we consider the quantity $B = (2C'T)^{1/((1+\beta)\alpha)}$ that serves as a cut-off between tail and bulk behaviors. Observe that

$$\left| \mathbb{E}\left[\max_{1 \leq i \leq T} X_i \right] - (TC)^{1/\alpha} \Gamma(1 - 1/\alpha) \right| \leq$$

$$\left| \int_0^\infty \left\{ 1 - \mathbb{P}\left(\max_{1 \leq i \leq T} X_i \leq x \right) - 1 + e^{-TCx^{-\alpha}} \right\} dx \right|$$

$$\leq \left| \int_0^B \left\{ \mathbb{P}\left(\max_{1 \leq i \leq T} X_i \leq x \right) - e^{-TCx^{-\alpha}} \right\} dx \right|$$

$$+ \left| \int_B^\infty \left\{ \mathbb{P}\left(\max_{1 \leq i \leq T} X_i \leq x \right) - e^{-TCx^{-\alpha}} \right\} dx \right|.$$

For $p \in \{2, \beta + 1\}$, we set $D_p = \Gamma(p - \frac{1}{\alpha})/\alpha$. Equipped with this notation, we may write

$$\left| \int_B^\infty \left\{ \mathbb{P}\left(\max_{1 \leq i \leq T} X_i \leq x \right) - e^{-TCx^{-\alpha}} \right\} dx \right| \leq \frac{4D_2 C^{1/\alpha}}{T^{1-1/\alpha}} + \frac{2C'D_{\beta+1}}{C^{\beta+1-1/\alpha}T^{\beta-1/\alpha}}.$$

Instead of loosely bounding the bulk term by B, we write

$$\left| \int_0^B \left\{ \mathbb{P}\left(\max_{1 \leq i \leq T} X_i \leq x \right) - e^{-TCx^{-\alpha}} \right\} dx \right| \leq B \, \mathbb{P}(X_1 \leq B)^T + \int_0^B e^{-TCx^{-\alpha}} dx.$$

$$\tag{4}$$

First, using (2) and the inequality $C'B^{-(1+\beta)\alpha} \leq CB^{-\alpha}/2$ (a direct consequence of Eq. (3)), we obtain

$$\mathbb{P}(X_1 \leq B)^T \leq \left(1 - CB^{-\alpha} + C'B^{-(1+\beta)\alpha}\right)^T$$

$$\leq \left(1 - \frac{1}{2}CB^{-\alpha}\right)^T \leq e^{-\frac{1}{2}TCB^{-\alpha}} = e^{-HT^{\beta/(\beta+1)}}.$$

Second, the integral in Eq. (4) can be bounded as follows:

$$\int_0^B e^{-TCx^{-\alpha}}\mathrm{d}x \leq Be^{-TCB^{-\alpha}} = (2C'T)^{1/((1+\beta)\alpha)}e^{-2HT^{\beta/(\beta+1)}}.$$

This concludes the proof.

3   The ExtremeHunter and ExtremeETC Algorithm

In this section, the tighter control provided by Theorem 1 is used in order to refine the analysis of the ExtremeHunter algorithm (Algorithm 1) carried out in Carpentier and Valko (2014). This theoretical analysis is also shown to be valid for ExtremeETC, a novel algorithm we next propose, that greatly improves upon ExtremeHunter, regarding computational efficiency.

3.1   Further Notations and Preliminaries

Throughout the paper, the indicator function of any event $\mathcal{E}$ is denoted by $\mathbb{1}\{\mathcal{E}\}$ and $\bar{\mathcal{E}}$ means the complementary event of $\mathcal{E}$. We assume that the reward related to each arm $k \in \{1, \ldots, K\}$ is drawn from a $(\alpha_k, \beta_k, C_k, C')$-second order Pareto distribution. Sorting the tail indices by increasing order of magnitude, we use the classical notation for order statistics: $\alpha_{(1)} \leq \cdots \leq \alpha_{(K)}$. We assume that $\alpha_{(1)} > 1$, so that the random rewards have finite expectations, and suppose that the strict inequality $\alpha_{(1)} < \alpha_{(2)}$ holds true. We also denote by $T_{k,t}$ the number of times the arm k is pulled up to time t. For $1 \leq k \leq K$ and $i \geq 1$, the r.v. $\widehat{X}_{k,i}$ is the reward obtained at the i-th draw of arm k if $i \leq T_{k,n}$ or a new r.v. drawn from ν_k independent from the other r.v.'s otherwise.

We start with a preliminary lemma supporting the intuition that the tail index α fully governs the extreme bandit problem. It will allow to show next that the algorithm picks the right arm after the exploration phase, see Lemma 2.

Lemma 1 (Optimal arm). *For n larger than some constant Q_4 depending only on $(\alpha_k, \beta_k, C_k)_{1 \leq k \leq K}$ and C', the optimal arm for the extreme bandit problem is given by:*

$$k^* = \operatorname*{arg\,min}_{1 \leq k \leq K} \alpha_k = \operatorname*{arg\,max}_{1 \leq k \leq K} V_k, \tag{5}$$

where $V_k = (nC_k)^{1/\alpha_k}\Gamma(1 - 1/\alpha_k)$.

Proof. We first prove the first equality. It follows from Theorem 1 that there exists a constant Q_2, depending only on $\{(\alpha_k, \beta_k, C_k)\}_{1 \leq k \leq K}$ and C', such that for any arm $k \in \{1, \ldots, K\}$, $|\mathbb{E}[G_n^{(k)}] - V_k| \leq V_k/2$. Then for $k \neq k^*$ we have, for all $n > Q_2$, $V_k/2 \leq \mathbb{E}[G_n^{(k)}] \leq \mathbb{E}[G_n^{(k^*)}] \leq 3V_{k^*}/2$. Recalling that V_k is proportional to n^{1/α_k}, it follows that $\alpha_{k^*} = \min_{1 \leq k \leq K} \alpha_k$. Now consider the following quantity:

$$Q_3 = \max_{k \neq k^*} \left[\frac{2C_k^{1/\alpha_k} \Gamma(1 - 1/\alpha_k)}{C_{k^*}^{1/\alpha_{k^*}} \Gamma(1 - 1/\alpha_{k^*})} \right]^{1/(1/\alpha_{k^*} - 1/\alpha_k)}. \tag{6}$$

For $n > Q_4 = \max(Q_2, Q_3)$, we have $V_{k^*} > 2V_k$ for any suboptimal arm $k \neq k^*$, which proves the second equality.

From now on, we assume that n is large enough for Lemma 1 to apply.

3.2   The EXTREMEHUNTER Algorithm (Carpentier and Valko 2014)

Before developing a novel analysis of the extreme bandit problem in Sect. 3.2 (see Theorem 2), we recall the main features of EXTREMEHUNTER, and in particular the estimators and confidence intervals involved in the indices of this optimistic policy.

Algorithm 1. ExtremeHunter (Carpentier and Valko 2014)

1: **Input:** K: number of arms, n: time horizon, $b > 0$ such that $b \leq \min_{1 \leq k \leq K} \beta_k$, N: minimum number of pulls of each arm (Eq. (9)).
2: **Initialize:** Pull each arm N times.
3: **for** $k = 1, \ldots, K$ **do**
4:    Compute estimators $\widehat{h}_{k,KN} = \widetilde{h}_k(N)$ (Eq. (8)) and $\widehat{C}_{k,KN} = \widetilde{C}_k(N)$ (Eq. (7))
5:    Compute index $B_{k,KN}$ (Eq. (12))
6: **end for**
7: Pull arm $I_{KN+1} = \arg\max_{1 \leq k \leq K} B_{k,KN}$
8: **for** $t = KN + 2, \ldots, n$ **do**
9:    Update estimators $\widehat{h}_{I_{t-1}, t-1}$ and $\widehat{C}_{I_{t-1}, t-1}$
10:    Update index $B_{I_{t-1}, t-1}$
11:    Pull arm $I_t = \arg\max_{1 \leq k \leq K} B_{k,t-1}$
12: **end for**

Theorem 1 states that for any arm $k \in \{1, \ldots, K\}$, $\mathbb{E}[G_n^{(k)}] \approx (C_k n)^{1/\alpha_k} \Gamma(1 - 1/\alpha_k)$. Consequently, the optimal strategy in hindsight always pulls the arm $k^* = \arg\max_{1 \leq k \leq K} \{(nC_k)^{1/\alpha_k} \Gamma(1 - 1/\alpha_k)\}$. At each round and for each arm $k \in \{1, \ldots, K\}$, EXTREMEHUNTER algorithm (Carpentier and Valko 2014) estimates the coefficients α_k and C_k (but not β_k, see Remark 2 in Carpentier and Valko (2014)). The corresponding confidence intervals are detailed below. Then, following the *optimism-in-the-face-of-uncertainty* principle (see (Auer et al.

2002) and references therein), the strategy plays the arm maximizing an optimistic plug-in estimate of $(C_k n)^{1/\alpha_k} \Gamma(1 - 1/\alpha_k)$. To that purpose, Theorem 3.8 in Carpentier and Kim (2014) and Theorem 2 in Carpentier et al. (2014) provide estimators $\widetilde{\alpha}_k(T)$ and $\widetilde{C}_k(T)$ for α_k and C_k respectively, after T draws of arm k. Precisely, the estimate $\widetilde{\alpha}_k(T)$ is given by

$$\widetilde{\alpha}_k(T) = \log \left(\frac{\sum_{t=1}^{T} \mathbb{1}\{X_t > e^r\}}{\sum_{t=1}^{T} \mathbb{1}\{X_t > e^{r+1}\}} \right),$$

where r is chosen in an adaptive fashion based on Lepski's method, see (Lepskiĭ 1990), while the estimator of C_k considered is

$$\widetilde{C}_k(T) = T^{-2b/(2b+1)} \sum_{i=1}^{T} \mathbb{1}\{\widetilde{X}_{k,i} \geq T^{\widetilde{h}_k(T)/(2b+1)}\}, \tag{7}$$

where

$$\widetilde{h}_k(T) = \min(1/\widetilde{\alpha}_k(T), 1). \tag{8}$$

The authors also provide finite sample error bounds for $T \geq N$, where

$$N = A_0 (\log n)^{2(2b+1)/b}, \tag{9}$$

with b a known lower bound on the β_k's ($b \leq \min_{1 \leq k \leq K} \beta_k$), and A_0 a constant depending only on $(\alpha_k, \beta_k, C_k)_{1 \leq k \leq K}$ and C'. These error bounds naturally define confidence intervals of respective widths Λ_1 and Λ_2 at level δ_0 defined by

$$\delta_0 = n^{-\rho}, \quad \text{where} \quad \rho = \frac{2\alpha_{k^*}}{\alpha_{k^*} - 1}. \tag{10}$$

More precisely, we have

$$\mathbb{P}\left(\left| \frac{1}{\alpha_k} - \widetilde{h}_k(T) \right| \leq \Lambda_1(T), \; \left| C_k - \widetilde{C}_k(T) \right| \leq \Lambda_2(T) \right) \geq 1 - 2\delta_0, \tag{11}$$

where

$$\Lambda_1(T) = D\sqrt{\log(1/\delta_0)} T^{-b/(2b+1)} \quad \text{and} \quad \Lambda_2(T) = E\sqrt{\log(T/\delta_0)} \log(T) T^{-b/(2b+1)},$$

denoting by D and E some constants depending only on $(\alpha_k, \beta_k, C_k)_{1 \leq k \leq K}$ and C'. When $T_{k,t} \geq N$, denote by $\widehat{h}_{k,t} = \widetilde{h}_k(T_{k,t})$ and $\widehat{C}_{k,t} = \widetilde{C}_k(T_{k,t})$ the estimators based on the $T_{k,t}$ observations for simplicity. EXTREMEHUNTER's index $B_{k,t}$ for arm k at time t, the optimistic proxy for $\mathbb{E}[G_n^{(k)}]$, can be then written as

$$B_{k,t} = \widetilde{\Gamma}\left(1 - \widehat{h}_{k,t} - \Lambda_1(T_{k,t})\right) \left(\left(\widehat{C}_{k,t} + \Lambda_2(T_{k,t})\right) n\right)^{\widehat{h}_{k,t} + \Lambda_1(T_{k,t})}, \tag{12}$$

where $\widetilde{\Gamma}(x) = \Gamma(x)$ if $x > 0$ and $+\infty$ otherwise.

On Computational Complexity. Notice that after the initialization phase, at each time $t > KN$, EXTREMEHUNTER computes estimators $\widehat{h}_{I_t,t}$ and $\widehat{C}_{I_t,t}$, each having a time complexity linear with the number of samples $T_{I_t,t}$ pulled from arm I_t up to time t. Summing on the rounds reveals that EXTREMEHUNTER's time complexity is quadratic with the time horizon n.

3.3 EXTREMEETC: A Computationally Appealing Alternative

In order to reduce the restrictive time complexity discussed previously, we now propose the EXTREMEETC algorithm, an *Explore-Then-Commit* version of EXTREMEHUNTER, which offers similar theoretical guarantees.

Algorithm 2. EXTREMEETC

1: **Input:** K: number of arms, n: time horizon, $b > 0$ such that $b \leq \min_{1 \leq k \leq K} \beta_k$, N: minimum number of pulls of each arm (Eq. (9)).
2: **Initialize:** Pull each arm N times.
3: **for** $k = 1, \ldots, K$ **do**
4:    Compute estimators $\widehat{h}_{k,KN} = \widetilde{h}_k(N)$ (Eq. (8)) and $\widehat{C}_{k,KN} = \widetilde{C}_k(N)$ (Eq. (7))
5:    Compute index $B_{k,KN}$ (Eq. (12))
6: **end for**
7: Set $I_{\text{winner}} = \arg\max_{1 \leq k \leq K} B_{k,KN}$
8: **for** $t = KN + 1, \ldots, n$ **do**
9:    Pull arm I_{winner}
10: **end for**

After the initialization phase, the *winner arm*, which has maximal index $B_{k,KN}$, is fixed and is pulled in all remaining rounds. Then EXTREMEETC's time complexity, due to the computation of $\widehat{h}_{k,KN}$ and $\widehat{C}_{k,KN}$ only, is $\mathcal{O}(KN) = \mathcal{O}((\log n)^{2(2b+1)/b})$, which is considerably faster than quadratic time achieved by EXTREMEHUNTER. For clarity, Table 1 summarizes time and memory complexities of both algorithms.

Table 1. Time and memory complexities required for estimating $(\alpha_k, C_k)_{1 \leq k \leq K}$ in EXTREMEETC and EXTREMEHUNTER.

Complexity	EXTREMEETC	EXTREMEHUNTER
Time	$\mathcal{O}\big((\log n)^{\frac{2(2b+1)}{b}}\big)$	$\mathcal{O}(n^2)$
Memory	$\mathcal{O}\big((\log n)^{\frac{2(2b+1)}{b}}\big)$	$\mathcal{O}(n)$

Due to the significant gain of computational time, we used the EXTREMEETC algorithm in our simulation study (Sect. 6) rather than EXTREMEHUNTER.

Controlling the Number of Suboptimal Rounds. We introduce a high probability event that corresponds to the favorable situation where, at each round, all coefficients $(1/\alpha_k, C_k)_{1 \leq k \leq K}$ simultaneously belong to the confidence intervals recalled in the previous subsection.

Definition 1. *The event ξ_1 is the event on which the bounds*

$$\left| \frac{1}{\alpha_k} - \tilde{h}_k(T) \right| \leq \Lambda_1(T) \quad and \quad \left| C_k - \tilde{C}_k(T) \right| \leq \Lambda_2(T)$$

hold true for any $1 \leq k \leq K$ and $N \leq T \leq n$.

The union bound combined with (11) yields

$$\mathbb{P}(\xi_1) \geq 1 - 2Kn\delta_0. \tag{13}$$

Lemma 2. *For $n > Q_5$, where Q_5 is the constant defined in (15), ExtremeETC and ExtremeHunter always pull the optimal arm after the initialization phase on the event ξ_1. Hence, for any suboptimal arm $k \neq k^*$, we have, on ξ_1:*

$$T_{k,n} = N \quad and \; thus \quad T_{k^*,n} = n - (K-1)N.$$

Proof. Here we place ourselves on the event ξ_1. For any arm $1 \leq k \leq K$, Lemma 1 in Carpentier and Valko (2014) provides lower and upper bounds for $B_{k,t}$ when $T_{k,t} \geq N$

$$V_k \leq B_{k,t} \leq V_k \left(1 + F \log n \sqrt{\log(n/\delta_0)} T_{k,t}^{-b/(2b+1)} \right), \tag{14}$$

where F is a constant which depends only on $(\alpha_k, \beta_k, C_k)_{1 \leq k \leq K}$ and C'. Introduce the horizon Q_5, which depends on $(\alpha_k, \beta_k, C_k)_{1 \leq k \leq K}$ and C'

$$Q_5 = \max \left(e^{\left(F\sqrt{1+\rho} A_0^{-b/(2b+1)} \right)^2}, Q_4 \right). \tag{15}$$

Then the following Lemma 3, proved in Appendix A, tells us that for n large enough, the exploration made during the initialization phase is enough to find the optimal arm, with high probability.

Lemma 3. *If $n > Q_5$, we have under the event ξ_1 that for any suboptimal arm $k \neq k^*$ and any time $t > KN$ that $B_{k,t} < B_{k^*,t}$.*

Hence the optimal arm is pulled at any time $t > KN$.

The following result immediately follows from Lemma 2.

Corollary 1. *For n larger than some constant depending only on $(\alpha_k, \beta_k, C_k)_{1 \leq k \leq K}$ and C' we have under ξ_1*

$$T_{k^*,n} \geq n/2.$$

Upper Bounding the Expected Extreme Regret. The upper bound on the expected extreme regret stated in the theorem below improves upon that given in Carpentier and Valko (2014) for ExtremeHunter. It is also valid for ExtremeETC.

Theorem 2. *For* EXTREMEETC *and* EXTREMEHUNTER, *the expected extreme regret is upper bounded as follows*

$$\mathbb{E}[R_n] = \mathcal{O}\left((\log n)^{2(2b+1)/b} n^{-(1-1/\alpha_{k^*})} + n^{-(b-1/\alpha_{k^*})}\right),$$

as $n \to +\infty$. *If* $b \geq 1$, *we have in particular* $\mathbb{E}[R_n] = o(1)$ *as* $n \to +\infty$.

The proof of Theorem 2 is deferred to Appendix A. It closely follows that of Theorem 2 in Carpentier and Valko (2014), the main difference being that their concentration bound (Theorem 1 therein) can be replaced by our tighter bound (see Theorem 1 in the present paper). Recall that in Theorem 2 in Carpentier and Valko (2014), the upper bound on the expected extreme regret for EXTREMEHUNTER goes to infinity when $n \to +\infty$:

$$\mathbb{E}[R_n] = \mathcal{O}\left(n^{\frac{1}{(1+b)\alpha_{k^*}}}\right). \tag{16}$$

In contrast, in Theorem 2 when $b \geq 1$, the upper bound obtained vanishes when $n \to +\infty$. In the case $b < 1$, the upper bound still improves upon Eq. (16) by a factor $n^{(\alpha_{k^*}b(b+1)-b)/((b+1)\alpha_{k^*})} > n^{b^2/(2\alpha_{k^*})}$.

4   Lower Bound on the Expected Extreme Regret

In this section we prove a lower bound on the expected extreme regret for EXTREMEETC and EXTREMEHUNTER in specific cases. We assume now that $\alpha_{(2)} > 2\alpha_{k^*}^2/(\alpha_{k^*} - 1)$ and we start with a preliminary result on second order Pareto distributions, proved in Appendix A.

Lemma 4. *If* X *is a r.v. drawn from a* (α, β, C, C')*-second order Pareto distribution and* r *is a strictly positive constant, the distribution of the r.v.* X^r *is a* $(\alpha/r, \beta, C, C')$*-second order Pareto.*

In order to prove the lower bound on the expected extreme regret, we first establish that the event corresponding to the situation where the highest reward obtained by EXTREMEETC and EXTREMEHUNTER comes from the optimal arm k^* occurs with overwhelming probability. Precisely, we denote by ξ_2 the event such that the bound

$$\max_{k \neq k^*} \max_{1 \leq i \leq N} \widetilde{X}_{k,i} \leq \max_{1 \leq i \leq n-(K-1)N} \widetilde{X}_{k^*,i}.$$

holds true. The following lemma, proved in Appendix A, provides a control of its probability of occurence.

Lemma 5. *For* n *larger than some constant depending only on* (α_k, β_k, C_k) *$_{1 \leq k \leq K}$ and* C', *the following assertions hold true.*

(i) We have:

$$\mathbb{P}(\xi_2) \geq 1 - K\delta_0,$$

where δ_0 *is given in Eq. (10).*

(ii) *Under the event $\xi_0 = \xi_1 \cap \xi_2$, the maximum reward obtained by* ExtremeETC *and* ExtremeHunter *comes from the optimal arm:*

$$\max_{1 \leq t \leq n} X_{I_t, t} = \max_{1 \leq i \leq n - (K-1)N} \widetilde{X}_{k^*, i}.$$

The following lower bound shows that the upper bound (Theorem 2) is actually tight in the case $b \geq 1$.

Theorem 3. *If $b \geq 1$ and $\alpha_{(2)} > 2\alpha_{k^*}^2/(\alpha_{k^*} - 1)$, the expected extreme regret of* ExtremeETC *and* ExtremeHunter *are lower bounded as follows*

$$\mathbb{E}[R_n] = \Omega\left((\log n)^{2(2b+1)/b} n^{-(1-1/\alpha_{k^*})}\right).$$

Proof. Here, π refers to either ExtremeETC or else ExtremeHunter. In order to bound from below $\mathbb{E}[R_n] = \mathbb{E}[G_n^{(k^*)}] - \mathbb{E}[G_n^{(\pi)}]$, we start with bounding $\mathbb{E}[G_n^{(\pi)}]$ as follows

$$\mathbb{E}\left[G_n^{(\pi)}\right] = \mathbb{E}\left[\max_{1 \leq t \leq n} X_{I_t, t}\right] = \mathbb{E}\left[\max_{1 \leq t \leq n} X_{I_t, t} \mathbb{1}\{\xi_0\}\right] + \mathbb{E}\left[\max_{1 \leq t \leq n} X_{I_t, t} \mathbb{1}\{\bar{\xi}_0\}\right]$$

$$\leq \mathbb{P}(\xi_0)\mathbb{E}\left[\max_{1 \leq t \leq n} X_{I_t, t} \mid \xi_0\right] + \sum_{k=1}^{K} \mathbb{E}\left[\max_{1 \leq i \leq T_{k,n}} \widetilde{X}_{k,i} \mathbb{1}\{\bar{\xi}_0\}\right], \qquad (17)$$

where $\widetilde{X}_{k,i}$ has been defined in Sect. 3.1. From (ii) in Lemma 5, we have

$$\mathbb{E}\left[\max_{1 \leq t \leq n} X_{I_t, t} \mid \xi_0\right] = \mathbb{E}\left[\max_{1 \leq i \leq n - (K-1)N} \widetilde{X}_{k^*, i} \mid \xi_0\right]. \qquad (18)$$

In addition, in the sum of expectations on the right-hand-side of Eq. (17), $T_{k,n}$ may be roughly bounded from above by n. A straightforward application of Hölder inequality yields

$$\sum_{k=1}^{K} \mathbb{E}\left[\max_{1 \leq i \leq T_{k,n}} \widetilde{X}_{k,i} \mathbb{1}\{\bar{\xi}_0\}\right] \leq \sum_{k=1}^{K} \left(\mathbb{E}\left[\max_{1 \leq i \leq n} \widetilde{X}_{k,i}^{\frac{\alpha_{k^*}+1}{2}}\right]\right)^{\frac{2}{\alpha_{k^*}+1}} \mathbb{P}\left(\bar{\xi}_0\right)^{\frac{\alpha_{k^*}-1}{\alpha_{k^*}+1}}. \qquad (19)$$

From (i) in Lemma 5 and Eq. (13), we have $\mathbb{P}(\bar{\xi}_0) \leq K(2n+1)\delta_0$. By virtue of Lemma 4, the r.v. $\widetilde{X}_{k,i}^{(\alpha_{k^*}+1)/2}$ follows a $(2\alpha_k/(\alpha_{k^*}+1), \beta_k, C_k, C')$-second order Pareto distribution. Then, applying Theorem 1 to the right-hand side of (19) and using the identity (18), the upper bound (17) becomes

$$\mathbb{E}\left[G_n^{(\pi)}\right] \leq \mathbb{E}\left[\max_{1 \leq i \leq n - (K-1)N} \widetilde{X}_{k^*, i} \mathbb{1}\{\xi_0\}\right]$$

$$+ \sum_{k=1}^{K} \left((nC_k)^{\frac{\alpha_{k^*}+1}{2\alpha_k}} \Gamma\left(1 - \frac{\alpha_{k^*}+1}{2\alpha_k}\right) + o\left(n^{\frac{\alpha_{k^*}+1}{2\alpha_k}}\right)\right)^{\frac{2}{\alpha_{k^*}+1}} (K(2n+1)\delta_0)^{\frac{\alpha_{k^*}-1}{\alpha_{k^*}+1}}$$

$$\leq \mathbb{E}\left[\max_{1 \leq i \leq n - (K-1)N} \widetilde{X}_{k^*, i}\right] + \mathcal{O}\left(n^{-(1-1/\alpha_{k^*})}\right), \qquad (20)$$

where the last inequality comes from the definition of δ_0. Combining Theorem 1 and (20) we finally obtain the desired lower bound

$$
\begin{aligned}
\mathbb{E}[R_n] &= \mathbb{E}\left[G_n^{(k^*)}\right] - \mathbb{E}\left[G_n^{(\pi)}\right] \\
&\geq \Gamma(1 - 1/\alpha_{k^*})C_{k^*}^{1/\alpha_{k^*}}\left(n^{1/\alpha_{k^*}} - (n - (K-1)N)^{1/\alpha_{k^*}}\right) + \mathcal{O}\left(n^{-(1-1/\alpha_{k^*})}\right) \\
&= \frac{\Gamma(1 - 1/\alpha_{k^*})C_{k^*}^{1/\alpha_{k^*}}}{\alpha_{k^*}}(K-1)Nn^{-(1-1/\alpha_{k^*})} + \mathcal{O}\left(n^{-(1-1/\alpha_{k^*})}\right),
\end{aligned}
$$

where we used a Taylor expansion of $x \mapsto (1+x)^{1/\alpha_{k^*}}$ at zero for the last equality.

5   A Reduction to Classical Bandits

The goal of this section is to render explicit the connections between the max K-armed bandit considered in the present paper and a particular instance of the classical Multi-Armed Bandit (MAB) problem.

5.1   MAB Setting for Extreme Rewards

In a situation where only the large rewards matter, an alternative to the max k-armed problem would be to consider the expected cumulative sum of the most 'extreme' rewards, that is, those which exceeds a given high threshold u. For $k \in \{1, \ldots, K\}$ and $t \in \{1, \ldots, n\}$, we denote by $Y_{k,t}$ these new rewards

$$
Y_{k,t} = X_{k,t}\mathbb{1}\{X_{k,t} > u\}.
$$

In this context, the classical MAB problem consists in maximizing the expected cumulative gain

$$
\mathbb{E}\left[G^{\mathrm{MAB}}\right] = \mathbb{E}\left[\sum_{t=1}^{n} Y_{I_t,t}\right].
$$

It turns out that for a high enough threshold u, the unique optimal arm for this MAB problem, $\arg\max_{1 \leq k \leq K}\mathbb{E}[Y_{k,1}]$, is also the optimal arm k^* for the max k-armed problem. We still assume second order Pareto distributions for the random variables $X_{k,t}$ and that all the hypothesis listed in Sect. 3.1 hold true. The rewards $\{Y_{k,t}\}_{1 \leq k \leq K, 1 \leq t \leq T}$ are also heavy-tailed so that it is legitimate to attack this MAB problem with the ROBUST UCB algorithm (Bubeck et al. 2013), which assumes that the rewards have finite moments of order $1 + \epsilon$

$$
\max_{1 \leq k \leq K}\mathbb{E}\left[|Y_{k,1}|^{1+\epsilon}\right] \leq v, \tag{21}
$$

where $\epsilon \in (0, 1]$ and $v > 0$ are known constants. Given our second order Pareto assumptions, it follows that Eq. (21) holds with $1 + \epsilon < \alpha_{(1)}$. Even if the knowledge of such constants ϵ and v is a strong assumption, it is still fair to compare

ROBUST UCB to EXTREMEETC/HUNTER, which also has strong requirements. Indeed, EXTREMEETC/HUNTER assumes that b and n are known and verify conditions depending on unknown problem parameters (e.g. $n \geq Q_1$, see Eq. (3)).

The following Lemma, whose the proof is postponed to Appendix A, ensures that the two bandit problems are equivalent for high thresholds.

Lemma 6

$$If \quad u > \max\left(1, \left(\frac{2C'}{\min_{1\leq k\leq K} C_k}\right)^{\frac{1}{\min_{1\leq k\leq K} \beta_k}}, \left(\frac{3\max_{1\leq k\leq K} C_k}{\min_{1\leq k\leq K} C_k}\right)^{\frac{1}{\alpha_{(2)}-\alpha_{(1)}}}\right),$$

(22)

then the unique best arm for the MAB problem is $arg\ min_{1\leq k\leq K} \alpha_k = k^*$.

Remark 1. Tuning the threshold u based on the data is a difficult question, outside our scope. A standard practice is to monitor a relevant output (e.g. estimate of α) as a function of the threshold u and to pick the latter as low as possible in the stability region of the output. This is related to the Lepski's method, see e.g. Boucheron and Thomas (2015); Carpentier and Kim (2014); Hall and Welsh (1985).

5.2   ROBUSTUCB Algorithm (Bubeck et al. 2013)

For the sake of completeness, we recall below the main feature of ROBUST UCB and make explicit its theoretical guarantees in our setting. The bound stated in the following proposition is a direct consequence of the regret analysis conducted by Bubeck et al. (2013).

Proposition 1. *Applying the* ROBUST UCB *algorithm of* (Bubeck et al. 2013) *to our MAB problem, the expected number of times we pull any suboptimal arm $k \neq k^*$ is upper bounded as follows*

$$\mathbb{E}[T_{k,n}] = \mathcal{O}(\log n).$$

Proof. See proof of Proposition 1 in Bubeck et al. (2013).

Hence, in expectation, ROBUST UCB pulls fewer times suboptimal arms than EXTREMEETC/HUNTER. Indeed with EXTREMEETC/HUNTER, $T_{k,n} \geq N = \Theta((\log n)^{2(2b+1)/b})$.

Remark 2. Proposition 1 may be an indication that the Robust UCB approach performs better than EXTREMEETC/HUNTER. Nevertheless, guarantees on its expected extreme regret require sharp concentration bounds on $T_{k,n}$ $(k \neq k^*)$, which is out of the scope of this paper and left for future work.

Algorithm 3. Robust UCB with truncated mean estimator (Bubeck et al. 2013)

1: **Input:** $u > 0$ s.t. Eq. (22), $\epsilon \in (0,1]$ and $v > 0$ s.t. Eq. (21).
2: **Initialize:** Pull each arm once.
3: **for** $t \geq K + 1$ **do**
4:    **for** $k = 1, \ldots, K$ **do**
5:       Update truncated mean estimator
$$\widehat{\mu_k} \leftarrow \frac{1}{T_{k,t-1}} \sum_{s=1}^{t-1} Y_{k,s} \mathbb{1}\left\{ I_s = k, Y_{k,s} \leq \left(\frac{vT_{k,s}}{\log(t^2)}\right)^{\frac{1}{1+\epsilon}} \right\}$$
6:       Update index
$$B_k \leftarrow \widehat{\mu_k} + 4v^{1/(1+\epsilon)} \left(\frac{\log t^2}{T_{k,t-1}}\right)^{\epsilon/(1+\epsilon)}$$
7:    **end for**
8:    Play arm $I_t = \arg\max_{1 \leq k \leq K} B_k$
9: **end for**

---

6   Numerical Experiments

In order to illustrate some aspects of the theoretical results presented previously, we consider a time horizon $n = 10^5$ with $K = 3$ arms and exact Pareto distributions with parameters given in Table 2. Here, the optimal arm is the second one (incidentally, the distribution with highest mean is the first one).

Table 2. Pareto distributions used in the experiments.

	Arms		
	$k = 1$	$k^* = 2$	$k = 3$
α_k	15	1.5	10
C_k	10^8	1	10^5
$\mathbb{E}[X_{k,1}]$	3.7	3	3.5
$\mathbb{E}[\max_{1 \leq t \leq n} X_{k,t}]$	7.7	$5.8 \cdot 10^3$	11

We have implemented ROBUST UCB with parameters $\epsilon = 0.4$, which satisfies $1 + \epsilon < \alpha_2 = 1.5$, v achieving the equality in Eq. (21) (ideal case) and a threshold u equal to the lower bound in Eq. (22) plus 1 to respect the strict inequality. EEXTREMEETC is runned with $b = 1 < +\infty = \min_{1 \leq k \leq K} \beta_k$. In this setting, the most restrictive condition on the time horizon, $n > KN \approx 7000$ (given by Eq. (9)), is checked, which places us in the validity framework of EXTREMEETC. The resulting strategies are compared to each other and to the random strategy pulling each arm uniformly at random, but not to THRESHOLD ASCENT algorithm (Streeter and Smith 2006) which is designed only for bounded rewards. Precisely, 1000 simulations have been run and Fig. 1 depicts the extreme regret (1) in each setting averaged over these 1000 trajectories. These experiments empirically support the theoretical bounds in Theorem 2: the expected extreme regret of EXTREMEETC converges to zero for large horizons. On the

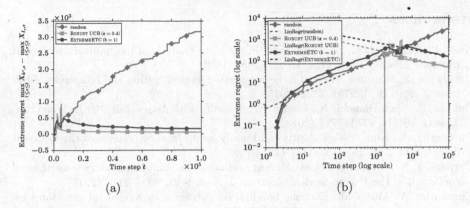

Fig. 1. Averaged extreme regret (over 1000 independent simulations) for EXTREMEETC, ROBUST UCB and a uniformly random strategy. Figure 1b is the log-log scaled counterpart of Fig. 1a with linear regressions computed over $t = 5 \cdot 10^4, \ldots, 10^5$.

log-log scale (Fig. 1b), EXTREMEETC's extreme regret starts linearly decreasing after the initialization phase, at $n > KN \approx 7000$, which is consistent with Lemma 2. The corresponding linear regression reveals a slope ≈ -0.333 (with a coefficient of determination $R^2 \approx 0.97$), which confirms Theorems 2 and 3 yielding the theoretical slope $-(1 - 1/\alpha_{k^*}) = -1/3$.

7 Conclusion

This paper brings two main contributions. It first provides a refined regret bound analysis of the performance of the EXTREMEHUNTER algorithm in the context of the max K-armed bandit problem that significantly improves upon the results obtained in the seminal contribution Carpentier and Valko (2014), also proved to be valid for EXTREMEETC, a computationally appealing alternative we introduce. In particular, the obtained upper bound on the regret converges to zero for large horizons and is shown to be tight when the tail of the rewards is sufficiently close to a Pareto tail (second order parameter $b \geq 1$). On the other hand, this paper offers a novel view of this approach, interpreted here as a specific version of a classical solution (*Robust UCB*) of the MAB problem, in the situation when only very large rewards matter.

Based on these encouraging results, several lines of further research can be sketched. In particular, future work will investigate to which extent the lower bound established for EXTREMEETC/HUNTER holds true for any strategy with exploration stage of the same duration, and whether improved performance is achievable with alternative stopping criteria for the exploration stage.

Acknowledgments. This work was supported by a public grant (*Investissement d'avenir* project, reference ANR-11-LABX-0056-LMH, LabEx LMH) and by the industrial chair *Machine Learning for Big Data* from Télécom ParisTech.

References

Auer, P., Cesa-Bianchi, N., Fischer, P.: Finite-time analysis of the multiarmed bandit problem. Mach. Learn. **47**(2–3), 235–256 (2002)

Boucheron, S., Thomas, M.: Tail index estimation, concentration and adaptivity. Electron. J. Stat. **9**(2), 2751–2792 (2015)

Bubeck, S., Cesa-Bianchi, N., Lugosi, G.: Bandits with heavy tail. IEEE Trans. Inf. Theory **59**(11), 7711–7717 (2013)

Carpentier, A., Kim, A.K.: Adaptive and minimax optimal estimation of the tail coefficient. Stat. Sin. **25**, 1133–1144 (2014)

Carpentier, A., Kim, A.K., et al.: Honest and adaptive confidence interval for the tail coefficient in the Pareto model. Electron. J. Stat. **8**(2), 2066–2110 (2014)

Carpentier, A., Valko, M.: Extreme bandits. In: Advances in Neural Information Processing Systems, vol. 27, pp. 1089–1097. Curran Associates Inc. (2014)

Carpentier, A., Valko, M.: Simple regret for infinitely many armed bandits. In: Proceedings of the 32nd International Conference on Machine Learning, pp. 1133–1141 (2015)

Cicirello, V.A., Smith, S.F.: The max k-armed bandit: a new model of exploration applied to search heuristic selection. In: The Proceedings of the Twentieth National Conference on Artificial Intelligence, vol. 3, pp. 1355–1361. AAAI Press (2005)

David, Y., Shimkin, N.: PAC lower bounds and efficient algorithms for the max k-armed bandit problem. In: Proceedings of The 33rd International Conference on Machine Learning (2016)

Hall, P., Welsh, A.H.: Adaptive estimates of parameters of regular variation. Ann. Stat. **13**(1), 331–341 (1985)

Lepskiĭ, O.V.: A problem of adaptive estimation in Gaussian white noise. Teor. Veroyatnost. i Primenen. **35**(3), 459–470 (1990)

Nishihara, R., Lopez-Paz, D., Bottou, L.: No regret bound for extreme bandits. In: Proceedings of the 19th International Conference on Artificial Intelligence and Statistics (AISTATS) (2016)

Resnick, S.: Heavy-Tail Phenomena: Probabilistic and Statistical Modeling, vol. 10. Springer, New York (2007). https://doi.org/10.1007/978-0-387-45024-7

Streeter, M.J., Smith, S.F.: A simple distribution-free approach to the max k-armed bandit problem. In: Benhamou, F. (ed.) CP 2006. LNCS, vol. 4204, pp. 560–574. Springer, Heidelberg (2006). https://doi.org/10.1007/11889205_40

Variational Thompson Sampling for Relational Recurrent Bandits

Sylvain Lamprier[1]([⊠]), Thibault Gisselbrecht[1,2,3], and Patrick Gallinari[1]

[1] Sorbonne Universités, UPMC Univ Paris 06, CNRS, LIP6 UMR 7606,
4 Place Jussieu, 75005 Paris, France
{sylvain.lamprier,thibault.gisselbrecht,patrick.gallinari}@lip6.fr
[2] IRT SystemX, 8 Avenue de la Vauve, 91120 Palaiseau, France
[3] SNIPS, 18 Rue Saint-Marc, 75002 Paris, France

Abstract. In this paper, we introduce a novel non-stationary bandit setting, called relational recurrent bandit, where rewards of arms at successive time steps are interdependent. The aim is to discover temporal and structural dependencies between arms in order to maximize the cumulative collected reward. Two algorithms are proposed: the first one directly models temporal dependencies between arms, as the second one assumes the existence of hidden states of the system behind the observed rewards. For both approaches, we develop a Variational Thompson Sampling method, which approximates distributions via variational inference, and uses the estimated distributions to sample reward expectations at each iteration of the process. Experiments conducted on both synthetic and real data demonstrate the effectiveness of our approaches.

1 Introduction

A multi-armed bandit problem is a real time sequential decision process in which, at each iteration, a learner is asked to select an action - called arm - among a set of K available ones. The aim of the forecaster is to maximize the cumulative reward over iterations by balancing exploitation (arms with higher observed rewards should be selected often) and exploration (all arms should be explored to improve the knowledge of their utility). The stochastic multi-armed bandit, originally introduced in [24], has been widely studied in the literature. In this instance, the agent assumes stationary distributions of rewards. As an alternative to the optimistic UCB algorithm [7], Thompson Sampling (TS) [32] is another well-known algorithm, based on posterior distributions for reward expectations, to deal with this kind of problem. Another area of research concerns the problem of time varying distributions, in which the expected value of each arm is allowed to change from iteration to iteration. Different scenarios regarding non-stationarity have been studied in the literature: expected rewards may vary either abruptly as in [19], stochastically as in [30] or [27], or with a budgeted total variation of the expected reward as in [9].

In this paper we introduce a new setup, called relational recurrent bandit, where rewards of actions are interdependent over time. This corresponds to problems where rewards of arms have both temporal and structural dependencies that

M. Ceci et al. (Eds.): ECML PKDD 2017, Part II, LNAI 10535, pp. 405–421, 2017.
https://doi.org/10.1007/978-3-319-71246-8_25

can be exploited to improve the efficiency of the reward collection. This is typically the case for tasks of sensors selection, where the aim is to collect useful data from streams under some budget constraints: for some reason (cost of sensor activation, restriction constraints, technical limits, etc.), data from every stream cannot be recorded simultaneously, the process must focus on the best current streams according to the task. In this context, dependencies can occur between streams, due to behavior correlations (similar reactions of sources to some external stimuli for instance, possibly with different delays), or to some kind of reward propagation (social influences between data sources for instance). To the best of our knowledge, such an instance of the multi-armed bandit problem has not been studied in the literature, although it corresponds to a realistic setting that can be met in several tasks, such as for instances dynamic sensor selection for climate modeling, useful source detection from social data streams, online information diffusion prediction and tracking, or even for advertising campaigns when the targeted communities have different reaction delays and/or influence relationships. This has to be distinguished from problems with structural dependencies between bandits, where reward distributions for connected situations are interdependent, for which there has been an increasing interest in the last few years [14, 17].

We investigate relational recurrent bandits for the multiple-plays scenario. While relational recurrent bandit could be defined for classical single-play process, considering temporal dependencies indeed becomes really attractive when more than one arm reveal their reward at each step. In this multiple-play problem, the agent selects $k > 1$ actions simultaneously, which leads to the collection of k rewards at each step. These observed rewards can be used to estimate reward expectations of every arm at the next time step, with the goal to select the most useful ones. Obviously, the temporal dependencies between arms' rewards are unknown a priori. They need to be learned online. The major difficulty in this problem is that the agent only knows the rewards of played arms, which leads to a problem of online learning with missing data.

Assuming linear dependencies between rewards of successive steps, we define a Bayesian model for the derivation of posterior distributions for unknown correlation parameters and the $K - k$ unobserved rewards at each time step. However, the exact computation of such posterior distributions is not analytically tractable. To overcome this difficulty, in order to build a TS procedure for relational recurrent bandits, a Variational Inference approach is developed to approximate true posterior distributions. In this context, two probabilistic models are considered: while a first one directly models temporal dependencies between arms, a second one is based on an underlying hidden Markov process that generates the rewards at each iteration. Although the first model better captures explicit relationships when some strong direct influences exist in the data, the second one allows one to greatly reduce the complexity and permits to encode more complex relationships.

To summarize, the contribution of this paper is three-fold:

- We propose a new instance of the bandit problem where distributions of rewards are defined recursively;
- We design a corresponding Thompson Sampling algorithm for two different formulations of the problem;
- We conduct experiments that assess the effectiveness of our approach on both synthetic and real data.

The paper is organized as follows. In Sect. 2, we review related state-of-the-art literature. In Sect. 3, we propose a first formulation of the relational recurrent bandit problem and an associated algorithm for this new bandit setting. Then, a second formulation for more complex dependencies is proposed in Sect. 4. Finally, Sect. 5 reports experimental results on both artificial and real data.

2   Related Works

Bandit problems have been extensively studied since the seminal paper of Lai and Robbins in 1958 [24]. For the case of stationary distributions of rewards, a huge variety of methods have been proposed to design efficient selection policies. The famous Upper Confidence Bound (UCB) algorithm proposed in [7] and many other UCB-based algorithms [4,5,20] have already proven to be efficient both empirically and theoretically. In these optimistic approaches, confidence intervals on the reward expectations are determined given uncertainty on the distribution estimates. The selected arm at each step is the one with the highest upper confidence bound, which allows one to guarantee logarithmic bounds of the regret[1]. On the other hand, the TS algorithm [32] introduces randomness on the exploration by sampling distribution parameters from their posterior at each time-step, before selecting the arm with the best reward expectation following the sampled distribution. It recently attracted increasing attention [15], due to its good performances and simplicity of use. It has also been proven to have a logarithmic regret bound in [2]. When more than one arm (say k) are selected at each iteration, the problem is called multiple-play multi-armed bandit, for which UCB-based [16] and TS policies [23] have been derived.

When K is very large, the problem becomes more challenging, since inducing sometimes intractable exploration spaces. Fortunately, in such scenarios, the rewards often exhibit some structural properties whose the agent can leverage. For instances, contextual versions of UCB are given in [1,25], for the case where rewards depend on some available contextual knowledge, according to unknown parameters to be learned. Arms with similar contexts at a given time t are supposed to lead to similar rewards, which helps the exploration process. Different cases have been investigated in the literature, for example linear [1,3,18], unimodal [29] or Lipschitz [10]. In every case, the structure reflects some relation between arms rewards.

---

[1] Where the regret corresponds to the expectation of what we missed with a given policy compared with an optimal strategy that knows exact distribution parameters.

Closer to our work, still quite far, are approaches that focus on bandits on graphs, first introduced in [26] for the adversarial bandit and then in [12] for the stochastic bandit. Later, in [11], the UCB-LP was proposed for an equivalent setting and improved the results of [12]. Note that in those cases, the graph structure is supposed to be known beforehand and is explicitly used by the algorithm. Basically, the main assumption is that when an arm is played, not only its reward is revealed, but the reward of all its neighbors in the graph is also shown to the learner. Another area of research concerns contextual bandit on graphs, for which each node is a contextual bandit and the edges of the graph are used to define the similarity between the weights of the different contextual bandits [14,17]. In [14] the edges of the graph are known, whereas in [17] the learner has to find the different clusters online. Our work differs from all of these approaches, by leveraging both temporal and structural dependencies between arms (and not between situations such as in [14,17]). Observing some good actions at some time step can provide information on the utility of the others ones for the following time step. In our setting, rewards at step $t - 1$ modify the distribution of all rewards at the next time step t. Capturing these dependencies may allow one to better handle non stationarity of rewards.

Non stationary bandits have been studied in some recent works, such as [19] which considers that the reward distribution can change during the process, but with a limited number of changes, and proposes two algorithms for this case: Discounted UCB which uses a discount factor to give more importance to recent observations than to older ones and Sliding Window UCB which uses a sliding window of size τ that restricts estimations to be performed only from observations in the last τ time steps. Rather than assuming radical distribution changes as in [19], *restless bandits* [33] consider that the state of each action evolves according to some random process[2]. Following this principle, [30] assume independent Brownian motions between consecutive steps, while [27], [6] or [31] consider that some hidden Markov process drives evolutions of reward expectations on each arm. These works differ from ours by the fact that reward expectations of different arms are independent. Moreover, considered states are usually discrete in existing works.

Finally, the recent work of [13] tackles the local influence maximization problem, in which authors do not assume any knowledge of the graph, but consider a setting where it can be gradually discovered. Indeed, the only information the learner has is a set of nodes each arm is currently influencing. Even if the temporal dependencies we aim at discovering in our approach can be seen as an influence graph, our proposal greatly differs from [13], which only focus on identifying the most influencial nodes of an unknown network without explicitly modeling the underlying dynamics of the successive rewards. Recurrent bandits stand as a novel instance of bandit problems, where temporal dependencies have to be extracted from incomplete data in an online fashion, in order to deal with the non-stationarity of the reward expectations.

---

[2] When only the state of the selected action changes at each iteration, the problem is called *rested bandit*.

3   Relational Recurrent Bandit

The bandit problem with multiple plays processes as follows: at each time step $t \in \{1, ..., T\}$, a subset $\mathcal{K}_t \subset \mathcal{K} = \{1, ..., K\}$ of k arms is selected and for every arm $i \in \mathcal{K}_t$ the agent receives the corresponding rewards $r_{i,t} \in \mathbb{R}$. The choice of $\mathcal{K}_t$ is done based on the historical decisions $\mathcal{H}_{t-1} = \{(i, r_{i,s}), i \in \mathcal{K}_s, s = 1..t-1\}$. The function that selects at each time-step t the subset $\mathcal{K}_t$ is called a bandit policy or algorithm.

In this section we propose to design a first TS algorithm for the relational recurrent bandit setting. The principle of TS is to produce a sample of the reward expectation according to its posterior distribution, and then to choose the arm with the best sampled expectation.

3.1   Probabilistic Model

In this section, we propose to consider direct recurrent relationships between arm's rewards from one time step to the following one. More specifically, we assume that the expected reward of an arm i at time step t is a linear combination of every rewards at time step $t-1$ plus a bias term, which we formalize as follows:

$$\forall i \in \{1, ..., K\}, \exists \theta_i \in \mathbb{R}^{K+1} \text{ such that } : \forall t \in \{2, ..., T\} : \mathbb{E}[r_{i,t} | R_{t-1}] = \theta_i^\top R_{t-1}^+ \tag{1}$$

where $R_t = (r_{1,t}, ..., r_{K,t})^\top \in \mathbb{R}^K$ is the vector of rewards at time step t and $R_t^+ = (R_t, r_{K+1,t})$ appends an additional constant bias term $r_{K+1,t}$ always equal to 1 at the end of R_t. We consider the following assumptions to derive our relational recurrent model:

- **Likelihood of data:** $\forall i \in \{1, ..., K\}, \exists \theta_i \in \mathbb{R}^{K+1}$ such that $\forall t \in \{2, ..., T\}$: $r_{i,t} = \theta_i^\top R_{t-1}^+ + \epsilon_{i,t}$, where $\epsilon_{i,t} \sim \mathcal{N}(0, \sigma^2)$ (Gaussian noise with mean 0 and variance σ^2).
- **Prior on parameters:** $\forall i \in \{1, ..., K\}$: $\theta_i \sim \mathcal{N}(0, \alpha^2 I)$ ($(K+1)$-dimensional Gaussian vector with mean 0 and covariance matrix $\alpha^2 I$).
- **Prior at time 1:** $\forall i \in \{1, ..., K\}$: $r_{i,1} \sim \mathcal{N}(0, \sigma^2)$.

For clarity, we consider simple covariance matrices for the different priors, however any prior can be used, depending on the knowledge of the agent. In the following, we propose a probabilistic analysis of the above model, which is then used to derive our bandit policy.

3.2   Algorithm

To perform TS, at each time step $t \geq 2$, we thus need to be able to sample a value $\tilde{r}_{i,t}$ from the expected reward posterior distribution $\theta_i^\top R_{t-1}^+$ for each action i. If the rewards of unplayed arms were revealed at each time step, i.e. if the full vector R_{t-1} was available, this would come down to a traditional contextual bandit problem. However, for every time step $s \in \{1, ..., t-1\}$ and $i \notin \mathcal{K}_s$, the reward $r_{i,s}$ is not available and must be treated as a random variable. Thus, TS must be performed from the following joint distribution:

$$P((r_{i,s})_{s=1..t-1,i\notin\mathcal{K}_s},(\theta_i)_{i=1..K}|(r_{i,s})_{s=1..t-1,i\in\mathcal{K}_s}) \tag{2}$$

However, due to the recurrent aspect of the problem, this distribution cannot be directly obtained. To overcome this difficulty, we propose to adopt a Variational Inference approach, which stands as an alternative to MCMC (Monte Carlo Markov Chain) methods, such as Gibbs Sampling, to approximate complex distributions. While MCMC methods provide numerical approximations via successive samplings following the true joint distribution, variational inference outputs an analytical locally optimal approximation of this distribution. In practice the use of MCMC approaches for relational recurrent bandits would induce too important computation costs at each step, and worst, too many steps to converge toward tight distributions.

The idea of Variational Inference is to approximate the true targeted distribution by a simpler distribution. We propose here to consider the following mean field approximation:

$$Q((r_{i,s})_{s=1..t-1,i\notin\mathcal{K}_s},(\theta_i)_{i=1..K}) = \prod_{i=1}^{K} q_{\theta_i}(\theta_i) \prod_{s=1}^{t-1}\prod_{i\notin\mathcal{K}_s} q_{r_{i,s}}(r_{i,s}) \tag{3}$$

with all q_{θ_i} and $q_{r_{i,s}}$ standing as independent variational distributions set for every factor of Q. The aim is then to find variational distributions for each factor, that minimize the Kullback-Leibler divergence (KL) from the true joint probability distribution P (as defined in (2)) to the variational joint distribution Q. This leads to an approximated distribution Q that is included in P: it can ignore some modes of P, but does not give posterior mass to regions where P has vanishing density. Following this, we obtain the variational distributions given in the two following distributions (all proofs are given in the supplementary material[3]).

Proposition 1. *Let $D_{t-1} = (R_s^{+\top})_{s=1..t-1}$ be the $(t-1)\times(K+1)$ matrix where row s corresponds to the rewards vector at time s concatenated with an additional component set to 1. We also note $D_{1..t-2}$ the $(t-2)\times(K+1)$ matrix of the $t-2$ first rows of D_{t-1} and $D_{i:2..t-1}$ the vector containing the $t-2$ last components of the i-th column of D_{t-1} (i.e., rewards obtained by i from 2 to $t-1$). Then, for all $t \geq 2$, and for $i \in \{1,...,K\}$, the best variational distribution of $q_{\theta_i}^*$ corresponds to a Gaussian $\mathcal{N}(A_{i,t-1}^{-1}b_{i,t-1}, A_{i,t-1}^{-1})$, with:*

- $A_{i,1} = \dfrac{I}{\alpha^2}$ *and* $b_{i,1} = 0$
- $A_{i,t-1} = \dfrac{\mathbb{E}[D_{1..t-2}^{\top}D_{1..t-2}]}{\sigma^2} + \dfrac{I}{\alpha^2}$ *and* $b_{i,t-1} = \dfrac{\mathbb{E}[D_{1..t-2}^{\top}]}{\sigma^2}\mathbb{E}[D_{i:2..t-1}]$ *for $t > 2$*

where $\mathbb{E}[D_{1..t-2}^{\top}D_{1..t-2}] = \sum_{s=1}^{t-2}\mathbb{E}[R_s^+]\mathbb{E}[R_s^+]^{\top} + Var(R_s^+)$, with values of $\mathbb{E}[R_s^+]$ and $Var(R_s^+)$ determined according to Proposition 2.

---

[3] Available at http://www-connex.lip6.fr/~lampriers/ECML2017-supMat.pdf.

Proposition 2. *We note Θ the $K \times (K+1)$ matrix where row i equals to $\theta_i^\top$ and β_j the j-th column of Θ. For $t \geq 2$ and $1 \leq s \leq t-1$, the best variational distribution $q_{r_{i,s}}^*$ is a Gaussian $\mathcal{N}(\mu_{i,s}, \sigma_{i,s}^2)$, with:*

- *If $s = 1$:* $\mu_{i,s} = \dfrac{\mathbb{E}[\beta_i]^\top \mathbb{E}[R_{s+1}] - \sum\limits_{j=1,j\neq i}^{K+1} \mathbb{E}[\beta_i^\top \beta_j] \mathbb{E}[r_{j,s}]}{1 + \mathbb{E}[\beta_i^\top \beta_i]}$, $\sigma_{i,s}^2 = \dfrac{\sigma^2}{1 + \mathbb{E}[\beta_i^\top \beta_i]}$

- *If $s = t-1$:* $\mu_{i,s} = \mathbb{E}[\theta_i]^\top \mathbb{E}[R_{s-1}^+]$, $\sigma_{i,s}^2 = \sigma^2$

- *Else:* $\mu_{i,s} = \dfrac{\mathbb{E}[\beta_i]^\top \mathbb{E}[R_{s+1}] - \sum\limits_{j=1,j\neq i}^{K+1} \mathbb{E}[\beta_i^\top \beta_j]\mathbb{E}[r_{j,s}] + \mathbb{E}[\theta_i]^\top \mathbb{E}[R_{s-1}^+]}{1 + \mathbb{E}[\beta_i^\top \beta_i]}$, $\sigma_{i,s}^2 = \dfrac{\sigma^2}{1 + \mathbb{E}[\beta_i^\top \beta_i]}$

where $\mathbb{E}[\beta_i^\top \beta_j] = \sum\limits_{l=1}^{K} Var(\theta_l)_{i,j} + \mathbb{E}[\theta_l]_i \mathbb{E}[\theta_l]_j$.

Variational distributions given in Propositions 1 and 2 are inter-dependent. Their estimation must therefore be performed via an iterative procedure, which is described in Algorithm 1. This algorithm uses a parameter $nbIt$ which corresponds to the number of iterations to achieve. Note that, when a reward is observed, one obviously uses the associated value rather than its expectation. That is, component i of $\mathbb{E}[R_s^+]$ (denoted $\mathbb{E}[R_s^+]_i$) equals $r_{i,s}$ in Propositions 1 and 2 if $i \in \mathcal{K}_s$, $\mathbb{E}[r_{i,s}]$ otherwise. Moreover, $Var(R_s)$ corresponds to a diagonal matrix where element (i,i) equals 0 if $i \in \mathcal{K}_s$, $\sigma_{i,s}^2$ otherwise. Note at last that $r_{K+1,s}$ is always known: $\mathbb{E}[r_{K+1,s}] = 1$ and $Var(r_{K+1,s}) = 0$ for any iteration s.

Algorithm 1.
Variational Inference

Input: $nbIt$

1 **for** $It = 1..nbIt$ **do**
2   **for** $i = 1..K$ **do**
3   Compute $A_{i,t-1}, b_{i,t-1}$
     w.r.t. proposition 1;
4   **for** $s = 1..t-1$ **do**
5   **if** $i \notin \mathcal{K}_s$ **then**
       Compute $\mu_{i,s}, \sigma_{i,s}^2$
       w.r.t. proposition 2 ;
6   **end**
7 **end**
8 **end**

Algorithm 2.
Recurrent Thompson Sampling

1 **for** $t = 1..T$ **do**
2   Perform Variational Inference;
3   **for** $i = 1..K$ **do**
4   Sample $\tilde{\theta}_i \sim q_{\theta_i}^*$;
5   **if** $i \notin \mathcal{K}_{t-1}$ **then**
6   Sample $\tilde{r}_{i,t-1} \sim q_{r_{i,t-1}}^*$;
7      $\tilde{r}_{i,t} = \tilde{\theta}_i^\top \tilde{R}_{t-1}^+$;
8   **end**
9    $\mathcal{K}_t \leftarrow \underset{\hat{\mathcal{K}} \subseteq \mathcal{K}, |\hat{\mathcal{K}}|=k}{\arg\max} \sum\limits_{i \in \hat{\mathcal{K}}} \tilde{r}_{i,t}$;
10   **for** $i \in \mathcal{K}_t$ **do** Collect $r_{i,t}$;
11 **end**

Algorithm 2 describes our recurrent relational TS procedure, which uses Algorithm 1 to estimate variational distributions for hidden variables at each iteration of the process. At each iteration t, the algorithm samples every hidden reward at time $t-1$ from $q_{r_{i,t-1}}^*$ and θ parameters from $q_{\theta_i}^*$ for every action i. It allows one to compute an expectation score $\tilde{r}_{i,t} = \tilde{\theta}_i^\top \tilde{R}_{t-1}^+$ for every action. The k actions with best $\tilde{r}_{i,t}$ scores are performed and the associated rewards are collected.

The complexity of the proposed algorithm increases linearly with the number of time steps from the beginning. At time t, we have $K(K+1) + (t-1)(K-k)$ random variables whose inter-dependent distributions have to be re-estimated at each iteration to include new observations (there are $(t-1)(K-k)$ missing rewards at time t). This is not compatible with the online nature of the bandit problem, since it might lead to memory overhead and complexity problems. To cope with this issue, we propose to use an approximated algorithm that restrains to a limited amount of time-steps in the past by introducing a sliding window of size S: instead of considering every missing reward from time 1 to time $t-1$ at each iteration t, the algorithm restricts distribution re-estimations to missing rewards from time $t-S-1$ to $t-1$, making its complexity constant with time. On the other hand both the $K(K+1)$ and the $(K-k)$ factors cannot be reduced easily with the proposed method. This comes from the fact that the algorithm tries to learn every weight of the model. When the number of arms becomes large, the model turns out to be very hard to learn. To cope with this issue, and also to deal with longer term dependencies, we propose a second model which considers transitions between hidden states of the system rather than explicit relationships.

4   State-Based Recurrent Bandit

In this section, we assume the existence of an underlying hidden state $h_t \in \mathbb{R}^d$ responsible for reward values at each iteration t. Moreover, the size d of this state is assumed to be smaller than K. On the other hand, unlike in the previous model, the recurrence is assumed to take place in the hidden state, such that there exists a linear transformation between h_{t-1} and h_t. Formally:

$$\exists \Theta \in \mathbb{R}^{d \times d}, \forall t \in \{2, ..., T\} : \mathbb{E}[h_t | h_{t-1}] = \Theta h_{t-1} \tag{4}$$

$$\forall i \in \{1, ..., K\}, \exists W_i \in \mathbb{R}^d, \exists b_i \in \mathbb{R}, \forall t \in \{1, ..., T\} : \mathbb{E}[r_{i,t} | h_t] = W_i^\top h_t + b_i \tag{5}$$

with $h_t \in \mathbb{R}^d$ the hidden state of the system at iteration t, Θ the transition matrix $d \times d$, $W_i \in \mathbb{R}^d$ the mapping vector from any state to the expected reward for i and b_i a bias term for every action i. The model is illustrated on 3 iterations in Fig. 1, where one observes temporal dependencies between states. The model is able to deal with long term dependencies thanks to recurrent relationships between continuous states.

We consider the following assumptions:

- **Likelihood 1:** $\exists \Theta \in \mathbb{R}^{d \times d}, \forall t \in \{2, ..., T\} : h_t = \Theta h_{t-1} + \epsilon_t$, where $\epsilon_t \sim \mathcal{N}(0, \delta^2 I)$.
- **Likelihood 2:** $\forall i \in \{1, ..., K\}, \exists W_i \in \mathbb{R}^d, \exists b_i \in \mathbb{R}$ such that : $\forall t \in \{1, ..., T\}$: $r_{i,t} = W_i^\top h_t + b_i \; \epsilon_{i,t} \sim \mathcal{N}(0, \sigma^2)$.
- **Prior on h_1:** $h_1 \sim \mathcal{N}(0, \delta^2 I)$, where I is the identity matrix of size d.

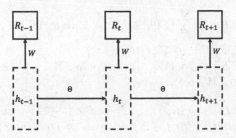

Fig. 1. Generative state-based recurrent model

- **Prior on** Θ: $\forall i \in \{1, ..., d\}$: $\theta_i \sim \mathcal{N}(0, \alpha^2 I)$, where θ_i is the i-th row of Θ.
- **Prior on** W: $\forall i \in \{1, ..., K\}$: $(W_i, b_i) \sim \mathcal{N}(0, \gamma^2 I)$, with (W_i, b_i) the $d + 1$ sized vector resulting from appending b_i to W_i, and I the identity matrix.

This defines a Linear Dynamic System for which parameters have to be estimated. With known transition Θ and emission $(W_i)_{i=1..K}$ matrices, this model would be similar to a Kalman Filter [22], usually designed to get the state of a physical system from noisy observations but with a known dynamical model. Our problem falls in Bayesian Linear Dynamical Systems, for which a variational approach has been proposed in [8]. However, the direct application of the very generic method proposed in this paper is complex. We now describe the resulting distributions for our specific case (proofs are given in the supplementary material).

Following the same approach as in the previous section we consider the following mean-field approximation:

$$Q(h_1, ..., h_{t-1}, \Theta, W, b) = \prod_{s=1}^{t-1} q_{h_s}(h_s) \prod_{i=1}^{K} q_{W_i, b_i}(W_i, b_i) \prod_{j=1}^{d} q_{\theta_j}(\theta_j) \quad (6)$$

Propositions 3, 4 and 5 give variational distributions based on this factorization.

Proposition 3. *Let W be a $K \times d$ matrix whose row i equals $W_i^\top$ and b a bias vector of size K. At step $t \geq 2$, for all s such that $1 \leq s \leq t - 1$, the best variational distribution $q_{h_s}^*$ is a Gaussian $\mathcal{N}(F_s^{-1} g_s, F_s^{-1})$ with:*

$$\begin{cases} \text{If } 1 < s < t - 1 : F_s = \mathcal{A}_s + \mathcal{B}_s + \mathcal{C}_s \text{ and } g_s = \mathcal{D}_s + \mathcal{E}_s + \mathcal{F}_s \\ \text{If } s = 1 : F_s = \mathcal{A}_s + \mathcal{B}_s + \mathcal{C}_s \text{ and } g_s = \mathcal{E}_s + \mathcal{F}_s \\ \text{If } s = t - 1 : F_s = \mathcal{A}_s + \mathcal{C}_s \text{ and } g_s = \mathcal{D}_s + \mathcal{E}_s \end{cases}$$

$$\mathcal{A}_s = \frac{I}{\delta^2}, \quad \mathcal{B}_s = \frac{\mathbb{E}[\Theta^\top \Theta]}{\delta^2}, \quad \mathcal{C}_s = \frac{\sum_{i \in \mathcal{K}_s} \mathbb{E}[W_i W_i^\top]}{\sigma^2}, \quad \mathcal{D}_s = \frac{\mathbb{E}[\Theta]\mathbb{E}[h_{s-1}]}{\delta^2},$$

$$\mathcal{E}_s = \frac{\sum_{i \in \mathcal{K}_s} \mathbb{E}[W_i] r_{i,s} - \mathbb{E}[W_i b_i]}{\sigma^2}, \quad \mathcal{F}_s = \frac{\mathbb{E}[\Theta]^\top \mathbb{E}[h_{s+1}]}{\delta^2};$$

$$\mathbb{E}[\Theta^\top \Theta] = \sum_{i=1}^{d} \mathbb{E}[\theta_i \theta_i^\top] = \sum_{i=1}^{d} (\mathbb{E}[\theta_i]\mathbb{E}[\theta_i]^\top + Var(\theta_i));$$

$$\mathbb{E}[W_i W_i^\top] = \mathbb{E}[W_i]\mathbb{E}[W_i]^\top + Var(W_i);$$

$\mathbb{E}[W_i b_i] = \mathbb{E}[W_i]\mathbb{E}[b_i] + Cov(W_i, b_i)$, with $Cov(W_i, b_i)$ the d first components of the last row of $Var((W_i, b_i))$.

Proposition 4. *Let $D_{t-1} = (h_s^\top)_{s=1..t-1}$ be the $(t-1) \times d$ matrix of states until $t-1$, $D_{1..t-2}$ the matrix of the $t-2$ first rows of D_{t-1} and $D_{i:2..t-1}$ the vector of the $t-2$ last components of the i-th column of D_{t-1}. Then, for $t \geq 2$ the best variational distribution $q_{\theta_i}^*$ is a Gaussian $\mathcal{N}(A_{i,t-1}^{-1} b_{i,t-1}, A_{i,t-1}^{-1})$ with:*

$$\begin{cases} A_{i,1} = \dfrac{I}{\alpha^2}; \; b_{i,1} = 0 \\ A_{i,t-1} = \dfrac{I}{\alpha^2} + \dfrac{\mathbb{E}[D_{1..t-2}^\top D_{1..t-2}]}{\delta^2}; \; b_{i,t-1} = \dfrac{\mathbb{E}[D_{1..t-2}]^\top}{\delta^2} \mathbb{E}[D_{i:2..t-1}] \; for \; t > 2 \end{cases}$$

where $\quad \mathbb{E}[D_{1..t-2}^\top D_{1..t-2}] = \sum_{s=1}^{t-2} (\mathbb{E}[h_s]\mathbb{E}[h_s]^\top + Var(h_s))$

Proposition 5. *For each action i, we note $\mathcal{T}_{i,t-1}$ the set of iterations where i has been played before iteration t, i.e., $\mathcal{T}_{i,t-1} = \{s, i \in \mathcal{K}_s \text{ for } 1 \leq s \leq t-1\}$. We also note $M_{i,t-1} = ((h_s, 1)^\top)_{s \in \mathcal{T}_{i,t-1}}$ and $c_{i,t-1} = (r_{i,s})_{s \in \mathcal{T}_{i,t-1}}$. For every i and $t \geq 1$, the best distribution $q_{(W_i, b_i)}^*$ is a Gaussian $\mathcal{N}(V_{i,t-1}^{-1} v_{i,t-1}, V_{i,t-1}^{-1})$ with:*

$$\begin{cases} V_{i,1} = \dfrac{I}{\gamma^2}; \; v_{i,1} = 0 \\ V_{i,t-1} = \dfrac{I}{\gamma^2} + \dfrac{\mathbb{E}[M_{i,t-1}^\top M_{i,t-1}]}{\sigma^2}; v_{i,t-1} = \dfrac{\mathbb{E}[M_{i,t-1}]^\top c_{i,t-1}}{\sigma^2} \end{cases}$$

where $\mathbb{E}[M_{i,t-1}^\top M_{i,t-1}] = \sum_{s \in \mathcal{T}_{i,t-1}} (\mathbb{E}[(h_s, 1)]\mathbb{E}[(h_s, 1)]^\top + Var((h_s, 1)))$ *and* $Var((h_s, 1))$ *is* $Var(h_s)$ *with an additional final row and column of 0.*

Based on Propositions 3, 4 and 5, a TS algorithm can be easily derived following a similar process as described in Algorithms 1 and 2: at each iteration t, a variational inference step allows one to estimate accurate distributions for the different factors. Then, distributions on h_{t-1}, Θ and each (W_i, b_i) allow one to sample a reward expectation score for each action $\tilde{r}_{i,t} = \tilde{W}_i^\top \tilde{\Theta} \tilde{h}_{t-1} + \tilde{b}_i$. A major benefit compared to the previous approach is that the complexity is much lower in the number of arms. Indeed here, at time t we only need to consider $d^2 + d(t-1) + K(d+1)$ random variables, which does not increase quadratically with the number of arms but with the dimension of the hidden space. Moreover, the same trick than before, which consists in restraining to a time window for learning, allows us to end up with a constant complexity. However, such a method might lead to forget important knowledge. To cope with this, we propose to introduce a memory, which consists in using values computed at time $t - 1 - S$ as new *priors*. For instance, for θ_i, rather than considering $A_{i,t-1} = I/\alpha^2 + \mathbb{E}[D_{t-2-S..t-2}^{\top}D_{t-2-S..t-2}]/\delta^2$, it comes down to set $A_{i,t-1} = A_{i,t-1-S} + \mathbb{E}[D_{t-2-S..t-2}^{\top}D_{t-2-S..t-2}]/\delta^2$. Similar operations can be performed on (W_i, b_i). In this setting, since complexity mostly arises from matrix inversions (which is in $\mathcal{O}(n^3)$ for a $n \times n$ matrix via Gauss-Jordan elimination), required updates at each step for our state-based recurrent bandit is in $\mathcal{O}(K * (d+1)^3 + S * d^3)$, with S the number of historical steps to consider. This appears reasonable compared with bandits with structural dependencies, such as [14] or [17], which are in $\mathcal{O}(K^2)$ in the first iterations ($d << K$ in most applications).

5   Experiments

In this section, we compare our algorithm with direct relationships defined in Sect. 3, hereafter called RelationalTS, and the state-based version defined in Sect. 4, hereafter denoted StateTS_d (with d the number of dimensions of the states). For RelationalTS, we set standard deviations on rewards to $\sigma = 0.1$ and standard deviations on parameters to $\alpha = 0.1$. Note that, to give more freedom to the bias parameter, we finally set the item $(K+1, K+1)$ to 1 in the prior covariance matrix of each θ_i). For each version of StateTS, $\gamma = \alpha = 1$ and $\sigma = \delta = 0.1$. These settings allowed us to observe the best average results. We consider a time window $S = 200$ and a number of variational inference iterations set to $nbIt = 10$. We also consider a memory for every version of our approach (see end of Sect. 4), as it allowed us to observe a slight gain of performances in every tested setting, while preventing from dramatic forgetting (e.g., if no activity is recorded during S iterations).

To the best of our knowledge, no algorithm already treats the recurrent setting we introduced. As baselines, we consider the following state-of-the-art policies, in addition to a random policy that chooses arms uniformly at each step:

- Algorithms for the stationary case: two combinatorial UCB policies CUCB [7] and CUCBV [21] and a Thompson Sampling algorithm TS designed for Gaussian rewards [2]. Those algorithms, while not designed to deal with time-varying rewards, could at least discover the bias part of the reward distributions.
- Algorithms for non-stationary rewards: non-stationary UCB-based policies D-UCB and SW-UCB [19] which can be adapted to the multiple-plays formulation by selecting the top-k arms with highest UCB scores. Note that D-UCB incorporates a discount factor while SW-UCB uses a sliding window in order to deal with the changes of expected rewards values through time. We tuned their parameter according to remarks 3 and 9 in [19], which allowed us to observe the best average results in our experiments.

Note that, while graph based approaches such as [17] could appear close to our work, they are not applicable in our setting. They are indeed usually designed for a finite set of decision situations (mostly users for which items have to be recommended), with relationships between situations (i.e., users with similar behaviors) and some observed features for the available actions (knowledge about the items to be recommended), which is not the case here.

5.1 Artificial Data

Two different sets of simulated data are considered:

- **XP1:** Rewards are generated following the relational model described in Sect. 3 (by taking $\sigma = \alpha = 1$ and $\mu_i = 0$ for every arm i. A matrix Θ for each action i and an initial rewards vector R_1 are randomly generated according to their prior distributions[4]. Then, rewards are iteratively generated following the relational model of Sect. 3.1.
- **XP2:** Rewards are generated following cycles defined for each action: the horizon of T iterations is split in periods of 25 iterations. For each action i, we uniformly sample 4 different means $\mu_{i,j} \in]0;1[$ (with $j \in \{0..3\}$). Then, at each iteration $t \in \{1, ..., T\}$ corresponds a period $per_t = (t \div 25)$. For each action i, a reward is sampled for iteration t from a Gaussian $\mathcal{N}(\mu_{i,x}, 1)$, with $x = per_t \mod 4$.

For both sets of experiments, a small version, with $K = 30$ arms and $T = 1000$ iterations, and a large version, with $K = 200$ and $T = 10000$, are considered. 100 datasets of each kind are generated, results are averages on these different datasets. While **XP1** aims at observing performances when direct temporal relationships exist between arms utility, **XP2** aims at assessing the approaches for settings with implicit structural dependencies, when some cycles exist in the data.

We present the results in term of cumulative reward at the last iteration as a function of k, in Fig. 2 for **XP1** (up) and **XP2** (bottom). In both pairs of plots, the curves on the left concern experiments on the small dataset ($K = 30$), while those on the right correspond to experiments on the larger one ($K = 200$). Note that, due to its complexity, RelationalTS has only been tested on the small versions of the datasets. Note also that the bell-shaped aspect of the curves of **XP1** is due to the fact that rewards can be negative. Hence, the final cumulative reward can decrease even with large k values. This is not the case for **XP2** where rewards are only positive.

For **XP1**, RelationalTS performs better than any other tested algorithm. While all baselines are only able to capture differences in average utilities (due to the bias component), our relational recurrent bandit efficiently discovers dependencies between arms to capture non-stationarity of rewards. On both datasets, there is no clear improvement with D-UCB and SW-UCB compared to traditional policies such as CUCB, except on the large dataset for D-UCB where this approach succeeds to capture some tendencies, but whose results remain significantly lower than the recurrent approaches proposed in this paper. On the other

---

[4] Note however that, to insure a non divergent model, Θ must be chosen such that $\lambda_{max}(\Theta^\top \Theta) \leq 1$, with $\lambda_{max}(A)$ the maximal eigenvalue of a matrix A (see the supplementary material for more details).

hand, despite it does not explicitly try to catch the model used for the data generation, our StateTS approach obtains very good results, while not as good as RelationalTS which exactly follows the underlying assumptions, but for a much lower complexity. This allows impressive results on the large dataset, where RelationalTS cannot be employed by encoding relational dependencies in a low-dimensional hidden space. We observe that the performance of our StateTS algorithm increases with the number of dimensions of the hidden space. However, on the large dataset, even if the rate of improvement between 5 and 10 or 10 and 20 dimensions is high, we notice that it then decreases and reaches a limit.

On the other hand, on **XP2**, RelationalTS is not able to outperform classical approaches. It has no mechanism to handle the non-stationarity of the dataset, where cycles of reward distributions are observed. However, our StateTS approach, which models utility distributions by hidden states of the system, is able to capture implicit dependencies and obtains interesting results on both small and large datasets. This highlights an additional capability of our state-based approach, which is able to adapt to various configurations of non-stationarity.

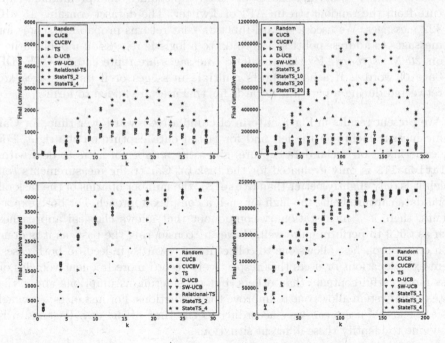

Fig. 2. Final cumulative reward w.r.t. k, for the small (on the left) and the large (on the right) version on the dataset for $XP1$ (up) and $XP2$ (bottom)

5.2   Real Data

Finally, two experiment sets on real data have been conducted:

- Car traffic measurements: we are given a total of $K = 30$ sensors to measure the traffic at different locations in the city of Paris[5] each hour in July 2015 ($T = 744$ time steps). The task is to efficiently select $k = 5$ sensors to monitor at each step in order to maximize the accumulated measured scores. This corresponds to a prediction of where traffic jams are the most likely to happen at each step, knowing some partial observation of the past. While rather artificial, this task simulates real-world settings where one has to secure a given area, but resources for observations and actions are limited and must be focused on areas requiring emergency responses.

- Social data capture: this task introduced in [21] aims at selecting users to follow on a social media such as Twitter to maximize the amount of useful collected data w.r.t. a need. At each iteration, the process has to choose $k = 50$ users to follow during a certain period of time (30 min in our experiments). All messages posted from these users during the period are collected, and their usefulness is determined by a reward function. We used a dataset collected from $K = 500$ users on Twitter during the Olympic games of 2016. The targeted users were those that were the first to use words "#Rio2016", "#Olympics", "#Olympics2016" or "#Olympicgames" in a preliminary capture from the random stream API of Twitter. The dataset contains 15 010 322 messages. We used a reward function that returns probability scores for messages to address politics, according to a logistic regression model learned on *20 Newsgroups*[6] for this thematic (messages are represented as TF-IDF bags of words). If a user i posts multiple messages or if its messages are retweeted during a period t, rewards for this user are added to form $r_{i,t}$.

We present results in Fig. 3 in term of cumulative reward over time, for Car traffic measurements on the left and for Social data capture on the right. For the experiment on social data capture, given the number of users to deal with, RelationalTS is only evaluated for the task on Car traffic measurements (on which it is only slightly better than StateTS), the number of arms of the task on social data capture being too high for such a complex approach. For both experiments, there is a high stationary component that allows classical approaches such as CUCB to perform quite well. However, considering the past, as it is done with our approaches, allows one to collect more rewards. Indeed, in both cases, there exist locations or users that respectively record more frequent vehicles or post more useful contents than others, but considering assumptions about the state of the system allows one to improve the predictions. For instances, crowded places or active users can vary according to the time of the day. Past rewards allow one to identify these different situations.

---

[5] http://opendata.paris.fr/.

[6] http://qwone.com/jason/20Newsgroups/.

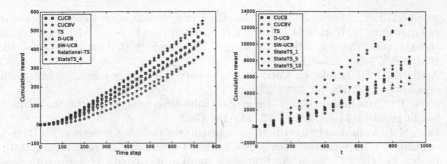

Fig. 3. Results on real data (car traffic on the left, social data capture on the right)

6   Conclusion

In this paper, we have proposed a new multi-armed bandit problem that considers relations between arms' rewards, based on a linear recurrent model, in a multiple-plays setting. In this case, not only the weights of the linear model are unknown to the learner, but a majority of the rewards component - which act as features - are hidden, since the agent is only allowed to play a restricted set of arms at each iteration. We proposed two new Thompson sampling algorithms that are able to leverage past observations via variational inference. While approximations performed avoid theoretical guarantees for the regret, they allowed us to obtain very interesting results on both synthetic and real-world datasets.

Future work concerns the introduction of non-linearity in our state-based model, notably by the use of Bayesian Recurrent Neural Networks (see [28] for instance), as transition functions between successive states and encoding/decoding functions of the rewards. Since they also use variational inference for obtaining posterior distributions of hidden variables, their application for relational recurrent exploration/exploitation problems appears really promising.

Acknowledgments. The work was supported by the IRT SystemX and the ANR project LOCUST (2015–2019, ANR-15-CE23-0027).

References

1. Abbasi-yadkori, Y., Pál, D., Szepesvári, C.: Improved algorithms for linear stochastic bandits. In: NIPS (2011)
2. Agrawal, S., Goyal, N.: Analysis of Thompson sampling for the multi-armed bandit problem. In: COLT (2012)
3. Agrawal, S., Goyal, N.: Thompson sampling for contextual bandits with linear payoffs. In: ICML (2013)
4. Audibert, J.Y., Bubeck, S.: Minimax policies for adversarial and stochastic bandits. In: COLT (2009)

5. Audibert, J.-Y., Munos, R., Szepesvári, C.: Tuning bandit algorithms in stochastic environments. In: Hutter, M., Servedio, R.A., Takimoto, E. (eds.) ALT 2007. LNCS (LNAI), vol. 4754, pp. 150–165. Springer, Heidelberg (2007). https://doi.org/10.1007/978-3-540-75225-7_15

6. Audiffren, J., Ralaivola, L.: Cornering stationary and restless mixing bandits with remix-ucb. In: NIPS, pp. 3339–3347 (2015)

7. Auer, P., Cesa-Bianchi, N., Fischer, P.: Finite-time analysis of the multiarmed bandit problem. Mach. Learn. **47**, 235–256 (2002)

8. Beal, M.J.: Variational algorithms for approximate Bayesian inference. Ph.D. thesis, Gatsby Computational Neuroscience Unit, University College London (2003)

9. Besbes, O., Gur, Y., Zeevi, A.: Stochastic multi-armed-bandit problem with non-stationary rewards. In: NIPS (2014)

10. Bubeck, S., Stoltz, G., Szepesvári, C., Munos, R.: Online optimization in x-armed bandits. In: NIPS (2009)

11. Buccapatnam, S., Eryilmaz, A., Shroff, N.B.: Stochastic bandits with side observations on networks. In: SIGMETRICS (2014)

12. Caron, S., Kveton, B., Lelarge, M., Bhagat, S.: Leveraging side observations in stochastic bandits. In: UAI (2012)

13. Carpentier, A., Valko, M.: Revealing graph bandits for maximizing local influence. In: AISTATS, Seville, Spain (2016)

14. Cesa-Bianchi, N., Gentile, C., Zappella, G.: A gang of bandits. In: NIPS (2013)

15. Chapelle, O., Li, L.: An empirical evaluation of Thompson sampling. In: NIPS. Curran Associates, Inc. (2011)

16. Chen, W., Wang, Y., Yuan, Y.: Combinatorial multi-armed bandit: general framework and applications. In: ICML (2013)

17. Claudio, G., Shuai, L., Giovanni, Z.: Online clustering of bandits. In: ICML (2014)

18. Dani, V., Hayes, T.P., Kakade, S.M.: Stochastic linear optimization under bandit feedback. In: COLT (2008)

19. Garivier, A., Moulines, E.: On upper-confidence bound policies for switching bandit problems. In: Kivinen, J., Szepesvári, C., Ukkonen, E., Zeugmann, T. (eds.) ALT 2011. LNCS (LNAI), vol. 6925, pp. 174–188. Springer, Heidelberg (2011). https://doi.org/10.1007/978-3-642-24412-4_16

20. Garivier, A.: The KL-UCB algorithm for bounded stochastic bandits and beyond. In: COLT (2011)

21. Gisselbrecht, T., Denoyer, L., Gallinari, P., Lamprier, S.: WhichStreams: a dynamic approach for focused data capture from large social media. In: ICWSM (2015)

22. Kalman, R.E.: A new approach to linear filtering and prediction problems. Trans. ASME-J. Basic Eng. **82**(Ser. D), 35–45 (1960)

23. Komiyama, J., Honda, J., Nakagawa, H.: Optimal regret analysis of Thompson sampling in stochastic multi-armed bandit problem with multiple plays. In: ICML (2015)

24. Lai, T., Robbins, H.: Asymptotically efficient adaptive allocation rules. Adva. Appl. Math. **6**(1), 4–22 (1985)

25. Li, L., Chu, W., Langford, J., Schapire, R.E.: A contextual-bandit approach to personalized news article recommendation. In: WWW (2010)

26. Mannor, S., Shamir, O.: From bandits to experts: on the value of side-observations. In: NIPS (2011)

27. Ortner, R., Ryabko, D., Auer, P., Munos, R.: Regret bounds for restless Markov bandits. Theor. Comput. Sci. **558**, 62–76 (2014)

28. Pczos, B., Lrincz, A., Ghahramani, Z.: Identification of recurrent neural networks by Bayesian interrogation techniques. JMLR **10**, 515–554 (2009)

29. Richard, C., Alexandre, P.: Unimodal bandits: regret lower bounds and optimal algorithms. In: ICML (2014)
30. Slivkins, A., Upfal, E.: Adapting to a changing environment: the Brownian restless bandits. In: COLT (2008)
31. Tekin, C., Liu, M.: Online learning of rested and restless bandits. IEEE Trans. Inf. Theory **58**(8), 5588–5611 (2012)
32. Thompson, W.: On the likelihood that one unknown probability exceeds another in view of the evidence of two samples. Am. Math. Soc. **25**, 285–294 (1933)
33. Whittle, P.: Restless bandits: activity allocation in a changing world. J. Appl. Probab. **25**, 287–298 (1988)

Subgroup Discovery

Explaining Deviating Subsets Through Explanation Networks

Antti Ukkonen[1], Vladimir Dzyuba[2], and Matthijs van Leeuwen[3(✉)]

[1] Department of Computer Science, University of Helsinki, Helsinki, Finland
antti.ukkonen@gmail.com
[2] Department of Computer Science, KU Leuven, Leuven, Belgium
vladimir.dzyuba@cs.kuleuven.be
[3] LIACS, Leiden University, Leiden, The Netherlands
m.van.leeuwen@liacs.leidenuniv.nl

Abstract. We propose a novel approach to finding explanations of deviating subsets, often called *subgroups*. Existing approaches for subgroup discovery rely on various quality measures that nonetheless often fail to find subgroup sets that are diverse, of high quality, and most importantly, provide good explanations of the deviations that occur in the data.

To tackle this issue we introduce *explanation networks*, which provide a holistic view on all candidate subgroups and how they relate to each other, offering elegant ways to select high-quality yet diverse subgroup sets. Explanation networks are constructed by representing subgroups by nodes and having weighted edges represent the extent to which one subgroup explains another. Explanatory strength is defined by extending ideas from database causality, in which interventions are used to quantify the effect of one query on another.

Given an explanatory network, existing network analysis techniques can be used for subgroup discovery. In particular, we study the use of Page-Rank for pattern ranking and seed selection (from influence maximization) for pattern set selection. Experiments on synthetic and real data show that the proposed approach finds subgroup sets that are more likely to capture the generative processes of the data than other methods.

1 Introduction

Within the field of exploratory data mining, subgroup discovery (SD) [7,21] is concerned with finding and explaining deviating subsets, i.e., regions in the data that stand out with respect to a given target. It has a number of closely related cousins, such as significant pattern mining [20] and emerging pattern mining [3], which all concern the discovery of patterns correlated with a Boolean target concept. The subgroup discovery task is more generic, as it is agnostic of the data and pattern types. For example, the target could be discrete or numeric [13], both of which we consider in this paper.

Since its introduction twenty years ago, many algorithms and quality measures have been proposed in the literature. While the initial focus was on devising

© Springer International Publishing AG 2017
M. Ceci et al. (Eds.): ECML PKDD 2017, Part II, LNAI 10535, pp. 425–441, 2017.
https://doi.org/10.1007/978-3-319-71246-8_26

more efficient algorithms, over time the focus has shifted towards redundancy elimination [9,10,14], (statistical) validation [5], and generalization of the task [12]. Nevertheless, existing approaches have several limitations, in particular where it concerns the core of the subgroup discovery task: *providing accurate explanations of the deviations that occur in the data.* This has several causes.

First of all, quality measures in subgroup discovery traditionally combine—often by multiplication—the size of the subgroup, i.e., the number of rows it covers, with its effect, i.e., the extent to which the target value for those rows deviates from the dataset average. The problem with this approach is that this results in a somewhat arbitrary trade-off between size and effect that has a very large impact on the scores (and thus ranking) of all patterns.

Second, most approaches that aim to eliminate redundancy take these individual qualities for granted and primarily consider the covers of subgroups to discard redundant patterns. This is true for, e.g., approaches based on relevancy/closedness [9] and generalization-aware pruning [14]. Furthermore, none of these methods explicitly considers redundancy among subgroups that do not share any attributes among their descriptions, neither do they explicitly consider the possibility that one description may be more interesting/relevant than another. That is, a subgroup is either kept or discarded and no alternatives are offered, where the exact choice between similar subgroups is pretty much random. Some approaches, such as DSSD [10], are heuristic and defined procedurally, which makes it even harder to assess the results.

As a consequence of the above, existing methods do not provide accurate *descriptions* for deviating subsets in practice, as we will empirically show in Sect. 6. That is, finding deviating subsets and descriptions that correspond to those subsets is doable, but identifying *accurate explanations*, i.e., descriptions that capture the data generating process, is a much harder task.

Approach and Contributions. We introduce *explanation networks*, i.e., networks in which the nodes represent subgroups and weighted, directed edges represent *explanations* between pairs of subgroups. Explanation networks offer a global perspective on all subgroups and their relationships, regardless of the branches of the search tree the subgroups happen to reside. As a result, the network naturally represents all information concerning relevancy and redundancy.

Technically, we build on ideas from *database causality* [15,16] to quantify to what extent subgroups "explain" each other. In particular, we use the notion of an *intervention* [19,22]: we say that a subgroup T explains a subgroup S if removing the cover of T from the cover of S results in a (much) smaller effect. Because a larger T is more likely to (partially) explain S by chance than a smaller T, we normalize its explanatory influence with its *expected* explanation. The result is an elegant formula that quantifies explanatory strength with a number of desirable properties. For example, it explicitly distinguishes effect from cover size and accounts for both in a principled way; others will be discussed later.

We demonstrate the strengths of explanation networks through two different pattern mining tasks. First, we show how subgroups can be ranked based on their global explanatory power, i.e., by considering all pairwise relationships.

We achieve this by observing that this setting is analogue to that of identifying relevant webpages on the World Wide Web and thus apply PageRank to explanation networks. Second, we observe another analogy to network analysis and show how the pattern set selection task, i.e., the task of selecting a small and diverse set of subgroups, can be formalized as a seed selection (Influence Maximization) problem on explanation networks.

The remainder of the paper is organized as follows. First, we discuss related work in Sect. 2, followed by preliminaries in Sect. 3. We then formally introduce explanation networks in Sect. 4 and describe the two tasks in Sect. 5. Section 6 presents experiment results, both on synthetic and real data, after which we conclude with conclusions and a brief outlook in Sect. 7.

2   Related Work

This section provides a high-level overview of two categories of related work aimed at discovering and explaining phenomena observed in data, namely (1) causality in databases, and (2) subgroup discovery.

Database Causality. Establishing "actual causality" requires controlled randomized experiments and thus cannot be accomplished using purely observational data [17]. Database research therefore uses a relaxed definition of causality, which originates from *database provenance* and focuses on identifying causal relations between tuples, i.e., which tuples in a database (input tuples) affect the results of a query (output tuples or columns thereof) [15,16]?

An important extension of this line of work replaces fine-grained "causes" described by (potentially large) collections of individual tuples by coarse-grained *explanations*, i.e., concise descriptions of those collections in a certain formal language, with an emphasis on aggregate queries [19,22]. The controlled experiment required to establish actual causality is approximated by a database *intervention*, i.e., by the removal of tuples that satisfy a certain description. Key challenges in this approach are (1) defining scoring functions for explanations, and (2) finding and returning the best explanations [18].

Subgroup Discovery. Subgroup discovery (SD) [7,21] is concerned with finding descriptions of subsets of a dataset that have a substantial deviation in a property of interest, relative to the entire dataset; see Atzmueller [1] for a recent overview. The property of interest, or *target*, is typically an aggregate of a chosen attribute, e.g., the mean of a numeric attribute. SD algorithms typically rely on bounds on the measure of deviation to prune the search space [13].

One of the crucial shortcomings in traditional SD is the *redundancy* of results, i.e., the situation wherein the subgroups with the highest quality contain many variations of the same theme and describe only few interesting subsets. Therefore, a wide range of approaches, including the one proposed in this paper, aim at eliminating redundancy in SD. Below we briefly discuss a number of existing methods; an empirical comparison is presented in Sect. 6.

Sequential covering schemes, e.g., CN2-SD [8], prune or penalize subgroups that overlap with higher-ranked subgroups. Likewise, in cascaded SD [4], subgroups that essentially improve regression accuracy for undescribed instances are incrementally added to the result set. Although we also compare subgroups by analyzing the subsets of the data that they describe (by means of a database intervention), we do not aim at incrementally constructing a single subgroup list. Impact rules [5] and generalization-aware SD [14] prune subgroups that do not improve on their (shorter) ancestors. Unlike these methods, we also relate and compare subgroups that do not share any part of their description. Skylines of subgroup sets [11] explicate the trade-off between quality and redundancy of a set of subgroups by building the Pareto front for the given dataset and target.

3 Preliminaries

In the following, we assume the data D to consist of n rows and $m+1$ attributes. There are m *description attributes* $x_1, \ldots, x_m$, and a single *target attribute* y. The domains of x_i need not be bounded; each domain can be either categorical (nominal or ordinal) or quantitative. In the definitions we assume y to be quantitative, i.e., $y \in \mathbb{R}$, but the results can be trivially extended to the common Boolean setting, $y \in \{\text{FALSE}, \text{TRUE}\}$, by considering the proportion of TRUEs instead of the mean when computing the quality or effect size of a subgroup.

In subgroup discovery, a *subgroup description* S is usually a conjunction of conditions on the description attributes, where every condition is of the form $x_i \odot v_i$, where $\odot$ is one of $<, >, \geq, \leq, =$, and v_i is some value from the domain of x_i. For example, $S = \{x_1 \leq 0.4 \text{ AND } x_3 = 1\}$. The set of all such descriptions constitutes the pattern language $\mathcal{L}$. However, the methods we discuss in this paper are agnostic of the particular type of description language and subgroup mining algorithm being used.

The *(subgroup) cover* of S, denoted $c_D(S)$, are the rows in data D that satisfy description S. One could also think of the subgroup description as a query, and the cover as the result set of this query. As a special case, we denote by $c_D(\emptyset)$ all rows of D, i.e., an empty description matches all rows in D. The (cover) size of a subgroup is defined as the number of data rows it covers, i.e., $|c_D(S)|$.

Define the *effect* of S in data D as

$$q_D(S) = \sum_{i \in c_D(S)} y_i. \tag{1}$$

The *average effect* of S in D is then defined as

$$\mu_D(S) = \frac{q_D(S)}{|c_D(S)|}. \tag{2}$$

Also, we denote by μ_D the mean of the target attribute in the entire data D.

Given the previous, the traditional subgroup discovery task is to find the top-k subgroup descriptions with regard to some quality measure $\phi : \mathcal{L} \to \mathbb{R}$. Quality measures typically combine the size of the subgroup cover with the observed deviation in the target attribute. Probably the best known quality measure is Weighted Relative Accuracy (WRAcc), which we will formally define when we need it in Sect. 6.

4 Explanation Networks

Before we formalize *explanation networks*, we first describe how to adapt ideas from database causality to define to what extent subgroups "explain" each other.

4.1 Interventions as Explanations

For the moment, we consider comparing two subgroups, S and T. Recent research [19,22] in the database community has developed methods that can be used to explain away outliers (deviations) in aggregate queries in relational databases. As the basic mechanism is that of an intervention and the goal is to explain deviations, this research area is often called *database causality*. We adopt a similar technique to quantify how much the effect of subgroup S can be explained by subgroup T.

The database causality approach is based on the simple principle where individual data items are the fundamental contributing factors to all observed effects. In an *intervention*, part of the data are removed, and we observe what happens to the deviation (of some target) in the data that remains. If this deviation is substantially changed after the removal of some data, we can conclude that the removed data play an important part in the observed deviation, and thus in part *explain* the observed deviation. Given a query that exhibits anomalous behavior, the goal is to find those queries that reduce this anomalous behavior the most, the idea being that those are likely to be (causal) explanations of the deviation.

This setting is strikingly similar to the subgroup discovery setting that we consider: subgroup descriptions can be interpreted as queries and we are also interested in the deviation in some target. Let us therefore translate Wu and Madden's [22] influence definition to our notation. First, we slightly abuse notation, and let $D \setminus T$ denote $D \setminus c_D(T)$ for short, i.e., $D \setminus T$ are those rows in data D that *do not belong* to the cover of subgroup description T. Then, the *(database) influence* of a subgroup T on S is defined as

$$\mathrm{infl}(S, T) = \frac{q_D(S) - q_{D \setminus T}(S)}{|c_D(S) \cap c_D(T)|}. \tag{3}$$

By Eq. 1, $q_{D \setminus T}(S)$ is the effect of S in data where subgroup T *is not true*. Informally, we compare the effect of S in data D to the effect of S in data from which $c_D(T)$ has been removed, normalized by the number of data rows that satisfy both S and T.

Considering the difference in effect of S with and without T is a natural choice, as it is this effect that we are trying to explain. Averaging over the number of affected data rows, however, causes a strong bias towards smaller explanations: the smaller $|c_D(S) \cap c_D(T)|$, the larger the influence. This is undesirable, in particular in the subgroup discovery setting, as subgroups with small covers tend to have long descriptions and do not generalize well. In practice, this results in many subgroups consisting of very few data rows that together 'explain' the larger subgroups.

No normalization at all, on the other extreme, is not an option either: in that case subgroups T having a large cover are more likely to have a large influence. In fact, without the denominator in Eq. 3 we would have $\mathrm{infl}(S, \emptyset) = q_D(S) - q_{D \setminus D}(S) = q_D(S)$, implying that all data $D = c_D(\emptyset)$ always has the largest possible influence on any subgroup S. (Recall that $\emptyset$ is the empty subgroup description that matches all of D.)

Motivated by the previous, the solution that we propose is to compare the observed influence of T on S to the *expected influence of a random subset of data rows having (roughly) the same size as* $c_D(T)$. The aim of this is to reduce the influence of subgroups that have a large cover, as their influence would otherwise be disproportionally strong. Let T^* denote a random subset of the rows of D, so that every row in D has an equal probability to belong to T^*, and we have $\mathbb{E}[|T^*|] = |c_D(T)|$. Notice that rather than requiring T^* to have exactly the same size as $c_D(T)$, we only constrain it to have the same size *in expectation*. This makes the resulting calculations much simpler, while achieving the same practical outcome of normalizing the influence with respect to cover size. The *expected effect* of S in data $D \setminus T^*$ is then given by

$$\mathbb{E}[q_{D \setminus T^*}(S)] = \sum_{i \in c_D(S)} \Pr[i \notin c_D(T^*)]y_i, \tag{4}$$

$$= \left(1 - \frac{|c_D(T)|}{n}\right) \sum_{i \in c_D(S)} y_i = \left(1 - \frac{|c_D(T)|}{n}\right) q_D(S). \tag{5}$$

Observe that this definition has two desirable properties: (1) it scales linearly with $q_D(S)$, meaning that it is potentially larger for subgroups that have a large deviation in the target attribute; and (2) the expected influence is smaller for larger $c_D(T)$ (and vice versa). The *explanation* of T on S, denoted $E[S, T]$, is then defined as the difference between the *expected* effect of S in $D \setminus T^*$ and the *observed* effect of S in $D \setminus T$, i.e.,

$$E[S, T] = \mathbb{E}[q_{D \setminus T^*}(S)] - q_{D \setminus T}(S) = \left(1 - \frac{|c_D(T)|}{n}\right) q_D(S) - q_{D \setminus T}(S). \tag{6}$$

This definition makes use of the desirable properties that expected effect has and has two desirable properties itself. First, it is 0 for both $c_D(T) = D$ (i.e., when $T = \emptyset$) and $c_D(T) = \emptyset$, meaning that neither the complete dataset nor the empty set is a good explanation of any subgroup S. Observe that the definition in Eq. 6 allows both positive as well as negative effects: the effect of S can both increase and decrease after the intervention, and the correction for expected effect does not exclude the possibility of either direction.

4.2   Defining the Network

Next we propose a novel concept that allows us to deal with the mutual relationships between multiple subgroups. While most traditional pattern mining approaches consider the pattern lattice—i.e., the search space, as defined by the pattern language, that search procedures typically traverse as a tree—we propose a holistic perspective instead and introduce a global *network* of patterns.

Specifically, we define a *weighted directed graph* G where individual subgroups are nodes, and two nodes, S and T, are connected with a directed edge (S, T) if S can be (partially) explained by T. The weight of edge (S, T), denoted $w(S, T)$, must be proportional to the amount with which T explains S. Clearly, we will use $w(S, T) = E[S, T]$. Formally, we have the following.

Definition 1 (Explanation Network). *Given data D and a set of subgroups $\mathcal{S}$, define the* explanation network G *as* $G = (V, W)$, *where* $V = \mathcal{S}$ *and* $W = \{(S, T) \mid S, T \in \mathcal{S}\}$. *Each* $(S, T) \in W$ *has weight* $w(S, T) = E[S, T]$.

A distinguishing feature of explanation networks is that they contain information regarding both the mutual relationships between patterns (1) from the same branch of the search tree; and (2) from *different branches* of the search tree. Especially the second property is unique in that it allows to discover and exploit overlap/redundancy across completely disjoint subgroup descriptions, which is achieved by the holistic view on the covers of all subgroups in $\mathcal{S}$. As such, explanation networks can be used for many different tasks; the next section will describe how they can be used for two tasks.

5   Using Explanation Networks

Explanation networks can be used for different purposes. In this section we describe how they can be used for two different mining tasks, i.e., (1) pattern ranking, and (2) pattern set selection.

5.1   Pattern Ranking

The explanations describe pairwise relationships between two subgroups, but in certain situations it may be of interest to provide a global score or ranking. To turn the pairwise relationships into a scoring method, we propose the following. Intuitively, *subgroups that are good at explaining other subgroups should have a high score*. Especially this should hold for subgroups that are good explanations of other high scoring subgroups. This can be expressed in the following recursive definition for $score(T)$:

$$score(T) \propto \sum_{S} score(S)E[S, T]. \tag{7}$$

This definition is analogous to that of PageRank [2], the well-known ranking method for web search that uses *random walks*, i.e., stochastic processes that move between a number of states, where the probability to move to any other state only depends on the current state of the walk. The PageRank score of a page is defined as its probability in the *stationary distribution*. Indeed, if this probability is high, the page must be easy to reach, and is thus of high quality.

We adapt this idea to the context of subgroups and the explanation network: if the stationary probability of a subgroup is high, then it must be easy to reach and therefore have high explanatory power. That is, we define a random walk where subgroups are states, and the transition probability from subgroup S to subgroup T is proportional to $w(S,T)$, i.e., the weight of the edge from S to T.

Formally, given the explanation network G with N nodes, we construct a random walk as follows. Let $\mathbf{A}$ denote a square matrix, the *transition probability matrix*, where element $\mathbf{A}[i,j]$ is equal to the transition probability from state i to state j, defined in two steps as follows.

1. Define the $N \times N$ matrix $\bar{\mathbf{A}}$ so that

$$\bar{\mathbf{A}}[S,T] = \begin{cases} w(S,T) & \text{if } w(S,T) > 0, \\ 0 & \text{otherwise} \end{cases}$$

   for all S and T. (Here we abuse notation slightly and denote the row and column indices that correspond to subgroups S and T simply by S and T.)
2. Define the $N \times N$ matrix $\mathbf{A}$ so that, for all S and T,

$$\mathbf{A}[S,T] = \bar{\mathbf{A}}[S,T] / \sum_S \bar{\mathbf{A}}[S,T].$$

For a random walk to have a stationary distribution, it must be *irreducible* and *aperiodic*. In PageRank this is commonly enforced by adding a *teleportation distribution*. Let $\mathbf{b}$ denote the teleportation distribution that satisfies $\sum_i \mathbf{b}_i = 1$, where $\mathbf{b}_i$ is equal to the probability to move from any state to state i.

Finally, the score of every subgroup is given by the PageRank vector $\mathbf{s} \in \mathbb{R}^N$, i.e., the stationary distribution of the random walk, defined by [2]:

$$\mathbf{s} = \alpha \mathbf{A}^\mathsf{T} \mathbf{s} + (1 - \alpha)\mathbf{b}. \tag{8}$$

If $\alpha = 1$ the teleportation distribution has no effect; in practice we usually set $\alpha = 0.7$, meaning that $\mathbf{s}$ is mainly affected by $\mathbf{A}$ (i.e., the $E[S,T]$ values). See also Subsect. 6.2 for a brief empirical study of the effect α has.

In the simplest case we can use a uniform distribution for $\mathbf{b}$, i.e., we let $\mathbf{b}_S = 1/N$ for all S. The teleportation distribution can also be used to bias the resulting scores based on some other criteria: subgroups S that have higher values of $\mathbf{b}_S$ also tend to have higher scores $\mathbf{s}_S$. This idea was used to define "personalized" variants of PageRank. [2] When scoring subgroups, we can define $\mathbf{b}$ so that $\mathbf{b}_S$ is proportional to, e.g., effect size $q_D(S)$, or cover size $|c_D(S)|$. PageRank thus allows to combine different ranking criteria using the same framework.

5.2   Pattern Set Selection

While a pattern ranking can be of interest in a wide range of scenarios, e.g., when using patterns as input for a next analysis phase, under certain circumstances it can be more useful to have a *set of non-redundant patterns*. When patterns are to be presented to domain experts, for example, the result set should be small. In these cases we can benefit from the explanation network by selecting *a set of patterns that explain numerous yet distinct patterns in the network*.

As with pattern ranking, we observe that this problem strongly resembles a well-known problem in network analysis: in this case, the *influence maximization (InfMax) problem* [6] in social networks. The InfMax problem concerns the selection of the k nodes in a network that are together the most influential, where influence is defined in terms of an influence *propagation model*.

We use the *Independent Cascade Model* (ICM) [6], because of its simplicity and nice theoretical properties (discussed below). The ICM assumes every node of the network to be either *active* or *inactive*. Initially all nodes are inactive, except a *seed set* of k nodes that are active. At every round, nodes that became active in the previous round (in the 1st round the seed nodes) attempt to activate their immediate neighbors. Each activation attempt is independent, and succeeds with probability $\mathbf{P}[T, S]$, where T is an active node and S is an inactive node. The process finishes when no activation attempt in a round is successful. The *influence of the seed set* is *the number of active nodes when the process finishes*.

To adapt this idea for subgroup set selection, we solve the InfMax problem on the explanation network G with appropriately defined activation probabilities $\mathbf{P}[T, S]$. Intuitively, as we want to find a pattern set that has a high explanation strength, we let $\mathbf{P}[T, S] \propto w(S, T)$. I.e., T activates S with a probability that is *directly proportional to the explanation of subgroup T on subgroup S*. In practice we let $\mathbf{P}[T, S] = \mathbf{A}[S, T]$, where $\mathbf{A}[S, T]$ is defined in the same way in Sect. 5.1. However, our approach is by no means tied to ICM; any propagation model that can be parametrized in terms of the network weights $w(S, T)$ can be used.

The *explanation maximization problem* is then defined as *finding those k subgroups that maximize influence when chosen as the seed set*. Kempe et al. [6] showed that solving the InfMax problem is in general NP-hard, but also that the ICM results in a *submodular* influence function. Therefore the problem can be solved efficiently by a greedy algorithm that one at a time selects the node that maximizes marginal gain in influence. This algorithm has a constant approximation ratio, i.e., it provides a solution of size k that has influence at least $(1 - 1/e)$ times the influence of the optimal solution. All experiments in this paper are carried out using this algorithm.

Finally, an important aspect that differentiates the explanation network from, e.g., social networks is its density, i.e., it contains substantially more edges in relation to the number of vertices. In practice the explanation network can be a single, large clique. As the complexity of seed selection algorithms mostly depends on the number of edges, it is very important to use an efficient algorithm despite the number of edges often being much lower than in social networks.

6 Experiments

In this section we evaluate how well the two tasks based on explanation networks that we introduced perform and empirically compare them to existing methods.

Baseline Methods: We consider the following four baseline methods, as they are well-known and the latter two are representative of the state of the art:

NRAcc: Rank patterns in decreasing order of Normalized Relative Accuracy, i.e., $(\mu_D(S) - \mu)/\sigma$, where μ and σ denote global mean and standard deviation of the target attribute.

WRAcc: Rank patterns in decreasing order of Weighted Relative Accuracy, defined as $\sqrt{|S|} \times$ NRAcc.

Generalization aware pruning (gap): Rank patterns in decreasing order of the *gap* score, defined as $\mu_D(S) - \max_{S'} \mu_D(S')$, where S' is a subgroup that is a *generalization* of S (i.e., its description consists of a subset of the conditions in the description of S).

Greedy-WRAcc: Sequential covering using WRAcc: iteratively select that subgroup that maximizes WRAcc (as defined above), remove the rows that belong to its cover from the data, and iterate until enough subgroups have been selected or until the data is exhausted.

Real Data: As datasets we use the Abalone (aba), Credit-G (cg), Mushroom (mush), Redwine (rw), and Wages (wag) from the UCI Machine Learning repository[1]. Further, we also include Elections (ele) as described in [10] and the Helsinki housing (hel) [23] dataset. cg and mush have a Boolean target, all others have a numeric target. On-the-fly discretisation of numeric description attributes was applied, meaning that 6 equal-size intervals were created upon pattern extension. Dataset sizes range up to 8337 rows (for hel) and 73 attributes (for ele).

Subgroup Candidates: In the experiments we assume a fixed candidate set of subgroups. The candidate set is obtained by mining subgroups using NRAcc (as defined above) using a support threshold of 10% and maximum search depth of 3. If there were more than 5000 candidates, these were initially ranked in terms of WRAcc, and the top-5000 were kept for the experiment.

Note on Running Times/Complexity: Constructing the explanation network requires computing intersections of $c_D(S)$ between all pairs of subgoups. Pagerank is computed using the basic power-iteration method, which converges rapidly in practice. For the seed selection task we use the greedy algorithm and some simple optimizations to speed up influence computations. Our methods[2] complete in approximately 15 min for the largest datasets (hel, elections), and in less than a minute for the smaller ones.

---

[1] http://archive.ics.uci.edu/ml/.

[2] Source code is available at: http://anttiukkonen.com/explanation-networks/.

6.1   Artificial Data Generation

We employ two approaches for creating artificial data with planted subgroups. The first one is based on a Bayesian network model, while the second one combines a real dataset with an artificial target attribute.

Bayesian Network. This generative model (illustrated in Fig. 1) consists of a number of independent *causal chains* with variables X that all end in the target attribute Y, and a number of random attributes R that are independent of everything, including Y. The idea is to model several different causes for observing $Y = 1$ in the underlying process. Ideally we find subgroups where the description does not contain any random attributes.

Here X_1^i is the *root cause* of the output Y in chain i, the other X_j^is are intermediary effects. The conditional probabilities of the X_j^i variables are adjusted so that $X_j^i = 1$ almost always when $X_{j-1}^i = 1$, and $X_j^i = 0$ almost always when $X_{j-1}^i = 0$. Also, $Y^i = 1$ almost always when $X_h^i = 1$, and $Y^i = 0$ almost always when $X_h^i = 0$. Finally, $Y = 1$ whenever at least one $Y^i = 1$, and $Y = 0$ otherwise. Data from the model is gen-

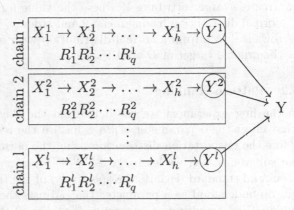

Fig. 1. A Bayesian network with l independent "chains", and a single target attribute Y.

erated by first drawing a dataset of size N separately from every chain. These are combined by concatenating the vectors and by introducing the global target variable Y that is simply the union of the Y^is from every chain. The original Y^is are removed. To generate data from one of the causal chains, we compute the exact joint distribution of the X_j^i and Y^i variables, and then draw N binary vectors from this distribution. In addition, we add q "non-causal" noise variables $R_1^i, \ldots, R_q^i$ by creating randomly permuted copies of some of the X_j^i.

Below we refer to data sampled from this model using the notation $\mathrm{BN}(h, q, l)$. For example, $\mathrm{BN}(2,4,5)$ refers to data from a model with five chains, each containing two causal variables and four noise variables (that is, 30 variables in total that serve as description attributes, plus one target).

Latent Cause Model. In a manner similar to the approach described above, we aim to simulate a scenario where the objective is to uncover the "true" cause of a phenomenon. In the Bayes network model this true cause was expressed by the observed X_j^i variables. Now we increase the level of difficulty and assume that the true cause is unobserved: it is only reflected in a noisy, numeric target attribute. Moreover, we assume that no perfect description of the true cause exists. In concrete, the true underlying cause is an unobserved binary attribute

Z. The observed target Y is generated from Z by drawing a random "low" value for those rows for which $Z = 0$, and a random "high" value for rows where $Z = 1$. The aim is to find subgroups the covers of which are a good match with $Z = 1$ (and do not necessarily have a high quality w.r.t. the observed target).

To maintain a realistic structure of the search space, we start from a given real dataset D and replace its original target attribute with an artificial one:

1. Select a random subset Z of the rows of D, s.t. $|Z| = 1/3|D|$. This corresponds to the "true" underlying cause.
2. Find a set of random subgroups. Select K of these s.t. their covers match the set Z as well as possible. (We used the F1-measure to calculate the goodness of the match.) These are the subgroups that we aim to find.
3. Create a target attribute Y where the value for rows in Z is drawn from a normal distribution having mean 2 and stdev 1, while the value for the other rows is drawn from a normal distribution with mean 0 and stdev 1. Replace the original target of D with Y.

6.2   Pattern Ranking

In the first experiment we use data from the Bayes network model to study what effect the α parameter of Eq. 8 has on the resulting pattern ranking. We define the teleportation distribution using the normalized relative accuracies of the subgroups, i.e., we let $\mathbf{b}_S \propto \frac{\mu_D(S) - \mu}{\sigma}$ for every S, where μ and σ are the mean and standard deviation, respectively, of the target attribute in D. Now α can be understood as a parameter that adjusts the effect between NRAcc and explanation strength of a subgroup. (For $\alpha = 0$ the ranking is only based on NRAcc, while for $\alpha = 1$ it is only based on the explanation strengths.)

With Bayes network data we can evaluate performance in terms of AUC by treating the pattern ranking problem as a classification problem where the objective is to separate those subgroups that have "non-causal" attributes in their description from those that only have "causal" attributes. We consider all subgroups that do not have *any noise* variables R_i in their descriptions as "causal". That is, we are very strict and want to find such subgroups that *only* describe phenomena that are associated with the target according to our model.

Results are shown in Fig. 2. From the two top-most panels we can observe that the PageRank-based pattern ranking approach performs well in terms of AUC. The lines are average AUCs over 20 independently drawn datasets (of 5000 rows) from the respective models, and the dashed lines show naïve confidence bands of $\pm$ 3 standard deviations. Especially when the target has several independent causes (BN(5,5,5)), we find the explanation network to show significant improvements over using WRAcc (or NRAcc) only. The panels in the bottom row of Fig. 2 show how explanation based pagerank is related to both WRAcc and subgroup size in subgroups mined from BN(5,5,5) (5000 rows) with $\alpha = 0.85$. Indeed, we can observe that pagerank separates the "causal" subgroups (shown in blue) from the "non-causal" subgroups (red) better than WRAcc. Moreover, while the explanation based ranking tends to favor large subgroups, simply ranking subgroups by size would not give the same result either.

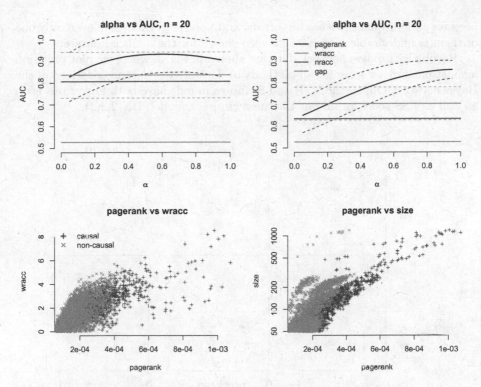

Fig. 2. Top-left: Mean AUC (solid) and naïve confidence bands (dashed) as a function of α using BN(3,6,3) data (see legend of top-right). Top-right: AUC as a function of α using BN(5,5,5) data. Bottom-left: Pagerank score vs. WRAcc for subgroups from BN(5,5,5). Bottom-right: Pagerank score vs. size for subgroups from BN(5,5,5). (Color figure online)

6.3 Pattern Set Selection

We continue with an experiment where the task is always to *retrieve a pattern set of 20 subgroups* from the given candidate set. Results with Bayes network data are evaluated in terms of *precision of retrieving "causal" subgroups*, i.e., the fraction of such subgroups in the 20 subgroups returned. Results with latent cause model are evaluated in terms of *precision of retrieving "correct" subgroups*, where a subgroup is considered as "correct" if it was chosen in step 2 of the target generation procedure ($K = 20$). Finally, results with real, unmodified data are evaluated in terms of a score function that is composed of four quantities: (1) average cover size (*avg.size*), (2) average quality (*avg.qual*), (3) entropy of the cover distribution (*cent*), and (4) fraction of data rows that are covered by at least one of the chosen subgroups (*ccov*). These are computed for all methods, and normalized to $[0, 1]$ by dividing with the value attained by the best performing method. The final score, denoted PSS (for "pattern set score"), is the *geometric mean* of these normalized numbers. We use the geometric mean because all four quantities are important and poor performance in even only one of them is undesirable. This score is also shown for the artificial datasets.

Results with Bayes network data are shown in Fig. 3, where we plot precision against pattern set score for all methods under different parametrizations of the Bayes network. Our methods (P and S, shown in red) have both higher precision as well as PSS score for all but the most simplest model, BN(2,2,2).

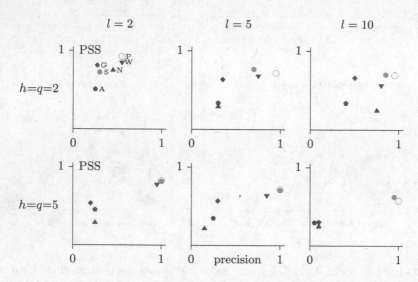

Fig. 3. Precision and pattern set scores for experiments with Bayes data. The proposed explanation based methods "pagerank" (P) and "seeds" (S) outperform the competitors in the hard settings with larger numbers of antecedents (h) and component chains (l). (Color figure online)

Results with latent cause data are shown in Table 1. This is a very hard task, as there are very few planted subgroups, and these are by definition not very well correlated with the noisy target. The explanation based approaches have the highest aggregate scores. Furthermore, they are the only methods that succeed in discovering some of the planted subgroups.

s Finally, we present average evaluation metric values over the real datasets in Table 2. We find that both algorithms that are based on a greedy selection heuristic, seeds and greedy-wracc, have the same average score. However, this score is composed differently for the two methods. Greedy-wracc has a higher entropy (cent) and cover (ccov) value, while seeds performs better in terms of average cover size and subgroup quality. Overall, evaluating subgroup sets in an objective, application independent manner, is difficult, and it is not obvious that the same method is appropriate for all tasks. However, the results of Table 2 suggest that the explanation based approaches (pagerank and seeds) find subgroups of reasonably high quality that have a fairly large cover, meaning the subgroups should generalize better to unseen data.

Table 1. Experiments with latent cause data (averages over 20 randomized runs).

	PSS				Precision			
	aba	cg	mush	wag	aba	cg	mush	wag
pagerank	**0.54**	**0.43**	**0.48**	**0.51**	0.00	**0.25**	**0.25**	0.00
seeds	**0.58**	**0.52**	**0.53**	**0.52**	0.00	**0.30**	**0.40**	**0.05**
gap	0.46	0.30	0.29	0.41	0.00	0.00	0.00	0.00
greedy-wracc	0.45	**0.39**	0.36	**0.50**	0.00	0.00	0.00	0.00
nracc	0.23	0.16	0.19	0.13	0.00	0.00	0.00	0.00
wracc	0.40	0.23	0.28	0.33	0.00	0.00	0.00	0.00

Table 2. Experiments with real data (averages over all seven real datasets).

	PSS	avg.size	ccov	cent	avg.qual
pagerank	0.56	**0.84**	0.49	0.33	0.76
seeds	**0.74**	**0.93**	**0.78**	0.67	0.66
gap	**0.68**	0.58	**0.75**	**0.77**	0.73
greedy-wracc	**0.74**	0.65	**0.98**	0.86	0.61
nracc	0.49	0.37	0.38	0.47	**1.00**
wracc	0.59	0.74	0.51	0.44	**0.82**

7 Conclusions

We introduced *explanation networks*, a novel, global perspective on subgroups and how they relate to each other. In particular, we used interventions to define the notion of explanation, which quantifies the explanatory influence of one subgroup on another and is normalized by the expected influence from a random subgroup of the same size. We showed how analogies with network analysis can be made and how they can lead to novel pattern mining methods. In this paper we have studied the use of PageRank for pattern ranking and the use of seed selection (influence maximization) for pattern set selection.

The experiments demonstrate that our explanation based approach provides advantages when it is of importance to select subgroup descriptions that capture the data generating process. Specifically, on artificial data we have shown—using very strict evaluation criteria—that our approach provides better rankings and pattern sets than the competitors.

Although these results clearly show the potential of explanation networks, there are also still many directions to be explored. For example, we will perform user studies in which the analyst is enabled to visually explore the network for alternative explanations. Further, it is of interest to investigate direct exploration and mining algorithms that avoid the need to materialise the full explanation network. As a third and final example, it would be interesting to develop a statistical test for assessing whether the influence of one subgroup on another is significant. In general, much is still to be gained from this novel network perspective on explanation and redundancy in pattern mining.

Acknowledgements. Antti Ukkonen was partially supported by Tekes (project Re:Know2) and Academy of Finland (decision 288814).

References

1. Atzmueller, M.: Subgroup discovery. Wiley Interdisc. Rev.: Data Mining Knowl. Discov. **5**(1), 35–49 (2015)
2. Brin, S., Page, L.: The anatomy of a large-scale hypertextual web search engine. Comput. Netw. **30**(1–7), 107–117 (1998)
3. Dong, G., Zhang, X., Wong, L., Li, J.: CAEP: classification by aggregating emerging patterns. In: Arikawa, S., Furukawa, K. (eds.) DS 1999. LNCS (LNAI), vol. 1721, pp. 30–42. Springer, Heidelberg (1999). https://doi.org/10.1007/3-540-46846-3_4
4. Grosskreutz, H.: Cascaded subgroups discovery with an application to regression. In: Proceedings of LeGo ECML/PKDD Workshop (2008)
5. Huang, S., Webb, G.I.: Discarding insignificant rules during impact rule discovery in large, dense databases. In: Proceedings of SDM, pp. 541–545 (2005)
6. Kempe, D., Kleinberg, J.M., Tardos, É.: Maximizing the spread of influence through a social network. In: Proceedings of KDD, pp. 137–146 (2003)
7. Klösgen, W.: Explora: a multipattern and multistrategy discovery assistant. In: Advances in Knowledge Discovery and Data Mining, pp. 249–271 (1996)
8. Lavrač, N., Kavšek, B., Flach, P., Todorovski, L.: Subgroup discovery with CN2-SD. J. Mach. Learn. Res. **5**(Feb), 153–188 (2004)
9. Lavrač, N., Gamberger, D.: Relevancy in constraint-based subgroup discovery. In: Boulicaut, J.-F., De Raedt, L., Mannila, H. (eds.) Constraint-Based Mining and Inductive Databases. LNCS (LNAI), vol. 3848, pp. 243–266. Springer, Heidelberg (2006). https://doi.org/10.1007/11615576_12
10. van Leeuwen, M., Knobbe, A.: Diverse subgroup set discovery. Data Mining Knowl. Discov. **25**(2), 208–242 (2012)
11. van Leeuwen, M., Ukkonen, A.: Discovering skylines of subgroup sets. In: Blockeel, H., Kersting, K., Nijssen, S., Železný, F. (eds.) ECML PKDD 2013. LNCS (LNAI), vol. 8190, pp. 272–287. Springer, Heidelberg (2013). https://doi.org/10.1007/978-3-642-40994-3_18
12. Leman, D., Feelders, A., Knobbe, A.: Exceptional model mining. In: Daelemans, W., Goethals, B., Morik, K. (eds.) ECML PKDD 2008. LNCS (LNAI), vol. 5212, pp. 1–16. Springer, Heidelberg (2008). https://doi.org/10.1007/978-3-540-87481-2_1
13. Lemmerich, F., Atzmueller, M., Puppe, F.: Fast exhaustive subgroup discovery with numerical target concepts. Data Min. Knowl. Discov. **30**(3), 711–762 (2016)
14. Lemmerich, F., Becker, M., Puppe, F.: Difference-based estimates for generalization-aware subgroup discovery. In: Blockeel, H., Kersting, K., Nijssen, S., Železný, F. (eds.) ECML PKDD 2013. LNCS (LNAI), vol. 8190, pp. 288–303. Springer, Heidelberg (2013). https://doi.org/10.1007/978-3-642-40994-3_19
15. Meliou, A., Gatterbauer, W., Halpern, J.Y., Koch, C., Moore, K.F., Suciu, D.: Causality in databases. IEEE Data Eng. Bull. **33**(3), 59–67 (2010)

16. Meliou, A., Roy, S., Suciu, D.: Causality and explanations in databases. Proc. VLDB Endow. **7**(13), 1715–1716 (2014)
17. Pearl, J.: Causality, 2nd edn. Cambridge University Press, Cambridge (2009)
18. Roy, S., Orr, L., Suciu, D.: Explaining query answers with explanation-ready databases. Proc. VLDB Endow. **9**(4), 348–359 (2015)
19. Roy, S., Suciu, D.: A formal approach to finding explanations for database queries. In: Proceedings of SIGMOD, pp. 1579–1590 (2014)
20. Terada, A., Okada-Hatakeyama, M., Tsuda, K., Sese, J.: Statistical significance of combinatorial regulations. Proc. Natl. Acad. Sci. **110**(32), 12996–13001 (2013)
21. Wrobel, S.: An algorithm for multi-relational discovery of subgroups. In: Komorowski, J., Zytkow, J. (eds.) PKDD 1997. LNCS, vol. 1263, pp. 78–87. Springer, Heidelberg (1997). https://doi.org/10.1007/3-540-63223-9_108
22. Wu, E., Madden, S.: Scorpion: explaining away outliers in aggregate queries. Proc. VLDB Endow. **6**(8), 553–564 (2013)
23. Zliobaite, I., Mathioudakis, M., Lehtiniemi, T., Parviainen, P., Janhunen, T.: Accessibility by public transport predicts residential real estate prices: a case study in Helsinki region. In: 2nd Workshop on Mining Urban Data at ICML 2015 (2015)

Flash Points: Discovering Exceptional Pairwise Behaviors in Vote or Rating Data

Adnene Belfodil[1]($\boxtimes$), Sylvie Cazalens[1], Philippe Lamarre[1], and Marc Plantevit[2]

[1] INSA Lyon, CNRS, LIRIS UMR 5205, 69621 Lyon, France
{adnene.belfodil,sylvie.cazalens,philippe.lamarre}@liris.cnrs.fr
[2] Université Lyon 1, CNRS, LIRIS UMR 5205, 69622 Lyon, France
marc.plantevit@liris.cnrs.fr

Abstract. We address the problem of discovering contexts that lead well-distinguished collections of individuals to change their pairwise agreement w.r.t. their usual one. For instance, in the European parliament, while in overall, a strong disagreement is witnessed between deputies of the far-right French party *Front National* and deputies of the left party *Front de Gauche*, a strong agreement is observed between these deputies in votes related to the thematic: *External relations with the union*. We devise the method *DSC* (*Discovering Similarities Changes*) which relies on exceptional model mining to uncover three-set patterns that identify contexts and two collections of individuals where an unexpected strengthening or weakening of pairwise agreement is observed. To efficiently explore the search space, we define some closure operators and pruning techniques using upper bounds on the quality measure. In addition of handling usual attributes (e.g. numerical, nominal), we propose a novel pattern domain which involves hierarchical multi-tag attributes that are present in many datasets. A thorough empirical study on two real-world datasets (i.e., European parliament votes and collaborative movie reviews) demonstrates the efficiency and the effectiveness of our approach as well as the interest and the actionability of the patterns.

Keywords: Exceptional model mining · Subgroup discovery

1 Introduction

The last decade has witnessed a huge growth in the collection of rating (e.g., Amazon, IMDb, Yelp, Foursquare) or vote (e.g., Parltrack, Voteview) data. Such data depict the opinion (i.e., review, or vote) of people (e.g., IMDb users, European parliament member) on an item (e.g., movie, restaurant, ballot) and need to be analyzed by leveraging contextual information to discover new actionable insights that cannot be obtained otherwise. There has been a rapid rise in the analysis of such data in many applications such as fact checking or lead finding in political journalism, and collaborative rating analysis.

Fact checking has become increasingly common in political journalism. It contributes to the quality of news provided by media[1]. For instance,

---

[1] *"increasing quality of journalism will lead to better decisions by citizens..."* [16].

© Springer International Publishing AG 2017
M. Ceci et al. (Eds.): ECML PKDD 2017, Part II, LNAI 10535, pp. 442–458, 2017.
https://doi.org/10.1007/978-3-319-71246-8_27

Truth-O-Meter[2] was extensively used during the 2016 US presidential debate. Delving deeply into the votes sessions makes it possible to enlighten some claims about consensus between politicians or finding some flashpoints (i.e., contexts that lead to strong disagreement). Average rating is not enough for an item. While some individuals are in agreement on many items, they can be in strong disagreement for certain types of items. Such information can directly be used for recommendation. For example, in Movielens dataset, while usually *middle-aged women* users are in agreement with *middle-aged men* users w.r.t. their overall ratings, these collections are in disagreement for *Comedy* movies released in 1998.

The discovery of descriptions that distinguish a group of objects given a target (class) has been widely studied in data mining and machine learning community under several vocables (subgroup discovery, emerging patterns, contrast sets) [14]. We consider here the well-established framework of subgroup discovery (SD) [22]. Given a set of objects taking a vector of attributes (of Boolean, nominal, or numerical type) as description, and a class label as a target, the goal is to efficiently discover subgroups of objects for which there is a high difference between the label distribution within the group compared to the distribution within the whole dataset. SD has been extended to a richer framework that handles more complicated target concepts, the so-called Exceptional Model Mining (EMM) [17]. A model is built over the labels from the objects in the subgroup and is compared to the model of the whole dataset using a quality measure. The more different is the model, the more exceptional is the subgroup. Many models have been investigated in the last decade [6–8,13,21]. However, no model in the EMM framework makes it possible to characterize collection of individuals whose pairwise agreement exceptionally deviates according to a subset of objects.

In this paper, we introduce the problem of discovering collections of individuals and particular contexts where their pairwise agreement exceptionally differs from their usual one as an instance of EMM. Figure 1 gives an overview of our approach. Based on an *aggregation level* set a priori, the method begins by constituting collections of individuals (1). Bi-sets of individuals are identified by a description (2) and their global pairwise behavior is computed (3). The method eventually aims to identify subset of reviewed items (4) for which the related pairwise behavior (5) substantially differs from the global one (6). To discover such patterns, we have to simultaneously explore the search space associated to the reviewed items and the search space associated to the reviewers. To this end, we devise the method *DSC* (*Discovering Similarities Changes*) to discover three-set patterns (*context, collection$_1$, collection$_2$*) that identify a context and two collections of individuals where an unexpected strengthening or weakening of pairwise agreement is observed. We define some closure operators and some effective pruning techniques based on the computation of tight upper bounds on the quality measure to efficiently explore the search space. *DSC* is able to handle numerical, nominal attributes and also hierarchical multi-tag attributes. The main contributions of this paper are manifold:

---

[2] http://www.politifact.com/truth-o-meter/.

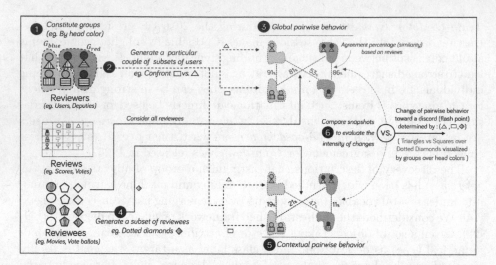

Fig. 1. Overview of *DSC*

Problem Formulation. We define the novel problem of exceptional pairwise behavior discovery in the EMM framework. This formulation makes it possible to consider several similarity measures to assess the pairwise agreement.

Algorithm and Analysis. We propose a branch-and-bound algorithm that efficiently exploits tight upper bounds and closure operators.

Evaluation. We report a thorough empirical study on real-world datasets that demonstrates the efficiency and the effectiveness of *DSC*.

The rest of the paper is organized as follows. Section 2 gives the formal definition of the exceptional pairwise behavior discovery problem. Section 3 presents the algorithms. Section 4 provides experimental results. Section 5 reviews the related work. Section 6 concludes and provides future directions.

2   Problem Definition

Data describing individuals outcomes about items are numerous, ranging from vote data to collaborative ratings through social-media platforms. We model such data as a triple $\langle E, U, R \rangle$ where E is a collection of objects (e.g., ballots, items, restaurants) and $\mathcal{A}_E = \{e_1, ..., e_n\}$ depicts the schema of the studied objects described by n attributes. U identifies the individuals (e.g., social network users, parliament members) described by m attributes over the schema $\mathcal{A}_U = \{u_1, ..., u_m\}$. Eventually, R represents the reviews (e.g., opinions, votes, ratings) of individuals over the objects. Each element of R is a triple $r = (e, u, o)$ where $o \in O$ is the outcome of a user $u \in U$ over an item $e \in E$. The function $o(e, u)$ returns the outcome o of u over an item e.

A description c over E defines a set of restrictions over the domains of the attributes $\mathcal{A}_E$. Such a description gives a context and identifies a subgroup of E

denoted E_c which is a collection of objects that fulfill the restrictions of c. We use the symbol $*$ to refer to the context that covers all the objects, therefore $E_* = E$. Similarly, a description g over U, which is a set of restrictions over the domains of the attributes $\mathcal{A}_U$, identifies a collection of individuals denoted U_g.

We aim to discover a context c and collections of individuals $U_{g'} \subseteq U$, $U_{g''} \subseteq U$ (*labeled respectively by their descriptions g', g'' over the attributes of U*) such that their pairwise agreement (similarities) differs exceptionally from the observed pairwise agreement over the whole objects. In other terms, we want to identify patterns (c, g', g'') that suggest an important change in pair-wise behavior between $U_{g'}$ and $U_{g''}$ within a context c. To this end, the outcomes of $U_{g'}$ and $U_{g''}$ have to be compared. Therefore, we need to define a similarity function between individuals over a given subgroup of objects. However, ratings data are generally sparse which limits the set of objects that have been rated by a pair of individuals. To overcome this issue, we have to consider aggregates of individuals and their *aggregated outcome*. The operator γ_L builds a parti-tion of U according to their values on the attributes $L \subseteq \mathcal{A}_U$. For instance, if U represents deputies affiliated to national parties depicted by the attribute np, $\gamma_{\{np\}}(U) = \{G_1, G_2, ...\}$ is a partition of U where each G_i represents a set of individuals affiliated to the same party.

We define an aggregated outcome operator $\theta : E \times 2^U \to O$ which maps an aggregate of individuals $G \subseteq U$ to its aggregated outcome w.r.t. an object e. For example, when dealing with movie ratings, aggregated outcome $\theta(e, G)$ can be defined as the mean of ratings given by some individuals of G to a movie e. We can compare the similarity between two sets of individuals based on their aggregated outcomes. The similarity measure is thus defined as: $sim : 2^E \times 2^U \times 2^U \longrightarrow [0, 1]$.

Our method relies on an EMM vision. Thus, we first need to determine a *model class* and a *quality measure* φ over this *model class*. We use a *simi-larity matrix* as a model to capture the pairwise agreement between pairs of user collections $(U_{g'}, U_{g''})$. Note that, contrary to common EMM approaches, there is *no unique base model* on the whole data but a model is *related to* a pair of descriptions (g', g'') identifying collections of individuals. The *base model* denoted $M_*^{g', g''}$, which represents the usual observed pairwise agreement over the whole objects between the candidate subgroups $U_{g'}, U_{g''}$, is defined as: $M_*^{g', g''} = \left(sim\left(E_*, i, j\right)\right)_{(i,j) \in \gamma_L(U_{g'}) \times \gamma_L(U_{g''})}$. The model built for a context c depicting a subgroup of objects E_c is: $M_c^{g', g''} = \left(sim\left(E_c, i, j\right)\right)_{(i,j) \in \gamma_L(U_{g'}) \times \gamma_L(U_{g''})}$.

The quality measure φ aims to quantify how much the model induced by the subgroup is different from the base model, i.e., how much the pairwise agreement observed over the whole objects differs from the one observed in a particular context between $U_{g'}$ and $U_{g''}$. Several quality measures can be defined according to the use case. For example, if we are interested in finding controversial contexts, we define $\varphi_{dissent}$ that captures the average similarity weakening between pairs of $\gamma_L(U_{g'}) \times \gamma_L(U_{g''})$:

$$\varphi_{dissent}(c, g', g'') = \frac{\sum_{(i,j)\in\gamma_L(U_{g'})\times\gamma_L(U_{g''})} max(sim(E_*,i,j)-sim(E_c,i,j),0)}{|\gamma_L(U_{g'})|\cdot|\gamma_L(U_{g''})|}$$

To find patterns (c, g', g'') that suggest an unexpected change of pairwise agreement, we rely on a well-known task, i.e., the discovery of *Top-k patterns* that fulfill a *minimum quality threshold* constraint σ_φ. Additional constraints can be taken into account (e.g. $\langle |E_c| \geq \sigma_E, |U_{g'}| \geq \sigma_U, |U_{g''}| \geq \sigma_U \rangle$).

3 'Discovery of Exceptional Pairwise Behaviors

In this section, we describe the enumeration principle based on closure operators, especially in the case of attributes whose domain is defined as a hierarchy. We then present different aggregates and similarities as well as the quality measures and their related tight upper and lower bounds. We eventually describe the algorithms to discover exceptional pairwise behaviors.

3.1   Candidate Descriptions Enumeration

Description Language. Let $\mathcal{G}$ be a generic collection of tuples which can be either E or U, and $\mathcal{A}_\mathcal{G} = (a_1, a_2, ..., a_n)$ its schema defined over n attributes. We denote by $dom(a_i)$ the domain of an attribute a_i. A *description* $d = \langle r_1, r_2, ..., r_n \rangle$ is a conjunction of restrictions over the attributes domains, where each restriction r_i corresponds to the attribute a_i. The restriction definition depends on the attribute type. If an attribute a_i is *nominal* then the corresponding restriction r_i is assimilated to a membership into a subset of $dom(a_i)$. Otherwise, if an a_i is *numeric* then the corresponding restriction r_i refers to a membership into an interval. The set of all possible descriptions is denoted $\mathcal{D}$. A description $d \in \mathcal{D}$ defines by intent a *subgroup* (*extent*) $\mathcal{G}_d \subseteq \mathcal{G}$ which contains the tuples of $\mathcal{G}$ verifying the restrictions of d. In order to bind the descriptions of $\mathcal{D}$ to subgroups in $\mathcal{G}$, we define a mapping function $\delta : \mathcal{G} \to \mathcal{D}$ that maps each tuple $g \in \mathcal{G}$ to its description in $\mathcal{D}$. To define this mapping function, we rely on the corresponding mappings $\delta_{a_i} : dom(a_i) \to \mathcal{D}_{a_i}$ that maps the values of an attribute a_i to its corresponding restriction $r_i \in \mathcal{D}_{a_i}$. Given a tuple g, an attribute a_i and its value a_i^g in g, if a_i is numeric, the restriction is an interval $\delta_{a_i}(a_i^g) = [a_i^g, a_i^g]$. Otherwise, if a_i is nominal, the restriction is a singleton $\delta_{a_i}(a_i^g) = \{a_i^g\}$. Finally, with the former definitions, for a tuple $g = (a_1^g, ..., a_n^g)$ we have $\delta(g) = \langle \delta_{a_1}(a_1^g), ..., \delta_{a_n}(a_n^g) \rangle$.

Description Space Structure. To enumerate candidate descriptions (*or candidate subgroups by extent*), we traverse the search space $\mathcal{D}$ in a bottom-up fashion. This search space is commonly depicted as a *meet-semi lattice* structured by an *infimum operator* denoted by $\sqcap$ [10] which simply allows to get the lowest common description of two given descriptions. The infimum operator definition relies on the n infimum operator $\sqcap_{a_i}$ corresponding each to the type of the attribute a_i. Let a be a *numeric* attribute, the corresponding infimum operator $\sqcap_a$ computes the minimum interval enclosing two intervals. In the other hand, if a is *nominal*, the corresponding infimum operator $\sqcap_a$ is represented by a set union operator. Thus the meet-semi lattice $(D, \sqcap)$ is the result of the cartesian product of the meet-semi lattices $(\mathcal{D}_a, \sqcap_a)$ each corresponding to an attribute $a \in \mathcal{A}_{\mathcal{G}}$. The infimum operator allows us to define a *partial order* denoted by $\sqsubseteq$ between descriptions. Given two descriptions c and d, we have $c \sqsubseteq d \Leftrightarrow c \sqcap d = c$.

Specialization and Neighborhood Relations. Let $c = \langle q_1, q_2, ..., q_n \rangle$ and $d = \langle r_1, r_2, ..., r_n \rangle$ be two descriptions of $\mathcal{D}$, r_i and q_i are two restrictions on the attribute a_i. r_i is a *specialization* of q_i iff $r_i \Rightarrow q_i$ which is equivalent to $q_i \sqsubseteq r_i \Leftrightarrow q_i \sqcap_{a_i} r_i = q_i$. A description d is a specialization of c (denoted $c \sqsubseteq d$) iff $\forall i \in [1..n] : q_i \sqsubseteq r_i$. Obviously, $c \sqsubseteq d \Longleftrightarrow \mathcal{G}_d \subseteq \mathcal{G}_c$ with $\mathcal{G}_d$ (*resp.* $\mathcal{G}_c$) the subgroup covered by d (*resp.* c). When traversing the search space we extend a description to more complex descriptions by atomic refinements. Thus, we define the *neighborhood relationship* $\prec$. We have $c \prec d$ iff $c \sqsubset d \wedge \nexists e \in \mathcal{D} : c \sqsubset e \sqsubset d$ and d is said to be an upper neighbor of c. To get the neighbors of a candidate description $c = \langle q_1, q_2, ..., q_n \rangle$, we rely on a similar neighborhood concept between restrictions. If a restriction q is over a nominal attribute a which is materialized by a subset $s_q \subseteq dom(a)$ membership, neighbors of q are candidates r which correspond to singletons of s_q. Similarly for a numeric attribute, candidate neighbors of a restriction r are the intervals q resulting from a left-minimal change or a right-minimal change on the interval bounds corresponding to r [12]. With these tools, we can easily define a refinement operator $\eta : \mathcal{D} \to 2^{\mathcal{D}}$ which maps to each description d its neighbors in $\mathcal{D}$ and we have:

$$\begin{aligned} \eta(c) &= \{d \in \mathcal{D} : d = \langle r_1, ..., r_n \rangle \succ c = \langle q_1, ..., q_n \rangle\} \\ &= \{d \in \mathcal{D} : \exists j \in [1..n] \mid r_j \succ q_j \text{ and } \forall i \in [1..n] \mid i \neq j \Rightarrow r_i = q_i\} \end{aligned} \quad (1)$$

Additionally we define η_f that computes the neighbors of a given description c by refining the fth restriction corresponding to the fth attribute as follows:

$$\eta_f(c) = \{d \in \mathcal{D} : r_f \succ q_f \text{ and } \forall i \in [1..n] \mid i \neq f \Rightarrow r_i = q_i\} \quad (2)$$

Closed Descriptions. We rely on the concept of *closed descriptions* to significantly decrease the number of explored descriptions by avoiding redundancy. A description c is said to be *closed iff* for every specialization d (*i.e.* $c \sqsubset d$) there is at least one object in $\mathcal{G}$ covered by c but not by d. More formally, $\forall d \in \mathcal{D} : c \sqsubset d \Rightarrow \mathcal{G}_d \subsetneq \mathcal{G}_c$. Two descriptions c and d are considered as equivalent (denoted $c \equiv d$) iff $\mathcal{G}_c = \mathcal{G}_d$. We can adapt the *CbO* (*Close-by-One*) algorithm [15] for our use in *DSC*.

To define the *closure operator* of a description d of $\mathcal{D}$, we need to introduce two derivation operators that create a Galois connection between $2^{\mathcal{G}}$ and $\mathcal{D}$: Given $S \subseteq \mathcal{G}$, the description $S^{\square} \in \mathcal{D}$ covering the subgroup S is:

$$S^{\square} := \sqcap_{g \in S} \delta(g) = \langle \sqcap_{g \in S} \delta_{a_1}(a_1^g), ..., \sqcap_{g \in S} \delta_{a_n}(a_n^g) \rangle$$

Given a description d, the subgroup $d^{\square}$ covered by d is:

$$G_d = d^{\square} = \{g \in \mathcal{G} \mid d \sqsubseteq \delta(g)\}$$

$(.)^{\square\square}$ is a closure operator and for every $d \in \mathcal{D}$ $d^{\square\square}$ is a closed description.

Canonicity Test. An important aspect in *CbO* enumeration is the *canonicity test*, which allows to determine if a description after closure was already generated and discard it, if appropriate. The canonicity test relies on a linear order $<$ between descriptions of $\mathcal{D}$. Given an arbitrary order between attributes $\mathcal{A}_{\mathcal{G}} = \{a_1, a_2, ..., a_n\}$, if $d = \langle r_1, ..., r_n \rangle$ comes from a closure after a refinement of the fth restriction of $c = \langle q_1, ..., q_n \rangle$ then we have: $c <_f d \iff \forall i \in [1..f-1] \mid q_i = r_i \wedge q_f <_{a_f} r_f$. Note that, in our case, the test part $q_f <_{a_f} r_f$ is always valid when the fth attribute is numeric or nominal. Although, the latter need to be assessed when the attribute is rather complex, such as for *HMT attributes* introduced in the next section.

3.2   Hierarchical Multi-tag Attribute (HMT)

Several votes and reviews datasets contain multi-tagged objects where each tag is a part of a hierarchical structure. For instance, the ballots in the European parliament can have multiple tags (e.g., the ballot *Gender mainstreaming in the work of the European Parliament* is tagged by *4.10.04-Gender equality* and *8.40.01-European Parliament*. Tag *4.10.04* itself identifies a hierarchy where tag *4.10* depicts *Social policy* which is a specialization of tag *4* that covers the ballots related to *Economic, social and territorial cohesion*). Let $\mathcal{G}$ be a set of tagged objects. For the sake of simplicity, each object g is described by a unique attribute *tags* which is a set of tags. Tags form a tree noted T.

We can define the partial order $\leq$ between tags as the same usual partial order in a tree structure where the tree root is the minimum (e.g. $* < 1 < 1.20$). This allows us to define the ascendants (resp. descendants) operator $\uparrow$ (resp. $\downarrow$) of a tag $t \in T$. We have $\uparrow t = \{u \in T | u \leq t\}$ and $\downarrow t = \{u \in T | u \geq t\}$. Let t and u be two tags, t is a lower neighbor of u denoted $t \prec u$ iff $\nexists e \in T \mid t < e < u$. Thus t is a parent of u denoted as $t = p(u)$.

A restriction over an *HMT* attribute is assimilated as a membership in a set of tags $\{t_1, ..., t_n\}$. We denote the description domain by $\mathcal{D}$ which is a subset of 2^T. Each object $g \in \mathcal{G}$ is mapped by $\delta(g)$ to its corresponding description in $\mathcal{D}$. Obviously if $\delta(g) = \{t_1, t_2\}$, the object g is tagged *explicitly* by the tags t_1 and t_2 but also *implicitly* by all their generalization $\uparrow t_1$ and $\uparrow t_2$ as shown in Fig. 2.

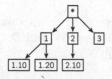

	tags
g_1	$\{1.20, 2.10\}$
g_2	$\{1, 3\}$
g_3	$\{1.10, 2.10, 3\}$
g_4	$\{2.10\}$
g_5	$\{1.20\}$

	*	1	1.10	1.20	2	2.10	3
g_1	×	×		×	×	×	
g_2	×	×					×
g_3	×	×	×		×		×
g_4	×				×	×	
g_5	×	×		×			

Fig. 2. A tags tree (left), a collection of tagged items (middle) and a vector representation (right)

To handle this attribute among the other attributes in the complex search space defined previously, we need to define the infimum operator $\sqcap_{HMT}$ between two descriptions of $\mathcal{D}$. Let $c = \{t_1, ..., t_n\}$ and $d = \{u_1, ..., u_m\}$ be two descriptions of $\mathcal{D}$, we define $\sqcap_{HMT}$ as: $c \sqcap_{HMT} d = max(\cup_{t \in c} \uparrow t \cap \cup_{u \in d} \uparrow u)$ where $max : 2^T \to 2^T$ is a function that maps each subset of tags $s \subseteq T$ to the leafs of the sub-tree compound of the tags of s: $max(s) = \{t \in s \mid (\downarrow t \setminus \{t\}) \cap s = \emptyset\}$.

Intuitively $c \sqcap_{HMT} d$ depicts the set of the maximum explicit or implicit tags shared by the two descriptions. For instance, if $c = \{1.10, 2\}$ and $d = \{1.20, 2.10\}$, $c \sqcap_{HMT} d = \{1, 2\}$. A description d is said to be a specialization of c denoted $c \sqsubseteq d$ iff $c \sqcap_{HMT} d = c$ which means $\forall t \in c \; \exists u \in d \mid u \in \downarrow t$. A description c is considered as a lower neighbor of d denoted $c \prec d$ iff:

$$
\begin{cases}
\exists! \, (t, u) \in c \times d : t \prec u \wedge \forall t' \in (c \setminus t) \; \exists u' \in d : t' = u' & \text{if} \quad |d| = |c| \\
\forall t \in c \; \exists u \in d : t - u \wedge \exists!(t, u) \in c \times d \; \exists t' \in \uparrow t : p(u) = p(t') & |d| = |c| + 1
\end{cases}
$$

Basically d is an upper neighbor of c, if either only one tag of d is refined in c by the neighborhood relation between tags or by adding a new tag in d that share parent with a tag in c or with one of its ascendants. The linear order between two conjunctions of tags $c = \{t_1, ..., t_n\}$ and $d = \{u_1, ..., u_n, ..., u_m\}$ given that d comes from a closure after refinement of the fth tag of c is defined as: $c <_f d \iff \forall i \in [1..f-1] : t_i = u_i \wedge t_f \leq u_f$. The linear order between tags can be provided by the depth first search on T.

Based on the definitions of $\sqcap_{HMT}$, neighborhood relation between two sets of tags and the linear order between them, the attribute HMT can be easily handled with the aforementioned attributes (numeric and nominal) in the complex search space dealing with n attributes.

3.3   Aggregations, Similarities and Quality Measures

An important aspect in DSC is the similarity measure between aggregates of individuals. Given $L \subseteq \mathcal{A}_U$ a set of individuals attributes on which we compute aggregates of individuals, a collection of individuals $U_g \subseteq U$ labeled by a description g, $\gamma_L(U_g) = \{G_1, G_2, ..., G_k\}$ is a partition of U_g. The aggregate outcome θ is defined according to the application domain. For example, the outcome of an aggregate of reviewers who give scores is defined as such: $\theta_{review}(e, G) = \frac{1}{|G|} \sum_{u \in G} o(e, u)$. The outcome of an aggregate G of European parliament members w.r.t a ballot is given by the vote of the majority[3] as $\theta_{votes}(e, G) = argmax_{v \in \mathcal{O}}\{count(v, \{o(e, u) \mid u \in G\})\}$ (See Fig. 3).

In this paper, we consider similarities that convey the average agreement proportion between two aggregates G_i, G_j based on their pairwise similarity $simobj$ over each object. We define such similarities over $2^E \times 2^U \times 2^U$:

$$
sim(E, G_i, G_j) = \frac{1}{|E|} \sum_{e \in E} simobj(E, G_i, G_j) \tag{3}
$$

---

[3] The same measure is used by **votewatch** to observe the voting behavior of deputies.

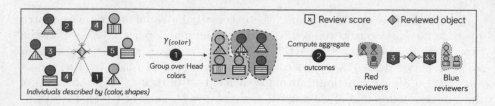

Fig. 3. Aggregates outcomes over one reviewed object

Indeed, the measure *simobj* which is defined over $E \times 2^U \times 2^U$ is adapted on the application domain. For example, if we want to compare deputies where vote decision can be either a *for*, *against* or *abstain*. we define:

$$simobj_{votes}(e, G_i, G_j) = \begin{cases} 1 & \text{if } \theta(e, G_i) = \theta(e, G_j) \\ 0 & \text{else} \end{cases} \tag{4}$$

For ratings ranging from 1 to 5, the similarity *simobj* is defined by how much the scores given by the two aggregates are close:

$$simobj_{review}(e, G_i, G_j) = 1 - \frac{1}{4}|\theta(e, G_i) - \theta(e, G_j)| \tag{5}$$

To discover interpretable patterns (c, g', g''), we define the two following quality measures $\varphi_{consent}$, $\varphi_{dissent}$ by relying on the defined similarities. $\varphi_{consent}$ makes it possible to consider a pattern as "*interesting*" if there is an important strengthening of similarities between individuals corresponding to g' and individuals corresponding to g'' for the context c. $\varphi_{dissent}$ aims to assess the weakening of similarities between individuals. We assume that the attributes $L \subseteq \mathcal{A}_U$ used to build partitions of individuals are given:

- $\varphi_{consent}(c, g', g'') = \dfrac{\sum_{(i,j) \in \gamma_L(U_{g'}) \times \gamma_L(U_{g''})} max(sim(E_c, i, j) - sim(E_*, i, j), 0)}{|\gamma_L(U_{g'})| \cdot |\gamma_L(U_{g''})|}$
- $\varphi_{dissent}(c, g', g'') = \dfrac{\sum_{(i,j) \in \gamma_L(U_{g'}) \times \gamma_L(U_{g''})} max(sim(E_*, i, j) - sim(E_c, i, j), 0)}{|\gamma_L(U_{g'})| \cdot |\gamma_L(U_{g''})|}$

3.4 Upper Bounds on Quality Measures

To early discard unpromising descriptions, we follow a branch-and-bound approach in which an upper bound on the quality measure φ is computed for a candidate description. We first define a generic upper bound UB_{sim} and a lower bound LB_{sim} on *sim*. Given a threshold σ_E that fix the minimum threshold on objects subgroup size, G_i and G_j two aggregates of individuals, we have:

- $LB_{sim}^1(E, G_i, G_j) = max\left(\frac{\sigma_E - |E|(1 - sim(E, G_i, G_j))}{\sigma_E}, 0\right)$
- $LB_{sim}^2(E, G_i, G_j) = \frac{1}{\sigma_E} smallest(\{simobj(e, G_i, G_j) \mid e \in E\}, \sigma_E)$
- $UB_{sim}^1(E, G_i, G_2) = min\left(\frac{|E| * sim(E, G_i, G_j)}{\sigma_E}, 1\right)$

$$- UB^2_{sim}(E, G_i, G_j) = \tfrac{1}{\sigma_E} largest(\{simobj(e, G_i, G_j) \mid e \in E\}, \sigma_E)$$

where $smallest(S, n)$ (resp. $largest(S, n)$) computes the sum of the n $minimum$ (resp. $maximum$) of given set S of real values. LB^1_{sim} (resp. UB^1_{sim}) is equivalent to LB^2_{sim} (resp. UB^2_{sim}) when $simobj$ gives binary results such as $simobj_{votes}$.

Given a description (c, g', g''), we define the following upper bounds[4] on the quality measure of every specialization d of c ($\forall d \mid c \sqsubseteq d$):

$$\varphi_{consent}(d, g', g'') \leq UB_{consent}(c, g', g'') \wedge \varphi_{dissent}(d, g', g'') \leq UB_{dissent}(c, g', g'')$$

$$- UB_{consent}(c, g', g'') = \frac{\sum_{(i,j) \in \gamma_L(U_{g'}) \times \gamma_L(U_{g''})} max(UB_{sim}(E_c, i, j) - sim(E_*, i, j), 0)}{|\gamma_L(U_{g'})| \cdot |\gamma_L(U_{g''})|}$$

$$- UB_{dissent}(c, g', g'') = \frac{\sum_{(i,j) \in \gamma_L(U_{g'}) \times \gamma_L(U_{g''})} max(sim(E_*, i, j) - LB_{sim}(E_c, i, j), 0)}{|\gamma_L(U_{g'})| \cdot |\gamma_L(U_{g''})|}$$

where $UB_{consent}$ (resp. $UB_{dissent}$) corresponds to $\varphi_{consent}$ (resp. $\varphi_{dissent}$).

3.5 Algorithms

Algorithm 1 called *EnumCC* (*Enumerate Closed Candidates*) describes the exploration of the search space over a collection of objects $\mathcal{G}$ defined by the attributes $\mathcal{A}_\mathcal{G} = \{a_1, ..., a_n\}$. *EnumCC* enumerates the *closed descriptions* c that verify the constraint $\sigma_\mathcal{G}$ on the size of its corresponding subgroup starting from a description d. Given a description d, *EnumCC* computes its corresponding subgroup S_c, if its size exceeds the threshold, the closure c of d is computed and the linear order between them is verified. If so c is returned as a valid candidate. The algorithm then generates the neighbors by refining the attributes $\{a_f, ..., a_n\}$. The flag f determines the attribute that was refined to generate the description d. Finally, a recursive call is done to explore the lattice structure formed by d in a DFS fashion. The parameter cnt is a Boolean that allows to prune the search space based on the computation of the upper bound on the quality of a candidate description. *EnumCC* is depicted as a generator.

Algorithm 2 depicts *DSC* method based on the use of the closure operator and a branch and bound exploration. It is related to the task of finding topk patterns with a minimum quality threshold σ_φ. The algorithm first generates the candidate pattern (c, g', g''), subsequently the upper bound of the candidate pattern is computed. If it does not exceed the threshold σ_φ, the search space is pruned. Otherwise the quality measure of the candidate is computed. If its quality exceeds the same threshold σ_φ then the *topk* set is updated. Subsequently, if the size of *topk* exceeds k, the worst pattern found w.r.t. φ is discarded and σ_φ is dynamically updated with the minimal quality of the current *topk* set. Note that E defines the objects on which individuals U_1 and U_2 (two subsets of U) gives outcomes. L determines the attributes on which the individuals are aggregated. Finally, σ_E, σ_U determines the thresholds of subgroups sizes of respectively E and U.

---

[4] Proofs of upperbounds are available in goo.gl/viQxhi.

Algorithm 1. $EnumCC(\mathcal{G}, d, \sigma_\mathcal{G}, f, cnt)$

---

1   $S_c \leftarrow d^\square$
2   **if** $|S_c| \geq \sigma_\mathcal{G}$ **then**
3     $c \leftarrow S_c^\square$
4   **if** $d \lessdot_f c$ **then**
5       $cnt\_c \leftarrow copy(cnt)$
6   **yield** (c, S_c, cnt_c) ;   // yield the results and wait for the next call
7   **if** cnt_c **then**
8   **foreach** $j \in [f, n]$ **do**
9   **foreach** $ngh \in \eta_j(c)$ **do**
10   **foreach** $(c_{ngh}, S_{ngh}, cnt_{ngh}) \in EnumCC(S_c, ngh, \sigma_\mathcal{G}, j, cnt\_c)$ **do**
11   **yield** $(c_{ngh}, S_{ngh}, cnt_{ngh})$

Algorithm 2. $DSC(E, U_1, U_2, L, \sigma_\mathcal{E}, \sigma_U, \sigma_\varphi, k)$

---

1   $\sigma_\varphi^{current} \leftarrow \sigma_\varphi$
2   $topk \leftarrow [\,]$
3   **foreach** $(g', U_{g'}, cont_{g'}) \in EnumCC(U_1, *, \sigma_{U_1}, 0, True)$ **do**
4   **foreach** $(g'', U_{g''}, cont_{g''}) \in EnumCC(U_2, *, \sigma_{U_2}, 0, True)$ **do**
5   **foreach** $(c, E_c, cont_c) \in EnumCC(\mathcal{E}, *, \sigma_\mathcal{E}, 0, True)$ **do**
6         $UB \leftarrow UB_{dissent}(c, g', g'')$;   // resp. $UB_{consent}$
7   **if** $UB < \sigma_\varphi^{current}$ **then**
8           $cont_c \leftarrow False$
9   **else**
10           $quality \leftarrow \varphi_{dissent}(c, g', g'')$;   // resp. $\varphi_{consent}$
11   **if** $quality \geq \sigma_\varphi^{current}$ **then**
12             $pattern \leftarrow (c, g', g'')$
13   **update** $topk$ **by** $\langle pattern, quality \rangle$ **limits by** k
14   **if** $|topk| = k$ **then**
15               $\sigma_\varphi^{current} \leftarrow min\_quality(topk)$
16   **output** $topk$

---

4   Empirical Study

In this section we report on both quantitative and qualitative experiments over the implemented algorithms. The algorithms were implemented in Python. The experiments were carried on an Intel Core i7-6700HQ 2.60 GHz machine with 16 GB RAM and were run by PyPy 5.4.1 For reproducibility purpose, the source code and the data are made available in our companion page[5]. These experiments aim to answer the following questions: *Q1* - Is the closure over an HMT attribute more effective than mining closed itemsets? *Q2* - Are the closing operator and the tights upper bounds effective and efficient? *Q3* - Does our algorithm scale w.r.t. different parameters? *Q4* - Does *DSC* provide actionable patterns?

---

[5] https://github.com/Adnene93/DiscoveringSimilarityChanges.

Experiments were carried out on two real-world datasets: a movie review dataset *Movielens*[6] and the European parliament dataset *EPD*[7]. The main characteristics of these datasets are reported in Table 1. In Movielens, 18 movie genres are organized through a flat hierarchy.

Table 1. Characteristics of the datasets

Characteristics	Movielens	EPD
#objects	1.681 movies	2.471 Ballots
#individuals	943	778
#outcomes	99.991	1.639.199
$\mathcal{A}_E$	1 HMT (18 tags), 1 Numeric	1 HMT (311 tags), 1 Numeric, 1 Nominal
$\mathcal{A}_U$	3 Nominal	3 Nominal

4.1 Performance Study

Q1 - We aim to study the performance of the closure operator in the presence of an HMT attribute. To this end, we compare it against the closure over itemsets (i.e., scaling) as illustrated in Fig. 2. A tree of tags is characterized by its *height* and its branching factor (*k-ary tree*). A dataset of multi-tagged object is described by the maximum number of tags (*maxtags*) that an object can have and also its size. Figure 4 reports the runtime and the number of explored candidates of the two closure operators when varying the branching factor, the tree height, the number of tags and the dataset size. For these experiments, we set the default values of these characteristics respectively to: 5, 3, 3 (*hierarchy of 125 tags*) and 5000 objects. HMTClosure exploits the structure of the tree and avoids exploring semantically equivalent descriptions (i.e.: $\{3, 3.10.05\}$ *is semantically equivalent to* $\{3.10.05\}$) whereas *ISClosure* explores them. In all configurations, HMTClosure outperforms ISClosure on both the execution time and the number of explored candidates. These experiments demonstrate that taking into account hierarchical relations makes the closure operator more efficient and effective.

Q2 - A baseline algorithm is obtained by deactivating the pruning techniques based on upper-bound and the closure operators . Thus, the baseline only pushes monotonic constraints. We compare *DSC* with the baseline and also with *closed* which is *DSC* without an upper bound computation on both Movielens and *EPD*. Notice that in *EPD* $UB^1_{dissent}$ and $UB^2_{dissent}$ are equivalent for the considered similarity. Therefore, we only report $UB^1_{dissent}$. We interrupt a method if its execution time exceeds one hour.

Figures 5 and 6 report the behavior (i.e., execution time and number of explored candidates) of the different methods when varying the characteristics

---

[6] https://grouplens.org/datasets/movielens/100k/.

[7] http://parltrack.euwiki.org/.

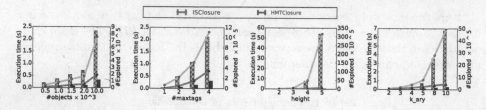

Fig. 4. Behavior of enumeration algorithms considering two closure operators for HMT attributes w.r.t. the number of objects, the height of the hierarchy, the number of tags and the branching factor which are set by default to respectively $5, 3, 3, 5000$.

of the datasets Movielens and *EPD*. Obviously, these experiments give evidence that each of the different optimizations of *DSC* are effective. For Movielens dataset, *DSC* is the most efficient when using $UB^2_{dissent}$ instead of $UB^1_{dissent}$. Indeed, $UB^2_{dissent}$ is more costly to compute than $UB^1_{dissent}$ but much tighter. The differences between the baseline and *DSC* are much more important on *EPD* because the HMT attribute is more complex than in Movielens. The experiments also demonstrate that the number of attributes used in a description of an object or a user heavily impacts the performance of the method as it increases the size of the search space.

Q3 - Figure 7 reports the behavior of *DSC* on *EPD* when varying the input parameters (i.e., the minimum thresholds σ_E and σ_U and the quality measure). Obviously, when the thresholds increase (i.e. become more stringent) the number of explored patterns and thus the execution time decrease. Nevertheless, we observe that when decreasing σ_E, *DSC* remains efficient thanks to its pruning abilities based on upper-bound computations and closure operators. The execution time increases in line with the number of dimensions $|L|$ on which are computed the group of individuals while the number of explored descriptions remains roughly the same. Indeed, the computation of the model is more costly. Finally, a greater σ_φ leads to an important reduction of the number of explored candidates and therefore a better execution time. This demonstrates the effectiveness of the pruning properties implemented in *DSC*. Even if the two quality measures behave similarly, $\varphi_{consent}$ performs slightly better than $\varphi_{dissent}$ as by default the relation between the parliament's deputies w.r.t. their voting behavior is rather consensual.

4.2   Qualitative Results (*Q4*)

Table 2 describes some patterns found by *DSC* when looking for contexts that weaken the pairwise agreement between collections of reviewers identified by gender and age group in Movielens. For instance, middle-aged females tend to be in discord with their peer males for 1998 comedy movies (*13 movies, e.g.: The Wedding Singer*) in the best pattern. This can be observed by a significant decrease of similarity (of 35%) between the two aggregates from 86% to 51%. The diver-

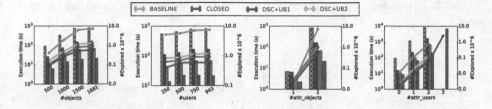

Fig. 5. Effectivness of *DSC* (*Top-5*) according to **Movielens** dataset characteristics which are set by default to $|E| = 1681$, $|U| = 943$, $\#attr_{objects} = 2$, $\#attr_{individuals} = 2$. The default thresholds are $\sigma_E = \sigma_U = 5$, $\sigma_\varphi = 0$, $|L| = 1$.

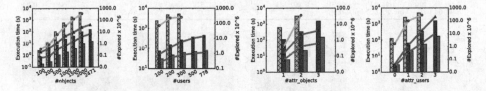

Fig. 6. Effectivness of *DSC* (*Top-5*) according to **EPD** dataset characteristics which are set by default to $|E| = 2471$, $|U| = 778$, $\#attr_{objects} = 3$, $\#attr_{individuals} = 2$. The default thresholds are $\sigma_E = \sigma_U = 15$, $\sigma_\varphi = 0$, $|L| = 1$.

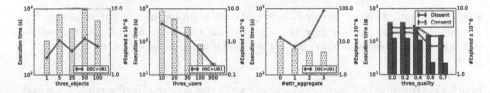

Fig. 7. Effectivness of *DSC* (top-5) over **EPD** according to constraints thresholds and quality measures. The default thresholds are $\sigma_E = \sigma_U = 15$, $\sigma_\varphi = 0$, $|L| = 1$

sification is done over the top-100 patterns. Two patterns are considered similar if one cover more than 50% of the reviewed objects contained in the second.

Figure 8 reports the patterns discovered suggesting flash points (particular contexts that lead European groups to important similarities weakening). These patterns allow us to explicit the differences between groups that usually share the same political line. For example, while PPE and S&D vote mostly the same (76% of the cases), the top pattern (1) uncovers the ballots (contextualized by their themes - *such as 3.40.16 Raw materials and 6.10.05 Peace preservation* - and their time period - Feb. 2015) on where the two groups strongly diverge. This is witnessed by a decrease of pairwise agreement from 76% to 0%. The heatmaps illustrated in Fig. 8. depict the overall pairwise agreement changes observed for the pattern (1). Such results can provide insights for both political analysts and journalists, where the analytic tool provided by *DSC* allows to help discover ideological idiosyncrasies when comparing deputies against their

Table 2. Diversified *Top-4* patterns discovered over Movielens by grouping on *age-groups*

i	context (c)	g'	g''	$|E_c|$	$|U_{g'}|$	$|U_{g''}|$	φ
1	[['Comedy'], [1998, 1998]]	[['middle-age'], ['F']]	[['middle-age'], ['M']]	13	119	228	0.35
2	[['Horror', 'Comedy'], [1992, 1996]]	[['middle-age'], ['F']]	[['old'], ['M']]	9	119	45	0.31
3	[['Drama'], [1949, 1950]]	[['middle-age'], ['F', 'M']]	[['young'], ['F', 'M']]	5	347	508	0.3
4	[['Romance'], [1998, 1998]]	[['middle-age'], ['M']]	[['young'], ['F']]	11	228	134	0.3

i	context (c)	g'	g''	$|E_c|$	$|U_{g'}|$	$|U_{g''}|$	φ
1	[['3.40.16', '6.10.05', '6.20.02', '6.30'], [Feb. 2015, Feb. 2015]]	[['PPE']]	[['S&D']]	29	227	191	0.76
2	[['2.40', '3.30.03.04', '3.30.05', '3.30.06', '3.30.20', '3.30.25', '4.60.06'], [July. 2015, July. 2015]]	[['S&D']]	[['Verts/ALE']]	13	191	51	0.75
3	[['2.50.08', '2.80', '3.45.04'], [Feb.2016, Feb.2016]]	[['ALDE']]	[['S&D']]	8	75	191	0.71
4	[['6.40.04.02'], [Mar.2015, Mar.2015]]	[['GUE/NGL']]	[['Verts/ALE']]	10	59	51	0.67

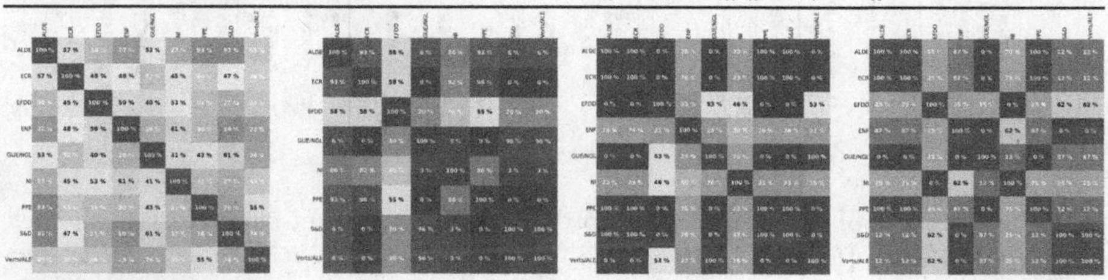

(1)   (2)   (3)   (4)

Fig. 8. Diversified *Top-4* patterns over EPD by grouping over political groups. (1) determine the usual pairwise observed between political groups and (2, 3 & 4) illustrate the heatmaps corresponding to the best 3 pattern found in top-k table

peers, determining red lines between political groups or exhibiting contexts where nations deputies coalesce against others in critical subjects.

5   Related Work

The problem of discovering exceptional subgroups based on the definition of a complex target model has been widely investigated in the recent years [7, 8, 13, 17, 18, 21]. Interestingly, de Sá et al. [6] use a similar matrix model to support the discovery of subgroups of individuals whose preference relation between ranked objects deviates from the norm. However, in the so-called exceptional preference mining, the dimensions of the model are fixed, i.e., the quality measure takes into account all objects and not dynamically a subset as in *DSC*. Dynamic EMM (i.e., EMM with a non-fixed model) has been recently investigated for different aims. Bosc et al. [4] propose a method to handle multi-label data where the number of labels per objects is much lower than the total number of labels which prevent the use of usual EMM model. Other dynamic EMM approaches aim to discover exceptional attributed sub-graphs [3, 13].

Thanks to open data policy, the analysis of political data has received much attention in the past decade. Most of them use basic data mining techniques. For instance, [11] uses clustering and PCA to identify cohesion blocs and dissimilarity blocs of voters within the US senate. Similar work was done on the Finnish [20] and the Italian [1] parliaments. An extensive tool was provided by [9] and applied to Swiss government datasets to detect opinion change of parliamentarians based on their expressed opinions before elections and votes cast afterwards.

Rating analysis has also received a wide interest in the last decade. In [5], the authors tackle the problem of rating interpretation by providing two methods (DEM, DIM). While the first one aims to discover groups of users that substantially agree for a given set of items, the second addresses the discovery of groups with an apparent inner discord. These two methods can be formalized as EMM instances with either a quality measure that assesses the average ratings of the identified subgroups or the average balance between positive and negative rating. While these methods consider a mono-objective measure (*rating average*), a similar work has been done to tackle multi-objective groups identification in [19]. It addresses a more complex statistical measure (rating distribution) and additionally coverage and diversity issues. In [2], the authors aim at using rating maps to identify subsets of reviews such that the distribution of rates observed is similar to the desired distributions.

6 Conclusion

In this paper, we introduced the novel problem of subjective exceptional pairwise behavior discovery in rating or vote data, rooted in the SD/EMM framework. We defined a branch-and-bound algorithm that exploits tight upper bounds and some closure operators to efficiently and effectively discover subgroups of interest. Experiments show that both quantitative and qualitative results are very satisfactory. We believe that this work opens new directions for future work. For example, the interactive discovery of exceptional pairwise behavior would make it possible to take into account prior knowledge. Such an exploration must be supported by instant mining algorithms.

Aknowledgement. This work has been partially supported by the project *ContentCheck* **ANR-15-CE23-0025** funded by the French National Research Agency.

References

1. Amelio, A., Pizzuti, C.: Analyzing voting behavior in Italian parliament: group cohesion and evolution. In: ASONAM, pp. 140–146. IEEE (2012)
2. Amer-Yahia, S., Kleisarchaki, S., Kolloju, N.K., Lakshmanan, L.V., Zamar, R.H.: Exploring rated datasets with rating maps. In: WWW 2017 (2017)
3. Bendimerad, A.A., Plantevit, M., Robardet, C.: Unsupervised exceptional attributed sub-graph mining in urban data. In: ICDM, pp. 21–30 (2016)
4. Bosc, G., Golebiowski, J., Bensafi, M., Robardet, C., Plantevit, M., Boulicaut, J.-F., Kaytoue, M.: Local subgroup discovery for eliciting and understanding new structure-odor relationships. In: Calders, T., Ceci, M., Malerba, D. (eds.) DS 2016. LNCS (LNAI), vol. 9956, pp. 19–34. Springer, Cham (2016). https://doi.org/10.1007/978-3-319-46307-0_2

5. Das, M., Amer-Yahiá, S., Das, G., Mri, C.Y.: Meaningful interpretations of collaborative ratings. PVLDB **4**(11), 1063–1074 (2011)
6. Rebelo de Sá, C., Duivesteijn, W., Soares, C., Knobbe, A.: Exceptional preferences mining. In: Calders, T., Ceci, M., Malerba, D. (eds.) DS 2016. LNCS (LNAI), vol. 9956, pp. 3–18. Springer, Cham (2016). https://doi.org/10.1007/978-3-319-46307-0_1
7. Duivesteijn, W., Feelders, A.J., Knobbe, A.: Exceptional model mining. Data Mining Knowl. Discov. **30**(1), 47–98 (2016)
8. Duivesteijn, W., Knobbe, A.J., Feelders, A., van Leeuwen, M.: Subgroup discovery meets Bayesian networks - an exceptional model mining approach. In: ICDM 2010 (2010)
9. Etter, V., Herzen, J., Grossglauser, M., Thiran, P.: Mining democracy. ACM (2014)
10. Ganter, B., Kuznetsov, S.O.: Pattern structures and their projections. In: Delugach, H.S., Stumme, G. (eds.) ICCS-ConceptStruct 2001. LNCS (LNAI), vol. 2120, pp. 129–142. Springer, Heidelberg (2001). https://doi.org/10.1007/3-540-44583-8_10
11. Jakulin, A., Buntine, W.: Analyzing the US Senate in 2003: similarities, networks, clusters and blocs (2004)
12. Kaytoue, M., Kuznetsov, S.O., Napoli, A., Duplessis, S.: Mining gene expression data with pattern structures in formal concept analysis. Inf. Sci. **181**(10), 1989–2001 (2011)
13. Kaytoue, M., Plantevit, M., Zimmermann, A., et al.: Exceptional contextual subgraph mining. Mach. Learn. **106**, 1171–1211 (2017). https://doi.org/10.1007/s10994-016-5598-0
14. Kralj Novak, P., Lavrač, N., Webb, G.I.: Supervised descriptive rule discovery: a unifying survey of contrast set, emerging pattern and subgroup mining. J. Mach. Learn. Res. **10**, 377–403 (2009)
15. Kuznetsov, S.O.: Learning of simple conceptual graphs from positive and negative examples. In: Żytkow, J.M., Rauch, J. (eds.) PKDD 1999. LNCS (LNAI), vol. 1704, pp. 384–391. Springer, Heidelberg (1999). https://doi.org/10.1007/978-3-540-48247-5_47
16. Lacy, S., Rosenstiel, T.: Defining and measuring quality journalism (2015)
17. Leman, D., Feelders, A., Knobbe, A.: Exceptional model mining. In: Daelemans, W., Goethals, B., Morik, K. (eds.) ECML PKDD 2008. LNCS (LNAI), vol. 5212, pp. 1–16. Springer, Heidelberg (2008). https://doi.org/10.1007/978-3-540-87481-2_1
18. Lemmerich, F., Becker, M., Atzmueller, M.: Generic pattern trees for exhaustive exceptional model mining. In: Flach, P.A., De Bie, T., Cristianini, N. (eds.) ECML PKDD 2012. LNCS (LNAI), vol. 7524, pp. 277–292. Springer, Heidelberg (2012). https://doi.org/10.1007/978-3-642-33486-3_18
19. Omidvar-Tehrani, B., Amer-Yahia, S., Dutot, P.-F., Trystram, D.: Multi-objective group discovery on the social web. In: Frasconi, P., Landwehr, N., Manco, G., Vreeken, J. (eds.) ECML PKDD 2016. LNCS (LNAI), vol. 9851, pp. 296–312. Springer, Cham (2016). https://doi.org/10.1007/978-3-319-46128-1_19
20. Pajala, A., Jakulin, A., Buntine, W.: Parliamentary group and individual voting behaviour in the finnish parliament in year 2003: a group cohesion and voting similarity analysis (2004)
21. van Leeuwen, M., Knobbe, A.J.: Diverse subgroup set discovery. Data Min. Knowl. Discov. **25**(2), 208–242 (2012)
22. Wrobel, S.: An algorithm for multi-relational discovery of subgroups. In: Komorowski, J., Zytkow, J. (eds.) PKDD 1997. LNCS, vol. 1263, pp. 78–87. Springer, Heidelberg (1997). https://doi.org/10.1007/3-540-63223-9_108

Time Series and Streams

A Multiscale Bezier-Representation for Time Series that Supports Elastic Matching

F. Höppner[✉] and T. Sobek

Department of Computer Science, Ostfalia University of Applied Sciences,
38302 Wolfenbüttel, Germany
f.hoeppner@ostfalia.de

Abstract. Common time series similarity measures that operate on the full series (like Euclidean distance or Dynamic Time Warping DTW) do not correspond well to the visual similarity as perceived by a human. Based on the interval tree of scale, we propose a multiscale Bezier representation of time series, that supports the definition of elastic similarity measures that overcome this problem. With this representation the matching can be performed efficiently as similarity is measured segment-wise rather than element-wise (as with DTW). We effectively restrict the set of warping paths considered by DTW and the results do not only correspond better to the analysts intuition but improve the accuracy in the standard 1NN time series classification.

1 Introduction

Time series analysis almost always starts with inspecting plots. But does a visually perceived similarity correspond to the similarity determined by common measures? We consider this as important whenever a domain expert, who has familiarized herself with the data earlier, wants to match her own knowledge and own expectations to the results of some data mining process or an exploratory analysis. Prominent approaches to time series similarity do not care that much about this correspondence and the results may thus be misinterpreted easily.

With this problem in mind, a new multiscale representation of time series is proposed, not to compress the series in the first place, but to capture its characteristic traits. It simplifies the definition of a notion of similarity that corresponds to our cognition. While we will showcase its usefulness in the context of classification, we see it as a versatile and helpful representation that enables new and carries over to existing approaches. As an example, a new efficient elastic similarity measure is derived directly from this representation. Although we focus on the *series as a whole* here – rather than subsequences as with shapelets or bag-of-words approaches – the representation may nevertheless be used as a starting point for a meaningful segmentation and subpattern-based approaches.

The main contribution of this paper is a new Bezier spline-based, multiscale time series representation, which (a) encodes many different, continuous representations of the same series, thereby capturing many possible alternative views

M. Ceci et al. (Eds.): ECML PKDD 2017, Part II, LNAI 10535, pp. 461–477, 2017.
https://doi.org/10.1007/978-3-319-71246-8_28

and mimicking the ambiguity in human perception. (b) It directly supports (no need to access the raw data) elastic similarity measures, which can be computed efficiently and finally (c) performs more than competitive to DTW.

The paper is organized as follows. In the next section we briefly review some existing time series representations and also dynamic time warping as the most prominent elastic similarity measure. We recall the so-called Interval Tree of Scales (IToS), which serves as a basis for the new time series representation. Then, in Sect. 3 we identify two drawbacks shared by most elastic measures and how we intend to use the IToS to overcome these problems. The new representation is presented in Sect. 4, including a proposal how to use it for an elastic similarity measure. The representation is evaluated, amongst others, on time series classification problems in Sect. 5.

2   Related Work

2.1   Similarity Measures and Time Series Representations

Similarity Measures. Given two series $\mathbf{x}$ and $\mathbf{y}$, Euclidean distance $d^2(\mathbf{x}, \mathbf{y}) = \sum_i (x_i - y_i)^2$ assumes a perfect alignment of both series as only values with the same time index are compared. If the series are not aligned, x_i might be better compared with some $y_{m(i)}$ where $m : \mathbb{N} \to \mathbb{N}$ is a monotonic index mapping. With dynamic time warping (DTW) [2] the optimal warping path m, that minimizes the Euclidean distance of $\mathbf{x}$ to a warped version of $\mathbf{y}$, is determined. An example warping path m is shown in Fig. 2(left): Two series are shown along the two axes; $\mathbf{x}$ on the left, $\mathbf{y}$ at the bottom. The matrix enclosed by both series encodes the warping path: an index pair (i, j) on the warping path denotes that x_i is mapped to $y_j = y_{m(i)}$. All pairs (i, j) of the warping path, starting at index pair $(0, 0)$ and leading to (m, m), contribute to the overall DTW distance. Despite the fact that DTW is rather old, recent studies [14] still recommend it as the best measure on average over a large range of datasets.

Time Series Representations. While a time series may consist of many values (high-dimensional), consecutive values typically correlate strongly. So we may seek for a more compact representation, such as piecewise polynomials. This requires a segmentation of the whole series, which may be adaptive (segments of varying length to approximate the data best) or static (subdivide the series into equal-length segments) [14]. Symbolic approaches may replace (short segments of) numeric values by a symbol, shifting the problem into the domain of strings (e.g. SAX [8]). *Bag of words* representations consider the series as a set of substrings as in text retrieval, but do not allow a reconstruction of an approximation of the original series as all positional information gets lost. While most approaches focus on approximation quality, compression rate, or bounding of the Euclidean distance, a few explicitly care about the perception of time series. The Landmark Model focuses on the extrema of time series to define similarity that is *consistent with human intuition* [11]. The extrema are also considered as *important points* in [12]. With noisy data, both approaches skip some extrema based on a priori defined thresholds. This is typical for smoothing operations, but it is difficult to come up with such a fixed threshold, because different degrees of smoothing may be advisable for different parts of the series. Too much smoothing bears the danger of smearing out important features, too little smoothing may draw off the attention from the relevant features.

Bezier Curves. Time series representations usually approximate the raw series. Polynomials are most frequently used, e.g. piecewise constant segments [8], linear segments [9], cubic segments or splines. The polynomials are defined as functions of time t or time series index i, such that $(t, f(t))$ resembles the approximation.

Bezier curves are common in computer graphics but not in data analysis. They are used to represent arbitrary curves in the plane and thus are not necessarily functional. A Bezier curve $(B_t(\tau), B_q(\tau))$ is a τ-parameterized 2D-curve, $\tau \in [0, 1]$, defined by two cubic polynomials (separately for the time and value domain). For given coefficients $\mathbf{c} = (c_0, c_1, c_2, c_3) \in \mathbb{R}^4$, $B_{\mathbf{c}}(\tau)$ is defined as

$$B_{\mathbf{c}}(\tau) = (1 - \tau)^3 c_0 + 3(1 - \tau)^2 \tau c_1 + 3(1 - \tau)\tau^2 c_2 + \tau^3 c_3 \qquad (1)$$

2.2 Interval Tree of Scales (IToS)

The relevance of extrema has been recognized early and by many authors. Extrema are, however, also introduced by noise, so we have to find means to distinguish extrema of different importance. Witkin was one of the first who recognized the usefulness of a scale-space representation of time series [15]. The *scale s* denotes the degree of smoothing (std. dev. of Gaussian filter) that is applied to the time series. The scale-space representation of a series depicts the location of extrema (or inflection points) as the scale s increases (cf. Fig. 1(left) for the time series shown at the bottom). An extremum can be tracked from the original series ($s \approx 0$) to the scale s at which it vanishes (where it gets smoothed away), indicating its persistence against smoothing.

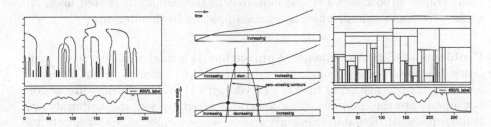

Fig. 1. Left: Depending on the variance of a Gaussian smoothing filter (vertical axis, logarithmic) the number and position of zero crossings in the first derivative varies. Mid: The zero-crossings of the first derivative (extrema in the original series) vanish pairwise. Right: Interval tree of scales obtained from left figure.

Zero-crossings vanish pairwise, three consecutive segments (e.g. increasing, decreasing, increasing) turn into a single segment (e.g. increasing), cf. Fig. 1(middle). The scale-space representation can thus be understood as a ternary tree of time series segments where the location of zero-crossings determine the temporal extent of the segment and the (dis-) appearance of zero-crossings limit the (vertical) extent or lifetime of a segment. By tracing the position of an extrema in the scale-space back to the position at $s \approx 0$ we can compensate the displacement caused by smoothing itself. We construct a so-called interval tree of scales [15] (cf. Fig. 1(right)), where the lifetime of a monotonic time series segment is represented by a box in the scale-space: its horizontal extent denotes the position of this segment in the series, the vertical extent denotes the stability or resistance against smoothing. The tree represents the time series at multiple scales and allows to take different views on the same series. It can be obtained by applying Gaussian filters or by means of the wavelet transform à trous [10].

We found the IToS already useful in a visualization technique to highlight discriminating parts of a class of time series [13]. But all comparisons had to be conducted directly on the raw data, which was time-consuming. Here, we propose a new general and efficient representation that supports various notions of elastic similarity (and also speeds up [13]). In the remainder of this work, we refer to the elements of the tessellation (rectangles) as *tiles*. A tile v covers a temporal range $[t_1^v, t_2^v]$ ($t_1^v < t_2^v$) and a scale range $[s_1^v, s_2^v]$ ($s_1^v < s_2^v$) and has an orientation $o \in \{increasing, decreasing\}$. By $\mathbf{x}|_v$ we denote the segment $\mathbf{x}|_{[t_1^v, t_2^v]}$ from the original series $\mathbf{x}$ covered by tile v.

3   Visual Perception of Series – Problems and Motivation

We want to draw the reader's attention to two problems with measures of the DTW-kind, the most prominent representative of elastic measures. In Fig. 2(middle) we have two similar series (linearly decreasing, incr., decr. segments), depicted in red and green. Both series are also shown on the x- and y-axis in the leftmost figure, together with the warping path. How the warping maps points of both series is also indicated in the middle by dotted lines. All series were standardized, which is common practice in the literature.

Problems of Existing Elastic Approaches. First, elastic time series similarity measures treat value and time differently: while a monotone but otherwise arbitrary transformation of time is allowed (with DTW), the values are usually not adapted to the comparison but are transformed a priori and uniformly (via standardization). But once we warp a time series locally, this also affects the value distribution and thus mean and variance: The local maximum of the red curve is mapped to many points of the green curve (cf. dotted assignment), so its local maximum occurs more often in the warped path than in the original red series. The initial mean and variance do not have much in common with the mean and variance of the warped curve, so why use the a priori values anyway?

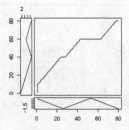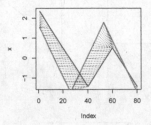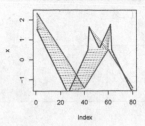

Fig. 2. Behaviour of time warping distances. Left: warping path of two time series (also shown in green and red in the middle). Mid and right: examples series; dotted lines indicate the DTW assignment. Right: Pair of blue and red series; structurally different, but with the same distance as green and red series. (Color figure online)

Second, elastic measures allow almost arbitrary (monotonic) warping, which may lead to some surprising results. If we would ask a human to align the red and green series in Fig. 2, an alignment of the local extrema would be natural, revealing the high similarity of both series as they behave identically between the local extrema. The local maximum m of the red curve (near $t = 60$), however, lies below the local maximum of the green curve, so all DTW approaches assign the red maximum to *all points of the green curve above* m. As a consequence, if we shuffle or reorder the green data above m (cf. rightmost subfigure, blue curve), neither the assignment nor the distance changes. This is in contrast to the human perception, we would never consider the blue series being as similar to the red series as the green.

Envisaged Use of the IToS. So we focus on two problems: (1) emphasis of local temporal scaling but ignorance of local vertical scaling, (2) arbitrary warping paths instead of warping that matches landmarks.

Regarding (2) we propose to utilize the IToS as a regularizer. We consider the IToS as a representation of all possible perceptions of a time series. Taking a finer or coarser look at some part of the series corresponds to selecting different tiles within the same time range but at different scales. Consider for example Fig. 3(right): If we intentionally ignore some fine-grained details of the series in the gray box we perceive it only as a single decreasing segment. By choosing tiles (from the IToS below the series) at different scales (b_4 rather than $b_3 b_5 b_6$) we adjust the level of detail and either ignore or consider local extrema. Any sequence of tiles (adjacent in time and covering the whole series) corresponds to a different perception of the series. Instead of allowing an arbitrary (monotonic) assignment of data points as with DTW, we propose to assign tiles to each other. In case of Fig. 3 we may choose perception a_0, a_1, a_2, a_4 for the left and b_1, b_4, b_7, b_8 for the right series and match them accordingly ($a_0 : b_1, a_1 : b_4, \ldots$).

Regarding (1) we propose to apply linear scaling to the temporal domain **and** the value domain. Continuing our example, we may match (decreasing) tiles a_1 and b_4. We align both start times and linearly rescale time to match the

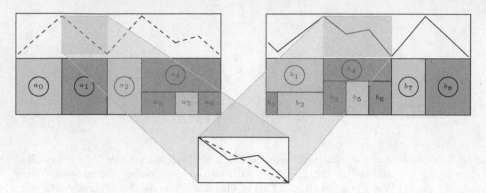

Fig. 3. The IToS for two time series. One possible perception of each time series is marked by circles. Matching time series is then accomplished by matching adjacent tiles subsequently.

temporal extent of both tiles. This is justified by the observations of different authors [5–7] that arbitrary warping is seldomly justified but linear warping suffices. In the same fashion we align the starting values and linearly rescale the vertical axis (replacing standardization) to have the same value range as illustrated at the bottom of Fig. 3. Once this alignment has taken place, we may measure the Euclidean distance between both series. This procedure limits the warping capabilities to *perceptually meaningful paths* and focuses on the shape of the tiles rather than (questionable) standardized absolute values. In contrast to other landmark approaches we want to keep the ambiguity in the perception of a time series by sticking to the multiscale representation.

4   A New Time Series Representation

The IToS just captures the temporal and vertical extent of monotone segments. In this section, we extend the IToS such that it fully represents the underlying series. The goal is to have all information readily available to measure similarity.

4.1   Multiscale Bezier-Representation of Time Series

We propose to attach a compact representation of an appropriately smoothed time series to each tile. This is shown in Fig. 4 for one series from the Beef dataset. At the top, the IToS is depicted. The light blue and red colors indicate increasing and decreasing segments. At the bottom, the original time series is shown, but this is hardly recognizable because for each tile from the IToS above we show an approximation of the corresponding segment (each in different color). Note that the segments connect seamlessly: for any sequence of adjacent tiles in the IToS we find a corresponding, smooth representation of the time series (a possible perception of the series, cf. Sect. 3).

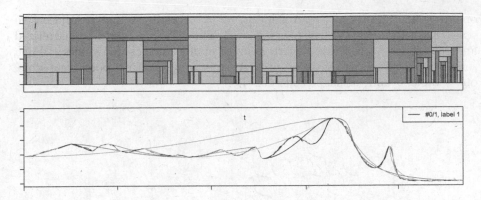

Fig. 4. Interval Tree of Scales for one series from the Beef dataset together with Bezier spline approximations. (Color figure online)

To achieve such a representation we cannot simply apply standard curve fitting to $\mathbf{x}|_v$, because that would ignore our background knowledge: By definition the tile boundaries mark local extrema of the series, so each segment has to be monotonic. From this alone a polynomial of degree 3 is uniquely determined (cf. [11]). However, one can actually think of many different shapes for an increasing segment, with a uniquely determined polynomial we would not be able to distinguish different shapes like those in Fig. 5 from one another. To better adapt to the data, we settle on Bezier curves, which are widely used in computer graphics. A Bezier curve (of degree 3) allows us to define a polynomial for both dimensions, the value domain as well as the time domain, which gives us the desired flexibility to adapt the approximations to the data as shown in Fig. 5.

These constraints are a strong regularizer of the fit. Note the last large peak on the right in Fig. 4 and the Bezier segments that correspond to the upmost two tiles in the IToS: Due to the constraints on monotonicity and start/end point the curve essentially ignores the peak to approximate the remaining data best. This is desired as the upmost tiles correspond to a smoothing level where the peak no longer exists. The associated Bezier curve does not contain this peak but still represent the original data well. This is quite different from a series we would obtain by actually smoothing the data (features would be smeared out and displaced).

The proposed time series representation is therefore as follows: Given a time series $\mathbf{x}$ and its IToS. The **Interval Tree of Bezier Segments (IToBS)** representation of $\mathbf{x}$ is a set V of tiles $v = (t_1, t_4, s_1, s_2, y_1, y_2, t_2, t_3, \sigma) \in V$, where $[t_1, t_4]$ denotes the temporal range of the tile, $[s_1, s_2]$ the scale range (both from the IToS) and $[y_1, y_1, y_2, y_2]$ / $[t_1, t_2, t_3, t_4]$ denote the parameters of the Bezier curve that approximates $\mathbf{x}|_v$ with standard deviation σ of the residuals. The reason why the coefficients of $B_{\mathbf{v}}(\cdot)$ require only the two values y_1 and y_2 will be explained in Sect. 4.3. The definition of a tile in the IToBS as a 9-tuple is convenient, but it is also highly redundant. Tiles v and w adjacent in time share one point in time $t_4^v = t_1^w$ and value $y_2^v = y_1^w$. For tiles v and w adjacent in scale (w is child-node of v) we have $s_2^v = s_1^w$. The values t_2, t_3 will be discretized and replaced by a code later (with a small number of possible values). When serializing the IToBS to disk, we need to store only one time point, one scale, one y-value, σ and the shape-code per tile.

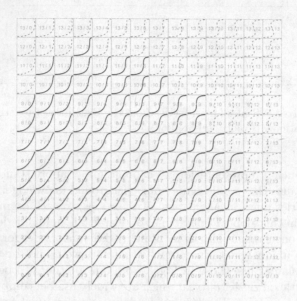

Fig. 5. Monotonically increasing Bezier curves with parameters $[0, 0, 1, 1]/[0, \frac{i}{10}, 1 - \frac{j}{10}, 1]$. For each spline the value of i/j is shown in the center. The dashed, red curves indicate parameter configurations that lead to non-monotonic or non-functional curves. (Color figure online)

4.2   Euclidean Tile Distance

The key idea of the tile distance (Sect. 3 and Fig. 3) was to rescale the content, i.e. the corresponding time series segments which will now be represented by Bezier segments, such that they fit into the same bounding box. We have not yet decided which size of the bounding box we actually want to use. But for any two functions $f(t)$ and $g(t)$ representing fitted segments, an affine transformation of values leads to a linearly scaled Euclidean distance (area between curves):

$$\int_{t_0}^{t_1} |(a \cdot f(t) + b) - (a \cdot g(t) + b)|\, dt = |a| \cdot \int_{t_0}^{t_1} |f(t) - g(t)|dt$$

The same holds for an affine transformation of time (integration by substitution). Thus, regardless of the bounding box we choose, whether we fit f to the original frame of g, the other way round or any other bounding box, we convert distances for different bounding boxes by applying linear scaling factors. So we determine *reference distances* using $[0,1]^2$ as the default bounding box (and use them to derive distances for any other bounding box).

We thus consider distances of Bezier segments rescaled to the unit square only. Applying such an affine transformation (in value and time) to a Bezier segment is simple: we have to apply the affine transformation to the Bezier curve coefficients. To rescale a monotonically increasing Bezier curve $[t_1, t_2, t_3, t_4]/[y_1, y_1, y_2, y_2]$ to the unit square, we obtain the coefficients as $[0, \tau_2, \tau_3, 1] / [0, 0, 1, 1]$ with $\tau_2 = \frac{t_2}{t_4 - t_1}$, $\tau_3 = \frac{t_3}{t_4 - t_1}$. For a decreasing segment, the value coefficients would be $[1, 1, 0, 0]$. From our intended application in Sect. 3 it makes only sense to compare tiles of the same orientation (both increasing or both decreasing), so the value coefficients of *both Bezier segments* to be compared will be either [0,0,1,1] or [1,1,0,0]. Switching between these two configurations again corresponds to an affine value transformation $1 - y$, the curve gets mirrored on the horizontal axis. The Euclidean distance between the curves is thus unaffected by this mirroring (scaling factor 1), so whenever we want to calculate the tile distance of two decreasing splines we could as well consider their mirrored, increasing counterparts and obtain the same distance.

At this point we know that it is sufficient to consider distances in the unit square for increasing curves. If we have efficient means to calculate tile distances for this case, we can provide arbitrary tile distances for arbitrary bounding boxes. While measuring Euclidean distance is a cheap operation on indexed time series (as x_i is aligned with y_i), it is an expensive operation with Bezier curves, because the temporal component $B_t(\tau)$ of tile v from series $\mathbf{x}$ is not necessarily aligned with that of tile w from series $\mathbf{y}$. Due to different cubic polynomials for $\mathbf{t}_v$ and $\mathbf{t}_w$ in the temporal dimension, we first have to identify parameters τ' and τ'' such that $B_{\mathbf{t}_v}(\tau')$ and $B_{\mathbf{t}_w}(\tau'')$ refer to the same point in time. We cannot afford to perform such expensive operations whenever we need to calculate a tile distance (it becomes a core operation in the similarity measure). We therefore suggest to discretize the possible values of τ_2 and τ_3 (say, consider values $0, 0.05, \ldots, 1$) and pre-calculate all possible tile distances offline. At a resolution of $R = 0.05$ we have about 500 possible pairs of τ_2/τ_3, so a lookup table that stores the tile distances between any two tiles consists of roughly 250.000 entries, which is small enough to fit into main memory. From this lookup table we can immediately return the distance of any two tiles on any chosen bounding box.

4.3   Determine the Bezier Segments

In this section we discuss how to obtain the Bezier parameters needed for the IToBS. While piecewise monotone approximations have been investigated for polynomials (e.g. [4]), this does not seem to be the case for Bezier curves. We assume that a monotonic time series segment with observations $\mathbf{s} = (x_i, y_i)_{1 \leq i \leq n}$ is given. We seek coefficients $\mathbf{t}, \mathbf{q} \in \mathbb{R}^4$ of a (cubic) Bezier curve $(B_t(\tau), B_q(\tau))_{\tau \in [0,1]}$ (cf. Eq. (1)) to approximate $\mathbf{s}$. First, we determine the Bezier coefficients $\mathbf{q} = (q_0, q_1, q_2, q_3)$ for the value domain. We want to preserve the extrema of the original function and thus require $B_\mathbf{q}(0) = y_1$ and $B_\mathbf{q}(1) = y_n$ and obtain $q_0 = y_1$ and $q_3 = y_n$. Furthermore we know that y_1 and

y_n are the extremá of the segment – so the gradient vanishes at $t = 0$ and $t = 1$:
$B_{\mathbf{q}}'(0) = 0 = B_{\mathbf{q}}'(1)$:

$$B_{\mathbf{q}}'(0) = 3(q_1 - q_0) = 0 \quad \Rightarrow \quad q_1 = q_0 \; (= y_1)$$
$$B_{\mathbf{q}}'(1) = 3(q_3 - q_2) = 0 \quad \Rightarrow \quad q_2 = q_3 \; (= y_n)$$

So $\mathbf{q}$ is fully determined as $\mathbf{q} = (y_1, y_1, y_n, y_n)$. (As already mentioned we have no degrees of freedom to adopt the Bezier spline $B_{\mathbf{q}}(\tau)$ to the data at hand.)

Second, we determine coefficients $\mathbf{t} = (t_0, t_1, t_2, t_3)$ for the time domain. Similar arguments as above lead us quickly to $t_0 = x_1$ and $t_3 = x_n$. By means of t_1 and t_2 we can adopt the shape of the Bezier curve to the time series segment. As $B_{\mathbf{q}}(\tau)$ is already fixed, we first identify a series τ_i such that $B_{\mathbf{q}}(\tau_i) = y_i$. The value τ_i may be found by solving a cubic equation or by a bisection method. There is a solution to this equation because for all i we have $y_i \in B_{\mathbf{q}}([0,1]) = [y_1, y_n]$. To optimize the fit we now seek a vector $\mathbf{t}$ that warps the temporal domain, that is, $B_{\mathbf{t}}(\tau_i)$ approximates x_i. So we deal with a regression problem that minimizes

$$f(\mathbf{t} \,|\, \mathbf{s}) = \sum_i (B_{\mathbf{t}}(\tau_i) - x_i)^2 \tag{2}$$

where $\mathbf{t}$ has only two degrees of freedom left (t_1 and t_2). Introducing the abbreviations $\delta_{ij} := t_i - t_j$, i.e., $t_1 = t_0 + \delta_{10}$, $t_2 = t_0 + \delta_{10} + \delta_{21}$, the optimal fit is then obtained from the linear equation system (with unknowns δ_{10}, δ_{21}):

$$\begin{aligned} 0 &= \sum_i 3(\tau_i^4 - 2\tau_i^5 + \tau_i^6)\delta_{21} + 3(\tau_i^3 - 2\tau_i^4 + \tau_i^5)\delta_{10} \\ &\quad + (\tau_i^5 - \tau_i^6)\delta_{30} + (\tau_i^2 - \tau_i^3)t_0 + (\tau_i^3 - \tau_i^2)x_i \\ 0 &= \sum_i 3(\tau_i^3 - 2\tau_i^4 + \tau_i^5)\delta_{21} + 3(\tau_i^2 - 2\tau_i^3 + \tau_i^4)\delta_{10} \\ &\quad + (\tau_i^4 - \tau_i^5)\delta_{30} + (\tau_i - \tau_i^2)t_0 + (\tau_i^2 - \tau_i)x_i \end{aligned} \tag{3}$$

Functional Curves. The regression problem provides us the global minimizer for the missing parameters δ_{10} and δ_{21}, but the solution to the regression problem may not fit our needs: In order to model a time series segment, the temporal component $B_{\mathbf{t}}(\tau)$ must be strictly increasing, otherwise the resulting Bezier spline $(B_{\mathbf{t}}(\tau), B_{\mathbf{q}}(\tau))$ might not be functional (cf. red cases in Fig. 5). Rather than an unconstrained optimization we actually have to impose a constraint on the monotonicity of $B_{\mathbf{t}}(\tau)$ within $[0,1]$. We require a non-negative first derivative

$$\frac{1}{3}B_{\mathbf{t}}'(\tau) = (\delta_{30} - 3\delta_{21}) \cdot \tau^2 + 2(\delta_{21} - \delta_{10}) \cdot \tau + \delta_{10} \geq 0 \quad \text{for } \tau \in [0,1].$$

We find the solution to this constrained problem by solving the unconstrained problem first (cf. Eq. (3)) and then, if it does not yield a monotonic $B_{\mathbf{t}}(\tau)$, find the minimizers of (2) among all boundary cases where $B_{\mathbf{t}}'(\tau)$ vanishes for some $\tau' \in [0,1]$. From the continuity of $B_{\mathbf{t}}'(\tau)$ and the fact that it must not become negative, we can conclude that either τ' has to be at the boundary of its valid range $[0,1]$, that is $\tau \in \{0,1\}$, or $B_{\mathbf{t}}'(\tau')$ must be a saddle point.

From $\tau \in \{0, 1\}$ we can conclude (by straightforward considerations, dropped due to lack of space) that either $\delta_{10} = 0$, $\delta_{21} = \delta_{30}/3$, or $\delta_{21} = \delta_{30} - 2\delta_{10}$, which yields three linear regression problems with a single parameter only. In case of a saddle point we can conclude $\delta_{21}^2 = \delta_{10}\delta_{30}$, which does unfortunately not lead to a linear regression problem. Instead of solving a quadratic optimization problem, we take the resolution at which the Bezier curves will be considered in the lookup-table into account, so we simply explore the possible values (according to the resolution R) for δ_{10}, set $\delta_{21} = \sqrt{\delta_{10}\delta_{30}}$ and finally pick the best boundary solution. This is done for all time series to turn the raw series into the IToBS representation, which is then stored to disk and for the subsequent operations we do not access the raw data.

4.4   A Tile Distance to Mimic Perceived Similarity

We can conveniently lookup tile distances on *arbitrary bounding boxes*, but to employ it in other contexts we have to settle on a *concrete distance*. Trying to mimic the human perception we take as much information into account as provided by the IToBS. But it is heuristic in nature and many other definitions are possible. The first aspect is the shape of both tiles: Let $d_{US}(v, w)$ denote the Euclidean distance between the tiles v and w in the *unit square* (retrieved from a lookup-table). To include the goodness-of-fit of both approximations we include their standard deviations $\sigma_v + \sigma_w$ (as obtained in the unit square). As the term $d_{US}(v, w) + \sigma_v + \sigma_w$ is independent of the tile dimensions (projection to the unit square), we choose a bounding box with the maximum of both tile durations and the maximum of both value ranges. This ensures that longer segments yield (potentially) larger distances than shorter segments and also establishes symmetry (by taking the maximum extent of both tiles). By $\max_{\Delta t}$ we denote the maximum temporal extent of both tiles, analogously for min and index y (value range) and s (scale). As already discussed, we obtain the distance in this bounding box by multiplying with the extents of the box ($\max_{\Delta t}, \max_{\Delta y}$):

$$\max_{\Delta t} \cdot \max_{\Delta y} \cdot (d_{US}(v, w) + \sigma_v + \sigma_w) \tag{4}$$

So far we account for the shape and approximation quality. But after rescaling, a flat and short segment may appear similar to a tall and long segment – but visually we would not perceive them as similar. Therefore we include penalty factors for differences in duration, vertical extent and *stability*, that is, their persistence against smoothing in the scale space. The latter is an aspect of similarity that is unique to multiscale approaches such as IToS. We penalize distances by a factor of 2 if one segment is twice as long, tall, or important (persistent) as the other:

$$TD(v, w) = \max_{\Delta t} \cdot \max_{\Delta y} \cdot (d_{US}(v, w) + \sigma_v + \sigma_w) \cdot \frac{\max_{\Delta t}}{\min_{\Delta t}} \cdot \frac{\max_{\Delta y}}{\min_{\Delta y}} \cdot \frac{\max_{\Delta s}}{\min_{\Delta s}}$$

4.5   Elastic Measure Based on IToBS

Once we have settled on a tile distance, we finally approach the definition of a distance for full time series. From a series $\mathbf{x}$ we obtain all of its tiles $V^{\mathbf{x}}$. (As long as only one time series is involved in the discussion, we drop the superscript $\mathbf{x}$.) Based on the tiles we can build a graph $G = (V, E)$ by connecting adjacent tiles (we have an edge $v \rightarrow w$, $v, w \in V$, iff $t_4^v = t_1^w$). We define a subset $V_S \subseteq V$ (resp. $V_E \subseteq V$) that contains all start-tiles (resp. end-tiles), that is, tiles which do not have a predecessor (resp. successor) in the graph. In the example of Fig. 6, the interval tree of the series on the left (resp. bottom) consists of 7 (resp. 9) tiles. The graphs are superimposed on the interval tree. Nodes belonging to V_S (or V_E) are connected to the virtual node S (or E). From these examples we find two perception for the left series $((a_0, a_1, a_2, a_4)$ and $(a_0, a_1, a_2, a_3, a_5, a_6))$ and several for the bottom series (e.g. $(b_0, b_2, b_4, b_7, b_8)$).

Among all perceptions of series $\mathbf{x}$ and $\mathbf{y}$ we want to identify those that lead to the smallest overall accumulated tile distance. As with DTW this requires a mapping $m(\cdot)$ of a tile $v^{\mathbf{x}}$ from the perception of series $\mathbf{x}$ to a tile $w^{\mathbf{y}}$ from a perception of series $\mathbf{y}$: $m(v^{\mathbf{x}}) = w^{\mathbf{y}}$. Unlike DTW we do not allow $m(\cdot)$ to skip tiles. When perceiving all details (path near the bottom of the interval tree), both series may not match structurally in the number of tiles: In Fig. 6 the series $\mathbf{x}$ (left) contains a landmark at the end (peak at a_5, a_6) while the other series $\mathbf{y}$ (bottom) does not. To perceive them as similar, we switch to a coarser scale for $\mathbf{x}$ (a_4 instead of a_3, a_5, a_6). Finding the path through both graphs $G_{\mathbf{x}}$ and $G_{\mathbf{y}}$, such that the tiles in both sequences correspond to each other best, corresponds to a structural comparison of the series.

The alignment-task is solved using dynamic programming (cf. Fig. 7), very much in the fashion of DTW. The classic DTW solution uses a distance matrix $D[i, j]$ to store the lowest accumulated cost of a warping path that matches $\mathbf{x}|_{1:i}$ to $\mathbf{y}|_{1:j}$. The subscripts refer to the time indices of the series. Now tiles take the role of points – but it is not possible to match arbitrary tiles, they have to have the same orientation and they must be adjacent in the graph. In Fig. 6 the hatched entries in the matrix denote illegal assignments due to incompatible tile orientation. An associative array L plays the role of a distance matrix in Fig. 7, line 2. To identify the minimal cost to reach an arbitrary pair $(v^{\mathbf{x}}, w^{\mathbf{y}})$ of tiles we have to consider how these tiles were reached: We need a predecessor $r^{\mathbf{x}}$ of $v^{\mathbf{x}}$ and a predecessor $s^{\mathbf{y}}$ of $w^{\mathbf{y}}$. Among all possible predecessor-pairs $(r^{\mathbf{x}}, s^{\mathbf{y}})$ leading to $(v^{\mathbf{x}}, w^{\mathbf{y}})$ we determine the one with minimal cost and add the tile distance between $v^{\mathbf{x}}$ and $w^{\mathbf{y}}$ (line 6). We have to make sure that all predecessors $(r^{\mathbf{x}}, s^{\mathbf{y}})$ have been evaluated earlier, which is easily achieved by sorting the tiles by their start-point in time (line 1). A predecessor of a tile must end before the new tile starts, so if we deal with a tile starting at t all possible predecessors must have started at $t' < t$ and were processed earlier. Finally, the matching of both series is complete if we reach the end of both series, so we find the minimal cost at pairs $(v^{\mathbf{x}}, w^{\mathbf{y}}) \in V_E^{\mathbf{x}} \times V_E^{\mathbf{y}}$ (line 9). The complexity of the algorithm is $O(n \cdot m)$, n and m being the number of tiles in the resp. graph (rather than number of points as in DTW). Only a fraction of the $n \cdot m$ tile combinations

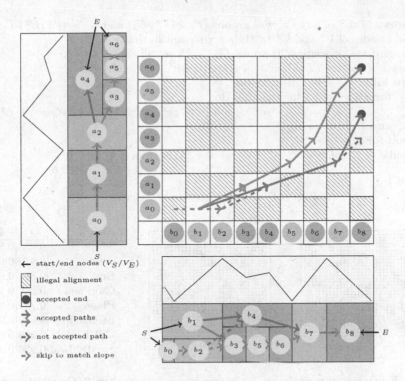

Fig. 6. Two series (left and bottom), represented by their interval tree and respective graph (slope is color-coded). The matrix in the center encodes the assignment of nodes from both graphs, a match of perceptions from both series is thus a path from the bottom left to top right edge of the matrix. (Color figure online)

will be considered (cf. hatched elements of Fig. 6). The tile distances need not be stored in D but may be calculated when needed as they are calculated in $O(1)$ thanks to the lookup-table of Sect. 4.2. We use the distance of Sect. 4.4, but the algorithm in Fig. 7 does not depend on a particular tile distance. To deal with series that start with different orientations ($\mathbf{x}$ starts only with increasing and $\mathbf{y}$ only with decreasing segments) we include virtual tiles of opposite orientation in V_S and V_E, which are considered as being of constant value y for distance calculations (where $y = y_1$ for virtual tiles in V_S and $y = y_n$ in V_E).

5 Evaluation

5.1 Sensitivity of Discretized Bezier Segments

The Bezier curves have been selected to keep the representation simple (polynomial of degree 3) but still flexible enough to adapt to the data. To increase efficiency, we consider only a discretized subset of all possible Bezier segments (cf. Fig. 5). Is this setting sensitive enough to capture differences in time series?

Require: IToBS of series $\mathbf{x}, \mathbf{y}$ as graphs $G^{\mathbf{x}} = (V^{\mathbf{x}}, E^{\mathbf{x}})$ and $G^{\mathbf{y}} = (V^{\mathbf{y}}, E^{\mathbf{y}})$

1: Sort nodes of $V^{\mathbf{x}}$ and $V^{\mathbf{y}}$ by their start point in time.
2: Instantiate associative 2D-arrays $L[v^{\mathbf{x}}, w^{\mathbf{y}}]$, set all entries to $+\infty$
3: Let $L[v^{\mathbf{x}}, w^{\mathbf{y}}] = TD(v^{\mathbf{x}}, w^{\mathbf{y}})$ for start tiles $v^{\mathbf{x}} \in V_S^{\mathbf{x}}$ and $w^{\mathbf{y}} \in V_S^{\mathbf{y}}$
4: **for all** $v^{\mathbf{x}} \notin V_S^{\mathbf{x}}$ in sort-order **do**
5:    **for all** $w^{\mathbf{y}} \notin V_S^{\mathbf{y}}$ in sort-order **do**
6:       $L[v^{\mathbf{x}}, w^{\mathbf{y}}] = TD(v^{\mathbf{x}}, w^{\mathbf{y}}) + \min\{L[r^{\mathbf{x}}, s^{\mathbf{y}}] \mid (r^{\mathbf{x}}, v^{\mathbf{x}}) \in E^{\mathbf{x}}, (s^{\mathbf{y}}, w^{\mathbf{y}}) \in E^{\mathbf{y}}\}$;
7:    **end for**
8: **end for**
9: find $d = \min\{L[v^{\mathbf{x}}, w^{\mathbf{y}}] \mid v^{\mathbf{x}} \in V_E^{\mathbf{x}}, v^{\mathbf{y}} \in V_E^{\mathbf{y}}\}$; **return** d

Fig. 7. Dynamic programming solution.

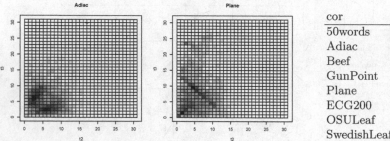

cor	1	2
50words	0.738	0.871
Adiac	0.979	0.975
Beef	0.828	0.985
GunPoint	0.907	0.909
Plane	0.797	0.942
ECG200	0.610	0.979
OSULeaf	0.908	0.965
SwedishLeaf	0.926	0.959

Fig. 8. Frequency of used shapes (as shown in Fig. 5) in the Adiac and Plane dataset.

Fig. 9. Correlation of distances: (1) raw vs BS distances, (2) raw vs BSN distances (cf. text).

The matrices of Fig. 8 show for each of the (discretized) segment types how often they occur in a specific dataset (Adiac and Plane [3]) at a resolution of 0.05. We can recognize that the frequency of segment types varies significantly in these datasets and the IToBS can thus capture differences between sets of time series.

5.2 Approximation of Euclidean Distance

Next we consider the approximation quality by the Bezier segments. We have calculated distances for valid pairs of tiles using (a) the Euclidean distance on the original time series segments without any approximation [raw], (b) the Euclidean distance using the approximated Bezier segments and the lookup table [BS] and (c) the Euclidean distance from a lookup-table including the standard deviation (as in Eq. (4)) [BSN]. Figure 9 shows how these distances typically correlate for some UCR datasets (cor. of (a) vs (b) in column 1, (a) vs (c) in column 2). The results show the (expected) loss in approximation quality when switching from the Euclidean distance on the raw data to the approximation by Bezier segments as the approximation eliminates details (column 1). But when including the approximation error in the tile distance the correlation coefficients increase and may get very close to 1 (column 2).

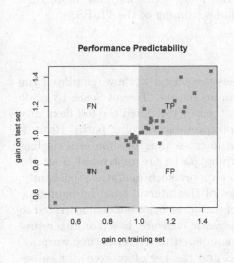

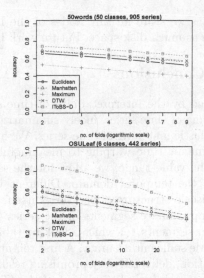

Fig. 10. Texas Sharp Shooter plot for the proposed elastic measure versus DTW.

Fig. 11. X-validated 1NN performance

5.3 Utility of Elastic Measure Based on IToBS

We demonstrate the utility of the IToBS/elastic measure in the standard 1-NN time series classification setting. We have performed cross-validated (CV) experiments with datasets from the UCR repository [3]. Figure 11 shows how the proposed approach (IToBS-D) compares against DTW [2] on two datasets. To eliminate the effect of the sample size in k-fold CV (one fold is used as training data, $k-1$ folds for testing), experiments were run over a full range of values for k. Both figures are examples where IToBS-D performs better than DTW for all k. Figure 10 shows the Texas Sharp Shooter plot [1] to demonstrate that the performance gain[1] over DTW obtained on the training set (x-axis) correlates with the gain for the test set (y-axis). Non-artificial UCR-datasets below 2 MB size were included in this figure. The number of folds was chosen such that the training set contained 5 examples from each class. For many cases near the center both measures perform comparable, but the top right quadrant (IToBS-D is better) contains more cases than the bottom left quadrant (DTW is better). The best performance is obtained for datasets where series from different classes have structural differences. If series are extremely similar across different classes, as it is the case with a few spectrogram datasets, very tiny differences must be exploited to distinguish the classes. These tiny differences may get lost in the Bezier approximation.

Compression was not our interest in the first place. Using a simple serialization, series with little noise use between $\frac{1}{3}$ and $\frac{1}{2}$ of the original disk space. As

---

[1] That is, how many times its accuracy is higher than that of DTW.

long as we store all tiny nodes of the IToBS, noisy series may use more than twice as much disk space. Future work includes pruning of the IToBS.

6   Conclusions

Guided by the interpretability of time series similarity, we have employed the extrema of time series, nicely summarized in an Interval Tree of Scale, to guide the elastic matching of time series. We have thereby restricted the full flexibility of dynamic time warping to meaningful, interpretable warping paths. Regarding the value range, we allow for the same linear scaling of time series values as in the temporal domain. Different warping paths are motivated from the multiscale structure of the series itself. To support such operations efficiently, we approximate segments of the series (tiles of the interval tree) by monotone Bezier segments, for which a look-up table of distances has been demonstrated to provide sufficient accuracy. A first tentative elastic measure, based on this representation, delivered very promising results compared to dynamic time warping. Future work includes pruning of the IToBS and the use of subtrees (or subsequences of tiles) as the basis for elastic shapelet- or bag of word approaches that directly connect to the visual perception.

References

1. Batista, G., Wang, X., Keogh, E.: A complexity-invariant distance measure for time series. In: Proceedings of International Conference on Data Mining, pp. 699–710 (2011)
2. Berndt, D.J., Clifford, J.: Finding patterns in time series: a dynamic programming approach. In: Advances in Knowledge Discovery and Data Mining, pp. 229–248 (1996). MIT Press. https://mitpress.mit.edu/books/advances-knowledge-discovery-and-data-mining
3. Chen, Y., Keogh, E., Hu, B., Begum, N., Bagnall, A., Mueen, A., Batista, G.: The UCR time series classification archive (2011)
4. Fritsch, F.N., Carlso, R.E.: Monotone picewise cubic interpolation. J. Numer. Anal. **17**(2), 238–246 (1980)
5. Höppner, F.: Less is more: similarity of time series under linear transformations. In: Proceedings of international Conference on Data Mining (SDM), pp. 560–568 (2014)
6. Keogh, E.: Efficiently finding arbitrarily scaled patterns in massive time series databases. In: Lavrač, N., Gamberger, D., Todorovski, L., Blockeel, H. (eds.) PKDD 2003. LNCS (LNAI), vol. 2838, pp. 253–265. Springer, Heidelberg (2003). https://doi.org/10.1007/978-3-540-39804-2_24
7. Keogh, E., Palpanas, T.: Indexing large human-motion databases. In: Proceedings of International Conference on Very Large Databases, pp. 780–791 (2004)
8. Lin, J., Keogh, E., Wei, L., Lonardi, S.: Experiencing SAX: a novel symbolic representation of time series. Data Min. Knowl. Discov. **15**(2), 107–144 (2007)
9. Malinowski, S., Guyet, T., Quiniou, R., Tavenard, R.: 1d-SAX: a novel symbolic representation for time series. In: Tucker, A., Höppner, F., Siebes, A., Swift, S. (eds.) IDA 2013. LNCS, vol. 8207, pp. 273–284. Springer, Heidelberg (2013). https://doi.org/10.1007/978-3-642-41398-8_24

10. Mallat, S.G.: A Wavelet Tour of Signal Processing. Elsevier, Amsterdam (2001)
11. Perng, C.-S., Wang, H., Zhang, S.R., Parker, D.S.: Landmarks: a new model for similarity-based pattern querying in time series databases. In: International Conference on Data Engineering (2000)
12. Pratt, K.B., Fink, E.: Search for patterns in compressed time series. Int. J. Image Graph. **02**(01), 89–106 (2002)
13. Sobek, T., Höppner, F.: Visual perception of discriminative landmarks in classified time series. In: Boström, H., Knobbe, A., Soares, C., Papapetrou, P. (eds.) IDA 2016. LNCS, vol. 9897, pp. 73–85. Springer, Cham (2016). https://doi.org/10.1007/978-3-319-46349-0_7
14. Wang, X., Mueen, A., Ding, H., Trajcevski, G., Scheuermann, P., Keogh, E.: Experimental comparison of representation methods and distance measures for time series data. Data Min. Knowl. Discov. **26**(2), 275–309 (2012)
15. Witkin, A.P.: Scale space filtering. In: Artifical Intelligence (IJCAI), Karlsruhe, Germany, pp. 1019–1022 (1983)

Arbitrated Ensemble for Time Series Forecasting

Vítor Cerqueira[1,2]([⊠]), Luís Torgo[1,2], Fábio Pinto[1,2], and Carlos Soares[1,2]

[1] University of Porto, Porto, Portugal
[2] INESC TEC, Porto, Portugal
{vmac,ltorgo,fhpinto}@inesctec.pt, csoares@fe.up.pt

Abstract. This paper proposes an ensemble method for time series forecasting tasks. Combining different forecasting models is a common approach to tackle these problems. State-of-the-art methods track the loss of the available models and adapt their weights accordingly. Metalearning strategies such as stacking are also used in these tasks. We propose a metalearning approach for adaptively combining forecasting models that specializes them across the time series. Our assumption is that different forecasting models have different areas of expertise and a varying relative performance. Moreover, many time series show recurring structures due to factors such as seasonality. Therefore, the ability of a method to deal with changes in relative performance of models as well as recurrent changes in the data distribution can be very useful in dynamic environments. Our approach is based on an ensemble of heterogeneous forecasters, arbitrated by a metalearning model. This strategy is designed to cope with the different dynamics of time series and quickly adapt the ensemble to regime changes. We validate our proposal using time series from several real world domains. Empirical results show the competitiveness of the method in comparison to state-of-the-art approaches for combining forecasters.

Keywords: Dynamic ensembles · Metalearning · Time series Numerical prediction · Reproducible research

1 Introduction

Time series forecasting is an important topic in the machine learning research community with several applications across different domains. Time series often comprise non-stationarities and time-evolving complex structures which difficult the forecasting process.

To cope with these issues a common approach is to combine several forecasting models. Combination strategies typically involve tracking the error of the available models and adaptively weigh them accordingly. Using stacking [30], a metalearning approach, to combine the available forecasting models is also a common approach. This method directly models inter-dependencies between models, which might be relevant to take into account the diversity among them [5].

In this paper we adopt a metalearning strategy to adaptively combine the available forecasting models. However, contrary to stacking, we model the individual expertise of each forecasting model and specialize them across the time

© Springer International Publishing AG 2017
M. Ceci et al. (Eds.): ECML PKDD 2017, Part II, LNAI 10535, pp. 478–494, 2017.
https://doi.org/10.1007/978-3-319-71246-8_29

series. Consequently, the forecasting models are combined in such a way that they are only picked for predicting in examples that they are good at. Moreover, as opposed to tracking the error on past instances, our combination approach is more proactive as it is based on predictions of future loss of models. This can result in·a faster adaptation to changes in the environment.

The motivation for our approach is that different learning models have different areas of expertise across the input space [5]. In time series forecasting there is evidence that forecasting models have a varying relative performance over time [2]. Moreover, it is also common for the underlying process generating the time series to have recurrent structures due to factors such as seasonality [12]. In this context, we hypothesize that the metalearning strategy enables the ensemble to better detect changes in the relative performance of models or changes between different regimes and quickly adapt itself to the environment.

Specifically, our proposed metalearning strategy, hereby denoted as ADE (Arbitrated Dynamic Ensemble), is based on an Arbitrating architecture [21]. In that sense, we build a meta-learner for each base-learner that is part of the ensemble. Each meta-learner is specifically designed to model how apt its base counterpart is to make a prediction for a given test example. This is accomplished by analysing how the error incurred by a given learning model relates to the characteristics of the data. At test time, the base-learners are weighted according to their degree of competence in the input observation, estimated by the predictions of the meta-learners. This is illustrated in Fig. 1.

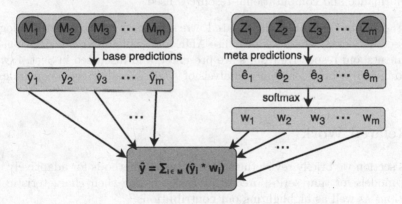

Fig. 1. Example scheme of the workflow of ADE for a new prediction. The base learners M produce the predictions $\hat{y}_i$, $i \in \{1, \ldots, m\}$ for the next value of the time series. In parallel, the meta learners Z produce the weights w of each base learner according to their predictions of error ($\hat{e}_i$). The final prediction $\hat{y}$ is computated using a weighted average of the predictions relative to the weights.

While a given base learner M_i is trained to model the future values of the time series, its metalearning associate Z_i is trained to model the error of M_i. Z_i then can make predictions regarding the error that M_i will incur when predicting the future values of the time series. In effect, the larger the estimates produced by Z_i (relative to the other models in the ensemble) the lower the weight of M_i will be in the combination rule. Although arbitrating models is not generally new [21], it was originally formulated for selecting classifiers. Moreover, we address several of its limitations which significantly improve its overall ability.

We validate the proposed method in 14 real-world time series. Empirical experiments suggest that our method is competitive against different adaptive methods for combining base-learners and other metalearning approaches such as stacking [30]. We note that all experiments are fully reproducible. Both the methods and time series data sets are publicly available as an R software package.

In summary, the contributions of this paper are:

- ADE, an arbitrated dynamic ensemble. Arbitrating was originally proposed for dynamic selection of classifiers [21], whereas we adapt it to time series forecasting tasks;
- We introduce a blocked prequential procedure in the Arbitrating approach to obtain out-of-bag predictions in the training set in order to increase the data used to train the metalearning model;
- A softmax function used to weight the forecasters. This function is commonly used in neural networks, however, to the best of our knowledge, it has not been applied yet to dynamically weight base-learners;
- An empirical study comparing different training strategies for the base and meta levels including a discussion about its implications in terms of predictive performance and computational resources used.

We start by outlining the related work in Sect. 2; the methodology is addressed in Sect. 3, where we formalise ADE and explain our contributions; the experiments and respective results are presented and discussed in Sect. 4, where we also include a brief scalability analysis of ADE. Finally, Sect. 5 concludes the paper.

2   Related Work

In this section we briefly revise the state-of-the-art methods for adaptively combining models for time series forecasting tasks, listing their characteristics and limitations as well as highlighting our contributions.

The simple average of the predictions of the available base-learners have been shown to be a robust combination method [7, 20, 26]. Nonetheless, other more sophisticated approaches have been proposed.

2.1   Windowing Strategies for Model Combination

AEC is a method for adaptively combining forecasters [25]. It uses an exponential re-weighting strategy to combine forecasters according to their past performance. It includes a forgetting factor to give more importance to recent values.

Timmermann argues that for the prediction of stock returns models have only short-lived periods of predictability [27]. He proposes an adaptive combination based on the recent R^2 of forecasters. If all models have poor explained variance (low R^2) in the recent observations then the forecast is set to the mean value of those observations. Otherwise, the base-learners are combined by averaging their predictions with the arithmetic mean.

Newbold and Granger proposed a method which weighs models in a linear way according to their performance in recent past data [19].

The outlined models are related to our work in the sense that they employ adaptive heuristics to combine forecasters. However, these heuristics are incremental or sliding summary statistics on relative past performance. Conversely, we explore differences among base-learners to specialise them across the data space. Moreover, we use a more proactive heuristic that is based on the prediction of relative future performance of individual forecasters.

2.2 Metalearning Strategies for Model Combination

Metalearning provides a way for modeling the learning process of a learning algorithm [4]. Several methods follow this approach to improve the combinination or selection of models [22, 30]. The most popular strategy to combine forecasters is similar to stacking [30]. Linear regression is used to adaptively estimate the weights of the base-learners [8].

Our proposal follows a metalearning strategy called arbitrating. This was introduced before for dynamic selection of classifiers [21] (Fig. 1). A prediction is made using a combination of different classifiers that are selected according to their expertise concerning the input data. The expertise of a model is learned using a meta-model, one for each available base classifier, which models the loss of its base counterpart. At runtime, the classifier with the highest confidence is selected to make a prediction. This work follows a first attempt on adapting the arbitration approach to numerical prediction tasks. This strategy was applied to solar radiation forecasting tasks, and the arbitration methodology is extended by weighing all forecasting models, as opposed to selecting the one with lowest predicted error [6]. In this paper we introduce other components to arbitrating. These address several of its drawbacks, such as its inefficient use of the available data, by using OOB samples from the training set; a more robust combination rule by using a committee of recent well performing models and weighing them using a softmax function; and the general translation to the time series forecasting tasks, which is fundamentally different than classification tasks.

3   Arbitrating for Time Series Forecasting

A time series Y is a temporal sequence of values $Y = \{y_1, y_2, \ldots, y_t\}$, where y_i is the value of Y at time i. We use time delay embedding to represent Y in an Euclidean space with embedding dimension K. Effectively, we construct a set of observations which are based on the past K lags of the time series.

This is accomplished by mapping the time series Y into the embedding vectors $V_{N-K+1} = \{\mathbf{v_1}, \mathbf{v_2}, \ldots \mathbf{v_{N-K+1}}\}$ where each $\mathbf{v_i} = \langle y_{i-(K-1)}, y_{i-(K-2)}, \ldots, y_i \rangle$.

The proposed methodology in ADE for time series forecasting settles on the three main steps: (i) An offline training step of M, the set of base-learners which are used to forecast future values of Y, and the online iterative steps: (ii) Training or updating of meta-learners Z, which model the expertise of base-learners, and (iii) prediction of y_{t+1} using M, dynamically weighed according to Z.

3.1   Learning Base-Level Models

In the offline learning phase we train m individual forecasters. Each $M^j, \forall j \in \{1, \ldots, m\}$ is built using the available time series Y_{tr}^K. The objective is to predict the next value of the series, y_{t+1}. We use the subscripts tr and ts to denote the training and testing sets, respectively.

By embedding the time series to an Euclidean space we are able to apply standard regression learning models. In this context, M is comprised by a set of heterogeneous models, such as Gaussian Processes and Neural Networks. Heterogeneous models have different inductive biases and assumptions regarding the process generating the data. Effectively, we expect models to have different expertise across the time series. The learning step is described in Algorithm 1.

3.2   Learning Meta-level Models

In the metalearning step of ADE the goal is to build algorithms capable of modelling the expertise of each model across the data-space. Our working hypothesis is that the ensemble can leverage individual learners with different inductive biases to better cope with the different regimes causing the time series.

Our assumption is that not all models will perform equally well at any given prediction point. This idea is in accordance with findings reported in prior work [2]. Systematic evidence was found that some models have varying relative performance over time and that other models are persistently good (or bad) throughout the time series. Furthermore, in many environments the dynamic concepts have a recurring nature, due to, for instance, seasonality. In effect, we use the meta learners to dynamically weigh base learners and adapt the combined model to changes in the relative performance of the base models, as well as for the presence of different regimes in the time series.

Algorithm 1. Learning M
___
Require: Available Time Series Y; Embedding Dimension K
1: Embed Y into Y_{tr}^K
2: **for all** $M^j \in M$ **do**
3:     train M^j using Y_{tr}^K
4: Return M

Our metalearning approach is based on an arbitrating architecture originally introduced in [21] (c.f. Sect. 2 for an explanation of arbitration). Specifically, a meta-learner Z^j, $\forall j \in \{1, \dots, m\}$ is trained to build the following model:

$$e^j = f(I) \tag{1}$$

where e^j is the absolute error incurred by M^j, I is the metafeature set and f is the regression function. The metafeatures are the embedding vectors, i.e., the past values of the time series.

We perform this regression analysis on a meta-level to understand how the error of a given model relates to the dynamics and the structure of the time series. Effectively, we can capitalise on this knowledge by dynamically combining base-learners according to the expectation of how they will perform.

Blocked Prequential for Out-of-Bag Predictions. In the original formulation of the Arbitrating strategy, the metalearning layer only starts at run-time, using only information from test observations [21]. This is motivated by the need for unbiased samples to build reliable metalearners. However, this means that at the beginning, few observations are available to train the meta learners, which might result in underfitting.

ADE uses the training set to produce out-of-bag (OOB) predictions which are then used to compute an unbiased estimate of the loss of each base learner. By retrieving OOB samples from the training set we are able to significantly increase the amount of data available to the meta learners. We hypothesize that this strategy improves the overall performance of the ensemble by improving the accuracy of each metalearner.

We produce OOB samples by running a blocked prequential procedure [9]. The available embedded time series used for training Y_{tr}^K is split into β equally-sized and sequential blocks of contiguous observations. In the first iteration, the first block is used to train the base learners M and the second is used to test them. Then, the second block is merged with the first one for training M and the third block is used for testing. This procedure continues until all blocks are tested. In summary, each metalearner uses all the information available up to the prediction point. However, each metalearner is trained every λ observations, so less computation is used to update the models.

Committee of Models for Prediction. As described earlier, the predictive performance of forecasting models has been reported to vary over a given time series. We address this issue with a committee of models, rather than selecting a single model [21], where we trim recently poor performing models from the combination rule for an upcoming prediction (e.g. [14]). Formally, we maintain the $\alpha\%$ base learners with lowest mean absolute error in the last Ω observations, trimming off the remaining ones. The predictions of the meta-level models are used to weigh the selected forecasters.

Additionally, if a base learner is consistently out of the committee, its meta-level pair is not trained. This saves computational resources. The meta-learning phase is described in Algorithm 2.

Algorithm 2. Metalearning Z

Require: Available observations at runtime Y_{ts}; Embedding Dimension K; Training
    Signal λ; meta-committee $^\alpha Z$
1: Embed Y_{ts} into $Y_{ts}^K \rightarrow I$
2: **for all** $Z^j \in {}^\alpha Z$ and not trained in the last training signal λ **do**
3:    train Z^j to model: $e^j = f(I)$
4: Return $^\alpha Z$

3.3 Predicting y_{t+1}

For a new observation y_{t+1}, ADE combines the base learners in M according to the meta information obtained from the models in Z to generate a prediction.

Initially we form the $^\alpha M$ committee with the $\alpha\%$ models in M with best performance in the last Ω observations. The corresponding metalearners are also discarded from the upcoming prediction so we also form the meta-committee $^\alpha Z$.

The weigh of a base learner M^j in $^\alpha M$ is given by the softmax of its negative predicted loss. This is formalised by the following equation:

$$w_{t+1}^j = \frac{exp(-\hat{e}^j)}{\sum_{j \in {}^\alpha M} exp(-\hat{e}^j)} \tag{2}$$

where $\hat{e}_{t+1}^j$ is the prediction made by $^\alpha Z^j$ for the absolute loss that $^\alpha M^j$ will incur in y_{t+1}, w_{t+1}^j is the weigh of M_j for observation y_{t+1} and exp denotes the exponential function. With the application of the softmax function, which is widely used in the modelling process of neural networks, the weigh of a given model decays exponentially as its predicted loss increases. We use this function (instead of a more traditional linear transformation) to further increase the influence of the predicted best performing models.

The final prediction is a weighted average of the predictions made by the base-learners $\hat{y}^j$ with respect to w_{t+1}^j: $\hat{y}_{t+1} = \sum_{j \in {}^\alpha M} \hat{y}_{t+1}^j \cdot w_{t+1}^j$.
The prediction step of ADE is described in Algorithm 3.

4 Experiments

In this section we present the experiments carried out to validate ADE. These address the following research questions:

Q1: How does the performance of the proposed method relates to the performance of the state-of-the-art methods for time series forecasting tasks and state-of-the-art methods for combining forecasting models?

Q2: Is it beneficial to use out-of-bag predictions from the training set to increase the data used to train the meta learners?

Q3: Is it beneficial to use a weighing scheme in our Arbitrating strategy instead of selecting the predicted best forecaster as originally proposed [21]?

Q4: How does the performance of ADE vary by the introduction of a committee, where poor recent base-learners are discarded from the upcoming prediction, as opposed to weighing all the models?

Q5: Does a non-linear weighting strategy using a softmax function produce better estimates instead of a traditional linear scaling?

Q6: How does the performance of ADE vary by using different updating strategies for the base and meta models?

To address these questions we used 14 real world time series from four different domains, briefly described in Table 1. In the interest of computational efficiency we truncated the time series to 2000 values.

Algorithm 3. Predicting y_{t+1}

Require: $K, M, Z, \alpha, \Omega, {}^{\alpha}M, {}^{\alpha}Z$
1: Embed the previous $K - 1$ values into Y_{t+1}^{K}
2: Get meta-predictions $\hat{e}_{t+1}^{j}$ from $Z^{j} \in {}^{\alpha}Z$
3: Compute weights $w_{t+1}^{j} = exp(-\hat{e}_{t+1}^{j})/\sum_{Z^{j} \subset {}^{\alpha}Z} exp(-\hat{e}_{t+1}^{j})$
4: Get predictions $\hat{y}_{t+1}^{j}$ from models $M^{j} \in {}^{\alpha}M$
5: Compute final prediction $\hat{y}_{t+1} = \sum_{j=1}^{m} \hat{y}_{t+1}^{j} \cdot w_{t+1}^{j}$
6: Add y_{t+1} to Y_{ts}
7: Update Committees ${}^{\alpha}M$ and ${}^{\alpha}Z$ according to α and Ω
8: Return $\hat{y}_{t+1}$ and go back to the metalearning step (Algorithm 2)

4.1   Experimental Setup

The methods used in the experiments were evaluated using the mean squared error (MSE) on 10 Monte Carlo repetitions. For each repetition, a random point in time is chosen from the full time window available for each series, and the previous window N consisting of 50% of the data set size is used for training the ensemble while the following window of size 30% is used for testing. The results obtained by the different methods were compared using the nonparametric Wilcoxon Signed Rank test. The experiments were carried out using performanceEstimation [28] R package.

We tested two different embedding dimensions by setting K to **7** and **15**.

4.2   Ensemble Setup and Baselines

The base-models M comprising the ensemble are the following:

Table 1. Datasets and respective summary

ID	Time series	Data source	Data characteristics
1	Electricity total load	Hospital energy loads [10]	Hourly values from Jan. 1, 2016 to Mar. 25, 2016
2	Equipment load		
3	Water heating load		
4	Gas energy		
5	Gas heat energy		
6	Ameal	Oporto water demand from different locations [1]	Half-hourly values from Feb. 6, 2013 to Mar. 19, 2013
7	Preciosa Mar		
8	Montes Burgos		
9	Global horiz. radiation	Solar radiation monitoring [3]	Hourly values from Feb. 16, 2016 to May 5, 2016
10	Direct normal radiation		
11	Diffuse horiz. radiation		
12	Sea level pressure	Ozone level detection [17]	Half-hourly from Feb. 6, 2013 to Mar. 13, 2013
13	Geo-potential height		
14	K index		

SVM: Support Vector Machines [15]; **GBR:** Generalized Boosted Regr. [24];
NN: Feed Forward Neural Nets [29]; **MARS:** MARS [18];
GP: Gaussian Processes [15];
GLM: Generalized Linear Models [11]; **RBR:** Rule-based Regression [16];
RF: Random Forests [31]; **PPR:** Projection Pursuit Regr. [23].

Different parameter settings are used for each of the individual learners, adding up to **40** models.

We use a Random Forest as meta-learner. The parameter λ was set to **10**, which means that at run-time the metalearners are re-trained every 10 observations. Each set contains the OOB samples from the training set and the observations from the test set up to the upcoming prediction. In the blocked prequential procedure used to obtain OOB samples we used 10 folds, i.e., β equal to **10**. The committee for each prediction contains the 50% forecasters with best performance in the last 50 observations (α and Ω values are set to **50**). We drop only half the models in the interest of keeping the combined model readily adaptable to changes in the environment. An average performing model may suddenly become important and the combined model should be able to capture these situations. By setting Ω to 50 we strive for estimates of recent performance that renders a robust committee. We compare the performance of ADE against the following 8 approaches:

Stacking: An adaptation of stacking [30] for times series, where a meta-model is learned using the base-level predictions as attributes. To preserve the temporal order of observations, the out-of-bag predictions used to train the metalearner (a random forest) are obtained using a blocked prequential procedure (c.f. Sect. 3.2). We tried different strategies for training the metalearner (e.g. holdout) but blocked prequential presented the best results;

Arbitrating: The original arbitrating approach [21], c.f. Sect. 2;

S: A static heterogeneous ensemble. All base learners are simply averaged using the arithmetic mean [26];

S-W: A weighted linear combination of the models, with weights according to their performance using all past information [14];

AEC: The adaptive combination procedure AEC [25], c.f. Sect. 2;

S-WRoll: The adaptive combination method, with a linear combination of the forecasters according to their recent performance [19];

ERP: The adaptive combination procedure proposed by Timmermann [27], c.f. Sect. 2;

ARIMA: A state-of-the-art method for time series forecasting. We use the implementation in the forecast R package [13], which automatically tunes ARIMA to an optimal parameter setting;

The following variants of ADE were tested:

ADE-Arb: A variant of ADE in which at each time point the best model is selected to make a prediction. Here best is the one with lowest predicted loss. This is in accordance with the original arbitrating architecture [21];

ADE-meta-runtime: A variant of ADE in which there is no blocked prequential procedure to obtain OOB samples to increase the data provided to the metalearners. In this scenario Z is trained in data obtained only at run-time, which is also in accordance with the original arbitrating strategy;

ADE-all-models: A variant of ADE, but without the formation of a committee. In this case, all forecasting models are weighed according to their expertise in the input data;

ADE-linear-committee: A variant of ADE, but using a linear transformation for weighting the base learners according to their predicted loss, instead of the proposed softmax;

ADE-meta-GP: A variant of ADE, but using a Gaussian Process with a linear kernel as meta-learner instead of a Random Forest;

All the variants of the proposed method use a random forest as meta-learner.

4.3   Results

Table 2 presents the paired comparisons between the proposed method, ADE, and the baselines. The numbers represent wins and losses of the proposed method. The numbers in parenthesis represent statistically significant wins and losses. The average rank for each model is also presented with the corresponding deviation. Figures 2 and 3 represent the critical difference diagrams for the

Table 2. Paired comparisons between the proposed method and the baselines for different embedding dimensions in the 14 time series. The Rank column stands for the average rank and respective standard deviation of each model. A rank of 1 in an experiment means that the model was the best method.

Method	K = 7			K = 15		
	ADE		Rank	ADE		Rank
	Looses	Wins		Looses	Wins	
S	3 (2)	**11** (11)	10.0 ± 3.3	3 (2)	**11** (10)	9.9 ± 3.3
S-W	5 (4)	**9** (8)	6.3 ± 3.0	6 (3)	**8** (8)	6.6 ± 3.8
ARIMA	3 (3)	**11** (11)	11.1 ± 5.5	3 (3)	**11** (11)	10.6 ± 5.3
Stacking	7 (5)	7 (5)	5.9 ± 5.2	3 (1)	**11** (8)	7.0 ± 4.2
Arbitrating	4 (0)	**10** (6)	7.8 ± 4.8	1 (0)	**13** (9)	9.0 ± 3.9
AEC	4 (2)	**10** (10)	8.6 ± 3.9	2 (0)	**12** (11)	9.3 ± 2.7
S-WRoll	5 (4)	**9** (7)	7.4 ± 3.4	4 (2)	**10** (9)	8.1 ± 3.6
ERP	3 (2)	**11** (11)	10.0 ± 3.3	3 (2)	**11** (11)	10.6 ± 3.8
ADE-Arb	1 (0)	**13** (7)	8.4 ± 3.1	2 (0)	**12** (8)	6.9 ± 3.9
ADE-meta-runt.	6 (0)	8 (5)	5.8 ± 3.9	3 (1)	**11** (6)	5.6 ± 3.0
ADE-all-models	5 (2)	**9** (3)	6.0 ± 2.3	3 (2)	**11** (3)	5.1 ± 3.3
ADE-linear-com.	5 (4)	**9** (6)	6.4 ± 3.3	5 (5)	**9** (7)	6.1 ± 3.2
ADE-meta-GP	6 (0)	8 (6)	6.1 ± 3.0	4 (3)	**10** (7)	6.0 ± 2.6
ADE	–	–	$\mathbf{5.2 \pm 2.8}$	–	–	$\mathbf{4.1 \pm 3.1}$

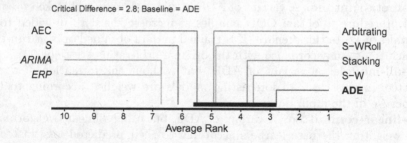

Fig. 2. Critical difference diagram for the post-hoc Bonferroni-Dunn test relative to baselines ($K = 7$)

post-hoc Bonferroni-Dunn test relative to the other baselines in the literature and baselines which are variants of ADE, respectively. In the interest of space we present the post-hoc test results only for K equal to 7. The analysis for K equal to 15 leads to similar conclusions.

The proposed method achieves competitive performance with respect to the baselines and ADE variants. Relative to **S**, **S-W**, **AEC**, **S-WRoll** and **ERP**, state of the art approaches for combining individual forecasters, and **ARIMA**, a state of the art method for time series forecasting, the difference is significant.

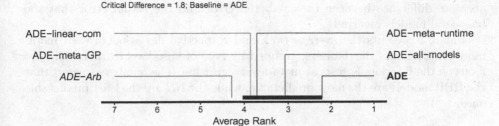

Fig. 3. Critical difference diagram for the post-hoc Bonferroni-Dunn test relative to ADE variants $(K = 7)$

These results answer the research question **Q1** regarding the performance of ADE relative to the state-of-the-art approaches for time series forecasting tasks.

Comparing the proposed method against **Stacking**, a widely used metalearning strategy for combining models, the performance is no significantly different, although our method presents a better average rank. Relative to the original arbitrating architecture, denoted as **Arbitrating**, the proposed method shows a systematic improvement, which results in a much better average rank. This proves that the introduced components are fundamental for the achieved performance, which answers question **Q3** regarding the comparison between ADE and the original arbitration approach. In the following, we discuss the effect of different components in the results.

ADE also shows consistent advantage over the performance of **ADE-meta-runtime** and **ADE-all-models**. This suggests that indeed it is worthwhile to get OOB predictions from the available data to improve the fit of the metalearners and to prune the ensemble for each prediction (as opposed to combining all the forecasters). These results answer our research questions **Q2** and **Q4** regarding the comparison of ADE to the respective components. The comparison between ADE and **ADE-linear-committee** shows that the softmax function renders a superior predictive performance relative to the linear transformation. In effect, a non-linear transformation indeed produces better estimates than a traditional linear scaling (question **Q5**). Comparing the proposed method against **ADE-Arb**, which used the predicted best model for forecasting, the difference is statistically significant, according to the post-hoc Bonferroni-Dunn test.

Regarding the embedding dimension, different values do not seem to alter the results significantly. Conversely, using Random Forests in the metalearning layer produced better results than with Gaussian Processes.

4.4   Further Analysis

Meta-Level Performance. We analysed the performance of the metalearning layer. The evaluation measure is mean absolute error. That is, the average absolute difference between the predicted error and the actual error that the base-level models incurred.

Figure 4 presents the average rank and respective deviation of each meta-model in Z, across the datasets, grouped by type of base-level learner. Here, we focus on the Random Forests as meta-level algorithm. The numbers suggest that the RBR models are the most predictable, while the NN are the least predictable ones.

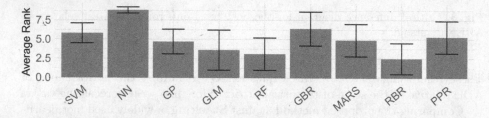

Fig. 4. Mean rank of the meta-models in terms of accuracy for predicting the performance of models obtained with several algorithms across the 14 datasets ($K = 7$) and corresponding standard-deviation.

Analyzing Training Strategies. In this section we address the research question **Q6**. In a dynamic environment it is common to update the model over time, either online or in chunks of observations. This is because time-dependent data is prone to changes in the underlying distribution and continuous training of models ensures that one has an up-to-date model. Since ADE settles on two layers of models we analysed different approaches for updating these and study their implications in predictive performance and computational resources used.

In the main experiments the base learners are not updated and meta-learners are updated in chunks of λ observations. Besides this strategy (which we denote as **A**) we try three other approaches:

B: M and Z are both trained only in the training set;
C: Both M and Z are re-trained every λ observations, which is particularly interesting if the models in M are typical online methods (e.g. ARIMA);
D: M is re-trained every λ observations but Z is trained only in the training data.

In these experiments we set λ to 1 meaning that re-training occurs at every prediction step. The test size is set to 15% of the time series total length. Other parameters follow the setup in the main experiments. We compare the four above-mentioned strategies in the 14 problems using the mean absolute scaled error (MASE). This is a scale invariant metric and is defined as follows:

$$\frac{\sum_{i=1}^{n} |e_i|}{\frac{n}{n-1} \sum_{i=2}^{n} |y_i - y_{i-1}|} \tag{3}$$

The results of this analysis are presented in Table 3. These show the average MASE and average runtime and corresponding deviations of ADE using the different retraining strategies across the 14 problems. The results for the predictive performance are comparable across the methods, although the ones that update the metalearners show a slightly lower error.

In terms of computational effort the strategies that update the base learners are significantly more expensive. Retraining the meta learners is faster than updating the base learners because of the formation of the committee in which only half the meta learners (according to the experimental setup) are retrained.

Table 3. Average MASE and respective deviation of the different retraining strategies in terms of predictive performance (a) and computational time spent in minutes (b) across the 14 problems.

(a)				(b)			
A	B	C	D	A	B	C	D
0.60 ± 0.18	0.62 ± 0.17	0.60 ± 0.18	0.62 ± 0.17	22.4 ± 6.6	2.1 ± 0.5	156.4 ± 38.2	130.3 ± 31.5

Finally, the results suggest that updating the meta learners and not updating the base learners (A) is better than the inverted strategy (D), both in predictive performance and runtime. In particular, the difference in computational time is mainly due to the selection of models included in the committee. In strategy A, the models outside the committee are not updated since they will not be used. Conversely, in strategy D, every base learner is updated.

4.5 Discussion

We empirically showed the advantages of the proposed method with respect to several state-of-the-art approaches for time series forecasting tasks.

One of the main limitations of ADE is that it is not able to directly model inter-dependencies between forecasters, which might be important to account for the diversity among models. We plan to address this issue in the future. Nonetheless, its performance is competitive with the widely used stacking method [30], which directly models these dependencies.

Some of the design decisions behind ADE are based on prior work regarding the variance in relative performance of forecasting models over a time series [2] and with potential recurring structures with the time series. However, there are cases in which time series change into new concepts and base learners may get outdated. Although we do not explicitly cover these scenarios, a simple strategy to address this issue is to track the loss of the ensemble. If its performance decreases beyond tolerance new base-learners are introduced (e.g. [12]) or existing ones are re-trained. Since an arbitration approach provides a modular architecture, models can be added (or removed) as needed.

5   Conclusions

We presented a new adaptive ensemble method for time series forecasting tasks. Our strategy for adaptively combining forecasters is based on metalearning, which provides a way of learning about the learning process of models [4]. Consequently, we are able to model their expertise in the different parts of the data and adapt the combined model to changes in the underlying environment and/or changes in relative performance of base learners. By analyzing each forecaster separately we specialize them in the sense that they will not be used for prediction – or they do but with a minor relevance – in observations that they are bad at.

ADE is motivated by the assumption that different forecasting models have varying relative performance over time [2]. Moreover, it is not uncommon for time series to show recurrent structures.

We proved the competitiveness of our approach in fourteen real-world time series against several baselines. These include state-of-the-art methods for adaptively combining forecasters and the most widely used metalearning strategy for model combination (stacking [30]). The effect of the different adaptations of arbitrating on the results was empirically analysed.

We argue that despite the competitive predictive performance, the proposed method does not directly model inter-dependencies between the available learners. We plan to investigate this issue in future work.

In the interest of reproducible science that the authors support, all methods and datasets are publicly available as an R software package[1].

Acknowledgements. This work is financed by the ERDF - European Regional Development Fund through the Operational Programme for Competitiveness and Internationalisation - COMPETE 2020 Programme within project POCI-01-0145-FEDER-006961, and by National Funds through the FCT - Fundação para a Ciência e a Tecnologia (Portuguese Foundation for Science and Technology) as part of project UID/EEA/50014/2013; Project "NORTE-01-0145-FEDER-000036" is financed by the North Portugal Regional Operational Programme (NORTE 2020), under the PORTUGAL 2020 Partnership Agreement, and through the European Regional Development Fund (ERDF). This work was partly funded by the ECSEL Joint Undertaking, program for research and innovation horizon 2020 (20142020) under grant agreement number 662189-MANTIS-2014-1.

References

1. ADDP: Oporto water consumption. http://addp.pt. Accessed 21 Nov 2016
2. Aiolfi, M., Timmermann, A.: Persistence in forecasting performance and conditional combination strategies. J. Econometr. **135**(1), 31–53 (2006)
3. Wilcox, S., Andreas, A.: Solar Radiation Monitoring Station (SoRMS): Humboldt State University, Arcata, California; NREL Rep DA-5500-56515 (2007)

---

[1] tsensembler: https://github.com/vcerqueira/tsensembler.

4. Brazdil, P., Carrier, C.G., Soares, C., Vilalta, R.: Metalearning: Applications to Data Mining. Springer Science & Business Media, Heidelberg (2008)
5. Brown, G., Kuncheva, L.I.: "Good" and "Bad" diversity in majority vote ensembles. In: El Gayar, N., Kittler, J., Roli, F. (eds.) MCS 2010. LNCS, vol. 5997, pp. 124–133. Springer, Heidelberg (2010). https://doi.org/10.1007/978-3-642-12127-2_13
6. Cerqueira, V., Torgo, L., Soares, C.: Arbitrated ensemble for solar radiation forecasting. In: Rojas, I., Joya, G., Catala, A. (eds.) IWANN 2017. LNCS, vol. 10305, pp. 720–732. Springer, Cham (2017). https://doi.org/10.1007/978-3-319-59153-7_62
7. Clemen, R.T., Winkler, R.L.: Combining economic forecasts. J. Bus. Econ. Stat. 4(1), 39–46 (1986)
8. Crane, D.B., Crotty, J.R.: A two-stage forecasting model: exponential smoothing and multiple regression. Manag. Sci. 13(8), B-501 (1967)
9. Dawid, A.P.: Present position and potential developments: some personal views: statistical theory: the prequential approach. J. Roy. Stat. Soc. Ser. A (Gener.) 278–292 (1984)
10. EERE: Commercial and residential hourly load profiles TMY3 location in the USA. http://en.openei.org/datasets/files/961/pub/. Accessed 21 Nov 2016
11. Friedman, J., Hastie, T., Tibshirani, R.: Regularization paths for generalized linear models via coordinate descent. J. Stat. Softw. 33(1), 1–22 (2010)
12. Gama, J., Kosina, P.: Tracking recurring concepts with meta-learners. In: Lopes, L.S., Lau, N., Mariano, P., Rocha, L.M. (eds.) EPIA 2009. LNCS (LNAI), vol. 5816, pp. 423–434. Springer, Heidelberg (2009). https://doi.org/10.1007/978-3-642-04686-5_35
13. Hyndman, R.J., with contributions from Athanasopoulos, G., Razbash, S., Schmidt, D., Zhou, Z., Khan, Y., Bergmeir, C., Wang, E.: forecast: Forecasting functions for time series and linear models (2014). R package version 5.6
14. Jose, V.R.R., Winkler, R.L.: Simple robust averages of forecasts: some empirical results. Int. J. Forecast. 24(1), 163–169 (2008)
15. Karatzoglou, A., Smola, A., Hornik, K., Zeileis, A.: kernlab - an S4 package for kernel methods in R. J. Stat. Softw. 11(9), 1–20 (2004)
16. Kuhn, M., Weston, S., Keefer, C., Coulter, N., C code for Cubist by Ross Quinlan: Cubist: Rule- and Instance-Based Regression Modeling (2014). R package version 0.0.18
17. Lichman, M.: UCI Machine Learning Repository (2013). archive.ics.uci.edu/ml
18. Milborrow, S.: earth: Multivariate Adaptive Regression Spline Models. Derived from mda:mars by Trevor Hastie and Rob Tibshirani (2012)
19. Newbold, P., Granger, C.W.: Experience with forecasting univariate time series and the combination of forecasts. J. Roy. Stat. Soc. Ser. A (Gener.) 131–165 (1974)
20. Oliveira, M., Torgo, L.: Ensembles for time series forecasting. In: ACML Proceedings of Asian Conference on Machine Learning, JMLR: Workshop and Conference Proceedings (2014)
21. Ortega, J., Koppel, M., Argamon, S.: Arbitrating among competing classifiers using learned referees. Knowl. Inf. Syst. 3(4), 470–490 (2001)
22. Pinto, F., Soares, C., Mendes-Moreira, J.: CHADE: metalearning with classifier chains for dynamic combination of classifiers. In: Frasconi, P., Landwehr, N., Manco, G., Vreeken, J. (eds.) ECML PKDD 2016. LNCS (LNAI), vol. 9851, pp. 410–425. Springer, Cham (2016). https://doi.org/10.1007/978-3-319-46128-1_26
23. R Core Team: R: A Language and Environment for Statistical Computing. R Foundation for Statistical Computing, Vienna, Austria (2013)

24. Ridgeway, G.: GBM: Generalized Boosted Regression Models (2015). R package version 2.1.1
25. Sánchez, I.: Adaptive combination of forecasts with application to wind energy. Int. J. Forecast. **24**(4), 679–693 (2008)
26. Timmermann, A.: Forecast combinations. In: Handbook of Economic Forecasting, vol. 1, pp. 135–196 (2006)
27. Timmermann, A.: Elusive return predictability. Int. J. Forecast. **24**(1), 1–18 (2008)
28. Torgo, L.: An Infra-structure for Performance Estimation and Experimental Comparison of Predictive Models (2013). R package version 0.1.1
29. Venables, W.N., Ripley, B.D.: Modern Applied Statistics with S, 4th edn. Springer, New York (2002). ISBN 0-387-95457-0
30. Wolpert, D.H.: Stacked generalization. Neural Netw. **5**(2), 241–259 (1992)
31. Wright, M.N.: ranger: A Fast Implementation of Random Forests (2015). R package

Cost Sensitive Time-Series Classification

Shoumik Roychoudhury[1], Mohamed Ghalwash[1,2,3], and Zoran Obradovic[1(✉)]

[1] Center for Data Analytics and Biomedical Informatics, Temple University,
Philadelphia, PA, USA
{shoumik.rc,mohamed.ghalwash,zoran.obradovic}@temple.edu
[2] IBM T. J. Watson Research Center, Cambridge, MA, USA
[3] Faculty of Science, Ain Shams University, Cairo, Egypt

Abstract. This paper investigates the problem of highly imbalanced time-series classification using shapelets, short patterns that best characterize the target time-series, which are highly discriminative. The current state-of-the-art approach learns generalized shapelets along with weights of the classification hyperplane via a classical cost-insensitive loss function. Cost-insensitive loss functions tend to treat different misclassification errors equally and thus, models are usually biased towards examples of majority class. The rare class (which will be referred to as positive class) is usually the important class and a false negative is always costlier than a false positive. Traditional 0–1 loss functions fail to differentiate between these two types of misclassification errors. In this paper, the generalized shapelets learning framework is extended and a cost-sensitive learning model is proposed. Instead of incorporating the misclassification cost as a prior knowledge, as was done by other published methods, we formulate a constrained optimization problem to *learn* the unknown misclassification costs along with the shapelets and their weights. First, we demonstrated the effectiveness of the proposed method on two case studies, with the objective to detect true alarms from life threatening cardiac arrhythmia dataset from Physionets MIMIC II repository. The results show improved true alarm detection rates over the current state-of-the-art method. Next, we compared to the state-of-the-art learning shapelet method on 16 balanced dataset from UCR time-series repository. The results show evidence that the proposed method outperforms the state-of-the-art method. Finally, we performed extensive experiments across additional 18 imbalanced time-series datasets. The results provide evidence that the proposed method achieves comparable results with the state-of-the-art sampling/non-sampling based approaches for highly imbalanced time-series datasets. However, our method is highly interpretable which is an advantage over many other methods.

Keywords: Cost sensitive · Time-series classification · Shapelets

1 Introduction

Research on time-series classification has garnered importance among practitioners in the data mining community. A major reason behind the ever increasing

© Springer International Publishing AG 2017
M. Ceci et al. (Eds.): ECML PKDD 2017, Part II, LNAI 10535, pp. 495–511, 2017.
https://doi.org/10.1007/978-3-319-71246-8_30

interest among data-miners is the plethora of time-series data available from a wide range of real-life domains. Temporal ordered data from areas such as financial forecasting, medical diagnosis, weather prediction etc. provide classification challenges more akin to real-world scenarios. Thus, building more robust time-series classification models is imperative.

One of the key sources of performance degradation in the field of time-series classification is the class imbalance problem [18] where the minority class (we call it the positive class) is outnumbered by abundant negative class instances. Models built using standard classification algorithms on such imbalanced datasets, which generally have minimum classification error as a criterion for classifier design often, are biased towards the majority class; and therefore, have higher misclassification error for the minority class examples. Moreover, in real-world scenarios such as object detection, medical diagnosis etc., the positive class is usually the more important class and false negatives are always costlier than false positives. Traditional 0–1 loss function classifiers fail to differentiate between these two types of errors and final outcomes are naturally biased towards the abundant negative class. Thus, a cost-sensitive classifier is preferred when dealing with datasets where examples from different classes carry different misclassification costs.

Recently, in the realm of time-series classification, Grabocka et al. [10] proposed a novel framework known as Learning Time-series Shapelets (LTS) to directly learn generalized short time-series subsequences known as shapelets [23] along with weights of a classifier hyperplane to differentiate temporal instances in a binary classification framework. Shapelets are local discriminative patterns (or subsequences) that can be used to characterize the target class, for determining the time-series class membership. Shapelets have been proven to have high predictive powers as they provide local variation information within the time-series as well as high interpretability of predictions due to easier visualizations. LTS formulates an optimization problem where a cost-insensitive 0–1 logistic loss function is minimized in order to learn generalized shapelets. The minimum Euclidean distances of the learned shapelets to the time-series can be used to linearly separate the time-series examples from different classes.

However, LTS uses cost-insensitive loss function that treats false positive and false negative errors equally, which limits its applicability on balanced datasets. In this paper, we propose a cost-sensitive time-series classification framework (henceforth known as CS-LTS) by extending the LTS model. A cost-sensitive logistic loss function is minimized to enhance the modeling capability of LTS. The cost-sensitive logistic loss function uses variable misclassification costs for false positive and false negative errors. Generally, these misclassification cost values are available from the cost matrix provided by domain experts which is often a cumbersome procedure. Instead of using fixed cost parameters, this paper *learns* the variable misclassification costs from the training data via a constrained optimization problem. Thus, the main contribution of this paper is summarized as the following.

1. The proposed method learns the misclassification costs from the training data thus nullifying the need for predetermination of cost values for misclassification errors. To the best of our knowledge, the proposed model is the first algorithmic approach to solve highly imbalanced time-series classification problem.
2. A constrained optimization problem is proposed which jointly learns shapelets (highly interpretable patterns), their weights, and most importantly misclassification costs, while other cost-sensitive approaches mainly consider misclassification costs are given a priori.
3. The effectiveness of the method is demonstrated on life-threatening cardiac arrhythmia dataset from Physionets MIMIC II repository showing improved true alarm detection rates over the current state-of-the-art method for false alarm suppression.
4. Finally, the method is evaluated extensively on 34 real-world time series datasets with varied degree of imbalances and compared to a large set of baseline methods previously proposed in the realm of imbalance time-series classification problems.

In Fig. 1(a), we show all time series examples for the blue and red classes. The blue class has only 3 time series, while the red class has 10 time series. Since LTS does not handle imbalance dataset, the learned hyperplane is very biased. This is clear from Fig. 1(b) that shows the distance between the two learned shapelets using LTS and the training time series. CS-LTS learns a hyperplane that is aware about the imbalance in the data, as shown in Fig. 1(c).

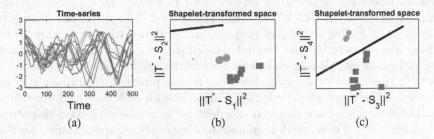

Fig. 1. An illustration of the proposed CS-LTS model (c) compared to LTS (b) using 2 shapelets learned on an imbalanced version of BirdChicken dataset (a). (Color figure online)

Next, we present a short literature review for time-series classification using shapelets and cost-sensitive time-series classification.

2   Related Work

Time-Series Classification via Shapelets. In the field of time-series classification, the concept of shapelets have received a lot of attention [8,10,16,23,24].

Shapelets are local discriminative patterns (or subsequences) that characterize the target class and maximally discriminate instances of time-series from various classes. Discovering the most discriminative subsequences is crucial for the success of time-series classification using shapelets. The primary approach, based on search-based techniques, proposed by Ye and Keogh [23], exhaustively search for all possible subsequences and a decision tree was constructed based on information gain criterion. The information gain accuracy was ranked based on the minimum distance of the candidate subsequences to the entire time-series training set. Hills et al. [15] perceived this minimum distance of the set of shapelets to a time-series dataset as a data transformation to a shapelet-transformed space where standard classifiers could be used to achieve high classification accuracy using the shapelet-transformed data as predictors. Recently, Grabocka et al. [10] proposed a novel framework known as Learning Time-series Shapelets (LTS) to jointly learn generalized shapelets along with weights of a logistic regression model using the minimum Euclidean distances of shapelets to time-series dataset as predictors. The method discovered optimal shapelets and reported statistically significant improvements in accuracy compared to other shapelet-based time-series classification models. However, a major drawback is low true positive rate in case of highly imbalanced time-series datasets. The logistic loss used in the LTS framework is a cost-insensitive loss function which treats false positive and false negative misclassifications errors equally. Classification models built using such loss functions suffer from the class imbalance problem.

Cost-Sensitive Classification. Classification techniques for handling imbalanced data-sets can broadly be divided into two kinds of approaches, data-level approaches [2–5, 12, 13, 17] and algorithmic-level [22] approaches. Data-level methods are sampling techniques that act as a pre-processing steps prior to the learning algorithm to balance the imbalanced datasets either through oversampling of the minority class or under sampling of the majority class or combination of both. Algorithmic-level approaches directly manipulate the learning algorithm by incorporating a predefined misclassification cost for each class to the loss function. These methods have reported excellent performance with good theoretical guarantees [14]; however, predetermination of optimal class misclassification cost or data-space weighting is required which can vary on a case-by-case basis among different datasets and also require domain expertise.

In this paper, an algorithmic approach is followed to directly manipulate the learning procedure by minimizing a cost-sensitive logistic loss function. An additive asymmetric learning function is fitted to the training data. In addition to learning the shapelets and weight parameters of the classification hyperplane, the cost parameters are also estimated from the training data. A constrained optimization problem is formulated that is optimized to jointly learn shapelets, weights of the classification hyperplane and misclassification cost parameters nullifying the need for predetermination of cost values for misclassification errors.

3   Model Description

Preliminaries: A binary class time-series dataset composed of I training examples denoted as $\mathbf{T} \in \mathbb{R}^{I \times Q}$ is considered where, each T_i $(1 \leq i \leq I)$ is of length Q and the label for each time-series instance is a nominal variable $Y \in \{0,1\}^I$. Candidate shapelets are segments of length L from a time-series starting from j-th time point inside the ith time-series. The objective is to learn k shapelets $\mathbf{S}$, each of length L, that are most discriminative in order to characterize the target class. The shapelets are denoted as $S \in \mathbb{R}^{K \times L}$.

The minimum distance $M_{i,k}$ between the ith series T_i and the kth shapelet S_k is the distance between the segment and time-series. This is defined as

$$M_{i,k} = \min_{j=1,\ldots,J} \frac{1}{L} \sum_{l=1}^{L} (T_{i,j+l-1} - S_{k,l})^2 \tag{1}$$

Given a set of I time-series training examples and K shapelets, a shapelet-transformed matrix [15] $\mathbf{M} \in \mathbb{R}^{I \times K}$ can be constructed which is composed of minimum distances $M_{i,k}$ between the ith series T_i and the kth shapelet S_k. The minimum distance M matrix is a representation in the shapelet transformed space and acts as predictors for each target time-series. However, the function in Eq. (3) is not continuous and thus non-differentiable. Grabocka et al. [10] defined a soft-minimum function (shown in Eq. (2)), which is an approximation for $M_{i,k}$.

$$M_{i,k} \approx \hat{M}_{i,k} = \frac{\sum_{j=1}^{J} D_{i,k,j} \exp(\alpha D_{i,k,j})}{\sum_{\bar{j}=1}^{J} \exp(\alpha D_{i,k,\bar{j}})} \tag{2}$$

where $D_{i,k,j}$ is defined as the distance between the jth segment of series i and the kth shapelet given by the formula

$$D_{i,k,j} = \frac{1}{L} \sum_{l=1}^{L} (T_{i,j+l-1} - S_{k,l})^2 \tag{3}$$

Learning Model: A linear learning model (shown in Eq. (4)) was proposed by [10] using the minimum distances M as predictors in the transformed shapelet space.

$$\hat{Y}_i = W_0 + \sum_{k=1}^{K} M_{i,k} W_k \quad \forall i \in \{1,\ldots,I\} \tag{4}$$

The learning function (Eq. (4)) is extended by incorporating C_{FN} and C_{FP} for false negative and false positive misclassifications cost respectively. The new asymmetric learning model is defined as Eq. (5).

$$Z_i = \frac{1}{C_{FN} + C_{FP}} ln \frac{\sigma(\hat{Y}) C_{FN}}{1 - \sigma(\hat{Y}) C_{FP}} = \frac{1}{C_{FN} + C_{FP}} (\hat{Y} + ln \frac{C_{FN}}{C_{FP}}) \tag{5}$$

$\sigma()$ is the logistic function and $\sigma(\hat{Y})$ represents the posterior probability of $P(Y = 1 \,|\, X)$.

Additionally, a cost-sensitive loss function (Eq. (6)) is proposed which is a differential cost-weighted logistic loss between the actual targets Y and the estimated targets Z.

$$\mathcal{L}(Y, Z) = -Y ln\sigma(C_{FN}Z) - (1 - Y)ln(1 - \sigma(C_{FP}Z)) \qquad (6)$$

A regularized cost-sensitive logistic loss function defined by Eq. (7) is the regularized objective function denoted by $\mathcal{F}$.

$$\underset{S,W,C}{\operatorname{argmin}} \mathcal{F}(S, W, C) = \underset{S,W,C}{\operatorname{argmin}} \sum_{i=1}^{I} \mathcal{L}(Y_i, Z_i) + \lambda_W \|W\|^2 \qquad (7)$$

where $C \in \{C_{FN}, C_{FP}\}$. The problem is formulated as a constrained optimization problem since the misclassification costs should always be positive. The misclassification cost denotes the loss incurred when a wrong prediction occurs. The constraints ensure both costs are positive and also the fact that cost of false negative is at least θ times greater than cost of false positive. These conditions ensure the loss function to be penalized more in the event of an error in the positive class than an error in the negative class.

$$\underset{S,W,C}{\operatorname{argmin}} \mathcal{F}(S, W, C)$$
$$\text{subject to } C_{FN} > 0, \ C_{FP} > 0 \qquad (8)$$
$$C_{FN} > \theta C_{FP}$$

Similar to [10], a Stochastic gradient descent (henceforth SGD) approach is adopted to solve the optimization problem. The SGD algorithm optimizes the parameters to minimize the loss function by updating through per instance of the training data. Thus, the per-instance decomposed objective function $\mathcal{F}_i$ (denoted by Eq. (9)) shows the division of Eq. (7) into per-instance losses for each time-series.

$$\mathcal{F}_i = \mathcal{L}(Y_i, Z_i) + \frac{\lambda_W}{I} \sum_{k=1}^{K} W_k^2 \qquad (9)$$

The objective of the learning algorithm is to learn the optimal shapelet S_k, the weights W for the hyperplane and the misclassification costs C which minimizes the loss function (Eq. (7)).

The SGD algorithm requires definitions of gradients of the objective function with respect to shapelets, hyperplane weights and misclassification costs. Eq. (10) shows the point gradient of objective function for the ith time-series with respect to shapelet S_k.

$$\frac{\partial \mathcal{F}_i}{\partial S_{k,l}} = \frac{\partial \mathcal{L}(Y_i, Z_i)}{\partial Z_i} \frac{\partial Z_i}{\partial \hat{Y}_i} \frac{\partial \hat{Y}_i}{\partial \hat{M}_{i,k}} \sum_{j=1}^{J} \frac{\partial \hat{M}_{i,k}}{\partial D_{i,k,j}} \frac{\partial D_{i,k,j}}{\partial S_{k,l}} \qquad (10)$$

Furthermore, the gradient of the cost-sensitive loss function with respect to the learning function Z_i is defined in Eq. (11). Also the gradient of the cost-sensitive learning function with respect to the estimated target variable $\hat{Y}_i$ is shown in Eq. (12)

$$\frac{\partial \mathcal{L}(Y_i, Z_i)}{\partial Z_i} = (1 - Y_i)\sigma(C_{FP}Z_i)C_{FP} - Y_i(1 - \sigma(C_{FN}Z_i))C_{FN} \qquad (11)$$

$$\frac{\partial Z_i}{\partial \hat{Y}_i} = \frac{1}{C_{FN} + C_{FP}} \qquad (12)$$

Equation (13) shows the gradient of the estimated target variable with respect to the minimum distance. The gradient of the over all minimum distance with respect to the segment distance and the gradient of the segment distance with respect to a shapelet point is defined by Eqs. (14) and (15) respectively.

$$\frac{\partial \hat{Y}_i}{\partial \hat{M}_{i,k}} = W_k \qquad (13)$$

$$\frac{\partial \hat{M}_{i,k}}{\partial D_{i,k,j}} = \frac{\exp(\alpha D_{i,k,j}(1 + \alpha(D_{i,k,j} - \hat{M}_{i,k})))}{\sum_{\bar{j}=1}^{J} \exp(\alpha D_{i,k,\bar{j}})} \qquad (14)$$

$$\frac{\partial D_{i,k,j}}{\partial S_{k,l}} = \frac{2}{L}(S_{k,l} - T_{i,j+l-1}) \qquad (15)$$

The hyperplane weights W are learned by minimizing the objective function 7 via SGD. The gradients for updating the weights W_k is shown in Eqs. (16) and (17) shows the gradient for update of the bias term W_0.

$$\frac{\partial \mathcal{F}_i}{\partial W_k} = \frac{\partial \mathcal{L}(Y_i, Z_i)}{\partial Z_i}\frac{\partial Z_i}{\partial \hat{Y}_i}\hat{M}_{i,k} + \frac{2\lambda_W}{I}W_k \qquad (16)$$

$$\frac{\partial \mathcal{F}_i}{\partial W_0} = \frac{\partial \mathcal{L}(Y_i, Z_i)}{\partial Z_i}\frac{\partial Z_i}{\partial \hat{Y}_i} \qquad (17)$$

The learning procedure for estimating the misclassification cost values in the proposed framework is a constrained optimization problem because we need to guarantee that $C_{FN} > 0$, $C_{FP} > 0$ and $C_{FN} > \theta C_{FP}$, where $\theta \in \mathbb{Z}$. However, Stochastic Gradient Descent algorithm can only be applied to solve unconstrained optimization problems. Thus, we convert the constrained optimization into an unconstrained optimization similar to [19] and apply SGD algorithm to solve the optimization problem for learning the optimal misclassification costs.

$$C_{FN} = \theta C_{FP} + \mathcal{D} \qquad (18)$$

The false negative misclassification cost (C_{FN}) is first written in terms of false positive misclassification cost as shown in Eq. (18) and replaced in Eq. (6) changing the optimization problem to Eq. (19).

Algorithm 1. Cost-sensitive learning time-series shapelets

1: **procedure** CS-LTS
2: **Input**: $T \in \mathcal{R}^{I \times Q}$, Number of shapelets K, length of a shapelet L, Regularization parameter λ_W, Learning rate η, maxIter
3: **Initialize**: Shapelets $S \in \mathbb{R}^{K \times L}$, classification hyperplane weights $W \in \mathbb{R}^K$, Bias $W_0 \in \mathbb{R}$, Misclassification cost $C_{FP} \in \mathbb{R}$, $\theta \in \mathbb{Z}$, $\mathcal{D} \in \mathbb{R}$
4:     **for** iterations $= \mathbb{N}_1^{maxIter}$ **do**
5:         **for** $i = 1, ..., I$ **do**
6:             **for** $k = 1, ..., K$ **do**
7:                 $W_k^{new} \leftarrow W_k^{old} - \eta \frac{\partial \mathcal{F}_i}{\partial W_k}$
8:                 **for** $l = 1, ..., L$ **do**
9:                     $S_{k,l}^{new} \leftarrow S_{k,l}^{old} - \eta \frac{\partial \mathcal{F}_i}{\partial S_{k,l}}$
10:         $W_0^{new} \leftarrow W_0^{old} - \eta \frac{\partial \mathcal{F}_i}{\partial W_0}$
11:         $\log C_{FP}^{new} \leftarrow \log C_{FP}^{old} - \eta \frac{\partial \mathcal{F}_i}{\partial \log C_{FP}}$
12:         $\mathcal{D}^{new} \leftarrow \mathcal{D}^{old} - \eta \frac{\partial \mathcal{F}_i}{\partial \mathcal{D}}$
        Return S, W, W_0, C_{FP}

$$\operatorname*{argmin}_{S,W,C_{FP},\mathcal{D}} \mathcal{F}(S, W, C_{FP}, \mathcal{D})$$
$$\text{subject to } C_{FP} > 0 \tag{19}$$

$\mathcal{D}$ is a regularization term for the misclassification cost. The objective function is then minimized with respect to $\log C_{FP}$ instead of C_{FP}. As a result, the new optimization problem becomes unconstrained. Derivatives of objective function with respect to $\log C_{FP}$ and $\mathcal{D}$ in gradient descent are computed as:

$$\frac{\partial \mathcal{F}_i}{\partial \log c_{FP}} = c_{FP} \frac{\partial \mathcal{L}(Y_i, Z_i)}{\partial c_{FP}} \tag{20}$$

$$\frac{\partial \mathcal{L}(Y_i, Z_i)}{\partial c_{FP}} = \frac{\partial \mathcal{L}(Y_i, Z_i)}{\partial Z_i} \frac{\partial Z_i}{\partial c_{FP}} \tag{21}$$

$$\frac{\partial \mathcal{L}(Y_i, Z_i)}{\partial \mathcal{D}} = \frac{\partial \mathcal{L}(Y_i, Z_i)}{\partial Z_i} \frac{\partial Z_i}{\partial \mathcal{D}} \tag{22}$$

The steps of the proposed cost-sensitive time-series classification method (CS-LTS, henceforth) are shown in Algorithm 1. The pseudocode shows that the procedure updates all K shapelets and the weights W, W_0, false positive cost C_{FP} and parameter $\mathcal{D}$ by a learning rate η.

4   Experimental Evaluation

In this section, we evaluate the effectiveness of the proposed method on different setting represented by different datasets. The objective function in Eq. (7) is a non-convex function with respect to parameters and solving it via SGD requires a good initialization of the parameters. The initialization step is very important in this scenario as it influences whether the optimization reaches the region of global minimum.

Model Parameter Initializations: Shapelets were initialized using K-means centroids of all segments similar to [10]. First we set the minimum length (L_{min}) of a shapelet to be 10% of the length of the time-series examples. Then the total number of shapelets was computed as L_{min} multiplied by number of training time series. The number of shapelets used as input for the optimization function was determined using $K = log(total\ number\ of\ segments)$. Three scales $\{L_{min}, 2 \times L_{min}, 3 \times L_{min}\}$ of subsequence lengths were investigated.

The weight parameters W_k and W_0 were initialized randomly around 0. C_{FP} was initially set to 1. The values for θ and initial value of $\mathcal{D}$ were determined through a grid search approach using internal cross-validations over the training data. The values for θ were searched from the set $\{1, 5, 10, 25, 50, 100\}$ and the initial values for $\mathcal{D}$ was chosen from $\{0.001, 0.01, 0.1, 10, 100, 1000\}$. The best parameter value was identified via internal cross-validation on training data. Once the best parameter value was identified, the methods were trained on the entire training set using the best chosen parameters, and the learned model was tested on the test set which was completely separate from the training procedure. The learning rate η was initialized to a small value of 0.01. The $maxIter$ for the optimization was set to 5000 iterations.

Evaluation Measures: We report F_β score for $\beta \in \{1, 2, 3\}$ since this is a commonly used performance metric for imbalanced learning. These are simple functions of the precision and recall. The traditional F-score or F_1 score is the harmonic mean of precision and recall that is considered a balanced measure between precision and recall. For $\beta > 1$ the evaluation metric rewards higher true positive rates. We also consider the sensitivity and specificity evaluation metrics, as the objective is to achieve lower false negative with minimum increase in false positive rates.

4.1   Cost Sensitive Cardiac Arrhythmia Alarms Detection

In this set of experiments, we demonstrate the effectiveness of the proposed method on two cost-sensitive applications from PhysioNets MIMIC II version 3 repository [9,21]. The objective is to detect true alarms while suppressing false alarms, where missing true alarms (positive class) is more severe than missing false alarms (negative class), since missing true alarm could lead to serious consequences and risk patients' lives.

The database is a multi-parameter ICU repository containing patients' records of up to eight signals from bedside monitors in Intensive Care Units (ICU). The extracted datasets contain human-annotated true and false cardiac arrhythmia alarms. We extracted a subset of patients' records that contained signal from lead ECG II, because it was identified as the sensor that contained the least number of missing values across the patients. For each alarm event, a 20-s window prior to the alarm event was extracted similar to [20].

We partition the dataset into four distinct cross-validation datasets, where we train the model on 3 folds and test on the fourth one. In addition to the cross validation experiment, we repeat the entire process of cross-validation for 10 independent trials (each trial has 4 distinct partitions on true alarm instances) which results in 40 different combination of training data. The mean and standard deviation of the evaluation metrics is then reported.

The two datasets selected are VTACH and CHALLENGE. VTACH consists of true and false Ventricular Tachycardia alarms from the ICU patients. CHALLENGE dataset is a mixture of different true and false arrhythmia alarms. The alarms categories are Asystole, Extreme Bradycardia, Extreme Tachycardia, Ventricular Tachycardia and Ventricular Flutter/Fibrillation. This dataset was presented at a competition in 2015 organized by PhysioNet to encourage the development of algorithms to reduce the incidence of false alarms in the Intensive Care Unit (ICU).

Achieving high true alarm detection rate (TAD) or high sensitivity is important when suppressing high false alarm rates from bedside monitors in ICU. High false alarm rates cause desensitization among care providers, thus risking patients' lives [7]. The objective of the prediction task is to provide high false alarms suppression (FAS) rates (achieve high specificity) while keeping TAD (sensitivity) high. In the two datasets, (Fig. 2) CS-LTS (circle) achieves higher TAD (Y-axis) than LTS (diamond) and the current state-of-the-art baseline BEHAR [1] (star) in the field of critical alarm detection. FAS (X-axis) is better for LTS (diamond) on both datasets compared to CS-LTS (circle). However, improving TAD by decreasing FAS is acceptable as missing true alarms may result in patient fatality. CS-LTS (circle) beats BEHAR (star) in terms of true alarm detection rate on both the datasets. In terms of false alarm suppression, CS-LTS achieves comparable performance on VTACH dataset. BEHAR (star) achieves 100% FAS for CHALLENGE dataset, however, true alarm detection rate is 0%. Figure 3 shows the comparison of F_β scores for VTACH and CHALLENGE datasets. In both datasets CS-LTS outperforms LTS with respect to $\beta = 2$ and $\beta = 3$. This proves that CS-LTS improves the TAD score on both datasets when compared to LTS.

4.2   Balanced Time Series Datasets

In this set of experiments, we highlight that the proposed model attains comparable or better classification accuracy when compared to state-of-the-art LTS on balanced datasets. So, incorporating cost sensitive learning does not hurt the optimization algorithm because it automatically learns the cost sensitive parameters. This is very useful if the intrinsic sensitivity of the data is not known a priori.

Sixteen binary-class datasets were selected from UCR time-series repository [6]. In order to ensure fair comparison with LTS, the default train and test splits were used. Ten independent runs (with different initialization for both

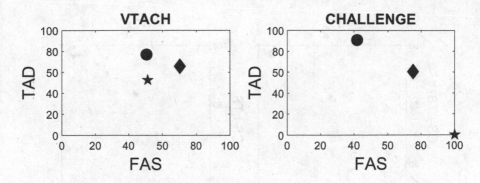

Fig. 2. CS-LTS[●] vs. LTS[♦] vs. BEHAR[★] in terms of true alarm detection (TAD) and false alarm suppression (FAS) rates over 2 critical alarm datasets. CS-LTS achieves higher TAD on both datasets compared to LTS and BEHAR.

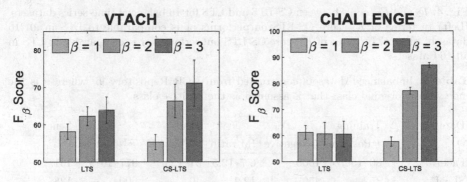

Fig. 3. Comparison of CS-LTS vs. LTS in terms F_1, F_2 and F_3 scores over 2 false alarm suppression datasets.

LTS and CS-LTS) were conducted and the average and standard deviation of the evaluation metric are reported.

The results of comparing CS-LTS to LTS on the 16 datasets are shown in Fig. 4. It is observed that CS-LTS outperforms or comparable to LTS on all 16 datasets. This set of experiments highlights the fact that the CS-LTS model provides a good alternative to LTS as it can handle balanced datasets quite effectively. The proposed method attains higher sensitivity with little loss of specificity when compared to LTS.

4.3   Imbalanced Time Series Datasets

In order to highlight the advantage of cost-sensitive learning over cost-insensitive learning, in this set of experiments, we extensively evaluate the model on 18 highly imbalanced datasets and compare it with LTS and different over-sampling and under-sampling methods. The imbalanced time series datasets were

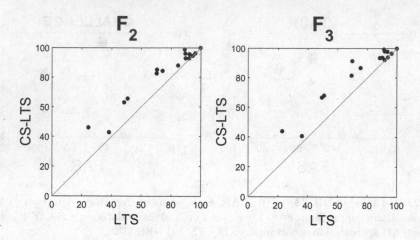

Fig. 4. F_2 and F_3 scores between CS-LTS and LTS for 16 balanced time-series datasets. (Left) In terms of F_2 score CS-LTS outperforms or is comparable to LTS in all 16 datasets. (Right) In terms of F_3 score CS-LTS outperforms or is comparable to LTS in all 16 datasets.

Table 1. Imbalanced datasets constructed from UCR Repository [6] where $*$ is the index of the original class that is assumed as the positive class

Dataset	Training			Test		Length
	#Positive	#Negative	IM ratio	#Positive	#Negative	
FaceAll*	80–150	1000	6.7–12.5	91–123	977–1079	131
SLeaf*	35	450	12.9	40	600	128
TwoPatterns*	200	180	9	1001–1106	1894–1999	128
Wafer*	200	380–3000	1.9–15	562–6220	392–3402	152
Yoga*	200	800–900	4–4.5	1300–1570	730–870	426

constructed by Cao et al. [4] from 5 multi-class datasets from the UCR time-series repository and the details are shown in Table 1.

The main advantage of CS-LTS over LTS is its superior performance in case of imbalanced datasets. In Fig. 5, it is shown that CS-LTS comfortably outperforms LTS on all 18 imbalanced datasets in terms of both F_1 and F_2 scores.

Moreover, in comparison to the state-of-the-art methods for imbalanced time-series classification, CS-LTS is very competitive. As shown in Table 2 in terms of F_1 score. The best method per dataset is shown in bold. The proposed CS-LTS method attains the highest number of absolute wins (5.86 wins) where a point is awarded to a method if it attains the highest F_1 score among the rest of the baseline methods for that particular dataset. In case of draws, the point is split into equal fractions and awarded to each method having the highest F_1 for a particular dataset.

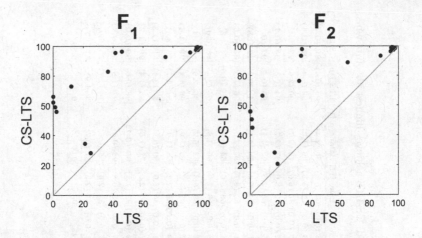

Fig. 5. F_1 and F_2 score between CS-LTS and LTS for 18 imbalanced time-series datasets. (Left) In terms of F_1 score CS-LTS achieves very high accuracy compared to LTS on 15 datasets and is comparable to LTS in 3. (Right) In terms of F_2 score CS-LTS outperforms or is comparable to LTS in all 18 datasets.

5   Discussion

Amongst the baselines, SPO [2], SMOTE [5], BORSMOTE [12], ADASYN [13], DB [11] and MoGT [4] are over-sampling techniques which mostly act as a preprocessing technique to over sample the rare class examples in order to construct balanced datasets. Easy [17] and Balanced [17] are under-sampling methods which reduces the number of examples from the majority class via under-sampling the majority class to balance the datasets.

From Table 2, we can infer that CS-LTS beats LTS and Easy across all datasets except 1 dataset (TwoPatterns3) in case of LTS which is a draw. Comparing with other baseline methods we see that CS-LTS has achieved similar accuracy as baseline methods on more than one datasets (such as wafer0 and wafer1). CS-LTS achieves comparable results with almost all of the over-sampling methods except for sleaf1 and TwoPatterns3 dataset. Results of CS-LTS on Sleaf1 and TwoPatterns3 certainly outperform LTS by huge margins; however, due to overlapping data-points in the feature space, it is hard for a linear model to achieve high classification accuracy in these two datasets. Compared to under-sampling methods (Easy and Balanced), CS-LTS is better than these baseline methods on most of the datasets. Another comparable method is the 1-Nearest Neighbor method (1-NN) which is known to be a good classifier for time-series classification problems. However, 1-NN computationally suffers from high dimensionality, hence it is time consuming compared to our method. Moreover, CS-LTS is an easier-to-interpret method as compared to 1-NN which makes it more desirable to domain experts. CS-LTS is an algorithmic approach to solve the imbalanced time-series classification problem whereas the state-of-the-art methods in this field are data manipulation methods that use over-sampling

Table 2. Comparison of mean F_1 scores for various baseline methods against proposed method. CS-LTS achieves highest absolute wins.

Dataset	SPO [2]	Repeat	SMOTE [5]	BORSMOTE [12]	ADASYN [13]	DB [11]	1MoGT [4]	2MoGT [4]	1 NN	Easy [17]	Balanced [17]	LTS [10]	CS-LTS
FaceAll1	96 (0.9)	94 (0.0)	95 (0.6)	95 (0.5)	95 (0.5)	95 (0.8)	96 (0.5)	97 (0.5)	98 (0.0)	67 (5.9)	86 (2.4)	98 (0.4)	**99 (0.2)**
FaceAll2	93 (1.0)	83 (0.0)	88 (0.5)	88 (0.7)	88 (0.8)	92 (0.4)	90 (0.5)	86 (0.8)	83 (0.0)	76 (3.2)	93 (1.3)	93 (0.4)	**95 (0.4)**
FaceAll3	95 (0.6)	**97 (0.0)**	96 (0.6)	**97 (0.2)**	96 (0.4)	91 (0.4)	95 (0.6)	94 (0.6)	**97 (0.0)**	60 (6.6)	73 (2.7)	90 (2.6)	92 (0.4)
FaceAll4	94 (0.5)	96 (0.0)	95 (0.6)	96 (0.5)	96 (0.5)	90 (1.0)	95 (0.5)	95 (0.5)	96 (0.0)	72 (3.0)	87 (2.7)	94 (0.3)	**98 (0.1)**
FaceAll5	96 (0.4)	**97 (0.0)**	**97 (0.1)**	**97 (0.1)**	**97 (0.2)**	95 (0.3)	**97 (0.2)**	95 (0.3)	95 (0.0)	85 (2.5)	92 (1.1)	95 (0.4)	**97 (0.1)**
SLeaf1	83 (0.8)	81 (0.0)	79 (1.4)	79 (1.6)	79 (1.6)	81 (1.6)	**87 (2.1)**	83 (1.7)	57 (0.0)	54 (5.1)	50 (4.4)	4 (20.2)	49 (1.8)
SLeaf2	96 (1.0)	94 (0.0)	95 (0.7)	96 (0.0)	96 (0.4)	96 (0.0)	**98 (0.7)**	95 (0.3)	91 (0.0)	85 (6.7)	87 (3.9)	96 (1.5)	**98 (0.5)**
SLeaf3	**88 (1.6)**	83 (0.0)	83 (1.0)	83 (1.1)	83 (1.1)	82 (0.5)	84 (0.7)	84 (1.4)	66 (0.0)	66 (5.6)	54 (6.6)	0.0 (0.0)	84 (1.6)
SLeaf4	**93 (1.0)**	61 (0.0)	72 (2.4)	71 (0.7)	73 (0.4)	89 (1.5)	83 (2.9)	88 (1.9)	68 (0.0)	56 (7.9)	66 (4.8)	66 (36.7)	88 (0.0)
SLeaf5	**90 (1.1)**	88 (0.0)	89 (0.7)	89 (0.7)	89 (0.6)	87 (0.8)	89 (1.0)	89 (0.8)	71 (0.0)	59 (8.3)	52 (5.2)	36 (3.9)	82 (1.4)
TwoPatterns1	92 (0.3)	71 (0.0)	77 (0.2)	77 (0.2)	78 (0.3)	89 (0.2)	84 (0.6)	84 (0.6)	92 (0.0)	95 (4.0)	75 (1.6)	96 (1.4)	**99 (1.4)**
TwoPattern2	78 (0.7)	65 (0.0)	68 (0.3)	68 (0.1)	68 (0.2)	73 (0.2)	75 (0.5)	81 (0.6)	**89 (0.0)**	31 (2.4)	68 (1.2)	51 (1.7)	72 (3.4)
Twopattern3	86 (0.3)	65 (0.0)	70 (0.4)	71 (0.5)	71 (0.7)	57 (0.2)	82 (0.6)	89 (0.6)	**91 (0.0)**	36 (3.0)	69 (0.9)	5 (13.1)	51 (11.3)
TwoPattern4	90 (0.5)	68 (0.0)	73 (0.2)	73 (0.2)	73 (0.2)	73 (0.2)	82 (0.7)	87 (0.4)	87 (0.0)	35 (2.5)	71 (1.5)	96 (1.4)	**99 (1.2)**
Wafer0	**99 (0.0)**	**99 (0.0)**	**99 (0.0)**	**99 (0.0)**	**99 (0.0)**	**99 (0.0)**	**99 (0.0)**	**99 (0.0)**	**99 (0.0)**	93 (1.1)	**99 (0.1)**	98 (0.6)	**99 (0.0)**
Wafer1	**99 (0.1)**	**99 (0.0)**	**99 (0.1)**	**99 (0.0)**	**99 (0.1)**	**99 (0.1)**	**99 (0.1)**	**99 (0.1)**	98 (0.0)	93 (0.8)	98 (0.6)	97 (1.6)	**99 (0.1)**
Yoga1	89 (0.2)	88 (0.0)	90 (0.1)	90 (0.2)	90 (0.2)	88 (0.0)	88 (0.1)	88 (0.2)	83 (0.0)	59 (2.5)	85 (0.6)	24 (1.7)	70 (0.9)
Yoga2	**91 (0.2)**	90 (0.0)	91 (0.1)	91 (0.1)	91 (0.1)	**91 (0.0)**	90 (0.1)	**91 (0.1)**	86 (0.0)	61 (2.5)	87 (0.7)	5 (0.0)	81 (1.7)
Absolute Wins	3.36	0.69	0.85	1.18	0.85	0.36	1.86	0.36	2.42	0	0.09	0	5.86

and under-sampling techniques, which act as a preprocessing step to solve the high imbalance time-series classification problem. Figure 6 shows the critical difference diagram amongst all the baseline methods and CS-LTS.

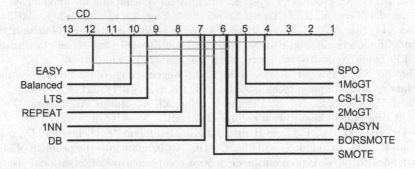

Fig. 6. Critical difference diagram showing average rank of CS-LTS against all baseline methods on 18 imbalanced datasets.

6   Conclusion

In this paper, we adapt the novel perspective of learning generalized shapelets for time-series classification via a logistic loss minimization, and extend the time-series classification framework to a cost-sensitive framework that can handle highly imbalanced time-series datasets. In contrast to the baseline model, whose prediction accuracy is biased towards the abundant negative class, the proposed CS-LTS does not suffer from class imbalance problem. Extensive experiments on 36 real-world time-series datasets reveal the proposed method is a good alternative to the baseline model. It can handle both balanced and imbalanced time-series datasets and achieve better or comparable results against the current state-of-the-art methods. In future, we plan to extend the cost-sensitive learning framework for multivariate time-series datasets in order to improve the performance of the model.

Acknowledgments. This research was supported in part by DARPA grant No. FA9550-12-1-0406 negotiated by AFOSR, the National Science Foundation grants NSF-SES-1447670, NSF-IIS-1636772 and Temple University Data Science Targeted Funding Program. Simulations were performed on the OwlsNest HPC cluster at Temple University, which is supported in part by the National Science Foundation through NSF grant CNS-1625061 and Pennsylvania Department of Health CURE grant.

References

1. Behar, J., Oster, J., Li, Q., Clifford, G.: ECG signal quality during arrhythmia and its application to false alarm reduction. IEEE Trans. Biomed. Eng. **60**(6), 1660–1666 (2013)

2. Cao, H., Li, X., Woon, D.Y., Ng, S.: SPO: structure preserving oversampling for imbalanced time series classification. In: 11th IEEE International Conference on Data Mining, ICDM 2011, Vancouver, BC, Canada, 11–14 December 2011, pp. 1008–1013 (2011)

3. Cao, H., Li, X., Woon, D.Y., Ng, S.: Integrated oversampling for imbalanced time series classification. IEEE Trans. Knowl. Data Eng. **25**(12), 2809–2822 (2013)

4. Cao, H., Tan, V.Y.F., Pang, J.Z.F.: A parsimonious mixture of Gaussian trees model for oversampling in imbalanced and multimodal time-series classification. IEEE Trans. Neural Netw. Learn. Syst. **25**(12), 2226–2239 (2014)

5. Chawla, N.V., Bowyer, K.W., Hall, L.O., Kegelmeyer, W.P.: Smote: synthetic minority over-sampling technique. J. Artif. Int. Res. **16**(1), 321–357 (2002)

6. Chen, Y., Keogh, E., Hu, B., Begum, N., Bagnall, A., Mueen, A., Batista, G.: The UCR time series classification archive, July 2015

7. Drew, B.J., Harris, P., Zgre-Hemsey, J.K., Mammone, T., Schindler, D., Salas-Boni, R., Bai, Y., Tinoco, A., Ding, Q., Hu, X.: Insights into the problem of alarm fatigue with physiologic monitor devices: a comprehensive observational study of consecutive intensive care unit patients. PLoS ONE **9**(10), e110274 (2014)

8. Ghalwash, M., Radosavljevic, V., Obradovic, Z.: Utilizing temporal patterns for estimating uncertainty in interpretable early decision making. In: Proceedings of ACM SIGKDD International Conference on Knowledge Discovery and Data Mining. pp. 402–411 (2014)

9. Goldberger, A.L., Amaral, L.A.N., Glass, L., Hausdorff, J.M., Ivanov, P.C., Mark, R.G., Mietus, J.E., Moody, G.B., Peng, C.K., Stanley, H.E.: PhysioBank, PhysioToolkit, and PhysioNet: components of a new research resource for complex physiologic signals. Circulation **101**(23), e215–e220 (2000)

10. Grabocka, J., Schilling, N., Wistuba, M., Schmidt-Thieme, L.: Learning time-series shapelets. In: Proceedings of 20th ACM SIGKDD International Conference on Knowledge Discovery and Data Mining, KDD 2014, pp. 392–401. ACM (2014)

11. Guo, H., Viktor, H.L.: Learning from imbalanced data sets with boosting and data generation: the DataBoost-IM approach. SIGKDD Explor. Newsl. **6**(1), 30–39 (2004)

12. Han, H., Wang, W.-Y., Mao, B.-H.: Borderline-SMOTE: a new over-sampling method in imbalanced data sets learning. In: Huang, D.-S., Zhang, X.-P., Huang, G.-B. (eds.) ICIC 2005. LNCS, vol. 3644, pp. 878–887. Springer, Heidelberg (2005). https://doi.org/10.1007/11538059_91

13. He, H., Bai, Y., Garcia, E.A., Li, S.: ADASYN: adaptive synthetic sampling approach for imbalanced learning. In: Proceedings of International Joint Conference on Neural Networks, IJCNN 2008, Part of the IEEE World Congress on Computational Intelligence, WCCI 2008, Hong Kong, China, 1–6 June 2008, pp. 1322–1328 (2008)

14. He, H., Garcia, E.A.: Learning from imbalanced data. IEEE Trans. Knowl. Data Eng. **21**(9), 1263–1284 (2009)

15. Hills, J., Lines, J., Baranauskas, E., Mapp, J., Bagnall, A.: Classification of time series by shapelet transformation. Data Min. Knowl. Discov. **28**(4), 851–881 (2014)

16. Hou, L., Kwok, J.T., Zurada, J.M.: Efficient learning of timeseries shapelets. In: Proceedings of 30th AAAI Conference on Artificial Intelligence, 12–17 February 2016, Phoenix, Arizona, USA, pp. 1209–1215 (2016)

17. Liu, X.Y., Wu, J., Zhou, Z.H.: Exploratory undersampling for class-imbalance learning. Trans. Sys. Man Cyber. Part B **39**(2), 539–550 (2009)

18. Lpez, V., Fernndez, A., Garca, S., Palade, V., Herrera, F.: An insight into classification with imbalanced data: empirical results and current trends on using data intrinsic characteristics. Inf. Sci. **250**, 113–141 (2013)

19. Radosavljevic, V., Vucetic, S., Obradovic, Z.: Continuous conditional random fields for regression in remote sensing. In: Proceedings of 2010 Conference on ECAI 2010: 19th European Conference on Artificial Intelligence, pp. 809–814. IOS Press, Amsterdam (2010)

20. Roychoudhury, S., Ghalwash, M.F., Obradovic, Z.: False alarm suppression in early prediction of cardiac arrhythmia. In: 2015 IEEE 15th International Conference on Bioinformatics and Bioengineering (BIBE), pp. 1–6, November 2015

21. Saeed, M., Villarroel, M., Reisner, A., Clifford, G., Lehman, L.W., Moody, G., Heldt, T., Kyaw, T., Moody, B., Mark, R.: Multiparameter intelligent monitoring in intensive care II: a public-access intensive care unit database. Crit. Care Med. **39**(5), 952–960 (2011)

22. Sun, Y., Kamel, M.S., Wong, A.K.C., Wang, Y.: Cost-sensitive boosting for classification of imbalanced data. Pattern Recogn. **40**(12), 3358–3378 (2007)

23. Ye, L., Keogh, E.: Time series shapelets: a new primitive for data mining. In: Proceedings of 15th ACM SIGKDD International Conference on Knowledge Discovery and Data Mining, KDD 2009, pp. 947–956. ACM, New York (2009)

24. Zhang, Q., Wu, J., Yang, H., Tian, Y., Zhang, C.: Unsupervised feature learning from time series. In: Proceedings of 25th International Joint Conference on Artificial Intelligence, IJCAI 2016, 9–15 July 2016, New York, NY, USA, pp. 2322–2328 (2016)

Cost-Sensitive Perceptron Decision Trees for Imbalanced Drifting Data Streams

Bartosz Krawczyk[✉] and Przemysław Skryjomski

Department of Computer Science, Virginia Commonwealth University,
Richmond, VA 23284, USA
bkrawczyk@vcu.edu

Abstract. Mining streaming and drifting data is among the most popular contemporary applications of machine learning methods. Due to the potentially unbounded number of instances arriving rapidly, evolving concepts and limitations imposed on utilized computational resources, there is a need to develop efficient and adaptive algorithms that can handle such problems. These learning difficulties can be further augmented by appearance of skewed distributions during the stream progress. Class imbalance in non-stationary scenarios is highly challenging, as not only imbalance ratio may change over time, but also relationships among classes. In this paper we propose an efficient and fast cost-sensitive decision tree learning scheme for handling online class imbalance. In each leaf of the tree we train a perceptron with output adaptation to compensate for skewed class distributions, while McDiarmid's bound is used for controlling the splitting attribute selection. The cost matrix automatically adapts itself to the current imbalance ratio in the stream, allowing for a smooth compensation of evolving class relationships. Furthermore, we analyze characteristics of minority class instances and incorporate this information during the model update process. It allows our classifier to focus on most difficult instances, while a sliding window keeps track of changes in class structures. Experimental analysis carried out on a number of binary and multi-class imbalanced data streams indicate the usefulness of the proposed approach.

Keywords: Machine learning · Data streams · Imbalanced data
Concept drift · Online learning · Multi-class imbalance

1 Introduction

Modern machine learning systems must take into account the phenomenon of data in motion, a scenario in which instances arrive rapidly and continuously. This has given birth to the notion of data streams, potentially unbounded and ordered data collections. They impose new challenges on learning systems, due to their ever-growing size, speed of incoming instances and difficulties or latencies for obtaining true class labels. Additionally, data characteristics may change over time, leading to a phenomenon known as concept drift [2]. When dealing

© Springer International Publishing AG 2017
M. Ceci et al. (Eds.): ECML PKDD 2017, Part II, LNAI 10535, pp. 512–527, 2017.
https://doi.org/10.1007/978-3-319-71246-8_31

with non-stationary data, the previously trained classifier may lose its competence over time, as the new concepts are vastly different from the previously seen ones. By using only incremental learning we will accommodate new instances, but do not take into account the changes in the relevance of older cases. Therefore, mining data streams require methods that are able to detect the potential presence of drift in order to reset the learning model, or smoothly adapt to incoming data with properly tuned forgetting mechanism. This is built on the top of need for high responsiveness, low computational resource consumption and highly limited data storage.

The discussed scenario becomes even more complex when we consider the potential presence of difficulty known as class imbalance [4]. Skewed distributions pose significant challenge for classifiers, leading to their bias towards the majority class. At the same time in many real-life applications minority instances are usually of higher significance. While learning from imbalanced data has gained broad attention from the research community over last two decades, online class imbalance is still an emerging topic [12]. Here the imbalance ratio may change dynamically with the stream progress. Furthermore, the concept drift may lead to changes in class relationships, allowing for minority class to become the majority one and vice versa. While most works in online imbalance consider binary cases, one must be aware that new classes may appear over time and old ones disappear, causing further complications for the learning process. As the number of classes may change over time, so may their mutual relationships. Multi-class imbalance, while difficult on itself in static scenarios, is an important direction to be addressed in order to obtain robust data stream classifiers. Additionally, we need proper performance metrics that can take into account streaming and multi-class skewed nature of analyzed data.

In this paper we propose a novel decision tree learning approach for handling imbalanced and drifting data streams. As a base for our model, we use fast perceptron trees and improve them to become skew-insensitive by using a moving threshold solution. It aims at re-balancing the supports for each class during the decision making step, thus alleviating the skew bias with almost no additional computational cost. This is achieved by weighting support functions for each class according to a specified cost function. In our solution the cost matrix evolves over time and adapts to the current state of the stream. This allows us to propose an adaptive cost-sensitive solution that is able to learn from both binary and multi-class imbalanced data streams. We augment it with drift detection and use McDiarmids bound for controlling the splitting attribute selection. Additionally, we show how to analyze the structure of minority classes in an online manner by using a sliding window. This allows us to estimate the difficulty of incoming minority class instances, giving an additional insight into the current state of the stream. We propose an efficient method of incorporating this background information into the update process of the proposed decision tree in order to better capture the minority class characteristics. Experimental study carried out on a number of drifting and imbalanced binary and multi-class data streams shows the usefulness of the proposed learning algorithm.

2   Learning from Imbalanced Data Streams

In this section, we will discuss the necessary background for this paper. This includes the area of mining drifting data streams, the problem of skewed class distributions and online learning in the presence of class imbalance.

2.1   Data Stream Mining

Let us define data stream as an ordered sequence of instances that arrive over time and can be of unbounded size. This leads to a set of learning characteristics specific to this problem that must be accounted for when designing data stream mining algorithms. Here, we do not have a predefined training set, but the instances become available at various time intervals sequentially with the stream progress. Additionally, due to the unknown and potentially massive size of the stream we cannot store it in memory and must use each instance a limited number of times before discarding it to limit the computational resources being used. Furthermore, characteristics of the stream are subject to change and we must accommodate this fact during the continuous learning process [2].

We will assume that data stream is composed of a set of states $S = \{S_1, S_2, \cdots, S_n\}$, where given state S_i comes from a distribution D_i. We may deal with online case (each state is a separate instance) or chunk case (each state is a set of instances). The simplest learning scenario is a stationary data stream, where transition between states $S_j \to S_{j+1}$ holds $D_j = D_{j+1}$. In most real-life scenarios the stream characteristics evolve over time, leading to the notion of non-stationary data stream and concept drift. It may affect various aspects of incoming data, thus leading to a number of views on the discussed phenomenon. From the point of view of influencing the existing decision boundaries, we may distinguish real and virtual concept drifts. The former has effect on posterior probabilities and may impact unconditional probability density functions. This forces the learning system to adapt to change in order not to lose the competence. The latter drift does not have any effect on posterior probabilities, but only on conditional probability density functions. It may still cause difficulties for the learning system, leading to false alarms and unnecessary computational expenses on rebuilding the classifier. Another view on concept drift comes form the severity of ongoing changes. Sudden concept drift appears when S_j is being suddenly replaced by S_{j+1}, where $D_j \neq D_{j+1}$. Gradual concept drift is a transition phase where examples in S_{j+1} are generated by a mixture of D_j and D_{j+1} with their proportions continuously changing. Incremental concept drift is characterized by a smooth and slow transition between distributions, where the differences between D_j and D_{j+1} are not significant. Additionally, we may face recurring concept drift, in which a state from k-th previous iteration may suddenly reemerge $D_{j+1} = D_{j-k}$, which may take place once or periodically.

In order to tackle the presence of concept drift one may chose among three main approaches: (a) rebuilding classifier from a scratch whenever new instances become available; (b) use a specific tool to detect changes and guide the model reconstruction; and (c) use adaptive classifier that will naturally follow the changes in the stream. The first approach is completely unsuitable due to prohibitive computational requirements, especially in case of online stream processing. Following the remaining two directions, we may distinguish four main approaches to handling drifting data streams. First one relies on using a concept drift detector - an external tool that monitors the characteristics and informs when a change is expected to appear. This allows for rebuilding the classifier only when it is deemed as necessary. Second one uses a sliding window in order to keep a track of most recent instances, as they should be most representative to the current state of the stream. Such a window follows the stream, discarding old instances and replacing them with most recent ones. However, the size of the window is a crucial factor affecting the performance of this approach. Third solution uses online or incremental classifiers that are able to incorporate new instances by updating the classification model without a need for a complete retraining. A forgetting mechanism is required in order to allow for better adaptation to changes and reduced model complexity. Finally, ensemble solutions are popularly used for mining drifting data streams. Here, new instances may be used to control diversity of the base learners, allowing them to better adapt to changes, while offering improved predictive capabilities.

2.2   Online Class Imbalance

Learning from imbalanced data is continuously challenging topic despite over two decades of developments in this domain [4]. Skewed distributions pose challenges to most of classifiers, as their will lead to a bias towards the majority class, while minority is usually the more important one. This has lead to a number of solutions that aims at alleviating this disproportion that can be grouped into three categories: data-level, algorithm-level and hybrid solutions. The first one uses preprocessing algorithms to balance class distributions. It involves oversampling the minority class, undersampling the majority one, or both at the same time. Second group focuses on identifying what causes a given classifier to fail in an imbalanced scenario and modifying its learning procedure in order to make it skew-insensitive. Third solution is a combination of one of the two previous ones with another learning paradigm, most commonly ensemble solution [14].

While there is a plethora of works devoted to binary imbalanced problems, its multi-class version still requires a significant attention [9]. One cannot view it as a simple extension from two to many classes, as the complexities go far beyond it. In binary cases the relationships between classes are easily defined and the bias is easy to be identified. In multi-class scenario we deal with much more complex dynamics among classes, leading to a notion of multi-minority and multi-majority cases. Using a simple decomposition into a set of pairwise relations leads to loss of useful information, as it is easy to gain performance on some of classes, while losing it on others. Furthermore, difficulties embedded in the nature of imbalanced data, such as noisy instances and class overlapping, become much more difficult to tackle.

The problem of imbalance becomes even more challenging when being considered from online and non-stationary perspective [12]. Here not only we must deal with skewed data distributions, but also with the fact that the underlying imbalance ratio is not know from the beginning and is subject to continuous change during the stream progress. As instances arrive one by one there is a need to monitor the relationship between classes and update the learning model accordingly. In this scenario two types of changes are bound to appear. First one is evolving imbalance ratio and class properties. Here incoming instances may influence the distribution skew, either strengthening or weakening it. The role of classes in no longer stationary and in time minority and majority distributions may swap places. Therefore, online class imbalance learning requires algorithms that are not fixed on a given minority class, but can adapt themselves to constantly evolving class dynamics. This may be accompanied with the concept drift, where class boundaries may be affected. A proper solution to this problem should be able to react to both types of changes in order to achieve good adaptability, generalization and minority-majority concept description. There is a number of works for imbalanced data streams that work under a much simpler assumption that the role of classes do not change over time or that data arrives in chunks and we must handle only local skewness. There is still little works devoted to actual online class imbalance and the most efficient approaches include a neural network-based solutions [12], combination of Hoeffding decision tree with Hellinger distance splitting criterion [5], and a Bagging-based ensemble solution [12].

Multi-class online imbalance is even more difficult to handle, as classes may swap their multi-minority and multi-majority roles [11]. Therefore, not only we need to model the evolving multi-distribution imbalance ratio, changing class relationships (together with overlapping levels or noisy instances), but also take into account the fact that number of minority and minority classes may change at any stage of processed data stream. So far there are only two solutions discussed in the literature to this problem, based on ensemble of neural networks and multi-class oversampling/undersampling with online Bagging ensemble.

3   Cost-Sensitive Perceptron Decision Trees

In this section, we will discuss in details the proposed cost-sensitive perceptron decision trees with adaptive threshold for online mining of imbalanced drifting data streams, as well as the usage of McDiarmid's bound for controlling the splitting attribute selection and online analysis of the minority class structures for gaining additional information to improve the performance of classifiers.

3.1   McDiarmid's Perceptron Decision Tree

We propose to build our learning algorithm for imbalanced and drifting data streams on top of the Fast Perceptron Decision Tree [1], as it provides both high accuracy and update speed, making it highly suitable for the task at hand. Its main advantage lies in using a linear perceptron at each leaf. This allows to speed-up the decision making process, as well as improve the overall accuracy. This hybrid solution combines the advantages of trees and neural models, allowing for efficient processing of data streams.

We use an online perceptron approach with sigmoid activation function (as suggested by Bifet et al. [1]) with squared error optimization. Let us assume that instances from the stream arrive in a form of $\langle \mathbf{x_i}, y_i \rangle$, where x_i is a feature vector for i-th instance and y_i is a class label associated with it. The perceptron learning scheme aims at minimizing the number of misclassified instances. We will annotate the learning hypothesis function of given perceptron for i-th instance as $h_\mathbf{w}(\mathbf{x_i})$. To evaluate the learning process, we apply the mean-square error defined as $J(\mathbf{w}) = \frac{1}{2} \sum (y_i - h_\mathbf{w}(\mathbf{x_i}))^2$. Bifet et al. [1] proposed to use sigmoid activation function instead of a traditional threshold and we follow this approach. The sigmoid activation function for hypothesis $h_\mathbf{w} = \sigma(\mathbf{w}^T \mathbf{x_i})$ is expressed as $\sigma(x) = 1/(1 = e^{-x})$, being differentiable as $\sigma'(x) = \sigma(x)(1 - \sigma(x))$. This allows us to calculate the error function gradient as follows:

$$\bigtriangledown J = - \sum_i (y_i - h_\mathbf{w}(\mathbf{x_i})) \bigtriangledown h_\mathbf{w}(\mathbf{x_i}), \tag{1}$$

where sigmoid hypothesis:

$$\bigtriangledown h_\mathbf{w}(\mathbf{x_i}) = h_\mathbf{w}(\mathbf{x_i})(1 - h_\mathbf{w}(\mathbf{x_i})), \tag{2}$$

which allows us to compute the following weight update rule:

$$\mathbf{w} = \mathbf{w} + \eta \sum_i (y_i - h_\mathbf{w}(\mathbf{x_i})) \, h_\mathbf{w}(\mathbf{x_i})(1 - h_\mathbf{w}(\mathbf{x_i}))\mathbf{x_i}. \tag{3}$$

As we deal with online learning, a stochastic gradient descent is being used with weights updated after each instance [1]. A single perceptron is trained per each class, making it suitable for both binary and multi-class problems. To obtain a final prediction regarding the class of new instance we select the highest value of support functions returned by each perceptron $\arg\max_k (h_{\mathbf{w_1}}(\mathbf{x}), \cdots , h_{\mathbf{w_K}}(\mathbf{x}))$, where K is the number of classes.

Original implementation of Fast Perceptron Decision Tree used Hoeffding inequality to determine the amount of instances needed for conducting a split [1]. However, recent study discussed flaws in the Hoeffding bound [8]. In this work, we propose to modify the underlying base of the original Fast Perceptron Decision Tree and use a McDiarmid's inequality for controlling the splitting criteria. It is a generalization of the Hoeffding's inequality, being applicable to both numerical and non-numerical data, as well as better describing the split measures. Let us now present the McDiarmid's theorem.

Theorem 1 (McDiarmid's Theorem). *Let $X_1, \cdots , X_n$ be a set of independent random variables and $f(x_1, \cdots , x_n)$ be a function fulfilling the following inequality:*

$$\sup_{x_1, \cdots , x_i, \cdots , x_n, \hat{x_i}} |f(x_1, \cdots , x_i, \cdots , x_n) - f(x_1, \cdots , \hat{x_i}, \cdots , x_n)| \leq c_i, \forall_{i=1, \cdots , n}.$$

$$\tag{4}$$

Then for any given $\epsilon > 0$ the following inequality is true:

$$Pr\left(f(X_1, \cdots, X_n) - E\left[f(X_1, \cdots, X_n)\right] \geq \epsilon\right) \leq \exp\left(-\frac{2\epsilon^2}{\sum_{i=1}^n c_i^2}\right) = \delta. \quad (5)$$

One may apply McDiarmid's inequality to any split measure. We use it in combination with the popular Gini gain in order to estimate the minimal number of instances n to conduct a split during data stream processing [8]. One may defined the Gini gain as follows:

$$\Delta g_i^G(S) = g^G(S) - \sum_{q \in \{L,R\}} \frac{n_{q,i}(S)}{n(S)}\left(1 - \sum_k^K \left(\frac{n_{q,i}^k(S)}{n_{q,i}(S)}\right)^2\right), \quad (6)$$

where S is a set of instances in analyzed node, L and R stand for children left and right nodes, $n_{q,i}(S)$ is the number of elements in given node that will be passed to q-th child node for split made on i-th attribute, and $n_{q,i}(S)$ is the number of instances belonging to k-th class that will be passed to q-th child node for split made on i-th attribute.

This allows us to formulate McDiarmid's inequality for comparing Gini gains for any two features.

Theorem 2 (McDiarmid's Inequality for Gini Gain). *Let $S = s_1, \cdots, s_n$ be a set of instances and let $\Delta g_i^G(S)$ and $\Delta g_j^G(S)$ be the Gini gain values (see Eq. 6) for i-th and j-th feature. If the following condition is satisfied by them:*

$$\Delta g_i^G(S) - \Delta g_j^G(S) > \sqrt{\frac{8\ln(1/\delta)}{n(S)}}, \quad (7)$$

then the following inequality holds with probability of $1 - \delta$ or higher:

$$E[\Delta g_i^G(S)] > E[\Delta g_j^G(S)]. \quad (8)$$

Corollary 1 (McDiarmid's Splitting Criterion for Gini Gain). *Let us assume that $\Delta g_{i_1}^G(S)$ and $Deltag_{i_2}^G(S)$ are the metric values for features with respectively highest and second highest Gini gain. If the following condition is satisfied:*

$$\Delta g_{i_1}^G(S) - \Delta g_{i_2}^G(S) > \sqrt{\frac{8\ln(1/\delta)}{n(S)}}, \quad (9)$$

then following Theorem 2, with the probability equal to $(1 - \delta)^{d-1}$ the following statement is true:

$$i_1 = \arg\max_{i=1,\cdots,d}\left\{E[g_i^G(S)]\right\}, \quad (10)$$

where d is the number of features and i_1-th feature is selected to split the current node.

3.2   Cost-Sensitive Modification with Adaptive Output

The decision tree learning algorithm discussed above will be further modified in order to make it suitable for learning from online imbalanced data. While decision trees are popular both in static imbalanced or balanced streaming data mining areas [13], for online skewed data there exists only a modification of Hoeffding Tree using Hellinger distance for conducting splits [5]. This metric, although skew insensitive, may still fail for difficult imbalanced datasets with complex class structures. On the other hand, it imposes minimal additional computational cost on the classifier - a highly desirable property in data stream mining. In non-stationary scenarios using data preprocessing is challenging and may lead to a prohibitively increased computational complexity. Therefore, algorithm-level solutions are worth pursuing and we will concentrate on them in this paper.

We propose to take an advantage of using perceptrons in leafs of the decision tree and enhance them with cost-sensitive approach. This will be achieved by modifying the output of each perceptron, instead of changing the structure of the training data or the training algorithm.

We will introduce the cost-sensitive modification in the prediction step. In the previous section for a K-class problem we denoted the continuous output of k-th perceptron for object $\mathbf{x}$ as $h_{\mathbf{w_k}}(\mathbf{x})$. In the proposed cost-sensitive approach, we will calculate the output of k-th perceptron in a leaf of our decision tree as:

$$h^*_{\mathbf{w_k}}(\mathbf{x}) = \sum_{l=1}^{K} h_{\mathbf{w_k}}(\mathbf{x}) \cdot cost[k, l], \tag{11}$$

where $cost[k, l]$ is the misclassification cost between k-th and l-th class, provided by the user.

Output modification approaches for neural classifiers have proved themselves to be efficient in tackling stationary imbalanced data [15], yet this is the first work on their usage for online class imbalance. This solution is highly compatible with data stream mining requirements, as it does not impose significant additional computational needs, do not rely on data preprocessing and can be easily included in the proposed decision tree learning scheme, taking the advantage of McDiarmid's inequality. Additionally, it is easily applicable for both binary and multi-class data streams, making it a versatile approach.

3.3   Adaptive Online Cost Matrix

The used misclassification costs have significant influence on the performance of this method. Too low cost would not alleviate the bias, while too high cost would degrade the performance over the majority classes. We need to strive for a balanced performance. In optimal scenario the cost would be provided by a domain expert or embedded in the nature of the problem. However, in most of real-life cases we do not have a supplied cost matrix and thus a manual setting is required. There exist some automatic and semi-automatic methods applicable to stationary data, yet they cannot be used for data streams.

As the imbalanced stream will evolve over time, it is to be expected that it will affect relationships among classes. Therefore, a static cost matrix will quickly become outdated, not being able to properly reflect the current concept. On the other hand tuning it for each incoming set of instances would impose additional computational requirements, as well as a need for validation instances to select the best settings. Therefore, a lightweight solution is needed that will be able to keep a track of changes in the stream.

We propose a simple, yet effective approach of monitoring the current imbalance ratio among classes and setting the cost according to local pairwise imbalance ratios. This will allow for an easy modeling of multi-minority and multi-majority cases. The costs will change with the progress of the stream, as labels of incoming instances will be recorded and used to update the current skewness levels. Please note that this solution does not require to keep the instances in memory, as only counters for each class are needed. However, as the stream evolve over time one cannot keep all of previous information regarding class relationships. Therefore, we propose to use a fixed time threshold, as well as time stamps with each recorded label and use them to remove outdated cases from imbalance ratio counting. This allows for dynamically adapting our cost matrix to changes and drifts in the data stream.

3.4   Drift and Imbalance Detection

In order to efficiently learn from imbalanced and drifting data streams, we require tools that will be able to monitor the imbalance ratio and the appearance of concept drifts. We propose to combine our Cost-Sensitive Perceptron Decision Tree with Drift Detection Method for Online Class Imbalance (DDM-OCI) [10]. It is based on monitoring the recall in the minority class and if there is a significant drop in it (as evaluated by drift detector), it reports a drift. Following other works in drift detection, it may also be used to output a two-stage decision: drift warning and drift detection. This will be very useful for the next solution to online class imbalance proposed in this paper. DDM-OCI was proposed for binary online imbalance, but can be easily extended to multi-class cases. Here, we monitor the averaged recall over all of minority classes. In our case that means all of classes with an exception to the current most frequent one, allowing for taking into account that in multi-class online imbalance roles of classes are also subject to change.

3.5   Online Analysis of Minority Class Structure

Imbalance ratio among classes is not the sole source of learning difficulty. The underlying class structures, overlapping and noisy instances have significant impact on the decision boundaries being estimated. Therefore, one may assume that minority class instances may pose a different level of difficulty to the learning procedure. Recent works for static imbalanced data propose to take this factor into account and analyze the types of minority instances [6]. However, there is still a need for approaches that will directly incorporate this information into training procedures. Additionally, no such analysis have been done for data streams and online class imbalance.

In this work, we propose to analyze the difficulty of incoming minority class objects, while taking into account the evolving structure of classes. Firstly, let us define the types of minority instances. For their identification, we will use a neighborhood search with $k = 5$, similar as in works dealing with static data. Based on that, we propose six levels of difficulty λ that can be assigned to each new minority instance based on how contaminated is its neighborhood. This is measured by parameter ρ that states how many of k neighbors belong to the same minority class. Details are presented in Table 1. This analysis may be extended to multi-class scenario by considering each multi-minority class separately, as we have shown for static scenarios in our previous work [9].

Table 1. Six levels of instance difficulty for minority class.

	safe	borderline	borderline+	rare	rare+	outlier
ρ	5	4	3	2	1	0
λ	1	2	3	4	5	6

We propose to label each new minority class instance based on this analysis. As we deal with an online scenario, we cannot nor want to keep the entire stream in the memory. Therefore, we propose to analyze the types of minority instances using a small sliding window that will keep only the most recent instances, allowing for a fast neighborhood search within it. Additionally, we will incorporate the information from the drift detector. When a warning signal is being raised, the window will be reduced to 1/4 of its original size. This will allow to accommodate the change that starts to appear by taking into account a reduced subset of recent instances. When a drift is being detected, we reset the window in order to not include instances from the previous concepts into the analysis after the change. When minority classes switch places with majority, we also reset the window.

Another issue lies in how to utilize this information regarding minority class structure during the online learning process. We propose to take advantage of perceptrons in the introduced cost-sensitive decision tree model. Their learning procedure can be influenced by the number of iterations they are allowed to spend on each instance. Therefore, each new minority instance will be presented to the cost-sensitive perceptron tree λ times during online learning, where λ is the difficulty level associated with this instance. This will shift our classifier towards concentrating on difficult instances, which in turn should lead to a better predictive performance.

4 Experimental Study

This experimental study was designed in order to answer the following research questions:

- Does the proposed cost-sensitive perceptron decision tree is able to efficiently learn from online binary and imbalanced drifting data streams?
- Does the introduced online analysis of minority instance difficulties can lead to better understanding the learning difficulties in online imbalance and thus improving the performance of an underlying classifier?
- Do the proposed modifications significantly influence the memory and time requirements of the learning model?

Following subsections will describe used datasets, experimental set-up, as well as present obtained results with their discussion.

4.1 Datasets

As there are no standard benchmarks for online class imbalance learning, we selected a diverse set of both artificially generated and real-life datasets with various levels of class imbalance. Let us now describe them shortly.

- **Binary data streams.** We have created six artificial datasets with varying proportions of minority instance types. They are generated by mixing one of five predefined states (described in Table 2), each consisting of 10000 instances and 7 features created using Clover data generator [7]. Therefore, each artificial dataset has 50000 instances. Artificial1 is $S_1 \to S_2 \to S_3 \to S1 \to S2$. Artificial1 is $S_1 \to S_2 \to S_3 \to S1 \to S2$. Artificial2 is $S_1 \to S_2 \to S_3 \to S2 \to S3$. Artificial3 is $S_2 \to S_3 \to S_4 \to S1 \to S3$. Artificial4 is $S_3 \to S_5 \to S_2 \to S3 \to S4$. Artificial6 is $S_1 \to S_5 \to S_3 \to S3 \to S4$. Additionally, we include Twitter dataset as described in [12]. We use MOA benchmarks, including RBF (10 classes), Hyperplane (5 classes), LED (10 classes), Random Tree (3 classes), and Poker (10 classes), generated using standard settings with 50000 instances each. They were transformed into binary problems by randomly selecting one class as minority and merging remaining ones as majority. After each 10000 instances minority class is swapped with one of the remaining ones.
- **Multi-class data streams.** Here, we used Chess and Tweet datasets, as described in [11]. Additionally, we used three MOA generators. RBF dataset consisted of 50000 instances, 20 features and 10 classes with random proportions and gradual drift. Hyperplane dataset consisted of 50000 instances, 10 features and 5 classes with proportions 1:5:10:20:50:100 and incremental concept drift. Random Tree dataset consisted of 50000 instances, 10 features and 3 classes with proportions 10:30:100 and sudden concept drift. Additionally, we use Yeast dataset with 8 features and 10 classes due to its difficult multi-class structure [9]. It has been copied and randomly shuffled to create 50000 instances.

Table 2. Five predefined states for generating artificial binary data streams with respect to composition [in %] of different minority instance types.

	safe	borderline	borderline+	rare	rare+	outlier
S_1	100	0	0	0	0	0
S_2	50	30	20	0	0	0
S_3	30	30	20	15	5	0
S_4	10	20	40	10	10	10
S_5	0	10	20	30	20	20

4.2   Set-up

There are but few methods for online class imbalance learning for both binary and multi-class cases. Most of them are either rooted in neural networks or ensemble approaches [12], thus not making a fair nor suitable reference for our single tree learning procedure. Therefore, as reference we have selected a standard Fast Perceptron Decision Tree (PDT) [1], to evaluate how our modifications increase its skew-insensitivity and Hellinger Hoeffding Decision Tree (HHT) [5], as this is another decision tree algorithm for imbalanced data streams. Furthermore, we evaluate the performance of the proposed cost-sensitive algorithm without taking into account the types of minority instances (CSPT) and with this extension included (CSPT+). Our algorithm is directly applicable for both binary and multi-class data streams, while for multi-class cases we modify HHT to conduct binary decompositions in each split in an identical fashion as its static version [3].

As evaluation metric, we use prequential G-mean for binary streams and prequential Averaged Recall for multi-class ones (where $AvRec = \frac{\sum_{k=1}^{K} TPR_k}{K}$). Decay factor is set to 0.995 for both metrics.

Experiments were conducted in R language using RMOA package on a machine equipped with an Intel Core i7-4700MQ Haswell @ 2.40 GHz processor and 32.00 GB of RAM.

4.3   Experiment 1: Binary Imbalanced Data Streams

Firstly, let us analyze the performance of our method on binary imbalanced data streams. We wanted to check how does the size of sliding window influence the performance of our method. Results for window size $\in [25, \cdots , 300]$ instances are reported in Table 3. From it one can see that in most cases 100 instances are enough to efficiently calculate both current imbalance ratio and minority instances types. Smaller windows cut-off too many instances, thus preventing us from gaining a more global view on the current state of the stream. Bigger windows do not contribute to the predictive power, yet significantly increase the computational complexity. Therefore, we may conclude that for estimating

Table 3. Averaged prequential G-mean [%] for varying sizes of sliding windows used to analyze the difficulty of minority instances over binary data streams. Best trade-off between size and predictive performance bolded.

	25	50	75	100	125	150	175	200	225	250	275	300
Artificial1	77.94	78.14	81.34	**82.59**	82.59	82.59	82.59	82.59	82.59	82.59	80.43	78.67
Artificial2	78.60	78.80	81.86	**83.18**	83.18	83.18	83.18	83.18	83.18	83.18	81.37	79.61
Artificial3	78.88	79.17	82.23	**83.59**	83.59	83.59	83.59	83.46	83.42	83.60	81.65	79.94
Artificial4	79.16	79.41	82.42	**84.18**	84.18	83.78	83.83	83.60	83.60	83.93	81.79	80.13
Artificial5	78.22	78.61	81.29	**82.99**	82.99	82.99	82.99	82.99	82.99	82.99	80.52	78.86
Artificial6	78.50	78.94	81.67	**83.94**	83.94	83.94	83.94	83.94	83.94	83.23	80.85	79.19
Twitter	46.43	48.92	**50.27**	50.27	50.27	50.27	50.27	50.27	50.27	50.27	46.47	45.18
RBF	90.08	92.34	94.18	**94.51**	94.51	94.51	94.51	94.51	94.51	94.51	94.51	92.01
Hyperplane	77.54	79.89	81.02	**81.87**	81.87	81.87	81.87	81.87	81.87	81.87	80.18	79.83
LED	53.19	55.88	**56.11**	56.11	56.11	56.11	56.11	56.11	56.11	56.11	54.36	52.19
RTree	49.75	**52.12**	52.12	52.12	52.12	52.12	52.12	52.12	52.12	52.12	52.12	52.12
poker	61.34	62.67	66.03	**67.69**	67.69	67.69	67.69	67.69	67.69	67.69	67.69	67.69

the imbalance ratios and minority class structures sliding window of size 100 is sufficient. This setting will be used in following experimental comparison.

Table 4 presents comparison with reference decision tree algorithms for binary online data streams. Standard PDT cannot tackle skewed distributions and becomes easily biased towards the majority class. HHT performs much better, yet CSPT and CSPT+ outperform it on 10 out of 12 data streams. This can be explained by Hellinger split criterion not being enough to counter severe class imbalance and difficult minority class structures, which is especially visible in case of six artificial datasets. In most cases CSPT+ returns the superior performance, showing that the proposed online analysis of minority instances difficulty can be beneficial to the learning process. It is interesting to notice that this happens on most of datasets, not only on six artificial ones that had explicitly generated such structures. We may conclude that difficult minority instances are bound to happen in online learning scenarios, especially when the minority class structure is constantly evolving. Therefore, it is worthwhile to incorporate such information during online classifier updating.

When taking into account both time and memory resources being used, one can see that perceptron-based solutions are faster than HHT. CSPT displays almost identical resource usage as the native PDT, proving that the proposed cost-sensitive modification and adaptive cost matrix does not impose any significant additional costs. CSPT+ displays slightly higher computational requirements, which was to be expected as for each new minority instance it analyzes its type and needs to store instances in a sliding window. However, this search is conducted only for minority instances, leading only to a slight increase in overall resource consumption which is far from being prohibitive.

Table 4. Averaged prequential G-mean [%], together with update time [s.] and memory consumption [RAM] per 1000 instances for binary data streams.

Dataset	PDT			HHT			CSPT			CSPT+		
	G-mean	Time	Memory	G-mean	Time	Memory	G-mean	Time	Memory	G-mean	Time	Memory
Artificial1	72.34	0.98	2.13	81.32	2.00	2.33	**82.59**	1.04	2.15	**82.59**	1.20	2.40
Artificial2	69.17	1.03	2.17	80.14	1.95	2.40	82.34	1.10	2.19	**83.18**	1.27	2.47
Artificial3	65.28	1.05	2.25	79.85	1.56	2.37	81.52	1.12	2.28	**83.59**	1.28	2.42
Artificial4	61.03	1.06	2.28	78.02	2.09	2.50	80.05	1.17	2.32	**84.18**	1.26	2.46
Artificial5	58.51	1.11	2.30	76.98	2.31	2.48	78.83	1.20	2.31	**82.99**	1.30	2.52
Artificial6	52.18	1.17	2.36	73.97	2.24	2.53	75.33	1.25	2.38	**83.94**	1.34	2.48
Twitter	37.37	0.99	1.78	47.24	1.43	1.98	48.42	1.04	1.82	**50.27**	1.13	2.00
RBF	70.38	1.36	2.03	87.04	2.56	2.21	92.18	1.42	2.05	**94.51**	1.51	2.19
Hyperplane	65.19	1.28	2.78	**81.26**	2.09	3.01	79.58	1.34	2.79	80.87	1.45	3.07
LED	21.89	2.17	3.01	**56.89**	3.05	3.23	54.29	2.23	3.04	56.11	2.31	3.32
RTree	19.76	2.02	2.32	47.51	2.74	2.49	51.15	2.08	2.35	**52.12**	2.18	2.46
poker	39.04	0.78	1.34	62.98	1.95	1.44	65.66	0.83	1.37	**67.69**	1.00	1.67

Table 5. Averaged prequential AvRec for varying sizes of sliding windows used to analyze the difficulty of minority instances over multi-class data streams. Best trade-off between size and predictive performance bolded.

	25	50	75	100	125	150	175	200	225	250	275	300
Chess	23.19	23.21	23.29	23.97	24.28	**25.72**	25.72	25.72	25.72	25.14	24.73	24.28
Tweet	28.54	28.93	28.99	29.36	30.42	**31.93**	31.93	31.93	31.93	31.93	31.74	31.02
RBF	37.56	37.87	37.89	40.05	**41.25**	41.25	41.25	41.25	41.25	41.25	40.51	39.99
Hyperplane	46.45	46.65	46.39	47.54	47.89	47.92	**48.31**	48.31	48.31	48.31	48.31	47.69
RTree	67.34	68.39	**70.18**	70.18	70.18	70.18	70.18	70.18	70.18	69.17	68.92	68.38
Yeast	78.23	78.78	79.02	79.43	79.58	**80.98**	80.98	80.98	80.98	80.02	78.93	78.75

4.4 Experiment 2: Multi-class Imbalanced Data Streams

We will now switch to multi-class imbalanced data streams. Let us once again check how does the size of sliding window influence the performance of our method, this time when higher number of classes is being taken into consideration. Results for window size $\in [25, \cdots , 300]$ instances are reported in Table 5. From it one can see that multi-class scenarios prefer slightly bigger sliding windows. This can be contributed the need for capturing more complex relationships among a number of distributions. Thus, more instances are needed as they will be divided among a number of classes. We will use the window size of 150 in the following experiments.

Table 6 presents comparison with reference decision tree algorithms for multi-class online data streams. Once again PDT fails to deliver satisfactory performance. However, we can see much bigger discrepancies between HHT and CSPT/CSPT+. As Hellinger distance is a binary metric, to adapt it for multi-class problems one must use a binary decomposition at each node and ten average the metric results when conducting splits. Our experiments show that this fails for multi-class imbalanced data streams. CSPT+ always returns the superior

Table 6. Averaged prequential AvRec [%], together with update time [s.] and memory consumption [RAM] per 1000 instances for multi-class data streams.

Dataset	PDT			HHT			CSPT			CSPT+		
	AvRec	Time	Memory	AvRec	Time	Memory	AvRec	Time	Memory	AvRec	Time	Memory
Chess	2.74	1.18	3.03	14.98	2.31	3.26	23.43	1.24	3.17	25.72	1.44	4.44
Tweet	3.19	1.39	2.87	22.12	2.25	2.97	28.71	1.48	2.99	31.93	1.59	4.33
RBF	3.89	3.03	4.03	33.91	3.68	4.22	40.59	3.08	4.15	41.25	3.19	5.02
Hyperplane	5.19	3.89	5.48	37.19	4.48	5.69	47.31	3.94	5.63	48.31	4.23	6.62
RTree	20.98	2.15	2.78	58.13	2.59	3.02	67.28	2.21	2.88	70.18	2.45	3.90
Yeast	21.48	4.02	6.48	61.93	5.01	6.71	72.19	4.14	6.58	80.98	4.25	8.05

performance, showing that taking into account minority class structures in multi-class online imbalance plays a very important role for the learning process.

When analyzing the resource usage, we can see that perceptron-based solutions increased their costs. This is due to higher number of perceptrons being trained at each leaf. Additionally, CSPT+ needs to store more instances in the sliding window and conduct more instance difficulty analyses, as minority instances may arrive from multiple classes. However, the displayed complexity does is not prohibitive and shows that CSPT+ can be used in real-life scenarios with multi-class imbalanced data streams.

5   Conclusions and Future Works

In this paper, we have introduced a novel decision tree learning algorithm for online learning from binary and multi-class data streams in presence of class imbalance and concept drift, using McDiarmid's inequality. A cost-sensitive improvement to Fast Perceptron Decision Trees was introduced. It modified the outputs of perceptrons in each leaf that were used to predict a class for new instances. Cost-sensitive weighting of support functions allowed to alleviate the bias towards the majority class without introducing additional computational costs associated with data preprocessing techniques. We proposed a simple, yet effective method for calculating cost matrix dynamically using pairwise imbalance ratios measured over most recent examples. This allowed for our cost matrix to swiftly adapt to changes in class distributions during the stream progress. Furthermore, we proposed to incorporate information regarding the types and difficulties of minority class instances into the learning process. We used a sliding window solution to store a small batch of most recent instances and use them to label types of incoming minority class instances by measuring how contaminated was their neighborhood. We proposed six levels of difficulty that were used to determine how many times a given instance is used by our cost-sensitive perceptron tree during the learning process. This allowed for more difficult instances to have a greater influence over the formed decision boundaries.

Obtained results encourage us to further pursue this direction. We envision a potential of using the proposed decision tree in ensemble set-up to improve its predictive power and drift handling capacities, as well as a need for evaluating alternative approaches to analyzing structure of minority classes.

References

1. Bifet, A., Holmes, G., Pfahringer, B., Frank, E.: Fast perceptron decision tree learning from evolving data streams. In: Zaki, M.J., Yu, J.X., Ravindran, B., Pudi, V. (eds.) PAKDD 2010. LNCS (LNAI), vol. 6119, pp. 299–310. Springer, Heidelberg (2010). https://doi.org/10.1007/978-3-642-13672-6_30
2. Gama, J., Zliobaite, I., Bifet, A., Pechenizkiy, M., Bouchachia, A.: A survey on concept drift adaptation. ACM Comput. Surv. **46**(4), 44:1–44:37 (2014)
3. Hoens, T.R., Qian, Q., Chawla, N.V., Zhou, Z.-H.: Building decision trees for the multi-class imbalance problem. In: Tan, P.-N., Chawla, S., Ho, C.K., Bailey, J. (eds.) PAKDD 2012. LNCS (LNAI), vol. 7301, pp. 122–134. Springer, Heidelberg (2012). https://doi.org/10.1007/978-3-642-30217-6_11
4. Krawczyk, B.: Learning from imbalanced data: open challenges and future directions. Prog. AI **5**(4), 221–232 (2016)
5. Lyon, R.J., Brooke, J.M., Knowles, J.D., Stappers, B.W.: Hellinger distance trees for imbalanced streams. In: 22nd International Conference on Pattern Recognition, ICPR 2014, 24–28 August 2014, Stockholm, Sweden, pp. 1969–1974 (2014)
6. Napierala, K., Stefanowski, J.: Types of minority class examples and their influence on learning classifiers from imbalanced data. J. Intell. Inf. Syst. **46**(3), 563–597 (2016)
7. Napierała, K., Stefanowski, J., Wilk, S.: Learning from imbalanced data in presence of noisy and borderline examples. In: Szczuka, M., Kryszkiewicz, M., Ramanna, S., Jensen, R., Hu, Q. (eds.) RSCTC 2010. LNCS (LNAI), vol. 6086, pp. 158–167. Springer, Heidelberg (2010). https://doi.org/10.1007/978-3-642-13529-3_18
8. Rutkowski, L., Pietruczuk, L., Duda, P., Jaworski, M.: Decision trees for mining data streams based on the McDiarmid's bound. IEEE Trans. Knowl. Data Eng. **25**(6), 1272–1279 (2013)
9. Sáez, J.A., Krawczyk, B., Woźniak, M.: Analyzing the oversampling of different classes and types of examples in multi-class imbalanced datasets. Pattern Recogn. **57**, 164–178 (2016)
10. Wang, S., Minku, L.L., Ghezzi, D., Caltabiano, D., Tiño, P., Yao, X.: Concept drift detection for online class imbalance learning. In: The 2013 International Joint Conference on Neural Networks, IJCNN 2013, 4–9 August 2013, Dallas, TX, USA, pp. 1–10 (2013)
11. Wang, S., Minku, L.L., Yao, X.: Dealing with multiple classes in online class imbalance learning. In: Proceedings of 25th International Joint Conference on Artificial Intelligence, IJCAI 2016, 9–15 July 2016, New York, NY, USA, pp. 2118–2124 (2016)
12. Wang, S., Minku, L.L., Yao, X.: A systematic study of online class imbalance learning with concept drift. CoRR abs/1703.06683 (2017). http://arxiv.org/abs/1703.06683
13. Wozniak, M.: A hybrid decision tree training method using data streams. Knowl. Inf. Syst. **29**(2), 335–347 (2011)
14. Woźniak, M., Graña, M., Corchado, E.: A survey of multiple classifier systems as hybrid systems. Inf. Fusion **16**, 3–17 (2014)
15. Zhou, Z., Liu, X.: Training cost-sensitive neural networks with methods addressing the class imbalance problem. IEEE Trans. Knowl. Data Eng. **18**(1), 63–77 (2006)

Efficient Temporal Kernels Between Feature Sets for Time Series Classification

Romain Tavenard[1(✉)], Simon Malinowski[2], Laetitia Chapel[3], Adeline Bailly[1],
Heider Sanchez[4], and Benjamin Bustos[4]

[1] Univ. Rennes 2 / LETG-Rennes COSTEL, IRISA, Rennes, France
`romain.tavenard@univ-rennes2.fr`
[2] Univ. Rennes 1 / IRISA, Rennes, France
[3] Univ. Bretagne Sud / IRISA, Vannes, France
[4] Department of Computer Science, University of Chile, Santiago, Chile

Abstract. In the time-series classification context, the majority of the
most accurate core methods are based on the Bag-of-Words framework,
in which sets of local features are first extracted from time series. A
dictionary of words is then learned and each time series is finally represented by a histogram of word occurrences. This representation induces
a loss of information due to the quantization of features into words as
all the time series are represented using the same fixed dictionary. In
order to overcome this issue, we introduce in this paper a kernel operating directly on sets of features. Then, we extend it to a time-compliant
kernel that allows one to take into account the temporal information. We
apply this kernel in the time series classification context. Proposed kernel
has a quadratic complexity with the size of input feature sets, which is
problematic when dealing with long time series. However, we show that
kernel approximation techniques can be used to define a good trade-off
between accuracy and complexity. We experimentally demonstrate that
the proposed kernel can significantly improve the performance of time
series classification algorithms based on Bag-of-Words.
Code related to this chapter is available at:
https://github.com/rtavenar/SQFD-TimeSeries
Data related to this chapter are available at:
http://www.timeseriesclassification.com

1   Introduction

Time series classification has many real-life applications in various domains such
as biology, medicine or speech recognition [17,24,27] and has received a large
interest over the last decades within the data mining and machine learning communities. Three main families of methods can be found in the literature in this
context: similarity-based methods, that make use of similarity measures between

Electronic supplementary material The online version of this chapter (https://
doi.org/10.1007/978-3-319-71246-8_32) contains supplementary material, which is
available to authorized users.

M. Ceci et al. (Eds.): ECML PKDD 2017, Part II, LNAI 10535, pp. 528–543, 2017.
https://doi.org/10.1007/978-3-319-71246-8_32

raw time series, shapelet-based methods aiming at extracting small subsequences that are discriminant of class membership, and feature-based methods, that rely on a set of feature vectors extracted from time series. The interested reader can refer to [1] for an extensive presentation of time series classification methods.

The most used dissimilarity measures are the euclidean distance (ED) and the dynamic time warping (DTW). The computational cost of ED is lower than the one of DTW, but ED is not able to deal with temporal distortions. The combination of DTW and k-Nearest-Neighbors (k-NN) is one of the seminal approaches for time series classification thanks to its good performance. Cuturi [9] introduces the Global Alignment Kernel that takes into account all possible alignments in order to produce a reliable similarity measure to be used at the core of standard kernel methods such as Support Vector Machines (SVM).

Ye and Keogh [29] introduce shapelets which are sub-sequences of time series that have a high discriminating power between the different classes. In this framework, classification is done with respect to the presence of absence of such shapelets in tested time series. Hills et al. [15] use shapelets to transform the time series into feature vectors representing distances from the time series to the extracted shapelets. Grabocka et al. [13] propose a new classification objective function (applied to the shapelet transform) to learn the shapelets, that improves accuracy and reduces the need to search for too many candidates.

Feature-based methods rely on extracting, from each time series, a set of feature vectors that describe it locally. These feature vectors are quantized into words, using a learned dictionary. Every time series is finally represented by a histogram of word occurrences that then feeds a classifier. Many feature-based approaches for time series classification can be found in the literature [2–4, 19, 25–27] and they mostly differ in the features they use.

This Bag-of-Word (BoW) framework has been shown to be very efficient. However, it suffers from a major drawback: it implies a quantization step that is done via a fixed partitioning of the feature space. Indeed, for a given dataset, words are obtained by clustering the whole set of features and this fixed clustering might not reflect very accurately the distribution of features for every individual time series. This problem has been studied in the computer vision domain for image retrieval for instance. To overcome this drawback, similarity measures operating directly on feature sets have been considered [5, 6, 16]: every instance is represented by its own raw feature sets, which models more accurately every single distribution of features. These measures have been shown to improve accuracy in the image classification context. However, their associated computational cost is quadratic with the size of the feature sets, which is a strong limitation for their direct use in real-world applications. Moreover, they have never been applied to time series classification purposes to the best of our knowledge.

In this paper, we propose a novel temporal kernel that takes as input a set of feature vectors extracted from the time series and their timestamps. Unlike standard Bag-of-Word approaches, this kernel takes feature localization into account, which leads to significant improvement in accuracy. The distance between feature sets is computed using the signature quadratic form distance (SQFD for short [5]) that is a very powerful tool for feature set comparison but has a high computing cost. We hence introduce an efficient variant of our feature set kernel that relies on kernel approximation techniques.

The rest of this paper is organized as follows. An overview of time series classification approaches based on BoW is given in Sect. 2, together with alternatives to BoW mainly used in the image community. The Signature Quadratic Form Distance [5] is detailed in Sect. 3. In Sect. 4, we introduce a kernel based on this distance together with a temporal variant of the latter kernel that enables taking temporal information into account. We also propose an approximation scheme that allows efficient computation of both kernels. In the experimental section, we evaluate the proposed kernel using SIFT features adapted to time series [2,7] and show that it significantly outperforms the original algorithm (based on quantized features) on the UCR datasets [8].

2   Related Work

In this section, we first give an overview of state-of-the-art techniques for time series classification that are based on the BoW framework, as the approach proposed in this paper aims at going beyond this framework. We focus on core classifiers, as the one proposed here. Such classifiers can be integrated into ensemble classifiers in order to build more accurate overall classifiers. More information about the use of ensemble classifiers in this context can be found in [1]. Then, we give an insight about similarity measures defined on feature sets for object comparison purposes. Such measures have been widely used in the image community, but never for time series classification to the best of our knowledge.

2.1   Bag-of-Words Methods for Time Series Classification

Inspired by the text mining and computer vision communities, recent works in time series classification have considered the use of Bag-of-Words [2–4,19,25–27]. In a BoW approach, time series are first converted into a histogram of word occurrences and then a classifier is built upon this representation. In the following, we focus on explaining how the conversion of time series into BoW is performed in the literature.

Baydogan et al. [4] propose Time Series Bag-of-Features (TSBF), a BoW approach for time series classification where local features such as mean, variance and extrema are extracted on sliding windows. A codebook learned by a class probability estimate distribution is then used to quantize the features into words. In [27], discrete wavelet coefficients are computed on sliding windows and then quantized into words by a k-means algorithm. A similar approach denoted BOSS using quantized Fourier coefficients is proposed in [25]. The SAX representation introduced in [18] can also be used to construct words. Histograms of n-grams of SAX symbols are computed in [19] to form the Bag-of-Patterns (BoP) representation. In [26], they propose the SAX-VSM method, which combines SAX with Vector Space Models. SMTS, a symbolic representation of Multivariate Time Series (MTS) is designed in [3]. This method works as follows: a feature matrix is built from MTS and the rows of this matrix are feature vectors composed of a time index, values and gradients of the time series on all dimensions at this time index. A dictionary of words is computed by giving random samples of this matrix to different decision trees. Xie and Beigi [28] extract keypoints from time series and describe them by scale-invariant features that characterize the shapes around the keypoints. Bailly et al. [2] compute multi-scale descriptors based on the SIFT framework and quantize them using a k-means algorithm. Features are either computed at regular time locations or at specific temporal locations discovered by a saliency detector.

2.2   Alternatives to BoW Quantization for Feature Set Classification

Such BoW approaches usually include a quantization step based on a fixed partitioning of the whole set of features. This step might prevent from accurately modeling the distribution of features for every single instance. To overcome this drawback, several similarity measures defined directly on feature sets have been designed. Huttenlocher et al. [16] propose the Hausdorff distance that measures the maximum nearest neighbor distance among features of different instances. The Fisher Kernel [22] relies on a parametric estimation of the feature distribution using a Gaussian Mixture Model. Beecks et al. [5] propose the Signature Quadratic Form distance (SQFD) that is based on the computation of cross-similarities between the features of different sets called *feature signatures*. As we will see later in this paper, SQFD is closely related to match kernels proposed in [6]. Beecks et al. [5] show that SQFD is able to reach higher accuracy than other above classical measures for image retrieval purposes. This makes SQFD a good candidate for being an alternative to BoW quantization in the time series classification context.

3   Signature Quadratic Form Distance

In the following, we assume that a time series S is represented as a set of n feature vectors $\{x_i \in \mathbb{R}^d\}$. Any algorithm presented in Sect. 2.1 can be used to extract such a set of feature vectors from the time series and one should note that considered time series may have different number of features extracted. SQFD is a distance that enables the comparison of instances (time series in our case) represented by weighted sets of features called *feature signatures*. In this section, we review the SQFD distance.

The feature signature of an instance is defined as follows:

$$\mathcal{F} = \{(x_i, w_i)\}_{i=1,\dots,n}, \text{ with } \sum_{i=1}^{n} w_i = 1. \tag{1}$$

$\mathcal{F}$ can either be composed of the full set of features $\{x_i\}$, in which case weights $\{w_i\}$ are all set to $1/n$, or by the result of a clustering of the full set of feature vectors from the instance into n clusters. In the latter case, $\{x_i\}$ are the centroids obtained after clustering and $\{w_i\}$ are the weights of the corresponding clusters. We explain here how the SQFD measure is defined (adopting the time series point of view).

Let $\mathcal{F}^1 = \{(x_i^1, w_i^1)\}_{i=1,\dots,n}$ and $\mathcal{F}^2 = \{(x_i^2, w_i^2)\}_{i=1,\dots,m}$ be two feature signatures associated with two time series $\mathcal{S}^1$ and $\mathcal{S}^2$. Let $k : \mathbb{R}^d \times \mathbb{R}^d \to \mathbb{R}$ be a similarity function defined between feature vectors. The SQFD between $\mathcal{F}^1$ and $\mathcal{F}^2$ is defined as:

$$\text{SQFD}(\mathcal{F}^1, \mathcal{F}^2) = \sqrt{w_{1-2}\, A\, w_{1-2}^T}, \tag{2}$$

where $w_{1-2} = (w_1^1, \dots, w_n^1, -w_1^2, \dots, -w_m^2)$ is the concatenation of the weights of $\mathcal{F}^1$ and the opposite of the weight of $\mathcal{F}^2$, and A is a square matrix of dimension $(n + m)$:

$$A = \begin{pmatrix} A^1 & A^{1,2} \\ \hline A^{2,1} & A^2 \end{pmatrix}, \tag{3}$$

where A^1 (resp. A^2) is the similarity matrix (computed using k) between features from $\mathcal{F}^1$ (resp. $\mathcal{F}^2$), $A^{1,2}$ is the cross-similarity matrix between features from $\mathcal{F}^1$ and those from $\mathcal{F}^2$, and $A^{2,1}$ is the transpose of $A^{1,2}$. It is shown in [5] that the RBF kernel

$$k_{\text{RBF}}(x_i, x_j) = e^{-\gamma_f \|x_i - x_j\|^2}, \tag{4}$$

where γ_f is called the kernel bandwidth, is a good choice for computing local similarity between two features.

SQFD is hence the square root of a weighted sum of local similarities between features from sets $\mathcal{F}^1$ and $\mathcal{F}^2$. When no clustering is used, pairwise local similarities are all taken into account, resulting in a very fine grain estimation of the similarity between series that we will refer to as *exact SQFD* in the following. However, its calculation has a high cost, as it requires the computation of a number of local similarities that is quadratic in the size of the sets. A reasonable alternative presented in [5] consists in first quantizing *each* set using a different k-means and then computing SQFD between feature signatures representing the quantized version of the sets. In the following, we will refer to this latter alternative as the k-means approximation of SQFD (SQFD-k-means for short).

4   Efficient Temporal Kernel Between Feature Sets

In this section, we first derive a kernel from the SQFD distance, considering an RBF kernel as the local similarity, and extend it to a time-sensitive kernel. We then propose a way to alleviate its computational burden.

4.1   Feature Set Kernel

Let us consider the equal weight case in the SQFD formulation, which is the one considered when no pre-clustering is performed on the feature sets. By expanding

Eq. (2) in this specific case, we get:

$$\text{SQFD}(\mathcal{F}^1, \mathcal{F}^2)^2 = \frac{1}{n^2} \sum_{i=1}^{n} \sum_{j=1}^{n} k_{\text{RBF}}(x_i^1, x_j^1) + \frac{1}{m^2} \sum_{i=1}^{m} \sum_{j=1}^{m} k_{\text{RBF}}(x_i^2, x_j^2)$$

$$- \frac{2}{n \cdot m} \sum_{i=1}^{n} \sum_{j=1}^{m} k_{\text{RBF}}(x_i^1, x_j^2). \tag{5}$$

Note that SQFD then corresponds to a biased estimator of the squared difference between the mean of the samples $\mathcal{F}^1$ and $\mathcal{F}^2$ which is classically used to test the difference between two distributions [14]. One can also recognize in Eq. (5) the match kernel [6] (also known as set kernel [11]). By denoting K the match kernel associated with k_{RBF} (in what follows, we will always denote with capital letters kernels that operate on sets whereas kernels operating in the feature space will be named with lowercase k), we have:

$$\text{SQFD}(\mathcal{F}^1, \mathcal{F}^2)^2 = K(\mathcal{F}^1, \mathcal{F}^1) + K(\mathcal{F}^2, \mathcal{F}^2) - 2K(\mathcal{F}^1, \mathcal{F}^2). \tag{6}$$

In other words, SQFD is the distance between feature sets embedded in the Reproducing Kernel Hilbert Space (RKHS) associated with K. Finally, we build a feature set kernel, denoted K_{FS} by embedding SQFD into an RBF kernel:

$$K_{\text{FS}}(\mathcal{F}^1, \mathcal{F}^2) = e^{-\gamma_K \, \text{SQFD}(\mathcal{F}^1, \mathcal{F}^2)^2}, \tag{7}$$

where γ_K is the bandwith of K_{FS}. This kernel can then be used at the core of standard kernel methods such as Support Vector Machines (SVM).

4.2 Time-Sensitive Feature Set Kernel

Kernel K_{FS} as defined in Eq. (7) ignores the temporal location of the features in the time series, only taking into account cross-similarities between the features. In order to integrate temporal information into k_{RBF}, let us augment the features with the time index at which they are extracted: for all $1 \leq i \leq n$, we denote $x_i^t = (x_i, t_i)$. The time-sensitive kernel between features associated with their time of occurrence, denoted k_{tRBF}, is defined as:

$$k_{\text{tRBF}}(x_i^t, x_j^t) = e^{-\gamma_t \, (t_j - t_i)^2} \cdot k_{\text{RBF}}(x_i, x_j). \tag{8}$$

In practice, t_i and t_j are relative timestamps ranging from 0 (beginning of the considered time series) to 1 (end of the time series) so that features extracted from time series of different lengths can easily be compared. As the product of two positive semi-definite kernels, k_{tRBF} is itself a positive semi-definite kernel. It can be seen as a temporal adaptation of the convolutional kernel for images introduced in [20]. Figure 1 illustrates the impact of the parameter γ_t of k_{tRBF} on the resulting similarity matrices A for SQFD: k_{RBF} ($\gamma_t = 0$) takes all matches into account without considering their temporal locations, whereas our time-sensitive kernel favors diagonal matches, γ_t controlling the rigidity of this process.

Denoting $x_i = (x_{i1}, x_{i2}, \ldots, x_{id})$, Eq. (8) can be re-written as:

$$k_{\text{tRBF}}(x_i^t, x_j^t) = e^{-\gamma_f \left(\sum_{l=1}^{d} (x_{jl} - x_{il})^2 + \frac{\gamma_t}{\gamma_f} (t_j - t_i)^2 \right)}, \tag{9}$$

where γ_f is the scale parameter. By defining $g(x_i, t_i) = \left(x_{i1}, \ldots, x_{id}, \sqrt{\frac{\gamma_t}{\gamma_f}} \, t_i \right)$, we get:

$$k_{\text{tRBF}}(x_i^t, x_j^t) = e^{-\gamma_f \, \|g(x_i, t_i) - g(x_j, t_j)\|^2}. \tag{10}$$

In other words, if we build time-sensitive features $g(x, t)$ by concatenating rescaled temporal information and the raw features, the time-sensitive kernel in Eq. (10) can be seen as a standard RBF kernel of scale parameter γ_f operating on these augmented features. Finally, replacing k_{RBF} with k_{tRBF} in Eq. (7) defines a time-sensitive variant for our feature set kernel.

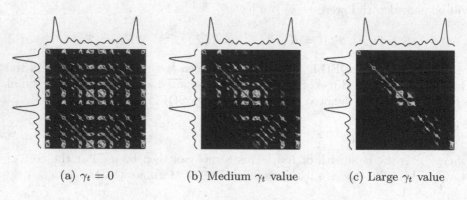

(a) $\gamma_t = 0$ (b) Medium γ_t value   (c) Large γ_t value

Fig. 1. Impact of the k_{tRBF} kernel on similarity matrices $A^{1,2}$. From left to right, growing γ_t values are used from $\gamma_t = 0$ (*i.e.* the k_{RBF} kernel case) to a large γ_t value that ignores almost all non-diagonal matches. Blue colors indicate low similarity whereas red colors represent high similarity (Matrices $A^{1,2}$ are shown but similar observations hold for matrices A^1, A^2 and $A^{2,1}$). Best viewed in color.

4.3   Temporal Kernel Normalization

As can be seen in Fig. 1, when γ_t grows, the number of zero entries in A increases, hence the norm of A decreases. It is valuable to normalize the resulting match kernel K so that, if all local kernel responses are equal to 1, the resulting match kernel evaluation is also equal to 1. The corresponding normalization factor is:

$$s^2 = \left(\sum_{i=1}^{n} \sum_{j=1}^{m} \exp^{-\gamma_t \left(\frac{i}{n} - \frac{j}{m} \right)^2} \right)^{-1}. \tag{11}$$

When $\gamma_t = 0$, this is equivalent to the $\frac{1}{n \cdot m}$ normalization term in the match kernels. When $\gamma_t > 0$, this can be computed either through a double for-loop or approximated by its limit when n and m tend towards infinity:

$$\hat{s}^2 = \left(\int_0^1 \int_0^1 e^{-\gamma_t (t_1 - t_2)^2} dt_1 dt_2 \right)^{-1} \tag{12}$$

$$= \left(\sqrt{\frac{\pi}{\gamma_t}} \left[2F\left(\sqrt{2\gamma_t}\right) - 1 \right] + \frac{e^{-\gamma_t} - 1}{\gamma_t} \right)^{-1} \tag{13}$$

where F is the cumulative distribution function of a centered-reduced Gaussian. In practice, we observe that even for very small feature sets ($n = m = 5$), the relative approximation error done when using $\hat{s}$ normalization instead of s normalization is less than 2%, which is sufficient to efficiently rescale kernels.

4.4   Efficient Computation of Feature Set Kernels

As mentioned earlier, exact SQFD computation is demanding as it requires filling the A matrix (Eq. (3)), leading to the evaluation of $(n + m)^2$ local kernels $k_{\mathrm{RBF}}(x_i, x_j)$. To lower the computational cost of kernel K_{FS} (Eq. (7)), two standard approaches can be considered. The first one, presented in Sect. 3, relies on a k-means quantization of *each* feature set, leading to a time complexity of $O(k^2)$ where k is the chosen number of centroids extracted per set. Another approach is to build an explicit finite-dimensional approximation of the RKHS associated with the match kernel K and approximate SQFD as the Euclidean Distance in this space, as explained below.

Kernel functions compute the inner product between two feature vectors embedded on a feature space thanks to a feature map Φ:

$$k(x_i, x_j) = \langle \Phi(x_i), \Phi(x_j) \rangle. \tag{14}$$

In the case of an RBF kernel, the associated feature map Φ_{RBF} is infinite dimensional. Let us now assume that one can embed feature vectors in a space of dimension D such that the dot product in this space is a good approximation of the RBF kernel on features. In other words, let us assume that there exists a finite mapping ϕ_{RBF} such that:

$$k_{\mathrm{RBF}}(x_i, x_j) \approx \langle \phi_{\mathrm{RBF}}(x_i), \phi_{\mathrm{RBF}}(x_j) \rangle. \tag{15}$$

Then, the match kernel K becomes:

$$K(\mathcal{F}^1, \mathcal{F}^2) = \frac{1}{n \cdot m} \sum_{i=1}^n \sum_{j=1}^m k_{\mathrm{RBF}}(x_i^1, x_j^2) \tag{16}$$

$$\approx \frac{1}{n \cdot m} \sum_{i=1}^n \sum_{j=1}^m \langle \phi_{\mathrm{RBF}}(x_i^1), \phi_{\mathrm{RBF}}(x_j^2) \rangle \tag{17}$$

$$\approx \left\langle \underbrace{\frac{1}{n} \sum_{i=1}^n \phi_{\mathrm{RBF}}(x_i^1)}_{\phi(\mathcal{F}^1)}, \underbrace{\frac{1}{m} \sum_{j=1}^m \phi_{\mathrm{RBF}}(x_j^2)}_{\phi(\mathcal{F}^2)} \right\rangle \tag{18}$$

Hence, approximating k_{RBF} using ϕ_{RBF} is sufficient to approximate the match kernel itself and the explicit feature map ϕ for K is the barycenter of explicit feature maps ϕ_{RBF} for features in the set.

Using Eq. (6), we can derive:

$$\text{SQFD}(\mathcal{F}^1, \mathcal{F}^2)^2 \approx \langle \phi(\mathcal{F}^1), \phi(\mathcal{F}^1) \rangle + \langle \phi(\mathcal{F}^2), \phi(\mathcal{F}^2) \rangle - 2 \langle \phi(\mathcal{F}^1), \phi(\mathcal{F}^2) \rangle \quad (19)$$

$$\approx \|\phi(\mathcal{F}^1) - \phi(\mathcal{F}^2)\|^2 \quad (20)$$

In other words, once feature sets are projected in this finite-dimensional space, approximate SQFD computation is performed through a Euclidean distance computation in $O(D)$ time, where D is the dimension of the feature map.

In this piece of work, we use the Random Fourier Features [23] to build a finite mapping that approximates the local kernel k_{RBF}. Other kernel approximation techniques [10] could be used, but our experience showed that Random Fourier Features reached very good performance for our problem, as showed in the next Section. Random Fourier Features represent each datapoint as its projection on a Fourier basis. The inverse Fourier transform of the RBF kernel is a Gaussian distribution $p(u) = \mathcal{N}(0, \sigma^{-2}I)$ and the Random Fourier Features are obtained by projecting each original feature into a set of sampling Fourier components $p(u)$, before passing through a sinusoid:

$$\phi_{RBF}(x_i) = \sqrt{\frac{2}{D}} \left[\cos(u_1^T x_i + b_1), \ldots, \cos(u_D^T x_i + b_D) \right]^T \quad (21)$$

where coefficients b_i are drawn from a uniform distribution in $[-\pi, \pi]$.

Overall, building on a kernelized version of SQFD, we have presented a way to incorporate time in the representation as well as a scheme for efficient computation of the kernel. In the following Section, we will show experimentally that these improvements help reaching very competitive performance for a wide range of time-series classification problems.

5   Experimental Results with Temporal SIFT Features

Our feature set kernel (with and without temporal information) can be implemented in any algorithm in lieu of a BoW approach. To evaluate the performances of this kernel, we use it in conjunction with time series-based Scale-Invariant Feature Transform (SIFT) features as it has been demonstrated in [2] that they significantly outperform most state-of-the-art local-feature-based time series classification algorithms applied on UCR datasets. In this section, we first recall the temporal SIFT features. We then evaluate the impact of kernel approximation in terms of both efficiency and accuracy. We also analyze the impact of integrating temporal information in the feature set kernel. Finally, we compare our proposed approach with state-of-the-art time series classifiers.

5.1   Dense Extraction of Temporal SIFT Features

In [2], time series-derived SIFT features are used in a BoW approach for time series classification. We review here how these features are computed (the interested reader can refer to [2] for more details). A time series S is described by keypoints extracted every τ_{step} time instants. Each keypoint is composed of a set of features that gives a description of the time series at different scales. More formally, let $L(S, \sigma) = S * G(t, \sigma)$ be the convolution of a time series with a Gaussian filter $G(t, \sigma) = \frac{1}{\sqrt{2\pi}\,\sigma}\,e^{-t^2/2\sigma^2}$ of width σ. A keypoint at time instant t is described at scale σ as follows. n_b blocks of size a are selected around the keypoint. At each point of these blocks, the gradient magnitudes of $L(S, \sigma)$ are computed and weighted so that points that are farther in time from the keypoint have less influence. Then, each block is described by two values: the sum of the positive gradients and the sum of negative gradients in the block. The feature vectors of all the keypoints computed at all scales compose the feature set describing the time series S.

5.2   Experimental Setting

For the sake of reproducibility, all presented experiments are conducted on public datasets from the UCR archive [8] and the Python source code used to produce the results (which heavily relies on `sklearn` [21] package) is made available for download[1]. All experiments are run on dense temporal SIFT features extracted using the publicly available software presented in [2]. For SIFT feature extraction, we choose to use fixed parameters for all datasets. Features are extracted at every time instant ($\tau_{step} = 1$), at all scales, with the block size $a = 4$ and the number of blocks per feature $n_b = 12$, resulting in 24-dimensional feature vectors. By using such a parameter set that achieves robust performance across datasets, we restrict the numbers of parameters to be tuned during cross-validation, without severely degrading performance. Finally, all experiments presented here are repeated 5 times and medians over all runs are reported.

5.3   Impact of Kernel Approximation

We analyze here the impact of the kernel approximation (in terms of trade-off between complexity and accuracy) on the proposed kernels. Timings are reported for execution on a laptop with 2.9 GHz dual core CPU and 8 GB RAM.

Effectiveness. Figure 2 presents the trade-off between accuracy and execution time obtained for the *ECG200* dataset of the UCR archive. Two methods are considered: SQFD-k-means is the approximation scheme that was proposed in [5] and SQFD-Fourier is the one used in this paper. First, this figure confirms that the assumption made in Eq. (15) is safe: one can obtain good approximation

---

[1] https://github.com/rtavenar/SQFD-TimeSeries: contains code and supplementary material.

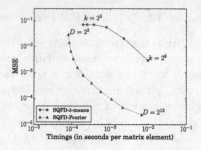

Fig. 2. Mean Squared Error (MSE) *vs* timings of the approximated kernel matrix (*ECG200*). Timings are reported in seconds per matrix element. As a reference, exact computation of feature set kernel takes 0.082 s per matrix element.

of an RBF kernel using finite dimension mapping. Then, for our approach, we observe that the use of larger dimensions leads to better kernel matrix estimation at the cost of larger execution time. The same applies for SQFD-k-means when varying the k parameter. In order to compare approximation methods, Fig. 2 can be read as follows: for a given MSE level (on the y-axis), the lower the timing, the better. This comparison leads to the conclusion that our proposed approximation scheme reaches better trade-offs for a wide range of MSE values. Note that this behaviour is observed on most of the datasets we have experimented on.

Sensitivity to the Amount of Training Data. In this section, we study the evolution of both training and testing times as a function of the amount of training time series. To do so, we compare both efficient approximations of K_{FS} listed above with a standard RBF kernel operating on BoW representations of the feature sets. In Fig. 3, all considered methods exhibit linear dependency between the training set size and the computation time for training. For BoW, this dependency comes from the k-means computation that has $O(nkd)$ time complexity, where n is the number of features used for quantization, k is the

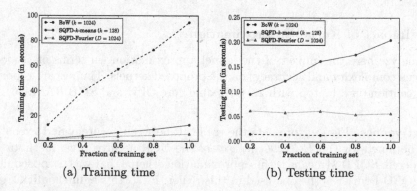

(a) Training time   (b) Testing time

Fig. 3. Training and testing times as a function of the amount of training data (*ECG200*). Training timings correspond to full training of the method for a given parameter set whereas test timings are reported per test time series.

number of clusters and d is the feature dimension. The same argument holds for SQFD-k-means, and this explains the observed difference in slope, as lower values of k are typically used in this context. SQFD-Fourier present an even lower slope, which correspond to the computation of projected features (one per time series). Concerning testing times, BoW as well as SQFD-Fourier have almost constant computation needs, whereas SQFD-k-means computation time is linearly dependent in the number of training time series. Note that this comparison is done on *ECG200* dataset for which the number of training time series is small. In this context, computation of the feature set representation (k-means quantization or feature map) dominates the processing time for both training and testing. In other settings where the number of training time series is large, processing time will be dominated by the computation of pairwise similarities which is, as stated above, linear in k for BoW, quadratic in k for SQFD-k-means, and linear in D for SQFD-Fourier. Once again, our proposed approximation scheme tends to better approximate the exact kernel matrix with lower timings (both in the training and the testing phase) than its competitor.

5.4 Impact of the Temporal Information

Let us now turn our focus to the impact of temporal information on the classification performance. To do so, we use K_{FS} in an SVM classifier. In order to have a fair comparison of the performances, all parameters (except the dimension D of the feature map) are set through cross-validation on the training set. The same applies for experiments presented in the following subsections and the range of tested parameter values are provided in the Supplementary Material. As a reference, BoW performance with RBF kernel (using the same temporal SIFT features) is also reported. To compute this baseline, the number k of codewords is also cross-validated. Figure 4 shows the error rates for dataset *ECG200* as a function of the dimension D of the feature map, considering (i) our feature set kernel without temporal information, (ii) its equivalent with temporal information and finally (iii) the normalized temporal kernel. One should first notice that in all cases, a higher dimension D tends to lead to better performance. This figure also illustrates the importance of the temporal kernel normalization:

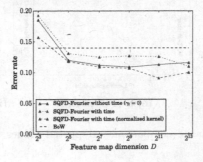

Fig. 4. Error rates as a function of the feature map dimension (*ECG200*).

the normalized temporal kernel reaches better performance than the feature set kernel with no time information, whereas the performance of the non-normalized one is worse. Indeed, when γ_t increases, the non-normalized version suffers from a bad scaling of kernel responses that impairs the learning process of the SVM.[2]

In the following, we use the same parameter ranges as above, and we cross-validate the parameter related to the time ($\gamma_t \in \{0\} \cup \{10^0 - 10^6\}$). By doing so, we offer the possibility for our method to learn (during training) whether time information is of interest or not for a given dataset. We present experiments run on the 85 datasets from the UCR Time Series Classification archive and observe that in more than 3/4 of our experiments, the temporal variant of our kernel is selected by cross-validation (*i.e.* $\gamma_t > 0$), which confirms the superiority of temporal kernels for such applications. Corresponding datasets are marked with a star in the full result table provided as Supplementary material.

5.5   Pairwise Comparisons on UCR Datasets

Pairwise comparisons of methods are presented in Fig. 5, in which green dots (data points lying below the diagonal) correspond to cases where the x-axis method has higher classification error rates, black ones represent cases where both methods share the same performance and red dots stand for cases for which the y-axis method has higher error rates. In these plots, Win/Tie/Lose scores are

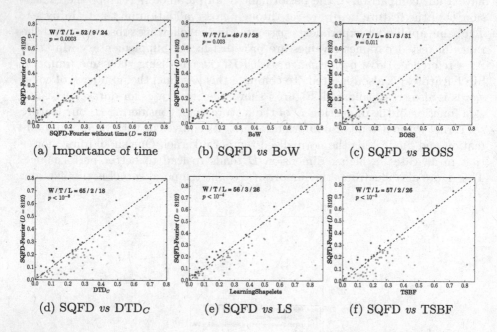

(a) Importance of time   (b) SQFD *vs* BoW   (c) SQFD *vs* BOSS

(d) SQFD *vs* DTD$_C$ (e) SQFD *vs* LS   (f) SQFD *vs* TSBF

Fig. 5. Pairwise performance comparisons. Reported values are error rates. (Color figure online)

---

[2] See Supplementary material for experiments on more datasets.

also reported where "Win" indicates the number of times the y-axis method outperforms the x-axis one. Finally, p-values corresponding to one-sided Wilcoxon signed rank tests are provided to assess statistical significance of observed differences and our significance level is set to 5%.

First, Fig. 5a confirms our observation made for *ECG200* dataset: incorporating temporal information allows, for many datasets, to improve the classification performances. We then compare the efficient version of our temporal kernel on feature sets with standard BoW approach running on the same feature sets (Fig. 5b). This figure shows improvement for a wide range of datasets: when testing the statistical differences between both methods, one can observe that our feature set kernel significantly outperforms the BoW approach. Finally, when considering other state-of-the-art competitors[3], our feature set kernel for time series show state-of-the-art performance, significantly outperforming BOSS [25], DTD_C [12], LearningShapelets (LS) [13] and TSBF [4]. This high accuracy is achieved with reasonable classification time (*e.g.* 300 ms per test time series on NonInvasiveFetalECG1 dataset, one of the largest UCR datasets).

6 Conclusion

Many local features have been designed for time-series representations and used in a BoW framework for classification purposes. To improve these approaches, we introduce in this paper a new temporal kernel between feature sets that gets rid of quantized representations. More precisely, we propose to kernelize SQFD for time-series classification purposes. We also derive a temporal feature set kernel, allowing one to take into account the time instant at which the features are extracted. In order to alleviate the high computational burden of this kernel and make it tractable for large datasets, we propose an approximation technique that allows fast computation of the kernel. Extensive experiments show that the temporal information helps improving classification accuracy. The temporal information is taken into account thanks to a simple RBF kernel, and we believe that the performance could be further improved by exploring other ways to incorporate time information into the local kernels. This is the main direction for our future work. Finally, our temporal feature set kernel significantly outperforms the initial BoW-based method and leads to competitive results *w.r.t* state-of-the-art time series classification algorithms. This kernel is likely to improve performance of any time series classification approach based on BoW.

Acknowledgments. Supported by the Millennium Nucleus Center for Semantic Web Research under Grant NC120004, the ANR through the ASTERIX project (ANR-13-JS02-0005-01), École des docteurs de l'UBL as well as by the Brittany Region.

---

[3] For the sake of brevity, we focus on standalone classifiers that are shown in [1] to outperform competitors in their categories.

References

1. Bagnall, A., Lines, J., Bostrom, A., Large, J., Keogh, E.: The great time series classification bake off: a review and experimental evaluation of recent algorithmic advances. Data Min. Knowl. Discov. **31**(3), 606–660 (2017). https://link.springer.com/article/10.1007/s10618-016-0483-9
2. Bailly, A., Malinowski, S., Tavenard, R., Chapel, L., Guyet, T.: Dense bag-of-temporal-SIFT-words for time series classification. In: Douzal-Chouakria, A., Vilar, J.A., Marteau, P.-F. (eds.) AALTD 2015. LNCS (LNAI), vol. 9785, pp. 17–30. Springer, Cham (2016). https://doi.org/10.1007/978-3-319-44412-3_2
3. Baydogan, M.G., Runger, G.: Learning a symbolic representation for multivariate time series classification. Data Min. Knowl. Discov. **29**(2), 400–422 (2015)
4. Baydogan, M.G., Runger, G., Tuv, E.: A bag-of-features framework to classify time series. IEEE Trans. Pattern Anal. Mach. Intell. **35**(11), 2796–2802 (2013)
5. Beecks, C., Uysal, M.S., Seidl, T.: Signature quadratic form distance. In: Proceedings of ACM International Conference on Image and Video Retrieval, pp. 438–445 (2010)
6. Bo, L., Sminchisescu, C.: Efficient match kernel between sets of features for visual recognition. Adv. Neural Inf. Process. Syst. **22**, 135–143 (2009)
7. Candan, K.S., Rossini, R., Sapino, M.L.: sDTW: computing DTW distances using locally relevant constraints based on salient feature alignments. In: Proceedings of International Conference on Very Large DataBases, vol. 5, pp. 1519–1530 (2012)
8. Chen, Y., Keogh, E., Hu, B., Begum, N., Bagnall, A., Mueen, A., Batista, G.: The UCR time series classification archive (2015). www.cs.ucr.edu/~eamonn/time-series_data/
9. Cuturi, M.: Fast global alignment kernels. In: Proceedings of International Conference on Machine Learning, pp. 929–936 (2011)
10. Drineas, P., Mahoney, M.W.: On the nyström method for approximating a gram matrix for improved kernel-based learning. J. Mach. Learn. Res. **6**, 2153–2175 (2005)
11. Gärtner, T., Flach, P.A., Kowalczyk, A., Smola, A.J.: Multi-instance kernels. In: Proceedings of International Conference on Machine Learning (2002)
12. Górecki, T., Łuczak, M.: Non-isometric transforms in time series classification using DTW. Knowl.-Based Syst. **61**, 98–108 (2014)
13. Grabocka, J., Schilling, N., Wistuba, M., Schmidt-Thieme, L.: Learning time-series shapelets. In: Proceedings of ACM SIGKDD International Conference on Knowledge Discovery and Data Mining, pp. 392–401 (2014)
14. Gretton, A., Borgwardt, K.M., Rasch, M., Schölkopf, B., Smola, A.J.: A kernel method for the two-sample-problem. In: Advances in Neural Information Processing Systems, pp. 513–520 (2006)
15. Hills, J., Lines, J., Baranauskas, E., Mapp, J., Bagnall, A.: Classification of time series by shapelet transformation. Data Min. Knowl. Discov. **28**(4), 851–881 (2014)
16. Huttenlocher, D.P., Klanderman, G.A., Rucklidge, W.J.: Comparing images using the Hausdorff distance. IEEE Trans. Pattern Anal. Mach. Intell. **15**(9), 850–863 (1993)
17. Le Cun, Y., Bengio, Y.: Convolutional networks for images, speech, and time series. In: The Handbook of Brain Theory and Neural Networks, vol. 3361, pp. 255–258 (1995)

18. Lin, J., Keogh, E., Lonardi, S., Chiu, B.: A symbolic representation of time series, with implications for streaming algorithms. In: Proceedings of ACM SIGMOD Workshop on Research Issues in Data Mining and Knowledge Discovery, pp. 2–11 (2003)

19. Lin, J., Khade, R., Li, Y.: Rotation-invariant similarity in time series using bag-of-patterns representation. Int. J. Inf. Syst. **39**, 287–315 (2012)

20. Mairal, J., Koniusz, P., Harchaoui, Z., Schmid, C.: Convolutional kernel networks. In: Advances in Neural Information Processing Systems, pp. 2627–2635 (2014)

21. Pedregosa, F., Varoquaux, G., Gramfort, A., Michel, V., Thirion, B., Grisel, O., Blondel, M., Prettenhofer, P., Weiss, R., Dubourg, V., Vanderplas, J., Passos, A., Cournapeau, D., Brucher, M., Perrot, M., Duchesnay, E.: Scikit-learn: machine learning in Python. J. Mach. Learn. Res. **12**, 2825–2830 (2011)

22. Perronnin, F., Dance, C.: Fisher kernels on visual vocabularies for image categorization. In: Proceedings of IEEE Conference on Computer Vision and Pattern Recognition, pp. 1–8 (2007)

23. Rahimi, A., Recht, B.: Random features for large-scale kernel machines. In: Advances in Neural Information Processing Systems, pp. 1177–1184 (2007)

24. Sakoe, H., Chiba, S.: Dynamic programming algorithm optimization for spoken word recognition. IEEE Trans. Acoust. Speech Sig. Process. **26**(1), 43–49 (1978)

25. Schäfer, P.: The BOSS is concerned with time series classification in the presence of noise. Data Min. Knowl. Discov. **29**(6), 1505–1530 (2014)

26. Senin, P., Malinchik, S.: SAX-VSM: interpretable time series classification using SAX and vector space model. In: Proceedings of IEEE International Conference on Data Mining, pp. 1175–1180 (2013)

27. Wang, J., Liu, P., She, M.F.H., Nahavandi, S., Kouzani, A.: Bag-of-words representation for biomedical time series classification. Biomed. Sig. Process. Control **8**(6), 634–644 (2013)

28. Xie, J., Beigi, M.: A scale-invariant local descriptor for event recognition in 1D sensor signals. In: Proceedings of IEEE International Conference on Multimedia and Expo, pp. 1226–1229 (2009)

29. Ye, L., Keogh, E.: Time series shapelets: a new primitive for data mining. In: Proceedings of ACM SIGKDD International Conference on Knowledge Discovery and Data Mining, pp. 947–956 (2009)

Forecasting and Granger Modelling
with Non-linear Dynamical Dependencies

Magda Gregorová[1,2(✉)], Alexandros Kalousis[1,2],
and Stéphane Marchand-Maillet[2]

[1] Geneva School of Business Administration,
HES-SO University of Applied Sciences of Western Switzerland,
Geneva, Switzerland
magda.gregorova@hesge.ch
[2] University of Geneva, Geneva, Switzerland

Abstract. Traditional linear methods for forecasting multivariate time series are not able to satisfactorily model the non-linear dependencies that may exist in non-Gaussian series. We build on the theory of learning vector-valued functions in the reproducing kernel Hilbert space and develop a method for learning prediction functions that accommodate such non-linearities. The method not only learns the predictive function but also the matrix-valued kernel underlying the function search space directly from the data. Our approach is based on learning multiple matrix-valued kernels, each of those composed of a set of input kernels and a set of output kernels learned in the cone of positive semi-definite matrices. In addition to superior predictive performance in the presence of strong non-linearities, our method also recovers the hidden dynamic relationships between the series and thus is a new alternative to existing graphical Granger techniques.

1 Introduction

Traditional methods for forecasting stationary multivariate time series from their own past are derived from the classical linear ARMA modelling. In these, the prediction of the next point in the future of the series is constructed as a linear function of the past observations. The use of linear functions as the predictors is in part based on the Wold representation theorem (e.g. [6]) and in part, probably more importantly, on the fact that the linear predictor is the best predictor (in the mean-square-error sense) in case the time series is Gaussian.

The Gaussian assumption is therefore often adopted in the analysis of time series to justify the simple linear modelling. However, it is indeed a simplifying assumption since for non-Gaussian series the best predictor may very well be a non-linear function of the past observations. A number of parametric non-linear models has been proposed in the literature, each adapted to capture specific sources of non-linearity (for example multiple forms of regime-switching models, e.g. [20]).

© Springer International Publishing AG 2017
M. Ceci et al. (Eds.): ECML PKDD 2017, Part II, LNAI 10535, pp. 544–558, 2017.
https://doi.org/10.1007/978-3-319-71246-8_33

In this paper we adopt an approach that does not rely on such prior assumptions for the function form. We propose to learn the predictor as a general vector-valued function **f** that takes as input the past observations of the multivariate series and outputs the forecast of the unknown next value (vector).

We have two principal requirements on the function **f**. The first is the standard prediction accuracy requirement. That is, the function **f** shall be such that we can expect its outputs to be close (in the squared error sense) to the true future observations of the process. The second requirement is that the function **f** shall have a structure that will enable the analysis of the relationships amongst the subprocesses of the multivariate series. Namely, we wish to understand how parts of the series help in forecasting other parts of the multivariate series, a concept known in the time-series literature as graphical Granger modelling [9,11].

To learn such a function **f** we employ the framework of regularised learning of vector-valued functions in the reproducing kernel Hilbert space (RKHS) [17]. Learning methods based on the RKHS theory have previously been considered for time series modelling (e.g. [10,15,19]). Though, as Pillonetto et al. note in their survey [18], their adoption for the dynamical system analysis is not a commonplace.

A critical step in kernel-based methods for learning vector-valued functions is the specification of the operator-valued kernel that exploits well the relationships between the inputs and the outputs. A convenient and well-studied class of operator-valued kernels (e.g. in [7,8,12]) are those decomposable into a product of a scalar kernel on the input space (input kernel) and a linear operator on the output space (output kernel).

The kernel uniquely determines the function space within which the function **f** is learned. It thus has significant influence on both our objectives described above. Instead of having to choose the input and the output kernels a priori, we introduce a method for learning the input and output kernels from the data together with learning the vector-valued function **f**.

Our method combines in a novel way the multiple-kernel learning (MKL) approach [14] with learning the output kernels within the space of positive semidefinite linear operators on the output space [12]. MKL methods for operator-valued kernels have recently been developed in [13,19]. The first learns a convex combination of a set of operator-valued kernels fixed in advance, the second combines a fixed set of input kernels with a single learned output kernel. To the best of our knowledge, ours is the first method in which the operator-valued kernel is learned by combining a set of input kernels with a set of multiple learned output kernels.

In accordance with our second objective stated above, we impose specific structural constraints on the function search space so that the learned function supports the graphical Granger analysis. We achieve this by working with matrix-valued kernels operating over input partitions restricted to single input scalar series (similar input partitioning has recently been used in [19]).

We impose diagonal structure on the output kernels to control the model complexity. Though this has a cost in the inability to model contemporaneous relationships, it addresses the strong over-parametrisation in a principled manner. It also greatly simplifies the final structure of the problem, which, in result, suitably decomposes into a set of smaller independent problems solvable in parallel.

We develop two forms of sparsity-promoting regularisation approaches for learning the output kernels. These are based on the ℓ_1 and ℓ_1/ℓ_2 norms respectively and are motivated by the search for Granger-causality relationships. As to our knowledge, the latter has not been previously used in the context of MKL.

Finally, we confirm on experiments the benefits our methods can bring to forecasting non-Gaussian series in terms of improved predictive accuracy and the ability to recover hidden dynamic dependency structure within the time series systems. This makes them valid alternatives to the state-of-the-art graphical Granger techniques.

Notation. We use bold upper case and lower case letters for matrices and vectors respectively, and the plain letters with subscripts for their elements. For any matrix or vector the superscript ^{T} denotes its transpose. Vectors are by convention column-wise so that $\mathbf{x} = (x_1, \ldots, x_n)^T$ is the n-dimensional vector $\mathbf{x}$. $\mathbb{R}, \mathbb{R}^n, \mathbb{R}^{m \times n}$ are the sets of real scalars, n-dimensional vectors, and $m \times n$ dimensional matrices. $\mathbb{R}_+^{m \times n}$ is the set of non-negative matrices, $\mathbb{S}_+^m$ the set of positive semi-definite $m \times m$ matrices and $\mathbb{D}_+^m$ the set of non-negative diagonal matrices. $\mathbb{N}_m$ is the set of positive integers $\{1, \ldots, m\}$. For any vectors $\mathbf{x}, \mathbf{y} \in \mathbb{R}^n$, $\langle \mathbf{x}, \mathbf{y} \rangle, ||\mathbf{x}||_1, ||\mathbf{x}||_2$ are the standard inner product, ℓ_1 and ℓ_2 norms in the real Hilbert spaces. For any square matrix $\mathbf{A}$, $\text{Tr}(\mathbf{A})$ denotes the trace. For any two matrices $\mathbf{A}, \mathbf{B} \in \mathbb{R}^{m \times n}$, $\langle \mathbf{A}, \mathbf{B} \rangle_F := \text{Tr}(\mathbf{A}^T \mathbf{B})$ is the Frobenius inner product and $||\mathbf{A}||_F := \sqrt{\langle \mathbf{A}, \mathbf{A} \rangle_F}$ the Frobenius norm. $\langle ., . \rangle_{\mathcal{F}}$ and $||.||_{\mathcal{F}}$ are the inner product and norm in the Hilbert space $\mathcal{F}$.

2 Problem Formulation

Given a realisation of a discrete stationary multivariate time series process $\{\mathbf{y}_t \in \mathcal{Y} \subseteq \mathbb{R}^m : t \in \mathbb{N}_n\}$, our goal is to learn a vector-valued function $\mathbf{f} : \mathcal{Y}^p \to \mathcal{Y}$ that takes as input the p past observations of the process and predicts its future vector value (one step ahead). The function $\mathbf{f}$ shall be such that

1. we can expect the prediction to be near (in the Euclidean distance sense) the unobserved future value
2. its structure allows to analyse if parts (subprocesses) of the series are useful for forecasting other subprocesses within the series or if some subprocesses can be forecast independently of the rest; in short, it allows Granger-causality analysis [9,11].

For notational simplicity, from now on we indicate the output of the function $\mathbf{f}$ as $\mathbf{y} \in \mathcal{Y} \subseteq \mathbb{R}^m$ and the input as $\mathbf{x} \in \mathcal{X} \subseteq \mathbb{R}^{mp}$ (bearing in mind that $\mathcal{X} = \mathcal{Y}^p$ is in fact the p-th order Cartesian product of $\mathcal{Y}$ and that the inputs $\mathbf{x}$ and outputs $\mathbf{y}$ are the past and future observations of the same m-dimensional series). We also align the time indexes so that our data sample consists of input-output data pairs $\{(\mathbf{y}_t, \mathbf{x}_t) : t \in \mathbb{N}_n\}$.

Following the standard function learning theory, we will learn $\mathbf{f} \in \mathcal{F}$ by minimising the regularised empirical squared-error risk (with a regularization parameter $\lambda > 0$)

$$\widehat{\mathbf{f}} = \arg \min_{\mathbf{f} \in \mathcal{F}} R(\mathbf{f})$$
$$R(\mathbf{f}) := \sum_{t=1}^{T} \|\mathbf{y}_t - \mathbf{f}(\mathbf{x}_t)\|_2^2 + \lambda \|\mathbf{f}\|_{\mathcal{F}}^2. \tag{1}$$

Here $\mathcal{F}$ is the reproducing kernel Hilbert space (RKHS) of $\mathbb{R}^m$-valued functions endowed with the norm $\|.\|_{\mathcal{F}}$ and the inner product $\langle ., . \rangle_{\mathcal{F}}$. The RKHS is uniquely associated with a symmetric positive-semidefinite matrix-valued kernel $\mathbf{H} : \mathcal{X} \times \mathcal{X} \to \mathbb{R}^{m \times m}$ with the reproducing property

$$\langle \mathbf{y}, \mathbf{g}(\mathbf{x}) \rangle = \langle \mathbf{H}_{\mathbf{x}} \mathbf{y}, \mathbf{g} \rangle_{\mathcal{F}} \quad \forall (\mathbf{y}, \mathbf{x}, \mathbf{g}) \in (\mathcal{Y}, \mathcal{X}, \mathcal{F}),$$

where the map $\mathbf{H}_{\mathbf{x}} : \mathcal{X} \to \mathbb{R}^{m \times m}$ is the kernel section of $\mathbf{H}$ centred at $\mathbf{x}$ such that $\mathbf{H}_{\mathbf{x}_i}(\mathbf{x}_j) = \mathbf{H}(\mathbf{x}_i, \mathbf{x}_j)$ for all $(\mathbf{x}_i, \mathbf{x}_j) \in (\mathcal{X}, \mathcal{X})$. From the classical result in [17], the unique solution $\widehat{\mathbf{f}}$ of the variational problem (1) admits a finite dimensional representation

$$\widehat{\mathbf{f}} = \sum_{t=1}^{T} \mathbf{H}_{\mathbf{x}_t} \mathbf{c}_t, \tag{2}$$

where the coefficients $\mathbf{c}_t \in \mathcal{Y}$ are the solutions of the system of linear equations

$$\sum_{t=1}^{T} (\mathbf{H}(\mathbf{x}_s, \mathbf{x}_t) + \lambda \delta_{st}) \mathbf{c}_t = \mathbf{y}_s, \quad \forall s \in \mathbb{N}_n, \tag{3}$$

where $\delta_{st} = 1$ if $s = t$ and is zero otherwise.

2.1   Granger-Causality Analysis

To study the dynamical relationships in time series processes, Granger [11] proposed a practical definition of causality based on the accuracy of least-squares predictor functions. In brief, for two time series processes $\mathbf{y}$ and $\mathbf{z}$, $\mathbf{y}$ is said to Granger-cause $\mathbf{z}$ ($\mathbf{y} \to \mathbf{z}$) if given all the other relevant information we can predict the future of $\mathbf{z}$ better (in the mean-square-error sense) using the history of $\mathbf{y}$ than without it.

Though the concept seems rather straightforward, there are (at least) three points worth considering. First, the notion is purely technical based on the predictive accuracy of functions with differing input sets; it does *not* seek to understand the underlying forces driving the relationships. Second, in practice the conditioning set of information needs to be reduced to all the *available* information instead of all the *relevant* information. Third, it only considers relationships between pairs of (sub-)processes and not the interactions amongst a set of series.

Eichler [9] extended the concept to multivariate analysis through graphical models. The discussion in the paper focuses on the notion of Granger non-causality rather than causality and describes the specific Markov properties (conditional non-causality) encoded in the graphs of Granger-causal relationships. In this sense, the absence of a variable in a set of inputs is more informative of the Granger (non-)causality than its presence. In result, graphical Granger methods are typically based on (structured) sparse modelling [4].

3   Function Space and Kernel Specification

The function space $\mathcal{F}$ within which $\mathbf{f}$ is learned is fully determined by the reproducing kernel $\mathbf{H}$. Its specification is therefore critical for achieving the two objectives for the function $\mathbf{f}$ defined in Sect. 2. We focus on the class of matrix-valued kernels decomposable into the product of input kernels, capturing the similarities in the inputs, and output kernels, encoding the relationships between the outputs.

To analyse the dynamical dependencies between the series, we need to be able to discern within the inputs of the learned function $\mathbf{f}$ the individual scalar series. Therefore we partition the elements of the input vectors according to the source scalar time series. In result, instead of a single kernel operating over the full vectors, we work with multiple partition-kernels, each of them operating over a single input series. We further propose to learn the partition-kernels by combining the MKL techniques with output kernel learning within the cone of positive semi-definite matrices.

More formally, the kernel we propose to use is constructed as a sum of kernels $\mathbf{H} = \sum_j^m \mathbf{H}^{(j)}$, where m is the number of the individual scalar-valued series in the multivariate process (dimensionality of the output space $\mathcal{Y}$). Each $\mathbf{H}^{(j)}$: $\mathcal{X}^{(j)} \times \mathcal{X}^{(j)} \to \mathbb{R}^{m \times m}$ is a matrix-valued kernel that determines its own RKHS of vector-valued functions. The domains $\mathcal{X}^{(j)} \subseteq \mathbb{R}^p$ are sets of vectors constructed by selecting from the inputs $\mathbf{x}$ only the p coordinates $i^{(j)} \in \mathbb{N}_{mp}$ that correspond to the past of a single scalar time series j.

$$\mathcal{X}^{(j)} = \{\mathbf{x}^{(j)} : x_i^{(j)} = x_{i(j)} \, \forall i, \, \mathbf{x} \in \mathcal{X}\}, \quad \cup_j \mathcal{X}^{(j)} = \mathcal{X}$$

Further, instead of choosing the individual matrix-valued functions $\mathbf{H}^{(j)}$, we propose to learn them. We construct each $\mathbf{H}^{(j)}$ again as a sum of kernels $\mathbf{H}^{(j)} = \sum_i^{s_j} \mathbf{H}^{(ji)}$ of possibly uneven number of summands s_j of matrix-valued kernels $\mathbf{H}^{(ji)} : \mathcal{X}^{(j)} \times \mathcal{X}^{(j)} \to \mathbb{R}^{m \times m}$. For this lowest level $\mathbf{H}^{(ji)}$ we focus on the family of decomposable kernels $\mathbf{H}^{(ji)} = k^{(ji)} \mathbf{L}^{(ji)}$. Here, the input kernels $k^{(ji)} : \mathcal{X}^{(j)} \times \mathcal{X}^{(j)} \to \mathbb{R}$ capturing the similarity between the inputs are fixed in advance from a dictionary of valid scalar-valued kernels (e.g. Gaussian kernels with varying scales). The set $\mathcal{L} = \{\mathbf{L}^{(ji)} : j = \mathbb{N}_m, i = \mathbb{N}_{s_j}, \sum_j^m s_j = l\}$ of output kernels $\mathbf{L}^{(ji)} : \mathcal{Y} \to \mathcal{Y}$ encoding the relations between the outputs is learned within the cone of symmetric positive semidefinite matrices $\mathbb{S}_+^m$.

$$\mathbf{H} = \sum_{j=1}^{m} \mathbf{H}^{(j)} = \sum_{j=1}^{m} \sum_{i=1}^{s_j} \mathbf{H}^{(ji)} = \sum_{j=1}^{m} \sum_{i=1}^{s_j} k^{(ji)} \mathbf{L}^{(ji)} \tag{4}$$

3.1 Kernel Learning and Function Estimation

Learning all the output kernels $\mathbf{L}^{(ji)}$ as full PSD matrices implies learning more than m^3 parameters. To improve the generalization capability, we reduce the complexity of the problem drastically by restricting the search space for $\mathbf{L}$'s to PSD diagonal matrices $\mathbb{D}_+^m$. This essentially corresponds to the assumption of no contemporaneous relationships between the series. We return to this point in Sect. 5.

As explained in Sect. 2.1, Granger (non-)causality learning typically searches for sparse models. We bring this into our methods by imposing a further sparsity inducing regularizer $Q : (\mathbb{R}^{m \times m})^l \rightarrow \mathbb{R}$ on the set of the output kernels $\mathcal{L}$. We motivate and elaborate suitable forms of Q in Sect. 3.2.

The joint learning of the kernels and the function can now be formulated as the problem of finding the minimising solution $\mathbf{f} \in \mathcal{F}$ and $\mathbf{L}$'s $\in \mathbb{D}_+^m$ of the regularised functional

$$J(\mathbf{f}, \mathcal{L}) := R(\mathbf{f}) + \tau Q(\mathcal{L}), \quad \tau > 0, \tag{5}$$

where $R(\mathbf{f})$ is the regularised risk from (1). By calling on the properties of the RKHS, we reformulate this as a finite dimensional problem that can be addressed by conventional finite-dimensional optimisation approaches. We introduce the gram matrices $\mathbf{K}^{(ji)} \in \mathbb{S}_+^n$ such that $K_{ts}^{(ji)} = k^{(ji)}(\mathbf{x}_t^{(j)}, \mathbf{x}_s^{(j)})$ for all $t, s \in \mathbb{N}_n$, the output data matrix $\mathbf{Y} \in \mathbb{R}^{n \times m}$ such that $\mathbf{Y} = (\mathbf{y}_1, \dots \mathbf{y}_n)^T$, and the coefficient matrix $\mathbf{C} \in \mathbb{R}^{n \times m}$ such that $\mathbf{C} = (\mathbf{c}_1, \dots \mathbf{c}_n)^T$.

Using these and (2) it is easy to show that the minimisation of the regularised risk $R(\mathbf{f})$ in (1) with respect to $\mathbf{f} \subset \mathcal{F}$ is equivalent to the minimisation with respect to $\mathbf{C} \in \mathbb{R}^{n \times m}$ of the objective

$$\widetilde{R}(\mathbf{C}) := ||\mathbf{Y} - \sum_{ji} \mathbf{K}^{(ji)} \mathbf{C} \mathbf{L}^{(ji)}||_F^2 + \lambda \sum_{ji} \langle \mathbf{C}^T \mathbf{K}^{(ji)} \mathbf{C}, \mathbf{L}^{(ji)} \rangle_F. \tag{6}$$

The finite dimensional equivalent of (5) is thus the joint minimisation of

$$\widetilde{J}(\mathbf{C}, \mathcal{L}) := \widetilde{R}(\mathbf{C}, \mathcal{L}) + \tau Q(\mathcal{L}). \tag{7}$$

3.2 Sparse Regularization

The construction of the kernel $\mathbf{H}$ and the function space $\mathcal{F}$ described in Sect. 3 imposes on the function $\mathbf{f}$ the necessary structure that allows the Granger-causality analysis (as per our 2nd objective set-out in Sect. 2). As explained in Sect. 2.1, the other ingredient we need to identify the Granger non-causalities is sparsity within the structure of the learned function.

In our methods, the sparsity is introduced by the regularizer Q. By construction of the function space, we can examine the elements of the output kernels $\mathbf{L}^{(ij)}$ (their diagonals) to make statements about the Granger non-causality. We say the j-th scalar time series is non-causal for the s series (given all the remaining series in the process) if $\mathbf{L}_{ss}^{(ji)} = 0$ for all $i \in \mathbb{N}_{s_j}$.

Essentially, any of the numerous regularizers that exist for sparse or structured sparse learning [3] could be used as Q, possibly based on some prior knowledge about the underlying dependencies within the time-series process.

We elaborate here two cases that do not assume any special structure in the dependencies as the base scenarios. The first is the entry-wise ℓ_1 norm across all the output kernels so that

$$Q_1(\mathcal{L}) = \sum_{ji} \|\mathbf{L}^{(ji)}\|_1 = \sum_{ji} \sum_{s}^{m} |L_{ss}^{(ji)}| \ . \tag{8}$$

The second is the ℓ_1/ℓ_2 grouped norm

$$Q_{1/2}(\mathcal{L}) = \sum_{js} \sqrt{\sum_{i} \left(L_{ss}^{(ji)}\right)^2}. \tag{9}$$

After developing the learning strategy for these in Sects. 4.1 and 4.2, we provide some more intuition of their effects on the models and link to some other known graphical Granger techniques in Sect. 5.

4   Learning Strategy

First of all, we simplify the final formulation of the problem (7) in Sect. 3.1. Rather than working with a set of diagonal matrices $\mathbf{L}^{(ji)}$, we merge the diagonals into a single matrix $\mathbf{A}$. We then re-formulate the problem with respect to this single matrix in place of the set and show how this reformulation can be suitably decomposed into smaller independent sub-problems.

We develop fit-to-purpose approaches for our two regularisers in Sects. 4.1 and 4.2. The first - based on the decomposition of the kernel matrices into the corresponding empirical features and on the variational formulation of norms [3] - shows the equivalence of the problem with group lasso [22,23]. The second proposes a simple alternating minimisation algorithm to obtain the two sets of parameters.

We introduce the non-negative matrix $\mathbf{A} \in \mathbb{R}_+^{l \times m}$ such that

$$\mathbf{A} = \left(diag(\mathbf{L}^{11}), \ldots, diag(\mathbf{L}^{ms_m})\right)^T \tag{10}$$

(each row in $\mathbf{A}$ corresponds to the diagonal of one output kernel; if $s_j = 1$ for all j we have $A_{js} = \mathbf{L}_{ss}^{(j1)}$). Using this change of variable, the optimisation problem (7) can be written equivalently as

$$\arg\min_{\mathbf{A},\mathbf{C}} \ddot{J}(\mathbf{C}, \mathbf{A})$$
$$\ddot{J}(\mathbf{C}, \mathbf{A}) := \ddot{R}(\mathbf{C}, \mathbf{A}) + \tau \ddot{Q}(\mathbf{A}), \tag{11}$$

where

$$\ddot{R}(\mathbf{C}, \mathbf{A}) = \sum_s^m \left(||\mathbf{Y}_{:s} - \sum_{ji} A_{(ji)s} \mathbf{K}^{ij} \mathbf{C}_{:s}||_2^2 + \lambda \sum_{ji} A_{(ji)s} \mathbf{C}_{:s}^T \mathbf{K}^{(ji)} \mathbf{C}_{:s} \right)$$

$$= \sum_s^m \left(\ddot{R}_s(\mathbf{C}_{:s}, \mathbf{A}_{:s}) \right), \tag{12}$$

and $\ddot{Q}(\mathbf{A})$ is the equivalent of $Q(\mathcal{L})$ so that

$$\ddot{Q}_1(\mathbf{A}) = ||\mathbf{A}||_1 = \sum_{rs} |A_{rs}| \tag{13}$$

and

$$\ddot{Q}_{1/2}(\mathbf{A}) = \sum_{js} \sqrt{\sum_i \left(A_{(ji)s} \right)^2} \tag{14}$$

In Eqs. (12) and (14) we somewhat abuse the notation by using $\sum_{ji} A_{(ji)s}$ to indicate the sum across the rows of the matrix $\mathbf{A}$.

From (12)–(14) we observe that, with both of our regularizers, problem (11) is conveniently separable along s into the sum of m smaller independent problems, one per scalar output series. These can be efficiently solved in parallel, which makes our method scalable to very large multivariate systems. The final complexity depends on the choice of the regulariser Q and the appropriate algorithm. The overhead cost can be significantly reduced by precalculating the gram matrices $\mathbf{K}^{(ij)}$ in a single preprocessing step and sharing these in between the m parallel tasks.

4.1 Learning with ℓ_1 Norm

To unclutter notation we replace the bracketed double superscripts (ij) by a single superscript $d = 1, \ldots, l$. We also drop the regularization parameter τ (fix it to $\tau = 1$) as it is easy to show that any other value can be absorbed into the rescaling of λ and the $\mathbf{C}$ and $\mathbf{A}$ matrices. For each of the s parallel tasks we indicate $\mathbf{A}_{:s} = \mathbf{a}$, $\mathbf{C}_{:s} = \mathbf{c}$ and $\mathbf{Y}_{:s} = \mathbf{y}$ so that the individual problems are the minimisations with respect to $\mathbf{a} \in \mathbb{R}_+^l$ and $\mathbf{c} \in \mathbb{R}^n$ of

$$P(\mathbf{c}, \mathbf{a}) := ||\mathbf{y} - \sum_d a_d \mathbf{K}^d \mathbf{c}||_2^2 + \lambda \sum_{ji} a_d \mathbf{c}^T \mathbf{K}^d \mathbf{c} + \sum_d a_d. \tag{15}$$

We decompose (for example by eigendecomposition) each of the gram matrices as $\mathbf{K}^d = \mathbf{\Phi}^d (\mathbf{\Phi}^d)^T$, where $\mathbf{\Phi}^d \in \mathbb{R}^{n \times n}$ is the matrix of the empirical features, and we introduce the variables $\mathbf{z}^d = a_d (\mathbf{\Phi}^d)^T \mathbf{c} \in \mathbb{R}^n$ and the set $\mathcal{Z} = \{\mathbf{z}^d : \mathbf{z}^d \in \mathbb{R}^n, d \in \mathbb{N}_l\}$. Using these we rewrite[1] Eq. (15)

$$\tilde{P}(\mathcal{Z}, \mathbf{a}) := ||\mathbf{y} - \sum_d \mathbf{\Phi}^d \mathbf{z}^d||_2^2 + \sum_d \left(\frac{\lambda ||\mathbf{z}^d||_2^2}{a_d} + a_d \right). \tag{16}$$

---

[1] We extend the function $x^2/y : \mathbb{R} \times \mathbb{R}_+ \to \mathbb{R}_+$ to the point $(0,0)$ by taking the convention $0/0 = 0$.

We first find the closed form of the minimising solution for $\mathbf{a}$ as $a_d = \sqrt{\lambda}||\mathbf{z}^d||_2$ for all d. Plugging this back to (16) we obtain

$$\min_{\mathbf{a}} \widetilde{P}(\mathcal{Z}, \mathbf{a}) = ||\mathbf{y} - \sum_d \mathbf{\Phi}^d \mathbf{z}^d||_2^2 + 2\sqrt{\lambda} \sum_d ||\mathbf{z}^d||_2. \tag{17}$$

Seen as a minimisation with respect to the set $\mathcal{Z}$ this is the classical group-lasso formulation with the empirical features $\mathbf{\Phi}^d$ as inputs. Accordingly, it can be solved by any standard method for group-lasso problems such as the proximal gradient descent method, e.g. [3], which we employ in our experiments. After solving for $\mathcal{Z}$ we can directly recover $\mathbf{a}$ from the above minimising identity and then obtain the parameters $\mathbf{c}$ from the set of linear equations

$$\left(\sum_d a_d \mathbf{K}^d + \lambda \mathbf{I}_n\right) \mathbf{c} = \mathbf{y}. \tag{18}$$

The algorithm outlined above takes advantage of the convex group-lasso reformulation (17) and has the standard convergence and complexity properties of proximal gradient descent. The empirical features $\mathbf{\Phi}^d$ can be pre-calculated and shared amongst the m tasks to reduce the overhead cost.

4.2 Learning with ℓ_1/ℓ_2 Norm

For the ℓ_1/ℓ_2 regularization, we need to return to the double indexation (ji) to make clear how the groups are created. As above, for each of the s parallel tasks we use the vectors $\mathbf{a}, \mathbf{c}$ and $\mathbf{y}$. However, for vector $\mathbf{a}$ we will keep the (ji) notation for its elements. The individual problems are the minimisations with respect to $\mathbf{a} \in \mathbb{R}_+^l$ and $\mathbf{c} \in \mathbb{R}^n$ of

$$P(\mathbf{c}, \mathbf{a}) := ||\mathbf{y} - \sum_{ji} a_{(ji)} \mathbf{K}^{(ji)} \mathbf{c}||_2^2 + \lambda \sum_{ji} a_{(ji)} \mathbf{c}^T \mathbf{K}^{(ji)} \mathbf{c} + \sum_j \sqrt{\sum_i a_{(ji)}^2} \tag{19}$$

We propose to use the alternating minimisation with a proximal gradient step. At each iteration, we alternatively solve for $\mathbf{c}$ and $\mathbf{a}$. For fixed $\mathbf{a}$ we obtain $\mathbf{c}$ from the set of linear equations (18). With fixed $\mathbf{c}$, problem (19) is a group lasso for $\mathbf{a}$ with groups defined by the sub-index j within the double (ji) indexation of the elements of $\mathbf{a}$. Here, the proximal gradient step takes place to move along the descend direction for $\mathbf{a}$. Though convex in $\mathbf{a}$ and $\mathbf{c}$ individually, the problem (19) is jointly non-convex and therefore can converge to local minima.

5   Interpretation and Crossovers

To help the understanding of the inner workings of our methods and especially the effects of the two regularizers, we discuss here the crossovers to other existing methods for MKL and Granger modelling.

ℓ_1 *Norm.* The link to group-lasso demonstrated in Sect. 4.1 is not in itself too surprising. The formulation in (15) can be recognised as a sparse multiple kernel learning problem which has been previously shown to relate to group-lasso (e.g. [2,21]). We derive this link in Sect. 4.1 using the empirical feature representation to (i) provide better intuition for the structure of the learned function $\widehat{\mathbf{f}}$, (ii) develop an efficient algorithm for solving problem (15).

The re-formulation in terms of the empirical features Φ^d creates an intuitive bridge to the classical linear models. Each Φ^d can be seen as a matrix of features generated from a subset $\mathcal{X}^{(j)}$ of the input coordinates relating to the past of a single scalar time series j. The group-lasso regularizer in Eq. (17) has a sparsifying effect at the level of these subsets zeroing out (or not) the whole groups of parameters $\mathbf{z}^d$. In the context of linear methods, this approach is known as the grouped graphical Granger modelling [16].

Within the non-linear approaches to time series modelling, Sindhwani et al. [19] recently derived a similar formulation. There the authors followed a strategy of multiple kernel learning from a dictionary of input kernels combined with a single learned output kernel (as opposed to our multiple output kernels). They obtain their IKL model, which is in its final formulation equivalent to problem (15), by fixing the output kernel to identity.

Though we initially formulate our problem quite differently, the diagonal constraint we impose on the output kernels essentially prevents the modelling of any contemporaneous relationships between the series (as does the identity output kernel matrix in IKL). What remains in our methods are the diagonal elements, which are non-constant and sparse, and which can be interpreted as the weights of the input kernels in the standard MKL setting.

ℓ_1/ℓ_2 *Norm.* The more complex ℓ_1/ℓ_2 regularisation discussed in Sect. 4.2 is to the best of our knowledge novel in the context of multiple kernel learning. It has again a strong motivation and clear interpretation in terms of the graphical Granger modelling. The norm has a sparsifying effect not only at the level of the individual kernels but at the level of the groups of kernels operating over the same input partitions $\mathcal{X}^{(j)}$. In this respect our move from the ℓ_1 to the ℓ_1/ℓ_2 norm has a parallel in the same move in linear graphical Granger techniques. The ℓ_1 norm Lasso-Granger method [1] imposes the sparsity on the individual elements of the parameter matrices in a linear model, while the ℓ_1/ℓ_2 of the grouped-Lasso-Granger [16] works with groups of the corresponding parameters of a single input series across the multiple lags p.

6   Experiments

To document the performance of our method, we have conducted a set of experiments on real and synthetic datasets. In these we simulate real-life forecasting exercise by splitting the data into a training and a hold-out set which is unseen by the algorithm when learning the function $\widehat{\mathbf{f}}$ and is only used for the final performance evaluation.

We compare our methods with the output kernel ℓ_1 regularization (NVARL1) and ℓ_1/ℓ_1 (NVARL12) with simple baselines (which nevertheless are often hard to beat in practical time series forecasting) as well as with the state-of-the-art techniques for forecasting and Granger modelling. Namely, we compare with simple mean and univariate linear autoregressive models (LAR), multivariate linear vector autoregressive model with ℓ_2 penalty (LVARL2), the group-lasso Granger method [16] (LVARL1), and a sparse MKL without the $\mathcal{X}^{(j)}$ input partitioning (NVAR). Of these, the last two are the most relevant competitors. LVARL1, similarly to our methods, aims at recovering the Granger structure but is strongly constrained to linear modelling only. NVAR has no capability to capture the Granger relationships but, due to the lack of structural constraints, it is the most flexible of all the models.

We evaluate our results with respect to the two objectives for the function $\widehat{\mathbf{f}}$ defined in Sect. 2. We measure the accuracy of the one-step ahead forecasts by the mean square error (MSE) for the whole multivariate process averaged over 500 hold-out points. The structural objective allowing the analysis of dependencies between the sub-processes is wired into the method itself (see Sects. 3 and 2.1) and is therefore satisfied by construction. We produce adjacency matrices of the graphs of the learned dependencies, compare these with the ones produced by the linear Granger methods and comment on the observed results.

6.1   Technical Considerations

For each experiment we preprocessed the data by removing the training sample mean and rescaling with the training sample standard deviation. We fix the number of kernels for each input partition to six ($s_j = 6$ for all j) and use the same kernel functions for all experiments: a linear, 2nd order and 3rd polynomial, and Gaussian kernels with width 0.5, 1 and 2. We normalise the kernels so that the training Gram matrices have trace equal to the size of the training sample.

We search for the hyper-parameter λ by a 5-fold cross-validation within a 15-long logarithmic grid $\lambda \in \{10^{-3}, \ldots, 10^4\}\sqrt{n}l$, where n is the training sample size and l is the number of kernels or groups (depending on the method). In each grid search, we use the previous parameter values as warm starts. We do not perform an exhaustive search for the optimal lag for each of the scalar input series by some of the classical testing procedures (based on AIC, BIC etc.). We instead fix it to $p = 5$ for all series in all experiments and rely on the regularization to control any excess complexity.

We implemented our own tools for all the tested methods based on variations of proximal gradient descent with ISTA line search [5]. The full Matlab code is available at https://bitbucket.org/dmmlgeneva/nonlinear-granger.

6.2   Synthetic Experiments

We have simulated data from a five dimensional non-Gaussian time-series process generated through a linear filter of a 5-dimensional i.i.d. exponential white noise $\mathbf{e}_t$ with identity covariance matrix (re-centered to zero and re-scaled to unit variance). The matrix $\boldsymbol{\Psi} = [0.7, 1.3, 0, 0, 0; 0, 0.6, -1.5, 0, 0;$ $0, -1.2, 1.46, 0, 0; 0, 0, 0, 0.6, 1.4; 0, 0, 0, 1.3, -0.5]$ in the filter $\mathbf{y}_t = \mathbf{e}_t + \boldsymbol{\Psi}\mathbf{e}_{t-1}$ is such that the process consists of two independent internally interrelated subprocesses, one composed of the first 3 scalar series, the other of the remaining two series. This structural information, though known to us, is unknown to the learning methods (not considered in the learning process).

We list in Table 1 the predictive performance of the tested methods in terms of the average hold-out MSE based on training samples of varying size. Our methods clearly outperform all the linear models. The functionally strongly constrained linear LVARL1 performs roughly on par with our methods for the small sample sizes. But for larger sample sizes, the higher flexibility of the function space in our methods yields significantly more accurate forecasts (as much as 10% MSE improvement).

Table 1. Synthetic experiments: MSE (std) for 1-step ahead forecasts (hold-out sample average)

Train size	300		700		1000	
Mean	0.925	(0.047)	0.923	(0.047)	0.923	(0.047)
LAR	0.890	(0.045)	0.890	(0.044)	0.890	(0.044)
LVAR	0.894	(0.045)	0.836	(0.041)	0.763	(0.035)
LVARL1	0.787	(0.037)	0.737	(0.031)	0.722	(0.030)
NVAR	0.835	(0.041)	0.735	(0.032)	0.719	(0.030)
NVARL1	*0.754*	(0.034)	0.706	(0.030)	**0.679**	(0.028)
NVARL12	0.808	(0.040)	0.710	(0.031)	0.684	(0.029)

Train size	1500		2000		3000	
Mean	0.923	(0.047)	0.922	(0.047)	0.922	(0.047)
LAR	0.888	(0.045)	0.889	(0.045)	0.888	(0.045)
LVAR	0.751	(0.034)	0.741	(0.033)	0.687	(0.028)
LVARL1	0.710	(0.029)	0.701	(0.028)	0.693	(0.028)
NVAR	0.699	(0.028)	0.682	(0.027)	0.662	(0.026)
NVARL1	*0.654*	(0.026)	*0.640*	(0.025)	**0.626**	(0.025)
NVARL12	**0.659**	(0.027)	0.685	(0.028)	**0.657**	(0.027)

In brackets is the average standard deviation (std) of the MSEs. Results for NVARL1 and NVARL12 in bold are significantly better than all the linear competitors, in *italics* are signicantly better than the non-linear NVAR (using one-sided paired-sample t-test at 10% significance level).

The structural constraints in our methods also help the performance when competing with the unstructured NVAR method, which has mostly less accurate forecasts. At the same time, as illustrated in Fig. 1, our methods are able to correctly recover the Granger-causality structure (splitting the process into the two independent subprocesses by the zero off-diagonal blocks), which NVAR by construction cannot.

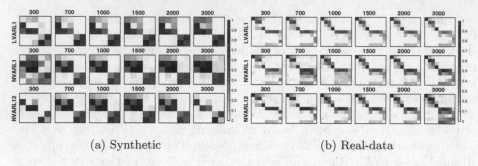

(a) Synthetic   (b) Real-data

Fig. 1. Schematics of the learned adjecency matrices of Granger-causality graphs for the three sparse learning methods across varying training sample size. A scalar time series y_i does not Granger-cause series y_j (given all the other series) if the element e_{ij} in the adjacency matrix is zero (white). The displayed adjacency matrices were derived from the learned matrices **A** by summing the respective elements across individual kernels. The values are rescaled so that the largest element in each matrix is equal to 1 (black).

6.3   Real Data Experiments

We use data on water physical discharge publicly available from the website of the Water Services of the US geological survey (http://www.usgs.gov/). Our dataset consists of 9 time series of daily rates of year-on-year growth at measurement sites along the streams of Connecticut and Columbia rivers.

The prediction accuracy of the tested methods is listed in Table 2. Our non-linear methods perform on par with the state-of-the-art linear models. This on one hand suggests that for the analysed dataset the linear modelling seems sufficient. On the other hand, it confirms that our methods, which in general have the ability to learn more complex relationships by living in a richer functional space, are well behaved and can capture simpler dependencies as well. The structure encoded into our methods, however, benefits the learning since the unstructured NVAR tends to perform less accurately.

The learned dynamical dependence structure of the time series is depicted in Fig. 1. In the dataset (and the adjacency matrices), the first 4 series are the Connecticut measurement sites starting from the one highest up the stream and moving down to the mouth of the river. The next 5 our the Columbia measurement sites ordered in the same manner.

From inspecting the learned adjacency matrices, we observe that all the sparse methods recover similar Granger-causal structures. Since we do not know the ground truth in this case, we can only speculate about the accuracy of the structure recovery. Nevertheless, it seems plausible that there is little dynamical cross-dependency between the Connecticut and Columbia measurements as the learned graphs suggest (the two rivers are at the East and West extremes of the US).

Table 2. Real-data experiments: MSE (std) for 1-step ahead forecasts (hold-out sample average)

Train size	300	700	1000
Mean	0.780 (0.053)	0.795 (0.054)	0.483 (0.026)
LAR	0.330 (0.023)	0.340 (0.024)	0.152 (0.013)
LVARL2	0.302 (0.021)	0.311 (0.022)	0.140 (0.012)
LVARL1	0.310 (0.022)	0.310 (0.023)	0.140 (0.012)
NVAR	0.328 (0.023)	0.316 (0.023)	0.148 (0.012)
NVARL1	0.308 (0.023)	0.317 (0.024)	0.140 (0.012)
NVARL12	0.321 (0.023)	0.322 (0.024)	0.141 (0.012)

Train size	1500	2000	3000
Mean	0.504 (0.03)	0.464 (0.027)	0.475 (0.017)
LAR	0.181 (0.015)	0.179 (0.013)	0.187 (0.008)
LVARL2	0.167 (0.014)	0.164 (0.013)	0.170 (0.007)
LVARL1	0.165 (0.014)	0.163 (0.013)	0.170 (0.008)
NVAR	0.169 (0.014)	0.166 (0.012)	0.173 (0.007)
NVARL1	0.164 (0.014)	0.161 (0.013)	0.167 (0.007)
NVARL12	0.162 (0.014)	0.160 (0.012)	0.166 (0.007)

In brackets is the average standard deviation (std) of the MSEs.

7 Conclusions

We have developed a new method for forecasting and Granger-causality modelling in multivariate time series that does not rely on prior assumptions about the shape of the dynamical dependencies (other than being sparse). The method is based on learning a combination of multiple operator-valued kernels in which the multiple output kernels are learned as sparse diagonal matrices. We have documented on experiments that our method outperforms linear competitors in the presence of strong non-linearities and is able to correctly recover the Granger-causality structure which non-structured kernel methods cannot do.

Acknowledgements. This work was partially supported by the research projects HSTS (ISNET) and RAWFIE #645220 (H2020). We thank Francesco Dinuzzo for helping to form the initial ideas behind this work through fruitful discussions while visiting in IBM Research, Dublin.

References

1. Arnold, A., Liu, Y., Abe, N.: Temporal causal modeling with graphical granger methods. In: Proceedings of 13th ACM SIGKDD International Conference on Knowledge Discovery and Data Mining - KDD 2007 (2007)
2. Bach, F.: Consistency of the group lasso and multiple kernel learning. J. Mach. Learn. Res. **9**, 1179–1225 (2008)

3. Bach, F., Jenatton, R., Mairal, J., Obozinski, G.: Optimization with sparsity-inducing penalties. Found. Trends Mach. Learn. **4**, 1–106 (2012)
4. Bahadori, M., Liu, Y.: An examination of practical granger causality inference. In: SIAM Conference on Data Mining (2013)
5. Beck, A., Teboulle, M.: Gradient-based algorithms with applications to signal recovery. In: Convex Optimization in Signal Processing and Communications (2009)
6. Brockwell, P.J., Davis, R.A.: Time Series: Theory and Methods, 2nd edn. Springer Science+Business Media, LLC, New York (2006). https://doi.org/10.1007/978-1-4899-0004-3
7. Caponnetto, A., Micchelli, C.A., Pontil, M., Ying, Y.: Universal multi-task kernels. Mach. Learn. Res. **9**, 1615–1646 (2008)
8. Dinuzzo, F., Ong, C.: Learning output kernels with block coordinate descent. In: International Conference on Machine Learning (ICML) (2011)
9. Eichler, M.: Graphical modelling of multivariate time series. Probab. Theory Relat. Fields **153**, 233–268 (2012)
10. Franz, M.O., Schölkopf, B.: A unifying view of wiener and volterra theory and polynomial kernel regression. Neural Comput. **18**, 3097–3118 (2006)
11. Granger, C.W.J.: Investigating causal relations by econometric models and cross-spectral methods. Econometrica **37**(3), 424–438 (1969)
12. Jawanpuria, P., Lapin, M., Hein, M., Schiele, B.: Efficient output kernel learning for multiple tasks. In: NIPS (2015)
13. Kadri, H., Rakotomamonjy, A., Bach, F., Preux, P.: Multiple operator-valued kernel learning. In: NIPS (2012)
14. Lanckriet, G.G.R., Cristianini, N., Bartlett, P., Ghaoui, L.E., Jordan, M.I.: Learning the kernel matrix with semidefinite programming. J. Mach. Learn. Res. **5**, 27–72 (2004)
15. Lim, N., D'Alché-Buc, F., Auliac, C., Michailidis, G.: Operator-valued Kernel-based vector autoregressive models for network inference. Mach. Learn. **99**, 489 (2015). https://doi.org/10.1007/s10994-014-5479-3
16. Lozano, A.C., Abe, N., Liu, Y., Rosset, S.: Grouped graphical Granger modeling for gene expression regulatory networks discovery. Bioinformatics **25**, i110–i118 (2009). (Oxford, England)
17. Micchelli, C.A., Pontil, M.: On learning vector-valued functions. Neural Comput. **17**, 177–204 (2005)
18. Pillonetto, G., Dinuzzo, F., Chen, T., De Nicolao, G., Ljung, L.: Kernel methods in system identification, machine learning and function estimation: a survey. Automatica **50**, 657–682 (2014)
19. Sindhwani, V., Minh, H.Q., Lozano, A.: Scalable matrix-valued kernel learning for high-dimensional nonlinear multivariate regression and granger causality. In: UAI (2013)
20. Turkman, K.F., Scotto, M.G., de Zea Bermudez, P.: Non-linear Time Series. Springer, Cham (2014). https://doi.org/10.1007/978-3-319-07028-5
21. Xu, Z., Jin, R., Yang, H., King, I., Lyu, M.R.: Simple and efficient multiple kernel learning by group lasso. In: International Conference on Machine Learning (ICML) (2010)
22. Yuan, M., Lin, Y.: Model selection and estimation in regression with grouped variables. J. Roy. Stat. Soc.: Ser. B (Stat. Methodol.) **68**, 49–67 (2006)
23. Zhao, P., Rocha, G.: Grouped and hierarchical model selection through composite absolute penalties (2006)

Learning TSK Fuzzy Rules from Data Streams

Ammar Shaker[✉], Waleri Heldt, and Eyke Hüllermeier

Department of Computer Science, Paderborn University, Paderborn, Germany
{ammar.shaker,eyke}@upb.de, heldt@mail.upb.de

Abstract. Learning from data streams has received increasing attention in recent years, not only in the machine learning community but also in other research fields, such as computational intelligence and fuzzy systems. In particular, several rule-based methods for the incremental induction of regression models have been proposed. In this paper, we develop a method that combines the strengths of two existing approaches rooted in different learning paradigms. Our method induces a set of fuzzy rules, which, compared to conventional rules with Boolean antecedents, has the advantage of producing smooth regression functions. To do so, it makes use of an induction technique inspired by AMRules, a very efficient and effective learning algorithm that can be seen as the state of the art in machine learning. We conduct a comprehensive experimental study showing that a combination of the expressiveness of fuzzy rules with the algorithmic concepts of AMRules yields a learning system with superb performance.

1 Introduction

Learning from data streams has been a topic of active research in recent years [11]. In this branch of machine learning, systems are sought that learn incrementally, and maybe even in real-time, on a continuous and potentially unbounded stream of data, and which is able to properly adapt themselves to changes of environmental conditions or properties of the data generating process. Systems with these properties have already been developed for different machine learning and data mining tasks, such as clustering and classification.

An extension of machine learning methods to the setting of data streams comes with a number of challenges. In particular, the standard batch mode of learning, in which the entire data as a whole is provided as an input to the learning algorithm, is no longer applicable. Correspondingly, the data must be processed in a single pass, which implies an incremental mode of learning and model adaptation.

Domingos and Hulten [9] list a number of properties that an ideal stream mining system should exhibit, and suggest corresponding design decisions: the system uses only a limited amount of memory; the time to process a single record is short and ideally constant; the data is volatile and a single data record accessed only once; the model produced in an incremental way is equivalent to the model that would have been obtained through common batch learning (on

M. Ceci et al. (Eds.): ECML PKDD 2017, Part II, LNAI 10535, pp. 559–574, 2017.
https://doi.org/10.1007/978-3-319-71246-8_34

all data records so far); the learning algorithm should react to concept change (i.e., any change of the underlying data generating process) in a proper way and maintain a model that always reflects the current concept.

Rule-based learning is a specifically popular approach in the realm of data streams, not only in the machine learning but also in the computational intelligence community, where it has been studied under the notion of "evolving fuzzy systems" [21]. In this paper, we develop a method that combines the strengths of two existing approaches for regression on data streams rooted in different learning paradigms. Our method induces a set of fuzzy rules, which, compared to conventional rules with Boolean antecedents, has the advantage of producing smooth regression functions. To do so, it makes use of an induction technique inspired by AMRules, a very efficient and effective learning algorithm that yields state-of-the-art performance in machine learning.

The rest of the paper is organized as follows. Following a review of related work, we introduce our method in Sect. 3. A comprehensive experimental study, in which the method is compared to several competitors, is presented in Sect. 4, prior to concluding the paper in Sect. 5.

2 Related Work

In the past ten years, learning from data streams has been considered for different learning tasks. Approaches to supervised learning have mostly focused on classification. Here, the Hoeffding tree method [8] has gained a lot of attention, and meanwhile, many modifications and improvements of the original method have been proposed [6]. In addition to the induction of decision trees, the learning of systems of decision rules is supported by several approaches, such as the Adaptive Very Fast Decision Rules (AVFDR) classifier [17]. AVFDR can be seen as an extension of the Very Fast Decision Rules (VFDR) classifier [12] that incrementally induce a compact set of decision rules from a data stream.

Regression on data streams has gained less attention than classification, with a few notable exceptions. AMRules [1] can be seen as an extension of AVFDR to the case of numeric target values. Another approach is based on the induction of model trees [15]. Besides, regression on data streams has been studied quite extensively in the computational intelligence and fuzzy systems community [2, 4, 21]. Specifically relevant for us is FLEXFIS [20], which learns a system of so-called Takagi-Sugeno-Kang (TSK) rules [25]. In the following, we describe these methods in some more detail.

FIMTDD (Fast Incremental Model Trees with Drift Detection) is a tree-based approach for inducing model trees for regression on data streams. Similar to Hoeffding trees, it uses Hoeffding's inequality [14] for choosing the best splitting attribute. Since FIMTDD tackles regression problems, attributes are evaluated in terms of the reduction of the target attribute's standard deviation. Each leaf node of the induced tree is associated with a linear function, which is learned (using stochastic gradient descent) in the subspace covering the instances that fall into that leaf node.

AMRules (Adaptive Model Rules) learns rules that are specified by a conjunction of literals on the input attributes in the premise part, and a linear function of the attributes in the consequent. The latter is chosen so as to maximize predictive accuracy in the sense of minimizing the root mean squared error. Adaptive statistical measures are maintained in each rule in order to describe the instance subspace covered by that rule. Each rule is initialized with a single literal and successively expanded with new literals. The best literal to be added, if any, is chosen on the basis of Hoeffding's bound, in a manner that is similar to the expansion of a Hoeffding tree. In their paper [1], the authors distinguish between decision lists and unordered rule sets and, correspondingly, propose two different update and prediction schemes. The first one sorts the set of rules in the order in which they were learned. Only the first rule covering an example is used for prediction and updated afterward. The second strategy updates all the rules that cover an example, and combines these rules' predictions by a weighted sum.[1] The authors also show that the latter strategy outperforms the former one, and hence used it for the rest of their study.

FLEXFIS (Fexible Fuzzy Inference Systems) induces a set of fuzzy rules, making use of fuzzy logic as a generalization of conventional (Boolean) logic [20]. Mores specifically, it uses so-called Takagi-Sugeno-Kang (TSK) rules that are defined by a fuzzy predicate in the premise part and a linear function of the input features in the consequent. As a result, an instance can be covered by a rule to a certain degree, reflecting a degree of relevance of the rule in the corresponding part of the instance space. Correspondingly, the prediction of a TSK system is produced by a weighted average of the outputs of the individual rules. The regions covered by the rules in the input space are defined by means of clustering methods: The instances in the training data are first clustered, and the fuzzy-logical predicates in the rule antecedents are obtained by projecting the clusters to the individual dimensions of this space. For learning on data streams, clustering is done in an incremental way. Moreover, the functions in the rule consequents are adapted using recursive weighted least squares (RWLS) estimation [19].

Our approach essentially seeks to combine the increased expressiveness of fuzzy rules as used by methods such as FLEXFIS, which allows for approximating a regression function is a smoother and much more flexible way, with the efficiency and effectivity of rule induction techniques such as AMRules. In fact, existing methods for learning fuzzy rules, including FLEXFIS and eTS+ [3], are usually slow and computationally inefficient. The complexity is mainly caused by the use of clustering methods, which have the additional disadvantage of producing rules that always contain all input attributes, as well as costly matrix operations (such as inversion) required by RWLS.

---

[1] In their paper [1], it is not mentioned how the weight of a rule is derived. Based on the delivered implementation, it seems a rule is weighted by the errors it has committed in the past.

3   The TSK-Streams Learning Algorithm

Our method, called TSK-Streams, is an adaptive incremental rule induction algorithm for regression on data streams. The model produced by TSK-Streams is a so-called Takagi-Sugeno-Kang (TSK) fuzzy system [25], a type of rule-based system that is widely used in the fuzzy logic community.

3.1   TSK Fuzzy Systems

A TSK rule R_i is a fuzzy rule of the following form:

$$\text{IF} \quad (x_1 \text{ IS } A_{i,1}) \text{ AND } \quad \ldots \quad \text{AND } (x_d \text{ IS } A_{i,d})$$
$$\text{THEN} \quad l_i(\boldsymbol{x}) = w_{i,0} + w_{i,1}x_1 + w_{i,2}x_2 + \ldots + w_{i,d}x_d, \tag{1}$$

where $(x_1, \ldots, x_d)^\top$ is the feature representation of an instance $\boldsymbol{x} \in \mathbb{R}^d$, and $A_{i,j}$ defines the jth antecedent of R_i in terms of a soft constraint. The coefficients $w_{i,0}, \ldots, w_{i,d} \in \mathbb{R}$ in the consequent part of the rule specify an affine function of the features (input attributes).

Modeling the soft constraint $A_{i,j}$ in terms of a fuzzy set with membership function $\mu_j^{(i)} : \mathbb{R} \longrightarrow [0,1]$, the truth degree of the predicate $(x_j \text{ IS } A_{i,j})$ is given by $\mu_j^{(i)}(x_j)$, that is, the degree of membership of x_j in $\mu_j^{(i)}$. Moreover, modeling the logical conjunction in terms of a triangular norm $\top$ [16], i.e., an associative, commutative, non-decreasing binary operator $\top : [0,1]^2 \longrightarrow [0,1]$ with neutral element 1 and absorbing element 0, the overall degree to which an instance $\boldsymbol{x}$ satisfies the premise of the rule R_i is given by

$$\mu_i(\boldsymbol{x}) = \top \left(\mu_1^{(i)}(x_1), \ldots, \mu_d^{(i)}(x_d) \right). \tag{2}$$

In the following, we will adopt the simple product norm, i.e., $\top(u,v) = uv$. Note that $A_{i,j}$ could be an empty constraint, which is modeled by $\mu_j^{(i)} \equiv 1$; this means that the jth attribute x_j does effectively not occur as part of the premise of the rule (1).

Now, consider a TSK system consisting of C rules $RS = \{R_1, \ldots, R_C\}$. Given an instance $\boldsymbol{x}$ as an input, each rule R_i is supposed to "fire" with the (activation) degree (2). Correspondingly, the output produced by the system is defined in terms of a weighted average of the outputs produced by the individual rules:

$$\hat{y} = \sum_{i=1}^{C} \Psi_i(\boldsymbol{x}) \cdot l_i(\boldsymbol{x}), \tag{3}$$

where

$$\Psi_i(\boldsymbol{x}) = \frac{\mu_i(\boldsymbol{x})}{\sum_{j=1}^{C} \mu_j(\boldsymbol{x})}. \tag{4}$$

3.2   Online Rule Induction

TSK-Streams learns rules incrementally, starting with a default rule. This rule
has an empty premise and covers the entire input space.

For each rule R_i, TSK-Streams continuously checks whether one of its extensions may improve the performance of the current system. Here, expanding a rule
R_i means splitting it into two new rules, which are obtained by adding, respec-
tively, a new predicate $(x_j \text{ IS } A_{i,j})$ and $(x_j \text{ IS } \neg A_{i,j})$ as an additional antecedent.
Considering the current rule as the default, the former defines a specialization,
while the latter can be seen as what remains of this default. $A_{i,j}$ is modeled in
terms of a fuzzy set with membership function $\mu_{j,l}^{(i)}$, which is chosen from a fuzzy
partition $\{\mu_{j,1}, \ldots, \mu_{j,k}\}$ of the domain of feature x_j (cf. Sect. 3.3 below), and
its negation $\neg A_{i,j}$ is characterized by the membership function $\bar{\mu}_{j,l}^{(i)} = 1 - \mu_{j,l}^{(i)}$.
We denote the corresponding expansions by $R_i \oplus \mu_{j,l}^{(i)}$ and $R_i \oplus \bar{\mu}_{j,l}^{(i)}$, respectively.

We distinguish between features x_j that are included by a positive literal
$(x_j \in \mu_{j,l}^{(i)})$ and those included by a negative literal $(x_j \in \bar{\mu}_{j,l}^{(i)})$, collecting the
indices of the former in the index set I and those of the latter in $\bar{I}$. In a single
rule, each attribute is only allowed to occur in a single positive literal. Negative
literals are allowed to be added as long as the conjunction of the constraints
on x_j does not become too restrictive, thus suggesting a kind of inconsistency
(there is not a single value of x_j satisfying the rule premise to a high degree).
Details of the rule expansion procedure in pseudocode are given in Algorithm 1.

3.3   Online Discretization and Fuzzification

As a basis for rule expansion, TSK-Streams maintains a fuzzy partition for each
feature x_j, i.e., a discretization of the domain of x_j into a finite number of
(overlapping) fuzzy sets $\{\mu_{j,1}, \ldots, \mu_{j,k}\}$.

The discretization process is based on the Partition Incremental Discretization (PID) proposed by Gama and Pinto [13]. PID is a technique that builds
histograms on data streams in an adaptive manner. In a first layer, continuous
input values produced by the data stream are grouped into intervals. A second layer then uses the intervals of the first layer to build histograms, using
either equal frequency or equal width binning. In this work, we extend the PID
approach as follows:

- Layer 1: This layer discretizes and summarizes the values observed for one
  input feature into an intial set of intervals.
- Layer 2: This layer merges or splits intervals of the first layer, with the goal
  to create intervals of equal frequencies.
- Layer 3: This layer transforms the second layer's intervals, which are of the
  form $X_{j,l} = [b, c]$, into fuzzy sets $\mu_{j,l}$. We employ fuzzy sets with a core $[b, c]$,
  in which the degree of membership is 1, and support $[a, d] \supset [b, c]$; outside the
  support, the membership is 0. The boundary of the fuzzy set $\mu_{j,l}$ is modeled in
  terms of a smooth "S-shaped" transition between full and zero membership:

Algorithm 1. GenerateExtendedRules

Input: $R = (I, M, \bar{I}, \bar{M}, \boldsymbol{\omega})$:
I the index set of the features considered in the premise.
M the set of fuzzy sets conjugated in premise.
$\bar{I}$ the index set of the features negated in the premise.
$\bar{M}$ the set of fuzzy sets whose negated form is conjugated in premise.
$\boldsymbol{\omega}$ the vector of coefficients of the linear function.
$P = \{\mu_{i,j}\}$: the set of all available fuzzy sets. $\mu_{i,j}$ is the jth fuzzy set on the ith feature.
υ the overlapping threshold.
Result: $S = \{(R_1, R_2)\}$: Set of expanded rules

1 **for** $i \in \{1, \ldots, d\} \setminus I$ **do**
2   **for** $\mu_{i,j} \in P$ **do**
3   **if** $i \notin \bar{I}$ **then**
4           $R_1 = (I \cup \{i\}, M \cup \{\mu_{i,j}\}, \bar{I}, \bar{M}, \boldsymbol{\omega})$
5           $R_2 = (I, M, \bar{I} \cup \{i\}, \bar{M} \cup \{\mu_{i,j}\}, \boldsymbol{\omega})$
6           $S \cup \{(R_1, R_2)\}$
7   **else**
8   find $m_{i.} \in M$
9   **if** $\mu_{i,j} \cap m_{i.} < \upsilon$ **then**
10             $R_1 = (I \cup \{i\}, M \cup \{\mu_{i,j}\}, \bar{I} \setminus \{i\}, \bar{M} \setminus \{m_{i.}\}, \boldsymbol{\omega})$
11             $m\prime_{i.} = m_{i.} \cup \mu_{i,j}$
12             $R_2 = (I, M, \bar{I}, \bar{M} \cup \{m\prime_{i.} \cup \mu_{i,j}\} \setminus \{m_{i.}\}, \boldsymbol{\omega})$
13             $S \cup \{(R_1, R_2)\}$

$$\mu_{j,l}(x) = \begin{cases} 0 & \text{if } x < a \\ 2\left(\frac{x-a}{b-a}\right)^2 & \text{if } a \leq x < (a+b)/2 \\ 1 - 2\left(\frac{x-b}{b-a}\right)^2 & \text{if } (a+b)/2 \leq x < b \\ 1 & \text{if } b \leq x \leq c \\ 1 - 2\left(\frac{c-x}{d-c}\right)^2 & \text{if } c < x \leq (c+d)/2 \\ 2\left(\frac{d-x}{d-c}\right)^2 & \text{if } (c+d)/2 < x \leq d \\ 0 & \text{if } x > d \end{cases} \tag{5}$$

For the fuzzy set $\mu_{j,l}$ associated with $X_{j,l}$, we set $a = b - \alpha \cdot |X_{j,l-1}|$ and $d = c + \alpha \cdot |X_{j,l+1}|$, where $|X_{j,l-1}|$ and $|X_{j,l+1}|$ denote, respectively, the lengths of the left and right neighbor interval of $X_{j,l}$, and $\alpha \in]0, 1[$ is an overlap degree. For the leftmost (rightmost) interval of the partition, we set $a = b$ $(d = c)$.

3.4   Learning Rule Consequents

FLEXFIS fits linear functions in the consequent parts of the rules via recursive weighted least squares estimation (RWLS) [19]. Since this approach requires

---

Algorithm 2. First Layer: Discretization

Input: x: newly observed value
k: the initial number of intervals
λ: exponential weighting factor, $\lambda \in [0, 1]$
Result: $L1Intervals, L1Counts, L1Time$: arrays for the produced intervals,
their counts and timestamps of the last update

1 **if** *first call* **then**
2   initialize $L1Intervals$ with k intervals of equal width

3 $sum = \sum_i L1Counts[i]$
4 **let** t be the current time
5 **let** i be the index s.t $x \in L1Intervals[i]$
6 $sum = sum * \lambda + 1$
7 $L1Counts[i] = L1Counts[i] * \lambda^{t-L1Time[i]} + 1$
8 $L1Time[i] = t$
9 **if** $(L1Counts[i]/sum) > threshold$ **then**
10   Split($L1Intervals[i], L1Counts[i]$)

---

multiple matrix inversions, it is computationally expensive. Therefore, inspired by AMRules, we instead apply a gradient method to learn consequents more efficiently.

Upon arrival of a new training instance $(\boldsymbol{x}_t, y_t)$, the squared error of the prediction $\hat{y}_t$ produced by TSK-Streams can be computed as follows:

$$E_t = (y_t - \hat{y}_t)^2 = \left(y_t - \left(\sum_{R_i \in RS} \frac{\Psi_i(\boldsymbol{x}_t)}{\sum_{R_k \in RS} \Psi_k(\boldsymbol{x}_t)} \sum_{j=0}^{d} \omega_{i,j} x_{t,j} \right) \right)^2 \quad (6)$$

Invoking the principle of stochastic gradient descent, the coefficients $\omega_{i,j}$ are then shifted into the negative direction of the gradient:

$$\boldsymbol{\omega} \leftarrow \boldsymbol{\omega} - \eta \nabla E_t, \quad (7)$$

where η is the learning rate. Component-wise, this yields the following update rule:

$$\omega_{i,j} \leftarrow \omega_{i,j} - 2\eta(y_t - \hat{y}_t) \left(\sum_{R_i \in RS} \frac{\Psi_i(\boldsymbol{x}_t)}{\sum_{R_k \in RS} \Psi_k(\boldsymbol{x}_t)} x_{t,j} \right) \quad (8)$$

The process of updating the rule consequents is summarized in Algorithm 4.

3.5 Adaptation of the Model Structure

As outlined above, TSK-Streams continuously adapts the rule system through the adaptation of fuzzy sets used as rule antecedents and linear functions in the rule consequents. While these are adaptations of the system's parameters, the decision to replace a rule by one of its expansions can be seen as a structural change.

Algorithm 3. Second Layer: Histograms

Input:

$L1Intervals, L1Counts$: arrays for the Layer1's intervals and their counts

k: number of intervals

Result: $L2Intervals$: the resulting intervals of the 2nd Layer

1   $sum = \sum_i L1Intervals[i]$

2   $maxCap = sum/k$

3   $currentCap = 0$

4   $interval.min = L1Intervals[1].min$

5   **while** $i <= length(L1Intervals)$ **do**

6   **while** $currentCap + L1Counts[i] < maxCap$ **do**

7       $currentCap = currentCap + L1Counts[i]$

8       $interval.max = intervalsL1[i].max$

9       $i++$

10    $newMax = L1Intervals[i].min + \frac{maxCap-currentCap}{L1Counts[i]}.L1Intervals[i].width$

11    $interval.max = newMax$

12    $L2Intervals.add(interval)$

13    $L1Intervals[i].min = newMax$

14    $L1Counts[i] = L1Counts[i] - \frac{maxCap-currentCap}{L1Counts[i]}$

15    $interval.min = newMax$

16    $currentCap = 0$

17   **if** $length(L1Intervals) > 1.5K$ **then**

18    $L1Intervals = L2Intervals$

19   **for** $i \in \{1, \ldots, k\}$ **do**

20      $L1Counts[i] = maxCap$

Needless to say, structural changes should generally be handled with care, especially when increasing the complexity of the model. Therefore, learning methods typically stick to the current model until being sufficiently convinced of a potential improvement through an expansion; to this end, the estimated difference in performance needs to be *significant* in a statistical sense.

Similar to Hoeffding trees [8], AMRules [10], and FIMT-DD [15], we apply Hoeffding's inequality in order to support these decisions. The Hoeffding inequality probabilistically bounds the difference between the expected value $E(X)$ of a random variable X with support $[a, b] \subset \mathbb{R}$ and its empirical mean $\bar{X}$ on an i.i.d. sample of size n in terms of

$$P\left(|\bar{X} - \mathrm{E}(X)| > \epsilon\right) \leq \exp\left(-\frac{2n\epsilon^2}{(b-a)^2}\right). \tag{9}$$

More specifically, we decide to split a rule R_i, i.e., to replace the rule with two rules $R_i \oplus \mu_{j,l}^{(i)}$ and $R_i \oplus \bar{\mu}_{j,l}^{(i)}$, by considering the reduction in the sum of squared errors (SSE). To this end, the SSE of the current system (rule set RS) is compared to the SSE of all alternative systems $(RS \setminus R_i) \cup \{R_i \oplus \mu_{j,l}^{(i)}, R_i \oplus \bar{\mu}_{j,l}^{(i)}\}$.

Algorithm 4. UpdateConsequent

Input: $RS = \{(R, S)\}$: the set of all rules and their extensions
$(\boldsymbol{x}_t, y_t)$: the new instance to train on
/* A rule takes the form $R = (I, M, \bar{I}, \bar{M}, \omega)$ */
/* A rule extension is $S = \{(R_1, R_2, SSE)\}$, where SSE is the sum of
   squared errors committed by this extension   */

1 $m_1 = \sum_{R_i \in RS} \mu_i(\boldsymbol{x}_t)$

2 $m_2 = \sum_{R_i \in RS} \mu_i(\boldsymbol{x}_t) l_i(\boldsymbol{x}_t)$

3 **for** $R_i \in RS$ **do**

4    $\mu_i(\boldsymbol{x}_t) = \top(\bigotimes_{\mu \in M_i} \mu(\boldsymbol{x}_t), \bigotimes_{\mu \in \bar{M}_i} 1 - \mu(\boldsymbol{x}_t))$

5   **if** $\mu_i(\boldsymbol{x}_t) > 0$ **then**

6   **for** $(R_1, R_2, SSE) \in S_i$ **do**

7        $\mu_{i1}(\boldsymbol{x}_t) = \top(\bigotimes_{\mu \in M_1} \mu(\boldsymbol{x}_t), \bigotimes_{\mu \in \bar{M}_1} 1 - \mu(\boldsymbol{x}_t))$

8        $\mu_{i2}(\boldsymbol{x}_t) = \top(\bigotimes_{\mu \in M_2} \mu(\boldsymbol{x}_t), \bigotimes_{\mu \in \bar{M}_2} 1 - \mu(\boldsymbol{x}_t))$

9        $m_1' = m_1 - \mu_i(\boldsymbol{x}_t) + \mu_1(\boldsymbol{x}_t) + \mu_2(\boldsymbol{x}_t)$

10       $m_2' = m_2 - \mu_i(\boldsymbol{x}_t) l_i(\boldsymbol{x}_t) + \mu_1(\boldsymbol{x}_t) l_1(\boldsymbol{x}_t) + \mu_2(\boldsymbol{x}_t) l_2$

11       $\omega_1 = \omega_1 + \eta(y_t - \frac{m_2'}{m_1'})\left(\frac{\mu_1(\boldsymbol{x}_t)}{m_1'}\boldsymbol{x}\right)$

12       $\omega_2 = \omega_2 + \eta(y_t - \frac{m_2'}{m_1'})\left(\frac{\mu_2(\boldsymbol{x}_t)}{m_1'}\boldsymbol{x}\right)$

13    $\omega_i = \omega_i + \eta(y_t - \frac{m_2}{m_1})\left(\frac{\mu_i(\boldsymbol{x}_t)}{m_1}\boldsymbol{x}\right)$

Let SSE_{best} and $SSE_{2ndbest}$ denote, respectively, the expansion with the lowest and the second lowest error. The best expansion is then adopted whenever

$$\frac{SSE_{best}}{SSE_{2ndbest}} < 1 - \epsilon, \tag{10}$$

or when ϵ becomes smaller than a tie-breaking constant τ. The constant ϵ is derived from (9) by setting the probability to a desired degree of confidence $1 - \delta$, i.e., setting the righ-hand side to $1 - \delta$ and solving for ϵ; noting that the ratio (10) is bounded in $]0, 1]$, $b - a$ is set to 1.[2] Refer to Algorithm 5 for details.

Instead of looking for a global improvement of the entire system, an alternative is to monitor the performance of individual rules and to base decisions about rule expansion on this performance. In this case, Hoeffding's bound is applied to the sum of weighted squared errors (SWSE), where the weighted squared error of a rule R_i on a training example $(\boldsymbol{x}_t, y_t)$ is given by

$$WSE_t = \Psi(\boldsymbol{x})(y_t - \hat{y}_t)^2 = \left(\frac{\mu(\boldsymbol{x}_t)}{\sum_{R_j \in RS} \mu_j(\boldsymbol{x}_t)}\right)(y_t - \hat{y}_t)^2.$$

---

[2] We are aware of theoretical issues caused by the use of the Hoeffding bound, the assumptions of which are normally not all satisfied [23]. Yet, the bound is commonly applied, in spite of these problems, and proved very useful in practice.

This error is then compared with the weighted error of the system in which the rule is replaced by extensions $R_i \oplus \mu_{j,l}^{(i)}$ and $R_i \oplus \bar{\mu}_{j,l}^{(i)}$. The usefulness of such extensions can be checked using the same kind of hypothesis testing as above.

To avoid an excessive increase in the number of rules, also coming with a danger of overfitting, we propose a penalization mechanism that consists of adding a complexity term C to ϵ. For the global variant, we set $C = \frac{1 - \log(2)/\log(|RS|)}{\sqrt{d}}$, where RS is the current set of rules and d the number of features. For the local variant, we use $C = \frac{1 - \log(2)/\log(|I \cup \bar{I}e|)}{\sqrt{d}}$ when comparing the extensions of a rule $R = (I, M, \bar{I}, \bar{M}, \omega)$.

Algorithm 5. ExpandSystem

Input: $RS = \{(R, S)\}$: the set of all rules and their extensions
δ: confidence level
τ: tie-breaking constant
n: number of examples seen so far by the current system
/* A rule takes the form $R = (I, M, \bar{I}, \bar{M}, \omega)$ */
/* A rule extension is $S = \{(R_1, R_2, SSE)\}$, where SSE is the sum of
   squared errors committed by this extension   */
1 $n = n + 1$
2 let $SSE_{current}$ be the sum of squared errors for the current system
3 let $(R, S)_{best}$ be the best performing extension with the lowest achieved sum of
   squared errors SSE_{best}
   /* $S_{best} = \{(R_1, R_2, SSE)_{best}\}$ */
4 let $(R, S)_{2ndbest}$ be the second best performing extension with $SSE_{2ndbest}$
   /* $S_{2ndbest} = \{(R_1, R_2, SSE)_{2ndbest}\}$ */
5 $\epsilon = \sqrt{\frac{\ln\left(\frac{1}{\delta}\right)(R)^2}{2n}}$
6 $\overline{X} = \frac{1}{n}(SSE_{best}/SSE_{2ndbest})$
7 $\overline{Y} = \frac{1}{n}(SSE_{best}/SSE_{current})$
8 **if** $((\overline{Y} + \epsilon) < 1)$ *AND* $((\overline{X} + \epsilon) < 1$ *OR* $\epsilon < \tau)$ **then**
9      $RS = RS \setminus \{(R, S) : R = R_{best}\}$
10     $RS = RS \cup \{(R_{best,1}, GenerateExtendedRules(R_{best,1})),$
11     $(R_{best,2}, GenerateExtendedRules(R_{best,2}))\}$
12     $n = 0, SSE_{current} = 0$
13   **for** $R \in RS$ **do**
14   reinitialize R

3.6 Change Detection

To detect a drop in a rule's performance, possibly caused by a concept drift, we employ the adaptive windowing (ADWIN) [5] drift detection method. The advantage of this technique, compared to the Page-Hinkely test (PH) [22] used by AMRules, is that ADWIN is non-parametric (it makes no assumptions about the observed random variable). Moreover, it has only one parameter δ_{adwin}, which represents the tolerance towards false alarms. We apply this change detection method locally in each rule on the absolute error committed by a rule on an example, given that the example is covered by this rule.

Upon detecting a drift in the rule $R_p = (I_p, M_p, \bar{I}_p, \bar{M}_p, \omega_p)$, we find its sibling rule $R_q = (I_q, M_q, \bar{I}_q, \bar{M}_q, \omega_q)$, from which it differs by only one single literal, i.e., there is a fuzzy set $\mu_{i,j}$ that satisfies one of the following criteria: $(\mu_{i,j} \in M_p) \wedge (i \in I_p) \wedge (\bar{\mu}_{i,j} \in \bar{M}_q) \wedge (i \in \bar{I}_q)$ or $(\bar{\mu}_{i,j} \in \bar{M}_p) \wedge (i \in \bar{I}_p) \wedge (\mu_{i,j} \in M_q) \wedge (i \in I_q)$. Removing the rule R_p can simply be achieved by removing it from the rule set and accordingly updating its sibling R_q by either removing $(i, \mu_{i,j})$ from (I_q, M_q) or removing $(i, \bar{\mu}_{i,j})$ from $(\bar{I}_q, \bar{M}_q)$, depending on which of the previous criteria was satisfied. If the sibling rule R_q was already extended before detecting the drift, one simply applies the same procedure to its children.

4   Empirical Evaluation

In this section, we conduct experiments in order to study the performance of TSK-Streams in comparison to other algorithms. More precisely, we analyze predictive accuracy and runtime of the algorithms, the size of the models they produce, as well as their ability to recover in the presence of a concept drift.

4.1   Setup

Our proposed fuzzy learner, TSK-Streams, is implemented under the MOA[3] (Massive Online Analysis) [7] framework, an open source software for mining and analyzing large data sets in a stream-like manner.

In the following evaluations, we compare TSK-Streams with the three methods introduced before: AMRules, FIMTDD, and FLEXFIS. Both AMRules and FIMTDD are implemented in MOA's distribution, and we use them in their default settings with $\delta = 0.01$ and $\tau = 0.05$ for the Hoeffding bound. Regarding the parametrization of TSK-Streams, we use the same values δ, τ, so as to assure maximal comparability with AMRules and FIMTDD. For the discretization, we use the following parameters: the number of intervals $k = 5$, the overlapping threshold $\upsilon = 0.2$, the exponential weighting factor $\lambda = 0.999$, and the overlapping degree $\alpha = 0.15$. FLEXFIS is implemented in Matlab and offers a function for finding optimal parameter values. We used this function to tune all parameters except the so-called "forgetting parameter", for which we manually found the value 0.999 to perform best.

All experiments are conducted using the test-then-train evaluation procedure; this procedure uses each instance for both training and testing. First, the model is evaluated on the instance, and then a single incremental learning step is carried out.

---

[3] http://moa.cms.waikato.ac.nz.

4.2   Results

In the first part of the evaluation, we preform experiments on standard synthetic and real benchmark data sets collected from the UCI repository[4] [18] and other repositories[5]; Table 1 provides a summary of the type, the number of attributes and instances of each data set.[6] Table 2 shows the average RMSE and the corresponding standard error on ten rounds for each data set. In this table, the winning approach on each data set is highlighted in bold font, and our approach is marked with an asterisk whenever it outperforms the three competitors. As can be seen, our fuzzy rule learner, both in its global and local variant, is superior to the other methods in terms of generalization performance. In a pairwise comparison, the global variant of TSK-Streams outperforms AMRules and FLEXFIS on 11 of the 14 data sets, and performs better than FIMTDD on 13; the local variant outperforms AMRules, FLEXFIS, and FIMTDD on 8, 11, and 13 data sets, respectively. Using a Wilcoxon signed-rank test, the global variant of our method thus outperforms AMRules, FLEXFIS, and FITDD with p-values 0.067, 0.041 and 0.0008, and the local variant outperforms FITDD with p-value 0.0008.

Table 1. Data sets

#	Name	Synthetic	Instances	Attributes
1	2dplanes	yes	40768	11
2	ailerons	no	13750	41
3	bank8FM	yes	8192	9
4	calHousing	no	20640	8
5	elevators	no	8752	19
6	fried	yes	40769	11
7	house16H	no	22784	16
8	house8L	no	22784	8
9	kin8nm	-	8192	9
10	mvnumeric	yes	40768	10
11	pol	no	15000	49
12	puma32H	yes	8192	32
13	puma8NH	yes	8192	9
14	ratingssweetrs	-	17903	2

---

[4] http://archive.ics.uci.edu/ml/.

[5] https://github.com/renatopp/arff-datasets/tree/master/regression, http://tunedit.org/repo/UCI/numeric.

[6] These are the same data sets as those used in the AMRules paper [1].

Table 2. Performance of the algorithms in terms of RMSE.

	AMRules	FIMTDD	FLEXFIS	TSK-streams (global)	TSK-streams (local)
2dplanes	$1.40E+00(0)$	$2.67E+00(0)$	$2.39E+00(0)$	$\mathbf{1.04E+00^*(0)}$	$1.05E+00^*(0)$
ailerons	$5.83E-04(0)$	$8.00E-04(0)$	$1.91E-04(0)$	$1.81E-04^*(0)$	$\mathbf{1.77E-04^*(0)}$
bank8FM	$3.81E-02(0)$	$1.25E-01(0)$	$\mathbf{3.64E-02(0)}$	$4.44E-02(0)$	$4.29E-02(0)$
calhousing	$7.00E+04(97)$	$8.44E+04(121)$	$\mathbf{6.72E+04(71)}$	$7.28E+04(90.8)$	$7.26E+04(10.3)$
elevators	$5.80E-03(0)$	$7.94E-03(0)$	$3.64E-03(0)$	$3.79E-03(0)$	$\mathbf{3.42E-03^*(0)}$
fried	$2.43E+00(0)$	$3.55E+00(0.02)$	$2.64E+00(0)$	$2.33E+00^*(0.01)$	$\mathbf{2.29E+00^*(0)}$
house16h	$4.74E+04(427)$	$5.39E+04(435)$	$4.84E+04(15)$	$4.46E+04^*(104)$	$\mathbf{4.44E+04(52.3)}$
house8	$4.07E+04(118)$	$4.67E+04(97.8)$	$4.04E+04(249)$	$\mathbf{4.02E+04^*(142)}$	$4.13E+04(106)$
kin8nm	$2.05E-01(0)$	$2.84E-01(0)$	$2.02E-01(0)$	$1.96E-01^*(0)$	$\mathbf{1.96E-01^*(0)}$
mvnumeric	$2.73E+00(0.02)$	$4.91E+00(0.01)$	$3.35E+00(0.05)$	$1.03E+00^*(0.01)$	$\mathbf{9.19E-01^*(0.01)}$
pol	$1.96E+01(0.17)$	$3.40E+01(0.24)$	$5.87E+01(0.23)$	$\mathbf{1.90E+01^*(0.11)}$	$2.02E+01(0.11)$
puma32H	$2.17E-02(0)$	$4.81E-02(0)$	$2.98E-02(0)$	$1.89E-02^*(0)$	$\mathbf{1.86E-02^*(0)}$
puma8NH	$4.18E+00(0.01)$	$6.30E+00(0)$	$4.47E+00(0)$	$\mathbf{4.07E+00^*(0.01)}$	$4.27E+00(0)$
sweetrs	$1.54E+00(0)$	$\mathbf{1.53E+00(0)}$	$1.61E+00(0)$	$1.60E+00(0)$	$1.61E+00(0)$
Average rank	2.85	4.64	3.28	2.14	2.07

Table 3. Performance of the algorithms in terms of the runtime and model size.

	AMRules	FIMTDD	FLEXFIS	TSK-Streams (global)	TSK-Streams (local)
	Execution time in seconds				
2dplanes	4.80 (0)	0.621 (0)	54.2 (0.13)	5.25 (0.03)	4.49 (0.02)
ailerons	3.23 (0)	0.814 (0)	28.6 (0.04)	18.7 (0.50)	16.3 (0.28)
bank8FM	1.70 (0)	0.321 (0)	14.5 (0.17)	0.829 (0)	1.05 (0)
calhousing	2.93 (0.01)	0.501 (0)	30.8 (0.79)	2.05 (0.01)	1.85 (0.01)
elevators	2.89 (0)	0.544 (0)	25.0 (0.03)	6.64 (0.20)	5.81 (0.03)
fried	4.85 (0.01)	1.13 (0)	56.6 (1.42)	6.54 (0.05)	8.18 (0.12)
house16h	3.67 (0)	0.881 (0)	33.4 (0.74)	9.05 (0.22)	6.05 (0.04)
house8	3.19 (0.01)	0.614 (0)	40.4 (2.35)	2.77 (0.04)	3.23 (0.02)
kin8nm	1.60 (0)	0.318 (0)	15.9 (0.36)	0.854 (0)	1.12 (0)
mvnumeric	5.04 (0)	0.885 (0)	89.4 (5.9)	7.24 (0.12)	13.3 (0.24)
pol	3.51 (0)	0.860 (0)	37.3 (0.39)	75.6 (1.2)	49.7 (0.65)
puma32H	2.47 (0)	0.724 (0)	19.8 (0.07)	8.08 (0.08)	8.89 (0.17)
puma8NH	1.63 (0)	0.308 (0)	21.0 (3.13)	0.745 (0)	1.04 (0)
sweetrs	2.35 (0.01)	0.338 (0)	45.5 (10.3)	0.384 (0)	0.819 (0.01)
Average rank	2.85	1	4.85	3.07	3.21
	Model size				
2dplanes	31.9 (0.19)	140.9 (0.23)	1 (0)	3 (0)	2 (0)
ailerons	5.4 (0.05)	24 (0.25)	1 (0)	3.3 (0.04)	3.1 (0.06)
bank8FM	8.1 (0.10)	26.8 (0.16)	1 (0)	2.9 (0.02)	2 (0)
calhousing	10 (0.06)	64.5 (0.26)	1.7 (0.12)	3 (0)	1.1 (0.02)
elevators	5.1 (0.02)	40.4 (0.39)	1 (0)	3.4 (0.04)	2.1 (0.02)
fried	18.1 (0.12)	119.6 (0.46)	1.6 (0.25)	3.6 (0.05)	3.2 (0.11)
house16h	6 (0.03)	64.9 (0.40)	1 (0)	5.1 (0.13)	2.1 (0.02)
house8	6.4 (0.05)	71.8 (0.29)	3.1 (0.67)	4.8 (0.11)	2.2 (0.03)
kin8nm	4.9 (0.02)	24 (0.22)	1.6 (0.22)	3 (0)	2 (0)
mvnumeric	24.5 (0.17)	130.2 (0.34)	3.6 (0.47)	6.1 (0.10)	6.5 (0.09)
pol	7.8 (0.08)	43 (0.28)	1 (0)	4.8 (0.06)	4.1 (0.07)
puma32H	11.1 (0.12)	28.3 (0.11)	1 (0)	3.3 (0.04)	3.9 (0.10)
puma8NH	6.7 (0.09)	28.1 (0.17)	1.1 (0.08)	3.1 (0.02)	2.2 (0.03)
sweetrs	9.9 (0.06)	63.2 (0.31)	7.8 (4.20)	3 (0)	2 (0)

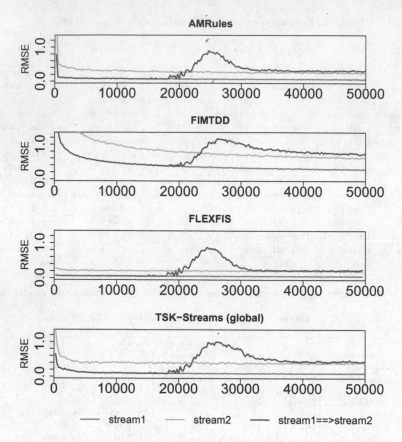

Fig. 1. Performance curves (RMSE, averaged over ten runs) on the distance to hyperplane data, with a drift from squared (red curve) to the cubed distance (green curve) in the middle of the episode. The recovery curve is plotted in blue. Ideally, this curve quickly reaches the performance level of the second stream (green curve). (Color figure online)

Table 3 shows the performance in terms of runtime and model size. TSK-Streams often remains a bit slower than AMRules and FIMTDD. At the same time, however, it is significantly faster than FLEXFIS, reducing runtime by a factor of around 10. Regarding the model size, we report the number of rules (leaves for FITDD) just to give an indicator of model complexity and without implying specific claims.

In the second part of the evaluation, we study the ability of our approach to recover from a performance drop in the presence of a concept drift. To this end, we make use of so-called *recovery analysis* as introduced in [24]. Recovery analysis aims at assessing a learner's ability to maintain its generalization performance in the presence of concept drift; it provides an idea of how quickly a drift is recognized, to what extent it affects the prediction performance, and how quickly the learner manages to adapt its model to the new condition. The main idea of recovery analysis is to employ three streams in parallel, two "pure streams" and one "mixture", instead of using a single data stream. The mixture stream resembles the first pure stream at the beginning and the second stream at the end, thus it contains a concept drift as a result of modeling the sampling probability as a sigmoidal function. Due to lack of space, we refer the reader to [24] for details of the methodology.

In general, we find that TSK-Streams recovers quite well in comparison to the other methods. As an illustration, we plot the recovery curves (blue lines) for the distance to hyperplane data set in Fig. 1. As can be seen, FLEXFIS exhibits a relatively large drop in performance. FIMTDD does not even manage to recover completely till the end of the stream. Compared to this, TSK-Streams and AMRules recover quite well.

5 Conclusion

In this paper, we proposed TSK-Streams, an evolving fuzzy rule learner for regression that meets the requirements of incremental and adaptive learning on data streams. Our method combines the expressivity and flexibility of TSK fuzzy rules with the efficiency and effectivity of concepts for rule induction as implemented in algorithms such as AMRules.

In an experimental study, we compared TSK-Streams with AMRules, FIMTDD, and FLEXFIS, the state-of-the-art regression algorithms for learning from data streams, on real and synthetic data. The results we obtained show that our learner compares very favorably and achieves superior performance. Moreover, it manages to adapt and recover well after a concept drift.

In future work, we plan to elaborate on extensions and variants of TSK-Rules that may lead to further improvements in performance. These developments will be accompanied by additional experiments and case studies.

Acknowledgments. This work was supported by the Competence Center for Cyber Physical Systems (CPS.HUB NRW).

References

1. Almeida, E., Ferreira, C., Gama, J.: Adaptive model rules from data streams. In: Blockeel, H., Kersting, K., Nijssen, S., Železný, F. (eds.) ECML PKDD 2013. LNCS (LNAI), vol. 8188, pp. 480–492. Springer, Heidelberg (2013). https://doi.org/10.1007/978-3-642-40988-2_31
2. Angelov, P.P.: Evolving Rule-based Models: A Tool for Design of Flexible Adaptive Systems. Springer, London (2002). https://doi.org/10.1007/978-3-7908-1794-2
3. Angelov, P.P.: Evolving Takagi-Sugeno fuzzy systems from data streams (eTS+). In: Angelov, P.P., Filev, D.P., Kasabov, N. (eds.) Evolving Intelligent Systems: Methodology and Applications. Wiley, Hoboken (2010)
4. Angelov, P.P., Filev, D.P., Kasabov, N. (eds.): Evolving Intelligent Systems: Methodology and Applications. Wiley, Hoboken (2010)

5. Bifet, A., Gavaldà, R.: Learning from time-changing data with adaptive windowing. In: Proceedings of 7th SIAM International Conference on Data Mining, Minneapolis, MN, USA, pp. 443–448 (2007)
6. Bifet, A., Gavaldà, R.: Adaptive learning from evolving data streams. In: Proceedings of IDA 2009, 8th International Symposium on Intelligent Data Analysis, Lyon, France, pp. 249–260 (2009)
7. Bifet, A., Holmes, G., Kirkby, R., Pfahringer, B.: MOA: massive online analysis. J. Mach. Learn. Res. **11**, 1601–1604 (2010)
8. Domingos, P., Hulten, G.: Mining high-speed data streams. In: Proceedings of 6th ACM SIGKDD International Conference on Knowledge Discovery and Data Mining, Boston, MA, USA, pp. 71–80 (2000)
9. Domingos, P., Hulten, G.: A general framework for mining massive data streams. J. Comput. Graph. Stat. **12**(4), 945–949 (2003)
10. Duarte, J., Gama, J., Bifet, A.: Adaptive model rules from high-speed data streams. ACM Trans. Knowl. Discov. Data **10**(3), 30:1–30:22 (2016)
11. Gama, J.: A survey on learning from data streams: current and future trends. Prog. Artif. Intell. **1**(1), 45–55 (2012)
12. Gama, J., Kosina, P.: Learning decision rules from data streams. In: Proceedings of 22nd International Joint Conference on Artificial Intelligence, Barcelona, Catalonia, Spain (2011)
13. Gama, J., Pinto, C.: Discretization from data streams: applications to histograms and data mining. In: Proceedings of 2006 ACM Symposium on Applied Computing, Dijon, France, pp. 662–667 (2006)
14. Hoeffding, W.: Probability inequalities for sums of bounded random variables. J. Am. Stat. Assoc. **58**(301), 13–30 (1963)
15. Ikonomovska, E., Gama, J., Dzeroski, S.: Learning model trees from evolving data streams. Data Min. Knowl. Discov. **23**(1), 128–168 (2011)
16. Klement, E.P., Mesiar, R., Pap, E.: Triangular Norms. Kluwer Academic Publishers, Dordrecht (2000)
17. Kosina, P., Gama, J.: Handling time changing data with adaptive very fast decision rules. In: Flach, P.A., De Bie, T., Cristianini, N. (eds.) ECML PKDD 2012. LNCS (LNAI), vol. 7523, pp. 827–842. Springer, Heidelberg (2012). https://doi.org/10.1007/978-3-642-33460-3_58
18. Lichman, M.: UCI machine learning repository (2013)
19. Ljung, L.: System Identification: Theory for the User, 2nd edn. Prentice Hall PTR, Upper Saddle River (1999)
20. Lughofer, E.: FLEXFIS: a robust incremental learning approach for evolving Takagi-Sugeno fuzzy models. IEEE Trans. Fuzzy Syst. **16**(6), 1393–1410 (2008)
21. Lughofer, E.: Evolving Fuzzy Systems: Methodologies, Advanced Concepts and Applications. Springer, Berlin (2011). https://doi.org/10.1007/978-3-642-18087-3
22. Page, E.S.: Continuous inspection schemes. Biometrika **41**(1–2), 100–115 (1954)
23. Rutkowski, L., Pietruczuk, L., Duda, P., Jaworski, M.: Decision trees for mining data streams based on the McDiarmid's bound. IEEE Trans. Knowl. Data Eng. **25**(6), 1272–1279 (2013)
24. Shaker, A., Hüllermeier, E.: Recovery analysis for adaptive learning from nonstationary data streams: experimental design and case study. Neurocomputing **141**, 97–109 (2014)
25. Takagi, T., Sugeno, M.: Fuzzy identification of systems and its applications to modeling and control. IEEE Trans. Syst. Man Cybern. **15**(1), 116–132 (1985)

Non-parametric Online AUC Maximization

Balázs Szörényi[1,2(✉)], Snir Cohen[1], and Shie Mannor[1]

[1] Technion, Haifa, Israel
szorenyi.balazs@gmail.com, snirc@cs.technion.ac.il, shie@ee.technion.ac.il
[2] Research Group on AI, Hungarian Academy of Sciences, University of Szeged,
Szeged, Hungary

Abstract. We consider the problems of online and one-pass maximization of the area under the ROC curve (AUC). AUC maximization is hard even in the offline setting and thus solutions often make some compromises. Existing results for the online problem typically optimize for some proxy defined via surrogate losses instead of maximizing the real AUC. This approach is confirmed by results showing that the optimum of these proxies, over the set of all (measurable) functions, maximize the AUC. The problem is that—in order to meet the strong requirements for per round run time complexity—online methods typically work with restricted hypothesis classes and this, as we show, corrupts the above compatibility and causes the methods to converge to suboptimal solutions even in some simple stochastic cases. To remedy this, we propose a different approach and show that it leads to asymptotic optimality. Our theoretical claims and considerations are tested by experiments on real datasets, which provide empirical justification to them.

1 Introduction

The *area under the ROC curve* (AUC) [16] measures how well a mapping h of the instance space to the reals respects the partial order defined by some "ideal" score function s; in the special case of bipartite ranking, s is simply a 0–1 valued function. As such, it has important applications in bioinformatics, information retrieval, anomaly detection, and many other areas.

Maximizing the AUC requires an approach different from maximizing the accuracy, even though there are some connections between the two [3,5,11]. Over the last decade, several approaches have been proposed and analyzed, guaranteeing consistency [9] and even optimal learning rates in some restricted cases [19]. Subsequently [22], followed by [13,21], considered AUC maximization in an online setting, while [14] introduced a one-pass AUC maximization framework.

In this paper we first point out two important shortcomings of the existing methods proposed for online and one-pass AUC optimization:

(A) None of them guarantees an optimal solution (not even asymptotically).
(B) They all need to store the whole data. The reason for this is that they require parameter tuning, and thus also multiple passes over the data.

© Springer International Publishing AG 2017
M. Ceci et al. (Eds.): ECML PKDD 2017, Part II, LNAI 10535, pp. 575–590, 2017.
https://doi.org/10.1007/978-3-319-71246-8_35

In contrast to (A), the k nearest neighbor method (k-NN), as we show, is guaranteed to converge to the optimum. This superiority of k-NN is also supported by the results of our empirical investigations. What is more, even though it clearly requires storing the whole data, it is not more demanding in terms of space complexity then previous algorithms, according to (B). Finally, one could argue that k-NN must perform poorly in terms of running time. This is not true, however: efficient solutions exist and, in fact, our experiments suggest that k-NN is competitive also in this regard. Additionally, dimensionality-originated issues can be taken care of using PCA or related methods.

The rest of the paper is structured as follows. First we introduce the formal framework and the definitions, then we show (A) formally, provide the theoretical justification for the k-NN method, present our experimental results, sum up the most important results from the literature, and finally we conclude with a short discussion.

2   Formal Setup

Given a set of n samples $(x_1, y_1), \ldots, (x_n, y_n) \in \mathcal{X} \times \mathcal{Y}$, where $\mathcal{X} \subseteq \mathbb{R}^d$ for some positive integer d and $\mathcal{Y} = \{-1, 1\}$, and given some mapping $h : \mathcal{X} \to \mathbb{R}$, the *area under the ROC curve* (AUC) [16] is the empirical mean

$$\mathrm{AUC}(h; \mathcal{X}^+, \mathcal{X}^-) = \sum_{x^+ \in \mathcal{X}^+} \sum_{x^- \in \mathcal{X}^-} \left(\frac{\mathbb{I}\left[h(x^+) > h(x^-)\right]}{T_+ T_-} + \frac{\mathbb{I}\left[h(x^+) = h(x^-)\right]}{2T_+ T_-} \right),$$

where $\mathcal{X}^+ = \{x_t : y_t = 1, 1 \le t \le n\}$, $\mathcal{X}^- = \{x_t : y_t = -1, 1 \le t \le n\}$, $T_+ = |\mathcal{X}^+|$, $T_- = |\mathcal{X}^-|$, and where $\mathbb{I}[\cdot]$ denotes the indicator function; i.e., $\mathbb{I}[E] = 1$ when event E holds and $\mathbb{I}[E] = 0$ otherwise. The regret hypothesis h with respect to some hypothesis set $\mathcal{H} \subseteq \mathbb{R}^{\mathcal{X}}$ is defined as

$$\mathrm{Regret}^{\mathcal{H}}(h; \mathcal{X}^+, \mathcal{X}^-) = \sup_{h' \in \mathcal{H}} \mathrm{AUC}(h'; \mathcal{X}^+, \mathcal{X}^-) - \mathrm{AUC}(h; \mathcal{X}^+, \mathcal{X}^-) \quad (1)$$

We also denote by $\mathrm{AUC}^*(\mathcal{X}^+, \mathcal{X}^-)$ the supremum of $\mathrm{AUC}(h'; \mathcal{X}^+, \mathcal{X}^-)$ over the set of all measurable functions h', and introduce the notation

$$\mathrm{Regret}(h; \mathcal{X}^+, \mathcal{X}^-) = \mathrm{AUC}^*(\mathcal{X}^+, \mathcal{X}^-) - \mathrm{AUC}(h; \mathcal{X}^+, \mathcal{X}^-).$$

Maximizing the AUC is also equivalent to minimizing the empirical risk

$$\mathrm{Risk}(h; \mathcal{X}^+, \mathcal{X}^-) = 1 - \mathrm{AUC}(h; \mathcal{X}^+, \mathcal{X}^-) = \sum_{t,t'=1}^{n} \frac{\ell^{\mathrm{AUC}}(h; (x_t, y_t), (x_{t'}, y_{t'}))}{2T_+ T_-} \quad (2)$$

of the loss function

$$\ell^{\mathrm{AUC}}(h; (x, y), (x', y')) = \mathbb{I}\left[(h(x) - h(x'))(y - y') < 0\right] + \mathbb{I}\left[h(x) = h(x'), y \ne y'\right].$$

One notorious problem with AUC is that it is non-convex and non-continuous, which makes it hard to work with. Especially in the online and one-pass settings, where having a low (typically constant or logarithmic, but at least sublinear) per round run time complexity is essential. To resolve this issue, papers that aim for maximizing AUC online [13,14,21,22], choose to replace ℓ^{AUC} in (2) by some surrogate loss function $\ell : \mathcal{H} \times (\mathcal{X} \times \mathcal{Y}) \to \mathbb{R}$, and instead of maximizing the AUC, they minimize the surrogate risk

$$\mathrm{Risk}^\ell(h; \mathcal{X}^+, \mathcal{X}^-) = \sum_{t,t'=1}^{t} \frac{\ell(h; (x_t, y_t), (x_{t'}, y_{t'}))}{2T_+ T_-},$$

and derive bounds for $\mathrm{Regret}^{\ell, \mathcal{H}}(h; \mathcal{X}^+, \mathcal{X}^-)$, which is obtained by replacing AUC in (1) by $1 - \mathrm{Risk}^\ell$.

If $(x, y_1), (x_2, y_2), \ldots$ are i.i.d. samples from some probability distribution $\mathbf{P}$ over $\mathcal{X} \times \mathcal{Y}$, then one can define

$$\mathrm{Risk}(h) = \mathbf{E}\left[\ell^{\mathrm{AUC}}(h; (X, Y), (X', Y')) \,\middle|\, Y > Y'\right] \tag{3}$$

and $\mathrm{AUC}(h) = 1 - \mathrm{Risk}(h)$. Similarly as above, replacing ℓ^{AUC} in (3) by some other loss function ℓ one obtains surrogate measures $\mathrm{Risk}^\ell(h)$ and $\mathrm{Regret}^{\ell, \mathcal{H}}(h)$. Along the same analogy, we also use the notation AUC^* and $\mathrm{Regret}(h)$.

One-Pass and Online Setting
The one-pass and the online settings both have the same underlying protocol: in each round t, the learner proposes some hypothesis $h_t : \mathcal{X} \to \mathbb{R}$ based on its previous experience, and then it observes the sample (x_t, y_t). The two frameworks only differ in their objectives:

- In the *online setting* we are concerned with the evolution of the empirical AUC—that is, with
$$\mathrm{AUC}_t = \mathrm{AUC}(h_t; \mathcal{X}_t^+, \mathcal{X}_t^-)$$
for $t = 1, \ldots, n$, where $\mathcal{X}_t^+ = \{x_i : y_i = 1, 1 \leq i \leq t\}$ and $\mathcal{X}_t^- = \{x_i : y_i = -1, 1 \leq i \leq t\}$.
- In the *one-pass setting* the generalization ability of the learner is tested after the whole data had been processed. More precisely, the measure of performance is $\mathrm{AUC}(h_n)$.

3   Surrogate Measures and Restricted Classes

Existing results for the online and one-pass AUC maximization problem optimize for some surrogate risk, instead of working with the AUC directly. In particular, many of them work with the square loss ℓ_2 (see Example 1). This approach is also confirmed by results showing *consistency* i.e., that $\mathrm{Regret}(h_t)$ converges to 0 whenever $\mathrm{Regret}^\ell(h_t)$ does for some sequence $h_1, h_2, \ldots$ of functions, as t goes to infinity. (See more about this in the section about the related work.)

These important results require, however, careful interpretation. And this is the starting point of our investigations: we claim that utilization of consistency is only legitimate when the hypotheses class $\mathcal{H}$ of our interest contains a global optimizer of the surrogate loss; that is, if $\sup_h \text{AUC}^\ell(h) = \sup_{h\in\mathcal{H}} \text{AUC}^\ell(h)$. When working with the set $\mathcal{H}_{\text{lin}} = \{h^w(x) = w^\top x : w \in \mathbb{R}^d\}$ of linear functions—which is the case for *all* existing results for online and one-pass AUC maximization—this criterion is not fulfilled. Indeed, even though square loss is consistent (see [14]), in Example 1 below it holds that $\text{AUC}(h') \ll \sup_{h\in\mathcal{H}_{\text{lin}}} \text{AUC}(h)$ for any $h' \in \text{argmin}_{h\in\mathcal{H}_{\text{lin}}} \text{Risk}^{\ell_2}(h)$.

The hinge loss $\ell^\gamma(h; (x,y), (x', y')) = \mathbb{I}\left[y \neq y'\right]\left[\gamma - \frac{1}{2}(y - y')(h(x) - h(x'))\right]_+$ was also used in algorithmic solutions, but that does not even satisfy consistency [15].

3.1   Square Loss with Linear Hypotheses

This section presents the example that existing results can fail completely in maximizing the real AUC even in a simple case. This is demonstrated by the following example.

Example 1. *Consider the setting when $\mathcal{X} = \mathbb{R}^2$, and $\mathbf{P}[X = (-\epsilon, -1 + \epsilon)|Y = 1] = 1$ and $\mathbf{P}[X = (0, -1 - \epsilon)|Y = -1] = \mathbf{P}[X = (0, 1)|Y = -1] = \mathbf{P}[X = (1, 0)|Y = -1] = 1/3$ for some small $\epsilon > 0$.*

[14] first shows that the square loss $\ell_2(h; (x,y), (x', y')) = (1 - \frac{y-y'}{2}(h(x) - h(x'))^2$ is consistent with AUC, and then use ℓ_2 as a surrogate loss to find the best linear score function $h^w(x) = w^\top x$. However, in the case above, AUC is maximized at $h^{\overline{w}^*}$ where $w^* = (-1, 0)$, where it actually takes value 1 (one has a very small freedom though, depending on the size of ϵ), and thus $\text{Risk}(h^{w^*}) = 0$ for the corresponding linear function h^{w^*}. On the other hand, the surrogate measure for h^{w^*} is

$$\text{Risk}^{\ell_2}(h^{w^*}) = \mathbf{E}[\ell_2(h^{w^*}; X, X')|Y > Y']$$

$$= \tfrac{1}{3}\left(1 + (w^*)^\top(\epsilon, 2\epsilon)\right)^2 + \tfrac{1}{3}\left(1 + (w^*)^\top(\epsilon, 2 - \epsilon)\right)^2 + \tfrac{1}{3}\left(1 + (w^*)^\top(1 + \epsilon, 1 - \epsilon)\right)^2$$

which evaluates approximately to 2/3. At the same time, the actual optimum w' of this surrogate measure is around $(-1/2, -1/2)$, where it takes the value:

$$\text{Risk}^{\ell_2}(h^{w'}) = \mathbf{E}[\ell_2(h; X, X')|Y > Y'] \approx 1/3,$$

whereas $\text{AUC}^*(h^{w'}) \approx 2/3$, which is very far from the true optimum.

That is, when $\mathcal{H}$ consists of the linear hypotheses (as is the case in [13,14]), then

$$\text{Regret}^{\ell_2, \mathcal{H}}(h^{w'}) = 0 ,$$

implying that

$$\text{Regret}^{\mathcal{H}}(h^{w'}) = \text{Regret}(h^{(-1/2, -1/2)}) \approx 1/3 .$$

Furthermore, adding a term $\|w\|^2$ to regularize the surrogate measure does not change on this.

Note that this does not contradict the consistency of the square loss. The reason is that consistency requires Regret to vanish as Regret^ℓ approaches 0, but as Regret^ℓ is huge for all linear hypotheses, this sets no restrictions on how Regret should behave in this example.

4   Conditional Probability as Rank Function

In this section we start the investigation of finding alternative algorithmic solutions. With that in mind, we reach back to the fundamentals of AUC, and show that good estimates of the conditional probability function $\eta(x) = \mathbf{P}[Y = 1|X = x]$ perform well at AUC maximization too.

The particular estimates that we consider here are of the form $\widehat{\eta} : \mathcal{X} \times \mathcal{Z} \to \mathbb{R}$ for some domain $\mathcal{Z}$. Here $\mathcal{Z}$ is the domain of a variable that is used to encode prior information (e.g., random samples) and internal randomization (used e.g., for tie breaking) of the learner. (In accordance with that, in some cases it will be more convenient to use the notation $\widehat{\eta}_z$ for $\widehat{\eta}(\cdot, z)$.) For example, given some series $\{k_n\}_n$ of stepsizes, the k_n−NN estimate of [12] makes use of some i.i.d. samples $U_1, \ldots, U_n, U$ drawn from the uniform distribution over $[0, 1]$. Putting $Z_n = (X_1, Y_1, \ldots, X_n, Y_n, U_1, \ldots, U_n, U)$, their k_n−NN estimate maps an instance x to

$$\widehat{\eta}_{Z_n}^{\text{DGKL}}(x) = \frac{1}{k_n} \sum_{i=1}^{k_n} Y_{\sigma(Z_n, x, i)}, \tag{4}$$

where $\sigma(Z_n, x, \cdot)$ is the permutation for which $(\|X_{\sigma(Z_n,x,1)} - x\|, \|U_{\sigma(Z_n,x,1)} - U\|), \ldots, (\|X_{\sigma(Z_n,x,n)} - x\|, \|U_{\sigma(Z_n,x,n)} - U\|)$ is in lexicographic order.

Given such an estimate, we show the following result (for the proof see Appendix A).

Theorem 2. *Let Z be some random variable over some domain $\mathcal{Z}$, and let $\widehat{\eta} : \mathcal{X} \times \mathcal{Z} \to \mathbb{R}$ be an estimate of the conditional distribution function $\eta(x) := \mathbf{E}[Y|X = x]$ as described above. Then $\mathbf{E}_Z[\text{Regret}(\widehat{\eta}_Z)] \leq \frac{3\sqrt{\epsilon}}{\mathbf{P}[Y=1]\mathbf{P}[Y=0]}$, where $\epsilon = \mathbf{E}_{X,Z}[|\widehat{\eta}(X, Z) - \eta(X)|]$.*

Similar result has also appeared in [1, 10, 19]. However, this particular estimator requires some small but essential differences in the analysis. Most importantly, kernel estimators are completely determined by the samples, whereas k_n−NN needs tie breaking. This requires additional randomness and complicates the analysis slightly.

5   AUC Maximization Using k-NN

In the previous section we have shown guarantees for the AUC performance of estimators of the conditional probability function η. In this section we review some of the results on estimating η using k_n−NN, and show what they give combined with Theorem 2.

First of all, Devroye et al. [12] have shown that the k_n–NN version presented as Algorithm 1 converges under any distribution, assuming some standard restrictions on k_n.[1]

Theorem 3 (Theorem 1 in [12]). *If stepsize k_n satisfies $\lim_{n\to\infty} k_n = \infty$ and $\lim_{n\to\infty} k_n/n = 0$, then $\mathbf{E}\left[|\widehat{\eta}_{Z_n}^{\mathrm{DGKL}}(X) - \eta(X)|\right] \to 0$, where $\widehat{\eta}_{Z_n}^{\mathrm{DGKL}}$ is defined as in (4).*

Plugging this into Theorem 2 we immediately obtain the following result on $\mathrm{AUC}(\widehat{\eta}_{Z_n}^{\mathrm{DGKL}})$.

Corollary 4. *Let $\widehat{\eta}_{Z_n}^{\mathrm{DGKL}}$ be defined as in (4). Then $\mathbf{E}_{Z_n}\left[\mathrm{Regret}(\widehat{\eta}_{Z_n}^{\mathrm{DGKL}})\right] \to 0$ if $\lim_{n\to\infty} k_n = \infty$ and $\lim_{n\to\infty} k_n/n = 0$, where $\widehat{\eta}_{Z_n}^{\mathrm{DGKL}}$ is defined as in (4).*

KNNOAM is thus guaranteed to converge in case of i.i.d. samples.

One can, in fact, derive results also for the rate of convergence based on the work by Chaudhuri and Dasgupta [8]. This would not hold uniformly though, only for some restricted distributions.

Algorithm 1. KNNOAM($\{k_t\}_t$)

1: Draw a random sample U_1 uniformly at random from $(0,1)$
2: **for** round $t = 1, \ldots, n$ **do**
3:     Observe sample (x_t, y_t)
4:     Draw a random sample U_{t+1} uniformly at random from $(0,1)$
5:     $Z_t = (x_1, y_1, x_2, y_2, \ldots, x_t, y_t, U_1, \ldots, U_t, U_{t+1})$
6:     Construct hypotheses $h_t : \mathcal{X} \to \mathbb{R}$, mapping x to $h_t(x) = \widehat{\eta}_{Z_t}^{\mathrm{DGKL}}(x)$ {As in (4)}
7: **end for**

Efficient Implementation

An important feature of k-NN-methods is that it can be implemented efficiently. For example, the Cover Tree structure [6] makes it possible to insert a new instance into an existing cover tree or to remove an old one from it in time $O(\log t)$, and also to find the k_n nearest neighbor of some arbitrary point in time $O(k_n \log t)$.

Choosing k

Choosing the right k for k-NN is a hard question. $k > \log\log n$ is recommended for pointwise convergence (see Remark 1 in 4), but the common practice is to use $\log n < k < n^{1/2}$. One can also think about using k that changes with context; i.e., depends on the particular instance that is queried. See [4] for further details.

For a given dataset, one can also use cross validation or some Bayesian approach to find the best k. This was used by all the linear methods mentioned in the Introduction, but in a real online setting this is not applicable.

---

[1] They actually show an even stronger equivalence result.

6   Dimensionality Reduction

In general, any learning method can be applied that approximates η with arbitrary accuracy. Note, however, that all these methods suffer from dimensionality issues; including k-NN. One way to deal with it is to apply first dimensionality reduction methods. More specifically, the idea is to first feed the obtained sample into some online PCA algorithm (like SGA or CCIPCA—see [7] for a thorough discussion), and then use its output as input for k-NN, Parzen-Rosenblatt kernels, etc. This way one maintains the good AUC performance guaranteed by the learning algorithms but prevents dimensionality-originated run-time issues thanks to the guarantees of the PCA methods.

The property this task requires from the aforementioned techniques to preserve the good AUC performance guarantees of k-NN is a kind of stability. More precisely, denoting by Φ_t the mapping they induce from the data from the first t rounds, it should fulfill the property

$$\|\Phi_t(x_t) - \Phi_T(x_t)\| \le \epsilon(t) \qquad \forall\, T \ge t$$

for some $\epsilon(t)$ converging to 0 as t goes to infinity. Maintaining the convergence of k-NN does not seem possible otherwise.

7   Experimental Results

In this section, we evaluate the empirical performance of the proposed KNN Online AUC Maximization (KNNOAM) algorithm on benchmark datasets.

Compared Algorithms

We compare the proposed KNNOAM algorithm with state-of-the-art online AUC optimization algorithms. Specifically, the compared algorithms in our experiments include:

- **OAM$_{seq}$:** the OAM algorithm with reservoir sampling and sequential updating [22];
- **OAM$_{gra}$:** the OAM algorithm with reservoir sampling and online gradient updating method [22];
- **OPAUC:** the one-pass algorithm AUC optimization algorithm proposed in [14];
- **AdaOAM:** the adaptive gradient AUC optimization algorithm proposed in [13];
- **KNNOAM:** our proposed KNN based algorithm.

General Experimental Setup

We conduct our experiments on sixteen benchmark datasets that have been used in previous studies on AUC optimization. The details of the datasets are summarized in Table 1. All these datasets can be downloaded from LIBSVM[2]

---

[2] https://www.csie.ntu.edu.tw/~cjlin/libsvmtools/.

and UCI Machine Learning Repository[3]. Note that half of the datasets (segment, satimage, vowel, letter, poker, usps, connect-4, acoustic and vehicle) are originally multi-class. These multi-class datasets have been converted into class-imbalanced binary datasets by choosing one class, setting its label to $+1$ and the rest to -1. This class has been chosen so that the ratio T_-/T_+ is below 50 and its cardinality has been minimized. (The ratio is kept below 50 to obtain conclusive results.) In case two or more classes have the same size, one has been chosen randomly. Previous studies use similar conversion methods. In addition, the features have been rescaled linearly to $[-1, 1]$ for all datasets. All experiments are performed with Matlab on a computer workstation with $3.40\,\text{GHz}$ CPU and $32\,\text{GB}$ memory.

Table 1. Details of the benchmark datasets used in the experiments. $T_+ = |\mathcal{X}^+|$ and $T_- = |\mathcal{X}^-|$.

Datasets	# instances	# features	T_-/T_+	Datasets	# instances	# features	T_-/T_+
fourclass	862	2	1.8078	segment	2310	19	6.0000
svmguide1	3089	4	1.8365	ijcnn1	141691	22	9.4453
magic04	19020	10	1.8439	connect-4	67557	126	9.4756
german	1000	24	2.3333	satimage	4435	36	9.6867
a9a	32561	123	3.1527	vowel	528	10	10.0000
svmguide3	1243	22	3.1993	usps	9298	256	12.1328
vehicle	846	18	3.2513	letter	20000	16	25.7380
acoustic	78823	50	3.3028	poker	25010	10	47.7524

Experimental Setup of the Online Setting

Our main goal is to compare the performance of the above algorithms in the online setting. In this setting, each algorithm receives a random sample from the dataset, suffers loss and updates its classifier according to this sample. Note that our proposed KNNOAM does not require any parameter tuning, as opposed to the other four existing algorithms. Clearly, parameter tuning requires multiple passes over the data, which is inconsistent with this setting. Although KNNOAM must store at time t all the samples up to time $t - 1$, the rest of the algorithms must store all of the samples in advance for the parameter tuning. Despite the fact that it gives the rest of the algorithms an unfair advantage, we follow the parameter tuning procedures for each algorithm suggested in [13,14,22] and use the best obtained parameters for each dataset and algorithm. We apply five-fold cross-validation on the training set to find the best learning rate $\eta \in 2^{[-10,10]}$ and the regularization parameter $\lambda \in 2^{[-10,2]}$ for both OPAUC and AdaOAM. For OAM_{seq} and OAM_{gra}, we apply five-fold cross-validation to tune the penalty parameter $C \in 2^{[-10:10]}$, and fix the buffer at 100 as suggested in [22]. For

---

[3] www.ics.uci.edu/~mlearn/MLRepository.html.

KNNOAM we only had to choose k_n that goes to infinity and is of order $o(n)$. We choose $k_n = 2 \log_2 n$.[4] For every dataset, we average over 20 runs and plot a graph showing the experimental AUC loss as a function of the number of samples. This gives us an opportunity to examine the evolution of the performance of the algorithms. The results of this experiment are presented in Appendix B. The code is available at https://bitbucket.org/snir/auc2017.

Experimental Setup of the One-Pass Setting
We also compare the performance of the algorithm in the one-pass setting. We follow the experimental setup suggested in previous studies [13,14,22]. The performance of the compared algorithms is evaluated by four trials of five-fold cross-validation using the parameters received by the parameter tuning procedure explained in the previous section about the online experimental setup. The AUC values are the average of these 20 runs. The results are summarized in Fig. 1.

7.1 Evaluation on Benchmark Datasets

In the online setting, KNNOAM outperforms the four other, state-of-the-art online AUC algorithms considered in our experiments in 12 out 16 datasets. What is more, in 10 out of those 12 datasets, the improvement by KNNOAM is significant. For example, in ijcnn1 and acoustic datasets, KNNOAM converges to a much lower empirical AUC loss than the rest of the algorithms. These are outstanding results for this setting, especially when recalling that the compared algorithms have been tuned before running the experiments. Our graphs demonstrate that KNNOAM has a much smoother convergence. The rest of the algorithms' performance is unstable: after receiving new samples it may improve but might also significantly deteriorate. This is a highly undesirable property.

In the german, svmguide3 and vehicle datasets, KNNOAM does not outperform all the compared algorithms. However, these datasets are very small and therefore the results are inconclusive.

In the a9a dataset, OPAUC and AdaOAM outperform KNNOAM. Although this dataset cannot be considered small, KNNOAM still behaves as if it were, because the samples are sparse and the dimensionality is relatively big. To strengthen this claim, we can examine the performance of KNNOAM on connect-4 dataset. This dataset has roughly the same dimensionality as the a9a dataset and the samples are as sparse, but has more than twice the samples, and KNNOAM outperforms the rest of the algorithms significantly.

Many of the graphs show that OAM_{seq}, OAM_{gra} are not guaranteed to converge at all.[5] Some of them also show that OPAUC and AdaOAM does not converge smoothly. This behavior is observed when the learning rate η is too high and a single sample could change the classifier drastically and cause a significant

---

[4] This choice guarantees low time complexity and also turned out to result in a competitive method. (We did try other choices; the results were similar.).

[5] It should be mentioned that the theoretical gurantees of OAM is also doubtful, according to [17].

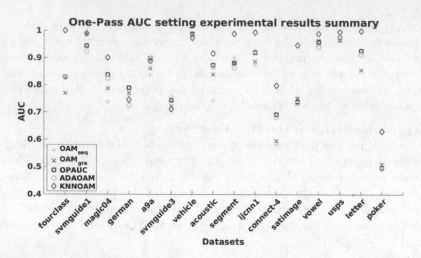

Fig. 1. Experimental results comparing the one-pass AUC performance summary

local deterioration of the performance as can be seen from the performance of OPAUC on svmguide1 dataset for example. On the other hand, if the chosen learning rate η is too small, the performance might be poor, as can be seen from the performance of AdaOAM on the same dataset.

The results for the one-pass setting are similar to the ones for the online setting: KNNOAM is better, or at least competitive in comparison with its competitors. For this setting we also present the running times (see Appendix C) KNNOAM is usually the fastest algorithm and is never the slowest among the compared algorithms.

8   Related Work

Reducing Ranking to Binary Classification
[3,5] considers the problem of ranking a finite random sample, and reduced this to a related binary classification problem. In particular, they bound the risk of the ranking (which is closely related to AUC of the ranker over this sample) in terms of the classification performance. However, the risk and the classification performance they use is only representative to that particular sample, and they do not consider prediction or generalization bounds. ([5] comment on the case when the rankings are drawn from some distribution, but does not imply any result in our setting.)

Offline Algorithmic Solutions
[9] aims to minimize (3), and actually obtain asymptotic optimality. Their algorithm constructs a tree in an iterative fashion by solving subsequent challenging optimization problems.

[19] uses kernel estimates of the conditional probability function $\eta(x) = \mathbf{E}[Y|X = x]$ based on the Parzen-Rosenblatt kernel, and have shown fast and

superfast convergence rates assuming Tsybakov-style noise. They also comple-
ment their results by showing lower bounds on the best convergence rate in some
situations.

Consistency of Surrogate Measures
The investigation of consistency with respect to AUC was initiated by [18] show-
ing consistency of a balanced version of the exponential and the logistic loss.
Later on [1,14,15,20] investigated the consistency of other loss functions like the
exponential, logistic, squared, and q-normed hinge loss, and variants of them.
Finally, [15] shows that hinge loss is not consistent.

One-Pass and Online Solutions
As mentioned, [22] was the first to analyze online AUC maximization. They
have defined the setup, presented algorithmic solutions optimizing for the hinge
loss, and provided regret bounds. [21] also uses the hinge loss with a perceptron-
like algorithm which, in round t, achieves regret $O(1/\sqrt{t})$. [17] works with the
same setting as [21], and achieves several improvements in terms of different
parameters.

[14] uses square loss in the one-pass setting, and obtains a convergence rate
of order $O(1/t)$ for the linearly separable case and $O(\sqrt{1/t})$ for the general one.
[13] obtains similar results, but using the Adaptive Gradient method.

All these papers work with the set of linear hypothesis.

Uniform Convergence Bounds
Uniform convergence bounds like the ones in [2] show how fast the empirical
AUC-risk converges to the actual AUC over a given class of hypothesis. Consequently, they do not provide any practical guidance on how one could acquire
some hypothesis with small risk.

9   Concluding Remarks

We have shown that existing methodology for maximizing AUC in an online or
one-pass setting can fail already in very simple situations. To remedy this, we
have proposed to reach back to the fundamentals of AUC, and suggested an
algorithmic solution based on the celebrated k-NN-estimate of the conditional
probability function. This has guarantees in the stochastic setting, has efficient
implementations, and outperforms previous methods on several real datasets.
The latter is even more surprising in view of the fact that, unlike its competitors,
it requires no parameter tuning.

Nevertheless, we feel that this should not be considered as an ultimate solution, but rather as an encouragement for future research to explore further alternative solutions. To mention a few:

- Combining KNNOAM with metric learning arises naturally, and could extend
  its applicability to more exotic domains.
- Maximizing the objective function AUC_n in the adversarial setting is another
  important question, which existing results do not tell anything about.

Acknowledgements. This research was supported in part by the European Communities Seventh Framework Programme (FP7/2007-2013) under grant agreement 306638 (SUPREL).

A   Proof of Theorem 2

Let us first introduce the notation $\mathcal{X}_z = \{x : |\widehat{\eta}(x, z) - \eta(x)| < \sqrt{\epsilon}\}$ for $z \in \mathcal{X}$ and define for $h : \mathcal{X} \to \mathbb{R}$ measurable and $x, x' \in \mathcal{X}$

$$a(h, x, x') = \tfrac{1}{2}\mathbb{I}\left[h(x) = h(x')\right]\left[\eta(x)(1 - \eta(x'))\right] + \mathbb{I}\left[h(x) > h(x')\right]\left[\eta(x)(1 - \eta(x'))\right]$$

and its symmetrization

$$b(h, x, x') = \tfrac{1}{2}a(h, x, x') + \tfrac{1}{2}a(h, x', x).$$

It then holds that

$$\mathbf{E}_{(X,Y)}\mathbf{E}_{(X',Y')}\left[\tfrac{1}{2}\mathbb{I}\left[h(X) > h(X')\right]\mathbb{I}\left[Y = 1, Y' = 0\right]\right.$$

$$\left. + \mathbb{I}\left[h(X) > h(X')\right]\mathbb{I}\left[Y = 1, Y' = 0\right]\right]$$

$$= \mathbf{E}_{X,X'}\left[\tfrac{1}{2}\mathbb{I}\left[h(X) = h(X')\right]\mathbf{E}_{Y,Y'}\left[\mathbb{I}\left[Y = 1, Y' = 0\right]\right]\right.$$

$$\left. + \mathbb{I}\left[h(X) > h(X')\right]\mathbf{E}_{Y,Y'}\left[\mathbb{I}\left[Y = 1, Y' = 0\right]\right]\right]$$

$$= \mathbf{E}_{X,X'}\left[a(h, X, X')\right]$$

$$= \mathbf{E}_{X,X'}\left[b(h, X, X')\right],$$

where the last equation follows because X and X' are i.i.d. This then gives

$$\mathrm{AUC}(h) = \frac{\mathbf{E}_{X,X'}[b(h, X, X')]}{\mathbf{P}[Y = 1]\mathbf{P}[Y = 0]}. \tag{5}$$

Now, note that $x, x' \in \mathcal{X}_z$ implies $|\eta(x)(1 - \eta(x')) - \eta(x')(1 - \eta(x))| \le 2\sqrt{\epsilon}$ because of the $\alpha\beta - \alpha'\beta' = (\alpha - \alpha')\beta + \alpha'(\beta - \beta')$ equality. Combining this with the fact that $\mathbb{I}\left[h(x) = h(x')\right] + \mathbb{I}\left[h(x) < h(x')\right] + \mathbb{I}\left[h(x) > h(x')\right] = 1$ for any $h : \mathcal{X} \to \mathbb{R}$ and any $x, x' \in \mathcal{X}$, it follows that

$$b(\eta, x, x') - b(\widehat{\eta}_z, x, x')) \le \sqrt{\epsilon}, \ z \in \mathcal{Z}, \ x, x' \in \mathcal{X}_z.$$

Accordingly,

$$\mathbf{E}_{X,X'}[b(\eta, X, X')] - \mathbf{E}_{X,X',z}[b(\widehat{\eta}_z, X, X')]$$

$$\le \sqrt{\epsilon} + \mathbf{P}_{X,X',z}[X \notin \mathcal{X}_Z \text{ or } X' \notin \mathcal{X}_Z]$$

$$\le 3\sqrt{\epsilon}, \tag{6}$$

where the last inequality is true because X and X' are i.i.d. and because $\mathbf{P}_{X,z}[X \notin \mathcal{X}_Z] = \mathbf{P}_{X,z}[|\widehat{\eta}(X, Z) - \eta(X)| \ge \sqrt{\epsilon}] \le \sqrt{\epsilon}$ by the definition of ϵ.

Finally, according to [10], $\mathrm{AUC}(h)$ is maximized when $h = \eta$. The theorem thus follows by combining (5) and (6).

B   Figures for Benchmark Datasets

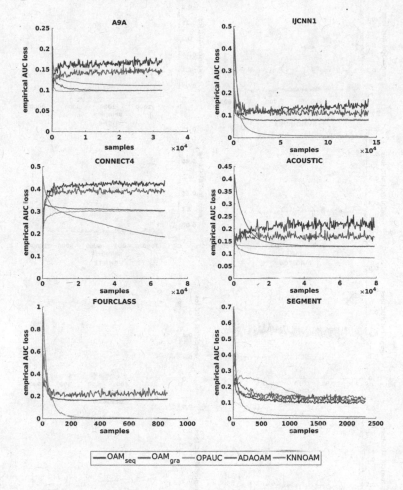

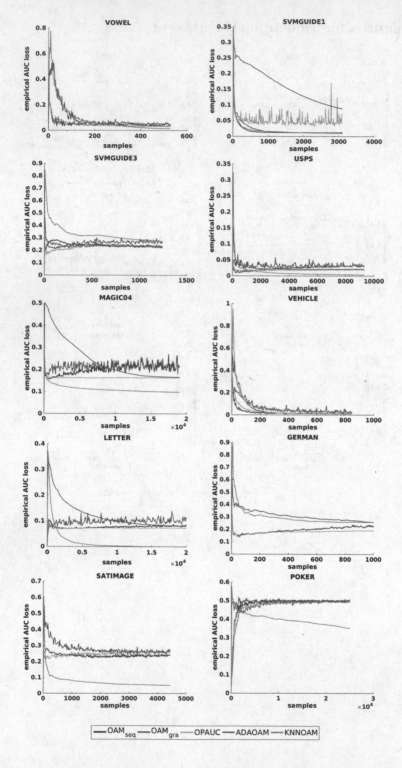

C Experimental Results for the One-Pass AUC Setup

See Table 2.

Table 2. Experimental results comparing the one-pass AUC performance (i.e., AUC(h_n)) of OAM$_{seq}$, OAM$_{gra}$, OPAUC, ADAOM and KNNOAM

Algorithm	fourclass		Algorithm	segment	
	AUC	Time (s)		AUC	Time (s)
OAM$_{seq}$	0.8253 ± 0.0290	0.5125	OAM$_{seq}$	0.9010 ± 0.0205	1.4031
OAM$_{gra}$	0.7694 ± 0.0518	0.1797	OAM$_{gra}$	0.8811 ± 0.0262	0.4719
OPAUC	0.8281 ± 0.0308	0.0422	OPAUC	0.8808 ± 0.0237	0.1289
AdaOAM	0.8278 ± 0.0269	0.0641	AdaOAM	0.8629 ± 0.0337	0.1922
KNNOAM	**1.0000 ± 0.0000**	**0.0070**	KNNOAM	**0.9891 ± 0.0087**	**0.0227**
Algorithm	svmguide1		Algorithm	ijcnn1	
	AUC	Time (s)		AUC	Time (s)
OAM$_{seq}$	0.9885 ± 0.0026	2.0953	OAM$_{seq}$	0.8733 ± 0.0253	101.9695
OAM$_{gra}$	0.9884 ± 0.0039	0.6022	OAM$_{gra}$	0.8862 ± 0.0345	34.1859
OPAUC	0.9440 ± 0.0175	0.1516	OPAUC	0.9201 ± 0.0080	**10.3078**
AdaOAM	0.9229 ± 0.0338	0.2336	AdaOAM	0.9237 ± 0.0036	14.2531
KNNOAM	**0.9900 ± 0.0041**	**0.0156**	KNNOAM	**0.9945 ± 0.0005**	66.9656
Algorithm	magic04		Algorithm	connect-4	
	AUC	Time (s)		AUC	Time (s)
OAM$_{seq}$	0.7370 ± 0.0910	13.4914	OAM$_{seq}$	0.5817 ± 0.0205	50.7984
OAM$_{gra}$	0.7875 ± 0.0530	4.6008	OAM$_{gra}$	0.5948 ± 0.0232	**17.4992**
OPAUC	0.8372 ± 0.0063	1.1969	OPAUC	0.6918 ± 0.0077	117.2625
AdaOAM	0.8240 ± 0.0085	1.7000	AdaOAM	0.6824 ± 0.0166	144.8133
KNNOAM	**0.9003 + 0.0061**	**0.3875**	KNNOAM	**0.7977 ± 0.0053**	77.9141
Algorithm	german		Algorithm	satimage	
	AUC	Time (s)		AUC	Time (s)
OAM$_{seq}$	0.7603 ± 0.0395	0.6086	OAM$_{seq}$	0.7520 ± 0.0250	2.9531
OAM$_{gra}$	0.7704 ± 0.0368	0.2172	OAM$_{gra}$	0.7504 ± 0.0262	1.1398
OPAUC	**0.7890 ± 0.0274**	0.0578	OPAUC	0.7380 ± 0.0166	0.4508
AdaOAM	0.7238 ± 0.0691	0.0859	AdaOAM	0.7328 ± 0.0267	0.5719
KNNOAM	0.7454 + 0.0268	**0.0086**	KNNOAM	**0.9466 ± 0.0108**	**0.1172**
Algorithm	a9a		Algorithm	vowel	
	AUC	Time (s)		AUC	Time (s)
OAM$_{seq}$	0.8368 ± 0.0199	25.3867	OAM$_{seq}$	0.9503 ± 0.0329	0.1195
OAM$_{gra}$	0.8611 ± 0.0108	**8.3141**	OAM$_{gra}$	0.9654 ± 0.0157	0.0547
OPAUC	**0.8989 ± 0.0036**	54.7289	OPAUC	0.9592 ± 0.0245	0.0273
AdaOAM	0.8967 ± 0.0036	67.2898	AdaOAM	0.9410 ± 0.0290	0.0422
KNNOAM	0.8879 ± 0.0040	17.4187	KNNOAM	**0.9895 ± 0.0092**	**0.0047**
Algorithm	svmguide3		Algorithm	usps	
	AUC	Time (s)		AUC	Time (s)
OAM$_{seq}$	0.7453 ± 0.0410	0.7477	OAM$_{seq}$	0.9634 ± 0.0121	7.3531
OAM$_{gra}$	0.7217 ± 0.0428	0.2703	OAM$_{gra}$	0.9650 ± 0.0097	**2.3188**
OPAUC	0.7433 ± 0.0384	0.0703	OPAUC	0.9759 ± 0.0066	35.9055
AdaOAM	**0.7519 ± 0.0442**	0.1047	AdaOAM	0.9752 ± 0.0081	43.8656
NNOAM	0.7112 ± 0.0339	**0.0102**	KNNOAM	**0.9957 ± 0.0029**	3.0055
Algorithm	vehicle		Algorithm	letter	
	AUC	Time (s)		AUC	Time (s)
OAM$_{seq}$	**0.9898 ± 0.0067**	0.4680	OAM$_{seq}$	0.9200 ± 0.0088	13.1813
OAM$_{gra}$	0.9872 ± 0.0084	0.1617	OAM$_{gra}$	0.8548 ± 0.0554	4.3805
OPAUC	0.9886 ± 0.0060	0.0469	OPAUC	0.9267 ± 0.0062	1.2367
AdaOAM	0.9838 ± 0.0101	0.0711	AdaOAM	0.9120 ± 0.0315	1.7828
KNNOAM	0.9745 ± 0.0074	**0.0063**	KNNOAM	**0.9993 ± 0.0003**	1.0063
Algorithm	acoustic		Algorithm	poker	
	AUC	Time (s)		AUC	Time (s)
OAM$_{seq}$	0.7421 ± 0.0587	58.3867	OAM$_{seq}$	0.4916 ± 0.0266	15.6133
OAM$_{gra}$	0.8379 ± 0.0277	19.5711	OAM$_{gra}$	0.5099 ± 0.0318	5.6508
OPAUC	0.8729 ± 0.0036	**8.0313**	OPAUC	0.4982 ± 0.0368	1.4664
AdaOAM	0.8653 ± 0.0120	10.4055	AdaOAM	0.5001 ± 0.0321	2.1391
KNNOAM	**0.9166 ± 0.0023**	42.7477	KNNOAM	**0.6299 ± 0.0309**	**1.1422**

References

1. Agarwal, S.: Surrogate regret bounds for bipartite ranking via strongly proper losses. J. Mach. Learn. Res. **15**(1), 1653–1674 (2014)
2. Agarwal, S., Graepel, T., Herbrich, R., Har-Peled, S., Roth, D.: Generalization bounds for the area under the ROC curve. JMLR **6**, 393–425 (2005)
3. Ailon, N., Mohri, M.: An efficient reduction of ranking to classification. In: COLT 2008, Helsinki, Finland, 9–12 July 2008, pp. 87–98 (2008)
4. Anava, O., Levy, K.: k^*-nearest neighbors: from global to local. In: Lee, D.D., Sugiyama, M., Luxburg, U.V., Guyon, I., Garnett, R. (eds.) Advances in Neural Information Processing Systems, vol. 29, pp. 4916–4924. Curran Associates Inc., Red Hook (2016)
5. Balcan, M.F., Bansal, N., Beygelzimer, A., Coppersmith, D., Langford, J., Sorkin, G.B.: Robust reductions from ranking to classification. Mach. Learn. **72**(1), 139–153 (2008)
6. Beygelzimer, A., Kakade, S., Langford, J.: Cover trees for nearest neighbor. In: ICML, pp. 97–104. ACM, New York (2006)
7. Cardot, H., Degras, D.: Online principal component analysis in high dimension: which algorithm to choose? CoRR abs/1511.03688 (2015). http://arxiv.org/abs/1511.03688
8. Chaudhuri, K., Dasgupta, S.: Rates of convergence for nearest neighbor classification. In: NIPS 2014, pp. 3437–3445 (2014)
9. Clémençon, S., Vayatis, N.: Tree-based ranking methods. IEEE Trans. Inf. Theory **55**(9), 4316–4336 (2009)
10. Clémençon, S., Lugosi, G., Vayatis, N.: Ranking and empirical minimization of U-statistics. Ann. Stat. **36**(2), 844–874 (2008)
11. Cortes, C., Mohri, M.: AUC optimization vs. error rate minimization. In: Thrun, S., Saul, L., Schölkopf, B. (eds.) NIPS, pp. 313–320. MIT Press, Cambridge (2004)
12. Devroye, L., Győrfi, L., Krżyzak, A., Lugosi, G.: On the strong universal consistency of nearest neighbor regression function estimates. Ann. Stat. **22**(3), 1371–1385 (1994)
13. Ding, Y., Zhao, P., Hoi, S.C.H., Ong, Y.: An adaptive gradient method for online AUC maximization. In: AAAI, pp. 2568–2574 (2015)
14. Gao, W., Jin, R., Zhu, S., Zhou, Z.: One-pass AUC optimization. In: ICML 2013, pp. 906–914 (2013)
15. Gao, W., Zhou, Z.: On the consistency of AUC pairwise optimization. In: IJCAI 2015, pp. 939–945 (2015)
16. Hanley, J.A., Mcneil, B.J.: The meaning and use of the area under a receiver operating characteristic (ROC) curve. Radiology **143**, 29–36 (1982)
17. Kar, P., Sriperumbudur, B.K., Jain, P., Karnick, H.: On the generalization ability of online learning algorithms for pairwise loss functions. In: 30th ICML 2013, 16–21 June 2013, Atlanta, GA, USA, pp. 441–449 (2013)
18. Kotlowski, W., Dembczynski, K., Hüllermeier, E.: Bipartite ranking through minimization of univariate loss. In: ICML, pp. 1113–1120. Omnipress (2011)
19. Robbiano, S., Clémençon, S.: Minimax learning rates for bipartite ranking and plug-in rules. ICML 2011, pp. 441–448 (2011)
20. Uematsu, K., Lee, Y.: On theoretically optimal ranking functions in bipartite ranking. Technical report 863, Department of Statistics, The Ohio State University, December 2011
21. Wang, Y., Khardon, R., Pechyony, D., Jones, R.: Generalization bounds for online learning algorithms with pairwise loss functions. In: COLT, pp. 13.1-13.22 (2012)
22. Zhao, P., Hoi, S.C.H., Jin, R., Yang, T.: Online AUC maximization. In: ICML, pp. 233–240 (2011)

On-Line Dynamic Time Warping
for Streaming Time Series

Izaskun Oregi[1]([✉]), Aritz Pérez[2], Javier Del Ser[1,2,3], and José A. Lozano[2,4]

[1] TECNALIA, 48160 Derio, Spain
{izaskun.oregui,javier.delser}@tecnalia.com
[2] Basque Center for Applied Mathematics (BCAM), 48009 Bilbao, Spain
{aperez,jdelser}@bcamath.org
[3] Department of Communications Engineering, University of the Basque Country
UPV/EHU, 48013 Bilbao, Spain
javier.delser@ehu.eus
[4] Department of Computer Science and Artificial Intelligence, University of the
Basque Country UPV/EHU, 20018 Donostia-San Sebastián, Spain
ja.lozano@ehu.eus

Abstract. Dynamic Time Warping is a well-known measure of dissimilarity between time series. Due to its flexibility to deal with non-linear distortions along the time axis, this measure has been widely utilized in machine learning models for this particular kind of data. Nowadays, the proliferation of streaming data sources has ignited the interest and attention of the scientific community around on-line learning models. In this work, we naturally adapt Dynamic Time Warping to the on-line learning setting. Specifically, we propose a novel on-line measure of dissimilarity for streaming time series which combines a warp constraint and a weighted memory mechanism to simplify the time series alignment and adapt to non-stationary data intervals along time. Computer simulations are analyzed and discussed so as to shed light on the performance and complexity of the proposed measure.

Keywords: Time series · On-line learning · Dynamic Time Warping

1   Introduction

In many fields such as manufacturing industry, energy, finance or health, time series are one of the most common forms under which data are captured and processed towards extracting valuable information. For this purpose, time series classification has played a central role in time series analysis: the goal is to build a predictive model based on labeled time series so as to use it to predict the label of previously unseen, unlabeled time series. In the presence of labeled data, k-Nearest Neighbor (k-NN) classification models have been extensively utilized with time series data due to their conceptual simplicity, efficiency and ease of implementation. In essence, k-NN algorithms consist of assigning a

© Springer International Publishing AG 2017
M. Ceci et al. (Eds.): ECML PKDD 2017, Part II, LNAI 10535, pp. 591–605, 2017.
https://doi.org/10.1007/978-3-319-71246-8_36

label to an unseen example according to the class distribution over its k most similar (*nearest*) data instances within the training set. It is obvious that the accuracy of nearest-neighbor techniques is closely related to the measure of similarity between examples. In this regard, research in pattern recognition for time series has originated a diverse collection of measures including the Euclidean Distance (ED), Elastic Similarity Measures (ESM) and Longest Common Subsequence (LCSS), each featuring properties and limitations that should match the requirements of the application at hand.

To the best of our knowledge, no attention has been paid in the literature to distance-based on-line classification models for time series data streams that build upon a proper design of elastic measures of similarity. In response to this lack of research, this work elaborates on an on-line DTW (ODTW) dissimilarity measure. The fundamental ingredient of the ODTW is given by a spotted DTW property that is exploited to avoid unnecessary dissimilarity recalculations. Moreover, computational resources (time and memory) are also controlled by virtue of a Sakoe-Chiba bounding approach. Finally, by under-weighting the influence of past events using a weighted memory mechanism, we make it possible to adapt the ODTW to non-stationarities in the data stream, in clear connection with the well-known stability-plasticity dilemma in on-line learning models. In order to assess the practical performance of the proposed ODTW dissimilarity measure under changing classification concepts, extensive experiments using 1-NN classifiers will be discussed over different public datasets. The ODTW accuracy rate as new data samples arrive will be compared to that of the DTW measure, showing that ODTW can be at least as accurate as DTW. The efficiency of the ODTW in terms of complexity will be also analyzed.

The remainder of the paper is organized as follows: Sect. 2 provides background information on time series similarity measures. Section 3 formulates the definition of the conventional DTW measure and introduces the Sakoe-Chiba band technique. Section 4 gives a detailed description of the proposed ODTW dissimilarity measure. Section 5 delves into the obtained experimental results and, finally, Sect. 6 summarizes the contributions and outlines future research lines, leveraging our findings in this work.

2   Background

Despite its neat advantages – low complexity and simplicity – ED is overly inflexible to deal with time series distortions. In classification problems where the learning process usually focuses on the shape of sequence, this limitation might pose a severe problem. To overcome this issue, Dynamic Time Warping (DTW), an elastic measure of similarity, has proved to be extremely effective to align sequences that are similar in shape but undergo non-linear variations in the time dimension. Along with DTW, the ESM family is completed by the so-called Edit distance [18], the Edit Distance for Real sequences (EDR, [3]) and the Edit distance with Real Penalty (EPR, [2]), among other DTW-based distances [12]. In general, the most important characteristic of all ESMs is their ability to shrink or stretch the time axis in order to find the alignment between the time series under comparison yielding the smallest distance. The ground difference among them, conversely, lies in the selected point-wise distance. Similarly, LCSS is a variation of ESM techniques that allows instances to be unmatched, i.e., a global sequence alignment is not required. Several studies have shown that the use of ESM with 1-NN classifiers outperforms results which are very accurate and hardly beaten in several classification problems [5,15]. Standing on this empirical evidence, DTW-based models have consolidated as the reference for shape-based classification tasks over time series data [9,16,17,23].

An important point to keep in mind when dealing with the DTW similarity measure is its quadratic computational complexity, which makes its computation prohibitive when tackling long time series. In order to avoid this drawback, techniques such as Itakura's Parallelogram [8] and Sakoe-Chiba band [20] are widely utilized to reduce the DTW complexity. These simple methods speed up the DTW computation just by limiting the flexibility of the measure when accommodating time-axis distortions. Similarly to these constraint-based methods, Salvador and Chan [22] estimate the DTW measure by means of a multi-level approach that recursively refines its resolution. Likewise, Keogh and Pazzani [11] propose a modified DTW approach which uses Piecewise Aggregate Approximation (PAA) in order to reduce the length of the time series under comparison and speed up the final computation. Indexing time series to accelerate the performance of different learning methods where DTW is involved is another solution to alleviate its computation [10,14,21].

The complexity issue of the DTW measure noted above is particularly challenging when time series are generated continuously along time, producing endless data streams potentially produced by non-stationary distributions. In many scenarios, data produced by systems and/or processes evolve over time, not necessarily in a stationary manner, making conventional classification methods unsuitable to handle data produced by time-varying generation processes. These stringent conditions under which stream data must be processed have motivated a recent upsurge of *on-line* classification models [6,7,13,24]. In order to identify changes in time series data generator models, Cavalcante et al. [1] have recently proposed a concept drift detector method coined as FEDD. Based on the feature vector similarity given by Pearson correlation distance (or cosine distance), this method monitors the evolution of sequence features in order to test whether a concept change has occurred. In [19] an incremental clustering system for time series data streams is presented: On-line Divisive-Agglomerative Clustering is a tree-like grouping technique that evolves with data based on a criterion to merge and split clusters using a correlation-based dissimilarity measure.

3   Dynamic Time Warping

The DTW measure between two time series, $X^m = (x_1, \ldots, x_i, \ldots, x_m)$ and $Y^n = (y_1, \ldots, y_j, \ldots, y_n)$, is given by the minimum cumulative distance resulting from the best point-wise alignment between both time series.

We represent an alignment of two time series X^m and Y^n by a **path** $p = \{(i_1, j_1), ..., (i_Q, j_Q)\}$ that goes from $(1,1)$ to (m,n) in a $[1,m] \times [1,n]$ lattice. Each pair $(i,j) \in p$ represents the alignment of the points x_i and y_j. We say that the path p is **allowed** if it satisfies that $(i_q, j_q) - (i_{q-1}, j_{q-1}) \in \{(1,0), (1,1), (0,1)\}$ for $q = 2, ..., Q$. That is, allowed paths are formed by $\uparrow, \rightarrow$ and $\nearrow$ steps. Figure 1a shows three possible paths (alignments) between time series X^m and Y^n. From here on, we will consider only allowed paths and, therefore, we will omit the term allowed, for the sake of brevity.

The **weight** of a path p is given by

$$w(p) = \sum_{(i,j) \in p} d_{i,j} \tag{1}$$

where $d_{i,j} = |x_i - y_j|$ is the point-wise distance. Next, we present the definition of the DTW measure:

Definition 1. *The **DTW** measure between time series X^m and Y^n is given by*

$$D(X^m, Y^n) = \min_{p \in \mathcal{P}} w(p)$$

where $\mathcal{P}$ is the set of all allowed paths in the $[1,m] \times [1,n]$ lattice.

When it is clear from the context, we will denote $D(X^m, Y^n)$ simply by $D_{m,n}$. The DTW value corresponds to the weight of the **optimal path**, i.e., the minimum weighted path.

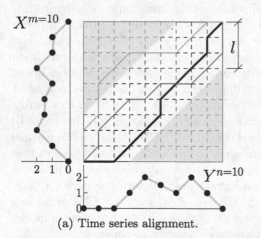

11	7	11	4	5	3.5	2	3	1	0
11	11	11	3	4	2	1	2	0	1
10	10	10	3	3	1.5	1	1	0	1
9	9	9	3	2	1	1	0	1	3
7	7	7	2	2	0.5	0	1	1	2
6	6	6	2	1	0	0.5	1	1.5	3
4.5	4.5	4.5	1.5	0.5	0	0.5	1	1.5	3
3	3	3	1	0	0.5	1.5	1.5	2.5	4.5
1	1	1	0	1	1.5	1.5	2.5	2.5	3.5
0	0	0	1	3	4.5	5.5	7.5	8.5	8.5

(a) Time series alignment.   (b) Cumulative distance matrix.

Fig. 1. Example of the computation of $D(X^m, Y^n)$ for two time series. It can be observed that the optimal path $\mathbf{w}^*$ (—) is that yielding the minimum cumulative distance among those paths connecting points $(1,1)$ to $(10,10)$. (Color figure online)

Since the number of allowed paths grows exponentially with the length of the time series X^m and Y^n, an exhaustive enumeration of all of them with the aim of finding the optimal path is computationally unfeasible, even for small values of n and m. Fortunately, by using dynamic programming, we can compute the DTW measure between two time series using the following recursion:

$$D_{m,n} = d_{m,n} + \min\{D_{m-1,n}, D_{m-1,n-1}, D_{m,n-1}\}, \tag{2}$$

where $D_{0,0} = 0$ and $D_{i,0} = D_{0,j} = \infty$ for $i = 1, \ldots, m$ and $j = 1, \ldots, n$, respectively (initial conditions). Using this recurrence, the complexity for computing the DTW measure between time series X^m and Y^n is $\mathcal{O}(m \cdot n)$. This is still unaffordable in many practical situations and, therefore, several techniques have been proposed to approximate the DTW measure by decreasing its computational complexity. Many of these techniques are based on reducing the number of paths by imposing additional constraints. Among these, we would like to highlight the Sakoe-Chiba bound and Itakura's parallelogram approaches due to their effective constraints. These constraints allow (i) discarding the subset of paths of higher lengths and (on average) with higher weights, and (ii) computing the (approximated) DTW measure in linear time with respect to the length of the time series considered.

In particular, the paths considered by the Sakoe-Chiba bound approach are composed by pairs (i_q, j_q) that satisfy the constraints $|i_q - j_q| \leq l$ for $1 \leq q \leq Q$, where l refers to the so-called **band width**. We call these additional constraints to the paths as the **Sakoe-Chiba constraints**. As a consequence, the DTW can be computed in $\mathcal{O}(l \cdot \max\{m, n\})$. The areas shadowed in Fig. 1a illustrate the forbidden (i_q, j_q) pairs for $l = 3$. In this case, the red path is not allowed while the blue and black paths remain allowed.

4   On-Line Dynamic Time Warping

In this section, we propose the on-line DTW (**ODTW**). The ODTW measure combines (i) an incremental computation of the DTW measure, (ii) the Sakoe-Chiba bound for limiting the computational and space complexities and (iii) a weighted memory mechanism. By the combination of these ideas, ODTW can be computed efficiently (i and ii) and it can control the contribution of the past values to the measure. Next, we introduce the three ideas in order and, finally, we combine them into the novel ODTW.

4.1   Controlling the Computational Complexity

When computing $D_{m,n}$ using Eq. (4), we also compute $D_{i,j}$ for $i = 1, ..., m$ and $j = 1, ..., n$. These values correspond to the DTW measures for time series X^i and Y^j for $1 \leq i \leq m$ and $1 \leq j \leq n$.

We call $M^{m,n} = \{D_{i,j} : i = 1, ..., m$ and $j = 1, ..., n\}$ the **measure matrix**. Figure 1b shows the measure matrix obtained when computing $D_{10,10}$. The set of values of the measure matrix shaded in gray corresponds to the set of DTW measures in the optimal path p^* associated to $D_{10,10}$, i.e., $\{D_{i,j} : (i,j) \in p^*\}$.

We call $F^{m,n} = \{D_{i,n}, D_{m,j} : 1 \leq i \leq m, 1 \leq j \leq n\}$ the **frontier** of the measure matrix $M^{m,n}$. In Fig. 1b the frontier $F^{7,8}$ corresponds to the set of matrix values shaded in green (plus the value at position $(7,8)$, shaded in gray).

Let us assume that we know the frontier $F^{r,s}$ for a given $1 \leq r < m$ and $1 \leq s < n$ and that we want to calculate $D_{m,n}$. Interestingly, in order to compute $D_{m,n}$, we can apply Eq. (4) recursively until a value from the frontier $F^{r,s}$ needs to be computed. At this point, the recursion can be stopped, which can avoid unnecessary calculations (see Fig. 2a). This simple idea is the basis for an incremental computation of the DTW in an on-line scenario.

Now, imagine a (general) on-line scenario where the time series arrive in chunks, sequentially. For instance, at time t_0, we can have the time series X^r and Y^s and, then, at time t_1, we can receive $X^{r+1,m} = (x_{r+1}, ..., x_m)$ and $Y^{s+1,n} = (y_{s+1}, ..., y_m)$. We want to compute $D_{m,n}$ incrementally. At time t_0, we can compute $D_{r,s}$, store the frontier $F^{r,s}$, and the time series X^r and Y^s. Then, at time t_1, we can compute $D_{m,n}$ using the stored frontier according to the previously described mechanism (see Fig. 2a). This incremental process can be repeated when a new chunk arrives.

Given the frontier $F^{r,s}$, the computational complexity for obtaining $D_{m,n}$ and $F^{m,n}$ using the proposed incremental procedure is $\mathcal{O}(m \cdot n - r \cdot s)$. We would like to highlight that, when a single point arrives for both time series ($m = r+1$ and $n = s+1$), the computational complexity is linear with respect to the length of the time series, $\mathcal{O}(\max\{m, n\})$. In addition, the described procedure requires X^m and Y^n to be stored, which leads to a space complexity of $\mathcal{O}(\max\{m, n\})$.

Unfortunately, the computational and space complexities of the proposed incremental computation of the DTW measure are impractical for most of the challenging on-line scenarios. To overcome this issue, Sakoe-Chiba constraints (see end of Sect. 3) can be imposed to the incremental computation, drastically reducing both the memory store and the computational complexity.

By using a band width l, after computing $D_{r,s}$, we store only $\mathcal{O}(l)$ values of the frontier $F^{r,s}$, because some values could correspond to a pair of points that do not fulfill the Sakoe-Chiba constraints. In addition, we only need to store the last l points of the time series X^r and Y^s, which leads to a space complexity of $\mathcal{O}(l)$. In addition, the computational complexity for calculating incrementally an approximation to $D_{m,n}$ according to Sakoe-Chiba constraints is linear in l, i.e., $\mathcal{O}((m-r) \cdot (n-s) + l \cdot (m+n-r-s))$. Therefore, by choosing an appropriate band width l, the proposed constrained and incremental DTW can effectively control the trade-off between (i) the required computational and memory resources, and (ii) the flexibility of DTW with respect to the distortions of the time series along the time axis.

4.2   Forgetting the Past

One of the most extended assumptions in off-line learning is that data samples are drawn from a stationary distribution. However, in on-line learning scenarios this assumption may not hold as stationarity of the data streaming can evolve over time. In consequence, we conceive a streaming time series as being divided

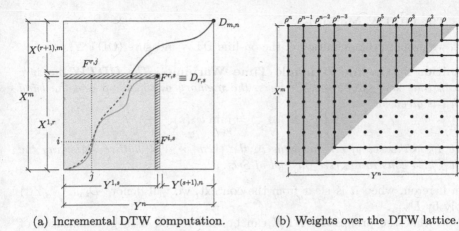

(a) Incremental DTW computation.   (b) Weights over the DTW lattice.

Fig. 2. (a) Example of the incremental computation of the DTW measure, and (b) exponential weights applied over a $[1, m] \times [1, n]$ lattice.

into a sequence of stationary intervals (concepts) of varying length and with a similar periodic shape. We say that a **concept drift** has occurred when the time series changes from one stationary interval to another.

In order to adapt the DTW measure to the concept drift phenomenon, we propose a simple yet efficient weighted memory mechanism that modulates the contribution of the point-wise distance $d_{i,j}$ in the weight of a path (see Eq. (1)). This mechanism leverages the assumption that recent points are more likely to have been produced in the last stationary interval. To this end, we propose the use of a memory parameter $\rho \in (0, 1]$ and we define the weight with memory of a path p as follows:

$$w_\rho(p) = \sum_{(i,j) \in p} \rho^{\max\{m-i, n-j\}} d_{i,j} \tag{3}$$

Note that $\max\{m - i, n - j\}$ corresponds to the Chebyshev distance between the points (i, j) and (m, n). In this manner, points belonging to the last stationary interval of a time series will contribute more significantly to the weight of a given path and to a measure that minimizes this weight with memory.

Figure 2b illustrates an example of the weighting function given in Eq. 3 for $m = n - 2$. All the pairs (i, j) in the lattice $[1, m] \times [1, n]$ at the same Chebyshev distance from (m, n) are connected with a black line.

We would like to highlight that, contribution of a pair of points (x_i, y_j) to the memory weight of any path is equal. In other words, the contribution of a pair of points to the memory weight does not depend on the path, which avoids favoring the longest paths. Intuitively, an appropriate memory parameter ρ should be selected according to the length of the stationary intervals or pattern period (see Sect. 5)

4.3   The ODTW Measure

Next, we propose the definition of the on-line DTW measure (ODTW).

Definition 2 (On-line Dynamic Time Warping). *The **ODTW** measure between time series X^m and Y^n given the memory parameter ρ and the band width l is given by*

$$D_{l,\rho}(X^m, Y^n) = \min_{p \in P_l} w_\rho(p)$$

where P_l is the set of all the paths in the $[1, m] \times [1, n]$ lattice satisfying the Sakoe-Chiba constraints (see the end of Sect. 3).

From here on, when it is clear from the context, we will denote $D_{l,\rho}(X^m, Y^n)$ simply by $D_{m,n}$.

Note that the proposed ODTW can be understood as a generalization of the DTW measure. For instance, if $\rho = 1$ and $l = \infty$ – without Sakoe-Chiba constraints – then ODTW corresponds to DTW. Additionally, when $n = m$, $\rho = 1$ and $l = 0$ ODTW corresponds to the Euclidean distance.

Again, by using dynamic programming, it is possible to compute the ODTW measure given ρ and l between X^m and Y^n using the following recursion

$$D_{m,n} = d_{m,n} + \min \left\{ \rho^{\mathbb{I}(m>n)} \cdot D_{m-1,n}, \rho \cdot D_{m-1,n-1}, \rho^{\mathbb{I}(m<n)} D_{m,n-1} \right\}, \quad (4)$$

where $D_{i,j} = \infty$ for any pair (i,j) not satisfying the Sakoe-Chiba additional constraints, and $\mathbb{I}(\cdot)$ is an auxiliary function taking value 1 if its argument is true, and 0 otherwise.

At this point, we would like to mention that Definition 2 and its recursive computation shown in Eq. (4) have been given in their simplest form, for the sake of brevity and readability. The simplest form is appropriate for on-line scenarios where the time series arrive in chunks consisting of a single point. However, both the definition and the recursive computation of ODTW can be easily generalized in order to deal with chunks of arbitrary sizes. Figure 3 illustrates the on-line DTW recursive computation in the general form, that is, when the time series arrive in chunks of arbitrary sizes. The green areas represent the stored l-frontiers.

Note: The source code in Python of the proposed ODTW measure (in its most general form) has been made available on-line at http://bitbucket.org/izaskun_oregui/ODTW.

5   Experimental Study

We explore the practical performance of the proposed ODTW measure by running several computer experiments aimed at two different yet related goals:

1. To show that ODTW is an efficient method and, hence, a suitable measure of dissimilarity, in terms of the computational complexity, for on-line classification scenarios.

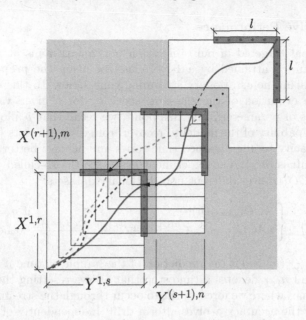

Fig. 3. Computation of the proposed ODTW measure in its general form. (Color figure online)

2. To provide practical evidence of the capacity of the ODTW measure and its memory mechanism to react and accommodate concept changes in the processed streaming time series.

5.1 Efficiency

As for the first goal, we compare the running time of ODTW and conventional DTW methods. In this experiment two streaming time series have been produced by drawing one sample at a time from a uniform distribution with support [0,1] from an initial length of $n = m = 3$ samples to a maximum of $n = m = 70$ samples. For the sake of fairness, the same value of the Sakoe-Chiba band width $l = 50$ has been used to compute both DTW and ODTW dissimilarities. Under these modeling assumptions, the complexity of the DTW measure is expected to be quadratic, $\mathcal{O}(n^2)$, as long as the length of the time series is less than the band width, i.e., $n \leq l$. However, for $n > l$ the DTW complexity is $\mathcal{O}(ln)$. As addressed in Sect. 4, the computational complexity of the proposed ODTW is $\mathcal{O}(n)$ when $n \leq l$ and $\mathcal{O}(l)$ when $n > l$.

Figure 4 shows the running time (in seconds when implemented natively in Python 2.7 on a single i7 core at 3.10 GHz) required to compute the ODTW (black) and DTW (gray) measures. A red dashed vertical line is included in the plot to indicate the value of the Sakoe-Chiba band width l. The empirical results shown in this plot support the hypothesis 1 discussed above, and show that ODTW is a suitable measure, in computational complexity terms, to quantify the dissimilarity between two streaming time series.

5.2 Predictive Performance

The second goal targeted in our simulation benchmark aims at assessing the predictive accuracy attained by a 1-NN classifier using the proposed ODTW metric when facing non-stationary streaming time series. To this end, we have monitored the evolution of the classifier accuracy for a given value of l and different values of parameter ρ. At this point, we recall that ρ allows balancing between the capability of the model to recover from drift concepts (low values of ρ) and its capacity to align warped time series and achieve better performance scores (high values of ρ). The experiment evaluates the so-called **prequential accuracy** $pACC(n)$ over different non-stationary datasets:

$$
pACC(n) = \begin{cases} ACC_{sample}(n) & \text{if } n = n_{ref}, \\ pACC(n-1) + \frac{ACC_{sample}(n) - pACC(n-1)}{n - n_{ref} + 1} & \text{otherwise,} \end{cases} \tag{5}
$$

where $ACC_{sample}(n)$ is 0 if the prediction of the sample at time n is wrong and 1 if correct; and n_{ref} denotes a time step that allows resetting the prequential accuracy at times where we force a drift to occur through the stream. This allows analyzing how the accuracy evolves after a drift, independently of the previous behavior of the classifier.

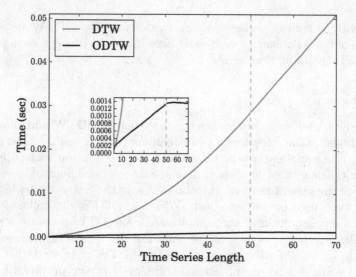

Fig. 4. Experimental running times required for the computation of the ODTW and DTW measures between two streaming time series of increasing length. (Color figure online)

In order to analyze the prequential accuracy in a systematic way, synthetic datasets for on-line classification of stream time series have been designed and utilized for this second set of experiments. Such artificial datasets are built upon several publicly available time series datasets commonly used in DTW-based time series analysis, which can all be retrieved from the UCR Time Series Classification Archive [4]. In particular, we will use the datasets listed in Table 1.

Given different values of the weight parameter ρ, the designed datasets should allow analyzing the ability of the ODTW measure to adjust to time series stationarity. Therefore, streaming time series should have at least two different stationary parts and be periodic in each stationary interval. We generalize this intuitive premise to build the reference and query streams for each dataset in Table 1. In particular, query streams are composed by an endless repetition of stationary intervals (concepts), each formed by P time series of the same class drawn uniformly at random from the corresponding subset. In order to simulate a non-stationarity in the stream, the generation process avoids repeating the same class between every two consecutive stationary periods. Reference streams are composed for every label in the dataset by concatenating uniformly sampled time streams for every label in the dataset. Consequently, each query sequence in the dataset presents a recurrent concept change occurring every P time series. For instance, query class labels in a binary classification problem with stationary periods of $P = 3$ time series of length 2 samples each would be given by $\{0, 0, 0, 0, 0, 0, 1, 1, 1, 1, 1, 1, 0, 0, 0, 0, 0, 0, 1, 1, 1, 1, 1, 1, 0, 0, \ldots\}$.

Table 1. Main characteristics of the utilized UCR datasets

Name	Train/TestSize	Time series length (n)	No. Classes	1-NN score	Sakoe-Chiba band l
Gun Point	50/150	150	2	0.913	0
Italy Power Demand	67/1029	24	2	0.955	0
Two Lead ECG	23/1139	82	2	0.868	4
Face Four	24/88	350	4	0.886	7
CBF	30/900	128	3	0.964	14
Toe Segmentation 2	36/130	343	2	0.908	17

We construct a total of 50 different test instances for each dataset in Table 1 to obtain an estimate of the average prequential accuracy achieved by an ODTW-based 1-NN classifier when predicting the label associated to the stream upon the arrival of every sample. Results are collected in Fig. 5 for every dataset with its optimal value of l (Table 1) and different choices of the memory parameter ρ. Vertical dashed lines indicate the time at which the end of a time series meets the beginning of the next time series (periodicity of the stationary interval), being highlighted in bold black if the transition involves a label change. In these experiments, a class label change is produced every $P = 3$ time series. The horizontal dashed line indicates the best DTW-based 1-NN accuracy rate reported in [4]

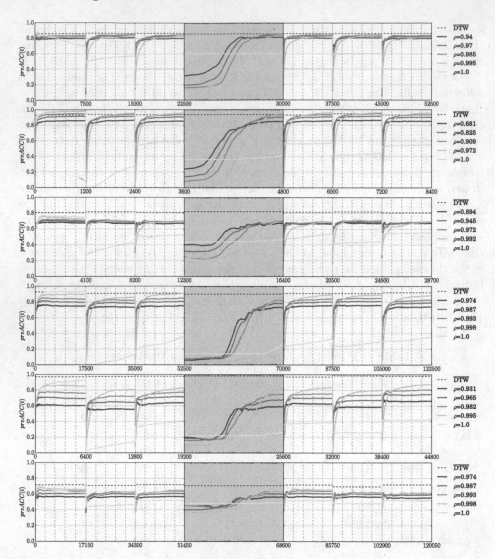

Fig. 5. Evolution of the prequential accuracy $pACC(n)$ over 7 stationary intervals of $P = 50$ concatenated time series each achieved by a 1-NN model using the proposed ODTW measure with the optimal l for each case and different values of ρ. The grey shadowed area corresponds to the 4-th stationary interval where the horizontal axis is represented in logarithmic scale. From top to bottom, plotted results correspond to Gun point, Italy Power Demand, Two Lead ECG, Face Four, CBF and Toe Segmentation 2 datasets.

and listed in Table 1. The values taken by ρ have been chosen according to the length of the original time series. Particularly, the chosen values correspond to $\rho^m \in \{0.0001, 0.01, 0.1, 0.5, 1\}$. These parameters represent very-short-, short-, middle-, long- and full-range memory. The value of the band width l – designed to sacrifice the ODTW flexibility for a computationally efficient computation – has been set equal to the optimal Sakoe-Chiba band width found for every dataset shown in Table 1.

Several conclusions can be drawn from the experiments in Fig. 5:

- As the value of ρ decreases (yielding a lower influence of past observations in the ODTW computation), the 1-NN model reacts more quickly to stationary changes, hence the prequential accuracy increases faster. However, in certain cases this involves a penalty in accuracy once the concept has become stable: the reason lies in the fact that the weighted memory of the ODTW measure fails to exploit past distance information of relevance for a discriminative alignment between the time series.
- As the values of ρ increase (a higher influence of past observations in the ODTW computation), even though the 1-NN model requires more time to recover from the stationary changes, its prequential accuracy tends to be better once the concept has been learned. In addition, as ρ increase the variance of the obtained prequential accuracy decreases. When $\rho = 1$, 1-NN performs worst due to the lack of a forgetting mechanism, which makes the value of the ODTW measure strongly biased by past alignments not linked to the concept to be predicted.

In light of the experimental results discussed above, we conclude that ODTW has both the reduced complexity and the flexibility to adapt to non-stationary environments needed for efficiently dealing with streaming time series.

6   Conclusions and Future Work

In this work, we have presented the On-line Dynamic Time Warping (ODTW), a natural adaptation of the popular DTW to the streaming time series setting. ODTW can be computed efficiently in an incremental way by avoiding unnecessary recalculations (see Sect. 4.1). It includes two parameters, l and ρ, that can be used in order to adapt the proposed measure to the particularities of the streaming time series under analysis. On the one hand, the band width parameter l is inspired by the Sakoe-Chiba band approach and can be used to control the trade-off between the complexity of the incremental computation of ODTW (linear in l) and the ability to shrink or stretch the time axis in order to align two time series (see Sect. 4.1). On the other hand, ρ is the forgetting parameter and it can be used to control the memory of ODTW by giving less importance to past values. By controlling the memory of the proposed measure, we can adjust the ability of ODTW to react to drift changes in streaming time series (see Sect. 4.2).

Due to the efficiency and flexibility of ODTW for dealing with streaming time series, we plan to extend its principles to other popular Elastic Similarity Measures such as the Edit distance, EDR and ERP. In addition, we will extend the experimentation by incorporating other on-line problems with streaming time series. Similar to DTW in off-line learning tasks, ODTW can also be used in on-line supervised and unsupervised problems such as on-line clustering, classification, outlier detection, etc. which is very useful in many different real world applications; for example, gesture classification, load profiling in energy grids and fraud detection.

Acknowledgments. This work has been supported by the Basque Government through the ELKARTEK program (ref. BID3ABI KK-2016/00096). Aritz Pérez is partially supported by the ELKARTEK program from the Basque Government, and by the Spanish Ministry of Economy and Competitiveness MINECO: BCAM Severo Ochoa excellence accreditation SVP-2014-068574 and SEV-2013-0323.

References

1. Cavalcante, R.C., Minku, L.L., Oliveira, A.L.: FEDD: feature extraction for explicit concept drift detection in time series. In: 2016 International Joint Conference on Neural Networks (IJCNN), pp. 740–747. IEEE (2016)
2. Chen, L., Ng, R.: On the marriage of Lp-norms and edit distance. In: Proceedings of the Thirtieth International Conference on Very Large Data Bases, vol. 30, pp. 792–803. VLDB Endowment (2004)
3. Chen, L., Özsu, M.T., Oria, V.: Robust and fast similarity search for moving object trajectories. In: Proceedings of the 2005 ACM SIGMOD International Conference on Management of Data, pp. 491–502. ACM (2005)
4. Chen, Y., Keogh, E., Hu, B., Begum, N., Bagnall, A., Mueen, A., Batista, G.: The UCR time series classification archive, July 2015. www.cs.ucr.edu/~eamonn/time_series_data/
5. Ding, H., Trajcevski, G., Scheuermann, P., Wang, X., Keogh, E.: Querying and mining of time series data: experimental comparison of representations and distance measures. Proc. VLDB Endow. $1(2)$, 1542–1552 (2008)
6. Ditzler, G., Roveri, M., Alippi, C., Polikar, R.: Learning in nonstationary environments: a survey. IEEE Comput. Intell. Mag. $10(4)$, 12–25 (2015)
7. Faisal, M.A., Aung, Z., Williams, J.R., Sanchez, A.: Data-stream-based intrusion detection system for advanced metering infrastructure in smart grid: a feasibility study. IEEE Syst. J. $9(1)$, 31–44 (2015)
8. Itakura, F.: Minimum prediction residual principle applied to speech recognition. IEEE Trans. Acoust. Speech Signal Process. $23(1)$, 67–72 (1975)
9. Jeong, Y.S., Jeong, M.K., Omitaomu, O.A.: Weighted dynamic time warping for time series classification. Pattern Recogn. $44(9)$, 2231–2240 (2011)
10. Keogh, E.: Exact indexing of dynamic time warping. In: Proceedings of the 28th International Conference on Very Large Data Bases, pp. 406–417. VLDB Endowment (2002)
11. Keogh, E.J., Pazzani, M.J.: Scaling up dynamic time warping for datamining applications. In: Proceedings of the Sixth ACM SIGKDD International Conference on Knowledge Discovery and Data Mining, pp. 285–289. ACM (2000)
12. Keogh, E.J., Pazzani, M.J.: Derivative dynamic time warping. In: Proceedings of the 2001 SIAM International Conference on Data Mining, pp. 1–11. SIAM (2001)
13. Krawczyk, B., Minku, L.L., Gama, J., Stefanowski, J., Woźniak, M.: Ensemble learning for data stream analysis: a survey. Inf. Fusion 37, 132–156 (2017)

14. Lin, J., Keogh, E., Lonardi, S., Chiu, B.: A symbolic representation of time series, with implications for streaming algorithms. In: Proceedings of the 8th ACM SIG-MOD Workshop on Research Issues in Data Mining and Knowledge Discovery, pp. 2–11. ACM (2003)

15. Lines, J., Bagnall, A.: Time series classification with ensembles of elastic distance measures. Data Min. Knowl. Discov. **29**(3), 565–592 (2015)

16. Rakthanmanon, T., Campana, B., Mueen, A., Batista, G., Westover, B., Zhu, Q., Zakaria, J., Keogh, E.: Searching and mining trillions of time series subsequences under dynamic time warping. In: Proceedings of the 18th ACM SIGKDD International Conference on Knowledge Discovery and Data Mining, pp. 262–270. ACM (2012)

17. Rath, T.M., Manmatha, R.: Word image matching using dynamic time warping. In: Proceedings of 2003 IEEE Computer Society Conference on Computer Vision and Pattern Recognition, vol. 2, p. II. IEEE (2003)

18. Ristad, E.S., Yianilos, P.N.: Learning string-edit distance. IEEE Trans. Pattern Anal. Mach. Intell. **20**(5), 522–532 (1998)

19. Rodrigues, P.P., Gama, J., Pedroso, J.: Hierarchical clustering of time-series data streams. IEEE Trans. Knowl. Data Eng. **20**(5), 615–627 (2008)

20. Sakoe, H., Chiba, S.: Dynamic programming algorithm optimization for spoken word recognition. IEEE Trans. Acoust. Speech Signal Process. **26**(1), 43–49 (1978)

21. Sakurai, Y., Yoshikawa, M., Faloutsos, C.: FTW: fast similarity search under the time warping distance. In: Proceedings of the Twenty-Fourth ACM SIGMOD-SIGACT-SIGART Symposium on Principles of Database Systems, pp. 326–337. ACM (2005)

22. Salvador, S., Chan, P.: Toward accurate dynamic time warping in linear time and space. Intell. Data Anal. **11**(5), 561–580 (2007)

23. Yu, D., Yu, X., Hu, Q., Liu, J., Wu, A.: Dynamic time warping constraint learning for large margin nearest neighbor classification. Inf. Sci. **181**(13), 2787–2796 (2011)

24. Zhao, X., Li, X., Pang, C., Zhu, X., Sheng, Q.Z.: Online human gesture recognition from motion data streams. In: Proceedings of the 21st ACM International Conference on Multimedia, pp. 23–32. ACM (2013)

PowerCast: Mining and Forecasting Power Grid Sequences

Hyun Ah Song[1(✉)], Bryan Hooi[2], Marko Jereminov[3], Amritanshu Pandey[3], Larry Pileggi[3], and Christos Faloutsos[1]

[1] School of Computer Science, Carnegie Mellon University, Pittsburgh, USA
{hyunahs,christos}@cs.cmu.edu
[2] Department of Statistics, Carnegie Mellon University, Pittsburgh, USA
bhooi@andrew.cmu.edu
[3] Deptartment of Electrical and Computer Engineering, Carnegie Mellon University, Pittsburgh, USA
{mjeremin,amritanp,pileggi}@andrew.cmu.edu

Abstract. What will be the power consumption of our institution at 8am for the upcoming days? What will happen to the power consumption of a small factory, if it wants to double (or half) its production? Technologies associated with the smart electrical grid are needed. Central to this process are algorithms that accurately model electrical load behavior, and forecast future electric power demand. However, existing power load models fail to accurately represent electrical load behavior in the grid. In this paper, we propose POWERCAST, a novel domain-aware approach for forecasting the electrical power demand, by carefully incorporating domain knowledge. Our contributions are as follows: 1. Infusion of **domain expert knowledge**: We represent the time sequences using an equivalent circuit model, the "BIG" model, which allows for an *intuitive interpretation* of the power load, as the BIG model is derived from physics-based first principles. 2. **Forecasting of the power load**: Our POWERCAST uses the BIG model, and provides (a) *accurate* prediction in multi-step-ahead forecasting, and (b) *extrapolations*, under *what-if* scenarios, such as variation in the demand (say, due to increase in the count of people on campus, or a decision to half the production in our factory etc.) 3. **Anomaly detection**: POWERCAST can spot and, even explain, anomalies in the given time sequences. The experimental results based on two real world datasets of up to three weeks duration, demonstrate that POWERCAST is able to forecast several steps ahead, with **59%** error reduction, compared to the competitors. Moreover, it is fast, and scales linearly with the duration of the sequences.

1 Introduction

The goal of the smart electrical grid is to manage the demand and supply of electricity while maintaining both efficiency and reliability. Indeed, [1] finds that improving the reliability of the U.S. grid could provide savings of around $49

© Springer International Publishing AG 2017
M. Ceci et al. (Eds.): ECML PKDD 2017, Part II, LNAI 10535, pp. 606–621, 2017.
https://doi.org/10.1007/978-3-319-71246-8_37

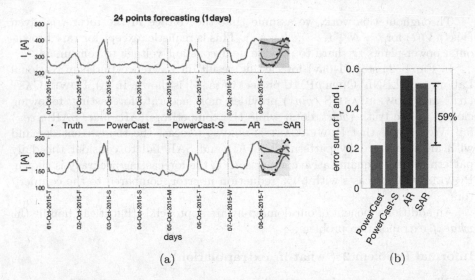

(a)   (b)

Fig. 1. PowerCast forecasts accurately. (a) POWERCAST (*red*) and POWERCAST-S (*pink*) forecasts 24 steps (1-day) on I_r (top row), and I_i (bottom row) more accurately compared to competitors (AR: *blue*; SAR: *green*; ground truth *black circles*). (b) RMSE comparison of the methods (Color figure online)

billion per year and provide a 12% to 18% reduction in emissions, while improving efficiency could save an additional \$20 *billion*. Toward this goal, monitoring systems have been put in place, including Phasor Measurement Units (PMUs) and newer high-precision micro-PMUs [34]. Using these data sources to accurately model load behavior in the grid, as well as to forecast future power load, is important in protecting the grid from failure and for maintaining reliability.

Our main goal is to understand how a specific service area consumes power (a university campus, a small factory, a village or neighborhood), by studying its past behavior. Once we have a good model for the power-consumption behavior, then we can do forecasting (*how much power will our campus/factory need tomorrow*), spot anomalies (when our forecast is too far from what actually happened), and answer "what-if" scenarios, like *how much power will we need, during spring-break on campus*; or *during a heat-wave, in our neighborhood*.

We focus on these two problems: forecasting, and 'what-if' scenarios. The informal definitions, are as follows. Note that alternating current (AC) (I) and voltage (V) are modeled as complex numbers.

Informal Problem 1 (Multi-step forecasting on power grid)

- ***Given:*** real and imaginary current ($I_r(t), I_i(t)$) and voltage ($V_r(t), V_i(t)$) of previous N time points ($t = 1, \cdots, N$),
- ***Forecast:*** the electric current demand, for N_f steps in the future (i.e., guess $I_r(t), I_i(t)$, for $t = N + 1, \cdots, N + N_f$)

Throughout this work, we assume that the voltage in the future, is given $(V_r(t), V_i(t)$ for $t = N+1, \cdots, N+N_f)$. This is realistic: except for rare brownouts, power-plants try hard to provide near-constant voltage to consumers.

In Fig. 1, 24 step (1-day) forecasting result on Lawrence Berkeley National Laboratory (LBNL) Open μPMU project data [34] is shown. In (a), POWERCAST (*red*) and POWERCAST-S (*pink*) provides more accurate forecasting, following closely to the truth (*black* dots) while the competitors (AR: *blue*, SAR:*green*) fail. We observe that POWERCAST is able to forecast the the current demand with an accurate daily pattern while AR and SAR fail to consider the daily pattern. In (b), a quantitative comparison on the forecasting accuracy is shown. POWERCAST forecasts with **59%** reduction in error, compared to the competitors.

An additional benefit of our domain-aware approach is that it can handle the *what-if* extrapolation problem:

Informal Problem 2 ('what-if' extrapolation)

- **Given:** *the historical data, as above (real and imaginary current $I_r(t)$, $I_i(t)$, and voltage $V_r(t)$, $V_i(t)$, $t = 1, \ldots, N$*
- **Guess:** *what will be the power demand in the future (currents $I_r(t)$, $I_i(t)$, $t > N$), if, say, the student population doubles on our campus (or our factory cuts production in half, etc.).*

Handling what-if scenarios is beyond the reach of black-box methods (ARIMA etc.), exactly because it demands domain knowledge, which we carefully infuse, using the established "BIG" model.

Our contributions are as follows:

1. **Infusion of domain knowledge**: Our method is domain-aware: among the many aggregated load models (BIG, "PQ", "ZIP"), we chose the first, because it allows for an accurate and *intuitive interpretation* of the power load, being derived from physics-based first principles.
2. **Forecasting/What-if**: The proposed POWERCAST (a) leads to *more accurate* forecasts (up to **59%** lower error) compared to textbook, black-box methods, and (b) it can answer *what-if* scenarios that black-box methods can not.
3. **Anomaly detection**: POWERCAST can spot, and even explain, anomalies in the given time sequences. (see Sect. 4.3, Fig. 6)

Reproducibility: Our code is open-sourced at: www.cs.cmu.edu/~hyunahs/code/PowerCast.zip; the LBNL dataset is at: powerdata.lbl.gov/.

The structure of the paper is typical: We give the background and related works (Sect. 2), our proposed method (Sect. 3), experiments (Sect. 4), and conclusions (Sect. 5).

2 Background and Related Work

2.1 Related Works

Load Model for Aggregated Electrical Demand: The BIG Model. Electrical load modeling for power system analysis has been traditionally done using the constant power *constant PQ* and *ZIP* models [28]. However, industry experience has shown that these models incorrectly characterize load behavior [22]. Recent advances in steady-state power system simulations [6,27] have introduced the use of physics based state variables, i.e. currents and voltages, and shown that load behavior at a given time instance can be accurately described by a linear relationship between current and voltage. From circuit theory, this can be represented by a parallel or series combination of reactance (B) and conductance (G). This model captures both voltage magnitude and angle information, in contrast to existing traditional load models [28]. Further adding a current source (I) results in the BIG load model, which accurately captures load sensitivities to voltage variations over a period of time [13]. The complete description of the BIG load model is given in Sect. 3 (Table 1).

Table 1. PowerCast captures all of the listed properties. $AR++$ = ARIMA, seasonal ARIMA etc. LDS = Linear Dynamical Systems. '*Pattern discovery*' = concept/latent-variable discovery.

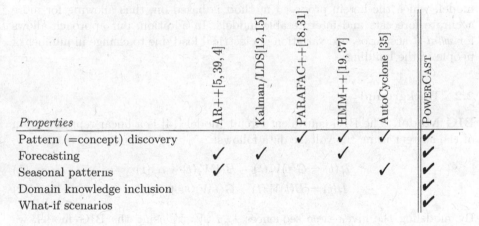

Properties	AR++[5, 39, 4]	Kalman/LDS[12,15]	PARAFAC++[18,31]	HMM++[19, 37]	AutoCyclone [35]	PowerCast
Pattern (=concept) discovery			✓	✓	✓	✓
Forecasting	✓	✓		✓		✓
Seasonal patterns	✓				✓	✓
Domain knowledge inclusion						✓
What-if scenarios						✓

Time Series Forecasting. Classical methods for time-series forecasting include the family of autoregression (AR)-based methods, including ARMA, exponential smoothing [7], ARIMA [5], ARMAX [39] and ARFIMA [4] models. Seasonal ARIMA [5] incorporates pre-determined, constant periods allowing the model to capture seasonal patterns. More recent generalizations include TBATS [9] which allows for more complex seasonality patterns. Other methods for time series forecasting include Kalman filtering [15], Linear Dynamical Systems (LDS) [12], Hidden Markov Models (HMMs) [19], wavelet-based methods such as AWSOM [29] and non-linear dynamical systems such as RegimeCast [23].

In the area of modeling of aggregated load in the power grid, autoregression is an extremely common approach due to its simplicity and interpretability [8, 11, 32, 33]. [30] uses ARMA models and exponential smoothing for short-term load forecasting. [26] uses a two-step procedure of seasonality removal, followed by ARMA with hyperbolic noise.

Tensor-Based Time Series Analysis. Tensor decomposition methods, including PARAFAC decomposition, Tucker decomposition [17, 18], multilinear principal components analysis [21], and Bayesian tensor analysis [36] are powerful tools to understand latent factors of a target dataset. Recent work such as Marble [10] and Rubik [38] applies tensor factorization for concept analysis with domain knowledge. In terms of time series applications, tensors have been used for modeling multiple coevolving time series for epidemiology [25], community discovery [20], and concept discovery [16]. [24] uses a tensor-based approach for forecasting based on complex sequences of timestamped events. AutoCyclone [35] models seasonality in tensor datasets by 'folding' the data into a higher-order tensor.

However, all of the aforementioned methods are typically not based on electrical load models derived from first principles, as is the case for the BIG load model, which the herein proposed method is based on, thus allowing for more accurate forecasts and interpretable models. In addition, our approach allows for *what-if* scenarios, e.g. variation of electrical load due to change in number of people in the building.

2.2 Background

BIG Model. The BIG equivalent circuit model [14] is a linear representation of the current using the voltage data follows:

$$I_r(t) = G(t)V_r(t) - B(t)V_i(t) + \alpha_r(t) \tag{1}$$
$$I_i(t) = B(t)V_r(t) + G(t)V_i(t) + \alpha_i(t) \tag{2}$$

By modeling the given time sequences I_r, I_i, V_r, V_i using the BIG model, we learn the BIG parameters G, B, α_r, α_i that provides us with an intuitive interpretation of the aggregated conductance (G) and susceptance (B) of the power grid as it varies over time. In POWERCAST, we convert the given time sequences to the BIG domain (G, B, α_r, α_i) and further proceed with pattern analysis and forecasting. As we will demonstrate in our analysis with experimental results in Sect. 4, working in the BIG domain provides more stable and interpretable analysis of the power systems: understanding the power system (Sect. 4.1), accurate forecasting and exploration of various what-if scenarios (Sect. 4.2), and anomaly detection (Sect. 4.3).

3 Proposed Method

In this work, our interest is in: (1) including the domain knowledge of the intrinsic behavior of the campus (G, B), (2) capture the latent periodic patterns in the sequences, and (3) do forecasting.

The core parts of POWERCAST algorithm is: (1) to convert the given time sequences into the BIG domain, and (2) forecasting in tensor structure. By converting the time sequences into the BIG domain, we try to understand the power system $(G, B$ value range, dynamics, ratio, etc.) that generated the sequences I_r, I_i, V_r, V_i that we observe, which enables us to do forecasting under *what-if* scenarios of change in the power systems on campus. Working with tensors tells us the long-term, and daily patterns. By performing forecasting based on the learned patterns from the past, POWERCAST provides more stable forecasting result (Sect. 4.2). In this section, we will explain how we convert the time sequences into the BIG domain and how we analyze tensors to do forecasting in more details.

A table of symbols that is used throughout this paper is shown in Table 2.

Table 2. Symbols and definitions

Symbols	Definitions
I_r, I_i, V_r, V_i	Given time sequences. Real and imaginary part of complex current and voltage
N	Total number of timeticks in the given time sequences. $N = N_d \times N_{dt}$
N_d	Number of days in the given time sequences
N_{dt}	Number of time points for each day
G, B, α_r, α_i	BIG parameters. (conductance, susceptance, offset to I_r and I_i)
$\mathcal{X}$	A tensor of given time sequences I_r, I_i, V_r, V_i. $\mathcal{X} \in \mathbb{R}^{N_d \times N_{dt} \times 4}$
$\mathcal{B}_a$	A tensor of BIG parameters G, B, α_r, α_i. $\mathcal{B}_a \in \mathbb{R}^{N_d \times N_{dt} \times 4}$
$\mathcal{B}_b$	A tensor of BIG parameters G, B, α_r, α_i, after extension. $\mathcal{B}_b \in \mathbb{R}^{(N_d + N_{fd}) \times N_{dt} \times 4}$
N_{fd}	Number of days for forecasting
N_f	Number of time steps for forecasting. $N_f = N_{fd} \times N_{dt}$
P_{ar}	Auto-regression parameter
R	Rank for tensor decomposition
N_w, σ	Parameters for Gaussian filter. (window size, standard deviation of Gaussian)

3.1 PowerCast Algorithm

Pseudocode of POWERCAST is described in Algorithm 1, where each function will be explained in more detail. In Fig. 2, a flowchart of the Algorithm 1 is illustrated.

Step 1: Seq2Tensor (Tensor construction): Given four time sequences $I_r(1 : N), I_i(1 : N), V_r(1 : N), V_i(1 : N)$, we construct a tensor $\mathcal{X} \in \mathbb{R}^{N_d \times N_{dt} \times 4}$ by cutting the sequences into daily unit as described in Fig. 2 *"Seq2Tensor."* The

Algorithm 1. POWERCAST.

Data: $I_r(1:N), I_i(1:N), V_r(1:N+N_f), V_i(1:N+N_f)$
Result: $\hat{I}_r(N+1:N+N_f), \hat{I}_i(N+1:N+N_f)$
Construct a tensor from given sequences:;
$[\mathcal{X}] = \text{Seq2Tensor}(I_r(1:N), I_i(1:N), V_r(1:N), V_i(1:N))$;
Covert the data to BIG domain (fit the data to BIG model):;
$[\mathcal{B}_a] = \text{X2B}(\mathcal{X})$;
Extend the tensor for forecasting:;
$[\mathcal{B}_b] = \text{Bextension}(\mathcal{B}_a)$;
Recover the data from BIG domain:;
$[\hat{I}_r(N+1:N+N_f), \hat{I}_i(N+1:N+N_f)] =$
$\text{B2Seq}(\mathcal{B}_b, V_r(N+1:N+N_f), V_i(N+1:N+N_f))$;

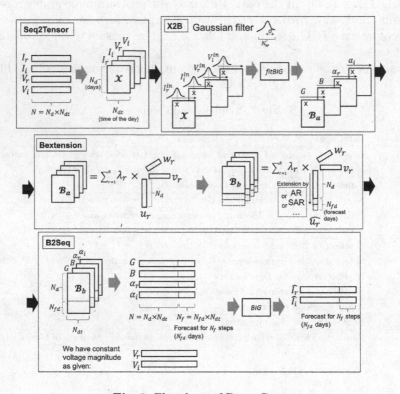

Fig. 2. Flowchart of PowerCast.

first, second, and third mode of the tensor corresponds to *days, hour of the day,* and *given time sequences,* respectively. For example, if the time sequences are hourly samples (24 points per day) for 7 days, then the dimensions of the first and second modes become 7 (days) and 24 (samples per day) respectively.

Step 2: X2B (Convert the data into BIG domain): Then we convert the data tensor ($\mathcal{X}$) to BIG domain ($\mathcal{B}_a$) by fitting the data to BIG model as described in "*X2B*" in Fig. 2. For fitting the BIG parameters, as we cannot fit the BIG parameters to a single point, we consider multiple points before and after the current timetick by apply a moving Gaussian filter to a window size of N_w, standard deviation σ to the data.

Function *fitBIG* takes in $[I_r^{in}, I_i^{in}, V_r^{in}, V_i^{in}]$ and fits the BIG models in Eqs. (1) and (2) to the given window of data, using weighted least squares fitting, using this Gaussian filter as weights. Then we construct a new tensor $\mathcal{B}_a$ consisting of BIG parameters.

We next combine these fitted values into a tensor in the BIG domain whose first, second, and third mode of the tensor corresponds to *days*, *hour of the day*, and *BIG parameters*, respectively.

Step 3: Bextension (Extend the tensor for forecasting): After constructing a tensor from the given data in the BIG domain, we decompose the tensor using canonical polyadic alternating least squares (CP_ALS) algorithm [17,18]: $\mathcal{B}_a = \sum_{r=1}^{R} \lambda_r \times \mathbf{u}_r \times \mathbf{v}_r \times \mathbf{w}_r$. Here, $\mathbf{u}, \mathbf{v}, \mathbf{w}$ corresponds to the hidden variables across multiple days, within a day, and among the BIG parameters. We will refer to these hidden variables as *long-term-concepts*, *daily-concepts*, *users-profile-concepts*, respectively.

After learning the tensor components, we extend the first mode of the tensor ($\mathbf{u}$, corresponding to *days*) for N_{fd} more points for forecasting - we conduct auto-regression on the $\mathbf{u}$ vector and forecast/extend it. We learn the AR parameters for P_{ar}^{th} order auto-regression $AR(P_{ar})$ using least squares on $\mathbf{u}$. Then we use AR parameters for the extension of the $\mathbf{u}$ vector to get an extended vector $\hat{\mathbf{u}}$: $\hat{\mathbf{u}}_r(N_d + f) = c_0 + \sum_{p=1}^{P_{ar}} c_p \hat{\mathbf{u}}_r(N_d + f - p)$

From the extended components $\hat{\mathbf{u}}$, we can reconstruct an *extended* tensor, $\mathcal{B}_b = \sum_{r=1}^{R} \lambda_r \times \hat{\mathbf{u}}_r \times \mathbf{v}_r \times \mathbf{w}_r$ with next N_{fd} days worth of data. A detailed steps are illustrated in Fig. 2 "*Bextension*" part.

Step 4: B2Seq (Recover the data from BIG domain): Now that we have a extended tensor $\mathcal{B}_b$ on BIG domain, we need to recover the N_{fd} days of forecast data back into the original data domain. A detailed description of steps is illustrated in Fig. 2 "*B2Seq*" part. We matricize the tensor into four matrices and reshape each matrix into a vector in row-wise manner. We apply the BIG Eqs. (1), and (2) to the assumed voltage data and forecast corresponding $I_r(t), I_i(t)$ for the next $t = N + 1, \cdots, N + N_f$ steps.

3.2   PowerCast-S

POWERCAST-S is a variation of POWERCAST. In function **Bextension**, we can use any type of forecasting function in exchange of *AR*.

We tried applying seasonal periodicity on the *long-term-concepts* ($\mathbf{u}$ vector), but on our dataset (which does not span over long enough period of time to capture the seasonal pattern) POWERCAST-S did not work well. Thus we recommend users to use plain POWERCAST unless: (1) you have a long enough history of the time sequences to capture the weekly pattern, and (2) you strongly believe that there is a seasonality in the data (weekly, monthly, seasonal, etc.).

4   Experiments

In this section we conduct experiments on the real world data to answer the following research questions: **Q1: Interpretability**, **Q2: Forecasting**, and **Q3: Anomaly detection**.

Dataset description: We next give a brief explanation of the datasets we used for experiments. All datasets are 5-tuples of the form (V_r, V_i, I_r, I_i, t).

- **CMU data:** The voltage and current measurements were recorded for the Carnegie Mellon University (CMU) campus for 23 days, from July 29, 2016 to August 20, 2016. The time sequences were sampled every hour ($N_{dt} = 24$).
- **LBNL data:** This is from the Lawrence Berkeley National Laboratory (LBNL) Open μPMU project[1] [34]. It spans 3 months, with sampling rate of 120 Hz; but we used only the interval (October 1, 2015 to October 8, 2015) and we down-sampled it to hourly samples ($N_{dt} = 24$). Moreover, we de-noised it using moving averages.

Experimental setup: For the parameters, we used $P_{ar} = 1$ for $AR(P_{ar})$ for extension of the **u** vector, $R = 2$ for the tensor decomposition, and $N_w = 5$, $\sigma = 0.5$ for the moving Gaussian filter. We used tensor tool box from [2,3].

Baseline methods: As baseline methods, we conducted experiments using autoregression (AR) and seasonal auto-regression (SAR) for comparison with our POWERCAST. For the baseline experiments, we assume that we do not have the domain knowledge on the power systems such as BIG model, etc. Thus we run AR and SAR methods on the given time sequences of I_r, I_i directly. The autoregression order for AR and SAR are determined via AIC criterion.

4.1   Q1: Interpretability

In this section we analyze the hidden variables in CMU data, and interpret the result using our domain knowledge.

As a reminder, G can be interpreted as a component of electrical load that contributes to real power consumption (for e.g. *light bulbs*), while B can be interpreted as a component of electric load that contributes to reactive power (for e.g. lagging reactive power ($+Q$) is absorbed by the *motors* where leading reactive power ($-Q$) is supplied by the capacitor) In our analysis below, we include images of *"light bulbs"* and *"motors"* to represent one example of the sources of G and B for more intuitive understanding.

Observations 1 (Weekly pattern in Fig. 3(a))

- The *long-term-concepts* (**u** vector) shows weekly periodicity accounting for the activities across the days (drops during weekends)

---

[1] http://powerdata.lbl.gov/.

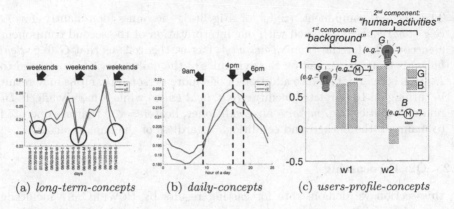

(a) *long-term-concepts*   (b) *daily-concepts*   (c) *users-profile-concepts*

Fig. 3. (a) Weekly periodicity in *long-term-concepts* (**u**), (b) Daily pattern in *daily-concepts* (**v**), (c) *users-profile-concepts* (**w**), on CMU data

– Both of the two components capture similar weekly pattern.

Observations 2 (Daily pattern in Fig. 3(b))

– The *daily-concepts* (**v** vector) captures daily activity patterns throughout the day, lowest during the night (midnight to 9am), and grows from 9am, when peaking start coming to campus, peaking at 4pm.
– The first component, $\mathbf{v}_1$, shows smoother varying pattern throughout the day. We can think of this component as the representation of the *"background"* activities, as it is insensitive to the dynamics of the human activities throughout the day.
– The second component, $\mathbf{v}_2$, shows more dynamic changes with (1) deeper drop during the night, (2) faster growth in early stage (logarithmic), and (3) rapid decay after the peak. This pattern closely follows the dynamics of the expected human activities throughout the day. We can think of this component as the representation of the *"human-activity"* factor.

Observations 3 (G and B factor analysis in Fig. 3(c))

– The *users-profile-concepts* (**w** vector) explains how much G and B account for each of the component.
– The first component - hidden variable (left of Fig. 3(a)) accounts for both G (e.g.*"light bulbs"*) and B (e.g. *"motors"*) in a close to 50-50 ratio. Combined with our interpretation in the *daily-concepts* for plot (b), we can interpret the first component as the *background* component that explains the background activities of the G and B (e.g. *light bulbs* and *motors*) factors that are operated independently of the dynamics of the major human factors. (basic facilities such as air conditioning units, *light bulbs*, etc., that runs throughout the day to maintain the minimum required conditions in the buildings)

- The second component (right of Fig. 3(a)) accounts dominantly for G (e.g. "*light bulbs*"). Aligned with our interpretation of the second component as accounting for the "*human-activity*" factor, this tells us that G (e.g. *light bulbs*) is more dependent on the dynamics of the daily activities compared to B. This is reasonable since addition of one more person on campus may result in turning on 10 G-related facilities (e.g. *light bulbs*) while it merely affects B-related facilities (e.g. *motors*- air-conditioner, heaters, etc.) that are operated to maintain the background conditions regardless of the human factors.

4.2   Q2: Forecasting

In this section we demonstrate forecasting results by POWERCAST including various *what-if* scenarios.

Multi-step forecasting. We start by showing multi-step ahead forecasting by POWERCAST in comparison with other competitors on two different real world datasets.

- **CMU data**: Comparisons on the forecasting by POWERCAST and competitors are shown for $N_f = 24$ steps ($N_{fd} = 1$-day) (Fig. 4).
  In (a), we observe that POWERCAST and POWERCAST-S are able to demonstrate the daily periodic pattern while competitors fails to correctly demonstrate daily periodicity. In (b), a quantitative comparison is given by root mean square error (RMSE) sum of I_r, I_i forecast results. The RMSE is computed as follows: $RMSE(x, \hat{x}) = \sqrt{\frac{\sum_{t=1}^{N_f}(x(t)-\hat{x}(t))^2}{\sum_{t=1}^{N_f}(x(t))^2}}$
- **LBNL data**:
  Figure 1 shows a similar quantitative comparison on the LBNL dataset, where POWERCAST also outperforms its competitors.

What-if scenario. If we know that the number of people on campus will increase by 10% tomorrow due to a big event, how much more power do we expect? Naive forecasting methods cannot tell us the future demand under various scenarios.

In this section, we illustrate how POWERCAST can handle forecasting under various *what-if* scenarios by adjusting G, B accordingly. We created three scenarios, assuming that we will have $\{10, 20, 30\}\%$ more activities on campus for next 2 days (thus increased values for G and B); and performed forecasting of the power demand under our scenarios. In Fig. 5, 48 step (2-day) ahead forecasting on CMU data under our scenarios are shown for (a) real current demand and (b) imaginary current demand. We see that I_r and I_i increase, according to the "BIG" model (see (a) and (b) respectively), allowing us to plan ahead. (Red, blue, green, and cyan, correspond to $\{0, 10, 20, 30\}\%$ increases).

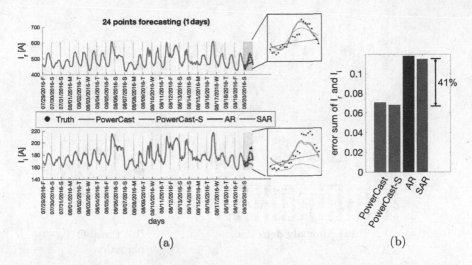

(a)   (b)

Fig. 4. PowerCast forecasts 24 steps (1-day) accurately on CMU data (a) POWERCAST (*red*) and POWERCAST-S (*pink*) forecasts 24 steps (I_r on the top row, and I_i on the bottom row) more accurately compared to the competitors (AR: *blue* and SAR: *green*). (b) RMSE comparison of the methods. POWERCAST achieves **41%** error reduction. (Color figure online)

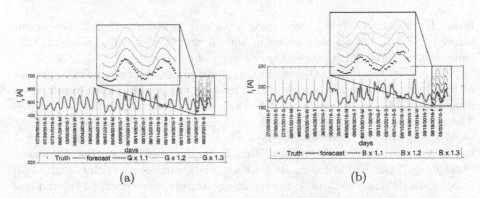

(a)   (b)

Fig. 5. PowerCast can handle forecasting under various what-if scenarios on demands for (a) I_r and (b) I_i on CMU data (Color figure online)

4.3   Q3: Anomaly Detection

In Fig. 6(a) $N_f = 144$ step forecasting ($N_{fd} = 6$ days) on CMU data is shown in solid red along with the actual reading in black circles. We report the highest deviation from our POWERCAST forecast, with a red vertical line. This happened on Tuesday, August 16, which was in the middle of the 3-day new graduate student orientation, probably ≈1,000 people.

The natural follow up question is: Was the extra population the reason for this anomaly? The answer seems 'no', for two reasons: extra people would mainly

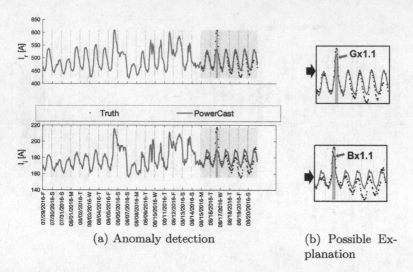

(a) Anomaly detection

(b) Possible Explanation

Fig. 6. PowerCast spots anomalies: (a) August 16, 2017 (spotted, and marked with red stripe) was during the new graduate student orientation. (b) PowerCast shows that a 10% increase in activities ('B' and 'G'), could explain this spike. (Color figure online)

boost I_r, that is 'G' (e.g. "*light bulbs*"), as our "*human-activity*" component showed in Fig. 3, but we have spikes in both I_r and I_i; the second reason is that there was no spike during the other two days of orientation.

The only explanation is that something boosted the "*background*" component of Fig. 3 (e.g., "*motors*", like elevators, heaters, air-conditioners), resulting in roughly equal boost to both I_r and I_i. Does reality corroborate the conjecture? The answer is '*yes*': the temperature on Aug. 16 spiked, to a scorching 88 °F ($\approx$30.5 °C), while the surrounding days was closer to 80 °F. This is exactly what we show in (b): PowerCast shows that for a 10% increase in the component "*background*" (air-conditioners, etc.), we get a 10% increase in both G (e.g. "*light bulbs*") and B (e.g. "*motors*"); the resulting answers correspond to the red line in (b), which is very close to the ground-truth black circles.

In short, PowerCast can not only spot anomalies, but also give hints about their cause.

4.4 Scalability

In Fig. 7, the wall clock time for PowerCast to run on a dataset of different number of timeticks (N) is plotted in black solid dots, along with a linear line in blue. The time was measured for 24 step (1-day) forecast given past time sequences of length $N = 240$–528 timeticks ($N_d = 10$–22 days) on CMU data.

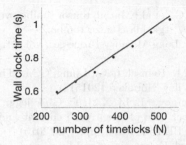

Fig. 7. PowerCast scales linearly with respect to the number of timeticks (N) (Color figure online)

5   Conclusions

We proposed POWERCAST, a novel domain-aware method to mine, understand, and forecast the power demand of an enterprise (university, factory, neighborhood). Our contributions are as follows:

1. **Infusion of domain knowledge**: We carefully picked a successful electric-power model, "BIG", which is derived from first-principles, and allows for *intuitive interpretation* of the power system of interest.
2. **Forecasting/What-if**: Our domain-aware approach, coupled with the BIG model provides (a) *accurate* prediction in multi-step ahead forecasting with up to 59% lower error, and (b) answers *what-if* scenarios, like "what will be our power demands when 80% of our students leave campus for March-break"
3. **Anomaly detection**: POWERCAST can spot anomalies and explain as we show in Sect. 4.3, Fig. 6.

Reproducibility: Our code is open-sourced at: www.cs.cmu.edu/~hyunahs/code/PowerCast.zip, and the LBNL dataset is at: powerdata.lbl.gov.

Acknowledgments. We would like to thank Dr. Tsubasa Takahashi, for discussions and feedback.

This material is based upon work supported by the National Science Foundation under Grant No. IIS-1247489, and by the Army Research Laboratory under Cooperative Agreement Number W911NF-09-2-0053. Any opinions, findings, and conclusions or recommendations expressed in this material are those of the author(s) and do not necessarily reflect the views of the National Science Foundation, or other funding parties. The U.S. Government is authorized to reproduce and distribute reprints for Government purposes notwithstanding any copyright notation here on.

References

1. Amin, S.M.: U.S. grid gets less reliable [The Data]. IEEE Spectr. **48**(1), 80 (2011)
2. Bader, B.W., Kolda, T.G.: Algorithm 862: MATLAB tensor classes for fast algorithm prototyping. ACM Trans. Math. Softw. **32**(4), 635–653 (2006)

3. Bader, B.W., Kolda, T.G., et al.: Matlab tensor toolbox version 2.6, February 2015. http://www.sandia.gov/~tgkolda/TensorToolbox/

4. Beran, J.: Statistics for Long-Memory Processes, vol. 61. CRC Press, Boca Raton (1994)

5. Box, G.E., Jenkins, G.M., Reinsel, G.C., Ljung, G.M.: Time Series Analysis: Forecasting and Control. Wiley, Hoboken (2015)

6. Bromberg, D.M., Jereminov, M., Li, X., Hug, G., Pileggi, L.: An equivalent circuit formulation of the power flow problem with current and voltage state variables. In: PowerTech, 2015 IEEE Eindhoven, pp. 1–6 (2015)

7. Brown, R.G.: Statistical Forecasting for Inventory Control. McGraw/Hill, New York City (1959)

8. Bunn, D., Farmer, E.D.: Comparative Models for Electrical Load Forecasting (1985)

9. De Livera, A.M., Hyndman, R.J., Snyder, R.D.: Forecasting time series with complex seasonal patterns using exponential smoothing. J. Am. Stat. Assoc. **106**(496), 1513–1527 (2011)

10. Ho, J.C., Ghosh, J., Sun, J.: Marble: high-throughput phenotyping from electronic health records via sparse nonnegative tensor factorization. In: ACM SIGKDD, pp. 115–124 (2014)

11. Hyndman, R.J., Fan, S.: Density forecasting for long-term peak electricity demand. IEEE Trans. Power Syst. **25**(2), 1142–1153 (2010)

12. Jain, A., Chang, E.Y., Wang, Y.F.: Adaptive stream resource management using kalman filters. In: ACM SIGMOD, pp. 11–22 (2004)

13. Jereminov, M., Bromberg, D.M., Li, X., Hug, G., Pileggi, L.: Improving robustness and modeling generality for power flow analysis. In: 2016 IEEE/PES Transmission and Distribution Conference and Exposition (T&D), pp. 1–5 (2016)

14. Jereminov, M., Pandey, A., Song, H.A., Hooi, B., Faloutsos, C., Pileggi, L.: Linear load model for robust power system analysis. In: IEEE PES Innovative Smart Grid Technologies (2017, p. (submitted))

15. Kalman, R.E.: A new approach to linear filtering and prediction problems. J. Basic Eng. **82**(1), 35–45 (1960)

16. Kang, U., Papalexakis, E., Harpale, A., Faloutsos, C.: Gigatensor: scaling tensor analysis up by 100 times-algorithms and discoveries. In: ACM SIGKDD, pp. 316–324 (2012)

17. Kolda, T.G., Bader, B.W.: Tensor decompositions and applications. SIAM Rev. **51**(3), 455–500 (2009)

18. Kolda, T.G., Sun, J.: Scalable tensor decompositions for multi-aspect data mining. In: ICDM, pp. 363–372 (2008)

19. Letchner, J., Re, C., Balazinska, M., Philipose, M.: Access methods for markovian streams. In: ICDE, pp. 246–257 (2009)

20. Lin, Y.R., Sun, J., Sundaram, H., Kelliher, A., Castro, P., Konuru, R.: Community discovery via metagraph factorization. ACM Trans. Knowl. Discov. Data (TKDD) **5**(3), 17 (2011)

21. Lu, H., Plataniotis, K.N., Venetsanopoulos, A.N.: Multilinear principal component analysis of tensor objects for recognition. In: ICPR, vol. 2, pp. 776–779 (2006)

22. Martí, J.R., Ahmadi, H., Bashualdo, L.: Linear power-flow formulation based on a voltage-dependent load model. IEEE Trans. Power Deliv. **28**(3), 1682–1690 (2013)

23. Matsubara, Y., Sakurai, Y.: Regime shifts in streams: Real-time forecasting of co-evolving time sequences. In: ACM SIGKDD, pp. 1045–1054 (2016)

24. Matsubara, Y., Sakurai, Y., Faloutsos, C., Iwata, T., Yoshikawa, M.: Fast mining and forecasting of complex time-stamped events. In: ACM SIGKDD, pp. 271–279 (2012)
25. Matsubara, Y., Sakurai, Y., Van Panhuis, W.G., Faloutsos, C.: Funnel: automatic mining of spatially coevolving epidemics. In: ACM SIGKDD, pp. 105–114 (2014)
26. Nowicka-Zagrajek, J., Weron, R.: Modeling electricity loads in california: ARMA models with hyperbolic noise. Sig. Process. **82**(12), 1903–1915 (2002)
27. Pandey, A., Jereminov, M., Hug, G., Pileggi, L.: Improving power flow robustness via circuit simulation methods. In: PES General Meeting (2017)
28. Pandey, A., Jereminov, M., Li, X., Hug, G., Pileggi, L.: Aggregated load and generation equivalent circuit models with semi-empirical data fitting. In: 2016 IEEE Green Energy and Systems Conference (IGSEC), pp. 1–6 (2016)
29. Papadimitriou, S., Brockwell, A., Faloutsos, C.: Adaptive, hands-off stream mining. In: VLDB, pp. 560–571 (2003)
30. Park, J., Park, Y., Lee, K.: Composite modeling for adaptive short-term load forecasting. IEEE Trans. Power Syst. **6**(2), 450–457 (1991)
31. Rogers, M., Li, L., Russell, S.J.: Multilinear dynamical systems for tensor time series. In: NIPS, pp. 2634–2642 (2013)
32. Smith, M.: Modeling and short-term forecasting of new south wales electricity system load. J. Bus. Econ. Stat. **18**(4), 465–478 (2000)
33. Soliman, S., Persaud, S., El-Nagar, K., El-Hawary, M.: Application of least absolute value parameter estimation based on linear programming to short-term load forecasting. Int. J. Electr. Power Energy Syst. **19**(3), 209–216 (1997)
34. Stewart, E.M., Liao, A., Roberts, C.: Open μPMU: a real world reference distribution micro-phasor measurement unit data set for research and application development, October 2016
35. Takahashi, T., Hooi, B., Faloutsos, C.: Autocyclone: automatic mining of cyclic online activities with robust tensor factorization. In: WWW, pp. 213–221 (2017)
36. Tao, D., Song, M., Li, X., Shen, J., Sun, J., Wu, X., Faloutsos, C., Maybank, S.J.: Bayesian tensor approach for 3-D face modeling. IEEE Trans. Circuits Syst. Video Technol. **18**(10), 1397–1410 (2008)
37. Wang, P., Wang, H., Wang, W.: Finding semantics in time series. In: ACM SIGMOD, pp. 385–396 (2011)
38. Wang, Y., Chen, R., Ghosh, J., Denny, J.C., Kho, A., Chen, Y., Malin, B.A., Sun, J.: Rubik: knowledge guided tensor factorization and completion for health data analytics. In: ACM SIGKDD, pp. 1265–1274 (2015)
39. Yang, H.T., Huang, C.M.: A new short-term load forecasting approach using self-organizing fuzzy ARMAX models. IEEE Trans. Power Syst. **13**(1), 217–225 (1998)

UAPD: Predicting Urban Anomalies from Spatial-Temporal Data

Xian Wu[1], Yuxiao Dong[1,2], Chao Huang[1], Jian Xu[1], Dong Wang[1],
and Nitesh V. Chawla[1(✉)]

[1] Department of Computer Science and Engineering, and iCeNSA,
University of Notre Dame, Notre Dame, USA
{xwu9,chuang7,jxu5,dwang5,nchawla}@nd.edu, yuxdong@microsoft.com
[2] Microsoft Research Redmond, Redmond, USA

Abstract. Urban city environments face the challenge of disturbances, which can create inconveniences for its citizens. These require timely detection and resolution, and more importantly timely preparedness on the part of city officials. We term these disturbances as anomalies, and pose the problem statement: if it is possible to also predict these anomalous events (proactive), and not just detect (reactive). While significant effort has been made in detecting anomalies in existing urban data, the prediction of future urban anomalies is much less well studied and understood. In this work, we formalize the future anomaly prediction problem in urban environments, such that those can be addressed in a more efficient and effective manner. We develop the Urban Anomaly PreDiction (*UAPD*) framework, which addresses a number of challenges, including the dynamic, spatial varieties of different categories of anomalies. Given the urban anomaly data to date, *UAPD* first detects the change point of each type of anomalies in the temporal dimension and then uses a tensor decomposition model to decouple the interrelations between the spatial and categorical dimensions. Finally, *UAPD* applies an autoregression method to predict which categories of anomalies will happen at each region in the future. We conduct extensive experiments in two urban environments, namely New York City and Pittsburgh. Experimental results demonstrate that *UAPD* outperforms alternative baselines across various settings, including different region and time-frame scales, as well as diverse categories of anomalies. Code related to this chapter is available at: https://bitbucket.org/xianwu9/uapd.

1 Introduction

Timely resolution of urban environment and infrastructure problems is an essential component of a well-functioning city. Cities have developed various reporting systems, such as 311 services [1], for the citizens to report, track, and comment on the urban anomalies that they encounter in their daily lives. These systems enable the city government institutions to respond on a timely basis to disturbances or anomalies such as noise, blocked driveway and urban infrastructure malfunctions [14]. However, these systems and resulting actions rely on accurate and timely reporting by the citizens, and are by nature reactive. That leads us to

© Springer International Publishing AG 2017
M. Ceci et al. (Eds.): ECML PKDD 2017, Part II, LNAI 10535, pp. 622–638, 2017.
https://doi.org/10.1007/978-3-319-71246-8_38

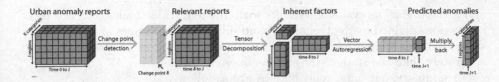

Fig. 1. The Urban Anomaly PreDiction (*UAPD*) Framework.

the question: what if we could *predict* which regions of a city will observe certain categories of anomalies in advance? We posit that this would enhance resource allocation and budget planning, and also result in more timely and efficient resolution of issues, thereby minimizing the disruption to the citizens' lives [19]. However, the development of such an urban anomaly prediction system faces several challenges:

(1) *Anomaly Dynamics*: The factors underlying urban anomalies may change over time. For example, anomalies in the winter may stem from winter related severe weather (e.g., snow), and it may be infeasible to train the predictive model by using historical data between spring and fall. (2) *Coupled Multi-Dimensional Correlations*: There may exist strong signals among the locations, time, and categories of occurred anomalies, that is, certain categories of anomalies occur at specific locations and/or specific time. Traditional time-series based models such as Autoregressive Moving-Average (ARMA) [28] and Gaussian Processing (GP) [8] largely rely on heuristic temporal features to make predictions. However, these approaches fail to capture the spatial-temporal factors that drive different categories of anomalies occur.

Contributions. To achieve a comprehensive solution that addresses these challenges and meets the goal of prediction, we develop a unified three-phase *Urban Anomaly PreDiction (UAPD)* framework to predict urban anomalies from spatial-temporal data—urban anomaly reports. As illustrated in Fig. 1, at the first phase, we propose a probabilistic model, whose parameters are inferred via Markov chain Monte Carlo, to detect the change point of the historical anomaly records of a city, such that only most relevant reports are used for the prediction of future anomalies. At the second phase, we model the anomaly data starting from the detected change point of all regions in the city with a three-dimensional tensor, where each of three dimensions stands for the regions, time slots, and categories of occurred anomalies, respectively. We then decompose the tensor to incorporate the underlying relationships between each dimension into their corresponding inherent factors in the tensor. Subsequently, the prediction of anomalies in next time slot is furthered to a time-series prediction problem. At the third phase, we leverage Vector Autoregression to capture the inter-dependencies of inherent factors among multiple time series, generating the prediction results.

We evaluate the performance of *Urban Anomaly PreDiction (UAPD)* by using two real-world datasets collected from the 311 service in New York City and Pittsburgh. The evaluation results with different region scales (i.e., precincts and regions divided by road segments) and different time-frames of training data show that our framework can predict different categories of anomalies more accurately than the state-of-the-art baselines.

In summary, our main contributions are as follows:

- We formalize the urban anomaly prediction problem from spatial-temporal data and develop a unified *UAPD* framework to predict future occurrences of different anomalies at different urban areas.
- In *UAPD*, we propose a change point detection solution to address the challenge of anomaly dynamics and leverage tensor decomposition to model the interrelations among multiple dimensions of urban anomaly data.
- We conduct extensive experiments with different scenarios in distinct urban environments to demonstrate the effectiveness of our presented framework.

2   Related Work

Anomaly Detection. Our work is related to urban sensing-based anomaly detection [5–7,18,23]. For example, Doan et al. presented a set of new clustering and anomaly detection techniques using a large-scale urban pedestrian dataset [6]. Chawla et al. used the Principal Component Analysis (PCA) scheme to discover events which may have caused anomalous behavior to appear in road traffic flow [5]. Pan et al. developed a novel system to detect anomalies according to drivers' routing behavior [23]. Zheng et al. proposed a probability-based anomaly detection method to discover collective anomalies from multiple spatial-temporal datasets [32]. Le et al. proposed an online algorithm to detect anomaly occurrence probability given the data from various sensors in highly dynamic contexts [18]. However, all the above solutions identified urban anomalies *after* they happen. In contrast, this paper develops a principled approach to *predict* urban anomalies in different regions of a city. Although there exists one recent work on anomaly prediction [15] by considering the dependency of anomaly occurrence between regions, two significant limitations exist: (i) it ignored the fact that anomalies of different categories can be correlated; (ii) it assumed that some portion of ground truth data from the predicted time slot are known. However, such assumptions do not always hold in the practical scenario for a couple of reasons. Firstly, dependencies between anomalies with different categories are ubiquitous in urban sensing. Secondly, it is difficult to know the ground truth information from the predicted time slot beforehand. To overcome those limitations, this paper develops a new urban anomaly prediction framework to explicitly explore the inter-category dependencies and model the time-evolving inherent factors with respect to different categories, without requiring any ground truth information from the predicted time slot.

Time Series Prediction. Our work is related to the literature that focus on time series prediction [2,8,25,28]. In particular, Vallis et al. [25] proposed a novel statistical technique to automatically detect long-term anomalies in cloud data.

A non-parametric approach Gaussian Processing (GP) has been developed to solve the general time series prediction problem [8]. Additionally, Wiesel et al. [28] proposed a time varying Autoregression Moving Average (ARMA) model for co-variance estimation. Bao et al. [2] proposed a Particle Swarm Optimization (PSO)-based strategy to determine the number of sub-models in a self-adaptive mode with varying prediction horizons. Inspired by the work above, we develop a new scheme to explicitly consider the anomaly distribution dependency between regions and incorporate it into the time series prediction model.

Bayesian Inference. Bayesian inference has been widely used in the data analytics communities [20, 21, 30]. For example, Yang et al. studied the problem of learning social knowledge graphs by developing a multi-modal Bayesian embedding model [30]. Lian et al. proposed a sparse Bayesian content-aware collaborative filtering approach for implicit feedback recommendation [21]. Lee et al. proposed a Bayesian nonparametric model for medical risks prediction by exploring phenotype topics from electronic health records [20]. In this paper, we study the problem of urban anomaly prediction by proposing a Bayesian inference model to capture the evolving relationships of anomaly sequences.

3   Problem Definition

Given the historical sequences of anomalies within a city's geographical regions, the objective is to predict whether certain categories of anomalies will happen at certain places in the future. We use a three-dimensional tensor $\mathcal{X} \in \mathbb{R}^{I \times J \times K}$ to represent the anomaly sequences of all regions in a city. I, J, K denote the number of regions (e.g., geographical areas divided by the city's road network), time slots, and anomaly categories (e.g., noise, blocked driveway, illegal parking, etc.) in the data, respectively. Each element $\mathcal{X}_{i,j,k}$ represents the number of the k-category anomalies that happened at region i at time j.

Urban Anomaly Prediction Problem: Given the historical anomaly data $\mathcal{X} \in \mathbb{R}^{I \times J \times K}$ of a city, the goal is to learn a predictive function $f(\mathcal{X}) \to \mathcal{X}_{:,J+1,:}$ to infer future occurrences of each category k of anomalies at each region i at time $J + 1$.

The urban anomaly prediction solution needs to incorporate additional factors as noted in Sect. 1. First, there may exist periodic and temporary patterns in urban anomalies. For example, the temporary road construction serves as the inherent factors for noise related anomalies. This pattern makes the model trained using data collected during the construction period ineffective for predictions over the following time-frame. Second, the urban anomaly data covers multiple aspects of coupled information—the location, time, and category of anomalies. It is unclear whether certain types of anomalies occur at correlated locations or time, and if so, how we can model the multidimensional data. The key notations are shown in Table 1.

Table 1. Symbols and definitions

Symbol	Interpretation
i, j, k	The indices of regions, time slots, anomaly categories
I, J, K	The number of regions, time slots, anomaly categories
M	The anomaly sequence matrix
M_j^k	The element in the k-th category, j-th time slot of the abnormal region matrix
$\mathcal{X}$	The anomaly tensor
R, T, C	The inherent factor matrix with regions, time slots, anomaly categories, respectively
L	The number of inherent factors (i.e., the rank of tensor)
δ	The change point
α, β	Parameters in Poisson distributions before/after change point
a, b	Shape/Scale parameters in Gamma distributions
p	Probability of each time slot being the change point

4   The *UAPD* Framework

Figure 1 illustrates the *UAPD* framework, with the following steps: (1) *UAPD* leverages Bayesian inference to detect the change point in the anomaly sequence; (2) *UAPD* incorporates the spatial-temporal anomaly data into a three-dimensional tensor, which can be decomposed to learn the latent correlations between all dimensions; (3) *UAPD* applies the Vector Autoregression model to predict the occurrence of different categories of anomalies in each region of a city. We will explain these three steps in detail in the following subsections.

4.1   Change Point Detection

Given an anomaly sequence, conventional prediction methods usually train a learning model by using all available data between time 0 and J to predict the future anomalies at time $J + 1$ [22]. However, the occurrence of urban anomalies can be largely influenced by temporary events and periodic patterns. For example, the road construction in a certain area can result in many noise reports during the construction period. As a result, the anomaly sequence of this area may be dramatically changed after the construction, i.e., a significant decrease in noise reports, leading to an ineffective prediction for post-construction anomalies by training models on the anomaly sequence that covers the construction period. To address this issue, we propose a Bayesian inference based method to detect the change point δ in an anomaly sequence between time 0 and J. The sequence follows two different data distributions before and after the change point. Thus, *UAPD* learns the anomaly predictive function, over the time sequences, by leveraging the most relevant latent factors that caused the anomalies.

Specifically, we aim to detect the change point of the anomaly sequence matrix M with regions that have k categories of anomalies. To do so, we first model M_j^k that reports the k-th category of anomalies at time j as a Poisson distribution. Before and after the change point δ, M_j^k follows Poisson distributions with different parameter configurations, α_k and β_k, respectively. Following the standard Bayesian inference, we set the conjugate prior of the Poisson distribution as the Gamma distribution with a and b as its shape and scale parameters, respectively. Finally, the change point δ is set to obey a multinomial distribution with parameters $p = (p_0, \cdots, p_j, \cdots, p_J)$, where p_j denotes the probability of time j to be the change point. The generative process behind our model can be summarized as follows:

$$\delta \sim Multinomial(\delta; p)$$
$$\alpha_k \sim Gamma(\alpha_k; a_k^\alpha, b_k^\alpha)$$
$$\beta_k \sim Gamma(\beta_k; a_k^\beta, b_k^\beta) \tag{1}$$
$$M_j^k \sim \begin{cases} Poisson(M_j^k; \alpha_k), \ 1 \leq j \leq \delta \\ Poisson(M_j^k; \beta_k), \ \delta + 1 \leq j \leq J \end{cases}$$

where M_j^k is the number of regions which have anomalies of k-th category at time j. The variable on the left side of the semi-colon is assigned a probability under the parameter on the right side. The posterior distribution is formally defined as follow:

$$\Pr(\alpha, \beta, \delta | M_{1:J}) \propto \Pr(M_{1:\delta} | \alpha)\Pr(M_{\delta+1:J} | \beta)\Pr(\alpha)\Pr(\beta)\Pr(\delta) \tag{2}$$
$$= \prod_{k=1}^{K} \left(\prod_{j=1}^{\delta} \Pr(M_j^k | \alpha_k) \prod_{j=\delta+1}^{J} \Pr(M_j^k | \beta_k)\Pr(\alpha_k)\Pr(\beta_k) \right) \Pr(\delta)$$

We could estimate the change point δ and distribution parameters α, β by maximizing the full joint distribution. However, to avoid fine parameter tuning and make accurate detection on change point, we apply the Markov Chain Monte Carlo (MCMC) method to infer parameter. In addition, since we have multiple random variables in this problem, we utilize the Gibbs sampling algorithm to execute the MCMC method. The basic idea in Gibbs sampling is that we sample the new value of each variable in each iteration accordingly by fixing other variables [17,24]. Algorithm 1 presents the outline of Gibbs sampling for change point detection. The conditional distribution of each variable and parameter derivations are given in the Appendix.

4.2   Tensor Decomposition

The inherent factors of anomalies among different regions, time slots and anomaly categories are revealed by tensor decomposition on a three dimensional tensor representing the anomaly records. Since the anomaly distributions of regions are often subject to changes in different time periods, the objective of tensor decomposition is to model the temporal effects of anomalies distribution in each region by learning the inherent factors as well as adopting these factors to different time periods.

Algorithm 1. Gibbs Sampling for Change Point

Initialize model parameter $\{ \alpha^{(1)}, \beta^{(1)}, \delta^{(1)} \}$;
for $itr = 1, ..., iterations$ **do**

    for $k = 1, ..., K$ **do**
       | Sample the $\alpha_k^{(itr+1)}$ according to $\Pr(\alpha_k|\delta, \beta^{itr}, M_{1:J})$ in Eq. (7);
    end

    for $k = 1, ..., K$ **do**
       | Sample the $\beta_k^{(itr+1)}$ according to $\Pr(\beta_k|\delta, \alpha^{(itr+1)}, M_{1:J})$ in Eq. (8);
    end

    for $\delta' = 1, ..., J$ **do**
       | Calculate parameter $p_{\delta'}^{(itr+1)}$ according to Eq. (9);
    end

    Sample the δ according to $\Pr(\delta|\alpha^{(itr+1)}, \beta^{(itr+1)}, M_{1:J})$;
end

We use the CANDECOMP/PARAFAC (CP) decomposition approach [3] to factorize the tensor into three different matrices $R \in \mathbb{R}^{I \times L}$, $T \in \mathbb{R}^{J \times L}$ and $C \in \mathbb{R}^{K \times L}$. Here L is the number of inherent factors and is indexed by l. R, T and C are the inherent factor matrices with respect to I regions, J time slots and K anomaly categories, respectively. We can express the three-way tensor factorization of $\mathcal{X}$ as:

$$\mathcal{X} \approx \sum_{l=1}^{L} R_{:,l} \circ T_{:,l} \circ C_{:,l} \tag{3}$$

where $R_{:,l}$, $T_{:,l}$ and $C_{:,l}$ represent the l-th column of R, T and C. $\circ$ denotes the vector outer product. Each entry $\mathcal{X}_{i,j,k}$ in tensor $\mathcal{X}$ can be computed as the inner-product of three L-dimensional vectors as follows:

$$\mathcal{X}_{i,j,k} \approx < R_i, T_j, C_k > \equiv \sum_{l=1}^{L} R_{i,l} T_{j,l} C_{k,l} \tag{4}$$

Although many techniques can be applied to CP decomposition, we utilize the *Alternative Least Square (ALS)* algorithm [4,12], which has been shown to outperform other algorithms in terms of solution quality [9].

4.3 Vector Autoregression

Using the tensor decomposition model discussed above, we can learn the inherent factors of anomalies from different time slots. Based on the three matrices generated by CP decomposition, we formulate the anomaly prediction problem as the time series prediction task on $T \in \mathbb{R}^{J \times L}$, since the inherent factors in other two dimensions remain constant. We use *Vector Autoregression (VAR)* [11] to capture the linear inter dependencies of inherent factors among multiple time series. In particular, VAR model formulates the evolution of a set of L inherent factors over the same sample period (i.e., time slot $j = 1, 2, ..., J$) as a linear function of their past values. We define the order S of VAR to represent the time series in the previous S time slots. Formally, it can be expressed as follows:

$$T_{J+1} = d + \sum_{s=1}^{S} B_s T_{J-s} + \epsilon_j \tag{5}$$

where d is a $L \times 1$ vector of constants, B_s is a time-invariant $L \times L$ matrix and ϵ_j is a $L \times 1$ vector of errors. Many techniques have been proposed for VAR order selection, such as Bayesian Information Criterion (BIC) [27], Final Prediction Error (FPE) [26] and Akaike information criteria (AIC) [10]. In this paper, we utilize the lowest AIC value to decide the order of VAR. Hence, the number of anomalies with the k-th category in region R_i in the next time slot can be derived as:

$$\mathcal{X}_{i,J+1,k} = < R_i, T_{J+1}, C_k > \equiv \sum_{l=1}^{L} R_{i,l} T_{J+1,l} C_{k,l} \tag{6}$$

5   Evaluation

In this section, we conduct experiments on two real-world datasets collected from 311 Service in New York City (NYC) and Pittsburgh, respectively. In particular, we answer the following research questions:

- **Q1:** How does *UAPD* perform as compared to the state-of-the-art solutions in predicting different categories of urban anomalies?
- **Q2:** How does *UAPD* perform in anomaly prediction with respect to different geographical region scales?
- **Q3:** How does *UAPD* perform in anomaly prediction with respect to different time-frames (i.e., #time-slots J)?
- **Q4:** Can *UAPD* effectively capture the change point of an anomaly sequence? Is change point detection effective for urban anomaly prediction?
- **Q5:** Is *UAPD* stable with regard to the rank of tensor L (i.e., the number of inherent factors)?

5.1   Experimental Setup

Datasets. We evaluated *UAPD* on real-world urban anomaly reports datasets collected from New York City (NYC) OpenData portal[1] and Pittsburgh Open-Data portal[2], respectively. Those datasets are collected from 311 Service that is an urban anomaly report platform, which allows citizens to report complaints about urban anomalies such as blocked driveway by texting, phone call or mobile

---

[1] https://data.cityofnewyork.us/.
[2] http://www.wprdc.org/.

Table 2. Data statistics

City	Anomaly category	Number of instances
New York city	Noise	151,174
	Blocked driveway	92,335
	Illegal parking	69,100
	Building/Use	27,724
Pittsburgh	Potholes	3,361
	Snow/Ice removal	2,504
	Building maintenance	2,352
	Weeds/Debris	2,023
	Refuse violations	1,196
	Abandoned vehicle	924
	Replace/Repair a sign	620

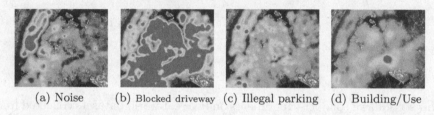

(a) Noise   (b) Blocked driveway   (c) Illegal parking   (d) Building/Use

Fig. 2. Geographical distribution of anomalies with different categories.

app. Each complaint record contains the timestamp, coordinates and category of anomaly. Different cities may have different anomaly categories due to different urban properties [29]. For the New York City datasets, we focus on 4 key anomaly categories (e.g., Blocked Driveway, Illegal Parking) which are selected in [15,31,32]. For the Pittsburgh datasets, we select the top frequently reported anomaly categories (e.g., Potholes, Refuse Violations) as the target predictive anomaly categories. The New York city data was collected from Jan 2014 to Dec 2014 and the Pittsburgh data was collected from Jan 2016 to Dec 2016.

The statistics of these datasets are summarized in Table 2. Due to space limit, we only show the geographical distribution of different categories of anomalies in NYC in Fig. 2. In those figures, firstly for the same anomaly category, we can observe that different regions have different number of anomalies. Secondly for the same region, we can observe that different anomaly categories have different distributions, which is the main motivation to use tensor decomposition model to identify the inherent factors by considering the correlations between different regions and different categories. Similar geographical distributions can be observed in Pittsburgh datasets.

Baselines and Evaluation Metrics. We compare our proposed scheme to the following state-of-the-art techniques:

- *CUAPS*: it predicts the anomaly by exploring the dependency of anomaly occurrence between different regions and use some portion of ground truth data from the predicted time slot [15].
- *Random Forest (RF)*: it incorporates the spatial-temporal features of anomaly sequence into the classification model.
- *Adaptive Boosting (AdaBoost)*: it predicts the anomaly occurrence of each region by maintaining a distribution of weights over the training examples.
- *Gaussian Processing (GP)*: it is a nonparametric approach that predicts the abnormal state of each region by exploring temporal features [8].
- *Autoregressive Moving-Average Model (ARMA)*: it is a time varying model that predicts the abnormal state of each region by exploring anomaly traces features [28].
- *Logistic Regression (LR)*: it is a regression model that estimates the abnormal state of each region by exploring temporal features [13].

5.2   Performance Validation

We use the following metrics to evaluate the estimation performance of the *UAPD* scheme: *F1-measure*, *Precision* and *Recall*.

Effectiveness of *UAPD* (Q1, Q2, Q3). In this subsection, we present the performance of *UAPD* on two real-world datasets with different geographical region scales and different time slots J. The source code of UAPD is publicly available[3]. In our evaluation, we consider two versions of our scheme: (i) *UAPD-CP*: a simplified version of *UAPD* framework which does not include the change point detection step; (ii) *UAPD*: the full version of the proposed scheme. We compare the *UAPD* scheme and its variant with the state-of-the-art algorithms.

Results on New York City Dataset. To investigate the effect of geographical scale, we partition the New York City into different geographical regions using the following methods:

- *High-Level Region*: New York City is divided into 76 geographical areas based on the political and administrative districts information[4]. We refer to each geographical region as a *high-level region*.
- *Fine-Grained Region*: we use major roads (road segments with levels from L_1 to L_5) to partition the entire city, which results in 862 geographical area [32]. We refer to each geographical area as a *fine-grained region*.

Based on the above region partition methods, each anomaly can be mapped into a particular region.

---

[3] https://bitbucket.org/xianwu9/uapd.
[4] https://data.cityofnewyork.us/Public-Safety/Police-Precincts/78dh-3ptz/data.

Table 3. Prediction results on June with high-level region in NYC

Anomaly	Noise			Blocked driveway			Illegal parking			Building/Use		
Algorithm	F1	Precision	Recall	F1	Precision	Recall	F1	Precision	Recall	F1	Precision	Recall
UAPD	0.966	0.944	0.989	0.878	0.838	0.923	0.867	0.776	0.980	0.679	0.602	0.779
UAPD-CP	0.952	0.940	0.964	0.837	0.809	0.867	0.867	0.776	0.982	0.628	0.549	0.733
CUAPS	0.930	0.941	0.920	0.824	0.826	0.825	0.808	0.794	0.822	0.576	0.620	0.539
RF	0.958	0.956	0.960	0.841	0.864	0.819	0.822	0.807	0.837	0.594	0.65	0.547
AdaBoost	0.950	0.947	0.953	0.841	0.827	0.856	0.812	0.783	0.843	0.584	0.572	0.596
GP	0.926	0.936	0.916	0.793	0.814	0.774	0.778	0.782	0.774	0.588	0.550	0.631
ARMA	0.926	0.962	0.892	0.818	0.946	0.720	0.688	0.866	0.571	0.408	0.804	0.274
LR	0.913	0.942	0.886	0.771	0.902	0.673	0.837	0.794	0.886	0.646	0.665	0.628

Table 4. Prediction results on Dec with high-level region in NYC

Anomaly	Noise			Blocked driveway			Illegal parking			Building/Use		
Algorithm	F1	Precision	Recall	F1	Precision	Recall	F1	Precision	Recall	F1	Precision	Recall
UAPD	0.892	0.838	0.954	0.894	0.880	0.908	0.873	0.779	0.992	0.724	0.639	0.836
UAPD-CP	0.893	0.834	0.962	0.894	0.891	0.898	0.873	0.779	0.992	0.733	0.655	0.833
CUAPS	0.881	0.826	0.945	0.869	0.904	0.837	0.805	0.762	0.853	0.613	0.642	0.587
RF	0.875	0.866	0.885	0.855	0.868	0.842	0.810	0.819	0.8	0.607	0.673	0.552
AdaBoost	0.880	0.849	0.914	0.848	0.853	0.844	0.833	0.828	0.839	0.623	0.625	0.621
GP	0.795	0.819	0.772	0.834	0.855	0.814	0.802	0.805	0.798	0.583	0.567	0.599
ARMA	0.823	0.897	0.761	0.817	0.959	0.712	0.708	0.861	0.601	0.454	0.813	0.315
LR	0.773	0.844	0.714	0.805	0.891	0.734	0.834	0.786	0.887	0.605	0.710	0.527

The evaluation results on NYC datasets with high-level region are shown in Tables 3 (June) and 4 (Dec), respectively. The evaluation results on NYC datasets with fine-grained region are shown in Tables 5 (June) and 6 (Dec), respectively. To investigate the effect of the number of time slots J, we study the performance of all compared schemes with different values of J. In Tables 3 and 4, we set J to be 150 (from January to May). In Tables 5 and 6, we set J to 330 (from January to November). The evaluation results are the average across the 10 consecutive days for prediction. From the results, we can observe that $UAPD$ consistently outperforms the state-of-the-art baselines in most cases on different categories of anomalies. For example, $UAPD$ outperforms the best baseline (AdaBoost) for high-level region-Building/Use case by 16.2% on F1-Score and the best baseline (RF) for fine-grained region-Blocked Driveway case by 34.7% on F1-Score. In the occasional case that $UAPD$ misses the best performance, it still generates very competitive results. The results are consistent with different geographical region scales and historical time slots.

Results on Pittsburgh Dataset. We repeated the same experiments on Pittsburgh dataset. Considering the space limit, we only present the evaluation results for high-level regions in June. In particular, we partition the Pittsburgh city based on its council districts information[5]. The evaluation results on different categories of anomalies are shown in Table 7. The reported results are also the

---

[5] http://pittsburghpa.gov/council/maps.

Table 5. Prediction results on June with fine-grained region in NYC

Anomaly	Noise			Blocked driveway			Illegal parking			Building/Use		
Algorithm	F1	Precision	Recall	F1	Precision	Recall	F1	Precision	Recall	F1	Precision	Recall
UAPD	0.630	0.534	0.768	0.555	0.491	0.639	0.421	0.404	0.441	0.262	0.268	0.257
UAPD-CP	0.635	0.593	0.684	0.511	0.550	0.478	0.373	0.468	0.310	0.122	0.367	0.073
CUAPS	0.486	0.599	0.409	0.415	0.478	0.366	0.220	0.187	0.268	0.170	0.118	0.309
RF	0.553	0.582	0.526	0.412	0.466	0.369	0.306	0.357	0.267	0.166	0.221	0.133
AdaBoost	0.528	0.522	0.535	0.397	0.370	0.429	0.301	0.264	0.350	0.183	0.155	0.222
GP	0.551	0.500	0.613	0.393	0.349	0.452	0.326	0.271	0.409	0.216	0.181	0.269
ARMA	0.404	0.787	0.272	0.199	0.631	0.118	0.104	0.517	0.058	0.028	0.412	0.014
LR	0.607	0.630	0.587	0.382	0.615	0.278	0.275	0.453	0.197	0.098	0.346	0.057

Table 6. Prediction results on Dec with fine-grained region in NYC

Anomaly	Noise			Blocked driveway			Illegal parking			Building/Use		
Algorithm	F1	Precision	Recall	F1	Precision	Recall	F1	Precision	Recall	F1	Precision	Recall
UAPD	0.521	0.466	0.589	0.558	0.570	0.546	0.403	0.455	0.362	0.171	0.406	0.109
UAPD-CP	0.373	0.366	0.380	0.530	0.537	0.522	0.373	0.389	0.359	0.210	0.247	0.183
CUAPS	0.423	0.437	0.410	0.452	0.508	0.407	0.242	0.270	0.218	0.111	0.096	0.133
RF	0.438	0.545	0.366	0.429	0.522	0.364	0.293	0.384	0.237	0.116	0.198	0.082
AdaBoost	0.380	0.338	0.433	0.393	0.367	0.422	0.290	0.257	0.332	0.160	0.124	0.227
GP	0.415	0.346	0.517	0.453	0.415	0.499	0.355	0.297	0.441	0.219	0.188	0.262
ARMA	0.358	0.715	0.239	0.253	0.663	0.156	0.151	0.641	0.086	0.023	0.286	0.012
LR	0.520	0.641	0.438	0.504	0.598	0.435	0.338	0.516	0.251	0.077	0.286	0.044

Table 7. Prediction results on June with high-level regions in pittsburgh

Anomaly	Potholes			Weeds/Debris			Building maintenance		
Algorithm	F1	Precision	Recall	F1	Precision	Recall	F1	Precision	Recall
UAPD	0.848	0.736	1.000	0.909	0.844	0.984	0.624	0.531	0.756
UAPD-CP	0.839	0.769	0.923	0.887	0.851	0.926	0.698	0.577	0.882
CUAPS	0.772	0.736	0.811	0.724	0.730	0.718	0.523	0.466	0.595
RF	0.793	0.842	0.750	0.794	0.862	0.735	0.585	0.649	0.533
AdaBoost	0.760	0.754	0.766	0.596	0.739	0.500	0.558	0.585	0.533
GP	0.733	0.716	0.750	0.812	0.800	0.824	0.615	0.542	0.711
ARMA	0.560	0.778	0.437	0.631	0.813	0.514	0.451	0.615	0.616
LR	0.778	0.790	0.766	0.821	0.833	0.808	0.494	0.525	0.467
Anomaly	Refuse violations			Abandoned vehicle			Replace/Repair a sign		
Algorithm	F1	Precision	Recall	F1	Precision	Recall	F1	Precision	Recall
UAPD	0.615	0.522	0.750	0.642	0.515	0.854	0.622	0.509	0.800
UAPD-CP	0.535	0.451	0.657	0.628	0.551	0.730	0.487	0.404	0.613
CUAPS	0.076	0.500	0.041	0.394	0.441	0.357	0.133	0.666	0.074
RF	0.490	0.522	0.462	0.560	0.618	0.512	0.364	0.526	0.278
AdaBoost	0.490	0.522	0.462	0.624	0.558	0.707	0.382	0.406	0.361
GP	0.393	0.342	0.461	0.494	0.478	0.512	0.459	0.447	0.472
ARMA	0.133	0.500	0.077	0.280	0.533	0.195	0.052	0.500	0.027
LR	0.324	0.545	0.231	0.500	0.543	0.463	0.217	0.500	0.139

average across the 10 consecutive days for prediction. In Table 7, we can observe that our scheme *UAPD* still outperforms other baselines in most of the evaluation metrics. Additionally, we can observe that *UAPD* outperforms *UAPD-CP* (without detecting any change point on anomaly sequence) in most cases. Since there do not exist reports of Snow/Ice removal category of anomaly on June, we do not consider this category in this evaluation.

The performance improvements of *UAPD* are achieved by (i) carefully considering dynamic causes of anomalies; (ii) explicitly exploiting inherent factors of different categories of anomalies; (iii) carefully handling the evolving relationship among different anomaly categories and regions, with Bayesian inference and tensor decomposition, which are missing from the state-of-the-art solutions.

Analysis of Change Point Detection (Q4). Additionally, we conducted experiments to analyze the results of change point detection (i.e., the first component of the proposed framework) and further visualize the result. The detection results on Pittsburgh datasets are shown in Fig. 3. We kept the value of J the same as the above experiments. In Fig. 3, we can observe that the detected starting point of anomaly sequences is Feb 18, 2016 instead of the actual beginning time (i.e., Jan 1, 2016). Moreover, the supplementary figures in both left and right side show that the anomaly distributions before the detected change point significantly differs from that after the change point, which demonstrates the effectiveness of our change point detection on anomaly sequences.

Impact of Rank Parameter (Q5). To investigate the effect of the only parameter (i.e., rank parameter L) in our framework, we studied the performance of our proposed scheme by varying the value of L. Particularly, we vary the value of *rank* from 2 to 9. The evaluation results on NYC datasets are shown in Figs. 4 and 5. We can observe that the performance of *UAPD* is stable with the value

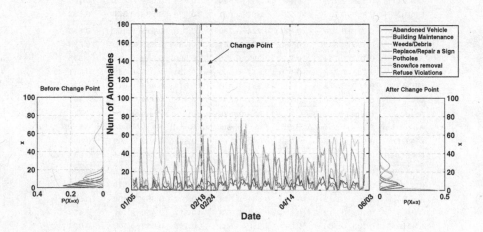

Fig. 3. Analysis of change point detection.

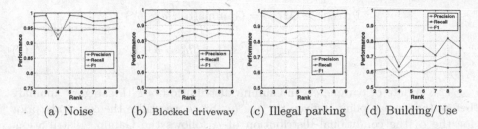

(a) Noise   (b) Blocked driveway   (c) Illegal parking   (d) Building/Use

Fig. 4. Performance w.r.t rank parameter L on Jun. with high-level regions in NYC

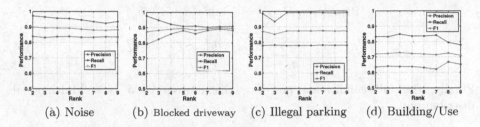

(a) Noise   (b) Blocked driveway   (c) Illegal parking   (d) Building/Use

Fig. 5. Performance w.r.t rank parameter L on Dec. with high-level regions in NYC

of *rank* from 5 to 8 which satisfy the value selection rubric of rank parameter L in tensor decomposition [16]. In our experiments, we set the value of *rank* to 8. Considering the space limit, we only present the results on NYC datasets. In the results on Pittsburgh datasets, similar results can be observed.

6   Conclusion

In this paper, we develop a Urban Anomaly PreDiction(*UAPD*) framework to predict urban anomalies from spatial-temporal data. *UAPD* enables the accurate prediction of the occurrences of different anomalies at each region of a city in the future. *UAPD* explicitly detects the change point of the anomaly sequences and also explores the time-evolving inherent factors and their relationships with each dimension tensor (i.e., regions and anomaly categories). We evaluate our presented framework on two sets of urban anomaly reports collected from 311 Service in New York City and Pittsburgh, respectively. The results show that *UAPD* significantly outperforms state-of-the-art baselines, and provides a framework for being predictive about disturbances, enabling the city officials to be more proactive in their preparation and response.

7   Appendix

In this section, we give the specific form of conditional posterior functions used in Algorithm 1. According to the posterior function in Eq. (3), we can obtain the full conditional posterior of α given β, δ as follows:

$$\Pr(\alpha_k|\delta, \boldsymbol{\beta}, M_{1:J}) = \text{Gamma}(\alpha_k; {}^*a_k^\alpha, {}^*b_k^\alpha)$$

$$^*a_k^\alpha = a_k^\alpha + \sum_{j=1}^{\delta} M_j^k; \quad {}^*b_k^\alpha = \delta + b_k^\alpha \tag{7}$$

The update of parameters in Gamma distribution of k-th category anomaly before the change point is given by ${}^*a_k, {}^*b_k$. Since we utilize the conjugate prior for the α, the conditional distribution of α follows the Gamma distribution. Similarly, the conditional posterior of β given α, δ has the same form as:

$$\Pr(\beta_k|\delta, \boldsymbol{\alpha}, M_{1:J}) = \text{Gamma}(\beta_k; {}^*a_k^\beta, {}^*b_k^\beta)$$

$$^*a_k^\beta = a_k^\beta + \sum_{j=\delta+1}^{J} M_j^k; \quad {}^*b_k^\beta = J - \delta + b_k^\beta \tag{8}$$

Finally, the full conditional posterior distribution of δ is a Multinomial distribution:

$$\Pr(\delta|\boldsymbol{\alpha}, \boldsymbol{\beta}, M_{1:J}) = \text{Multinomial}(\delta; \boldsymbol{p})$$

$$p_{\delta'} = \exp\Big(\sum_{k=1}^{K} \big(\sum_{j=1}^{\delta'} M_j^k \log\alpha_k + \sum_{j=\delta'}^{J} M_j^k \log\beta_k - \delta'\alpha_k - (J - \delta')\beta_k \big) + \log\sigma \Big) \tag{9}$$

where σ is the normalization constant of $p_{\delta'}$. $\boldsymbol{p} = (p_1, ..., p_{\delta'}, ..., p_J)$.

Acknowledgments. Research was in part sponsored by the Army Research Laboratory and was accomplished under Cooperative Agreement Number W911NF-09-2-0053 (Network Science CTA). The views and conclusions contained in this document are those of the authors and should not be interpreted as representing the official policies, either expressed or implied, of the Army Research Laboratory or the U.S. Government. The U.S. Government is authorized to reproduce and distribute reprints for Government purposes notwithstanding any copyright notation here on.

References

1. 3-1-1. https://en.wikipedia.org/wiki/3-1-1. Accessed July 2017
2. Bao, Y., Xiong, T., Hu, Z.: PSO-MISMO modeling strategy for multistep-ahead time series prediction. IEEE Trans. Cybern. **44**, 655–668 (2014)
3. Carroll, J.D., Chang, J.J.: Analysis of individual differences in multidimensional scaling via an N-way generalization of "Eckart-Young" decomposition. Psychometrika **35**(3), 283–319 (1970)
4. Carroll, J.D., Pruzansky, S., Kruskal, J.B.: Candelinc: a general approach to multidimensional analysis of many-way arrays with linear constraints on parameters. Psychometrika **45**(1), 3–24 (1980)
5. Chawla, S., Zheng, Y., Hu, J.: Inferring the root cause in road traffic anomalies. In: ICDM, pp. 141–150. IEEE (2012)

6. Doan, M.T., Rajasegarar, S., Salehi, M., Moshtaghi, M., Leckie, C.: Profiling pedestrian distribution and anomaly detection in a dynamic environment. In: CIKM, pp. 1827–1830. ACM (2015)
7. Dong, Y., Pinelli, F., Gkoufas, Y., Nabi, Z., Calabrese, F., Chawla, N.V.: Inferring unusual crowd events from mobile phone call detail records. In: Appice, A., Rodrigues, P.P., Santos Costa, V., Gama, J., Jorge, A., Soares, C. (eds.) ECML PKDD 2015. LNCS (LNAI), vol. 9285, pp. 474–492. Springer, Cham (2015). https://doi.org/10.1007/978-3-319-23525-7_29
8. Esling, P., Agon, C.: Time-series data mining. ACM Comput. Surv. (CSUR) 45, 12 (2012)
9. Faber, N.K.M., Bro, R., Hopke, P.K.: Recent developments in CANDE-COMP/PARAFAC algorithms: a critical review. Chemom. Intell. Lab. Syst. 65(1), 119–137 (2003)
10. Gelman, A., Hwang, J., Vehtari, A.: Understanding predictive information criteria for bayesian models. Stat. Comput. 24, 997–1016 (2014)
11. Hamilton, J.D.: Time Series Analysis, vol. 2. Princeton University Press, Princeton (1994)
12. Harshman, R.A.: Foundations of the parafac procedure: models and conditions for an "explanatory" multi-modal factor analysis (1970)
13. Hosmer Jr., D.W., Lemeshow, S., Sturdivant, R.X.: Applied Logistic Regression, vol. 398. Wiley, Hoboken (2013)
14. Hsieh, H.P., Lin, S.D., Zheng, Y.: Inferring air quality for station location recommendation based on urban big data. In: KDD, pp. 437–446. ACM (2015)
15. Huang, C., Wu, X., Wang, D.: Crowdsourcing-based urban anomaly prediction system for smart cities. In: CIKM. ACM (2016)
16. Kolda, T.G., Bader, B.W.: Tensor decompositions and applications. SIAM Rev. 51(3), 455–500 (2009)
17. Koller, D., Friedman, N.: Probabilistic Graphical Models: Principles and Techniques. MIT press, Cambridge (2009)
18. Le, V.D., Scholten, H., Havinga, P.: Flead: online frequency likelihood estimation anomaly detection for mobile sensing. In: Ubicomp, pp. 1159–1166. ACM (2013)
19. Lee, J.Y., Kang, U., Koutra, D., Faloutsos, C.: Fast anomaly detection despite the duplicates. In: WWW, pp. 195–196. ACM (2013)
20. Lee, W., Lee, Y., Kim, H., Moon, I.C.: Bayesian nonparametric collaborative topic poisson factorization for electronic health records-based phenotyping. In: IJCAI (2016)
21. Lian, D., Ge, Y., Yuan, N.J., Xie, X., Xiong, H.: Sparse Bayesian content-aware collaborative filtering for implicit feedback. In: IJCAI (2016)
22. Lv, Y., Duan, Y., Kang, W., Li, Z., Wang, F.Y.: Traffic flow prediction with big data: a deep learning approach. IEEE Trans. Intell. Transp. Syst. 16(2), 865–873 (2015)
23. Pan, B., Zheng, Y., Wilkie, D., Shahabi, C.: Crowd sensing of traffic anomalies based on human mobility and social media. In: SIGSPATIAL, pp. 344–353. ACM (2013)
24. Resnik, P., Hardisty, E.: Gibbs sampling for the uninitiated. Technical report, DTIC Document (2010)
25. Vallis, O., Hochenbaum, J., Kejariwal, A.: A novel technique for long-term anomaly detection in the cloud. In: Proceedings of the 6th USENIX Workshop on Hot Topics in Cloud Computing (HotCloud 2014) (2014)
26. Wang, Y.: On efficiency properties of an R-square coefficient based on final prediction error. Stat. Probab. Lett. 83(10), 2276–2281 (2013)

27. Watanabe, S.: A widely applicable bayesian information criterion. J. Mach. Learn. Res. **14**(Mar), 867–897 (2013)
28. Wiesel, A., Bibi, O., Globerson, A.: Time varying autoregressive moving average models for covariance estimation. IEEE Trans. Signal Process. **61**, 2791–2801 (2013)
29. Wikipedia: https://en.wikipedia.org/wiki/3-1-1#availability (2017)
30. Yang, Z., Tang, J., Cohen, W.: Multi-modal Bayesian embeddings for learning social knowledge graphs. In: IJCAI (2016)
31. Zheng, Y., Liu, T., Wang, Y., Zhu, Y., Liu, Y., Chang, E.: Diagnosing New York city's noises with ubiquitous data. In: Ubicomp, pp. 715–725. ACM (2014)
32. Zheng, Y., Zhang, H., Yu, Y.: Detecting collective anomalies from multiple spatio-temporal datasets across different domains. In: SIGSPATIAL. ACM (2015)

Transfer and Multi-task Learning

LKT-FM: A Novel Rating Pattern Transfer Model for Improving Non-overlapping Cross-Domain Collaborative Filtering

Yizhou Zang[✉] and Xiaohua Hu

Drexel University, Philadelphia, PA 19104, USA
yizhouzang@gmail.com

Abstract. Cross-Domain Collaborative Filtering (CDCF) has attracted various research works in recent years. However, an important problem setting, i.e., "users and items in source and target domains are totally different", has not received much attention yet. We coin this problem as Non-Overlapping Cross-Domain Collaborative Filtering (NOCDCF). In order to solve this challenging CDCF task, we propose a novel 3-step rating pattern transfer model, i.e. low-rank knowledge transfer via factorization machines (LKT-FM). Our solution is able to mine high quality knowledge from large and sparse source matrices, and to integrate the knowledge without losing much information contained in the target matrix via exploiting Factorization Machine (FM). Extensive experiments on real world datasets show that the proposed LKT-FM model outperforms the state-of-the-art CDCF solutions.

1 Introduction

Cross-Domain Collaborative Filtering (CDCF) is an emerging research topic in recommender systems. It aims to improve recommendations in an individual domain by drawing upon the knowledge acquired from related domains. Most CDCF models transfer knowledge based on explicit correspondence among entities of target and source domains [1–3]. However, few works have studied a more practical problem setting, in which "users and items in source and target domains are totally different". This setting is called Non-Overlapping CDCF (NOCDCF), which is the most challenging problem in cross-domain recommendation.

For the NOCDCF problem, the most well-known solutions may be the RPT-based (Rating Pattern Transfer) methods [5,9]. This set of methods shares across domains a group-level preference, which is referred to as a rating pattern. For example, suppose a newly opened book-selling website would like to build a recommender system. Due to the lack of visiting at the beginning, very few ratings are available for collaborative filtering. Fortunately, in the meanwhile, there is a popular movie review website shares its rating data to the public. Since movie domain is correlated to book domain in some aspects (First, they have correspondence in genres, e.g. comedy movies corresponds to humorous books. Second, the user sets of two domains may be sampled from the same population and reflect

© Springer International Publishing AG 2017
M. Ceci et al. (Eds.): ECML PKDD 2017, Part II, LNAI 10535, pp. 641–656, 2017.
https://doi.org/10.1007/978-3-319-71246-8_39

similar social aspects [5,17] even though they don't overlap), similar user-item rating pattern is deemed to exist in both domains. RPT methods extract such rating pattern by co-clustering the rows (users) and columns (movies) of the source matrix. The knowledge of source domain (movie domain) is then transferred to the target domain (book domain) via sharing the co-clusters.

However, existing NOCDCF approaches have their limitations:

1. All these methods are based on the assumption that the source rating matrix is dense. Unfortunately, as we know, canonical CF datasets such as Amazon, Netflix and MovieLens rating sets are mostly sparse. The sparseness may considerably degrade the performance of existing methods. The reason for this is that current methods need to impute the missing values in the source matrix in order to apply co-clustering algorithms, which is undefined when the matrix is incomplete [4]. For sparse matrices, the imputation may easily distort the data and further affect the clustering quality. As a result, a lot of noise will be introduced into the co-clusters and thus lower the recommendation performance. Current methods try to avoid the impact of sparseness by exploiting a dense but small portion of the original source matrix [5,6]. However, this does not solve the whole problem: using only a dense subset might lose the useful knowledge contained in the remaining large though sparse portion of the original matrix. The decrease of recommendation accuracy of existing NOCDCF methods with the increase of data sparseness is also observed in our empirical studies in Sect. 4. Therefore, it raises a demand to devise a novel model that can extract knowledge from sparse matrices.
2. Current NOCDCF methods do not integrate the shared knowledge very well. An earlier work [19] observes that existing NOCDCF methods cannot transfer knowledge under some conditions. We argue that this is caused by the knowledge integration method they use. Most of these methods utilize Direct-Expansion (DE) for knowledge integration [6,9]. However, the DE approach relies too heavily on the shared knowledge and may miss useful knowledge contained in the target matrix itself. As a result, the transferred knowledge would hurt the recommendation performance. Such negative transfer is also observed in our empirical studies in Sect. 4. We find that current NOCDCF methods perform even worse than some single domain recommendation methods while using DE. Therefore, it brings a new challenge of how to integrate the shared knowledge while avoiding negative transfer.

To overcome the above limitations, we propose a novel 3-step RPT model, i.e. low-rank knowledge transfer via factorization machines (LKT-FM). In our first step, we factorize the source matrix into two low-rank matrices (i.e. user-factor and item-factor matrices), and then generate clusters from them. In the second step, we follow the idea in [5] to map users/items of target domain to corresponding clusters. In the third step, we expand the design matrix of Factorization Machine (FM), which incorporates the shared knowledge into the target data in a seamless manner. We then conduct extensive empirical studies on several real world datasets and achieve considerably better results than the state-of-the-art methods. Note that we also compare our LKT-FM with ordinary

CDCF methods, which requires overlapping information as additional input, and achieve competitive results. Therefore, our model is a generic solution to CDCF problems, not limited to NOCDCF.

The contribution of this paper can be summarized as follows:

- We devise a novel rating pattern transfer solution LKT-FM for solving NOCDCF problem.
- We perform experiments on real world datasets and show that our LKT-FM method is considerably better than the state-of-the-art NOCDCF methods in terms of both knowledge extraction and knowledge integration, and is also competitive with ordinary CDCF methods even in an overlapping setting.

2 Background

2.1 Problem Definition

As mentioned in Sect. 1, in NOCDCF, there is no overlap between users and items across domains. More formally, assume we have a source domain D_S and a target domain D_T. Respectively for such domains, let U_S, U_T be their sets of users, I_S, I_T be their sets of items, and $\boldsymbol{X}_S$, $\boldsymbol{X}_T$ be their user-item rating matrices. Our goal is to predict the unobserved entries in $\boldsymbol{X}_T$ by taking advantage of the knowledge in $\boldsymbol{X}_S$ with the restriction that $U_S \cap U_T = \emptyset$ and $I_S \cap I_T = \emptyset$.

2.2 Non-Overlapping CDCF (NOCDCF)

The seminal paper [5] proposed one of the first NOCDCF methods that exploit rating pattern (a.k.a. codebook) to transfer knowledge. Its name is CodeBook Transfer (CBT). CBT is an adaptive knowledge transfer approach. It consists of 3 steps: (1) rating pattern construction, (2) cluster membership mapping and (3) rating pattern integration. In the first step, a rating pattern is constructed via applying co-clustering on the source matrix $\boldsymbol{X}_S$ so as to obtain a rating pattern $\boldsymbol{B}$. In the second step, each user/item is assigned to the cluster identified in the rating pattern $\boldsymbol{B}$. In the third step, the filled target matrix $\widehat{\boldsymbol{X}}_T$ is obtained by expanding the rating pattern $\boldsymbol{B}$.

In a later work, [9] extended the same idea and proposed a probabilistic approach that transfers knowledge in a collective manner. More recently, [6] believed that a rating pattern consists of two substructures, a domain-specific rating pattern and a common rating pattern. Each domain has its own domain-specific rating pattern, while all correlated domains share the common rating pattern. Furthermore, [10] learned the relatedness between different source domains and target domain, and then integrated appropriate amount of knowledge (i.e. rating pattern) from each source domain to the target domain. However, as mentioned in Sect. 1, all the existing NOCDCF methods depend on co-clustering the imputed matrix, thus do not work well in a sparse setting. In addition, all these methods do not leverage MF techniques for knowledge integration, which may lead to negative transfer due to the loss of knowledge contained in the target matrix.

2.3   Matrix Factorization (MF)

In this work, we apply MF as a pre-processing method to obtain a low-rank representation of users and items (i.e. U and V), in order for further clustering. While there are several variants of MF [11–13], we only review the basic MF in this paper.

Matrix factorization (MF) may be the most common and successful technique for single-domain recommendation tasks [7]. The basic idea of MF is to approximate the rating matrix using the product of two low-rank latent factor matrices:

$$X_{ij} \approx \widehat{X}_{ij} = S_{ij} + U_i V_j^T. \tag{1}$$

where $X \in \mathbb{R}^{m \times n}$ represents the rating matrix (m is number of users and n is number of items), S indicates the bias matrix [7], $U \in \mathbb{R}^{m \times d}$ is the user-factor latent matrix and $V \in \mathbb{R}^{n \times d}$ is the item-factor latent matrix. The system learns by minimizing the squared error function as follows, only considering the observed ratings:

$$\min \Sigma_{i=1}^m \Sigma_{j=1}^n I_{ij} (X_{ij} - \widehat{X}_{ij})^2 + \lambda_1 ||U||_F^2 + \lambda_2 ||V||_F^2. \tag{2}$$

where I_{ij} is the indicator function that equals 1 if user rated item, and equals 0 otherwise, λ_1 and λ_2 are constants controlling the extent of regularization, and $|| \cdot ||_F^2$ denotes Frobenius norm.

2.4   Factorization Machine (FM)

In this work, we apply FM [8] for knowledge integration. FM is a generic predictive model that allows to mimic most collaborative filtering (CF) models by feature engineering. More specifically, in a rating prediction problem, let S denote the set of tuples (x, y) where $x = (x_1, ..., x_k) \in \mathbb{R}^k$ is a k-dimensional feature vector and y is corresponding class label. FM models all possible interactions between variables in x using factorized interactions. The FM model considering pairwise interactions can be represented as follows:

$$\widehat{y}(x) = w_0 + \Sigma_{i=1}^k w_i x_i + \Sigma_{i=1}^k \Sigma_{i'=i+1}^k v_i v_{i'} x_i x_{i'} \tag{3}$$

where $w_0 \in \mathbb{R}$ is the global bias, $w_i \in \mathbb{R}$ are the biases of feature i, vector $v_i \in \mathbb{R}^{1 \times f}$ are interaction parameter vectors of feature i. In FM, the original interaction parameters $w_{ii'}$ are replaced by the product of v_i and $v_{i'}$. By doing this, the number of parameters decreases significantly and thus the interactions can be estimated even under high data sparsity. In practice, FM performs prominently for various CF tasks [8] and thus is a very strong baseline for single-domain recommendation evaluation. Although FM works remarkably for single-domain recommendation problems, how to use it for CDCF problems still remains open. To the best of our knowledge, there is only one previous work using FM for CDCF. However, that work [15] solved a much easier CDCF task in which the users in different domains totally overlap. Thus, it is different from our goal of solving the NOCDCF problem.

3   LKT-FM

Our proposed LKT-FM solution follows the 3-step framework of CBT model. We illustrate our model in Fig. 1. Specifically, in the first step, we aim to extract high quality rating pattern via low-rank clustering; in the second step, we assign each user/item in target domain to corresponding cluster; and in the third step, we propose to integrate the extracted rating pattern through feature expansion of factorization machines (FM). These three steps are described in detail in the following three subsections.

3.1   Low-Rank Rating Pattern Construction

As discussed in Sect. 1, constructing rating patterns through user-item co-clustering has potential issues when the source matrix is sparse. Thus, we propose a new construction method to alleviate the sparseness problem.

We first preprocess the source matrix $X_S \in \mathbb{R}^{m_S \times n_S}$ by applying basic MF [7]. As shown in the upper half part of Fig. 1, the user-item rating matrix is factorized into two low-rank matrices $U_S \in \mathbb{R}^{m_S \times d}$ and $V_S \in \mathbb{R}^{n_S \times d}$, $d \ll m_S, n_S$. U_S is user latent factor matrix and V_S is item latent factor matrix. Secondly, we apply K-means clustering on the row vectors of U_S and V_S respectively to generate user/item clusters. Note that, since U_S and V_S are both complete matrices (i.e. no missing values), no imputation is needed before clustering. Thus, it effectively avoids the impact of imputation, which may introduce much noise when source matrix is sparse. Fig. 1 shows the obtained user-cluster membership matrix $P_S \in \{0,1\}^{m_S \times p}$ and item-cluster membership matrix $Q_S \in \{0,1\}^{n_S \times q}$, where p, q are given cluster numbers. Then the clustering-level rating pattern B is constructed as follows:

$$B = [P_S^T X_S Q_S] \oslash [P_S^T \mathbf{1}\mathbf{1}^T Q_S] \tag{4}$$

where $\oslash$ denotes the entry-wise division. Equation 4 means averaging the ratings of each user-item co-cluster as an entry in B.

3.2   Cluster Membership Mapping

In the second step, source and target domains are bridged by mapping users/items in the source domain to the clusters identified in B. We adopt the mapping method proposed in [5]. We learn such mapping by minimizing the following quadratic loss function:

$$\mathcal{L} = ||[X_T - P_T B X_T^T] \circ W||_F^2, \quad \text{s.t.} P_T \mathbf{1} = \mathbf{1}, Q_\mathbf{T} \mathbf{1} = \mathbf{1} \tag{5}$$

where $X_T \in \mathbb{R}^{m_T \times n_T}$ is the target matrix, $P_T \in \{0,1\}^{m_T \times p}$ and $Q_T \in \{0,1\}^{n_T \times q}$ are cluster membership matrices, $\circ$ denotes the entry-wise product, and W denotes a binary weighting matrix where $W_{ij} = 1$ if user i rated item j and $W_{ij} = 0$ otherwise. We learn the parameters P_T and Q_T by applying Algorithm 2 in [5].

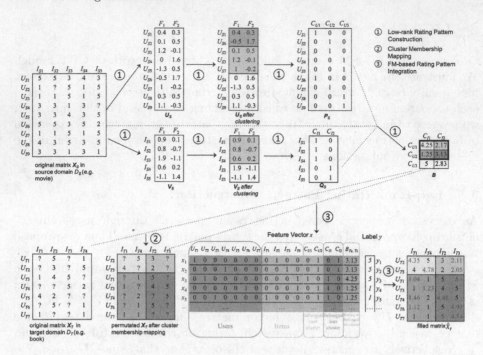

Fig. 1. Illustration of LKT-FM

3.3   FM-based Rating Pattern Integration

Once the rating pattern B and the membership matrices P_T and Q_T are obtained, existing NOCDCF approaches [5,6] usually incorporate the transferred knowledge in B by expanding B directly. More specifically, target matrix X_T is reconstructed by duplicating the rows and columns of B using $P_T B Q_T^T$. Figure 2 illustrates this incorporation process. More formally, the target matrix X_T is approximated by $\widehat{X}_T$, which is defined as follows:

$$\widehat{X}_T = W \circ X_T + [1 - W] \circ [P_T B Q_T^T] \tag{6}$$

We call this method Direct Expansion (DE), which may miss useful knowledge in the target matrix itself according to the previous analysis. Thus, in this work, we propose a new knowledge integration approach to solve this problem. We treat rating pattern as the side information of collaborative filtering tasks and incorporate it by using Factorization Machines (FM). The new approach is superior to the conventional solution as it takes advantage of both RPT and matrix factorization techniques.

First, assume U_T and I_T to be the sets of users and items in the target domain D_T. The rating prediction problem in D_T can be modeled by a target function $f : U_T \times I_T \to \mathbb{R}$. According to FM, each user-item interaction $(u, i) \in U_T \times I_T$ in X_T is represented by a feature vector $x \in \mathbb{R}^k$, $k = |U_T| + |I_T|$. The feature

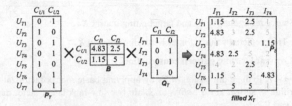

Fig. 2. Rating pattern incorporation by Direct Expansion (DE)

vector x consists of binary variables indicating which user rated which item. In other words, for each non-zero entry x_{ui} in X_T, its corresponding feature vector x can be represented as:

$$x = (\underbrace{0, ..., 0, 1, 0, ..., 0}_{|U_T|}, \underbrace{0, ..., 0, 1, 0, ..., 0}_{|I_T|}) \tag{7}$$

where the first $|U_T|$ binary indicator variables represent user, and the following $|I_T|$ binary indicator variables represent item.

Given the rating pattern B, cluster membership matrices P_T and Q_T, we can then incorporate the rating pattern by adding more features into x. There are various possible ways of extending the vector x. We first provide a straightforward solution, which adds three types of information, as follows:

$$x = (\underbrace{0, ..., 0, 1, 0, ..., 0}_{|U_T|}, \underbrace{0, ..., 0, 1, 0, ..., 0}_{|I_T|}, \underbrace{0, ..., 0, 1, 0, ..., 0}_{p}, \underbrace{0, ..., 0, 1, 0, ..., 0}_{q}, B_{C_u, C_i}) \tag{8}$$

where p, q denote the numbers of user clusters and item clusters. In the added part, the first p binary indicator variables represent which cluster user u belongs to, the next q binary indicator variables refer to which cluster item i belongs to, and B_{C_u, C_i} denotes the cluster-level rating of user cluster C_u to item cluster C_i. The extended feature vector x then serves as the input for Eq. 3, and the output y is the predicted rating of user u gives to item i.

In our empirical studies, we also tried other six solutions for expanding design matrix. Unlike Eq. 8 that integrates all the three new features, those six solutions make use of partial information extracted. The details of these six expansions are given in Appendix. We compare and discuss the differences of these expansions in Sect. 4.

3.4 Algorithm

We depict the above three steps of *low-rank rating pattern construction*, *cluster membership mapping* and *FM-based rating pattern integration* in Fig. 3, which contains five components of *dimension reduction*, *clustering*, *mapping*, *incorporation* and *factorization*. In specific, we first employ dimension reduction technique to obtain the low-rank user-factor and item-factor matrices, which are then used

Input: Source rating matrix X_S and target rating matrix X_T.

Output: The predicted value r for each missing entry in X_T

Step 1. Low-rank rating pattern construction

> *Step 1.1.* Reduce dimension of X_S by applying basic matrix factorization algorithm as shown in Eq.1 so as to obtain two low-rank latent factor matrices U_S and V_S
>
> *Step 1.2.* Cluster U_S and V_S to obtain user-cluster membership matrix P_S and item-cluster membership matrix Q_S, based on which the rating pattern B is constructed according to Eq.4

Step 2. Cluster membership mapping

> *Step 2.1.* Map each user and item in X_T to the cluster identified in B using Eq.5

Step 3. FM-based rating pattern integration

> *Step 3.1.* Incorporate rating pattern B into the target matrix X_T via expanding design matrix as shown in Eq.8
>
> *Step 3.2.* Factorize design matrix using factorization machines as shown in Eq.3

Fig. 3. The algorithm of LKT-FM (low-rank knowledge transfer via factorization machines)

by clustering algorithms to extract a group-level rating matrix. After that, we map each user and item into its corresponding cluster. In the end, we transfer the rating pattern via incorporating it into the design matrix and factorize the expanded design matrix using factorization machines.

Note that our algorithm is generic and flexible because we may find alternatives to each of these five components to derive a new solution according to the requirements of real-world applications. For example, we may employ another matrix factorization algorithm rather than basic MF, or another clustering algorithm rather than the simple K-means.

4   Experiments

4.1   Experiment Goal

The goal of our experiments is to answer the following research questions:

- **RQ1 Evaluation of Knowledge Extraction:** Can the proposed low-rank knowledge transfer solution (LKT) extract higher quality rating pattern than the traditional co-clustering method?
- **RQ2 Evaluation of Knowledge Integration:** Can factorization machines (FM) better integrate the rating pattern than traditional Direct-Expansion (DE) methods?

- **RQ3 Overall Performance**: Overall, how is the performance of our LKT-FM model compared with state-of-the-art NOCDCF techniques?
- **RQ4 Performance in Overlapping Setting**: How is the performance of our LKT-FM model compared with ordinary CDCF techniques, when users/items are overlapped across domains?

Note that RQ1, RQ2, RQ3 aim at evaluating our model in a NOCDCF setting from different aspects, while RQ4 aims at exploring its generality. We all know that NOCDCF has much wider range of applications in real world. And this is in fact the major motivation of this paper to solve this problem. But we are also interested in evaluating our model on overlapping CDCF datasets. If our model performs better than current NOCDCF approaches and in the meanwhile is competitive with existing Overlapping-CDCF methods, we can safely say our model is a generic solution to all CDCF tasks. And that is the reason why we evaluate RQ4. In the following sub-sections, we will describe datasets, baselines and experiment setups for these two settings respectively.

4.2   Datasets

For RQ1, RQ2, RQ3, we use three benchmark real-world datasets for evaluation:

- MovieLens 1M dataset[1]: A movie rating dataset contains 1,000,209 ratings of 3,900 movies made by 6,040 users (rating ratio 4.2%). Since we want to explore the effect of the sparseness of source matrix on the performance of NOCDCF algorithms, besides using the original dataset, we also use three sub-matrices with different sparseness as source matrix: 224,745 ratings by 500 users on 1000 movies (rating ratio 44.9%); 476,409 ratings by 1000 users on 2000 movies (rating ratio 23.8%); 634,680 ratings by 1500 users on 3000 movies (rating ratio 14.1%). We use the same sub-matrix extraction method in [5,6].
- Book-Crossing dataset[2]: A book rating data set contains more than 1.1 million ratings (scales 0-9) by 278,858 users on 271,379 books. Following [5,6], we obtain target matrix by randomly choosing 500 users with at least 20 ratings, and 1000 movies (rating ratio 3.03%).
- EachMovie dataset[3]: A movie rating dataset contains 2.8 million ratings (scales 1-6) by 72,916 users on 1682 movies. We still randomly choose 500 users with at least 20 ratings, and 1000 movies for experiment (rating ratio 12.4%).

Note that, in order for rating scale consistency, we normalize the rating scales from 1 to 5 for Book-Crossing and EachMovie dataset.

For RQ4, we use Amazon dataset for evaluation:

- Amazon dataset [20]: A diverse product rating data set contains 7,593,243 ratings provided by 1,555,170 users over 548,552 different products including 393,558 books, 103,144 music CDs, 19,828 DVDs and 26,132 VHS video tapes.

---

[1] http://www.grouplens.org/node/73.

[2] http://www.informatik.uni-freiburg.de/?cziegler/BX/.

[3] http://www.cs.cmu.edu/?lebanon/IR-lab.htm.

4.3   Baseline

For RQ1, RQ2, RQ3, we consider three state-of-the-art techniques as baselines: Codebook Transfer (CBT) [5], Cluster-Level Latent Factor Model (CLFM) [6] and Factorization Machines (FM) [8]. Note that, this work focuses on solving the single-domain knowledge-transfer problem [1], in which only a target domain and a source domain are considered. There are also some models designed for multi-domain knowledge-transfer problem, such as TALMUD [10], which however are identical to CBT when only two domains are involved. Therefore, we didn't adopt these methods for evaluation.

- CBT is a widely used RPT method in previous works [6,10]. It adopts orthogonal nonnegative matrix tri-factorization (ONMTF) algorithm [4] for codebook (i.e. rating pattern) extraction and is also one of the first approaches that can handle NOCDCF problem.
- CLFM is a more recent work that extends the idea of CBT. It not only learns a common rating pattern shared by all domains, but also learns domain-specific rating pattern for each individual domain.
- FM is a generic predictive model for single domain recommendations. To the best of our knowledge, no previous work adopted FM as a baseline algorithm for NOCDCF problem. This may be because cross-domain approaches were deemed to perform better than single domain approaches. However, cross-domain recommendation may easily have negative transfer issue. Thus, including a prominent single-domain CF approach is a necessity for the evaluation of CDCF algorithms. Thus in this work, we adopt FM as a baseline to explore whether NOCDCF methods can really improve the recommendation.

For RQ4, we consider two state-of-the-art CDCF methods as baselines: PF2-CDTF [18] and CDFM [15].

- PF2-CDTF is a tensor-based factorization model that can capture triadic relation between users, items and domains to improve CDCF recommendation.
- CDFM is a Factorization-Machine-based CDCF method that incorporates different domains by expanding the design matrix of FM. This method is relevant to our model but requires user correspondence information as additional input, which makes it only applicable in the Overlapping-CDCF setting.

4.4   Experimental Setup

For RQ1, RQ2, RQ3, we choose MovieLens as source domain, Book-Crossing and EachMovie as target domains. Following the work in [6], we evaluate our method under different configurations. For each target dataset, 300 users are randomly selected as the training set, and the remaining users for testing. For each test user, three different sizes of observed rating (Given5, Given10, Given15) are provided to avoid cold-start and the remaining ratings are for evaluation. Note that, as mentioned in Sect. 4.1, we obtained three sub-matrices of MovieLens

1M dataset with different levels of sparseness. In our experiments, all the three sub-matrices were used as source matrix, besides the original one, in order to explore the impact of data sparseness.

For RQ4, we choose Amazon-Music and Amazon-Book as target domain, the rests as source domains. We build the training and test set in two different ways similar to [15, 18] to allow comparison with them. In the first setup, TR75, 75% of data is considered as training set and the rest as test set, and in the second setup, TR20, only 20% of data is considered as training set and the rest as test set.

We adopt Mean Absolute Error (MAE) as evaluation metric. MAE is computed as $\text{MAE} = (\Sigma_{i \in T} |r_i - \widehat{r_i}|)/|T|$, where T means the test set, r_i is true value and $\widehat{r_i}$ is the predicted rating.

4.5 Experimental Results

RQ1 Evaluation of Knowledge Extraction

Since the performance of recommendation depends on both knowledge construction and integration methods, to evaluate the quality of knowledge extraction method, we need to make sure all the model adopt the same integration method. Thus, we use Direct-Expansion (DE) instead of Factorization Machines (FM) for our model in this part. We re-denoted our model as LKT-DE and compare it with two baselines CBT and CLFM.

Note that, in Fig. 4, "ML (224,745, 44.9%)" denotes that the source is a subset of MovieLens, containing 224,745 ratings and the rating ratio is 44.9%. "BX-5" denotes the target is Book-Crossing dataset and 5 ratings are given for each test user. Similarly, "EM-10" denotes the target is EachMovie dataset and 10 ratings are given for each test user. We use these denotations in the rest of the paper.

From Fig. 4 we can see that our method LKT-DE outperforms CBT and CLFM all the time, which means our knowledge extraction method behaves better than the co-clustering methods used by baselines. It is interesting to see that, in general, as the source matrix ML becomes sparser, the performances of two baselines degrade (i.e. MAE becomes larger). For example, in the first chart of Fig. 4, the blue line, representing baseline method CBT, goes up from 0.56 to 0.63 while the rating radio of source matrix decrease from 44.9% to 4.3%. This supports the analysis in Sect. 1 that current NOCDCF methods do not work well in sparse settings.

On the contrary, our method is not affected by the sparseness. As the source matrix becomes sparser, the performance becomes even better. This might be because as the source matrix ML becomes sparser, number of ratings in it also increases (from 224,745 to 1,000,209). Thus, although the sparseness makes it harder to extract knowledge for baseline methods, our method takes advantage of the increased ratings, and thus extract even more knowledge from them.

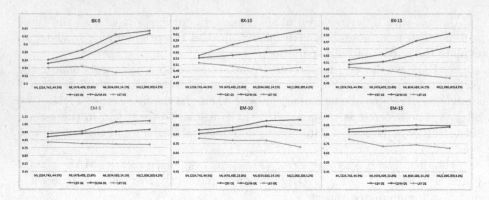

Fig. 4. MAE comparison of different knowledge integration methods (Color figure online)

RQ2 Evaluation of Knowledge Integration

Before comparing our FM-based integration method with the traditional Direct-Expansion (DE) approach, we first compare the seven different expansions of design matrix discussed in Sect. 3.3. Similar to RQ1, we make sure all the integration methods, including our seven solutions and the baseline DE, use the same rating pattern constructed by LKT.

Figure 5 shows the result on EM (EachMovie) dataset using ML (476,409, 23.8%) as source matrix. All the other seven lines are below the black line representing FM, which only uses target matrix for recommendation. This indicates that all the three types of information, item cluster index, user cluster index and co-cluster rating in source domain are useful for enhancing recommendation tasks in the target. Among the seven, FM $(U + R + I)$ which includes all the three types of information U, R, I, performs the best.

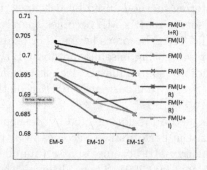

Fig. 5. MAE comparison of different design matrix expansions when source is ML (476,409, 23.8%) and target is EM

Note that, in our future work, we are planning to improve the current approach by allowing automatic selection of feature expansion for FM, but in this work, we choose the simple but effective FM $(U + R + I)$ expansion for further comparisons.

To save space, we don't show the comparison results when using other two source matrices, ML (224,745, 44.9%) and ML (634,680, 14.1%). We report the final results in the following.

Table 1 shows the comparison of our FM-based method with the baseline DE, which is now re-denoted as LKT-DE.

Table 1. MAE comparison of different rating pattern incorporation methods

	ML(224,745, 44.9%)		ML (476,409, 23.8%)		ML (634,680, 14.1%)		ML (1,000,209, 4.2%)	
	LKT-DE	LKT-FM	LKT-DE	LKT-FM	LKT-DE	LKT-FM	LKT-DE	LKT-FM
BX-5	0.540	**0.527**	0.543	**0.526**	0.527	**0.503**	0.530	**0.510**
BX-10	0.516	**0.494**	0.505	**0.480**	0.490	**0.468**	0.501	**0.475**
BX-15	0.496	**0.470**	0.488	**0.461**	0.475	**0.449**	0.464	**0.431**
EM-5	0.821	**0.705**	0.800	**0.701**	0.792	**0.695**	0.788	**0.683**
EM-10	0.807	**0.696**	0.785	**0.684**	0.774	**0.669**	0.763	**0.662**
EM-15	0.798	**0.694**	0.775	**0.681**	0.768	**0.672**	0.744	**0.660**

It is clear that our method always performs better than the baseline DE. We also find that the improvement on EM dataset is prominent. It may be because EM and ML are both movie-rating dataset while BX is book-rating dataset. The relatedness between ML and EM is higher than that between ML and BX. Thus, more useful knowledge is transferred from ML to EM and thus improves recommendation accuracy more.

RQ3 Overall Performance

To answer this question, three baselines CBT, CLFM and FM are compared with the proposed LKT-FM method. Note that, FM generates the prediction only with target-domain data. Thus, the MAE values for FM do not change when the source matrix changes.

The experimental results are shown in Table 2. We can see that the proposed method outperforms all the other models. It is very surprising to find that FM behaves better than the two state-of-the-art NOCDCF methods (CBT and CLFM), especially on the EachMovie dataset. This indicates that existing NOCDCF methods have negative transfer issue in this case. As analyzed in Sect. 1, existing NOCDCF methods incorporate rating pattern via direct expansion, which may rely too excessively on the knowledge from source domain and miss knowledge in the target domain itself. When the loss of knowledge from target domain exceeds the gain from source domain, current NOCDCF methods become even less effective than single-domain algorithms. On the contrary, our model overcomes this by striking a good balance between getting knowledge from the source and the target, and thus outperforms both current single-domain (FM) and cross-domain algorithms (CBT, CLFM).

Table 2. MAE comparison of different NOCDCF methods

	ML (224,745, 44.9%)				ML (476,409, 23.8%)			
	CBT	CLFM	FM	LKT-FM	CBT	CLFM	FM	LKT-FM
BX-5	0.560	0.551	0.541	**0.527**	0.585	0.566	0.541	**0.526**
BX-10	0.540	0.532	0.508	**0.494**	0.576	0.541	0.508	**0.480**
BX-15	0.517	0.504	0.491	**0.470**	0.534	0.512	0.491	**0.461**
EM-5	0.927	0.889	**0.703**	0.705	0.957	0.927	0.703	**0.691**
EM-10	0.897	0.856	0.701	**0.696**	0.925	0.893	0.701	**0.684**
EM-15	0.906	0.877	0.701	**0.694**	0.938	0.886	0.701	**0.681**
	ML (634,680, 14.1%)				ML(1,000,209, 4.2%)			
	CBT	CLFM	FM	LKT-FM	CBT	CLFM	FM	LKT-FM
BX-5	0.624	0.606	0.541	**0.503**	0.633	0.626	0.541	**0.510**
BX-10	0.601	0.551	0.508	**0.468**	0.621	0.559	0.508	**0.475**
BX-15	0.573	0.532	0.491	**0.449**	0.534	0.525	0.491	**0.431**
EM-5	1.071	0.950	0.703	**0.695**	0.998	0.901	0.703	**0.683**
EM-10	0.996	0.937	0.701	**0.669**	0.957	0.894	0.701	**0.662**
EM-15	0.952	0.904	0.701	**0.672**	0.942	0.929	0.701	**0.660**

RQ4 Performance in Overlapping Setting

Figures 6 and 7 show the comparison of different CDCF methods on Amazon
dataset. In these two figures we can see that CDFM, which is a FM-based CDCF
method, outperforms other methods, including our LKT-FM. This is not surprising because both LKT-FM and CDFM adopt FM for knowledge integration,
while CDFM also utilizes user correspondence information as addition input. But
except for CDFM, we can see that our LKT-FM outperforms other CDCF methods including the prominent PRE2-CDTF algorithm. Note that PRE2-CDTF
performs even worse than FM when 20% of data is used as training set (TR20),
which means negative transfer happens. On the contrary, our LKT-FM outperforms FM consistently. These results indicate that our LKT-FM is competitive
with other methods even in the overlapping setting. Thus, we can argue that
our LTK-FM model is an appropriate approach for solving various CDCF tasks,
not limited to NOCDCF problems.

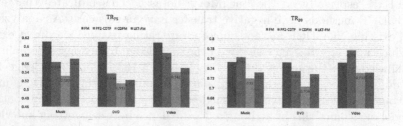

Fig. 6. MAE comparison of different CDCF methods on Amazon dataset (target: Book)

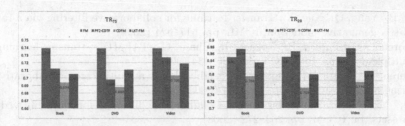

Fig. 7. MAE comparison of different CDCF methods on Amazon dataset (target: Music)

5 Conclusion and Future Work

In this paper we presented LKT-FM, a novel rating pattern transfer model, which aims at addressing the Non-Overlapping CDCF (NOCDCF) problem. The proposed model consists of 3 components: (1) a low-rank clustering method that enables knowledge extraction on large and sparse matrices; (2) a membership mapping algorithm which assigns users and items into clusters identified in the rating pattern; (3) a FM-based knowledge integration method that incorporates the shared knowledge into the target data in a seamless manner. Our experimental results showed LKT-FM outperforms state-of-the-art single-domain and NOCDCF approaches. In addition, LKT-FM is competitive with ordinary CDCF methods even in overlapping settings, which makes it a generic solution to all CDCF tasks.

References

1. Fernndez-Tobas, I., Cantador, I., Kaminskas, M., Ricci, F.: Cross-domain recommender systems: a survey of the state of the Art. In: CERI, pp. 187–198 (2012)
2. Berkovsky, S., Kuflik, T., Ricci, F.: Cross-domain mediation in collaborative filtering. In: Conati, C., McCoy, K., Paliouras, G. (eds.) UM 2007. LNCS (LNAI), vol. 4511, pp. 355–359. Springer, Heidelberg (2007). https://doi.org/10.1007/978-3-540-73078-1_44
3. Tang, J., Wu, S., Sun, J., Su, H.: Cross-domain collaboration recommendation. In: KDD, pp. 1285–1293 (2012)
4. Ding, C.H.Q., Li, T., Peng, W., Park, H.: Orthogonal nonnegative matrix t-factorizations for clustering. In: KDD, pp. 126–135 (2006)
5. Li, B., Yang, Q., Xue, X.: Can movies and books collaborate? Cross-domain collaborative filtering for sparsity reduction. In: IJCAI, pp. 2052–2057 (2009)
6. Gao, S., Luo, H., Chen, D., Li, S., Gallinari, P., Guo, J.: Cross-domain recommendation via cluster-level latent factor model. In: Blockeel, H., Kersting, K., Nijssen, S., Železný, F. (eds.) ECML PKDD 2013. LNCS (LNAI), vol. 8189, pp. 161–176. Springer, Heidelberg (2013). https://doi.org/10.1007/978-3-642-40991-2_11
7. Koren, Y., Bell, R., Volinsky, C.: Matrix factorization techniques for recommender systems. Computer **42**, 30–37 (2009)
8. Rendle, S.: Factorization machines with libFM. ACM Trans. Intell. Syst. Technol. **3**, 57:1–57:22 (2012)

9. Li, B., Yang, Q., Xue, X.: Transfer learning for collaborative filtering via a rating-matrix generative model. In: ICML, pp. 617–624 (2009)
10. Moreno, O., Shapira, B., Rokach, L., Shani, G.: TALMUD: transfer learning for multiple domains. In: CIKM, pp. 425–434 (2012)
11. Koren, Y.: Factorization meets the neighborhood: a multifaceted collaborative filtering model. In: KDD, pp. 426–434 (2008)
12. Hu, Y., Koren, Y., Volinsky, C.: Collaborative filtering for implicit feedback datasets. In: ICDM, pp. 263–272 (2008)
13. Koren, Y.: Collaborative filtering with temporal dynamics. In: KDD, pp. 426–434 (2009)
14. Ji, K., Sun, R., Li, X., Shu, W.: Improving matrix approximation for recommendation via a clustering-based reconstructive method. Neurocomputing **173**, 912–920 (2016)
15. Loni, B., Shi, Y., Larson, M., Hanjalic, A.: Cross-domain collaborative filtering with factorization machines. In: de Rijke, M., Kenter, T., de Vries, A.P., Zhai, C.X., de Jong, F., Radinsky, K., Hofmann, K. (eds.) ECIR 2014. LNCS, vol. 8416, pp. 656–661. Springer, Cham (2014). https://doi.org/10.1007/978-3-319-06028-6_72
16. Shi, Y., Larson, M., Hanjalic, A.: Tags as bridges between domains: improving recommendation with tag-induced cross-domain collaborative filtering. In: Konstan, J.A., Conejo, R., Marzo, J.L., Oliver, N. (eds.) UMAP 2011. LNCS, vol. 6787, pp. 305–316. Springer, Springer (2011). https://doi.org/10.1007/978-3-642-22362-4_26
17. Coyle, M., Smyth, B.: (Web search)[shared]: social aspects of a collaborative, community-based search network. In: Nejdl, W., Kay, J., Pu, P., Herder, E. (eds.) AH 2008. LNCS, vol. 5149, pp. 103–112. Springer, Heidelberg (2008). https://doi.org/10.1007/978-3-540-70987-9_13
18. Hu, L., Cao, J., Xu, G., Cao, L., Gu, Z., Zhu, C.: Personalized recommendation via cross-domain triadic factorization. In: WWW, pp. 595–606 (2013)
19. Cremonesi, P., Quadrana, M.: Cross-domain recommendations without overlapping data: myth or reality? In: RecSys, pp. 297–230 (2014)
20. Leskovec, J., Adamic, L.A., Huberman, B.A.: The dynamics of viral marketing. ACM Trans. Web **1**, 5 (2007)

Distributed Multi-task Learning for Sensor Network

Jiyi Li[1]([⊠]), Tomohiro Arai[1], Yukino Baba[1],
Hisashi Kashima[1], and Shotaro Miwa[2]

[1] Department of Intelligence Science and Technology, Kyoto University,
Yoshida-honmachi, Sakyo-ku, Kyoto 606-8501, Japan
{jyli,baba,kashima}@i.kyoto-u.ac.jp, arai@ml.ist.i.kyoto-u.ac.jp
[2] Mitsubishi Electric Corporation, Tokyo, Japan
Miwa.Shotaro@bc.mitsubishielectric.co.jp

Abstract. A sensor in a sensor network is expected to be able to make prediction or decision utilizing the models learned from the data observed on this sensor. However, in the early stage of using a sensor, there may be not a lot of data available to train the model for this sensor. A solution is to leverage the observation data from other sensors which have similar conditions and models with the given sensor. We thus propose a novel distributed multi-task learning approach which incorporates neighborhood relations among sensors to learn multiple models simultaneously in which each sensor corresponds to one task. It may be not cheap for each sensor to transfer the observation data from other sensors; broadcasting the observation data of a sensor in the entire network is not satisfied for the reason of privacy protection; each sensor is expected to make real-time prediction independently from neighbor sensors. Therefore, this approach shares the model parameters as regularization terms in the objective function by assuming that neighbor sensors have similar model parameters. We conduct the experiments on two real datasets by predicting the temperature with the regression. They verify that our approach is effective, especially when the bias of an independent model which does not utilize the data from other sensors is high such as when there is not plenty of training data available.

Keywords: Sensor network · Multi-task learning
Distributed approach

1 Introduction

Traditional data integration and analysis for sensor data collects the observation data from all sensors to construct *centralized* global models at a data center (Fig. 1(a)). With the advances in the device technology, the computing power and storage capacity of sensors have been improved. Meanwhile, with the large scale of sensor network growth, transferring all observation data in the network and learning the centralized model utilizing them require a huge amount of

M. Ceci et al. (Eds.): ECML PKDD 2017, Part II, LNAI 10535, pp. 657–672, 2017.
https://doi.org/10.1007/978-3-319-71246-8_40

computation. Therefore, a sensor in a sensor network is expected to be able to make prediction or decision by itself utilizing the decentralized local models learned from the data observed on this sensor.

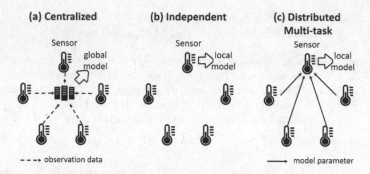

Fig. 1. Different types of solutions for learning the models of sensors in sensor network: (a) Centralized global model, (b) Independent local model, (c) Distributed multi-task local model (Our proposal).

A naïve solution of learning a local model by a sensor is only utilizing the observation data on this sensor. We define it as *independent* model (Fig. 1(b)). One of the issues that need to be solved for learning such local models is the limitation of available data for training. In the early stage of using a sensor, there may be not a lot of data available to train the local model for this sensor. A solution is to leverage the observation data from other sensors which have similar conditions and thus may have similar models with the given sensor. However, sensors can be overloaded to transfer the observation data to other sensors, and broadcasting the observation data of each sensor in the entire network has the privacy leakage problem. In addition, in the prediction stage, we hope each sensor is able to make prediction independently from the observation data from neighbor sensors.

To overcome these problems, based on the assumption that neighbor sensors have similar model parameters, we propose sharing the model parameters among the sensors instead of transferring the observation data. We treat the problems of learning the local models for a set of sensors as a *multitask* learning problem in which each sensor corresponds to one task. In the learning process, the proposed method incorporates the neighborhood relations among sensors to learn multiple models simultaneously. We then present a *distributed* learning algorithm for our multi-task model. In this algorithm, at each iteration, the local model of each sensor is computed simultaneously, and then the model parameters are shared among all the sensors (Fig. 1(c)). The parameters are incorporated in the regularization terms of the objective function of each sensor, thus we can use the information from the other sensors to update each local model. Although a few algorithms for distributed multi-task learning algorithms for client-server models have been proposed [6, 17], we propose a novel distributed algorithm for multi-task learning in peer-to-peer models.

In this paper, we focus on the time series regression tasks to verify our approach. We utilize several real datasets collected from real sensor networks in which the sensors recorded the temperature information for different applications. The task is to predict the temperature at a given place with a sensor based on the historical temperature observation data. The experiments show that our approach is effective, especially when the volume of training data is small which is one of the reasons of the under-fitting of an independent model learned based on the data on a sensor.

Furthermore, although a centralized model generally is expected to have better performance than local models because it can utilize much more observation data for learning, in some cases, because the environmental conditions and data characteristics of the sensors may be somewhat different, a local model for a sensor may be able to provide better presentation of this sensor when the training data is not a lot.

The contributions of this paper are as follows.

- We focus on the problem of predictive modeling in sensor networks in which each sensor learns a local model by utilizing its own data and the data from neighbor sensors. We propose a distributed approach which the sensors learn multiple models simultaneously.
- We propose a novel approach to solve the problem in a manner of multi-task learning by treating each sensor as a task. To avoid transferring observation data of neighbor sensors in the network, our approach only leverages the model parameters from neighbor sensors.
- We use real datasets to verify that our approach is effective especially when the bias of an independent model which does not utilize the data from other sensors is high such as when there is not plenty of training data available.

2   Related Work

We review the existing work on the following four issues, i.e., multi-task learning, distributed learning, prediction of sensor data and time-series data mining.

Multi-task Learning: Rather than implementing multiple related learning tasks separately, multi-task learning leverages the information from different tasks to train these tasks simultaneously so that it can raise the model performance of each single task [4]. In this paper, because learning the model of each sensor can be regarded as one learning task, and modeling all sensors can be simultaneously performed. We thus solve our problem with multi-task learning.

Existing work on multi-task learning models the relatedness and model parameters of multiple tasks in different manners. Argyriou et al. [2] assumed the common underlying representation shared across multiple related tasks and learned a low-dimensional representation. Evgeniou et al. [8] modeled the relation between tasks in terms of a kernel function that used a task coupling parameter. Evgeniou et al. [7] also leveraged the similarity between tasks as prior knowledge, assumed that similar tasks have similar parameters and utilized as regularization. In our work, we model the task relatedness and model parameters of multiple tasks by using the manner proposed in [7]. In our scenario, the similarity between tasks can be given by the neighborhood graph based on the physical distance between sensors. In other words, the sensors which are near to each other are regarded to have similar parameters, and we utilize this similarity to perform regularization.

Regarding the implementation of multi-task learning in a distributed environment, Wang et al. [17] proposed distributed learning method followed the assumption of the common underlying representation like [2]. The manner of the common underlying representation requires the distributed learning method having a client-server manner. Our approach and [17] differ in the regularization schemes. [17] incorporated shared-sparsity among local models, while ours makes local model parameters close to each other. Dinuzzo et al. [6] also proposed another method of distributed multi-task learning with a client-server manner. However, the client-server manner is not suitable for our scenario in which a server does not exist. In contrast, in our distributed approach, the tasks learn simultaneously in a peer-to-peer manner. In addition, Vanhaesebrouck et al. [16] proposed asynchronous algorithms for collaborative peer-to-peer network based on label propagation which jointly learned and propagated their model based on both their local dataset and the behavior of their neighbors.

Distributed Learning: The idea of conducting learning in a distributed manner has also been used for various areas related to machine learning and data mining. For example, in the area of reinforcement learning, multi-agent learning [14] performs learning of agents in distributed cooperative manner. [12] proposed parallel and distributed learning approach in which a number of processors cooperated together for a single learning task. In contrast, our work concentrates on the area of sensor data and multiple tasks.

Prediction of Sensor Data: With the development of technology such as Internet of Things, sensors have reach miniaturization, high performance, and low price. The sensor networks in which the sensors can communicate to each other and have some computation capability have become possible. In this research, we focus on the prediction tasks for sensor observation data. There are several previous studies on time series modeling of sensor data, e.g., [15], while they do not deal with collaborative learning among sensors. [10] is the existing work which is most similar to our work. It proposed distributed regression model based on divided local region. However, it needs the observation data from neighbor sensors in the prediction stage. In addition, Ahmadi et al. [1] proposed a solution with a centralized clustering on entire area and an independent method for each cluster. Ceci et al. [5] used neighbor distances as features for a centralized model. In contrast, in our approach, the predictions of a sensor can be made independently from the observation data of neighbor sensors.

Time Series Data Mining: There is much existing work on learning the prediction model from time series data. The regression model is just one of the alternatives which may be not the best model in the existing work. For example, the approaches based on RNN with LSTM [9,11,13] may have better prediction performance on non-stationary time series data. Our work does not exactly focus on the prediction task for time series data. We focus on the distributed multi-task learning in a sensor network by utilizing the regression task to illustrate our idea. Although the sensor datasets we use are time series data, our proposal can actually also be utilized to the other kinds of regression tasks.

3   Our Approach

We describe our approach in details in this section. Our purpose is to learn decentralized local models for different sensors in the network. The learning process thus needs to be carried out at each corresponding sensor, instead of a data center which collects all observation data from the entire sensor network. We propose a distributed multi-task learning approach in which each sensor corresponds to one task to solve this problem.

The type of learning task on a sensor can be diverse based on the practical requirements. In this paper, we focus on time series regression tasks, in which we are given observed data in a specific time window and aim to predict the value of the observations in future timestamps. Such time-series prediction tasks are very useful and widely required for sensor networks, e.g., predicting the temperature at a place with a weather sensor. Our proposal is not limited to such specific prediction task. Our approach can be directly utilized to other kinds of regression tasks. Our idea of distributed multi-task learning can also be adapted to other types of learning tasks such as classification.

3.1   Preliminary

For a given sensor network $\mathcal{S}$ ($|\mathcal{S}| = n$) which generates observation data based on time series, we learn a local regression model f_k for predicting the value of observation data at timestamp i on each sensor $s_k \in \mathcal{S}$ based on the observation data at a time window before i with a fixed window size t. k is the index of a sensor. The feature data on a sensor s_k is $\mathcal{X}_k = (\mathbf{x}_{k1}, \mathbf{x}_{k2}, \ldots, \mathbf{x}_{kn_k})^\top$ which contains both the observation data in a specific time window on this sensor and an intercept term with a value of 1, $\mathbf{x}_{ki} = (x_{k(i-t)}, \ldots, x_{k(i-2)}, x_{k(i-1)}, 1)$. n_k is the number of instances on sensore s_k. We denote the target value as $\mathbf{y}_k = (y_{k1}, y_{k2}, \ldots, y_{kn_k})^\top$ in which $y_{ki} = x_{ki}$ is the predicted observation value at timestamp i.

Given feature data $\{\mathcal{X}_k\}_k$ and target data $\{\mathbf{y}_k\}_k$ from a sensor network $\mathcal{S} = \{s_k\}_k$, our task is to learn a local regression model for each sensor s_k with model parameters $\mathbf{w}_k$.

The basic linear regression model f_k and the object function $\mathcal{L}_0$ which are used for the independent approach in this paper is as follows. Because we focus

on our idea of the distributed multi-task learning, we thus utilize this fundamental regression model without any additional settings such as regularization
to highlight the effectiveness of our proposal.

$$f_k(\mathcal{X}_k, \mathbf{w}_k) = \mathbf{w}_k^\top \mathcal{X}_k, \quad \mathcal{L}_0(\mathbf{w}_k) = ||\mathbf{w}_k^\top \mathcal{X}_k - \mathbf{y}_k||_2^2, \quad \mathbf{w}_k = (\mathcal{X}_k^\top \mathcal{X}_k)^{-1} \mathcal{X}_k^\top \mathbf{y}_k. \quad (1)$$

3.2   Multi-task Model

To learn the local model on a given sensor, one of the problems that need to
be solved is the limited available data for training. In the early stage of using
a sensor, there may be not a lot of data available to train the local model for
this sensor. A solution is to leverage the observation data from other sensors.
The other sensors in the sensor network may have similar but somewhat different models with the given sensor. In the learning process, all these sensors
can communicate with each other to learn multiple models simultaneously. The
problem of learning the local models for a set of sensors can thus be regarded as
a multi-task learning problem in which each sensor corresponds to one task. We
thus propose a multi-task learning approach for learning the decentralized local
models of the sensors.

In the training stage, sensors are allowed to transfer the data with other
sensors. In a sensor network, there may be huge amount of sensors installed in
an immense geographical range. For a given sensor, it costs too much to transfer
the data from all other sensors. It would be better to constrain that each sensor
can only communicate with the neighbor sensors which are physically near to
the given sensor. Considering that the neighbor sensors are more possible to
have more similar conditions with the given sensor because of shorter physically
distance, this settings of using neighbor sensors only is rational and cost-effective.
We denote the set of neighbor sensors of a given sensor s_k as $\mathcal{S}_k$.

Which kinds of data are transferred with neighbor sensors and used for learning models is another issues needs to be solved. Intuitively, we can transfer and
use the observation data. However, there are three main problems in this manner.
First, it may be not cheap for each sensor to keep transferring the observation
data with a number of other sensors; Second, broadcasting the observation data
of an sensor in the entire network have the privacy leakage problem; Third, each
sensor needs to be able to make computation in the prediction stage without
waiting for using the observation data from neighbor sensors. To overcome these
problems, instead of transferring the observation data in the network, our distributed approach in the network transfers the model parameters with neighbor
sensors. This solution is based on the assumption that neighbor sensors have similar model parameters. To leverage the model parameters from neighbor sensors,
our approach shares them as the regularization terms in the objective function.
Note that all sensors are not necessary to have the same model.

Based on the above discussion, we propose a novel distributed multi-task
learning approach which incorporates neighborhood relations among sensors to
learn multiple models simultaneously in which each sensor corresponds to one

task. Following [7], the overall object function of all local regression models for all sensors can be written as follows.

$$\mathcal{L}(\{\mathbf{w}_k\}_k) = \sum_{k=1}^{n} ||\mathbf{w}_k^\top \mathcal{X}_k - \mathbf{y}_k||_2^2 + \lambda \sum_{k=1}^{n} \sum_{s^{k'} \in \mathcal{S}_k} ||\mathbf{w}_k - \mathbf{w}_{k'}||_2^2 \qquad (2)$$

λ is the hyperparameter to control the regularization term in the object function. In this paper, we use same λ to all neighbor sensors. An option is to set or learn different λ for a different pair of sensors. For the purpose of clarifying our proposal of distributed multi-task learning of our problem settings, we use this simpler solution with single λ hyperparameter.

The object function $\mathcal{L}_k(\mathbf{w}_k)$ of a local regression model for a sensor $\mathbf{s}_k$ can thus be written as follows.

$$\mathcal{L}_k(\mathbf{w}_k) = ||\mathbf{w}_k^\top \mathcal{X}_k - \mathbf{y}_k||_2^2 + \lambda \sum_{s_{k'} \in \mathcal{S}_k} ||\mathbf{w}_k - \mathbf{w}_{k'}||_2^2 \qquad (3)$$

We can utilize formulation (3) for each sensor respectively instead of utilizing formulation (2) for all sensors. Because this formulation (3) for the original formulation (2) can be regarded as a block-coordinate descent for a convex function, the convergence of the formulation (2) can be guaranteed [3].

The parameters $\mathbf{w}_k$ in our multi-task model which minimizes this object functions can be computed as follows, in which m_k is the size of $\mathcal{S}_k$.

$$\mathbf{w}_k = (\mathcal{X}_k^\top \mathcal{X}_k + m_k \lambda \mathcal{I})^{-1}(\mathcal{X}_k^\top \mathbf{y}_k + \lambda \sum_{s_{k'} \in \mathcal{S}_k} \mathbf{w}_{k'}) \qquad (4)$$

3.3 Distributed Learning

This multi-task formulation in the previous sub-section shows that it requires integrating the $\mathbf{w}_k$ for the given sensors $\mathbf{s}_k$ and its neighbor sensors $\mathcal{S}_k$ to compute together for solving the optimal values of $\{\mathbf{w}_k\}_k$. Generally, it can be implemented at a data center which collects the observation data from all sensors. However, based on our scenario discussed in this paper, we need to distribute the computation to each sensor. We thus propose a distributed approach for our multi-task model.

In our distributed approach, the local model parameters of the sensors are computed on multiple sensors simultaneously. On a given sensor $\mathbf{s}_k$, in each iteration r, the model parameter $\mathbf{w}_k^{(r)}$ is updated based on the values of model parameters $\mathbf{w}_{k'}^{(r-1)}$ of other neighbor sensors $s_{k'} \in \mathcal{S}_k$ at iteration $r-1$. The formulation of the multi-task model can be revised in the following manner.

$$\mathbf{w}_k^{(r)} = (\mathcal{X}_k^\top \mathcal{X}_k + m_k \lambda \mathcal{I})^{-1}(\mathcal{X}_k^\top \mathbf{y}_k + \lambda \sum_{s_{k'} \in \mathcal{S}_k} \mathbf{w}_{k'}^{(r-1)}) \qquad (5)$$

In the distributed computation, a sensor s_k sends $\mathbf{w}_k^{(r-1)}$ to other neighbor sensors $s_{k'}$ with $s_k \in \mathcal{S}_{k'}$, and waits for receiving the current model parameters

---

Algorithm 1: DISTRIBUTED MULTI-TASK LEARNING

Input: Feature Data $\{\mathcal{X}_k\}_k$; Target Data $\{\mathbf{y}_k\}_k$
Output: Model Parameters $\{\mathbf{w}_k\}_k$;
   // Initialization
1 **forall** the $s_k \in \mathcal{S}$ **do**
2     $\mathbf{w}_k^{(0)} = (\mathcal{X}_k^\top \mathcal{X}_k)^{-1} \mathcal{X}_k^\top \mathbf{y}_k$;
3 **end**
4 $r = 1$;
   // Learning
5 **forall** the $s_k \in \mathcal{S}$ **do**
6   **while** $r \leq r_{max}$ **do**
        // broadcast the parameters to neighbor sensors
7   **forall** the $s_{k'}$, $s_k \in \mathcal{S}_{k'}$ **do**
8   Send($\mathbf{w}_k^{(r-1)}$);
9   **end**
        // collect the parameters from neighbor sensors
10   **forall** the $s_{k'} \in \mathcal{S}_k$ **do**
11   Receive($\mathbf{w}_{k'}^{(r-1)}$);
12   **end**
        // update parameters at iteration r based on the parameters of
           neighbor sensors at iteration r-1
13      $\mathbf{w}_k^{(r)} = (\mathcal{X}_k^\top \mathcal{X}_k + m_k \lambda \mathcal{I})^{-1}(\mathcal{X}_k^\top \mathbf{y}_k + \lambda \sum_{s_{k'} \in \mathcal{S}_k} \mathbf{w}_{k'}^{(r-1)})$;
14      $r = r + 1$;
15   **if** $||\mathbf{w}_k^{(r)} - \mathbf{w}_k^{(r-1)}||_2^2 < \theta$ **then**
16   break;
17   **end**
18   **end**
19 **end**
20 **return** $\{\mathbf{w}_k\}_k$

---

at iteration $r - 1$ from the neighbor sensors $s_k \in \mathcal{S}_k$. After that, it updates the model parameters $\mathbf{w}_k^{(r)}$ at iteration r based on its own observation data ($\mathcal{X}_k$, $\mathbf{y}_k$) and the model parameters from neighbor sensors. When all updated model parameters from neighbor sensors in $\mathcal{S}_k$ have reached the sensor $\mathbf{s}_k$, it starts the iteration $r + 1$ of computation. There are two stop criteria of the iterations, i.e., the difference of model parameters between two continue iterations are smaller than a threshold θ or the maximum number of iterations has been reached. Algorithm 1 lists the computation of our distributed approach.

4   Experiments

In this section, we verify our approach with data collected from real sensor networks. All these data are time-series temperature observation data, while the sensor networks are from different applications. We utilize our approach to solve the problem of temperature prediction of a given place with a sensor in a sensor network at a timestamp.

4.1   Dataset: Weather

The weather observation data such as temperature, precipitation, sunshine duration, wind and so on are collected by the sensors installed at specific places by the national meteorological agency. These collected data then can be used for various purposes of data analysis such as predicting the weather in the future, analyzing weather conditions in the past and so on. These sensors construct a huge weather sensor network around a country.

In this paper, we collect the weather observation data from the official website of Japan Meteorological Agency[1]. We select 11 places in a local region in the area of Gifu province. The geographical coordinates of these places are listed in Table 1. The temperature observation data is provided hour by hour. The number of timestamps is thus 24 in one day. The range of days is from the year 2005 to the year 2014 (ten years) and September 1 to September 30th (one month in each year). There are 7200 timestamps in total.

We set the location of the place $s0$ as $(0,0)$ and create a Cartesian coordinate system, and assign the locations of the nearby places by converting the latitude and longitude into geometric distances in the coordinate system. The converted coordinates system is shown in Fig. 2. The neighborhood relation is decided by a distance threshold (i.e., 30000) and is also shown in Fig. 2. Table 2 lists the number of neighbor sensors of each sensor.

Table 1. Weather dataset: geographical coordinates of places

ID	City name	(North latitude, East longitude)
s0	Kanayama	$(35°39.7',137°9.6')$
s1	Hagiwara	$(35°53.5',137°12.4')$
s2	Miyachi	$(35°45.8',137°17.3')$
s3	Kurogawa	$(35°35.9',137°19.1')$
s4	Nakatsugawa	$(35°28.6',137°29.2')$
s5	Ena	$(35°26.8',137°24.2')$
s6	Tajimi	$(35°20.8',137°6.5')$
s7	Minokamo	$(35°26.7',137°0.3')$
s8	Mino	$(35°33.3',136°54.6')$
s9	Yawata	$(35°45.4',136°58.7')$
s10	Nagataki	$(35°55.4',136°49.9')$

---

[1] http://www.data.jma.go.jp/gmd/risk/obsdl/index.php.

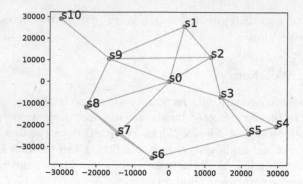

Fig. 2. Weather dataset: places and neighborhood relations

Table 2. Weather dataset: number of neighbor sensors

Sensors	s0	s1	s2	s3	s4	s5	s6	s7	s8	s9	s10
Number of neighbors	6	3	4	4	2	3	3	3	4	5	1

4.2   Dataset: HVACS Evaluation House

This dataset is collected from the House-type HVACS (Heating, Ventilation and Air Conditioning System) Evaluation Facility built by Mitsubishi Electric R&D Centre Europe[2]. As one aspect of the smart house technology, we can install the sensors with various functions in a house. Figure 3 shows the outside and inside view of the smart house from which we collect the data. We install many temperature sensors in the whole house. In this paper, we use two groups of temperature sensors in the living room as two separated subsets, i.e. the sensors in the floor (HEH-Floor) and the sensors in the ceiling (HEH-Ceiling). There are eight sensors in each subset. The sensors record the temperature every 10 s. The data is collected in one day. There are 8619 records in the HEH-Floor dataset and 8618 records in the HEH-Ceiling dataset. Figure 4 shows the positions of sensors and the neighborhood relations among the sensors. The related positions of the sensors in the floor or ceiling are same (Table 3).

Table 3. HVACS evaluation house dataset: number of neighbor sensors

Dataset	Sensors	s1	s2	s3	s4	s5	s6	s7	s8
HEH-Floor	Number of neighbors	3	4	5	7	4	3	5	3
Dataset	Sensors	s1	s2	s3	s4	s5	s6	s7	s8
HEH-Ceiling	Number of neighbors	3	4	5	7	4	3	5	3

---

[2] http://www.mitsubishielectric.com/brief/randd/index.html.

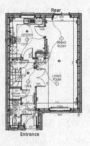

Fig. 3. Outside and inside view of the HVACS evaluation house

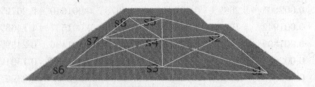

Fig. 4. HVACS evaluation house datasets (floor or ceiling): positions and neighborhood relations

4.3 Experimental Settings

We utilize the following two evaluation metrics for evaluating the performance, i.e., Root Mean Square Error (RMSE) and Mean Absolute Error (MAE). The formulas of these two metrics for a sensor s_k are as follows. We evaluate the average RMSE and MAE on all sensors. $\hat{y}_{ki}$ is the estimated value.

$$RMSE_k = \sqrt{\frac{1}{n_k} \sum_i (y_{ki} - \hat{y}_{ki})^2}, \quad MAE_k = \frac{1}{n_k} \sum_i |y_{ki} - \hat{y}_{ki}|$$

We compare our *multitask* approach with the *independent* approach which is based on formulation (1) and only utilizes its own observation data as the features for learning, in the scenario of learning decentralized local models. In addition, we also show the performance of an ideal *centralized* approach as a reference. It is also based on formulation (1) but utilizes the observation data of all sensors as the features for learning.

We verify the performance of our approach on different window sizes. The window size t is set as $\{2,3,5,11,23\}$ respectively for all datasets. Especially, for the weather dataset, $t = 23$ means using the observation data of previous hours in the length of one day to predict the temperature.

For our approach, the regularization hyperparameter λ is tuned by grid search with five-fold cross validation on the training set; the candidate values is in $\{0, 10^{-6}, 10^{-5}, 10^{-4}, 5 \times 10^{-4}, 10^{-3}, 5 \times 10^{-3}, 10^{-2}\}$. The stop criterion θ is set to 10^{-6}. The maximum iteration number is set to 100.

The feature and target data of each instance is generated with overlap which can increase the number of instances and avoid the learned model overfitting

Table 4. Experimental results of each dataset; (10% training, 90% testing)

Win.	RMSE				MAE			
Size t	Independ	Multitask	Improve	Central	Independ	Multitask	Improve	Central
(a). Weather dataset								
2	0.029693	0.029684	0.0312%	0.024120	0.021661	0.021631	0.1401%	0.017166
3	0.029296	0.029273	0.0779%	0.024309	0.021745	0.021709	0.1662%	0.017242
5	0.030002	0.029955	0.1578%	0.024574	0.022568	0.022508	0.2653%	0.017549
11	0.030982	0.030927	0.1748%	0.025636	0.023448	0.023384	0.2728%	0.018689
23	0.026703	0.026678	0.0924%	0.027869	0.019213	0.019179	0.1781%	0.020831
(b). HEH-Floor dataset								
2	0.002112	0.002065	2.2689%	0.067664	0.001678	0.001640	2.2672%	0.053601
3	0.001965	0.001942	1.1907%	0.060207	0.001555	0.001537	1.1463%	0.047720
5	0.001904	0.001899	0.2744%	0.055196	0.001489	0.001486	0.2439%	0.043977
11	0.002088	0.002084	0.2027%	0.058846	0.001569	0.001567	0.1391%	0.047019
23	0.002821	0.002819	0.0603%	0.060079	0.002004	0.002003	0.0396%	0.047240
(c). HEH-Ceiling dataset								
2	0.004019	0.003937	2.0369%	0.084777	0.003162	0.003095	2.1184%	0.065068
3	0.003765	0.003727	1.0020%	0.075965	0.002947	0.002916	1.0325%	0.058240
5	0.003615	0.003612	0.0718%	0.064683	0.002807	0.002805	0.0751%	0.049914
11	0.003770	0.003769	0.0110%	0.065543	0.002865	0.002865	−0.0009%	0.050197
23	0.004500	0.004490	0.2242%	0.064775	0.003280	0.003274	0.1743%	0.050013

to a chronological cycle. For example, assuming that window size $t = 2$, an instance is $(\mathbf{x}_{ki} = (x_{k(i-2)}, x_{k(i-1)}, 1), \; y_{ki} = x_{ki})$, then the next instance is $(\mathbf{x}_{k(i+1)} = (x_{k(i-1)}, x_{k(i)}, 1), \; y_{k(i+1)} = x_{k(i+1)})$.

We split the training and testing sets with different rates of the entire data. In this paper, because we especially focus on the scenario that there is not lots of training data available, when using the testing data, we ignore the real timestamp of the instances. In other words, for an instance in the testing data, even if there are some observation data in the time interval between the training data and this instance, we do not use the data in this time interval to improve the model.

4.4   Experimental Results

Table 4 lists the experimental results. To facilitate the presentation and highlight our contribution, we separately show the experimental results with the 10% split rate for the training set in the table. Considering the comparison between our multi-task approach and the independent approach, the *improve* column shows the improvement of our multi-task approach to the independent approach. It shows that out approach has better performance than the independent one on all three datasets when the split rate for training set is low, which means that there is not lots of training data available.

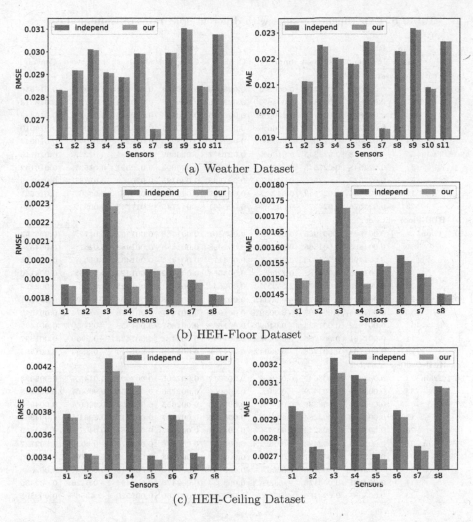

(a) Weather Dataset

(b) HEH-Floor Dataset

(c) HEH-Ceiling Dataset

Fig. 5. Performance of each sensor. Window size $t = 3$; (10% training, 90% testing)

Figure 5 shows the performance of each sensor respectively when the split rate for training is set to 10%. The window size is set to 3. It shows the robustness of our approach on each sensor, i.e., our approach not only improves the average performance of all sensors but also improves the performance of each sensor. Therefore, each sensor can obtain profits by using our approach.

Table 5 shows the experimental results on other split rate for training and testing sets. One of the observations is that in the cases of the high split rate for training set, e.g., 90% split rate for the training set, there are somewhat differences between the performance results of the weather dataset and the HVACS Evaluation House datasets. In the weather dataset, our approach still always outperforms the independent approach; while our approach is not better than

Table 5. Experimental results with different split rates of training data

	Win.	RMSE				MAE			
	Size t	Independ	Multitask	Improve	Central	Independ	Multitask	Improve	Central
(a). Weather dataset									
50% Train	2	0.029746	0.029744	0.0065%	0.023633	0.020934	0.020929	0.0230%	0.016556
	3	0.028881	0.028880	0.0048%	0.023585	0.020407	0.020403	0.0173%	0.016513
	5	0.028874	0.028872	0.0060%	0.023535	0.020453	0.020449	0.0198%	0.016482
	11	0.028719	0.028717	0.0066%	0.023420	0.020401	0.020397	0.0204%	0.016510
	23	0.026070	0.026069	0.0018%	0.022627	0.018233	0.018231	0.0135%	0.016044
90% Train	2	0.028348	0.028345	0.0130%	0.021877	0.020699	0.020695	0.0223%	0.016116
	3	0.027548	0.027545	0.0082%	0.021866	0.020094	0.020091	0.0153%	0.016142
	5	0.027578	0.027566	0.0414%	0.021823	0.020125	0.020109	0.0773%	0.016119
	11	0.027554	0.027542	0.0438%	0.021629	0.020198	0.020181	0.0860%	0.016091
	23	0.024576	0.024575	0.0035%	0.020461	0.017745	0.017743	0.0093%	0.015139
(b). HEH-Floor dataset									
50% Train	2	0.001967	0.001966	0.0776%	0.005801	0.001568	0.001567	0.0777%	0.005297
	3	0.001865	0.001865	0.0278%	0.005098	0.001488	0.001488	0.0267%	0.004624
	5	0.001793	0.001793	0.0055%	0.004175	0.001431	0.001431	0.0053%	0.003734
	11	0.001795	0.001782	0.7256%	0.003554	0.001432	0.001422	0.7018%	0.003110
	23	0.001826	0.001779	2.5525%	0.002644	0.001461	0.001423	2.5784%	0.002217
90% Train	2	0.002011	0.002011	−0.0004%	0.002066	0.001608	0.001608	−0.0003%	0.001651
	3	0.001895	0.001895	−0.0089%	0.001932	0.001517	0.001518	−0.0074%	0.001547
	5	0.001822	0.001823	−0.0469%	0.001833	0.001457	0.001458	−0.0659%	0.001469
	11	0.001781	0.001786	−0.2562%	0.001746	0.001424	0.001428	−0.2930%	0.001397
	23	0.001764	0.001768	−0.2804%	0.001739	0.001410	0.001414	−0.2854%	0.001393
(c). HEH-Ceiling dataset									
50% Train	2	0.003647	0.003647	0.0124%	0.006317	0.002905	0.002905	0.0145%	0.005473
	3	0.003459	0.003459	0.0000%	0.006060	0.002756	0.002756	0.0000%	0.005238
	5	0.003336	0.003336	0.0000%	0.005793	0.002655	0.002655	0.0000%	0.004986
	11	0.003298	0.003284	0.4034%	0.005135	0.002627	0.002617	0.3951%	0.004364
	23	0.003264	0.003243	0.6477%	0.004621	0.002602	0.002585	0.6407%	0.003859
90% train	2	0.003494	0.003494	−0.0022%	0.003411	0.002776	0.002776	−0.0029%	0.002727
	3	0.003277	0.003277	0.0000%	0.003247	0.002603	0.002603	0.0000%	0.002595
	5	0.003160	0.003160	0.0000%	0.0031 30	0.002516	0.002516	0.0000%	0.002508
	11	0.003105	0.003126	−0.6876%	0.003034	0.002478	0.002494	−0.6426%	0.002430
	23	0.003087	0.003160	-2.3520%	0.002991	0.002460	0.002520	-2.4356%	0.002395

the independent approach in the HVACS Evaluation House datasets. The reason is that the weather dataset is collected from a very large time period (ten years), while the HVACS Evaluation House datasets are collected from a short time period (one day). The time series data in the weather dataset is thus possible to be much more non-stationary. Even the split rate of the training set is high, the independent regression model is still under-fitting to the weather data. In contrast, the independent regression model can fit the HVACS Evaluation House datasets well when there are many training data. In other words, our approach can outperform the independent approach when the bias of the independent model is high. Lack of enough training data is one of the reasons which can cause the under-fitting of the independent model.

The centralized approach is ideal and can use all observation data from all sensors for learning the models. Therefore, when there are plenty of training data, generally we expect that it can reach better performance than the independent approach and our distributed multi-task approach (e.g., weather dataset results in Tables 4 and 5). The advantages and disadvantages of the centralized approach and the reasons for proposing decentralized approaches have been discussed at the beginning of the introduction section. In addition, because the candidate sensors are not selected and the model is not specialized for a given sensor, it is possible to have worse performance than our approach when the training data is not large enough for it. (e.g., HVACS Evaluation House datasets results in Tables 4 and 5).

5   Conclusion

In this paper, we focus on the problem of learning decentralized local models for each sensor in a sensor network. We propose a novel distributed multi-task learning approach which incorporates neighborhood relations among sensors learn multiple models simultaneously in which each sensor corresponds to one task. Instead of broadcasting the observation data of a sensor in the entire network which is not satisfied for the reason of cost and privacy protection, this approach shares the model parameters as regularization terms in the objective function by assuming that neighbor sensors have similar model parameters. We conduct the experiments on three real datasets by predicting the temperature with the regression. They verify that our approach is effective, especially when the bias of an independent model which does not utilize the data from other sensors is high such as when there is not plenty of training data available.

In this paper, we select linear prediction because it is commonly used in practical cases. It would be useful to consider the solution for other types of complex models rather than linear prediction. On one hand, for many complex models which solve an optimization problem, we can add a regularization term of parameters in the object function like our approach. On the other hand, for complex models which cannot be solved·in such manner, other specific topics need to be proposed, e.g., for clustering task and so on. We will address them in the future work. We assume that neighbor sensors have similar model parameters. If some neighbor sensors show a significant different behavior, e.g., the regularization term is high. It can be a useful clue for detecting fault sensors or unordinary places. These issues can be separated topics in the future work.

References

1. Ahmadi, M., Huang, Y., John, K.: Application of spatio-temporal clustering for predicting ground-level ozone pollution. In: Griffith, D., Chun, Y., Dean, D. (eds.) Advances in Geocomputation. Advances in Geographic Information Science, pp. 153–167. Springer, Heidelberg (2017). https://doi.org/10.1007/978-3-319-22786-3_15

2. Argyriou, A., Evgeniou, T., Pontil, M.: Multi-task feature learning. In: Advances in Neural Information Processing Systems, vol. 19, p. 41 (2007)

3. Beck, A., Tetruashvili, L.: On the convergence of block coordinate descent type methods. SIAM J. Optim. **23**(4), 2037–2060 (2013)

4. Caruana, R.: Multitask learning. Mach. Learn. **28**(1), 41–75 (1997). http://dx.doi.org/10.1023/A:1007379606734

5. Ceci, M., Corizzo, R., Fumarola, F., Malerba, D., Rashkovska, A.: Predictive modeling of PV energy production: how to set up the learning task for a better prediction? IEEE Trans. Indus. Inform. **13**(3), 956–966 (2017)

6. Dinuzzo, F., Pillonetto, G., De Nicolao, G.: Client-server multitask learning from distributed datasets. IEEE Trans. Neural Netw. **22**(2), 290–303 (2011)

7. Evgeniou, T., Micchelli, C.A., Pontil, M.: Learning multiple tasks with kernel methods. J. Mach. Learn. Res. **6**(Apr), 615–637 (2005)

8. Evgeniou, T., Pontil, M.: Regularized multi-task learning. In: Proceedings of the Tenth ACM SIGKDD International Conference on Knowledge Discovery and Data Mining, pp. 109–117. ACM (2004)

9. Gers, F.A., Eck, D., Schmidhuber, J.: Applying LSTM to time series predictable through time-window approaches. In: Dorffner, G., Bischof, H., Hornik, K. (eds.) ICANN 2001. LNCS, vol. 2130, pp. 669–676. Springer, Heidelberg (2001). https://doi.org/10.1007/3-540-44668-0_93

10. Guestrin, C., Bodik, P., Thibaux, R., Paskin, M., Madden, S.: Distributed regression: an efficient framework for modeling sensor network data. In: Third International Symposium on Information Processing in Sensor Networks, IPSN 2004, pp. 1–10. IEEE (2004)

11. Hochreiter, S., Schmidhuber, J.: Long short-term memory. Neural Comput. **9**(8), 1735–1780 (1997)

12. Low, Y., Gonzalez, J.E., Kyrola, A., Bickson, D., Guestrin, C.E., Hellerstein, J.: Graphlab: a new framework for parallel machine learning. arXiv preprint arXiv:1408.2041 (2014)

13. Sak, H., Senior, A.W., Beaufays, F.: Long short-term memory recurrent neural network architectures for large scale acoustic modeling. In: Interspeech, pp. 338–342 (2014)

14. Shoham, Y., Powers, R., Grenager, T.: Multi-agent reinforcement learning: a critical survey. Technical report, Stanford University (2003)

15. Tulone, D., Madden, S.: PAQ: time series forecasting for approximate query answering in sensor networks. In: Römer, K., Karl, H., Mattern, F. (eds.) EWSN 2006. LNCS, vol. 3868, pp. 21–37. Springer, Heidelberg (2006). https://doi.org/10.1007/11669463_5

16. Vanhaesebrouck, P., Bellet, A., Tommasi, M.: Decentralized collaborative learning of personalized models over networks. In: Artificial Intelligence and Statistics, pp. 509–517 (2017)

17. Wang, J., Kolar, M., Srebro, N., et al.: Distributed multi-task learning. In: Proceedings of the 19th International Conference on Artificial Intelligence and Statistics (AISTATS), pp. 751–760 (2016)

Learning Task Clusters via Sparsity Grouped Multitask Learning

Meghana Kshirsagar[1(✉)], Eunho Yang[2], and Aurélie C. Lozano[3]

[1] Memorial Sloan Kettering Cancer Center,
1275 York Avenue, New York, NY, USA
meghana.ksagar@gmail.com

[2] School of Computing, Korea Advanced Institute of Science and Technology,
Daejeon, South Korea
eunhoy@kaist.ac.kr

[3] IBM T. J. Watson Research, Yorktown Heights, New York, NY, USA
aclozano@us.ibm.com

Abstract. Sparse mapping has been a key methodology in many high-dimensional scientific problems. When multiple tasks share the set of relevant features, learning them jointly in a group drastically improves the quality of relevant feature selection. However, in practice this technique is used limitedly since such grouping information is usually hidden. In this paper, our goal is to recover the group structure on the sparsity patterns and leverage that information in the sparse learning. Toward this, we formulate a joint optimization problem in the task parameter and the group membership, by constructing an appropriate regularizer to encourage sparse learning as well as correct recovery of task groups. We further demonstrate that our proposed method recovers groups and the sparsity patterns in the task parameters accurately by extensive experiments.

1 Introduction

Humans acquire knowledge and skills by categorizing the various problems/tasks encountered, recognizing how the tasks are related to each other and taking advantage of this organization when learning a new task. Statistical machine learning methods also benefit from exploiting such similarities in learning related problems. Multitask learning (MTL) (Caruana 1997) is a paradigm of machine learning, encompassing learning algorithms that can share information among related tasks and help to perform those tasks together more efficiently than in isolation. These algorithms exploit task relatedness by various mechanisms. Some works enforce that parameters of various tasks are close to each other in some geometric sense (Evgeniou and Pontil 2004; Maurer 2006). Several works leverage the existence of a shared low dimensional subspace (Argyriou et al. 2008; Liu et al. 2009; Jalali et al. 2010; Chen et al. 2012) or manifold (Agarwal et al. 2010) that contains the task parameters. Some bayesian MTL methods assume the same prior on parameters of related tasks (Yu et al. 2005; Daumé III 2009), while neural networks based methods share some hidden units (Baxter 2000).

This work was done while MK and EY were at IBM T. J. Watson research.

M. Ceci et al. (Eds.): ECML PKDD 2017, Part II, LNAI 10535, pp. 673–689, 2017.
https://doi.org/10.1007/978-3-319-71246-8_41

A key drawback of most MTL methods is that they assume that all tasks are equally related. Intuitively, learning unrelated tasks jointly may result in poor predictive models; i.e. tasks should be coupled based on their relatedness. While the coupling of task parameters can sometimes be controlled via hyper-parameters, this is infeasible when learning several hundreds of tasks. Often, knowing the *task relationships* themselves is of interest to the application at hand. While these relationships might sometimes be derived from domain specific intuition (Kim and Xing 2010; Widmer et al. 2010; Rao et al. 2013), they are either not always known apriori or are pre-defined based on the knowledge of $P(X)$ rather than $P(Y|X)$. We aim to automatically learn these task relationships, while simultaneously learning individual task parameters. This idea of jointly learning task groups and parameters has been explored in prior works. For instance Argyriou et al. (2008) learn a set of kernels, one per group of tasks and Jacob et al. (2009) cluster tasks based on similarity of task parameters. Others (Zhang and Yeung 2010; Gong et al. 2012) try to identify "outlier" tasks. Kumar and Daumé III (2012); Kang et al. (2011) assume that task parameters within a group lie in a shared low dimensional subspace. Zhang and Schneider (2010) use a matrix-normal regularization to capture task covariance and feature covariance between tasks and enforce sparsity on these covariance parameters and Fei and Huan (2013) use a similar objective with a structured regularizer. Their approach is however, not suited for high dimensional settings and they do not enforce any sparsity constraints on the task parameters matrix W. A Bayesian approach is proposed in Passos et al. (2012), where parameters are assumed to come from a nonparametric mixture of nonparametric factor analyzers.

Here, we explore the notion of *shared sparsity* as the structure connecting a group of related tasks. More concretely, we assume that tasks in a group all have similar relevant features or analogously, the same zeros in their parameter vectors. Sparsity inducing norms such as the ℓ_1 norm capture the principle of parsimony, which is important to many real-world applications, and have enabled efficient learning in settings with high dimensional feature spaces and few examples, via algorithms like the Lasso (Tibshirani 1996). When confronted by several tasks where sparsity is required, one modeling choice is for each task to have its' own sparse parameter vector. At the other extreme is the possibility of enforcing shared sparsity on all tasks via a structured sparsity inducing norm such as ℓ_1/ℓ_2 on the task parameter matrix: $\|W\|_{1,2}$ (Bach et al. 2011)[1]. We choose to enforce sparsity at a group level by penalizing $\|W_g\|_{1,2}$, where $\|W_g\|$ is the parameter matrix for all tasks in group g, while learning group memberships of tasks.

To see why this structure is interesting and relevant, consider the problem of transcription factor (TF) binding prediction. TFs are proteins that bind to the DNA to regulate expression of nearby genes. The binding specificity of a TF to an arbitrary location on the DNA depends on the pattern/sequence of nucleic acids (A/C/G/T) at that location. These sequence preferences of TFs

---

[1] Note: this cross-task structured sparsity is different from the Group Lasso (Yuan and Lin 2006), which groups covariates within a task ($\min_{w \in \mathbb{R}^d} \sum_g \|w_g\|$, where w_g is a group of parameters).

have some similarities among related TFs. Consider the task of predicting TF binding, given segments of DNA sequence (these are the examples), on which we have derived features such as n-grams (called k-mers)[2]. The feature space is very high dimensional and a small set of features typically capture the binding pattern for a single TF. Given several tasks, each representing one TF, one can see that the structure of the ideal parameter matrix is likely to be group sparse, where TFs in a group have similar binding patterns (i.e. similar important features but with different weights). The applicability of task-group based sparsity is not limited to isolated applications, but desirable in problems involving billions of features, as is the case with web-scale information retrieval and in settings with few samples such as genome wide association studies involving millions of genetic markers over a few hundred patients, where only a few markers are relevant.

The main contributions of this work are:

- We present a new approach towards learning task group structure in a multi-task learning setting that simultaneously learns both the task parameters W and a clustering over the tasks.
- We define a regularizer that divides the set of tasks into groups such that all tasks within a group share the same sparsity structure. Though the ideal regularizer is discrete, we propose a relaxed version and we carefully make many choices that lead to a feasible alternating minimization based optimization strategy. We find that several alternate formulations result in substantially worse solutions.
- We evaluate our method through experiments on synthetic datasets and two interesting real-world problem settings. The first is a regression problem: QSAR, quantitative structure activity relationship prediction (see (Ma et al. 2015) for an overview) and the second is a classification problem important in the area of regulatory genomics: transcription factor binding prediction (described above). On synthetic data with known group structure, our method recovers the correct structure. On real data, we perform better than prior MTL group learning baselines.

1.1   Relation to Prior Work

Our work is most closely related to Kang et al. (2011), who assume that each group of tasks shares a latent subspace. They find groups so that $\|W_g\|_*$ for each group g is small, thereby enforcing sparsity in a *transformed* feature space. Another approach, GO-MTL (Kumar and Daumé III 2012) is based on the same idea, with the exception that the latent subspace is shared among all tasks, and a low-rank decomposition of the parameter matrix $W = LS$ is learned. Subsequently, the coefficients matrix S is clustered to obtain a grouping of tasks. Note that, learning group memberships is not the goal of their approach, but rather a post-processing step upon learning their model parameters.

To understand the distinction from prior work, consider the weight matrix W^* in Fig. 4(a), which represents the true group sparsity structure that we wish

---

[2] e.g.: GTAATTNC is an 8-mer ('N' represents a wild card).

to learn. While each task group has a low-rank structure (since s of the d features are non-zero, the rank of any $\boldsymbol{W}_g$ is bounded by s), it has an additional property that $(d-s)$ features are zero or irrelevant for all tasks in this group. Our method is able to exploit this additional information to learn the correct sparsity pattern in the groups, while that of Kang et al. (2011) is unable to do so, as illustrated on this synthetic dataset in Fig. 5 (details of this dataset are in Sect. 5.1). Though Kang et al. (2011) recovers some of the block diagonal structure of W, there are many non-zero features which lead to an incorrect group structure. We present a further discussion on how our method is sample efficient as compared to Kang et al. (2011) for this structure of W in Sect. 3.1.

We next present the setup and notation, and lead to our approach by starting with a straight-forward combinatorial objective and then make changes to it in multiple steps (Sects. 2–4). At each step we explain the behaviour of the function to motivate the particular choices we made; present a high-level analysis of sample complexity for our method and competing methods. Finally we show experiments (Sect. 5) on four datasets.

2   Setup and Motivation

We consider the standard setting of multi-task learning in particular where tasks in the same group share the sparsity patterns on parameters. Let $\{T_1, \ldots, T_m\}$ be the set of m tasks with training data $\mathcal{D}_t$ $(t = 1 \ldots m)$. Let the parameter vectors corresponding to each of the m tasks be $\boldsymbol{w}^{(1)}, \boldsymbol{w}^{(2)}, \ldots, \boldsymbol{w}^{(m)} \in \mathbb{R}^d$, d is the number of covariates/features. Let $\mathcal{L}(\cdot)$ be the loss function which, given $\mathcal{D}_t$ and $\boldsymbol{w}^{(t)}$ measures deviation of the predictions from the response. Our goal is to learn the task parameters where (i) each $\boldsymbol{w}^{(t)}$ is assumed to be sparse so that the response of the task can be succinctly explained by a small set of features, and moreover (ii) there is a partition $\mathcal{G}^* := \{G_1, G_2, \ldots, G_N\}$ over tasks such that all tasks in the same group G_i have the same sparsity patterns. Here N is the total number of groups learned. If we learn every task independently, we solve m independent optimization problems:

$$\underset{\boldsymbol{w}^{(t)} \in \mathbb{R}^d}{\text{minimize}} \ \mathcal{L}(\boldsymbol{w}^{(t)}; \mathcal{D}_t) + \lambda \|\boldsymbol{w}^{(t)}\|_1$$

where $\|\boldsymbol{w}^{(t)}\|_1$ encourages sparse estimation with regularization parameter λ. However, if $\mathcal{G}^*$ is given, jointly estimating all parameters together using a group regularizer (such as ℓ_1/ℓ_2 norm), is known to be more effective. This approach requires fewer samples to recover the sparsity patterns by sharing information across tasks in a group:

$$\underset{\boldsymbol{w}^{(1)}, \ldots, \boldsymbol{w}^{(m)}}{\text{minimize}} \ \sum_{t=1}^{m} \mathcal{L}(\boldsymbol{w}^{(t)}; \mathcal{D}_t) + \sum_{g \in \mathcal{G}^*} \lambda_g \|\boldsymbol{W}_g\|_{1,2} \tag{1}$$

where $\boldsymbol{W}_g \in \mathbb{R}^{d \times |g|}$, where $|g|$ is the number of tasks in the group g and $\| \cdot \|_{1,2}$ is the sum of ℓ_2 norms computed over row vectors. Say $t_1, t_2 \ldots t_k$ belong to group

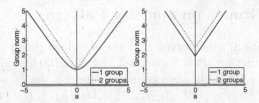

Fig. 1. Toy examples with two fixed parameter vectors: (left) $\boldsymbol{w}^{(1)} = (1, 0, 0)^\top$, $\boldsymbol{w}^{(2)} = (a, 0, 0)^\top$, and (right) $\boldsymbol{w}^{(1)} = (1/2, 1, 0)^\top$, $\boldsymbol{w}^{(2)} = (0, 1, a)^\top$, where we only vary one coordinate a fixing all others to visualize the norm values in 2-d space. Functions show the group norms, $\sum_g \|\boldsymbol{W}\boldsymbol{U}_g\|_{1,2}$ in (2) where two tasks belong to a single group (solid) or to separate groups (dotted). In both cases, this group regularizer favors the case with a single group.

g, then $\|\boldsymbol{W}_g\|_{1,2} := \sum_{j=1}^d \sqrt{(w_j^{(t_1)})^2 + (w_j^{(t_2)})^2 + \ldots + (w_j^{(t_k)})^2}$. Here $w_j^{(t)}$ is the j-th entry of vector $\boldsymbol{w}^{(t)}$. Note that here we use ℓ_2 norm for grouping, but any ℓ_α norm $\alpha \geq 2$ is known to be effective.

We introduce a membership parameter $u_{g,t}$: $u_{g,t} = 1$ if task T_t is in a group g and 0 otherwise. Since we are only allowing a hard membership without overlapping (though this assumption can be relaxed in future work), we should have exactly one active membership parameter for each task: $u_{g,t} = 1$ for some $g \in \mathcal{G}$ and $u_{g',t} = 0$ for all other $g' \in \mathcal{G} \setminus \{g\}$. For notational simplicity, we represent the group membership parameters for a group g in the form of a matrix $\boldsymbol{U}_g$. This is a diagonal matrix where $\boldsymbol{U}_g := \text{diag}(u_{g,1}, u_{g,2}, \ldots, u_{g,m}) \in \{0, 1\}^{m \times m}$. In other words, $[\boldsymbol{U}_g]_{ii} = u_{g,i} = 1$ if task T_i is in group g and 0 otherwise. Now, incorporating $\boldsymbol{U}$ in (1), we can derive the optimization problem for learning the task parameters $\{\boldsymbol{w}^{(t)}\}_{t=1,\ldots,m}$ and $\boldsymbol{U}$ simultaneously as follows:

$$\underset{\boldsymbol{W},\boldsymbol{U}}{\text{minimize}} \sum_{t=1}^m \mathcal{L}(\boldsymbol{w}^{(t)}; \mathcal{D}_t) + \sum_{g \in \mathcal{G}} \lambda_g \|\boldsymbol{W}\boldsymbol{U}_g\|_{1,2}$$

$$\text{s.t.} \sum_{g \in \mathcal{G}} \boldsymbol{U}_g = \boldsymbol{I}^{m \times m}, \quad [\boldsymbol{U}_g]_{ii} \in \{0, 1\}. \tag{2}$$

where $\boldsymbol{W} \in \mathbb{R}^{d \times m} := [\boldsymbol{w}^{(1)}, \boldsymbol{w}^{(2)}, \ldots, \boldsymbol{w}^{(m)}]$. Here $\boldsymbol{I}^{m \times m}$ is the $m \times m$ identity matrix. After solving this problem, $\boldsymbol{U}$ encodes which group the task T_t belongs to. It turns out that this simple extension in (2) fails to correctly infer the group structure as it is biased towards finding a smaller number of groups. Figure 1 shows a toy example illustrating this. The following proposition generalizes this issue.

Proposition 1. *Consider the problem of minimizing (2) with respect to $\boldsymbol{U}$ for a fixed $\widehat{\boldsymbol{W}}$. The assignment such that $\widehat{\boldsymbol{U}}_g = \boldsymbol{I}^{m \times m}$ for some $g \in \mathcal{G}$ and $\widehat{\boldsymbol{U}}_{g'} = \boldsymbol{0}^{m \times m}$ for all other $g' \in \mathcal{G} \setminus \{g\}$, is a minimizer of (2).*

Proof: Please refer to the appendix.

3   Learning Groups on Sparsity Patterns

In the previous section, we observed that the standard group norm is beneficial when the group structure $\mathcal{G}^*$ is known but not suitable for inferring it. This is mainly because it is basically aggregating groups via the ℓ_1 norm; let $\boldsymbol{v} \in \mathbb{R}^N$ be a vector of $(\|\boldsymbol{W}\boldsymbol{U}_1\|_{1,2}, \|\boldsymbol{W}\boldsymbol{U}_2\|_{1,2}, \ldots, \|\boldsymbol{W}\boldsymbol{U}_N\|_{1,2})^\top$, then the regularizer of (2) can be understood as $\|\boldsymbol{v}\|_1$. By the basic property of ℓ_1 norm, $\boldsymbol{v}$ tends to be a sparse vector, making $\boldsymbol{U}$ have a small number of active groups (we say some group g is active if there exists a task T_t such that $u_{g,t} = 1$).

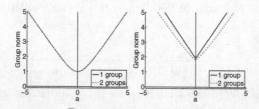

Fig. 2. For the two toy examples from Fig. 1, we show the behaviour of $\left(\sum_g (\|\boldsymbol{W}\boldsymbol{U}_g\|_{1,2})^2\right)^{0.5}$ (the group regularizer in (3) with $\alpha = 0.5$). See the caption of Fig. 1 for the choice of $\boldsymbol{W} = [\boldsymbol{w}^{(1)}, \boldsymbol{w}^{(2)}]$. In the example on the right, the regularizer now favors putting the tasks in separate groups.

Based on this finding, we propose to use the ℓ_α norm ($\alpha \geq 2$) for summing up the regularizers from different groups, so that the final regularizer as a whole forces most of $\|\boldsymbol{W}\boldsymbol{U}_g\|_{1,2}$ to be non-zeros:

$$\operatorname*{minimize}_{\boldsymbol{W},\boldsymbol{U}} \sum_{t=1}^m \mathcal{L}(\boldsymbol{w}^{(t)}; \mathcal{D}_t) + \sum_{g \in \mathcal{G}} \lambda_g \left(\|\boldsymbol{W}\boldsymbol{U}_g\|_{1,2}\right)^\alpha$$

$$\text{s.t.} \sum_{g \in \mathcal{G}} \boldsymbol{U}_g = \boldsymbol{I}^{m \times m}, \quad [\boldsymbol{U}_g]_{ik} \in \{0, 1\}. \tag{3}$$

Note that strictly speaking, $\|\boldsymbol{v}\|_\alpha$ is defined as $(\sum_{i=1}^N |v_i|^\alpha)^{1/\alpha}$, but we ignore the relative effect of $1/\alpha$ in the exponent. One might want to get this exponent back especially when α is large. ℓ_α norms give rise to exactly the opposite effects in distributions of $u_{g,t}$, as shown in Fig. 2 and Proposition 2.

Proposition 2. *Consider a minimizer $\widehat{\boldsymbol{U}}$ of (3), for any fixed $\widehat{\boldsymbol{W}}$. Suppose that there exist two tasks in a single group such that $\widehat{w}_i^{(s)} \widehat{w}_j^{(t)} \neq \widehat{w}_j^{(s)} \widehat{w}_i^{(t)}$. Then there is no empty group g such that $\widehat{\boldsymbol{U}}_g = \boldsymbol{0}^{m \times m}$.*

Proof: Please refer to the appendix.

Figure 3 visualizes the unit surfaces of different regularizers derived from (3) (i.e. $\sum_{g \in \mathcal{G}} (\|\boldsymbol{W}\boldsymbol{U}_g\|_{1,2})^\alpha = 1$ for different choices of α.) for the case where we

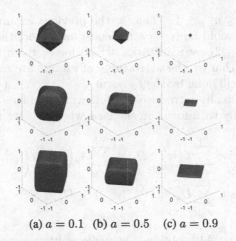

(a) $a = 0.1$ (b) $a = 0.5$ (c) $a = 0.9$

Fig. 3. Unit balls of the regularizer in (3) for different values of α. Suppose we have $\boldsymbol{w}^{(1)} = (x, y)^\top$ in group G_1 and $\boldsymbol{w}^{(2)} = (z, a)^\top$ in G_2. In order to visualize in 3-d space, we vary 3 variables x, y and z, and fix $w_2^{(2)}$ to some constant a. The first row is using ℓ_1 norm for summing the groups: $(|x| + |y|) + (|z| + a)$, the second row is using ℓ_2 norm: $\sqrt{(|x| + |y|)^2 + (|z| + a)^2}$, and the last row is using ℓ_5 norm: $\left((|x| + |y|)^5 + (|z| + a)^5\right)^{0.2}$. As a increases (from the first column to the third one), $\boldsymbol{w}_1$ quickly shrinks to zero in case of ℓ_1 summation. One the other hand, in case of ℓ_2 summation, x and y in $\boldsymbol{w}^{(1)}$ are allowed to be non-zero, while z shrinks to zero. This effect gets clearer as α increases.

have two groups, each of which has a single task. It shows that a large norm value on one group (in this example, on G_2 when $a = 0.9$) does not force the other group (i.e. G_1) to have a small norm as α becomes larger. This is evidenced in the bottom two rows of the third column (compare it with how ℓ_1 behaves in the top row). In other words, we see the benefits of using $\alpha \geq 2$ to encourage more active groups.

While the constraint $[\boldsymbol{U}_g]_{ik} \in \{0, 1\}$ in (3) ensures hard group memberships, solving it requires integer programming which is intractable in general. Therefore, we relax the constraint on $\boldsymbol{U}$ to $0 \leq [\boldsymbol{U}_g]_{ik} \leq 1$. However, this relaxation along with the ℓ_α norm over groups prevents both $\|\boldsymbol{W}\boldsymbol{U}_g\|_{1,2}$ and also individual $[\boldsymbol{U}_g]_{ik}$ from being zero. For example, suppose that we have two tasks (in $\mathbb{R}^2$) in a single group, and $\alpha = 2$. Then, the regularizer for any g can be written as $\left(\sqrt{(w_1^{(1)})^2\, u_{g,1}^2 + (w_1^{(2)})^2\, u_{g,2}^2} + \sqrt{(w_2^{(1)})^2\, u_{g,1}^2 + (w_2^{(2)})^2\, u_{g,2}^2}\right)^2$. To simplify the situation, assume further that all entries of $\boldsymbol{W}$ are uniformly a constant w. Then, this regularizer for a single g would be simply reduced to $4w^2(u_{g,1}^2 + u_{g,2}^2)$, and therefore the regularizer over all groups would be $4w^2(\sum_{t=1}^m \sum_g u_{g,t}^2)$. Now it is clearly seen that the regularizer has an effect of grouping over the group membership vector $(u_{g_1,t}, u_{g_2,t}, \ldots, u_{g_N,t})$ and encouraging the set of membership parameters for each task to be uniform.

To alleviate this challenge, we re-parameterize $u_{g,t}$ with a new membership parameter $u'_{g,t} := \sqrt{u_{g,t}}$. The constraint does not change with this reparameterization: $0 \le u'_{g,t} \le 1$. Then, in the previous example, the regularization over all groups would be (with some constant factor) the sum of ℓ_1 norms, $\|(u_{g_1,t}, u_{g_2,t}, \dots, u_{g_N,t})\|_1$ over all tasks, which forces them to be sparse. Note that even with this change, the activations of groups are not sparse since the sum over groups is still done by the ℓ_2 norm.

Toward this, we finally introduce the following problem to jointly estimate U and W (specifically with focus on the case when α is set to 2):

$$\underset{W,U}{\text{minimize}} \sum_{t=1}^{m} \mathcal{L}(w^{(t)}; \mathcal{D}_t) + \sum_{g \in \mathcal{G}} \lambda_g \left(\left\| W \sqrt{U_g} \right\|_{1,2} \right)^2$$

$$\text{s.t.} \sum_{g \in \mathcal{G}} U_g = I^{m \times m}, \quad 0 \le [U_g]_{ik} \le 1. \tag{4}$$

where $\sqrt{M}$ for a matrix M is obtained from element-wise square root operations of M. Note that (3) and (4) are not equivalent, but the minimizer $\widehat{U}$ given any fixed W usually is actually binary or has the same objective with some other binary $\widehat{U}'$ (see Theorem 1 of Kang et al. (2011) for details). As a toy example, we show in Fig. 4, the estimated U (for a known W) via different problems presented so far (2), (4).

Fusing the Group Assignments. The approach derived so far works well when the number of groups $N << m$, but can create many singleton groups when N is very large. We add a final modification to our objective to encourage tasks to have similar group membership wherever warranted. This makes the method more robust to the mis-specification of the number of groups, 'N' as it prevents the grouping from becoming too fragmented when $N >> N^*$. For each task

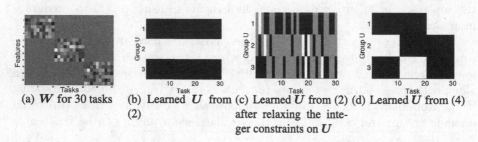

(a) W for 30 tasks   (b) Learned U from (2)   (c) Learned U from (2) after relaxing the integer constraints on U (d) Learned U from (4)

Fig. 4. Comparisons on different regularizers. (a) W for 30 tasks with 21 features, which is assumed to be *known and fixed*. Three groups are clearly separated: (T_1–T_{10}), (T_{11}–T_{20}), (T_{21}–T_{30}) whose nonzero elements are block diagonally located (black is negative, white is positive). (b) Learned U from (2) (with a relaxing of discrete constraints, and the square root reparameterization). All tasks are assigned to a single group. (c) Estimation on U using (2) after relaxing the integer constraints on U. All there groups are active (have tasks in them), but most of $u_{g,t}$ are not binary. (d) Estimation on U using (4). All group memberships are recovered correctly. Note that the order of groups does not matter. For (b)–(d), white is 1 and black is 0.

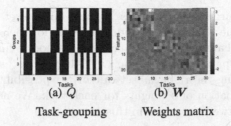

(a) Q (b) W

Task-grouping   Weights matrix

Fig. 5. Comparison of the regularizer from Kang et al. (2011) when learned on our synthetic dataset (set-1). Figure 4(a) shows the W^*. (a) the learned Q, analogous to U in our notation (white is 1 and black is 0) and (b) $\widehat{W}$

$t = 1, \ldots, m$, we define $N \times N$ matrix $V_t := \mathrm{diag}(u_{1,t}, \ldots, u_{N,t})$. Note that the V_t are entirely determined by the U_g matrices, so no actual additional variables are introduced. Equipped with this additional notation, we obtain the following objective where $\| \cdot \|_F$ denotes the Frobenius norm (the element-wise ℓ_2 norm), and μ is the additional regularization parameter that controls the number of active groups.

$$\underset{W,U}{\text{minimize}} \sum_{t=1}^m \mathcal{L}(w^{(t)}; \mathcal{D}_t) + \sum_{g \in \mathcal{G}} \lambda_g \left(\left\| W \sqrt{U_g} \right\|_{1,2} \right)^2$$

$$+ \mu \sum_{t<t'} \left\| V_t - V_{t'} \right\|_F^2$$

$$\text{s.t.} \sum_{g \in \mathcal{G}} U_g = I^{m \times m}, \quad [U_g]_{ik} \in [0,1] \tag{5}$$

3.1 Theoretical Comparison of Approaches

It is natural to ask whether enforcing the shared sparsity structure, when groups are unknown leads to any efficiency in the number of samples required for learning. In this section, we will use intuitions from the high-dimensional statistics literature in order to compare the sample requirements of different alternatives such as independent lasso or the approach of Kang et al. (2011). Since the formal analysis of each method requires making different assumptions on the data X and the noise, we will instead stay intentionally informal in this section, and contrast the number of samples each approach would require, assuming that the desired structural conditions on the x's are met. We evaluate all the methods under an idealized setting where the structural assumptions on the parameter matrix W motivating our objective (2) hold exactly. That is, the parameters form N groups, with the weights in each group taking non-zero values only on a common subset of features of size at most s. We begin with our approach.

Complexity of Sparsity Grouped MTL. Let us consider the simplest inefficient version of our method, a generalization of subset selection for Lasso which searches over all feature subsets of size s. It picks one subset S_g for each group g and then estimates the weights on S_g independently for each task in group g. By a simple union bound argument, we expect this method to find the right support sets, as well as good parameter values in $O(Ns \log d + ms)$ samples. This is the complexity of selecting the right subset out of $\binom{d}{s}$ possibilities for each group, followed by the estimation of s weights for each task. We note that there is no direct interaction between m and d in this bound.

Complexity of Independent Lasso per Task. An alternative approach is to estimate an s-sparse parameter vector for each task independently. Using standard bounds for ℓ_1 regularization (or subset selection), this requires $O(s \log d)$ samples per task, meaning $O(ms \log d)$ samples overall. We note the multiplicative interaction between m and $\log d$ here.

Complexity of Learning All Tasks Jointly. A different extreme would be to put all the tasks in one group, and enforce shared sparsity structure across them using $\| \cdot \|_{1,2}$ regularization on the entire weight matrix. The complexity of this approach depends on the sparsity of the union of all tasks which is Ns, much larger than the sparsity of individual groups. Since each task requires to estimate its own parameters on this shared sparse basis, we end up requiring $O(msN \log d)$ samples, with a large penalty for ignoring the group structure entirely.

Algorithm 1. SG-MTL (Eq. (4))

Input: $\{\mathcal{D}_t\}_{t=1}^m$
Initialize W using single task learning
Initialize U to random matrix
repeat
    Update U by solving a projected gradient descent
    for all tasks $t = 1, 2, \ldots, m$ **do**
        Update $w^{(t)}$ using a coordinate descent for all features $j = 1, 2, \ldots, d$
    end for
until stopping criterion is satisfied

Complexity of Kang et al. (2011). As yet another baseline, we observe that an s-sparse weight matrix is also naturally low-rank with rank at most s. Consequently, the weight matrix for each group has rank at most s, plausibly making this setting a good fit for the approach of Kang et al. (2011). However, appealing to the standard results for low-rank matrix estimation (see e.g. Negahban and Wainwright (2011)), learning a $d \times n_g$ weight matrix of rank at most s requires $O(s(n_g + d))$ samples, where n_g is the number of tasks in the group g. Adding up across tasks, we find that this approach requires a total of $O(s(m + md))$, considerably higher than all other baselines even if the groups are already provided. It is easy to see why this is unavoidable too. Given a group, one requires $O(ms)$ samples to estimate the entries of the s linearly independent rows. A method utilizing sparsity information knows that the rest of the columns are filled with zeros, but one that only knows that the matrix is low-rank assumes that the remaining $(d-s)$ rows all lie in the linear span of these s rows, and the coefficients of that linear combination need to be estimated giving rise to the additional sample complexity. In a nutshell, this conveys that estimating a sparse matrix using low-rank regularizers is sample inefficient, an observation hardly surprising from the available results in high-dimensional statistics but important in comparison with the baseline of Kang et al. (2011).

For ease of reference, we collect all these results in Table 1 below.

Table 1. Sample complexity estimates of recovering group memberships and weights using different approaches

	SG-MTL	Lasso	Single group	Kang et al. (2011)
Samples	$O(Ns \log d + ms)$	$O(ms \log d)$	$O(msN \log d)$	$O(s(m+md))$

4 Optimization

We solve (4) by alternating minimization: repeatedly solve one variable fixing the other until convergence (Algorithm 1) We discuss details below.

Solving (4) w.r.t U: This step is challenging since we lose convexity due the reparameterization with a square root. The solver might stop with a premature U stuck in a local optimum. However, in practice, we can utilize the random search technique to get the minimum value over multiple re-trials. Our experimental results reveal that the following projected gradient descent method performs well.

Given a fixed W, solving for U only involves the regularization term i.e. $R(U) = \sum_{g \in \mathcal{G}} \lambda_g \left(\sum_{j=1}^{d} \left\| W_j \sqrt{U_g} \right\|_2 \right)^2$ which is differentiable w.r.t U. The derivative is shown in the appendix along with the extension for the fusion penalty from (5). Finally after the gradient descent step, we project $(u_{g_1,t}, u_{g_2,t}, \ldots, u_{g_N,t})$ onto the simplex (independently repeat the projection for each task) to satisfy the constraints on it. Note that projecting a vector onto a simplex can be done in $O(m \log m)$ (Chen and Ye 2011).

Solving (4) w.r.t W. This step is more amenable in the sense that (4) is convex in W given U. However, it is not trivial to efficiently handle the complicated regularization terms. Contrast to U which is bounded by $[0,1]$, W is usually unbounded which is problematic since the regularizer is (not always, but under some complicated conditions discovered below) non-differentiable at 0.

While it is challenging to directly solve with respect to the entire W, we found out that the coordinate descent (in terms of each element in W) has a particularly simple structured regularization.

Consider any $w_j^{(t)}$ fixing all others in $\boldsymbol{W}$ and $\boldsymbol{U}$; the regularizer $R(\boldsymbol{U})$ from (4) can be written as

$$\sum_{g \in \mathcal{G}} \lambda_g \left\{ u_{g,t}(w_j^{(t)})^2 + 2 \left(\sum_{j' \neq j} \sqrt{\sum_{t'=1}^{m} u_{g,t'}(w_{j'}^{(t')})^2} \right) \sqrt{\sum_{t'=1}^{m} u_{g,t'}(w_j^{(t')})^2} \right\} + C(j,t) \tag{6}$$

where $w_j^{(t)}$ is the only variable in the optimization problem, and $C(j,t)$ is the sum of other terms in (4) that are constants with respect to $w_j^{(t)}$.

For notational simplicity, we define $\kappa_{g,t} := \sum_{t' \neq t} u_{g,t'}(w_j^{(t')})^2$ that is considered as a constant in (6) given $\boldsymbol{U}$ and $\boldsymbol{W} \setminus \{w_j^{(t)}\}$. Given $\kappa_{g,t}$ for all $g \in \mathcal{G}$, we also define $\mathcal{G}^0$ as the set of groups such that $\kappa_{g,t} = 0$ and $\mathcal{G}^+$ for groups s.t. $\kappa_{g,t} > 0$. Armed with this notation and with the fact that $\sqrt{x^2} = |x|$, we are able to rewrite (6) as

$$\sum_{g \in \mathcal{G}} \lambda_g u_{g,t}(w_j^{(t)})^2 + 2 \sum_{g \in \mathcal{G}^+} \lambda_g \left(\sum_{j' \neq j} \sqrt{\sum_{t'=1}^{m} u_{g,t'}(w_{j'}^{(t')})^2} \right) \sqrt{\sum_{t'=1}^{m} u_{g,t'}(w_j^{(t')})^2} \tag{7}$$

$$+ 2 \sum_{g \in \mathcal{G}^0} \lambda_g \left(\sum_{j' \neq j} \sqrt{\sum_{t'=1}^{m} u_{g,t'}(w_{j'}^{(t')})^2} \right) \sqrt{u_{g,t}} |w_j^{(t)}|$$

where we suppress the constant term $C(j,t)$. Since $\sqrt{x^2 + a}$ is differentiable in x for any constant $a > 0$, the first two terms in (7) are differentiable with respect to $w_j^{(t)}$, and the only non-differentiable term involves the absolute value of the variable, $|w_j^{(t)}|$. As a result, (7) can be efficiently solved by proximal gradient descent followed by an element-wise soft thresholding. Please see appendix for the gradient computation of $\mathcal{L}$ and soft-thresholding details.

5   Experiments

We conduct experiments on two synthetic and two real datasets and compare with the following approaches.

(1) Single task learning (STL): Independent models for each task using elastic-net regression/classification.
(2) AllTasks: We combine data from all tasks into a single task and learn an elastic-net model on it.
(3) Clus-MTL: We first learn STL for each task, and then cluster the task parameters using k-means clustering. For each task cluster, we then train a multitask lasso model.
(4) GO-MTL: group-overlap MTL (Kumar and Daumé III 2012).

(5) Kang et al. (2011): nuclear norm based task grouping.
(6) SG-MTL: our approach from Eq. 4.
(7) Fusion SG-MTL: our model with a fusion penalty (see Sect. 3).

Table 2. Synthetic datasets: (upper table) Average MSE from 5 fold CV. (lower table) Varying group sizes and the corresponding average MSE. For each method, lowest MSE is highlighted.

Dataset	STL	ClusMTL	Kang	SG-MTL	Fusion SG-MTL
set-1 (3 groups)	1.067	1.221	1.177	0.682	0.614
set-2 (5 groups)	1.004	1.825	0.729	0.136	0.130

Synthetic data-2 with 30% feature overlap across groups					
Method	*Number of groups, N*				
	2	4	$N^* = 5$	6	10
ClusMTL	1.900	1.857	1.825	1.819	**1.576**
Kang et al. (2011)	**0.156**	0.634	0.729	0.958	1.289
SG-MTL	0.145	**0.135**	0.136	0.137	0.137
Fusion SG-MTL	0.142	0.139	**0.130**	0.137	0.137

5.1 Results on Synthetic Data

The first setting is similar to the synthetic data settings used in Kang et al. (2011) except for how W is generated (see Fig. 4(a) for our parameter matrix and compare it with Sect. 4.1 of Kang et al. (2011)). We have 30 tasks forming 3 groups with 21 features and 15 examples per task. Each group in W is generated by first fixing the zero components and then setting the non-zero parts to a random vector w with unit variance. Y_t for task t is $X_t W_t + \epsilon$. For the second dataset, we generate parameters in a similar manner as above, but with 30 tasks forming 5 groups, 100 examples per task, 150 features and a 30% overlap in the features across groups. In Table 2, we show 5-fold CV results and in Fig. 6 we show the groups (U) found by our method.

How Many Groups? Table 2 (Lower) shows the effect of increasing group size on three methods (smallest MSE is highlighted). For our methods, we observe a dip in MSE when N is close to N^*. In particular, our method with the fusion penalty gets the lowest MSE at $N^* = 5$. Interestingly, Kang et al. (2011) seems to prefer the smallest number of clusters, possibly due to the low-rank structural assumption of their approach, and hence cannot be used to learn the number of clusters in a sparsity based setting.

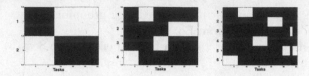

Fig. 6. Groupings found for the case where $N^* = 5$. We show results of a typical run of our method with $N = 2, 4, 6$. On the x-axis are the 30 tasks and on the y-axis are the group ids.

Table 3. QSAR prediction: average MSE and R^2 over 10 train: test splits with 100 examples per task in the training split (i.e. $n = 100$). The standard deviation of MSE is also shown. For all group learning methods, we use $N = 5$.

Table 4. Avg. MSE of our method (Fusion SG-MTL) as a function of the number of clusters and training data size. The best average MSE is observed with 7 clusters and $n = 300$ training examples (green curve with squares). The corresponding R^2 for this setting is 0.401.

	μ_{MSE}	σ_{mse}	μ_{R^2}	σ_{R2}
STL	0.811	0.02	0.223	0.01
AllTasks	0.908	0.01	0.092	0.01
ClusMTL	0.823	0.02	0.215	0.02
GOMTL	0.794	0.01	0.218	0.01
Kang	1.011	0.03	0.051	0.03
Fusion SG-MTL	**0.752**	0.01	**0.265**	0.01

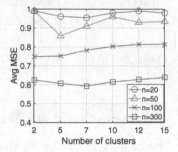

5.2 Quantitative Structure Activity Relationships (QSAR) Prediction: Merck Dataset

Given features generated from the chemical structures of candidate drugs, the goal is to predict their molecular activity (a real number) with the target. This dataset from Kaggle consists of 15 molecular activity data sets, each corresponding to a different target, giving us 15 tasks. There are between 1500 to 40000 examples and ≈5000 features per task, out of which 3000 features are common to all tasks. We create 10 train: test splits with 100 examples in training set (to represent a setting where $n \ll d$) and remaining in the test set. We report R^2 and MSE aggregated over these experiments in Table 3, with the number of task clusters N set to 5 (for the baseline methods we tried $N = 2, 5, 7$). We found that Clus-MTL tends to put all tasks in the same cluster for any value of N. Our method has the lowest average MSE.

In Fig. 4, for our method we show how MSE changes with the number of groups N (along x-axis) and over different sizes of the training/test split. The dip in MSE for $n = 50$ training examples (purple curve marked 'x') around $N = 5$

Table 5. TFBS prediction: average AUC-PR, with training data size of 200 examples, test data of ≈1800 for number of groups $N = 10$.

Method	AUC-PR
STL	0.825
AllTasks	0.709
ClusMTL	**0.841**
GOMTL	0.749
Kang	0.792
Fusion SG-MTL	**0.837**

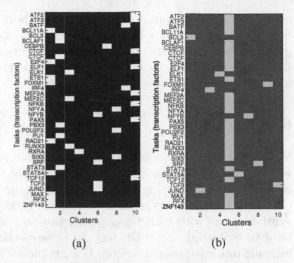

(a)   (b)

Fig. 7. Matrix indicating the groups learned by two methods. The rows show the task names and columns are cluster-ids. A white entry at position (i, j) indicates that task i belongs to cluster j. (a) Groups learned by SG-MTL on the TFBS problem (b) Groups learned by the Clus-MTL baseline show that it tends to put all tasks in the same cluster.

suggests there are 5 groups. The learned groups are shown in the appendix in Table 7 followed by a discussion of the groupings.

5.3 Transcription Factor Binding Site Prediction (TFBS)

This dataset was constructed from processed ChIP-seq data for 37 transcription factors (TFs) downloaded from ENCODE database (Consortium et al. 2012). Training data is generated in a manner similar to prior literature (Setty and Leslie 2015). Positive examples consist of 'peaks' or regions of the DNA with binding events and negatives are regions away from the peaks and called 'flanks'. Each of the 37 TFs represents a task. We generate all 8-mer features and select 3000 of these based on their frequency in the data. There are ≈2000 examples per task, which we divide into train:test splits using 200 examples (100 positive, 100 negative) as training data and the rest as test data. We report AUC-PR averaged over 5 random train:test splits in Table 5. For our method, we found the number of clusters giving the best AUC-PR to be $N = 10$. For the other methods, we tried $N = 5, 10, 15$ and report the best AUC-PR.

Though our method does marginally better (not statistically significant) than the STL baseline, which is a ridge regression model, in many biological applications such as this, it is desirable to have an interpretable model that can produce biological insights. Our MTL approach learns groupings over the TFs which are shown in Fig. 7(a). Overall, ClusMTL has the best AUC-PR on this dataset however it groups too many tasks into a single cluster (Fig. 7(b)) and forces each group to have at least one task. Note how our method leaves some groups empty (column 5 and 7) as our objective provides a trade-off between adding groups and making groups cohesive.

6   Conclusion

We presented a method to learn group structure in multitask learning problems, where the task relationships are unknown. The resulting non-convex problem is optimized by applying the alternating minimization strategy. We evaluate our method through experiments on both synthetic and real-world data. On synthetic data with known group structure, our method outperforms the baselines in recovering them. On real data, we obtain a better performance while learning intuitive groupings. Code is available at: https://github.com/meghana-kshirsagar/treemtl/tree/groups.

Full paper with appendix is available at: https://arxiv.org/abs/1705.04886.

Acknowledgements. We thank Alekh Agarwal for helpful discussions regarding Sect. 3.1. E.Y. acknowledges the support of MSIP/NRF (National Research Foundation of Korea) via NRF-2016R1A5A1012966 and MSIP/IITP (Institute for Information & Communications Technology Promotion of Korea) via ICT R&D program 2016-0-00563, 2017-0-00537.

References

Agarwal, A., Gerber, S., Daumé III, H.: Learning multiple tasks using manifold regularization. In: Advances in Neural Information Processing Systems, pp. 46–54 (2010)

Argyriou, A., Evgeniou, T., Pontil, M.: Convex multi-task feature learning. Mach. Learn. **73**, 243–272 (2008)

Bach, F., Jenatton, R., Mairal, J., Obozinski, G., et al.: Convex optimization with sparsity-inducing norms. Optim. Mach. Learn. **5**, 19–53 (2011)

Baxter, J.: A model of inductive bias learning. J. Artif. Intell. Res. (JAIR) **12**, 149–198 (2000)

Caruana, R.: Multitask learning. Mach. Learn. **28**(1), 41–75 (1997). ISSN 0885-6125

Chen, J., Liu, J., Ye, J.: Learning incoherent sparse and low-rank patterns from multiple tasks. ACM Trans. Knowl. Discov. Data (TKDD) **5**(4), 22 (2012)

Chen, Y., Ye, X.: Projection onto a simplex. arXiv preprint arXiv:1101.6081 (2011)

ENCODE Project Consortium, et al.: An integrated encyclopedia of DNA elements in the human genome. Nature **489**(7414), 57–74 (2012)

Daumé III, H.: Bayesian multitask learning with latent hierarchies. In: Proceedings of the Conference on Uncertainty in Artificial Intelligence, pp. 135–142. AUAI Press (2009)

Evgeniou, T., Pontil, M.: Regularized multi-task learning. In: ACM SIGKDD (2004)

Fei, H., Huan, J.: Structured feature selection and task relationship inference for multi-task learning. Knowl. Inf. Syst. **35**(2), 345–364 (2013)

Gong, P., Ye, J., Zhang, C.: Robust multi-task feature learning. In: Proceedings of the 18th ACM SIGKDD International Conference on Knowledge Discovery and Data Mining, pp. 895–903. ACM (2012)

Jacob, L., Vert, J.P., and Bach, F.R.: Clustered multi-task learning: a convex formulation. In: Advances in Neural Information Processing Systems (NIPS), pp. 745–752 (2009)

Jalali, A., Sanghavi, S., Ruan, C., Ravikumar, P.K.: A dirty model for multi-task learning. In: Advances in Neural Information Processing Systems, pp. 964–972 (2010)

Kang, Z., Grauman, K., Sha, F.: Learning with whom to share in multi-task feature learning. In: International Conference on Machine learning (ICML) (2011)

Kim, S., Xing, E.P.: Tree-guided group lasso for multi-task regression with structured sparsity. In: The Proceedings of the International Conference on Machine Learning (ICML) (2010)

Kumar, A., Daumé III, H.: Learning task grouping and overlap in multi-task learning. In: The Proceedings of the International Conference on Machine Learning (ICML) (2012)

Liu, J., Ji, S., Ye, J.: Multi-task feature learning via efficient $l_{2,1}$-norm minimization. In: Proceedings of the Twenty-Fifth Conference on Uncertainty in Artificial Intelligence (UAI), pp. 339–348 (2009)

Ma, J., Sheridan, R.P., Liaw, A., Dahl, G.E., Svetnik, V.: Deep neural nets as a method for quantitative structure-activity relationships. J. Chem. Inf. Model. **55**(2), 263–274 (2015)

Maurer, A.: Bounds for linear multi-task learning. J. Mach. Learn. Res. **7**, 117–139 (2006)

Negahban, S., Wainwright, M.J.: Estimation of (near) low-rank matrices with noise and high-dimensional scaling. Ann. Stat. **39**, 1069–1097 (2011)

Passos, A., Rai, P., Wainer, J., Daumé III, H.: Flexible modeling of latent task structures in multitask learning. In: The Proceedings of the International Conference on Machine Learning (ICML) (2012)

Rao, N., Cox, C., Nowak, R., Rogers, T.T.: Sparse overlapping sets lasso for multitask learning and its application to fMRI analysis. In: Advances in Neural Information Processing Systems, pp. 2202–2210 (2013)

Setty, M., Leslie, C.S.: SeqGL identifies context-dependent binding signals in genome-wide regulatory element maps. PLoS Comput. Biol. **11**(5), e1004271 (2015)

Tibshirani, R.: Regression shrinkage and selection via the lasso. J. R. Stat. Soc. Ser. B (Methodol.) **58**, 267–288 (1996)

Widmer, C., Leiva, J., Altun, Y., Rätsch, G.: Leveraging sequence classification by taxonomy-based multitask learning. In: Berger, B. (ed.) RECOMB 2010. LNCS, vol. 6044, pp. 522–534. Springer, Heidelberg (2010). https://doi.org/10.1007/978-3-642-12683-3_34

Yu, K., Tresp, V., Schwaighofer, A.: Learning Gaussian processes from multiple tasks. In: Proceedings of the 22nd International Conference on Machine Learning, pp. 1012–1019. ACM (2005)

Yuan, M., Lin, Y.: Model selection and estimation in regression with grouped variables. J. R. Stat. Soc.: Ser. B (Stat. Methodol.) **68**(1), 49–67 (2006)

Zhang, Y., Schneider, J.G.: Learning multiple tasks with a sparse matrix-normal penalty. In: Advances in Neural Information Processing Systems, pp. 2550–2558 (2010)

Zhang, Y., Yeung, D.-Y.: A convex formulation for learning task relationships in multitask learning (2010)

Lifelong Learning with Gaussian Processes

Christopher Clingerman and Eric Eaton[(✉)]

Department of Computer and Information Science,
University of Pennsylvania, Philadelphia, PA 19104, USA
{chcl,eeaton}@seas.upenn.edu

Abstract. Recent developments in lifelong machine learning have demonstrated that it is possible to learn multiple tasks consecutively, transferring knowledge between those tasks to accelerate learning and improve performance. However, these methods are limited to using linear parametric base learners, substantially restricting the predictive power of the resulting models. We present a lifelong learning algorithm that can support non-parametric models, focusing on Gaussian processes. To enable efficient online transfer between Gaussian process models, our approach assumes a factorized formulation of the covariance functions, and incrementally learns a shared sparse basis for the models' parameterizations. We show that this lifelong learning approach is highly computationally efficient, and outperforms existing methods on a variety of data sets.

Keywords: Lifelong machine learning · Online multi-task learning Gaussian process

1 Introduction

Recent advances in lifelong machine learning [3,23] have shown that it is now possible to learn tasks consecutively and obtain equivalent model performance to batch multi-task learning (MTL) [14,15] with dramatic computational speedups. Lifelong learning methods have been developed for either classification and regression [23] or reinforcement learning [3,6] domains, including support for autonomous cross-domain mapping of the learned knowledge [5]. These multi-task and lifelong learning methods maintain a repository of learned knowledge that acts as a basis over the model space, and is shared between the task models to facilitate transfer between tasks. Although effective, these methods are currently limited to using linear parametric base learners, substantially limiting the predictive power of the resulting models.

We present the first lifelong learning algorithm that supports non-parametric models through Gaussian process (GP) regression. GPs have been used successfully in MTL [1,2,4,21,28,29], but lifelong learning with GPs has not yet been explored. To enable transfer between the GP task models, we assume a factorized formulation of the models' parameterizations using a shared sparse basis, and incrementally learn a repository of shared knowledge that acts as that basis

© Springer International Publishing AG 2017
M. Ceci et al. (Eds.): ECML PKDD 2017, Part II, LNAI 10535, pp. 690–704, 2017.
https://doi.org/10.1007/978-3-319-71246-8_42

to underly the covariance functions. This shared knowledge base is updated with each new task, serving both to accelerate learning of GP models for future tasks via knowledge transfer and to refine the models of known tasks. This process is computationally efficient, and we provide theoretical analysis of the convergence of our approach. We demonstrate the effectiveness of GP lifelong learning on a variety of data sets, outperforming existing GP MTL methods and lifelong learning with linear models.

2 Related Work

Gaussian processes (GP) have proven to be effective tools for modeling spatial, time-varying, or nonlinear data [22], providing substantially more predictive power than linear models. However, the model complexity of a standard GP places restrictions on the amount of data that can be processed in a batch multitask setting. Typical (albeit naïve) computation requires $O(n^3)$ time, where n is the number of training data instances. In single-task settings, many interesting approaches such as input sparsification [24], hierarchical GPs with input clustering [18], and distributed GPs for large-scale regression [9] have been investigated for reducing computation time while maintaining prediction accuracy. Other work has focused on scaling the amount of data that GPs can reasonably handle [7,20]. While these single-task GP learning methods do not consider MTL or lifelong learning scenarios, their approaches could be used as the base learners in our framework to (1) further scale the amount of data that can be handled for individual tasks, and (2) reduce the data storage requirements for previous tasks.

Several works have tackled the MTL setting with GP predictors [1,2,4,21, 28,29]. These methods attempt to learn some form of shared knowledge among related tasks in a batch setting. However, the manner in which these models form connections among tasks increases complexity and, incidentally, computation time. Consequently, these methods are inappropriate for the lifelong learning setting, in which tasks arrive consecutively and the models must be repeatedly updated. Our approach, in contrast, considers the case where tasks are observed online and utilizes a factorized model that is more computationally efficient. Also, in direct comparison with batch MTL using GPs [4], we show that our method requires significantly less computation time while maintaining comparable accuracy.

Lifelong machine learning [8] has similarly seen much interest and development over the past several years in the aforementioned settings of classification, regression, and reinforcement learning [3,6,10,19,23,27], with applications to robotics [12,25], user modeling [13], and learning of structured information [16]. A popular choice for the foundation of these algorithms is linear parametric models. While linear models are simple and computationally efficient, they lack the predictive power of GPs. As a natural yet non-trivial extension of this prior work, we have devised a novel approach to lifelong learning using GPs as the base learning algorithm to merge, in some sense, the best of both worlds—our approach combines the predictive power of GPs in the multi-task setting while utilizing the computational efficiency and longevity of lifelong learning.

3   Background on Gaussian Processes

A Gaussian process (GP) is a distribution over functions $g(\mathbf{x})$, where the distribution is determined solely by mean and covariance functions [22]:

$$g(\mathbf{x}) \sim \mathcal{GP}(m(\mathbf{x}), \mathbf{K}) \;, \tag{1}$$

with mean $m(\cdot)$ and covariance matrix $\mathbf{K}$. GPs can be used for modeling labeled data by assuming that the data are samples drawn from a multivariate Gaussian distribution. Given a set of labeled training instances $\{(\mathbf{x}_i, y_i)\}_{i=1}^n$ with $\mathbf{x}_i \in \mathcal{X} \subseteq \mathbb{R}^d$ and $y_i \in \mathbb{R}$, GP regression models the likelihood as a Gaussian with $\mathrm{P}(\mathbf{y} \mid \mathbf{X}, g) = \mathcal{N}(g, \sigma^2 \mathbf{I})$, where $g : \mathcal{X} \mapsto \mathbb{R}$ maps from instances to their corresponding labels. The prior on g is then defined as $\mathrm{P}(g(\mathbf{x}_i) \mid \boldsymbol{\theta}) = \mathcal{GP}(m(\mathbf{x}_i), \mathbf{K})$, where $m(\mathbf{x}_i)$ is the mean function on the input datum $\mathbf{x}_i$, $\mathbf{K}$ is a covariance matrix $[\mathbf{K}]_{i,j} = \kappa(\mathbf{x}_i, \mathbf{x}_j)$ between all pairs of data instances $\mathbf{x}_i, \mathbf{x}_j \in \mathbf{X}$ based on a chosen covariance function, and $\boldsymbol{\theta}$ is a set of parameters of the chosen covariance function. Given a new input datum $\mathbf{x}_*$, the GP predicts the distribution of the corresponding label value y_* as $\mathrm{P}(y_* \mid \mathbf{x}_*, \mathbf{X}, \mathbf{y}) = \mathcal{N}(\mu_*, \sigma_*)$, where

$$\mu_* = m(\mathbf{x}_*) + \mathbf{k}_*^\mathsf{T}[\mathbf{K} + \sigma^2 \mathbf{I}]^{-1}(\mathbf{y} - m(\mathbf{x}_*)) \;,$$

$$\sigma_* = \kappa(\mathbf{x}_*, \mathbf{x}_*) + \sigma^2 - \mathbf{k}_*^\mathsf{T}[\mathbf{K} + \sigma^2 \mathbf{I}]^{-1}\mathbf{k}_* \;,$$

and $\mathbf{k}_*$ is a vector of covariance values $[\mathbf{k}_*]_i = \kappa(\mathbf{x}_*, \mathbf{x}_i)$ for all $\mathbf{x}_i \in \mathbf{X}$. In other words, GP regression models:

$$\begin{bmatrix} \mathbf{y} \\ y_* \end{bmatrix} \sim \mathcal{N}\left(\begin{bmatrix} m(\mathbf{X}) \\ m(\mathbf{x}_*) \end{bmatrix}, \begin{bmatrix} \mathbf{K} & \mathbf{k}_*^\mathsf{T} \\ \mathbf{k}_* & k_{**} \end{bmatrix} \right) \;. \tag{2}$$

4   Lifelong Learning with Gaussian Processes

After first summarizing the lifelong learning problem, we extend GP regression to the multi-task setting and derive our approach for lifelong GP learning. To ensure consistency with the existing literature, we adopt the notational conventions of Ruvolo and Eaton [23].

4.1   The Lifelong Learning Problem

The lifelong learning agent (Fig. 1) faces a series of consecutive learning tasks $\mathcal{Z}^{(1)}, \mathcal{Z}^{(2)}, \ldots, \mathcal{Z}^{(T_{\max})}$. In our setting, each task is a supervised or semi-supervised learning problem $\mathcal{Z}^{(t)} = (id^{(t)}, \hat{f}^{(t)}, \mathbf{X}^{(t)}, \mathbf{y}^{(t)})$, where $id^{(t)}$ is a unique task identifier, $\hat{f}^{(t)} : \mathcal{X}^{(t)} \mapsto \mathcal{Y}^{(t)}$ defines an unknown mapping from an instance space $\mathcal{X}^{(t)} \subseteq \mathbb{R}^d$ to the label space $\mathcal{Y}^{(t)}$. Typically, $\mathcal{Y}^{(t)} = \{-1, +1\}$ for classification tasks and $\mathcal{Y}^{(t)} = \mathbb{R}$ for regression tasks. Each task t has n_t training instances $\mathbf{X}^{(t)} \in \mathbb{R}^{d \times n_t}$ with corresponding labels $\mathbf{y}^{(t)} \in \mathcal{Y}^{(t)^{n_t}}$ given by $\hat{f}^{(t)}$. A priori, the lifelong learner does not know the total number of tasks $T_{\max}$, the task distribution, or their order.

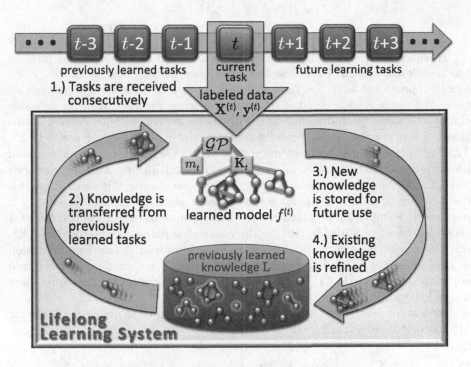

Fig. 1. The lifelong learning process with GPs (adapted from [23]).

Each time step, the agent receives a batch of labeled training data for some task t, either a new task or as additional data for a previously learned task. Let T denote the number of unique tasks the agent has encountered so far ($0 \leq T \leq T_{\max}$). At any time, the agent may be asked to make predictions on instances from any previously learned task. Its goal is to construct task models $f^{(1)}, \ldots, f^{(T)}$ where each $f^{(t)} : \mathbb{R}^d \mapsto \mathcal{Y}^{(t)}$ will approximate $\hat{f}^{(t)}$. Each $f^{(t)}$ must be able to be rapidly updated as the agent encounters additional training data for known tasks, and new $f^{(t)}$'s must be added efficiently as the agent encounters new tasks. We assume that the total number of tasks $T_{\max}$ will be large, and so the algorithm must have a computational complexity to update the task models that scales favorably to numerous tasks.

4.2   Gaussian Process Regression for Multi-task Learning

To extend the GP framework (Sect. 3) to the multi-task and lifelong learning setting, we assume that each task t corresponds to training an individual GP model $f^{(t)}$ from training data $\mathbf{D}^{(t)} = \left\{ (\mathbf{x}_i^{(t)}, y_i^{(t)}) \right\}_{i=1}^{n_t}$ with its own mean $m_t(\cdot)$ and covariance function $\mathbf{K}_t$. Given T tasks, the goal is to learn each task model $f^{(t)}$ such that

$$P(\mathbf{y}^{(t)} \mid \mathbf{X}^{(t)}, g^{(t)}) \sim \mathcal{N}(g^{(t)}, \sigma_t^2 \mathbf{I})$$

$$g^{(t)} \sim \mathcal{GP}(m_t, \mathbf{K}_t) \quad \forall t \in \{1, \dots, T\} .$$

To share knowledge between task models, we assume a factorized form of the covariance kernels, such that the covariance kernel parameters $\boldsymbol{\theta}^{(t)} \in \mathbb{R}^d$ can be represented as $\boldsymbol{\theta}^{(t)} = \mathbf{Ls}^{(t)}$. The matrix $\mathbf{L} \in \mathbb{R}^{d \times k}$ is shared between all tasks and forms a basis over the space of covariance kernels, capturing reusable chunks of knowledge. Individual tasks' covariance kernels are then defined in this shared basis via the sparse coefficient vectors $\{\mathbf{s}^{(1)}, \dots, \mathbf{s}^{(T)}\}$, with $\mathbf{s}^{(t)} \in \mathbb{R}^k$. This factorized sparse representation has shown success for transfer between linear parameterized models in previous MTL and lifelong learning methods [3,14,15,23], and here we adapt it to the non-parametric GP setting. In this manner, our approach is similar to these other methods in that we employ a parametric formulation in which the prediction function $f^{(t)}(\mathbf{x}) = f(\mathbf{x}; \boldsymbol{\theta}^{(t)})$ for each task t is determined in part by the covariance parameter vector $\boldsymbol{\theta}^{(t)} \in \mathbb{R}^d$, but the model itself is non-parametric. This factorized sparse representation is also somewhat related to techniques for fast GP training using factorized covariance matrices [11,21], but with a different factorization designed to share latent knowledge between task models.

We assume that the tasks are drawn i.i.d., allowing us to define the following MTL objective function for GPs:

$$\min_{\mathbf{L}} \frac{1}{T} \sum_{t=1}^{T} \min_{\mathbf{s}^{(t)}} \left\{ \mathcal{L}\left(f\left(\mathbf{X}^{(t)}; \mathbf{Ls}^{(t)}\right), \mathbf{y}^{(t)}\right) + \mu \|\mathbf{s}^{(t)}\|_1 \right\} + \lambda \|\mathbf{L}\|_{\mathsf{F}}^2 , \qquad (3)$$

where $(\mathbf{x}_i^{(t)}, y_i^{(t)})$ is the ith labeled training instance for task t, the L_1 norm $\| \cdot \|_1$ is used as a convex approximation to the true vector sparsity of the $\mathbf{s}^{(t)}$ vectors, $\| \cdot \|_{\mathsf{F}}$ is the Frobenius norm to control the complexity of $\mathbf{L}$, and μ and λ are regularization coefficients. The loss function $\mathcal{L}$ optimizes the fit of each task model to the training data. For GP models, we define this loss function as the negative log-marginal likelihood of a standard Gaussian process, which must be minimized to fit the GP:

$$\mathcal{L}\left(f\left(\mathbf{X}^{(t)}; \mathbf{Ls}^{(t)}\right), \mathbf{y}^{(t)}\right) = {\mathbf{y}^{(t)}}^{\mathsf{T}} \left[\mathbf{K}_t(\mathbf{Ls}^{(t)}) + \sigma_t^2 \mathbf{I}\right]^{-1} \mathbf{y}^{(t)}$$

$$+ \log\left|\mathbf{K}_t(\mathbf{Ls}^{(t)}) + \sigma_t^2 \mathbf{I}\right| + n_t \log(2\pi) . \qquad (4)$$

Note that Eq. 3 is not jointly convex in $\mathbf{L}$ and the $\mathbf{s}^{(t)}$ vectors, so most MTL methods [14,15] solve related forms of this objective using alternating optimization, repeatedly solving for $\mathbf{L}$ while holding $\mathbf{s}^{(t)}$'s fixed and then optimizing the $\mathbf{s}^{(t)}$'s while holding $\mathbf{L}$ fixed. While effective in determining a locally optimal solution, these MTL approaches are computationally expensive and would require re-optimization as tasks were added incrementally, making them unsuitable for the lifelong learning setting.

Our approach to optimizing Eq. 3 is based upon the Efficient Lifelong Learning Algorithm (ELLA) [23], which provides a computationally efficient method for learning the $\mathbf{s}^{(t)}$'s and $\mathbf{L}$ online over multiple consecutive tasks. Although ELLA can support a variety of parametric linear models, it cannot natively support non-parametric models, limiting its predictive power. In the next section, we develop a lifelong learning approach for the (non-parametric) GP framework, and show that the resulting algorithm provides an efficient method for learning consecutive GP task models.

4.3   Efficient Updates for Lifelong GP Learning

To solve Eq. 3 efficiently in a lifelong learning setting, we first eliminate the inner recomputation of the loss function by approximating it via a sparse-coded solution, following Ruvolo and Eaton [23]. We approximate the loss function via a second-order Taylor expansion around $\boldsymbol{\theta} = \boldsymbol{\theta}^{(t)}$ of $\mathcal{L}\left(f\left(\mathbf{X}^{(t)}; \mathbf{Ls}^{(t)}\right), \mathbf{y}^{(t)}\right)$, where $\boldsymbol{\theta}^{(t)} = \arg\min_{\boldsymbol{\theta}} \mathcal{L}\left(f\left(\mathbf{X}^{(t)}; \boldsymbol{\theta}, \mathbf{y}^{(t)}\right)\right)$. Substituting this expansion into Eq. 3, we obtain

$$\min_{\mathbf{L}} \frac{1}{T} \sum_{t=1}^{T} \min_{\mathbf{s}^{(t)}} \left\{ \|\boldsymbol{\theta}^{(t)} - \mathbf{Ls}^{(t)}\|_{\mathbf{H}^{(t)}}^2 + \mu\|\mathbf{s}^{(t)}\|_1 \right\} + \lambda\|\mathbf{L}\|_{\mathsf{F}}^2 , \tag{5}$$

where $\mathbf{H}^{(t)}$ is the Hessian matrix given by

$$\mathbf{H}^{(t)} = \frac{1}{2} \nabla_{\boldsymbol{\theta},\boldsymbol{\theta}}^2 \mathcal{L}\left(f\left(\mathbf{X}^{(t)}; \boldsymbol{\theta}\right), \mathbf{y}^{(t)}\right)\bigg|_{\boldsymbol{\theta}=\boldsymbol{\theta}^{(t)}}, \tag{6}$$

and $\|\mathbf{v}\|_{\mathbf{A}}^2 = \mathbf{v}^\top \mathbf{A} \mathbf{v}$.

The second inefficiency in Eq. 3 involves the need to recompute the $\mathbf{s}^{(t)}$'s whenever we evaluate a new $\mathbf{L}$. This dependency can be simplified by only recomputing the $\mathbf{s}^{(t)}$ for the current task t, leaving all other coefficient vectors fixed. Essentially we have removed the minimization over all $\mathbf{s}^{(t)}$'s in place of a minimization over only the current task's $\mathbf{s}^{(t)}$. Later, we provide convergence guarantees that show that this choice to update $\mathbf{s}^{(t)}$ only when training on task t does not significantly affect the quality of model fit as T grows large. With these simplifications, we can solve the optimization in Eq. 5 incrementally via the following update equations:

$$\mathbf{s}^{(t)} \leftarrow \arg\min_{\mathbf{s}^{(t)}} \ell(\mathbf{L}_m, \mathbf{s}^{(t)}, \boldsymbol{\theta}^{(t)}, \mathbf{H}^{(t)}) \tag{7}$$

$$\mathbf{L}_{m+1} \leftarrow \arg\min_{\mathbf{L}} \hat{g}_m(\mathbf{L}) \tag{8}$$

$$\hat{g}_m(\mathbf{L}) = \lambda\|\mathbf{L}\|_{\mathsf{F}}^2 + \frac{1}{T}\sum_{t=1}^{T} \ell\left(\mathbf{L}, \mathbf{s}^{(t)}, \boldsymbol{\theta}^{(t)}, \mathbf{H}^{(t)}\right) , \tag{9}$$

where

$$\ell\left(\mathbf{L}, \mathbf{s}, \boldsymbol{\theta}, \mathbf{H}\right) = \mu\|\mathbf{s}\|_1 + \|\boldsymbol{\theta} - \mathbf{Ls}\|_{\mathbf{H}}^2 \tag{10}$$

and $\mathbf{L}_m$ corresponds to $\mathbf{L}$ at the algorithm's m-th iteration.

To apply these update equations in practice, given a new task t, we first compute $\boldsymbol{\theta}^{(t)}$ via single-task GP learning on $\mathbf{D}^{(t)}$. Then, we compute the Hessian $\mathbf{H}^{(t)}$, as described in detail in Appendix A. Next, we compute $\mathbf{s}^{(t)}$ and $\mathbf{L}_m$ in a single gradient-descent step. For each iteration of gradient descent, we compute $\mathbf{s}^{(t)}$ using the current $\mathbf{L}_m$ by solving an instance of LASSO. Then, we use $\mathbf{s}^{(t)}$ to find $\nabla\mathbf{L}$ and recompute $\mathbf{L}_m$. This process is repeated until either convergence or a set maximum number of iterations has taken place; typically, this process requires only a few iterations to converge in practice. The equation for $\nabla\mathbf{L}$ is found by deriving Eq. 9 with respect to $\mathbf{L}$, similar to the derivation by Bou Ammar et al. [5], yielding:

$$\nabla\mathbf{L} = \lambda\mathbf{L} + \frac{1}{T}\sum_{t=1}^{T}\left(-\mathbf{H}^{(t)}\boldsymbol{\theta}^{(t)}\mathbf{s}^{(t)^\mathsf{T}} + \mathbf{H}^{(t)}\mathbf{L}\mathbf{s}^{(t)}\mathbf{s}^{(t)^\mathsf{T}}\right). \tag{11}$$

For computational efficiency, the sum in Eq. 11 is computed incrementally (by updating it for the current task) and stored; it is not necessary to recompute it via summing over all previous tasks. As a final step, we reinitialize any unused columns of $\mathbf{L}_m$. Algorithm 1 presents our complete algorithm for lifelong GP learning, which we refer to as GP-ELLA.

In contrast to linear methods for lifelong learning [3,23], GP-ELLA must explicitly store training data for each individual task to compute the non-parametric kernel values for new instances. To further improve its scalability for lifelong learning with large amounts of data, our approach could also be adapted to employ sparse GPs [24] as the base learners, which store a reduced number of instances per task, or use approximations to the kernel matrix. We started exploring this direction, but found that the adaptation is nontrivial, well beyond the scope of this paper, and so leave it to future work.

4.4 Details of Label Prediction: Recovering σ_f^2 and Prediction Smoothing

In practice, we minimize the negative log-marginal likelihood in training the task-based GP model over all parameters, including σ_f^2. However, our model of $\boldsymbol{\theta}^{(t)}$ does not include a representation of σ_f^2, so in the label prediction step, we hold out a portion of the training data as a verification test set and perform line search to determine the best possible σ_f^2 parameter given the generated $\boldsymbol{\theta}^{(t)} = \mathbf{L}\mathbf{s}^{(t)}$.

For our label predictions, we also utilize an idea from Nguyen-Tuong et al. [17], where the training data are partitioned into subsets and the final predictions are smoothed using a weighted average of the local model predictions. In our predictions for a task t, we first generate predictions with respect to $\boldsymbol{\theta}^{(t)} = \mathbf{L}\mathbf{s}^{(t)}$. For all other known tasks t', we also generate $\boldsymbol{\theta}^{(t')} = \mathbf{L}\mathbf{s}^{(t')}$ and its corresponding predictions for the current task t. Then, we compute a weighted average of all predictions based on similarities in the $\boldsymbol{\theta}$ vectors, using an exponentially decaying

---

Algorithm 1. GP-ELLA$(d, k, \lambda, \mu, \gamma)$

$T \leftarrow 0, \quad \mathbf{L} \sim \mathcal{N}(0,1)_{d,k}$
while isMoreTrainingDataAvailable() **do**
$\quad (\mathbf{X}_{\text{new}}, \mathbf{y}_{\text{new}}, t) \leftarrow$ getNextTrainingData()
$\quad$ **if** isNewTask(t) **then**
$\quad\quad T \leftarrow T+1, \quad \mathbf{X}^{(t)} \leftarrow \mathbf{X}_{\text{new}}, \quad \mathbf{y}^{(t)} \leftarrow \mathbf{y}_{\text{new}}$
$\quad$ **else**
$\quad\quad \mathbf{X}^{(t)} \leftarrow \left[\mathbf{X}^{(t)} \; \mathbf{X}_{\text{new}}\right], \quad \mathbf{y}^{(t)} \leftarrow \left[\mathbf{y}^{(t)}; \mathbf{y}_{\text{new}}\right]$
$\quad$ **end if**
$\quad \left(\boldsymbol{\theta}^{(t)}, \mathbf{H}^{(t)}\right) \leftarrow$ GPLearner$(\mathbf{X}^{(t)}, \mathbf{y}^{(t)})$
$\quad$ **while** $\mathbf{s}^{(t)}$ and $\mathbf{L}$ have not converged **do**
$\quad\quad \mathbf{s}^{(t)} \leftarrow$ Equation [7]
$\quad\quad \nabla \mathbf{L} \leftarrow$ Equation [11]
$\quad\quad \mathbf{L} \leftarrow \mathbf{L} - \gamma \nabla \mathbf{L}$
$\quad$ **end while**
$\quad \mathbf{L} \leftarrow$ reinitializeAllZeroColumns($\mathbf{L}$)
end while

L_2-norm weight function. This additional label generation step can be thought of as further smoothing predictions among similar known tasks, as measured by the similarity of their $\boldsymbol{\theta}^{(t)}$'s. In the case where the tasks are different, this additional smoothing step will not detrimentally affect the predictions, since their corresponding $\boldsymbol{\theta}^{(t)}$'s will have low similarity.

4.5 Theoretical Guarantees

This section provides theoretical results that show that GP-ELLA converges and that the simplifications to enable efficient updates have an asymptotically negligible effect on model performance. First, recall that $\hat{g}_T(\mathbf{L})$ (as defined by Eq. 9 with $m = T$) represents the cost of $\mathbf{L}$ under the current choice of the $\mathbf{s}^{(t)}$'s after GP-ELLA observes T tasks. Let $e_T(\mathbf{L}_T)$ be the MTL objective function as defined by Eq. 3, but for the given specific choice of $\mathbf{L}$ (instead of the optimization over all $\mathbf{L}$). Ruvolo and Eaton [23] showed that:

Proposition 1: The latent basis becomes more stable over time at a rate of $\mathbf{L}_{T+1} - \mathbf{L}_T = O\left(\frac{1}{T}\right)$.

Proposition 2:

(a) $\hat{g}_T(\mathbf{L}_T)$ converges almost surely (a.s.); and
(b) $\hat{g}_T(\mathbf{L}_T) - e_T(\mathbf{L}_T)$ converges a.s. to 0.

Proposition 1 shows that $\mathbf{L}$ becomes increasingly stable as T increases. Proposition 2 shows that the algorithm converges to a fixed per-task loss on the approximate objective function $\hat{g}_T$, and that the approximate objective function converges to the same value as the MTL objective.

In order for these propositions to apply to GP-ELLA, we must show that it satisfies the following assumptions:

1. The tuples $\left(\mathbf{H}^{(t)}, \boldsymbol{\theta}^{(t)}\right)$ are drawn *i.i.d.* from a distribution with compact support.
2. For all L, $\mathbf{H}^{(t)}$, and $\boldsymbol{\theta}^{(t)}$, the smallest eigenvalue of $L_\gamma^\top \mathbf{H}^{(t)} L_\gamma$ is at least κ (with $\kappa > 0$), where γ is the subset of non-zero indices of the vector $\mathbf{s}^{(t)} = \arg\min_\mathbf{s} \|\boldsymbol{\theta}^{(t)} - L\mathbf{s}\|_{\mathbf{H}^{(t)}}^2$. In this case the non-zero elements of the unique minimizing $\mathbf{s}^{(t)}$ are given by: $\mathbf{s}^{(t)}{}_\gamma = \left(L_\gamma^\top \mathbf{H}^{(t)} L_\gamma\right)^{-1}\left(L_\gamma^\top \mathbf{H}^{(t)} \boldsymbol{\theta}^{(t)} - \mu\boldsymbol{\epsilon}_\gamma\right)$, where the vector $\boldsymbol{\epsilon}_\gamma$ contains the signs of the non-zero entries of $\mathbf{s}^{(t)}$.

To verify the first assumption, we must show that the entries of $\mathbf{H}^{(t)}$ and $\boldsymbol{\theta}^{(t)}$ are bounded with compact support. We can show easily that the Hessian and $\boldsymbol{\theta}^{(t)}$ are contained within a compact region by examining their form, thus verifying the first assumption. The second assumption is a condition upon the sparse coding solution being unique, which holds true under the standard sparse coding assumptions. Therefore, the propositions above apply to GP-ELLA. In particular, since Proposition 2 holds, this verifies that the simplifications made in the optimization process (Sect. 4.3) do not cause GP-ELLA to incur any penalty in terms of the average per-task loss.

Computational Complexity: Each GP-ELLA update requires running a single-task GP on n_t d-dimensional data instances; let $M(d, n_t)$ be the complexity of this operation. Then, the algorithm iteratively optimizes $\mathbf{s}^{(t)}$ and L at a cost of $O(k^2 d^3)$ per iteration [23]. This process typically requires very few iterations i, or we can limit the number of iterations i, which works well in practice. Together, this gives GP-ELLA a per-task cost of $O(M(d, n_t) + ik^2 d^3)$.

5   Evaluation

To evaluate GP-ELLA, we analyze its prediction accuracy and computation time in comparison to four alternatives: task-independent GP learners ("Indiv. GPs"), a single GP trained over the entire data set ("Shared GP"), Bonilla et al's [4] batch MTL GP algorithm ("Batch MTGP"), and Ruvolo and Eaton's [23] ELLA with linear base learners. We also considered comparing to the more recent MTL GP method by Rakitsch et al. [21], but their approach cannot handle our problem setting since it requires that each instance be represented in all tasks (with varying labels).

5.1   Data Sets

We used four data sets in our evaluation:

Synthetic Regression Tasks: This set of 100 synthetic tasks was used previously to evaluate lifelong learning [23]. Each task has $n_t = 100$ instances with

$d = 13$ features, with labels that were generated via a linear factorized model on six latent basis vectors with added Gaussian noise.

London Schools: The London schools data set has been used extensively for MTL evaluation. It consists of examination scores from 15,362 students in 139 schools from the Inner London Education Authority. The goal is to predict the exam score for each student, treating each school as a separate task. We employ the same feature encoding as previous studies [14], yielding $d = 27$ features.

Robot Arm Kinematics: We generated this data set by simulating the forward kinematics of synthetic 8-DOF robot arms of various link lengths and configurations. The lengths of each arm link were kept consistent for each task, while the joint angles (all DOFs being revolute joints) varied in each data instance. Each task's goal is to predict the distance of the arm's end-effector from the $\mathbb{R}^3$ point $[0.1, 0.1, 0.1]$. We generated 113 tasks, each with a minimum of 50 joint configurations.

Parkinson's Vocal Tests: This data set consists of vocal signal tests recorded from patients with Parkinson's disease [26]. We split the data set into tasks based on the 42 unique patient IDs and trained on the 16 vocal signal features. The data set had two labels applied to each instance: the linearly interpolated clinician's motor score ("Parkinson's Motor") and the total UPDRS score ("Parkinson's UPDRS"), which we evaluated in separate experiments.

5.2   Methodology

Each data set was evaluated separately. Results were averaged over 10 runs for all data sets. For each trial, the data features and labels were mean-centered, and the data randomly divided into equal-sized training and testing sets. In the parameter verification step, described below, the training set was further divided into a sub-training set and a validation set. The tasks were presented in a random order to the lifelong learners and in batch to the other learners.

The covariance matrix self-noise parameter σ^2 was set to $1.0e-6$ for all data sets except London schools, where it was set to 1.0. The variance used in the weight function for comparing $\theta^{(t)}$ values in the prediction step was set to 100 for all data sets. Also, the number of iterations of gradient descent used to learn the covariance function hyperparameters, across all algorithms, was set to 50.

The parameters k, μ, and λ for GP-ELLA and ELLA were tuned on a 50% subset of the training data for the first five tasks, evaluating those parameter values on the validation portion of the training data for those tasks. Then, those tuned parameter values were used for retraining the models for the first five tasks with all available training data, and for learning the remaining tasks. Values of k used in this tuning step were all even numbers from 2 to $2d$, and μ and λ were both set to values in $\{e^{-12}, e^{-8}, e^{-4}, e^0\}$. Performance was measured on the held-out test sets after all tasks had been learned.

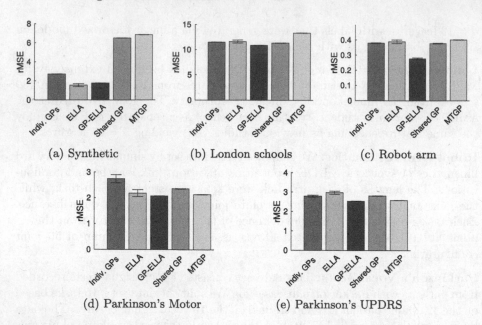

(a) Synthetic   (b) London schools   (c) Robot arm

(d) Parkinson's Motor   (e) Parkinson's UPDRS

Fig. 2. Prediction accuracy reported in root mean squared error (rMSE), where lower scores are better. The whiskers are standard error bars. Our approach achieves consistently superior performance across all data sets, with the sole exception of the synthetic data. On the synthetic data, ELLA is slightly superior since the data was generated using a linear task model.

For Batch MTGP, we use the rank-1 matrix approximation to achieve a training time that is competitive with the other algorithms. However, using this approximation, instead of the more accurate but significantly more computationally intense rank-2 approximation or full-rank matrix (as reported by [4]), caused Batch MTGP to perform worse in some cases than individual GPs.

All experiments were performed on a Mac Pro with dual 6-core Xeon 2.67 GHz processors with 24 threads total; computation for each experiment was limited to four threads. Our algorithm was implemented in MATLAB and makes use of two external packages: MTIMESX, for fast MEX-based computation of the Hessian matrix, and SPAMS, for running LASSO. We also use GPML [22] for GP computations, including minimizing the log-marginal likelihood/fitting the covariance function hyperparameters for a given training data set.

5.3   Results

As shown in Fig. 2 and Table 1, our approach compares favorably to competing methods, including multi-task GP. In terms of prediction accuracy, our method is superior to all competing methods over all data sets, with one exception. The one exception is on the synthetic data that was generated according to a linear task model (the same as used by ELLA), for which ELLA slightly outperforms

Table 1. Computation time in seconds. GP-ELLA shows less than an order of magnitude increase in time over individual GPs, while the Shared GP and MTGP methods show approximately 2–3 orders of magnitude increase in time. Due to its use a linear model, ELLA is the fastest by far, as expected. The standard error of each value is given after the $\pm$.

	Synthetic	London schools	Robot arm	Parkinson's Motor	Parkinson's UPDRS
Indiv. GPs	25.4 ± 0.7	56.9 ± 0.5	10.5 ± 0.2	12.7 ± 0.5	13.1 ± 0.2
ELLA	0.14 ± 0.016	0.22 ± 0.029	0.11 ± 0.002	0.05 ± 0.002	0.05 ± 0.004
GP-ELLA	63.8 ± 1.1	489.8 ± 18.7	38.3 ± 0.2	113.6 ± 4.6	109.4 ± 4.2
Shared GP	$3,474.9 \pm 313.7$	$13,070.5 \pm 1,784.9$	$2,023.3 \pm 85.5$	$1,972.7 \pm 871.0$	$1,150.5 \pm 113.9$
MTGP	$21,603.8 \pm 756.5$	$72,338.6 \pm 7,361.6$	$20,124.1 \pm 507.3$	$4,449.3 \pm 79.2$	$4,538.7 \pm 63.9$

our approach. On all other data sets, GP-ELLA significantly outperforms ELLA, which learns linear models, demonstrating the increased predictive power of nonparametric lifelong learning with GPs.

In terms of computation time, our method is slower than the Individual GPs or ELLA, as expected. Since task-independent GPs are a subroutine of the GP-ELLA algorithm, GP-ELLA will undoubtedly have a higher computation time than the independent GPs, but our results show less than an order of magnitude increase in computation time. Additionally, these results show that GP-ELLA is significantly faster than the competing GP methods, while obtaining lower prediction error. These competing multi-task GP methods are slower than individual GPs by approximately 2–3 orders of magnitude. Please note that the reported computation times do not include parameter tuning for any algorithm, but only the training and evaluation. Our results are consistent across all data sets and clearly demonstrate the benefits of GP-ELLA.

6   Conclusion

Given the recent advances in lifelong learning and batch GP multi-task learning, it is natural to combine the advantages of both paradigms to enable nonparametric lifelong learning. Our algorithm, GP-ELLA, constitutes the first nonparametric lifelong learning method, providing substantially improved predictive power over existing lifelong learning methods that rely on linear parametric models. GP-ELLA also has favorable performance in terms of both prediction accuracy and computation time when compared to multi-task GP methods, with guarantees on convergence in a lifelong learning setting.

Acknowledgements. This research was supported by ONR grant #N00014-11-1-0139, AFRL grant #FA8750-14-1-0069, and AFRL grant #FA8750-16-1-0109. We would like to thank Paul Ruvolo and the anonymous reviewers for their helpful feedback.

Appendix A: Computing the Hessian $\mathbf{H}^{(t)}$

To compute the Hessian $\mathbf{H}^{(t)}$ for the GP loss function, we combine Eqs. 4 and 6. Letting $\mathbf{K}_\sigma = \mathbf{K}_t(\boldsymbol{\theta}^{(t)}) + \sigma_t^2 I$,

$$[\mathbf{H}^{(t)}]_{ab} = \frac{1}{2}\mathbf{y}^{(t)\mathsf{T}}\left(\mathbf{K}_\sigma^{-1}\frac{\partial \mathbf{K}_\sigma}{\partial \theta_a}\mathbf{K}_\sigma^{-1}\frac{\partial \mathbf{K}_\sigma}{\partial \theta_b}\mathbf{K}_\sigma^{-1} - \mathbf{K}_\sigma^{-1}\frac{\partial^2 \mathbf{K}_\sigma}{\partial \theta_a \partial \theta_b}\mathbf{K}_\sigma^{-1}\right) \tag{12}$$

$$+ \mathbf{K}_\sigma^{-1}\frac{\partial \mathbf{K}_\sigma}{\partial \theta_b}\mathbf{K}_\sigma^{-1}\frac{\partial \mathbf{K}_\sigma}{\partial \theta_a}\mathbf{K}_\sigma^{-1}\bigg)\mathbf{y}^{(t)} \tag{13}$$

$$+ \frac{1}{2}\mathrm{tr}\left(\mathbf{K}_\sigma^{-1}\frac{\partial^2 \mathbf{K}_\sigma}{\partial \theta_a \partial \theta_b} - \mathbf{K}_\sigma^{-1}\frac{\partial \mathbf{K}_\sigma}{\partial \theta_b}\mathbf{K}_\sigma^{-1}\frac{\partial \mathbf{K}_\sigma}{\partial \theta_a}\right). \tag{14}$$

As an example, consider the squared exponential (SE) covariance kernel function, defined as

$$k_{SE}\left(\mathbf{x}_i^{(t)}, \mathbf{x}_j^{(t)}\right) = \sigma_f^2 \exp\left(-\frac{1}{2}\left(\mathbf{x}_i^{(t)} - \mathbf{x}_j^{(t)}\right)^{\mathsf{T}}\mathbf{M}\left(\mathbf{x}_i^{(t)} - \mathbf{x}_j^{(t)}\right)\right).$$

If we let $\mathbf{M} = \mathrm{diag}(l_1^{-2}, l_2^{-2}, \ldots, l_d^{-2})$, we now have the ARD variant of k_{SE}. Additionally, if we set $\sigma_f^2 = 1$ and $\boldsymbol{\theta}^{(t)} \triangleq \{\theta_1^{-2} = l_1^{-2}, \theta_2^{-2} = l_2^{-2}, \ldots, \theta_d^{-2} = l_d^{-2}\}$, we have a kernel with $\boldsymbol{\theta}^{(t)} \in \mathbb{R}^d$. Sample values of the first- and second-order derivatives of $\mathbf{K}_\sigma$ are

$$\left[\frac{\partial \mathbf{K}_\sigma}{\partial \theta_a}\right]_{ij} = \frac{\partial k_{SE}(\mathbf{x}_i^{(t)}, \mathbf{x}_j^{(t)})}{\partial \theta_a}$$

$$= k_{SE}(\mathbf{x}_i^{(t)}, \mathbf{x}_j^{(t)})([\mathbf{x}_i^{(t)}]_a - [\mathbf{x}_j^{(t)}]_a)^2\theta_a^{-3},$$

$$\left[\frac{\partial^2 \mathbf{K}_\sigma}{\partial \theta_a^2}\right]_{ij} = \frac{\partial^2 k_{SE}(\mathbf{x}_i^{(t)}, \mathbf{x}_j^{(t)})}{\partial \theta_a^2}$$

$$= k_{SE}(\mathbf{x}_i^{(t)}, \mathbf{x}_j^{(t)})\left(-3([\mathbf{x}_i^{(t)}]_a - [\mathbf{x}_j^{(t)}]_a)^2\theta_a^{-4}\right.$$

$$\left.+ ([\mathbf{x}_i^{(t)}]_a - [\mathbf{x}_j^{(t)}]_a)^4\theta_a^{-6}\right),$$

$$\left[\frac{\partial^2 \mathbf{K}_\sigma}{\partial \theta_a \partial \theta_b}\right]_{ij} = \frac{\partial^2 k_{SE}(\mathbf{x}_i^{(t)}, \mathbf{x}_j^{(t)})}{\partial \theta_a \partial \theta_b}$$

$$= k_{SE}(\mathbf{x}_i^{(t)}, \mathbf{x}_j^{(t)})\left(([\mathbf{x}_i^{(t)}]_a - [\mathbf{x}_j^{(t)}]_a)^2\theta_a^{-3}\right)$$

$$\times \left(([\mathbf{x}_i^{(t)}]_b - [\mathbf{x}_j^{(t)}]_b)^2\theta_b^{-3}\right).$$

References

1. Álvarez, M.A., Lawrence, N.D.: Computationally efficient convolved multiple output Gaussian processes. J. Mach. Learn. Res. **12**, 1459–1500 (2011)
2. Álvarez, M.A., Luengo, D., Titsias, M.K., Lawrence, N.D.: Efficient multioutput Gaussian processes through variational inducing kernels. In: Proceedings of the 13th International Conference on Artificial Intelligence and Statistics (AISTATS), pp. 25–32 (2010)
3. Bou Ammar, H., Eaton, E., Ruvolo, P., Taylor, M.E.: Online multi-task learning for policy gradient methods. In: Proceedings of the 31st International Conference on Machine Learning (ICML), June 2014
4. Bonilla, E.V., Chai, K.M., Williams, C.: Multi-task Gaussian process prediction. In: Advances in Neural Information Processing Systems (NIPS), pp. 153–160 (2008)
5. Bou Ammar, H., Eaton, E., Luna, J.M., Ruvolo, P.: Autonomous cross-domain knowledge transfer in lifelong policy gradient reinforcement learning. In: Proceedings of the International Joint Conference on Artificial Intelligence (IJCAI), July 2015
6. Bou Ammar, H., Tutunov, R., Eaton, E.: Safe policy search for lifelong reinforcement learning with sublinear regret. In: Proceedings of the 32nd International Conference on Machine Learning (ICML), July 2015
7. Chalupka, K., Williams, C.K., Murray, I.: A framework for evaluating approximation methods for Gaussian process regression. J. Mach. Learn. Res. **14**(1), 333–350 (2013)
8. Chen, Z., Liu, B.: Lifelong Machine Learning. Synthesis Lectures on Artificial Intelligence and Machine Learning. Morgan & Claypool Publishers, San Rafael (2016)
9. Deisenroth, M., Ng, J.W.: Distributed Gaussian processes. In: Proceedings of the 32nd International Conference on Machine Learning (ICML), pp. 1481–1490 (2015)
10. Fei, G., Wang, S., Liu, B.: Learning cumulatively to become more knowledgeable. In: Proceedings of the 22nd ACM SIGKDD International Conference on Knowledge Discovery and Data Mining (2016)
11. Flaxman, S., Gelman, A., Neill, D., Smola, A., Vehtari, A., Wilson, A.G.: Fast hierarchical Gaussian processes. Manuscript in preparation (2016). http://sethrf.com/files/fast-hierarchical-GPs.pdf
12. Isele, D., Luna, J.M., Eaton, E., Gabriel, V., Irwin, J., Kallaher, B., Taylor, M.E.: Lifelong learning for disturbance rejection on mobile robots. In: Proceedings of the IEEE/RSJ International Conference on Intelligent Robots and Systems (IROS) (2016)
13. Kamar, E., Kapoor, A., Horvitz, E.: Lifelong learning for acquiring the wisdom of the crowd. In: Proceedings of the 23rd International Joint Conference on Artificial Intelligence (IJCAI) (2013)
14. Kumar, A., Daumé III, H.: Learning task grouping and overlap in multi-task learning. In: Proceedings of the 29th International Conference on Machine Learning (ICML) (2012)
15. Maurer, A., Pontil, M., Romera-Paredes, B.: Sparse coding for multitask and transfer learning. In: Proceedings of the 30th International Conference on Machine Learning (ICML), pp. 343–351, May 2013
16. Mitchell, T.M., Cohen, W.W., Talukdar, P.P., Betteridge, J., Carlson, A., Gardner, M., Kisiel, B., Krishnamurthy, J., et al.: Never ending learning. In: Proceedings of the 29th AAAI Conference on Artificial Intelligence (2015)

17. Nguyen-Tuong, D., Peters, J.R., Seeger, M.: Local Gaussian process regression for real time online model learning. In: Advances in Neural Information Processing Systems (NIPS), pp. 1193–1200 (2009)
18. Park, S., Choi, S.: Hierarchical Gaussian process regression. In: Proceedings of the 2nd Asian Conference on Machine Learning (ACML), pp. 95–110 (2010)
19. Pentina, A., Urner, R.: Lifelong learning with weighted majority votes. In: Advances in Neural Information Processing Systems (NIPS), pp. 3612–3620 (2016)
20. Quinonero-Candela, J., Rasmussen, C.E.: A unifying view of sparse approximate Gaussian process regression. J. Mach. Learn. Res. **6**, 1939–1959 (2005)
21. Rakitsch, B., Lippert, C., Borgwardt, K., Stegle, O.: It is all in the noise: efficient multi-task Gaussian process inference with structured residuals. In: Advances in Neural Information Processing Systems (NIPS), pp. 1466–1474 (2013)
22. Rasmussen, C.E., Williams, C.K.I.: Gaussian Processes for Machine Learning (Adaptive Computation and Machine Learning). The MIT Press, Cambridge (2005)
23. Ruvolo, P., Eaton, E.: ELLA: an efficient lifelong learning algorithm. In: Proceedings of the 30th International Conference on Machine Learning (ICML), June 2013
24. Snelson, E., Ghahramani, Z.: Sparse Gaussian processes using pseudo-inputs. In: Advances in Neural Information Processing Systems (NIPS), pp. 1257–1264 (2005)
25. Thrun, S.: A lifelong learning perspective for mobile robot control. In: Proceedings of the IEEE/RSJ/GI International Conference on Intelligent Robots and Systems (IROS) 'Advanced Robotic Systems and the Real World' (1994)
26. Tsanas, A., Little, M.A., McSharry, P.E., Ramig, L.O.: Accurate telemonitoring of Parkinson's disease progression by noninvasive speech tests. IEEE Trans. Biomed. Eng. **57**(4), 884–893 (2010)
27. Wang, B., Pineau, J.: Generalized dictionary for multitask learning with boosting. In: Proceedings of the International Joint Conference on Artificial Intelligence (IJCAI) (2016)
28. Wang, Y., Khardon, R.: Sparse Gaussian processes for multi-task learning. In: Flach, P.A., De Bie, T., Cristianini, N. (eds.) ECML PKDD 2012. LNCS (LNAI), vol. 7523, pp. 711–727. Springer, Heidelberg (2012). https://doi.org/10.1007/978-3-642-33460-3_51
29. Yu, K., Tresp, V., Schwaighofer, A.: Learning Gaussian processes from multiple tasks. In: Proceedings of the 22nd International Conference on Machine Learning (ICML), pp. 1012–1019 (2005)

Personalized Tag Recommendation for Images Using Deep Transfer Learning

Hanh T. H. Nguyen[✉], Martin Wistuba, and Lars Schmidt-Thieme

Information Systems and Machine Learning Lab, University of Hildesheim,
Universitätsplatz 1, 31141 Hildesheim, Germany
{nthhanh,wistuba,schmidt-thieme}@ismll.de

Abstract. Image tag recommendation in social media systems provides the users with personalized tag suggestions which facilitate the users' tagging task and enable automatic organization and many image retrieval tasks. Factorization models are a widely used approach for personalized tag recommendation and achieve good results. These methods rely on the user's tagging preferences only and ignore the contents of the image. However, it is obvious that especially the contents of the image, such as the objects appearing in the image, colors, shapes or other visual aspects, strongly influence the user's tagging decisions.

We present a personalized content-aware image tag recommendation approach that combines both historical tagging information and image-based features in a factorization model. Employing transfer learning, we apply state of the art deep learning image classification and object detection techniques to extract powerful features from the images. Both, image information and tagging history, are fed to an adaptive factorization model to recommend tags. Empirically, we can demonstrate that the visual and object-based features can improve the performance up to 1.5% over the state of the art.

Keywords: Image tagging · Convolutional neural networks · Personalized tag recommendation · Factorization models

1 Introduction

A large number of digital resources are stored, shared and accessed by users around the world everyday. To assist the organization and retrieval of images, social media services allow users to annotate their resources with their own keywords, called tags. Even though tagging is a relatively simple task, it is tedious, time-consuming and thus discourages the users from tagging their images. A study by Sigurbjörnsson and Van Zwol revealed that most images uploaded to Flickr have only few or even no tags [17]. They analyzed photos uploaded between February 2004 and June 2007 and reported that around 64% of them have 1 to 3 tags and around 20% have no tags at all.

Tag recommendation is used to save the user's time by suggesting relevant tags for the uploaded content. These suggestions are preferably based on the

© Springer International Publishing AG 2017
M. Ceci et al. (Eds.): ECML PKDD 2017, Part II, LNAI 10535, pp. 705–720, 2017.
https://doi.org/10.1007/978-3-319-71246-8_43

user's tag preferences and the contents of the uploaded resource. However, in practice the tag recommendation systems are often solely based on the user's tagging history, often ignoring the content of the uploaded items [2,13,16].

One disadvantage of the narrow folksonomy systems, which allow one or few people providing tags for a given resource, is the item cold-start problem. Most images uploaded to platforms such as Flickr are tagged only by few users, i.e. the owner of the image and other users with permissions granted by the owner. Hence, personalized tag recommendation models that are solely based on the user's preferences have tremendous problems providing useful predictions, especially for images that just have been uploaded. Thus, these recommendation models are often predicting the most popular tags.

According to Sigurbjörnsson and Van Zwol [17], people usually choose words related to the contents or contexts such as location or time to annotate images. Image features could be used to solve the cold-start problem. The low-level features such as color histograms have been often used in the personalized content-aware tag recommendation to overcome the problem.

In this paper, we propose a personalized tag recommendation which uses various deep learning methods and publicly available data sets for image classification and object recognition to extract powerful image features. These image features are combined with factorization models in order to boost the prediction performance. We propose to train a convolutional neural network on the famous ImageNet data set which is able to extract useful features from images on our image data set. Furthermore, we are training a convolutional neural network to detect 80 different objects on the MS COCO data set. Both these tasks (classification and object detection) are different to our task (tag recommendation) and use different data sets. However, we will show that we can use these networks to extract useful features from images that will help us recommending better tags. The extracted visual features are finally used by factorization machines (FM) [13] and pairwise interaction tensor factorization (PITF) [16] in order to give final recommendations. Our experiments are conducted on a real world data set, namely NUS-WIDE, and we can show that our proposed way of extracting image features improves the accuracy of the tag recommender by at least 1%.

The motivation for our approach can be explained easily and follows the way how human beings tag images. Lets have a look at Fig. 1. The user tagged this image with "urban", "motorcycle" and "downtown". While the COCO data set only allows us to distinguish 80 different objects which are completely unrelated to our task, we can nevertheless detect a person, a motorbike, a car and few further objects. This is also what a human being does and the appearance of the motorbike likely resulted into the tag "motorcycle". However, object detection can also help to recommend tags such as "urban". This tag is obviously no object which you see on the image but the recommender system can learn that whenever motorbikes, cars and people are detected on an image that an urban area or city is pictured. In a similar way the classification algorithm can extract image features such as specific shapes, colors and so on.

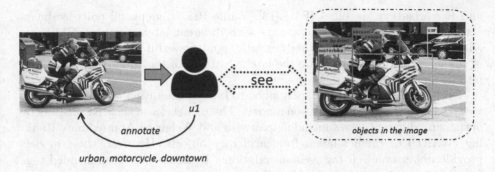

Fig. 1. When tagging an image, the user is highly influenced by what she sees on the image. In this example the user chooses "motorcycle" due to its occurrence. Furthermore, the tag "urban" is chosen since the image shows a street, people, cars and other things typical for cities. Our idea is to create an automatic system that uses object detection as a part of it to improve the tag recommendation performance.

2   Related Work

Tag recommendation can be based on different information such as the user's tagging behavior, the image contents, the time and location when the image was taken. A large number of approaches have been proposed which target various of these information. Li and Wang [5] extracted color and texture features and learned a mapping between these features and semantic concepts described by several keywords. The recommended tags were obtained based on the profiling models constructed from the concepts and the visual features. Li [6] also focused on using visual features in a neighbor voting model. The relevant tags for a given image are retrieved by the votes of similar images. The users' vocabularies approach also searches all neighbors of a given image according to location, time and visual features from the tagging history of the image's owner. A tag list is generated from tags of these neighbors and the most frequent tags are recommended [9].

An other approach is based on collective knowledge [17]. Tags correlated with the user-provided tags having higher co-occurrence scores are recommended to the given user. The approach proposed by Garg and Weber [2] also depends on the co-occurrence metric to get global and personal candidate tags correlated to the initial tags.

The correlated scores of tags retrieved from different contexts such as the personal or social tagging history are aggregated to compute the final scores of tags [10]. The social features extracted from users' social activities are combined with the textual features derived from tags, titles, contents and comments to represent tags. A predictor such as logistic regression or Naïve Bayes is employed to compute the scores of tags.

The relation between users, items and tags is mostly used in factorization models that provide a great performance for tag recommendation. Two of the state-of-the-art models are Pairwise Interaction Tensor Factorization (PITF) [16] and Factorization Machines (FM) [13]. While PITF models all pairwise inter-actions between users, items and tags with different latent features, FM takes advantage of feature engineering flexibility and powerful predicting capability of the factorization which share the latent features of tags between all pairwise interactions.

Deep learning approaches are applied to image annotation [3,19] that can be viewed as multi-label classification models. The models learn their parameters by optimizing different losses including pairwise and Weighted Approximate Rank-ing (WARP) or predict labels from arbitrary objects. However, these models provide unpersonalized tag recommendations. It means the recommended tags for similar images are the same for all users.

Factorization models, which do not use image features, cannot recommend tags which are related to the image's contents. They perform worse when recom-mending tags for new images and they merely recommend the most popular tags by users. In contrast, neural networks that are able to suggest the content-based tags to images will miss personal tags during the recommendation process. We propose a novel approach which combines the best of both worlds. Our model can catch both, tags which are related to the image itself and those which are user-specific, and is able to recommend the most relevant tags to the user.

3   Problem Formulation

The personalized tag recommender will suggest a ranked list of tags to a given user and image. The set of historical tagging assignments represented as $\mathcal{A}$ is a relation between the set of user U, images I and tags T. If user u assigns tag t to image i, the value of $a_{u,i,t} = 1$, or otherwise $a_{u,i,t} = 0$ [7].

The observed tagging set is defined as

$$S := \{(u, i, t) | a_{u,i,t} = 1\}$$

and all observed pairs (u, i), called posts, are grouped in a set [14] that are defined as

$$P_S := \{(u, i) \mid \exists t \in T : (u, i, t) \in S\}$$

Our content-aware recommendation extracts various image features from color images in the set $R := \{R_i \mid i \in I\}$. Visual features of an image $i \in I$ are extracted by the image classification network and denoted as $z_i \in \mathbb{R}^m$. Object detection features are represented by a vector $o_i \in \mathbb{R}^n$.

The scoring function $\hat{y}(u, z_i, o_i, t)$ of the image-based recommendation model computes the scores of tags for a given post $p_{u,i}$ which are used to rank tags. If the score $\hat{y}_{u,z_i,o_i,t_a}$ is larger than the score $\hat{y}_{u,z_i,o_i,t_b}$, the tag t_a is more relevant to the post $p_{u,i}$ than the tag t_b.

A content-aware tag recommendation model is expected to provide a top-K tag list $\hat{T}_{u,i}$ that is ranked in descending order of tags' scores for a post $p_{u,i}$.

$$\hat{T}_{u,i} := \operatorname*{argmax}_{t \in T}^{K} \hat{y}(u, z_i, o_i, t) \tag{1}$$

4 The Proposed Architecture

Our personalized image-aware tag recommendation aims at taking benefit of deep learning methods to extract visual information and improve the recommendation capability of factorization models. Our proposed architecture is illustrated in Fig. 2.

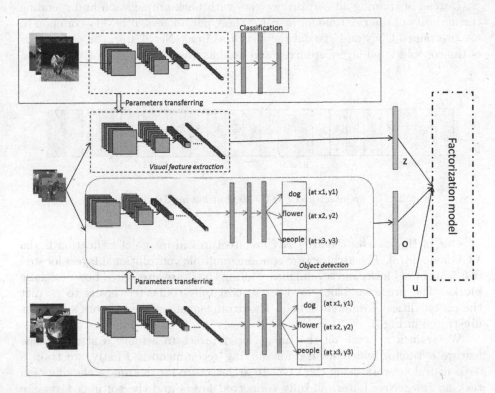

Fig. 2. The proposed architecture for personalized content-aware tag recommendation. We train one network for the task of image classification and one network for the task of object detection on two different data sets. These networks are finally used to extract image features or detect objects. These features and predictions are used as visual features in order to train a factorization model.

A deep neural network is trained on the ImageNet data set for the task of classification and another network is trained on the COCO data set for the task of object detection. The parameters of the networks are transferred to the tag recommender system and used to build the feature extractor.

The image features and the historical tagging assignments are fed to an adapted factorization model to compute the tag scores.

4.1   Visual Feature Extraction

Convolutional neural networks (CNNs) have recently achieved a great success in image classification. They can be used as strong extractors to achieve valuable visual features for images. In these networks, one or more convolutional layers are deployed to generate feature maps by sliding kernel windows across images. Several pooling layers can follow the convolutional layers.

Instead of training all network weights with back-propagation and spending the majority of the run time for learning these parameters, it is very common to use pretrained CNN on large data sets such as ImageNet. Later, the parameters of the convolutional layers are fixed and used as given feature extraction layers.

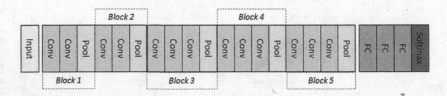

Fig. 3. The architecture of VGG-16 that having 16 weighted layers

One of the state-of-the-art CNN architectures in image classification is the VGG model [18]. The architecture contains multiple convolutional layers located in 5 sequential blocks and several max pooling layers are alternated between these blocks. The predictor block involves several fully-connected layers to predict the probabilities of different labels. The arrangement of the network's layers is illustrated in Fig. 3.

We train a network on the ImageNet data set to achieve a strong image feature extractor which we use for our tag recommender. Firstly, we train a deep neural network using the VGG-16 architecture for the image classification task on ImageNet. Later, all fully connected layers and the softmax layer are removed from the network and a global average pooling replaces these layers in the network. Finally, the network is used as the feature extractor in the tag recommender system and the output of the network is used as the new representation of the image. The extracting features process is formulated as:

$$z_i := f_{\text{vgg16}}(R_i) : \mathbb{R}^{224 \times 224 \times 3} \to \mathbb{R}^m$$

4.2   Object Detection

Deep learning does not only achieve state-of-the-art performance for image classification but is also applied successfully for the task of object detection. One of the state-of-the-art system that works fast and effective is YOLOv2 [12]. It is based on the DarkNet19 architecture that is described in Table 1. It is an improved version of YOLO (You Only Look Once) [11]. YOLO uses a single

Table 1. YOLOv2 is a fully convolutional network and is based on the Darknet-19 architecture sketched below.

Type	Filters	Size/stride	Output
Convolutional	32	3×3	224×224
Maxpool		$2 \times 2/2$	112×112
Convolutional	64	3×3	112×112
Maxpool		$2 \times 2/2$	56×56
Convolutional	128	3×3	56×56
Convolutional	64	1×1	56×56
Convolutional	128	3×3	56×56
Maxpool		$2 \times 2/2$	28×28
Convolutional	256	3×3	28×28
Convolutional	128	1×1	28×28
Convolutional	256	3×3	28×28
Maxpool		$2 \times 2/2$	14×14
Convolutional	512	3×3	14×14
Convolutional	256	1×1	14×14
Convolutional	512	3×3	14×14
Convolutional	256	1×1	14×14
Convolutional	512	3×3	14×14
Maxpool		$2 \times 2/2$	7×7
Convolutional	1024	3×3	7×7
Convolutional	512	1×1	7×7
Convolutional	1024	3×3	7×7
Convolutional	512	1×1	7×7
Convolutional	1024	3×3	7×7
Convolutional	1000	1×1	7×7
Avgpool		Global	1000
Softmax			

convolutional network in order to predict multiple bounding boxes and the label probabilities for these boxes.

The network comprises multiple convolutional layers mostly having 3×3 filters and the number of feature maps are doubled after each pooling step.

Our proposed architecture uses the probabilities of detected objects as features. If one object has been detected multiple times, we are using the maximum probability of this object. The information of bounding boxes is not used in the models and it is ignored during the extracting process. We train YOLOv2 on the COCO data set. Then, the network is used to extract the object representation for images in tag recommendation. The output of the network is a sparse vector representing for the detected probabilities of 80 categories (one for each object in the COCO data set) and it is denoted as:

$$o_i := f_{\text{YOLO}}(R_i) : \mathbb{R}^{448 \times 448 \times 3} \rightarrow \mathbb{R}^n$$

4.3   Factorization Models

Two state-of-the-art factorization models applied widely for tag recommendation are factorization machines (FMs) [13] and pairwise interaction tensor factorization (PITF) [16] that model the interaction between different elements of tag assignments. While PITF distinguishes latent features of tags for different pairs of interaction, FM shares these features between all pairs of interaction. In more detail, the input of these models is defined as the following,

$$x_{u,i,t} = \Big(\underbrace{0,\ldots, \overset{u}{1} ,\ldots,0}_{|U|}, \underbrace{0,\ldots, \overset{i}{1} ,\ldots,0}_{|I|}, \underbrace{0,\ldots, \overset{t}{1} ,\ldots,0}_{|T|} \Big) \qquad (2)$$

The scoring function in FM models is denoted as

$$\hat{y}(u,i,t) = b + \sum_{j=1}^{p} x_j w_j + \sum_{j=1}^{p-1} \sum_{j'=j+1}^{p} x_j x_{j'} \langle \mathrm{v}_j, \mathrm{v}_{j'} \rangle \qquad (3)$$

where $p = |U| + |I| + |T|$ and $\mathrm{v}_j \in \mathbb{R}^k$ are the latent features of the j-th feature. Moreover, $\langle \mathrm{v}_j, \mathrm{v}_{j'} \rangle$ is computed as

$$\langle \mathrm{v}_j, \mathrm{v}_{j'} \rangle = \sum_k v_{j,k} \cdot v_{j',k}$$

Because exactly one x_u, x_i and x_t are one and all others are zero, and we are applying a pair-wise loss function, the prediction function of the FM can be simplified to

$$\hat{y}(u,i,t) = w_t + \sum_{j=1}^{k} (v_{u,j}^U + v_{i,j}^I) v_{t,j}^T \qquad (4)$$

where k is the number of latent features, $V^U \in \mathbb{R}^{|U| \times k}$, $V^I \in \mathbb{R}^{|I| \times k}$ and $V^T \in \mathbb{R}^{|T| \times k}$ are the latent features of users, images and tags.

Similarly, the PITF prediction model simplifies to

$$\hat{y}(u,i,t) = \sum_{j=1}^{k} v_{u,j}^U \cdot v_{t,j}^{T^U} + v_{i,j}^I \cdot v_{t,j}^{T^I} \qquad (5)$$

where model parameters are denoted as $V^U \in \mathbb{R}^{|U| \times k}$, $V^I \in \mathbb{R}^{|I| \times k}$, $V^{T^U} \in \mathbb{R}^{|T| \times k}$ and $V^{T^I} \in \mathbb{R}^{|T| \times k}$.

The models are plainly based on the relation between different elements and use the index of all elements as their input. We cannot directly apply these models to content-aware recommendation where the input contains information of images representing in feature vectors.

4.4   Factorization Models for Image-Aware Tag Recommendation

The aforementioned factorization models focus on using users' preferences, instead of using contents of images. To feed image-based features to the factorization models, the part representing the image in Eq. (2) is replaced by its image-based features. If the features are the combination of visual and object features, it is denoted as:

$$x_{u,z_i,o_i,t} = \Big(\underbrace{\ldots, \overset{u}{1}, \ldots,}_{|U|} \underbrace{z_{i_1}, \ldots, z_{i_m}}_{m}, \underbrace{o_{i_1}, \ldots, o_{i_n}}_{n}, \underbrace{\ldots, \overset{t}{1}, \ldots}_{|T|} \Big)$$

Depending on the types of features used to predict the tags' scores, the part of unused features is removed from the input.

Based on the description of the input, we propose different factorization models based on FM and PITF to generate the scoring functions.

If both types of features are used to predict the relevant tags, the scoring functions are formulated as:

– The FM-based formula is:

$$\hat{y}(u, z_i, o_i, t) = w_t + \sum_{j=1}^{k} \left(v_{u,j}^U + \sum_{a=1}^{m} z_{i_a} \cdot v_{a,j}^Z + \sum_{a=1}^{n} o_{i_a} \cdot v_{a,j}^O \right) v_{t,j}^T \qquad (6)$$

– The PITF-based function is:

$$\hat{y}(u, z_i, o_i, t) = w_t + \sum_{j=1}^{k} v_{u,j}^U \cdot v_{t,j}^{T^U} + \Big(\sum_{a=1}^{m} z_{i_a} \cdot v_{a,j}^Z\Big) v_{t,j}^{T^Z} + \Big(\sum_{a=1}^{n} o_{i_a} \cdot v_{a,j}^O\Big) v_{t,j}^{T^O} \quad (7)$$

If the input contains one type of features, the parameters associated with the unused features are removed from the formula.

Depending on the types of image-based features and the scoring function, the models are named differently. In detail, **FM-OD** and **PITF-OD** use only the object detection features while **FM-IC** and **PITF-IC** use the feature extraction obtained by the image classification knowledge. **FM-IC-OD** and **PITF-IC-OD** use all image-based features.

4.5   Optimization

The criterion of the optimization used is Bayesian Personalized Ranking (BPR) optimization criterion [15]. The parameters found satisfy that the difference between the relevant and irrelevant tags are maximal.

The stochastic gradient descent applied to BPR is in respect of quadruples (u, i, t^+, t^-); i.e., for each $(u, i, t^+) \in S_{train}$ and an unobserved tag of $p_{u,i}$ drawn at random t^-, the loss is computed and is used to update the model's parameters.

$$\text{BPR}(u, z_i, o_i, t^+, t^-) := \ln \sigma(\hat{y}'(u, z_i, o_i, t^+, t^-)) \qquad (8)$$

where

$$\sigma(\psi) = \frac{1}{1 + e^{-\psi}}$$

The tag assigned by the user u for image i, called t^+ and the unobserved tag t^- of the pair (u, i) are denoted as

$$t^+ \in T_{u,i}^+ := \{t \in T \mid (u, i, t) \in S_{train}\}; \quad t^- \in T_{u,i}^- := \{t \in T \mid (u, i, t) \notin S_{train}\}$$

Moreover, the difference between two types of tags is defined as

$$\hat{y}'(u, z_i, o_i, t^+, t^-) = \hat{y}'(u, z_i, o_i, t^+) - \hat{y}'(u, z_i, o_i, t^-)$$

The learning algorithm is described in Algorithm 1. For each random post, a relevant tag and an irrelevant tag are sampled and the scores of these tags are computed. The gradients of the cost function in Eq. (8) with respect to the model's parameters are obtained as follows:

$$\frac{\partial \text{BPR}}{\partial \Theta} = \frac{e^{-\hat{y}'(u, z_i, o_i, t^+, t^-)}}{1 + e^{-\hat{y}'(u, z_i, o_i, t^+, t^-)}} \times \left(\frac{\partial \hat{y}'(u, z_i, o_i, t^+)}{\partial \Theta} - \frac{\partial \hat{y}'(u, z_i, o_i, t^-)}{\partial \Theta} \right) \quad (9)$$

Algorithm 1. Learning BPR

1: **Input:** P_S, S, Z, O, α
2: **Output:** Θ

3: Initialize $\Theta \leftarrow \mathcal{N}(0, 0.1)$
4: **repeat**
5:     Pick $(u, i) \in P_{S_{train}}$, $z_i \in Z$, $o_i \in O$
6:     Get $t_{u,i}^+ \in T$ and $(u, i, t) \in S$
7:     Pick $t_{u,i}^- \in T$ randomly whereas $(u, i, t) \notin S$
8:     Compute $\hat{y}'(u, z_i, o_i, t^+)$ and $\hat{y}'(u, z_i, o_i, t^-)$
9:     Update $\Theta \leftarrow \Theta + \alpha \left(\frac{\partial \text{BPR}(u, z_i, o_i, t^+, t^-)}{\Theta} \right)$
10: **until** convergence
11: **return** Θ

In order to learn the model, the gradients $\frac{\partial \hat{y}'(u, z_i, o_i, t^+)}{\partial \Theta}$ and $\frac{\partial \hat{y}'(u, z_i, o_i, t^-)}{\partial \Theta}$ have been computed. For examples, from Eq. (6), the derivatives with respect to parameters of tags are computed as:

$$\frac{\partial \hat{y}'(u, z_i, o_i, t)}{\partial v_{t,j}^T} = v_{u,j}^U + \sum_{a=1}^{m} z_{i_a} \cdot v_{a,j}^Z + \sum_{a=1}^{n} o_{i_a} \cdot v_{a,j}^O$$

5   Evaluation

5.1   Dataset

We obtained experiments on subsets of the publicly available multilabel data set NUS-WIDE [1] that contains 269,648 images. We preprocessed the first subset by keeping available images tagged by the 100 most popular tags, sampling 1.000 users, refining to get 10-core dataset referring to users and tags where each user or tag occurs at least in 10 posts [4]. Later we remove tags assigning more than 50% of images by one user to avoid the case that users tag all their images by the same words.

In a similar way, the second subset is obtained after several steps. First, tags are filtered by matching to WordNet [8] and only English tags are kept. Later, the data set is refined to get 20-core regarding to users, 100-core to tags and removing tags annotating more than 50% of images by one user.

Table 2. Dataset characteristics

Dataset	Users $|U|$	Images $|I|$	Tags $|T|$	Triples $|S|$	Posts $|P_S|$	Training posts $|P_{S_{train}}|$	Test posts $|P_{S_{test}}|$
NUS-WIDE-1	1000	27.662	100	81.263	27.858	25.858	2.000
NUS-WIDE-2	1.999	90.483	1.661	634.739	95.130	76.842	18.288

We created our train/test split using leave-one-post-out [7]: for each user in NUS-WIDE-1, 2 posts are randomly picked and put into its test set. Similarly, 20% of NUS-WIDE-2 posts for each user are sampled to put into the test set. These data sets are described with respect to users, images, tags, triples and posts as in Table 2. The color images used to extract features are crawled from Flickr and rescaled into 224×224 dimension. The distribution of posts per tag in NUS-WIDE-1 is more balanced than in NUS-WIDE-2 which has more than 50% of tags appearing less than 500 times.

5.2   Experimental Setup

The visual features extracted are combined in a 512-dimension vector while the object recognition probabilities of a given image are appended in a 80-dimension vector.

The factor dimension for both factorization architecture is fixed in 128. The evaluation metric used in this paper is the F1-measure in top K tag lists where K is in the range of 1 to 10.

$$\text{F1@K} = \frac{2 \cdot \text{Prec@K} \cdot \text{Recall@K}}{\text{Prec@K} + \text{Recall@K}} \tag{10}$$

where

$$\text{Prec@K} = \underset{(u,i) \in S_{test}}{\text{avg}} \frac{|\hat{T}_{u,i} \cap T_{u,i}|}{K} \qquad \text{Recall@K} = \underset{(u,i) \in S_{test}}{\text{avg}} \frac{|\hat{T}_{u,i} \cap T_{u,i}|}{|T_{u,i}|}$$

$$\hat{T}_{u,i} = \text{Top}(u, z_i, o_i, K) = \underset{t \in T}{\text{argmax}}^K \hat{y}(u, z_i, o_i, t)$$

The best learning rate α are searched within the range $\{10^{-2}, 10^{-3}, 10^{-4}\}$ and the best L2-regularization λ are found from the range $\{10^{-5}, 10^{-6}, 10^{-7}\}$. The proposed models **FM-IC-OD** and **PITF-IC-OD** are compared to following personalized tag recommendation approaches that are based only on the users' preference: **PITF** [16] and **FM** [13].

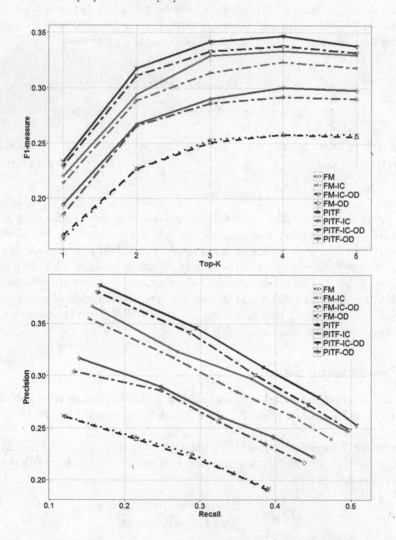

Fig. 4. F1-measure and Precision-Recall for NUS-WIDE-1

Moreover, these models are also compared to the factorization models using visual features or object detection features: **FM-OD**, **PITF-OD**, **FM-IC** and **PITF-IC**.

5.3 Results

As shown in Figs. 4 and 5, the personalized models **FM** and **PITF** which do not consider content information have the worst performance. They solely depend on the users' preferences and their power in catching the interaction between new images with other elements is not effective. In the NUS data set, most images in the test set do not appear in the training set and their latent parameters are not learned.

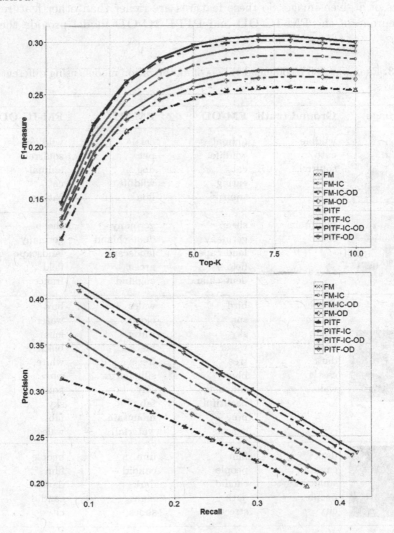

Fig. 5. F1-measure and Precision-Recall for NUS-WIDE-2

The claim that image-features improve the prediction quality is clearly shown in these figures. These features boost the performance from 1% to more than 3%. The object detection features are less effective than other features while the combination of image-based features helps improving accuracy the most effectively. Because the most popular tags in these data set are related to color such as blue or green, the object detection cannot capture these information and the models using them will miss these tags.

Otherwise, the visual features capture more unique information of a given images. For this reason, the performance of the models using visual features is better than the models using only the information of object detection. The combination of image-based features prove the powerful abilities in boosting performance. They can capture general object information and unique visual features of a given image. So these features are richer than other features and the accuracy of the **FM-IC-OD** and **PITF-IC-OD** model provide the best results.

Table 3. Examples top recommended tags of factorization models using different types of features

Image	Ground truth	FM-OD	FM-IC	FM-IC-OD
	wildlife cute squirrel	animal wildlife cat eating squirrel	cat pet dog wildlife lion	wildlife squirrel animal cat eating
	mountain sheep	sheep germany landscape field deutschland	germany deutschland landscape green england	sheep germany landscape field france
	clouds gothic stone bird dark castle wall	bird sun sky water tree blue silhouette beautiful prey sunset	wales cloud water black fresh silhouette sun fab mountain waterfall	bird water lens black white sun aqua sky fab wales
	dark candid sunglasses people city	film people candid dark street	film candid dark night shoes	people film dark candid city

Moreover, the PITF-based models generally work better than the FM-based in most cases. They separate the latent features of tags depending on the elements that they interact with. So they can capture the different representative of tags and combine the scores computing for each interaction into the final score. The difference between the PITF-based and FM-based approaches is clearly in the models using visual or object features while the performances of the models using both features are nearly compatible.

Examples in Table 3 show that the proposed models can capture the visual-based tags and object-tags compared to the models that are purely based on one type of image features. For example, **FM-IC-OD** recommends to a given user the object-based tag as "bird" and the visual-based tag as "black" in the fourth images.

6 Conclusion

In this paper, we showed how to extract image features using transfer learning. We used two different data sets in order two train two different neural networks for the task of image classification and object detection. We used these networks to extract powerful image features in order to improve the performance of the current state of the art for tag recommendation. Our proposed approach is able to recommend tags related to objects in images, tags representing image attributes and tags which are typically chosen by a user. For this reason, the performance of the models has been improved at least up to 1%. The experiments show that different types of image-based features improve the accuracy of tag recommendation in different levels. In the future, the contents used in the recommendation are not only limited on the information of images but are also broadened to contents of users such as vocabularies of users or their social activities.

References

1. Chua, T.S., Tang, J., Hong, R., Li, H., Luo, Z., Zheng, Y.: NUS-WIDE: a real-world web image database from national university of Singapore. In: Proceedings of the ACM International Conference on Image and Video Retrieval, p. 48 (2009)
2. Garg, N., Weber, I.: Personalized, interactive tag recommendation for flickr. In: Proceedings of the 2008 ACM Conference on Recommender Systems, pp. 67–74 (2008)
3. Gong, Y., Jia, Y., Leung, T., Toshev, A., Ioffe, S.: Deep convolutional ranking for multilabel image annotation. arXiv preprint arXiv:1312.4894 (2013)
4. Jäschke, R., Marinho, L., Hotho, A., Schmidt-Thieme, L., Stumme, G.: Tag recommendations in folksonomies. In: Kok, J.N., Koronacki, J., Lopez de Mantaras, R., Matwin, S., Mladenič, D., Skowron, A. (eds.) PKDD 2007. LNCS (LNAI), vol. 4702, pp. 506–514. Springer, Heidelberg (2007). https://doi.org/10.1007/978-3-540-74976-9_52
5. Li, J., Wang, J.Z.: Real-time computerized annotation of pictures. IEEE Trans. Pattern Anal. Mach. Intell. **30**(6), 985–1002 (2008)

6. Li, X., Snoek, C.G., Worring, M.: Learning tag relevance by neighbor voting for social image retrieval. In: Proceedings of the 1st ACM International Conference on Multimedia Information Retrieval, pp. 180–187 (2008)

7. Marinho, L.B., Hotho, A., Jäschke, R., Nanopoulos, A., Rendle, S., Schmidt-Thieme, L., Stumme, G., Symeonidis, P.: Recommender Systems for Social Tagging Systems. Springer Science & Business Media, Heidelberg (2012). https://doi.org/10.1007/978-1-4614-1894-8

8. Miller, G.A.: Wordnet: a lexical database for English. Commun. ACM **38**, 39–41 (1995)

9. Qian, X., Liu, X., Zheng, C., Du, Y., Hou, X.: Tagging photos using users' vocabularies. Neurocomputing **111**, 144–153 (2013)

10. Rae, A., Sigurbjörnsson, B., van Zwol, R.: Improving tag recommendation using social networks. In: Adaptivity, Personalization and Fusion of Heterogeneous Information, pp. 92–99 (2010)

11. Redmon, J., Divvala, S., Girshick, R., Farhadi, A.: You only look once: unified, real-time object detection. In: Proceedings of the IEEE Conference on Computer Vision and Pattern Recognition, pp. 779–788 (2016)

12. Redmon, J., Farhadi, A.: YOLO9000: Better, faster, stronger. arXiv preprint arXiv:1612.08242 (2016)

13. Rendle, S.: Factorization machines. In: 2010 IEEE 10th International Conference on Data Mining (ICDM), pp. 995–1000 (2010)

14. Rendle, S., Balby Marinho, L., Nanopoulos, A., Schmidt-Thieme, L.: Learning optimal ranking with tensor factorization for tag recommendation. In: Proceedings of the 15th ACM SIGKDD International Conference on Knowledge Discovery and Data Mining, pp. 727–736 (2009)

15. Rendle, S., Freudenthaler, C., Gantner, Z., Schmidt-Thieme, L.: BPR: Bayesian personalized ranking from implicit feedback. In: Proceedings of the Twenty-Fifth Conference on Uncertainty in Artificial Intelligence, pp. 452–461 (2009)

16. Rendle, S., Schmidt-Thieme, L.: Pairwise interaction tensor factorization for personalized tag recommendation. In: Proceedings of the Third ACM International Conference on Web Search and Data Mining, pp. 81–90 (2010)

17. Sigurbjörnsson, B., Van Zwol, R.: Flickr tag recommendation based on collective knowledge. In: Proceedings of the 17th International Conference on World Wide Web, pp. 327–336 (2008)

18. Simonyan, K., Zisserman, A.: Very deep convolutional networks for large-scale image recognition. arXiv preprint arXiv:1409.1556 (2014)

19. Wei, Y., Xia, W., Huang, J., Ni, B., Dong, J., Zhao, Y., Yan, S.: CNN: single-label to multi-label. arXiv preprint arXiv:1406.5726 (2014)

Ranking Based Multitask Learning of Scoring Functions

Ivan Stojkovic[1,2], Mohamed Ghalwash[1,3,4], and Zoran Obradovic[1(✉)]

[1] Center for Data Analytics and Biomedical Informatics, Temple University,
Philadelphia, PA 19122, USA
zoran.obradovic@temple.edu
[2] School of Electrical Engineering, University of Belgrade, 11120 Belgrade, Serbia
[3] IBM T. J. Watson Research Center, Cambridge, MA, USA
[4] Faculty of Science, Ain Shams University, 11566 Cairo, Egypt

Abstract. Scoring functions are an important tool for quantifying properties of interest in many domains; for example, in healthcare, a disease severity scores are used to diagnose the patient's condition and to decide its further treatment. Scoring functions might be obtained based on the domain knowledge or learned from data by using classification, regression or ranking techniques - depending on the type of supervised information. Although learning scoring functions from collected data is beneficial, it can be challenging when limited data are available. Therefore, learning multiple distinct, but related, scoring functions together can increase their quality as shared regularities may be easier to identify. We propose a multitask formulation for ranking-based learning of scoring functions, where the model is trained from pairwise comparisons. The approach uses mixed-norm regularization to impose structural regularities among the tasks. The proposed regularized objective function is convex; therefore, we developed an optimization approach based on alternating minimization and proximal gradient algorithms to solve the problem. The increased predictive accuracy of the presented approach, in comparison to several baselines, is demonstrated on synthetic data and two different real-world applications; predicting exam scores and predicting tolerance to infections score.

Keywords: Sparse learning · Multitask · Mixed-norm regularization

1 Introduction

Quantifying the properties of interest is an integral part in many domains, e.g., assessing the condition of a patient [27], estimating the risk of an investment [1], or predicting binding affinity of a ligand [4] when developing new drugs. Various measuring technologies and sensors are devised to quantify such properties of interest, which would in turn be utilized for informing decisions and making appropriate actions. However, the properties of interest are often not easy to obtain, whether they are difficult to measure directly or completely unobservable. This is usually the case when the properties are conceptual, i.e. they are

© Springer International Publishing AG 2017
M. Ceci et al. (Eds.): ECML PKDD 2017, Part II, LNAI 10535, pp. 721–736, 2017.
https://doi.org/10.1007/978-3-319-71246-8_44

latent constructs, such as health, satisfaction, and even intelligence. Under these circumstances, other measurable characteristics, considered related and informative of the true target, are observed and used as surrogate variables. In clinical settings, variables like temperature, blood pressure and various biomarkers measured from tissues are commonly tracked and considered when determining the health of the patient.

Typically, some heuristic rules are decided to map these surrogate variables into the desired score. The process of deciding these heuristic rules (or scoring functions) is usually long and tedious. For example, disease severity scores that are needed in clinical practices for patient diagnostics require years of effort and consensus of the medical community before the scoring functions can become part of the protocols. Fortunately, developments in machine learning and increasing amounts of collected data allows an alternative and complementary way for engineering the scoring functions by extracting rules automatically from the data, which facilitates and complements traditional approaches.

Algorithms for learning scoring functions from data were previously proposed, mainly in the medical domain, with the objective to learn disease severity scores [11,12,21,28,31]. Initial approaches posed the problem as traditional supervised learning tasks of classification [21,28] and regression [31]. However, classification and regression approaches require scores to be already accessible up front, which limits their applicability to problems with a good surrogate. The approach in [11,12] suggests the very appealing idea that there is a more convenient alternative form of supervised information to learn the scoring function from. Namely, ranked pairs are much easier to obtain than direct score estimates, and moreover, learning from pairs of ranked examples may result in more reliable and robust scoring functions.

In this work, we extend the suggested ranking-based approach [11] for score learning in multitask settings. The efforts are motivated by the applications in which there are multiple related tasks, with a limited amount of data for each task. Related tasks commonly share underlying regularities which could be learned more accurately by modeling all tasks together. For example, in education, scores on different subjects (e.g. Math and English) are dependent on the same characteristics of a particular student and a particular school. In the medical domain, disease severity scores for related illnesses (e.g. various respiratory viral infections) are expected to share common underlying biological mechanisms.

Consequently, we propose a novel multitask formulation for learning scoring functions from pairwise comparisons, by enforcing structural regularities on joint parameter space, using a matrix norm regularizations. In addition, we provide another contribution by developing an optimization algorithm in the form of an alternate minimization scheme based on a proximal gradient method. We evaluated the proposed approach on a synthetic data and two real-world applications. The objective of the first application is learning exam scores of elementary school pupils, while the objective of the second application is learning the tolerance to respiratory viral infections in humans. The results showed increased prediction accuracy of the proposed approach over individual tasks.

2   Related Work

Early efforts to learn scoring functions were dependent on complete supervised information (e.g. classification and regression tasks). In the classification settings, where the discrete class labels are provided, the classification methods were used to estimate the probability of a sample belonging to a certain class; these probabilities were used as a scoring function. For example, the method in [28] uses sparsity inducing L_1 norm in combination with a classical logistic loss function to learn the disease severity scoring function for assessing the abnormality of the skull in craniosynostosis cases.

Another similar approach is to learn the scoring function in a regression manner from the continuous outcome. In [31], Alzheimer's disease severity, as measured by cognitive scores, is modeled as a (temporal) multi-task regression problem using the fused sparse group lasso approach. The approach was more concerned with the progression of the disease; hence, the multi-task problem was formulated considering each time-step as a separate task. In contrast, we are interested in multiple score mapping from a single time-point set of measurements. There is also work on multitask learning to rank in the context of web search results ranking [6], where the ranking function is learned using the gradient boosted trees from the ranking scores provided by the human experts.

The problem with such completely-supervised methods is the necessity of providing direct values of scores for training purposes, which render the approaches as less powerful in settings where characteristics of interest are latent and not directly accessible. However, rather than giving direct estimates of the score, the easier task seems to be comparing two samples and asserting whether one has a higher score than the other. Ranking SVM [18] was the first approach that recognized the benefits of learning from ordered pairs of samples. This method was applied to learn an improved relevance function for documents retrieval from click-through data. Main insight was that clicked links are definitely more relevant for the search, as compared to non-clicked ones. And such kind of data is much more abundant than the user provided rankings. Recently, the ranking SVM-based method was adopted for Sepsis severity score learning [11] and extended for temporal applications by introducing a term that ensures gradual score change over consecutive time points.

Multitask learning is based on the idea that generalization (predictive performance) can be increased by accounting for the intrinsic relationships among multiple tasks. Multitask approach is found particularly effective when the number of samples per task is small. To the best of our knowledge, there are no published multitask formulations for ranking-based scoring functions, that is, for methods that learn from pairwise comparisons. The closest approaches are the previously mentioned multitask regression-based models for Alzheimer's disease progression [31] and search results ranking [6]. Other multitask regression methods exist that learn the structure among the tasks using norm regularization [30], or methods that utilize fixed relatedness structure [23] obtained from domain knowledge [25] or learned from a statistical correlation [24]. However, since they are not directly proposed for ranking-based learning of the scoring functions, we will not consider them, nor will compare with them in this work.

The main problem in multi-task learning is finding the most appropriate assumption on how the tasks are related and incorporating such assumption into the model. Typically, in linear models, such structural assumptions are imposed on the joint parameter matrix, where rows correspond to features and columns to different tasks. Kernel methods assume that all tasks are related and similar [13], but some methods enforce tasks to be grouped into clusters [16]. For example, "Dirty method" [17] encourages block-structured row-sparsity in the joint parameter matrix by $\|.\|_{1,1}$ norm, and element-wise sparsity with $\|.\|_{1,\infty}$. The robust approach [14] selects sparse rows of features for related tasks with $\|.\|_{2,1}$ and dense columns for outlier tasks with $\|.\|_{1,2}$, in order to discern between related and unrelated tasks. Other approaches assume some shared common set of features [3] or shared common subspace [2,9]. The approach proposed in [10] attempts to learn such relatedness subspace with trace (nuclear) norm $\|.\|_*$ by encouraging the parameter matrix to have low rank, and finding outlier tasks with additional sparse group norm $\|.\|_{1,2}$.

In this work we use regularization composed of trace norm [10], and grouped Lasso penalty [3] to jointly learn multiple ranking based scoring tasks, from temporal data.

3   Model

Let us assume that we have N samples (examples), where each sample i is represented as $X_i \in \mathbb{R}^d$, and where X_{ij} is the value (measurement) of the feature $j = \{1, 2, \ldots, d\}$ for the sample $i = \{1, 2, \ldots, N\}$. Let us assume that $y_i \in \mathbb{R}$ represents the property of interest (outcome variable) for the sample i. Scoring function $score : \mathbb{R}^d \to \mathbb{R}$ is then a mapping $X_i \mapsto y'_i$ that provides a close estimate y'_i of the true score y_i.

However, in many cases the values of the true scoring function are difficult to obtain. In such situations, it is easier to assess the ranking between the scores of two samples p and q, i.e. to assert that one has perceived higher score than the other: $score(X_p) > score(X_q)$. Therefore, a set of multiple such ordered pairs can be used to find a projection in the space of measured features, that will preserve the orders in the best possible way, and that might be used as a scoring function.

Moreover, measurements collected on multiple occasions over time might belong to the same subject; In this case, the measurements at each time step will be considered as a sample. We assume that the outcome variable changes gradually (smoothly) over time for the same subject, e.g. the disease severity score changes smoothly over consecutive time points for the same patient. This assumption will lead to improving the quality of the scoring function. We assume that X_p represents the feature vector for the sample p (which could be one particular subject at one particular time point).

In this work, we constrain such functional mapping *score* to the linear case, where the score estimate is computed as a weighted sum of the measured characteristics: $score(X) = w^T X$. Therefore, the problem of learning the scoring function becomes finding the appropriate weight (or parameter) vector $w \in \mathbb{R}^d$.

3.1   Single Task Model Formulation

Maximizing the number of correctly ordered training pairs can be performed using the soft max-margin framework expressed in a Hinge loss form (1), as suggested in [18].

$$max(0, 1 - (X_p - X_q)w) \tag{1}$$

If sample p should have higher score compared to sample q, the formulation (1) will favor the weighted difference $(X_p - X_q)w$ that is positive and greater than 1, thus even achieving some margin in the score difference.

The L_2 norm on the weight vector $||w||^2$, is introduced to regularize the magnitude of the weights, and to turn the problem into simultaneous maximization of correct ordering and maximization of normalized margin.

Gradual (smooth) change of the scoring function over time can be obtained by penalizing high changes of the score (e.g. for two samples X_{i+1}^s, X_i^s of the same subject s), over short time intervals. In [12] such effect is achieved by using the temporal smoothness term:

$$\left(\frac{(X_{i+1}^s - X_i^s)w}{(t_{i+1}^s - t_i^s)} \right)^2, \tag{2}$$

which essentially ensures that squared magnitude in difference, normalized with the time interval length, is kept low.

Therefore, for single task formulation of ranking-based scoring function learning, we adopted the Linear Disease Severity Score Learning formulation [11] which combines attractive properties of ranking SVM [18], with temporal smoothness term (2) that enforces the gradual change of the scoring function over time:

$$\hat{w} = \underset{w}{\operatorname{argmin}} \frac{1}{2} \|w\|_2^2 + c \sum_{\{p,q\} \in O} max(0, 1 - (X_p - X_q)w)$$

$$+ b \sum_{\{i,i+1\}_s \in S} \left(\frac{(X_{i+1}^s - X_i^s)w}{(t_{i+1}^s - t_i^s)} \right)^2 \tag{3}$$

Every measurement (row) vector X_i, $i = \{1, 2, \ldots, N\}$ has associated timestamp t, while $\hat{w} \in \mathbb{R}^d$ denotes the solution of the objective 3.

Set O is composed of ordered pairs $\{p, q\}$, where p has a higher rank than q (p is perceived to have a higher score than q), and which corresponds to the measurement vectors X_p and X_q, respectively. Sum of the Hinge loss terms over all pairs from the O set, serves to reduce the extent of incorrectly ordered pairs.

Set of all consecutive pairs in all subjects is denoted S and the sum of the Temporal smoothness terms in Eq. (3) penalizes high rates of change in score values in consecutive time steps t_i and t_{i+1} for all subjects $s \in S$. Scalar constants c and b are hyperparameters that determine the cost of the respective loss terms, the Hinge loss and the Temporal loss.

We aggregate the differences of measurements in the Hinge loss term into a single data matrix $D_{k \times d}$, where k is the number of pairs in the comparison set O. Similarly, measurement and temporal difference ratios in the Temporal loss term we write as matrix $R_{l \times d}$, where l is a number of pairs in the consecutive measurements set S. We aggregate the L_2 norm and temporal smoothness terms (they are essentially weighting the square of optimization parameters) into a single weighted quadratic term $\frac{1}{2} w^T Q w$, where Q is constant square matrix defined in Eq. (4):

$$Q = I + 2b R^T R, \tag{4}$$

I being the d-dimensional identity matrix.

The formulation (3) can now be rewritten more concisely as (5):

$$\hat{w} = \underset{w}{\operatorname{argmin}} \frac{1}{2} w^T Q w + c \sum_i max(0, 1 - D^i w) \tag{5}$$

3.2   Multitask Formulation

As mentioned before, in case of a limited amount of data for training the scoring function for a single task (5), it is beneficial to exploit the relatedness among the multiple similar tasks, by learning them together, as illustrated in Fig. 1.

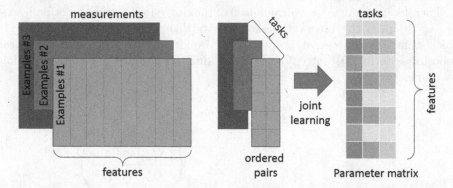

Fig. 1. Illustration of joint training of multiple ranking based score learning tasks. Three distinct task are depicted, where measured data in combination with supervision in form of ordered pairs, are jointly optimized to obtain the scoring function parameters, represented as parameter matrix. Parameter matrix is typically regularized to encode the structural assumptions regarding the task relatedness.

For m different tasks, individual parameter vectors w_i are aligned into a matrix $W_{d \times m}$, and a joint objective is obtained as a superposition of individual losses (Eq. (5)) over the multiple tasks $i \in \{1, 2, ..., m\}$:

$$\underset{W}{\operatorname{argmin}} \sum_{i=1}^{m} \left(\frac{1}{2} W_i^T Q_i W_i + c \sum_j max(0, 1 - D_i^j W_i) \right) \tag{6}$$

Instead of the non-smooth Hinge loss $L(a) = max(0, a)$ in Eq. (6), we work with the twice differentiable approximation in the form of Huber loss [11]:

$$L_h(a) = \begin{cases} 0 & \text{, if } a < -h \\ \frac{(a+h)^2}{4h} & \text{, if } |a| \leq h \\ a & \text{, if } a > h. \end{cases} \tag{7}$$

where the approximation threshold h can be chosen arbitrarily small.

Further, we regularize the objective in Eq. (6) with a joint norm on parameter matrix $\|W\|_{p,q} = (\sum_i ((\sum_j (W_{ij}^q)^{\frac{1}{q}})^p)^{\frac{1}{p}}$. For $p = 2$ and $q = 1$, this approach is known as a group Lasso penalty on the row groups (of W), which forces sparsity in the parameter weights corresponding to certain features [3]. Additionally, we introduce the trace norm L_* in order to get the low rank component, or in other words, the parameter weight pattern common among all the tasks. To accommodate such a setup, which will be further clarified in the Optimization section, the parameter matrix W was split into two distinct matrices A and B, where $W = A + B$.

Multitask Ranking Based Scoring Function Learning (MultiRBSFL) objective is now given in Eq. (8), and it takes as an input two matrices (per task i) obtained from the data: $Q_{d \times d}^i$ and $D_{k \times d}^i$; hyperparameters b, c, λ_1 and λ_2 weighting the influence of Temporal loss, Huber loss, trace norm and sparse group norm, respectively.

$$\underset{W=A+B}{\operatorname{argmin}} \mathcal{L}_1 + \lambda_1 \|A\|_* + \lambda_2 \|B\|_{2,1} \tag{8}$$

where

$$\mathcal{L}_1 = \frac{1}{m} \sum_{i=1}^{m} \left(\frac{1}{2}(A^i + B^i)^T Q^i (A^i + B^i) + c \sum_{j=1}^{k} L_h(1 - D_j^i (A^i + B^i)) \right) \tag{9}$$

A^i and B^i are column vectors $\mathbb{R}^{d \times 1}$, and D_j^i is $\mathbb{R}^{1 \times k}$ row-vector.

4   Optimization

The optimization (8) is composed of smooth and non-smooth terms. However, although the regularaization terms are separable in A and B, the loss term $\mathcal{L}_1$ is not separable. Therefore, we solve the problem by using the alternative minimization scheme, where, in each iteration, we fix A and minimize (8) with respect to B, and then fix B and minimize (8) w.r.t A. In this case, each subproblem can be decomposed into two different optimizations. This will be explained in the next section.

Fix A

$$\underset{B}{\operatorname{argmin}} \ \mathcal{L}_1 + \lambda_2 \|B\|_{2,1} \tag{10}$$

Fix B

$$\underset{A}{\operatorname{argmin}} \ \mathcal{L}_1 + \lambda_1 \|A\|_* \tag{11}$$

In general, problem (10) and (11) can be written as:

$$\underset{\Theta}{\operatorname{argmin}} \ \mathcal{L}_1 + \gamma \|\Theta\|_p, \tag{12}$$

where $\Theta = \{A, B\}$ and $p = \{*, \{2, 1\}\}$.

The optimization (12) is convex. The expression $\mathcal{L}_1$ is smooth and the regulariation term (either group lasso or trace norm) is non-smooth. Therefore, we solve (12) using the proximal methods.

4.1   Proximal Algorithm

We solve (12) using the proximal gradient method [20].

$$\Theta^{k+1} := \mathbf{prox}_{\lambda\|\Theta\|_p}(\Theta^k - \lambda\nabla\mathcal{L}_1(\Theta^k))$$

$$= \underset{\Theta}{\operatorname{argmin}} \left(\|\Theta\|_p + \frac{1}{2\lambda} \|\Theta - (\Theta^k - \lambda\nabla\mathcal{L}_1(\Theta^k))\|_2^2 \right), \tag{13}$$

where $\mathbf{prox}_{\lambda\|\Theta\|_p}$ is the proximal operator of the scaled function $\|\Theta\|_p$, and $\lambda \in (0, 1/L]$ is a *constant* step size, and L is a Lipschitz constant of $\nabla\mathcal{L}_1$. Problem (12) can be solved analytically, where the proximal operator associated with the norm can be obtained as in [5].

Trace norm. Let us assume that $M = U\Sigma V$ is the singular value decompoistion of M, where Σ is a diagonal matrix and its entries σ_i are the singluar values of the matrix M. The proximal operator of the trace norm is defined as [8]:

$$\mathbf{prox}_{\lambda\|.\|_*}(M) = U\mathbf{diag}(\mathbf{prox}_{\lambda\|.\|_1}(\sigma(M)))V$$

i.e., the proximal operator of $\|.\|_*$ can be calculated by carrying out a singular value decomposition of Z and evaluating the proximal operator of the corresponding absolutely symmetric function at the singular values $\sigma(M)$. Therefore,

$$\mathbf{prox}_{\lambda\|.\|_*}(M) = U\mathbf{diag}(\overline{\sigma}_1, \overline{\sigma}_2, \ldots, \overline{\sigma}_n)V, \tag{14}$$

where:

$$\overline{\sigma}_i = \begin{cases} \sigma_i - \lambda & \sigma_i \geq \lambda \\ 0 & -\lambda \leq \sigma_i \leq \lambda \\ \sigma_i + \lambda & \sigma_i \leq -\lambda \end{cases}$$

Equation (14) is sometimes called the singular value thresholding operator.

Group lasso norm. The proximal operator associated with the group lasso norm is defined as:

$$\left[\mathbf{prox}_{\lambda\|.\|_{1,2}}(u)\right]_g = \begin{cases} (1 - \frac{\lambda}{\|u_g\|_2})u_g & \|u_g\|_2 > \lambda \\ 0 & \text{otherwise} \end{cases}$$

4.2 Step Size

In order to find an adaptive step size λ^k in each iteration k, we employ the backtracking line search algorithm [7], which requires computing an upper bound for $\mathcal{L}_1$. Since $\mathcal{L}_1$ is convex and smooth, and $\nabla \mathcal{L}_1$ is L-Lipschitz continuous, it follows that:

$$\mathcal{L}_1(\boldsymbol{\Theta}) \leq \underbrace{\mathcal{L}_1(\boldsymbol{\Theta}^k) + \nabla \mathcal{L}_1(\boldsymbol{\Theta}^k)^T(\boldsymbol{\Theta} - \boldsymbol{\Theta}^k) + \frac{L}{2}\left\|\boldsymbol{\Theta} - \boldsymbol{\Theta}^k\right\|_2^2}_{\widehat{\mathcal{L}_1}_{\frac{1}{L}}(\boldsymbol{\Theta}, \boldsymbol{\Theta}^k)} \tag{15}$$

By utilizing (15), it can be shown that the optimization (13) is equivalent to [20]:

$$\boldsymbol{\Theta}^{k+1} := \operatorname*{argmin}_{\boldsymbol{\Theta}} \widehat{\mathcal{L}_1}_{\lambda^k}(\boldsymbol{\Theta}, \boldsymbol{\Theta}^k) + \|\boldsymbol{\Theta}\|_p \tag{16}$$

where $\lambda^k = \frac{1}{L}$. So at each iteration, the function $\mathcal{L}_1$ is linearized around the current point and the problem (16) is solved. The final fast proximal gradient method with backtracking is shown in Algorithm 1. The final alternative minimization algorithm is shown in Algorithm 2.

5   Empirical Evaluation

The proposed approach for multitask learning of ranking-based scoring functions is tested on one synthetic and two real-world datasets. We compared our MultiRBSFL approach against the following baseline approaches:

1. L_2 - independently learning scoring functions for each task (objective (3));
2. L_1 - independently learning sparse (L_1 regularized) scoring functions for each task;
3. L_* - learning multiple scoring functions by imposing low rank regularization on their joint parameter matrix (L_* regularized objective (6));
4. $L_{2,1}$ - joint objective (6), regularized by mixed $\|.\|_{2,1}$ norm.

Algorithm 1. Fast Gradient Proximal Method with Backtracking Step Size

---

1: **Input:** Θ^0 (random), η (usually $1/2$), $L > 0$
2: $\lambda = \frac{1}{L}$, $\mathbf{z}^1 = \Theta^0$, $t_1 = 1$, $k = 0$
3: **repeat**
4:     $k \longleftarrow k + 1$
5:     **while** true **do**
6:         $\mathbf{z} \longleftarrow$ Solve (12)   ▷ use λ and $\mathbf{z}^k$
7:         **if** $\mathcal{L}_1(\mathbf{z}) \leq \widehat{\mathcal{L}}_1(\mathbf{z}, \mathbf{z}^k)$ **then**
8:             break
9:         **end if**
10:        $\lambda \longleftarrow \eta\lambda$
11:    **end while**
12:    $\Theta^k \longleftarrow \mathbf{z}$
13:    $t_{k+1} = \frac{1+\sqrt{1+4t_k^2}}{2}$
14:    $\mathbf{z}^{k+1} = \Theta^k + (\frac{t_k-1}{t_{k+1}})(\Theta^k - \Theta^{k-1})$
15: **until** Convergence

Algorithm 2. Alternative Minimization

1: **Input:** A^0, B^0 (random)
2: **repeat**
3:     Fix A, solve (10) using Algorithm (1).
4:     Fix B, solve (11) using Algorithm (1).
5: **until** Convergence

Our MultiRBSFL approach, which uses composite low rank and mixed norm regularized joint objective (8), we will denote as $L_* + L_{2,1}$ for consistency in naming the alternative approaches.

We measured the predictive performance in terms of accuracy, which is the number of correctly ordered test pairs. As the pairwise ranking relation is antisymmetric, it is sufficient to use only the positive training instances (i.e. where the first sample in a pair has the larger score). Test pairs are exclusively generated from examples not contained in the training set. Accuracy values that we report in this study are obtained by doing 5-fold cross-validation experiments.

5.1   Experiments on Synthetic Data

In this settings, a Gaussian processes model with an exponential kernel was used to generate the temporal data. We compiled 250 processes to mimic $d = 250$ measured variables (features) per subject. Each single process was used to generate a time series with 10 time points (10 samples). We followed the same principle to generate 10 different multivariate time series (subjects) for training and 10 subjects for test, resulting in 100 samples $X^{train}_{100\times250}$ for training, and 100 samples $X^{test}_{100\times250}$ for test.

Four different tasks were created by randomly generating the weight matrix $W_{250 \times 4}$, with only 5 nonzero rows, which corresponds to the $L_{2,1}$ assumption (row-sparsity). This row-wise sparse matrix was then superimposed with a dense rank-1 matrix, generated by multiplication of two random vectors, which suits the L_* trace norm part of the objective. True underlying scores on four tasks, for each of the 250-dimensional samples (one time point of one patient), are calculated as the weighted sum of the feature values $X * W$. Zero mean random vector was subsequently superimposed to input X data to model the measurement noise.

A training set is then obtained by making pairs out of samples whose scores are sufficiently different (in our case we set the threshold to 1). Pairs of examples were generated independently for each task based on their scores, totaling 14,187 pairs for all four tasks jointly. Test set pairs were generated in the same fashion, but with a smaller threshold and consisted out of 19,390 pairs. Training pairs were used to learn the weight matrix $\hat{W}$, which was used to estimate the testing scores from the test samples. The obtained estimates were used to infer the relative order of the testing pairs. The accuracy (percentage of correct guesses) is reported in the Table 1. It is no surprise that the proposed $L_* + L_{1,2}$ approach achieves the highest accuracy on all four tasks, as the underlying assumptions were explicitly built into the synthetic example.

Table 1. Comparison of accuracy indicators (fraction of correctly ordered pairs) for alternative score learning methods on the synthetic data of four related tasks.

Task	L_2	L_1	L_*	$L_{1,2}$	$L_* + L_{1,2}$
TASK1	0.538	0.745	0.680	0.744	**0.757**
TASK2	0.556	0.707	0.763	0.782	**0.795**
TASK3	0.592	0.765	0.744	0.821	**0.837**
TASK4	0.466	0.864	0.700	0.874	**0.885**
AVG	0.538	0.770	0.722	0.805	**0.818**

5.2   School Exam Score

Intelligence as well as the capacity for understanding and using mathematics or languages are all examples of properties that are latent - yet important and often evaluated (estimated). We have tested the multitask score learning framework on data from an elementary school study [19], which contains longitudinal data on performance in Math and English language for pupils in 50 inner London schools[1]. In total there are scores for 3,236 exams (Math and English each), taken by 1,402 students over three consecutive school years. The goal is to rank the students' performances on Math and English test based on known score from Ravens ability test and additional information like demographics, social status, gender, class and school type. Distributions of scores for two tasks are given in the Fig. 2.

---

[1] http://www.bristol.ac.uk/cmm/media/migrated/jsp.zip.

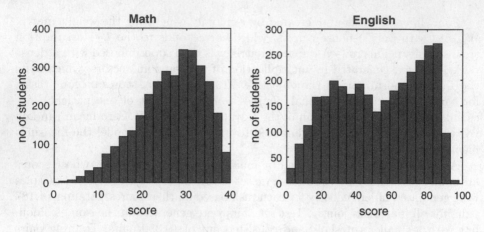

Fig. 2. Distributions of test scores for Math and English tasks, respectively.

According to results depicted in Table 2, our $L_* + L_{1,2}$ approach achieved the best predictive performance in both tasks.

Table 2. Comparison of accuracy indicators (fraction of correctly ordered pairs) for alternative score learning methods on the task of learning the performance on Math and English tests.

Task	L_2	L_1	L_*	$L_{1,2}$	$L_* + L_{1,2}$
MATH	0.780	0.794	0.725	0.789	**0.812**
ENGLISH	0.820	0.863	0.717	0.857	**0.870**
AVG	0.800	0.828	0.721	0.823	**0.841**

5.3   Tolerance to Infections Score

Tolerance is the host's behavior that arises from interactions with a pathogen, which describes the ability of the host to preserve fitness despite the presence of a large amount of pathogen. Therefore, it is defined as changes in host fitness (health) with respect to changes in pathogen load [22]. However, tolerance is a very understudied topic, where there is no established scoring function, despite the necessity.

We analyzed three publicly available datasets[2] that allows characterization of the tolerance behavior in humans. The data comes from the human viral challenge studies [29] where human volunteers were infected with H3N2 influenza, rhinovirus (HRV) and respiratory syncytial virus (RSV), respectively. For all subjects in each dataset, symptoms were recorded twice a day and quantified by the modified Jackson Score [15]. Thereafter, subjects were classified

---

[2] http://people.ee.duke.edu/~lcarin/reproduce.html

based on the modified Jackson Score values into "symptomatic" and "asymptomatic" groups. In addition, viral load temporal measurements are available for 28 "symptomatic" subjects, given in Table 3. Gene expression measurements (for 12,023 genes) were collected temporally, starting at a baseline (24 hours prior to inoculation with virus) and measured at certain time points following the experimental procedure described in detail in [29], making a total of 16, 14 and 21 time-point measurements for H3N2, HRV and RSV datasets, respectively. Table 3 shows the viral shedding and symptom scores for subjects who developed clinically relevant symptoms from H3N2, HRV and RSV datasets.

Temporal measurements about symptoms (proxy for fitness) and viral (pathogen) load for each subject were used to derive tolerance scores according to the definition given in [22]. In particular, the tolerance score for each subject was calculated by dividing the maximum viral load with the maximum severity of symptoms observed for that subject (Table 3). Gene expression measurements were used as an explanatory variables in our ranking task.

Table 3. Tolerance scores (Ratio) derived by dividing maximum viral load (Max V) with maximum severity score (Max S).

H3N2				HRV				RSV			
Sub ID	Max S	Max V	Ratio	Sub ID	Max S	Max V	Ratio	Sub ID	Max S	Max V	Ratio
FLU05	12.00	5.45	0.45	HRV06	8.00	2.72	0.34	RSV01	11.00	0.00	0.00
FLU08	10.00	4.70	0.47	HRV19	2.00	0.95	0.47	RSV20	6.00	0.00	0.00
FLU01	9.00	4.25	0.47	HRV04	8.00	3.94	0.49	RSV07	20.00	4.46	0.22
FLU07	12.00	6.25	0.52	HRV15	7.00	3.45	0.49	RSV02	20.00	5.10	0.26
FLU06	7.00	5.00	0.71	HRV07	7.00	4.44	0.63	RSV12	4.00	2.50	0.62
FLU10	5.00	3.75	0.75	HRV20	6.00	4.44	0.74	RSV06	9.00	5.65	0.63
FLU12	4.00	5.00	1.25	HRV16	6.00	4.69	0.78	RSV14	6.00	4.54	0.76
FLU15	2.00	4.50	2.27	HRV09	3.00	2.46	0.82	RSV11	5.00	3.85	0.77
FLU13	2.00	5.45	2.70	HRV11	3.00	2.47	0.83	RSV03	6.00	4.70	0.78
				HRV03	4.00	3.45	0.86				

Biological rationale behind the task relatedness is that the three infections are viruses that cause similar respiratory symptoms (runny nose, fever, cough) and are quantified by the same Jackson score, suggesting that some shared genetic mechanisms might be responsible for the disease manifestations. Consequently, we sought to learn the tolerance scoring functions jointly.

The tolerance scores were used to compile a set of ranked pairs, and the objective was to learn the scoring functions for tolerance to H3N2, HRV and RSV viruses (3 tasks), from high-dimensional gene expression data. Since 12,023 dimensions is very computationally expensive to optimize, we reduced the dimensionality of the data to the 100 most informative genes according to the correlation with the target. The results of learning the scoring functions with different approaches are summarized in the Table 4.

The results from the Table 4 show that the HRV task is the most difficult one in the described formulation. Although some alternative approaches achieved

Table 4. Comparison of accuracy indicators (fraction of correctly ordered pairs) for alternative score learning methods on the tolerance to three viruses learning task.

Task	L_2	L_1	L_*	$L_{1,2}$	$L_* + L_{1,2}$
FLU	0.766	0.980	0.809	0.988	**0.996**
HRV	0.344	0.122	0.389	**0.500**	0.400
RSV	0.806	**0.972**	0.861	0.306	0.861
AVG	0.638	0.692	0.686	0.598	**0.752**

better accuracy in two of the tasks, the proposed approach achieved the best generalization trade-off as can be concluded from the highest average (overall) accuracy.

6   Discussion and Conclusions

We proposed the method that jointly learns multiple scoring functions from a set of ranked examples. The approach utilizes composite regularization consisting of the trace norm and row-wise grouped Lasso penalty, to impose the structural regularity among the model parameters of different tasks. We also provide optimization algorithm, based on the alternate minimization and proximal gradient techniques, for solving the proposed convex MultiRBSFL objective.

Presented empirical evaluations in one synthetic and two real world datasets suggest the benefits of utilizing the multitask approach for learning related ranking based scoring functions. According to the results, the model with only L_* performs worse than $L_{1,2}$, probably because sparsity in features seems to be the more dominant pattern in the data than the low-rank component. However, utilizing both L_* and $L_{1,2}$ in the same model turned out to be most beneficial for studied applications.

The proposed proximal gradient algorithm with alternating minimization for optimization of the multitask objective proved valuable for applications with low to moderate dimensionality of the feature space. However, as the contemporary applications have ever increasing number of measured variables, more efficient optimization approaches and with better scalability would be required. One potential way to accelerate the proximal gradient algorithm is to adopt the approach proposed in [26].

Acknowledgments. This research was supported in part by DARPA grant W911NF-16-C-0050 and in part by DARPA grant No. FA9550-12-1-0406 negotiated by AFOSR. Computations were performed on the OwlsNest HPC cluster at Temple University, which is supported in part by the National Science Foundation through NSF grant NSF-CNS-1625061 and Pennsylvania Department of Health CURE grant.

References

1. Anderson, R.: The Credit Scoring Toolkit: Theory and Practice for Retail Credit Risk Management and Decision Automation. Oxford University Press, Oxford (2007)
2. Ando, R.K., Zhang, T.: A framework for learning predictive structures from multiple tasks and unlabeled data. J. Mach. Learn. Res. **6**(Nov), 1817–1853 (2005)
3. Argyriou, A., Evgeniou, T., Pontil, M.: Convex multi-task feature learning. Mach. Learn. **73**(3), 243–272 (2008)
4. Ashtawy, H.M., Mahapatra, N.R.: Machine-learning scoring functions for identifying native poses of ligands docked to known and novel proteins. BMC Bioinform. **16**(6), S3 (2015)
5. Bach, F., Jenatton, R., Mairal, J., Obozinski, G.: Optimization with sparsity-inducing penalties. Found. Trends Mach. Learn. **4**, 1–106 (2012)
6. Bai, J., Zhou, K., Xue, G., Zha, H., Sun, G., Tseng, B., Zheng, Z., Chang, Y.: Multi-task learning for learning to rank in web search. In: Proceedings of the 18th ACM Conference on Information and Knowledge Management, pp. 1549–1552. ACM (2009)
7. Beck, A., Teboulle, M.: Gradient-based algorithms with applications to signal recovery. In: Convex Optimization in Signal Processing and Communications, pp. 42–88 (2009)
8. Cai, J.F., Candès, E.J., Shen, Z.: A singular value thresholding algorithm for matrix completion. SIAM J. Optim. **20**(4), 1956–1982 (2010)
9. Chen, J., Tang, L., Liu, J., Ye, J.: A convex formulation for learning shared structures from multiple tasks. In: Proceedings of the 26th Annual International Conference on Machine Learning, pp. 137–144. ACM (2009)
10. Chen, J., Zhou, J., Ye, J.: Integrating low-rank and group-sparse structures for robust multi-task learning. In: Proceedings of the 17th ACM SIGKDD International Conference on Knowledge Discovery and Data Mining, pp. 42–50. ACM (2011)
11. Dyagilev, K., Saria, S.: Learning (predictive) risk scores in the presence of censoring due to interventions. Mach. Learn. **102**(3), 1–26 (2015)
12. Dyagilev, K., Saria, S.: Learning severity score for sepsis: a novel approach based on clinical comparisons. In: AMIA Annual Symposium Proceedings, pp. 1890–1898 (2015)
13. Evgeniou, T., Micchelli, C.A., Pontil, M.: Learning multiple tasks with kernel methods. J. Mach. Learn. Res. **6**(Apr), 615–637 (2005)
14. Gong, P., Ye, J., Zhang, C.: Robust multi-task feature learning. In: Proceedings of the 18th ACM SIGKDD International Conference on Knowledge Discovery and Data Mining, pp. 895–903. ACM (2012)
15. Jackson, G.G., Dowling, H.F., Spiesman, I.G., Board, A.V.: Transmission of the common cold to volunteers under controlled conditions: I. the common cold as a clinical entity. AMA Arch. Intern. Med. **101**(2), 267–278 (1958)
16. Jacob, L., Vert, J.P., Bach, F.R.: Clustered multi-task learning: a convex formulation. In: Advances in Neural Information Processing Systems, pp. 745–752 (2009)
17. Jalali, A., Sanghavi, S., Ruan, C., Ravikumar, P.K.: A dirty model for multi-task learning. In: Advances in Neural Information Processing Systems, pp. 964–972 (2010)
18. Joachims, T.: Optimizing search engines using clickthrough data. In: Proceedings of the Eighth ACM SIGKDD International Conference on Knowledge Discovery and Data Mining, pp. 133–142. ACM (2002)

19. Mortimore, P., Sammons, P., Stoll, L., Lewis, D., Ecob, R.: School Matters: The Junior Years. Open Books (1988)
20. Parikh, N., Boyd, S.: Proximal algorithms. Found. Trends Optim. **1**(3), 127–239 (2014)
21. Santolino, M., Boucher, J.P.: Modelling the disability severity score in motor insurance claims: an application to the spanish case. IREA-Working Papers, 2009, IR09/002 (2009)
22. Simms, E.L.: Defining tolerance as a norm of reaction. Evol. Ecol. **14**(4–6), 563–570 (2000)
23. Stojkovic, I., Jelisavcic, V., Milutinovic, V., Obradovic, Z.: Distance based modeling of interactions in structured regression. In: Procedeengs of the 25th International Joint Conference on Artificial Intelligence IJCAI 2016, pp. 2032–2038 (2016)
24. Stojkovic, I., Jelisavcic, V., Milutinovic, V., Obradovic, Z.: Fast sparse Gaussian Markov random fields learning based on cholesky factorization. In: Proceedings of the 26th International Joint Conference on Artificial Intelligence, IJCAI 2017, pp. 2758–2764 (2017)
25. Stojkovic, I., Obradovic, Z.: Predicting sepsis biomarker progression under therapy. In: Proceedings of the 30th IEEE International Symposium on Computer-Based Medical Systems, CBMS 2017, pp. 19–24. IEEE (2017)
26. Toh, K.C., Yun, S.: An accelerated proximal gradient algorithm for nuclear norm regularized linear least squares problems. Pac. J. Optim. **6**(615–640), 15 (2010)
27. Vincent, J.L., Moreno, R., Takala, J., Willatts, S., De Mendonça, A., Bruining, H., Reinhart, C., Suter, P., Thijs, L.: The sofa (sepsis-related organ failure assessment) score to describe organ dysfunction/failure. Intensive Care Med. **22**(7), 707–710 (1996)
28. Yang, S., Shapiro, L., Cunningham, M., Speltz, M., Birgfeld, C., Atmosukarto, I., Lee, S.-I.: Skull retrieval for craniosynostosis using sparse logistic regression models. In: Greenspan, H., Müller, H., Syeda-Mahmood, T. (eds.) MCBR-CDS 2012. LNCS, vol. 7723, pp. 33–44. Springer, Heidelberg (2013). https://doi.org/10.1007/978-3-642-36678-9_4
29. Zaas, A.K., Chen, M., Varkey, J., Veldman, T., Hero, A.O., Lucas, J., Huang, Y., Turner, R., Gilbert, A., Lambkin-Williams, R., et al.: Gene expression signatures diagnose influenza and other symptomatic respiratory viral infections in humans. Cell Host Microbe **6**(3), 207–217 (2009)
30. Zhou, J., Chen, J., Ye, J.: Malsar: Multi-task learning via structural regularization. Arizona State University 21 (2011)
31. Zhou, J., Liu, J., Narayan, V.A., Ye, J.: Modeling disease progression via fused sparse group lasso. In: Proceedings of the 18th ACM SIGKDD International Conference on Knowledge Discovery and Data Mining, pp. 1095–1103. ACM (2012)

Theoretical Analysis of Domain Adaptation with Optimal Transport

Ievgen Redko[1]([✉]), Amaury Habrard[2], and Marc Sebban[2]

[1] Univ. Lyon, INSA-Lyon, Université Claude Bernard Lyon 1, UJM-Saint Etienne,
CNRS, Inserm, CREATIS UMR 5220, U1206, 69266 Lyon, France
`ievgen.redko@creatis.insa-lyon.fr`
[2] Univ. Lyon, UJM-Saint-Etienne, CNRS, Lab. Hubert Curien UMR 5516,
42023 Saint-Étienne, France
{`amaury.habrard,marc.sebban`}`@univ.st-etienne.fr`

Abstract. Domain adaptation (DA) is an important and emerging field
of machine learning that tackles the problem occurring when the distributions of training (source domain) and test (target domain) data are
similar but different. This kind of learning paradigm is of vital importance for future advances as it allows a learner to generalize the knowledge across different tasks. Current theoretical results show that the
efficiency of DA algorithms depends on their capacity of minimizing the
divergence between source and target probability distributions. In this
paper, we provide a theoretical study on the advantages that concepts
borrowed from optimal transportation theory [17] can bring to DA. In
particular, we show that the Wasserstein metric can be used as a divergence measure between distributions to obtain generalization guarantees
for three different learning settings: (i) classic DA with unsupervised
target data (ii) DA combining source and target labeled data, (iii) multiple source DA. Based on the obtained results, we motivate the use of
the regularized optimal transport and provide some algorithmic insights
for multi-source domain adaptation. We also show when this theoretical analysis can lead to tighter inequalities than those of other existing
frameworks. We believe that these results open the door to novel ideas
and directions for DA.

Keywords: Domain adaptation · Generalization bounds
Optimal transport

1 Introduction

Many results in statistical learning theory study the problem of estimating the
probability that a hypothesis chosen from a given hypothesis class can achieve

Electronic supplementary material The online version of this chapter (https://
doi.org/10.1007/978-3-319-71246-8_45) contains supplementary material, which is
available to authorized users.

a small true risk. This probability is often expressed in the form of generalization bounds on the true risk obtained using concentration inequalities with respect to (w.r.t.) some hypothesis class. Classic generalization bounds make the assumption that training and test data follow the same distribution. This assumption, however, can be violated in many real-world applications (e.g., in computer vision, language processing or speech recognition) where training and test data actually follow a related but different probability distribution. One may think of an example, where a spam filter is learned based on the abundant annotated data collected for one user and is further applied for newly registered user with different preferences. In this case, the performance of the spam filter will deteriorate as it does not take into account the mismatch between the underlying probability distributions. The need for algorithms tackling this problem has led to the emergence of a new field in machine learning called domain adaptation (DA), subfield of transfer learning [18], where the source (training) and target (test) distributions are not assumed to be the same. From a theoretical point of view, existing generalization guarantees for DA are expressed in the form of bounds over the target risk involving the source risk, a divergence between domains and a term λ evaluating the capability of the considered hypothesis class to solve the problem, often expressed as a joint error of the ideal hypothesis between the two domains. In this context, minimizing the divergence between distributions is a key factor for the potential success of DA algorithms. Among the most striking results, existing generalization bounds based on the H-divergence [3] or the discrepancy distance [15] have also an interesting property of being able to link the divergence between the probability distributions of two domains w.r.t. the considered class of hypothesis.

Despite their advantages, the above mentioned divergences do not directly take into account the geometry of the data distribution. Recently, [6,7] has proposed to tackle this drawback by solving the DA problem using ideas from optimal transportation (OT) theory. Their paper proposes an algorithm that aims to reduce the divergence between two domains by minimizing the Wasserstein distance between their distributions. This idea has a very appealing and intuitive interpretation based on the transport of one domain to another. The transportation plan solving OT problem takes into account the geometry of the data by means of an associated cost function which is based on the Euclidean distance between examples. Furthermore, it is naturally defined as an infimum problem over all feasible solutions. An interesting property of this approach is that the resulting solution given by a joint probability distribution allows one to obtain the new projection of the instances of one domain into another directly without being restricted to a particular hypothesis class. This independence from the hypothesis class means that this solution not only ensures successful adaptation but also influences the capability term λ. While showing very promising experimental results, it turns out that this approach, however, has no theoretical guarantees. This paper aims to bridge this gap by presenting contributions covering three DA settings: (i) classic unsupervised DA where the learner has only access to labeled source data and unsupervised target instances, (ii) DA where one has access to labeled data from both source and target domains, (iii) multi-source DA where labeled instances for a set of distinct source domains (more than 2) are available. We provide new theoretical guarantees in the form of generalization bounds for these three settings based on the Wasserstein distance thus justifying its use in DA. According to [26], the Wasserstein distance is rather strong and can be combined with smoothness bounds to obtain convergences in other distances. This important advantage of Wasserstein distance leads to tighter bounds in comparison to other state-of-the-art results and is more computationally attractive.

The rest of this paper is organized as follows: Sect. 2 is devoted to the presentation of optimal transport and its application in DA. In Sect. 3, we present the generalization bounds for DA with the Wasserstein distance for both single- and multi-source learning scenarios. Finally, we conclude our paper in Sect. 4.

2 Definitions and Notations

In this section, we first present the formalization of the Monge-Kantorovich [13] optimization problem and show how optimal transportation problem found its application in DA.

2.1 Optimal Transport

Optimal transportation theory was first introduced in [17] to study the problem of resource allocation. Assuming that we have a set of factories and a set of mines, the goal of optimal transportation is to move the ore from mines to factories in an optimal way, i.e., by minimizing the overall transport cost. More formally, let $\Omega \subseteq \mathbb{R}^d$ be a measurable space and denote by $\mathcal{P}(\Omega)$ the set of all probability measures over Ω. Given two probability measures $\mu_S, \mu_T \in \mathcal{P}(\Omega)$, the Monge-Kantorovich problem consists in finding a probabilistic coupling γ defined as a joint probability measure over $\Omega \times \Omega$ with marginals μ_S and μ_T for all $x, y \in \Omega$ that minimizes the cost of transport w.r.t. some function $c : \Omega \times \Omega \to \mathbb{R}_+$:

$$\arg\min_{\gamma} \int_{\Omega_1 \times \Omega_2} c(\boldsymbol{x}, \boldsymbol{y})^p d\gamma(\boldsymbol{x}, \boldsymbol{y})$$
$$\text{s.t. } \boldsymbol{P}^{\Omega_1} \# \gamma = \mu_S, \boldsymbol{P}^{\Omega_2} \# \gamma = \mu_T,$$

where $\boldsymbol{P}^{\Omega_i}$ is the projection over Ω_i and $\#$ denotes the pushforward measure. This problem admits a unique solution γ_0 which allows us to define the Wasserstein distance of order p between μ_S and μ_T for any $p \in [1; +\infty]$ as follows:

$$W_p^p(\mu_S, \mu_T) = \inf_{\gamma \in \Pi(\mu_S, \mu_T)} \int_{\Omega \times \Omega} c(\boldsymbol{x}, \boldsymbol{y})^p d\gamma(\boldsymbol{x}, \boldsymbol{y}),$$

where $c : \Omega \times \Omega \to \mathbb{R}^+$ is a cost function for transporting one unit of mass $\boldsymbol{x}$ to $\boldsymbol{y}$ and $\Pi(\mu_S, \mu_T)$ is a collection of all joint probability measures on $\Omega \times \Omega$ with marginals μ_S and μ_T.

Remark 1. In what follows, we consider only the case $p = 1$ but all the obtained results can be easily extended to the case $p > 1$ using Hölder inequality implying for every $p \leq q \Rightarrow W_p \leq W_q$.

In the discrete case, when one deals with empirical measures $\hat{\mu}_S = \frac{1}{N_S} \sum_{i=1}^{N_S} \delta_{x_S^i}$ and $\hat{\mu}_T = \frac{1}{N_T} \sum_{i=1}^{N_T} \delta_{x_T^i}$ represented by the uniformly weighted sums of N_S and N_T Diracs with mass at locations x_S^i and x_T^i respectively, Monge-Kantorovich problem is defined in terms of the inner product between the coupling matrix γ and the cost matrix C:

$$W_1(\hat{\mu}_S, \hat{\mu}_T) = \min_{\gamma \in \Pi(\hat{\mu}_S, \hat{\mu}_T)} \langle C, \gamma \rangle_F$$

where $\langle \cdot, \cdot \rangle_F$ is the Frobenius dot product, $\Pi(\hat{\mu}_S, \hat{\mu}_T) = \{\gamma \in \mathbb{R}_+^{N_S \times N_T} | \gamma \mathbf{1} = \hat{\mu}_S, \gamma^T \mathbf{1} = \hat{\mu}_T\}$ is a set of doubly stochastic matrices and C is a dissimilarity matrix, i.e., $C_{ij} = c(x_S^i, x_T^j)$, defining the energy needed to move a probability mass from x_S^i to x_T^j. Figure 1 shows how the solution of optimal transport between two point-clouds can look like.

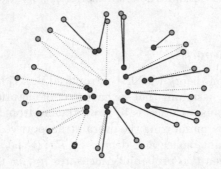

Fig. 1. Blue points are generated to lie inside a square with a side length equal to 1. Red points are generated inside an annulus containing the square. Solution of the regularized optimal transport problem is visualized by plotting dashed and solid lines that correspond to the large and small values given by the optimal coupling matrix γ. (Color figure online)

It turns out that the Wasserstein distance has been successfully used in various applications, for instance: computer vision [22], texture analysis [21], tomographic reconstruction [12] and clustering [9]. The huge success of algorithms based on this distance is due to [8] who introduced an entropy-regularized version of optimal transport that can be optimized efficiently using matrix scaling algorithm. We are now ready to present the application of OT to DA below.

2.2   Domain Adaptation and Optimal Transport

The problem of DA is formalized as follows: we define a domain as a pair consisting of a distribution μ_D on Ω and a labeling function $f_D : \Omega \rightarrow [0, 1]$. A hypothesis class H is a set of functions so that $\forall h \in H, h : \Omega \rightarrow \{0, 1\}$.

Definition 1. *Given a convex loss-function l, the probability according to the distribution μ_D that a hypothesis $h \in H$ disagrees with a labeling function f_D (which can also be a hypothesis) is defined as*

$$\epsilon_D(h, f_D) = \mathbb{E}_{x \sim \mu_D} \left[l(h(x), f_D(x)) \right].$$

When the source and target error functions are defined w.r.t. h and f_S or f_T, we use the shorthand $\epsilon_S(h, f_S) = \epsilon_S(h)$ and $\epsilon_T(h, f_T) = \epsilon_T(h)$. We further denote by $\langle \mu_S, f_S \rangle$ the source domain and $\langle \mu_T, f_T \rangle$ the target domain. The ultimate goal of DA then is to learn a good hypothesis h in $\langle \mu_S, f_S \rangle$ that has a good performance in $\langle \mu_T, f_T \rangle$.

In unsupervised DA problem, one usually has access to a set of source data instances $\boldsymbol{X_S} = \{\boldsymbol{x}_S^i \in \mathbb{R}^d\}_{i=1}^{N_S}$ associated with labels $\{y_S^i\}_{i=1}^{N_S}$ and a set of unlabeled target data instances $\boldsymbol{X_T} = \{\boldsymbol{x}_T^i \in \mathbb{R}^d\}_{i=1}^{N_T}$. Contrary to the classic learning paradigm, unsupervised DA assumes that the marginal distributions of $\boldsymbol{X_S}$ and $\boldsymbol{X_T}$ are different and given by $\mu_S, \mu_T \in \mathcal{P}(\Omega)$.

For the first time, optimal transportation problem was applied to DA in [6,7]. The main underlying idea of their work is to find a coupling matrix that efficiently transports source samples to target ones by solving the following optimization problem:

$$\gamma_o = \underset{\gamma \in \Pi(\hat{\mu}_S, \hat{\mu}_T)}{\arg \min} \langle C, \gamma \rangle_F.$$

Once the optimal coupling γ_o is found, source samples $\boldsymbol{X_S}$ can be transformed into target aligned source samples $\hat{\boldsymbol{X}}_S$ using the following equation

$$\hat{\boldsymbol{X}}_S = \text{diag}((\gamma_o \boldsymbol{1})^{-1}) \gamma_o \boldsymbol{X_T}.$$

The use of Wasserstein distance here has an important advantage over other distances used in DA (see Sect. 5) as it preserves the topology of the data and admits a rather efficient estimation as mentioned above. Furthermore, as shown in [6,7], it improves current state-of-the-art results on benchmark computer vision data sets and has a very appealing intuition behind.

3   Generalization Bounds with Wasserstein Distance

In this section, we introduce generalization bounds for the target error when the divergence between tasks' distributions is measured by the Wasserstein distance.

3.1   A Bound Relating the Source and Target Error

We first consider the case of unsupervised DA where no labelled data are available in the target domain. We start with a lemma that relates the Wasserstein metric with the source and target error functions for an arbitrary pair of hypothesis. Then, we show how the target error can be bounded by the Wasserstein distance for empirical measures. We first present the Lemma that introduces Wasserstein distance to relate the source and target error functions in a Reproducing Kernel Hilbert Space.

Lemma 1. *Let $\mu_S, \mu_T \in \mathcal{P}(\Omega)$ be two probability measures on $\mathbb{R}^d$. Assume that the cost function $c(\boldsymbol{x}, \boldsymbol{y}) = \|\phi(\boldsymbol{x}) - \phi(\boldsymbol{y})\|_{\mathcal{H}_{k_l}}$, where $\mathcal{H}_{k_l}$ is a Reproducing Kernel Hilbert Space (RKHS) equipped with kernel $k_l : \Omega \times \Omega \to \mathbb{R}$ induced by $\phi : \Omega \to \mathcal{H}_{k_l}$ and $k_l(\boldsymbol{x}, \boldsymbol{y}) = \langle \phi(\boldsymbol{x}), \phi(\boldsymbol{y}) \rangle_{\mathcal{H}_{k_l}}$. Assume further that the loss function $l_{h,f} : x \longrightarrow l(h(x), f(x))$ is convex, symmetric, bounded, obeys the triangular equality and has the parametric form $|h(x) - f(x)|^q$ for some $q > 0$. Assume also that kernel k_l in the RKHS $\mathcal{H}_{k_l}$ is square-root integrable w.r.t. both μ_S, μ_T for all $\mu_S, \mu_T \in \mathcal{P}(\Omega)$ where Ω is separable and $0 \leq k_l(\boldsymbol{x}, \boldsymbol{y}) \leq K, \forall \, \boldsymbol{x}, \boldsymbol{y} \in \Omega$. Then the following holds*

$$\epsilon_T(h, h') \leq \epsilon_S(h, h') + W_1(\mu_S, \mu_T)$$

for every hypothesis h', h.

Proof. As this Lemma plays a key role in the following sections, we give its proof here. We assume that $l_{h,f} : x \longrightarrow l(h(x), f(x))$ in the definition of $\epsilon(h)$ is a convex loss-function defined $\forall h, f \in \mathcal{F}$ where $\mathcal{F}$ is a unit ball in the RKHS $\mathcal{H}_k$. Considering that $h, f \in \mathcal{F}$, the loss function l is a non-linear mapping of the RKHS $\mathcal{H}_k$ for the family of losses $l(h(x), f(x)) = |h(x) - f(x)|^{q}$[1]. Using results from [23], one may show that $l_{h,f}$ also belongs to the RKHS $\mathcal{H}_{k_l}$ admitting the reproducing kernel k_l and that its norm obeys the following inequality:

$$\|l_{h,f}\|_{\mathcal{H}_{k_l}}^2 \leq \|h - f\|_{\mathcal{H}_k}^{2q}.$$

This result gives us two important properties of $l_{f,h}$ that we use further:

- $l_{h,f}$ belongs to the RKHS that allows us to use the reproducing property;
- the norm $\|l_{h,f}\|_{\mathcal{H}_{k_l}}$ is bounded.

For simplicity, we can assume that $\|l_{h,f}\|_{\mathcal{H}_{k_l}}$ is bounded by 1. This assumption can be verified by imposing the appropriate bounds on the norms of h and f and is easily extendable to the case when $\|l_{h,f}\|_{\mathcal{H}_{k_l}} \leq M$ by scaling as explained in [15, Proposition 2]. We also note that q does not necessarily have to appear in the final result as we seek to bound the norm of l and not to give an explicit expression for it in terms of $\|h\|_{\mathcal{H}_k}, \|f\|_{\mathcal{H}_k}$ and q. Now the error function defined above can be also expressed in terms of the inner product in the corresponding Hilbert space, i.e[2]:

$$\epsilon_S(h, f_S) = \mathbb{E}_{x \sim \mu_S}[l(h(x), f_S(x))] = \mathbb{E}_{x \sim \mu_S}[\langle \phi(x), l \rangle_{\mathcal{H}}].$$

We define the target error in the same manner:

$$\epsilon_T(h, f_T) = \mathbb{E}_{y \sim \mu_T}[l(h(y), f_T(y))] = \mathbb{E}_{y \sim \mu_T}[\langle \phi(y), l \rangle_{\mathcal{H}}].$$

---

[1] If $h, f \in \mathcal{H}$ then $h - f \in \mathcal{H}$ implying that $l(h(x), f(x)) = |h(x) - f(x)|^q$ is a nonlinear transform for $h - f \in \mathcal{H}$.

[2] For the sake of simplicity, we will further write $\mathcal{H}$ meaning $\mathcal{H}_{k_l}$ and l meaning $l_{f,h}$.

With the definitions introduced above, the following holds:

$$\epsilon_T(h, h') = \epsilon_T(h, h') + \epsilon_S(h, h') - \epsilon_S(h, h')$$
$$= \epsilon_S(h, h') + \mathbb{E}_{y \sim \mu_T}[\langle \phi(y), l \rangle_{\mathcal{H}}] - \mathbb{E}_{x \sim \mu_S}[\langle \phi(x), l \rangle_{\mathcal{H}}]$$
$$= \epsilon_S(h, h') + \langle \mathbb{E}_{y \sim \mu_T}[\phi(y)] - \mathbb{E}_{x \sim \mu_S}[\phi(x)], l \rangle_{\mathcal{H}}$$
$$\leq \epsilon_S(h, h') + \|l\|_{\mathcal{H}} \|\mathbb{E}_{y \sim \mu_T}[\phi(y)] - \mathbb{E}_{x \sim \mu_S}[\phi(x)]\|_{\mathcal{H}}$$
$$\leq \epsilon_S(h, h') + \| \int_\Omega \phi d(\mu_S - \mu_T) \|_{\mathcal{H}}.$$

Second line is obtained by using the reproducing property applied to l, third line follows from the properties of the expected value. Fourth line here is due to the properties of the inner-product while fifth line is due to $\|l_{h,f}\|_{\mathcal{H}} \leq 1$. Now using the definition of the joint distribution we have the following:

$$\| \int_\Omega \phi d(\mu_S - \mu_T) \|_{\mathcal{H}} = \| \int_{\Omega \times \Omega} (\phi(x) - \phi(y)) d\gamma(x, y) \|_{\mathcal{H}}$$
$$\leq \int_{\Omega \times \Omega} \|\phi(x) - \phi(y)\|_{\mathcal{H}} d\gamma(x, y).$$

As the last inequality holds for any γ, we obtain the final result by taking the infimum over γ from the right-hand side, i.e.:

$$\int_\Omega \phi d(\mu_S - \mu_T) \|_{\mathcal{H}} \leq \inf_{\gamma \in \Pi(\mu_S, \mu_T)} \int_{\Omega \times \Omega} \|\phi(x) - \phi(y)\|_{\mathcal{H}} d\gamma(x, y).$$

which gives

$$\epsilon_T(h, h') \leq \epsilon_S(h, h') + W_1(\mu_S, \mu_T).$$

$\square$

Remark 2. We note that the functional form of the loss-function $l(h(x), f(x)) = |h(x) - f(x)|^q$ is just an example that was used as the basis for the proof. According to [23, Appendix 2], we may also consider more general nonlinear transformations of h and f that satisfy the assumption imposed on $l_{h,f}$ above. These transformations may include a product of hypothesis and labeling functions and thus the proposed results is valid for hinge-loss too.

This lemma makes use of the Wasserstein distance to relate the source and target errors. The assumption made here is to specify that the cost function $c(x, y) = \|\phi(x) - \phi(y)\|_{\mathcal{H}}$. While it may seem too restrictive, this assumption is, in fact, not that strong. Using the properties of the inner-product, one has:

$$\|\phi(x) - \phi(y)\|_{\mathcal{H}} = \sqrt{\langle \phi(x) - \phi(y), \phi(x) - \phi(y) \rangle_{\mathcal{H}}}$$
$$= \sqrt{k(x, x) - 2k(x, y) + k(x, y)}.$$

Now it can be shown that for any given positive-definite kernel k there is a distance c (used as a cost function in our case) that generates it and vice versa (see Lemma 12 from [24]).

In order to prove our next theorem, we present first an important result showing the convergence of the empirical measure $\hat{\mu}$ to its true associated measure w.r.t. the Wasserstein metric. This concentration guarantee allows us to propose generalization bounds based on the Wasserstein distance for finite samples rather than true population measures. Following [4], it can be specialized for the case of W_1 as follows[3].

Theorem 1 ([4], **Theorem 1.1**). *Let μ be a probability measure in $\mathbb{R}^d$ so that for some $\alpha > 0$, we have that $\int_{\mathbb{R}^d} e^{\alpha \|x\|^2} d\mu < \infty$ and $\hat{\mu} = \frac{1}{N} \sum_{i=1}^N \delta_{x_i}$ be its associated empirical measure defined on a sample of independent variables $\{x_i\}_{i=1}^N$ drawn from μ. Then for any $d' > d$ and $\varsigma' < \sqrt{2}$ there exists some constant N_0 depending on d' and some square exponential moment of μ such that for any $\varepsilon > 0$ and $N \geq N_0 \max(\varepsilon^{-(d'+2)}, 1)$*

$$\mathbb{P}\left[W_1(\mu, \hat{\mu}) > \varepsilon\right] \leq \exp\left(-\frac{\varsigma'}{2} N \varepsilon^2\right),$$

where d', ς' can be calculated explicitly.

The convergence guarantee of this theorem can be further strengthened as shown in [11] but we prefer this version for the ease of reading. We can now use it in combination with the previous Lemma to prove the following theorem.

Theorem 2. *Under the assumptions of Lemma 1, let $\mathbf{X_S}$ and $\mathbf{X_T}$ be two samples of size N_S and N_T drawn i.i.d. from μ_S and μ_T respectively. Let $\hat{\mu}_S = \frac{1}{N_S} \sum_{i=1}^{N_S} \delta_{x_S^i}$ and $\hat{\mu}_T = \frac{1}{N_T} \sum_{i=1}^{N_T} \delta_{x_T^i}$ be the associated empirical measures. Then for any $d' > d$ and $\varsigma' < \sqrt{2}$ there exists some constant N_0 depending on d' such that for any $\delta > 0$ and $\min(N_S, N_T) \geq N_0 \max(\delta^{-(d'+2)}, 1)$ with probability at least $1 - \delta$ for all h the following holds:*

$$\epsilon_T(h) \leq \epsilon_S(h) + W_1(\hat{\mu}_S, \hat{\mu}_T) + \sqrt{2 \log\left(\frac{1}{\delta}\right) / \varsigma'} \left(\sqrt{\frac{1}{N_S}} + \sqrt{\frac{1}{N_T}}\right) + \lambda,$$

where λ is the combined error of the ideal hypothesis h^* that minimizes the combined error of $\epsilon_S(h) + \epsilon_T(h)$.

Proof.

$$\begin{aligned}
\epsilon_T(h) &\leq \epsilon_T(h^*) + \epsilon_T(h^*, h) = \epsilon_T(h^*) + \epsilon_S(h, h^*) + \epsilon_T(h^*, h) - \epsilon_S(h, h^*) \\
&\leq \epsilon_T(h^*) + \epsilon_S(h, h^*) + W_1(\mu_S, \mu_T) \\
&\leq \epsilon_T(h^*) + \epsilon_S(h) + \epsilon_S(h^*) + W_1(\mu_S, \mu_T) \\
&= \epsilon_S(h) + W_1(\mu_S, \mu_T) + \lambda \\
&\leq \epsilon_S(h) + W_1(\mu_S, \hat{\mu}_S) + W_1(\hat{\mu}_S, \mu_T) + \lambda
\end{aligned}$$

---

[3] We present the original version of this Theorem in the Supplementary material.

$$\leq \epsilon_S(h) + \sqrt{2\log\left(\frac{1}{\delta}\right)/N_S\varsigma'} + W_1(\hat{\mu}_S, \hat{\mu}_T) + W_1(\hat{\mu}_T, \mu_T) + \lambda$$

$$\leq \epsilon_S(h) + W_1(\hat{\mu}_S, \hat{\mu}_T) + \lambda + \sqrt{2\log\left(\frac{1}{\delta}\right)/\varsigma'}\left(\sqrt{\frac{1}{N_S}} + \sqrt{\frac{1}{N_T}}\right).$$

Second and fourth lines are obtained using the triangular inequality applied to the error function. Third inequality is a consequence of Lemma 1. Fifth line follows from the definition of λ, sixth, seventh and eighth lines use the fact that Wasserstein metric is a proper distance and Theorem 1. □

A first immediate consequence of this theorem is that it justifies the use of the optimal transportation in DA context. However, we would like to clarify the fact that the bound does not suggest that minimization of the Wasserstein distance can be done independently from the minimization of the source error nor it says that the joint error given by the lambda term becomes small. First, it is clear that the result of W_1 minimization provides a transport of the source to the target such as W_1 becomes small when computing the distance between newly transported sources and target instances. Under the hypothesis that class labeling is preserved by transport, i.e. $P_{\text{source}}(y|x_s) = P_{\text{target}}(y|\text{Transport}(x_s))$, the adaptation can be possible by minimizing W_1 only. However, this is not a reasonable assumption in practice. Indeed, by minimizing the W_1 distance only, it is possible that the obtained transformation transports one positive and one negative source instance to the same target point and then the empirical source error cannot be properly minimized. Additionally, the joint error will be affected since no classifier will be able to separate these source points. We can also think of an extreme case where the positive source examples are transported to negative target instances, in that case the joint error λ will be dramatically affected. A solution is then to regularize the transport to help the minimization of the source error which can be seen as a kind of joint optimization. This idea was partially implemented as a class-labeled regularization term added to the original optimal transport formulation in [6,7] and showed good empirical results in practice. The proposed regularized optimization problem reads

$$\min_{\gamma \in \Pi(\hat{\mu}_S, \hat{\mu}_T)} \langle C, \gamma \rangle_F - \frac{1}{\lambda}E(\gamma) + \eta \sum_j \sum_{\mathcal{L}} \|\gamma(I_{\mathcal{L}}, j)\|_q^p.$$

Here, the second term $E(\gamma) = -\sum_{i,j}^{N_S, N_T} \gamma_{i,j} \log(\gamma_{i,j})$ is the regularization term that allows one to solve optimal transportation problem efficiently using Sinkhorn-Knopp matrix scaling algorithm [25]. Second regularization term $\eta \sum_j \sum_c \|\gamma(I_c, j)\|_q^p$ is used to restrict source examples of different classes to be transported to the same target examples by promoting group sparsity in the matrix γ thanks to $\|\cdot\|_q^p$ with $q = 1$ and $p = \frac{1}{2}$. In some way, this regularization term influences the capability term by ensuring the existence of a good hypothesis that will be able to be discriminant on both source and target domains data. Another recent paper of [28] also suggests that transport regularization is important for the use of OT in domain adaptation tasks. Thus, we conclude that the regularized transport formulations such as the one of [6,7] can be seen as algorithmic solutions for controlling the trade-off between the terms of the bound.

Assuming that $\epsilon_S(h)$ is properly minimized, only λ and the Wasserstein distance between empirical measures defined on the source and target samples have an impact on the potential success of adaptation. Furthermore, the fact that the Wasserstein distance is defined in terms of the optimal coupling used to solve the DA problem and is not restricted to any particular hypothesis class directly influences λ as discussed above. We now proceed to give similar bounds for the case where one has access to some labeled instances in the target domain.

3.2   A Learning Bound for the Combined Error

In semi-supervised DA, when we have access to an additional small set of labeled instances in the target domain, the goal is often to find a trade-off between minimizing the source and the target errors depending on the number of instances available in each domain and their mutual correlation. Let us now assume that we possess βn instances drawn independently from μ_T and $(1-\beta)n$ instances drawn independently from μ_S and labeled by f_S and f_T, respectively. In this case, the empirical combined error [2] is defined as a convex combination of errors on the source and target training data:

$$\hat{\epsilon}_\alpha(h) = \alpha\hat{\epsilon}_T(h) + (1-\alpha)\hat{\epsilon}_S(h),$$

where $\alpha \in [0,1]$.

The use of the combined error is motivated by the fact that if the number of instances in the target sample is small compared to the number of instances in the source domain (which is usually the case in DA), minimizing only the target error may not be appropriate. Instead, one may want to find a suitable value of α that ensures the minimum of $\hat{\epsilon}_\alpha(h)$ w.r.t. a given hypothesis h. We now prove a theorem for the combined error similar to the one presented in [2].

Theorem 3. *Under the assumptions of Theorem 2 and Lemma 1, let D be a labeled sample of size n with βn points drawn from μ_T and $(1-\beta)n$ from μ_S with $\beta \in (0,1)$, and labeled according to f_S and f_T. If $\hat{h}$ is the empirical minimizer of $\hat{\epsilon}_\alpha(h)$ and $h_T^* = \min_h \epsilon_T(h)$ then for any $\delta \in (0,1)$ with probability at least $1-\delta$ (over the choice of samples),*

$$\epsilon_T(\hat{h}) \leq \epsilon_T(h_T^*) + c_1 + 2(1-\alpha)(W_1(\hat{\mu}_S, \hat{\mu}_T) + \lambda + c_2),$$

where

$$c_1 = 2\sqrt{\frac{2K\left(\frac{(1-\alpha)^2}{1-\beta} + \frac{\alpha^2}{\beta}\right)\log(2/\delta)}{n}} + 4\sqrt{K/n}\left(\frac{\alpha}{n\beta\sqrt{\beta}} + \frac{(1-\alpha)}{n(1-\beta)\sqrt{1-\beta}}\right),$$

$$c_2 = \sqrt{2\log\left(\frac{1}{\delta}\right)/\varsigma'}\left(\sqrt{\frac{1}{N_S}} + \sqrt{\frac{1}{N_T}}\right).$$

Proof

$$\epsilon_T(\hat{h}) \leq \epsilon_\alpha(\hat{h}) + (1-\alpha)(W_1(\mu_S, \mu_T) + \lambda)$$

$$\leq \hat{\epsilon}_\alpha(\hat{h}) + \sqrt{\frac{2K\left(\frac{(1-\alpha)^2}{1-\beta} + \frac{\alpha^2}{\beta}\right)\log(2/\delta)}{n}} + (1-\alpha)(W_1(\mu_S, \mu_T) + \lambda)$$

$$+ 2\sqrt{K/n}\left(\frac{\alpha}{n\beta\sqrt{\beta}} + \frac{(1-\alpha)}{n(1-\beta)\sqrt{1-\beta}}\right)$$

$$\leq \hat{\epsilon}_\alpha(h_T^*) + \sqrt{\frac{2K\left(\frac{(1-\alpha)^2}{1-\beta} + \frac{\alpha^2}{\beta}\right)\log(2/\delta)}{n}} + (1-\alpha)(W_1(\mu_S, \mu_T) + \lambda)$$

$$+ 2\sqrt{K/n}\left(\frac{\alpha}{n\beta\sqrt{\beta}} + \frac{(1-\alpha)}{n(1-\beta)\sqrt{1-\beta}}\right)$$

$$\leq \epsilon_\alpha(h_T^*) + 2\sqrt{\frac{2K\left(\frac{(1-\alpha)^2}{1-\beta} + \frac{\alpha^2}{\beta}\right)\log(2/\delta)}{n}} + (1-\alpha)(W_1(\mu_S, \mu_T) + \lambda)$$

$$+ 4\sqrt{K/n}\left(\frac{\alpha}{n\beta\sqrt{\beta}} + \frac{(1-\alpha)}{n(1-\beta)\sqrt{1-\beta}}\right)$$

$$\leq \epsilon_T(h_T^*) + 2\sqrt{\frac{2K\left(\frac{(1-\alpha)^2}{1-\beta} + \frac{\alpha^2}{\beta}\right)\log(2/\delta)}{n}} + 2(1-\alpha)(W_1(\mu_S, \mu_T) + \lambda)$$

$$+ 4\sqrt{K/n}\left(\frac{\alpha}{n\beta\sqrt{\beta}} + \frac{(1-\alpha)}{n(1-\beta)\sqrt{1-\beta}}\right)$$

$$\leq \epsilon_T(h_T^*) + c_1 + 2(1-\alpha)(W_1(\hat{\mu}_S, \hat{\mu}_T) + \lambda + c_2).$$

The proof follows the standard theory of uniform convergence for empirical risk minimizers where lines 1 and 5 are obtained by observing that $|\epsilon_\alpha(h) - \epsilon_T(h)| = |\alpha\epsilon_T(h) + (1-\alpha)\epsilon_S(h) - \epsilon_T(h)| = |(1-\alpha)(\epsilon_S(h) - \epsilon_T(h))| \leq (1-\alpha)(W_1(\mu_T, \mu_S) + \lambda)$ where the last inequality comes from line 4 of the proof of Theorem 2, line 3 follows from the definition of $\hat{h}$ and h_T^* and line 6 is a consequence of Theorem 1. Finally, lines 2 and 4 are obtained based on the concentration inequality obtained for $\epsilon_\alpha(h)$. Due to the lack of space, we put this result in the Supplementary material. □

This theorem shows that the best hypothesis that takes into account both source and target labeled data (i.e., $0 \leq \alpha < 1$) performs at least as good as the best hypothesis learned on target data instances alone ($\alpha = 1$). This result agrees well with the intuition that semi-supervised DA approaches should be at least as good as unsupervised ones.

4   Multi-source Domain Adaptation

We now consider the case where not one but many source domains are available during the adaptation. More formally, we define N different source domains (where T can either be or not a part of this set). For each source j, we have a labelled sample S_j of size $n_j = \beta_j n$ $\left(\sum_{j=1}^{N} \beta_j = 1, \sum_{j=1}^{N} n_j = n\right)$ drawn from the associated unknown distribution μ_{S_j} and labelled by f_j. We now consider the empirical weighted multi-source error of a hypothesis h defined for some vector $\boldsymbol{\alpha} = \{\alpha_1, \ldots, \alpha_N\}$ as follows:

$$\hat{\epsilon}_{\boldsymbol{\alpha}}(h) = \sum_{j=1}^{N} \alpha_j \hat{\epsilon}_{S_j}(h),$$

where $\sum_{j=1}^{N} \alpha_j = 1$ and each α_j represents the weight of the source domain S_j.

In what follows, we show that generalization bounds obtained for the weighted error give some interesting insights into the application of the Wasserstein distance to multi-source DA problems.

Theorem 4. *With the assumptions from Theorem 2 and Lemma 1, let S be a sample of size n, where for each $j \in \{1, \ldots, N\}$, $\beta_j n$ points are drawn from μ_{S_j} and labelled according to f_j. If $\hat{h}_{\boldsymbol{\alpha}}$ is the empirical minimizer of $\hat{\epsilon}_{\boldsymbol{\alpha}}(h)$ and $h_T^* = \min_h \epsilon_T(h)$ then for any fixed $\boldsymbol{\alpha}$ and $\delta \in (0,1)$ with probability at least $1 - \delta$ (over the choice of samples),*

$$\epsilon_T(\hat{h}_{\boldsymbol{\alpha}}) \le \epsilon_T(h_T^*) + c_1 + 2 \sum_{j=1}^{N} \alpha_j \left(W_1(\hat{\mu}_j, \hat{\mu}_T) + \lambda_j + c_2 \right),$$

where

$$c_1 = 2\sqrt{\frac{2K \sum_{j=1}^{N} \frac{\alpha_j^2}{\beta_j} \log(2/\delta)}{n}} + 2\sqrt{\sum_{j=1}^{N} \frac{K\alpha_j}{\beta_j n}},$$

$$c_2 = \sqrt{2 \log\left(\frac{1}{\delta}\right) / \varsigma'} \left(\sqrt{\frac{1}{N_{S_j}}} + \sqrt{\frac{1}{N_T}} \right),$$

where $\lambda_j = \min_h \left(\epsilon_{S_j}(h) + \epsilon_T(h) \right)$ represents the joint error for each source domain j.

Proof. The proof of this Theorem is very similar to the proof of Theorem 4. The final result is obtained by applying the concentration inequality for $\epsilon_{\boldsymbol{\alpha}}(h)$ (instead of those used for $\epsilon_{\boldsymbol{\alpha}}(\hat{h})$ in the proof of Theorem 4) and by using the following inequality that can be obtained easily by following the principle of the proof of [2, Theorem 4]:

$$|\epsilon_{\boldsymbol{\alpha}}(h) - \epsilon_T(h)| \le \sum_{j=1}^{N} \alpha_j \left(W_1(\mu_j, \mu_T) + \lambda_j \right),$$

where $\lambda_j = \min\limits_{h} (\epsilon_{S_j}(h) + \epsilon_T(h))$. For the sake of completness, we present the concentration inequality for $\epsilon_\alpha(h)$ in the Supplementary material.   □

While the results for multi-source DA may look like a trivial extension of the theoretical guarantees for the case of two domains, they can provide a very fruitful implication on their own. As in the previous case, we consider that the potential term that should be minimized in this bound by a given multi-source DA algorithm is the term $\sum_{j=1}^{N} \alpha_j W_1(\hat\mu_j, \hat\mu_T)$.

Assume that $\hat\mu$ is an arbitrary unknown empirical probability measure on $\mathbb{R}^d$. Using the triangle inequality and bearing in mind that $\alpha_j \leq 1$ for all j, we can bound this term as follows:

$$\sum_{j=1}^{N} \alpha_j W_1(\hat\mu_j, \hat\mu_T) \leq (\sum_{j=1}^{N} \alpha_j W_1(\hat\mu_j, \hat\mu)) + N W_1(\hat\mu, \hat\mu_T).$$

Now, let us consider the following optimization problem

$$\inf_{\hat\mu \in \mathcal{P}(\Omega)} \frac{1}{N} \sum_{j=1}^{N} \alpha_j W_1(\hat\mu_j, \hat\mu) + W_1(\hat\mu, \hat\mu_T). \tag{1}$$

In this formulation, the first term $\frac{1}{N} \sum_{j=1}^{N} \alpha_j W_1(\hat\mu_j, \hat\mu)$ corresponds exactly to the problem known in the literature as the Wasserstein barycenters problem [1] that can be defined for W_1 as follows.

Definition 2. *For N probability measures $\mu_1, \mu_2, \ldots, \mu_N \in \mathcal{P}(\Omega)$, an empirical Wasserstein barycenter is a minimizer $\mu_N^* \in \mathcal{P}(\Omega)$ of $J_N(\mu) = \min_\mu \frac{1}{N} \sum_{i=1}^{N} a_i W_1(\mu, \mu_i)$, where for all i, $a_i > 0$ and $\sum_{i=1}^{N} a_i = 1$.*

The second term $W_1(\hat\mu, \hat\mu_T)$ of Eq. 1 finds the probability coupling that transports the barycenter to the target distribution. Altogether, this bound suggests that in order to adapt in the multi-source learning scenario, one can proceed by finding a barycenter of the source probability distributions and transport it to the target probability distribution.

On the other hand, the optimization problem related to the Wasserstein barycenters is closely related to the Multimarginal optimal transportation problem [19] where the goal is to find a probabilistic coupling that aligns N distinct probability measures. Indeed, as shown in [1], for a quadratic Euclidean cost function the solution μ_N^* of the barycenter problem in the Wasserstein space is given by the following equation:

$$\mu_N^* = \sum_{k \in \{k_1, \ldots, k_N\}} \gamma_k \delta_{A_k(x)},$$

where $A_k(x) = \sum_{j=1}^{N} \gamma_j x_{k_j}$ and $\boldsymbol{\gamma} \in \mathbb{R}^{\Pi_{j=1}^{N} n_j}$ is an optimal coupling solving for all $k \in \{1, \ldots, N\}$ the multimarginal optimal transportation problem with the following cost:

$$c_k = \sum \frac{a_j}{2} \|x_{k_j} - A_k(x)\|^2.$$

We note that this reformulation is particularly useful when the source distributions are assumed to be Gaussians. In this case, there exists a closed form solution for the multimarginal optimal transportation problem [14] and thus for Wasserstein barycenters problem too. Finally, it is also worth noticing that the optimization problem Eq. 1 has already been introduced to solve the multiview learning task [12]. In their formulation, the second term is referred to as an a priori knowledge about the barycenter which, in our case, is explicitly given by the target probability measure simultaneously.

5   Comparison to Other Existing Bounds

As mentioned in the introduction, there are numerous papers that proposed DA generalization bounds. The main difference between them lies in the distance used to measure the divergence between source and target probability distributions. The seminal work of [3] considered a modification of the total variation distance called H-divergence given by the following equation:

$$d_H(p,q) = 2 \sup_{h \in H} |p(h(x) = 1) - q(h(x) = 1)|.$$

On the other hand, [5,15] proposed to replace it with the discrepancy distance:

$$\text{disc}(p,q) = \max_{h,h' \in H} |\epsilon_p(h,h') - \epsilon_q(h,h')|.$$

The latter one was shown to be tighter in some plausible scenarios. A more recent work on generalization bounds using integral probability metric

$$D_{\mathcal{F}}(p,q) = \sup_{f \in \mathcal{F}} | \int f dp - \int f dq |$$

and Rényi divergence

$$D_\alpha(p\|q) = \frac{1}{\alpha - 1} \log \left(\sum_{i=1}^n \frac{p_i^\alpha}{q_i^{\alpha-1}} \right)$$

were presented in [16,27], respectively. [27] provides a comparative analysis of discrepancy and integral metric based bounds and shows that the former are less tight. [16] derives the domain adaptation bounds in multisource scenario by assuming that the good hypothesis can be learned as a weighted convex combination of hypothesis from all the sources available. Considering a reasonable amount of previous work on the subject, a natural question about the tightness of the DA bounds based on the Wasserstein metric introduced above arises in spite of the Theorem 3.

The answer to this question is partially given by the Csiszàr-Kullback-Pinsker inequality [20] defined for any two probability measures $p, q \in \mathcal{P}(\Omega)$ as follows:

$$W_1(p,q) \leq \text{diam}(\Omega)\|p - q\|_{\text{TV}} \leq \sqrt{2\text{diam}(\Omega)\text{KL}(p\|q)},$$

where $\text{diam}(\Omega) = \sup_{x,y \in \Omega}\{d(x,y)\}$ and $\text{KL}(p\|q)$ is the Kullback-Leibler divergence.

A first consequence of this inequality shows that the Wasserstein distance not only appears naturally and offers algorithmic advantages in DA but also gives tighter bounds than total variation distance (L1) used in [2, Theorem 1]. On the other hand, it is also tighter than bounds presented in [16] as the Wasserstein metric can be bounded by the Kullback-Leibler divergence which is a special case of Rényi divergence when $\alpha \to 1$ as shown in [10]. Regarding the discrepancy distance and omitting the hypothesis class restriction, one has $d_{min}\text{disc}(p,q) \le W_1(p,q)$, where $d_{min} = \min_{x \ne y \in \Omega}\{d(x,y)\}$. This inequality, however, is not very informative as minimum distance between two distinct points can be dramatically small thus making it impossible to compare the considered distances directly.

Regarding computational guarantees, we note that the H-divergence used in [3] is defined as the error of the best hypothesis distinguishing between the source and target domain samples pseudo-labeled with 0's and 1's and thus presents an intractable problem in practice. For the discrepancy distance, authors provided a linear time algorithm for its calculation in 1D case and showed that in other cases it scales as $O(N_S^2 d^{2.5} + N_T d^2)$ when the squared loss is used [15]. In its turn, the Wasserstein distance with entropic regularization can be calculated based on the linear time Sinkhorn-Knopp algorithm regardless the choice of the cost function c that presents a clear advantage over the other distances considered above.

Finally, none of the distances previously introduced in the generalization bounds for DA take into account the geometry of the space meaning that the Wasserstein distance is a powerful and precise tool to measure the divergence between domains.

6   Conclusion

In this paper, we studied the problem of DA in the optimal transportation context. Motivated by the existing algorithmic advances in domain adaptation, we presented the generalization bounds for both single and multi-source learning scenarios where the distance between source and target probability distributions is measured by the Wasserstein metric. Apart from the distance term that taken alone justifies the use of optimal transport in domain adaptation, the obtained bounds also included the capability term depicting the existence of a good hypothesis for both source and target domains. A direct consequence of its appearance in the bounds is the need to regularize optimal transportation plan in a way that allows to ensure efficient learning in the source domain once the interpolation was done. This regularization, achieved in [6,7] by the means of the class-based regularization, thus can be also viewed as an implication of the obtained results. Furthermore, it explains the superior performance of both class-based and Laplacian regularized optimal transport in domain adaptation compared to it simple entropy regularized form. On the other hand, we also showed that the use of the Wasserstein distance leads to tighter bounds compared to the bounds based on the total variation distance and Rényi divergence and is more computationally attractive than some other existing results. From the analysis of the bounds obtained for the multi-source DA, we derived a new algorithmic idea that suggests the minimization of two terms: first term corresponds to the Wasserstein barycenter problem calculated on the empirical source measures while the second one solves the optimal transport problem between this barycenter and the empirical target measure.

Future perspectives of this work are many and concern both the derivation of new algorithms for domain adaptation and the demonstration of new theoretical results. First of all, we would like to study the extent to which the cost function used in the derivation of the bounds can be used on actual real-world DA problems. This distance, defined as a norm of difference between two feature maps, can offer a flexibility in the calculation of the optimal transport metric due to its kernel representation. Secondly, we aim to produce new concentration inequalities for the λ term that will allow to bound the true best joint hypothesis by its empirical counter-part. These concentration inequalities will allow to access the adaptability of two domains from the given labelled samples while the speed of convergence may show how many data instances from the source domains is needed to obtain a reliable estimate of λ. Finally, the introduction of the Wasserstein distance to the bounds means that new DA algorithms can be designed based on the other optimal coupling techniques. These include, for instance, Knothe-Rosenblatt coupling and Moser coupling.

Acknowledgments. This work was supported in part by the French ANR project LIVES ANR-15-CE23-0026-03.

References

1. Agueh, M., Carlier, G.: Barycenters in the Wasserstein space. SIAM J. Math. Anal. **43**(2), 904–924 (2011)
2. Ben-David, Sh., Blitzer, J., Crammer, K., Kulesza, A., Pereira, F., Vaughan, J.: A theory of learning from different domains. Mach. Learn. **79**, 151–175 (2010)
3. Ben-David, Sh., Blitzer, J., Crammer, K., Pereira, O.: Analysis of representations for domain adaptation. In: NIPS (2007)
4. Bolley, Fr., Guillin, Ar., Villani, C.: Quantitative concentration inequalities for empirical measures on non-compact spaces. Prob. Theory Relat. Fields **137**(3–4), 541–593 (2007)
5. Cortes, C., Mohri, M.: Domain adaptation and sample bias correction theory and algorithm for regression. Theoret. Comput. Sci. **519**, 103–126 (2014)
6. Courty, N., Flamary, R., Tuia, D.: Domain adaptation with regularized optimal transport. In: Calders, T., Esposito, F., Hüllermeier, E., Meo, R. (eds.) ECML PKDD 2014. LNCS (LNAI), vol. 8724, pp. 274–289. Springer, Heidelberg (2014). https://doi.org/10.1007/978-3-662-44848-9_18
7. Courty, N., Flamary, R., Tuia, D., Rakotomamonjy, A.: Optimal transport for domain adaptation. IEEE Trans. Pattern Anal. Mach. Intell. **39**(9), 1853–1865 (2017)
8. Cuturi, M.: Sinkhorn distances: lightspeed computation of optimal transport. In: NIPS, pp. 2292–2300 (2013)

9. Cuturi, M., Doucet, A.: Fast computation of Wasserstein barycenters. In: ICML, pp. 685–693 (2014)
10. Ding, Y.: Wasserstein-divergence transportation inequalities and polynomial concentration inequalities. Stat. Probab. Lett. **94**(C), 77–85 (2014)
11. Fournier, N., Guillin, A.: On the rate of convergence in Wasserstein distance of the empirical measure. Probab. Theory Relat. Fields **162**(3–4), 707 (2015)
12. Bergounioux, M., Abraham, I., Abraham, R., Carlier, G.: Tomographic reconstruction from a few views: a multi-marginal optimal transport approach. Appl. Math. Optim. **75**(1), 1–19 (2016)
13. Kantorovich, L.: On the translocation of masses. C.R. (Doklady) Acad. Sci. URSS (N.S.) **37**(10), 199–201 (1942)
14. Knott, M., Smith, C.S.: On a generalization of cyclic-monotonicity and distances among random vectors. Linear Algebra Appl. **199**, 363–371 (1994)
15. Mansour, Y., Mohri, M., Rostamizadeh, A.: Domain adaptation: learning bounds and algorithms. In: COLT (2009)
16. Mansour, Y., Mohri, M., Rostamizadeh, A.: Multiple source adaptation and the rényi divergence. In: UAI, pp. 367–374 (2009)
17. Monge, G.: Mémoire sur la théorie des déblais et des remblais. In: Histoire de l'Académie Royale des Sciences, pp. 666–704 (1781)
18. Pan, S.J., Yang, Q.: A survey on transfer learning. IEEE Trans. Knowl. Data Eng. **22**(10), 1345–1359 (2010)
19. Pass, B.: Uniqueness and monge solutions in the multimarginal optimal transportation problem. SIAM J. Math. Anal. **43**(6), 2758–2775 (2011)
20. Pinsker, M.S.: Information and Information Stability of Random Variables and Processes. Holden-Day, San Francisco (1964)
21. Rabin, J., Peyré, G., Delon, J., Bernot, M.: Wasserstein barycenter and its application to texture mixing. In: Bruckstein, A.M., ter Haar Romeny, B.M., Bronstein, A.M., Bronstein, M.M. (eds.) SSVM 2011. LNCS, vol. 6667, pp. 435–446. Springer, Heidelberg (2012). https://doi.org/10.1007/978-3-642-24785-9_37
22. Rubner, Y., Tomasi, C., Guibas, L.J.: The earth mover's distance as a metric for image retrieval. Int. J. Comput. Vis. **40**(2), 99–121 (2000)
23. Saitoh, S.: Integral Transforms, Reproducing Kernels and their Applications. Pitman Research Notes in Mathematics Series (1997)
24. Sejdinovic, D., Sriperumbudur, B.K., Gretton, A., Fukumizu, K.: Equivalence of distance-based and RKHS-based statistics in hypothesis testing. Ann. Stat. **41**(5), 2263–2291 (2013)
25. Sinkhorn, R., Knopp, P.: Concerning nonnegative matrices and doubly stochastic matrices. Pac. J. Math. **21**, 343–348 (1967)
26. Villani, C.: Optimal Transport: Old and New. Grundlehren der mathematischen Wissenschaften. Springer, Berlin (2009). https://doi.org/10.1007/978-3-540-71050-9
27. Zhang, C., Zhang, L., Ye, J.: Generalization bounds for domain adaptation. In: NIPS (2012)
28. Perrot, M., Courty, N., Flamary, R., Habrard, A.: Mapping estimation for discrete optimal transport. In: NIPS, pp. 4197–4205 (2016)

TSP: Learning Task-Specific Pivots
for Unsupervised Domain Adaptation

Xia Cui[✉], Frans Coenen, and Danushka Bollegala

University of Liverpool, Liverpool, UK
xia.cui@liverpool.ac.uk

Abstract. Unsupervised Domain Adaptation (UDA) considers the problem of adapting a classifier trained using labelled training instances from a source domain to a different target domain, without having access to any labelled training instances from the target domain. Projection-based methods, where the source and target domain instances are first projected onto a common feature space on which a classifier can be trained and applied have produced state-of-the-art results for UDA. However, a critical pre-processing step required by these methods is the selection of a set of common features (aka. *pivots*), this is typically done using heuristic approaches, applied prior to performing domain adaptation. In contrast to the one of heuristics, we propose a method for learning Task-Specific Pivots (TSPs) in a systematic manner by considering both the labelled and unlabelled data available from both domains. We evaluate TSPs against pivots selected using alternatives in two cross-domain sentiment classification applications. Our experimental results show that the proposed TSPs significantly outperform previously proposed selection strategies in both tasks. Moreover, when applied in a cross-domain sentiment classification task, TSP captures many sentiment-bearing pivots.

1 Introduction

Domain Adaptation (DA) [5,7,30] considers the problem of adapting a model trained on one domain (*the source*) to a different domain (*the target*). DA is useful when we do not have sufficient labelled training data for a novel target domain to which we would like to apply a model that we have already trained using the labelled data for an existing source domain. If the source and target domains are sufficiently similar, then DA methods can produce accurate classifiers for the target domain [3]. DA methods have been widely applied for NLP tasks such as Part-Of-Speech (POS) tagging [32], sentiment classification [4], named entity recognition [22], and machine translation [24]. For example, in cross-domain sentiment classification, we might like to apply a sentiment classifier trained using labelled reviews on *books* for classifying reviews written on *Laptops*[1].

---

[1] Here, a collection of reviews on a particular product category is considered as a *domain*.

M. Ceci et al. (Eds.): ECML PKDD 2017, Part II, LNAI 10535, pp. 754–771, 2017.
https://doi.org/10.1007/978-3-319-71246-8_46

Table 1. Positive (+) and negative (−) sentiment reviews on *Books* and *Laptops*. Sentiment-bearing features are shown in bold, whereas selected pivots are underlined.

Books	*Laptops*
+	I think that this is an **excellent** book	This is a **powerful**, yet **compact** laptop An **excellent** choice for a travelling businessman!
−	This book is a **disappointment**, definitely **not recommended**	It's the **worst** laptop I have ever had. It is **slow** and forever **crashing**
−	Found myself skipping most of it found it **boring**	**Pricy** laptop and does not deliver. A big **disappointment** **Loud** fan, **noisy** hard drive. Never buy again

In Unsupervised Domain Adaptation (UDA), we assume the availability of unlabelled data from both source and target domains, and labelled data only from the source domain. In contrast, Supervised Domain Adaptation (SDA) assumes the availability of a small labelled dataset from the target domain [12,13]. UDA is a significantly harder task compared to SDA because of the unavailability of any labelled data from the target domain. In this paper, we focus on UDA. The main challenge of UDA is *feature mismatch* [6,8] – the features that appear in the source domain training instances are different from that in the target domain test instances. Because of the feature mismatch problem, even if we learn a highly accurate model from the source domain, most of those features will not affect in the target domain, resulting in a lower accuracy.

Current state-of-the-art methods for UDA first learn an embedding between the source and target domain feature spaces, and then learn a classifier for the target task (e.g. sentiment classification) in this embedded space [5,6,30]. In order to learn this embedding, these methods must select a subset of common features (here onwards referred to as *pivots*) to the two domains. For example, consider the user reviews shown in Table 1 for *Books* and *Laptops* selected from Amazon[2]. Sentiment-bearing features, such as *powerful, loud, crashing, noisy* are specific to *laptops*, whereas *boring* would often be associated with a *book*. On the other hand, features such as *disappointment* and *excellent* are likely to be domain-independent, hence suitable as pivots for UDA.

As detailed later in Sect. 2, all existing strategies for selecting pivots are based on heuristics, such as selecting the top-frequent features that are common to both source and target domains [5]. In addition to frequency, Mutual Information (MI) [4], Pointwise Mutual Information (PMI) [6], and Positive Pointwise Mutual Information (PPMI) [9] have been proposed in prior work on UDA as pivot selection strategies.

There are two fundamental drawbacks associated with all existing heuristic-based pivot selection strategies. First, existing pivot selection strategies focus on

---

[2] www.amazon.com.

either (a) selecting a subset of the common features to source and target domains as pivots, or (b) selecting a subset of task-specific features (eg. sentiment-bearing features selected based on source domain's labelled training instances) as pivots. However, as we see later in our experiments, to successfully adapt to a new domain, the pivots must be both domain-independent (thereby capturing sufficient information for the knowledge transferring from the source to the target), as well as task-specific (thereby ensuring the selected pivots are accurately related to the labels). Second, it is non-trivial as to how we can combine the two requirements (a) and (b) in a consistent manner to select a set of pivots. Pivot selection can be seen as an optimal subset selection problem in which we must select a subset of the features from the intersection of the feature spaces of the two domains. Optimal subset selection is an NP-complete problem [14], and subset enumeration methods are practically infeasible considering that the number of subsets of a feature set of cardinality n is $2^n - 1$, where $n > 10^4$ in typical NLP tasks.

To overcome the above-mentioned limitations of existing pivot selection strategies, we propose **TSP** – Task-Specific Pivot selection for UDA. Specifically, we define two criteria for selecting pivots: one based on the *similarity between the source and target domains under a selected subset of features* (Sect. 3.2), and another based on *how well the selected subset of features capture the information related to the labelled instances in the source domain* (Sect. 3.3). We show that we can model the combination of the two criteria as a single constrained quadratic programming problem that can be efficiently solved for large feature spaces (Sect. 3.4). The reduction of pivot selection from a subset selection problem to a feature ranking problem, based on *pivothood* (α) scores learn from the data, enables us to make this computation feasible. Moreover, the salience of each criteria can be adjusted via a *mixing parameter* (λ) enabling us to gradually increase the level of task-specificity in the selected pivots, which is not possible with existing heuristic-based pivot selection methods.

We compare the proposed TSP selection method against existing pivot selection strategies using two UDA methodssps Spectral Feature Alignment (SFA) [30], and Structural Correspondence Learning (SCL) [4] on a benchmark dataset for cross-domain sentiment classification. We see that TSP, when initialised with pre-trained word embeddings trained using Continuous Bag-of-Words (CBOW) and Global Vector Prediction (GloVe) significantly outperforms previously proposed pivot selection strategies for UDA with respect to most domain pairs. Moreover, analysing the top-ranked pivots selected by TSP for their sentiment polarity, we see that more sentiment-bearing pivots are selected by TSP when we increase the mixing parameter. More importantly, our results show that we must consider both criteria discussed above when selecting pivots to obtain an optimal performance in UDA, which cannot be done using existing pivot selection strategies.

2   Related Work

In this section, we summarise previously proposed pivot selection strategies for UDA. For a detailed overview of DA methods, the intended reader is referred to [20].

Selecting common high-frequent features (FREQ) in source and target domains was proposed as a pivot selection strategy for cross-domain POS tagging by Blitzer et al. [5]. This strategy was used to select pivots for SCL to adapt a classification-based POS tagger. However, in their follow-up work Blitzer et al. [4] observed that for sentiment classification, where it is important to consider the polarity of the pivots, frequency is an inadequate criteria for pivot selection. To select pivots that behave similarly in the target domain as in the source domain, they computed the MI between a feature and source domain's positive vs. negative sentiment labelled reviews. If a particular feature was biased towards positive or negative labelled reviews, then it was likely that the feature contained information related to sentiment, which is useful for adapting a sentiment classifier for the target domain. Blitzer et al. [4] empirically showed MI to be a better pivot selection strategy than FREQ for cross-domain sentiment classification.

Pan et al. [30] proposed a modified version of MI where they weighted the MI between a feature and a domain by the probability of the feature in the domain to encourage domain-independent, as well as high-frequent, features to be selected as pivots. However, their pivot selection strategy used only the unlabelled data from the source and target domains. Unfortunately, there is no guarantee that the pivots selected purely based on unlabelled data will be related to the supervised target task (eg. sentiment classification or POS tagging) for which we plan to apply DA.

Bollegala et al. [6] proposed PMI [11] as a pivot selection strategy for UDA. Empirically, PMI gives a better normalisation result by considering individual feature occurrences. PPMI, which simply sets negative PMI values to zero has also been used as a pivot selection strategy for UDA [9]. Negative PMI values are often caused by noisy and unreliable co-occurrence counts; by ignoring these PPMI can select a more reliable set of pivots.

Overall, the above-mentioned heuristic-based pivot selection strategies can be seen as evaluating the dependence of a feature on: (a) source vs. target domain unlabelled training instances, or (b) source domain positive vs. negative labelled training instances. The above-mentioned four pivot selection strategies FREQ, MI, PMI and PPMI can be computed using either unlabelled data (corresponding to setting (a)) to derive four unlabelled versions of the pivot selection strategies $FREQ_U$, MI_U, PMI_U, and $PPMI_U$, or labelled data (corresponding to setting (b)) to derive four labelled versions of the pivot selection strategies $FREQ_L$, MI_L, PMI_L, and $PPMI_L$. In our experiments, we conducted an extensive comparison over all eight combinations, to the best of our knowledge, ours is the first paper to do so.

Although we focus on *feature-based* DA methods where the objective is to learn a common embedding between the source and target domain feature spaces, we note that there exist numerous *instance-based* DA methods that operate directly on training instances [21, 22]. In instance-based DA, the goal is to select a subset of source domain labelled instances that is similar to the target domain unlabelled instance, and train a supervised classifier for the target task using only those selected instances. Sampling-based approaches [18] and weighting-based approaches have been proposed for instance-based DA [21].

Recently, deep learning approaches [10, 16, 17, 27, 33, 34] have been proposed for further improving instance-based DA. Glorot et al. [17] used Stacked Denoising Auto-encoders (SDAs) to learn non-linear mappings for the data variations. Chen et al. [10] improved SDAs by marginalizing random corruptions under a specific network structure. Ganin and Lempitsky [16] proposed to regularise the immediate layers to perform feature learning, domain adaptation and classifier learning jointly using backpropagation for cross-domain image classification tasks.

Pivot selection does not apply for instance-based DA methods, hence not considered in the remainder of the paper.

3   Methods

3.1   Outline

Let us consider a source domain $\mathcal{S}$ consisting of a set of labelled instances $\mathcal{D}_L^{(S)}$ and unlabelled instances $\mathcal{D}_U^{(S)}$. Without loss of generality we assume the target adaptation task is binary classification, where we have a set of positively labelled instances $\mathcal{D}_+^{(S)}$ and a set of negatively labelled instances $\mathcal{D}_-^{(S)}$. Although we limit our discussion to the binary classification setting for simplicity, we note that the proposed method can be easily extended to other types of DA setting, such as multi-class classification and regression. For the target domain, denoted by $\mathcal{T}$, under UDA we assume the availability of only a set of unlabelled instances $\mathcal{D}_U^{(T)}$.

Given a pair of source and target domains, we propose a novel pivot selection method for UDA named **TSP** – Task Specific Pivot Selection. TSP considers two different criteria when selecting pivots.

First, we require that the *selected set of pivots must minimise the distance between the source and target domains*. The exact formulation of distance between domains and the corresponding optimisation problem are detailed in Sect. 3.2.

Second, we require that the *selected set of pivots be task-specific*. For example, if the target task is sentiment classification, then the selected set of pivots must contain sentiment bearing features. Given labelled data (annotated for the target task) from the source domain, we describe an objective function in Sect. 3.3 for this purpose.

Although the above-mentioned two objectives could be optimised separately, by jointly optimising for both objectives simultaneously we can obtain task-specific pivots that are also transferrable to the target domain. In Sect. 3.4, we formalise this joint optimisation problem and provide a quadratic programming-based solution. For simplicity, we present TSP for the pairwise UDA case, where we attempt to adapt from a single source domain to a single target domain. However, TSP can be easily extended to multi-source and multi-target UDA settings following the same optimisation procedure.

3.2   Pivots are Common to the Two Domains

Theoretical studies in UDA show that the upper bound on the classification accuracy on a target domain depends on the similarity between the source and the target domains [3]. Therefore, if we can somehow make the source and target domains similar by selecting a subset of the features from the intersection of their feature spaces, then we can learn better UDA models.

To formalise this idea into an objective we can optimise for, let us denote the possibility of a feature w_k getting selected as a pivot by $\alpha_k \in [0, 1]$. We refer to α_k as the *pivothood* of a feature $w_k \in \mathcal{W}$ ($|\mathcal{W}| = K$). Higher pivothood values indicate that those features are more appropriate as pivots. The features could be, for example in sentiment classification, n-grams of words, POS tags, dependencies or any of their combinations. For simplicity of the disposition, let us assume that w_k are either unigrams or bigrams of words, and that we have pre-trained word embeddings $\boldsymbol{w}_k \in \mathbb{R}^D$ obtained using some word embedding learning algorithm. Our proposed method does not assume any specific properties of the word embeddings used; any fixed-dimensional representation of the features is sufficient, not limited to word embeddings.

Given source domain unlabelled data, we can compute, $\boldsymbol{c}^{(S)}$, the centroid for the source domain as follows:

$$\boldsymbol{c}^{(S)} = \frac{1}{\left|\mathcal{D}_U^{(S)}\right|} \sum_{d \in \mathcal{D}_U^{(n)}} \sum_{w_k \in d} \alpha_k \phi(w_k, d) \boldsymbol{w}_k \tag{1}$$

Here, $\phi(w_k, d)$ indicates the salience of w_k as a feature representing an instance d, such as the tf-idf measure popularly used in text classification. Likewise, we can compute the $\boldsymbol{c}^{(T)}$ centroid for the target domain using the unlabelled data from the target domain as follows:

$$\boldsymbol{c}^{(T)} = \frac{1}{\left|\mathcal{D}_U^{(T)}\right|} \sum_{d \in \mathcal{D}_U^{(T)}} \sum_{w_k \in d} \alpha_k \phi(w_k, d) \boldsymbol{w}_k \tag{2}$$

The centroid can be seen as a representation for the domain consisting all of the instances (reviews in the case of sentiment classification). If two domains are similar under some representation, then it will be easier to adapt from one domain to the other. Different distance measures can be used to compute the distance (or alternatively the similarity) between two domains under a particular representation such as ℓ_1 distance (Manhattan distance) or ℓ_2 distance (Euclidean distance). In this work, we use Euclidean distance for this purpose.

The problem of selecting a set of pivots can then be formulated as minimising the squared Euclidean distance between $\boldsymbol{c}^{(S)}$ and $\boldsymbol{c}^{(T)}$ given by,

$$\text{min.}_{\alpha} \left\| \boldsymbol{c}^{(S)} - \boldsymbol{c}^{(T)} \right\|_2^2 \tag{3}$$

Here, $\boldsymbol{\alpha} = (\alpha_1, \ldots, \alpha_K)^\top$ is the pivothood indicator vector.

Assuming $\phi(w, d) = 0$ for $w \notin d$, and by substituting (1) and (2) in (3), we can re-write the objective as follows:

$$\text{min.}_{\alpha} \left\| \sum_{k=1}^{K} \alpha_k f(w_k) \boldsymbol{w}_k \right\|_2^2 \tag{4}$$

Here, $f(w_k)$ can be pre-computed and is given by,

$$f(w_k) = \frac{1}{\left| \mathcal{D}_U^{(S)} \right|} \sum_{d \in \mathcal{D}_U^{(S)}} \phi(w_k, d) - \frac{1}{\left| \mathcal{D}_U^{(T)} \right|} \sum_{d \in \mathcal{D}_U^{(T)}} \phi(w_k, d). \tag{5}$$

Minimisation of (4) can be trivially achieved by setting $\alpha_k = 0, \forall k = 1, \ldots, K$. To avoid this trivial solution and to make the objective scale invariant, we introduce the constraint $\sum_k \alpha_k = 1$.

Further, if we define $\boldsymbol{u}_k = f(w_k)\boldsymbol{w}_k$, then (4) reduces to the constrained least square regression problem given by (6).

$$\text{min.}_{\alpha} \left\| \sum_{k=1}^{K} \alpha_k \boldsymbol{u}_k \right\|_2^2$$
$$\text{s.t.} \sum_{k=1}^{K} \alpha_k = 1 \tag{6}$$

3.3   Pivots are Task Specific

The objective defined in the previous Section is computed using only unlabelled data, hence agnostic to the target task. However, prior work on UDA [4,6] has shown that we must select pivots that are specific to the target task in order to be accurately adapted to cross-domain prediction tasks. Labelled data can be used to measure the task specificity of a feature. However, in UDA, the only labelled data we have is from the source domain. Therefore, we define the task specificity γ_k of a feature w_k as follows:

$$\gamma_k = \left(h(w_k, \mathcal{D}_+^{(S)}) - h(w_k, \mathcal{D}_-^{(S)}) \right)^2 \tag{7}$$

Here, $h(w_k, \mathcal{D}_+^{(S)})$ denotes the association between the feature w_k and the source domain positive labelled instances. A wide range of association measures such as MI, PMI, PPMI, χ^2, log-likelihood ratio can be used as h [28]. PMI [11] between a feature w_k and a set of training instances $\mathcal{D}$ is given by:

$$\text{PMI}(w_k, \mathcal{D}) = \log\left(\frac{p(w_k, \mathcal{D})}{p(w_k)p(\mathcal{D})}\right), \tag{8}$$

We compute the probabilities p in (8) using frequency counts. PPMI [26] is a variation of PMI and follows $\text{PMI}(w_k, \mathcal{D})$ defined by (8), PPMI can be given by,

$$\text{PPMI}(w_k, \mathcal{D}) = \max(\text{PMI}(w_k, \mathcal{D}), 0) \tag{9}$$

We use PPMI as h in our experiments to reduce the noise caused by negative PMI values (Sect. 2).

Given the task specificity of features, we can formulate the problem of pivot selection as follows:

$$\min._{\boldsymbol{\alpha}} -\frac{\sum_k \alpha_k \gamma_k}{\sum_k \gamma_k} \tag{10}$$

3.4 Joint Optimisation

Ideally, we would like to select pivots that: (a) make source and target domain closer, as well as (b) are specific to the target task. A natural way to enforce both of those constraints is to linearly combine the two objectives given by (6) and (10) into the joint optimisation problem (11).

$$\min._{\boldsymbol{\alpha}} \left\|\sum_{k=1}^{K} \alpha_k \boldsymbol{u}_k\right\|_2^2 - \lambda \frac{\sum_k \alpha_k \gamma_k}{\sum_k \gamma_k}$$

$$\text{s.t.} \sum_{k=1}^{K} \alpha_k = 1 \tag{11}$$

Here, the mixing parameter $\lambda \geq 0$ is a hyperparameter that controls the level of task specificity of the selected pivots. For example, by setting $\lambda = 0$ we can select pivots purely using unlabelled data. When we gradually increase λ, the source domain labelled data influences the pivot select process, making the selected pivots more task specific.

To further simplify the optimisation problem given by (11), let us define $\mathbf{U} \in \mathbb{R}^{D \times K}$ to be a matrix where the K columns correspond to $\boldsymbol{u}_k \in \mathbb{R}^D$, and $\mathbf{c}$ to be a vector whose k-th element is set to

$$c_k = -\lambda \gamma_k / \sum_k \gamma_k.$$

Then, (11) can be written as the following quadratic programming problem:

$$\min._{\boldsymbol{\alpha}} \boldsymbol{\alpha}^T \mathbf{U}^\top \mathbf{U} \boldsymbol{\alpha} + \mathbf{c}^T \boldsymbol{\alpha}$$

$$\text{s.t.} \ \boldsymbol{\alpha}^T \mathbf{1} = 1, \tag{12}$$

$$\boldsymbol{\alpha} \geqslant 0.$$

We solve the quadratic programming problem given by (11) using the conjugate gradient (CG) method [29]. In practice, $D \ll K$, for which $\mathbf{U}^\top \mathbf{U} \in \mathbb{R}^{K \times K}$ is rank deficient and is not positive semidefinite. However, performing a small diagonalised Gaussian noise perturbation is sufficient in practice to obtain locally optimal solutions via CG that are sufficiently accurate for our datasets. We use CVXOPT[3] to solve (12) to obtain pivothoods α_k. We rank the features w_k in the descending order of their corresponding α_k and select the top-ranked features as pivots for an UDA method.

The computational complexity of the quadratic programming problem is dominated by the eigenvalue decomposition of $\mathbf{U}^\top \mathbf{U}$, which is of the dimensionality $K \times K$. Recall that K is the total number of features in the feature space representing the source and target domain instances. Computing the eigenvalue decomposition of a $K \times K$ square matrix is $\mathcal{O}(K^3)$. However, we can use randomised truncated Eigensolvers to compute the top-ranked eigenvalues (corresponding to the best pivots), without having to perform the full eigenvalue decomposition [19].

4   Experiments

Because the purpose of selecting pivots is to perform UDA, and because we cannot determine whether a particular feature is suitable as a pivot by manual observation, the most direct way to evaluate a pivot selection method is to use the pivots selected by that method in a state-of-the-art UDA method and evaluate the relative increase/decrease in performance in that DA task. For this purpose, we select SCL and SFA, which are UDA methods and perform a cross-domain sentiment classification task. The task here is to learn from labelled reviews from the source domain (a collection of reviews about a particular product) and predict sentiment for a different target domain (a collection of reviews about a different product). It is noteworthy that sentiment classification is used here purely as an evaluation task, and our goal is not to improve the performance of sentiment classification itself but to evaluate pivot selection methods. Therefore, we keep all other factors other than the pivots used by the UDA methods fixed during our evaluations.

We use the multi-domain sentiment dataset [4], which contains Amazon user reviews for the four product categories *Books* (**B**), *Electronic appliances* (**E**), *Kitchen appliances* (**K**), and *Dvds* (**D**). For each domain we have 1000 positive and 1000 negative reviews, and ca. 15,000 unlabelled reviews. We use the standard split of $800 \times 2 = 1600$ train and $200 \times 2 = 400$ test reviews. We generate 12 DA tasks by selecting one of the four domains as the source and another as the target. We represent a review using a bag-of-features consisting of unigrams and bigrams, excluding a standard stop words list. We drop features that occur less than 5 times in the entire dataset to remove noise. A binary logistic regression classifier with an ℓ_2 regulariser is trained in each setting to develop a binary sentiment classifier. The regularisation coefficient is tuned using the

---

[3] http://cvxopt.org/.

music domain as a development domain, which is not part of the train/test data. Although different classifiers could be used in place of logistic regression, by using a simpler classifier we can readily evaluate the effect of the pivots on the UDA performance. The classification accuracy on the target domain's test labelled reviews is used as the evaluation measure.

On average, for a domain pair we have $K = 2648$ features. We set the number of pivots selected to the top-ranked 500 pivots in all domain pairs. Later in Sect. 4.4 we study the effect of the number of pivots on the performance of the proposed method. We use the publicly available $D = 300$ dimensional GloVe[4] (trained using 42B tokens from the Common Crawl) and CBOW[5] (trained using 100B tokens from Google News) embeddings as the word representations required by TSP. For bigrams not listed in the pretrained models, we sum the embeddings of constituent unigrams following prior work on compositional approaches for creating phrase (or sentence) embeddings using word embeddings [1,23]. We denote TSP trained using GloVe and CBOW embeddings respectively by **T-GloVe** and **T-CBOW**. By using two different types of word embeddings we can evaluate the dependance (if any) of TSP on a particular type of word embedding learning algorithm. Later in Sect. 4.4, we evaluate the effect of counting-based and prediction-based word embeddings on the performance of TSP when used for computing source and target centroids respectively in (1) and (2). We use the inverse document frequency (IDF) [31] as the feature salience ϕ and PPMI as h in our experiments[6].

4.1   Cross-Domain Sentiment Classification

To evaluate the effect on performance of an UDA method when we select pivots using the proposed TSP and previously proposed pivots selection methods, we compare the performance of SCL and SFA in 12 different adaptation tasks from a source **S** to a target domain **T** as shown in Tables 2 and 3. We use the binomial exact test at two significance levels to evaluate statistical significance. The *no-adapt* (**NA**) lower-baselines, which simply applies a model trained using the source domain labelled data on the target domain's test data without performing DA, produced a near random-level performance indicating the importance of performing DA for obtaining good performance.

In SCL, T-CBOW reports the best performance in 8 domain-pairs, whereas T-GloVe in 3. Although in B-E and K-D pairs respectively PMI_U and PMI_L report the best results, the difference in performance with T-CBOW is not significant. Overall, T-CBOW is the best method in SCL closely followed by T-GloVe. For each of the previously proposed pivot selection methods except frequency, we see that the performance is always better when we use labelled data (denoted by subscript L) than unlabelled data (denoted by subscript U) to select pivots.

---

[4] http://nlp.stanford.edu/projects/glove/.

[5] https://code.google.com/archive/p/word2vec/.

[6] Performance of TSP was found to be robust over a wide-range of ϕ and h combinations.

Table 2. Classification accuracy of SCL using pivots selected by TSP (T-CBOW & T-GloVe) and prior methods. For each domain pair, the best results are given in bold font. The last row is the average across all the domain pairs. Statistically significant improvements over the $FREQ_U$ baseline according to the binomial exact test, are shown by "*" and "**" respectively at $p = 0.01$ and $p = 0.001$ levels.

S-T	NA	SCL									
		$FREQ_L$	$FREQ_U$	MI_L	MI_U	PMI_L	PMI_U	$PPMI_L$	$PPMI_U$	T-CBOW	T-GloVe
B-E	52.03	69.75	68.25	68.75	65.75	69.50	**75.75***	69.50	67.50	73.25	75.25*
B-D	53.51	70.25	73.25	74.25	59.75	76.50	72.00	76.50	70.50	**77.50**	77.00
B-K	51.63	76.25	74.25	78.25	63.50	80.00*	79.50	80.00*	77.00	**83.25****	82.00**
E-B	51.02	60.50	65.25	66.25	55.75	64.75	63.00	64.25	60.50	**68.75**	67.25
E-D	50.94	68.00	67.75	68.00	66.25	70.50	67.00	71.50	65.50	73.25	**74.00***
E-K	56.00	81.00	80.50	82.50	80.50	86.25*	77.50	85.75*	77.25	**87.50****	**87.50****
D-B	52.50	72.00	69.25	72.00	56.25	74.75	68.50	75.75*	69.50	**76.75***	75.50*
D-E	53.25	71.75	70.50	74.25	66.00	74.25	65.25	74.00	65.25	**76.25**	75.75
D-K	54.39	70.75	75.25	74.00	57.25	80.50	77.25	80.25	79.75	83.00**	**84.00****
K-B	51.29	66.75	67.75	68.50	56.00	74.00*	70.00	74.00*	69.25	**75.25***	74.25*
K-E	54.86	74.00	74.25	75.50	78.00	80.00*	72.25	80.00*	71.75	**82.00****	81.25*
K-D	50.94	67.00	65.75	68.00	60.00	**71.50**	67.50	**71.50**	68.75	70.75	70.75
AVG	52.70	70.67	71.00	72.52	63.75	75.21	71.30	75.25	70.21	**77.30***	77.04*

Table 3. Classification accuracy of SFA using pivots selected by TSP (T-CBOW & T-GloVe) and prior methods. For each domain pair, the best results are given in bold font. The last row is the average across all the domain pairs. Statistically significant improvements over $FREQ_U$ baseline according to the binomial exact test are shown by "*" and "**" respectively at $p = 0.01$ and $p = 0.001$ levels.

S-T	NA	SFA									
		$FREQ_L$	$FREQ_U$	MI_L	MI_U	PMI_L	PMI_U	$PPMI_L$	$PPMI_U$	T-CBOW	T-GloVe
B-E	52.03	70.50	74.00	73.25	66.00	74.00	71.00	74.00	70.50	**74.75**	73.75
B-D	53.51	71.50	**78.00**	69.50	60.00	72.75	74.75	72.75	73.00	77.75	**78.00**
B-K	51.63	72.75	74.25	73.00	66.50	78.50	75.75	78.50	75.00	**81.00***	80.75*
E-B	51.02	64.75	64.50	64.00	57.25	65.75	59.00	64.00	57.75	64.50	**67.00**
E-D	50.94	67.50	**74.50**	63.25	60.75	71.50	65.00	71.50	61.50	73.50	71.50
E-K	56.00	81.00	82.50	78.25	71.75	85.50	79.00	85.25	78.00	**87.00**	85.50
D-B	52.50	74.25	**79.00**	69.50	62.00	73.50	73.00	73.75	67.75	**79.00**	78.00
D-E	53.25	72.50	75.50	71.75	65.75	69.00	68.75	69.00	66.50	74.25	**76.75**
D-K	54.39	73.75	76.75	74.75	56.50	81.00	79.75	81.00	79.00	81.50	**83.25***
K-B	51.29	67.75	70.00	69.00	58.00	66.50	**71.25**	66.50	**71.25**	69.00	70.00
K-E	54.86	80.50	**84.50**	79.25	70.25	77.25	71.75	77.25	72.50	80.50	79.75
K-D	50.94	67.25	**77.75**	67.75	60.50	68.00	71.00	68.00	71.00	72.25	72.50
AVG	52.70	72.00	75.94	71.10	62.94	73.60	71.67	73.46	70.31	76.25	**76.40**

On the other hand in SFA, T-GloVe reports the best performance in 5 domain-pairs, whereas T-CBOW in 3. Among the previously proposed pivot selection methods, $FREQ_U$ performs best in 5 domain-pairs. However, we see that overall, T-GloVe and T-CBOW outperform all the other pivot selection methods. Moreover, the difference between performance reported by T-CBOW and T-GloVe is not significant, indicating that TSP is relatively robust against the actual word embedding method being used.

Table 4. Top 5 pivots selected for adapting from **E** to **K** by T-CBOW and T-GloVe for different mixing parameter values. Bigrams are denoted by "+".

λ	T-CBOW	T-GloVe
0	unlike complaint fair i+havent name	wont whether yes complete so+good
10^{-5}	sent+it of+junk im+very fast+and an+excellent	go+wrong of+junk im+very i+called an+excellent
10^{-4}	of+junk value+for i+called fast+and it+comes	is+excellent sent+it im+very an+excellent good+price
10^{-3}	an+excellent of+junk i+called fast+and pleased+with	sent+it stopped+working of+junk is+perfect very+happy
10^{-2}	i+love of+junk dont+buy i+called very+happy	i+love of+junk i+called a+great very+happy
1.0	return stopped of+junk dont+buy i+called	stopped of+junk dont+buy i+called very+happy

Overall, we see that SCL benefits more from accurate pivot selection than SFA. SCL learns separate linear predictors for each pivot using non-pivots (i.e. features other than pivots) as features, whereas SFA builds a bi-partite graph between pivots and non-pivots on which eigenvalue decomposition is applied to obtain a lower-dimensional embedding of pivots. Therefore, the effect of pivots on SCL is more direct and critical than in SFA.

4.2 Effect of the Mixing Parameter

To evaluate the effect of the mixing parameter on the pivots selected by TSP, we use the pivots selected by TSP for different λ values with SCL and SFA, and measure the classification accuracy on a target domain's sentiment labelled test reviews. Hyperparameters such as the number of singular vectors used in SCL and the number of latent dimensions used in SFA are tuned using the *music* development domain as the target for each of the four source domains. Due to the space limitations we show the results for adapting from the **D** to **K** pair with SCL and SFA in Fig. 1. We see that when we do not require pivots to be task-specific (i.e. $\lambda = 0$) the accuracy is low. However, when we gradually increase λ thereby enforcing the criterion that the selected pivots must be also task-specific, we see a steady improvement in accuracy, which then saturates. SFA [30] constructs a bipartite graph based on the co-occurrence relationship between domain-independent (pivots) and domain-specific features (non-pivots) after pivot selection. FREQ$_U$ uses a larger dataset (S_U and T_U) compared to TSP (S_L and T_U), such that decreases the chance of rare co-occurrence may happen for a smaller dataset. For the same reason, adding γ that computes from an even smaller dataset (S_L only) does not help much in SFA.

This result shows the importance of considering both objectives (6) and (10) appropriately when selecting pivots, instead of focusing on only one of the two.

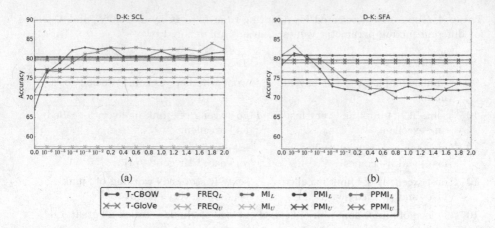

Fig. 1. Accuracy of SCL and SFA when adapting from **D** source to **K** target (x-axis not to scale).

Moreover, performance of TSP is relatively robust for $\lambda > 10^{-3}$, which is attractive because we do not have to fine-tune the mixing-parameter for each UDA setting.

4.3 Sentiment Polarity of the Selected Pivots

To evaluate whether the pivots selected by TSP are task-specific, we conducted the following experiment. For a particular value of λ, we sort the features in the descending order of their α values, and select the top-ranked 500 features as pivots. Because our task in this case is sentiment classification, we would expect the selected pivots to be sentiment-bearing (i.e. containing words that express sentiment). Next, we compare the selected pivots against the sentiment polarity ratings provided in SentiWordNet [15]. SentiWordNet classifies each synset in the WordNet into positive, negative or neutral sentiment polarities such that any word with a positive sentiment will have a positive score, a negative sentiment a negative score, and zero otherwise. For bigrams not listed in the SentiWordNet, we compute the average polarity of the two constituent unigrams as the sentiment polarity of the bigrams. We consider a pivot to be task-specific if it has a non-zero polarity score.

Figure 2 shows the percentage of the task-specific (sentiment-bearing) pivots among the top-500 pivots selected by TSP for different λ values when adapting between **E** and **K**. We see that initially, when λ is small, the percentage of task-specific pivots is small. However, when we increase λ, thereby encouraging TSP to consider the task-specificity criterion more, we end up with pivots that are more sentiment-bearing. Moreover, when $\lambda > 10^{-3}$ the percentage of the task-specific pivots remains relatively stable, indicating that we have selected all pivots that are sentiment-bearing for the particular domain-pair.

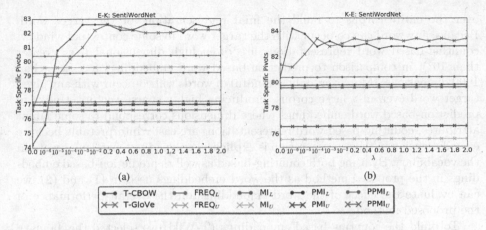

Fig. 2. Task specific pivots for adapting between **E** and **K** by TSP with $\lambda \in [0, 2]$ (x-axis not to scale).

As a qualitative example, we show the top-5 pivots ranked by their pivothood scores by T-CBOW and T-GloVe when adapting between **E** and **K** in Table 4. At $\lambda = 0$, we see that most of the top-ranked pivots are not sentiment-bearing. However, when we increase λ, we see that more and more sentiment-bearing pivots appear among the top-ranks. This result re-confirms that TSP can accurately capture task-specific pivots by jointly optimising for the two criteria proposed in Sect. 3. Interestingly, we see that many pivots are selected by both T-CBOW and T-GloVe such as *an+excellent*, *i+love*, and *very+happy*. This is encouraging because it shows that TSP depends weakly on the word representation method we use, thereby enabling us to potentially use a wide-range of existing pre-trained word embeddings with TSP.

4.4   Number of Selected Pivots and Effect of Word Embeddings

The number of pivots selected and the word embeddings are external inputs to TSP. In this section, we experimentally study the effect of the number of pivots selected by our proposed TSP and the word embeddings on the performance of the cross-domain sentiment classification. Specifically, we use TSP to select different k numbers of pivots, with three types of word embeddings: pre-trained word embeddings using CBOW (**T-CBOW**), pre-trained word embeddings using GloVe (**T-GloVe**), and counting-based word embeddings computed from Wikipedia (referred to as **T-Wiki**).

Counting-based word embeddings differ from prediction-based word embeddings such as CBOW and GloVe in several important ways [2,25]. In counting-based word embeddings we represent a target word by a vector where the elements correspond to words that co-occur with the target word in some contextual window. Entire sentences or windows of a fixed number of tokens can be considered as the co-occurrence window. Next, co-occurrences are aggregated over the entire corpus to build the final representation for the target word. Because any word can co-occur with the target word in some contextual window, counting-based word representations are often high dimensional (e.g. greater than 10^5), in comparison to prediction-based word embeddings (e.g. less than 1000 dimensions). Moreover, only a handful of words will co-occur with any given target word even in a large corpus, producing highly sparse vectors. Unlike in prediction-based word embeddings where dimensions correspond to some latent attributes, counting-based word representations are easily interpretable because each dimension in the representation is explicitly assigned to a particular word in the vocabulary. By using both counting-based as well as prediction-based embeddings in the proposed method as the word embeddings used in (1) and (2), we can evaluate the effect of the word embeddings on the overall performance of the proposed method.

To build the counting-based embeddings (T-Wiki) we selected the January 2017 dump of English Wikipedia[7], and processed it using a Perl script[8] to create a corpus of 4.6 billion tokens. We select unigrams occurring at least 1000 times in this corpus amounting to a vocabulary of size 73, 954. We represent each word by a vector whose elements correspond to the PPMI values computed from the co-occurrence counts between words in this Wikipedia corpus.

To study the effect of the number of pivots k on the performance of TSP, we fix the mixing parameter $\lambda = 10^{-3}$ for all domain pairs and vary $k \in [100, 1000]$. We show the performance of SCL and SFA for the B-D pair respectively in Figs. 3a and b. From Fig. 3a, we see that, overall for SCL, T-CBOW outperforms the other two embeddings across a wide range of pivot set sizes. Moreover, its performance increases with k, which indicates that with larger pivot sets it can better represent a domain using the centroid. T-Wiki on the other hand reports

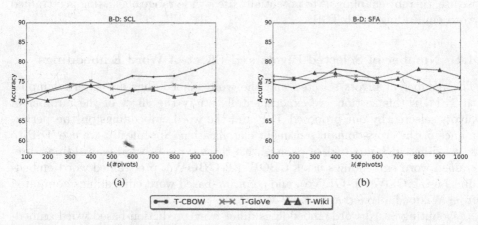

Fig. 3. Accuracy of SCL and SFA when adapting from **B** source to **D** target, from k (number of pivots) in the range $[100, 1000]$. The mixing parameter is set to $\lambda = 10^{-3}$.

---

[7] https://dumps.wikimedia.org.
[8] http://mattmahoney.net/dc/textdata.html.

lowest accuracies across all k values. Prior work evaluating word embedding learning algorithms has shown that prediction-based word embeddings outperform counting-based word embeddings for a wide-range of language processing tasks [2]. On the other hand, for SFA (Fig. 3b) we do not see much difference in performance among the different word embeddings. Overall, the best performance is reported using **T-CBOW** with $k = 400$ pivots. This observation is in agreement with the recommendation by Pan et al. [30] to use 500 domain independent (pivot) features for this dataset. Similar trends were observed for all 12 domain pairs.

5   Conclusion

We proposed TSP, a task-specific pivot selection method that simultaneously requires pivots to be both similar in the two domains as well as task-specific. TSP jointly optimises the two criteria by solving a single quadratic programming problem. The pivots selected by TSP improve the classification accuracy in multiple cross-domain sentiment classification tasks, consistently outperforming previously proposed pivot selection methods. Moreover, comparisons against SentiWordNet reveal that indeed the top-ranked pivots selected by TSP are task-specific. We conducted a series of experiments to study the behaviour of the proposed method with various parameters and the design choices involved such as the mixing parameter, number of pivots used, UDA method where the selected pivots are used, and the word embeddings used for representing features when computing the domain centroids. Our experimental results shows that TSP can find pivots for various pairs of domains and improve the performance of both SCL and SFA when compared to the performance obtained by using pivots selected by prior heuristics. Moreover, our analysis shows that it is important to use both unlabelled data (available for both source and target domains) as well as labelled data (available only for the source domain) when selecting pivots.

References

1. Arora, S., Liang, Y., Ma, T.: A simple but tough-to-beat baseline for sentence embeddings. In: Proceedings of ICLR (2017)
2. Baroni, M., Dinu, G., Kruszewski, G.: Don't count, predict! a systematic comparison of context-counting vs. context-predicting semantic vectors. In: Proceedings of the 52nd Annual Meeting of the Association for Computational Linguistics, Baltimore, Maryland, vol. 1: Long Papers, pp. 238–247. Association for Computational Linguistics, June 2014
3. Ben-David, S., Blitzer, J., Crammer, K., Kulesza, A., Pereira, F., Vaughan, J.W.: A theory of learning from different domains. Mach. Learn. **79**, 151–175 (2009)
4. Blitzer, J., Dredze, M., Pereira, F.: Biographies, bollywood, boom-boxes and blenders: domain adaptation for sentiment classification. In: Proceedings of ACL, pp. 440–447 (2007)

5. Blitzer, J., McDonald, R., Pereira, F.: Domain adaptation with structural corre-spondence learning. In: Proceedings of EMNLP, pp. 120–128 (2006)
6. Bollegala, D., Mu, T., Goulermas, J.Y.: Cross-domain sentiment classification using sentiment sensitive embeddings. IEEE Trans. Knowl. Data Eng. **28**(2), 398–410 (2015)
7. Bollegala, D., Weir, D., Carroll, J.: Using multiple sources to construct a sentiment sensitive thesaurus for cross-domain sentiment classification. In: Proceedings of ACL, pp. 132–141 (2011)
8. Bollegala, D., Weir, D., Carroll, J.: Cross-domain sentiment classification using a sentiment sensitive thesaurus. IEEE Trans. Knowl. Data Eng. **25**(8), 1719–1731 (2013)
9. Bollegala, D., Weir, D., Carroll, J.: Learning to predict distributions of words across domains. In: Proceedings of ACL, pp. 613–623 (2014)
10. Chen, M., Xu, Z., Weinberger, K., Sha, F.: Marginalized denoising autoencoders for domain adaptation. arXiv preprint arXiv:1206.4683 (2012)
11. Church, K.W., Hanks, P.: Word association norms, mutual information, and lexi-cography. Comput. Linguist. **16**(1), 22–29 (1990)
12. Daumé III, H.: Frustratingly easy domain adaptation. In: Proceedings of ACL, pp. 256–263 (2007)
13. Daumé III, H., Kumar, A., Saha, A.: Frustratingly easy semi-supervised domain adaptation. In: Proceedings of the Workshop on Domain Adaptation for Natural Language Processing, pp. 53–59 (2010)
14. Davies, S., Russell, S.: NP-completeness of searches for smallest possible feature sets. In: Proceedings of AAAI (1994)
15. Esuli, A., Sebastiani, F.: SentiWordNet: a publicly available lexical resource for opinion mining. In: Proceedings of LREC, pp. 417–422 (2006)
16. Ganin, Y., Lempitsky, V.: Unsupervised domain adaptation by backpropagation. In: Proceedings of ICML, pp. 1180–1189 (2015)
17. Glorot, X., Bordes, A., Bengio, Y.: Domain adaptation for large-scale sentiment classification: a deep learning approach. In: ICML 2011 (2011)
18. Gong, B., Grauman, K., Sha, F.: Connecting the dots with landmarks: Discrim-inatively learning domain-invariant features for unsupervised domain adaptation. In: Proceedings of ICML (2013)
19. Halko, N., Martinsson, P.G., Tropp, J.A.: Finding structure with randomness: probabilistic algorithms for constructung approximate matrix decompositions. SIAM Rev. **53**(2), 217–288 (2010)
20. Jiang, J.: A literature survey on domain adaptation of statistical classifiers. Tech-nical report, UIUC (2008)
21. Jiang, J., Zhai, C.: Instance weighting for domain adaptation in NLP. In: Proceed-ings of ACL, pp. 264–271 (2007)
22. Jiang, J., Zhai, C.: A two-stage approach to domain adaptation for statistical classifiers. In: Proceedings of CIKM, pp. 401–410 (2007)
23. Kenter, T., Borisov, A., de Rijke, M.: Siamese CBOW: optimizing word embeddings for sentence representations. In: Proceedings of the 54th Annual Meeting of the Association for Computational Linguistics, Berlin, Germany, vol. 1: Long Papers, pp. 941–951. Association for Computational Linguistics, August 2016
24. Koehn, P., Schroeder, J.: Experiments in domain adaptation for statistical machine translation. In: Proceedings of the Second Workshop on Statistical Machine Trans-lation, pp. 224–227 (2007)
25. Levy, O., Goldberg, Y., Dagan, I.: Improving distributional similarity with lessons learned from word embeddings. Trans. Assoc. Comput. Linguist. **3**, 211–225 (2015)

26. Lin, D.: Automatic retrieval and clustering of similar words. In: Proceedings of the 17th International Conference on Computational Linguistics, vol. 2, pp. 768–774. Association for Computational Linguistics (1998)
27. Long, M., Wang, J.: Learning transferable features with deep adaptation networks. In: Proceedings of ICML (2015)
28. Manning, C.D., Schutze, H.: Foundations of Statistical Natural Language Processing. MIT Press, Cambridge (1999)
29. Nocedal, J., Wright, S.J.: Numerical Optimization. Springer, New York (1999). https://doi.org/10.1007/b98874
30. Pan, S.J., Ni, X., Sun, J.T., Yang, Q., Chen, Z.: Cross-domain sentiment classification via spectral feature alignment. In: Proceedings of WWW, pp. 751–760 (2010)
31. Salton, G., Buckley, C.: Introduction to Modern Information Retreival. McGraw-Hill Book Company, New York (1983)
32. Schnabel, T., Schütze, H.: Towards robust cross-domain domain adaptation for part-of-speech tagging. In: Proceedings of IJCNLP, pp. 198–206 (2013)
33. Yosinski, J., Clune, J., Bengio, Y., Lipson, H.: How transferable are features in deep neural networks? In: Advances in Neural Information Processing Systems, pp. 3320–3328 (2014)
34. Ziser, Y., Reichart, R.: Neural structural correspondence learning for domain adaptation. arXiv preprint arXiv:1610.01588 (2016)

Unsupervised and Semisupervised Learning

k^2-means for Fast and Accurate Large Scale Clustering

Eirikur Agustsson[1], Radu Timofte[1,2(✉)], and Luc Van Gool[1,3]

[1] Computer Vision Lab, D-ITET, ETH Zurich, Zürich, Switzerland
radu.timofte@vision.ee.ethz.ch
[2] Merantix GmbH, Berlin, Germany
[3] KU Leuven, Leuven, Belgium

Abstract. We propose k^2-means, a new clustering method which efficiently copes with large numbers of clusters and achieves low energy solutions. k^2-means builds upon the standard k-means (Lloyd's algorithm) and combines a new strategy to accelerate the convergence with a new low time complexity divisive initialization. The accelerated convergence is achieved through only looking at k_n nearest clusters and using triangle inequality bounds in the assignment step while the divisive initialization employs an optimal 2-clustering along a direction. The worst-case time complexity per iteration of our k^2-means is $O(nk_nd + k^2d)$, where d is the dimension of the n data points and k is the number of clusters and usually $n \gg k \gg k_n$. Compared to k-means' $O(nkd)$ complexity, our k^2-means complexity is significantly lower, at the expense of slightly increasing the memory complexity by $O(nk_n + k^2)$. In our extensive experiments k^2-means is order(s) of magnitude faster than standard methods in computing accurate clusterings on several standard datasets and settings with hundreds of clusters and high dimensional data. Moreover, the proposed divisive initialization generally leads to clustering energies comparable to those achieved with the standard k-means++ initialization, while being significantly faster.

1 Introduction

The k-means algorithm in its standard form (Lloyd's algorithm) employs two steps to cluster n data points of d dimensions and k initial cluster centers [19]. The *expectation* or *assignment step* assigns each point to its nearest cluster while the *maximization* or *update step* updates the k cluster centers with the mean of the points belonging to each cluster. The k-means algorithm repeats the two steps until convergence, that is the assignments no longer change in an iteration i.

k-means is one of the most widely used clustering algorithms, being included in a list of top 10 data mining algorithms [27]. Its simplicity and general applicability vouch for its broad adoption. Unfortunately, its $O(ndki)$ time complexity depends on the product between number of points n, number of dimensions d, number of clusters k, and number of iterations i. Thus, for large such values even a single iteration of the algorithm is very slow.

© Springer International Publishing AG 2017
M. Ceci et al. (Eds.): ECML PKDD 2017, Part II, LNAI 10535, pp. 775–791, 2017.
https://doi.org/10.1007/978-3-319-71246-8_47

Table 1. Notations

n	Number of data points to cluster
k	Number of clusters
k_n	Number of nearest clusters
d	Number of dimensions of the data points
X	The data $(x_i)_{i=1}^n, x_i \in \mathbb{R}^d$
C	Cluster centers $C = (c_j)_{j=1}^k, c_j \in \mathbb{R}^d$
a	Cluster assignments $\{1, \cdots, n\} \rightarrow \{1, \cdots, k\}$
$a(x_i)$	Cluster assignment of x_i, i.e. $a(i)$
$a(X')$	Cluster assignment of some set of points X'
X_j	Points assigned to cluster j, $(x_i \in X \vert a(i) = j)$
$\mu(X_j)$	The mean of X_j: $\frac{1}{\vert X_j \vert} \sum_{x \in X_j} x$
$\Vert x \Vert$	l_2 norm of $x \in \mathbb{R}^d$
$\phi(X_j)$	Energy of X_j: $\sum_{x \in X_j} \Vert x - \mu(X_j) \Vert^2$
$\mathcal{N}_{k_n}(c_l)$	k_n nearest neighbours of c_l in C (including c_l)

The simplest way to handle larger datasets is parallelization [28,29], however this requires more computation power as well. Another way is to process the data online in batches as done by the MiniBatch algorithm of Sculley [23], a variant of the Lloyd algorithm that trades off quality (i.e. the converged energy) for speed.

To improve both the speed and the quality of the clustering results, Arthur and Vassilvitskii [1] proposed the k-means++ initialization method. The initialization typically results in a higher quality clustering and fewer iterations for k-means, than when using the default random initialization. Furthermore, the expected value of the clustering energy is within a $8(\ln k + 2)$ factor of the optimal solution. However, the time complexity of the method is $O(ndk)$, i.e. the same as a single iteration of the Lloyd algorithm - which can be too expensive in a large scale setting. Since k-means++ is sequential in nature, Bahman et al. [2] introduced a parallel version k-means|| of k-means++, but did not reduce the time complexity of the method.

Another direction is to speed up the actual k-means iterations. Elkan [8], Hamerly [11] and Drake and Hamerly [7] go in this direction and use the triangle inequality to avoid unnecessary distance computation between cluster centers and the data points. However, these methods still require a full Lloyd iteration in the beginning to then gradually reduce the computation of progressive iterations. The recent Yinyang k-means method of Ding et al. [6] is a similar method, that also leverages bounds to avoid redundant distance calculations. While typically performing 2–3× faster than Elkan method, it also requires a full Lloyd iteration to start with.

Philbin *et al.* [22] introduce an approximate k-means (AKM) method based on kd-trees to speed up the assignment step, reducing the complexity of each k-means iteration from $O(nkd)$ to $O(nmd)$, where $m < k$. In this case m, the distance computations performed per each iteration, controls the trade-off between a fast and an accurate (i.e. low energy) clustering. Wang *et al.* [26] use cluster closures for further 2.5× speedups.

Mazzeo *et al.* [20] introduce a centroid-based method that combines divisive and agglomerative clustering, obtaining quickly high quality clusters as measured by the CH-index [5].

In this paper we propose k^2-means, a method aiming at both fast and accurate clustering. Following the observation that usually the clusters change gradually and affect only local neighborhoods, in the assignment step we only consider the k_n nearest neighbours of a center as the candidates for the clusters members. Furthermore we employ the triangle inequality bounds idea as introduced by Elkan [8] to reduce the number of operations per each iteration. For initializing k^2-means, we propose a divisive initialization method, which we experimentally prove to be more efficient than k-means++.

Our k^2-means gives a significant algorithmic speedup, i.e. reducing the complexity to $O(nk_nd)$ per iteration, while still maintaining a high accuracy comparable to methods such as k-means++ for a chosen $k_n < k$. Similar to m in AKM, k_n also controls a trade-off between speed and accuracy. However, our experiments show that we can use a significantly lower k_n when aiming for a high accuracy.

The paper is structured as follows. In Table 1 we summarize the notations used in this paper. In Sect. 2 we introduce our proposed k^2-means method and our divisive initialization. In Sect. 3 we describe the experimental benchmark and discuss the results obtained, while in Sect. 4 we draw conclusions.

2 Proposed k^2-means

In this section we introduce our k^2-means method and motivate the design decisions. The pseudocode of the method is given in Algorithm 1.

Given some data $X = (x_i)_{i=1}^n, x_i \in \mathbb{R}^d$, the k-means clustering objective is to find cluster centers $C = (c_j)_{j=1}^k, c_j \in \mathbb{R}^d$ and cluster assignments $a : \{1, \cdots, n\} \to \{1, \cdots, k\}$, such that the cluster energy

$$\sum_{j=1}^k \sum_{x \in X_j} \|x - c_j\|^2 \tag{1}$$

is minimized, where $X_j := (x_i \in X | a(i) = j)$ denotes the points assigned to a cluster j. For a data point x_i, we sometimes write $a(x_i)$ instead of $a(i)$ for the cluster assignment. Similarly, for a subset X' of the data, $a(X')$ denotes the cluster assignments of the corresponding points.

Algorithm 1. k^2-means

1: **Given**: k, data X, neighbourhood size k_n
2: Initialize centers C
3: Initialize assignments $a : \{1 \cdots n\} \to \{1, \cdots, k\}$.
4: **while** Not converged **do**
5:    Build k_n-NN graph of C:
6:    $\mathcal{N}_{k_n} : C \to \{1 \cdots k\}^{k_n}$
7:    **for** $x \in X$ **do**
8:       Get current center for x:
9:       $l \leftarrow a(x)$
10:      Assign x to nearest candidate center:
11:      $a(x) \leftarrow \arg\min_{l' \in \mathcal{N}_{k_n}(c_l)} \|x - c_{l'}\|$
12:   **end for**
13:   **for** $j \in \{1 \cdots k\}$ **do**
14:      $c_j \leftarrow \mu(X_j)$ {Update center}
15:   **end for**
16: **end while**
17: **return** C, a

Standard Lloyd obtains an approximate solution by repeating the following until convergence: (i) In the *assignment step*, each x is assigned to the nearest center in C. (ii) For the *update step*, each center is recomputed as the mean of its members.

The assignment step requires $O(nk)$ distance computations, i.e. $O(nkd)$ operations, and dominates the time complexity of each iteration. The update step requires only $O(nd)$ operations for mean computations.

To speed up the assignment step, an approximate nearest neighbour method can be used, such as kd-trees [21, 22] or locality sensitive hashing [13]. However, these methods ignore the fact that the cluster centers *are moving* across iterations and often this movement is *slow*, affecting a *small neighborhood* of points. With this observation, we obtain a very simple fast nearest neighbour scheme:

Suppose at iteration i, a data point x was assigned to a nearby center, $l = a(x)$. After updating the centers, we still expect c_l to be close to x. Therefore, the centers nearby c_l are likely candidates for the nearest center of x in iteration $i + 1$. To speed up the assignment step, we thus only consider the k_n nearest neighbours of c_l, $\mathcal{N}_{k_n}(c_l)$, as candidate centers for the points $x \in X_l$. Since for each point we only consider k_n centers in the assignment step (in line 11 of Algorithm 1), the complexity is reduced to $O(nk_nd)$. In practice, we can set $k_n \ll k$.

We also use inequalities as in [8] to avoid redundant distance computations in the assignment step (in line 11 of Algorithm 1). We use the exact same triangle inequalities as described in the Elkan paper [8], but only maintain the nk_n lower bounds, for the neighbourhood of each point, instead of nk for the Elkan method. It is easy to see that this modification is valid as an exact speed up of the assignment step within then neighbourhood. When a point is assigned to a new cluster, we however need to update the k_n lower bounds of the point, since the neighbourhood changes in this case. We refer to the original Elkan paper [8] for a detailed discussion on triangle inequalities and bounds.

As for standard Lloyd, the total energy can only decrease in each iteration of the algorithm. In the assignment step, points are only moved to closer centers, reducing their contribution to the total energy. In the update step, the center z of a cluster S is updated as the mean of its members, $\mu(S)$. As Lemma 1 shows, this clearly reduces the energy of the cluster since the second right hand side term is positive. Thus, the total energy is monotonically decreasing as a function of iterations, which guarantees convergence.

As shown by Arthur and Vassilvitskii [1], a good initialization, such as k-means++, often leads to a higher quality clustering compared to random sampling. Since the $O(ndk)$ complexity of k-means++ would negate the benefits of the k^2-means computation savings, we propose an alternative fast initialization scheme, which also leads to high quality clustering solutions.

2.1   Greedy Divisive Initialization (GDI)

For the initialization of our k^2-means, we propose a simple hierarchical clustering method named Greedy Divisive Initialization (GDI), detailed in Algorithm 2. Similarly to other divisive clustering methods, such as [4,24], we start with a single cluster and repeatedly split the highest energy cluster until we reach k clusters.

To efficiently split each cluster, we use Projective Split (Algorithm 3), a variant of k-means with $k = 2$, that is motivated by the following observation: Suppose we have points X' and centers (c_1, c_2) in the k-means method. Let H be the hyperplane with normal vector $c_2 - c_1$, going through $\mu(c_1, c_2)$ (see e.g. the top left corner of Fig. 1). When we perform the standard k-means assignment step, we greedily assign each point to its closest centroid to get a solution with a lower energy, thus assigning the points on one side of H to c_1, and the other side of H to c_2.

Although this is the best assignment choice for the current centers c_1 and c_2, this may not be a good split of the data. Therefore, we depart from the standard assignment step and consider instead *all* hyperplanes along the direction $c_2 - c_1$ (i.e. with normal vector $c_2 - c_1$). We project X' onto $c_2 - c_1$ and "scan" a hyperplane through the data to find the split that gives the lowest energy (lines 4–8 in Algorithm 3). To efficiently recompute the energy of the cluster splits as the hyperplane is scanned, we use the following Lemma:

Lemma 1 [14, Lemma 2.1]. *Let S be a set of points with mean $\mu(S)$. Then for any point $z \in \mathbb{R}^d$*

$$\sum_{x \in S} \|x - z\|^2 = \sum_{x \in S} \|x - \mu(S)\|^2 + |S|\|z - \mu(S)\|^2 \tag{2}$$

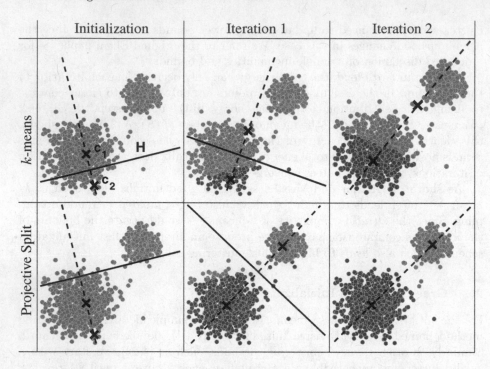

Fig. 1. Example of two iterations of Projective Split and standard k-means with $k = 2$ using the same initialization. The dashed line shows the direction defined by two centers $(c_2 - c_1)$. The solid line shows where the algorithms split the data in each iteration. The splitting line of k-means always goes through the midpoint of the two centers, while Projective Split picks the minimal energy split along the dashed line. Even though the initial centers start in the same cluster, Projective Split can almost separate the clusters in a single iteration.

We can now compute

$$\phi(S \cup \{y\}) = \sum_{x \in S \cup \{y\}} \|x - \mu(S \cup \{y\})\|^2 \tag{3}$$

$$= \sum_{x \in S} \|x - \mu(S \cup \{y\})\|^2 + \|y - \mu(S \cup \{y\})\|^2 \tag{4}$$

$$= \phi(S) + |S| \|\mu(S \cup \{y\}) - \mu(S)\|^2 + \|y - \mu(S \cup \{y\})\|^2, \tag{5}$$

where we used Lemma 1 in (4). Equipped with (5) we can efficiently update energy terms in line 8 in Algorithm 3 as we scan the hyperplane through the data X_j (after sorting it along $c_a - c_b$ in line 5–6), using in total only $O(|X_j|)$ distance computations and mean updates. Note that $\mu(S \cup \{y\})$ is easily computed with an add operation as $(|S|\mu(S) + y)/(|S| + 1)$.

Algorithm 2. Greedy Divisive Initialization (GDI)

1: **Given:** k, data X
2: Assign all points to one cluster
3: $C = \{\mu(X)\}$, $a(X) = 1$
4: **while** $|C| < k$ **do**
5:     Pick highest energy cluster:
6:     $j \leftarrow \arg\max_l \phi(X_l)$
7:     Split the cluster:
8:     $X_a, c_a, X_b, c_b \leftarrow \text{ProjectiveSplit}(X_j)$
9:     $c_j \leftarrow c_a$
10:     $c_{|C|+1} \leftarrow c_b$
11:     $a(X_b) \leftarrow |C| + 1$
12:     $C \leftarrow C \cup \{c_{|C|+1}\}$
13: **end while**
14: **return** C, a

Algorithm 3. Projective Split

1: **Given:** data $X_j = (x_i)_{i=1}^{n_j}$
2: Pick two random samples c_a, c_b from X_j
3: **while** Not Converged **do**
4:     Sort X_j along $c_a - c_b$:
5:     $P_j \leftarrow (x_i \cdot (c_a - c_b) | x_i \in X_j)$
6:     $X_j \leftarrow X_j$ sorted by P_j
7:     Find minimum-energy split:
8:     $l_{min} = \arg\min_l \phi((\tilde{x}_i)_{i=1}^l) + \phi((\tilde{x}_i)_{i=l+1}^{n_j})$
9:     $X_a \leftarrow (\tilde{x}_i)_{i=1}^{l_{min}}$
10:     $X_b \leftarrow (\tilde{x}_i)_{i=l_{min}+1}^{n_j}$
11:     $c_a, c_b \leftarrow \mu(X_a), \mu(X_b)$
12: **end while**
13: **return** X_a, c_a, X_b, c_b

Compared to standard k-means with $k = 2$, our Projective Split takes the optimal split along the direction $c_2 - c_1$ but greedily considers only this direction. In Fig. 1 we show how this can lead to a faster convergence.

2.2 Time Complexity

Table 2 shows the time and memory complexity of Lloyd, Elkan, MiniBatch, AKM, and our k^2-means.

The time complexity of each k^2-means iteration is dominated by two factors: building the nearest neighbour graph of C (line 6), which costs $O(k^2)$ distance computations, as well as computing distances between points and candidate centers (line 11), which initially costs nk_n distance computations. Elkan and k^2-means use the triangle inequality to avoid redundant distance calculations and

Table 2. Time and memory complexity per iteration for Lloyd, Elkan, MiniBatch, AKM and our k^2-means.

Method	Time complexity	Memory complexity
Lloyd	$O(nkd)$	$O((n+k)d)$
Elkan [8]	$O(nkd + k^2d) \sim O(nd + k^2d)$	$O((n+k)d + nk + k^2)$
MiniBatch [23]	$O(bkd)$	$O((b+k)d)$
AKM [22]	$O(nmd)$	$O((n+k)d)$
k^2-means (ours)	$O(nk_nd + k^2d) \sim O(nd + k^2d)$	$O((n+k)d + nk_n + k^2)$

empirically we observe the $O(nkd)$ and $O(nk_nd)$ terms (respectively) gradually reduce down to $O(nd)$ at convergence.

In MiniBatch k-means processes only b samples per iteration (with $b \ll n$) but needs more iterations for convergence. AKM limits the number of distance computations to m per iteration, giving a complexity of $O(nmd)$.

Table 3 shows the time and memory complexity of random, k-means++ and our GDI initialization. For the GDI, the time complexity is dominated by calls to Projective Split. If we limit Projective Split to maximum $O(1)$ iterations (2 in our experiments) then a call to ProjectiveSplit(X_j) costs $O(|X_j|)$ distance computations and vector additions, $O(|X_j|)$ inner products and $O(|X_j| \log |X_j|)$ comparisons (for the sort), giving in total $O(|X_j|(\log |X_j| + d))$ complexity. However, the resulting time complexity of GDI depends on the data.

Table 3. Time and memory complexity for initialization.

Initialization	Time complexity	Memory complexity
Random	$O(k)$	$O(k)$
k-means++ [1]	$O(nkd)$	$O(n+k)$
GDI (ours)	$O(n(\log k)(d + \log n)) \sim O(nk(d + \log n))$	$O(n + kd)$

For pathological datasets, it could happen for each call to Projective Split(X'), that the minimum split is of the form $\{y\}, X' \setminus \{y\}$, i.e. only one point y is split off. In this case, for $|X| = n$, the total complexity will be $O(n(\log n + d) + (n-1)(\log(n-1) + d) + \cdots + (n-k)(\log(n-k) + d)) = O(nk(d + \log n))$.[1]

A more reasonable case is when at each call ProjectiveSplit(X') splits each cluster into two similarly large clusters, i.e. the minimum split is of the form (X'_a, X'_b) where $|X_a| \approx |X_b|$. In this case the worst case scenario is when in each

---

[1] A simple example of such a pathological dataset is $X = (x_i)_{i=1}^n \subset \mathbb{R}$ where $x_1 = 0$, $x_2 = 1$, $x_3 = \phi(x_1, x_2)$, $x_4 = \phi(x_1, x_2, x_3)$ and $x_n = \phi(x_1, \cdots, x_n)$. The size of x_n grows extremely fast though, e.g. $x_{10} \approx 1581397605569$ and x_{14} has 195 digits.

split the highest energy cluster is the largest cluster (in no. of samples), resulting a total complexity of $O(n \log k(d + \log n))$.[2] Therefore the time complexity of GDI is somewhere between $O(n \log k(d + \log n)) \sim O(n(d + \log n)k)$.

In our experiments we count vector operations for simplicity (i.e. dropping the $O(d)$ factor), as detailed in the next section. To fairly account for the $O(|X_j| \log |X_j|)$ complexity of the sorting step in ProjectiveSplit, we artificially count it as $|X_j| \log_2(|X_j|)/d$ vector operations.

3 Experiments

For a fair comparison between methods implemented in various programming languages, we use the number of vector operations as a measure of complexity, i.e. distances, inner products and additions. While the operations all share an $O(d)$ complexity, the distance computations are most expensive accounting for the constant factor. However, since the runtime of all methods is dominated by distance computations (i.e. more than 95% of the runtime), for simplicity we count all vector operations equally and refer to them as "distance computations", using the terminology from [8].

3.1 Datasets

In our experiments we use datasets with 2414–150000 samples ranging from 50 to 32256 dimensions as listed in Table 5. The datasets are diverse in content and feature representation.

To create **cnnvoc** we extract 4096-dimensional CNN features [16] for 15662 bounding boxes, each belonging to 20 object categories, from PASCAL VOC 2007 [9] dataset. **covtype** uses the first 150000 entries of the Covertype dataset [3] of cartographic features. From the **mnist** database [17] of handwritten digits we also generate **mnist50** by random projection of the raw pixels to a 50-dimensional subspace. For **tinygist10k** we use the first 10000 images with extracted gist features from the 80 million tiny images dataset [25]. **cifar** represents 50000 training images from the CIFAR [15] dataset. **usps** [12] has scans of handwritten digits (raw pixels) from envelopes. **yale** contains cropped face images from the Extended Yale B Database [10,18].

3.2 Methods

We compare our k^2**-means** with relevant clustering methods: Lloyd (standard k-means), Elkan [8] (accelerated Lloyd), MiniBatch [23] (web-scale online clustering), and AKM [22] (efficient search structure).

Aside from our **GDI** initialization, we also use **random** initialization and k**-means++** [1] in our experiments. For k-means++ we use the provided Matlab implementation. We Matlab implement MiniBatch k-means according to

---

[2] If we split all clusters of approximately equal size simultaneously, we need $O(\log k)$ passes and perform $O(n(d + \log n))$ computations in each pass.

Algorithm 1 in [23] and use the provided codes for Elkan and AKM. **Lloyd++**
and **Elkan++** combine k-means++ initialization with Lloyd and Elkan, respectively.

We run all methods, except MiniBatch, for a maximum of 100 iterations. For
MiniBatch k-means we use $b = 100$ samples per batch and $t = n/2$ iterations.
For the Projective Split, Algorithm 3, we perform only 2 iterations.

3.3   Initializations

We compare k-means++, random and our GDI initialization by running 20 trials
of k-means (Lloyd) clustering with $k \in \{100, 200, 500\}$ on the datasets. Table 4
reports minimum and average cluster energy as well as the average number of
distance computations, relative to k-means++, averaged over 20 seeds.

Our GDI gives a (slightly) better average and minimum convergence energy
than the other initializations, while its runtime complexity is an order of magnitude smaller than in the case of k-means++ initialization. Notably, the speedup
of GDI over k-means++ improves as k grows, and at $k = 500$ is typically more
than an order of magnitude. This makes GDI a good choice for the initialization
of k^2-means.

Table 4. Comparison of energy and runtime complexity for random, k-means++, and
our GDI initialization. The results are displayed relative to k-means++, averaged over
20 seeds. Random initialization does not require distance computations. GDI is an
order of magnitude faster while giving comparable energies to k-means++.

Dataset	k	Average convergence energy			Minimum convergence energy			Average runtime complexity	
		Random	k-means++	GDI	Random	k-means++	GDI	k-means++	GDI
cnnvoc	100	1.00	1.00	1.00	1.00	1.00	1.00	1.00	0.16
	200	1.00	1.00	0.99	1.00	1.00	0.99	1.00	0.09
	500	1.00	1.00	0.99	1.00	1.00	0.99	1.00	0.04
covtype	100	1.51	1.00	0.99	1.47	1.00	0.99	1.00	0.19
	200	1.58	1.00	0.98	1.38	1.00	0.99	1.00	0.11
	500	1.43	1.00	0.99	1.30	1.00	0.99	1.00	0.05
mnist	100	1.00	1.00	1.00	1.00	1.00	1.00	1.00	0.15
	200	1.00	1.00	1.00	1.00	1.00	1.00	1.00	0.09
	500	1.00	1.00	1.00	1.00	1.00	1.00	1.00	0.04
mnist50	100	1.00	1.00	1.00	1.00	1.00	1.00	1.00	0.19
	200	1.00	1.00	1.00	1.00	1.00	1.00	1.00	0.11
	500	1.00	1.00	1.00	1.00	1.00	1.00	1.00	0.05
tinygist10k	100	1.00	1.00	1.00	1.00	1.00	1.00	1.00	0.16
	200	1.00	1.00	0.99	1.00	1.00	1.00	1.00	0.09
	500	1.00	1.00	0.99	1.00	1.00	1.00	1.00	0.04
usps	100	1.01	1.00	0.99	1.01	1.00	1.00	1.00	0.16
	200	1.01	1.00	0.99	1.01	1.00	0.99	1.00	0.09
	500	1.04	1.00	1.00	1.04	1.00	1.00	1.00	0.05
yale	100	1.01	1.00	1.00	1.00	1.00	0.99	1.00	0.16
	200	1.02	1.00	1.00	1.02	1.00	1.00	1.00	0.10
	500	1.05	1.00	1.03	1.05	1.00	1.02	1.00	0.05
Average		1.078	1.000	0.996	1.061	1.000	0.997	1.000	0.103

Table 5. Algorithmic speedup in reaching an energy within 1% from the final Lloyd++ energy. (-) marks failure in reaching the target of 1% relative error. For each method, the parameter(s) that gave the highest speedup at 1% error is used.

Dataset	k	AKM	Elkan++	Elkan	Lloyd++	Lloyd	k^2-means
cifar	50	1.0	2.6	3.7	1.0	1.0	**9.5**
$n = 50000$	200	1.9	3.0	4.6	1.0	1.1	**26.2**
$d = 3072$	1000	4.9	3.0	5.1	1.0	1.2	**86.7**
cnnvoc	50	**13.8**	2.1	2.9	1.0	1.4	9.0
$n = 15662$	200	**22.6**	2.0	2.8	1.0	1.2	19.2
$d = 4096$	1000	3.3	1.9	2.8	1.0	0.9	**20.2**
covtype	50	-	6.1	-	1.0	-	**35.1**
$n = 150000$	200	-	6.3	-	1.0	-	**78.7**
$d = 54$	1000	-	8.5	-	1.0	-	**176.6**
mnist	50	7.3	3.6	5.3	1.0	1.5	**12.3**
$n = 60000$	200	1.9	3.7	5.7	1.0	1.2	**24.6**
$d = 784$	1000	4.7	3.6	5.9	1.0	0.8	**43.4**
mnist50	50	**12.7**	3.7	5.4	1.0	1.3	8.8
$n = 60000$	200	1.9	4.2	6.7	1.0	1.2	**22.3**
$d = 50$	1000	3.1	4.1	6.6	1.0	0.8	**38.0**
tinygist10k	50	**16.2**	2.4	3.6	1.0	1.4	11.7
$n = 10000$	200	12.8	2.3	3.5	1.0	1.3	**22.3**
$d = 384$	1000	1.5	2.1	-	1.0	-	**13.6**
usps	50	5.3	4.1	-	1.0	-	**11.8**
$n = 7291$	200	16.8	4.4	-	1.0	-	**23.6**
$d = 256$	1000	**18.5**	2.7	-	1.0	-	-
yale	50	2.1	4.2	6.3	1.0	0.6	**17.9**
$n = 2414$	200	**21.9**	2.9	-	1.0	-	13.9
$d = 32256$	1000	-	**1.9**	-	1.0	-	-
Avg. speedup		8.7	3.6	4.7	1.0	1.1	**33.0**

3.4 Performance

Our goal is *fast accurate clustering*, where the cluster energy differs only slightly from Lloyd with a good initialization (such as k-means++) at convergence. Therefore, we measure the runtime complexity needed to achieve a clustering energy that is within 1% of the energy obtained with Lloyd++ at convergence.

For a given budget i.e. the maximum number of iterations and parameters such as m for AKM and k_n for k^2 means, it is not known beforehand how well the algorithms approximate the targeted Lloyd++ energy. For a fair comparison we use an oracle to select the best parameters and the number of iterations for each

Table 6. Algorithmic speedup in reaching the same energy as the final Lloyd++ energy. (-) marks failure in reaching the target of 0% relative error. For each method, the parameter(s) that gave the highest speedup at 0% error is used.

Dataset	k	AKM	Elkan++	Elkan	Lloyd++	Lloyd	k^2-means
cifar	50	-	17.8	-	1.0	-	**37.9**
	200	1.2	24.2	-	1.0	-	**139.8**
	1000	11.3	17.5	28.2	1.0	2.6	**373.6**
cnnvoc	50	2.4	9.3	-	1.0	-	**26.2**
	200	3.7	9.3	-	1.0	-	**59.7**
	1000	5.8	**8.1**	-	1.0	-	-
covtype	50	-	28.9	-	1.0	-	**172.0**
	200	-	40.2	-	1.0	-	**442.4**
	1000	-	**44.5**	-	1.0	-	-
mnist	50	1.1	17.3	26.6	1.0	2.9	**39.3**
	200	-	25.8	-	1.0	-	**81.0**
	1000	9.0	29.8	-	1.0	-	**141.1**
mnist50	50	-	18.7	-	1.0	-	**31.0**
	200	2.1	26.6	-	1.0	-	**80.3**
	1000	5.1	22.7	-	1.0	-	**94.1**
tinygist10k	50	12.5	12.5	20.1	1.0	3.6	**50.1**
	200	4.7	11.0	-	1.0	-	**71.8**
	1000	2.6	**7.8**	-	1.0	-	-
usps	50	-	12.6	-	1.0	-	**31.7**
	200	3.4	14.6	-	1.0	-	**54.4**
	1000	-	**9.4**	-	1.0	-	-
yale	50	2.8	9.5	-	1.0	-	**32.5**
	200	**20.8**	6.5	-	1.0	-	18.7
	1000	-	**4.0**	-	1.0	-	-
Avg. speedup		5.9	17.9	25.0	1.0	3.0	**104.1**

method, i.e. the ones that give the highest speedup but still reach the reference error. In practice, one can use a rule of thumb or progressively increase k, m and the number of iterations until a desired energy has been reached.

To measure performance we run AKM, Elkan++, Elkan, Lloyd++, Lloyd, MiniBatch, and k^2-means with $k \in \{50, 200, 1000\}$ on various datasets, with 3 different seeds and report average speedups over Lloyd++ when the energy reached is within 1% from Lloyd++ at convergence in Table 5.

Each method is stopped once it reaches the reference energy and for AKM and k^2-means, we use the parameters m and k_n from $\{3, 5, 10, 20, 30, 50, 100, 200\}$ that give the highest speedup.

Table 5 shows that for most settings, our k^2-means has the highest algorithmic speedup at 1% error. It benefits the most when both the number of clusters and the number of points are large, e.g. for $k = 200$ at least $19\times$ speedup for all datasets with $n \geq 7000$ samples. We do not reach the target energy for usps and yale with $k = 1000$, because k_n was limited to 200.

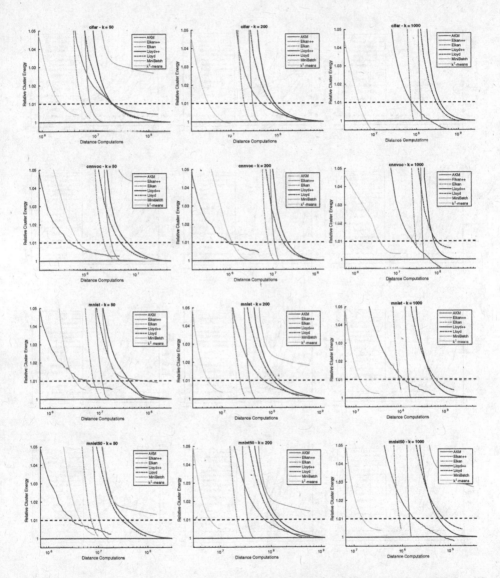

Fig. 2. Cluster Energy (relative to best Lloyd++ energy) vs distance computations on cifar, cnnvoc, mnist and mnist50 for $k \in \{50, 200, 1000\}$. For AKM and k^2-means, we use the parameter with the highest algorithmic speedup at 1% error.

Figure 2 show the convergence curves corresponding to cifar, cnnvoc, mnist and mnist50 entries in Table 5. Figure 3 shows the convergence curves of AKM and k^2 means under same settings, when varying the parameters m and k_n. On cifar the benefit of k^2-means is clear since it reaches the reference error significantly faster than the other methods. On mnist50 k^2-means is considerably

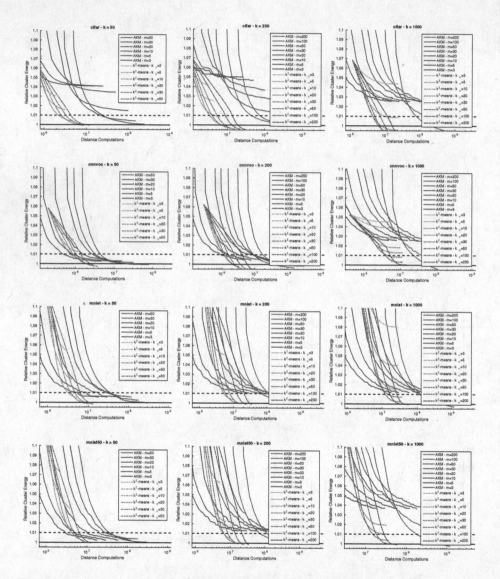

Fig. 3. Cluster Energy (relative to best Lloyd++ energy) vs distance computations on cifar, cnnvoc, mnist and mnist50 for $k \in \{50, 200, 1000\}$. For AKM and k^2-means, we use the parameter with the highest algorithmic speedup at 1% error.

faster than AKM for $k = 1000$ but AKM reaches the 1% reference faster for $k = 50$.

In all settings of Table 5, Elkan++ gives a consistent up to 8.5× speedup (since it is an exact acceleration of Lloyd++). For some settings Elkan is faster than Elkan++ in reaching the desired accuracy. This is due to the faster initialization. MiniBatch fails in all but one case (mnist, $k = 50$) to reach the reference error of 1% and is thus not shown. In 2/40 cases, we do not reach the 1% reference error - since the maximum k_n employed is $k_n = 200$.

For accurate clustering, when the reference energy is the Lloyd++ convergence energy (i.e. 0% error), Table 6 shows that the speedups of k^2-means are even higher. This is partially because in 87.5% of the cases (35/40) we obtain a lower energy than Lloyd++ since our proposed GDI initialization is comparable or better than k-means++ (see Table 4). For this setting, the second fastest method is Elkan++, which is designed for accelerating the exact Lloyd++.

4 Conclusions

We proposed k^2-means, a simple yet efficient method ideally suited for fast and accurate large scale clustering ($n > 10000$, $k > 10$, $d > 50$). k^2 means combines an efficient divisive initialization with a new method to speed up the k-means iterations by using the k_n nearest clusters as the new set of candidate centers for the cluster members as well as triangle inequalities. The algorithmic complexity of our k^2-means is sublinear in k for $n \gg k$ and experimentally shown to give a high accuracy on diverse datasets. For accurate clustering, k^2-means requires an order of magnitude fewer computations than alternative methods such as the fast approximate k-means (AKM) clustering. Moreover, our efficient divisive initialization leads to comparable clustering energies and significantly lower runtimes than the k-means++ initialization under the same conditions.

Acknowledgments. This work was supported by the ETH Zurich General Fund OK and by the Google 2017 Faculty Research Award.

References

1. Arthur, D., Vassilvitskii, S.: K-means++: the advantages of careful seeding. In: Proceedings of the Eighteenth Annual ACM-SIAM Symposium on Discrete Algorithms, pp. 1027–1035. Society for Industrial and Applied Mathematics (2007)
2. Bahmani, B., Moseley, B., Vattani, A., Kumar, R., Vassilvitskii, S.: Scalable K-means++. Proc. VLDB Endow. **5**(7), 622–633 (2012)
3. Blake, C., Keogh, E., Merz, C.: UCI repository of machine learning databases (1998). http://www.ics.uci.edu/mlearn/MLRepository.html
4. Boley, D.: Principal direction divisive partitioning. Data Min. Knowl. Disc. **2**(4), 325–344 (1998)
5. Caliński, T., Harabasz, J.: A dendrite method for cluster analysis. Commun. Stat.-Theory Methods **3**(1), 1–27 (1974)

6. Ding, Y., Zhao, Y., Shen, X., Musuvathi, M., Mytkowicz, T.: Yinyang K-means: a drop-in replacement of the classic K-means with consistent speedup. In: Proceedings of the 32nd International Conference on Machine Learning (ICML 2015), pp. 579–587 (2015)
7. Drake, J., Hamerly, G.: Accelerated K-means with adaptive distance bounds. In: Proceedings of the 5th NIPS Workshop on Optimization for Machine Learning (2012)
8. Elkan, C.: Using the triangle inequality to accelerate K-means. In: ICML, vol. 3, pp. 147–153 (2003)
9. Everingham, M., Van Gool, L., Williams, C.K., Winn, J., Zisserman, A.: The pascal visual object classes (VOC) challenge. Int. J. Comput. Vis. **88**(2), 303–338 (2010)
10. Georghiades, A., Belhumeur, P., Kriegman, D.: From few to many: illumination cone models for face recognition under variable lighting and pose. IEEE Trans. Pattern Anal. Mach. Intell. **23**(6), 643–660 (2001)
11. Hamerly, G.: Making K-means even faster. In: SDM, pp. 130–140. SIAM (2010)
12. Hull, J.J.: A database for handwritten text recognition research. IEEE Trans. Pattern Anal. Mach. Intell. **16**(5), 550–554 (1994)
13. Indyk, P., Motwani, R.: Approximate nearest neighbors: towards removing the curse of dimensionality. In: Proceedings of the Thirtieth Annual ACM Symposium on Theory of Computing, pp. 604–613. ACM (1998)
14. Kanungo, T., Mount, D.M., Netanyahu, N.S., Piatko, C.D., Silverman, R., Wu, A.Y.: A local search approximation algorithm for k-means clustering. In: Proceedings of the Eighteenth Annual Symposium on Computational Geometry, pp. 10–18. ACM (2002)
15. Krizhevsky, A., Hinton, G.: Learning multiple layers of features from tiny images. Masters thesis, Department of Computer Science, University of Toronto (2009)
16. Krizhevsky, A., Sutskever, I., Hinton, G.E.: Imagenet classification with deep convolutional neural networks. In: Advances in Neural Information Processing Systems, pp. 1097–1105 (2012)
17. LeCun, Y., Cortes, C., Burges, C.J.: The MNIST database of handwritten digits (1998)
18. Lee, K., Ho, J., Kriegman, D.: Acquiring linear subspaces for face recognition under variable lighting. IEEE Trans. Pattern Anal. Mach. Intell. **27**(5), 684–698 (2005)
19. Lloyd, S.P.: Least squares quantization in PCM. IEEE Trans. Inf. Theory **28**(2), 129–137 (1982)
20. Mazzeo, G.M., Masciari, E., Zaniolo, C.: A fast and accurate algorithm for unsupervised clustering around centroids. Inf. Sci. **400-401**, 63–90 (2017)
21. Muja, M., Lowe, D.G.: Fast approximate nearest neighbors with automatic algorithm configuration. In: VISAPP, vol. 1, p. 2 (2009)
22. Philbin, J., Chum, O., Isard, M., Sivic, J., Zisserman, A.: Object retrieval with large vocabularies and fast spatial matching. In: 2007 IEEE Conference on Computer Vision and Pattern Recognition, CVPR 2007, pp. 1–8. IEEE (2007)
23. Sculley, D.: Web-scale K-means clustering. In: Proceedings of the 19th International Conference on World Wide Web, pp. 1177–1178. ACM (2010)
24. Su, T., Dy, J.: A deterministic method for initializing K-means clustering. In: 2004 16th IEEE International Conference on Tools with Artificial Intelligence, ICTAI 2004, pp. 784–786. IEEE (2004)
25. Torralba, A., Fergus, R., Freeman, W.T.: 80 million tiny images: a large data set for nonparametric object and scene recognition. IEEE Trans. Pattern Anal. Mach. Intell. **30**(11), 1958–1970 (2008)

26. Wang, J., Wang, J., Ke, Q., Zeng, G., Li, S.: Fast approximate k-means via cluster closures. In: 2012 IEEE Conference on Computer Vision and Pattern Recognition (2012)
27. Wu, X., Kumar, V., Quinlan, J.R., Ghosh, J., Yang, Q., Motoda, H., McLachlan, G.J., Ng, A., Liu, B., Philip, S.Y., et al.: Top 10 algorithms in data mining. Knowl. Inf. Syst. **14**(1), 1–37 (2008)
28. Xu, Y., Qu, W., Li, Z., Min, G., Li, K., Liu, Z.: Efficient K-means++ approximation with mapreduce. IEEE Trans. Parallel Distrib. Syst. **25**(12), 3135–3144 (2014)
29. Zhao, W., Ma, H., He, Q.: Parallel K-means clustering based on mapreduce. In: Jaatun, M.G., Zhao, G., Rong, C. (eds.) CloudCom 2009. LNCS, vol. 5931, pp. 674–679. Springer, Heidelberg (2009). https://doi.org/10.1007/978-3-642-10665-1_71

A Simple Exponential Family Framework for Zero-Shot Learning

Vinay Kumar Verma and Piyush Rai[(✉)]

Department of Computer Science and Engineering, IIT Kanpur, Kanpur, India
{vkverma,piyush}@cse.iitk.ac.in

Abstract. We present a simple generative framework for learning to predict previously unseen classes, based on estimating *class-attribute-gated* class-conditional distributions. We model each class-conditional distribution as an exponential family distribution and the parameters of the distribution of each seen/unseen class are defined as functions of the respective observed class attributes. These functions can be learned using only the seen class data and can be used to predict the parameters of the class-conditional distribution of each unseen class. Unlike most existing methods for zero-shot learning that represent classes as fixed embeddings in some vector space, our generative model naturally represents each class as a probability distribution. It is simple to implement and also allows leveraging additional unlabeled data from unseen classes to improve the estimates of their class-conditional distributions using transductive/semi-supervised learning. Moreover, it extends seamlessly to few-shot learning by easily updating these distributions when provided with a small number of additional labelled examples from unseen classes. Through a comprehensive set of experiments on several benchmark data sets, we demonstrate the efficacy of our framework.

1 Introduction

The problem of learning to predict unseen classes, also popularly known as Zero-Shot Learning (ZSL), is an important learning paradigm which refers to the problem of recognizing objects from classes that were not seen at training time [13,26]. ZSL is especially relevant for learning "in-the-wild" scenarios, where new concepts need to be discovered on-the-fly, without having access to labelled data from the novel classes/concepts. This has led to a tremendous amount of interest in developing ZSL methods that can learn in a robust and scalable manner, even when the amount of supervision for the classes of interest is relatively scarce.

A large body of existing prior work for ZSL is based on embedding the data into a semantic vector space, where distance based methods can be applied to find the most likely class which itself is represented as a point in the same semantic space [20,26,33]. However, a limitation of these methods is that each class is represented as a fixed point in the embedding space which does not adequately

© Springer International Publishing AG 2017
M. Ceci et al. (Eds.): ECML PKDD 2017, Part II, LNAI 10535, pp. 792–808, 2017.
https://doi.org/10.1007/978-3-319-71246-8_48

account for intra-class variability [2,18]. We provide a more detailed overview of existing work on ZSL in the Related Work section.

Another key limitation of most of the existing methods is that they usually lack a proper generative model of the data. Having a generative model has several advantages [19]. For example, (1) data of different types can be modeled in a principled way using appropriately chosen class-conditional distributions; (2) unlabeled data can be seamlessly integrated (for both seen as well as unseen classes) during parameter estimation, leading to a transductive/semi-supervised estimation procedure, which may be useful when the amount of labeled data for the seen classes is small, or if the distributions of seen and unseen classes are different from each other [11]; and (3) a rich body of work, both frequentist and Bayesian, on learning generative models [19] can be brought to bear during the ZSL parameter estimation process.

Motivated by these desiderata, we present a generative framework for zero-shot learning. Our framework is based on modelling the class-conditional distributions of seen as well as unseen classes using exponential family distributions [3], and further conditioning the parameters of these distributions on the respective class-attribute vectors via a linear/nonlinear regression model of one's choice. The regression model allows us to predict the parameters of the class-conditional distributions of *unseen* classes using only their class attributes, enabling us to perform zero-shot learning.

In addition to the generality and modelling flexibility of our framework, another of its appealing aspects is its simplicity. In contrast with various other state-of-the-art methods, our framework is very simple to implement and easy to extend. In particular, as we will show, parameter estimation in our framework simply reduces to solving a linear/nonlinear regression problem, for which a closed-form solution exists. Moreover, extending our framework to incorporate unlabeled data from the unseen classes, or a small number of labelled examples from the unseen classes, i.e., performing few-shot learning [17,23] is also remarkably easy under our framework which models class-conditional distributions using exponential family distributions with conjugate priors.

2   A Generative Framework for ZSL

In zero-shot learning (ZSL) we assume there is a total of S seen classes and U unseen classes. Labelled training examples are only available for the seen classes. The test data is usually assumed to come only from the unseen classes, although in our experiments, we will also evaluate our model for the setting where the test data could come from both seen and unseen classes, a setting known as generalised zero-shot learning [6].

We take a generative modeling approach to the ZSL problem and model the class-conditional distribution for an observation x from a seen/unseen class c ($c = 1, \ldots, S + U$) using an exponential family distribution [3] with natural parameters $\boldsymbol{\theta}_c$

$$p(\boldsymbol{x}|\boldsymbol{\theta}_c) = h(\boldsymbol{x}) \exp \left(\boldsymbol{\theta}_c^\top \phi(\boldsymbol{x}) - A(\boldsymbol{\theta}_c) \right) \tag{1}$$

where $\phi(x)$ denotes the sufficient statistics and $A(\theta_c)$ denotes the log-partition function. We also assume that the distribution parameters θ_c are given conjugate priors

$$p(\theta_c|\tau_0, \nu_0) \propto \exp(\theta_c^\top \tau_0 - \nu_0 A(\theta_c)) \tag{2}$$

Given a test example x_*, its class y_* can be predicted by finding the class under which x_* is most likely (i.e., $y_* = \arg\max_c p(x_*|\theta_c)$), or finding the class that has the largest posterior probability given x_* (i.e., $y_* = \arg\max_c p(\theta_c|x_*)$). However, doing this requires first estimating the parameters $\{\theta_c\}_{c=S+1}^{S+U}$ of all the unseen classes.

Given labelled training data from any class modelled as an exponential family distribution, it is straightforward to estimate the model parameters θ_c using maximum likelihood estimation (MLE), maximum-a-posteriori (MAP) estimation, or using fully Bayesian inference [19]. However, since there are no labelled training examples from the unseen classes, we cannot estimate the parameters $\{\theta_c\}_{c=S+1}^{S+U}$ of the class-conditional distributions of the unseen classes.

To address this issue, we learn a model that allows us to predict the parameters θ_c for any class c using the *attribute vector* of that class via a *gating* scheme, which is basically defined as a linear/nonlinear regression model from the attribute vector to the parameters. As is the common practice in ZSL, the attribute vector of each class may be derived from a human-provide description of the class or may be obtained from an external source such as Wikipedia in form of word-embedding of each class. We assume that the class-attribute of each class is a vector of size K. The class-attribute of all the classes are denoted as $\{a_c\}_{c=1}^{S+U}$, $a_c \in \mathbb{R}^K$.

2.1 Gating via Class-Attributes

We assume a regression model from the class-attribute vector a_c to the parameters θ_c of each class c. In particular, we assume that the class-attribute vector a_c is mapped via a function f to generate the parameters θ_c of the class-conditional distribution of class c, as follows

$$\theta_c = f_\theta(a_c) \tag{3}$$

Note that the function f_θ itself could consist of multiple functions if θ_c consists of multiple parameters. For concereteness, and also to simplify the rest of the exposition, we will focus on the case when the class-conditional distribution is a D dimensional Gaussian, for which θ_c is defined by the mean vector $\mu_c \in \mathbb{R}^D$ and a p.s.d. covariance matrix $\Sigma_c \in \mathcal{S}_+^{D \times D}$. Further, we will assume Σ_c to be a diagonal matrix defined as $\Sigma_c = \mathrm{diag}(\sigma_c^2)$ where $\sigma_c^2 = [\sigma_{c1}^2, \ldots, \sigma_{cD}^2]$. Note that one can also assume a full covariance matrix but it will significantly increase the number of parameters to be estimated. We model μ_c and σ_c^2 as functions of the attribute vector a_c

$$\mu_c = f_\mu(a_c) \tag{4}$$

$$\sigma_c^2 = f_{\sigma^2}(a_c) \tag{5}$$

Note that the above equations define two regression models. The first regression model defined by the function f_μ has a_c as the input and μ_c as the output. The second regression model defined by f_{σ^2} has a_c as the input and σ^2 as the output. The goal is to learn the functions f_μ and f_{σ^2} from the available training data. Note that the form of these functions is a modelling choice and can be chosen appropriately. We will consider both linear as well as nonlinear functions.

2.2   Learning the Regression Functions

Using the available training data from all the seen classes $c = 1, \ldots, S$, we can form empirical estimates of the parameters $\{\hat{\mu}_c, \hat{\sigma}_c^2\}_{c=1}^S$ of respective class-conditional distributions using MLE/MAP estimation. Note that, since our framework is generative, both labeled as well as unlabeled data from the seen classes can be used to form the empirical estimates $\{\hat{\mu}_c, \hat{\sigma}_c^2\}_{c=1}^S$. This makes our estimates of $\{\hat{\mu}_c, \hat{\sigma}_c^2\}_{c=1}^S$ reliable even if each seen class has very small number of labeled examples. Given these estimates for the seen classes

$$\hat{\mu}_c = f_\mu(a_c) \qquad c = 1, \ldots, S \tag{6}$$

$$\hat{\sigma}_c^2 = f_{\sigma^2}(a_c) \qquad c = 1, \ldots, S \tag{7}$$

We can now learn f_μ using "training" data $\{a_c, \hat{\mu}_c\}_{c=1}^S$ and learn f_{σ^2} using training data $\{a_c, \hat{\sigma}^2{}_c\}_{c=1}^S$. We consider both linear and nonlinear regression models for learning these.

The Linear Model. For the linear model, we assume $\hat{\mu}_c$ and $\hat{\sigma}_c^2$ to be linear functions of the class-attribute vector a_c, defined as

$$\hat{\mu}_c = \mathbf{W}_\mu a_c \qquad c = 1, \ldots, S \tag{8}$$

$$\hat{\rho}_c = \log \hat{\sigma}_c^2 = \mathbf{W}_{\sigma^2} a_c \qquad c = 1, \ldots, S \tag{9}$$

where the regression weights $\mathbf{W}_\mu \in \mathbb{R}^{D \times K}$, $\mathbf{W}_{\sigma^2} \in \mathbb{R}^{D \times K}$, and we have re-parameterized $\hat{\sigma}_c^2 \in \mathbb{R}_+^D$ to $\hat{\rho}_c \in \mathbb{R}^D$ as $\hat{\rho}_c = \log \hat{\sigma}_c^2$.

We use this re-parameterization to map the output space of the second regression model f_{σ^2} (defined by $\mathbf{W}_{\sigma^2}$) to real-valued vectors, so that a standard regression model can be applied (note that $\hat{\sigma}_c^2$ is positive-valued vector).

Estimating Regression Weights of Linear Model: We will denote $\mathbf{M} = [\hat{\mu}_1, \ldots, \hat{\mu}_S] \in \mathbb{R}^{D \times S}$, $\mathbf{R} = [\hat{\rho}_1, \ldots, \hat{\rho}_S] \in \mathbb{R}^{D \times S}$, and $\mathbf{A} = [a_1, \ldots, a_S] \in \mathbb{R}^{K \times S}$. We can then write the estimation of the regression weights $\mathbf{W}_\mu$ as the following problem

$$\hat{\mathbf{W}}_\mu = \arg\min_{\mathbf{W}_\mu} ||\mathbf{M} - \mathbf{W}_\mu \mathbf{A}||_2^2 + \lambda_\mu ||\mathbf{W}_\mu||_2^2 \tag{10}$$

This is essentially a multi-output regression [7] problem $\mathbf{W}_\mu : a_s \mapsto \hat{\mu}_s$ with least squares loss and an ℓ_2 regularizer. The solution to this problem is given by

$$\hat{\mathbf{W}}_\mu = \mathbf{M}\mathbf{A}^\top(\mathbf{A}\mathbf{A}^\top + \lambda_\mu \mathbf{I}_K)^{-1} \tag{11}$$

Likewise, we can then write the estimation of the regression weights $\mathbf{W}_{\sigma^2}$ as the following problem

$$\hat{\mathbf{W}}_{\sigma^2} = \arg\min_{\mathbf{W}_{\sigma^2}} ||\mathbf{R} - \mathbf{W}_{\sigma^2}\mathbf{A}||_2^2 + \lambda_{\sigma^2}||\mathbf{W}_{\sigma^2}||_2^2 \qquad (12)$$

The solution of the above problem is given by

$$\hat{\mathbf{W}}_{\sigma^2} = \mathbf{R}\mathbf{A}^\top(\mathbf{A}\mathbf{A}^\top + \lambda_{\sigma^2}\mathbf{I}_K)^{-1} \qquad (13)$$

Given $\hat{\mathbf{W}}_\mu$ and $\hat{\mathbf{W}}_{\sigma^2}$, parameters of the class-conditional distribution of each unseen class $c = S+1, \ldots, S+U$ can be easily computed as follows

$$\hat{\boldsymbol{\mu}}_c = \hat{\mathbf{W}}_\mu \boldsymbol{a}_c \qquad (14)$$

$$\hat{\boldsymbol{\sigma}}_c^2 = \exp(\hat{\boldsymbol{\rho}}_c) = \exp(\hat{\mathbf{W}}_{\sigma^2}\boldsymbol{a}_c) \qquad (15)$$

The Nonlinear Model: For the nonlinear case, we assume that the inputs $\{\boldsymbol{a}_c\}_{c=1}^S$ are mapped to a kernel induced space via a kernel function k with an associated nonlinear mapping ϕ. In this case, using the representer theorem [24], the solution for the two regression models f_μ and f_{σ^2} can be written as the spans of the inputs $\{\phi(\boldsymbol{a}_c)\}_{c=1}^S$. Note that mappings $\phi(\boldsymbol{a}_c)$ do not have to be computed explicitly since learning the nonlinear regression model only requires dot products $\phi(\boldsymbol{a}_c)^\top\phi(\boldsymbol{a}_{c'}) = k(\boldsymbol{a}_c, \boldsymbol{a}_{c'})$ between the nonlinear mappings of two classes c and c'.

Estimating Regression Weights of Nonlinear Model: Denoting $\mathbf{K}$ to be the $S \times S$ kernel matrix of the pairwise similarities of the attributes of the seen classes, the nonlinear model f_μ is obtained by

$$\hat{\boldsymbol{\alpha}}_\mu = \arg\min_{\boldsymbol{\alpha}_\mu} ||\mathbf{M} - \boldsymbol{\alpha}_\mu\mathbf{K}||_2^2 + \lambda_\mu||\boldsymbol{\alpha}_\mu||_2^2 \qquad (16)$$

where $\hat{\boldsymbol{\alpha}}_\mu$ is a $D \times S$ matrix consists of the coefficients of the span of $\{\phi(\boldsymbol{a}_c)\}_{c=1}^S$ defining the nonlinear function f_μ.

Note that the problem in Eq. 16 is essentially a multi-output *kernel* ridge regression [7] problem, which has a closed form solution. The solution for $\hat{\boldsymbol{\alpha}}_\mu$ is given by

$$\hat{\boldsymbol{\alpha}}_\mu = \mathbf{M}(\mathbf{K} + \lambda_\mu\mathbf{I}_S)^{-1} \qquad (17)$$

Likewise, the nonlinear model f_{σ^2} is obtained by solving

$$\hat{\boldsymbol{\alpha}}_{\sigma^2} = \arg\min_{\boldsymbol{\alpha}_{\sigma^2}} ||\mathbf{M} - \boldsymbol{\alpha}_{\sigma^2}\mathbf{K}||_2^2 + \lambda_{\sigma^2}||\boldsymbol{\alpha}_{\sigma^2}||_2^2 \qquad (18)$$

where $\hat{\boldsymbol{\alpha}}_{\sigma^2}$ is a $D \times S$ matrix consists of the coefficients of the span of $\{\phi(\boldsymbol{a}_c)\}_{c=1}^S$ defining the nonlinear function f_{σ^2}. The solution for $\hat{\boldsymbol{\alpha}}_{\sigma^2}$ is given by

$$\hat{\boldsymbol{\alpha}}_{\sigma^2} = \mathbf{R}(\mathbf{K} + \lambda_\mu\mathbf{I}_S)^{-1} \qquad (19)$$

Given $\hat{\boldsymbol{\alpha}}_\mu$, $\hat{\boldsymbol{\alpha}}_{\sigma^2}$, parameters of class-conditional distribution of each unseen class $c = S + 1, \ldots, S + U$ will be

$$\hat{\boldsymbol{\mu}}_c = \hat{\boldsymbol{\alpha}}_\mu \boldsymbol{k}_c \tag{20}$$

$$\hat{\boldsymbol{\sigma}}_c^2 = \exp(\hat{\boldsymbol{\rho}}_c) = \exp(\hat{\boldsymbol{\alpha}}_{\sigma^2} \boldsymbol{k}_c) \tag{21}$$

where $\boldsymbol{k}_c = [k(\boldsymbol{a}_c, \boldsymbol{a}_1), \ldots, k(\boldsymbol{a}_c, \boldsymbol{a}_S)]^\top$ denotes an $S \times 1$ vector of kernel-based similarities of the class-attribute of unseen class c with the class-attributes of all the seen classes.

Other Exponential Family Distributions: Although we illustrated our framework taking the example of Gaussian class-conditional distributions, our framework readily generalizes to the case when these distributions are modelled using any exponential family distribution. The estimation problems can be solved in a similar way as the Gaussian case with the basic recipe remaining the same: Form empirical estimates of the parameters $\boldsymbol{\Theta} = \{\hat{\boldsymbol{\theta}}_c\}_{c=1}^S$ for the seen classes using all the available seen class data and then learn a linear/nonlinear regression model from the class-attributes $\mathbf{A}$ (or their kernel representation $\mathbf{K}$ in the nonlinear case) to $\boldsymbol{\Theta}$.

In additional to its modeling flexibility, an especially remarkable aspect of our generative framework is that it is very easy to implement, since both the linear model as well as the nonlinear model have closed-form solutions given by Eqs. 11, 13, 17 and 19, respectively (the solutions will be available in similar closed-forms in the case of other exponential family distributions). A block-diagram describing our framework is shown in Fig. 1. Note that another appealing aspect of our framework is its modular architecture where each of the blocks in Fig. 1 can make use of a suitable method of one's choice.

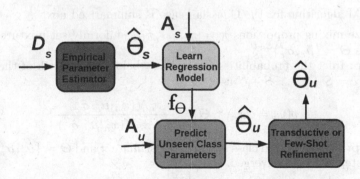

Fig. 1. Block-diagram of our framework. $\mathcal{D}_s$ denotes the seen class data (can be labeled (and optionally also unlabeled); $\mathbf{A}_s$ denotes seen class attributes; $\mathbf{A}_u$ denotes unseen class attributes; $\hat{\boldsymbol{\Theta}}_s$ denotes the estimated seen class parameters; $\hat{\boldsymbol{\Theta}}_u$ denotes the estimated unseen class parameters. The last stage - transductive/few-shot refinement - is optional (Sects. 2.3 and 4.2)

2.3 Transductive/Semi-supervised Setting

The procedure described in Sect. 2.2 relies only on the seen class data (labeled and, optionally, also unlabeled). As we saw for the Gaussian case, the seen class data is used to form empirical estimates of the parameters $\{\hat{\mu}_c, \hat{\sigma}^2_c\}^S_{c=1}$ of the class-conditional distributions of seen classes, and then these estimates are used to learn the linear/nonlinear regression functions f_μ and f_{σ^2}. These functions are finally used to compute the parameters $\{\hat{\mu}_c, \hat{\sigma}^2_c\}^{S+U}_{c=S+1}$ of class-conditionals of unseen classes. We call this setting the *inductive* setting. Note that this procedure does not make use of any data from the unseen classes. Sometimes, we may have access to unlabeled data from the unseen classes.

Our generative framework makes it easy to leverage such *unlabeled* data from the *unseen* classes to further improve upon the estimates $\{\hat{\mu}_c, \hat{\sigma}^2_c\}^{S+U}_{c=S+1}$ of their class-conditional distributions. In our framework, this can be done in two settings, *transductive* and *semi-supervised*, both of which leverage unlabeled data from unseen classes, but in slightly different ways. If the unlabeled data is the unseen class test data itself, we call it the *transductive* setting. If this unlabeled data from the unseen classes is different from the actual unseen class test data, we call it the *semi-supervised* setting.

In either setting, we can use an Expectation-Maximization (EM) based procedure that alternates between inferring the labels of unlabeled examples of unseen classes and using the inferred labels to update the estimates of the parameters $\{\hat{\mu}_c, \hat{\sigma}^2_c\}^{S+U}_{c=S+1}$ of the distributions of unseen classes.

For the case when each class-conditional distribution is a Gaussian, this procedure is equivalent to estimating a Gaussian Mixture Model (GMM) using the unlabeled data $\{x_n\}^{N_u}_{n=1}$ from the unseen classes. The GMM is initialized using the estimates $\{\hat{\mu}_c, \hat{\sigma}^2_c\}^{S+U}_{c=S+1}$ obtained from the inductive procedure of Sect. 2.2. Note that each of the U mixture components of this GMM corresonds to an unseen class.

The EM algorithm for the Gaussian case is summarized next

1. Initialize mixing proportions $\pi = [\pi_1, \ldots, \pi_U]$ uniformly set mixture parameters as $\Theta = \{\hat{\mu}_c, \hat{\sigma}^2_c\}^{S+U}_{c=S+1}$
2. **E Step:** Infer the probabilities for each x_n belonging to each of the unseen classes $c = S+1, \ldots, S+U$ as

$$p(y_n = c | x_n, \pi, \Theta) = \frac{\pi_c \mathcal{N}(x_n | \hat{\mu}_c, \hat{\sigma}^2_c)}{\sum_c \pi_c \mathcal{N}(x_n | \hat{\mu}_c, \hat{\sigma}^2_c)}$$

3. **M Step:** Use to inferred class labels to re-estimate π and $\Theta = \{\hat{\mu}_c, \hat{\sigma}^2_c\}^{S+U}_{c=S+1}$.
4. Go to step 2 if not converged.

Note that the same procedure can be applied even when each class-conditional distribution is some exponential family distribution other than Gaussian. The E and M steps in the resulting mixture model are straightforward in that case as well. The E step will simply require the Gausian likelihood to be replaced by the corresponding exponential family distribution's likelihood. The M step will require doing MLE of the exponential family distribution's parameters, which has closed-form solutions.

2.4 Extension for Few-Shot Learning

In few-shot learning, we assume that a very small number of labeled examples may also be available for the unseen classes [17,23]. The generative aspect of our framework, along with the fact the the data distribution is an exponential family distribution with a conjugate prior on its parameters, makes it very convenient for our model to be extended to this setting. The outputs $\{\hat{\boldsymbol{\mu}}_c, \hat{\boldsymbol{\sigma}}_c^2\}_{c=S+1}^{S+U}$ of our generative zero-shot learning model can naturally serve as the hyper-parameters of a conjugate prior on parameters of class-conditional distributions of unseen classes, which can then be updated given a small number of labeled examples from the unseen classes. For example, in the Gaussian case, due to conjugacy, we are able to update the estimates $\{\hat{\boldsymbol{\mu}}_c, \hat{\sigma}_c^2\}_{c=S+1}^{S+U}$ in a straightforward manner when provided with such labeled data. In particular, given a small number of labeled examples $\{\boldsymbol{x}_n\}_{n=1}^{N_c}$ from an unseen class c, $\hat{\boldsymbol{\mu}}_c$ and $\hat{\sigma}_c^2$ can be easily updated as

$$\boldsymbol{\mu}_c^{(FS)} = \frac{\hat{\boldsymbol{\mu}}_c + \sum_{n=1}^{N_c} \boldsymbol{x}_n}{1 + N_c} \tag{22}$$

$$\sigma_c^{2(FS)} = \left(\frac{1}{\hat{\sigma}_c^2} + \frac{N_c}{\sigma^2} \right)^{-1} \tag{23}$$

where $\sigma^2 = \frac{1}{N_c} \sum_{n=1}^{N_c} (\boldsymbol{x}_n - \hat{\boldsymbol{\mu}}_c)^2$ denotes the empirical variance of the N_c observations from the unseen class c.

A particularly appealing aspect of our few-shot learning model outlined above is that it can also be updated in an online manner as more and more labelled examples become available from the unseen classes, without having to re-train the model from scratch using all the data.

3 Related Work

Some of the earliest works on ZSL are based on predicting attributes for each example [13]. This was followed by a related line of work based on models that assume that the data from each class can be mapped to the class-attribute space (a shared semantic space) in which each seen/unseen class is also represented as a point [1,26,33]. The mapping can be learned using various ways, such as linear models or feed forward neural networks or convolutional neural networks. Predicting the label for a novel unseen class example then involves mapping it to this space and finding the "closest" unseen class. Some of the work on ZSL is aimed at improving the semantic embeddings of concepts/classes. For example, [29] proposed a ZSL model to incorporate relational information about concepts. In another recent work, [4] proposed a model to improve the semantic embeddings using a metric learning formulation. A complementary line of work to the semantic embedding methods is based on a "reverse" mapping, i.e., mapping the class-attribute to the observed feature space [32,37].

In contrast to such semantic embedding methods that assume that the classes are collapsed onto a single point, our framework offers considerably more flexibility by modelling each class using its own distribution. This makes our model more suitable for capturing the intra-class variability, which the simple point-based embedding models are incapable of handling.

Another popular approach for ZSL is based on modelling each unseen class as a linear/convex combination of seen classes [20] or of a set of "abstract" or "basis" classes [5,22]. The latter class of methods, in particular, can be seen as a special case of our framework since, for our linear model, we can view the columns of the $D \times K$ regression weights as representing a set of K basis classes. Note however that our model has such regression weights for each parameter of the class-conditional distribution, allowing it to be considerably more flexible. Moreover, our framework is also significantly different in other ways due to its fully generative framework, due to its ability to incorporate unlabeled data, performing few-shot learning, and its ability to model different types of data using an appropriate exponential family distribution.

A very important issue in ZSL is the *domain shift* problem which may arise if the seen and unseen class come from very different domains. In these situations, standard ZSL models tend to perform badly. This can be somewhat alleviated using some additional unlabeled data from the unseen classes. To this end, [11] provide a dictionary learning based approach for learning unseen class classifiers in which the dictionary is adapted to the unseen class domain. The dictionary adaptation is facilitated using unlabeled data from the unseen classes. In another related work, [8] leverage unlabeled data in a transductive ZSL framework to handle the domain shift problem. Note that our framework is robust to the domain shift problem due to its ability to incorporate unlabeled data from the unseen classes (the transductive setting). Our experimental results corroborate this.

Semi-supervised learning for ZSL can also be used to improve the semantic embedding based methods. [16] provide a semi-supervised method that leverages prior knowledge for improving the learned embeddings. In another recent work, [37] present a model to incorporate unlabeled unseen class data in a setting where each unseen class is represented as a linear combination of seen classes. [34] provide another approach, motivated by applications in computer vision, that jointly facilitates the domain adaptation of attribute space and the visual space. Another semi-supervised approach presented in [15] combines a semisupervised classification model over the observed classes with an unsupervised clustering model over unseen classes together to address the zero-shot multi-class classification.

In contrast to these models for which the mechanism for incorporating unlabeled data is model-specific, our framework offers a general approach for doing this, while also being simple to implement. Moreover, for large-scale problems, it can also leverage more efficient solvers (e.g., gradient methods) for estimating the regression coefficients associated with class-conditional distributions.

4   Experiments

We evaluate our generative framework for zero-shot learning (hereafter referred to as **GFZSL**) on several benchmark data sets and compare it with a number of state-of-the-art baselines. We conduct our experiments on various problem settings, including standard *inductive* zero-shot learning (only using seen class labeled examples), *transductive* zero-shot learning (using seen class labeled examples and unseen class unlabeled examples), and few-shot learning (using seen class labeled examples and a very small number of unseen class labeled examples). We report our experimental results on the following benchmark data sets:

- **Animal with Attribute (AwA):** The AwA data set contains 30475 images with 40 seen classes (training set) and 10 unseen classes (test set). Each class has a human-provided binary/continuous 85-dimensional class-attribute vector [12]. We use continuous class-attributes since prior works have found these to have more discriminative power.
- **Caltech-UCSD Birds-200-2011 (CUB-200):** The CUB-200 data set contains 11788 images with 150 seen classes (training set) and 50 unseen class (test set). Each image has a binary 312-dimensional class-attribute vector, specifying the presence or absence of various attribute of that image [28]. The attribute vectors for all images in a class are averaged to construct its continuous class-attribute vector [2]. We use the same train/test split for this data set as used in [2].
- **SUN attribute (SUN):** The SUN data set contains 14340 images with 707 seen classes (training set) and 10 unseen classes (test set). Each image is described by a 102-dimensional binary class-attribute vector. Just like the CUB-200 data set, we average the attribute vectors of all images in each class to get its continuous attribute vector [10]. We use the same train/test split for this data set as used in [10].

For image features, we considered both GoogleNet features [27] and VGG-19(4096) fc7 features [25] and found that our approach works better with VGG-19. All of the state-of-the-art baselines we compare with in our experiments use VGG-19 fc7 features or GoogleNet features [27]. For the nonlinear (kernel) variant of our model, we use a quadratic kernel. Our set of experiments include:

- **Zero-Shot Learning:** We consider both inductive ZSL as well as transductive ZSL.
  - **Inductive ZSL:** This is the standard ZSL setting where the unseen class parameters are learned using only seen class data.
  - **Transductive ZSL:** In this setting [34], we also use the unlabeled test data while learning the unseen class parameters. Note that this setting has access to more information about the unseen class; however, it is only through unlabeled data.
- **Few-Shot Learning:** In this setting [17,23], we also use a small number of labelled examples from each unseen class.

– **Generalized ZSL:** Whereas standard ZSL (as well as few-shot learning) assumes that the test data can only be from the unseen classes, generalized ZSL assumes that the test data can be from unseen as well as seen classes. This is usually a more challenging setting [6] and most of the existing methods are known to be biased towards predicting the seen classes.

We use the standard train/test split as given in the data description section. For selecting the hyperparameters, we further divide the train set further into train and validation set. In our model, we have two hyper-parameter λ_μ and λ_{σ^2}, which we tune using the validation dataset. For AwA, from the 40 seen classes, a random selection of 30 classes are used for the training set and 10 classes are used for the validation set. For CUB-200, from the 150 seen classes, 100 are used for the training set and rest 50 are used for the validation set. Similarly, for the SUN dataset from the 707 seen classes, 697 are used for the training set and rest 10 is used for the validation set. We use cross-validation on the validation set to choose the best hyperparameter $[\lambda_\mu, \lambda_{\sigma^2}]$ for the each data set and use these for testing on the unseen classes.

4.1 Zero-Shot Learning

In our first set of experiments, we evaluate our model for zero-shot learning and compare with a number of state-of-the-art methods, for the inductive setting (which uses only the seen class labelled data) as well as the transductive setting (which uses the seen class data and the unseen class unlabeled data).

Inductive ZSL: Table 1 shows our results for the inductive ZSL setting. The results of the various baselines are taken from the corresponding papers. As shown in the Table 1, on CUB-200 and SUN, both of our models (linear and nonlinear) perform better than all of the other state-of-the-art methods. On AwA, our model has only a marginally lower test accuracy as compared to the best performing baseline [34]. However, we also have an average improvement 5.67% on all the 3 data sets as compared to the overall best baseline [34]. Among baselines using VGG-19 features (bottom half of Table 1), our model achieves a 21.05% relative improvement over the best baseline on the CUB-200 data, which is considered to be a difficult data set with many fine-grained classes.

In contrast to other models that embed the test examples in the semantic space and then find the most similar class by doing a Euclidean distance based nearest neighbor search, or models that are based on computing the similarity scores between seen and unseen classes [33], for our models, finding the "most probable class" corresponds to computing the distance of each test example from a *distribution*. This naturally takes into account the shape and spread of the class-conditional distribution. This explains the favourable performance of our model as compared to the other methods.

Table 1. Accuracy (%) of different type of images features. Top: Deep features like AlexNet, GoogleNet, etc. Bottom: Deep VGG-19 features. The '-' indicates that this result was not reported.

Method	AwA	CUB-200	SUN	Average
Akata et al. [2]	66.70	50.1	–	–
Qiao et al. [21]	66.46 ± 0.42	29 ± 0.28	–	–
Xian et al. [31]	71.9	45.5	–	–
Changpimyo et al. [5]	72.9	54.7	62.7	63.43
Wang et al. [29]	75.99	33.48	–	–
Lampert et al. [14]	57.23	–	72.00	–
Romera and Torr [22]	75.32 ± 2.28	–	82.10 ± 0.32	–
Bucher et al. [4]	77.32 ± 1.03	43.29 ± 0.38	84.41 ± 0.71	68.34
Zhang and Saligrama [35]	79.12 ± 0.53	41.78 ± 0.52	83.83 ± .29	68.24
Wang and Chen [30]	79.2 ± 0.0	46.7 ± 0.0	–	–
Zhang and Saligrama [34]	**81.03 ± 0.88**	46.48 ± 1.67	84.10 ± 1.51	70.53
GFZSL: Linear	79.90	52.09	86.50	72.23
GFZSL: Nonlinear	80.83	**56.53**	**86.50**	**74.59**

Transductive Setting: For transductive ZSL setting [9,35,36], we follow the procedure described in Sect. 2.3 to estimate parameters of the class-conditional distribution of each unseen class. After learning the parameters, we find the most probable class for each test example by evaluating its probability under each unseen class distribution and assign it to the class under which it has the largest probability. Tables 2 and 3 compare our results from the transductive setting with other state-of-the-art baselines designed for the transductive setting. In addition to accuracy, we also report precision and recall results of our model and the other baselines (wherever available). As we can see from Tables 2 and 3, both of our models (linear and kernel) outperform the other baselines on all the 3 data sets. Also comparing with the inductive setting results presented in Table 1, we observe that our generative framework is able to very effectively leverage unlabeled data and significantly improve upon the results of a purely inductive setting.

4.2 Few-Shot Learning (FSL)

We next perform an experiment with the few-shot learning setting [17,23] where we provide each model with a small number of labelled examples from each of the unseen classes. For this experiment, we follow the procedure described in Sect. 4.2 to learn the parameters of the class-conditional distributions of the unseen classes. In particular, we train the inductive ZSL model (using only the seen class training data) and the refine the learned model further using a very small number of labelled examples from the unseen classes (i.e., the few-shot learning setting).

Table 2. ZSL accuracy (%) obtained in the transductive setting: results reported using the VGG-19 feature. Average Precision and recall for the all dataset with its standard deviation over the 100 iteration. The '-' indicates that this result was not reported in the original paper.

Method	AwA	CUB-200	SUN	Average
Guo et al. [9]	78.47	–	82.00	–
Romera and Torr [22] + Zhang and Saligrama [36]	84.30	–	37.50	–
Zhang and Saligrama [35] + Zhang and Saligrama [36]	92.08 ± 0.14	55.34 ± 0.77	86.12 ± 0.99	77.85
Zhang and Saligrama [34] + Zhang and Saligrama [36]	88.04 ± 0.69	55.81 ± 1.37	85.35 ± 1.56	76.40
GFZSL: Linear	94.20	57.14	87.00	79.45
GFZSL: Kernel	**94.25**	**63.66**	**87.00**	**80.63**

Table 3. ZSL precision and recall scores obtained in the transductive setting: results reported using the VGG-19 features. Average precision and recall for the all dataset with its standard deviation over the 100 iteration. Note: Precision and recall scores not available for Guo et al. [9] and Romera et al. [22] + Zhang et al. [36]

	Average Precision			Average Recall		
Method	AwA	CUB-200	SUN	AwA	CUB-200	SUN
Zhang et al.[35]+Zhang et al. [36]	91.37 ± 14.75	57.09 ± 27.91	85.96 ± 10.15	90.28 ± 8.08	55.73 ± 31.80	86.00 ± 13.19
Zhang et al.[34]+Zhang et al. [36]	89.19 ± 15.09	57.20 ± 25.96	86.06 ± 12.36	86.04 ± 9.82	55.77 ± 26.54	85.50 ± 13.68
GFZSL: Linear	93.70	57.90	87.40	92.20	57.40	87.00
GFZSL: Kernel	**93.80**	**64.09**	**87.40**	**92.30**	**63.96**	**0.87**

To see the effect of knowledge transfer from the seen classes, we use a multiclass SVM as a baseline that is provided with the same number of labelled examples from each unseen class. In this experiment, we vary the number of labelled examples of unseen classes from 2 to 20 (for SUN we only use 2, 5, and 10 due to the small number of labelled examples). In Fig. 2, we also compare with standard (inductive) ZSL which does not have access to the labelled examples from the unseen classes. Our results are shown Table 4 and Fig. 2.

As shown in Table 4 (all data sets) and Fig. 2, the classification accuracy on the unseen classes shows a significant improvement over the standard inductive ZSL, even with as few as 2 or 5 additional labelled examples per class. We also observe that the few-shot learning method outperform multiclass SVM which only relies on the labelled data from the unseen classes. This demonstrates the advantage of the knowledge transfer from the seen class data.

Table 4. Accuracy (%) in the few-shot learning setting: For each data set, the accuracies are reported using $2, 5, 10, 15, 20$ labeled examples for each unseen class

Dataset	Method	2	5	10	15	20
AwA	GFZSL	87.96 ± 1.47	91.64 ± 0.81	93.31 ± 0.50	$94.01 \pm .36$	94.30 ± 0.33
	SVM	74.81	83.19	90.44	91.22	92.04
CUB-200	GFZSL	60.84 ± 1.39	64.81 ± 1.14	68.44 ± 1.21	70.11 ± 0.93	71.23 ± 0.87
	SVM	46.19	59.33	68.75	73.87	75.42
SUN	GFZSL	75.57 ± 4.79	83.05 ± 3.60	82.09 ± 3.30	–	–
	SVM	56.00	77.00	78.00	–	–

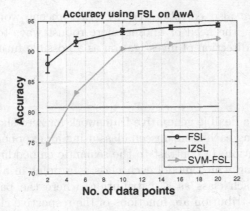

Fig. 2. (On AwA data): a comparison on classification accuracies of the few-shot learning variant of our model with multi-class SVM (training on labeled examples from seen classes) and the inductive ZSL

4.3 Generalized Few-Shot Learning (GFSL)

We finally perform an experiment on the more challenging generalized few-shot learning setting [6]. This setting assumes that test examples can come from seen as well as unseen classes. This setting is known to be notoriously hard [6]. In this setting, although the ZSL models tend to do well on predicting test examples from seen classes, the performance on correctly predicting the unseen class example is poor [6] since the trained models are heavily biased towards predicting the seen classes.

One way to mitigate this issue could be to use some labelled examples from the unseen classes (akin to what is done in few-shot learning). We, therefore, perform a similar experiment as in Sect. 4.2. In Table 5, we show the results of our model on classifying the unseen class test examples in this setting.

As shown in Table 5, our model's accuracies on the generalized FSL task improve as it gets to see labelled examples from unseen classes. However, it is still outperformed by a standard multiclass SVM. The better performance of SVM can be attributed to the fact that it is not biased towards the seen classes since the classifier for each class (seen/unseen) is learned independently.

Table 5. Accuracies (%) in the generalized few-shot learning setting.

Dataset	Method	2	5	10	15	20
AwA	GFZSL	25.32 ± 2.43	37.42 ± 1.60	43.20 ± 1.39	45.09 ± 1.17	45.96 ± 1.09
	SVM	40.84	60.81	75.36	77.00	77.10
CUB-200	GFZSL	6.64 ± 0.87	15.12 ± 1.17	22.02 ± 0.76	25.03 ± 0.71	26.47 ± 0.83
	SVM	25.97	37.98	47.10	53.87	54.42
SUN	GFZSL	1.17 ± 1.16	4.20 ± 1.77	9.48 ± 2.22	–	–
	SVM	9.94	20.00	27.00	–	–

Our findings are also corroborated by other recent work on generalized FSL [6] and suggest the need of finding more robust ways to handle this setting. We leave this direction of investigation as a possible future work.

5   Conclusion

We have presented a flexible generative framework for zero-shot learning, which is based on modelling each seen/unseen class using an exponential family class-conditional distribution. In contrast to the semantic embedding based methods for zero-shot learning which model each class as a point in a latent space, our approach models each class as a distribution, where the parameters of each class-conditional distribution are functions of the respective class-attribute vectors. Our generative framework allows learning these functions easily using seen class training data (and optionally leveraging additional unlabeled data from seen/unseen classes).

An especially appealing aspect of our framework is its simplicity and modular architecture (cf., Fig. 1) which allows using a variety of algorithms for each of its building blocks. As we showed, our generative framework admits natural extensions to other related problems, such as transductive zero-shot learning and few-shot learning. It is particularly easy to implement and scale to a large number of classes, using advances in large-scale regression. Our generative framework can also be extended to jointly learn the class attributes from an external source of data (e.g., by learning an additional embedding model with our original model). This can be an interesting direction of future work. Finally, although we considered a point estimation of the parameters of class-conditional distributions, it is also possible to take a fully Bayesian approach for learning these distributions. We leave this possibility as a direction for future work.

Acknowledgements. This work is supported by a grant from Tower Research CSR, Dr. Deep Singh and Daljeet Kaur Fellowship, and Research-I Foundation, IIT Kanpur. Vinay Verma acknowledges support from Visvesvaraya Ph.D. fellowship.

References

1. Akata, Z., Perronnin, F., Harchaoui, Z., Schmid, C.: Label-embedding for attribute-based classification. In: CVPR (2013)
2. Akata, Z., Reed, S., Walter, D., Lee, H., Schiele, B.: Evaluation of output embeddings for fine-grained image classification. In: CVPR (2015)
3. Brown, L.D.: Fundamentals of statistical exponential families. Institute of Mathematical Statistics (1986)
4. Bucher, M., Herbin, S., Jurie, F.: Improving semantic embedding consistency by metric learning for zero-shot classification. arXiv preprint arXiv:1607.08085 (2016)
5. Changpinyo, S., Chao, W.-L., Gong, B., Sha, F.: Synthesized classifiers for zero-shot learning. In: CVPR (2016)

6. Chao, W.-L., Changpinyo, S., Gong, B., Sha, F.: An empirical study and analysis of generalized zero-shot learning for object recognition in the wild. In: Leibe, B., Matas, J., Sebe, N., Welling, M. (eds.) ECCV 2016. LNCS, vol. 9906, pp. 52–68. Springer, Cham (2016). https://doi.org/10.1007/978-3-319-46475-6_4
7. Friedman, J., Hastie, T., Tibshirani, R.: The Elements of Statistical Learning. Springer Series in Statistics, vol. 1. Springer, Berlin (2001). https://doi.org/10.1007/978-0-387-21606-5
8. Fu, Y., Hospedales, T.M., Xiang, T., Gong, S.: Transductive multi-view zero-shot learning. PAMI 37(11), 2332–2345 (2015)
9. Guo, Y., Ding, G., Jin, X., Wang, J.: Transductive zero-shot recognition via shared model space learning. In: AAAI (2016)
10. Jayaraman, D., Grauman, K.: Zero-shot recognition with unreliable attributes. In: NIPS (2014)
11. Kodirov, E., Xiang, T., Fu, Z., Gong, S.: Unsupervised domain adaptation for zero-shot learning. In: ICCV (2015)
12. Krizhevsky, A., Hinton, G.: Learning multiple layers of features from tiny images (2009)
13. Lampert, C.H., Nickisch, H., Harmeling, S.: Learning to detect unseen object classes by between-class attribute transfer. In: CVPR (2009)
14. Lampert, C.H., Nickisch, H., Harmeling, S.: Attribute-based classification for zero-shot visual object categorization. PAMI 36(3), 453–465 (2014)
15. Li, X., Guo, Y.: Max-margin zero-shot learning for multi-class classification. In: AISTATS (2015)
16. Li, X., Guo, Y., Schuurmans, D.: Semi-supervised zero-shot classification with label representation learning. In: CVPR (2015)
17. Mensink, T., Gavves, E., Snoek, C.G.: Costa: co-occurrence statistics for zero-shot classification. In: CVPR (2014)
18. Mukherjee, T., Hospedales, T.: Gaussian visual-linguistic embedding for zero-shot recognition. In: EMNLP (2016)
19. Murphy, K.P.: Machine Learning: A Probabilistic Perspective. MIT Press, Cambridge (2012)
20. Norouzi, M., Mikolov, T., Bengio, S., Singer, Y., Shlens, J., Frome, A., Corrado, G.S., Dean, J.: Zero-shot learning by convex combination of semantic embeddings. In: ICLR (2014)
21. Qiao, R., Liu, L., Shen, C., van den Hengel, A.: Less is more: zero-shot learning from online textual documents with noise suppression. In: CVPR (2016)
22. Romera-Paredes, B., Torr, P.: An embarrassingly simple approach to zero-shot learning. In: ICML (2015)
23. Salakhutdinov, R., Tenenbaum, J.B., Torralba, A.: Learning with hierarchical-deep models. PAMI 35(8), 1958–1971 (2013)
24. Schölkopf, B., Smola, A.J.: Learning with Kernels: Support Vector Machines, Regularization, Optimization, and Beyond. MIT Press, Cambridge (2001)
25. Simonyan, K., Zisserman, A.: Very deep convolutional networks for large-scale image recognition. arXiv preprint arXiv:1409.1556 (2014)
26. Socher, R., Ganjoo, M., Manning, C.D., Ng, A.: Zero-shot learning through cross-modal transfer. In: NIPS (2013)
27. Szegedy, C., Liu, W., Jia, Y., Sermanet, P., Reed, S., Anguelov, D., Erhan, D., Vanhoucke, V., Rabinovich, A.: Going deeper with convolutions. In: CVPR (2015)
28. Wah, C., Branson, S., Welinder, P., Perona, P., Belongie, S.: The Caltech-UCSD Birds-200-2011 Dataset (2011)

29. Wang, D., Li, Y., Lin, Y., Zhuang, Y.: Relational knowledge transfer for zero-shot learning. In: AAAI (2016)
30. Wang, Q., Chen, K.: Zero-shot visual recognition via bidirectional latent embedding. arXiv preprint arXiv:1607.02104 (2016)
31. Xian, Y., Akata, Z., Sharma, G., Nguyen, Q., Hein, M., Schiele, B.: Latent embeddings for zero-shot classification. In: CVPR (2016)
32. Zhang, L., Xiang, T., Gong, S.: Learning a deep embedding model for zero-shot learning. arXiv preprint arXiv:1611.05088 (2016)
33. Zhang, Z., Saligrama, V.: Zero-shot learning via semantic similarity embedding. In: ICCV (2015)
34. Zhang, Z., Saligrama, V.: Learning joint feature adaptation for zero-shot recognition. arXiv preprint arXiv:1611.07593 (2016)
35. Zhang, Z., Saligrama, V.: Zero-shot learning via joint latent similarity embedding. In: CVPR (2016)
36. Zhang, Z., Saligrama, V.: Zero-shot recognition via structured prediction. In: Leibe, B., Matas, J., Sebe, N., Welling, M. (eds.) ECCV 2016. LNCS, vol. 9911, pp. 533–548. Springer, Cham (2016). https://doi.org/10.1007/978-3-319-46478-7_33
37. Zhao, B., Wu, B., Wu, T., Wang, Y.: Zero-shot learning via revealing data distribution. arXiv preprint arXiv:1612.00560 (2016)

DeepCluster: A General Clustering Framework Based on Deep Learning

Kai Tian[1], Shuigeng Zhou[1(✉)], and Jihong Guan[2]

[1] Shanghai Key Lab of Intelligent Information Processing,
School of Computer Science, Fudan University, Shanghai 200433, China
{ktian14,sgzhou}@fudan.edu.cn
[2] Department of Computer Science and Technology, Tongji University,
Shanghai 201804, China
jhguan@tongji.edu.cn

Abstract. In this paper, we propose a general framework DeepCluster to integrate traditional clustering methods into deep learning (DL) models and adopt Alternating Direction of Multiplier Method (ADMM) to optimize it. While most existing DL based clustering techniques have separate feature learning (via DL) and clustering (with traditional clustering methods), DeepCluster simultaneously learns feature representation and does cluster assignment under the same framework. Furthermore, it is a general and flexible framework that can employ different networks and clustering methods. We demonstrate the effectiveness of DeepCluster by integrating two popular clustering methods: K-means and Gaussian Mixture Model (GMM) into deep networks. The experimental results shown that our method can achieve state-of-the-art performance on learning representation for clustering analysis. Code and data related to this chapter are available at: https://github.com/JennyQQL/DeepClusterADMM-Release.

1 Introduction

Clustering is one of the most important techniques for analyzing data in an unsupervised manner, it has a wide range of applications including computer vision [11,14,23], natural language processing [1,2,26] and bioinformatics [22,28]. In the past decades, a large number of algorithms have been proposed to handle clustering problems [6,15]. However, there is no algorithm that fits all problems. Clustering method choosing depends on the data to handle and the specific task. Roughly, there are two sets of approaches, the feature-based clustering algorithms and the similarity-based clustering algorithms. Most of them try to find the intrinsic data structure from the original feature space or the underlying subspace.

Among the existing algorithms, K-means [12] and Gaussian Mixture Models (GMM) [4] are two popular feature-based methods. K-means makes hard clustering that assigns each sample to its nearest cluster center. GMM assumes that data are generated from several independent Gaussian distributions and tries to

© Springer International Publishing AG 2017
M. Ceci et al. (Eds.): ECML PKDD 2017, Part II, LNAI 10535, pp. 809–825, 2017.
https://doi.org/10.1007/978-3-319-71246-8_49

infer these distributions from the data. Thus, it makes soft assignments. However, they both do clustering in the original feature space. Spectral clustering [15] is a representative algorithm of similarity-based clustering or subspace clustering methods. Most of those approaches start with building an affinity matrix and project the original data to a linear subspace. Finally, clustering is done in the subspace.

One problem with most feature-based clustering methods is that they cannot scale well to high-dimensional data due to the curse of dimensionality. In high-dimensional data analysis, it is more reasonable to consider some compact and representative features instead of the whole feature space. Recently, deep learning (DL) has been developed and with a great success in many areas, such as image classification and speech recognition [19]. DL aims to learn a powerful representation from the raw data through high-level non-linear mapping [3]. Recently, how to use deep representation to improve clustering performance becomes a hot research topic.

Basically, there are mainly two ways to use deep features for clustering. One is clustering the hidden features that are extracted from a well-trained deep network [21, 26]. However, these approaches cannot fully exploit the power of deep neural network for clustering. The other is to embed an existing clustering method into DL models, which is an end-to-end approach. For example, [18] integrates K-means algorithm into deep autoencoders and does cluster assignment on the middle layers. It alternatively updates the network parameters and cluster centers. [25] proposes a clustering objective to learn non-linear deep features, which minimizes the KL divergence between the auxiliary target distribution and the model assignments.

In this paper, we propose a new and general framework to integrate traditional clustering methods into deep learning models and develop an algorithm to optimize the underlying non-convex and non-linear objective based on Alternating Direction of Multiplier Method (ADMM) [5]. Concretely, we can use a deep autoencoder to reconstruct the data, and associate deep features with clustering methods by introducing a dummy variable (say $\mathbf{Y}$). We combine deep models and clustering methods with the constraint $\mathbf{Y} = f_{\theta_1}(\mathbf{X})$ where $\mathbf{X}$ is the data and $f_{\theta_1}(\cdot)$ is the encoder of deep network. In the optimization process, we optimize each part of the model's parameters alternatively. Our experimental results shown that instead of directly clustering the hidden features, our framework works better.

The novelties and contributions of our work are as follows:

- A general clustering framework base on DL, where clustering parameters (except for the network parameters) can be represented in closed form.
- Both network and clustering method are configurable according to user requirements, which make it a flexible framework.
- Based on ADMM, relaxation is introduced to the model by doing clustering on dummy variable $\mathbf{Y}$.
- Experiments on real datasets show that the new method can achieve state-of-the-art clustering performance.

2   Related Work

Clustering is an extensively studied area, and up to now many clustering methods have been developed. Here, we review mainly on the clustering methods that employ DL techniques, and briefly highlight the advantages/differences of our work over/from the most-related existing ones.

Among the popular clustering methods, K-means and GMM are widely used in many applications. However, they have two drawbacks: one is that they mainly work in the original feature space; the other is that they cannot handle large and high-dimensional data sets well.

Spectral clustering and its variants are extensively popular among subspace clustering methods. [13] develops a distributed framework to solve sparse subspace clustering via ADMM. However, it considers only linear subspaces. To address this problem, another approach [16] was proposed to incorporate non-linearity into subspace clustering. The objective is to minimize the data reconstruction error and add a sparsity prior to the deep features.

To make use of deep learning features, some works first train a network and then cluster the hidden features. One of them is to learn a deep autoencoder on a graph and then run K-means algorithm on the non-linear embedding to get cluster assignments [21]. [26] introduces a novel method for short text clustering by first training a Convolutional Neural Network (CNN). The target to train CNN is spectral hashing code, and after network training it extracts deep features on which K-means is run. These approaches have separate feature learning and clustering.

To conduct end-to-end clustering in deep networks, [18] proposes a model to simultaneously learn the deep representations and the cluster centers. It makes hard assignment to each sample and directly does clustering on the hidden features of deep autoencoder. A recent attempt is the Deep Embedding Clustering (DEC) method [25], which achieves stat-of-the-art results on many datasets, but it may fails when closely related clusters exist.

Different from the above works, our DeepCluster is a general DL based clustering framework that can embrace different clustering methods and network structures such as DAEs, CNNs and RNNs. It provides a flexible mechanism to fit a clustering method to a deep network for a specific clustering task. Concretely, the most-related existing methods are DAEC [18] and DEC [25].

Though DAEC is the first work to explore deep feature learning and clustering simultaneously, it does clustering directly on the feature space, which is not flexible. DEC uses only the encoder part of DAE, and assumes that the model can correctly make high confidence predictions, which if not satisfied, it performs badly. DeepCluster introduces a copy of features $\mathbf{Y}$ and does cluster assignment on it, which makes feature learning and clustering independent from each other given this $\mathbf{Y}$. On the one hand, DeepCluster is able to fully take advantages of deep representation for clustering; On the other hand, recent studies have shown that ADMM can be used to train deep neural network without backpropagation, which means that our DeepCluster is parallelizable and can be used in asynchronous systems [20]. In summary, DeepCluster provides the feasibility of combining the most suitable network and clustering method for a specific clustering task.

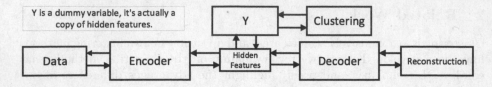

Fig. 1. The DeepCluster architecture.

3   Preliminaries

3.1   Deep Autoencoder (DAE)

A single autoencoder is a three layer neural network with one hidden layer, its output is to reconstruct the input x. An autoencoder is composed of an encoder and a decoder. The encoder can be formalized as

$$a_1 = f(W_1 x + b_1) \tag{1}$$

where W_1 is the weight and b_1 is the bias of encoder, a_1 means the hidden features of x. The decoder is formulated as

$$\hat{x} = g(W_2 a_1 + b_2) \tag{2}$$

where W_2 is the weight and b_2 is the bias of decoder, $\hat{x}$ is the reconstruction of input.

Deep autoencoder is to stack several autoencoders to build a deep neural network, where the hidden features learned by a lower level autoencoder are fed to a higher level autoencoder as input. The first layer autoencoder takes raw data as input.

3.2   Alternating Directed Method of Multipliers (ADMM)

Let us consider a general optimization problem as suggested in [7], which can be formulated as follows:

$$\min_{x \in \mathbb{R}^n} h(x) + o(Dx) \tag{3}$$

where D is an $m \times n$ matrix, which is often assumed to have full column rank. h and o are supposed to be convex functions on $\mathbb{R}^n$ and $\mathbb{R}^m$, respectively. To solve Eq. (3), we can rewrite it by introducing an additional dummy variable $z \in \mathbb{R}^m$:

$$\begin{aligned} \min h(x) + o(z) \\ s.t. \ \ Dx = z. \end{aligned} \tag{4}$$

This is a constrained convex problem that can be solved by the classical augmented Lagrangian algorithm (ALM). However, it is not decomposable due to the constraints, and the subproblems are unlikely to be easier to solve than the original one. The alternating directed method of multipliers (ADMM) is proposed to overcome the drawbacks of ALM. It is robust and supports decomposition.

To solve Eqs. (3) and (4), ADMM uses the following forms with a scalar parameter $\rho > 0$:

$$x^{k+1} \in \arg\min_{x \in \mathbb{R}^n} \left\{ h(x) + o(z^k) + \langle \lambda^k, Dx - z^k \rangle + \frac{\rho}{2} \parallel Dx - z^k \parallel^2 \right\}$$

$$z^{k+1} \in \arg\min_{z \in \mathbb{R}^m} \left\{ h(x^{k+1}) + o(z) + \langle \lambda^k, Dx^{k+1} - z \rangle + \frac{\rho}{2} \parallel Dx^{k+1} - z \parallel^2 \right\}$$

$$\lambda^{k+1} = \lambda^k + \rho(Dx^{k+1} - z^{k+1})$$

$$(5)$$

From Eq. (5), we can see that ADMM essentially decouples the functions h and o, and makes it possible to exploit the individual structures of h and o. Thus, the optimization procedure can be efficient and parallelizable.

Although the ADMM method was proposed to solve convex problems, many studies have shown that this approach can be used in non-convex cases, such as nonnegative matrix factorization [5] and network lasso [8].

4   The DeepCluster Framework

In this section, we give the architecture, formal formulation and algorithm of our DeepCluster framework.

Figure 1 is the architecture of DeepCluster. The encoder learns hidden feature representations and $\mathbf{Y}$ is a dummy variable that is required to be equal to the features. We do clustering on $\mathbf{Y}$ instead of the features. The constraint controls the interaction between autoencoder and the clustering method. During the clustering process, $\mathbf{Y}$ is adjusted to minimize the objective, meanwhile the constraint requires the encoder to learn better representations for clustering. This framework is flexible and robust, it can be seen as a multi-task learning model or the clustering part can be seen as a regularizer of deep autoencoder (DAE). By introducing a copy of deep features as $\mathbf{Y}$ and requiring it equal to the hidden features, we essentially introduce a kind of relaxation to the optimization procedure, which makes the framework decomposable to two components.

DeepCluster is formulated as follows:

$$\min : \parallel \mathbf{X} - \hat{\mathbf{X}} \parallel_F^2 + \lambda * \mathcal{G}_\mathbf{w}(\mathbf{Y})$$
$$\text{s.t.} \quad \mathbf{Y} = f_{\theta_1}(\mathbf{X}) \tag{6}$$

where $\mathbf{X}$ is the raw data to be clustered and $\hat{\mathbf{X}}$ is the reconstruction learned by the deep network. $\mathcal{G}_w(\mathbf{Y})$ can be any specific clustering objective function such as K-means. $\mathbf{Y}$ is the dummy variable in order to make parameters decomposable. λ defines a trade-off between the network objective and the clustering objective. When $\lambda = 0$, the problem degrades to deep network optimization. We add the constraint on $\mathbf{Y}$ to make it close to the features learned from $\mathbf{X}$.

Algorithm 1. The general learning algorithm of DeepCluster

Input: input data $\mathbf{X}$, hyperparameters ρ, λ, learning rate η
Output: a well-trained autoencoder, cluster assignments
1: Initialize parameters of DAE $\boldsymbol{\theta}$, parameters of clustering model $\mathbf{w}$
2: **while** not converged **do**
3:     update $\boldsymbol{\theta}$: $\boldsymbol{\theta} = \boldsymbol{\theta} - \eta * d\boldsymbol{\theta}$
4:     update $\mathbf{Y}$:
5:     $\mathbf{Y} = \mathrm{argmin}_{\mathbf{Y}} \; \lambda * \mathcal{G}_{\mathbf{w}}(\mathbf{Y}) + \frac{\rho}{2}\|\mathbf{Y} - f_{\boldsymbol{\theta}_1}(\mathbf{X}) + \mathbf{U}\|_F^2$
6:     update $\mathbf{U}$: $\mathbf{U} = \mathbf{U} + \mathbf{Y} - f_{\boldsymbol{\theta}_1}(\mathbf{X})$
7:     update $\mathbf{w}$: $\mathbf{w} = \mathrm{argmin}_{\mathbf{w}} \; \mathcal{G}_{\mathbf{w}}(\mathbf{Y})$
8: **return** $\boldsymbol{\theta}, \mathbf{w}$

The first part of the above objective is non-convex as deep neural networks contain multilayer non-linear transformations. It is hard to optimize this model by gradient-based optimization methods directly, because some clustering methods cannot be solved by them. In this paper, we adopt ADMM to optimize it. First, we derive its augmented Lagrangian [9] as follows:

$$\mathcal{L}_{\rho}(\boldsymbol{\theta}, \mathbf{Y}, \mathbf{U}, \mathbf{w}) = \|\mathbf{X} - \hat{\mathbf{X}}\|_F^2 + \lambda * \mathcal{G}_{\mathbf{w}}(\mathbf{Y}) + \frac{\rho}{2}\|\mathbf{Y} - f_{\boldsymbol{\theta}_1}(\mathbf{X}) + \mathbf{U}\|_F^2 \tag{7}$$

where $\mathbf{U}$ is the scaled dual variable (or the reciprocal of $\boldsymbol{\lambda}$ in Eq. (5)) and $\rho > 0$ is the penalty parameter. ρ is a very important parameter to control how close between $\mathbf{Y}$ and $f_{\boldsymbol{\theta}_1}(\mathbf{X})$. Then, we solve the equation by alternatively optimizing some parameters while keeping the others fixed. For a deep autoencoder that consists of an encoder and a decoder, we denote $\boldsymbol{\theta} = \{\boldsymbol{\theta}_1, \boldsymbol{\theta}_2\}$ as the total network parameters, where $\boldsymbol{\theta}_1$ indicates the encoder parameters and $\boldsymbol{\theta}_2$ means the decoder parameters. To optimize $\boldsymbol{\theta}$, we actually optimize $\|\mathbf{X}-\hat{\mathbf{X}}\|_F^2 + \alpha \cdot \frac{\rho}{2}\|\mathbf{Y} - f_{\boldsymbol{\theta}_1}(\mathbf{X}) + \mathbf{U}\|_F^2$. Here, α is to control the gradient influence of the constraint, which is set to 1 by default.

Algorithm 1 outlines the general optimization process. Note that by introducing $\mathbf{Y}$, we can decompose the parameters into two parts and keep the parameter update formulations of clustering methods $\mathcal{G}_{\mathbf{w}}(\mathbf{Y})$ remain the same as before. The challenge is how to optimize $\mathbf{Y}$ that depends on the specific clustering model. This framework is flexible because we can select the most appropriate clustering algorithms for any specific clustering task. Moreover, the deep network is also configurable. There are many variants of deep autoencoders such as denoise autoenocders, convolutional autoencoders and variational autoenocders.

In what follows, we give a brief convergence and complexity analysis on Deep-Cluster. DeepCluster's objective function consists of three parts. The first part is the reconstruction error of DAE, the parameter update rules here are the same as stochastic gradient descent (SGD) or its variants. The second part is the clustering objective that is independent from DAE when $\mathbf{Y}$ is given. The last part is the constraint imposed on $\mathbf{Y}$, which is a convex function of $\mathbf{Y}$. Following the ADMM optimization procedure, as DAE is nonconvex function of $\boldsymbol{\theta}$, it is hard to prove DeepCluster's global convergence directly. Although there are some works to prove the convergence of ADMM on non-convex problems under some specific assumptions, they do not suit for our case [10,24]. However, our experiments have shown that DeepCluster can converge to a good local minima when the value of ρ is properly chosen.

The computational complexity of DeepCluster consists of three parts: the complexity of clustering algorithm, the complexity of DAE, and the computation of $\mathbf{Y}$. The complexity of $\mathbf{Y}$ optimization is related to the clustering algorithm. For example, it takes $O(TND)$ in DC-Kmeans where T, N and D are the number of iterations, the number of samples and the hidden feature dimensionality respectively. This term can often be omitted in the complexity of clustering algorithm. Thus, DeepCluster has similar time complexity to that of DAEC. However, DeepCluster needs additional $O(ND)$ space to store $\mathbf{Y}$ and $\mathbf{U}$.

5   Two Specific DeepCluster Implementations

Here we give two specific implementations of integrating commonly used clustering algorithms into deep autoencoder (DAE). We choose K-means and GMM as examples.

5.1   DC-Kmeans: Integrating K-means into DAE

K-means tries to find the nearest cluster center for each sample. That is, if the j-th cluster center is the closest to $\mathbf{x}_i$, it assigns $\mathbf{x}_i$ to cluster j with 100% confidence. K-means is a very simple and has no tunable parameter except K. In the following, we embed K-means into DeepCluster, and refer this method as DC-Kmeans for convenience.

Following Eq. (6), we have the objective function of DC-Kmeans as follows:

$$\min : \frac{1}{N} \sum_{i=1}^{N} \|\mathbf{x}_i - \hat{\mathbf{x}}_i\|^2 + \frac{\lambda}{2} * \|\mathbf{y}_i - \mathbf{c}_i^*\|^2 \tag{8}$$

$$s.t. \quad \mathbf{y}_i = f_{\theta_1}(\mathbf{x}_i) \quad i = 1, ..., N$$

where $\mathbf{c}_i^* = \mathrm{argmin}_{\mathbf{c}_j} \|\mathbf{y}_i - \mathbf{c}_j\|^2, j = 1, ..., K$ is the closest centroid to $\mathbf{y}_i$. Besides, N is the total number of samples and K is the number of clusters. And following Eq. (7), we have the corresponding augmented Lagrangian of Eq. (8):

$$\mathcal{L}_\rho(\boldsymbol{\theta}, \mathbf{Y}, \mathbf{U}, \mathbf{C}) = \frac{1}{N} \sum_{i=1}^{N} \|\mathbf{x}_i - \hat{\mathbf{x}}_i\|^2 + \frac{\lambda}{2} * \|\mathbf{y}_i - \mathbf{c}_i^*\|^2 + \frac{\rho}{2} \|\mathbf{y}_i - f_{\theta_1}(\mathbf{x}_i) + \mathbf{u}_i\|^2 \tag{9}$$

Algorithm 2. The Learning Algorithm of DC-Kmeans

Input: input data $\mathbf{X}$, hyperparameters ρ, λ, learning rate η
Output: a well-trained autoencoder, cluster assignments
1: Initialize parameters of DAE $\boldsymbol{\theta}$, parameters of clustering model $\mathbf{C}$
2: **while** not converged **do**
3:    update $\boldsymbol{\theta}$: $\boldsymbol{\theta} = \boldsymbol{\theta} - \eta * d\boldsymbol{\theta}$
4:    update $\mathbf{Y}$: $\mathbf{y}_i = \frac{\lambda * \mathbf{c}_i^* + \rho * (f_{\boldsymbol{\theta}_1}(\mathbf{x}_i) - \mathbf{u}_i)}{\lambda + \rho}$, $i = 1, ..., N$
5:    update $\mathbf{U}$: $\mathbf{u}_i = \mathbf{u}_i + \mathbf{y}_i - f_{\boldsymbol{\theta}_1}(\mathbf{x}_i)$, $i = 1, ..., N$
6:    update $\mathbf{C}$: $\mathbf{c}_j = \frac{1}{N_j} \sum_{\mathbf{x}_i \in \mathcal{C}_j} \mathbf{y}_i$, $j = 1, ..., K$

7: **return** $\boldsymbol{\theta}, \mathbf{C}$

To solve the above equation, we treat an autoencoder as a non-linear and non-convex function of $\boldsymbol{\theta}$. Inspired by ADMM, we alternatively optimize each part of the variables in this objective function. Concretely, we use gradient-based optimization algorithm to find a good candidate for the network parameters $\boldsymbol{\theta}$. As for $\mathbf{Y}$, $\mathbf{C}$ and $\mathbf{U}$, closed form solutions are available as follows (detailed derivations are omitted):

$$
\begin{aligned}
\mathbf{y}_i^{\text{new}} &= \frac{\lambda * \mathbf{c}_i^* + \rho * (f_{\boldsymbol{\theta}_1^{\text{new}}}(\mathbf{x}_i) - \mathbf{u}_i)}{\lambda + \rho} \\
\mathbf{u}_i^{\text{new}} &= \mathbf{u}_i + f_{\boldsymbol{\theta}_1^{\text{new}}}(\mathbf{x}_i) - \mathbf{y}_i^{\text{new}} \\
\mathbf{c}_j^{\text{new}} &= \frac{1}{N_j} \sum_{\mathbf{x}_j \in \mathcal{C}_j} \mathbf{y}_i^{\text{new}}
\end{aligned}
\tag{10}
$$

where N_j is the size of the j-th cluster, $\mathcal{C}_j$ denotes the j-th cluster data set. We set $\lambda = 1$ in our experiments for simplicity. We outline the learning algorithm in Algorithm 2.

5.2   DC-GMM: Integrating GMM into DAE

Another widely-used feature-based clustering method is Gaussian mixture model (GMM). GMM assumes that all data samples are generated from multiple independent Gaussian distributions. GMM can be seen as generalized K-means clustering by incorporating the covariance structure of each cluster. GMM is a probabilistic model and its objective function is to maximize likelihood $P(\mathbf{Y})$. Let π_i be the mixing coefficient of each Gaussian distribution component, $\mathbf{z}_i$ be a K-dimensional binary random variable with $\sum_k z_{ik} = 1$. We denote the posterior probability of z_{ik} as $p(z_{ik} = 1|\mathbf{y}_i) = \gamma(z_{ik})$. Each cluster is assumed a multivariate Gaussian distribution with mean $\boldsymbol{\mu}_k$ and covariance $\boldsymbol{\Sigma}_k$.

We can simply treat this model as adding mixture-of-Gaussian prior to the dummy variables $\mathbf{Y}$, rather than to deep features. The log likelihood of $\mathbf{Y}$ is defined as below:

$$\ln p(\mathbf{Y}|\boldsymbol{\pi}, \boldsymbol{\mu}, \boldsymbol{\Sigma}) = \sum_{i=1}^{N} \ln \left[\sum_{k=1}^{K} \pi_k \mathcal{N}(\boldsymbol{y}_i | \boldsymbol{\mu}_k, \boldsymbol{\Sigma}_k) \right] \tag{11}$$

where $\mathcal{N}(\boldsymbol{y}_i | \boldsymbol{\mu}_k, \boldsymbol{\Sigma}_k)$ is a multivariate Gaussian distribution with $\boldsymbol{\mu}_k$ and $\boldsymbol{\Sigma}_k$ as its parameters. In this case, our objective function is:

$$\min : \frac{1}{N} \sum_{i=1}^{N} \left\{ \|\boldsymbol{x}_i - \hat{\boldsymbol{x}}_i\|^2 - \lambda * \ln \left[\sum_{k=1}^{K} \pi_k \mathcal{N}(\boldsymbol{y}_i | \boldsymbol{\mu}_k, \boldsymbol{\Sigma}_k) \right] \right\} \tag{12}$$

$$s.t. \quad \boldsymbol{y}_i = f_{\boldsymbol{\theta}_1}(\boldsymbol{x}_i) \quad i = 1, ..., N$$

And the augmented Lagrangian of this model is:

$$\mathcal{L}_\rho(\boldsymbol{\theta}, \mathbf{Y}, \mathbf{U}, \boldsymbol{\mu}, \boldsymbol{\Sigma}, \boldsymbol{\pi}) = \frac{1}{N} \sum_{i=1}^{N} \left\{ \|\boldsymbol{x}_i - \hat{\boldsymbol{x}}_i\|_2^2 \right.$$

$$- \lambda * \ln \left[\sum_{k=1}^{K} \pi_k p(\boldsymbol{y}_i | \boldsymbol{\mu}_k, \boldsymbol{\Sigma}_k) \right] \tag{13}$$

$$\left. + \frac{\rho}{2} \|\boldsymbol{y}_i - f_{\boldsymbol{\theta}_1}(\boldsymbol{x}_i) + \boldsymbol{u}_i\|^2 \right\}$$

We can derive the update equation of $\boldsymbol{y}_i$ in closed form as follows:

$$\boldsymbol{y}_i^{\text{new}} = \left[\lambda \sum_{k=1}^{K} \gamma(z_{ik}) \boldsymbol{\Sigma}_k^{-1} + \rho \mathbf{I} \right]^{-1} \left[\rho * (f_{\boldsymbol{\theta}_1}(\boldsymbol{x}_i) - \boldsymbol{u}_i) \right.$$

$$\left. + \lambda \sum_{k=1}^{K} \gamma(z_{ik}) \boldsymbol{\Sigma}_k^{-1} \boldsymbol{\mu}_k \right] \tag{14}$$

where $\mathbf{I}$ is the identity matrix. The other parameters remains the same as in standard GMM algorithm:

$$\mathbf{u}_i^{\text{new}} = \mathbf{u}_i + \boldsymbol{y}_i^{\text{new}} - f_{\boldsymbol{\theta}_1^{\text{new}}}(\mathbf{x}_i)$$

$$\boldsymbol{\mu}_k^{\text{new}} = \frac{1}{N_k} \sum_{i=1}^{N} \gamma(z_{ik}) \mathbf{y}_i^{\text{new}}$$

$$\boldsymbol{\Sigma}_k^{\text{new}} = \frac{1}{N_k} \sum_{k=1}^{K} \gamma(z_{ik}) (\mathbf{y}_i^{\text{new}} - \boldsymbol{\mu}_k^{\text{new}})(\mathbf{y}_i^{\text{new}} - \boldsymbol{\mu}_k^{\text{new}})^T \tag{15}$$

$$\pi_k^{\text{new}} = \frac{N_k}{N}$$

where $N_k = \sum_{i=1}^{N} \gamma(z_{ik})$.

As in DC-Kmeans, we set $\lambda = 1$ in this model too. The learning algorithm of DC-GMM is given in Algorithm 3.

Algorithm 3. The Learning Algorithm of DC-GMM

Input: input data $\mathbf{X}$, hyperparameters ρ, λ, learning rate η

Output: a well-trained autoencoder, cluster assignments

1: Initialize parameters of DAE $\boldsymbol{\theta}$, parameters of GMM $\boldsymbol{\mu}, \boldsymbol{\Sigma}, \boldsymbol{\pi}$

2: **while** not converged **do**

3:     update $\boldsymbol{\theta}$: $\boldsymbol{\theta} = \boldsymbol{\theta} - \eta * d\boldsymbol{\theta}$

4:     update Y:

5:

6:     $\boldsymbol{y}_i = \left[\lambda \sum_{k=1}^{K} \gamma(z_{ik})\boldsymbol{\Sigma}_k^{-1} + \rho\mathbf{I}\right]^{-1}\left[\rho * (f_{\theta_1}(\boldsymbol{x}_i) - \boldsymbol{u}_i)\right.$

7:         $\left. + \lambda \sum_{k=1}^{K} \gamma(z_{ik})\boldsymbol{\Sigma}_k^{-1}\boldsymbol{\mu}_k\right], \quad i = 1, ..., N$

8:

9:     update U: $\boldsymbol{u}_i = \boldsymbol{u}_i + \boldsymbol{y}_i - f_{\theta_1}(\boldsymbol{x}_i), \quad i = 1, ..., N$

10:    update μ: $\boldsymbol{\mu}_k = \frac{1}{N_k}\sum_{i=1}^{N} \gamma(z_{ik})\boldsymbol{y}_i, \quad k = 1, ..., K$

11:    update Σ:

12:

13:    $\boldsymbol{\Sigma}_k = \frac{1}{N_k}\sum_{k=1}^{K} \gamma(z_{ik})(\boldsymbol{y}_i - \boldsymbol{\mu}_k)(\boldsymbol{y}_i - \boldsymbol{\mu}_k)^T,$

14:

15:        $k = 1, ..., K$

16:    update π: $\pi_k = \frac{N_k}{N}, k = 1, ..., K$

17: **return** θ, μ, Σ, π

---

6   Performance Evaluation

To evaluate our framework, we use three real-world datasets and compare our methods against several existing clustering methods. As our aim is to demonstrate the effectiveness of our framework, instead of pursuing the best performance on each dataset, we choose the vanilla deep autoencoder for simplicity. There are many choices that can be exploited to achieve better results, including substituting the network or the clustering algorithm by a better one.

6.1   Experimental Settings

Baseline algorithms: We compare DeepCluster with K-means and GMM method on the original data space as well as deep feature space of a fine-tuned autoencoder. We denote them as DAE+Kmeans and DAE+GMM. We also compare DeepCluster with DAEC [18] that is closely related to DC-Kmeans and DEC [25] that is a state-of-the-art unsupervised clustering model. To evaluate the effectiveness of DeepCluster, we simply set the number of clusters to the true number of classes for all experiments.

Datasets: Three real datasets are used to do evaluation and comparison, including two handwritten digits datasets and one text datasets[1].

---

[1] USPS can be downloaded from: http://www.cs.nyu.edu/~roweis/data.html.

Fig. 2. Sample images of MNIST (left) and USPS (right).

- **MNIST:** a benchmark dataset for many machine learning tasks. It has 60,000 handwritten digit images in the training dataset and 10,000 images in the test dataset. Each image is of 28 * 28 pixel size. Some sample images are shown in Fig. 2 (left).
- **USPS:** also a handwritten digits (0-9) dataset and each class have 1100 samples. These images have been deslanted and size normalized. So the total number of images in this dataset is 11000, and each image is of 16 * 16 pixel size. Some sample images are shown in Fig. 2 (right).
- **Reuters10k:** there are 810000 English news stories in the Reuters dataset. Here, following [25], we consider the four root categories: corprate/industrial, government/social, markets, economics and computer, and use TF-IDF features of the top-2000 frequently used word stems. Finally, we randomly select 10000 examples as in DEC because some methods do not scale well.

Table 1 summarizes the major statistics of the datatsets.

Table 1. Dataset information

Dataset	#Classes	#Dims	#Samples
MNIST	10	784	70,000
USPS	10	256	11,000
Reuters10k	4	2000	10,000

Evaluation Measures: To measure clustering performance, we adopt four metrics, *i.e.*, *Accuracy*, Normalized Mutual Information (*NMI*), *Purity* and Adjusted Rand Index (*ARI*). High value of these metrics indicates better performance. These measures are defined as follows:

$$ACC = \max_m \frac{\sum_{i=1}^{N} \mathbf{1}(r_i = m(c_i))}{N}$$

$$NMI = \frac{I(\boldsymbol{r}, \hat{c})}{(H(\boldsymbol{r}) + H(\hat{c}))/2}$$

$$Purity = \sum_{k=1}^{K} \frac{max_i(n_k^i)}{N} \tag{16}$$

$$ARI = \frac{\sum_{ij} \binom{n_{ij}}{2} - [\sum_i \binom{a_i}{2} \sum_j \binom{b_j}{2}]/\binom{n}{2}}{\frac{1}{2}[\sum_i \binom{a_i}{2} + \sum_j \binom{b_j}{2}] - [\sum_i \binom{a_i}{2} \sum_j \binom{b_j}{2}]/\binom{n}{2}}$$

where $\mathbf{1}(\cdot)$ is an indicator function, r_i is the ground-truth label, c_i is the cluster assignment and $m(\cdot)$ denotes all possible one-to-one mapping between clusters and labels. $\boldsymbol{r}$ denotes the ground truth labels and $\hat{c}$ is the cluster assignments. $I(\cdot)$ is the mutual information metric and H is the entropy. n_k^i is the number of samples of class i but assigned to cluster k. ARI is quite complex, n_{ij}, a_i, b_j are values from the contingency table (see details in [17]).

6.2   Implementation Details

In order to conduct a fair comparison, we choose the same network structure as in [25]. In other words, we stack four autoencoders to build a deep autoencoder and the encoder structure is d-500-500-2000-10, where d is the dimension of original data. To obtain a good initialization for our methods, we also do layer-wise pre-training and fine-tuning.

Note that the features of Reuters10k are very sparse and feature cardinality ranges from 0 to 38, which is quite different from the other datasets. We conduct two different sets of initialization steps and activation functions. For Reuters10k, we use exactly the same initialization and the same activation functions as DEC. The pre-training of the two image datasets are done by layer-wise training of RBMs, and the fine-tuning are done as DAEC [18] with sigmoid activation functions. We optimize $\boldsymbol{\theta}$ by the AdaDelta algorithm, which is a variant of the stochastic gradient decent algorithm [27]. However, we find SGD is more suitable for the Reuters10k dataset as suggested in DEC.

In the training stage of our methods, we use warm start by initializing $\mathbf{Y}$ to $f_{\boldsymbol{\theta}_1}(\mathbf{X})$ and $\mathbf{U} = 0$. As mentioned before, we set $\lambda = 1$ for all experiments. We set $\alpha = 0.01$ for DC-GMM models on image datasets and DC-Kmeans on Reuters10K dataset. We carry out linear search to find the best ρ. The convergence threshold of DC-Kmeans is set to 0.1% and the max iteration is set to 200.

For K-means and GMM-based methods, we run each method 10 times and report the best performance. For DAEC, we vary λ in $\{0.2 * i\}, i = 1, ..., 5$ and report the best performance. We give the best results of DEC by running its code and do the hyper-parameter selection as suggested[2].

---

[2] https://github.com/piiswrong/dec.

Table 2. Performance comparison on the three real datasets. The highest values of all metrics on the datasets are in bold.

Method	MNIST				USPS				Reuters10K			
	Accuracy	NMI	Purity	ARI	Accuracy	NMI	Purity	ARI	Accuracy	NMI	Purity	ARI
K-means	0.5618	0.5163	0.5997	0.3837	0.4585	0.4503	0.4767	0.3063	0.6018	0.3863	0.6595	0.3271
GMM	0.3505	0.2836	0.3672	0.1811	0.29	0.2107	0.2917	0.1077	0.494	0.1964	0.4954	0.1048
DAE+Kmeans	0.6903	0.6469	0.7171	0.5325	0.5955	0.5203	0.5985	0.4053	0.6648	0.4456	0.7499	0.4283
DAE+GMM	0.7853	0.7525	0.7896	0.6718	0.6422	0.5967	0.6422	0.475	0.6349	0.3576	0.7096	0.1884
DAEC	0.734	0.6615	0.7383	0.6093	0.6111	0.5449	0.6255	0.4368	0.7019	0.342143	0.7096	0.3247
DEC	0.8496	0.8273	0.8496	0.7721	0.6246	0.6191	0.651	0.4692	0.6945	**0.5124**	0.7726	**0.4963**
DC-Kmeans	0.8015	0.7448	0.8015	0.689	0.6442	0.5737	0.6546	0.4559	**0.7301**	0.4447	0.7663	0.4892
DC-GMM	**0.8555**	**0.8318**	**0.8555**	**0.7823**	**0.6476**	**0.6939**	**0.6713**	**0.4913**	0.6906	0.4458	**0.7765**	0.404

6.3 Experimental Results

Effect of Deep Features: We compare the performance of different K-means based algorithms and GMM based algorithms on all datasets, the results are shown in Fig. 3. Obviously, the performance on the raw data is worse than on learnt feature spaces. When features learnt by a fine-tuned deep autoencoder are used, better performance can be achieved. Moreover, if we embed K-means algorithm into deep models as in DAEC, the results are much better, as shown in Fig. 3(a), (b), (c). DC-Kmeans outperforms DAEC on all metrics because of the introduced relaxation. Figure 3(d), (e), (f) also show that feature learning and clustering jointly is better than doing them separately. Thus, we can conclude that a more compact and representative feature space is important and effective for clustering, and deep learning can learn such features.

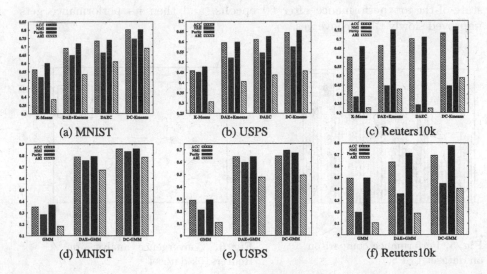

(a) MNIST   (b) USPS   (c) Reuters10k

(d) MNIST   (e) USPS   (f) Reuters10k

Fig. 3. Performance comparison in different feature spaces on all datasets. Kmeans-based methods (a) (b) (c). GMM-based methods (d) (e) (f).

Effectiveness of DeepCluster Framework: As we know, the digits images of USPS are more illegible than MNIST. We reconstruct the cluster centers by DAEC, DC-Kmeans and DC-GMM, and the results are shown in Fig. 4. We can see that the centers learned by our methods are more reasonable than DAEC as there exist duplicated digits in the reconstructed centroids by DAEC.

DAEC

DC-Kmeans

DC-GMM

Fig. 4. The reconstruction of cluster centroids learned by different methods.

To demonstrate the effectiveness of our methods, we present the performance results of all methods on the USPS dataset in Fig. 5, and give the performance results on all datasets in Table 2. We can see that DC-GMM outperforms the other methods in all metrics on USPS and MNIST. Both DC-Kmeans and DC-GMM outperform DAEC, which indicates that clustering the dummy variable is better than clustering the deep features directly. Certainly, DEC outperforms the other methods in *NMI* and performs slightly better than DC-Kmeans in *ARI* on Reuters10K. But DC-Kmeans outperforms DEC in *ACC*. DC-GMM achieves comparable performance to DEC, however it underperforms DEC and DC-Kmeans in *ARI*.

As shown in Fig. 6, DeepCluster's convergence is fast and stable. It achieves state-of-the-art performance after 60 epcohs, and then its performance gets improved slowly till it converges.

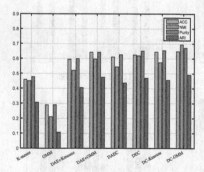

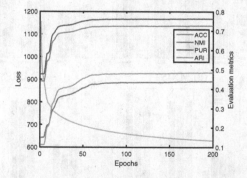

Fig. 5. Performance comparison on dataset USPS.

Fig. 6. Convergence analysis on Reuters10k dataset

Effect of Hyperparameter Choice: Here we check our method's sensitivity to the hyperparameter ρ. We use different values of ρ to evaluate the model's sensitivity on all our experimental settings, the results are shown in Fig. 7. We can see that our method is robust to ρ. This is very important as it is not possible to do cross-validation in real-world applications [25].

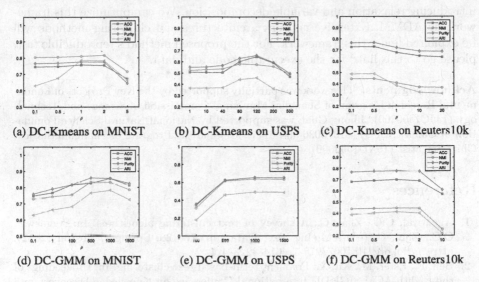

(a) DC-Kmeans on MNIST   (b) DC-Kmeans on USPS   (c) DC-Kmeans on Reuters10k

(d) DC-GMM on MNIST   (e) DC-GMM on USPS   (f) DC-GMM on Reuters10k

Fig. 7. Parameter sensitivity analysis of DeepCluster algorithms on different datasets.

Time Analysis: We present the time cost of our method in Table 3. DeepCluster algorithms spend approximately the same time as DAEC. DEC spend the least time because it works on GPU in Caffe, while DeepCluster and DAEC run on CPU in MATLAB code.

Table 3. Experimental time information. (In seconds)

Methods	DAEC	DEC	DC-Kmeans	DC-GMM
MNIST	18323	*1802*	20778	19234
USPS	2896	*1037*	2657	2715
Reuters10k	2025	*1994*	2527	2332

7 Conclusion

This paper presents a deep learning based clustering framework that simultaneously learns hidden features and does cluster assignment. Thanks to employing the ADMM algorithm, we can optimize our models in an end-to-end manner. We demonstrate the effectiveness of this framework by embedding K-means and GMM into DAE. Experimental results validate the effectiveness and advantage of our framework, which can achieve state-of-the-art performance on real-world datasets. Compared to DAEC, our framework runs clustering algorithms on the dummy variable while constraining the variable close to the learned features. By introducing relaxation and variable decomposition. We can optimize this framework by ADMM. Extensive network architectures and clustering methods will be exploited under this framework. For the proposed method's reproducible test, please go to this link[3] for the executable code and data.

Acknowledgments. This work was partially supported by the Key Projects of Fundamental Research Program of Shanghai Municipal Commission of Science and Technology (14JC1400300). Jihong Guan was supported by National Natural Science Foundation of China (NSFC) (61373036 and 61772367) and the Program of Shanghai Subject Chief Scientist (15XD1503600).

References

1. Aggarwal, C.C., Zhai, C.: A survey of text clustering algorithms. In: Aggarwal, C.C., Zhai, C. (eds.) Mining Text Data, pp. 77–128. Springer, Boston (2012). https://doi.org/10.1007/978-1-4614-3223-4_4
2. Beil, F., Ester, M., Xu, X.: Frequent term-based text clustering. In: Proceedings of the Eighth ACM SIGKDD International Conference on Knowledge Discovery and Data Mining, pp. 436–442. ACM (2002)
3. Bengio, Y., Courville, A., Vincent, P.: Representation learning: a review and new perspectives. IEEE Trans. Pattern Anal. Mach. Intell. **35**(8), 1798–1828 (2013)
4. Bishop, C.M.: Pattern recognition. Mach. Learn. **128**, 1–58 (2006)
5. Boyd, S., Parikh, N., Chu, E., Peleato, B., Eckstein, J.: Distributed optimization and statistical learning via the alternating direction method of multipliers. Found. Trends® Mach. Learn. **3**(1), 1–122 (2011)
6. Dueck, D., Frey, B.J.: Non-metric affinity propagation for unsupervised image categorization. In: IEEE 11th International Conference on Computer Vision, ICCV 2007, pp. 1–8. IEEE (2007)
7. Eckstein, J., Yao, W.: Understanding the convergence of the alternating direction method of multipliers: theoretical and computational perspectives. Pac. J. Optim. **11**(4), 619–644 (2015)
8. Hallac, D., Leskovec, J., Boyd, S.: Network lasso: clustering and optimization in large graphs. In: Proceedings of the 21th ACM SIGKDD International Conference on Knowledge Discovery and Data Mining, pp. 387–396. ACM (2015)
9. Hestenes, M.R.: Multiplier and gradient methods. J. Optim. Theory Appl. **4**(5), 303–320 (1969)
10. Hong, M., Luo, Z.Q., Razaviyayn, M.: Convergence analysis of alternating direction method of multipliers for a family of nonconvex problems. SIAM J. Optim. **26**(1), 337–364 (2016)

---

[3] https://github.com/JennyQQL/DeepClusterADMM-Release.

11. Joulin, A., Bach, F., Ponce, J.: Discriminative clustering for image co-segmentation. In: 2010 IEEE Conference on Computer Vision and Pattern Recognition (CVPR), pp. 1943–1950. IEEE (2010)
12. Kanungo, T., Mount, D.M., Netanyahu, N.S., Piatko, C.D., Silverman, R., Wu, A.Y.: An efficient k-means clustering algorithm: analysis and implementation. IEEE Trans. Pattern Anal. Mach. Intell. **24**(7), 881–892 (2002)
13. Liu, B., Yuan, X.T., Yu, Y., Liu, Q., Metaxas, D.N.: Decentralized robust subspace clustering. In: Proceedings of the Thirtieth AAAI Conference on Artificial Intelligence, pp. 3539–3545. AAAI Press (2016)
14. Liu, H., Shao, M., Li, S., Fu, Y.: Infinite ensemble for image clustering. In: Proceedings of ACM SIGKDD International Conference on Knowledge Discovery and Data Mining (2016)
15. Ng, A.Y., Jordan, M.I., Weiss, Y., et al.: On spectral clustering: analysis and an algorithm. In: Advances in Neural Information Processing Systems, vol. 2, pp. 849–856 (2002)
16. Peng, X., Xiao, S., Feng, J., Yau, W.Y., Yi, Z.: Deep subspace clustering with sparsity prior In: The 25th International Joint Conference on Artificial Intelligence (2016)
17. Santos, J.M., Embrechts, M.: On the use of the adjusted rand index as a metric for evaluating supervised classification. In: Alippi, C., Polycarpou, M., Panayiotou, C., Ellinas, G. (eds.) ICANN 2009. LNCS, vol. 5769, pp. 175–184. Springer, Heidelberg (2009). https://doi.org/10.1007/978-3-642-04277-5_18
18. Song, C., Liu, F., Huang, Y., Wang, L., Tan, T.: Auto-encoder based data clustering. In: Ruiz-Shulcloper, J., Sanniti di Baja, G. (eds.) CIARP 2013. LNCS, vol. 8258, pp. 117–124. Springer, Heidelberg (2013). https://doi.org/10.1007/978-3-642-41822-8_15
19. Szegedy, C., Liu, W., Jia, Y., Sermanet, P., Reed, S., Anguelov, D., Erhan, D., Vanhoucke, V., Rabinovich, A.: Going deeper with convolutions. In: Proceedings of the IEEE Conference on Computer Vision and Pattern Recognition, pp. 1–9 (2015)
20. Taylor, G., Burmeister, R., Xu, Z., Singh, B., Patel, A., Goldstein, T.: Training neural networks without gradients: a scalable ADMM approach. In: International Conference on Machine Learning (2016)
21. Tian, F., Gao, B., Cui, Q., Chen, E., Liu, T.Y.: Learning deep representations for graph clustering. In: AAAI, pp. 1293–1299 (2014)
22. Tian, K., Shao, M., Wang, Y., Guan, J., Zhou, S.: Boosting compound-protein interaction prediction by deep learning. Methods **110**, 64–72 (2016)
23. Wang, R., Shan, S., Chen, X., Gao, W.: Manifold-manifold distance with application to face recognition based on image set. In: IEEE Conference on Computer Vision and Pattern Recognition, CVPR 2008, pp. 1–8. IEEE (2008)
24. Wang, Y., Yin, W., Zeng, J.: Global convergence of ADMM in nonconvex nonsmooth optimization. arXiv preprint arXiv:1511.06324 (2015)
25. Xie, J., Girshick, R., Farhadi, A.: Unsupervised deep embedding for clustering analysis. In: International Conference on Machine Learning (ICML) (2016)
26. Xu, J., Wang, P., Tian, G., Xu, B., Zhao, J., Wang, F., Hao, H.: Short text clustering via convolutional neural networks. In: Proceedings of NAACL-HLT, pp. 62–69 (2015)
27. Zeiler, M.D.: Adadelta: an adaptive learning rate method. arXiv preprint arXiv:1212.5701 (2012)
28. Zhang, R., Cheng, Z., Guan, J., Zhou, S.: Exploiting topic modeling to boost metagenomic reads binning. BMC Bioinform. **16**, S2 (2015)

Multi-view Spectral Clustering on Conflicting Views

Xiao He[1,2](✉), Limin Li[3], Damian Roqueiro[1,2], and Karsten Borgwardt[1,2](✉)

[1] Department of Biosystems Science and Engineering, ETH Zurich,
Basel, Switzerland
{xiao.he,damian.roqueiro,karsten.borgwardt}@bsse.ethz.ch
[2] Swiss Institute of Bioinformatics, Lausanne, Switzerland
[3] School of Mathematics and Statistics, Xi'an Jiaotong University, Xi'an, China
liminli@mail.xjtu.edu.cn

Abstract. In a growing number of application domains, multiple feature representations or *views* are available to describe objects. *Multi-view clustering* tries to find similar groups of objects across these views. This task is complicated when the corresponding clusterings in each view show poor agreement (*conflicting* views). In such cases, traditional multi-view clustering methods will not benefit from using multi-view data. Here, we propose to overcome this problem by combining the ideas of multi-view spectral clustering with alternative clustering through kernel-based dimensionality reduction. Our method automatically determines feature transformations in each view that lead to an optimal clustering w.r.t to a new proposed objective function for conflicting views. In our experiments, our approach outperforms state-of-the-art multi-view clustering methods by more accurately detecting the ground truth clustering supported by all views.

Keywords: Multi-view clustering · Alternative clustering
Conflicting views · Kernel dimensionality reduction

1 Introduction

In many application domains, it is commonplace that a single object can be described by multiple feature representations or *views*. We will expect to obtain a clustering of better quality if information on all views is taken into account. *Multi-view clustering* tries to find similar groups of objects across different views, with a number of methods having been proposed in the literature, including Multi-view EM [1], Canonical Correlation Analysis for multi-view clustering [2], Multi-view spectral clustering [3–5], Multi-view clustering with unsupervised feature selection [6,7] and Nonnegative Matrix Factorization [8]. All these methods share the assumption of a common clustering structure across views, which is interpreted as having the corresponding clusterings in each view in agreement with a ground truth partitioning.

© Springer International Publishing AG 2017
M. Ceci et al. (Eds.): ECML PKDD 2017, Part II, LNAI 10535, pp. 826–842, 2017.
https://doi.org/10.1007/978-3-319-71246-8_50

However, in real-world datasets, certain views may contain subsets of features with varying degrees of relatedness which may lead to multiple non-redundant alternative clustering solutions in each view. For example, while clustering university webpages by text features (words), some words such as 'major', 'position' or 'homework' will lead to a partitioning of webpages into categories such as 'student', 'faculty' and 'course'. Alternatively, other words (e.g. 'biology', 'cell', 'computer science', 'code' etc.) will lead to a partitioning of webpages by their department of affiliation, which is independent of the categories described before. The final clustering will be closer to one of the partitionings but contaminated by the other one.

Several methods [9–11] were proposed to tackle this alternative clustering problem, but they only focus on single-view data. The problem exists in the multi-view setting as well but it has received little attention. When the corresponding clusterings in each view show poor agreement, we say that we have *conflicting* views. The users may only be in favor of the underlying clustering that is closer to the partitioning on one view. In such cases, traditional multi-view clustering methods [1–3,5,6,8] will fail to be beneficial or may even be harmful when using multi-view data.

Going back to our example of university webpages, if we consider webpages comprised of two views: (a) text (words) and (b) hyperlinks, a clustering on the text view will cluster webpages into a partitioning by categories ('student', 'faculty', 'course', etc.) since more word features are related to this partition. Suppose users are interested in finding this partition. However, a clustering on the hyperlink view will mainly partition webpages by their department of affiliation. This is due to the fact that, for example, students' webpages may link to the courses for which they are registered while webpages of faculty members are linked to the pages of the courses which they teach. Therefore, the two views conflict and their corresponding partitions are likely to disagree. In addition, as mentioned before, the clustering on the text view is contaminated by word features that lead to a partitioning of webpages by their department of affiliation (partitioned on the hyperlinks view). We consider such an underlying structure (e.g. department) which unduly influences the partitioning to be a *confounder*. There might be useful information in the hyperlinks view, but this is masked by the confounder. To unveil the desired clustering structure across conflicting views, we need to find agreement between patterns across views and correct for the confounder.

Our goal in this article is, therefore, to perform multi-view clustering on conflicting views and to correct for possible confounders. We define a novel objective function that combines ideas of multi-view spectral clustering and alternative clustering and propose a new algorithm MvKDR to solve it. More specifically, we project each view onto two different subspaces where two alternative clusterings can be found based on kernel dimensionality reduction [12,13]. With the prior knowledge of which view is more informative, we then try to find a consensus partition by maximizing the agreement between clusterings on one subspace from each view and minimize the agreement between this consensus partition with the alternative clustering on the other subspaces from all views. The motivation behind our method MvKDR is that conflicting views lead to disagreement (statistical independence) between the clusterings across different views. We aim to maximize agreement (statistical dependence) between clusterings across all views on the reduced subspaces and correct for possible confounders through the process of finding an alternative clustering in each view.

Our motivating example for solving this problem of conflicting views comes from cancer genomics. Here, patients are described by different views which consist of molecular tests performed on their tissues. These are the expression level of (a) genes and (b) DNA methylation. The aim is to discover cancer subtypes by clustering the patients. DNA methylation is known to be a mechanism that the cell uses to control gene expression, and so it is reasonable to expect an intrinsic disease-related clustering across both views. However, we found in our experiments that independent clusterings on the gene expression view and the DNA methylation view show little agreement. This may be due to the fact that not all genes or DNA methylation mechanisms are disease-related. Some biological processes that affect both gene expression and DNA methylation may act as confounders. In our experiments, we perform a survival analysis on gene and DNA methylation expression data of cancer patients to show the effectiveness of our proposed method MvKDR.

In regards to related work, and as mentioned before, most of the multi-view clustering techniques in the literature [1–8] do not consider confounding factors and conflicting views, which are the focus of this paper. Christoudias et al. [14] studied the multi-view problem in the presence of view disagreement but in a supervised manner. The most related work are Pareto [15] and MVUFS [16]. Pareto is a multi-objective spectral clustering method based on pareto optimization. However, Pareto performs on the Laplacian matrix in the full feature space, which may fail to detect clustering structure that can only be found in subspaces. MVUFS tries to do feature selection on the second view when it conflicts with the first one. The proposed MvKDR differs from MVUFS by considering the correction of confounding factors across views. In our experiments, we compare to Pareto and MVUFS, as well as other state-of-the-art multi-view clustering approaches on both synthetic and real-world data. Our goal is to show the advantages offered by our proposed method MvKDR by more accurately detecting the underlying ground truth clusterings.

The remainder of this paper is organized as follows: in the following section we describe the proposed multi-view spectral clustering model and describe the algorithm in detail. Section 3 contains an extensive experimental evaluation. Section 4 concludes the paper with a summary of our work and future direction.

2   Multi-view Spectral Clustering on Conflicting Views

In this section, we first review the co-regularized multi-view spectral clustering framework. We then extend it to our new model MvKDR with confounder correction by applying the technique of kernel dimensionality reduction. We finally provide the optimization algorithm for solving the model.

2.1   Co-regularized Spectral Clustering

Suppose we are given an m-view dataset of $\{X_1, \cdots, X_m\}$, where $X_v \in \mathbb{R}^{n \times d_v}$, n and d_v are the number of samples and features in view v. Suppose the number of true clusters k is known.

The kernel matrix of X_v is denoted as K_v and D_v is the degree matrix of K_v. The normalized graph Laplacian for view v can be written as $L_v = D_v^{-\frac{1}{2}} K_v D_v^{-\frac{1}{2}}$. Based on this definition, co-regularized spectral clustering CoReg [3] was proposed as:

$$\max_{\text{All } U_v} \sum_{v=1}^{m} tr(U_v^T L_v U_v) + \lambda \sum_{v \neq w} R(U_v, U_w),$$
$$s.t. \ U_v^T U_v = I \tag{1}$$

where $U_v \in \mathbb{R}^{n \times k}$ with $v \in 1, ..., m$, $R(U_v, U_w) = tr(U_v U_v^T U_w U_w^T)$ is a regularizer that measures the agreement between the embeddings U_v and U_w and λ trades-off the spectral clustering objective and the embedding agreement. The problem can be solved by alternating maximization cycling over the views with all but one U_v fixed. Since alternating maximization converges to a local maximum, CoReg usually starts with an informative view and performs k-means on the final embedding of that view, with the assumption that we have prior knowledge about views.

2.2   Multi-view Spectral Clustering with Kernel Dimensionality Reduction

In the scenario of conflicting views, the agreement of embeddings from Laplacians obtained in the full space will not be useful, or even worse, it can be harmful. As mentioned before, our idea is to first project each view to a low-dimensional subspace, and then maximize the agreement of embeddings from the Laplacians calculated in the subspaces. To this effect, we propose the following model:

$$\max_{\text{All } U_v, W_v} \sum_{v=1}^{m} tr(U_v^T L_v U_v) + \lambda_1 \sum_{v \neq w} R(U_v, U_w),$$
$$s.t. \ U_v^T U_v = I, W_v^T W_v = I \tag{2}$$
$$L_v = D_v^{-\frac{1}{2}} K_{W_v^T X_v} D_v^{-\frac{1}{2}}$$

where $W_v \in \mathbb{R}^{d_v \times k}$ is the projection matrix, $K_{W_v^T X_v}$ is the kernel matrix on the projected subspace, D_v is the degree matrix of $K_{W_v^T X_v}$ and L_v is the corresponding normalized graph Laplacian with $v = 1, \ldots, m$.

The model in (2) seeks a low-dimensional subspace for each view, where the clustering structures are strong (as described by the first term) and the dependence between these clustering embeddings (second term) is maximized. By integrating dimensionality reduction into multi-view clustering, we can find useful information in the projected subspace and maximize the agreement of clusterings there.

However, when there are confounders in the conflicting views, only searching for agreement of spectral clustering in the reduced subspace is not enough. In such cases, the low-dimensional subspace will still be affected by confounders even after the dimensionality reduction.

To tackle this problem, assuming we have prior knowledge that the first view has more discriminatory power with respect to the samples for users' interest, we try to find two non-redundant alternative clustering embeddings U_v and U'_v for each of all the other views. In addition to the regularizer that measures the agreement between desired embeddings, we introduce another regularizer for confounding correction. We propose our MvKDR model as:

$$
\begin{aligned}
\max_{\text{All } U,W} & \sum_{v=1}^{m} tr(U_v^T L_v U_v) + \sum_{v=1}^{m} tr(U_v'^T L_v' U_v') \\
& + \lambda_1 \sum_{v \neq w} R(U_v, U_w) - \lambda_2 \sum_{v,w} R(U_v, U_w'), \\
s.t. \ & U^T U = I, W^T W = I, \\
& L = D^{-\frac{1}{2}} K_{W^T X_v} D^{-\frac{1}{2}}
\end{aligned}
\tag{3}
$$

where $U \in \{U_v, U_v'\}$, $W \in \{W_v, W_v'\}$, $L \in \{L_v, L_v'\}$, and $D \in \{D_v, D_v'\}$ for $v = 1, \ldots, m$. W_v, U_v and L_v are, respectively, the projection, embedding and Laplacian matrix corresponding to the desired clustering in view v, and W_v', U_v' and L_v' are the projection, embedding and Laplacian matrix corresponding to alternative clustering in view v, respectively.

The model in (3) corrects for confounders at the clustering stage which, as discussed before, fits well in many real applications. It helps find strong clustering structures, through dimensionality reduction, in the consensus embedding U_v (first term in the equation) and in the alternative embedding U_v' (second term). In addition, it maximizes the agreement/dependence between desired embeddings U_v and U_w from different views (third term), and minimizes the agreement/dependence between consensus embedding U_v and alternative embedding U_w' in the other views to correct for confounders (fourth term).

The optimization problem in (3) can be solved by the technique of alternating optimization and kernel dimensionality reduction, which is discussed in Sect. 2.3.

2.3 Optimization Algorithm

In this section, we propose an algorithm to solve the optimization problem in (3). We take the alternating maximization strategy in the same way as in co-regularized spectral clustering [3].

We first optimize for U_v by assuming all other variables fixed. For each U_v, the optimization problem of (3) becomes that of (4):

$$
\begin{aligned}
\max_{U_v} \ & tr(U_v^T (L_v + \lambda_1 \sum_{v \neq w} U_w U_w^T - \lambda_2 \sum_w U_w' U_w'^T) U_v) \\
s.t. \ & U_v^T U_v = I
\end{aligned}
\tag{4}
$$

The objective function in (4) is the same as the one in spectral clustering with a modified Laplacian matrix. We optimize each U_v with (4) by using eigenvalue decomposition. U_v' can be solved in the same way.

We then optimize for W_v by assuming that all other variables are fixed. Note that the optimization for each W_v and W_v' is independent with U_v and U_v' fixed. Then, with all but one W_v fixed, the optimization problem becomes:

$$\max_{W_v} \; tr(D_v^{-\frac{1}{2}} U_v U_v^T D_v^{-\frac{1}{2}} K_{W_v^T X_v}) \\ s.t. \; W_v^T W_v = I, \tag{5}$$

This optimization problem can be solved by gradient ascent. For simplification, we assume that D_v and D_v' are also fixed. Otherwise we can treat $D_v^{-\frac{1}{2}} K_v D_v^{-\frac{1}{2}}$ as the kernel function and apply the chain rule to get the gradient. In practice, we found that this strategy yields similar results compared to fixed D_v and D_v'.

We use the *kernel dimensionality reduction* (KDR) technique [12,13] to solve the problem in (5) with an input kernel matrix $G = D_v^{-\frac{1}{2}} U_v U_v^T D_v^{-\frac{1}{2}}$. Following the scheme of gradient ascent, in each step we calculate the derivative of (5) with a fixed kernel function, i.e. Gaussian kernel. We describe KDR in more detail later and give an example of performing KDR with a Gaussian kernel.

Finally, we repeat these two steps alternatively until convergence. We obtain the clustering by performing k-means on the resulting embedding of the first view (the most informative one). Algorithm 1 provides a summary of our approach MvKDR.

Algorithm 1: MvKDR

Data: $X_1, ..., X_m$, k, λ_1, λ_2
Result: U_1
// Initialization
1 **for** $v \in 1, ..., m$ **do**
2     $K_v, D_v, L_v, U_v = \text{SpectralClustering}(X_v)$;
3   Update G_v with (5);
4     $W_v = \text{KDR}(X_v, G_v)$, $W_v' = W_v$;
5 **end**
6 **repeat**
    // Step 1: Given W, solve U
7   **for** $v \in 1, ..., m$ **do**
8   Update K_v, D_v, L_v with $W_v^T X_v$;
9   Update K_v', D_v', L_v' with $W_v'^T X_v$;
10   **end**
11   **for** $v \in 1, ..., m$ **do**
12   Solve U_v, U_v' with (4);
13   **end**
    // Step 2: Given U, solve W
14   **for** $v \in 1, ..., m$ **do**
15   Update G_v, G_v' with (5);
16        $W_v = \text{KDR}(X_v, G_v)$, $W_v' = \text{KDR}(X_v, G_v')$;
17   **end**
18 **until** *Converge*;

Kernel dimensionality reduction. We propose to use *kernel dimensionality reduction* (KDR) to solve (5). KDR was first introduced by Fukumizu et al. [12] for the purpose of regression. Given data X and response Y, KDR aims to find the projection of X onto a subspace $W^T X$ that captures the dependency of X on Y in the Reproducing kernel Hilbert space (RKHS) via two semidefinite kernels $K_{W^T X}$ and K_Y.

Wang et al. [13] extended it to the unsupervised case and employed another kernel-based measure of independence, the *Hilbert-Schmidt Independence Criterion* (HSIC) [17]. HSIC is the Hilbert-Schmidt norm of the cross-covariance operator on two random variables. Its empirical estimate is given by $\text{HSIC}(X, Y) = tr(HK_X HK_Y)$, where H is a centering matrix [17]. The objective function of the HSIC version of KDR can be written as:

$$\max_{W^T W = I} tr(GK_{W^T X}), \tag{6}$$

where G is the centralized input kernel matrix. This is exactly the same as our optimization problem in (5).

Equation (6) can be solved by the steepest gradient ascent method with line search. To fulfill the orthogonal constraints, the gradient is projected onto the tangent space after each update.

With a Gaussian kernel, the function is defined as:

$$K(W^T x_i, W^T x_j) = \exp(-\frac{\|W^T x_i - W^T x_j\|^2}{2\sigma^2}), \tag{7}$$

where x_i is the i_{th} sample of X_v. To simplify the formula, we write $K_{W^T X_v}$ as K. The derivative of (6) is shown in (8), where $z_i = W^T x_i$ is the i_{th} sample in the projected space. With the orthogonality constraint of W, the problem is non-convex. But as shown in [12,13], the gradient based method works well in practice.

$$\begin{aligned}
\frac{\partial(tr(GK))}{\partial W} &= \sum_{i,j=1}^{n} G_{i,j} \frac{\partial K_{ij}}{\partial W} \\
&= \sum_{i,j=1}^{n} (-\frac{1}{\sigma^2} G_{i,j} K_{i,j}(x_i - x_j)(x_i - x_j)^T W) \\
&= \sum_{i,j=1}^{n} (-\frac{1}{\sigma^2} G_{i,j} K_{i,j}(x_i - x_j)(z_i - z_j)^T),
\end{aligned} \tag{8}$$

Computational complexity. The computational runtime complexity of MvKDR consists of two parts: $O(n^3)$ in general for eigen-decomposition of the Laplacian matrix and $O(n^2 dkt_1 t_2)$ for the derivative calculation, where n is the number of samples, d is the largest number of features in all views, k is the dimension of the embedding and t_1 and t_2 are the numbers of iterations for the gradient ascent in KDR and outer loops in Algorithm 1 respectively. Both iterations converge fast in our experiments. Therefore the complexity of MvKDR is empirically in the same order as the multi-view spectral clustering method CoReg [3] when $d \ll n$.

3   Experiments

In this section, we report our empirical clustering results by comparing the proposed method MvKDR to a number of baseline methods on both synthetic and real-world multi-view datasets. In addition, we use our method to perform a survival analysis of cancer patients on two genomic datasets.

3.1   Baseline Algorithms and Setting

To demonstrate how clustering performance can be improved by our proposed approach, we compared MvKDR with the following algorithms:

1. **Single view** (SPV1 and SPV2): Consists in running a spectral clustering on each view separately.
2. **Kernel addition** (KerAdd): Adds kernels from different views and performs spectral clustering.
3. **Co-regularized spectral clustering** (CoReg): Adopts the co-regularization framework to spectral clustering, pairwise version as described in [3].
4. **Multi-view unsupervised feature selection** (MVUFS): Integrates sparse unsupervised feature selection and non-negative matrix factorization into a multi-view clustering framework [16].
5. **Multi-view multi-objective spectral clustering** (Pareto): Finds multiple alternative cuts across views with multi-objective Pareto optimization [15].
6. **Multiple non-redundant spectral clustering views** (mSC): Finds multiple non-redundant clustering solutions on a single view [11].

All the comparison methods need the number of clusters k to be predetermined. We set k to be the true number of clusters when this is known as ground truth. We choose Gaussian RBF kernel for all the methods and fix the parameter σ using the median of pairwise distances of each view [17]. For methods using KDR we project each view to $k - 1$ dimensional subspaces. For methods with a regularization parameter, we set it to $\{10^{-2}, 10^{-1}, 10^{0}, 10^{1}, 10^{2}\}$, and determine the best parameters with the smallest objective function of k-means. For all methods, we initialize k-means with 10 random re-starts and record the average of the objective function for parameter selection and to report the results. For datasets with class labels, we measure the performance of the clustering methods based on the accuracy (ACC) and the normalized mutual information (NMI), which are widely used for evaluating clusterings. Please refer to [18] for detailed definitions. The source code for Pareto and MVUFS was provided by the authors [15,16]. We implemented all the rest of the methods. The source code of MvKDR can be found online[1].

---

[1] https://github.com/BorgwardtLab/MvKDR.

3.2   Datasets

We evaluate the performance of all the above mentioned methods on three synthetic datasets and on two real datasets: UCI and WebKB.

- **Synthetic datasets:** We generated three synthetic datasets containing views with varying degrees of confounders to compare the performance of the methods listed in Sect. 3.1. The data generation process was as following. We randomly drew 300 samples in 2D Euclidean space with three clusters in order, each with 100 samples. Suppose the samples are represented by two 300-dimensional column vectors $[a_1\ a_2]$, where $a_i = [a_{i,1}^T\ a_{i,2}^T\ a_{i,3}^T]^T$ and $a_{i,j} \in \mathbb{R}^{100}$ for feature $i = 1, 2$, and cluster $j = 1, 2, 3$. We generated $[a_{1,j}\ a_{2,j}]$ from three Gaussian distributions to get three clusters. These three clusters form the main clustering of these 300 samples. We randomly drew two more vectors a_3 and a_4 in the same way but with random order of the samples, such that the alternative clusterings by a_3 and a_4 are independent of the main clustering. For these 300 samples with four features, we constructed View1 with a_1 and αa_3 and, View2 with a_2 and βa_4. Three synthetic datasets were generated with different α and β representing different amount of conflicting between views.
- **UCI datasets**[2]: We chose six UCI benchmark datasets [19] for evaluation as in [15], namely Hepatitis, Iris, Wine, Glass, Ionosphere, and Wdbc. To construct the two views, we divided the features into two disjoint subsets where the first view contains the first half of the features and the second view the contains the remaining features. The divisions are performed on the data after standardization.
- **WebKB datasets**[3]: This dataset contains information of webpages from four universities in the US. We obtained a preprocessed dataset from a previous work [20]. Webpages from each university are document-samples, represented as 0/1-valued word vectors (View 1) and hyperlinks between documents (View 2). These webpages are classified into one of five classes: course, faculty, student, project and staff. We performed clustering of samples from each university as well as on all the samples. Similarly to the processing of documents described in [3], we first reduced the dimensionality of both views to 100 by Latent Semantic Analysis.

3.3   Results

Synthetic datasets. The clustering results of NMI/ACC on synthetic datasets are reported in Table 1. The numbers in parentheses indicate the differences between the obtained result and the best single-view clustering (SPV1). The result is highlighted in bold if the improvement is at least 0.01.

From Table 1 we can see that the clusterings of View 1 and View 2 have little agreement. Our proposed method MvKDR improves significantly over SPV1

---

[2] https://archive.ics.uci.edu/ml/datasets.html.
[3] http://linqs.umiacs.umd.edu/projects/projects/lbc/.

Table 1. NMI/ACC in % for synthetic data and difference to that of SPV1 in parentheses, the result is highlighted in bold if the improvement to SPV1 is more than 1%.

		SPV1	SPV2	KerAdd	CoReg	mSC	Pareto	MVUFS	MvKDR
NMI	Syn1	71.5	0.80	40.6 (−30)	**75.8 (+4.3)**	71.3 (−0.2)	46.1 (−25)	57.9 (−13)	**80.1 (+8.6)**
	Syn2	71.5	0.70	33.3 (−38)	71.8 (+0.3)	71.3 (−0.2)	52.7 (−18)	57.9 (−13)	**74.6 (+3.1)**
	Syn3	53.9	0.70	5.60 (−48)	15.5 (−38)	54.2 (+0.3)	49.0 (−4.9)	45.5 (−8.4)	**71.5 (+17)**
ACC	Syn1	91.9	38.2	62.9 (−29)	**93.2 (+1.3)**	91.7 (−0.2)	72.9 (−19)	82.8 (−9.1)	**94.7 (+2.8)**
	Syn2	91.9	37.8	60.1 (−31)	91.9 (+0.0)	91.7 (−0.2)	76.6 (−15)	82.8 (−9.1)	**93.0 (+1.1)**
	Syn3	81.5	37.8	44.2 (−37)	52.1 (−29)	81.8 (+0.4)	74.6 (−6.9)	74.3 (−7.2)	**91.3 (+9.8)**

and provides the best results in all scenarios with an average improvement in NMI/ACC of +9.7%/4.6%. KerAdd, Pareto and MVUFS severely degrade the clustering performance compared to SPV1 in all three datasets, with a reduction of NMI/ACC of −39.1%/32.7% for KerAdd, −18.7%/15.8% for Pareto and −11.6%/8.4% for MVUFS. mSC performs almost the same as SPV1. CoReg can improve the clustering quality on Syn1, but shows no benefit in Syn2, and its performance degrades severely on Syn3 (−38.4%/29.4%).

UCI dataset. Table 2 shows the clustering results on the UCI datasets and, in parentheses, the difference of NMI/ACC to the result of the best single-view clustering (SPV1). As before, the result is highlighted in bold if the improvement is at least 0.01. The Hepa and Iono datasets stand for Hepatitis and Ionosphere respectively.

From Table 2 we can see that the clusterings on the two constructed views for Hepatitis, Iris and Wdbc are not in agreement as the results of spectral clustering on View 1 are much closer to the ground truth than those on View 2. For Wine and Glass we see the opposite behavior: the clusterings on each view partly reflect the ground truth. For Ionosphere it is hard to draw any conclusion because clustering on each view performs badly.

MvKDR improves the best single-view clustering on all six datasets, whether there are conflicting views or not, with an average improvement in NMI/ACC of +6.5%/2.7%. In addition, MvKDR provides the best results on one dataset, second best results on two datasets, and third best results on the other three datasets.

CoReg can improve the clustering quality on four out of six datasets since many of them have clusterings that agree across views, with an average improvement in NMI/ACC of +4.8%/2.3%, which is inferior to the one obtained with MvKDR. In addition, it cannot gain from the second view on Iris and its performance degrades on Hepatitis.

The clustering performance of KerAdd is also degraded on datasets with conflicting views such as Hepatitis, Iris and Wdbc, with an average degradation of NMI/ACC of −27.8%/6.3%.

Pareto provides the best results on the Iris dataset, but its performance degrades severely on the other five datasets with an average degradation in NMI/ACC of −20.0%/11.1%. The reason might be that Pareto only considers binary alternative cuts. It is not clear how to merge these binary cuts into higher number of clusters.

Table 2. NMI/ACC in % for UCI data and difference to that of SPV1 in parentheses, the result is highlighted in bold if the improvement to SPV1 is more than 1%.

		SPV1	SPV2	KerAdd	CoReg	mSC	Pareto	MVUFS	MvKDR
NMI	Hepa	25.9	7.40	19.3 (−6.6)	25.0 (−0.9)	21.8 (−4.1)	5.70 (−20)	25.1 (−0.8)	26.6 (+0.7)
	Iris	67.3	10.7	32.9 (−34)	67.3 (+0.0)	67.7 (+0.4)	**80.8 (+13)**	3.80 (−63)	**75.8 (+8.5)**
	Wine	75.7	54.7	**88.4 (+12)**	**91.1 (+15)**	60.6 (−15)	42.4 (−33)	9.40 (−66)	**88.5 (+12)**
	Glass	37.9	25.2	**43.1 (+5.2)**	**43.4 (+5.5)**	**44.0 (+6.1)**	11.0 (−26)	24.1 (−13)	**43.7 (+5.8)**
	Iono	9.50	9.60	**12.0 (+2.5)**	**11.7 (+2.2)**	10.0 (+0.5)	4.60 (−4.9)	**16.8 (+7.3)**	**11.7 (+2.2)**
	Wdbc	51.3	3.50	49.9 (−1.4)	**57.9 (+6.6)**	**58.6 (+7.3)**	2.80 (−48)	46.5 (−4.8)	**57.3 (+6.0)**
ACC	Hepa	77.4	61.9	72.9 (−4.5)	76.6 (−0.8)	71.0 (−6.4)	58.9 (−18)	76.9 (−0.5)	78.1 (+0.7)
	Iris	94.0	69.0	82.0 (−12)	94.0 (+0.0)	94.0 (+0.0)	**97.0 (+3.0)**	51.0 (−43)	**96.0 (+2.0)**
	Wine	92.1	84.3	**97.2 (+4.9)**	**97.8 (+5.7)**	83.4 (−8.7)	70.8 (−21)	49.7 (−42)	**96.9 (+4.7)**
	Glass	71.5	56.5	**76.2 (+4.7)**	**76.6 (+5.5)**	**78.5 (+7.0)**	71.5 (+0.0)	59.3 (−12)	**75.7 (+4.2)**
	Iono	68.4	69.2	70.1 (+0.9)	69.8 (+0.6)	**69.5 (+1.5)**	65.0 (−3.4)	**75.1 (+6.7)**	69.8 (+0.6)
	Wdbc	89.6	63.3	87.2 (−2.4)	**91.7 (+2.1)**	**92.1 (+2.5)**	63.1 (−26)	87.9 (−1.7)	**91.7 (+2.1)**

MVUFS also provides the best results on Ionosphere, but is not effective on the rest of datasets with an average degradation in NMI/ACC of −27.8%/15.3%. This may be due to the fact that the dimensionality of the UCI datasets is rather small, and this affects the method as it is based on feature selection.

As mentioned before, mSC tries to find multiple clustering solutions of a single view. It is clear from the table that mSC improves over SPV1 on Glass and Wdbc by only correcting possible confounders in a single view. This is a sign that correction of confounders may indeed improve the clustering. However, mSC performs much worse on the rest of the datasets because it only uses single view information.

Figure 1 depicts the mean difference of NMI values of different methods with regard to the best-performing technique on each dataset (i.e. the largest NMI value obtained in any run). Similar observations can be seen for the ACC value,

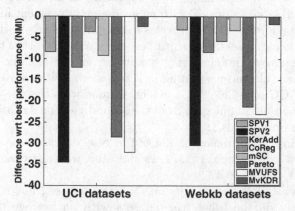

Fig. 1. The mean difference of NMI of different methods with respect to the best-performing technique on each dataset, grouped by two cases: UCI and WebKB datasets.

Table 3. NMI/ACC in % for Webkb data and difference to that of SPV1 in parentheses, the result is highlighted in bold if the improvement to SPV1 is more than 1%.

		SPV1	SPV2	KerAdd	CoReg	mSC	Pareto	MVUFS	MvKDR
NMI	Cor	41.4	6.60	27.0 (−14)	41.7 (+0.3)	40.0 (−1.)	14.1 (−27)	11.8 (−29)	**45.1 (+3.7)**
	Tex	36.7	16.0	24.2 (−12)	25.6 (−11)	**38.2 (+1.5)**	11.5 (−25)	27.7 (−9.0)	**37.9 (+1.2)**
	Was	34.9	7.80	33.4 (−1.5)	35.0 (+0.1)	32.9 (−2.0)	27.7 (−7.2)	20.7 (−14)	34.6 (−0.3)
	Wis	33.0	6.50	**40.2 (+7.2)**	33.0 (+0.0)	34.4 (+1.4)	20.1 (−12)	6.30 (−26)	33.8 (+0.8)
	All	16.5	5.70	10.7 (−5.8)	16.8 (+0.3)	15.9 (−0.6)	2.70 (−13)	11.6 (−4.9)	**17.9 (+1.4)**
ACC	Cor	55.9	38.5	46.0 (−9.9)	54.4 (−1.5)	53.8 (−2.1)	38.5 (−17)	37.9 (−18)	**57.2 (+1.3)**
	Tex	46.6	49.9	49.5 (−0.4)	47.1 (−2.8)	48.3 (−1.6)	35.9 (−10)	**53.4 (+3.5)**	48.0 (−1.9)
	Was	50.5	39.1	48.8 (−0.7)	50.6 (+0.1)	49.3 (−1.2)	27.7 (−22)	42.3 (−8.2)	50.5 (+0.0)
	Wis	50.9	37.4	**59.9 (+9.9)**	50.9 (+0.0)	**52.0 (+1.1)**	41.0 (−9.9)	40.1 (−10)	**53.3 (+2.4)**
	All	41.9	28.3	34.7 (−7.9)	41.7 (−0.2)	41.3 (−0.6)	29.5 (−12)	38.9 (−3.0)	**43.2 (+1.3)**

which is not shown due to the space limitations. From the figure it is clear that MvKDR outperforms all other methods on the UCI datasets.

WebKB dataset. Table 3 depicts the clustering results on the WebKB datasets. Cor, Tex, Was and Wis stand for the dataset from University of Cornell, Texas, Washington and Wisconsin. As it was the case for the UCI datasets, the improvement/degradation compared to SPV1 is shown in parentheses, numbers are highlighted in bold if the improvement is at least 0.01.

In Table 3 we see that the clustering on the link view (View 2) hardly agrees with the ground truth labels (on average, NMI of only 8.5%). The proposed method MvKDR improves the best single-view clustering on all but one dataset and provides the best results on two datasets and second best on the other two, with an average improvement in NMI/ACC of +1.4%/0.6%. KerAdd performs poorly on four out of five datasets compared to the best single view, with an average of degradation in NMI/ACC of −5.4%/1.8%. CoReg outputs nearly the same results as the best single view clustering on four datasets and has worse performance on the other one. The clustering performance of Pareto degrades on all five datasets (average degradation in NMI/ACC of −17.3%/12.2%), again the reason is probably that it only considers alternative binary cuts and there are five clusters in WebKB datasets. The performance of MVUFS also degrades significantly on four out of five WebKB datasets (degradation in NMI/ACC of −17.3%/12.2%). The reason might be the existence of confounders in the conflicting views.

The overall NMI performance of different methods with regard to the best-performing technique on WebKB datasets is also shown in Fig. 1. MvKDR outperforms all other methods and is the only method to improve the clustering only on View 1. These experiments show the effectiveness of our proposed method MvKDR when clustering with conflicting views.

3.4   Cancer Patients Survival Analysis

We conducted additional analyses of cancer genomics data from The Cancer Genome Atlas (TCGA) Research Network [21]. The data were preprocessed and provided by Wang et al. [22] and it includes five cancer types: glioblastoma multiforme (GBM), breast invasive carcinoma (BRCA), kidney renal clear cell carcinoma (KIRC), colon adenocarcinoma (COAD) and lung squamous cell carcinoma (LUSC). For each type of cancer, we used gene expression and DNA methylation expression data as two separate views for clustering. Our goal is to identify clusters in which patients can be considered to have a specific cancer subtype. This is a discovery process as there are no ground truth labels. Yet, we have information about certain drugs that some patients took. We performed a survival analysis on the stratified clusters from MvKDR and from the other methods and investigated if patients within a cluster had the same response to the drug treatment (here response is measured as survival time). We expect to find clusters of patients that respond the same to a drug treatment. We performed a two-sample t-test within each cluster to compare whether the survival time is significantly different between the patients that received the drug versus those that did not.

Patients without drug treatment (or with missing drug information) were removed from the analysis. We end up with 141 samples for GBM cancer, 76 samples for BRCA and 0, 20, 27 samples for COAD, KIRC and LUSC respectively. Due to small sample sizes of other three cancer types, we perform the survival analysis only on GBM and BRCA. We select the drug that is used by most of patients in each cancer type, Temozolomide for GBM and Cytoxan for BRCA. Of the 141 GBM patients, 95 were treated with Temozolomide and 52 out of the 76 BRCA patients were treated with Cytoxan.

To verify our assumption of conflicting views on this dataset, we first conducted a spectral clustering on each view and measured the NMI value between them. The NMI between clustering results from gene expression and DNA methylation is only 0.021 and 0.051 for number of clusters $k = 2$ and $k = 3$, respectively. This experiment shows that the gene expression view and DNA methylation view conflict with each other.

Next we compare the our method MvKDR to the baseline method of Single View spectral clustering SPV1 and SPV2 and state-of-the-art multi-view spectral clustering method CoReg. Gene expression profiles are more often used to define cancer subtypes, therefore we used this view as the more informative one for CoReg and MvKDR. Similarly to the analyses on the UCI and WebKB datasets, the parameters of CoReg and MvKDR were determined by the objective function of k-means. We performed clustering with increasing number of clusters k, starting from $k = 2$. For each cluster we conducted a two-sample t-test with different variance of survival time for patients with and without the drug treatment. In our analysis we found that when $k > 6$ the number of samples in a cluster is too small and the clustering tends to be more correlated to the previous one with smaller k.

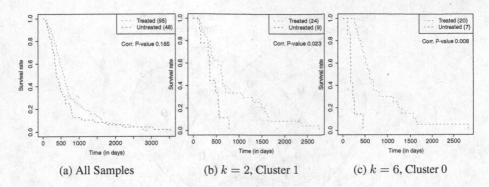

(a) All Samples   (b) $k = 2$, Cluster 1   (c) $k = 6$, Cluster 0

Fig. 2. Survival analysis of GBM patients for treatment with Temozolomide in the whole dataset and two significant clusters found by MvKDR. The numbers in parentheses denote the number of patients in the group; the p-values are corrected for multiple testing using the Bonferroni method.

MvKDR detects two significant clusters in regards to the survival analysis of patients treated with the drug Temozolomide versus those not treated with it. Figure 2 depicts the comparison of survival time for all patients and of patients in the two discovered clusters. In cluster 0 with $k = 6$, patients with the drug treatment had a significantly increased survival time with a p-value < 0.01 after Bonferroni correction. In cluster 1 with $k = 2$ we make a similar observation that treated patients live longer than untreated ones with a p-value < 0.05 after Bonferroni correction. For patients in the whole GBM dataset and in other clusters found by MvKDR, we did not observe significant differences of survival time between treated and untreated patients. This experiment shows that MvKDR can discover meaningful subgroups of patients based on their genomic profiles, where the drug treatment of Temozolomide can significantly increase the survival time. These findings will be useful for recommendations of Temozolomide treatment to patients with genomic profiles similar to those found by our model.

We further performed spectral clustering on each data view and CoReg on both views, where SPV1 corresponds to the gene expression view and SPV2 to the DNA methylation view. Both SPV1 and CoReg found one significant cluster regarding survival time of treated and untreated patients of the drug after Bonferroni corection. Figure 3 depicts the number of overlapping patients of all four significant clusters found by SPV1, CoReg and MvKDR. From the figure we can see that Cluster 2 found by MvKDR, Cluster 3 found by SPV1 and Cluster 4 found by CoReg are essentially identical, with the majority of the patients overlapping. Further, Cluster 2 found by MvKDR achieves the smallest p-value compared to the other two. Cluster 1 found by MvKDR overlaps with the other three in only four patients. This analysis shows that MvKDR is able to find useful clusters that can be also detected by state-of-the-art methods. In addition, MvKDR discovers a novel cluster that was missed by the comparison methods. One possible reason for this could be that MvKDR unveils the masked information in the DNA methylation view by using kernel dimensionality reduction and confounding correction.

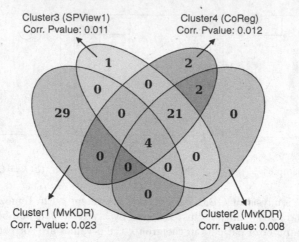

Fig. 3. Overlapping comparison of all four significant clusters of survival analysis on GBM data from different methods. MvKDR detects Cluster 2 that is almost the same as Cluster 3 (SPV1) and Cluster 4 (CoReg) with a smaller P-value of 0.008. In addition, MvKDR finds a novel significant cluster (Cluster 1) missed by other methods.

We performed survival analysis on the BRCA dataset as well, where we have similar observations as on GBM dataset. Specifically, MvKDR detects one significant cluster of 51 patients regarding survival analysis of the drug Cytoxan with a p-value of 0.036 after Bonferroni correction. SPV2 finds one significant cluster of 50 patients with p-value of 0.005. The number of overlap of patients between two groups is 33. This shows again that MvKDR can detect meaningful and novel cluster compared to the baseline algorithms.

4   Conclusion

Most existing multi-view learning approaches suffer on conflicting views and confounders. In this work, we propose a new approach named MvKDR to find the desired consensus clustering across different views, which is normally hidden or masked by confounders in conflicting views. With prior knowledge about the most informative view, our main idea is to extract two kinds of independent information from each of all the other views: the first is consistent with the desired consensus structure and the second is independent of it. The consensus clustering can be obtained by the consistent information across all the views. Our experiments on synthetic and real datasets show that the MvKDR significantly improves the clustering. In our model, we assume that we have prior knowledge about the most informative view, which in certain cases may be difficult to obtain in practice. If so, it is challenging to distinguish which clustering is more interesting. The problem may be solved in a semi-supervised manner with a few user inputs as constraints that guide the direction of the dimensionality reduction. Extending our approach in this direction may be a topic of future work.

Acknowledgments. This work was funded in part by the SNSF Starting Grant Significant Pattern Mining (XH, KB), the Marie Curie Initial Training Network MLPM2012, Grant No. 316861 (KB), the NSFC projects 11471256 and 11631012 (LL). The authors greatly acknowledge Dean Bodenham for helpful discussions and proof-reading of the manuscript.

References

1. Bickel, S., Scheffer, T.: Multi-view clustering. In: ICDM, pp. 19–26 (2004)
2. Chaudhuri, K., Kakade, S.M., Livescu, K., Sridharan, K.: Multi-view clustering via canonical correlation analysis. In: ICML, pp. 129–136 (2009)
3. Kumar, A., Rai, P., III, H.D.: Co-regularized multi-view spectral clustering. In: NIPS, pp. 1413–1421 (2011)
4. Kumar, A., III, H.D.: A co-training approach for multi-view spectral clustering. In: ICML, pp. 393–400 (2011)
5. Xia, R., Pan, Y., Du, L., Yin, J.: Robust multi-view spectral clustering via low-rank and sparse decomposition. In: AAAI, pp. 2149–2155 (2014)
6. Tang, J., Hu, X., Gao, H., Liu, H.: Unsupervised feature selection for multi-view data in social media. In: SDM, pp. 270–278 (2013)
7. Wang, H., Nie, F., Huang, H.: Multi-view clustering and feature learning via structured sparsity. In: ICML, pp. 352–360 (2013)
8. Gao, J., Han, J., Liu, J., Wang, C.: Multi-view clustering via joint nonnegative matrix factorization. In: SDM, pp. 252–260 (2013)
9. Cui, Y., Fern, X.Z., Dy, J.G.: Non-redundant multi-view clustering via orthogonalization. In: ICDM, pp. 133–142 (2007)
10. Gondek, D., Hofmann, T.: Non-redundant data clustering. In: ICDM, pp. 75–82 (2004)
11. Niu, D., Dy, J.G., Jordan, M.I.: Multiple non-redundant spectral clustering views. In: ICML, pp. 831–838 (2010)
12. Fukumizu, K., Bach, F.R., Jordan, M.I.: Kernel dimension reduction in regression. Ann. Stat. **37**(4), 1871–1905 (2009)
13. Wang, M., Sha, F., Jordan, M.I.: Unsupervised kernel dimension reduction. In: NIPS, pp. 2379–2387 (2010)
14. Christoudias, C.M., Urtasun, R., Darrell, T.: Multi-view learning in the presence of view disagreement. In: Proceedings of the Twenty-Fourth Conference on Uncertainty in Artificial Intelligence, pp. 88–96. UAI (2008)
15. Davidson, I., Qian, B., Wang, X., Ye, J.: Multi-objective multi-view spectral clustering via pareto optimization. In: SDM, pp. 234–242 (2013)
16. Qian, M., Zhai, C.: Unsupervised feature selection for multi-view clustering on text-image web news data. In: CIKM, pp. 1963–1966 (2014)
17. Gretton, A., Bousquet, O., Smola, A., Schölkopf, B.: Measuring statistical dependence with hilbert-schmidt norms. In: Jain, S., Simon, H.U., Tomita, E. (eds.) ALT 2005. LNCS (LNAI), vol. 3734, pp. 63–77. Springer, Heidelberg (2005). https://doi.org/10.1007/11564089_7
18. Xu, W., Liu, X., Gong, Y.: Document clustering based on non-negative matrix factorization. In: SIGIR, pp. 267–273 (2003)
19. Lichman, M.: UCI Machine Learning Repository (2013)

20. Sen, P., Namata, G.M., Bilgic, M., Getoor, L., Gallagher, B., Eliassi-Rad, T.: Collective classification in network data. AI Mag. **29**(3), 93–106 (2008)
21. TCGA Network: The Cancer Genome Atlas (2006). http://cancergenome.nih.gov/
22. Wang, B., Mezlini, A.M., Demir, F., Fiume, M., Tu, Z., Brudno, M., Haibe-Kains, B., Goldenberg, A.: Similarity network fusion for aggregating data types on a genomic scale. Nat. Methods **11**(3), 333–337 (2014)

Pivot-Based Distributed K-Nearest Neighbor Mining

Caitlin Kuhlman[1]([⊠]), Yizhou Yan[1], Lei Cao[2], and Elke Rundensteiner[1]

[1] Worcester Polytechnic Institute, Worcester, MA 01609, USA
{cakuhlman,yyan2,rundenst}@wpi.edu
[2] Massachusetts Institute of Technology, Cambridge, MA 02139, USA
lcao@csail.mit.edu

Abstract. k-nearest neighbor (kNN) search is a fundamental data mining task critical to many data analytics methods. Yet no effective techniques to date scale kNN search to large datasets. In this work we present PkNN, an exact distributed method that by leveraging modern distributed architectures for the first time scales kNN search to billion point datasets. The key to the PkNN strategy is a *multi-round kNN search* that exploits pivot-based data partitioning at each stage. This includes an *outlier-driven partition adjustment* mechanism that effectively minimizes data duplication and achieves a balanced workload across the compute cluster. Aggressive data-driven bounds along with a tiered support assignment strategy ensure correctness while limiting computation costs. Our experimental study on multi-dimensional real-world data demonstrates that PkNN achieves significant speedup over the state-of-the-art and scales effectively in data cardinality. Code and data related to this chapter are available at: http://solar-10.wpi.edu/cakuhlman/PkNN.

Keywords: K-nearest neighbor search · Distributed computing
MapReduce

1 Introduction

Detecting the nearest neighbors of all points in a dataset is a ubiquitous task common to numerous data mining endeavors. Many techniques, including classification and regression [7], clustering [3,10], and outlier detection [4,20], require a k-nearest neighbor (kNN) search to be performed for every point in the dataset. As real-world applications increasingly rely on the analysis of very large datasets, they require resources beyond what is feasible on a single machine. Therefore, the development of highly distributed solutions for kNN search is no longer an option, but a necessity.

Several distributed exact kNN search methods have been proposed in the literature [5,15,24], using a *support set strategy*. First, the data is divided into disjoint partitions by grouping nearby points together. Then these *core partitions* are each augmented by additional sets of *support points* which could potentially

M. Ceci et al. (Eds.): ECML PKDD 2017, Part II, LNAI 10535, pp. 843–860, 2017.
https://doi.org/10.1007/978-3-319-71246-8_51

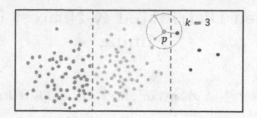

Fig. 1. kNN partitioning problem.

be neighbors of the points in the core partition. The support sets ensure that a local kNN search over core points will return exact results without requiring access to data stored on any other machines.

Although this approach successfully distributes kNN computation across machines, existing methods are not scalable beyond million point datasets, as confirmed by our experiments (Sect. 5). Major shortcomings include the use of worst-case estimation for support sets [15] leading to excessive data duplication, or computationally intensive support determination leading to excessive run-times [5]. Worse yet, the *outlier problem* in skewed data sets has largely been overlooked. Data points far from all others can disproportionately inflate support sets and overload machines, resulting in job failures.

Challenges. A key property of kNN search is that the distance from each point to its neighbors may vary considerably across a dataset. The distance tends to be small in dense areas, while very large for points in sparse areas, as illustrated in Fig. 1. Therefore the distance from points to potential neighbors in the support set varies greatly across partitions. To locate the exact set of all necessary support points for each partition a priori would essentially require a kNN search be conducted even before the data partitioning is formed. Alternatively, if the distance to support points is estimated given only limited information about the data distribution, it must be very conservative to ensure correctness. Unfortunately, this inevitably leads to prohibitively high data duplication as shown above. Designing a strategy that safely bounds the support set for each partition while introducing minimum data duplication thus represents an open problem.

Second, load balancing across the partitions is critical, since a single long-running analytics task will determine the response time of the entire job. Furthermore, if the workload assigned to one machine exceeds its accommodation capacity, it will cause job failures. In distributed kNN search, the workload of each machine is determined not only by the cardinality of each core data partition, but also the size of its associated support set. The strongly interdependent nature of the partitioning and support point estimation problems complicates load balancing. On the one hand, to ensure load balance the partitioning has to take the support points into consideration. On the other hand, the support points required for a partition are determined by the distribution of data within. This raises the proverbial chicken and egg problem.

Proposed Approach. In this work we propose a distributed method that for the first time scales exact kNN search to billion point datasets. Our approach fully exploits a pivot-based partitioning strategy, so called **Pivot-Based kNN Search (PkNN)**.

PkNN overturns the common understanding that a distributed analytics task should be completed in as few rounds as possible. We show that decomposing the kNN search into multiple rounds reveals opportunities to solve the problems of generating balanced partitions and accurately predicting support points. The rich knowledge learned in the kNN search itself can be fully explored to derive a data-driven bound on the support set much tighter than bounds based on worst-case estimation. In particular PkNN features a multi-granularity pruning strategy to accurately yet efficiently locate support points. By leveraging the learned knowledge it is able to identify and prune the set of *partitions* that cannot possibly contain any support points. Then at a finer individual data point granularity it introduces the concept of a boundary hyperplane to quickly discover support points based on their position relative to the hyperplane.

The support set bound and learned knowledge together are then leveraged to dynamically adjust the imperfect initial partitioning to ensure load balancing for the next round of kNN search. Our key observation here is that severe load imbalance is caused by a few far-located points in the dataset, so called outliers. By dynamically re-distributing these points, our *outlier-driven partition adjustment* mechanism ensures that no partition will overwhelm the resources of its assigned machine. Furthermore, it is shown to tighten the support point bound – minimizing data duplication to a rate close to 1.

Contributions. The key contributions include:

1. We present PkNN, the first distributed solution that scales kNN search to billion point datasets.
2. Our multiple round kNN search strategy successfully minimizes the data duplication rate and balances the workload across different machines.
3. Our outlier-driven partition adjustment strategy overcomes the *outlier problem* in kNN search, adapting partitioning to the data distribution and available computing resources of each machine.
4. Experimental evaluation on large real-world datasets shows that PkNN significantly outperforms the state-of-the-art in data duplication rate and runtime.

2   Problem Statement

Here we formalize the distributed kNN search problem. Frequently used symbols are listed in Table 1. Given a multi-dimensional dataset D, the task is to find for each point p in D a set of k other points which are most similar to p. We assume this similarity measure to be a well defined distance metric $|\cdot|$ such as any L^p norm.

Definition 1. *kNN Search.* $\forall$ *points* $p \in D$, *find the set of points* $kNN(p)$ *where* $kNN(p) = \{q_1, q_2, ..., q_k \in D \mid \forall r \in D \setminus kNN(p), |p, q_i| \leq |p, r|, \ r \neq p\}$.

Table 1. Table of symbols

Symbol	Definition
kNN(p)	The k nearest neighbors of point p
V_i	Voronoi cell
v_i	Pivot corresponding to V_i
$V_i.core$	Core points of a Voronoi cell $\{p \in D \mid |p, v_i| \leq |p, v_j| \ i \neq j\}$
$V_i.support$	Support set of Voronoi cell V_i
$core\text{-}dist(V_i)$	$max(|p, q|) \ \forall p, q \in V_i.core \ q \in kNN(p) \in V_i.core$
H_{ij}	Hyperplane boundary between Voronoi cells V_i and V_j
$hp\text{-}dist(q, V_i)$	The distance from a point $q \in V_j$ to H_{ij}
$support\text{-}dist(V_i)$	$max(|p, v_i| + |p, q|) \ \forall p, q \in V_i.core \ q \in kNN(p) \in V_i.core$

The naïve solution for centralized kNN search is to simply compare each point with every other point in the dataset and choose the k closest. This requires access to all points in the dataset, i.e. it is quadratic. For today's huge datasets, the time to complete this exhaustive search is prohibitive, and furthermore, the data may be too large to be accommodated on a single machine. Therefore the kNN search task must be completed in a distributed fashion without local access to the entire dataset. We target shared-nothing platforms such as Hadoop [22] and Spark [25] where unrestricted pairwise exchange of data is not allowed between machines. Therefore our goal is to complete the kNN search autonomously for the portion of the dataset processed on each machine. To meet stringent response time requirements of big data applications, this distributed approach must effectively minimize end-to-end execution time.

Definition 2. *Distributed kNN Search.* *Given a dataset D stored across n machines in a distributed compute infrastructure, perform kNN Search by processing n subsets (called cells) C_i such that $C_1 \cup C_2 \cup ... \cup C_n = D$ on separate machines independently in parallel while minimizing end-to-end execution time.*

2.1   Data Partitioning

Data partitioning is the first step of any distributed analytics task. If data is distributed among machines randomly, then it is likely that the neighbors of a given point will be sent to a different machine. Therefore a data partitioning strategy which preserves data proximity is desired. Here we adopt *pivot-based data partitioning*. First a small set of n initial points, or *pivots*, is chosen and then all data points are grouped based on their distances to the pivots. The result is a data partitioning known as a Voronoi Diagram which determines a unique division of the metric space into Voronoi cells.

Definition 3. *Voronoi Cell.* *Given a dataset D and set of n pivots $P = \{v_1, v_2, \ldots, v_n\}$ we have n corresponding Voronoi cells $\{V_i \mid V_1 \cup V_2 \cup ... \cup V_n = D, V_i \cap V_j = \emptyset\}$ and $\forall p \in V_i, distance(p, v_i) \leq distance(p, v_j), i \neq j$.*

Although pivot-based partitioning preserves the locality of the data, for points that lie along the boundaries of cells, their neighbors may still be sent to other machines. The use of a *support set* for each data partition ensures that the neighbors of all points in a partition can be found locally. Each cell is augmented by additional points which may be neighbors of those points near the boundary of the cell [5, 15, 24]. The points inside each cell C_i are denoted as $C_i.core = \{p \mid p \in C_i\}$. The support set of the cell, denoted $C_i.support$, must be sufficient to guarantee that the kNN of all core points in each cell C_i can be found among $C_i.core$ and $C_i.support$.

Definition 4. Support Set. *The support set of a cell C_i contains at least all data points which satisfy the following two conditions: (1) $\forall q \in C_i.support$, $q \notin C_i.core$, and (2) there exists at least one point $p \in C_i.core$ such that $q \in kNN(p)$.*

This suppport set strategy categorizes data points into two classes, namely *core points* and *support points*. Each data point will be assigned as a core point to exactly one cell, and possibly many support sets. Support points must be duplicated and transmitted multiple times. A large number of support points increases the computation costs per partition since many more points must be searched. Worst yet, it also introduces heavy communication costs which often dominate the cost of distributed methods [2]. We model these costs by the *duplication rate* in Definition 5. To minimize the duplication rate, the support set should contain as few points as possible.

Definition 5. *Given a dataset D and a distributed algorithm A for computing the kNN of all points in D, the **duplication rate** $dr(D, A) = \frac{|Rec(D,A)|}{|D|}$, where $\mid D \mid$ represents the cardinality of D and $\mid Rec(D, A) \mid$ the number of data records produced by the partitioning of Algorithm A.*

3 PkNN: Pivot-Based Distributed kNN Search

3.1 The Overall PkNN Approach

Our Pivot-based Distributed kNN Search approach or in short PkNN not only ensures autonomous kNN computation for each data partition, but also minimizes the data duplication caused by the use of support sets. PkNN is based on our critical observation about the **data dependent property of kNN distance** A common understanding in the distributed analytics literature is that in shared-nothing distributed infrastructures an analytics task should be completed in as few rounds (or MapReduce jobs) as possible – ideally in one round. This avoids reading and writing the data many times, as well as high communication costs incurred from repeated shuffling phases, which are often the dominant costs of a distributed algorithm [21]. Although widely adopted in the literature, this approach does not hold in the distributed kNN search context due to the data dependent property of kNN distance. In order to complete the kNN search in one round, the distance to the support points has to be estimated a priori. However, the distance from each point to its kNN may vary considerably across a dataset. It may be small in dense areas, while very large in sparse areas. Therefore it is difficult to predict the distance to support points and in turn accurately estimate the support set of each data partition without precisely knowing the underlying data distribution characteristics. Given only limited information about the data distribution, estimates must be very conservative to ensure correctness [15]. This inevitably leads to prohibitively high data duplication mitigating the benefit of the one round solution.

The Multi-round Approach. Inspired by the *data dependent property* observation, PkNN no longer aims to complete the kNN search in one round. Instead it decomposes the kNN search process into two steps, namely **core kNN search** and **support kNN search**. The core kNN search step performs an initial kNN search *only over the core points in each cell*. The kNN produced for a point p in this step is called the core kNN of p. The key insight is that by examining the points in each cell directly we learn essential information about the characteristics of the underlying data in each Voronoi cell V_i. Specifically, we determine the *core-distance* and *support-distance* of each cell to derive a tight upper bound on the distance from any core point to its support points.

After the support set of each Voronoi cell is determined, the support kNN search step then completes the kNN search over *only the support points* of each cell. We observe that given a core point $p \in V_i$, its true kNN are guaranteed to be located among its core kNN and $V_i.support$. In other words, no repeated computation ensues in our multi-round kNN search strategy. Next we detail how to utilize the core-distance and support-distance bounds to locate the support set.

3.2 Support Set Discovery

Our support set discovery method features a *multi-granularity pruning strategy* to accurately yet efficiently locate support points. Given a cell V_i, by utilizing the *support-distance*(V_i) and the distances between the pivots, it is able to quickly identify and prune the set of Voronoi cells V_j that do not have any chance to contain the support points of V_i. Furthermore, at a finer individual data point granularity, by introducing the concept of boundary hyperplane, it quickly prunes the points based on their distances from the hyperplane and the *core-distance*(V_i). First, we formally define the concepts of core-distance and support-distance.

Definition 6. *The core-distance of a Voronoi cell V_i.* core-distance(V_i) = $max(|p, q|) \; \forall \; p, q \in V_i.core$ *where* $q \in kNN(p) \in V_i.core$.

The core-distance of a given Voronoi cell V_i represents the maximum distance from a core point p to its kth nearest *core* neighbor q. The core-distance effectively defines an upper bound on the distance between any core point of V_i and the possible support points. In other words, given a point q outside V_i, it is guaranteed not to be a support point of V_i if its distance to any core point of V_i is larger than the *core-distance*(V_i).

The support-distance takes the pivot v_i of cell V_i into consideration. It captures the maximum distance of a possible support point of V_i to the pivot of V_i.

Definition 7. *The support-distance of a Voronoi cell. support-distance* $(V_i) = max(|v_i, p| + |p, q|)$ *where q is the kth nearest neighbor of p $\forall p, q \in$ $V_i.core$.*

Cell Granularity Pruning. For each Voronoi cell V_i, we first utilize the *support-distance*(V_i) to determine a set of *candidate support cells*, as shown in Fig. 2. This is equivalent to pruning the cells which could not possibly contain support points for V_i, as stated in Lemma 1.

Lemma 1. *Cell Granularity Pruning. Given Voronoi cells V_i, V_j and their corresponding pivots v_i, v_j $i \neq j$, if the support-distance$(V_i) \leq |v_i, v_j|/2$, then V_j does not contain any support points of V_i.*

Proof. Recall that by Definition 3, since q is in V_j, it is closer to the pivot v_j than to any other pivot. Therefore, $|q, v_i| \geq |v_i, v_j|/2$. We give a proof by contradiction: Let some point $q \in V_j$ be a necessary support point of cell V_i, i.e., $q \in kNN(p)$ for $p \in V_i$. Then by Definition 4, $|p, q| < |p, r|$ where $r \in V_i.core$ is the kth nearest neighbor of p out of all the core points in V_i.

Assume that Theorem 1 is not true. Then we have:

$$|r, p| + |p, v_i| < |v_i, v_j|/2 \qquad \text{by assumption}$$
$$|q, p| + |p, v_i| < |v_i, v_j|/2 \qquad \text{def 4}$$
$$|q, v_i| \leq |q, p| + |p, v_i| \qquad \text{triangle inequality}$$
$$|q, v_i| < |v_i, v_j|/2 \qquad \text{transitivity}$$

This results in a contradiction. If $|q, v_i| < |v_i, v_j|/2$, then $q \in V_i$, which violates the original assumption.

Based on Lemma 1, we can quickly determine whether a cell V_j is a support cell of V_i by only computing the distance between their pivots. Performing this pruning at the cell level, we not only avoid unnecessary data duplication, but also reduce the number of cells each point must be checked against when mapping points to support sets.

Point Granularity Pruning. Even if a cell V_j is determined to be a support cell of V_i, not every point in V_j is a necessary support point of V_i. Our point granularity pruning strategy is based on the concept of a boundary hyperplane. Given two adjacent Voronoi cells V_i and V_j, the boundary hyperplane is a hyperplane H_{ij} that contains the midpoint of V_i and V_j. Given a point q in V_j, the distance from q to H_{ij} – denoted as *hp-distance*(q, H_{ij}) (*simplified as hp-distance*(q) *when no ambiguity arises*) – and the core-distance of V_i together can determine whether q is a support point of V_i.

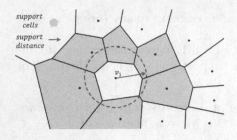
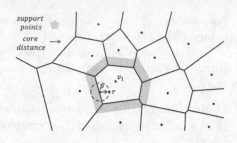

Fig. 2. Support-distance(V_1) used to define support cells.

Fig. 3. Core-distance (V_1) used to define support points.

Lemma 2. _Point Granularity Pruning._ _Given any point $p \in V_i$, $q \in V_j$, $i \neq j$, if hp-distance$(q) \geq$ core-distance(V_i), then $q \notin kNN(p)$_

Proof. Let $r \in V_i$ be the furthest core kNN of $p \in V_i$. By Definition 6, $|p, r| \leq$ _core-distance_(V_i). Since $|q, p| >$ _hp-distance_(q) and _hp-distance_$(q) \geq$ _core-distance_(V_i), we get $|q, p| >$ _core-distance_(V_i). Therefore, $|q, p| > |p, r|$. This proves that q is guaranteed to be not a kNN of p.

Figures 2 and 3 illustrate the intuition behind these core and support bounds in a 2D space. In Fig. 2 the _support-distance_ of a cell V_i is used to determine a set of shaded candidate cells which may possibly contain support points. Then in Fig. 3 the _core-distance_(V_i) is determined by the max distance from a point to its kth nearest neighbor – in this case $|p, r|$. Points in the candidate support cells which fall within this distance of the linear boundaries of V_i are assigned to $V_i.support$.

hp-distance Computation. When the commonly employed Euclidean or Mahalanobis distances are used to define the space over which the Voronoi Diagram is constructed, the boundaries of Voronoi cells are comprised of piecewise linear hyperplanes described by Theorem 1. In this case, the exact hp-distance between a point and a cell can be computed [19].

Theorem 1. _The boundary between two adjacent Voronoi cells V_i and V_j is a hyperplane H_{ij} containing their midpoint. In $\mathbf{R}^d$ the hyperplane is given as $H_{ij} = y : y^T n + p = 0$ where n is the normal vector orthogonal to the plane. $\forall y \in H_{ij}, |y, v_i| = |y, v_j|$. The distance of any point $p \in \mathbf{R}^d$ to the plane is $hp - distance(q) = \frac{|s^T n + p|}{\|n\|_2}$._

In arbitrary metric spaces the exact computation of the hp-distance may not be straightforward. Fortunately, the lower bound on _hp-distance_(q) in Theorem 2 is shown in [13] via triangle inequality to hold. This lower bound can be used in place of the exact value of _hp-distance_(q) in Lemma 2. The proof of this conclusion is straightforward. If the lower bound of _hp-distance_(q) is larger than core-distance(V_i), then the exact $hp\text{-}distance(q)$ is guaranteed to be larger than core-distance(V_i).

Theorem 2. *Given two cells V_i and V_j with corresponding pivots v_i and v_j and a point $q \in V_j$, $hp\text{-}distance(q) \geq \frac{|q,v_i| - |q,v_j|}{2}$.*

Point-granularity support assignment along with the cell-based pruning strategy introduced in Lemma 1 together ensure that PkNN quickly discovers the support set of each cell V_i, while minimizing the data duplication rate by only including a small set of potential neighbors.

4   Outlier Driven Partition Adjustment

Our multi-round PkNN approach reduces the duplication rate dramatically and significantly outperforms the alternatives in both end-to-end execution time and scalability to the size of the dataset as confirmed in our experiments. However, our experimental study and theoretical analysis demonstrate that it still cannot handle datasets with billion point cardinality due to load imbalance. It is well known that load imbalance causes job failures if the workload assigned to one machine exceeds its processing capacity.

Support-Aware Load Balancing. Intuitively, pivot-based partitioning tends to generate partitions with a balanced workload if pivots are uniformly randomly sampled from the data utilizing for example a reservoir sampling technique [23]. More pivots are selected from dense areas and fewer from sparse areas, naturally reflecting the distribution of the data and tending to evenly partition even data with skewed distribution. Unfortunately, the workload of each machine is not only determined by the cardinality of each data partition, but also by the size of its support set. This makes the pivot-based partitioning problem and the support point estimation problem strongly interdependent. The support points are determined by the distribution characteristics of each data partition. Therefore without knowing the data partitioning it is impossible to estimate the support points.

The remedy again comes from our multi-round PkNN approach. Since PkNN decomposes kNN search into two steps, core kNN search and support kNN search, it gives us the opportunity to *adjust* the partitions formed at the core kNN search step and avoid an unbalanced workload during the support kNN search caused by the addition of support points. In other words, it is not necessary to solve the 'mission impossible' of producing partitions with perfect balanced workload before conducting the kNN search. First, uniform sampling is sufficient to produce partitions balanced enough for core kNN search. Next, based on the support set discovered via the method introduced in Sect. 3.2, we adjust partitions that could potentially overload the compute nodes.

Outlier Driven Partition Adjustment (ODA). Our partitioning adjustment method is based on the key observation that given a data partition V_i, some particular points in V_i lead to a large number of support points and in turn cause severe load imbalance. These particular points, called outliers, are points that have large distance from their kth nearest core neighbor. By Lemma 2, a point q is assigned to $V_i.support$ if the distance from q to the boundary hyperplane is smaller than the $core\text{-}distance(V_i)$. Therefore the larger the $core\text{-}distance(V_i)$, the more points $V_i.support$ contains. Since the $core\text{-}distance(V_i)$ is determined by the point p in V_i that is furthest from its core kNN, the outliers will significantly increase the number of the support points. Leveraging this observation, our outlier driven partition adjustment, or in short ODA not only ensures no machine will be overloaded, but also minimizes the data duplication rate close to 1.

4.1 ODA: Outlier Driven Partitioning Adjustment

Definition 8. *Given a compute cluster* $\mathbb{C}$ *with a constraint* m *on the number of points that can be processed by a single machine* M_i *in* $\mathbb{C}$ *(mapped from the resource constraint of M such as memory), the goal of ODA is to ensure that each cell* V_i *of dataset* D *satisfies the constraint:* $|V_i.core| + |V_i.support| <= m, \forall\ V_i \in D.$

Outlier Discovery. ODA first discovers outliers by translating the data cardinality constraint m put on each single machine M_i to a distance constraint α_i based on the distribution of the core points in cell V_i. A point o is considered as an outlier if the distance to its kth core neighbor is larger than α_i.

ODA first maps the irregular shaped Voronoi cell V_i generated at core kNN search step into regular hypersphere centered on the pivot v_i. Assuming the core points $V_i.core$ are uniformly distributed in this hypersphere, then the radius r of V_i can be represented utilizing the mean distance $|v_i, p|\ \forall p \in V_i.core$. The number of core points c assigned to V_i can be naturally considered as the "mass" of V_i. Then the "density" of V_i can be modeled as $density(V_i) = \frac{c}{Volume(V_i)}$. Based on the well known mathematical principle that the "volume" of a hypersphere in n dimensional Euclidean space, is proportional to the nth power of the radius r by a constant factor, $Volume(V_i)$ can be represented by $x \times r^n$, therefore $density(V_i) = \frac{c}{x \times r^n}$.

Assume that the distribution of the data close to a Voronoi cell V_i will be similar to that contained within the cell, then a hypersphere $\hat{V}_i$ centered also on the pivot v_i but with a larger radius will have the similar density to V_i. Suppose $\hat{V}_i$ contains m points, where m represent the data cardinality constraint of machine M_i. Then we have

$$\frac{c}{r^n} = \frac{m}{(r + \alpha_i)^n} \tag{1}$$

Here α_i denotes the allowable distance to expand V_i to a sphere $\hat{V}_i$ that contains no more than m points. This value α_i can be computed as follows.

$$\alpha_i = \left(\frac{m \times r^n}{c}\right)^{1/n} - r \tag{2}$$

For each Voronoi cell V_i, α_i gives a customized max threshold on the distance to points in V_i.support subject to resource constraints and the data distribution of V_i.core.

Lemma 3. *If* $|p, q| < \alpha_i \; \forall \; p, q \in V_i.core$ *where* $q \in kNN(p) \in V_i.core$, *then* $|V_i.core| + |V_i.support| < m$.

Proof. If the core k-distance of $\forall$ point p $\in V_i$.core is smaller than α_i, then all points in V_i.support are covered in hypersphere $\hat{V}_i$. Since $\hat{V}_i$ covers at most m points, this proves $|V_i.core| + |V_i.support| < m$.

Based on Lemma 3, points $p \in V_i.support$ are identified as outliers of cell V_i if their distance to their core kNN cause *core-distance* (V_i) to be larger than α_i.

Partition Adjustment. To ensure that outliers cannot skew a cell V_i and cause job failure, V_i will be adjusted before conducting the support kNN search step by redistributing the outliers within the compute cluster. This eliminates the influence of the outliers on our data-driven bounds, while still leaving the original partitioning intact.

Algorithm 1. ODA

1: **function** ODA(*pivList*)
2:     avgDistList = calcAvgDistance(*pivList*)
3:     outlierList = detectOutliers(*pivList, avgDistList*)
4:     concat(*pivList, outlierList*)
5:     supList = calcSupportList(*pivList*)
6:     supportKNNSearch(*pivList, supList*)

The overall outlier-driven partition adjustment approach (ODA) is shown in Algorithm 1. First, for each cell V_i the information necessary to compute α_i, namely the average distance from each point to the pivot v_i can be easily collected during the initial pivot-based data partitioning phase (Line 2). Then outliers are identified during core kNN search. After the core k-distance of a point p is calculated, p is marked as an *outlier* if its core k-distance exceeds the threshold α_i (Line 3). Next when determining the support set of V_i, these outliers are treated as additional "special" pivots. These special pivots differ from regular pivots in that no additional core points are assigned to their partitions (Line 4). Since an outlier o would have already found their kth nearest neighbor from among the core points (core kNN) in their original partition V_i, the distance from o to its kth core kNN determines the core-distance of its new special partition. This in turn is utilized to determine the support points of o (Line 5). In the support kNN search step, kNN search is conducted only for the single core outlier point of each of these "special" partitions (Line 6). Since by their nature the outliers are far from all other points, it is unlikely that many additional support points will be mapped to the newly formed special partitions.

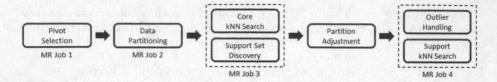

Fig. 4. PkNN framework.

By discovering and handling the outliers in this way, ODA ensures that no partition V_i will exceed the·processing capacity of each machine. Furthermore, the overall data duplication rate of the whole dataset is further reduced to be close to 1 as confirmed in our experiments. The overall structure of the full-fledged PkNN framework is shown in Fig. 4.

5 Experimental Evaluation

5.1 Experimental Setup

All experiments are conducted on a shared-nothing cluster with one master node and 28 slave nodes. Each node consists of 16 core AMD 3.0 GHz processors, 32 GB RAM, 250 GB disk, and nodes are interconnected with 1Gbps Ethernet. Each server runs CentOS Linux (kernel version 2.6.32), Java 1.7, Hadoop 2.4.1. Each node is configured to run up to 4 map and 4 reduce tasks concurrently. Speculative execution is disabled to boost performance. The replication factor is set to 3.

Datasets. We evaluate PkNN on real world data using *OpenStreetMap* [1], one of the largest real datasets publicly available, which has been used in similar research work [26]. It contains geolocation information for physical landscape features such as a buildings and roads all over the world. Two attributes are used, longitude and latitude. Hierarchical datasets (Shown in Table 2) evaluate the scalability of PkNN with regard to the data size. The datasets grow from a Massachusetts dataset of 10 million points to over 1 billion points covering more than the western hemisphere.

Real data are also used to evaluate the performance of PkNN in higher dimensions. The Sloan Digital Sky Survey (*SDSS*) [8] is one of the largest astronomical catalogs publicly accessible. We extracted a dataset containing 100 million records from the thirteenth release of SDSS data. In this experiment we utilize seven numerical attributes including Right Ascension, Declination, three Unit Vectors, Galactic longitude and Galactic latitude. The *TIGER* [9] dataset contains 70 million line segments, representing GIS features of the United States. Four numerical attributes giving the longitude and latitude of two endpoints of line segments are used.

Methods. We compare PkNN[1] against two state-of-the-art distributed solutions for kNN search in MapReduce. The first called PBJ [15] uses pivot-based partitioning to perform kNN Join. The authors of the [15] generously shared their code, which we adapted to run on a single dataset. The second called Spitfire [5] uses grid-based partitioning to perform kNN search on a message-passing architecture. We have implemented their method and adapted it for MapReduce.

Metrics. *End-to-end execution time* is measured, which is common for the evaluation of distributed algorithms. Furthermore, the *execution time for the key stages* of the MapReduce workflow is broken down to evaluate the performance of different stages of computation. The other key metric is the *duplication rate* (Definition 5). This measure, calculated as the total number of core and support points processed divided by the number of points in the input dataset, captures how effectively supporting areas are bounded.

5.2 Evaluating Data Duplication Rates

We first evaluate the data duplication rate of the three methods with respect to the number of data partitions. For a truly scalable solution, the benefit of adding partitions (necessary as the data size grows) should not be outweighed by the increased communication costs. Experiments are conducted on the Massachusetts dataset containing 10 million records. This data set has areas of varying density throughout the domain space. k is fixed at 5 and the number of pivots is varied from 50 to 1100.

To perform a fair comparison of PkNN and PBJ, they are required to use the same pivot set, as the choice of pivots impacts performance significantly. Therefore we omit the grouping step introduced in [15], which starts with a very large number of pivots and then groups them into a small number of final partitions. As their own evaluation shows, this grouping step only results in a modest decrease in execution time [15]. The number of partitions in Spitfire is tuned to a similar amount, although the equi-depth partitioning algorithm [5] does not allow us to set an exact number of partitions. Both the end-to-end runtime as well as the data duplication rate are measured for all methods.

Figure 5a shows PkNN and Spitfire clearly attain much lower data duplication than PBJ (shown on the left), and consequently execution time (shown on the right). Even using a small number of pivots, the minimum duplication rate for PBJ is around 30. PkNN and Spitfire on the other hand remain close to the optimal rate of 1. The PBJ method shows a quadratic increase in running time. Figure 5b compares the overall execution time for PkNN and Spitfire. For these methods, adding more data partitions improves runtime. Both methods perform similarly on this small data size. However, as we demonstrate in Sect. 5.3, PkNN outperforms Spitfire due to its computational complexity as the data size increases, despite both methods achieving low data duplication rates.

---

[1] PkNN source code available at http://solar-10.wpi.edu/cakuhlman/PkNN.

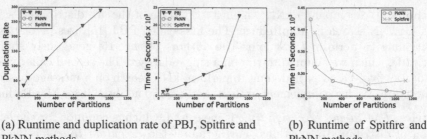

(a) Runtime and duplication rate of PBJ, Spitfire and PkNN methods.

(b) Runtime of Spitfire and PkNN methods.

Fig. 5. Impact of number of partitions on Massachusetts dataset.

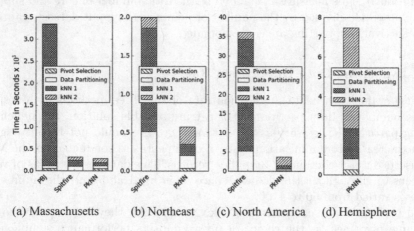

(a) Massachusetts   (b) Northeast   (c) North America   (d) Hemisphere

Fig. 6. Varying dataset size from 10 million to over 1 billion data points.

5.3   Evaluating Scalability

In this section the scalability of PBJ, Spitfire, and PkNN with regard to the data size is evaluated. For this set of experiments we use 2d data from the Open Street Map dataset and fix k at 5. Figure 6 shows the end-to-end execution time of each method on datasets of varying size, with the time for each key stage of the MapReduce workflow broken down. In Fig. 6a we observe that the single round kNN search approach of the PBJ method is almost 20 times slower than the two kNN search rounds in PkNN. Attempts to evaluate PBJ on data larger than 10 million points were not successful due to job failures caused by the high data duplication rate.

Figure 6b and c show PkNN outperforms Spitfire by a factor of 3 and a factor of 7 respectively. The first kNN job in Spitfire performs the support discovery step using the concept of point hiding [5]. These experiments clearly show the computational complexity of this approach hinders performance when the data contains hundreds of millions of points. Figure 6d evaluates the performance of PkNN on a billion point dataset, which neither competing method could handle. This demonstrates that PkNN truly scales to handle modern data quantities.

Table 2 shows the dataset sizes and parameters used for evaluation. To choose the number of pivots, it was observed that a ratio of 1: 30,000 points performed well on the small dataset in initial experiments. Fewer pivots were experimentally determined to perform better on larger data. Attempts to run PkNN on datasets larger than 10 million without the ODA partition adjustment mechanism lead to job failures. However, by identifying a small fraction of the data as outliers in each dataset ODA achieves a more balanced workload allowing for the processing of these large datasets, as well as data duplication rates close to 1.

Table 2. Dataset sizes.

Dataset	Points	Pivots	Mean cell size	Outliers	Duplication rate
Massachusetts	10,000,000	300	33,333	3	1.93
Northeast	80,464,841	2400	33,527	56	1.66
North America	812,233,510	24,000	33,848	361	1.51
Western Hemisphere	1,185,194,762	11,000	107,744	1857	1.49

5.4   Evaluating the Impact of the Number of Dimensions

Finally we show that PkNN effectively handles multidimensional data at scale. The PBJ method cannot accommodate data beyond the 10 million points, and Spitfire only supports 2-dimensional data. Table 3 shows the performance of PkNN on the Tiger and SDSS datasets which contain 70 and 100 million records and four and seven dimensions respectively. We can see PkNN continues to achieve low data duplication rates under 2 on real multidimensional data as well as fast runtime.

Table 3. PkNN performance on multidimensional data.

Dataset	Cardinality	Dimension	End-to-end execution time (sec)				Duplication rate
			Pivot selection	Partitioning	kNN 1	kNN 2	
Tiger	72,729,686	4	41	148	167	203	1.18
SDSS	100,000,000	7	41	494	237	230	1.59

6   Related Work

kNN search is a well-studied problem with many solutions to mitigate its high computational complexity $(O(n^2))$. In centralized algorithms, spatial indexing structures are typically used, such as grid-based indices for low dimensional data or hierarchical tree structures [6,11,14], as they achieve an $O(nlogn)$ expected complexity.

In recent years, a number of distributed solutions for kNN search have been proposed for message-passing and peer-to-peer distributed systems [5,12,18].

These approaches cannot be utilized on modern distributed architectures such as MapReduce [22] and Spark [25] which are the focus of our work due to their scalability, fault tolerance, and ability to run on commodity hardware. A recent methods for kNN classification have targeted these platforms [16,17], however they utilize a broadcast join technique, where the neighbors of a small test set of points are found among a larger training set distributed across a compute cluster. This technique is not applicable to our problem where we aim to find all neighbors of all points in large datasets (over 1 billion points) which cannot be processed in their entirety on a single machine.

A MapReduce based kNN search approach was proposed in [15]. Similar to PkNN this approach also utilizes pivot-based partitioning to divide the dataset. Worst-case estimation based on the distance from pivots to a small number of points in each partition bounds the size of the supporting sets. As shown in their experiments and confirmed by our results (Sect. 5.2), this leads to extremely high data duplication (upwards of 20x the size of the original dataset). Although an adaptation of the bound and pivot selection method is presented in [24] as part of system specific to spatial data, the worst case estimation of support sets still cannot scale to large datasets due to the duplication rate. Our multi-round PkNN approach effectively cuts the duplicate rate close to be 1 by adopting a multi-round strategy.

The Spitfire approach [5] effectively improves over the duplication rate of [15,24] by utilizing grid-based partitioning to divide the data and bound the support set. However, its point hiding concept to discover support sets leads to high computation costs as the data size increases, as confirmed by our experiments (Sect. 5.3). Spitfire is also specific to two-dimensional geo-location datasets while our PkNN approach is effective in supporting the general case of multi-dimensional datasets.

7   Conclusion

This work presents PkNN, an innovative distributed approach for kNN search over modern distributed architectures. We introduce a multi-round computation strategy along with data-driven bounds and a tiered support discovery technique which effectively limit data duplication. An outlier-driven partition adjustment mechanism ensures load balance. PkNN for the first time scales kNN search to billion-point real world datasets and gracefully handles multidimensional data.

References

1. Open Street Map. http://www.openstreetmap.org/. Accessed 23 Apr 2016
2. Afrati, F.N., Sarma, A.D., Salihoglu, S., Ullman, J.D.: Upper and lower bounds on the cost of a map-reduce computation. Proc. VLDB Endow. **6**(4), 277–288 (2013)
3. Ankerst, M., Breunig, M.M., Kriegel, H.P., Sander, J.: OPTICS: ordering points to identify the clustering structure, pp. 49–60. ACM Press (1999)

4. Breunig, M.M., Kriegel, H.P., Ng, R.T., Sander, J.: LOF: identifying density-based local outliers. In: ACM SIGMOD International Conference on Management of Data, pp. 93–104. ACM, New York (2000)
5. Chatzimilioudis, G., Costa, C., Zeinalipour-Yazti, D., Lee, W.C., Pitoura, E.: Distributed in-memory processing of all k nearest neighbor queries. IEEE Trans. Knowl. Data Eng. **28**(4), 925–938 (2016)
6. Ciaccia, P., Patella, M., Zezula, P.: M-tree: an efficient access method for similarity search in metric spaces. In: Proceedings of 23rd International Conference on Very Large Data Bases, pp. 426–435 (1997)
7. Cover, T., Hart, P.: Nearest neighbor pattern classification. IEEE Trans. Inf. Theory **13**(1), 21–27 (1967)
8. Dawson, K.S.: The SDSS-IV extended baryon oscillation spectroscopic survey: overview and early data. Astron. J. **151**(2), 44 (2016). http://stacks.iop.org/1538-3881/151/i=2/a=44
9. Eldawy, A., Mokbel, M.F.: SpatialHadoop: a MapReduce framework for spatial data. In: International Conference on Data Engineering, pp. 1352–1363. IEEE (2015)
10. Ester, M., Kriegel, H.P., Sander, J., Xu, X.: A density-based algorithm for discovering clusters in large spatial databases with noise. In: KDD, vol. 96, pp. 226–231 (1996)
11. Guttman, A.: R-trees: a dynamic index structure for spatial searching, vol. 14. ACM (1984)
12. Haghani, P., Michel, S., Cudré-Mauroux, P., Aberer, K.: LSH at large-distributed KNN search in high dimensions. In: International Workshop on the Web and Databases (2008)
13. Hjaltason, G.R., Samet, H.: Index-driven similarity search in metric spaces (survey article). ACM Trans. Database Syst. **28**(4), 517–580 (2003)
14. Lin, K.I., Jagadish, H.V., Faloutsos, C.: The TV-tree: an index structure for high-dimensional data. Int. J. Very Large Data Bases **3**(4), 517–542 (1994)
15. Lu, W., Shen, Y., Chen, S., Ooi, B.C.: Efficient processing of k nearest neighbor joins using mapreduce. Proc. VLDB Endow. **5**(10), 1016–1027 (2012)
16. Maillo, J., Ramírez, S., Triguero, I., Herrera, F.: kNN-IS: an iterative spark-based design of the k-nearest neighbors classifier for big data. Knowl.-Based Syst. **117**, 3–15 (2017)
17. Maillo, J., Triguero, I., Herrera, F.: A mapreduce-based k-nearest neighbor approach for big data classification. In: Trustcom/BigDataSE/ISPA, vol. 2, pp. 167–172. IEEE (2015)
18. Novak, D., Zezula, P.: M-Chord: a scalable distributed similarity search structure. In: Proceedings of the 1st International Conference on Scalable Information Systems, p. 19. ACM (2006)
19. Ramaswamy, S., Rose, K.: Adaptive cluster distance bounding for high-dimensional indexing. IEEE Trans. Knowl. Data Eng. **23**(6), 815–830 (2011)
20. Ramaswamy, S., Rastogi, R., Shim, K.: Efficient algorithms for mining outliers from large data sets, vol. 29, pp. 427–438. ACM (2000)
21. Sarma, A.D., Afrati, F.N., Salihoglu, S., Ullman, J.D.: Upper and lower bounds on the cost of a map-reduce computation. In: Proceedings of the VLDB Endowment, vol. 6, pp. 277–288. VLDB Endowment (2013)
22. Shvachko, K., Kuang, H., Radia, S., Chansler, R.: The hadoop distributed file system. In: IEEE 26th Symposium on Mass Storage Systems and Technologies, pp. 1–10. IEEE (2010)

23. Vitter, J.S.: Random sampling with a reservoir. ACM Trans. Math. Softw. **11**(1), 37–57 (1985)
24. Xie, D., Li, F., Yao, B., Li, G., Zhou, L., Guo, M.: Simba: efficient in-memory spatial analytics. In: ACM SIGMOD International Conference on Management of Data, pp. 1071–1085 (2016)
25. Zaharia, M., Chowdhury, M., Franklin, M.J., Shenker, S., Stoica, I.: Spark: cluster computing with working sets. HotCloud **10**, 10 (2010)
26. Zhang, C., Li, F., Jestes, J.: Efficient parallel KNN joins for large data in mapreduce. In: Proceedings of the 15th International Conference on Extending Database Technology, pp. 38–49. ACM (2012)

Author Index

Printed in the United States
By Bookmasters